W9-BBF-740

International Encyclopedia of STATISTICS

International Encyclopedia of STATISTICS

Edited by

WILLIAM H. KRUSKAL and **JUDITH M. TANUR**
University of Chicago

*State University of New York
at Stony Brook*

VOLUME 2

THE FREE PRESS
A Division of Macmillan Publishing Co., Inc.
NEW YORK

Collier Macmillan Publishers
LONDON

THE *International Encyclopedia of Statistics* includes articles on
statistics and articles relevant to statistics that were first published in
the *International Encyclopedia of the Social Sciences* (1968). These
articles have been brought up to date by the addition of postscripts or by
revision in whole or in part and have been supplemented by new articles
and biographies listed in the Introduction.

THE FREE PRESS
A Division of Macmillan Publishing Co., Inc.
866 Third Avenue, New York, N.Y.   10022

COLLIER MACMILLAN CANADA, LTD.

Library of Congress Catalog Card Number: 78-17324

PRINTED IN THE UNITED STATES OF AMERICA

printing number

   3 4 5 6 7 8 9 10

Library of Congress Cataloging in Publication Data
Main entry under title:

International encyclopedia of statistics.

   Includes index.
   1. Statistics—Dictionaries. 2. Social sciences—
Statistical methods—Dictionaries. I. Kruskal,
William H.   II. Tanur, Judith M.
HA17.I63   001.4′22′03   78-17324
ISBN 0-02-917960-2 (set)
ISBN 0-02-917970-X (Vol. I)
ISBN 0-02-917980-7 (Vol. II)

ODDS RATIO

See COUNTED DATA; SURVEY ANALYSIS, *article on* METHODS OF SURVEY ANALYSIS.

OFFICIAL STATISTICS

See CENSUS; GOVERNMENT STATISTICS; PUBLIC POLICY AND STATISTICS; VITAL STATISTICS.

OPERATIONS RESEARCH

The roots of "operations research" (commonly referred to as OR) go back at least to the industrial revolution, which brought with it the mechanization of production, power generation, transportation, and communication. Machines replaced man as a source of power and made possible the development of the large industrial, military, and governmental complexes that we know today. These developments were accompanied by the continuous subdivision of industrial, commercial, military, and governmental management into more and more specialized functions and eventually resulted in the kind of multilevel structure that today characterizes most organizations in our culture.

As each new type of specialized manager appeared, a new specialized branch of applied science or engineering developed to provide him with assistance. For example, in industry this progression began with the emergence of mechanical and chemical engineering to serve production management and has continued into more recent times with the development of such specialties as industrial engineering, value analysis, statistical quality control, industrial psychology, and human engineering. Today, no matter how specialized the manager, at least one relevant type of applied science or engineering is available to him.

Whenever a new layer of management is created, a new managerial function, that of the *executive*, is also created at the next higher level. The executive function consists of coordinating and integrating the activities of diverse organizational units so that they serve the interests of the organization as a whole, or at least the interests of the unit that contains them. The importance of the executive function has grown steadily with the increase in the size and complexity of industrial, military, and governmental organizations.

The executive function in business and industry has developed gradually. The executive was not subjected to violent stimuli from new technology as was, for example, the manager of production. Consequently the executive "grew" into his problems, and these appeared to him to require for their solution nothing but good judgment based on relevant past experience. The executive, therefore, felt no need for a more rigorous scientific way of looking at his problems. However, the demands on his time grew, and he sought aid from those who had more time for, and more experience with, the problems that he faced. It was this need that gave rise to management consulting in the 1920s. Management consulting, however, was based on experience and qualitative judgment rather than on experimentation and quantitative analysis. The executive function was left without a scientific arm until World War II.

The major difference between the development of military executives and of their industrial counterparts is to be found in the twenty-year gap between the close of World War I and the opening of World

667

War II. Because there was little opportunity to use military technology under combat conditions during this period, this technology developed too rapidly for effective absorption into military tactics and strategy. Thus, it is not surprising that British military executives turned to scientists for aid when the German air attack on Britain began. Initially they sought aid in incorporating the then new radar into the tactics and strategy of air defense. Small teams of scientists, drawn from any disciplines from which they could be obtained, worked on such problems with considerable success in 1939 and 1940. Their success bred further demand for such services, and their use spread to Britain's allies—the United States, Canada, and France. These teams of scientists were usually assigned to the executive in charge of operations, and their work came to be known in the United Kingdom as "operational research" and in the United States by a variety of names: operations research, operations analysis, operations evaluation, systems analysis, systems evaluation, and management science. The name operations research was and is the most widely used in the United States.

At the end of the war very different things happened to OR in the United Kingdom and in the United States. In the United Kingdom expenditures on defense research were reduced. This led to the release of many OR workers from the military at a time when industrial managers were confronted with the need to reconstruct much of Britain's manufacturing facilities that had been damaged during the war and to update obsolete equipment. In addition the British Labour party, which had come into power, began to nationalize several major and basic industries. Executives in these industries in particular sought and received assistance from the OR men coming out of the military. Coal, iron and steel, transport, and many other industries began to create industrial OR.

In contrast to the situation in Great Britain, defense research in the United States was increased at the end of the war. As a result military OR was expanded, and most of the war-experienced OR workers remained in the service of the military. Industrial executives did not ask for help because they were slipping back into a familiar peacetime pattern that did not involve either major reconstruction of plant or nationalization of industry.

During the late 1940s, however, the electronic computer became available and confronted the industrial manager with the possibility of automation —the replacement of man by machines as a source of control. The computer also made it possible for a man to control more effectively widely spread and large-scale activities because of its ability to process large amounts of data accurately and quickly. It provided the spark that set off what has sometimes been called the second industrial revolution. In order to exploit the new technology of control, industrial executives began to turn to scientists for aid as the military leaders had done before them. They absorbed the OR workers who trickled out of the military and encouraged academic institutions to educate additional men for work in this field.

Within a decade there were at least as many OR workers in academic, governmental, and industrial organizations as there were in the military. More than half of the largest companies in the United States have used or are using OR, and there are now about 4,000 OR workers in the country. A national society, the Operations Research Society of America, was formed in 1953. Other nations followed, and in 1957 the International Federation of Operational Research Societies was formed. Books and journals on the subject began to appear in a wide variety of languages. Graduate courses and curricula in OR began to proliferate in the United States and elsewhere.

In short, after vigorous growth in the military, OR entered its second decade with continued growth in the military and an even more rapid growth in industrial, academic, and governmental organizations.

Essential characteristics of OR. The essential characteristics of OR are its systems (or executive) orientation, its use of interdisciplinary teams, and its methodology.

Systems approach to problems. The systems approach to problems is based on the observation that in organized systems the behavior of any part ultimately has some effect on the performance of every other part. Not all these effects are significant or even capable of being detected. Therefore the essence of this orientation lies in the systematic search for significant interactions when evaluating actions or policies in any part of the organization. Use of such knowledge permits evaluation of actions and policies in terms of the organization as a whole, that is, in terms of their over-all effect.

This way of approaching organizational problems is diametrically opposed to one based on "cutting a problem down to size." OR workers almost always enlarge the scope of a problem that is given to them by taking into account interactions that were not incorporated in the initial formulation of the problem. New research methods had to be developed to deal with these enlarged and more complicated problems. These are discussed below.

As an illustration of the systems approach to organizational problems, consider the case of a company which has 5 plants that convert a natural material into a raw material and 15 finishing plants that use this raw material to manufacture the products sold by the company. The finishing plants are widely dispersed and have different capacities for manufacturing a wide range of finished products. No single finishing plant can manufacture all the products in the line, but any one product may be produced in more than one plant.

Many millions of dollars are spent each year in shipping the output of the first group of plants to the second group. The problem that management presented to an OR group, therefore, was how to allocate the output of the raw-material plants to the finishing plants so as to minimize total between-plant transportation costs. So stated, this is a well-defined, self-contained problem for which a straight-forward solution can be obtained by use of one of the techniques of OR, linear programming [see PROGRAMMING].

In the initial phases of their work, the OR workers observed that whereas all the raw-material plants were operating at capacity, none of the finishing plants were. They inquired whether the unit-production costs at the finishing plants varied with the percentage of capacity in use. They found that this was the case and also that the costs varied in a different way in each plant. As a result of this inquiry the original problem was reformulated to include not only transportation costs, but also the increased costs of production resulting from shipping to a finishing plant less material than it required for capacity operation. In solving this enlarged problem it was found that increased costs of production outweighed transportation costs and that a solution to the original problem (as formulated by management) would have resulted in an increase in production costs that would have more than offset the saving in transportation costs.

The OR workers then asked whether the increased costs of production that resulted from unused capacity depended on how production was planned, and they discovered that it did. Consequently, another related study was initiated in an effort to determine how to plan production at each finishing plant so as to minimize the increase in unit-production costs that resulted from unused capacity. In the course of this study of production planning it also became apparent that production costs were dependent on what was held where in semifinished inventory. Therefore, another study was begun to determine at what processing stage semifinished inventories should be held and what

they should contain. Eventually the cost of shipping finished products to customers also had to be considered.

In the sequence of studies briefly described, it was not necessary to wait until all were completed before the results of the first could be applied. Solutions to each part of the total problem were applied immediately because precautions had been taken not to harm other operations. With each successive finding previous solutions were suitably adjusted. Eventually some change was made in every aspect of the organization's activities, but each with an eye on its over-all effect. This is the essence of the systems approach to organizational problems.

The interdisciplinary team. Although division of the domain of scientific knowledge into specific disciplines is a relatively recent phenomenon, we are now so accustomed to classifying scientific knowledge in a way that corresponds either to the departmental structure of universities or to the professional organization of scientists that we often act as though nature were structured in the same way. Yet we seldom find such things as pure physical problems, pure psychological problems, pure economic problems, and so on. There are only problems; the disciplines of science simply represent different ways of looking at them. Nearly every problem may be looked at through the eyes of every discipline, but, of course, it is not always fruitful to do so.

If we want to explain an automobile's being struck by a locomotive at a grade crossing, for example, we could do so either in terms of the laws of motion, or the engineering failure of warning devices, or the state of physical or mental health of the driver, or the social use of automobiles as an instrument of suicide, and so on. The way in which we look at the event depends on our purposes in doing so. A highway engineer and a driving instructor would look at it quite differently.

Though experience indicates a fruitful way of looking at most familiar problems, we tend to deal with unfamiliar and complicated situations in the way that is most familiar to us. It is not surprising, therefore, that given the problem, for example, of increasing the productivity of a manufacturing facility, a personnel psychologist will try to select better workers or improve the training that workers are given. A mechanical engineer will try to improve the machines. An industrial engineer will try to improve the plant layout, simplify the operations performed by the workers, or offer them more attractive incentives. The systems and procedures analyst will try to improve the flow of information into and through the plant, and so on. All may produce improvements, but which is best? For complicated problems we seldom can know in advance. Hence it is desirable to consider and evaluate as wide a range of approaches to the problem as possible. OR has greatly enlarged our capacity to deal with all the complexities of and the approaches to a given problem and has therefore expanded our opportunities to benefit from the use of interdisciplinary teams in solving problems.

Since more than a hundred scientific disciplines, pure and applied, have been identified, it is clearly not possible to incorporate each in most research projects. But in OR as many diverse disciplines are used on a team as possible, and the team's work is subjected to critical review by as many of the disciplines not represented on the team as possible.

Methodology. Experimentation lies at the heart of scientific method, but it is obvious that the kind of organized man–machine system with which industrial, military, and governmental managers are concerned can never be brought into the laboratory, and only infrequently can such systems be manipulated enough in their natural environment to experiment on them there. Consequently, the OR worker finds himself in much the same position as the astronomer, and he takes a way out of his difficulty much like that taken by the astronomer. If he cannot manipulate the system itself, he builds a representation of the system, a *model* of it, that he can manipulate. In OR such models are abstract (symbolic) representations that may be very complicated from a mathematical point of view. From a logical point of view, however, they are quite simple. In general they take the form of an equation in which the performance of the system, P, is expressed as a function, f, of a set of controlled variables, C, and a set of uncontrolled variables, U:

$$P = f(C, U).$$

The controlled variables represent the aspects of the system that management can manipulate, for example, production quantities, prices, range of product line, and so on. Such variables are often called decision variables since managerial decision making may be thought of as assigning values to these variables. The uncontrolled variables represent aspects of the system and its environment that significantly affect the system's performance but are not under the control of management, for example, product demand, competitors' prices, cost of raw material, and location of customers.

The measure of performance of the system may be very difficult to construct since it must reflect the relative importance of each relevant objective of the organization. This measure is sometimes called the criterion or objective function since it provides the basis for selecting the "best" or "better" courses of action.

Limitations or restrictions may be imposed on the possible values of the controlled variables. For example, in preparing a budget a limitation is normally placed on the total amount that may be allocated to different departments, or there may be legal constraints on the decision-making activities of managers. Such restrictions can usually be expressed mathematically as equations or inequalities and can be incorporated in the model.

Once the decision maker's choices and the system involved have been represented by a mathematical model, the researcher must find a set of values of the controlled variables that yields the best (or as close as possible to the best) performance of the system. These "optimizing" values may be found either by experimenting on the model (i.e., by *simulation*) or by mathematical analysis. In either case the result is a set of equations, one for each controlled variable, giving the value of that controlled variable relative to a particular set of values of the uncontrolled variables and other controlled variables that yields the best performance of the system as a whole [see SIMULATION].

If the problem is a recurrent one, then the values of the uncontrolled variables (for example, demand) may change from one decision-making period to another. In such cases a procedure must also be provided for determining when values of the uncontrolled variables have undergone significant change and for adjusting the solution appropriately. Such a procedure is called a solution-control system.

The output of an OR study, then, is usually a set of rules for determining the optimal values of the controlled variables together with a procedure for continuously checking the values of the uncontrolled variables. It must be borne in mind, however, that a single, unified, and comprehensive OR study is seldom possible in an organization of any appreciable size. Rather, what usually occurs is a sequence of interrelated studies, each of which is designed to be adjustable to the results of the others.

Ten years of constructing and working with models of managerial problems in industry have shown that, despite the fact that no two problems are ever exactly alike in content, most problems fall into one, or a combination, of a small number of basic types. These problem-types have now been studied extensively so that today we have considerable knowledge about how to construct and solve models that are relevant to them. Adequate definitions of these problem-types require more space than is available here, but the following brief characterizations indicate their nature.

Inventory problem—to determine the amount of a resource to be acquired or the frequency of acquisition when there is a penalty for having either too much or too little available.

Allocation problem—to determine the allocation of resources to a number of jobs where available resources do not permit each job to be done in the best possible way, so as to do all (or as many as possible) of the jobs in such a way as to achieve the best over-all performance, given criteria for measuring performance.

Queuing problem—to determine the amount of service facilities required or how to schedule arrival of tasks at service facilities so that losses associated with idle facilities, waiting, and turned-away tasks are minimized.

Sequencing problem—to determine the order in which a set of tasks should be performed in a multistage facility so as to minimize costs associated with the performance of the tasks and delays in completing them.

Routing problem—to determine which path or route through a network of points or locations is shortest (or longest), has maximum (or minimum) capacity, or is least (or most) costly to traverse subject to certain limitations on the paths or routes that are permissible.

Replacement problem—to determine when to replace instruments, tools, or facilities so that acquisition, maintenance, and operating and failure costs are minimized.

Competition problem—to determine the rule to be followed by a decision maker that yields the best results when the outcome of his decision depends in part on decisions made by others.

Search problem—to determine the amount of resources to employ and how to allocate them in seeking information to be used for a particular purpose so as to minimize the costs associated with the search and with the errors that can result from use of incorrect information.

The future of OR. OR has been primarily concerned with the executive's decision-making or control process. There are, of course, other approaches to improving the performance of organizations, for example, selecting better personnel, providing better personnel training, better motivating personnel, accelerating their operations through work study, changing equipment and materials, modifying communications, changing organizational structure. This multiplicity of available approaches presents the executive with the additional problem of selecting which approaches to pursue. He seldom has an objective basis for doing so. Clearly it would be desirable to develop an integrated and comprehensive approach to organizations, one that rationally selects from or combines different points of view. OR and other systems-oriented interdisciplinary research are taking steps to develop such an over-all approach to organizational problems. This is leading to mathematical descriptions of organizational structures and communications systems, thus providing the ultimate possibility of integrating studies of organizational structure, communication, and control.

Precise solutions of some limited problems of organizational structure have already been found. For example, given an organization's over-all objective and a description of its task and environment, it is possible to determine the number and types of units into which the organization should be divided and the objectives to be assigned to these units so as to minimize inefficiency arising from the organization's structure. This is a problem in structural design. Or, given an organization that has an inefficient structure, it is possible to determine the types of decentralized control to be applied to decentralized decision making so as to minimize inefficiency. This is a problem in structural control.

Such developments are leading to an integrated theory of, and generalized methodology for, research on organized systems. Since all of these systems are, in some sense, social systems, the participation of the social scientist in these interdisciplinary efforts is essential.

RUSSELL L. ACKOFF

[*See also* Gochman 1968; Kaplan 1968; Mitchell 1968; Parsons 1968; Rapoport 1968.]

BIBLIOGRAPHY

ACKOFF, RUSSELL L. (editor) 1961 *Progress in Operations Research.* Volume 1. Operations Research Society of America, Publications in Operations Research, No. 5. New York: Wiley.

▶ACKOFF, RUSSELL L. 1974 *Redesigning the Future: A Systems Approach to Societal Problems.* New York: Wiley.

ACKOFF, RUSSELL L.; and RIVETT, PATRICK 1963 *A Manager's Guide to Operations Research.* New York: Wiley.

CHURCHMAN, C. WEST; ACKOFF, RUSSELL L.; and ARNOFF, E. LEONARD 1957 *Introduction to Operations Research.* New York: Wiley.

DUCKWORTH, WALTER E. 1962 *A Guide to Operational Research.* London: Methuen.

EDDISON, R. T.; PENNYCUICK, K.; and RIVETT, PATRICK

1962 *Operational Research in Management.* New York: Wiley.

▶GOCHMAN, DAVID S. 1968 Systems Analysis: V. Psychological Systems. Volume 15, pages 486–495 in *International Encyclopedia of the Social Sciences.* Edited by David L. Sills. New York: Macmillan and Free Press.

▶KAPLAN, MORTON A. 1968 Systems Analysis: IV. International Systems. Volume 15, pages 479–486 in *International Encyclopedia of the Social Sciences.* Edited by David L. Sills. New York: Macmillan and Free Press.

MILLER, DAVID W.; and STARR, MARTIN K. 1960 *Executive Decisions and Operations Research.* Englewood Cliffs, N.J.: Prentice-Hall.

▶MITCHELL, WILLIAM C. 1968 Systems Analysis: III. Political Systems. Volume 15, pages 473–479 in *International Encyclopedia of the Social Sciences.* Edited by David L. Sills. New York: Macmillan and Free Press.

▶PARSONS, TALCOTT 1968 Systems Analysis: II. Social Systems. Volume 15, pages 458–473 in *International Encyclopedia of the Social Sciences.* Edited by David L. Sills. New York: Macmillan and Free Press.

▶RAPOPORT, ANATOL 1968 Systems Analysis: I. General Systems Theory. Volume 15, pages 452–458 in *International Encyclopedia of the Social Sciences.* Edited by David L. Sills. New York: Macmillan and Free Press.

SASIENI, MAURICE; YASPAN, ARTHUR; and FRIEDMAN, LAWRENCE 1959 *Operations Research: Methods and Problems.* New York: Wiley.

OPINION POLLS

See ERRORS, *article on* NONSAMPLING ERRORS; PUBLIC POLICY AND STATISTICS; SAMPLE SURVEYS; STATISTICS AS LEGAL EVIDENCE; SURVEY ANALYSIS.

OPTIONAL STOPPING

See SIGNIFICANCE, TESTS OF.

ORDER STATISTICS

See under NONPARAMETRIC STATISTICS.

ORGANIZATIONS: METHODS OF RESEARCH

This article was first published in IESS *with four companion articles less relevant to statistics.*

A formal organization consists of a set of people who are engaged in activities coordinated by the relatively consistent expectations that these people have about one another and about the purposes of the organization. Such expectations define a set of organizational statuses or offices, each status with a set of roles linking it to the other statuses with which it interacts. Organizational research examines how people behave in their organizational roles and how organizations behave as collective units. Here we will consider the types of data collection and measurement, and research design and analysis used in organizational research.

Data collection

Organizational research uses three main sources of data: qualitative observation and interviewing, surveys of organization members, and institutional records. Qualitative methods of data collection are described elsewhere (see Powdermaker 1968). Surveys within organizations take three forms: surveys of one stratum only; surveys of two or more strata within the organization (for instance, workers and management, or students and faculty); and "relational surveys" of linked pairs or sets of role partners. In surveys of two or more organizational strata ("multistratum surveys") individuals are usually sampled at random within status groups and are identified only by the general status each one occupies. In "relational surveys" each individual is identified by the specific others with whom he interacts, and the sample is set up to include these role partners. When a sample of individuals is asked to provide the information on their role partners, who are not themselves interviewed, we have a "pseudo-relational survey."

Institutional data include records made by organizations for their own use; information from directories, which can be turned into "data banks"; and "institutional questionnaires," sent to samples of organizations, to be filled out by one or more key informants in each organization.

Measurement

The formal aspect of measurement that is most significant for organizational research is the relation of the units from which the basic data are gathered to those which are characterized by the measurement. Information can be gathered on the organization as a whole or on component parts of it, such as individuals, interacting pairs and sets, and subgroups of members. Data on component parts can be aggregated in various ways to characterize the organization; and data on the organization can be considered as a "contextual characteristic" of the members. The following types of characteristics of collectives and their members have been distinguished by Lazarsfeld and Menzel (1961), and further discussed by Barton (1961, pp. 2–3 and appendix 1).

Integral or global characteristics of a unit do not derive from aggregation of members, pairs, or sets within that unit, but from properties of the unit as a whole. In the case of an organization, examples of such properties are its physical equipment, formal rules, budgets, programs, collective events, and collective outputs.

Relational characteristics derive from information concerning the relationship between a unit and other units. In the case of an individual, they include popularity, measured by number of choices received on sociometric questions; participation in an occupational community, measured by the number of friends who belong to the same occupation; supportiveness of political environment, measured by the proportion of his associates who vote the same way as the respondent; cosmopolitanism, measured by the number of extraorganizational contacts he has; and so on. For collective units, relational characteristics would include the amount of communication between one unit and various others in its environment, the frequency of cooperation or conflict with other units, the volume of economic transactions between them, and so on (Levine & White 1961; Litwak & Hylton 1962).

The underlying model for relational data is a "who-to-whom" matrix showing the value of the relationship for each pair within a group. There may be different types of relationships (who likes whom, who gives orders to whom, who talks shop with whom, etc.), each of which generates a matrix [see SOCIOMETRY]. Summary scores for individuals are found in the marginal totals of each matrix. Other operations can give us the number of second-order and higher-order connections which an individual has (through his immediate partners) and the characteristics of his interpersonal environment. Organizations likewise have higher-order connections and an interorganizational environment.

Aggregate measures characterize a collective in terms of distributions of data concerning its members. The simplest aggregate measures are additive properties, such as rates and means. Thus the morale of army units may be measured by the proportion of their men who feel strong loyalty to the unit; the political climate of a college by the proportion of its faculty members who are liberal; the ability level of a school by the mean IQ of its students. In these cases, as we would expect, there is a simple correspondence between the individual and the collective properties; relationships true on the individual level are duplicated on the organizational level. However, this kind of correspondence does not always exist, as discussions of the "ecological fallacy" have shown (Robinson 1950; Duncan & Davis 1953; Goodman 1953). Additive measures can also be used as indicators of more complex organizational characteristics that are not simply the sum of individual traits: for example, the proportion of an organization's total personnel that is in administrative positions may indicate how bureaucratized the organization is.

A peculiar type of aggregate property arises when researchers ask samples of members for their perceptions of the organization or parts of it, and add up their answers to obtain a kind of aggregate perception. For example, an army company may be classified as having authoritarian leadership if a large proportion of members report that the leader behaves in an authoritarian way (Selvin 1960). A problem here is that different groups within an organization may evaluate the same behavior differently: Halpin (1956) found school principals reporting themselves as high on "consideration" for teachers; the school boards agreed with them, but the teachers themselves did not.

The *variance* of the distribution of member data may also be used to characterize the collective. Thus the range or standard deviation of the incomes of an organization's members can be used as a measure of the organization's equalitarianism, and the same measures, when applied to the ability levels of a high school class or the values of members of a union local, can indicate the degree of heterogeneity in the class or of value consensus in the union. Equality, homogeneity, and consensus are emergent properties on the collective level, without individual counterparts.

Finally, the *correlations* between variables within a collective can be used as aggregate measures. For instance, the correlation of values with rank in the American army indicates its degree of "value stratification" (Speicr 1950). Similarly, the correlation of military rank with external social status measures the degree of ascription as compared with achievement in the organization's promotion system.

The members of the collective from which data are aggregated need not be individuals; they may be smaller collectives within the larger one. Thus an aggregate measure of the degree of local democracy within national unions might be the proportion of locals within a union that had contested elections.

When we have data from several status groups within a number of organizations, we can obtain aggregate measures for each stratum. An organization might thus be characterized as having high morale in the upper strata but low morale among the rank and file, as having authoritarian top managers and democratic foremen, and so on. The *differences* between different strata, in rates or mean scores on various attitudes, behaviors, background data, etc., are actually a form of correlational index.

Relational-pattern measures (sometimes termed *structural* or *sociometric* measures) are aggregations of relational data on the members of a collective unit. Group cohesion is often measured in this way by the ratio of in-group to out-group choices when each member of a group is asked whom he chooses or would choose as a close friend; group integration may be indicated by the over-all frequency with which group members communicate with one another (see Schachter 1968). More complex measures would include the extent to which relations within a group form self-contained cliques. Patterns of relationship among subunits—for example, the amount of conflict between subgroups in an organization—can also characterize a collective. Just as relational characteristics of individuals are indicated by their row or column scores in a who-to-whom matrix, relational pattern measures for a collective derive from the matrix as a whole—that is, from the distribution or patterning of pair relations.

Contextual measures arise when we characterize a member by the properties of the collective of which the member is a part. Being dissatisfied is an integral property of the individual; being a member of a work group where a high proportion are dissatisfied is a contextual property, derived from the aggregate of individual properties of members. Being a member of a conflict-ridden organization is a contextual property derived from a relational-pattern property of the collective; attending a college with a large library is a contextual property derived from an integral property of the collective.

In a multistratum survey of many organizations, contextual data can also be derived from different strata. For example, aggregate data on university faculty are contextual for students, and characteristics of foremen are contextual for the workers under them.

Organizations, too, have contexts, such as the larger organization of which they are part or the industry or community in which they are located. Locals may be part of a "democratic" or an "undemocratic" national union; business firms may be located in communities with growing or shrinking populations, or belong to competitive or concentrated industries.

In summary, individual members of an organization have integral, relational, and contextual properties. Collectives likewise have integral properties, relational properties (with regard to other organizations), and contextual properties (those of larger collectivities of which they are members). But in addition, collectives have aggregate properties derived from their members' integral characteristics, and relational-pattern properties derived from their members' relationships.

Research designs

We have distinguished five main sources of data: qualitative observation, one-stratum surveys, multistratum surveys, relational surveys, and institutional data; this is the first of three dimensions on which research designs can be classified. A second dimension is the number of organizational units studied: "case studies" of one organization, "comparative studies" of two or several organizations; and "large-sample studies" of enough organizations to make statistical analyses with organizations as units of analysis. (Either a whole organization or a formal subunit can be the focus of study; a study of 100 work groups within a large corporation would be a "large-sample study" of work groups.) The third dimension is whether data are gathered for one point in time or whether comparable data are gathered for each of two or more time periods. Studies over time permit analysis of change and inferences of causal relationships.

A typology of designs for organizational studies is generated if we run these three dimensions against one another (see Table 1, in which each row of six cells derives from a different source of data).

Analysis of data on organizations

The different types of research design outlined in Table 1 permit several basic types of analysis, some qualitative and some quantitative. Each type of analysis can be performed for units at different levels of aggregation—for individuals, for linked sets of individuals, or for organizations. We will examine each type of analysis in turn, indicating its limitations as well as the problems that it permits us to study.

Ideal-type comparison. A study of one organization at one point in time, using qualitative methods, provides a very limited scope for analysis. Such analysis usually involves comparing the case at hand with a theoretically derived ideal type—Weber's model of bureaucracy, for instance, or the economists' model of a firm composed of rational economic men. Thus, an analysis of bureaucratic patterns in the navy officer corps describes a number of striking behavior patterns (avoiding responsibility, ritualism, insulation from the outside world, and ceremonialism) and concludes: "The military variant of bureaucracy may thus be viewed

Table 1 — A typology of designs of organizational studies

TIME PERIODS	NUMBER OF UNITS STUDIED		
	One	Several	Many
1	Qualitative case study Davis 1948	Qualitative comparative study Coser 1958 Form & Nosow 1958	Many qualitative case studies Udy 1959
2+	Qualitative case study over time Selznick 1949 Gouldner 1954 Roethlisberger & Dickson 1939	Qualitative comparative study over time Guest 1962	Many qualitative case studies over time ———
1	One-stratum one-unit survey Walker & Guest 1952	One-stratum comparative survey Katz & Hyman 1947	One-stratum survey of a large sample of organizations Lazarsfeld & Thielens 1958 Bowers 1964
2+	One-stratum one-unit panel survey Stouffer et al. 1949	One-stratum comparative panel survey ———	One-stratum panel survey in a large sample of organizations
1	Multistratum one-unit survey Stouffer et al. 1949	Multistratum comparative survey Georgopoulos & Mann 1962 Illinois, University of 1954 Mann & Hoffman 1960	Multistratum survey of a large sample of organizations ———
2+	Multistratum one-unit panel survey Newcomb 1943 Lieberman 1956	Multistratum comparative panel survey ———	Multistratum panel survey of a large sample of organizations ———
1	One-unit relational survey Weiss 1956 Kahn 1964 Stogdill et al. 1956	Comparative relational survey ———	Relational survey of a large sample of organizations Kahn & Katz 1953 Gross et al. 1958
2+	One-unit relational panel survey W. Wallace 1964 A. F. C. Wallace & Rashkis 1959	Comparative relational panel survey Morse & Reimer 1956	Relational panel survey of a large sample of organizations ———
1	Case study using one-time institutional data ———	Comparative study using one-time institutional data Harbison et al. 1955	Study of a large sample of organizations using one-time institutional data Faunce 1962 Douglass 1926
2+	Case study using institutional data over time Brown 1956	Comparative study using institutional data over time Haire 1959	Study of a large sample of organizations using institutional data over time Lipset et al. 1956

as a skewing of Weber's ideal type by the situational elements of uncertainty and standing by" (Davis [1948] 1952, p. 384). Similarly, the authors of the famous Western Electric study (Roethlisberger & Dickson 1939) sought to discover, by means of qualitative interviewing and observation, whether factory workers behaved like a set of discrete and rational economic units and concluded, when faced with such phenomena as group production norms and informal leaders, that they did not.

Single case studies thus can disclose the existence of phenomena that raise problems for the theories from which ideal types have been generated. Such studies therefore inspire further research of more complex design to find the conditions under which the ideal–typical or the deviant phenomena occur.

Multivariate comparison of a few cases. When we have data on two or several organizations—whether these data are qualitative, aggregated from surveys, or compiled from institutional records—we can use some form of "quasi-experimental" analysis (also known as "quasi-correlational" analysis). The simplest form is "strategic paired comparison," a method that calls for two organizations that are formally similar but differ strongly in one respect. The presumed consequences of this difference are then explored in detail. Thus Coser (1958) compared a medical ward with a surgical ward in the same hospital; the wards, though formally similar, presented many differences in the actual exercise of authority and the informal relations of doctors and nurses. Mann and Hoffman (1960), using a survey of top managers, foremen, and workers, compared an automated power plant with a nonautomated one. Treating the results as a "quasi-experiment," they suggest that automation increased the men's sense of having an influence on plant operations, as well as their interest and satisfaction, but also aggravated tensions related to the job.

In these two examples, the researchers translated a gross institutional characteristic (medical versus surgical, automated versus nonautomated) into several more general sociological characteristics (time pressure on decision making, job enlargement, interdependence of components), which were capable of explaining the observed differences in the dependent variables in terms of general sociological propositions (for example, when there is more time pressure on decision making, there will be less group consultation). Having only two cases, however, they could not test the relative effects of the several proposed explanatory variables or control for the effects of other variables.

When researchers have somewhat more cases, they can locate each organization in a multidimensional classification of explanatory variables. Thus a study of six organizations' responses to disaster found seven characteristics that differentiated the more effective from the less effective ones; each organization represented a particular pattern of these seven attributes (Form & Nosow 1958). Such an analysis is suggestive but not conclusive, since there is only one case per cell.

Qualitative study of change. Most qualitative case studies that go beyond description of formal structure necessarily examine time sequences of the behavior of organization members. These may disclose the "microprocesses"—the exercise of informal pressures, the ways of getting around rules, the immediate causes of deviance—that maintain normal equilibrium in the organization. Many examples of such microprocesses were observed in the Western Electric study. A worker who went beyond the group norm of output was ridiculed or "binged" on the arm by fellow workers. A researcher entering the room was mistaken for a time-study man, and everyone slowed down. Changes in production often followed off-the-job personal problems (Roethlisberger & Dickson 1939). More recently, it was found that episodes of patient disturbance in a mental hospital followed staff disagreements

on the patients' treatment (Stanton & Schwartz 1954).

The logic of sequence analysis is *post hoc, ergo propter hoc;* but this kind of reasoning can be misleading in the absence of experimental or statistical controls to eliminate accidental or spurious factors. Researchers therefore try to test their interpretations by observing as many repeated sequences as they can and by locating changes precisely in the time sequence. This procedure can become a qualitative form of time-series experiment [*see* EXPERIMENTAL DESIGN, *article on* QUASI-EXPERIMENTAL DESIGN; *see also* Campbell & Stanley 1963].

Qualitative studies over long time periods, particularly if they cover major organizational changes such as succession, growth, and reorganization, permit derivation of relationships between organizational variables. Such studies also test hypotheses about the functions of various parts of the system by observing what happens when one part of it is changed. For instance, McCleery (1957) studied the process of change in a prison from an authoritarian to a more liberal regime. His main argument was that the arbitrary behavior of the old regime created such insecurity that the inmates accepted an exploitative elite of "old cons" who interpreted and negotiated with the authorities, and who controlled disorderly inmate behavior that might have jeopardized their own special privileges. At the same time, this administrative arbitrariness made the inmates hostile to the official programs of rehabilitation. When the new administration created fairer procedures and more communication with the inmates, this hostility was reduced, but the power of the inmate elite was reduced still faster, so that an outburst of disorder resulted. A new equilibrium was finally achieved without an exploitative elite and with more rehabilitative activity.

Several operations can be distinguished here: identifying the "anatomy" of the system—the major formal and informal status groups; identifying the key variables characterizing each group; reporting the values of these variables at several stages in the process of change; deriving causal relationships among them; and locating these causal relationships in a functional model of the whole system (see Figure 1). This model can be analyzed to account for the initial equilibrium, the sequence of changes, and the final equilibrium (Barton & Anderson 1961).

That such a system of relationships was found to exist in a single case is hardly conclusive evidence that it exists in all such cases. Comparative study of several organizations can begin to provide checks against alternative possibilities and to spe-

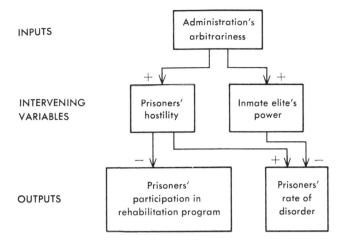

INPUTS

INTERVENING VARIABLES

OUTPUTS

Figure 1 — Causal relationships in prison disorder

Source: Adapted from Barton & Anderson 1961.

cify conditions under which the relationships hold. Thus, Guest (1962) compared two studies of succession of a top manager and proposed no fewer than seven variables in an effort to explain why the two men adopted opposite policies.

Multivariate analysis of individuals. A survey of individual attitudes and behavior within a single organization can analyze the correlates of job satisfaction, productivity, mobility aspirations, or any other organizationally related attitudes and behaviors, in exactly the manner of ordinary public-opinion studies, except that detailed information on organizational status and activities can be added to the usual limited background data. For instance, Stouffer and his colleagues (1949) analyzed the relation of rank, combat experience, length of service, and similar organizational-status variables to morale, mobility aspirations, etc., for the army as a whole and for various components.

Multivariate analysis of linked sets. Analysis of characteristics of pairs or sets of role partners is made possible by relational samples. This is easiest when the role set corresponds to an organizational subunit—as in studies of the correlation of supervisor behavior with worker morale or productivity (see, for instance, Kahn & Katz 1953). In Hall's study (1955) of 40 air crews, behavior of each commander was described for three variables by the aggregate perceptions of his crewmen; each crew's role prescriptions for their commander were measured for the same three variables; and a measure of the conformity of commander behavior to crew prescriptions was derived. His conformity could then be correlated with other attributes of his crew, such as its cohesion and degree of consensus.

Where the role set does not correspond to an

organizational subunit, "snowball sampling" can be used. For example, Kahn and his associates (*Organizational Stress . . .* 1964) took 53 managers as "focal persons" and interviewed two superiors, two subordinates, and three pairs of each. Tensions of the focal person could then be related to characteristics of his role set. The variety of measures that can be created with such samples is shown by a study by Stogdill and his colleagues (1956), who surveyed 47 role sets of a superior and one or two subordinates in a research agency, measuring role expectations and perceived role behavior for self and other for each of 45 role elements. From these basic data they derived measures of each pair's consensus on role expectations, conformity of behavior to own and partner's expectations, agreement on perceptions of behavior, and differentiations or division of labor between the two statuses, both perceived and expected.

Multivariate analysis of organizations. Whenever we have data on a large number of organizations or subunits within organizations, it is possible to apply the same methods of analysis that are used in survey analysis of individual data. The organizational data may derive from a large set of qualitative studies containing comparable data, from surveys that have sufficient samples within a large number of organizations to provide aggregate data, or from institutional data derived from records, "data banks," and surveys of informants by means of institutional questionnaires. Thus Udy (1959) classified 82 production organizations, using data from anthropological monographs compiled in the Human Relations Area Files. He found a high correlation between the degree of complexity of their technology and the number of levels of authority (see Table 2).

Blau and his colleagues (1966) studied a large sample of public bureaucracies with an institutional questionnaire which measured size, division of labor, professionalization, and centralization of authority. Quite complex conditional relationships were found; for example, in organizations with few professionals, division of labor was positively related to centralization of authority, while in highly professionalized organizations the relationship was the reverse.

Table 2 — Authority and technological complexity in 82 organizations

Number of levels of authority	INDEX OF TECHNOLOGICAL COMPLEXITY	
	High	Low
3 or more	28	3
2 or less	5	46

Source: Udy 1959, p. 584.

Very large samples of organizations can be studied by using published directories, data banks of institutional information, or questionnaires to informants in each unit. Thus, Faunce (1962) studied 753 union locals by passing out a questionnaire at the U.A.W. convention, and Douglass (1926) studied 1,000 city churches by using informant questionnaires.

Contextual analysis of individuals. When we have drawn samples of individuals within several or many organizations, contextual analysis becomes possible. The simplest form of this is an examination of the relationship between individual attitudes or behavior and the attributes of the organization of which the individuals in question are members. Thus, students report more cheating at colleges that are large, have many students per faculty member, have most students living off campus, have easy admissions policies, and are coeducational (Bowers 1964). Moreover, it can be shown by means of three-variable "contextual tables" that the relationship between two individual attributes is different in different organizational contexts, and thus that there is an organizational factor influencing the processes of individual behavior. For instance, Lipset and his colleagues, in a study of a large local in the International Typographical Union, surveyed two strata: chapel chairmen and rank-and-file members. Over-all, the chairmen were more interested, knowledgeable, and active in union politics than the rank and file were. However, when union members were separated according to size of the shop, it was found that this difference between chairmen and rank and file appeared only in the larger shops (Lipset et al. 1956, pp. 176–182). Similar effects were obtained in a study of three prisons by Berk (1966).

Contextual analysis of linked sets. Relationships between characteristics of pairs of role partners or between those of whole role sets may be modified by their organizational context. To study this requires comparing relational surveys in a number of organizations—for example, examining the relation of informal contact to consensus between role partners in organizations of different degrees of bureaucratization.

Contextual analysis of organizations. The larger setting—type of community, industry, or society—may affect the characteristics of organizations or the interrelations among them (Udy 1965). Cross-national comparisons show, for instance, that the ratio of supervisors to workers may be 1 to 15 in an American steel plant and 1 to 50 in a similar German steel plant (Harbison et al. 1955; see also Evan 1963). Presumably the relationship between plant size and the supervisor–worker ratio is also different in each case. A complex contextual effect is exemplified by the finding that the relationship between secular orientation and effectiveness is positive for YMCAs at nondenominational colleges, but negative for those at denominational colleges (Lucci 1960).

Time-series analysis. When information is available on several variables within one organizational unit for many points in time, it is possible to search for causal relationships by examining the sequence of changes in these variables. Brown (1956) examined union records covering a period of 50 years. Indicators of intraunion conflict— number of resolutions for constitutional change, challenges to convention delegates, failure of officers to be re-elected, jurisdictional disputes between locals—were found to be inversely related to size year by year during a cycle of growth, decline, and new growth. Tsouderos (1955) used time series of income, expenditure, capital, membership, and number of administrative employees to study the process of growth, bureaucratization, and decline in ten voluntary organizations. His finding was that loss of members after a period of growth led to increased administrative staff, which led to maintenance of income and activities but further loss of members and to eventual decline.

Time series could also be provided by survey data on several organizational strata repeated over many periods. Some large corporations collect data of this kind as a "morale barometer," but it does not seem to be systematically analyzed in the way that public-opinion time series have been.

Panel analysis of individuals. A design that obtains information on a sample of organization members at two or more points in time permits a much clearer isolation of causal relationships among individual characteristics than does a one-time survey. For instance, Stouffer and his associates found that higher-ranking enlisted men were more often in good spirits, accepted the soldier role, were more satisfied with their army job, and thought the army was well run. To test whether these conformist attitudes were a cause of promotion or only a result of being promoted, samples of privates were surveyed when newly recruited and subsequently followed up after several months during which some had been promoted. Those with more conformist attitudes early in their army careers were more likely to be promoted later on. The difference held good when various background factors related to both promotion and attitudes were held constant (Stouffer et al. 1949, vol. 1, pp. 147–154).

Another study dealt with workers who became foremen or shop stewards, repeating on them and on matched control groups a set of attitude questions given a year previously to all workers in the plant. A third wave was done two years later, when some of the foremen had been demoted because of layoffs and some of the stewards had been replaced in office. Many attitudes had changed considerably in response to changes in status (Lieberman 1956).

A panel study that included many organizations would permit comparison of social processes in different settings. Such studies do not appear to exist. (See, however, Miller 1958 for an example of a "pseudo-panel" study which compares students in different years of college for different types of institutions, suggesting major differences in the socialization process.)

Panel analysis of linked sets. If, as is often said, organizations are systems of interacting parts, the best design for studying them is a relational panel, since this would permit analysis of changes among related individuals and groups over time. But studies that make use of such panels are remarkably rare. We have mentioned the qualitative analysis by Stanton and Schwartz (1954) of patient disturbance and staff disagreement in a mental hospital. One quantitative panel study of staff and patients found no such relationship (Wallace & Rashkis 1959), but another found not only that staff disagreement was associated with subsequent patient disturbance but also that patient disturbance led to staff disagreement (Rashkis & Wallace 1959).

Walter Wallace (1964; 1965) obtained a complete sociogram of contacts among students at one college. He also obtained panel attitude data from freshmen at entry, at midsemester, and at the end of the first semester, and attitude data on all other students at midsemester. He constructed measures of the predominant attitude of each freshman's "interpersonal environment," which he related to change in freshman attitudes. Segments of the interpersonal environment, classified by sex and college class, were shown to have differential effects.

Relational panel analysis would be particularly useful in studying role relationships as a system, since it would permit examination of such phenomena as the effects on the whole network of expectations and behavior of a change in one partner's role expectations. But such studies do not appear to exist.

Panel analysis of organizations. If standardized data were gathered for a sample of organizations over time, the determinants of organizational change could be analyzed by the same methods that are used in analyzing panels of individuals. The basic data might be from any of the sources we have discussed—qualitative studies, surveys, or institutional records—provided they measure the same characteristics of a large number of organizations at several comparable points in time. A near example of this type of analysis is the finding by Stouffer and his colleagues that companies which before D-day had the highest willingness for combat subsequently had the lowest rates of nonbattle casualties. The relationship was much higher for veteran regiments than for nonveteran regiments (Stouffer et al. 1949, vol. 2, chapter 1).

A comparable approach was taken by Lipset and his colleagues, who analyzed time-series data on party voting in the printers' union, broken down by large versus small locals. They were able to show that a rising opposition normally gains strength first in the large locals, while the small ones support the party in power; when the party supported by the large locals wins, these locals are again the first to go into opposition (Lipset et al. 1956, pp. 373–382). But, once again, such studies are hardly ever undertaken.

For purposes of developing theories of organizational processes at the microlevel, relational panel studies obtaining data on role expectations and behavior would appear to be particularly appropriate. They would provide factual data on which to base development of simulation models and mathematical models, just as panel studies of voting behavior (including data on the *perceived* norms and behaviors of friends and associates and of various social groups) did for the development of simulation models of electoral processes [*see* SIMULATION, *article on* POLITICAL PROCESSES; *see also* Stokes 1968].

For testing theories of organizational change, panel data on organizational characteristics for large samples of organizations would be highly desirable. We now have a good many examples of studies of large samples of organizations, but because they deal with the organizations at only one point in time, they are severely limited in establishing causal relations.

ALLEN H. BARTON

[*Directly related are the entries* EXPERIMENTAL DESIGN; PANEL STUDIES; *see also* Becker 1968; Powdermaker 1968; Wax 1968. *Other relevant material may be found in* COUNTED DATA; INTERVIEWING IN SOCIAL RESEARCH; MULTIVARIATE ANALYSIS; *see also* Benoit-Guilbot 1968; Simon 1968; Smith 1968; Whyte 1968.]

BIBLIOGRAPHY

BARTON, ALLEN H. 1961 *Organizational Measurement and Its Bearing on the Study of College Environments.* New York: College Entrance Examination Board.

BARTON, ALLEN H.; and ANDERSON, BO 1961 Change in an Organizational System: Formalization of a Qualitative Study. Pages 400–418 in Amitai Etzioni (editor), *Complex Organizations: A Sociological Reader.* New York: Holt.

► BECKER, HOWARD S. 1968 Observation: I. Social Observation and Social Case Studies. Volume 11, pages 232–238 in *International Encyclopedia of the Social Sciences.* Edited by David L. Sills. New York: Macmillan and Free Press.

► BENOIT-GUILBOT, ODILE 1968 Industrial Relations: II. The Sociology of Work. Volume 7, Pages 230–240 in *International Encyclopedia of the Social Sciences.* Edited by David L. Sills. New York: Macmillan and Free Press.

BERK, BERNARD B. 1966 Organizational Goals and Inmate Organization. *American Journal of Sociology* 71: 522–534.

BLAU, PETER M. 1960 Structural Effects. *American Sociological Review* 25:178–193.

BLAU, PETER M.; HEYDEBRAND, WOLF V.; and STAUFFER, ROBERT E. 1966 The Structure of Small Bureaucracies. *American Sociological Review* 31:179–191.

BOWERS, WILLIAM J. 1964 Student Dishonesty and Its Control in College. Cooperative Research Project No. OE 1672. Unpublished manuscript, Columbia Univ., Bureau of Applied Social Research.

BROWN, JULIA S. 1956 Union Size as a Function of Intra-union Conflict. *Human Relations* 9:75–89.

CAMPBELL, DONALD T.; and STANLEY, JULIAN C. 1963 Experimental and Quasi-experimental Designs for Research on Teaching. Pages 171–246 in Nathaniel L. Gage (editor), *Handbook of Research on Teaching.* Chicago: Rand McNally.

COSER, ROSE L. 1958 Authority and Decision-making in a Hospital: A Comparative Analysis. *American Sociological Review* 23:56–63.

DAVIS, ARTHUR K. (1948) 1952 Bureaucratic Patterns in the Navy Officer Corps. Pages 380–395 in Robert K. Merton et al. (editors), *Reader in Bureaucracy.* Glencoe, Ill.: Free Press. → First published in Volume 27 of *Social Forces.*

DOUGLASS, H. 1926 *1000 City Churches: Phases of Adaptation to Urban Environment.* New York: Doran.

DUNCAN, OTIS DUDLEY; and DAVIS, BEVERLY 1953 An Alternative to Ecological Correlation. *American Sociological Review* 18:665–666.

EVAN, WILLIAM M. 1963 Indices of the Hierarchical Structure of Industrial Organizations. *Management Science* 9:468–478.

FAUNCE, WILLIAM A. 1962 Size of Locals and Union Democracy. *American Journal of Sociology* 68:291–298.

FORM, WILLIAM H.; and NOSOW, SIGMUND N. 1958 *Community in Disaster.* New York: Harper.

GEORGOPOULOS, BASIL S.; and MANN, FLOYD C. 1962 *The Community General Hospital.* New York: Macmillan.

GOODMAN, LEO A. 1953 Ecological Regressions and Behavior of Individuals. *American Sociological Review* 18:663–664.

GOULDNER, ALVIN W. 1954 *Patterns of Industrial Bureaucracy.* Glencoe, Ill.: Free Press.

GROSS, NEAL; MASON, WARD S.; and McEACHERN, ALEXANDER W. 1958 *Explorations in Role Analysis: Studies of the School Superintendency Role.* New York: Wiley.

GUEST, ROBERT H. 1962 Managerial Succession in Complex Organizations. *American Journal of Sociology* 68:47–53.

HAIRE, MASON 1959 Biological Models and Empirical Histories of the Growth of Organizations. Pages 272–306 in Foundation for Research on Human Behavior, *Modern Organization Theory: A Symposium.* Edited by Mason Haire. New York: Wiley.

HALL, ROBERT L. 1955 Social Influences on the Aircraft Commander's Role. *American Sociological Review* 20: 292–299.

HALPIN, ANDREW W. 1956 *The Leadership Behavior of School Superintendents: The Perceptions and Expectations of Board Members, Staff Members and Superintendents.* Columbus: Ohio State Univ., College of Education.

HARBISON, FREDERICK H. et al. 1955 Steel Management on Two Continents. *Management Science* 2:31–39.

ILLINOIS, UNIVERSITY OF, INSTITUTE OF LABOR AND INDUSTRIAL RELATIONS 1954 *Labor–Management Relations in Illini City,* by William E. Chalmers et al. Champaign, Ill.: The Institute.

KAHN, ROBERT L.; and KATZ, DANIEL (1953) 1960 Leadership Practices in Relation to Productivity and Morale. Pages 554–570 in Dorwin Cartwright and Alvin F. Zander (editors), *Group Dynamics: Research and Theory.* 2d ed. Evanston, Ill.: Row, Peterson.

KAHN, ROBERT L. et al. 1964 *Organizational Stress: Studies in Role Conflict and Ambiguity.* New York: Wiley.

KATZ, DANIEL; and KAHN, ROBERT L. 1966 *The Social Psychology of Organizations.* New York: Wiley.

KATZ, DAVID; and HYMAN, HERBERT H. 1947 Morale in War Industries. Pages 437–448 in Society for the Psychological Study of Social Issues, *Readings in Social Psychology.* New York: Holt.

LAWRENCE, PAUL R. 1958 *The Changing of Organizational Behavior Patterns: A Case Study of Decentralization.* Boston: Harvard Univ., Graduate School of Business Administration, Division of Research.

LAZARSFELD, PAUL F.; and MENZEL, HERBERT 1961 On the Relation Between Individual and Collective Properties. Pages 422–440 in Amitai Etzioni (editor), *Complex Organizations: A Sociological Reader.* New York: Holt.

LAZARSFELD, PAUL F.; and THIELENS, WAGNER JR. 1958 *The Academic Mind: Social Scientists in a Time of Crisis.* A report of the Bureau of Applied Social Research, Columbia University. Glencoe, Ill.: Free Press.

LEVINE, SOL; and WHITE, PAUL E. 1961 Exchange as a Conceptual Framework for the Study of Interorganizational Relationships. *Administrative Science Quarterly* 5:583–601.

LIEBERMAN, SEYMOUR 1956 The Effects of Changes in Roles on the Attitudes of Role Occupants. *Human Relations* 9:385–402.

LIPSET, SEYMOUR M.; TROW, MARTIN A.; and COLEMAN, JAMES S. 1956 *Union Democracy: The Internal Politics of the International Typographical Union.* Glencoe, Ill.: Free Press.

LITWAK, EUGENE; and HYLTON, LYDIA F. 1962 Interorganizational Analysis: A Hypothesis on Co-ordinating Agencies. *Administrative Science Quarterly* 6: 395–420.

Lucci, York 1960 The YMCA on the Campus. Unpublished manuscript, Columbia Univ., Bureau of Applied Social Research.

McCleery, Richard H. (1957) 1961 Policy Change in Prison Management. Pages 376–400 in Amitai Etzioni (editor), *Complex Organizations: A Sociological Reader*. New York: Holt.

Mann, Floyd C.; and Hoffman, L. Richard 1960 *Automation and the Worker: A Study of Social Change in Power Plants*. New York: Holt.

Miller, Norman 1958 Social Class and Value Differences Among American College Students. Ph.D. dissertation, Columbia Univ.

Morse, Nancy C.; and Reimer, Everett 1956 The Experimental Change of a Major Organizational Variable. *Journal of Abnormal and Social Psychology* 52: 120–129.

Newcomb, Theodore M. (1943) 1957 *Personality and Social Change: Attitude Formation in a Student Community*. New York: Dryden.

►Powdermaker, Hortense 1968 Field Work. Volume 5, pages 418–424 in *International Encyclopedia of the Social Sciences*. Edited by David L. Sills. New York: Macmillan and Free Press.

Rashkis, Harold A.; and Wallace, Anthony F. C. 1959 The Reciprocal Effect: How Patient Disturbance Is Affected by Staff Attitudes. *Archives of General Psychiatry* 1:489–498.

Robinson, W. S. 1950 Ecological Correlations and the Behavior of Individuals. *American Sociological Review* 15:351–357.

Roethlisberger, Fritz J.; and Dickson, William J. (1939) 1961 *Management and the Worker: An Account of a Research Program Conducted by the Western Electric Company, Hawthorne Works, Chicago*. Cambridge, Mass.: Harvard Univ. Press. → A paperback edition was published in 1964 by Wiley.

►Schachter, Stanley 1968 Cohesion, Social. Volume 2, pages 542–546 in *International Encyclopedia of the Social Sciences*. Edited by David L. Sills. New York: Macmillan and Free Press.

Selvin, Hanan C. 1960 *The Effects of Leadership*. Glencoe, Ill.: Free Press.

Selznick, Philip 1949 *TVA and the Grass Roots: A Study in the Sociology of Formal Organization*. University of California Publications in Culture and Society, Vol. 3. Berkeley: Univ. of California Press.

Sills, David L. 1957 *The Volunteers: Means and Ends in a National Organization*. Glencoe, Ill.: Free Press.

►Simon, Herbert A. 1968 Administration: III. Administrative Behavior. Volume 1, pages 74–79 in *International Encyclopedia of the Social Sciences*. Edited by David L. Sills. New York: Macmillan and Free Press.

►Smith, M. Brewster 1968 Stouffer, Samuel A. Volume 15, pages 277–280 in *International Encyclopedia of the Social Sciences*. Edited by David L. Sills. New York: Macmillan and Free Press.

Speier, Hans 1950 *The American Soldier* and the Sociology of Military Organization. Pages 106–132 in Robert K. Merton and Paul F. Lazarsfeld (editors), *Continuities in Social Research*. Glencoe, Ill.: Free Press.

Stanton, Alfred H.; and Schwartz, M. S. 1954 *The Mental Hospital: A Study of Institutional Participation in Psychiatric Illness and Treatment*. New York: Basic Books.

Stinchcombe, Arthur L. 1961 On the Use of Matrix Algebra in the Analysis of Formal Organization. Pages 478–484 in Amitai Etzioni (editor), *Complex Organizations: A Sociological Reader*. New York: Holt.

Stogdill, Ralph M.; Scott, Ellis L.; and Jaynes, William E. 1956 *Leadership and Role Expectations*. Bureau of Business Research, Monograph No. 86. Columbus: Ohio State Univ., College of Commerce and Administration.

►Stokes, Donald E. 1968 Voting. Volume 16, pages 387–395 in *International Encyclopedia of the Social Sciences*. Edited by David L. Sills. New York: Macmillan and Free Press.

Stouffer, Samuel A. et al. 1949 *The American Soldier*. Studies in Social Psychology in World War II, Vols. 1 and 2. Princeton Univ. Press. → Volume 1: *Adjustment During Army Life*. Volume 2: *Combat and Its Aftermath*.

Tsouderos, John E. 1955 Organizational Change in Terms of a Series of Selected Variables. *American Journal of Sociology* 20:206–210.

Udy, Stanley H. Jr. 1959 The Structure of Authority in Non-industrial Production Organizations. *American Journal of Sociology* 64:582–584.

Udy, Stanley H. Jr. 1965 The Comparative Analysis of Organizations. Pages 678–709 in James G. March (editor), *Handbook of Organizations*. Chicago: Rand McNally.

Walker, Charles R.; and Guest, Robert H. 1952 *The Man on the Assembly Line*. Cambridge, Mass.: Harvard Univ. Press.

Wallace, Anthony F. C.; and Rashkis, Harold A. 1959 The Relation of Staff Consensus to Patient Disturbance in Mental Hospital Wards. *American Sociological Review* 24:829–835.

Wallace, Walter L. 1964 Institutional and Life-cycle Socialization of College Freshmen. *American Journal of Sociology* 70:303–318.

Wallace, Walter L. 1965 Peer Influences and Undergraduates' Aspirations for Graduate Study. *Sociology of Education* 38:375–392.

►Wax, Rosalie Hankey 1968 Observation: II. Participant Observation. Volume 11, pages 238–241 in *International Encyclopedia of the Social Sciences*. Edited by David L. Sills. New York: Macmillan and Free Press.

Weiss, Robert S. 1956 *Processes of Organization*. Ann Arbor: Univ. of Michigan, Institute for Social Research, Survey Research Center.

►Whyte, William F. 1968 Mayo, Elton. Volume 10, pages 82–83 in *International Encyclopedia of the Social Sciences*. Edited by David L. Sills. New York: Macmillan and Free Press.

Postscript

Organizational research methods have progressed particularly in the area of multivariate analysis of organizational characteristics. Use of institutional questionnaires permitted Blau and Schoenherr (1971) to measure theoretically relevant structural variables for large samples of bureaucracies. Multiple regression analysis, used with reasonable assumptions about time-order, permitted the isolation of the effects of size, division of labor, shape of organizational pyramid, and decentralization. Meyer (1972; 1975) gathered data over time in a panel of organizations and used cross-lagged regression analysis to clarify the causal direction of relationships. These methods permit much more powerful analysis than research limited to published data on organizations or to researcher-collected data on small samples. However, these studies have been restricted to structural characteristics, and they exclude variables that would require surveys of individual members such as managerial style, attitudes of various strata of members, mutual role definitions in role sets, or the sociometric structure of relationships. A study by Zablocki (1977) of 60 urban communal groups combines structural information with attitude and sociometric data from members, gathered at two points in time, permitting analysis of the mutual effects of all three types of variables. It does not appear that progress has been made in developing longitudinal versions of the classic study of mutual role expectations in role sets by Gross et al. (1958).

The study of comparable organizations in different societal contexts (for example, in conservative capitalist, welfare-capitalist, centralized socialist, and decentralized socialist systems) has only just begun. For example the measurement by Tannenbaum (1968) of the distribution of control over strata has been applied to Yugoslav "worker-controlled" firms, Israeli kibbutzim, and Italian, Austrian, and American enterprises (Rus 1972; see also Tannenbaum et al. 1974).

The systematic analysis of existing case studies is exemplified in the review by Price (1968) of 50 highly diverse studies to produce a propositional inventory on organizational effectiveness, and the coding by Gamson (1975) of historical data on a sample of 53 social protest groups.

Improvements in measurement of organizational characteristics are exemplified in the comparison by Pennings (1973) of institutional and survey data on a set of ten organizations, and in Price's *Handbook of Organizational Measurement* (1972), which presents operational measures of 22 organizational variables used in important studies.

Future development of organizational research still requires the mutual contributions of qualitative case studies and quantitative comparative and longitudinal research (Sieber 1973). Statistical methods for analyzing mutual effects of variables in longitudinal studies, rather than simple one-way causation, need to be applied (Duncan et al. 1971). Furthermore, a broadening of the cultural contexts of research to include more egalitarian societies and non-Western cultures offers the chance to develop less culture-bound theories.

ALLEN H. BARTON

ADDITIONAL BIBLIOGRAPHY

BLAU, PETER M.; and SCHOENHERR, RICHARD A. 1971 *The Structure of Organizations.* New York: Basic Books.

DUNCAN, OTIS DUDLEY; HALLER, A. O.; and PORTES, A. 1971 Peer Influence on Aspirations: A Reinterpretation. Pages 219–244 in Hubert M. Blalock, Jr. (editor), *Causal Models in the Social Sciences.* Chicago: Aldine-Atherton.

GAMSON, WILLIAM A. 1975 *The Strategy of Social Protest.* Homewood, Ill.: Dorsey.

MEYER, MARSHALL W. 1972 Size and the Structure of Organizations: A Causal Analysis. *American Sociological Review* 37:434–440.

MEYER, MARSHALL W. 1975 Leadership and Organizational Structure. *American Journal of Sociology* 81:515–542.

PENNINGS, JOHANNES 1973 Measures of Organizational Structure: A Methodological Note. *American Journal of Sociology* 79:686–704.

PRICE, JAMES L. 1968 *Organizational Effectiveness: An Inventory of Propositions.* Homewood, Ill.: Irwin.

PRICE, JAMES L. 1972 *Handbook of Organizational Measurement.* Lexington, Mass.: Heath.

RUS, VELJKO 1972 The Limits of Organized Participation. Volume 2, pages 165–188 in International Sociological Conference on Participation and Self-management, First, Dubrovnik, Yugoslavia, 1972, *Participation and Self-management: Report.* Institute for Social Research, Univ. of Zagreb.

SIEBER, SAM D. 1973 The Integration of Field Work and Survey Methods. *American Journal of Sociology* 78:1335–1359.

TANNENBAUM, ARNOLD S. (editor) 1968 *Control in Organizations.* New York: McGraw-Hill.

TANNENBAUM, ARNOLD S. et al. (editors) 1974 *Hierarchy in Organizations: An International Comparison.* San Francisco: Jossey-Bass.

ZABLOCKI, BENJAMIN 1977 *Alienation and Charisma: American Communitarian Experiments.* New York: Free Press.

OUTLIERS

See under STATISTICAL ANALYSIS, SPECIAL PROBLEMS OF.

P

PANEL STUDIES

The potentials of panel analysis were first developed at Columbia University under the aegis of Paul F. Lazarsfeld. The first major panel study carried out under the techniques pioneered by Lazarsfeld was a study of voter decision making during the 1940 presidential election campaign (Lazarsfeld et al. 1944). Since then there have been numerous panel studies in a variety of sociological, political, and economic areas. Panel studies have been made of other elections in the United States and other countries, of the socialization of medical students to professional norms, and of social climates in high schools.

The idea behind panel analysis is deceptively simple. Instead of comparing *aggregates* over time, panel analysis compares repeated observations of *individuals*. Insofar as the study of social change is limited to net changes in a social aggregate, the absence of such net changes is often assumed to be indicative of social stability. But constancy in the aggregate may obscure considerable compensatory change among individuals. For example, the distribution of income in a society may show no net change over a ten-year period, but various processes of economic mobility may be at work. In short, panel analysis gives rise to the study of an aspect of social change that tends to be neglected in studies of aggregate trends.

Origins. The technique of using repeated interviews with a constant sample of people was first attempted by Stuart Rice (1928) during the 1924 presidential election campaign. Rice's panel consisted of students of sociology in the three upper classes at Dartmouth College. His purpose was to find out how preferences for candidates shifted during the campaign. Though his research design included few variables, Rice's analysis was quite modern in that he did differentiate between net change and gross change. Theodore Newcomb (1943), in his classic study of Bennington College students, conducted repeated interviews for the four-year period 1935 to 1939. His main interest was in learning what effect a liberal college environment had upon the attitudes of girls coming from well-to-do, conservative families.

In both of these early efforts at panel analysis, the researchers were interested in identifying students who had changed attitudes and in discerning the reasons for the changes. Neither Rice nor Newcomb, however, developed an appropriate technique for handling repeated data. Newcomb described in detail the girls who showed a marked shift from a conservative to a liberal outlook during their stay at Bennington and, conversely, those who were impervious to the liberalism of the faculty. But although he collected a wealth of statistical information in his interviews with students, Newcomb tended to rely on qualitative analysis of the extreme cases of change.

The first systematic statement on the technique of panel analysis was made by Lazarsfeld and Fiske (1938). These researchers reported their experiences and problems with the new panel technique.

Applications. Formally, panel analysis is a research technique for collecting and analyzing data through repeated interviews with a sample of individuals in a natural, rather than a laboratory, setting. Before examining the characteristic modes of panel analysis, it is useful to compare panel analysis with other research techniques in social science that involve over-time data.

Econometrics, for example, makes use of quantitative time series data. Monthly, quarterly, or annual series may cover hundreds of time points, whereas panel studies seldom exceed six interviews. In econometrics, the data typically involve a single geographic or political entity. Though data may be collected from cross-sectional samples of individuals, families, or firms, the interest is usually in the aggregate dynamics of a single economic system. Panel studies typically range from 300 to 3,000 cases, and the interest is in the molecular changes of these hundreds or thousands of individuals.

In experimental psychology, learning studies involve longitudinal data covering hundreds of observations, but the focus is usually on the change in one criterion. The research tends to be highly experimental and theoretical, and involves the use of mathematical models. Panel analysis, on the other hand, is nonexperimental and descriptive. It emphasizes the interrelationships of many changing variables and is statistical, though not highly mathematical. In such panel studies, a set of interlocking variables may be investigated because they are thought to be theoretically fruitful, and there is considerable theoretical improvisation during the analysis of the data.

In educational research, follow-up studies involve over-time data. The interest, however, is not in periodically reobserving the same variables but in correlating a set of predictors with a criterion, as when a battery of aptitude tests is correlated with subsequent vocational success.

In social psychology, research design usually takes the form of controlled experiment involving before–after data obtained from subjects in a laboratory setting. The researchers decide which subjects will be exposed to what stimulus. In panel studies, by contrast, the researchers have no control over the individuals in the study, nor do they manipulate any stimuli. Experiments, unlike panels, are not intended to uncover new concepts or descriptively explore new terrain but, rather, to verify or nullify hypotheses dictated by psychological theory and formulated before the start of the experiments. However, a type of panel study known as the "impact" panel does parallel controlled experiments; the similarities between these two techniques will be considered at greater length.

The turnover table

The starting point, or central concept, of panel analysis is the turnover table, showing a categorical variable at time 1 cross-tabulated with itself at time 2. For a variable with n possible response categories, a turnover table is an $n \times n$ table summarizing the responses of each individual in the panel on a particular item at two successive interviews. In essence, such a table shows not only *net change* but also *gross change*.

Since the turnover table is basic, let us first examine the simplest case: a dichotomy ($n = 2$) observed at two time points. An example comes from a study by the Bureau of Labor Statistics, which in February 1963 carried out a nation-wide survey of men between the ages of 16 and 21 who were no longer enrolled in school. Two years later the bureau resurveyed them to analyze their early work experiences and problems. Table 1, which includes only those who were in the labor force at both times, shows the employment status of the youths at the two times.

From the totals, we observe that the employment rate for these youths rose from approximately 81 per cent to almost 91 per cent. Although there

Table 1 — Employment status in February 1963 and 1965

	FEBRUARY 1965		
FEBRUARY 1963	Employed	Unemployed	1963 totals
Employed	76.6% (1,628,000)	4.7% (101,000)	81.3% (1,729,000)
Unemployed	14.0% (298,000)	4.7% (99,000)	18.7% (397,000)
1965 totals	90.6% (1,926,000)	9.4% (200,000)	100.0% (N = 2,126,000)

Source: Perrella & Waldman 1966, table A.

Table 2 — Net change equals gross change

	FEBRUARY 1965		
FEBRUARY 1963	Employed	Unemployed	1963 totals
Employed	1,729,000	0	1,729,000
Unemployed	197,000	200,000	397,000
1965 totals	1,926,000	200,000	2,126,000

Source: Perrella & Waldman 1966, table A.

Table 3 — Gross change equals almost three times net change

	FEBRUARY 1965		
FEBRUARY 1963	Employed	Unemployed	1963 totals
Employed	1,529,000	200,000	1,729,000
Unemployed	397,000	0	397,000
1965 totals	1,926,000	200,000	2,126,000

Source: Perrella & Waldman 1966, table A.

was a 10 per cent net change, there were actually 19 per cent whose employment status changed from working to being jobless, or the reverse. The information that the rate rose from 81 to 91 per cent is not enough to predict how many individuals experienced a change in employment status. Table 2 shows one possible situation—where net change and gross change are identical. Table 3 shows a contrasting possibility—where the gross change is almost three times as much as the net change. Thus, tables 1–3, with the same aggregate employment trend, represent very different economic situations.

I will return to this example in more depth when I come to discuss qualifier analysis. Meanwhile, it is useful to examine more complex multicategoried turnover tables.

The data in Table 4 come from a panel study of the 1948 presidential election campaign between President Truman and Governor Dewey (Berelson et al. 1954). The study was carried out in Elmira, New York, which in 1948 was a predominantly Republican community. Before examining the interior cells, we first examine the trend, as shown in Table 5.

Thus, as the campaign progressed, instead of more would-be voters reaching a decision, both parties appear to have lost adherents, with a corresponding increase in the undecided category. This information could of course have been obtained from two polls of separate samples. When we examine the interior cells of the turnover table, we see that there was more than a 6 per cent change. Dividing all of the entries in Table 4 by the grand total, we can obtain the distribution of changer types (see Table 6).

The percentages on the diagonal running from upper left to lower right represent those whose vote intention remained constant, the total being 81.8 per cent. The gross change is thus about three times the net change. The individual changes, largely compensatory, can be classified into three types: (1) crystallizers (4.5 per cent), or those who were undecided but reached a decision in October; (2) waverers (10.7 per cent), or those who were decided but became doubtful in October; (3) converters (3 per cent), or those who switched from Republican to Democrat or from Democrat to Republican.

Turnover tables are also frequently examined by dividing the entries in each row by the row total; these percentages, which provide estimates of "transition probabilities," enable us to examine change, controlling for initial position (see Table 7).

Table 4 — Vote intention, August and October 1948

AUGUST VOTE INTENTION	OCTOBER VOTE INTENTION			August totals
	Republican	Undecided	Democrat	
Republican	369	54	17	440
Undecided	22	102	12	136
Democrat	6	27	151	184
October totals	397	183	180	760

Source: Adapted from Berelson et al. 1954, p. 23.

Table 5 — Trend in vote intention, August to October 1948 (per cent)

VOTE INTENTION	AUGUST	OCTOBER	NET CHANGE
Republican	57.9	52.2	−5.7
Undecided	17.9	24.1	+6.2
Democrat	24.2	23.7	−0.5

Source: Adapted from Berelson et al. 1954, p. 23.

Table 6 — Distribution of vote intention (per cent)

AUGUST VOTE INTENTION	OCTOBER VOTE INTENTION			August totals
	Republican	Undecided	Democrat	
Republican	48.5	7.2	2.2	57.9
Undecided	2.9	13.4	1.6	17.9
Democrat	0.8	3.5	19.9	24.2
October totals	52.2	24.1	23.7	100 = 760

Source: Adapted from Berelson et al. 1954, p. 23.

Table 7 — Changes in vote intention (per cent)

AUGUST VOTE INTENTION	OCTOBER VOTE INTENTION			Totals[*]	Number of cases
	Republican	Undecided	Democrat		
Republican	83.9	12.3	3.9	100	440
Undecided	16.2	75.0	8.8	100	136
Democrat	3.3	14.7	82.1	100	184

[*] Details may not add to totals because of rounding.

Source: Adapted from Berelson et al. 1954, p. 23.

To predict what decisions will be made by those who are undecided in August, we might advance one of the following four theories. (1) In a predominantly Republican community, those who are undecided might tend to lean toward the Democrats, since those leaning toward the Republicans would not hesitate to say so. (2) There are presumably equal forces pulling the undecided toward the Republican camp and toward the Democratic camp. Hence, those who do decide for whom to vote will split their votes 50–50 for Republicans and Democrats. (3) The undecided will vote in the same proportion as the rest of the community—2.4 Republicans to 1 Democrat. (4) Each undecided will be exposed, on the average, to 2.4 Republican stimuli to 1 Democratic stimulus and will vote for Republicans in most cases—or, at any rate, having no better criterion, most of the undecided will vote with the majority. Conjectures of this kind can best be tested by panel studies.

In Table 7, the transition probabilities on the main diagonal show that the undecided group is the most volatile: 16.2 per cent switched to Republican and 8.8 per cent switched to Democrat, distributing their votes to Republican and Democrat in a ratio of 1.8 to 1. This is less than the August ratio of Republicans to Democrats (440/184 = 2.4); it is also less than the October ratio (397/180 = 2.2).

Examination of Table 4 revealed that of 138 individual changes, all but 23 involved either switching from undecided to a party preference or from a party preference to undecided. Analysis of turnover using the Republican–Undecided–Democrat trichotomy, as in Table 7, tends to focus on the voters who are least involved politically. As was shown by an earlier panel study of the 1940 presidential campaign (Lazarsfeld et al. 1944), the so-called independent voter who listened to both candidates and judiciously weighed the soundness of their respective programs before deciding on his vote was largely a mythical being. For the most part, those who oscillated from either Republican or Democrat to indecision, or who did not decide whom they would vote for until late in the campaign, were the politically uninterested who often were persuaded to join one or another political camp for the flimsiest of reasons. Most of the changers were politically uninvolved and uninformed. Their shifts were more easily explained by social determinants than by political considerations. As a result of these findings, some political scientists charged the authors with supplanting traditional concepts of the political process with a type of sociological determinism. Accordingly, when the 1940 election study was replicated in 1948, the researchers sought a deeper analysis of the decision-making processes of politically involved voters. By combining the variable of party preference with that of strength of party preference, a five-point vote intention scale was constructed. The turnover table with this as the criterion shows much more of the dynamics of the campaign than the trichotomous turnover tables above (see Table 8).

Although a moderate Republican's vote counts as much at the ballot box as a strong Republican's vote, this analytic device of transforming the vote

Table 8 — Change of vote intention from August to October (per cent)

VOTE INTENTION IN AUGUST 1948	VOTE INTENTION IN OCTOBER 1948					August totals	
	R+	R	?	D	D+	Per cent*	Number of cases
Strong Republican	75	13	10	2	1	100	241
Moderate Republican	32	47	16	4	2	100	199
Undecided	6	10	75	7	2	100	136
Moderate Democrat	0	4	16	59	21	100	122
Strong Democrat	1	0	13	15	71	100	62
October totals (number of cases)	254	143	183	102	78		760

* Details may not add to totals because of rounding.

Source: Adapted from Berelson et al. 1954, p. 23.

intention into a five-point scale allowed study of intraparty change and, hence, more study of the political issues. One further feature of the table is worth noting. The percentages on the main diagonal indicate for each political position the proportion who remained constant. Note that the moderate Republicans and the moderate Democrats are the least constant. They tend either to become more partisan or to become undecided. Thus, as the campaign progresses, there is an increase in the number of undecided, together with an increase in partisanship. In August, the ratio of strong to moderate Republicans is 241/199 = 1.2; by October the ratio rises to 1.8. Among the Democrats, the ratio is 0.51 in August and rises to 0.76 in October. It might be supposed that a minority which is outnumbered 2.4 to 1 would show intense partisanship, but this is clearly not the case here.

Qualifiers

Turnover tables, by showing all of the gross change with respect to some criterion, serve the function of raising questions about the different types of changers. What are their characteristics, the processes that impel them to change, and the psychological and social effects of different patterns of change? In order to answer such questions, researchers have stratified turnover tables by other variables, which are sometimes called qualifiers. In panel studies, analysis by means of qualifiers constitutes the great bulk of the work performed.

Qualifiers can be classified according to whether they are changing or constant. Constant qualifiers can be further classified by precedence: those which occur prior to the first interview are called antecedent qualifiers, and those which occur between interviews 1 and 2 are called intervening qualifiers.

The constant qualifiers that precede interview 1

are the conventional demographic characteristics, such as age, education, nationality, sex, and social class. Of course, in some instances such characteristics might change; whether or not they are categorized as constant depends on the interval between interviews in the particular study. An individual can change his marital status or primary group attachments or socioeconomic status when the period between interviews is sufficiently long. The function of such qualifiers is to elaborate the original turnover table by showing the conditions under which there is more or less change, just as in survey analysis the researcher starts with the association between two variables and introduces other variables to find conditions under which the original relationship is heightened or diminished.

Tables 1–3 above showed changes in employment status over a two-year period. In order to learn more about the effects of different characteristics on employability, the same researchers examined the employment turnover by the educational level of the youths. The results are shown in tables 9 and 10.

The trend is the same for both groups. Dropouts increased from 74 per cent employed to 84 per cent, and high school graduates increased from 88 per cent employed to 97 per cent. The transition probabilities reveal more of the situation. Among graduates who were employed at the time of the first survey, only 1.6 per cent were jobless at the time of the resurvey, but of those who were unemployed at the time of the first survey, 11.4 per

cent were still unemployed at the time of the resurvey. Among the dropouts, of those employed at the time of the first survey, 11.2 per cent were jobless at the time of the resurvey, whereas of those dropouts who were unemployed at the time of the first survey, fully 31.7 per cent were jobless at the time of the resurvey. These findings strongly suggest that there exists a hard-core group of jobless youth. It should also be noted that 26 per cent of the dropouts underwent some change in employment, compared with 12 per cent of the graduates.

To discover why some dropouts were employed at both times while some graduates were unemployed at both times, we might qualify the employment turnover table by *race*. There are income statistics which show that there is a considerable gap between the earnings of whites who did not graduate from high school and nonwhite high school graduates. In the course of a lifetime the average nonwhite high school graduate can expect to earn approximately $50,000 *less* than the white dropout. It would not be surprising to find that race accounts for some of the anomalous cases.

We might study other qualifiers. For example, among the graduates, did those who pursued a strictly academic program do better or worse than those who pursued a business education or vocational program? Were married youths more likely to hold their jobs? They might be more diligent or more motivated to work to pay for furniture or to accumulate savings for a prospective family. And there might be a tendency among employers to lay off unmarried workers first when work slackens. Undoubtedly, employment and marital status are related. In fact, more dropouts than graduates were unmarried at the time of the first survey. If we were interested in the social effects of employment, we could use marital status as the criterion and examine turnover of marital status as qualified by employment. Were those who were employed and unmarried in 1963 more likely to become married by 1965 than those who were unemployed? And were those unemployed and married in 1963 more likely than the employed to become separated or divorced? We could go further and analyze the joint change of employment and marital status, which, with dichotomous variables, would involve analysis of a sixteenfold table.

Impact panels. Frequently the panel technique is used for evaluating the effects of information campaigns. A public health agency, for example, might be interested in learning the effects of an information campaign on the recognition and reporting of cancer symptoms; the federal government might want to determine the effectiveness

Table 9 — Employment status of high school graduates (1963) in 1963 and 1965 (per cent)

	FEBRUARY 1965		1963 totals	
FEBRUARY 1963	Employed	Unemployed	Per cent	Number of cases
Employed	98.4	1.6	100	963,000
Unemployed	88.6	11.4	100	132,000
1965 totals	1,065,000	30,000		1,095,000

Source: Adapted from Perrella & Waldman 1966, table A.

Table 10 — Employment status of high school dropouts (1963) in 1963 and 1965 (per cent)

	FEBRUARY 1965		1963 totals	
FEBRUARY 1963	Employed	Unemployed	Per cent	Number of cases
Employed	88.8	11.2	100	766,000
Unemployed	68.3	31.7	100	265,000
1965 totals	861,000	170,000		1,031,000

Source: Adapted from Perrella & Waldman 1966, table A.

of a consumer education program among the poor; a state commission might want to determine the effectiveness of a campaign aimed at reducing discrimination against minority groups. These questions and others like them can best be answered by a type of panel study called an impact panel, in which the qualifiers intervene between the first and second series of interviews and refer to exposure to some stimulus or succession of stimuli. The impact study thus represents one equivalent of the controlled experiment when observations must be made in a natural social setting and the individuals being studied cannot be randomly assigned to experimental or control groups. [*See* EXPERIMENTAL DESIGN, *article on* QUASI-EXPERIMENTAL DESIGN; *see also* EVALUATION RESEARCH.]

One general observation should be made concerning the place of theory in experiments as compared with panel studies. In experiments virtually all of the theoretical thinking must be done in advance of the field work. In panel studies, however, theorizing takes place at both ends of the research process. A small set of interlocking variables must be decided upon at the outset; these variables determine the boundaries of the analysis. Even the most empirically oriented researcher must have some implicit theory of the relative fruitfulness of different variables. After the data have been coded and put on punch cards or tape, theory serves an organizing function, since without some notion about which tabulations to run, it would be possible to keep a computer busy for months before exhausting the astronomical number of possible cross tabulations.

Some problems connected with impact studies may be worth pointing out. The experiment seeks to measure the effects of a stimulus when the individual is exposed to it. The panel, on the other hand, seeks to explain what happens under nonlaboratory conditions; it aims to assess effects of exposure under conditions where audience self-selection is operative. Under experimental conditions a captive audience is exposed to some prepared stimulus, such as a lecture, and often shows significant before–after differences in information or attitude. Usually, the shift is among the least educated and least interested individuals. When the same program is transferred from the laboratory to a mass medium, a study of effects frequently reveals little or no change: those who are least educated, least interested, and least in agreement with the communication rarely expose themselves to the stimulus in the first place. The panel, unlike the experiment, deals with a noncaptive audience; hence, not only the effects of exposure but also the nature of the audience self-selection are studied (see Janis 1968; Klapper 1968).

Determination of exposure to mass media or to personal influence is difficult, for it generally involves the use of retrospective questions. This, of course, violates the central conception of panel analysis, because we cannot be sure to what extent selective recall is operative. For example, in the study of political behavior during a national election campaign, it would be important to assess the role of local political parties. What, for instance, is the effect of receiving campaign literature? What influence does a personal visit by a party worker have on the turnout and vote intentions of potential voters? But if we ask respondents whether they received literature or were contacted by a party worker, how can we be sure that we are not dealing with memory biases—that is, the voters were contacted randomly by workers but those interested in politics tended to remember the contact? If responses to the queries on being visited or on receiving political literature are analyzed, we do find that the more interested respondents report greater personal and impersonal contact.

To some extent this result may reflect selective contact. Party workers may have approached would-be voters who seemed more receptive to political discussion, or they may have selected names of contacts from previous voting lists or made their contacts in various other nonrandom ways.

One safeguard that can be built into panel studies to check against the bias of selective recall is stratification. When effects of party contact are analyzed, for instance, it is necessary to control by level of political interest. Another safeguard is to try to measure exposure independently by finding out from party workers whom they contacted.

Inherent in panel study design is the possibility of examining the differential processes through which the stimulus is related to its effects. Experiments ordinarily do not permit such detailed study, for the simple reason that the number of cases is usually not sufficient, although there are occasional exceptions. In panel studies, more refined analyses of the relations between the stimulus and its effects are carried out by determining the characteristics of those respondents who are most influenced by the stimulus—whether they are men or women, old or young, educated or uneducated, and so on. The stimulus may have had a marked effect in the desired direction on certain groups of respondents and no effect or a "boomerang" effect on other groups.

Particularly important in this respect is the consideration of effects in terms of the respondents' *initial position on the criterion.* In the impact panel study, unlike the experimental study, the respondents who have been exposed to the stimulus are quite likely to differ from unexposed respondents on the criterion prior to the exposure. If they do, it is necessary, in order to impute any effect to the stimulus, to control the criterion prior to the campaign, for there is every reason to expect that the effects of a stimulus will not be the same for respondents who are initially favorable, indifferent, or unfavorable on the criterion. By controlling for initial position on the criterion, we do not altogether approximate experimental matching, although we do reduce its urgency. By holding initial position constant, the subgroups to be compared are made more homogeneous on everything except exposure to the stimulus.

For other material bearing on impact panels and controlled experiments, see the discussions by Hovland (1959) and Cohen (1964, chapter 9).

Mutual effects

The most interesting part of panel analysis is the analysis of mutually interacting variables. Some examples of two interacting variables are the following. (1) In the investigation of psychosomatic disorders we find that worry and anxiety produce hypertension or ulcers, and these disabilities in turn raise the level of anxiety. (2) Studies have been carried out relating chronic illness with poverty. Chronic illness drains a family's resources, while low socioeconomic status reduces access to appropriate medical care, proper diet, and a congenial working environment. (3) In studies of family life, we often find that problem children and lack of parental love tend to be associated. But which causes which? Does lack of parental love induce aberrant behavior, or does aberrant behavior corrode parental affection? It seems likely that each reinforces the other. (4) The social psychologist may find that friendship and similarity in values feed into each other. People tend to select friends among those with similar values, and values tend to become convergent in the course of friendship. (5) The sociologist may observe that individuals who reveal greater conformity to company goals tend to have higher rates of promotion. Correspondingly, promotion tends to reinforce conformity with company goals. In each example, one cannot say which variable is the dependent variable and which the independent; each influences the other. The problem in research is to analyze their relative influence. Let us look at an example of the analysis of relative influence.

Suppose there is to be a referendum in some community on the issue of fluoridating the water, and we are studying how people influence each other on public health matters. We twice ask a panel of married couples: "As things stand now, how do you intend to vote on this issue?" The resultant sixteenfold table (Table 11) shows the hypothetical joint distribution of vote intention at two interviews of 671 hypothetical couples. Assuming these figures were empirical, what could we infer from them about the relative influence of the spouses?

Consider, on the one hand, those couples who at time 1 were in agreement but at time 2 were in disagreement. Altogether there were $b + c + n + o = 1 + 4 + 8 + 2 = 15$ such changes. In 12 of them,

Table 11 — Opinion of spouses at successive interviews on fluoridation issue*

HUSBAND	WIFE	TIME 2 Favor Husband — Favor	TIME 2 Favor Husband — Oppose	TIME 2 Oppose Husband — Favor	TIME 2 Oppose Husband — Oppose	Time 1 totals
TIME 1 Favor	Favor	*a* 300	*b* 1	*c* 4	*d* 3	308
Favor	Oppose	*e* 25	*f* 40	*g* 1	*h* 10	76
Oppose	Favor	*i* 5	*j* 2	*k* 45	*l* 25	77
Oppose	Oppose	*m* 3	*n* 8	*o* 2	*p* 197	210
Time 2 totals		333	51	52	235	671

* Hypothetical data.

it was the husband who "defected," so to speak, while in only 3 did the wife "defect." Thus, as far as preserving agreement between the spouses, the husband was about four times more influential than the wife.

Consider, on the other hand, the couples who were initially in disagreement but as a consequence of one spouse changing opinion were subsequently in agreement. Altogether there were $e + i + h + l = 25 + 5 + 10 + 25 = 65$ such cases. To what extent did each spouse influence the other to change in accord with his own opinion? Of the 65 cases, 50 agreements were generated as a result of the wife changing her opinion to accord with her husband's; only 15 agreements resulted from the husband changing to harmonize with his wife's opinion. Thus, as far as generating agreement, the husband was more than three times as influential as the wife. From the four agreement-to-disagreement cells of the sixteenfold table and the four disagreement-to-agreement cells, an index of relative influence might be devised. During the last twenty years, Lazarsfeld has proposed several indices of relative influence, all based on the single-change cells of the sixteenfold table. If the eight single-change cells are labeled as in Table 11, one of Lazarsfeld's indices is the following:

$$I_{H,W} = \left(\frac{e}{e+b} - \frac{o}{o+l} \right) - \left(\frac{i}{i+c} - \frac{n}{n+h} \right).$$

The terms in the first pair of parentheses represent H's influence on W; the terms in the second pair of parentheses represent W's influence on H.

In the hypothetical husband–wife sixteenfold table (Table 11),

$$I_{H,W} = \left(\frac{25}{26} - \frac{2}{27} \right) - \left(\frac{5}{9} - \frac{8}{18} \right) = .78$$

(in favor of husbands).

A second method of measuring relative influence is based on the relative magnitude of the cross-lagged correlations: $r_{H_1W_2}$ versus $r_{W_1H_2}$ (Campbell 1963; Pelz & Andrews 1964). If Yule's Q is used as a measure of correlation, the cross-lagged correlations in the husband–wife example are $Q_{H_1W_2} = .93 > Q_{W_1H_2} = .83$. Since $Q_{H_1W_2} > Q_{W_1H_2}$, the inference would be that the husband is more influential. Pelz and Andrews (1964) also consider as a measure the relative magnitude of the partial cross-lagged correlations: $r_{H_1W_2 \cdot W_1}$ versus $r_{W_1H_2 \cdot H_1}$. Generally, the Lazarsfeld index, the simple cross-lagged correlations, and the partial cross-lagged correlations will agree as far as indicating which of two variables is more influential. [*For a discussion of*

Yule's Q, see STATISTICS, DESCRIPTIVE, *article on* ASSOCIATION.]

In conclusion, it should be stated that much of the potential value of panel analysis for sociological inquiry remains untapped. The analysis of mutual effects has been directed almost entirely at two-variable interactions. The burgeoning literature on causal analysis in recent years attests to the readiness of the field to move beyond this. Moreover, panel methodology has developed around change of individuals, principally opinion change. Problems in which the unit of analysis is not the isolated individual, but individuals interlocked sociometrically or in a formal organization, have rarely been the subject of panel research.

BERNARD LEVENSON

[*See also* COUNTED DATA; ERRORS, *article on* NONSAMPLING ERRORS; STATISTICS, DESCRIPTIVE, *article on* ASSOCIATION; SURVEY ANALYSIS; *and the biography of* RICE; Kadushin 1968; Katz 1968.]

BIBLIOGRAPHY

ANDERSON, T. W. 1954 Probability Models for Analyzing Time Changes in Attitudes. Pages 17–66 in Paul F. Lazarsfeld (editor), *Mathematical Thinking in the Social Sciences.* Glencoe, Ill.: Free Press.

BERELSON, BERNARD; LAZARSFELD, PAUL F.; and McPHEE, WILLIAM N. 1954 *Voting: A Study of Opinion Formation in a Presidential Campaign.* Univ. of Chicago Press. → Tables 2, 3, 4, and 5 are based on data from this book, by permission of the authors and the University of Chicago Press. Copyright 1954 by the University of Chicago.

BLUMEN, ISADORE; KOGAN, MARVIN; and McCARTHY, PHILIP 1955 *The Industrial Mobility of Labor as a Probability Process.* Cornell Studies in Industrial and Labor Relations, Vol. 6. Ithaca, N.Y.: Cornell Univ. Press.

CAMPBELL, DONALD T. 1963 From Description to Experimentation: Interpreting Trends as Quasi-experiments. Pages 212–242 in Chester W. Harris (editor), *Problems in Measuring Change.* Madison: Univ. of Wisconsin Press.

COHEN, ARTHUR R. 1964 *Attitude Change and Social Influence.* New York and London: Basic Books.

COLEMAN, JAMES S. 1964a *Introduction to Mathematical Sociology.* New York: Free Press. → See especially pages 132–188, "Relations Between Attributes: Over-time Data."

COLEMAN, JAMES S. 1964b *Models of Change and Response Uncertainty.* Englewood Cliffs, N.J.: Prentice-Hall.

GLOCK, CHARLES Y. (1951) 1955 Some Applications of the Panel Method to the Study of Change. Pages 242–250 in Paul F. Lazarsfeld and Morris Rosenberg (editors), *The Language of Social Research: A Reader in the Methodology of Social Research.* Glencoe, Ill.: Free Press.

GLOCK, CHARLES Y. 1952 Participation Bias and Reinterview Effect in Panel Studies. Ph.D. dissertation, Columbia Univ.

GOODMAN, LEO A. 1962 Statistical Methods for Analyzing Processes of Change. *American Journal of Sociology* 68:57–78.

GOODMAN, LEO A. 1965 On the Statistical Analysis of Mobility Tables. *American Journal of Sociology* 70: 564–585.

HOVLAND, CARL I. 1959 Reconciling Conflicting Results Derived From Experimental and Survey Studies of Attitude Change. *American Psychologist* 14:8–17.

►JANIS, IRVING L. 1968 Persuasion. Volume 12, pages 55–65 in *International Encyclopedia of the Social Sciences.* Edited by David L. Sills. New York: Macmillan and Free Press.

►KADUSHIN, CHARLES 1968 Reason Analysis. Volume 13, pages 338–343 in *International Encyclopedia of the Social Sciences.* Edited by David L. Sills. New York: Macmillan and Free Press.

KATONA, GEORGE 1958 Attitude Change: Instability Response and Acquisition of Experience. *Psychological Monographs* 72, no. 10.

►KATZ, ELIHU 1968 Diffusion: III. Interpersonal Influence. Volume 4, pages 178–185 in *International Encyclopedia of the Social Sciences.* Edited by David L. Sills. New York: Macmillan and Free Press.

KENDALL, PATRICIA L. 1954 *Conflict and Mood: Factors Affecting Stability of Response.* Glencoe, Ill.: Free Press.

►KLAPPER, JOSEPH T. 1968 Communication, Mass: V. Effects. Volume 3, pages 81–90 in *International Encyclopedia of the Social Sciences.* Edited by David L. Sills. New York: Macmillan and Free Press.

LAZARSFELD, PAUL F.; BERELSON, BERNARD; and GAUDET, HAZEL (1944) 1960 *The People's Choice: How the Voter Makes Up His Mind in a Presidential Campaign.* 2d ed. New York: Columbia Univ. Press.

LAZARSFELD, PAUL F.; and FISKE, MARJORIE 1938 The "Panel" as a New Tool for Measuring Opinion. *Public Opinion Quarterly* 2:596–612.

LIPSET, SEYMOUR M. et al. (1954) 1959 The Psychology of Voting: An Analysis of Political Behavior. Volume 2, pages 1124–1175 in Gardner Lindzey (editor), *Handbook of Social Psychology.* Cambridge, Mass.: Addison-Wesley.

McDILL, EDWARD L.; and COLEMAN, JAMES S. 1963 High School Social Status, College Plans, and Academic Achievement: A Panel Analysis. *American Sociological Review* 28:905–918.

NEWCOMB, THEODORE M. (1943) 1957 *Personality and Social Change: Attitude Formation in a Student Community.* New York: Dryden.

PELZ, DONALD C.; and ANDREWS, F. M. 1964 Detecting Causal Priorities in Panel Study Data. *American Sociological Review* 29:836–848.

PERRELLA, VERA C.; and WALDMAN, ELIZABETH 1966 *Out-of-school Youth: Two Years Later.* U.S. Bureau of Labor Statistics, Division 7, Labor Force Studies, Special Labor Force Report No. 71. Washington: Government Printing Office.

RICE, STUART A. 1928 *Quantitative Methods in Politics.* New York: Knopf.

UNDERHILL, RALPH 1966 Values and Post-college Career Change. *American Journal of Sociology* 72:163–172.

WIGGINS, LEE M. 1955 Mathematical Models for the Interpretation of Attitude and Behavior Change: The Analysis of Multi-wave Panels. Ph.D. dissertation, Columbia Univ.

ZEISEL, HANS (1947) 1957 *Say It With Figures.* 4th ed., rev. New York: Harper. → See especially pages 215–254, "The Panel."

PARETO COEFFICIENT
See SIZE DISTRIBUTIONS IN ECONOMICS.

PARTIAL CORRELATION
See MULTIVARIATE ANALYSIS, *article on* CORRELATION METHODS; STATISTICS, DESCRIPTIVE, *article on* ASSOCIATION.

PATH ANALYSIS
See CAUSATION; LINEAR HYPOTHESES, *article on* REGRESSION; MULTIVARIATE ANALYSIS, *article on* CORRELATION METHODS; PREDICTION; SOCIAL MOBILITY; SURVEY ANALYSIS, *article on* METHODS OF SURVEY ANALYSIS.

PATTERN ANALYSIS
See CLUSTERING.

PEARSON, KARL

Karl Pearson, "founder of the science of statistics," was born in London in 1857 and died at Coldharbour in Surrey, England, in 1936.

Pearson's father, William Pearson, was a barrister, Queen's Counsel, and a leader in the chancery courts. He was a man of great ability, with exceptional mental and physical energy and a keen interest in historical research, traits which his son also exhibited.

An incident from Pearson's infancy, which Julia Bell, his collaborator, once related, contains in miniature many of the characteristics which marked his later life. She had asked him what was the first thing he could remember. He recalled that it was sitting in a highchair and sucking his thumb. Someone told him to stop sucking it and added that unless he did so, the thumb would wither away. He put his two thumbs together and looked at them a long time. "They look alike to me," he said to himself. "I can't see that the thumb I suck is any smaller than the other. I wonder if she could be lying to me." Here in this simple anecdote we have rejection of constituted authority, appeal to empirical evidence, faith in his own interpretation of the meaning of observed data, and, finally, imputation of moral obliquity to a person whose judgment differed from his own. These characteristics were prominent throughout his entire career. (The chief source of information about Pearson's early life is a 170-page memoir written immediately after his death by his son, Egon S. Pearson [1938], also a distinguished statistician.)

In Pearson's early educational history there are indications of a phenomenal range of interests, unusual intellectual vigor, delight in controversy, the determination to resist anything which he considered misdirected authority, an appreciation of scholarship, and the urge to self-expression, but there is almost no suggestion of any special leaning toward those studies for which he is now chiefly remembered.

In 1866 he was sent to University College School, London, but after a few years was withdrawn for reasons of health. At the age of 18 he obtained a scholarship at King's College, Cambridge, being placed second on the list.

In an autobiographical note entitled "Old Tripos Days at Cambridge" (1936), published shortly before his death, he wrote of those undergraduate days as some of the happiest of his life. "There was pleasure in the friendships, there was pleasure in the fights, there was pleasure in the coaches' teaching, there was pleasure in searching for new lights as well in mathematics as in philosophy and religion." His tutor was Edward J. Routh, considered by some the most successful tutor in the history of Cambridge, a man for whom he developed a real affection. Pearson used to speak of the stimulation received from his mathematics teachers, Routh, Burnside, and Frost, and described contacts with other distinguished persons. He gave an amusing account of an examination held on four days in the homes of the four examiners: George Gabriel Stokes, whom he venerated as the greatest mathematical physicist in England and one of the two best lecturers he had ever known; James Clerk Maxwell, another great physicist but a poor lecturer; Arthur Cayley, lawyer and mathematician, inventor of the theory of matrices and the geometry of *n*-dimensional space; and Isaac Todhunter, who had by that time published his *History of the Mathematical Theory of Probability*. The examination paper set by Todhunter provided a turning point in Pearson's career. A demonstration which he submitted in this examination was attached, with an approving comment, by Todhunter to the unfinished manuscript of his *History of the Theory of Elasticity* (1886–1893). After Todhunter's death Pearson was invited to finish and edit this *History*. This task was the beginning of his vital association with the Cambridge University Press, whose proofs were, for the next half century, rarely absent from his writing table.

Besides mathematics and the theory of elasticity, his interests during his Cambridge years included philosophy, especially that of Spinoza; the works of Goethe, Dante, and Rousseau, which he read in the original; the history of religious thought; and a search for a concept of the Deity that would be consistent with what he knew of science. Deeply concerned with religion but resenting coercion, he challenged the university authorities, first by his refusal to continue attendance at compulsory divinity lectures and then by his objection to compulsory chapel. He won both fights, and the university regulations were altered, but he continued to attend chapel on a voluntary basis.

After taking his degree with mathematical honors at Cambridge in 1879, he read law at Lincoln's Inn and was called to the bar in 1881. There followed travel in Germany and a period of study at the universities of Heidelberg and Berlin, where he balanced the study of physics with that of metaphysics; of Roman law with the history of the Reformation; of German folklore with socialism and Darwinism. After returning to England he was soon lecturing and writing on German social life and thought, on Martin Luther, Karl Marx, Maimonides, and Spinoza, contributing hymns to the Socialist Song Book, writing papers in the field of elasticity, teaching mathematics in King's College, London, and engaging in literary duels with Matthew Arnold and the librarians of the British Museum.

One of the friendships which had a deep influence on his life was with Henry Bradshaw, librarian of Cambridge University, to whom Pearson referred in "Old Tripos Days" as "the man who most influenced our generation." In a speech, he described Bradshaw as "the ideal librarian, but something greater—the guide of the young and foolish" and added that the librarian showed him what the essentials of true workmanship must be. So deep was their friendship that Bradshaw could reprove the younger man for excessive ardor and lack of wisdom in intellectual controversy.

His first publication, at 23 years of age, was a little book, which must have been largely autobiographical, entitled *The New Werther* (1880), written in the form of letters from a young man named "Arthur" to his fiancée. It foreshadows *The Ethic of Freethought* (1888) and *The Grammar of Science* (1892). Arthur writes:

I rush from science to philosophy, and from philosophy to our old friends the poets; and then, overwearied by too much idealism, I fancy I become practical in returning to science. Have you ever attempted to conceive all there is in the world worth knowing—that not one subject in the universe is unworthy of study? The giants of literature, the mysteries of many-dimensional space, the attempts of Boltzmann and Crookes to penetrate Nature's very laboratory, the Kantian theory of the universe, and the latest discoveries in embryology, with their wonderful tales of the development of life—what an immensity beyond our grasp! . . . Mankind seems on the verge of a new and glorious discovery. What Newton did to simplify the various planetary motions must now be done to unite in one whole the various isolated theories of mathematical physics.   (Quoted in Egon S. Pearson, 1938, p. 8)

All that the young writer wanted was a complete understanding of the universe. Thirty years later, the first issue of *Biometrika* carried as its frontispiece a picture of a statue of Charles Darwin with the words: *Ignoramus, in hoc signo laboremus.* Those five words ring out like the basic theme of Pearson's life: "We are ignorant; so let us work."

In 1884 Pearson became professor of applied mathematics and mechanics at University College, teaching mathematics to engineering students as well as courses on geometry. He occupied himself for the next few years with writing papers on elasticity, completing Todhunter's *History of the Theory of Elasticity,* lecturing on socialism and free thought, publishing *The Ethic of Freethought,* writing in German (*Die Fronica* 1887) a historical study of the Veronica legends concerning pictures of Christ, collecting material on the German passion play which later formed the substance of *The Chances of Death* (1897), completing a book called *The Common Sense of the Exact Sciences* (see 1885), which had been begun by W. K. Clifford, and taking an active part in a small club whose avowed purpose was to break down the conventional barriers which prevented free discussion of the relations between men and women.

In 1890 Pearson was invited to lecture on geometry at Gresham College, with freedom to choose the subject matter on which he would lecture. In March 1891 he delivered his first course of four lectures on "The Scope and Concepts of Modern Science." In 1892 he published the first edition of *The Grammar of Science,* and in 1893 he wrote an article on asymmetrical frequency curves (1894). It is apparent that a very important change had taken place in his concepts of scientific method, that he had reached a new conviction about the statistical aspects of the foundations of knowledge, that problems of heredity and evolution had acquired a new urgency, and in short, that a dramatic change had taken place in his professional life.

Influences on Pearson's thinking. In 1890 W. F. R. Weldon was appointed to the chair of biology at University College. Weldon was already acquainted with Francis Galton and engaged in statistical research. In 1890 Weldon published a paper on variations in shrimp, in 1892 one on correlated variations, and in 1893 a third paper which contains the sentence "It cannot be too strongly urged that the problem of animal evolution is essentially a statistical problem." In 1893 that point of view was heresy. The importance for science of the intense personal friendship which soon sprang up between Pearson and Weldon, then both in their early thirties, can scarcely be exaggerated. Weldon asked the questions that drove Pearson to some of his most significant contributions. Weldon's sudden death from pneumonia at the age of 46 was a heavy blow to science and a great personal tragedy to Pearson.

In 1889 Galton, then 67 years old, published *Natural Inheritance,* summarizing his researches between 1877 and 1885 on the subject of regression. This work moved Weldon to undertake his studies of regression in biological populations and moved Pearson to arithmetical researches that culminated in 1897 with the famous product-moment correlation coefficient r. The elaboration of correlations led, in other hands than Pearson's, to such diverse statistical inventions as factor analysis and the analysis of variance. The stimulation Pearson received from Galton and the devotion he felt toward the older man show on every page of the four volumes of *The Life, Letters and Labours of Francis Galton* (1914–1930), one of the world's great biographies. On this work of over 1,300 quarto pages and about 170 full-page plates, Pearson lavished some twenty years of work and much of his personal fortune.

Development of a science of statistics. The year 1890 represented not only a turning point in Pearson's career; it marked the beginning of the science of statistics. Antedating this development and preparing the way for it had been a long period of slowly increasing interest in the statistical way of thinking. In 1890 this interest was still sporadic, restricted in scope, and shared by very few people. It exhibited itself primarily in the collection of such public statistics as population data, vital statistics, and economic data. It was also evident in actuarial work and in the adjustment of observations in astronomy and meteorology, particularly least squares adjustment. Outside these areas, this development was hampered not only by lack of interest but also by paucity of data and the absence of adequate theory. Statistical theory was almost entirely that which had been developed by the great astronomers and mathematicians concerned with mathematical probability related to errors of observation. It related chiefly to the binomial distribution or the normal distribution of a single variable.

The gathering of public statistics by governments and semipublic agencies was well established. After about 1800 most of the industrialized countries had instituted the official national census. Several nongovernmental societies had been set up, chiefly for the purpose of improving the quality of public statistics, for example, the Statistical Society of London (now the Royal Statistical Society) in 1834 and the American Statistical Association in 1839. Actuarial work had become a fairly well-developed and respected profession. The 25-year period from 1853 to 1878 was the era of the great international statistical congresses. Economic statistics moved ahead greatly in this period, with notable improvements in methods of gathering data. Governments were beginning to take physical measurements of their soldiers and were making these data available to anthropometrists.

Among Pearson's predecessors were men who made significant contributions to the mathematical theory of probability in relation to gambling problems, but they never tested that theory on data and never proposed its application in any other area. Early theorists of this kind were Pierre de Fermat, Blaise Pascal, Christian Huygens, and Abraham de Moivre. Other mathematicians wrote about the possible application of probability theory to social phenomena, but they had no data: Jakob (Jacques) Bernoulli wrote on such possible application to economics; Daniel Bernoulli on inoculation as a preventive of small pox; and Niklaus (Nicholas) Bernoulli, Condorcet, and Poisson, among others, on the credibility of testimony and related legal matters. Before 1800 William Playfair had invented the statistical graph and published many beautiful statistical charts from quite dubious data.

None of these men, then, either cared to test his theories on data or had appropriate data to work with, and contrariwise, many other men worked in statistical agencies tabulating data with very little idea of how to analyze them.

Two groups of persons, the actuaries and the mathematical astronomers, possessed both mathematical acumen and relevant data for testing theory, but neither group proposed a general statistical approach outside of its own field. The great mathematical astronomers of the first half of the nineteenth century, notably Laplace and Gauss, did lay the foundations for modern statistical theory by developing the concept of errors of observation and an impressive accompanying mathe-

matical theory, and the ferment of ideas which they stimulated spread over Europe. Important contributions were made by Friedrich Wilhelm Bessel and Johann Franz Encke in Germany; Giovanni Plana in Italy; Adrien Marie Legendre, Poisson, Jean Baptiste Fourier, Auguste Bravais, and "Citizen" Kramp in France; Quetelet in Belgium; George Biddell Airy and Augustus De Morgan in England; and Thorwald Nicolai Thiele in Denmark. Only a very few persons before Pearson had thought of the statistical analysis of concrete data as a general method applicable to a wide range of problems; one such was Cournot, whose extensive writing, both on the theory of chance and on such matters as wealth and supply and demand, laid the foundations for mathematical economics. And more than any other person in the nineteenth century, Quetelet brought together mathematical theory, the collection of official statistics, and a concern for practical problems and fused the three into a single tool for studying the problems of life. Finally, of course, there was Galton, whose work had a great impact on Pearson and whose close friendship with him had an incalculable influence.

Although these men had put a high value on concrete data, the amount of data to which they had access was paltry beside what soon began to be collected by Pearson and his associates. Pearson always insisted on publishing the original data as well as the statistics derived from them. His primary aim was to develop a methodology for the exploration of life, not the refinement of mathematical theory. Whenever he developed a new piece of statistical theory, he immediately used it on data, and if his mathematics was cumbersome, this did not concern him.

Major contributions

Frequency curves. One of the problems on which Pearson spent a great deal of time and energy was that of deriving a system of generalized frequency curves based on a single differential equation, with parameters obtained by the method of moments. Quetelet seems to have believed that almost all social phenomena would show approximately normal distributions if the number of cases could be made large enough. Before 1890 J. P. Gram and Thiele in Denmark had developed a theory of skew frequency curves. After Pearson published his elaborate and extremely interesting system (1894; 1895), many papers were written on such related topics as the fitting of curves to truncated or imperfectly known distributions and tables of the probability distribution of selected curves.

Chi-square. Having fitted a curve to a set of observations, Pearson needed a criterion to indicate how good the fit was, and so he invented "chi-square" (1900). Quetelet and others who wanted to demonstrate the closeness of agreement between the frequencies in a distribution of observed data and frequencies calculated on the assumption of normal probability merely printed the two series side by side and said, in essence, "Behold!" They had no measure of discrepancy and were apparently not made uncomfortable by the lack of such a measure. Pearson not only devised the measure but he worked out its distribution and had it calculated. He himself never seems to have understood the concept of degrees of freedom, either in relation to chi-square or to his probable-error formulas. Yet chi-square is an enormously useful device with a range of applications far greater than the specific problem for which it was created, and it occupies an important position in modern statistical theory.

Correlation. The idea of correlation is due to Galton, who published a paper entitled "Co-relations and Their Measurement Chiefly From Anthropometric Data" in 1880 and another entitled "Regression Towards Mediocrity in Hereditary Stature" in 1885, and who gave a more widely read statement in *Natural Inheritance* (1889). The mathematics of the normal correlation surface had been derived earlier in connection with errors made in estimating the position of a point in space. In 1808 Adrain gave the first known derivation of the probability that two such errors will occur together but dealt with uncorrelated errors only. The density function for two related errors was given by Laplace in 1810 and for n related errors by Gauss in 1823 or perhaps earlier. Plana in 1812, studying the probability of errors in surveying, and Bravais in 1856, that of errors in artillery fire, each obtained an equation in which there is a term analogous to r. Being concerned about the probability of the occurrence of error and not with the strength of relationship between errors, these men all studied the density function and paid no attention to the product term in the exponent, which is a function of r. They applied their findings only to errors of observation, and the relation of their work to the correlation surface was noted only long after the important works on correlation had been written. In a study made in 1877 of the height, weight, and age of 24,500 Boston school children, Henry Pickering Bowditch published curves showing the relation of height to weight but missed discovering the correlation between two variables.

Galton had been seriously hampered in his study

of correlation by both lack of data and lack of an efficient routine of computation. He did have data on sweet peas and on the stature of parents and adult offspring for two hundred families. While Pearson began to lecture on correlation, Weldon began to make measurements on shrimp for correlation studies.

In Pearson's first fundamental paper on correlation, entitled "Regression, Heredity and Panmixia" (1896), he generalized Galton's conclusions and methods; derived the formula which we now call the "Pearson's product moment" and two other equivalent formulas; gave a simple routine for computation which could be followed by a person without much mathematical training; stated the general theory of correlation for three variables; and gave the coefficients of the multiple regression equation in terms of the zero-order correlation coefficients.

There followed a series of great memoirs on various aspects of correlation, some by students or associates of Pearson, such as G. Udny Yule and W. F. Sheppard, but most of them from his own hand. These dealt with such matters as correlation in nonnormal distributions, tetrachoric r, correlation between ranks, correlation when one or both variables are not scaled or regression is not linear. There were many papers presenting the results of correlation analysis in a great variety of fields. A large amount of labor went into the derivation of the probable error of each of these various coefficients and the tabulation of various probabilities related to correlation. It is fitting that the product-moment correlation coefficient is named the "Pearson r."

Individual variability. Variation among errors made in observations on the position of a heavenly body had been studied extensively by the great mathematical astronomers. The list of those who before 1850 had written on the "law of facility of error," derived the formula for the normal curve, and compiled probability tables would be a long one. The term "probable error" had come into widespread use within a few years after Bessel employed the term *der wahrscheinliche Fehler* in 1815 in a paper on the position of the polar star.

The concept of true variability among individuals is very different from the concept of chance variation among errors in the estimation of a single value. The idea of individual variability is prominent in the writings of Quetelet, Fechner, Ebbinghaus, Lexis, Edgeworth, Galton, and Weldon, but it was not commonly appreciated by other scientists of the nineteenth century. Pearson's emphasis upon this idea is one of his real contributions to the understanding of life. In that first great paper on asymmetrical frequency curves (1894) he introduced the term "standard deviation" and the symbol σ, and he consistently used this term and this symbol when discussing variation among individuals. However, when writing about sampling variability, he always used the term "probable error," thus clearly distinguishing variability due to individual differences from variability due to chance errors.

Probable errors of statistics. Pearson himself probably considered that one of his greatest contributions was the derivation of the probable errors of what he called "frequency constants" and various tables to facilitate the computation of such. His method, already well known in other connections, was to write the equation for a statistic, take the differential of both sides of that equation, square, sum, and reduce the result by any algebraic devices he could think of. The process of reduction was often formidable. Even though he was not much concerned with the distinction between a statistic and its parameter and he frequently used the former in place of the latter, these probable errors marked a great advance over the previous lack of any measure of the sampling variability of most statistics. In this era new statistics were being proposed on every side, and the amount of energy which went into the derivation of these probable errors was tremendous. With the successful search for exact sampling distributions that has been under way ever since the publication of Student's work in 1908 and R. A. Fisher's 1915 paper on the sampling distribution of the correlation coefficient, better methods than Pearson's probable error formulas could provide are now in many cases available.

Publication of tables. An editorial in the first number of *Biometrika* (unsigned, but always attributed to Pearson) referred to the urgent need for tables to facilitate the work of the statistician and biometrician and promised that such tables would be produced as rapidly as possible. Such tables as were then available were in widely scattered sources, some of them almost impossible to obtain. By 1900, tables of the binominal coefficients, the trigonometric functions, logs, and antilogs were readily available. A large table of the logs of factorials computed in 1824 by F. C. Degen was almost unknown. There were tables of squares, cubes, square roots, and reciprocals, of which Barlow's is the best known, and there were multiplication tables by Crelle and by Coatsworth. Legendre had published a table of logarithms of the gamma-function, but copies were very scarce. The normal probability function had been ex-

tensively tabulated but always with either the probable error ($.6745\sigma$) or the modulus ($\sigma/\sqrt{2}$) as argument, never the standard error. Poisson had not tabulated the distribution which bears his name, but Bortkiewicz had done so in 1898 (in his *Gesetz der kleinen Zahlen*).

A list of the tables which have been issued in *Biometrika* from its second issue in 1902 until the present or which have appeared in the separate volumes of *Tables for Statisticians and Biometricians* (1914) or in the *Drapers' Company Series of Tracts for Computers* would be a very long one. Some of these tables are no longer used, others appear to be timeless in value, even after the advent of the electronic computer. The *Tables of the Incomplete Beta-function* (1934) was among Pearson's last contributions to science, published when he was 78 years old.

Controversies

The frequent controversies in which Pearson was embroiled cannot be disregarded. In his youth he did battle for such unpopular radical ideas as socialism, the emancipation of women, and the ethics of free thought. A few years later he was involved in a long struggle for the unpopular idea that mathematics should be applied to the study of biology. Much bitterness arose over this question, and the Royal Society, while ready to accept papers dealing with either mathematics or biology, refused to accept papers dealing with both. That refusal was one of the circumstances that led to the founding of *Biometrika* in 1900, and this in turn gave great impetus to the young sciences of biometry and mathematical statistics: now there was a journal in which mathematical papers on the biological sciences could be published.

In 1904 Galton established the Eugenics Record Office to further the scientific study of eugenics. It became known as the Eugenics Laboratory two years later, when Galton turned it over to Pearson so that he might operate it in connection with his Biometric Laboratory. The Biometric Laboratory, whose existence Pearson dated back to 1895, was a center for training postgraduate workers in this new branch of exact science. In 1911 these two laboratories were united to form the department of applied statistics in University College, with Pearson as its first professor.

Beginning in 1907 the Eugenics Laboratory published numerous very substantial statistical papers on three of the most controversial issues of the day: pulmonary tuberculosis, alcoholism, and mental deficiency and insanity. These appeared in two series entitled "Studies in National Deterioration" and "Questions of the Day and of the Fray."

In contrast to the idea then current that tuberculosis could be eradicated by improving the environment, Pearson's statistical studies indicated that the predisposition to tuberculosis was more hereditary than environmental and that there was no clear evidence that patients treated in sanatoria had a higher recovery rate than those treated elsewhere.

Another common assumption at that time was that alcoholic parents produce children with mental and physical deficiencies. The first studies on this subject coming from the Eugenics Laboratory found no marked relation between parental alcoholism and the intelligence, physique, or disease of offspring (1910). Later papers concluded that alcoholism is more likely to be a consequence than a cause of mental defect. Pearson commented that "the time is approaching when real knowledge must take the place of energetic but untrained philanthropy in dictating the lines of feasible social reform" (quoted in Egon S. Pearson 1938, p. 61).

After the American Eugenics Record Office announced in 1912 that mental defect was almost certainly a recessive Mendelian character and advised that "weakness in any trait should marry strength in that trait and strength may marry weakness," Pearson or his associates marshaled statistical evidence to refute this pronouncement.

Each time Pearson took up such an issue, the reaction of medical authorities and public officials was angry and violent, and their personal attacks on Pearson were prolonged and vituperative; open conflict also developed between the more traditional Eugenics Education Society, of which Galton was honorary president, and the Eugenics Laboratory, of which he had been the founder.

The young sciences of biometry and statistics may well have profited from these major struggles with organized groups that allowed them to break the restraining bonds of apathy, of ignorance, of entrenched authority. Pearson was something of a crusader, and among the qualities a crusader needs are self-confidence, the courage to fight for his convictions, and a touch of intellectual intolerance. He was a perfectionist and had scant patience with ideas or work which he considered incorrect. Moreover, he was trained for a legal career and from childhood had in his father the example of a successful trial lawyer. However, his first thought was to get at the truth, and, if intellectually convinced of an error, Pearson was ready to admit it. He once published in *Biometrika* a paper called "Peccavimus" ("We Have Erred").

Although Pearson made contributions to statistical technique that now appear to be of enduring importance, these techniques are of less importance than what he did in rousing the scientific world from a state of sheer uninterest in statistical studies to one of eager effort by a large number of well-trained persons, who developed new theory, gathered and analyzed statistical data from every field, computed new tables, and re-examined the foundations of statistical philosophy. This is an achievement of fantastic proportions. His laboratory was a world center in which men from all countries studied. Few men in all the history of science have stimulated so many other people to cultivate and to enlarge the fields they themselves had planted. He provided scientists with the concept of a general methodology underlying all science, one of the great contributions to modern thought.

HELEN M. WALKER

[For the historical context of Pearson's work, see STATISTICS, article on THE HISTORY OF STATISTICAL METHOD; and the biographies of the BERNOULLI family; CONDORCET; COURNOT; FISHER, R. A.; GALTON; GAUSS; LAPLACE; MOIVRE; POISSON; QUETELET; for discussion of the subsequent development of his ideas, see GOODNESS OF FIT; MULTIVARIATE ANALYSIS, article on CORRELATION METHODS; NONPARAMETRIC STATISTICS, article on RANKING METHODS; STATISTICS, DESCRIPTIVE, article on ASSOCIATION; and the biography of YULE.]

WORKS BY PEARSON

1880   *The New Werther.* London: Kegan.
(1885) 1946   CLIFFORD, WILLIAM K. *The Common Sense of the Exact Sciences.* Edited, and with a preface, by Karl Pearson; newly edited by James R. Newman. New York: Knopf.
1886–1893   TODHUNTER, ISAAC *A History of the Theory of Elasticity and of the Strength of Materials From Galilei to the Present Time.* 2 vols. Edited and completed by Karl Pearson. Cambridge Univ. Press.
1887   *Die Fronica: Ein Beitrag zur Geschichte des Christusbildes im Mittelalter.* Strassburg (then Germany): Trübner.
(1888) 1901   *The Ethic of Freethought, and Other Addresses and Essays.* London: Black.
(1892) 1937   *The Grammar of Science.* 3d ed., rev. & enl. New York: Dutton. → A paperback edition was published in 1957 by Meridian.
(1894) 1948   Contributions to the Mathematical Theory of Evolution. I. Pages 1–40 in Karl Pearson, *Karl Pearson's Early Statistical Papers.* Cambridge Univ. Press. → First published as "On the Dissection of Asymmetrical Frequency Curves" in Volume 185 of the *Philosophical Transactions* of the Royal Society of London, Series A.
(1895) 1948   Contributions to the Mathematical Theory of Evolution. II: Skew Variation in Homogeneous Material. Pages 41–112 in Karl Pearson, *Karl Pearson's Early Statistical Papers.* Cambridge Univ. Press. → First published in Volume 186 of the *Philosophical Transactions* of the Royal Society of London, Series A.
(1896) 1948   Mathematical Contributions to the Theory

of Evolution. III: Regression, Heredity and Panmixia. Pages 113–178 in Karl Pearson, *Karl Pearson's Early Statistical Papers.* Cambridge Univ. Press. → First published in Volume 187 of the *Philosophical Transactions* of the Royal Society of London, Series A.

1897   *The Chances of Death, and Other Studies in Evolution.* 2 vols. New York: Arnold.

(1898) 1948   PEARSON, KARL; and FILON, L. N. G. Mathematical Contributions to the Theory of Evolution. IV: On the Probable Errors of Frequency Constants and on the Influence of Random Selection on Variation and Correlation. Pages 179–261 in Karl Pearson, *Karl Pearson's Early Statistical Papers.* Cambridge Univ. Press. → First published in Volume 191 of the *Philosophical Transactions* of the Royal Society of London, Series A.

(1900) 1948   On the Criterion That a Given System of Deviations From the Probable in the Case of a Correlated System of Variables Is Such That It Can Be Reasonably Supposed to Have Arisen From Random Sampling. Pages 339–357 in Karl Pearson, *Karl Pearsons' Early Statistical Papers.* Cambridge Univ. Press. → First published in Volume 50 of the *Philosophical Magazine,* Fifth Series.

1906   Walter Frank Raphael Weldon: 1860–1906. *Biometrika* 5:1–52.

1907   *A First Study of the Statistics of Pulmonary Tuberculosis.* Drapers' Company Research Memoirs, Studies in National Deterioration, No. 2. London: Dulau.

1910   ELDERTON, ETHEL M.; and PEARSON, KARL *A First Study of the Influence of Parental Alcoholism on the Physique and Ability of the Offspring.* Univ. of London, Francis Galton Laboratory for National Eugenics, Eugenics Laboratory, Memoirs, Vol. 10. London: Dulau.

(1914) 1930–1931   PEARSON, KARL (editor) *Tables for Statisticians and Biometricians.* 2 vols. London: University College, Biometric Laboratory. → Part 1 is the third edition; Part 2 is the first edition.

1914–1930   *The Life, Letters and Labours of Francis Galton.* 3 vols. in 4. Cambridge Univ. Press.

▶(1921–1933)   1978   *The History of Statistics in the Seventeenth and Eighteenth Centuries, Against the Changing Background of Intellectual, Scientific and Religious Thought.* Edited by Egon S. Pearson. London: Griffin; New York: Macmillan. → Lectures given at University College, London.

(1922) 1951   PEARSON, KARL (editor) *Tables of the Incomplete Gamma-function.* Computed by the staff of the Department of Applied Statistics, Univ. of London. London: Office of Biometrika.

1923   *On the Relationship of Health to the Psychical and Physical Characters in School Children.* Cambridge Univ. Press.

1934   PEARSON, KARL (editor) *Tables of the Incomplete Beta-function.* London: Office of Biometrika.

1936   Old Tripos Days at Cambridge, as Seen From Another Viewpoint. *Mathematical Gazette* 20:27–36.

Karl Pearson's Early Statistical Papers. Cambridge Univ. Press, 1948. → Contains papers published between 1894 and 1916.

SUPPLEMENTARY BIBLIOGRAPHY

Annals of Eugenics. → Published since 1925 by the Eugenics Laboratory. Karl Pearson was the editor until his death and also contributed articles.

Biometrika: A Journal for the Statistical Study of Biological Problems. → Published since 1900, the journal was founded by W. F. R. Weldon, Francis Galton, and Karl Pearson. Edited by Karl Pearson until his death, and since then by Egon S. Pearson.

Drapers' Company Research Memoirs Biometric Series. → Published since 1904. Contains a series of major contributions by Pearson and his associates. The series was edited by Pearson.

▶EISENHART, CHURCHILL   1974   Karl Pearson. Volume 10, pages 447–473 in *Dictionary of Scientific Biography.* Edited by Charles C. Gillispie. New York: Scribner's. → An excellent new biography.

○FILON, L. N. G.   1936   Karl Pearson as an Applied Mathematician. Royal Society of London, *Obituary Notices of Fellows* 2, no. 5:104–110.

FISHER, R. A.   1915   Frequency Distribution of the Value of the Correlation Coefficient in Samples From an Indefinitely Large Population. *Biometrika* 10:507–521.

HALDANE, J. B. S.   1957   Karl Pearson, 1857–1957: A Centenary Lecture Delivered at University College, London, on May 13, 1957. *Biometrika* 44:303–313.

MORANT, GEOFFREY (editor)   1939   *A Bibliography of the Statistical and Other Writings of Karl Pearson.* Cambridge Univ. Press.

PEARSON, EGON S.   1938   *Karl Pearson: An Appreciation of Some Aspects of His Life and Work.* Cambridge Univ. Press. → First published in 1936 and 1938 in Volumes 28 and 29 of *Biometrika.* Contains a partial bibliography of Karl Pearson's writings.

WALKER, HELEN M.   1958   The Contributions of Karl Pearson. *Journal of the American Statistical Association* 53:11–22.

WILKS, S. S.   1941   Karl Pearson: Founder of the Science of Statistics. *Scientific Monthly* 53:249–253.

▶YULE, G. UDNY   1936   Karl Pearson, 1857–1936. Royal Society of London, *Obituary Notices of Fellows* 2, no. 5:73–104.

PEARSON PRODUCT–MOMENT CORRELATION COEFFICIENT

See MULTIVARIATE ANALYSIS, *article on* CORRELATION METHODS.

PEIRCE, CHARLES SANDERS

Charles Sanders Peirce (1839–1914), the greatest of America's scientific philosophers, was born in Cambridge, Massachusetts, the second son of the famous Harvard mathematician and astronomer Benjamin Peirce (1809–1880). Peirce was coached by his father in mathematics, physics, and astronomy, and was later to revise his father's *Linear Associative Algebra* of 1870. After receiving his bachelor's degree from Harvard in 1859, and a master's degree *summa cum laude,* he joined the United States Coast and Geodetic Survey in 1861. During the thirty years he worked there, he became internationally famous for his pendular measurements of gravity and of starlight intensity (1878a). Between 1879 and 1884, he taught at the Johns Hopkins University, his only university teaching position. He was elected a member of the National Academy of Sciences in 1877 and, even before that, to the American Academy of Arts and Sciences.

Peirce published an improvement of Boole's logic (1867), which was later further developed by H. M. Sheffer in his work on the stroke function. He also improved on De Morgan's notation (1870) and developed further the logic of propositions, classes, and relations (1883). This was Peirce's chief contribution to logic and helped pave the way for twentieth-century developments in symbolic logic. Just as the work of Giuseppe Peano in Italy on the foundations of mathematics influenced such Italian philosophers as the "logical pragmatist" G. Vailati, so Peirce's research in logic and critical philosophy (especially his studies of Kant, Reid, Bain, Schelling, and Hegel) led to his own formulation of the pragmatic theory that the meaning of a conception lies in the sum total of the conceivable consequences which the object of that conception can possibly have on the conduct of "an indefinite community of investigators" (1877–1878, vol. 12, pp. 286–302). The final opinion of that ideal community would constitute the truth, and the object of that ultimate opinion would be "reality."

This social criterion of meaning and truth mistakenly has been thought by critics to reduce philosophy to mere public opinion and group prejudice. But Peirce clearly limited his ideal community to scientific investigators. He insisted, furthermore, on the *fallibilism* of all beliefs, even if common sense and science do require that *some* premises be taken tentatively as indubitable until experience or experiment shows that a disparity exists between the accepted consequences of these premises and observation. Applying the test of performing specifiable procedures for verifying the calculable consequences of hypotheses has become known as operationalism; Percy Bridgman advocated it in his study of the logic of physics, and John Dewey in his study of the methodology of the social sciences.

Peirce formulated his early statements of pragmatism in two essays, "The Fixation of Belief" and "How to Make Our Ideas Clear," the first two of six essays in "Illustrations of the Logic of Science" (1877–1878). Although in these early publications he did not call his philosophy "pragmatism," he did use the term in his discussions about it around Harvard. Many of these discussions (mainly on the significance of the Darwinian controversy for methods of reasoning in the physical and social sciences) took place in an informal club to which he belonged in the early 1870s and whose members also included William James, Chauncey Wright, John Fiske, Oliver Wendell Holmes, Jr., Nicholas St. John Green, and Joseph B. Warner—the last three being law students at the time (see Wiener 1949).

Peirce's survey of methodological theories in the first of the two essays was a critical analysis of these theories based on social psychology and intellectual history. He listed the four chief methods used to settle "doubts" (by which he meant externally caused disturbances of mind rather than the more subjective kind of Cartesian doubts): tenacity, authority, apriorism, and the scientific method. The scientific method is the most satisfactory in the long run because it is the only one of the four that is not in principle inflexible; it is self-corrective, whereas proponents of the other three methods can claim "infallibility" only by resting on premises other than those encountered in experience or experimental situations. Peirce also rejected intuitionism as being a form of a priori rationalism.

He believed in a kind of metaphysical evolutionism. Originally there was chaos, but as man continually used reason to order his experience, there was a "growth of concrete reasonableness" (1891; 1893). His three main metaphysical categories (that is, modes of representing all that is known) are exemplified by chance qualities (firstness), brute existence (secondness), and generality or order (thirdness) (1877–1878, vol. 12, pp. 604–615, 705–718; vol. 13, pp. 203–217). In his later works, Peirce insisted that generality is objectively real, that it exists independently of our beliefs, and that it is given in perception of the particular.

Although Peirce defended a frequency theory of inductive probability, he also proposed other meanings of probable hypotheses or likelihood (1883). He thought all reasoning processes could be divided into three types, depending on their consequences: purely explicative deduction (for example, mathematics and formal logic), ampliative induction (for example, empirical generalization), and conjectural abduction, or reasoning culminating in a probable hypothesis. This third type of reasoning (in his logic of hypothesis) is also called retroduction and is exemplified by cryptography, medical diagnosis, historical inference, and detective work. For all scientific theory he advocated (at the same time as Ernst Mach, but independently of him) a principle of economy, or practical simplicity in the consideration of hypotheses, and a statistical conception of the laws of nature as predictive, fallible, and subject to modification in time.

In social matters, Peirce was intensely opposed to the rugged individualism or "philosophy of greed" of the political economy of Simon Newcomb and of other social Darwinists of his day. He also deplored the lack of faith in the gospel of love on the part of those theologians who practiced the "higher criticism." In general, Peirce was concerned with the neglect of humanistic ethics in social or political matters.

During the last thirty years of his life he continued his studies in the logic and philosophy of the sciences, although he lived almost like a hermit at Milford, Pennsylvania, with his second wife, a French widow who spoke little English. His contributions to logic, to the philosophy of science, and to the theory of signs were, nevertheless, very influential in the development of the views of people who participated more actively in academic or political affairs, for example, William James, John Dewey, George Herbert Mead, C. I. Lewis, Charles E. Morris, Morris R. Cohen, F. P. Ramsey, Ernest Nagel, Sidney Hook, and others who call themselves pragmatists.

Recent commentators (Feibleman 1946; Thompson 1953; Wiener & Young 1952; Goudge 1950; Murphey 1961) have examined Peirce's philosophy "as a whole" and have discerned a latent system in his diverse writings, a system that, although incomplete, is architectonic in its professed aims. Kant built an architectonic system of categories that was based on what he regarded as a finished classical logic, but Peirce went far beyond the classical syllogistic logic in his "logic of relatives."

Murray G. Murphey, in *The Development of Peirce's Philosophy* (1961, p. 432) has outlined four major phases of Peirce's philosophy: a Kantian phase, 1857–1865; the development of the irreducibility of the three syllogistic figures corresponding to deduction, abduction, and induction, 1866–1869; the logic of relations, 1870–1884; and quantification and set theory, 1884–1912. These phases are not actually sharply demarcated in Peirce's work. Murphey is correct, nevertheless, in emphasizing the fact that Peirce's philosophical categories of firstness, secondness, and thirdness underwent changes as his logical theories developed; for like Kant, Peirce believed one's philosophy should follow one's logic.

PHILIP P. WIENER

[*For the historical context of Peirce's work, see* STA-TISTICS, *article on* THE HISTORY OF STATISTICAL METHOD; *see also* Kaplan 1968; Tax & Krucoff 1968. *For discussion of the subsequent development of his ideas, see* ERRORS, *article on* NONSAMPLING ERRORS; *see also* Frankel 1968; Konefsky 1968; Phelan 1968; Rosenfield 1968; Shibutani 1968; Weinreich 1968.]

BIBLIOGRAPHY

The most complete bibliography of Peirce's writings and of works on Peirce is contained in Volume 8 of Peirce's Collected Papers. *The best commentary is in* Murphey 1961. *For Peirce's posthumously published contributions to logic, see Volumes 2, 3, and 4 of the* Collected Papers.

WORKS BY PEIRCE

1867 On an Improvement in Boole's Calculus of Logic. American Academy of Arts and Sciences, *Proceedings* 7:250–261.

(1870) 1933 Description of a Notation for the Logic of Relatives, Resulting From an Amplification of the Conceptions of Boole's Calculus of Logic. Volume 3, pages 27–98 in Charles S. Peirce, *Collected Papers.* Cambridge, Mass.: Harvard Univ. Press.

1873 On the Theory of Errors of Observations. U.S. Coast and Geodetic Survey, *Report of the Superintendent* [1870]:200–224.

1877–1878 Illustrations of the Logic of Science. *Popular Science Monthly* 12:1–15, 286–302, 604–615, 705–718; 13:203–217, 470–482.

1878a Photometric Researches. Leipzig: Engelmann.

(1878b) 1956 The Probability of Induction. Volume 2, pages 1341–1354 in James R. Newman (editor), *The World of Mathematics: A Small Library of the Literature of Mathematics From A'h-mosé the Scribe to Albert Einstein.* New York: Simon & Schuster.

(1878c) 1956 The Red and the Black. Volume 2, pages 1334–1340 in James R. Newman (editor), *The World of Mathematics: A Small Library of the Literature of Mathematics From A'h-mosé the Scribe to Albert Einstein.* New York: Simon & Schuster.

1881 On the Logic of Number. *American Journal of Mathematics* 4:85–95.

1883 A Theory of Probable Inference. Pages 126–181 in Charles S. Peirce (editor), *Studies in Logic.* Boston: Little.

1884 The Numerical Measure of the Success of Predictions. *Science* 4:453–454.

1891 The Architecture of Theories. *Monist* 1:161–176.

1893 Evolutionary Love. *Monist* 3:176–200.

Collected Papers. Edited by Charles Hartshorne et al. 8 vols. Cambridge, Mass.: Harvard Univ. Press, 1931–1958. → Volume 1: *Principles of Philosophy.* Volume 2: *Elements of Logic.* Volume 3: *Exact Logic.* Volume 4: *The Simplest Mathematics.* Volume 5: *Pragmatism and Pragmaticism.* Volume 6: *Scientific Metaphysics.* Volume 7: *Science and Philosophy.* Volume 8: *Reviews, Correspondence, and Bibliography.*

SUPPLEMENTARY BIBLIOGRAPHY

AIRY, G. B. 1856 [Letter from Professor Airy, Astronomer Royal, to the editor.] *Astronomical Journal* 4:137–138. → On the work of Benjamin Peirce.

BUCHLER, JUSTUS 1939 *Charles Peirce's Empiricism.* New York: Harcourt.

FEIBLEMAN, JAMES 1946 *An Introduction to Peirce's Philosophy, Interpreted as a System.* New York: Harper.

►FRANKEL, CHARLES 1968 Dewey, John. Volume 4, pages 155–159 in *International Encyclopedia of the Social Sciences.* Edited by David L. Sills. New York: Macmillan and Free Press.

GOODMAN, LEO A.; and KRUSKAL, WILLIAM H. 1959 Measures of Association for Cross Classifications: II. Further Discussion and References. *Journal of the American Statistical Association* 54:123–163. → See especially pages 127–132, "Doolittle, Peirce, and Contemporary Americans."

GOUDGE, THOMAS A. 1950 *The Thought of C. S. Peirce.* Univ. of Toronto Press.

GOULD, B. A. JR. 1855 On Peirce's Criterion for the Rejection of Doubtful Observations, With Tables for Facilitating Its Application. *Astronomical Journal* 4:81–87. → On the work of Benjamin Peirce.

►KAPLAN, ABRAHAM 1968 Positivism. Volume 12, pages 389–395 in *International Encyclopedia of the Social Sciences*. Edited by David L. Sills. New York: Macmillan and Free Press.

►KONEFSKY, SAMUEL J. 1968 Holmes, Oliver Wendell. Volume 6, pages 491–493 in *International Encyclopedia of the Social Sciences*. Edited by David L. Sills. New York: Macmillan and Free Press.

LEWIS, C. I. 1918 *A Survey of Symbolic Logic*. Berkeley: Univ. of California Press.

MURPHEY, MURRAY G. 1961 *The Development of Peirce's Philosophy*. Cambridge, Mass.: Harvard Univ. Press.

NEWMAN, JAMES R. 1956 Commentary on Charles Sanders Peirce. Volume 3, pages 1767–1772 in James R. Newman (editor), *The World of Mathematics: A Small Library of the Literature of Mathematics From A'h-mosé the Scribe to Albert Einstein*. New York: Simon & Schuster.

PEIRCE, BENJAMIN 1852 Criterion for the Rejection of Doubtful Observations. *Astronomical Journal* 2:161–163.

►PHELAN, WILLIAM D. JR. 1968 James, William. Volume 8, pages 227–234 in *International Encyclopedia of the Social Sciences*. Edited by David L. Sills. New York: Macmillan and Free Press.

►ROSENFIELD, LEONORA COHEN 1968 Cohen, Morris R. Volume 2, pages 540–542 in *International Encyclopedia of the Social Sciences*. Edited by David L. Sills. New York: Macmillan and Free Press.

►SHIBUTANI, TAMOTSU 1968 Mead, George Herbert. Volume 10, pages 83–87 in *International Encyclopedia of the Social Sciences*. Edited by David L. Sills. New York: Macmillan and Free Press.

STEWART, R. M. 1920a Peirce's Criterion. *Popular Astronomy* 28:2–3. → On the work of Benjamin Peirce.

STEWART, R. M. 1920b The Treatment of Discordant Observations. *Popular Astronomy* 28:4–6. → On the work of Benjamin Peirce.

►TAX, SOL; and KRUCOFF, LARRY S. 1968 Social Darwinism. Volume 14, pages 402–406 in *International Encyclopedia of the Social Sciences*. Edited by David L. Sills. New York: Macmillan and Free Press.

THOMPSON, MANLEY H. 1953 *The Pragmatic Philosophy of C. S. Peirce*. Univ. of Chicago Press.

►WEINREICH, URIEL 1968 Semantics and Semiotics. Volume 14, pages 164–169 in *International Encyclopedia of the Social Sciences*. Edited by David L. Sills. New York: Macmillan and Free Press.

WIENER, PHILIP P. 1949 *Evolution and the Founders of Pragmatism*. Cambridge, Mass.: Harvard Univ. Press.

WIENER, PHILIP P.; and YOUNG, FREDERIC H. (editors) 1952 *Studies in the Philosophy of Charles Sanders Peirce*. Cambridge, Mass.: Harvard Univ. Press.

WILSON, EDWIN B.; and HILFERTY, MARGARET M. 1929 Note on C. S. Peirce's Experimental Discussion of the Law of Errors. National Academy of Science, *Proceedings* 15:120–125.

WINLOCK, JOSEPH 1856 On Professor Airy's Objections to Peirce's Criterion. *Astronomical Journal* 4:145–147. → On the work of Benjamin Peirce.

Postscript

Although best known today as a philosopher and logician, Peirce also produced work in statistics of depth and originality unmatched by any other nineteenth-century American. Much of this work went unappreciated (or even unpublished) during his lifetime and is today primarily of interest as evidence that an American genius could, in the 1870s and 1880s, anticipate some of the major conceptual advances usually associated with European work of the 1920s and 1930s. But in at least one instance, the design of experiments in experimental psychology, Peirce's work had a direct and lasting influence upon the development of statistical methodology.

Peirce's father and his early employment with the U.S. Coast and Geodetic Survey may have been responsible for his exposure to probability and statistics. Benjamin Peirce had, in 1852, published one of the earliest significance tests for the rejection of outliers. This paper excited international controversy (Gould 1855; Airy 1856; Winlock 1856; Stewart 1920a, 1920b; Stigler 1973), and his son's first major statistical publication (1873) included a defense of his father's criterion. This same paper (1873) presented an empirical investigation of the distribution of a person's reaction times that was both a pioneering work in psychophysics and an interesting early appearance of nonparametric density estimation. Peirce estimated densities by a method of repeated smoothing that is equivalent to a modern kernel estimate. Peirce's conclusion that the density was similar to a normal curve was reexamined and criticized much later by Wilson and Hilferty (1929).

Another early paper (1879) by Peirce provided a mathematical analysis of the problem of optimally allocating experimental observations between competing experiments, under a model with two components of variance. The solution, presented in the framework of utility theory, includes a derivation of the basic result of marginal utility theory that may have been arrived at independently of earlier work of Gossen, Jevons, and Walras. In a different vein, a later letter to the editor of *Science* on the possibility of predicting tornadoes derived a latent structure measure of association for 2×2 tables (1884; Goodman & Kruskal 1959).

Peirce's major conceptual contribution to statistics was his understanding of the role randomization could play in inference. In an unpublished paper (circa 1896), he endorsed mathematical randomization as being capable of serving as a basis for inference and in fact defined "induction" as "reasoning from a sample taken at random to the whole lot sampled." His understanding of randomization actually influenced the development of experimental psychology through a paper (1884) written with Joseph Jastrow. Peirce and Jastrow gave an empirical refutation to earlier claims concerning the existence of a "least perceptible difference" of sensations, by an experiment using a balanced randomization scheme with a shuffled deck of cards. Incidental to this investigation, Peirce found that an individual's subjective assessment of his confidence in his judgment varied roughly linearly with the logarithm of the odds that his judgment was in fact correct. Empirical assessments of subjective or personalistic probabilities are usually associated with much later workers (for example, Savage 1971). Jastrow's later development and advocacy of Peirce's methodology had an important influence on psychological research. Peirce's approach to probability was that of an objective frequentist, and he opposed subjective theories of probabilistic reasoning, although ironically his work in logic had a profound impact on that of F. P. Ramsey. [See PROBABILITY, *article on* INTERPRETATIONS.]

Peirce left a vast collection of unpublished manuscripts when he died (Robin 1967). Many are still to be published, and it will be many years before an accurate assessment of this extraordinary man will be possible.

STEPHEN M. STIGLER

OTHER WORKS BY PEIRCE

1879 Note on the Theory of the Economy of Research. U.S. Coast and Geodetic Survey, *Report of the Superintendent* [1876]:197–201. → Reprinted in Volume 7, pages 76–88 in Peirce's *Collected Papers*.

1884 PEIRCE, CHARLES S.; and JASTROW, JOSEPH On Small Differences of Sensation. National Academy of Sciences, *Memoirs* 3, part 1:75–83. → Read October 17, 1884. Reprinted in Volume 7, pages 13–34 in Peirce's *Collected Papers*.

(c. 1896) 1957 Lessons From the History of Science. Pages 195–234 in Charles S. Peirce, *Essays in the Philosophy of Science*. Edited by Vincent Tomas. New York: Liberal Arts. → From a manuscript of notes (c. 1896) for a projected but never completed history of science. Also in Volume 1, pages 19–49 in Peirce's *Collected Papers*.

ADDITIONAL BIBLIOGRAPHY

EISELE, CAROLYN 1976 The New Elements of Mathematics by Charles S. Peirce. Pages 111–121 in *Men and Institutions in American Mathematics*. Graduate Studies, Texas Tech Univ., No. 13. Lubbock: The University.

JASTROW, JOSEPH 1887–1888 A Critique of Psychophysic Methods. *American Journal of Psychology* 1:271–309.

ROBIN, RICHARD S. 1967 *Annotated Catalogue of the Papers of Charles S. Peirce*. Amherst: Univ. of Massachusetts Press.

SAVAGE, LEONARD J. 1971 Elicitation of Personal Probabilities and Expectations. *Journal of the American Statistical Association* 66:783–801.

STIGLER, STEPHEN M. 1973 Simon Newcomb, Percy Daniell, and the History of Robust Estimation 1885–1920. *Journal of the American Statistical Association* 68:872–879.

STIGLER, STEPHEN M. 1978 Mathematical Statistics in the Early States. *Annals of Statistics* 6, no. 2:239–265.

PERMUTATIONS

See PROBABILITY, *article on* FORMAL PROBABILITY.

PERSONAL PROBABILITY

See BAYESIAN INFERENCE; DECISION MAKING; PROBABILITY, *article on* INTERPRETATIONS; *and the biographies of* BAYES *and* SAVAGE.

PETTY, WILLIAM

William Petty (1623–1687), the English economist and first systematic exponent of "the art of political arithmetic," was a self-made man with a wide range of talents and immense mental energy. He was the son of a clothier, traditionally described by his biographers as "poor," but if, as is likely, Petty's father left him the good house and 8 acres of land that he owned in his birthplace, Romsey, in 1685, the family was comfortably above the poverty line. Certainly, when the young Petty went to sea as a cabin boy at the age of about 14, he was sufficiently literate in Latin and Greek to gain entry to the Jesuits' college at Caen, where he was put ashore on breaking a leg; there he studied Latin, Greek, French, and mathematics. Then, after a short spell in the Royal Navy, he went on to study medicine at the universities of Utrecht, Amsterdam, Paris, and Oxford. By 1651 he had taken his Oxford degree of Doctor of Physic and entered the London College of Physicians. At the age of 28, he became vice-principal of Brasenose College and professor of anatomy at Oxford.

Although one of the leading intellectuals of his day and a founding member of the Royal Society, Petty was a man of the world rather than a scholar. He wrote more than he read. He was a persistent inventor: he patented a double-writing device before he got his medical degree, and he pursued his design for a twin-hulled ship through four prototypes. He worked out a scheme for the rebuilding of London after the Great Fire of 1666. In 1652 he went to Ireland as physician-general to the army, which was no sinecure in a country ravaged by the plague and other lethal epidemic diseases. Within two years of his arrival in Ireland he had taken over the complex task of surveying the forfeited estates of the Irish rebels as a basis for their redistribution among the English conquerors. This task, too, he performed with his usual efficiency and drive and with the determination and pugnacity which earned him many enemies and carried him into countless lawsuits. He spent much of his working life defending his Irish survey and handling his own Irish estates, but he still managed to read a number of communications to the Royal Society on topics varying from dyeing practices in the clothing trade to the testing of mineral waters and to achieve a formidable and serious literary output on economic, demographic, naval, medical, and scientific subjects.

A good deal has been written about Petty, and the published opinions of his contemporaries leave no doubt of the respect and esteem with which he was regarded. It is unlikely that he was author, as has been claimed, of the "Natural and Political Observations Upon the Bills of Mortality," which was signed by his friend John Graunt (see Graunt [1662] 1963, vol. 2, pp. 314–435), but he could well have been; and the title page of his own "Observations Upon the Dublin-Bills of Mortality" ([1683] 1963, vol. 2, pp. 479–491) explicitly associates him with the earlier work, which was a pathbreaker in demographic analysis. John Aubrey, his first biographer, the diarists John Evelyn and Samuel Pepys, and the economist and statesman, Davenant, all men of distinction in their own right, regarded Petty as one of the outstanding men of their time.

Petty's claim to fame as an economist lies not so much in his originality or his theoretical ability as in his analytical skill. His insistence on measurement and his clear schematic view of the economy make him the first econometrician, and he was constantly evolving and using concepts and analytical methods that were in advance of his time. His evaluation of the gain from foreign trade in "Another Essay in Political Arithmetick Concerning the Growth of the City of London" ([1682] 1963, vol. 2, pp. 451–478) is based on a statement of the benefits of the division of labor and specialization and was written a century before Adam Smith's famous account. Petty put so much stress on the role of labor in creating wealth that he has been regarded (for instance, by Marx) as an early exponent of the labor theory of value. But, as is shown by a characteristic and frequently quoted passage from his "Treatise of Taxes and Contributions" ([1662] 1963, vol. 1, pp. 1–97)—"Labour is the Father and active principle of Wealth as Lands are the Mother"—his theory of production and value is based on the two original factors of production of the early economists. He was the author of the first known national income estimates, in "Verbum sapienti" ([1665] 1963, vol. 1, pp. 99–120), "Political Arithmetick" ([c. 1676] 1963, vol. 1, pp. 233–313), and "Treatise of Ireland" ([1687] 1963, vol. 2, pp. 545–621), although he did not

trouble to define or develop his concepts and was rough, even careless at times, in his use of figures. Some of the calculations in his "Treatise of Taxes and Contributions" and elsewhere are essentially exercises in what is now called "cost-benefit analysis." He was the first writer, so far as we know, to grasp the concept of the velocity of money, again in "Verbum sapienti," although in his "Quantulumcunque Concerning Money" ([1695] 1963, vol. 2, pp. 437–448) there is no trace of it. He was not above manipulating his data in ways that would justify his polemical arguments, and it would be rash to accept his statistics uncritically. But he was no slave to political prejudice, and his analysis of the economy of his time for England and Ireland —for example, "The Political Anatomy of Ireland" ([1671–1676] 1963, vol. 1, pp. 121–231)— and his various essays on "political arithmetick" are both shrewd and penetrating.

PHYLLIS DEANE

[*Other relevant material may be found in the biography of* GRAUNT.]

WORKS BY PETTY

(1662) 1963 Treatise of Taxes and Contributions. Volume 1, pages 1–97 in William Petty, *The Economic Writings* . . . New York: Kelley.

(1662–1695) 1963 *The Economic Writings of Sir William Petty.* 2 vols. Edited by Charles H. Hull. New York: Kelley. → Contains Petty's main writings and a general bibliography.

(1665) 1963 Verbum sapienti. Volume 1, pages 99–120 in William Petty, *The Economic Writings* . . . New York: Kelley. → Written in 1665; first published posthumously in 1691.

(1671–1676) 1963 The Political Anatomy of Ireland. Volume 1, pages 121–231 in William Petty, *The Economic Writings* . . . New York: Kelley. → Written between 1671 and 1676; first published posthumously in 1691.

(c. 1676) 1963 Political Arithmetick. Volume 1, pages 233–313 in William Petty, *The Economic Writings* . . . New York: Kelley. → First written c. 1676; first published surreptitiously in 1683 as "England's Guide to Industry." The first authorized edition was published posthumously in 1690 by Petty's son.

(1682) 1963 Another Essay in Political Arithmetick Concerning the Growth of the City of London, 1682. Volume 2, pages 451–478 in William Petty, *The Economic Writings* . . . New York: Kelley.

(1683) 1963 Observations Upon the Dublin-Bills of Mortality, 1681, and the State of That City. Volume 2, pages 479–491 in William Petty, *The Economic Writings* . . . New York: Kelley.

(1687) 1963 Treatise of Ireland, 1687. Volume 2, pages 545–621 in William Petty, *The Economic Writings* . . . New York: Kelley. → Written in 1687; first published posthumously in 1899.

(1695) 1963 Sir William Petty's Quantulumcunque Concerning Money, 1682. Volume 2, pages 437–448 in

William Petty, *The Economic Writings* . . . New York: Kelley. → Published posthumously.

The Double Bottom or Twin-hulled Ship of Sir William Petty. Edited by the Marquis of Lansdowne. Oxford: Roxburghe Club, 1931.

History of the Cromwellian Survey of Ireland: A.D. 1655–1656. Edited by Thomas A. Larcom. Irish Archaelogical and Celtic Society Publications, Vol. 15. Dublin Univ. Press, 1851. → Commonly called the "Down Survey."

The Petty Papers. 2 vols. Edited by the Marquis of Lansdowne. London: Constable, 1927.

The Petty–Southwell Correspondence: 1676–1687. Edited by the Marquis of Lansdowne. London: Constable, 1928.

SUPPLEMENTARY BIBLIOGRAPHY

AUBREY, JOHN (1898) 1957 Sir William Petty. Pages 237–241 in John Aubrey, *Brief Lives*. Edited from the original manuscripts and with a life of John Aubrey by Oliver L. Pick. Ann Arbor: Univ. of Michigan Press.

BEVAN, WILSON L. 1894 Sir William Petty: A Study in English Economic Literature. Volume 9, pages 370–472 in American Economic Association, *Publications*. Baltimore: The Association.

FITZMAURICE, EDMOND G. P. 1895 *The Life of Sir William Petty: 1623–1687*. London: Murray. → Contains Petty's autobiographical will.

GRAUNT, JOHN (1662) 1963 Natural and Political Observations Upon the Bills of Mortality. Volume 2, pages 314–435 in William Petty, *The Economic Writings* . . . New York: Kelley. → The authorship of this work is in doubt.

GREENWOOD, MAJOR 1928 Graunt and Petty. *Journal of the Royal Statistical Society* 91:79–85.

PASQUIER, MAURICE 1903 *Sir William Petty: Ses idées économiques*. Paris: Giard & Brière.

STRAUSS, EMIL 1954 *Sir William Petty: Portrait of a Genius*. London: Bodley Head; Glencoe, Ill.: Free Press.

PHILOSOPHY

See SCIENCE, *article on* THE PHILOSOPHY OF SCIENCE; *see also* CAUSATION; SCIENTIFIC EXPLANATION.

POINT ESTIMATION

See under ESTIMATION.

POISSON, SIMÉON DENIS

Siméon Denis Poisson (1781–1840) was born at Pithiviers. He entered the École Polytechnique in 1798 and did so well there that the school exempted him from his final examinations and immediately appointed him assistant in mathematical analysis in 1800. He became successively an associate professor and a full professor (this in 1806) there, and then was called to the Bureau des Longitudes and to the Institut de France in 1812. After the Bourbon restoration, he was appointed, in 1816, professor of mechanics in the Faculté des Sciences at Paris and was then appointed to the Royal Council of the University, where he took charge of the mathematics curriculum for all the secondary schools of France. In 1837 Poisson was ennobled, and in 1840 he succeeded Thénard as dean of the Faculté des Sciences.

Poisson's principal interests were mechanics, mathematical analysis, physics, and the theory of probability. Although we are primarily concerned here with his work on probability and statistics, it is important to mention the fundamental nature of his contributions on such subjects as the invariability of the major axes of planets (1811), the distribution of electricity on the surfaces of bodies, capillary phenomena (1831), and the mathematical theory of heat (1835a).

It was toward the end of his life that Poisson's interests focused on probability and statistics. His most important work in this field is *Recherches sur la probabilité des jugements en matière criminelle et en matière civile, précédées des règles générales du calcul des probabilités* (1837), but slightly earlier he had published several brief papers that were very important for their time, since they contained the bases of a precise statistical method for the social sciences.

Poisson, like Laplace before him, sought to establish the probability that a juror may err in arriving at a verdict of guilt. Laplace had tried to formulate a hypothesis that does not do violence to common sense; he assumed that the probability that the juror will not be mistaken is governed by a law of equal distribution between $\frac{1}{2}$ and 1. From this law he deduced the probability that the verdict is correct as a function of the majority obtained. Poisson sharply criticized this procedure (1835b) and laid down the principle that any chain of reasoning in this type of problem should be based on the observation of such facts as demonstrate the existence of a *law of large numbers*. He attacked the problem by going back to the sources and examining the texts of laws as well as legal statistical documents. Thus, he analyzed the records of the Cours d'Assises. The juries of the Cours d'Assises had 12 members, and although before 1831 the majority required for conviction was 7 to 5, from that date it was 8 to 4. Poisson noted that before 1831 the annual proportion of acquittals was 0.39 in the mean, always remaining between 0.38 and 0.40, and the proportion of convictions by a vote of 7 to 5 was 0.07. He then claimed that even before the required majority was changed in 1831, the effect of the new rules could have been predicted: "The proportion of convictions . . . had to be $0.61 - 0.07 = 0.54$. If a vote of 8 to 4 was required, the ratio of acquittals to trials would thus be 0.46. This is in fact what happened during the year 1831" (1836a).

This sort of reasoning seems simple and obvious to us today, but at the time it was received with skepticism. Thus, immediately after Poisson read his paper before the Academy of Sciences, Poinsot objected violently to "the idea that a calculus could be applied to the moral sciences . . ." (Poinsot 1836).

In Poisson's view, the ordinary law of large numbers (which he evidently knew in the case of Bernoulli's urn) was primarily an experimental fact that should be taken as the foundation of statistics. It is important to remember that at that time the axioms used to construct the calculus of probabilities were different from the axioms used today, probability being defined, after Laplace, as the ratio of favorable cases to the total number of cases, assuming all these cases to be equally probable. Such a definition certainly does not preclude successful work, and Poisson used this definition in discovering his eponymous distribution. But the definition cannot deal with statistical problems, so that it was necessary for Poisson to enunciate what might be called a physical law.

The Poisson distribution (or law). The probability of having r successes in the course of n independent trials for an event whose probability is p is given by the binomial term $Pr(r) = \binom{n}{r}p^r(1-p)^{n-r}$. If, in this formula, p approaches 0 and n approaches infinity, with $np \to \lambda$ finite and not 0, we find, by using Stirling's formula, that

$$Pr(r) = \left(\frac{n!}{(n-r)!\,r!}\right)\frac{\lambda^r}{n^r}\left(1-\frac{\lambda}{n}\right)^{n-r}$$

$$(1) \quad = \frac{\lambda^r}{r!}\left(1-\frac{\lambda}{n}\right)^n\left(1-\frac{\lambda}{n}\right)^{-r}(1)\left(1-\frac{1}{n}\right)$$

$$\cdots\left(1-\frac{r-1}{n}\right)$$

$$\sim \frac{\lambda^r}{r!}e^{-\lambda}.$$

We are thus led to consider the discrete distribution

$$Pr(0) = e^{-\lambda}$$
$$Pr(1) = \lambda e^{-\lambda}$$
$$Pr(2) = \frac{\lambda^2}{2!}e^{-\lambda}$$
$$\vdots \qquad \vdots$$
$$Pr(r) = \frac{\lambda^r}{r!}e^{-\lambda}$$
$$\vdots \qquad \vdots$$

which is called the *Poisson distribution* (better, *Poisson's law*) or *rare events distribution*. This last term has the advantage of indicating that a Poisson distribution may be applicable when a large group is considered, each member of which has only a small probability of incurring an accident (as, for example, the number of men killed every year by the kick of a horse in a regiment of the Prussian army). The term "rare" is, however, ambiguous, and one should avoid the error of supposing that the absolute number of events is necessarily small.

The limit proof sketched in (1) above is incomplete, and we are indebted to Cauchy, a colleague of Poisson's, for the notion of characteristic function, which makes it possible to establish rigorously the tendency of the binomial distribution to approach that of Poisson.

The characteristic function of the binomial law is

$$[1 - p + pe^{it}]^n = \left[1 + \frac{\lambda}{n}(e^{it} - 1)\right]^n.$$

This tends toward $\phi(t) = \exp[\lambda(e^{it} - 1)]$ as $n \to \infty$, and the latter is the characteristic function of the Poisson distribution.

The cumulant generating function for the Poisson distribution is clearly $\psi(t) = \lambda(e^{it} - 1)$, which shows that all the cumulants are equal to λ. The quantity λ is called the *parameter* of Poisson's law; in particular, both mean and variance are λ.

Poisson distribution tables. Various tables and charts of the frequencies of the Poisson distribution have been prepared. The cumulative distribution function for the Poisson distribution may readily be expressed in terms of the incomplete gamma function Γ given in separate tables (for example, Pearson & Hartley's *Biometrika Tables for Statisticians* 1954), for if X is the number of successes, then $Pr(X > r) = \Gamma[\lambda(r)]/\Gamma(r)$.

Generalized Poisson distribution. Let $x_0, x_1, \cdots, x_r, \cdots$ be quantities in arithmetical progression having the form $x_r = a + \mu r$, where μ is any positive number, and let X be a stochastic variable such that

$$Pr(X = x_k) = \frac{\lambda^k}{k!}e^{-\lambda};$$

we then say that X obeys a *generalized Poisson distribution (or law)*. Other, and deeper, directions of generalization are discussed in the next section [see also DISTRIBUTIONS, STATISTICAL, article on MIXTURES OF DISTRIBUTIONS].

Poisson process. Suppose that the number, $X(t)$, of events taking place between instant 0 and instant t obeys a Poisson law (thus, the number of calls received by a telephone exchange between 0 and t might obey a Poisson law). Clearly, $X(t)$ can be considered as a random variable with integral values which, as t varies, engenders a nondecreasing random function with nonnegative integral values, called a *Poisson process.*

Stationary Poisson process. Without knowing anything a priori about the distribution of events between instants 0 and t (for example, again, the number of calls received by a telephone exchange between specified times), we can grant that the number $\Delta X = X(t + \Delta t) - X(t)$ of events between t and $t + \Delta t$ is independent of $X(\tau)$ for $\tau \leqslant t$. This hypothesis, whatever its legitimacy, means that the occurrence of an event between t and $t + \Delta t$ is independent of the previous occurrence of a similar event; under these conditions $X(t)$ is a stochastic function with *independent increments.* Let us grant, further, that the events under consideration constitute a stationary phenomenon. Then, by dividing the interval $(0,t)$ into n partial intervals, $X(t)$ can be regarded as the sum of n independent random variables, all obeying the same law.

Finally, if it is assumed, as it is natural to do, that whatever the $(0,t)$ interval, there is a zero probability that the number of events is infinite and a zero probability that several events will occur at the same instant, it can be shown that, whatever t may be, the random variable $X(t)$ obeys a Poisson law having the parameter $\lambda = at$, where a is a positive constant.

The foregoing process is called the *stationary Poisson process* and is characterized by the constant a, known as the *density of the process.* It can be shown to have the following properties: (1) over a given time interval $(\tau, \tau + \lambda)$ the events are distributed at random and independently of each other; (2) if t is a given instant and $t + v$ the instant at which the first event subsequent to t occurs, the probability density of the stochastic variable v is $f(v) = ae^{-av}$.

R. FÉRON

[*Directly related are the entries* PROBABILITY; QUEUES. *Other relevant material may be found in* DISTRIBUTIONS, STATISTICAL; *and in the biographies of* BORTKIEWICZ *and* LAPLACE.]

WORKS BY POISSON

(1811) 1842 *A Treatise of Mechanics.* London: Longmans. → First published as *Traité de mécanique.*

(1815) 1816 Mémoire sur la théorie des ondes. Académie des Sciences, Paris, *Mémoires* 2d Series 1:71–186.

1831 *Nouvelle théorie de l'action capillaire.* Paris: Bachelier.

1835a *Théorie mathématique de la chaleur.* Paris: Bachelier. → A supplement consisting of notes and a memoir was published in 1837.

1835b Recherches sur la probabilité des jugements, principalement en matière criminelle. Académie des Sciences, Paris, *Comptes rendus hebdomadaires* 1:473–494.

1835 POISSON, SIMÉON DENIS et al. Recherches de statistique sur l'affection calculeuse: Rapport sur ces recherches. Académie des Sciences, Paris, *Comptes rendus hebdomadaires* 1:167–177.

1836a Note sur le calcul des probabilités. Académie des Sciences, Paris, *Comptes rendus hebdomadaires* 2:395–398.

1836b Formules relatives aux probabilités qui dépendent de très grands nombres. Académie des Sciences, Paris, *Comptes rendus hebdomadaires* 2:603–613.

1836c Note sur la loi des grands nombres. Académie des Sciences, Paris, *Comptes rendus hebdomadaires* 2:377–382. → Includes three pages of comment.

1837 *Recherches sur la probabilité des jugements en matière criminelle et en matière civile, précédées des règles générales du calcul des probabilités.* Paris: Bachelier.

SUPPLEMENTARY BIBLIOGRAPHY

ARAGO, DOMINIQUE FRANÇOIS JEAN 1854 Poisson. Pages 593–671 in Dominique François Jean Arago, *Oeuvres complètes de François Arago.* Volume 2: Notices biographiques. Paris: Gide & Baudry. → See pages 672–689 for a bibliography edited by Poisson and pages 690–698 for "Discours prononcé aux funérailles de Poisson, le jeudi 30 avril 1840."

BIENAYMÉ, JULES 1855 Communication sur un principe que M. Poisson avait cru découvrir et qu'il avait appelé loi des grands nombres. Académie des Sciences Morales et Politiques, *Séances et travaux* 31:379–389.

FAURE, FERNAND 1909 *Les précurseurs de la Société de Statistique de Paris.* Nancy (France): Berger-Levrault.

GINI, CORRADO 1941 Alle basi del metodo statistico: Il principio della compensazione degli errori accidentali e la legge dei grandi numeri. *Metron* 14:173–240.

GOURAUD, CHARLES M. C. 1848 *Histoire du calcul des probabilités depuis ses origines jusqu'à nos jours.* Paris: Durand.

HAIGHT, FRANK A. 1967 *Handbook of the Poisson Distribution.* New York: Wiley.

KEYNES, JOHN MAYNARD (1921) 1952 The Law of Great Numbers. Pages 332–366 in John Maynard Keynes, *Treatise on Probability.* London: Macmillan.

LAPLACE, PIERRE SIMON DE (1820) 1951 *A Philosophical Study on Probabilities.* New York: Dover. → First published as *Essai philosophique sur les probabilités.*

LOTTIN, JOSEPH 1909 La théorie des moyennes et son emploi dans les sciences d'observation. *Revue néo-scolastique de philosophie* 16:537–569. → Now called *Revue philosophique de Louvain.*

LOTTIN, JOSEPH 1910 Le calcul des probabilités et les régularités statistiques. *Revue néo-scolastique de philosophie* 17:23–52.

MOIVRE, ABRAHAM DE (1718) 1756 *The Doctrine of Chances: Or, a Method of Calculating the Probabilities of Events in Play.* 3d ed. London: Millar.

PEARSON, EGON S.; and HARTLEY, H. O. (editors) (1954) 1966 *Biometrika Tables for Statisticians.* 3d ed. Vol. 1. Cambridge Univ. Press.

POINSOT, LOUIS 1836 [Remarques sur la communication de Poisson.] Académie des Sciences, Paris, *Comptes rendus hebdomadaires* 2:398–399. → See pages 399–400 for Poisson's reply.

WESTERGAARD, HARALD L. 1932 *Contributions to the History of Statistics.* London: King. → See pages 149–150 for a discussion of Poisson's concern with medical statistics.

POISSON DISTRIBUTION

See DISTRIBUTIONS, STATISTICAL, *article on* SPECIAL DISCRETE DISTRIBUTIONS.

POLLS

See ERRORS, *article on* NONSAMPLING ERRORS; PUBLIC POLICY AND STATISTICS; SAMPLE SURVEYS; STATISTICS AS LEGAL EVIDENCE; SURVEY ANALYSIS.

POLYNOMIAL REGRESSION

See LINEAR HYPOTHESES, *article on* REGRESSION.

POPULATION

See DEMOGRAPHY, *article on* THE FIELD.

POSTERIOR DISTRIBUTION

See BAYESIAN INFERENCE; PROBABILITY, *article on* INTERPRETATIONS.

POWER

See HYPOTHESIS TESTING; SIGNIFICANCE, TESTS OF.

PREDICTION

In sociological writing the term "prediction" means a stated expectation about a given aspect of social behavior that may be verified by subsequent observation. Within this general meaning the term is used in two principal senses: for deductions from known to unknown events within a conceptually static system and for statements about future outcomes based on recurring sequences of events. This article is largely restricted to the latter sense, although the former usage, which covers significant forms of logical reasoning, is widely prevalent.

The estimate of a given variable from one or more concurrent variables as in regression analysis is conventionally referred to as a prediction. Similarly, the estimate of a population characteristic may be referred to as a prediction, although the sample from which the inference is drawn is not separated in time from the population that it represents. Still more common, especially in the writing on social systems, is the designation of the term y in the expression "If x, then y," as a prediction, even though the x and y are often regarded as of simultaneous occurrence. Although such usage has been questioned on linguistic grounds, it is well established, and its currency is not likely to be affected by such arguments.

In its second principal sense the term "prediction" refers to assertions about future outcomes based on the observed regularities among consecutive events of the past. This category contains its own distinction, depending on whether the statement holds for a single, concrete instance, with due regard for the accidents of time and place, or abstractly holds for any case in a class satisfying stated conditions. If the statement is concrete and necessarily bound to the calendar, it carries the label "forecast"; otherwise, when it is not so restricted, it carries the more general term "prediction." According to this distinction, the expected volume of crime in the United States in the next calendar year would constitute a forecast, whereas the expected success of the individual on parole under specified conditions would be a prediction. Although this terminology is useful for distinguishing between special and general formulations, it has not been consistently applied, even in technical writing, and many so-called predictions would have to be relabeled as forecasts if the distinction were to be strictly maintained.

Prediction as a social process. At least some form of prediction, in the broadest sense of the term, is practiced on all levels of culture (Tylor 1871, chapter 4). The contemporary emphasis in sociology, as described below, is thus consistent with traditional enterprise, answering to the same general purposes but differing in the process by which the foreknowledge is obtained. The social purpose of prediction, whether of physical or social events, is to secure a measure of control over what otherwise would be less manageable circumstances. The effects of such natural calamities as typhoons and floods may be mitigated, if not averted, by forehanded preparation; similarly, by stating the conditions under which a social upheaval can occur, steps may be taken to prevent the occurrence of one. Some of the Biblical prophecies were of this nature, urging the people to righteousness in order to avoid the wrath of God. At times such a statement may be more in the nature of a promise than a threat, setting forth the conditions to be met in order to achieve a desired objective, such as an annuity upon retirement. But however they differ in meaning, practically all predictions are potential instruments of social action, enabling the group either to facilitate a favorable outcome or to impede an unfavorable one.

The process of sociological prediction. Although all predictions are alike in broad social purpose, they differ in the process of their formulation, which will be more or less scientific according to the nature of the underlying analysis. The process of sociological prediction has in varying degree those elements common to all scientific prediction: some theory of behavior from which deductions may be drawn and some factual evidence that is relevant to the propositions of the theory. Sociological prediction has arisen naturally from the concerns of sociology itself, both theoretical and empirical, and thus sociologists now take prediction of the forms and processes of social life as one of their principal tasks. This commitment has its roots in the writings of Auguste Comte and has been regularly affirmed by leading representatives of the discipline since that time. Max Weber held that the purpose of sociology is to predict the patterns of social interaction, and Albion Small, one of the founders of American sociology, took very much the same position (1916).

Although sociological predictions ideally are to be drawn from theory, for the most part they have been little more than statistical projections based on compilations of empirical data within categories of perhaps little theoretical significance. But such compilations in the form of time series and actuarial tables have had their bearing on theory. For example, the hypothesis of "cultural lag" (Ogburn 1922) was derived in part from the empirical growth curve of inventions, and Edwin H. Sutherland's theory of "differential association" has been refined on the basis of parole prediction studies (see Glaser 1954). In this way the construction of statistical trends and experience tables and their corresponding projections have had some impact on theory, in both extending and recasting it. Nevertheless, it is not the statistical materials and their manipulation that give sociological prediction its special character, for comparable series and their analysis are part of the natural sciences; rather, it is the underlying categories from which sociological predictions are derived.

Prediction research. Although many sociological investigations have a bearing on the predictability of social and cultural events, relatively few studies have had prediction as their primary goal. For the purposes of outlining these more specialized studies and citing examples, prediction research will be classified here according to whether its focus is the collective characteristics of the group or the characteristics of its constituent members.

In those studies analyzing the collective aspects of the group, the prediction has in some cases extended over several classes of events, whereas in various others the prediction has been restricted to a single outcome. The prediction of a relatively wide range of events is perhaps best represented by the work of William F. Ogburn, a consistent

theme of which was the proposition that technological trends of the past provide a useful key to cultural trends of the future. This idea was developed in Ogburn's work during the 1930s, which included reviews of selected social and economic trends in the United States from 1900 to 1930 (President's Research Committee . . . 1933) and a government report on social and economic conditions affecting the rate of invention and the impact of invention on social life (U.S. National Resources Committee, Science Committee 1937), and it was later restated in *The Social Effects of Aviation* (Ogburn et al. 1946).

Not all studies of trends and cycles have been at the societal level; some have been concerned with the pattern of interaction and the sequence of its development within the small group. The work of Bales and his associates (summarized in Bales 1959) provides an example of this approach, and Bales's writings contain an assessment of its potential for predictive knowledge.

The prediction of a single social outcome may be illustrated by the election forecast, since its problems are well defined (Mosteller et al. 1949) and its operating procedures are well standardized. Such a forecast rests on a succession of carefully designed and drawn sample polls taken at regular intervals shortly before the election. Based on the trend of these results, with due allowance for sampling and measurement error, the percentage of the vote for each candidate is predicted, and thus the probable winning candidate can be named. The critical matters bearing on the accuracy of such prediction include the correspondence between the sampled population and the population actually voting on election day as well as the stability of the observed trend, at least through the day of the election. Notwithstanding these and related difficulties, scientific polling agencies have been quite successful in predicting the results of political elections held in the United States since 1952. Students of this process have noted that such predictions may affect the election itself.

With minor exceptions, the prediction of individual behavior has been limited to those forms of personal adjustment whose variation is thought to be largely due to differences in social background and circumstance: for example, adjustment in the armed forces (Star 1950), postwar adjustment (Cottrell 1949), adjustment on parole (Burgess 1928), and adjustment in marriage (Burgess & Cottrell 1939).

The device by which the prediction of personal adjustment is usually effected is the experience table, which in principle is no different from the actuary's life table that shows probabilities of death by age. Similarly, the table of social experience gives the odds of success for the several subclasses into which the population has been arranged, and it thereby yields a prediction for the individual case; obviously, the more nearly the probabilities approach zero or one, the more accurate is the prediction for each person. This method can be illustrated by a table from Burgess' early study of factors determining success on parole (see Table 1); this table, an adaptation of the original,

Table 1 — Frequency distribution of 1,000 parolees, by prediction score

Prediction score*	Number of men in each class interval	Per cent violators of parole
16–21	68	1.5
14–15	140	2.2
13	91	8.8
12	106	15.1
11	110	22.7
10	88	34.1
7–9	287	43.9
5–6	85	67.1
2–4	25	76.0

* Score for each parolee is the number of factors for which he scored above the group mean.

Source: Adapted from Burgess 1928, p. 248.

shows the percentage of failures (that is, parole violators) for each of the classes into which 1,000 parolees had been grouped according to scores based on 21 factors correlated with outcome on parole. This table, which may be regarded as stating a set of empirical probabilities, is characteristic of practically all studies seeking to predict the behavior of the individual.

Recent studies of social adjustment are somewhat distinctive in their employment of refined statistical methods to differentiate between successes and failures. An important innovation consists in their attention to the possible discrepancy between decisions maximizing predictive accuracy and those minimizing social cost. Ohlin (1951) discusses the relevance of prediction tables for parole selection and analyzes their differing purposes, and Cronbach and Gleser (1957) provide a comparable discussion of personnel selection based on psychological test scores.

Techniques of social prediction

The derivation of a prediction is usually accomplished by general statistical methods, none of which is restricted in its application to social data. Despite their general familiarity, these methods will be briefly reviewed here, primarily to illustrate representative applications in social prediction. Although not all statistical predictions are derived in the same way, all share the requirement that both predictor and criterion variables be subject to reliable measurement. Except for this scant reference, and despite its importance, the problem of measurement will be ignored in the following discussion.

Extrapolation. Roughly speaking, extrapolation is the process of predicting a variable from itself—for example, predicting the future growth of a population from its past growth. By this technique it is possible to obtain a succession of expected values that are arrayed in the future from least to most distant. Such expected values will materialize only if the underlying social process or causal system that the curve expresses is constant for the period over which the prediction extends. Prediction by extrapolation will be in error to the degree that a given process changes; accordingly, prediction by this method will generally be more accurate for shorter rather than longer durations. An example of extrapolation is provided by Hart's prediction of life expectancy on the basis of the trend in life expectancy during the last 75 years (1954). [For other examples of extrapolation see PREDICTION AND FORECASTING, ECONOMIC; see also Grauman 1968.]

Correlation methods. Reduced to its lowest terms, prediction by correlation is based on measured linkages between earlier and later events in a given sequence—for example, between scholastic performance in high school and that in college, between adjustment in childhood and that in marriage. It goes without saying that such obtained relationships will have no predictive utility beyond the specific population for which they hold. Thus, if the correlation between type of infant feeding and social maturity (to take a hypothetical example) obtains only in the middle class, the former category will be worthless as a predictor of social maturity in the lower class.

In basing our prediction on correlated events, the guiding principle is that the errors of prediction must be minimized in some well-defined sense. To illustrate the application of this principle and also to mark points of entry into the pertinent literature, we will consider in barest detail the most common procedures for predicting a variable from a set of attributes or variables and those for predicting an attribute from a set of attributes or variables. When the criterion consists of more than a single element (Hotelling 1935) and/or when attributes and variates are employed together as predictors (Mannheim & Wilkins 1955), the procedures will be more complicated but no different in principle.

In predicting a variable from one or more variables, the practice is to predict so that the sum of the squared errors around the fitted curve is a minimum. When the fitted curve is linear, as is usually the case, the product-moment coefficient of correlation r, or an adaptation of it (partial or multiple), serves to gauge the relative accuracy of comparable predictions. For example, a succession of similar studies of marriage obtained the following correlations between scores considered to be prognostic of marital adjustment and scores of actual adjustment in marriage: Burgess and Cottrell, .51 (1939); Burgess and Wallin, .50 (1953); Terman, .54 (see Terman et al. 1938).

In predicting a quantitative variable from one or more attributes, the rule is to determine the predicted values so as to maximize the sum of squares between groups and correspondingly to minimize the sum of squares within groups, as in the analysis of variance; hence, the coefficient of intraclass correlation, or an equivalent, may be taken as a measure of predictive accuracy. The fact that no instance of such a measure appears in the bibliography to this article is intended to suggest the possibly limited value of attributes for purposes of predicting quantitative variables.

Where the prediction is from one or more attributes to a single attribute, the rule is to predict repeatedly whichever attribute has the largest subclass frequency, or conditional probability (Guttman 1941). The accuracy of such prediction will vary according to the difference between the conditional and marginal probabilities. Since measures of association such as C, T, or phi reflect that difference, they have been widely used to gauge the predictive accuracy of attributes. The examples given in Table 2, which are from one of the Gluecks' early studies of recidivism, will serve to illustrate the form and usual magnitude of such coefficients.

Finally, where the prediction is from one or more variables to a single attribute the preferred procedure is to derive from the variables a composite variable that will yield the least overlap among the several within-class distributions and hence the least error in prediction. When the composite variable is a linear function of its components, the procedure is termed a *discriminant function* (Fisher 1936). The experience table for predicting recidivism among Borstal (i.e., reform school) lads (Mannheim & Wilkins 1955) was based on this method and is indicative of both its value and limitations; Kirby (1954) has also used this method, in a parole prediction study.

Markov chains. Although the Markov chain, as well as the general class of stochastic processes to which it belongs, is covered in most standard references on probability (Feller 1950–1966), its potentialities for sociological prediction have only recently been explored, and these efforts have been largely, if not exclusively, limited to the fitting of historical data to theoretical chains. Although such materials indicate whether conditions as observed at time t_k might have been accurately predicted by Markov methods beginning at time t_0, they contain no demonstration that a given process will repeat itself indefinitely and therefore give no indication whether it may be confidently used for predictive purposes. This remark, however, should not be construed as a criticism of the studies cited illustratively below, which purported to be more suggestive than conclusive.

The feasibility of predicting political attitudes has been considered by Anderson (1954) in a secondary analysis of panel data. His particular concern was with the probability that a person would hold the same political opinion at the end of a sequence that he held at the start. To determine whether such probabilities might be attained by Markovian methods, a comparison was drawn between distributions based on opinions expressed by a panel of voters in each of the six months immediately preceding the U.S. presidential election of 1940 and those expected on the basis of probability theory. Although the findings of this particular analysis were inconclusive, they do hint at the potentialities of the Markov chain as a device for analyzing the process of attitude change and as a tool for social prediction.

The correspondence between actual patterns of labor mobility and those obtained by treating the movement of workers as a Markov process has been examined by Blumen, Kogan, and McCarthy (1955). Although this investigation was concerned primarily with the dynamics of labor mobility, it has considerable relevance for prediction in its emphasis on statements expressing the probability that a worker in a given industrial group will be

Table 2 — Correlation between recidivism and selected social traits

Social trait	Coefficient of contingency (C)
Mental condition	.43
Work habits	.35
Economic responsibility	.28

Source: Adapted from Glueck & Glueck 1937, p. 135.

in that same group after k intervals of time. Apart from its substantive value for industrial sociology, this study is constructively important as a demonstration of the potential utility of Markov theory in the prediction of mobility, both social and geographical.

Current problems

During the past fifty years of American sociology, the problems, or problematics, of prediction have received at least as much attention as the predictions themselves, which, as suggested in the foregoing discussion, have been relatively few in number and circumscribed in content. The range of these problems is reflected in the kinds of issues that have been regularly debated in the sociological journals. Some of these issues are briefly presented here.

Actuarial approach versus case method. The issue of the actuarial approach versus the case method has arisen in the prediction of personal adjustment. This issue has two parts: whether it is possible to frame a prediction solely from case materials in total disregard of all probabilities, and whether such prediction, when there is no explicit reference to actuarial materials, is more accurate than that based on statistical averages, or rates. Most serious students of prediction would answer "no" to the first question and "possibly under some circumstances" to the second.

Prediction that claims to be wholly devoid of probabilities is regarded as logically impossible by some students (Social Science Research Council 1941). Briefly put, their argument is: notwithstanding the claims of the predictor to the contrary, the predicted case will be treated as a class member or located in a risk table although that table may be wholly based on the experience of the forecaster and may exist only in his mind. That the process is subjective and personal does not alter its essential nature as probabilistic; therefore—so runs the argument—all prediction is actuarial and, correspondingly, no prediction is wholly free of uncertainty.

The second answer, "possibly under some circumstances," shows a recognition that predictions from extensive case materials may in some instances be more accurate than those based on group averages, by virtue of the analyst's exceptional ability to assign cases to risk categories having probabilities very close to zero or one. However, such idiosyncratic accuracy embodies "procedures" that cannot be readily codified and transmitted to the public, as would be required of scientific methods. Hence, prediction from case materials, no matter how accurate, must be regarded as an expression of clinical insight and judgment rather than as the application of scientific law.

Prediction as feedback. Despite the antiquity of the idea that a stated prediction may affect its own fulfillment (Popper 1957) and despite the relevance of this idea to social prediction, few empirical studies have sought to measure the influence of predictions on subsequent events. However, an illustrative literature has emerged, including references to rumor, stereotyped expectation, the election forecast, and false prophecy (see especially Merton 1948; Simon 1954; Festinger et al. 1956). Students of voting behavior have noted the possible effects of a computer prediction of the election of specific candidates based on national election results in an earlier time zone on the pattern of voting in a later time zone.

Such materials lend themselves to systematic classification according to whether the outcome is favorable or unfavorable and according to whether the effect of prediction is positive or negative. Thus, the prognosis that the patient will recover may generate either a confidence that will hasten recovery or an overconfidence that will delay it; on the other hand, an unfavorable prognosis of chronic illness may create either an attitude of resignation or an attitude of defiance, with differing effects. Instances of this kind support the common opinion that by reason of the symbolic nature of human life, social prediction is reflexive and that in consequence the validity of social prediction is subject to greater uncertainty than physical prediction.

Efficiency of social prediction. Broadly speaking, the concept of the efficiency of prediction refers to the accuracy of a given method of prediction relative to that of an alternative that is taken as a standard or norm (Reiss 1951). If, for example, the chosen standard produces 20 errors per 100 trials and the alternative in question produces only 10, then the alternative may be said to be twice as efficient as the standard.

Usually the predictions compared are that derived from the joint distribution of a criterion and one or more predictors and that derived from the distribution of the criterion alone. Thus, given a male delinquency rate of 20 per cent, the prediction of nondelinquency on repeated trials would carry an error rate of 20 per cent; if, after introducing social class as a predictor, the error rate is reduced to 10 per cent, then the prediction from the two-way classification may be said to be twice as efficient as prediction from the single classification of children as delinquent or nondelinquent.

By and large, the measured efficiency of social prediction has been relatively low. For example, in criminological studies, where the identification of the prospective delinquent or recidivist would be of considerable practical importance, the gains resulting from the introduction of information thought to bear on such behavior have been negligible (Schuessler 1954). To achieve substantial gains in efficiency it will be necessary to identify and measure those variables that are closely correlated with the criterion.

Explanation versus prediction. Although explanation and prediction are not irreconcilable or even rival alternatives, there are differences of opinion about which of these should receive the greater emphasis. In general, this disagreement reflects the difference between a preference for theory and an emphasis on factual research.

In this disagreement, those placing greater emphasis on explanation usually argue as follows: Predictions (hypotheses) may be deduced from a general explanation (theory), but a collection of statistical predictions, no matter how accurate, does not constitute a theory; hence, explanation should be the first order of business. Moreover, obtained statistical regularities may be an accident of time or place and hence an unstable basis for prediction, whereas a scientific law is universal and therefore an unerring source of prediction. According to this line of reasoning, the poor showing of prediction in sociology is a reflection of crudities in general theory. Thus, the development of a set of valid explanations is considered to be a prerequisite for the improvement of predictions.

Those putting greater stress on prediction would probably not disagree with the premise of the foregoing argument, although they would emphasize the interplay between statistical association and theoretical explanation and the steady impact of the former on the latter. Furthermore, they would hold that if they are to frame hypotheses and to test them, they have no choice but to operate pragmatically within existing theory, reforming it as they proceed. Probably the rank and file of American sociologists are "agnostics" with regard to this issue, excepting those whose major interest lies in the philosophy of social science.

Limits of prediction. Another issue is the general question whether some classes of social events are inherently unpredictable. Although this question is valid, it necessarily admits of only a speculative answer. The judgment that the "future course of human affairs is unpredictable" (Toynbee 1934–1961, vol. 12) has to do with the broad question of cultural history or evolution and has little to do with prediction as a specialty within sociology. In general, sociologists have been primarily concerned with problems of much smaller compass: land use, migration, suburban growth, fertility rates of human populations, rates of assimilation, patterns of racial violence, and political movements. Nevertheless, the importance of predicting social mutations as well as recurring phenomena has been increasingly emphasized in sociological writings (e.g., Moore 1964), and it is to be anticipated that this problem will be studied empirically in the coming decades. Such factual studies will enable sociologists to set provisional limits to the range of social prediction (Bell 1965).

In conclusion, it should be noted that prediction and sociology have always been closely linked: sociology grew out of a concern with prediction (Comte's *savoir pour prévoir*) and has always had the securing of predictive knowledge as one of its express aims. However, relatively few empirical studies have been specifically concerned with producing it. Thus, the importance of recent prediction studies lies as much in the methodological understandings that have grown out of them as in the substantive findings themselves. As these understandings become systematized and gain in currency, a wider variety of factual investigations will probably appear as a sequel to the pioneering work of the last several decades.

KARL F. SCHUESSLER

[*Directly related are the entries* CAUSATION; LIFE TABLES; MARKOV CHAINS; PREDICTION AND FORECASTING, ECONOMIC. *Other relevant material may be found in* LINEAR HYPOTHESES; MULTIVARIATE ANALYSIS. *See also* Barnes 1968; Bendix 1968; Glaser 1968; Jaffe 1968; König 1968; Stokes 1968.]

BIBLIOGRAPHY

ANDERSON, T. W. 1954 Probability Models for Analyzing Time Changes in Attitudes. Pages 17–66 in Paul F. Lazarsfeld (editor), *Mathematical Thinking in the Social Sciences.* Glencoe, Ill.: Free Press.

BALES, ROBERT F. 1959 Small-group Theory Research. Pages 293–305 in American Sociological Society, *Sociology Today.* Edited by Robert K. Merton, Leonard Broom, and Leonard S. Cottrell. New York: Basic Books.

► BARNES, HARRY ELMER 1968 Small, Albion W. Volume 14, pages 320–322 in *International Encyclopedia of the Social Sciences.* Edited by David L. Sills. New York: Macmillan and Free Press.

BELL, DANIEL 1965 Twelve Modes of Prediction. Pages 96–127 in Julius Gould (editor), *Penguin Survey of the Social Sciences 1965.* Baltimore: Penguin.

► BENDIX, REINHARD 1968 Weber, Max. Volume 16, pages 493–502 in *International Encyclopedia of the Social Sciences.* Edited by David L. Sills. New York: Macmillan and Free Press.

BLUMEN, ISADORE; KOGAN, MARVIN; and McCARTHY, PHILIP 1955 *The Industrial Mobility of Labor as a Probability Process.* Cornell Studies in Industrial and Labor Relations, Vol. 6. Ithaca, N.Y.: Cornell Univ. Press.

BORGATTA, EDGAR F.; and WESTOFF, CHARLES F. 1954 The Prediction of Total Fertility. *Milbank Memorial Fund Quarterly* 32:383–419.

BOWERMAN, CHARLES E. 1964 Prediction Studies. Pages 215–246 in Harold T. Christensen (editor), *Handbook of Marriage and the Family.* Chicago: Rand McNally.

BURGESS, ERNEST W. 1928 Factors Determining Success or Failure on Parole. Pages 203–249 in Illinois, Committee on Indeterminate-sentence Law and Parole, *The Workings of the Indeterminate-sentence Law and the Parole System in Illinois.* Springfield, Ill.: Division of Pardons and Paroles.

BURGESS, ERNEST W.; and COTTRELL, LEONARD S. 1939 *Predicting Success or Failure in Marriage.* New York: Prentice-Hall.

BURGESS, ERNEST W.; and WALLIN, PAUL 1953 *Engagement and Marriage.* Philadelphia: Lippincott.

CANTRIL, HADLEY 1938 The Prediction of Social Events. *Journal of Abnormal and Social Psychology* 33:364–389.

CHAPIN, FRANCIS S. (1947) 1955 *Experimental Designs in Sociological Research.* Rev. ed. New York: Harper.

CLAUSEN, JOHN A. 1950 Studies of the Postwar Plans of Soldiers: A Problem in Prediction. Pages 568–708 in Samuel A. Stouffer et al., *Measurement and Prediction.* Studies in Social Psychology in World War II, Vol. 4. Princeton Univ. Press.

COTTRELL, LEONARD S. 1949 The Aftermath of Hostilities. Pages 549–595 in Samuel A. Stouffer et al., *The American Soldier.* Studies in Social Psychology in World War II, Vol. 2. Princeton Univ. Press.

CRONBACH, LEE J.; and GLESER, GOLDINE C. (1957) 1965 *Psychological Tests and Personnel Decisions.* 2d ed. Urbana: Univ. of Illinois Press.

DOLLARD, JOHN 1948 Under What Conditions Do Opinions Predict Behavior? *Public Opinion Quarterly* 12:623–632.

DUNCAN, OTIS DUDLEY et al. 1953 Formal Devices for Making Selection Decisions. *American Journal of Sociology* 58:573–584.

FELLER, WILLIAM 1950–1966 *An Introduction to Probability Theory and Its Applications.* 2 vols. New York: Wiley. → A second edition of the first volume was published in 1957.

FESTINGER, LEON; RIECKEN, H. W.; and SCHACHTER, STANLEY 1956 *When Prophecy Fails.* Minneapolis: Univ. of Minnesota Press.

FISHER, R. A. 1936 The Use of Multiple Measurements in Taxonomic Problems. *Annals of Eugenics* 7:179–188.

GLASER, DANIEL 1954 A Reconsideration of Some Parole Prediction Factors. *American Sociological Review* 19:335–341.

►GLASER, DANIEL 1968 Penology: II. Probation and Parole. Volume 11, pages 518–523 in *International Encyclopedia of the Social Sciences.* Edited by David L. Sills. New York: Macmillan and Free Press.

GLUECK, SHELDON; and GLUECK, ELEANOR 1937 *Later Criminal Careers.* New York: Commonwealth Fund.

GLUECK, SHELDON; and GLUECK, ELEANOR 1959 *Predicting Delinquency and Crime.* Cambridge, Mass.: Harvard Univ. Press.

GOODMAN, LEO 1952 Generalizing the Problem of Prediction. *American Sociological Review* 17:609–612.

►GRAUMAN, JOHN V. 1968 Population: VI. Population Growth. Volume 12, pages 376–381 in *International Encyclopedia of the Social Sciences.* Edited by David L. Sills. New York: Macmillan and Free Press.

GUTTMAN, LOUIS 1941 Mathematical and Tabulation Techniques. Pages 251–364 in Social Science Research Council, Committee on Social Adjustment, *The Prediction of Personal Adjustment,* by Paul Horst et al. New York: The Council.

HART, HORNELL 1954 Expectation of Life: Actual Versus Predicted Trends. *Social Forces* 33:82–85.

HOTELLING, HAROLD 1935 The Most Predictable Criterion. *Journal of Educational Psychology* 26:139–142.

►JAFFE, A. J. 1968 Ogburn, William Fielding. Volume 11, pages 277–281 in *International Encyclopedia of the Social Sciences.* Edited by David L. Sills. New York: Macmillan and Free Press.

KAPLAN, A.; SKOGSTAD, A. L.; and GIRSHICK, M. A. 1950 The Prediction of Social and Technological Events. *Public Opinion Quarterly* 14:93–110.

KIRBY, BERNARD C. 1954 Parole Prediction Using Multiple Correlation. *American Journal of Sociology* 59:539–550.

►KÖNIG, RENÉ 1968 Comte, Auguste. Volume 3, pages 201–206 in *International Encyclopedia of the Social Sciences.* Edited by David L. Sills. New York: Macmillan and Free Press.

LAVIN, DAVID E. 1965 *The Prediction of Academic Performance: A Theoretical Analysis and Review of Research.* New York: Russell Sage Foundation.

LAZARSFELD, PAUL F.; and FRANZEN, RAYMOND H. 1945 Prediction of Political Behavior in America. *American Sociological Review* 10:261–273.

McCORMICK, THOMAS C. 1952 Toward Causal Analysis in the Prediction of Attributes. *American Sociological Review* 17:35–44.

MANNHEIM, HERMANN; and WILKINS, LESLIE T. 1955 *Prediction Methods in Relation to Borstal Training.* London: H.M. Stationery Office.

MEEHL, PAUL E. 1954 *Clinical Versus Statistical Prediction.* Minneapolis: Univ. of Minnesota Press.

MERTON, ROBERT K. (1948) 1957 The Self-fulfilling Prophecy. Pages 421–436 in Robert K. Merton, *Social Theory and Social Structure.* Rev. & enl. ed. Glencoe, Ill.: Free Press.

MOORE, WILBERT E. 1964 Predicting Discontinuities in Social Change. *American Sociological Review* 29:331–338.

MOSTELLER, FREDERICK et al. 1949 *The Pre-election Polls of 1948.* New York: Social Science Research Council.

OGBURN, WILLIAM F. (1922) 1950 *Social Change, With Respect to Culture and Original Nature.* New edition with supplementary chapter. New York: Viking.

OGBURN, WILLIAM F. 1934 Studies in Prediction and the Distortion of Reality. *Social Forces* 13:224–229.

OGBURN, WILLIAM F.; ADAMS, JEAN L.; GILFILLAN, S. C. 1946 *The Social Effects of Aviation.* Boston: Houghton Mifflin. → Includes an extended discussion of the methodology of prediction and a selected bibliography.

OHLIN, LLOYD E. 1951 *Selection for Parole.* New York: Russell Sage Foundation.

POPPER, KARL R. 1957 *The Poverty of Historicism.* Boston: Beacon.

PRESIDENT'S RESEARCH COMMITTEE ON SOCIAL TRENDS

1933 *Recent Social Trends in the United States*. 2 vols. New York: McGraw-Hill. → Incorporates the results of thirty studies of selected social and economic trends in the United States from 1900 to 1930.

REISS, ALBERT J. JR. 1951 The Accuracy, Efficiency, and Validity of a Prediction Instrument. *American Journal of Sociology* 56:552–561.

SAWYER, JACK T. 1966 Measurement and Prediction, Clinical and Statistical. *Psychological Bulletin* 66:178–200.

SCHUESSLER, KARL F. 1954 Parole Prediction: Its History and Status. *Journal of Criminal Law, Criminology, and Police Science* 45:425–431.

SIMON, HERBERT A. (1954) 1957 Bandwagon and Underdog Effects of Election Predictions. Pages 79–87 in Herbert A. Simon, *Models of Man: Social and Rational; Mathematical Essays on Rational Human Behavior in a Social Setting*. New York: Wiley.

SMALL, ALBION W. 1916 Fifty Years of Sociology in the United States (1865–1915). *American Journal of Sociology* 21:721–864.

SOCIAL SCIENCE RESEARCH COUNCIL, COMMITTEE ON SOCIAL ADJUSTMENT 1941 *The Prediction of Personal Adjustment*, by Paul Horst et al. New York: The Council.

SOROKIN, PITIRIM A. (1941) 1962 *Social and Cultural Dynamics*. Volume 4: Basic Problems, Principles and Methods. Totowa, N.J.: Bedminster Press.

STAR, SHIRLEY A. 1950 The Screening of Psychoneurotics in the Army: Technical Development of Tests. Pages 486–567 in Samuel A. Stouffer et al., *Measurement and Prediction*. Studies in Social Psychology in World War II, Vol. 4. Princeton Univ. Press.

►STOKES, DONALD E. 1968 Voting. Volume 16, pages 387–395 in *International Encyclopedia of the Social Sciences*. Edited by David L. Sills. New York: Macmillan and Free Press.

TERMAN, L. M. et al. 1938 *Psychological Factors in Marital Happiness*. New York: McGraw-Hill.

THORNDIKE, EDWARD L. et al. 1934 *Prediction of Vocational Success*. New York: Commonwealth Fund.

TOYNBEE, ARNOLD J. 1934–1961 *A Study of History* 12 vols. Oxford Univ. Press.

TYLOR, EDWARD B. (1871) 1958 *Primitive Culture: Researches Into the Development of Mythology, Philosophy, Religion, Art and Custom*. Volume 1: Origins of Culture. Gloucester, Mass.: Smith.

U.S. NATIONAL RESOURCES COMMITTEE, SCIENCE COMMITTEE 1937 *Technological Trends and National Policy, Including the Social Implications of New Inventions*. Washington: Government Printing Office.

U.S. NATIONAL RESOURCES COMMITTEE, SCIENCE COMMITTEE 1938 *The Problems of a Changing Population*. Report of the Committee on Population Problems. Washington: Government Printing Office.

WESTOFF, CHARLES F.; SAGI, PHILIP C.; and KELLY, E. LOWELL 1958 Fertility Through Twenty Years of Marriage: A Study in Predictive Possibilities. *American Sociological Review* 23:549–556.

Postscript

Work on sociological prediction has changed little in either pattern or significance: intermittent effort and perhaps some progress but no critical breakthroughs. As before, sociologists seem to be as concerned with the form of prediction as with the substance, while still disclaiming an interest in prediction per se. In their writing they seldom make fine distinction between prediction in the sense of deduction from theory and prediction in the sense of forecast of the future.

Increased attention has, however, been given to forecasting the future, possibly because of the public's growing concern over the limits of social and economic growth, and partly because of the public's expectation that sociologists contribute something to public policy. To mark this trend, the scope of this review has been broadened to cover prediction in both of its principal meanings. However, as in the main article, discussion is limited largely to quantitative prediction.

Predicted differences. In analyzing patterns of racial discrimination, sociologists have given descriptions of what blacks would be like if they were more like whites. Since these statements hold for hypothetical conditions, they may be regarded as sociological predictions.

Cutright's work (1974, table 1, p. 6) supplies an example. His sample consisted of 1,439 black and 5,573 white males, ages 30–37; his procedure was to calculate the regressions of yearly earning on years of schooling, for blacks and whites separately. Extrapolating the regression of earnings on schooling for blacks gives their expected earnings if they had as much schooling as whites. These calculations, shown in Table 3, answer the question of what blacks would earn on the average if they had the same average schooling as whites.

Table 3 — Comparisons between black and white earnings

	White	Black	Difference
Average earnings	$6,077	$3,481	$2,596
Dollar value of school year (slope)	$ 476	$ 129	$ 347
Regression value for blacks at white average schooling and average earnings for whites	$6,077	$3,766	$2,311

Source: Cutright 1974, table 1, p. 6.

They are quoted here not for their substantive interest but rather to illustrate a form of prediction that is increasingly common. Such predictions generally provide the expected change in the average standing of the members of the minority on one scale (for example, income) after giving them statistically the benefit of the average standing of the majority on some other scale (education) or scales.

Markov projections. Sociologists, albeit in very small numbers, continue to be intrigued by the possibility that a population distribution at the end of a given period may be projected from the distribution at the start by means of transition probabilities. To date, however, this possibility has received little empirical testing. The available evidence suggests that social moving, in the sense of either territorial migration or social mobility, does not conform to a simple Markov chain and that projections based on this assumption will probably go awry. [See MARKOV CHAINS.] Since a simple Markov projection for the population assumes that transition probabilities are identical for individuals, and stable over time, and since these assumptions are unrealistic, it is natural that efforts have been directed mainly to making population projections under sociologically more realistic assumptions. McFarland (1970) and Spilerman (1972) are examples. [See SOCIAL MOBILITY.]

McFarland's proposed method, considerably oversimplified, involves averaging individual transition probabilities and entering these average values in the population transition matrix. Manipulating this matrix gives a population projection adjusted for individual differences in transition probabilities. Spilerman (1972) devised a procedure for estimating individual transition probabilities and their changes over time from categories of relevant sociological information. His rationale is that the individual's tendency to move or stay, either territorially or socially, will depend on such factors as age and race, and that projections will be made more accurate by taking such factors into account. His demonstration, based on migration data for the 1930s and 1940s, lends some support to his claim.

Structural equation models. Since the mid-1960s, sociologists in increasing numbers have engaged in a discussion of the potentiality of structural equation models for social research in general (Goldberger & Duncan 1973) and for social forecasting in particular (Spilerman 1975). One theme of this discussion is that it will be possible to predict social conditions at the end of some period from conditions at the start, provided that structural relations have been accurately specified, and provided that the system itself is stable over time.

Although this idea is very appealing in the abstract, its concrete application in sociology is likely to encounter obstacles. The concept of social system is itself an obstacle. Although economists often seem to be reasonably clear about the systems they wish to model, sociologists seem less certain about the social structures they wish to represent mathematically. If they disregard educational, economic, and political activities, they are reduced to modeling systems of residual categories; if their model includes these major social activities, they seem to do little more than assemble parts prefabricated by economists and political scientists. In any event, it is doubtful that modeling with structural equations will go very far in sociology in the absence of clear notions about the underlying social structures to be modeled. [See SURVEY ANALYSIS; MULTIVARIATE ANALYSIS, *article on* CORRELATION METHODS; SIMULTANEOUS EQUATION ESTIMATION.]

Sociological measurement constitutes another major obstacle. If the structural model specifies continuous variation, it will be necessary to represent concepts by continuous measures; if no continuous measures are available or if available measures are inferior, then it will be impossible to fit the model. This is the predicament in which sociologists commonly find themselves: they cannot fit their models either because requisite measures are unavailable or because available measures are of uncertain reliability and validity.

Social indicators. The effort to develop social indicators has been justified in part by its relevance for social forecasting. The term "social indicator" has come to mean a series of periodic measures indexing some significant aspect of people's social condition. The concept is not new and is implicit in the work of Sorokin and Ogburn, to say nothing of Quetelet. Only since the mid-1960s, however, have American sociologists made a concerted effort to sort out the most significant social indicators and to provide for systematic publication. Although social indicators are interrelated generally, they are usually separated into eight groups: health, public safety, education, employment, income, housing, leisure and recreation, and population. For a synopsis of this movement and pertinent bibliography, see Sheldon and Parke (1975).

Social indicators, in the sense of reliable time series, are essential for testing structural models about the social process and for making and testing social forecasts. For example, it will be impossible to confirm the presence of periodicities in a social process, and to take them into account in forecasting, unless a time series of significant duration has been compiled. Similarly, in extrapolating trend lines, it is necessary to have extensive time series in order to obtain reliable estimates of the change coefficients. Thus the availability of reliable and meaningful social indicators will permit the sociologist to arrive at social forecasts comparable in accuracy to population and economic market forecasts.

The sociology of the future. The sociological study of the future (Bell & Mau 1971) is presently more a possible than actual specialty; little or no research has been done under its auspices, and even its proposed subject matter is somewhat hazy, since there are no future facts, only possibilities, a point conceded by its exponents. It is mentioned here because of its concern with anticipating (predicting) the future. However, its aim is not merely to understand the future but also to shape it; and in this latter emphasis, it deviates from orthodox sociology, which limits itself largely to observing and explaining social events.

In shaping the future, it is necessary to consider not only alternative possibilities (scenarios) but also the likelihood of their ever materializing; and in assigning probabilities to alternative futures, it is necessary to rely on the experience of the past. In analyzing this experience, one will of necessity fall back on statistical time series and their generating mechanisms, which, if not misspecified, seem to afford the most secure basis for both anticipating and shaping the future. Social prognosis and treatment, no less than clinical prognosis, seem to be ultimately dependent on statistical knowledge of the past. This is not to question the legitimacy of the future as a focus of sociological study, but simply to note that the method of this specialty will probably be some form of time-series analysis. [See TIME SERIES.]

The conclusion of the main article is still apt: much methodological exploration but little in the way of confirmed prediction. Some sociologists believe that the method of sociometrics—equivalent by definition to that of econometrics—holds much promise for obtaining predictive knowledge; the work of Blau and Duncan (1967) on occupational mobility is cited to back up this claim. Whatever their opinion about sociometrics, most sociologists would concur in the requirements that the lines of influence among the elements (variables) comprising a system be precisely specified and that the magnitude of these effects be reliably estimated. Few sociologists would regard these requirements as impediments to attaining predictive knowledge in sociology.

KARL F. SCHUESSLER

ADDITIONAL BIBLIOGRAPHY

BELL, WENDELL; and MAU, JAMES A. (editors) 1971 *The Sociology of the Future.* New York: Russell Sage.

BLAU, PETER M.; and DUNCAN, OTIS DUDLEY 1967 *The American Occupational Structure.* New York: Wiley.

BURKE, PETER J.; and SCHUESSLER, KARL F. 1974 Alternative Approaches to Analysis-of-variance Tables. Pages 145–188 in Herbert L. Costner (editor), *Sociological Methodology, 1973–1974.* San Francisco: Jossey-Bass.

CUTRIGHT, PHILLIPS S. 1974 Academic Achievement, Schooling and the Earnings of White and Black Men. *Journal of Social and Behavioral Sciences* 20:1–18.

GOLDBERGER, ARTHUR S.; and DUNCAN, OTIS DUDLEY (editors) 1973 *Structural Equation Models in the Social Sciences.* New York: Academic Press.

McFARLAND, DAVID D. 1970 Intragenerational Social Mobility as a Markov Process: Including a Time-stationary Markovian Model That Explains Observed Declines in Mobility Rates Over Time. *American Sociological Review* 35:463–476.

SHELDON, ELEANOR B.; and PARKE, ROBERT 1975 Social Indicators. *Science* 188:693–699.

SPILERMAN, SEYMOUR 1972 The Analysis of Mobility Processes by the Introduction of Independent Variables Into a Markov Chain. *American Sociological Review* 37:277–294.

SPILERMAN, SEYMOUR 1975 Forecasting Social Events. Pages 381–403 in Kenneth C. Land and Seymour Spilerman (editors), *Social Indicator Models.* New York: Russell Sage.

PREDICTION AND FORECASTING, ECONOMIC

A forecast can be defined generally as a statement about an unknown and uncertain event—most often, but not necessarily, a future event. Such a statement may vary greatly in form and content: it can be qualitative or quantitative, conditional or unconditional, explicit or silent on the probabilities involved. A reasonable requirement, however, is that the forecast should be verifiable, at least in principle; trivial predictions that are so broad or vague that they could never be found incorrect merit no consideration, and the same applies to predictions that are rendered meaningless by relying on entirely improbable assumptions or conditions. As for the "event" that is being predicted, it too is to be interpreted very broadly. Thus the forecast may refer to one particular or several interrelated situations (single versus multiple predictions). It may identify a single value or a range of values likely to be assumed by a certain variable (point versus interval predictions). Various combinations of these categories are possible, and some are interesting; for example, an unconditional interval prediction can be viewed as a set of conditional point predictions (that is, the forecaster estimates the range of probable outcomes by setting limits to the variation in the underlying conditions).

In principle, the unknown event that is the target of a prediction could pertain to the past or the present, but it is the future that is of primary concern to the forecaster. Information about the past and present is often incomplete and inadequate, but it is as a rule far richer and firmer than whatever may be "knowable" about the future. In fact, inferences from the past are the sole basic source for the expectations of events that are beyond the forecaster's control. And the quality of these expectations (forecasts) is clearly a major determinant of the quality of plans and decisions that refer to factors over which the maker or user of the forecast does have substantial control.

Economic forecasts refer to the economic aspects of unknown events. Looking into the future, these predictions may be classified into short-run (with spans or distances to the target period of up to one or two years), intermediate (two to five years), and long-term (relating to more persistent developments or distant occurrences). Forecasts in all these categories are made for purposes of business planning, of aiding economic policies of governments, and of testing generalizations of economic theory. It is true that business forecasts deal largely with the near future, because this is both what is most needed and what stands a better chance of relative success in the reduction of avoidable business risks. Nonetheless, in certain areas, such as planning new industrial plant construction or acquisition of new businesses, prediction of rather long developments is needed, and lately business forecasters seem to have grown bolder in undertaking to project long trends in the economy. With the recent emphasis on growth objectives, long-range forecasts are also gaining ground as tools of governmental planning and decision making.

Important generalizations of economic theory typically imply qualitative conditional predictions; for example, the "law of demand" predicts that a decline in the price of a good will lead consumers to purchase more of that good. In the history of economic thought one also finds another type of prediction, in which the author presents as a forecast of things to come what is essentially an empirical hypothesis based on assumptions that may only have been valid at the time or may be questioned altogether. Predictions of secular developments by the classical economists provide several major examples, such as the law of historically diminishing returns and the Malthusian population principle. (Marx's projections of a falling rate of profit and increasing pauperization and crises belong in the same logical category.) History dealt harshly with some of these prognostications, while many others were left untested by events and must often be viewed as inconclusive or lacking in present interest. In any case, these so-called evolutionary laws and other predictions of such a general nature are not included in the subject matter of this article, which will concentrate on more limited, specific, and, in particular, quantitative forecasts.

Historical background

It is clear from the preceding that prediction in the general sense has always been one of the products of economic thought; indeed, the ideal aim of any scientific generalization is to establish regularities or relationships that would hold not only for the past but also for the future observations of the same phenomena. Specific quantitative prediction, however, is of a much more recent origin in economics, having had to await the development of empirically oriented research and its statistical and mathematical tools. Two contributions of enduring influence in this development were Ernst Engel's analysis of cross-section data on workers' household budgets, which appeared in 1883, and Clément Juglar's study of time series data on prices and finance, which appeared in 1862 and introduced the idea of observable "cycles" in business activity.

Growing interest in the persistent and disturbing phenomenon of business cycles gave considerable impetus to the collection and analysis of a variety of economic time series. A succession of significant studies of business cycles, with frequent references to historical and statistical materials, appeared between 1898 and 1925: works by Wicksell, Tugan-Baranovskii, Aftalion, Spiethoff, and Schumpeter in Europe, and by Mitchell in the United States. In the course of his later work on business cycles, Mitchell developed a strong (though healthily skeptical) concern about the possibilities of predicting the near-term fortunes of the economy. His 1938 paper on statistical indicators of cyclical revivals, written with Arthur F. Burns, initiated a series of several studies by the National Bureau of Economic Research (NBER), which produced tools that have recently been widely used in practical forecasting. (See Zarnowitz 1968.)

Another flow of important contributions to the present-day techniques of economic forecasting had its source in the development of new methods of statistical inference, which were first applied in the physical and biological sciences and soon attracted the attention of those interested in social and economic data. Early illustrations of such work are found in Henry L. Moore's studies on economic cycles and forecasting, which appeared in 1914 and 1917. Moore's work stimulated the use of regression methods in forecasting prices and production of individual (particularly agricultural) commodities. Irving Fisher's major achievements in monetary economics, index numbers, the study of distributed lags, etc., have long been acknowledged as early models of what came to be known as econometric research. Other pioneers in this approach include Paul H. Douglas (production functions, wages), Henry Schultz (demand functions), Ragnar Frisch (marginal utility measurement), Charles F. Roos (automobile and housing demand), and Jan Tinbergen (statistical tests of business cycle theories). Modern analysts and forecasters who use econometric models clearly owe a major debt to the work of these men.

More directly concerned with forecasting of short-term changes in general business conditions were the efforts of Warren M. Persons (1931) and the Harvard University Committee on Economic Research to identify time series that tend to move cyclically and would therefore help to anticipate what was ahead for the economy. The result was the Harvard Index Chart, consisting of three curves: (A) speculation (stock prices); (B) business (wholesale prices, later bank debits); and (C) money market (short-term interest rates). The Harvard chart was first published in 1919 for the last decade before World War I, then extended back to 1875; but the first extensive use of the underlying relationship was made by one of the early commercial business forecasting services, J. H. Brookmire, in 1911. The A–B–C sequence was found to have occurred with substantial regularity in the period before World War I. In the 1920s, the three-curve barometer proved less consistently successful, and it was considerably modified and often used in combination with other statistical devices. The Harvard service did not survive the great depression, although the index was published periodically in the *Review of Economic Statistics* until 1941. However, a recent unpublished evaluation shows that the sequence underlying the Harvard ABC curves persisted over a long period, not only in the interwar years but also in the post-World War II years. The Brookmire–Harvard method, for all its shortcomings, deserves to be acknowledged as an early forerunner of the present indicator techniques and also as a highly influential factor in the evolution of business cycle forecasting in Europe during the interwar period.

For men of affairs, prediction in some (not necessarily explicit) form must have always been unavoidable. Most decisions made in business and many decisions made in government imply or follow some forecasts of economic conditions. Economic and business forecasting as a specialized activity, however, is a relatively recent phenomenon which made its appearance largely in the twentieth century and developed rapidly only after World War I and, particularly, after World War II. In the latter postwar period forecasts have become both more abundant and more ambitious than ever before. Increasingly, forecasts of such comprehensive aggregates as the gross national product (GNP) and its major components, industrial production, and total employment are being made in numerical form not only for the next year but often also over a sequence of short periods, say the next four or six quarters. The spread of such predictions was stimulated by two recent developments: the rise of active interest in the application of macroeconomic theory, and the corresponding accumulation and improvement of aggregative data. The former can be attributed largely to the intensive work done by economists in the last thirty years— since Keynes's *General Theory*—on problems in the determination of aggregate income, employment, and the price level. The latter goes back to the development of concepts and data on national income accounts by Simon Kuznets and others at the NBER and in the government statistical agencies. Very recently, the rapid growth of electronic computer technology has greatly accelerated the rate at which economic data (the raw materials for the forecaster) are compiled and processed. The same factor also had some more direct effects —without the computer, for example, the large-scale econometric models could not have been produced, and hence output of forecasts of the econometric variety would have been severely limited.

Types and methods of forecasting

It is instructive to classify forecasts by several different criteria. First, one may distinguish between forecasts that are based solely on the past and current values of the variable to be predicted and forecasts that rely on postulated or observed relationships between the variable to be predicted and other variables. The former type of forecast will be referred to as an *extrapolation*. Second, the degree to which formalized methods are used establishes at least in principle a whole gradation ranging from informal judgment forecasts to predictions based on fully specified and strictly implemented econometric models. And third, one may distinguish between forecasts constructed by a single source, which could be an individual or a team (say, the staff of a business, or government agency), and forecasts derived as weighted or unweighted averages of different predictions made by a few or many individuals or organizations (the latter category includes opinion polls, surveys of businessmen's anticipations, etc.).

These various classes of forecasts overlap and can be combined in various ways. For example, a forecast of next year's GNP and its major components by a business economist may consist of any or all of the following ingredients: (1) extrapolation, of some kind, of the past behavior of the given series; (2) relation of the series to be predicted to known or estimated values of some other variables; (3) other external information considered relevant, such as a survey of investment intentions or a government budget estimate; and (4) the judgment of the forecaster. Also, it should be noted that a group forecast, say an opinion poll, will incorporate as many different techniques as are used by the different respondents. Any single classification of forecasting methods that cuts across the different criteria is likely to have only limited application to actual forecasts, which are built much more often on a combination of methods than on one particular technique.

Increasingly, starting with the early post-World War II forecasts, the analysis of factors affecting the course of the economy is being carried on in terms of the major components of GNP. This framework is thought to have the advantages of (*a*) ensuring that none of the main components of "aggregate effective demand" will be overlooked in the forecast, since they are all represented in the expenditure categories of the GNP accounts; (*b*) steering the forecaster to think in terms of the basic determinants of spending decisions by consumers, business, government, and foreign buyers; and (*c*) providing some safeguards that the various parts of the forecast, being constructed according to the internally consistent GNP system, will not be inconsistent with each other. However, there are also some disadvantages in an exclusive reliance on data such as those for GNP, which are highly aggregative, subject to frequent and often substantial revisions, and available only annually (with a relatively short dependable historical record) and quarterly (for the recent years).

A few mechanical forecasting techniques that were once fairly popular can be quickly disposed of as having little scientific basis or empirical soundness. For example, the method that assumes periodic cycles of given duration founders on the fact that cycles in business activity are far from being strictly periodic. The device of equalizing the areas below and above trend, or predicting that a business index will turn once it deviates from the "normal" by some critical amount, is unlikely to score well because trends are difficult to determine on a current basis and need not have the connotation of "normal" or "equilibrium" levels even when they are satisfactorily identified.

Judgment, in the broad sense of the word, is,

of course, a necessary ingredient of all types of prediction: the forecaster must "judge" what information and methods of analysis to use and how to interpret and evaluate the results. Thus, judgment is bound to enter the forecasting process at various stages, but its proper role is to be a complement to, not a substitute for, a competent economic and statistical analysis. Informed judgment can go far in making forecasts more consistent and dependable, while pure guesswork can only rely on luck for success. It is well to recognize, however, that the ability to reach "good judgments" is not a well-defined, technical, transferable skill but rather a puzzling function of personal talent, experience, and training.

Many economic forecasts, particularly from business sources, are not based on formal models and do not disclose the underlying assumptions and methods. Some are likely to be little more than products of intuition, yet most seem presently to take the specific form of numerical point forecasts. However, there is no general presumption that the informal judgmental forecasts are largely hunches; on the contrary, at least the better ones among them originate in the application of various analytical techniques as well as judgment to diverse and substantial bodies of information.

Judgmental inferences from data samples and from any other evidence that the forecaster may have and regard as pertinent involve probability distributions. It would be informative and helpful for the appraisal of predictions if forecasters stated the odds they attach to the expected outcomes (the practice is frequent in weather forecasts, for example). The step from interval to distribution forecasting is, in principle, short. However, probabilistic distribution predictions appear to be regrettably rare in practical business and economic forecasting.

After eliminating those forecasting procedures that have little relevance currently, the following seem to merit further discussion: (1) extrapolative techniques; (2) surveys of intentions or anticipations; (3) business cycle indicators; and (4) econometric models. Typically, forecasters are using one or another of these approaches, or more likely some combination of them, for the most part tempering their results with considerable doses of "judgment."

Extrapolations. The term "extrapolations" will be used here as a shorthand expression for "extrapolations of the past behavior of the series to be predicted." (The term is often applied in a broader sense to include projections of relationships between different variables, but such forecasts are more conveniently discussed under other headings—"econometric models" and to some extent also "business cycle indicators." To be sure, extrapolative elements can be combined with the others, as in an econometric model in which, say, x_t is associated with both x_{t-1} and y_{t-1}.) Being restricted to the history of just one variable or process, extrapolations make only minimal or no use of economic theory, which deals largely with relations between different factors. Technically, however, extrapolations can vary a great deal, from very simple to very complicated forms. The simplest "naive models" project the last-known level or the last-known change in the series, that is, they assume that next period's value of the series will equal this period's value or that it will equal this period's value plus the change from the preceding to this period. Few, if any, forecasters would cast their predictions in the form of such crude extrapolations, but the naive models are useful as minimum standards against which to measure the performance of forecasts proper.

Since trends are common in many economic time series because of the pervasive influence of growth in the economy, trend extrapolations usually provide more effective predictions and are therefore more demanding as criteria for forecast evaluation. Particularly in application to long-term forecasts, trend fittings and projections are widely used. The trends are usually conceived as smooth (often but not always monotonic) functions of time; the methods of describing them vary greatly, from visual freehand projections and long-period moving averages to diverse (for example, exponential, logistic) curves fitted by mathematical formulas. In the short-run context, the other typical components of economic time series become of primary importance, namely, the cyclical, seasonal, and purely "irregular" or random movements. (However, for some series, trends are substantial even over short periods, and in such cases it is especially rewarding for the forecaster to approximate them well.) [See TIME SERIES.]

Strictly periodic, repetitive fluctuations are, like persistent trends, relatively easy to extrapolate: stable seasonal movements would often be more or less of this type. Average ratios of raw (say, monthly) data to smoothed values of the series representing mainly the longer-term movements (a centered 12-month moving average is the simplest example) are most commonly employed as a set of "seasonal indexes." Extrapolations of these indexes then serve as forecasts of the seasonal movements. Since such movements are a major source of instability for business firms, their projection is particularly important in industrial and sales fore-

casting. Complications arise when the seasonal patterns are not very pronounced and not very stable, hence difficult to isolate from other component movements. Forecasters whose interest is mainly in the other movements often try to work around the seasonal effects by predicting changes in the seasonally adjusted series.

This leaves the cyclical and irregular components as the major objects of concern for the short-term forecaster. Looking forward, it is usually anything but easy to distinguish the cyclical from the random element in the movement of an economic time series, although retrospectively it is often possible to do so with fair results (by decomposing the time series and testing the residually obtained estimates of the irregular component for randomness).

The forecasting errors that are directly traceable to very short random movements must be accepted as unavoidable. The forecaster can hardly be expected to predict an event generally regarded as unforeseeable, such as an outbreak of a war or a strike started without advance warning. However, although such "shocks" cannot themselves be predicted individually with the tools of economics and statistics, their more significant effects on the economy are, of course, the proper concern of the forecaster. In probabilistic predictions, which aim at the distribution of unknown parameters and outcomes rather than at point forecasts of future events, the effects of shocks and other random errors would be taken into account as an important part of the system to be analyzed. The role of random impulses in propagating business cycles has been given considerable attention in recent simulation studies of aggregative econometric models. [See SIMULATION, *article on* ECONOMIC PROCESSES.]

For sequences of successive point predictions, which are the most common type of economic forecasts, the requirement of a good forecast is, in brief, that it predict well the systematic movements—trends and cycles—not that it predict perfectly the actual values of economic series, which, as a rule, contain random elements. Smoothing techniques can reveal the past patterns of systematic changes in the given series, and extrapolations can help the forecaster in his task to the extent that they preserve these patterns and to the extent that the patterns continue to apply. But, in regard to economic and related social events, the future seldom reproduces the past without significant modifications, and the historical "patterns" are often complex enough to elude efficient extrapolation.

In particular, extrapolations are by and large

incapable of signalizing the turning points in business. The turns in extrapolations will as a rule *lag* behind those in the actual values; the strength of a good projection lies almost entirely in that it may predict well the longer-term trends. This contributes to the fact that, along with the short random variations, it is the cyclical fluctuations, not the longer trends, that produce the greatest difficulties in short-term forecasting. These fluctuations are recurrent but nonperiodic; they vary greatly in duration and amplitude; and calling them cyclical should convey neither more nor less than that they reflect mainly the participation of the given economic factor in "the business cycle."

Important mathematical studies of smoothing and extrapolation of time series (by A. Kolmogoroff and by Norbert Wiener) appeared in the early 1940s. The method of autocorrelation, or the extrapolation of a series by means of a correlation of the series with itself at different points of time, was added and related to the methods of trend projection and harmonic analysis of the residuals from trend. It is an essential feature of dynamic process analysis in economics that variables at different points of time are functionally (usually stochastically) related; this approach may result in difference equations which generate their own solutions, as in the multiplier–accelerator models that yield a dependence of the current value of national income on the past values of national income. Thus, there is a strain of theoretical thought here that suggests the use of autoregressive extrapolation functions in aggregate income forecasting.

More recent developments have led to application of such functions in the analysis of how expectations are formed and how lagged adjustments are made. In particular, functions in which the weights decline geometrically as one goes back to progressively earlier past period values have been used in a variety of problems involving either expectation or partial adaptation to change. [*See* DISTRIBUTED LAGS.] Forecasts derived from such exponentially declining weighted averages were found to have certain desirable properties for a class of autoregressive time series. But these are "optimum" predictions only if the structure of the series is known to belong in the given class and only if the past values of the series are all that one has to go on. These conditions are seldom fulfilled or relevant. As one of the pioneers in econometric forecasting methods has observed: "Mathematical processing or analysis can never substitute for sophistication. And until one understands the forces that built a particular structure in an economic time series and how these forces are currently changing, he cannot forecast with confidence even

though by chance he scores a preponderance of successes" (Roos 1955, pp. 368–369).

Surveys of anticipations or intentions. The collection and evaluation of expectational data for the U.S. economy did not develop on a large scale until after World War II, but since then work in this area has proceeded at a rapid pace. The data relate to future consumer expenditures, planned or anticipated capital outlays of business firms, business expectations about "operating variables," and government budget estimates.

Reports dealing with purchases of household appliances and automobiles are published periodically by the Survey Research Center of the University of Michigan. They currently include quarterly measures of consumer attitudes and inclinations to buy as well as annual financial data. Since 1959, a quarterly household survey has also been conducted by the Bureau of the Census.

The surveys of business plans and anticipations relating to future expenditures on plant and equipment include one that is carried on annually and quarterly as a joint enterprise by the U.S. Department of Commerce and the Securities and Exchange Commission (SEC) and another that is conducted annually by the McGraw-Hill Publishing Company. Since 1955, a quarterly survey of new and unspent capital appropriations has been conducted by the National Industrial Conference Board.

The oldest of the current surveys of businessmen's expectations in the United States is the Railroad Shippers' Forecast, which has given quarterly anticipations for carloadings by commodity since 1927. The *Fortune* magazine program, which started in the 1950s, includes surveys of "business expectations and mood," retail sales, farm spending, homebuilding, inventories, and capital goods production (all of these are now semiannual, except for the annual farm and the quarterly inventory surveys). Data on manufacturers' sales expectations have been gathered since 1948 in the course of the U.S. Department of Commerce–SEC survey of investment anticipations; quarterly figures on manufacturers' sales and inventory expectations are now published regularly by the Department of Commerce in a program initiated in 1957.

Two other sources of data on business expectations about operating variables can be grouped together inasmuch as their output takes the special form of "diffusion indexes." Such data indicate, for each successive forecast period, the percentage of respondents in the sample who expect either rises or declines or no change in the given variable. The Dun and Bradstreet surveys of manufacturers, wholesalers, and retailers cover employment, inventories, prices, new orders, sales, and profits; they started in the 1950s, are quarterly, and refer in each case to businessmen's expectations for the impending six-month period. The monthly questionnaire of the National Association of Purchasing Agents, first issued in 1947, is addressed to participating members, covers production, new orders, commodity prices, inventories, and employment, and asks how the month ahead is going to compare with the preceding month (whether it will be better, worse, or the same).

Surveys of enterprise expectations have also spread in other countries, originating apparently with the IFO-Institute for Economic Research in Munich at the beginning of 1950. This institute sends monthly questionnaires to a large number of companies in West Germany and on the basis of replies from executives compiles diffusion data on the actual and expected directions of change in several important economic variables. By 1959, eight European countries, as well as Japan, South Africa, and Australia, adopted methods of entrepreneurial surveys similar to the IFO procedure.

Business expectations have been classified into intentions (plans for action where the firm can make binding decisions), market anticipations (relating to the interplay between the firm's actions and its environment), and outlook (expectations about conditions which the firm cannot significantly influence but which will affect the markets). To illustrate, new capital appropriations or plans regarding next year's outlays for plant and equipment fall into the first category, as does the scheduling over shorter periods of production and employment. The firm's sales forecast, which depends on customers' reaction to the terms offered, advertising efforts, etc., belongs to the second category, as do the expectations concerning financing, inventories, and selling prices (for firms that are not able to set their own prices). Finally, the class labeled "outlook" includes forecasts of the general situation of the economy or industry.

Consumer intentions to buy are in principle akin to business plans to acquire productive resources, but in practice they are often more vague and attitudinal and usually less firmly budgeted. Government budget estimates also represent intentions: they document what the central administration would undertake to spend for diverse specified purposes, subject to approval by the legislature. In government, as well as in many large business companies, the process of forming "expectations" or forecasts is often highly decentralized and complex, as is indeed the related process of reaching decisions.

The distinction between intentions, market anticipations, and outlook is of significance for the question of the predictive value of expectational data. It is plausible that accuracy will tend to be higher the greater the degree of control that those holding the expectation have over the variable concerned. This suggests that intentions should be on the whole more accurate than market anticipations, and the outlook estimates should be the least accurate. There is some evidence consistent with this view, notably the fact that business anticipations of plant and equipment expenditures have a much better forecasting record than business sales anticipations (as shown by the U.S. Department of Commerce–SEC sample surveys). However, there are other relevant factors which can modify such comparisons, such as the variabilities of the predicted series (forecasts of a very stable aggregate, classifiable as outlook, may be better than market anticipations for a variable which is highly volatile and therefore difficult to predict) and span of forecast (surveys looking far enough ahead, even for largely "controllable" variables, will be more in the nature of market anticipations and less of intentions, and they could well be less accurate than outlook surveys for the very near future).

Clearly, expectations of all kinds always involve some degree of uncertainty. They are presumably based in part on historical evidence, such as extrapolation of past behavior of the given series and inferences from observed relations with other series; but they are unlikely to incorporate only such evidence. As a result of expert insight or mere hunches, they may well include some additional information not contained in the patterns of the past. Hence, even where expectations are not very efficient when used alone as a direct forecast, they may still have a net predictive value as an ingredient in a forecasting process that combines expectational with other inputs.

In long-run forecasting, informed judgment or expectations play a major role along with extrapolative techniques. In large part, these forecasts are growth projections, which have been described as tools for exploring economic potentials. They are not intended to provide predictions of actual conditions in a distant year but rather estimates of likely conditions under some specified assumptions regarded as more or less reasonable. Attempts are often made to allow for the uncertainties of the future by constructing alternative projections that assume several different paths of economic developments within the range considered plausible. These forecasts are thus essentially concerned with trends of the economy at full employment. The variables projected are typically population, labor force, hours of work, and productivity on the supply side; expenditures on the major GNP components on the demand side; income, saving, and investment; and price level movements (which are often only implicitly considered, the projections being expressed in constant dollars).

Business cycle indicators. Business cycle indicators, which are used in analyzing and forecasting short-term economic developments, are time series selected for the relative consistency of their timing at cyclical revivals and recessions (other criteria being the economic significance of the series in relation to business cycles, their statistical adequacy, historical conformity to general movements of the economy, smoothness, and currency). The series are selected from large collections of quarterly and monthly economic series and then subjected to detailed analysis and repeated examinations of the quality of their performance as cyclical indicators. Such selections, based on studies of 500–800 series and successive reviews of the results, were made at the National Bureau of Economic Research in 1938 (by Mitchell and Burns), in 1950 and 1960 (by Geoffrey H. Moore), and in 1965 (by Moore and Shiskin). The first list included only indicators of cyclical revivals; the later ones covered indicators of both revivals and recessions, classified into those that tend to lead the turns in general business activity, those that tend to coincide roughly with these turns, and those that tend to lag. Further revisions and extensions of the list have been prompted by the appearance of new and improvement of old data, the accumulation of knowledge about the behavior of the series and the processes they represent, and the great increase in efficiency with which time series can be processed and analyzed. These reviews resulted in many significant changes, but the core of the list has not been essentially altered as to its composition in terms of the represented processes. In the most recent list, this core consists of the following:

Leading indicators (14 series)—average work week in manufacturing, nonagricultural job placements, new building permits for private housing, net business formation, new orders for durable goods, contracts for plant and equipment; change in unfilled orders for durables, change in manufacturers' and trade inventories; industrial materials prices, stock prices, corporate profits after taxes, ratio of price to unit labor costs in manufacturing; change in consumer installment debt, liabilities of business failures.

Roughly coincident indicators (8 series)—nonagricultural employment, unemployment rate, GNP in constant dollars, industrial production, personal income, retail sales, manufacturing and trade sales, wholesale price index.

Lagging indicators (6 series)—long-duration unemployment, book value of manufacturers' and trade inventories, labor cost per unit of output in manufacturing, business expenditures on new plant and equipment, bank rates on short-term business loans, commercial and industrial bank loans outstanding.

While the selection is based mainly on historical evidence, it is also broadly supported by general economic considerations and logic. Thus the aggregative series on production, employment, and income measure approximately the general level of business or economic activity whose major fluctuations are defined as the business cycle; hence, these series could hardly fail to be "roughly coincident." The leaders include series that anticipate production and employment, such as hours worked, job vacancies, and new orders and contracts. For example, an increase in demand calling for additional labor input is likely to be met first by lengthening the work week and only later, if still needed, by hiring new workers (the former adjustment is less binding and costly than the latter). New orders precede production and employment by sizable intervals for goods made largely in response to prior offers or commitments to buy. Most durable manufactured goods belong in this category and, particularly, the capital equipment items which are as a rule produced to fill advance orders; construction of industrial and commercial plant is similarly anticipated by building contracts. The execution of these new investment orders and contracts takes time, however, and so the expenditures on plant and equipment is a roughly coincident or slightly lagging series.

Another type of sequence arises from the fact that a stock series often undergoes retardation before reversal; hence the corresponding flow series (or rate of change in the stock) tends to turn ahead of the stock. Thus, inventory changes lead at business cycle turns, while total inventories lag. Still other sequences are recognized when downturns in some indicators are related to upturns in others. For example, the decline in inventories lags behind, and is a possible consequence of, the downturns in the comprehensive measures of economic activity (such as GNP, industrial production, and retail sales); but the downturn in inventories also leads, and may be contributing to, the later upturns in these and other series (as the need for the stocks to be ultimately replenished will stimulate orders and help to bring about the next business recovery). Such considerations suggest that the coinciders and laggers are not merely of value as confirming indicators; some of them also play an active role as links in the continuous round of business cycle developments.

In addition to the 28 series identified above, more than sixty other series are included in the full list of indicators of business expansions and contractions. Up-to-date charts and tabulations and various summary measures for the entire set of indicators have been published since 1961 by the Bureau of the Census in the monthly report *Business Cycle Developments*. The full list contains several related series for each type of economic process having significance for business cycle analysis and forecasting. Most of the series are monthly (less than 20 per cent are quarterly), and more than half of those classified by timing are leaders. Change in money supply, a series with long leads and rather pronounced irregular component movements, deserves a special reference in view of the hypothesis that this variable is a fundamental factor in initiating business downturns and upturns.

The indicators are used mainly to reduce the lag in the recognition of cyclical turning points. The lead time provided by the indicators is, on the whole, short; often, especially when the economy reverses its course in a relatively abrupt manner, the best obtainable result seems to be to recognize the cyclical turning point at about the time it is reached. Even this is not a negligible achievement, however, since revivals and recessions are generally not recognized as such until several months after they have occurred.

The greatest difficulty in using indicators for forecasting purposes arises from the need to establish on the current basis the direction in which these series are moving or the dates of their turning points. This is because the trend and cyclical movements in many indicators are typically overlaid and often obscured by other short-period variations, partly of a seasonal but mainly of an erratic nature. Seasonal adjustment and subsequent smoothing of the series can help to distinguish its cyclical from its shorter irregular movements, but these are essentially descriptive–historical procedures which are not very efficient and which cause losses in up-to-dateness when applied currently.

The leading indicators, in particular, are highly sensitive to all kinds of short-term influences and not only to the forces making for the general cyclical movements. They have anticipated marked retardations in business activity as well as the major recessions and revivals. Skilled (or lucky) judgment of the user could sometimes succeed in distinguishing between these different episodes, but

no mechanical, replicable method of applying the indicators has been able to do so.

Individual indicators occasionally fail to signalize the approach of a general business reversal, and their leads often vary a great deal in length from one recession or revival to another. The evidence of groups of indicators is considered to be on the whole more reliable than the evidence of any single indicator. Accordingly, the degree of consensus in the behavior of these series attracts considerable attention of business analysts and forecasters. Several measures of the consensus are in use, including diffusion and other composite indexes for groups of the indicator series.

Econometric model forecasts. Aggregate econometric models are systems of equations designed to represent the basic quantitative relationships among, and the behavior over time of, such major economic variables as national income and product, consumption, investment, employment, and price level. Such models are used for forecasting and also for other purposes (simulation of the likely effects of alternative fiscal and monetary policies, tests of hypotheses, etc.). In recent years, intensive work in this area has been done in several universities and government agencies in the United States and abroad, notably by Lawrence R. Klein and his associates.

The equations in econometric models describe the behavior of consumers, producers, investors, and other groups of economic agents; they also describe the market characteristics, institutional conditions, and technological requirements that guide and constrain economic action. They include variables selected to represent important systematic factors entering each of these functions and relate these variables by means of statistical estimates of their net marginal effects. The estimation of these unknown values (called parameters and taken, as a rule, to be constant) is based on the assumption that all the major factors affecting the relationship have been properly identified, leaving only random disturbances with expected values of zero. Ideally, the residual disturbance terms, which entail the net effect of all influences other than those of the specified explanatory variables, should be small, not serially correlated, and not associated with the systematic factors. In practice, the model builder hopes that the disturbances will be at least approximately random and relatively small, that is, that economic theory or other insights will lead him to a sufficiently good specification of a few principal determinants in each of these relationships.

In contrast to these "stochastic" equations, which are supposed to hold only approximately, the remaining equations of the model are accounting "identities," which are based on definitions and are therefore supposed to hold exactly. The stochastic equations and the identities together form a description of the structure of the economy.

The unknown variables that are to be determined by the model are called the jointly dependent or current endogenous variables; their number equals that of the equations in the complete system. To solve for these variables, each of them is expressed as a function of the estimated structural parameters, the disturbances, and the predetermined variables. The predetermined variables are inputs required by the model and consist of (a) the values of those variables (labeled exogenous) that are viewed as determined by factors outside the model; and (b) the lagged values of the endogenous variables, which are given by outside estimates or by the past operation of the system. It is the jointly determined estimates of the current endogenous variables, all of which are functions of the predetermined variables with disturbances typically assumed to be zero, that represent the forecasting output or the "reduced form" of the model.

If the predetermined variables are taken as given, say at their reported ex post values, forecasts made from the reduced form will be conditional upon these data. "Unconditional" forecasts of the jointly dependent variables will be obtained when the unknown future values of the predetermined variables are themselves predicted (which for the exogenous factors necessarily means prediction outside the model). A closely related distinction is between alternative hypothetical forecasts (for example, of GNP next year, assuming 5 or 7 or 10 per cent increases in government expenditures) and the single preferred forecast (that government expenditures will be up 7 per cent and GNP will be such and such).

It follows that unconditional forecasts could have substantial errors because of wrong projections of exogenous variables, even if the specification and solution of the model were essentially correct. The accuracy of the conditional forecasts, on the other hand, depends (apart from any effects of errors in the data to which the equations are fitted) only on the errors that occur in the construction and solution of the model. These may arise for several different reasons: (1) incorrect specification of the behavioral or other economic relationships, that is, failure to use the right explanatory variables in their proper form; (2) deficient methods of statistical inference, for example, inability to measure well the separate effects of several closely interrelated exogenous variables; (3) errors in the parameter estimates resulting from sampling fluctuations,

that is, from the presence of the disturbance terms which obscure the underlying relationships; and (4) discrepancies between predictions and realizations that result from the assumption that the disturbance terms vanish—in any particular period these terms may differ from zero, even though their expected values are zero.

Statistical inferences as to the probabilistic meaning of the parameter estimates, the goodness of fit to the data in the sample period, the presence or absence of autocorrelation in the disturbance terms, etc., can be used to evaluate the severity of the errors resulting from sampling variations and the inefficiency of the estimation methods employed. As far as the appraisal of the model per se is concerned, however, misspecifications are clearly the decisive sources of error, and these are more difficult to detect and evaluate. Theoretical and other a priori considerations can tell us something about the correctness of some of a model's specifications; but the correctness of the specifications for the forecast period and not just the sample period must be judged primarily from the quality of the conditional forecasts made with a model that proved satisfactory on the other statistical tests just mentioned.

The unit time period varies from a quarter to a year in different models; for the purpose of analyzing and forecasting near-term developments, short unit periods are desirable, and aggregate econometric models have recently progressed from annual to quarterly units. Models also vary greatly in size and complexity. One view is that simple small-scale models can be sufficient for forecasting the broad course of the economy in the near future and indeed that they may be preferable in this role to large models which present greater opportunities for error by taxing heavily the present inadequate knowledge and data. But, while small models with as few as five equations have recently been proposed, the trend appears to be in the direction of ever larger and more complex systems.

It should be noted that conditional forecasts of particular variables are sometimes obtained by econometricians from one-equation systems. If the equation is unlagged, it merely shifts the burden of prediction in that, to get an authentic ex ante forecast of the dependent variable, the independent variable(s) must somehow be predicted outside the model. If the dependent variable is taken with a lag, it can be predicted from the equation inasmuch as the earlier values of the explanatory factors are known or treated as known. The single-equation approach is strictly applicable only to cause–effect relations where one variable, say x, depends on others, $y_1, y_2, \cdots$, but the y's do not depend on x. If the variables depend on each other, a multiequation model should be applied.

An econometric forecasting model can in principle be so constructed and annotated as to be available for production of successive forecasts, replication by users other than the authors of the model, and continuous inspection and testing. This possibility holds out the promise of scientific advance, but it is difficult to achieve in practice for complex models with large requirements in terms of data and methods, especially when the data and methods are themselves subject to frequent and substantial changes. Moreover, some econometric forecasters wish to use their models in a flexible manner, modifying them repeatedly so as to take advantage of additional information. Thus, anticipatory data from surveys, indicators, etc., are introduced into the models as exogenous variables, or judgment about the probable effects of a recent event is used to alter the constant term in an individual equation. Such modifications are often informal and sometimes unrecorded; they may (although they need not) improve any particular forecast, but they certainly increase the difficulty of replicating and evaluating the models.

Input–output tables. Input–output analysis, which was developed by Wassily Leontief and first presented in 1941, involves a relatively detailed division of the commodity-producing sector of the economy into individual industries and estimation of the relations between these industries. In a statistical input–output table, each row shows the sales made by a given industry to every other industry, and each column shows what one industry purchased from every other industry. If the quantity of each input per unit of output is treated as a structural constant (which implies constant returns to scale and, a particularly drastic assumption, absence of substitution among inputs when relative prices change), then a system of linear equations can be set up describing the interdependence between the outputs of the different industries by means of these "technical coefficients" of production. Such a model can be used to make conditional predictions of the values of industry outputs, given the estimates or projections of the "bill of goods," that is, of purchases by the autonomous sectors— consumers, government, and foreign countries. (See Leontief 1968.)

Evaluaton of forecasts

Despite the widespread and apparently increasing use of economic forecasts in business, government, and research, surprisingly little has been done to test these forecasts in a systematic and comprehensive way. Yet it is clear that forecasts must be properly evaluated if their makers are to learn from past errors and if their users are to be able to discriminate intelligently among the available sources and methods. Without dependable assessments of forecasting accuracy, informed comparisons of costs and returns associated with forecasting are clearly impossible.

Conceivably, the effectiveness of forecasts could be such as to complicate seriously their evaluation. Forecasts may influence economic behavior and, in particular, the variables being predicted; to the extent that this happens, the forecasts may validate or invalidate themselves. For example, if almost everyone predicted better economic conditions in the period ahead, this very consensus of optimistic expectations could contribute to the stimulation of the economy. Or, conversely, if the government accepted the forecast that the economy is threatened by recession, it would probably adopt policies designed to avert or at least postpone the undesired outcome initially foreseen.

It is easy to exaggerate or misjudge such feedback effects. Of the limited theoretical work on the problem, some is highly abstract and speculative. The best result appears to negate the thesis that public announcements must *necessarily* invalidate an otherwise accurate prediction. Under some plausible assumptions (notably that the predicted variable has a lower and an upper bound), if unpublished forecasts can be accurate, so can published forecasts, because the reaction to them can conceptually be known and taken into account.

Actually, the changes in GNP and other macrovariables that are predicted by different sources for any given period show sufficiently large dispersion to suggest that no single forecast is generally accepted (the evidence comes from a recent NBER study of American forecasts referred to below). For groups of forecasters, average forecasts are in the long run typically more accurate than most of the forecasts of the individual members because of compensating errors among the latter. *Consistently* superior forecasts are evidently hard to find. Some predictions are much more influential than others (and the special significance of official government forecasts is widely recognized), but there is no single authoritative source that enjoys unquestioned leadership.

A careful appraisal of American business forecasts presented by Cox (1929) was one of the few early studies in this area. The years following World War II produced a number of particularly unsatisfactory forecasts (reflecting expectations of serious unemployment), which became the subject of some instructive reviews. Christ (see Conference on Business Cycles . . . 1951) published a test of an early econometric model constructed by Klein. Essays concerned with the development and appraisal of different forecasting approaches were collected in several volumes by the National Bureau of Economic Research (Conference on Models of Income Determination 1964; Conference on Research in Income and Wealth 1954; 1955; Universities–National Bureau Committee for Economic Research 1960; Moore 1961). In Europe, Henri Theil and his associates (1958) produced a major analysis of the methodology and quality of forecasts, directed to both business surveys and econometric models. A comprehensive study of accuracy and other properties of short-term forecasts of economic activity in the United States has been conducted since 1963 at the National Bureau of Economic Research, and several reports on methods and results of this evaluation have been prepared.

Judging from the over-all results of most of these studies, the record of economic forecasters in general leaves a great deal to be desired, although it also includes some significant achievements and may be capable of further improvements. According to the current NBER study, the annual GNP predictions for 1953–1963 made by some three hundred to four hundred forecasters (company staffs and groups of economists from various industries, government, and academic institutions) had errors averaging $10 billion. Although this amounts to only about 2 per cent of the average level of GNP, the errors were big enough to make the difference between a good and a bad business year. The average annual change in GNP in this period was approximately $22 billion. Hence, the errors were, according to absolute averages, not quite one-half the size of those errors that would be produced by assuming that next year's GNP will be the same as last year's (since the error in assuming no change is equal to the actual change). Had the forecasters assumed that GNP would advance next year by the average amount it had advanced in the preceding postwar years, the resulting average error would not have been greater than $12 billion. However, while it is true that in terms of the absolute average errors some forecasts of GNP have not been significantly better than simple trend extrapolations, the forecasts were typically superior in terms of correlations with actual changes. In fact, recent forecasts of aggregate economic activity in the coming year, whether measured by GNP or industrial production, have generally been more accurate than both the simple extrapolations of the last level or change and the considerably more effective models of trend projection and autoregression.

Forecasters, then, were able to make a net predictive contribution over and above what could be obtained by means of mechanical extrapolations alone. This is a favorable result that is by no means always attained—for example, simple trend projections would have done better than the recorded forecasts that envisaged a serious business depression in 1947–1948. It is not a major achievement, however, since the course of the economy has been relatively smooth in recent years and therefore, on the whole, probably less difficult to predict than the developments in the earlier part of the postwar period. In the decade after 1953, there were fewer and smaller exogenous "shocks" of the type represented, say, by the outbreak of the Korean War in 1950. Also, the timing of recent business downturns was early enough to make the presence of the recessions rather widely known by the ends of the peak years (1953, 1957, and 1960), and most forecasts are made at the end of the year. This, plus the presumption that the contractions would continue to be short, made predicting the direction of the year-to-year changes in aggregate economic activity relatively easy.

By and large, the business economists' forecasts covered by this analysis are informal, eclectic, and framed loosely in terms of the GNP accounts. The econometric model forecasts for GNP in the years since 1953 appear to have been about as accurate as the better business forecasts. It must be noted, however, that very few ex ante econometric forecasts are available for such tests, since most of the models now in use have been constructed only in the last few years and their accuracy beyond the sample periods cannot as yet be evaluated with any confidence. The econometric forecasts that do lend themselves to this appraisal come mainly from a series of closely related models used in a flexible manner, with many judgmental modifications.

Many forecasts show a bias of underrating the growth of the economy; the declines are much less frequently underestimated than the increases. The charge that forecasters tend to be too cautious or conservative seems, in this sense, to be justified. Since such series as GNP or industrial production have strong upward trends, their future levels as well as their changes have been understated most of the time (underestimation of increases typically results in underestimation of the ensuing levels).

Forecast errors are also affected by the cyclical characteristics of the forecast period. The levels of GNP and industrial production are underestimated most in the first year of expansion, when the increases in these series tend to be very large. Later in the expansions, when the increases are usually smaller, the levels are underestimated much less. In contractions, the predicted levels are as a rule too high, often because the downturn was missed and sometimes because the decline turned out to be larger than foreseen.

The forecasts of total GNP are often substantially better, in the sense of having smaller percentage change errors, than the forecasts of most major GNP expenditure components from the same source. Apparently, the over-all forecasts benefit from a partial cancellation of errors in forecasts of the components. This is definitely preferable to the opposite case of positively correlated and mutually reinforcing errors, but gross inaccuracies in the component forecasts are, of course, disturbing, even if these errors happen to be largely compensating.

Errors in predicting percentage changes in personal consumption expenditures are considerably smaller than those in corresponding forecasts of gross private domestic investment, while errors in predicting government spending are of intermediate size. Consumption forecasts have suffered from a pronounced tendency toward underestimation. In contrast, changes in series that fluctuate more and grow less strongly, notably the components of investment, have been as often overestimated as underestimated.

Although the errors of consumption forecasts are smaller than those for the other major GNP components when measured in deviations of percentage changes, they are large relative to the errors of some extrapolations. The consumption series, including those for nondurable goods and services, are smoothly growing series that could have been predicted very well in recent years by simple trend projections; and, indeed, the average errors of the latter have often been smaller than those of recorded consumption forecasts.

Aggregation of short-term expectations or plans of business concerns about their outlays on plant and equipment, as developed in periodic intentions surveys, results in better predictions of total business capital expenditures than those made independently for the entire economy. This can be inferred from comparisons between investment forecasts made before and after the McGraw-Hill investment intentions survey and also from comparisons involving the U.S. Department of Commerce–SEC anticipations data.

The average accuracy of short-term forecasts diminishes as they reach further into the future: the various series covered are all predicted better over the next three months than over the next six and better over six months than over nine or twelve. However, the errors increase less than proportionately to the extension of the span. Most extrapolations also tend to worsen with the lengthening of their span, and the simplest among them (the naive models) usually show the fastest deterioration. While forecasts of three to nine months (which includes the annual forecasts whose average spans are little more than six months) are generally superior to all kinds of extrapolations, the longer forecasts, which aim at targets 12 to 18 months ahead, are often worse than some of the relatively efficient types of trend projections and autoregressive extrapolations.

Conditional forecasts from formal models appear to produce errors that are similarly related to the time span of prediction. Thus, a recent analysis of input–output tables for the Netherlands during the period 1949–1958 concluded that predictions based on this method were on the whole better than extrapolations when the tables used were less than two or three years older than the extrapolation data. The input–output forecasts of industry output values for three or more years ahead (given the actual final demand schedules) proved inferior to the results of some fairly simple mechanical extrapolations.

Forecasts made frequently for sequences of two or more short intervals (for example, quarterly, for four successive quarters each) are more relevant for an appraisal of turning-point errors than are the annual forecasts, and they present a rather unfavorable picture. There are very few indications in the record of an ability to forecast turning dates several months ahead. Not only were actual turns missed, but some turns were predicted that did not occur. Reports by observers of the economic scene in business and financial periodicals are consistent with this finding in that they commonly show substantial lags in recognizing revivals and recessions. Business cycle indicators have at times demonstrably reduced these lags for competent users, but they also occasionally gave signals of retardations or minor declines that were misread as forewarning recession.

Multiperiod forecasts also provide data that lend themselves to an analysis of forecast revisions, that is, changes in predictions for a given target period made at dates successively closer to that period. These revisions do tend to improve the forecasts in most cases, and they appear to be related to errors in previous forecasts. Positive correlations between forecast revisions and errors of forecast have been interpreted as evidence of "adaptive behavior," or "learning from past errors." Such correlations, however, could also be due to autoregressive elements in some forecasts, since they are implied in fixed-weight autoregressive extrapolations.

Better statistical data and better utilization of the historical content of the predicted series could lead to significant improvements of the forecasts (as indicated, in particular, by the relative success of some types of extrapolations in predicting consumption and also price levels). Improvements in the record-keeping practices of forecasters are definitely needed. The records should include the estimates of the level of the series at the time the forecast was made, as errors in these base values are as a rule substantial and their measurability is important for the appraisal of forecasts.

VICTOR ZARNOWITZ

[*Directly related are the entries* BUSINESS CYCLES; ECONOMETRIC MODELS, AGGREGATE. *See also* Morgan 1968.]

BIBLIOGRAPHY

ADELMAN, IRMA; and ADELMAN, FRANK L. 1959 The Dynamic Properties of the Klein–Goldberger Model. *Econometrica* 27:596–625.

ALEXANDER, SIDNEY S. 1958 Rate of Change Approaches to Forecasting: Diffusion Indexes and First Differences. *Economic Journal* 68:288–301.

ALEXANDER, SIDNEY S.; and STEKLER, H. O. 1959 Forecasting Industrial Production: Leading Series Versus Autoregression. *Journal of Political Economy* 67:402–409.

ALMON, SHIRLEY 1965 The Distributed Lag Between Capital Appropriations and Expenditures. *Econometrica* 33:178–196.

BARGER, HAROLD; and KLEIN, LAWRENCE R. 1954 A Quarterly Model for the United States Economy. *Journal of the American Statistical Association* 49.413–437.

BASSIE, V. LEWIS 1958 *Economic Forecasting.* New York: McGraw-Hill.

BRATT, ELMER C. (1940) 1961 *Business Cycles and Forecasting.* 5th ed. Homewood, Ill.: Irwin.

¹BUTLER, WILLIAM F.; and KAVESH, ROBERT A. (editors) 1966 *How Business Economists Forecast.* Englewood Cliffs, N.J.: Prentice-Hall.

CHRIST, CARL F. 1956 Aggregate Econometric Models: A Review Article. *American Economic Review* 46:385–408.

CLARK, COLIN 1949 A System of Equations Explaining the United States Trade Cycle, 1921 to 1941. *Econometrica* 17:93–124.

COLM, GERHARD 1955 Economic Barometers and Economic Models. *Review of Economics and Statistics* 37:55–62.

COLM, GERHARD 1958 Economic Projections: Tools of Economic Analysis and Decision Making. *American Economic Review* 48, no. 2:178–187.

CONFERENCE ON BUSINESS CYCLES, NEW YORK, *1949* 1951 *Conference on Business Cycles.* New York: National Bureau of Economic Research. → See especially pages 35–129 for Carl F. Christ's article and comments by Milton Friedman et al.; see pages 339–374 for Ashley Wright's article.

CONFERENCE ON MODELS OF INCOME DETERMINATION, CHAPEL HILL, *1962* 1964 *Models of Income Determination.* Studies in Income and Wealth, Vol. 28. Princeton Univ. Press.

CONFERENCE ON RESEARCH IN INCOME AND WEALTH 1954 *Long-range Economic Projection.* Studies in Income and Wealth, Vol. 16. Princeton Univ. Press.

CONFERENCE ON RESEARCH IN INCOME AND WEALTH 1955 *Short-term Economic Forecasting.* National Bureau of Economic Research, Studies in Income and Wealth, Vol. 17. Princeton Univ. Press.

CORNFIELD, JEROME; EVANS, W. DUANE; and HOFFENBERG, MARVIN 1947 Full Employment Patterns, 1950. *Monthly Labor Review* 64:163–190, 420–432.

COX, GARFIELD V. (1929) 1930 *An Appraisal of American Business Forecasts.* Rev. ed. Univ. of Chicago Press.

DUESENBERRY, JAMES S.; ECKSTEIN, OTTO; and FROMM, GARY 1960 A Simulation of the United States Economy in Recession. *Econometrica* 28:749–809.

DUESENBERRY, JAMES S. et al. (editors) 1965 *The Brookings Quarterly Econometric Model of the United States.* Chicago: Rand McNally.

EISNER, ROBERT 1963 Investment: Fact and Fancy. *American Economic Review* 53:237–246.

FELS, RENDIGS 1963 The Recognition-lag and Semi-automatic Stabilizers. *Review of Economics and Statistics* 45:280–285.

[Forecasting.] 1954 *Journal of Business* 27, no. 1.

FOSS, MURRAY F.; and NATRELLA, VITO 1957 Ten Years' Experience With Business Investment Anticipations. *Survey of Current Business* 37, no. 1:16–24.

GORDON, ROBERT A. 1962 Alternative Approaches to Forecasting: The Recent Work of the National Bureau. *Review of Economics and Statistics* 44:284–291.

GRUNBERG, EMILE; and MODIGLIANI, FRANCO 1954 The Predictability of Social Events. *Journal of Political Economy* 62:465–478.

HART, ALBERT G. 1965 Capital Appropriations and the Accelerator. *Review of Economics and Statistics* 47:123–136.

JORGENSON, DALE W. 1963 Capital Theory and Investment Behavior. *American Economic Review* 53:247–259.

KLEIN, LAWRENCE R.; and GOLDBERGER, A. S. 1955 *An Econometric Model of the United States: 1929–1952.* Amsterdam: North-Holland Publishing.

KLEIN, LAWRENCE R. et al. 1961 *An Econometric Model of the United Kingdom.* Oxford: Blackwell.

►LEONTIEF, WASSILY 1968 Input–Output Analysis. Volume 7, pages 345–354 in *International Encyclopedia of the Social Sciences.* Edited by David L. Sills. New York: Macmillan and Free Press.

LEWIS, JOHN P. 1962 Short-term General Business Conditions Forecasting: Some Comments on Method. *Journal of Business* 35:343–356.

LIEBENBERG, MAURICE; HIRSCH, ALBERT A.; and POPKIN, JOEL 1966 A Quarterly Econometric Model of the United States: A Progress Report. *Survey of Current Business* 46, no. 5:13–39.

LIU, TA-CHUNG 1963 An Exploratory Quarterly Econometric Model of Effective Demand in the Postwar U.S. Economy. *Econometrica* 31:301–348.

MAHER, JOHN E. 1957 Forecasting Industrial Production. *Journal of Political Economy* 65:158–165.

MARQUARDT, WILHELM; and STRIGEL, WERNER 1959 *Der Konjunkturtest: Eine neue Methode der Wirtschaftsbeobachtung.* Berlin: Duncker & Humblot.

MODIGLIANI, FRANCO; and COHEN, KALMAN J. 1961 *The Role of Anticipations and Plans in Economic Behavior and Their Use in Economic Analysis and Forecasting.* Urbana: Univ. of Illinois.

MODIGLIANI, FRANCO; and WEINGARTNER, H. M. 1958 Forecasting Uses of Anticipatory Data on Investment and Sales. *Quarterly Journal of Economics* 72:23–54.

►MORGAN, JAMES N. 1968 Survey Analysis: III. Applications in Economics. Volume 15, pages 429–436 in *International Encyclopedia of the Social Sciences.* Edited by David L. Sills. New York: Macmillan and Free Press.

MOORE, GEOFFREY H. (editor) 1961 *Business Cycle Indicators.* 2 vols. National Bureau of Economic Research, Studies in Business Cycles, No. 10. Princeton Univ. Press.

NATIONAL PLANNING ASSOCIATION 1959 *Long-range Projections for Economic Growth: The American Economy in 1970.* Planning Pamphlet No. 107. Washington: The Association.

NERLOVE, MARC 1962 A Quarterly Econometric Model for the United Kingdom: A Review Article. *American Economic Review* 52:154–176.

NERLOVE, MARC 1966 A Tabular Survey of Macro-econometric Models. *International Economic Review* 7:127–175.

NEWBURY, FRANK D. 1952 *Business Forecasting: Principles and Practices.* New York: McGraw-Hill.

OKUN, ARTHUR M. 1959 A Review of Some Economic Forecasts for 1955–1957. *Journal of Business* 32:199–211.

OKUN, ARTHUR M. 1960 On the Appraisal of Cyclical Turning-point Predictions. *Journal of Business* 33:101–120.

OKUN, ARTHUR M. 1962 The Predictive Value of Surveys of Business Intentions. *American Economic Review* 52, no. 2:218–225.

PASHIGIAN, B. PETER 1964 The Accuracy of the Commerce–S.E.C. Sales Anticipations. *Review of Economics and Statistics* 46:398–405.

PERSONS, WARREN M. 1931 *Forecasting Business Cycles.* New York: Wiley.

PROCHNOW, HERBERT V. (editor) 1954 *Determining the Business Outlook.* New York: Harper.

ROOS, CHARLES F. 1955 Survey of Economic Forecasting Techniques. *Econometrica* 23:363–395.

SHISKIN, JULIUS 1961 *Signals of Recession and Recovery: An Experiment With Monthly Reporting.* National Bureau of Economic Research, Occasional Paper No. 77. New York: The Bureau.

SPENCER, MILTON H.; CLARK, C. G.; and HOGUET, P. W. 1961 *Business and Economic Forecasting: An Econometric Approach.* Homewood, Ill.: Irwin.

STEKLER, HERMAN O. 1961a Diffusion Index and First Difference Forecasting. *Review of Economics and Statistics* 43:201–208.

STEKLER, HERMAN O. 1961b Forecasting Industrial Production. *Journal of the American Statistical Association* 56:869–877.

STEKLER, HERMAN O. 1962 A Simulation of the Forecasting Performance of the Diffusion Index. *Journal of Business* 35:196–200.

SUITS, DANIEL B. 1962 Forecasting and Analysis With an Econometric Model. *American Economic Review* 52:104–132.

THEIL, HENRI (1958) 1961 *Economic Forecasts and Policy.* 2d ed. Amsterdam: North-Holland Publishing.

THEIL, HENRI 1966 *Applied Economic Forecasting.* Chicago: Rand McNally.

TINBERGEN, JAN 1938–1939 *Statistical Testing of Business-cycle Theories.* 2 vols. Geneva: League of Nations, Economic Intelligence Service. → Volume 1: *A Method and Its Application to Investment Activity.* Volume 2: *Business Cycles in the United States of America: 1919–1932.*

UNIVERSITIES–NATIONAL BUREAU COMMITTEE FOR ECONOMIC RESEARCH 1960 *The Quality and Economic Significance of Anticipations Data: A Conference.* Princeton Univ. Press.

ZARNOWITZ, VICTOR 1967 *An Appraisal of Short-term Economic Forecasts.* New York: National Bureau of Economic Research.

►ZARNOWITZ, VICTOR 1968 Mitchell, Wesley C. Volume 10, pages 373–378 in *International Encyclopedia of the Social Sciences.* Edited by David L. Sills. New York: Macmillan and Free Press.

Postscript

Turbulent developments in the United States and elsewhere continue to test severely the data and methods as well as the abilities and resources of economic forecasters. At the same time, research in forecasting continues at a fair pace, concentrating on the record and potential of different sources and techniques, with the ultimate aim to improve the anaytical tools of the trade.

Major errors of contemporary forecasts. Inflation, which became the principal economic feature and problem of the years 1967–1974, has been greatly underestimated in the large majority of forecasts. Its strength and persistence were unprecedented for the United States in a period without a major war. Moreover, the inflationary trend seemed to be but weakly affected by either the various policy efforts to contain it or the recurrent shortfalls of aggregate demand—the slowdown in 1967 and the recessions in 1970 and 1973–1975. Presumably, the tenacity of the inflation and its coexistence with considerable unemployment account for much of the large underestimation error in recent price-level forecasts.

Other likely sources of forecasting errors in general lie in political and other external events whose occurrence or timing were not predictable by means of the professional skills and tools of economists and statisticians, and whose consequences were difficult to anticipate. Examples of unpredictable events and the changes they create are numerous: the long, costly war in Vietnam; the Arab oil embargo and cartel pricing; environmental problems; the Watergate crisis, and so on. The domestic and international economic policies, created in an effort to deal with these events, themselves compounded the uncertainties and became suspect to many as inefficient or even perverse; for example, the intermittent actions to initiate, manipulate, discontinue, and revive price and wage controls. The abandonment of the system of fixed exchange rates, two devaluations, and the (mostly downward) "float" of the dollar also added to the complications facing the analysts of current economic trends.

Where a particularly great need for study and correction lies is often revealed by adversities of economic change and policy. Thus recent developments indicate that the determination of the general level of prices (and wages and interest rates) is far from being as well understood as good forecasting would require; and much the same applies to the related matter of the effectiveness of fiscal and monetary policies. The need for improvement of the corresponding equations in the forecasting models is not really new, but it is more urgent than ever and is being increasingly recognized as such.

The record shows that the forecasts of GNP and its major expenditure components neither improved nor deteriorated significantly in two periods for which some systematic comparisons were made: 1953–1963 and 1964–1969. The average errors in predicting changes in aggregate income and output tended to be much smaller for the economists' forecasts than for simple extrapolative models. On the whole, this can be viewed as a moderately creditable performance. Adding the latest years to these comparisons does not alter this general conclusion. On the other hand, forecasts of the rate of inflation have been very inaccurate and not much better than simple extrapolations, and these failures, although most conspicuous of late, were by no means limited to the most recent period only.

Comparative assessments. Studies of forecasting techniques and performance (for summaries, see Fiftieth Anniversary Colloquium 1971; Zarnowitz 1971) indicate that the large majority of forecasters use, in varying combinations, several approaches: models of economic relationships, anticipatory data from surveys, business cycle indicators, and judgments about the probable effects of recent and expected events. A broadly representative survey of the economic outlook, conducted quarterly since 1968 by the American Statistical Association and NBER, regularly collects information on the major premises and procedures incorporated in the members' forecasts; it leaves little doubt about the fact that none of the identified techniques enjoys a general preference over all others. This, significantly, suggests that none is expected to prove consistently superior. Indeed, when the survey participants were grouped according to the method they ranked first, the analysis of their errors revealed no systematic differences between

the accuracy of those who relied most on the "informal GNP model," those who favored more explicit and specific econometric models, and those who favored leading indicators.

Forecasts from different sources often show a great deal of diversity, especially near turning points in business activity and in other periods clouded by unusually high uncertainty about the near-term economic outlook. But at any given time they naturally draw upon much the same information, use similar techniques, and to some extent influence each other directly; hence, when addressed to the same target variables and periods, they are likely to have much in common, and do. Consequently, errors in many reputable forecasts show substantial positive correlations over time. This applies to predictions from private as well as government sources. The latter, despite their prompt and full access to data from official statistical agencies and the budgetary process, are not superior. The forecasts of the President's Council of Economic Advisers (CEA) certainly deserve and get much attention, notably in the context of governmental policy intentions, but they at times seemed (and later proved) too optimistic and met with considerable skepticism. On balance, the CEA predictions were about as accurate as the more successful ones of the private predictions—better in some years and for some variables, worse in other cases.

Econometric models as tools of prediction. Forecasts with econometric models require projections of exogenous variables, which are essentially judgmental (although often derived with the aid of various extrapolations); moreover, such forecasts typically incorporate judgmental adjustments of the computer solutions of the models. With such adjustments, the ex ante econometric forecasts are on the average about as accurate as the comparable noneconometric forecasts from professional sources —actually often better, but not systematically so and mostly by slight margins. Without the adjustments, the econometric model predictions are on the whole considerably worse. The evidence suggests that the interaction of the econometricians with their models can and frequently does produce good forecasts, but that a mechanical use of present-day models cannot be relied upon to do so.

Substantial work has been done on the appraisal of the predictive value of U.S. economic models, starting with the reports prepared for the 1969 Conference on Econometric Models of Cyclical Behavior (1972) and continuing with the projects of the Seminars on Evaluation and Comparison of Econometric Models sponsored by the National Science Foundation and NBER (1970——).The whole area of research in the interrelated subjects of the forecasting properties, business cycle simulations, and other dynamic analyses of the models is now definitely an active and promising one. It is clear that there is much need for improvement here, and it is likely that there is also much room for it. The process of building ever larger and more complex macroeconometric models may have been pressed forward too fast in the preceding years; it certainly left far behind the work on testing the analytical foundations and evaluating the performance of the models. The more recent studies should help restore the necessary balance between the complementary efforts in model construction and evaluation.

Time-series analysis and extrapolations. Other approaches to forecasting were also recognized to be in need of reappraisal and further analytical development. The best extrapolative models are now statistically sophisticated products of pure time-series analysis, as represented by the new techniques for optimal characterization of discrete linear processes [Box & Jenkins 1970; *see also* TIME SERIES]. Predictions with these models establish rather high standards of accuracy for the economists' forecasts. Furthermore, economists' forecasts may omit some of the information utilized in the time-series models, which opens up the possibility of (1) useful inferences about the structure of the forecasts (their apparent extrapolative and other components) and (2) improvement through appropriately weighted composite predictions. In principle, such ways of comparing, decomposing, and combining forecasts can be applied quite generally to any type of predictor, provided that the forecasts can be matched with the actual values and with each other over sufficiently long periods of time. In practice, the main arguments made in the literature on behalf of the integration of different forecasting methods refer to the need for having econometric models more efficiently incorporate the information from surveys of plans of economic decision-making units and other anticipatory data.

Cyclical indicators. A comprehensive review of business cycle indicators was initiated late in 1972 by the Bureau of Economic Analysis (BEA) in the U.S. Department of Commerce, and some of its results have recently been incorporated in the monthly BEA report, *Business Conditions Digest,* which now systematically covers several hundred time series of principal interest to business analysts and forecasters. The new evaluation of many of these and other indicators should significantly enhance the value of these data for predictive purposes. Such reviews become necessary from time to time for several reasons: changes in the data

(revisions, additions, deletions); new developments in economic theory and research; structural changes in the economy; effects of exogenous disturbances of all sorts. To illustrate, before the late 1960s, inflation in the United States had never in recorded history been so rapid and persistent as to cause current-dollar GNP (and other important nominal aggregates) to rise during a recession, but it did so in both 1970 and 1974, which distorted the cyclical movements in some of the indicators series and made a timely identification of the onset of each of the last two recessions more difficult. However, those of the historically leading indicators that are expressed in nonmonetary terms (physical units, ratios, etc.) and those that have been "deflated," i.e., adjusted for changes in prices, generally continued to show the useful property of early cyclical timing. The situation is similar for the rates of change in certain monetary aggregates, commodity stocks, and sensitive price indexes, which, when appropriately smoothed, clearly lead at turning points in business expansions, contractions, and also accelerations and slowdowns.

Prospects. In the final analysis, all remediable deficiencies of the tools and techniques of economic forecasting reflect lacking or incorrect information in the relevant areas of knowledge—mainly economics and statistics. Presumably, then, the reduction of such cognitive shortcomings would help improve the ability of experts to predict major changes in the economy. Yet it seems clear, too, that forecasts concern events that are not "knowable" in the same sense that the data of history and science are; hence such forecasts also are not "perfectible" in nearly the same degree that our understanding of the corresponding past events and relationships is. We are essentially ignorant about the limits to the accuracy of economic and other forecasts in our puzzlingly mixed stochastic–deterministic universe, where human (individual and group) behavior is variously constrained and yet in part voluntarily decided. Although these limits are probably more narrowly drawn than many people presume, it is difficult to believe that they are already approached by the present-day forecasts in economics. Richer, better, and prompter reporting on current conditions could, surely, contribute to a significant improvement of short-term forecasts, which in turn should help raise the quality of longer projections. Macroeconometric models are undoubtedly in parts misspecified and probably capable of being improved by better specification, as well as by better estimation methods. It may well be possible to get more information out of surveys of business and consumer expectations and use it more effectively in judgmental and model forecasting. However, progress on any of these fronts is likely to be costly, uneven, and time-consuming.

VICTOR ZARNOWITZ

ADDITIONAL BIBLIOGRAPHY

Because of space limitations only new books and summary articles of broad and particular interest are included, as well as a few older books that were inadvertently omitted from the bibliography of the main article. For further references, see the 1970 Fiftieth Anniversary Colloquium 1971; Zarnowitz 1971; Butler, Kavesh, & Platt 1974.

ALMON, CLOPPER 1966 *The American Economy to 1975: An Interindustry Forecast.* New York: Harper & Row.

BOX, GEORGE E. P.; and JENKINS, GWYLIM M. 1970 *Time Series Analysis: Forecasting and Control.* San Francisco: Holden-Day.

BURMEISTER, EDWIN; and KLEIN, LAWRENCE R. (editors) 1974 Econometric Model Performance: Comparative Simulation Studies of Models of the U.S. Economy. Symposium. *International Economic Review* 15:264–414, 539–653.

Business Conditions Digest. Monthly. → Published by the U.S. Department of Commerce, Bureau of Economic Analysis, since 1968. Previously (1961–1968) published by the Bureau of the Census as *Business Cycle Developments.*

ᴵBUTLER, WILLIAM F.; KAVESH, ROBERT A.; and PLATT, ROBERT B. (editors) (1966) 1974 *Methods and Techniques of Business Forecasting.* Englewood Cliffs, N.J.: Prentice-Hall. → First published as *How Business Economists Forecast.*

COLE, ROSANNE 1969 *Errors in Provisional Estimates of Gross National Product.* National Bureau of Economic Research, Studies in Business Cycles, No. 21. New York: The Bureau.

CONFERENCE ON ECONOMETRIC MODELS OF CYCLICAL BEHAVIOR, HARVARD UNIVERSITY, 1969 1972 *Econometric Models of Cyclical Behavior: Proceedings.* 2 vols. Edited by Bert G. Hickman. Conference on Research in Income and Wealth, Studies in Income and Wealth, No. 36. New York: National Bureau of Economic Research.

DHRYMES, PHOEBUS J. et al. 1972 Criteria for Evaluation of Econometric Models. *Annals of Economic and Social Measurement* 1:291–324.

EVANS, MICHAEL K. 1969 *Macroeconomic Activity: Theory, Forecasting, and Control; An Econometric Approach.* New York: Harper & Row.

EVANS, MICHAEL K.; and KLEIN, LAWRENCE R. (1967) 1968 *The Wharton Econometric Forecasting Model.* 2d enl. ed. Studies in Quantitative Economics, No. 2. Philadelphia: Economics Research Unit, Dept. of Economics, Wharton School of Finance and Commerce, Univ. of Pennsylvania.

FAIR, RAY C. 1971 *A Short-run Forecasting Model of the United States Economy.* Lexington, Mass.: Heath.

FELS, RENDIGS; and HINSHAW, C. ELTON 1968 *Forecasting and Recognizing Business Cycle Turning Points.* National Bureau of Economic Research, Studies in Business Cycles, No. 17. New York: The Bureau.

FIFTIETH ANNIVERSARY COLLOQUIUM, FIRST, NEW YORK, 1970 1971 *The Business Cycle Today.* Edited by Victor Zarnowitz. Fiftieth Anniversary Colloquium Series; National Bureau of Economic Research, General Series, No. 96. New York: The Bureau.

HAITOVSKY, YOEL et al. 1974 *Forecasts With Quarterly Macroeconometric Models.* National Bureau of Economic Research, Studies in Business Cycles, No. 23. New York: The Bureau.

HICKMAN, BERT G.; and COEN, ROBERT M. 1976 *An Annual Growth Model of the U.S. Economy.* Amsterdam and New York: North-Holland Publishing.

KLEIN, LAWRENCE R. (1968) 1970 *An Essay on the Theory of Economic Prediction.* Chicago: Markham.

LEWIS, JOHN P.; and TURNER, ROBERT C. (1959) 1967 *Business Conditions Analysis.* 2d ed. New York: McGraw-Hill.

MINCER, JACOB (editor) 1969 *Economic Forecasts and Expectations: Analyses of Forecasting Behavior and Performance.* National Bureau of Economic Research, Studies in Business Cycles, No. 19. New York: The Bureau.

MOORE, GEOFFREY H. 1969 Forecasting Short-term Economic Change. *Journal of the American Statistical Association* 64:1–22.

MOORE, GEOFFREY H.; and SHISKIN, JULIUS 1967 *Indicators of Business Expansions and Contractions.* National Bureau of Economic Research, Occasional Paper No. 103. New York: The Bureau.

MORGENSTERN, OSKAR 1928 *Wirtschaftsprognose: Eine Untersuchung ihrer Voraussetzungen und Möglichkeiten.* Vienna: Springer.

MORGENSTERN, OSKAR (1950) 1970 *On the Accuracy of Economic Observations.* 2d ed., completely rev. Princeton Univ. Press.

STEKLER, HERMAN O. 1970 *Economic Forecasting.* New York: Praeger.

ZARNOWITZ, VICTOR 1971 New Plans and Results of Research in Economic Forecasting. Pages 53–70 in National Bureau of Economic Research, *Fifty-first Annual Report.* New York: The Bureau.

PRICE, RICHARD

▶ *This article was specially written for this volume.*

Richard Price (1723–1791) was a major figure in the advance of British quantitative thought during the eighteenth century. A dissenting minister, he published important work in probability, statistics, demography, actuarial methods, welfare, philosophy, theology, and politics.

He is best known in statistical circles as the editor, annotator, and extender of Thomas Bayes' famous 1764 paper in which a special case of what is now called Bayes' theorem is developed, together with cautious suggestions for its application in statistical inference. [*Detailed references and discussions are given in the biography of* BAYES; BAYESIAN INFERENCE; *and* PROBABILITY, *article on* INTERPRETATIONS.]

Price applied Bayesian ideas in his treatment of the validity of testimony, especially as that relates to miracles. To my eye, he makes ironic, unfortunate errors by a confused treatment of conditional probabilities. In effect, the probabilities of A given B and of B given A are confused.

In a wholly different area—estimating the population of England before census taking began—Price did not apply Bayes' ideas. In 1780, Price gave one of the lowest estimates in a range that ran from 5 to 11 million and over which there was bitter controversy. In retrospect, looking back from the early British censuses (which began in 1801), Price's estimate was indeed much too low.

Richard Price was a founder of actuarial science in England. Actuarial work in his day was on a shaky basis, both as to factual background and as to theory. Price was led to actuarial thought at first by concern over pensions for ministers and insurance for their widows, and later by concern for old age support in general. He encouraged improvements in survival statistics, he clarified much actuarial mathematics and published an early and influential textbook on insurance (1771), and his was the major actuarial influence on the Equitable Life Assurance Society in its early days.

Price was an ardent supporter of the American colonists in their struggle for freedom from arbitrary British rule. He had extensive correspondence and a long friendship with Benjamin Franklin, as well as with other American leaders. Toward the end of his life, Price became an ardent supporter of the French Revolution, then just beginning. His 1789 sermon, *A Discourse on the Love of Our Country*, led to Burke's famous rebuttal, *Reflections on the Revolution in France* (1789). Perhaps it was, in a way, fortunate for Price that he died in 1791, before the bloody excesses in France. All observers agreed that Price was a man of sweetness of character, simplicity of taste, and great humility and candor.

In theology and philosophy, Price wrote on liberty, prayer, providence, and ethics. The work on ethics, described as a forerunner to Kant, has motivated a series of commentaries that continues to our time.

WILLIAM H. KRUSKAL

[*For Price and Bayes' theorem, surely his most important connection with statistics, see the biography of* BAYES *and the article on* BAYESIAN INFERENCE.]

WORKS BY PRICE

(1771) 1783 *Observations on Reversionary Payments: On Schemes for Providing Annuities for Widows, and for Persons in Old Age; On the Method of Calculating the Values of Assurances on Lives; and on the National Debt.* 2 vols. London: T. Cadell & W. Davis. → Many subsequent editions have been published; the seventh (1812) was edited and enlarged by William Morgan, Price's nephew.

(1789) 1790 *A Discourse on the Love of Our Country, Delivered on November 4, 1789, at the Meeting-house in the Old Jewry, to the Society for Commemorating the Revolution in Great Britain.* 6th ed. London: T. Cadell. → Includes additions to the appendix.

SUPPLEMENTARY BIBLIOGRAPHY

This bibliography lists the major general treatments of Price; they in turn provide bibliographical entry to the special treatments of various aspects of Price's life and work, as well as to his own publications.

BURKE, EDMUND (1789) 1790 *Reflections on the Revolution in France, and on the Proceedings of Certain Societies in London Relative to That Event.* London: Dodsley. → For a discussion of currently available editions, see Alexander M. Bickel, "Edmund Burke," *New Republic*, March 17 [1973]:30–35.

CONE, CARL B. 1952 *Torchbearer of Freedom: The Influence of Richard Price on Eighteenth Century Thought.* Lexington: Univ. Press of Kentucky.

GLASS, DAVID V. 1973 *Numbering the People.* Farnborough (England): Heath. → Contains a detailed description of the eighteenth-century population controversy in Britain and of Price's part in it.

HOLLAND, J. D. 1968 An Eighteenth-century Pioneer: Richard Price, D.D., F.R.S. (1723–1791). Royal Society of London, *Notes and Records* 23:43–64.

LABOUCHEIX, HENRI 1970 *Richard Price, théoricien de la révolution americaine, le philosophe et le sociologue, le pamphlétaire et l'orateur.* Paris: Didier.

MORGAN, WILLIAM 1815 *Memoirs of the Life of the Rev. Richard Price.* London: Hunter & Rees. → This biography, by Price's nephew, is unreliable in detail and has been roundly criticized by later writers.

OGBORN, MAURICE EDWARD 1962 *Equitable Assurances: The Story of Life Assurance in the Experience of the Equitable Life Assurance Society, 1762–1962.* London: Allen & Unwin. → Chapter 7 is on Price.

THOMAS, D. O. 1972 Richard Price, Apostle of Candour. *Études anglaises* 25:290–298. → This is in part a review of Laboucheix (1970).

THOMAS, ROLAND 1924 *Richard Price, Philosopher and Apostle of Liberty.* London: Humphrey Milford.

PRINCIPAL COMPONENTS

See FACTOR ANALYSIS AND PRINCIPAL COMPONENTS; MULTIVARIATE ANALYSIS.

PRIOR DISTRIBUTION

See BAYESIAN INFERENCE; PROBABILITY, *article on* INTERPRETATIONS.

PRIVACY

See COMPUTATION; ETHICAL ISSUES IN THE SOCIAL SCIENCES; PUBLIC POLICY AND STATISTICS.

PROBABILITY

I. FORMAL PROBABILITY   Gottfried E. Noether
II. INTERPRETATIONS   Bruno de Finetti

I

FORMAL PROBABILITY

This article deals with the mathematical side of probability, the *calculus of probabilities*, as it has sometimes been called. While there is sometimes disagreement on the philosophy and interpretation of probability, there is rather general agreement on its formal structure. Of course, when using the calculus of probabilities in connection with probabilistic models or statistical investigations, the social scientist must decide what interpretation to associate with his probabilistic statements.

The axiomatic approach

The sample space. Mathematically the most satisfactory approach to probability is an axiomatic one. Although a rigorous description of such an approach is beyond the scope of this article, the basic ideas are simple. The framework for any probabilistic investigation is a *sample space*, a set whose elements, or *points*, represent the possible outcomes of the "experiment" under consideration. An *event*, sometimes called a chance event, is represented by a subset of the sample space. For example, in a guessing-sequence experiment, test subjects are instructed to write down sequences of numbers, with each number chosen from among the digits 1, 2, 3. For sequences of length 2, an appropriate sample space is given by the nine number pairs (1,1), (1,2), (2,1), (1,3), (2,2), (3,1), (2,3), (3,2), (3,3). In this sample space the subset consisting of the three pairs (1,1), (2,2), (3,3) stands for the event *a run*, and the subset consisting of the pairs (1,3), (2,2), (3,1) stands for the event *sum equals 4*. For convenience of reference, these two events will be denoted more briefly by R and F.

The algebra of events. It is helpful to introduce some concepts and notation from the algebra of events. Events (or sets) are denoted by letters $A, B, \cdots$, with or without subscripts. The event consisting of every experimental outcome—that is, the sample space itself—is denoted by S. The event *not A* is denoted by A^c (alternate notations sometimes used include $\bar{A}$, CA, and A'), the event *A and B* by AB (or $A \cap B$ or $A \wedge B$), the event *A or B* by $A + B$ (or $A \cup B$ or $A \vee B$), with corresponding definitions for more than two events. The *or* in *A or B* is to be interpreted in the inclusive sense, implying either A or B alone or both A and B. Two events A and B are said to be *disjoint* (or *mutually exclusive*) if $AB = \phi$, where $\phi = S^c$ is the *impossible* event, an event consisting of none of the experimental outcomes. The events $A + B$ and AB are often called, respectively, the *union* and the *intersection* of A and B, and the event A^c is often called the *negation* of A. The events A and A^c are said to be *complementary* events. Figure 1 is a pictorial representation of two events (sets), A and B, by means of a *Venn diagram*. The sample space, S, is represented by a rectangle; events A and B partition S into four disjoint parts labeled according

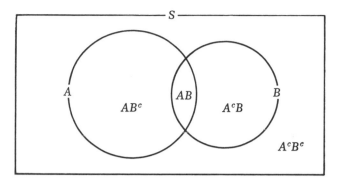

Figure 1 — Venn diagram

to their description in terms of A, A^c, B, and B^c. Thus, for example, A^cB^c is the part of the sample space that belongs both to A^c and B^c. By definition, $A + B = AB^c + A^cB + AB$.

The axioms. In an axiomatic treatment, the probability, $P(A)$, of the event A is a number that satisfies certain axioms. The following three axioms are basic. They may be heuristically explained in terms of little weights attached to each point, with the total of all weights one pound and with the probability of an event taken as the sum of the weights attached to the points of that event.

$$(i) \qquad P(A) \geq 0,$$
$$(ii) \qquad P(S) = 1,$$
$$(iii) \qquad P(A + B) = P(A) + P(B), \qquad \text{if } AB = \phi.$$

Thus, no probability should be negative, and the maximum value of any probability should be 1. This second requirement represents an arbitrary though useful normalization. Finally, the probability of the union of two disjoint events should be the sum of the individual probabilities. It turns out that for sample spaces containing only a finite number of sample points no additional assumptions need be made to permit a rigorous development of a probability calculus. In sample spaces with infinitely many sample points the above axiom system is incomplete; in particular, (iii) requires extension to denumerably many events. Mathematically more serious is the possibility that there may be events to which no probability can be assigned without violating the axioms; practically, this possibility is unimportant.

It should be noted that the axiomatic approach assumes only that *some* probability $P(A)$ is associated with an event A. The axioms do not say how the probability is to be determined in a given case. Any probability assignment that does not contradict the axioms is acceptable.

The problem of assigning probabilities is particularly simple in sample spaces containing a finite number of points, and the weight analogy applies particularly neatly in this case. It is then only necessary to assign probabilities to the individual sample points (the *elementary* events), making sure that the sum of all such probabilities is one. According to axiom (*iii*) it follows that the probability of an arbitrary event, A, is equal to the sum of the probabilities of the sample points in A. In particular, if there are n points in the sample space and *if* it is decided to assign the same probability, $1/n$, to each one of them, then the probability of an event A is equal to $m_A(1/n) = m_A/n$, where m_A is the number of sample points in A. For example, in the guessing-sequence experiment the same probability, $\frac{1}{9} = .111$, may be assigned to each one of the nine sample points if it is assumed that each number written down by a test subject is the result of a mental selection of one of three well-shuffled cards marked 1, 2, and 3, each successive selection being uninfluenced by earlier selections. This model will be called the dice model, since it is also appropriate for the rolling of two unconnected and unbiased "three-sided" dice. An actual guessing-sequence experiment in which 200 test subjects participated produced the results given in Table 1. For example, 13 of the 200 subjects produced sequences in which both digits were 1. Table 2 gives the proportion of cases in each category (for example, $13/200 = .065$). Inspection reveals important deviations from the theoretical dice model, deviations that are statistically significant [*see* COUNTED DATA; HYPOTHESIS TESTING]. For illustrative purposes an empirical model can be set up in which the entries in Table 2 serve as basic probabilities. In the dice model $P(R) = P(F) = \frac{3}{9} = \frac{1}{3}$, while in the empirical model $P_e(R) = .065 + .065 + .045 = .175$ and $P_e(F) = .130 + .065 + .160 = .355$. The subscript "$e$" distinguishes probabilities in the empirical model from those in the dice model.

Some basic theorems and definitions

Probabilities of complementary events. The probability of the event *not A* is 1 minus the probability of the event A, $P(A^c) = 1 - P(A)$; in particular, the probability of the impossible event is zero, $P(\phi) = 0$. This result may seem self-evident, but logically it is a consequence of the axioms.

Addition theorem. If A and B are any two events, the probability of the event $A + B$ is equal to the sum of the separate probabilities of events A and B minus the probability of the event AB, $P(A + B) = P(A) + P(B) - P(AB)$. For example, $P(R + F) = \frac{1}{3} + \frac{1}{3} - \frac{1}{9} = \frac{5}{9} = .556$, and $P_e(R + F) = .175 + .355 - .065 = .465$. If A and B are disjoint, $P(AB) = 0$, so the addition theorem is then just a restatement of axiom (*iii*).

Conditional probability. If $P(A)$ is not zero, the *conditional* probability of the event B given the event A is defined as

$$(1) \qquad P(B|A) = \frac{P(AB)}{P(A)}.$$

The meaning of this quantity is most easily interpreted in sample spaces with equally likely outcomes. Then $P(B|A) = (m_{AB}/n)/(m_A/n) = m_{AB}/m_A$, where m_{AB} is the number of sample points in AB. If A is considered a new sample space and new probabilities, $1/m_A$, are assigned to each of the m_A sample points in A, the probability of the event B in this new sample space is $m_{AB}/m_A = P(B|A)$, since m_{AB} of the m_A points in A are in B. The interpretation of $P(B|A)$ as the probability of B in the reduced sample space A is valid generally. Figure 2 shows the two ways of looking at conditional probability. The events AB and A can be considered in the sample space S, as in the first part of Figure 2. It is also possible to consider the event B in the sample space A, as in the second part of

Table 1 — Guessed digit pairs

		SECOND DIGIT			
		1	2	3	Total
	1	13	27	32	72
FIRST DIGIT	2	25	13	29	67
	3	26	26	9	61
Total		64	66	70	200

Source: Private communication from Frederick Mosteller.

Table 2 — Proportions of subjects guessing each digit pair

		SECOND DIGIT			
		1	2	3	Total
	1	.065	.135	.160	.360
FIRST DIGIT	2	.125	.065	.145	.335
	3	.130	.130	.045	.305
Total		.320	.330	.350	1.000

Source: Private communication from Frederick Mosteller.

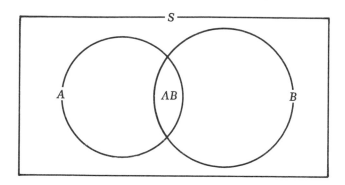

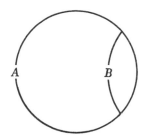

Figure 2 — Conditional probability P(B | A)

Figure 2. For example, $P_e(R|F) = P_e(RF)/P_e(F) = .065/.355 = .183$.

The multiplication theorem. Equation (1) can be rewritten as $P(AB) = P(A)P(B|A)$. This is the *multiplication theorem* for probabilities: The probability of the event AB is equal to the product of the probability of A and the conditional probability of B given A. The multiplication theorem is easily extended to more than two events. Thus, for three events A, B, C, $P(ABC) = P(A)P(B|A)P(C|AB)$.

Independent events. In general, the conditional probability $P(B|A)$ differs from the probability $P(B)$. Thus, in general, knowledge of whether A has occurred will change the evaluation of the probability of B. If the occurrence of A leaves the probability of B unchanged, that is, if $P(B|A) = P(B)$, B is said to be *independent of* A. It then follows that A is also independent of B, or simply that A and B are *independent*. A useful way of defining the independence of A and B is

(2) $$P(AB) = P(A)P(B);$$

that is, A and B are independent just when the probability of the joint occurrence of A and B is equal to the product of the individual probabilities. For example, in the dice model $P(RF) = \frac{1}{6} = P(R)P(F)$, while in the empirical model $P_e(RF) = .065 \neq P_e(R)P_e(F)$. Thus the events R and F are independent in the dice model but *dependent* in the empirical model.

The extension of the concept of independence from two to more than two events produces certain complications. For two independent events, equation (2) remains true if either A or B or both are replaced by the corresponding negations A^c and B^c. Put differently, if A and B are independent, so are A and B^c, A^c and B, A^c and B^c. If the concept of independence of three or more events is to imply a corresponding result—and such a requirement seems desirable—the formal definition of independence becomes cumbersome: The events $A_1, \cdots, A_n$ are said to be independent if the probability of the joint occurrence of any number of them equals the product of the corresponding individual probabilities. Thus, for the independence of three events A, B, C, *all* of the following four conditions must be satisfied: $P(ABC) = P(A)P(B)P(C)$, $P(AB) = P(A)P(B)$, $P(AC) = P(A)P(C)$, $P(BC) = P(B)P(C)$.

Bayes' formula. Let A be an event that can occur only in conjunction with one of k mutually exclusive and exhaustive events $H_1, \cdots, H_k$. If A is observed, the probability that it occurs in conjunction with H_i ($i = 1, \cdots, k$) is equal to

(3) $$\begin{aligned} &P(H_i|A) \\ &= \frac{P(H_iA)}{P(H_1A) + \cdots + P(H_kA)} \\ &= \frac{P(H_i)P(A|H_i)}{P(H_1)P(A|H_1) + \cdots + P(H_k)P(A|H_k)}, \end{aligned}$$

where $P(H_i)$ is the *prior* probability of the event H_i, the probability of H_i before it is known whether A occurs, and $P(A|H_i)$ is the (conditional) probability of the event A given H_i, so that the denominator of (3) is $P(A)$. The probability (3) is called the *posterior* probability of H_i. It is important to note that application of (3) requires knowledge of prior probabilities. For example, suppose that each of the 200 number pairs of the guessing-sequence experiment has been written on a separate card. In addition, two three-sided dice have been rolled 200 times and the results noted on additional cards. One of the 400 cards is selected at random and found to contain a run. The probability that this card comes from the guessing sequence is $(\frac{1}{2} \times .175)/[(\frac{1}{2} \times .175) + (\frac{1}{2} \times \frac{1}{3})] = .344$. Thus the appearance of a run changes the prior probability .500 to the posterior probability .344, reflecting the much smaller probability of a run in the empirical model than in the dice model.

Combinatorial formulas. The task of counting the number of points in the sample space S and in subsets of S can often be carried out more economically with the help of a few formulas from combinatorial analysis. A *permutation* is an arrangement of some of the objects in a set in which order is relevant. A *combination* is a selection in which order is irrelevant. Given n different objects, $m \leqslant n$ of them can be selected in

$$\binom{n}{m} = \frac{n!}{m!\,(n-m)!}$$
$$= \frac{n(n-1)\cdots(n-m+1)}{m(m-1)\cdots 1}$$

ways, when no attention is paid to the order. The symbol $\binom{n}{m}$ (called a binomial coefficient and sometimes written C_m^n) is read as "the number of combinations of n things taken m at a time." For a positive integer r, $r!$ (read "r-factorial") stands for the product of the first r positive integers, $r! = r(r-1)\cdots 1$. By definition, $0! = 1$. It is possible to arrange m out of n different objects in $(n)_m = m!\binom{n}{m} = n(n-1)\cdots(n-m+1) = n!/(n-m)!$ ways. The symbol $(n)_m$ is read as "the number of permutations of n things taken m at a time." If the n objects are not all different—in particular, if there are n_1 of a first kind, n_2 of a second kind, and so on up to n_k of a kth kind such that $n = n_1 + \cdots + n_k$—then there are $n!/(n_1!\cdots n_k!)$ ways in which all n objects can be permuted. For $k = 2$ this expression reduces to $\binom{n}{n_1} = \binom{n}{n_2}$. See Niven (1965) and Riordan (1958) for combinatorial problems.

Random variables

Definition. In many chance experiments the chief interest relates to numerical information furnished by the experiment. A numerical quantity whose value is determined by the outcome of a chance experiment—mathematically, a single-valued function defined for every point of the sample space—is called a *random variable* (abbreviated r.v.). The remainder of this article deals with such variables. For reasons of mathematical simplicity, the discussion will be in terms of r.v.'s that take only a finite number of values. Actually, with suitable modifications the results are valid generally. (Some details are given below.) It is customary to denote r.v.'s by capitals such as X, Y, Z. For example, consider the sample space consisting of the n students in a college, each student having the same probability, $1/n$, of being selected for an "experiment." A possible r.v., X, is the IQ score of a student. A second r.v., Y, is a student's weight to the nearest five pounds. A third r.v., Z, may take only the value 1 or 2, depending on whether a student is male or female. The example shows that many r.v.'s can be defined on the same sample space.

As a further example, consider the r.v. W, equal to the sum of the two numbers in a guessing sequence of length 2. Table 3 shows the relationship between sample points and values of W. To each point in S there corresponds exactly one value of W. Generally, the reverse is not true.

While it is possible to study the calculus of probabilities without specifically defining r.v.'s, formal introduction of such a concept greatly clarifies basic ideas and simplifies notation. If the sample space S in Table 3 actually refers to successive rolls of a three-sided die, possibly with unequal probabilities for the three sides, it is natural to consider only the sum of the two rolls as the event of interest [see SUFFICIENCY]. This can be done by defining separately the five events *sum equals 2*, *sum equals 3*, $\cdots$, *sum equals 6* or by considering vaguely a "wandering" variable that is able to take the values 2 or 3 or $\cdots$ or 6. The r.v. W defined above and illustrated in Table 3 combines both approaches in a simple and unambiguous way. Note that the event F can be expressed as the event $W = 4$.

Table 3 — Relation between sample points and values of a random variable

Sample space				Values of random variable, W		
(1,1)	(1,2)	(1,3)	W	2	3	4
(2,1)	(2,2)	(2,3)	$\rightarrow$	3	4	5
(3,1)	(3,2)	(3,3)		4	5	6

The advantages of a formal concept are even more pronounced when it is desirable to consider two or more measurements jointly. Thus, in the earlier example an investigator might not be interested in considering weight and sex of students each by itself but as they relate to one another. This is accomplished by considering for every point in the sample space the number pair (Y, Z) where Y and Z are the r.v.'s defined earlier. Perhaps most important, the concept of r.v. permits more concise probability statements than the ones associated with the basic sample space.

Frequency function of a random variable. Let X be a random variable with possible values $x_1, \cdots, x_k$. Define a function $f(x)$ such that for $x = x_i$ ($i = 1, \cdots, k$), $f(x_i)$ is the probability of the event that the basic experiment results in an outcome for which the r.v. X takes the value x_i, $f(x_i) = P(X = x_i)$. Clearly, $f(x_1) + \cdots + f(x_k) = 1$. The function $f(x)$ defined in this way is called the *frequency* function of X. For example, in the dice model the frequency function of W is

w	2	3	4	5	6
$f(w)$	$\frac{1}{9}$	$\frac{2}{9}$	$\frac{3}{9}$	$\frac{2}{9}$	$\frac{1}{9}$

Mean and variance of a random variable. The *expected value* of the r.v. X, which is denoted by $E(X)$ (or by EX or μ_x or simply μ if there is no ambiguity), is defined as the weighted average $E(X) = x_1 f(x_1) + \cdots + x_k f(x_k)$. $E(X)$ is also called the *mean* of X. More generally, if $H(x)$ is a function of x, the expected value of the r.v. $H(X)$ is given by

$$\text{(4)} \quad \begin{aligned} E[H(X)] &= H(x_1)f(x_1) + \cdots + H(x_k)f(x_k) \\ &= \mu_{H(X)} . \end{aligned}$$

The function $H(x) = (x - \mu)^2$ is of particular interest and usefulness. Its expected value is called the *variance* of X and is denoted by var X (or σ_X^2 or σ^2), var $X = E(X - \mu)^2 = (x_1 - \mu)^2 f(x_1) + \cdots + (x_k - \mu)^2 f(x_k)$. The positive square root of the variance is called the *standard deviation* and is denoted by $s.d.X$ (or σ_X or σ).

If the r.v. Y is a linear function of the r.v. X, $Y = a + bX$, where a and b are constants, then $\mu_Y = a + b\mu_X$, $\sigma_Y^2 = b^2 \sigma_X^2$, and $\sigma_Y = |b|\sigma_X$, where $|b|$ denotes the numerical value of b without regard to sign.

The mean and variance of a r.v. are two important characteristics. Their true theoretical significance emerges in connection with such advanced theorems as the law of large numbers and the central limit theorem (see below). The useful additive property of means and variances is also stated below. On an elementary level, the mean and the variance are useful descriptive or summary measures [see STATISTICS, DESCRIPTIVE, *article on* LOCATION AND DISPERSION].

The mean is but one of the "averages" that can be computed from the values $x_1, \cdots, x_k$ of the r.v. Another average is the median, Med, defined by the two inequalities $P(X \leqslant \text{Med}) \geqslant \frac{1}{2}$ and $P(X \geqslant \text{Med}) \geqslant \frac{1}{2}$. Thus the median—which may not be uniquely defined—is a number that cuts in half, as nearly as possible, the frequency function of X.

The standard deviation is a measure of variability or spread around the mean, a small standard deviation indicating little variability among the possible values of the r.v., a large standard deviation indicating considerable variability. This rather vague statement is made more precise by the Bienaymé–Chebyshev inequality. This inequality establishes a connection between the size of the standard deviation and the concentration of probability in intervals centered at the mean.

The Bienaymé–Chebyshev inequality. Let δ be an arbitrary positive constant. The probability that a r.v. X takes a value that deviates from its mean by less than δ standard deviations is at least $1 - 1/\delta^2$, $P(|X - \mu| < \delta\sigma) \geqslant 1 - 1/\delta^2$. This is one of the forms of the Bienaymé–Chebyshev inequality

(often called the Chebyshev inequality). Often the complementary result is more interesting. The probability that a r.v. X takes a value that deviates from its mean by at least δ standard deviations is at most $1/\delta^2$, $P(|X - \mu| \geqslant \delta\sigma) \leqslant 1/\delta^2$. Although there are r.v.'s for which the inequality becomes an equality, for most r.v.'s occurring in practice, the probability of large deviations from the mean is considerably smaller than the upper limit indicated by the inequality. An example is given below. The Chebyshev inequality illustrates the fact that for probabilistic purposes the standard deviation is the natural unit of measurement.

The binomial distribution. Consider an experiment that has only two possible outcomes, called *success* and *failure*. The word *trial* will be used to denote a single performance of such an experiment. A possible sample space for describing the results of n trials consists of all possible sequences of length n of the type $FFSSS \cdots F$, where S stands for success and F for failure. A natural r.v. defined on this sample space is the total number X of successes [see SUFFICIENCY]. If successive trials are independent and the probability of success in any given trial is a constant p, then

$$\text{(5)} \quad f(x) = P(X = x) = \binom{n}{x} p^x (1 - p)^{n-x}, \quad x = 0, 1, \cdots, n.$$

The r.v. X is called the *binomial* r.v. and the frequency function (5) the binomial distribution. [See DISTRIBUTIONS, STATISTICAL *for a discussion of the binomial distributions and other distributions mentioned below.*] The expected or mean value of X is np and the standard deviation is $\sqrt{np(1 - p)}$. According to the Chebyshev inequality the upper limit for the probability that a r.v. deviates from its mean by two standard deviations or more is .25. For the binomial variable with $n = 50$ and $p = \frac{1}{2}$, for example, the exact probability is only .033, considerably smaller than the Chebyshev limit.

Joint frequency functions. Let X and Y be two r.v.'s defined on the same sample space. Let X take the values $x_1, \cdots, x_k$, and let Y take the values $y_1, \cdots, y_h$. What then is the probability of the joint event $X = x_i$ *and* $Y = y_j$? The function $f(x,y)$ such that $f(x_i, y_j) = P(X = x_i \text{ and } Y = y_j)$ $(i = 1, \cdots, k; \ j = 1, \cdots, h)$ is called the *bivariate* frequency function of X and Y. In terms of $f(x,y)$ the *marginal* frequency function of X is given by $f(x_i) = P(X = x_i) = \sum_{j=1}^{h} f(x_i, y_j)$ and, similarly, the marginal frequency function of Y by $g(y_j) = P(Y = y_j) = \sum_{i=1}^{k} f(x_i, y_j)$. Marginal distributions are used to make probability statements that involve only one of the variables without regard to the value of the other variable. A different situation arises if the value of one of the variables becomes known. In that case, probability statements involving the other variable should be conditional on what is known. If in (1), A and B are defined as the events $X = x_i$ and $Y = y_j$, respectively, the conditional frequency function of the r.v. Y, given that X has the value x_i, is found to be $g(y_j|x_i) = P(Y = y_j|X = x_i) = f(x_i,y_j)/f(x_i)$, with a corresponding expression for the conditional frequency function of X given Y. For example, in the empirical model for the guessing-sequence experiment, associate a r.v. X with the first guess and a r.v. Y with the second guess. Then $f(x,y) = P_e$ (the first guess is x and the second guess is y), $x,y = 1,2,3$. In Table 2 the last column on the right represents the marginal distribution of X and the last row on the bottom the marginal distribution of Y. It is noteworthy that these marginal distributions do not differ very much from the marginal distributions for the dice model in which all probabilities equal $\frac{1}{3}$. As an example of a conditional probability, note that $g(1|1) = P_e(Y=1|X=1) = .065/.360 = .181$, a result that differs considerably from the corresponding probability $\frac{1}{3}$ for the dice model.

The two r.v.'s X and Y are said to be independent (or independently distributed) if the joint frequency function of X and Y can be written as the product of the two marginal frequency functions,

$$(6) \qquad f(x_i,y_j) = f(x_i)g(y_j).$$

If X and Y are independent r.v.'s, knowledge of the value of one of the variables, provided that $f(x,y)$ is known, furnishes no information about the other variable, since if (6) is true, the conditional frequency functions equal the marginal frequency functions, $f(x|y) = f(x)$ and $g(y|x) = g(y)$, whatever the value of the conditioning variable. The r.v.'s X and Y in the guessing-sequence example are independent in the dice model and dependent in the empirical model.

In addition to the means μ_X and μ_Y and the variances σ_X^2 and σ_Y^2 of the marginal frequency functions one defines the *covariance*,

$$
\begin{aligned}
\sigma_{XY} &= E(X - \mu_X)(Y - \mu_Y) \\
&= \sum_{i=1}^{k} \sum_{j=1}^{h} (x_i - \mu_X)(y_j - \mu_Y)f(x_i,y_j) \\
&= E(XY) - \mu_X\mu_Y,
\end{aligned}
$$

and the *correlation* coefficient, $\rho = \sigma_{XY}/\sigma_X\sigma_Y$. If X and Y are independent, their covariance and, consequently, their correlation are zero. The reverse, however, is not necessarily true; two r.v.'s, X and Y, can be uncorrelated (that is, can have correlation coefficient zero) without being independent.

The covariance of two r.v.'s is a measure of the "co-variability" of the two variables about their respective means; a positive covariance indicates a tendency of the two variables to deviate in the same direction, while a negative covariance indicates a tendency to deviate in opposite directions. Although the covariance does not depend on the zero points of the scales in which X and Y are measured, it does depend on the units of measurement. By dividing the covariance by the product of the standard deviations a normalization is introduced making the resulting correlation coefficient independent of the units of measurement as well. (If, for example, X and Y represent temperature measurements, the correlation between X and Y is the same whether temperatures are measured as degrees Fahrenheit or as degrees centigrade.) The concepts of covariance and correlation are closely tied to *linear* association. One may have very strong (even complete) nonlinear association and very small (even zero) correlation.

The concepts discussed in this section generalize from two to more than two variables.

Sums of random variables

Mean and variance of a sum of random variables. Let $X_1, \cdots, X_n$ be a set of n r.v.'s with means $E(X_i) = \mu_i$, variances $E(X_i - \mu_i)^2 = \sigma_i^2$, and covariances $E(X_i - \mu_i)(X_j - \mu_j) = \sigma_{ij}$ $(i,j = 1,\cdots, n; i \neq j)$. Some of the most fruitful studies in the calculus of probabilities are concerned with the properties of sums of the type $Z = c_1X_1 + \cdots + c_nX_n$ where the c_i are given constants. In this article only some of the simpler, although nevertheless highly important, results will be stated.

The mean and variance of Z are $\mu_Z = c_1\mu_1 + \cdots + c_n\mu_n$ and $\sigma_Z^2 = c_1^2\sigma_1^2 + \cdots + c_n^2\sigma_n^2 + 2c_1c_2\sigma_{12} + 2c_1c_3\sigma_{13} + \cdots + 2c_{n-1}c_n\sigma_{n-1,n}$. For the remainder assume that the r.v.'s $X_1, \cdots, X_n$ are independently and identically distributed. This is the mathematical model assumed for many statistical investigations. The common mean of the r.v.'s $X_1, \cdots, X_n$ is denoted by μ and the common variance by σ^2. Of particular interest are the two sums $S_n = X_1 + \cdots + X_n$ and $\bar{X} = (X_1 + \cdots + X_n)/n = S_n/n$. (The binomial r.v. X is of the form S_n if a r.v. X_i is associated with the ith trial, X_i taking the value 1 or 0, depending on whether the ith trial results in success or failure.) For the sum S_n, $\mu_{S_n} = n\mu$, $\sigma_{S_n}^2 = n\sigma^2$, and for $\bar{X}$, $\mu_{\bar{X}} = \mu$, $\sigma_{\bar{X}}^2 = \sigma^2/n$.

The law of large numbers. If Chebyshev's inequality is applied to $\bar{X}$, then for arbitrarily small positive δ

$$P(|\bar{X} - \mu| < \delta) \geqslant 1 - \frac{\sigma^2}{n\delta^2}.$$

By choosing n sufficiently large, the right side can be made to differ from 1 by as little as desired. It follows that the probability that $\bar{X}$ deviates from μ by more than some arbitrarily small positive quantity δ can be made as small as desired. In particular, in the case of the binomial variable X, the probability that the observed success ratio, $\bar{X} = X/n$, deviates from the probability, p, of success in a single trial by more than δ can be made arbitrarily small by performing a sufficiently large number of trials. This is the simplest version of the celebrated *law of large numbers*, more commonly known—and misinterpreted—as the "law of averages." The law of large numbers does *not* imply that the observed number of successes X necessarily deviates little from the expected number of successes np, only that the relative frequency of success X/n is close to p. Nor does the law of large numbers imply that, given $\bar{X} > p$ after n trials, the probability of success on subsequent trials is small in order to compensate for an excess of successes among the first n trials. Nature "averages out" by swamping, not by fluctuating.

In more advanced treatments this law is called the weak law of large numbers, to distinguish it from a stronger form.

The central limit theorem. The law of large numbers has more theoretical than practical significance, since it does not furnish precise or even approximate probabilities in any given situation. Such information is, however, provided by the *central limit theorem*: As n increases indefinitely, the distribution function of the standardized variable $(S_n - \mu_{S_n})/\sigma_{S_n} = (S_n - n\mu)/\sqrt{n\sigma^2}$ converges to the so-called standard normal distribution. (A general discussion of the normal distribution is given below.) For practical purposes the stated result means that for large n the probability that S_n takes a value between two numbers $a < b$ can be obtained approximately as the area under the standard normal curve between the two points $(a - n\mu)/\sqrt{n\sigma^2}$ and $(b - n\mu)/\sqrt{n\sigma^2}$. In particular, the probability that in a binomial experiment the number of successes is at least k_1 and at most k_2 (where k_1 and k_2 are integers) is approximately equal to the area under the normal curve between the limits $(k_1 - \frac{1}{2} - np)/\sqrt{np(1-p)}$ and $(k_2 + \frac{1}{2} - np)/\sqrt{np(1-p)}$, provided n is sufficiently large. Here a *continuity correction* of $\frac{1}{2}$ has been used in order to improve the approximation. For most practical purposes n may be assumed to be "sufficiently large" if $np(1-p)$ is at least 3.

The central limit theorem occupies a basic position not only in theory but also in application. The sample observations $x_1, \cdots, x_n$ drawn by the statistician may be looked upon as realizations of n jointly distributed random variables $X_1, \cdots, X_n$. It is customary to refer to a function of sample observations as a *statistic*. From this point of view, a statistic is a r.v. and its distribution function is called the *sampling distribution* of the statistic. The problem of determining the sampling distributions of statistics of interest to the statistician is one of the important problems of the calculus of probabilities. The central limit theorem states that under very general conditions the sampling distribution of the statistic S_n can, for sufficiently large samples, be approximated in a suitable manner by the normal distribution.

More complicated versions of the law of large numbers and the central limit theorem exist for the case of r.v.'s that are not identically distributed and are even dependent to some extent.

A more general view

A more general view of random variables and their distributions will now be presented.

Discrete and continuous random variables. For reasons of mathematical simplicity the discussion so far has been in terms of r.v.'s that take only a finite number of values. Actually this limitation was used explicitly only when giving such definitions as that of an expected value. Theorems like Chebyshev's inequality, the law of large numbers, and the central limit theorem were formulated without mention of a finite number of values. Indeed, they are true for very general r.v.'s. The remainder of this article will be concerned with such r.v.'s. Of necessity the mathematical tools have to be of a more advanced nature. (For infinite sample spaces, there arises the need for a concept called measurability in the discussion of events and of random variables. For simplicity such discussion is omitted here.)

By definition, a r.v. X is a single-valued function defined on a sample space. For every number x $(-\infty < x < \infty)$, the probability that X takes a value that is smaller than or equal to x can be determined. Let $F(x)$ denote this probability considered as a function of x, $F(x) = P(X \leq x)$. $F(x)$ is called the (cumulative) *distribution function* of the r.v. X. The following properties of a distribution function are consequences of the definition: $F(-\infty) = 0$; $F(\infty) = 1$; $F(x)$ is monotonically nondecreasing, that is, $F(x_1) \leq F(x_2)$ if $x_1 < x_2$. Furthermore, $F(x_2) - F(x_1) = P(x_1 < X \leq x_2)$. Such a function may be continuous or discontinuous. If discontinuous, it has at most a denumerable number of discontinuities, at each of which $F(x)$ has a simple jump, or saltus. The height of this jump is equal to the probability with which the r.v. X takes the value x where the discontinuity occurs. At the same time, if $F(x)$ is continuous at x, the probability of the event $X = x$ is zero.

Let $F(x)$ be discontinuous with discontinuities occurring at the points $x_1, x_2, \cdots, x_n, \cdots$. (If there are only a finite number of discontinuities, denote their number by n.) Let the size of the jump occurring at $x = x_i$ ($i = 1, 2, \cdots, n, \cdots$) be equal to $f(x_i)$. A particularly simple case occurs if $f(x_1) + f(x_2) + \cdots + f(x_n) + \cdots = 1$. In this case $F(x)$ is a "step function" and $x_1, x_2, \cdots, x_n, \cdots$ are the only values taken by the r.v. X. Such a r.v. is said to be *discrete*. Clearly the r.v.'s considered earlier are discrete r.v.'s with a finite number of values. As before, call $f(x)$ the frequency function of the r.v. X. In terms of $f(x)$, $F(x)$ is given by $F(x) = \sum f(x_i)$, where $-\infty < x < \infty$ and the summation extends over all x_i that are smaller than or equal to x.

If $F(x)$ is continuous for all x, X is said to be a *continuous* r.v. Consider the case where there exists a function $f(x)$ such that $F(x) = \int_{-\infty}^{x} f(t)\, dt$. (In statistical applications this restriction is of little importance.) The function $f(x) = dF(x)/dx$ is called the *density* function of the continuous r.v. X. Clearly $f(x) \geqslant 0$ and $\int_{-\infty}^{\infty} f(t)\, dt = 1$. Furthermore, $F(x_2) - F(x_1) = \int_{x_1}^{x_2} f(t)\, dt = P(x_1 < X < x_2) = P(x_1 \leqslant X \leqslant x_2)$ $= P(x_1 < X \leqslant x_2) = P(x_1 \leqslant X \leqslant x_2)$, since for a continuous r.v. X, $P(X = x) = 0$ for every x. It follows that for a continuous r.v. X with density function $f(x)$ the probability that X takes a value between two numbers x_1 and x_2 is given by the area between the curve representing $f(x)$ and the x-axis and bounded by the ordinates at x_1 and x_2 (Figure 3). Although r.v.'s that are part continuous and part discrete occur, they will not be considered here.

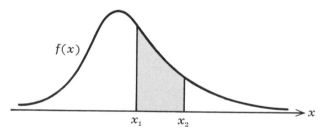

Figure 3 — P(x₁ ≤ X ≤ x₂) as area

For two continuous r.v.'s defined on the same sample space, stipulate the existence of a bivariate density function $f(x,y)$ such that

$$P(x_1 \leqslant X \leqslant x_2 \text{ and } y_1 \leqslant Y \leqslant y_2) = \int_{x_1}^{x_2} \int_{y_1}^{y_2} f(x,y)\, dx\, dy.$$

In terms of $f(x,y)$, marginal and conditional density functions can be defined as in the case of two discrete variables.

Let $H(x)$ be a function of x. An interesting and

important problem is concerned with the distribution of the derived random variable $H(X)$. Here only the expected value of $H(X)$ will be discussed. $E[H(X)]$ can be expressed in terms of the distribution function $F(x)$ by means of the Stieltjes integral

$$(7) \qquad E[H(X)] = \int_{-\infty}^{\infty} H(x)\, dF(x).$$

For discrete r.v.'s, (7) reduces to $E[H(X)] = \sum H(x_i) f(x_i)$. If X has a density function $f(x)$, (7) becomes $E[H(X)] = \int_{-\infty}^{\infty} H(x) f(x)\, dx$. One important new factor arises that was not present in (4), the question of existence. In (7) the expected value exists if and only if the corresponding sum or integral converges absolutely. This condition means that in the discrete case, for example, $E[H(X)]$ exists just when $\sum |H(x_i)| f(x_i)$ converges.

If $H(x) = x^k$, $k = 1, 2, \cdots$, the corresponding expected value is called the *kth moment about the origin* and is denoted by μ'_k, $\mu'_k = E(X^k)$. Of particular interest is the first moment, μ'_1, which was written earlier as μ. The *kth central moment*, μ_k, is defined as $\mu_k = E(X - \mu)^k = \mu'_k - k\mu\mu'_{k-1} \pm \cdots + (-1)^k \mu^k$. The first central moment is zero. The second central moment is the variance, denoted by σ^2, $\sigma^2 = E(X - \mu)^2 = \mu_2' - \mu^2$. The moments are of interest because of the information that they provide about the distribution function. Thus the Chebyshev inequality shows the kind of information provided by the first two moments. Additional moments provide more and more precise information. Finally, in many circumstances, knowledge of the moments of all orders uniquely determines the distribution function.

Generating functions. Rather than compute moments from their definitions, it is often simpler to make use of a generating function. The *moment-generating function*, $M(u)$, is defined as the expected value of the random variable e^{uX}, where u is a real variable. The *characteristic function* is defined as the expected value of e^{iuX}, where u again is real and $i = \sqrt{-1}$. The characteristic function has the advantage that it always exists. The moment-generating function exists only for r.v.'s that have moments of all orders. For $k = 1, 2, \cdots$, the kth moment can be found as the kth derivative of $M(u)$ evaluated at $u = 0$, $\mu'_k = d^k M(u)/du^k \big|_{u=0}$. A corresponding result holds for characteristic functions.

While the moment-generating property of the moment-generating function is useful, its main significance arises from the *uniqueness theorem*. A moment-generating function uniquely determines its distribution function, and it is often easier to find the moment-generating function of a r.v. than its distribution function. As an example, consider the r.v. $Z = X + Y$, where X and Y are two independently distributed r.v.'s. It follows from the definition of a moment-generating function that the moment-generating function of Z is the product of the moment-generating functions of X and Y. No such simple relationship exists between the distribution function of Z and those of X and Y. However, once the moment-generating function of Z is known, it is theoretically possible to determine its distribution function. These results also hold for characteristic functions.

The Poisson and normal distributions. In conclusion two examples will be given—one involving a discrete r.v., the other a continuous r.v. When considering random events occurring in time, one is often interested in the total number of occurrences in an interval of given length. An example is the number of suicides occurring in a community in a year's time. Then a discrete r.v. X with possible values $0, 1, 2, \cdots$ can be defined. Often an appropriate mathematical model is given by the *Poisson distribution*, according to which

$$f(x) = P(X = x) = e^{-\lambda} \frac{\lambda^x}{x!}, \qquad x = 0, 1, 2, \cdots,$$

where λ is a characteristic of the type of random event considered. The moment-generating function of X is $M(u) = \exp[\lambda(e^u - 1)]$, where $\exp w$ stands for e^w. Then $\mu = \lambda = \sigma^2$. Thus λ is the mean number of occurrences in the given time interval and, at the same time, is also the variance of the number of occurrences. Furthermore, the sum of two independent Poisson variables is again a Poisson variable whose parameter λ is the sum of the parameters of the two independent variables.

The Poisson distribution serves as an excellent approximation for binomial probabilities, if the probability of success, p, is small. More exactly, if np is set equal to λ, the binomial probability in (5) can be approximated by $e^{-\lambda}\lambda^x/x!$, provided n is sufficiently large. This approximation is particularly useful when only the product np is known but not the values of n and p separately.

A continuous r.v. with density function $f(x) = \exp[-(x-a)^2/2b^2]/\sqrt{2\pi b^2}$ is said to have the *normal distribution* with parameters a and $b > 0$. The moment-generating function of such a variable is $\exp(au + \frac{1}{2}b^2u^2)$, from which $\mu = a$, $\sigma^2 = b^2$. It is therefore customary to write the density as

$$f(x) = \frac{1}{\sqrt{2\pi\sigma^2}} \exp\left[-\frac{1}{2\sigma^2}(x-\mu)^2\right].$$

A particular normal density function is obtained by setting $\mu = 0$ and $\sigma^2 = 1$. This simple function is called the *unit* (or *standard*) *normal density* function; it appeared above in the discussion of the central limit theorem. The sum of two or more independent normal variables is again normally distributed with mean and variance equal to the sums of means and variances, respectively.

The normal distribution is often used as a mathematical model to describe populations such as that of scores on a test. Arguments in support of the normality assumption are customarily based on the central limit theorem. Thus it is argued that the value of a given measurement is determined by a large number of factors. It is less frequently realized that reference to the central limit theorem implies also that factors act in an *additive* fashion. Nevertheless, experience shows that the degree of nonnormality occurring in practice is often so small that the assumption of actual normality does not lead to erroneous conclusions.

GOTTFRIED E. NOETHER

[*For a discussion of the various distributions mentioned in the text, see also* DISTRIBUTIONS, STATISTICAL.]

BIBLIOGRAPHY

WORKS REQUIRING AN ELEMENTARY MATHEMATICAL BACKGROUND

CRAMÉR, HARALD (1951) 1955 *The Elements of Probability Theory and Some of Its Applications.* New York: Wiley. → First published as *Sannolikhetskalkylen och några av dess anwändningar.*

[1]HODGES, J. L. JR.; and LEHMANN, E. L. 1964 *Basic Concepts of Probability and Statistics.* San Francisco: Holden-Day.

[2]MOSTELLER, FREDERICK; ROURKE, ROBERT E. K.; and THOMAS, GEORGE B. JR. 1961 *Probability With Statistical Applications.* Reading, Mass.: Addison-Wesley.

NIVEN, IVAN 1965 *Mathematics of Choice.* New York: Random House.

WEAVER, WARREN 1963 *Lady Luck: The Theory of Probability.* Garden City, N.Y.: Doubleday.

WORKS OF A MORE ADVANCED NATURE

[3]FELLER, WILLIAM 1950–1966 *An Introduction to Probability Theory and Its Applications.* 2 vols. New York: Wiley. → The second edition of Volume 1 was published in 1957.

GNEDENKO, BORIS V. (1950) 1962 *The Theory of Probability.* New York: Chelsea. → First published as *Kurs teorii veroiatnostei.*

PARZEN, EMANUEL 1960 *Modern Probability Theory and Its Applications.* New York: Wiley.

RIORDAN, JOHN 1958 *An Introduction to Combinatorial Analysis.* New York: Wiley.

Postscript

New editions of several cited works have appeared: a second edition of Hodges and Lehmann (1964), a second edition of Mosteller et al. (1961), and further editions of both volumes of Feller (1950–1960). A new work falling into the category of "more advanced" is Chung (1974).

GOTTFRIED E. NOETHER

[See also the biography of RÉNYI.]

ADDITIONAL BIBLIOGRAPHY

CHUNG, KAI LAI 1974 Elementary Probability Theory With Stochastic Processes. Berlin and New York: Springer.

[3]FELLER, WILLIAM (1950–1960) 1968–1971 An Introduction to Probability Theory and Its Applications. 2 vols. New York: Wiley. → The third edition of Volume 1 was published in 1968, the second edition of Volume 2 in 1971.

[1]HODGES, J. L. JR.; and LEHMANN, E. L. (1964) 1970 Basic Concepts of Probability and Statistics. 2d ed. San Francisco: Holden-Day.

[2]MOSTELLER, FREDERICK; ROURKE, ROBERT E. K.; and THOMAS, GEORGE B. JR. (1961) 1970 Probability With Statistical Applications. 2d ed. Reading, Mass.: Addison-Wesley.

II
INTERPRETATIONS

Many disputes—what are they about? There are myriad different views on probability, and disputes about them have been going on and increasing for a long time. Before outlining the principal questions, and the main attitudes toward them, we note a seeming contradiction; it may be said with equal truth that the different interpretations alter in no substantial way the contents and applications of the theory of probability and, yet, that they utterly alter everything. It is important to have in mind precisely what changes and what does not.

Nothing changes for the mathematical theory [see PROBABILITY, article on FORMAL PROBABILITY]. Thus a mathematician not conceptually interested in probability can do unanimously acceptable work on its theory, starting from a merely axiomatic basis. And often nothing changes even in practical applications, where the same arguments are likely to be accepted by everyone, if expressed in a sufficiently acritical way (and if the validity of the particular application is not disputed because of preconceptions inherent in one view or another).

For example, suppose someone says that he attaches the probability one sixth to an ace at his next throw of a die. If asked what he means, he may well agree with statements expressed roughly thus: he considers $1 the fair insurance premium against a risk of $6 to which he might be exposed by occurrence of the ace; the six faces are equally likely and only one is favorable; it may be expected that every face will appear in about $\frac{1}{6}$ of the trials

in the long run; he has observed a frequency of $\frac{1}{6}$ in the past and adopts this value as the probability for the next trial; and so on. Little background is needed to see that each of these rough statements admits several interpretations (or none at all, if one balks at insufficient specification). Moreover, only one of the statements can express the very idea, or definition, of probability according to this person's language, while the others would be accepted by him, if at all, as consequences of the definition and of some theorems or special additional assumptions for particular cases. It would be a most harmful misappraisal to conclude that the differences in interpretation are meaningless except for pedantic hairsplitters or, even worse, that they do not matter at all (as when the same geometry is constructed from equivalent sets of axioms beginning with different choices of the primitive notions). The various views not only endow the same formal statement with completely different meanings, but a particular view also usually rejects some statements as meaningless, thereby restricting the validity of the theory to a narrowed domain, where the holders of that view feel more secure. Then, to replace the rejected parts, expedients aiming at suitable reinterpretation are often invented, which, naturally, are only misinterpretations for the adherents of other views.

A bit of history

The beginnings. It is an ambitious task even to make clear the distinctions and connections between the various schools of thought, so we renounce any attempt to enter far into their historical vicissitudes. A sketch of the main lines of evolution should be enough to give perspective. [See STATISTICS, article on THE HISTORY OF STATISTICAL METHOD.]

In an early period (roughly 1650–1800), the mathematical theory of probability had its beginnings and an extraordinarily rapid and fruitful growth. Not only were the fundamental tools and problems acquired, but also the cornerstones of some very modern edifices were laid: among others, the principle of utility maximization, by Daniel Bernoulli in 1738; the probabilistic approach to inductive reasoning and behavior, by Thomas Bayes in 1763; and even the minimax principle of game theory, by de Waldegrave in 1712 (Guilbaud 1961). But interest, in those days, focused on seemingly more concrete problems, such as card games; conceptual questions were merely foreshadowed, not investigated critically; utility theory remained unfruitful; Bayes' principle was misleadingly linked with Bayes' postulate of uniform initial distribution; de Waldegrave's idea went unnoticed [see BAYESIAN INFERENCE; GAME THEORY].

It happened thus that some applications of probability to new fields, such as judicial decisions, were bold and careless, that the Bayesian approach was often misused, and that ambiguity in interpretation became acute in some contexts. Particularly troublesome was the meaning of "equally likely," when—with Laplace in 1814—this notion came forward as the basis of an ostensible definition of probability.

A confining criticism. A bitter criticism arose that was prompt to cut away all possible causes of trouble rather than to analyze and recover sound underlying ideas. This attitude was dominant in the nineteenth century and is still strong. Concerning "equal probabilities," it has long been debated whether they ought to be based on "perfect knowledge that all the relevant circumstances are the same" or simply on "ignorance of any relevant circumstance that is different," whatever these expressions themselves may mean. To illustrate, is the probability of heads $\frac{1}{2}$ for a single toss only if we know that the coin is perfect, or even if it may not be but we are not informed which face happens to be favored? The terms *objective* and *subjective* —now used to distinguish two fundamentally different natures that probability might be understood to have—first appeared in connection with these particular, not very well specified, meanings of "equally probable."

It is apparent how narrow the field of probability becomes if restricted to cases arising from symmetric partitions. Sometimes, it was said, applications outside this field could be allowed by "analogy," but without any real effort to explain how far (and why, and in what sense) such an extension would be valid.

Authors chiefly interested in statistical problems were led to another confining approach that hinges on that property of probability most pertinent to their field, namely, the link with frequency, for example, Venn (1866), or the limit of frequency, for example, von Mises (1928). In any such theory, the limitation of the field of probability is less severe, in a sense, but is vaguely determined—or altogether undetermined if, as I believe, such theories are unavoidably circular.

A liberating criticism. There is today a vigorous revival of the current of thought (mentioned above in the subsection "The beginnings") that could not find its true development in the eighteenth century for want of full consciousness of its own implications. According to this outlook, an attempt to replace the familiar intuitive notion of probability by any, necessarily unsuccessful, imitation is by no means required, or even admissible. On the contrary, it suffices to make this intuitive notion neat and clear, ready to be used openly for what it is.

The most deliberate contributions to this program were along two convergent lines, one based on what may be called *admissibility*, by Frank P. Ramsey in 1926 (see 1923–1928) and Leonard J. Savage (1954), and one on *coherence*, or *consistency*, by Bruno de Finetti (1930) and B. O. Koopman (1940a; 1940b). In addition, the impact of authors supporting views concordant only in part with this one (like John Maynard Keynes 1921; Émile Borel 1924; and Harold Jeffreys 1939), as well as the impact of many concomitant circumstances, was no less effective. We can but list the following principal ones: the development of somewhat related theories (*games*, as by John von Neumann and Oskar Morgenstern 1944 [see GAME THEORY]; *decisions*, Abraham Wald 1950 [see DECISION THEORY]; *logical investigations*, Rudolf Carnap 1950; 1952); the detection of shortcomings in objectivistic, or frequentistic, statistics; the applications of probability to problems in economics and operations research; the survival of the "old–new" ideas in some spheres where common sense and practical needs were not satisfied by other theories, as in engineering (Fry 1934; Molina 1931), in actuarial science, and in credibility theory (Bailey 1950); and so on. Above all, the less rigid conception of scientific thought following the decline of rationalistic and deterministic dogmatism facilitated acceptance of the idea that a theory of uncertainty should find its own way.

The objective and the subjective

In logic. To make the step to probability easier, let us start with logic. The pertinent logic is that of sentences—more precisely, of sentences about the outside world, that is, sentences supposed to have some verifiable meaning in the outside world. We call such a sentence or the fact that it asserts—a harmless formal ambiguity—an *event*.

(There is a usage current in which *event* denotes what would here be called a class of events that are somehow homogeneous or similar, and *individual event*, or *trial*, denotes what is here called an event. The alternative usage is favored by some who emphasize frequencies in such classes of events. The nomenclature adopted here is the more flexible; while not precluding discussion of classes of events, it raises no questions about just what classes of events, if any, play a special role.)

Some examples of events—that is, of sentences,

assertions, or facts—are these: A = Australia wins the Davis Cup in 1960. B = the same for 1961. C = the same for 1960 *and* 1961.

What can be said of an event, objectively? Objectively it is either true or false (*tertium non datur*), irrespective of whether its truth or falsity is known to us or of the reason for our possible ignorance (such as that the facts are in the future, that we have not been informed, that we forget, and so on). However, a third term, *indeterminate*, is sometimes used to denote events that depend on future facts, or, in a narrower sense, on future facts other than those considered to be fully controlled by deterministic physical laws. In this sense, event A was indeterminate until the deciding ball of the 1960 Cup finals, and B and C were indeterminate until 1961, when they *became* true. Yet, in the strictest logical terminology, they *are* true in the atemporal sense; or, with reference to time, they were and will be true forever, irrespective of the time–space point of a possible observer.

Subjectively—for some person, at a given moment—an event may be certain, impossible, or dubious. But mistakes are not excluded; certain (or impossible) does not necessarily imply true (or false).

In logic restricted to the subjective aspect, that is, ignoring or disregarding reality, one can say only whether a person's assertions are consistent or not; including the objective aspect, one may be able to add that they are *correct, unmistaken,* or *mistaken*. In the Davis Cup example (if $\bar{A}$ means not-A, and a that A is dubious), then ABC, Abc, aBc, abc, $ab\bar{C}$, $\bar{A}B\bar{C}$, $A\bar{B}\bar{C}$, $\bar{A}\bar{B}\bar{C}$, $\bar{A}b\bar{C}$, $a\bar{B}\bar{C}$ are the only consistent assertions. For example, AbC is not, for there is no possible doubt about B if C is certain. Taking reality into account (namely, that A, B, C are actually all true), only the first, ABC, is correct; the next three are unmistaken, for no false event is considered as certain nor any true one as impossible; the remaining assertions are mistaken.

From logic to probability. Probability is the expedient devised to overcome the insufficiency of the coarse logical classification. To fill the questionable gap between true and false, instead of the generic term *indeterminate*, a continuous range of *objective probabilities* may be contemplated. And the unquestionable gap between certain and impossible, the *dubious*, is split into a continuous range of degrees of doubt, or degrees of belief, which are *subjective probabilities.*

The existence of subjective probabilities and of some kind of reasoning thereby is a fact of daily psychological experience; it cannot be denied, although it may be considered by some to be proba-

bilistically uninteresting. (In the most radical negation its relation to the true probability theory is compared to that between the energy of a person's will and energy as a physical notion.) To exploit this common experience, we need only a device to measure, and hence to define effectively, subjective probabilities in numerical terms and criteria to build up a theory. We consider three kinds of theories about subjective probability, SP, SC, and SR, which aim, respectively, to characterize psychological, consistent, and rational behavior under uncertainty.

Since the meaningfulness of objective probabilities is at least as questionable as the notion of indeterminateness, it is essential to ponder the various bases proposed for objective probability and the forces impelling many people to feel it important. Here, too, we shall consider three kinds of theories of objective probability, OS, OF, and OL, based, respectively, on symmetry (or equiprobable cases), on frequency, and on the limit of frequency.

Some remarks. The above classification into six kinds of theories is a matter of convenience. Some theories that do not fit the scheme must be mentioned separately. In the *axiomatic approach*, for example, probability has, by convention, the nature of a measure as defined by Kolmogorov ([1933] 1956, p. 2) and discussed by Savage (1954, p. 33); applications are made without specifying the meaning of probability or alluding to any of the interpretations discussed here. In other theories, probabilities may be noncomparable, and hence nonnumerical; there are such variants not only of SP but also of SC (Smith 1961) or even SR, as by Keynes.

Different theories are not necessarily incompatible. For instance, SC should be, and sometimes is, accepted by a supporter of SR as a preliminary weaker construction—as projective is preliminary to Euclidean geometry. Even less natural associations are current: Carnap admits two distinct notions, prob$_1$ and prob$_2$, that correspond to SR (or SC?) and OF (or OL?); if, like many psychologists, one stops at SP or at its variant with noncomparability (Ellsberg 1961), judging SC to be unreasonably stringent, one is often inclined to accept much more stringent assumptions, such as OS, in particular fields; and so on.

The order we shall follow is from the least to the most restrictive theory. In a sense, each requires the preceding ones together with further restrictions. Beginning with the subjectivistic interpretation, we have a good tool for investigating the special objectivistic constructions; they can in fact be imbedded into it, not only in form but also in substance. For the opinions a person holds as objectively right will of course also be adopted by him among those he considers subjectively right for himself.

Toward a subjectivistic definition. There is no difficulty in finding methods that attach to a subjective probability its numerical value in the usual scale from 0 to 1; what is difficult is weighing the pros and cons of various essentially equivalent measuring devices. Verbalistic comparisons or assertions are simple, but not suitable, at least at first, when we must grasp the real meaning from the measuring device. The device must therefore reveal preferences, by replies or by actions. The first way, by replies, is still somewhat verbalistic; the second is truly behavioristic but may be vitiated insofar as actions are often unpondered or dictated by caprice or by elusive side effects.

Obviously, the latter inconveniences are particularly troublesome in measuring subjective entities. It must be considered, for example, whether some opponent exists who may perhaps have, or be able to obtain, more information or be inclined to cheat. Such trouble is not, however, peculiar to probability. Measuring any physical quantity gives rise to the same sort of difficulties when its definition is based on a sufficiently realistic idealization of a measurement device.

Roughly speaking, the value p, given by a person for the probability $P(E)$ of an event E, means the price he is just willing to pay for a unit amount of money conditional on E's being true. That is, the preferences on the basis of which he is willing to behave (side effects being eliminated) are determined, with respect to gains or losses depending on E, by assuming that an amount S conditional on E is evaluated at pS. That must be asserted only for sufficiently small amounts; the general approach should deal in the same way with utility, a concept that, for want of space, is not discussed here (see Georgescu-Roegen 1968).

To construct a device that obliges a person to reveal his true opinions, it suffices to offer him a choice among a set of decisions, each entailing specified gains or losses according to the outcome of the event (or events) considered; such a set can be so arranged that any choice corresponds to a definite probability evaluation. The device can also be so simply constructed as to permit an easy demonstration that coherence requires the evaluation to obey the usual laws of the calculus of probability.

The subjectivistic approach

If we accept such a definition, the fundamental question of what should be meant by a theory of probability now arises. There seems to be agreement that it must lead to the usual relations between probabilities, to the rules of the probability calculus, and also to rules of behavior based on any evaluation of probabilities (still excluding decisions about substantial amounts of money, which require utilities). The three kinds of theory announced earlier differ in that they are looking for: a theory of actual behavior (SP); a theory of coherent behavior (SC); a theory of rational behavior (SR)—in other words, they are looking either for a theory of behavior under uncertainty as it is or for a theory of behavior as it ought to be, either in the weaker sense of avoiding contradictions or in the stronger sense of choosing the "correct" probability evaluations.

Subjective probability, psychological (SP). No one can deny the existence of actual behavior or the interest in investigating it in men, children, rats, and so on. Such experimental studies yield only descriptive theories, which cannot be expected to conform to the ordinary mathematics of probability. A descriptive theory may exist whether a corresponding normative one does or not. For example, when studying tastes, there are no questions of which tastes are intrinsically true or false. When studying responses to problems of arithmetic or logic, it is meaningful to distinguish, and important to investigate, whether the answers are correct and what mistakes are made.

One can be content to stop with the study of SP, accepting no normative theory—which SC, SR, OS, OF, and OL are, in a more or less complete sense. But one cannot object to a normative theory on the grounds that it does not conform to the actual behavior of men or rats or children. A normative theory states what behavior is good or bad. We may question whether any normative theory exists or whether a given one has any claims to be accepted. These questions have nothing to do with whether any beings do in fact behave according to the conclusions of the theory. It cannot therefore be confirmed or rejected on the basis of observational data, which can on the contrary say only whether there is more or less urgent need to teach people how to behave consistently or rationally.

Subjective probability, consistent (SC). The probability evaluations over any set of events whatever can be mathematically separated into the classes of those that are coherent and those that are not. *Coherence* means that no bet resulting in certain loss is considered acceptable; coherence is equivalent to *admissibility*, according to which no decision is preferable to another that, in every case, gives as good an outcome [see DECISION THEORY]. This is meant here with reference to maximizing expectations; it should ultimately be transformed into maximizing expected utility, which is the most general case of coherent behavior (Savage 1954).

As may be seen rather easily from the behavioristic devices on which personal probability is founded, coherence is equivalent to the condition that the whole usual calculus of probability be satisfied. For instance, if $C = AB$ (as in the example about the Davis Cup), the necessary and sufficient condition for coherence is

$$P(A) + P(B) - 1 \leqslant P(C) \leqslant \min [P(A), P(B)].$$

Properly subjectivistic authors (like the present one) think coherence is all that theory can prescribe; the choice of any one among the infinitely many coherent probability distributions is then free, in the sense that it is the responsibility of, and depends on the feelings of, whatever person is concerned.

Subjective probability, rational (SR). We denote as *rational*, or *rationalistic*, a theory that aims at selecting (at least in a partial field of events) just one of the coherent probability distributions, supposed to be prescribed by some principles of thought. (This is called a "necessary" view. Some writers use "rational" for what we call "coherent.")

In most cases, such a view amounts to presenting as cogent the feeling of symmetry that is likely to arise in many circumstances and with it the conclusion that some probabilities are equal, or uniformly distributed. But what really are the conditions where such an argument applies? A symmetry in somebody's opinion is the conclusion itself, not a premise, and the ostensible notion of absolute ignorance seems inappropriate to any real situation. Less strict assumptions, such as symmetry of syntactical structure of the sentences asserting a set of events, are likely to permit arbitrariness and to lead to misuses—in most radical form, to the d'Alembert paradox, according to which any dubious event has $p = \frac{1}{2}$, on the pretext that we are in a symmetric situation, being unable to deny either E or non-E, which means simply stopping at "dubious."

The objectivistic approach

The picture changes utterly in passing from the subjectivistic to the objectivistic approach, if due attention is paid to the underlying ideas. There are no longer people intent on weighing doubts. It is Nature herself who is facing the doubts, irresolute toward decisions, committing them to Chance or Fortune. The mythological expressions are but images; yet the expressions commonly used in objectivistic probability are, at bottom, equivalent to them, and the objectivistic language is so widespread that even attentive subjectivists sometimes lapse into it. It is considered meaningful to ask, for example, whether some effect is due to a cause or to chance (that is, is random); whether a fact modifies the probability of an event; whether a random variable is normal, or two are independent, or a process is Poisson; whether chance intervenes in a process (once for all, at one step, or at every instant); whether this or that phenomenon obeys the laws of probability or what their underlying chance mechanism is; and so on.

In fact, an objective probability is regarded as something belonging to Nature herself (like mass, distance, or other physical quantities) and is supposed to "exist" and have a determined value even though it may be unknown to anyone. Quite naturally, therefore, objectivistic theories do actually deal with unknown probabilities, which are of course meaningless in a subjectivistic theory. Furthermore, it can fairly be said that an objective probability is always unknown, although hypothetical estimates of its values are made in a not really specifiable sense. How can one hope to communicate with such a mysterious pseudoworld of objective probabilities and to acquire some insight on it?

Objective probability, symmetrical (OS). The first partial answer comes from the objectivistic interpretation of the "symmetry principle," or "the principle of cogent reason," according to which identical experiments repeated under identical conditions have the same probability of success. This applies also to the case of several symmetric possible outcomes of one experiment (such as the six faces of a die) and is often asserted also for combinations (such as the 2^n sequences of possible outcomes of tossing a coin n times). Accepting—perhaps on the basis of SC, admitting that objective probabilities must be consistent with subjective ones, or perhaps by convention—the rule of favorable divided by possible cases, probability is defined in a range and way very similar to those of SR; but, even apart from the change from subjective to objective interpretation, it is not so close as it may seem, because the role of information is now lost. We can no longer content ourselves with asserting something about symmetry of the real world as it is known to us but are compelled to entangle ourselves in asserting perfect symmetry of what is unknown if not indeed unknowable. Or we may switch from this supernatural attitude to the harmless one of regarding such assertions as merely "hypothetical," but then obtain only hypothetical knowledge of the objective probabilities, since the perfect symmetry is only hypothetical. For instance, what about differences arising from magnetism before magnetism was known? Strictly, the perfect symmetry is contradictory, unless unavoidable differences in time, place, past and current circumstances, etc. are bypassed as irrelevant.

Objective probability, frequency (OF).   Another answer is: "Objective probability is revealed by frequency." It is a property that somehow drives frequency (with respect to a sufficiently large number of "identical" events) toward a fixed number, p, that is the value of the probability of such events. Statistical data are, then, a clue to the ever unknowable p if the events concerned are considered identical or to an average probability if their probabilities differ.

OF actually presupposes OS whenever we intend to justify the necessity of arranging events in groups to get frequencies and to use these frequencies for other, as yet unknown, events of the group. Nonetheless, there may be a conflict if, for instance, the frequencies that occur with a die, accepted as perfect, are not almost equal, as can happen.

Objective probability, limit of frequency (OL). In order to remove the unavoidable indeterminateness of OF, the suggestion has been made to increase the large number to infinity and to define p as the limit of frequency. Whether one does or does not like this idea as a theoretical expedient, clearly no practical observations or practical questions do concern eternity; for dealing with real problems this theory is at best only an elusive analogy.

Critique of objectivistic theories.   Objectivistic theories are often preferred (especially by some practically oriented people, such as statisticians and physicists), because they seem to join the fundamental notions with practically useful properties by a direct short cut. But the short cut leaps unfathomable gulfs. It admits of no bridge between us, with our actual knowledge, and the imagined objectivistic realm, which can be turned into an innocent allegory for describing some models but has no proper claim to being complete and self-supporting. The needed bridge is supplied by subjective probability, whose role seems necessary and unchanged, whether or not we want also to make some use of the unnecessary notion of objective probability for our descriptions of the world. If this bridge is rejected, only recourse to expedients is left. It will shortly be explained why subjective probabilities are said to provide the natural bridge, while objectivistic criteria (such as the usual methods of estimating quantities, testing hypotheses, or defining inductive behavior) appear to be artificial and inadequate expedients.

Inductive reasoning and behavior

A feature that has been postponed, to avoid premature distraction, must now be dealt with. A subjective probability, $P(E)$, is of course conditional on the evidence, or state of information, currently possessed by the subject concerned; to make that explicit, we may write $P(E|A)$, where A is the current state of information, which is usually left implicit. Any additional information, real or hypothetical, consisting in learning that an event H is true (and H may be the joint assertion, or logical product, of any number and kind of "simpler" events) leads from the probabilities $P(E)$ to $P(E|H)$, conditional on H (or on AH, if A need be made explicit). The coherence condition of SC suffices for all rules in the whole field of conditional probabilities and hence specifies by implication what it means to reason and behave coherently, not only in a static sense, that is, in a given state of information, but also in a dynamic one, in which new information arises freely or may be had at some cost by experiment or request.

The passage from $P(E)$ to $P(E|H)$ is prescribed simply by the theorem of compound probabilities, or equivalently—in a slightly more elaborate and specific form—by Bayes' theorem. As for the decision after knowledge of H, it must obviously obey the same rules as before, except with $P(E|H)$ in the place of $P(E)$; the sole essentially new question is how best to spend time, effort, and money for more information and when to stop for final decision, but that too is settled by the same rules.

Coherence thus gives a complete answer to decision problems, including even induction, that is, the use of new information; no room is left for arbitrariness or additional conventions. Of course, the freedom in choosing $P(E)$, the initial distribution, is still allowed (unless we accept uniqueness from SR), and similarly for utility. The unifying conclusion is this: Coherence obliges one to behave *as if* he accepted some initial probabilities and utilities, acting then so as to maximize expected utility.

The particular case of *statistical* induction is simply that in which H expresses the outcome of several "similar" events or trials. The simplest condition, called *exchangeability*, obtains when only information about the number of successes and failures is relevant for the person, irrespective of just which events, or trials, are successes or failures; and the most important subcase is exchangeability of a (potentially) infinite class of events—like coin or die tossing, or drawings with replacement from an urn, or repetitions of an experiment under sufficiently similar conditions. The model of an infinite class of exchangeable events can be proved equivalent to the model that presents the events as independent conditional on the value of an unknown probability, whether the "unknown probability" is interpreted objectivistically or otherwise. In fact, to deal with the latter model consistently, one must—if not explicitly, at least in effect—start with an initial subjective distribution of the unknown probability; but exactly the same results are obtainable directly from the definition of exchangeability itself, without recourse to any probabilities other than those to which we have direct subjective access (Feller 1950–1966, vol. 2, p. 225). Whatever the approach, Bayes' rule acts here in a simple way (we cannot go into the details here) that explains how we all come to evaluate probabilities in statistical-induction situations according to the observed frequencies, insofar as our common sense induces approximate coherence.

The inconsistency just noted of using any but the Bayesian approach, even under an objectivistic formulation, should be discussed further. The initial subjective probability distribution of a person must be the same for all decision problems that depend on a given set of events. It cannot be chosen by criteria that, like the minimax rule, depend on the specific problem or on the instrument of observation, because such criteria, although coherent within each problem, are not coherent over-all. Also, there is no justification for calling the Bayesian method unreasonable because the needed initial distribution is "unknown." More accurately, it is of dubious choice (see, for example, Lindley 1965, vol. 2, pp. 19–21). Actually, any method proposed for avoiding the risk of such a choice is demonstrably even worse than a specific choice. The situation is as though someone were to estimate the center of gravity of some weighted points of a plane by a point outside the plane because he did not know the weights of the points sufficiently well; yet the projection of the estimate back onto the plane must improve it. This is not a mere analogy but a true picture in a suitable mathematical representation. This picture should be emphasized because it shows that, far from opposition, there is necessarily agreement between inductive behavior and inductive reasoning, which is contrary to an opinion current in objectivistic statistics.

A few nuances of the views

No sketch of some of the representative views can cover the opinions of all authors, if only because some might take a particular idea in earnest and push it to its extreme consequences while others might consider it merely as a suggestive abstraction to be taken with a grain of salt. Our sketches may therefore appear either as insufficient or as caricatures. Still, even mentioning a few of the nuances may create a more realistic impression.

Each notion changes its meaning with the theory; let us take, as an example, independence. Two drawings, with replacement, from an urn of unknown composition are called independent by an objectivist (since for him the probability is the unknown but constant proportion of white balls) but not by a subjectivist (for whom "independent" means "devoid of influence on my opinion"). For him, such events are only exchangeable; the observed outcomes, through their frequency, do alter the conditional probability of those not yet observed, whether in terms of an unknown probability —here, the urn composition—or directly.

In OF, on the contrary, independence and the law of large numbers are almost prerequisites for the definition of probability. To escape confusing circularity, duplicates of notions are invented. Thus, the preliminary form of the law of large numbers is called the "empirical law of chance," making a distinction based on a similar one between "highly probable" and "almost certain"—itself created expressly to be eliminated by the "principle of Cournot." This ostensible principle is one that seems to suggest that we can practically forget the "almost" and consider probability as a method of deriving certainties; in a safer version, it might offer a link between subjective and objective probabilities.

But the first prerequisite for OF is the grouping of events into classes, sometimes tacitly slipped in by the terminology that uses the word "event" for a sequence of events rather than for an individual event. What is intended by such a class raises confusing questions. Does maintaining that probability belongs to a class imply that all events, or trials, in the class are equally probable? Or is that question to be rejected and the probability of a single trial held to be meaningless? What information entitles us to assign a given single event to such a class? These questions are aspects of a more general one that has still other aspects. For any objectivist, frequentistic or not, which states of information allow us to regard a probability as known? Which are insufficient and leave it unknown? Is there only one specific state of information concerning a given event to which a probability corresponds? Does the whole information on which a probability is based consist of all relevant circumstances of the present or of the past? Or can nothing be excluded as irrelevant so that it consists of the whole present and past?

Actually, some conventional state of information, far from knowledge of the whole past and present, is usually considered appropriate. For instance, for extractions from an urn, when the proportion of white balls is known (and perhaps when it is also known that the balls are well mixed), objectivists say that the probability is known and equal to the proportion. But why not require fuller information? If, for example, we have noted that the child drawing generally chooses a ball near the top, then additional information about the position of the white balls in the urn would seem to be relevant.

Something between equal probabilities for events in classes and individual probabilities for each trial may seem to be offered by the precaution of speaking of the probability of an event with respect to a given set of trials (see Fréchet 1951, pp. 15–16). An example will illustrate the meaning and the implications not only of this attitude but more generally of all attempts to evade the main question above, namely, whether the probabilities of events in a class are equal, unequal, or nonsense. There is no proper premium for insurance on the life of Mr. Smith, unless we specify that he is to be insured as, for instance, a lawyer, a widower, a man forty years old, a blonde, a diabetic, an ex-serviceman, and so on. How then should an objectivistic insurer evaluate Mr. Smith's application for insurance?

Probability and philosophy

The view of a world ruled by Chance has also been opposed for philosophical reasons—not to mention theological and moral ones. Chance is incompatible with determinism, which, it was once said, is a prerequisite to science. By replying, "Chance is but the image of a set of many little causes producing large effects," faith in perfect determinism in the microcosm was reconciled with probability, but perhaps only with SR rather than with OS, which was the real point of conflict. At any rate, the advent of probabilistic theories in physics and elsewhere later showed that determinism is not the only possible basis for science.

Subjectivistic views have been charged with "idealism" by Soviet writers (for example, Gnedenko 1950), and "priesthood" by objectivistic statisticians (for example, van Dantzig 1957), inasmuch as these views draw their principles only from human understanding. For SR that may be partially justified; it should, however, be ascribed simply to misunderstanding when it is said of SC, which seems exposed to the opposite charge, if any. Namely, SC allows too absolute a freedom for a

person's evaluations, abstaining from any prescription beyond coherence. Is that not assent to arbitrariness? The answer is, Yes, in freedom from prefabricated schemes; but the definition calls on personal responsibility; and mathematical developments based on the coherence conditions show how and why the usual prescriptions—above all, those based on symmetry and frequency—ought to be applied, not as rigid artificial rules, but as patterns open to intelligence and discernment for proper interpretation in each case.

BRUNO DE FINETTI

[See also BAYESIAN INFERENCE; CAUSATION; SCIENCE, article on THE PHILOSOPHY OF SCIENCE; SCIENTIFIC EXPLANATION.]

BIBLIOGRAPHY

Contributions by Venn, Borel (about Keynes), Ramsey, de Finetti, Koopman, and Savage are collected and commented on in Kyburg & Smokler 1964.

BAILEY, ARTHUR L. 1950 Credibility Procedures. Casualty Actuarial Society, Proceedings 37:7–23.

○BAYES, THOMAS (1764) 1958 An Essay Towards Solving a Problem in the Doctrine of Chances. Biometrika 45:296–315. → First published in Volume 53 of the Royal Society of London, Philosophical Transactions. Reprinted in 1963 in Bayes' Facsimiles of Two Papers by Bayes, published by Hafner.

BERNOULLI, DANIEL (1738) 1954 Exposition of a New Theory on the Measurement of Risk. Econometrica 22:23–36. → First published as "Specimen theoriae novae de mensura sortis."

BOREL, ÉMILE (1924) 1964 Apropos of a Treatise on Probability. Pages 45–60 in Henry E. Kyburg, Jr. and Howard E. Smokler (editors), Studies in Subjective Probability. New York: Wiley. → First published in French in Volume 98 of the Revue philosophique.

CARNAP, RUDOLF (1950) 1962 Logical Foundations of Probability. 2d ed. Univ. of Chicago Press.

CARNAP, RUDOLF 1952 The Continuum of Inductive Methods. Univ. of Chicago Press.

DE FINETTI, BRUNO 1930 Fondamenti logici del ragionamento probabilistico. Unione Matematica Italiana, Bollettino Series A 9:258–261.

ELLSBERG, DANIEL 1961 Risk, Ambiguity, and the Savage Axioms. Quarterly Journal of Economics 75:643–669.

○FELLER, WILLIAM (1950–1960) 1968–1971 An Introduction to Probability Theory and Its Applications. 2 vols. New York: Wiley. → The third edition of Volume 1 was published in 1968, the second edition of Volume 2 in 1971.

FRÉCHET, MAURICE 1951 Rapport général sur les travaux du colloque de calcul des probabilités. Pages 3–21 in Congrès International de Philosophie des Sciences, Paris, 1949, Actes. Volume 4: Calcul des probabilités. Paris: Hermann.

FRY, THORNTON C. 1934 A Mathematical Theory of Rational Inference. Scripta mathematica 2:205–221.

►GEORGESCU-ROEGEN, NICHOLAS 1968 Utility. Volume 16, pages 236–267 in International Encyclopedia of the Social Sciences. Edited by David L. Sills. New York: Macmillan and Free Press.

GNEDENKO, BORIS V. (1950) 1962 The Theory of Prob-

ability. New York: Chelsea. → First published as *Kurs teorii veroiatnostei*.

GUILBAUD, GEORGES 1961 Faut-il jouer au plus fin? Pages 171–182 in Colloque sur la décision, Paris, 25–30 mai, 1960 [*Actes*]. France, Centre National de la Recherche Scientifique, Colloques Internationaux, Sciences Humaines. Paris: The Center.

HACKING, IAN 1965 *Logic of Statistical Inference.* Cambridge Univ. Press.

JEFFREYS, HAROLD (1939) 1961 *Theory of Probability.* 3d ed. Oxford: Clarendon.

KEYNES, JOHN MAYNARD (1921) 1952 *A Treatise on Probability.* London: Macmillan. → A paperback edition was published in 1962 by Harper.

KOLMOGOROV, ANDREI N. (1933) 1956 *Foundations of the Theory of Probability.* New York: Chelsea. → First published in German.

KOOPMAN, BERNARD O. 1940a The Axioms and Algebra of Intuitive Probability. *Annals of Mathematics* Second Series 41:269–292.

KOOPMAN, BERNARD O. (1940b) 1964 The Bases of Probability. Pages 159–172 in Henry E. Kyburg, Jr. and Howard E. Smokler (editors), *Studies in Subjective Probability.* New York: Wiley.

KYBURG, HENRY E. JR.; and SMOKLER, HOWARD E. (editors) 1964 *Studies in Subjective Probability.* New York: Wiley.

LAPLACE, PIERRE SIMON DE (1814) 1951 *A Philosophical Study on Probabilities.* New York: Dover. → First published as *Essai philosophique sur les probabilités.*

LINDLEY, DENNIS V. 1965 *Introduction to Probability and Statistics From a Bayesian Viewpoint.* 2 vols. Cambridge Univ. Press.

MOLINA, EDWARD C. 1931 Bayes' Theorem. *Annals of Mathematical Statistics* 2:23–37.

RAMSEY, FRANK P. (1923–1928) 1950 *The Foundations of Mathematics, and Other Logical Essays.* New York: Humanities.

○SAVAGE, LEONARD J. (1954) 1972 *The Foundations of Statistics.* Rev. ed. New York: Dover.

SMITH, CEDRIC A. B. 1961 Consistency in Statistical Inference and Decision. *Journal of the Royal Statistical Society* Series B 23:1–25.

VAN DANTZIG, DAVID 1957 Statistical Priesthood: Savage on Personal Probabilities. *Statistica neerlandica* 11:1–16.

VENN, JOHN (1866) 1962 *The Logic of Chance: An Essay on the Foundations and Province of the Theory of Probability, With Special Reference to Its Logical Bearings and Its Application to Moral and Social Science.* 4th ed. New York: Chelsea.

VON MISES, RICHARD (1928) 1961 *Probability, Statistics and Truth.* 2d ed., rev. London: Allen & Unwin.

○VON NEUMANN, JOHN; and MORGENSTERN, OSKAR (1944) 1953 *Theory of Games and Economic Behavior.* 3d ed., rev. Princeton Univ. Press. → A paperback edition was published in 1964 by Wiley.

WALD, ABRAHAM (1950) 1964 *Statistical Decision Functions.* New York: Wiley.

Postscript

This postscript includes topics that did not find a proper place in the framework of the main article, together with brief mention of developments and discussions since publication of the article. The two kinds of material are intermingled for unity.

Logical probability. Research along lines begun by Carnap (discussed in the main article), and centered chiefly on the problem of induction, has been carried out separately by several scholars, notably by Jaakko Hintikka (1969) and by Hintikka and Patrick Suppes (1966). Similar ideas have been discussed by Popper and his associates, for example, by Douglas A. Gillies (1973).

After Carnap's death in 1970, materials he had circulated privately and revised were published with commentary by Richard Jeffrey (Carnap & Jeffrey 1971). Jeffrey finds himself drawing near to the subjectivist position, and he maintains that Carnap held a subjectivist view in an underlying sense, although he had hoped to find a logical substitute through an abstract apparatus.

Subjective probability. A great figure in subjective probability, Leonard Jimmie Savage, also died [*see the biography of* SAVAGE]. I must express my admiration for and indebtedness to so invaluable a friend and thinker. Posthumous papers illuminate Savage's standpoint on two central topics: first (1971), on the elicitation of probabilities, and second (1973), on inductive reasoning.

One should mention the concept of plausible reasoning discussed by George Pólya (1954), mainly in connection with mathematical conjectures but also for all kinds of practical questions. Although Pólya avoids explicit use of numerical probabilities, confining himself to a qualitative approach, his treatment perfectly parallels that of subjective probability and Bayesian inference.

My own presentation of the theory of probability in all its aspects from the subjective viewpoint (de Finetti 1970) has appeared in English (1974–1975).

Bayesian statistics. Bayesian statistics, when probabilities are taken as subjective, is the necessary, unique, correct statistical process. Other methods are then but approximations, rough "ad-hoceries" (see Good 1950) or rules of thumb. (See also Cornfield 1967.)

A short, elementary, yet brilliant defense of this position is given by Dennis V. Lindley (1971); Harry V. Roberts (1965) shows how the usual parametric formulation may be avoided by direct consideration of so-called predictive distributions. I have a review paper on this topic (de Finetti 1973).

One example of application of the Bayesian approach to a fundamental scientific question is given by I. J. Good (1969) as he discusses the probabilities of theories of the origin of the solar system in terms of regularities (Bode's law) of the distances from the sun to the planets. The ensuing discussion shows the inevitable large margin for guessing, and also preconceived bitter opposition because of basic misconceptions (see de Finetti 1972).

Philosophers and psychologists. Philosophers, especially philosophers of science, appear to be giving increased attention to the foundations of probability, and even to its methods and application. I mention in particular an important work by Wolfgang Stegmüller (1973).

Psychologists also evince widespread interest in probability, its interpretations, and experimental investigations of its empirical forms. For example, significant questions about risk taking by pedestrians and drivers have been investigated by John Cohen (1960; 1964; see also Cohen & Hansel 1956). Again, there have been systematic studies of forecasts by probability assessors in particular fields—meteorology, sports, security prices, and so on. I cite publications by Winkler (1967); Winkler and Murphy (1968); Murphy and Winkler (1970), and Staël von Holstein (1970a; 1970b) as examples. Yet again, one should mention response-and-scoring methods to investigate partial knowledge on single test items, for example by C. H. Coombs (1953), Chernoff (1962), and Shuford, Arthur, and Massengill (1966).

Other important, interesting developments in the psychological investigation of probability are associated with Ward Edwards (for example, 1968) and Amos Tversky (for example, Tversky & Kahneman 1974).

Five international conferences on subjective probability have been held (Hamburg, 1969; Amsterdam, 1970; London, 1971; Rome, 1973; and Darmstadt, 1975) in order to bring together psychologists, mathematicians, statisticians, and others; the proceedings of the Amsterdam conference have been published (De Zeeuw et al. 1970). Yet progress toward mutual understanding seems slow and awkward.

BRUNO DE FINETTI

[See also the biography of SAVAGE.]

ADDITIONAL BIBLIOGRAPHY

CARNAP, RUDOLF; and JEFFREY, RICHARD C. (editors) 1971 Studies in Inductive Logic and Probability. Volume 1. Berkeley: Univ. of California Press.

CHERNOFF, HERMAN 1962 The Scoring of Multiple Choice Questionnaires. Annals of Mathematical Statistics 33:375–393.

COHEN, JOHN 1960 Chance, Skill and Luck: The Psychology of Guessing and Gambling. Baltimore: Penguin.

COHEN, JOHN 1964 Behaviour in Uncertainty, and Its Social Implications. London: Allen & Unwin.

COHEN, JOHN; and HANSEL, MARK 1956 Risk and Gambling: The Study of Subjective Probability. London: Longmans.

COOMBS, C. H. 1953 On the Use of Objective Examinations. Educational and Psychological Measurement 13:308–310.

CORNFIELD, JEROME 1967 Bayes' Theorem. Review of the International Statistical Institute 35:34–39.

DE FINETTI, BRUNO (1970) 1974–1975 Theory of Probability. 2 vols. New York and London: Wiley. → First published in Italian.

DE FINETTI, BRUNO 1972 Probability, Induction and Statistics: The Art of Guessing. New York: Wiley.

DE FINETTI, BRUNO 1973 Bayesianism: Its Unifying Role for Both the Foundations and the Applications of Statistics. International Statistical Institute, Bulletin 45:349–368. → Discussion on pages 369–376.

DE ZEEUW, G.; VLEK, C. A. J.; and WAGENAAR, W. A. (editors) 1970 Subjective Probability: Theory, Experiments, Applications. Amsterdam: North-Holland. → Reprinted in Acta Psychologica 34 (1970), nos. 2 and 3.

EDWARDS, WARD 1968 Conservatism in Human Information Processing. Pages 17–52 in Benjamin Kleinmuntz (editor), Formal Representation of Human Judgment. New York: Wiley.

GILLIES, DOUGLAS A. 1973 An Objective Theory of Probability. London: Methuen.

GOOD, I. J. 1950 Probability and the Weighing of Evidence. London: Griffin; New York: Hafner.

GOOD, I. J. 1969 A Subjective Evaluation of Bode's Law and an "Objective" Test for Approximate Numerical Rationality. Journal of the American Statistical Association 64:23–66. → Includes discussion.

HINTIKKA, JAAKO 1969 Statistics, Induction, and Law-likeness: Comments on Dr. Vetter's Paper. Synthese 20:72–83.

HINTIKKA, JAAKO; and SUPPES, PATRICK (editors) 1966 Aspects of Inductive Logic. Amsterdam: North-Holland.

LINDLEY, DENNIS V. 1971 Making Decisions. New York and London: Wiley.

MURPHY, ALLEN H.; and WINKLER, ROBERT L. 1970 Scoring Rules in Probability Assessment and Evaluation. Pages 273–286 in G. De Zeeuw, C. A. J. Vlek, and W. A. Wagenaar (editors), Subjective Probability: Theory, Experiments, Applications. Amsterdam: North-Holland. → Reprinted in Acta Psychologica 34 (1970): 273–286.

PÓLYA, GEORGE (1954) 1968 Mathematics and Plausible Reasoning. Volume 2: Patterns of Plausible Inference. Rev. ed. Princeton Univ. Press.

ROBERTS, HARRY V. 1965 Probabilistic Prediction. Journal of the American Statistical Association 60:50–62.

SAVAGE, LEONARD J. (1961) 1964 The Foundations of Statistics Reconsidered. Pages 173–186 in Henry E. Kyburg, Jr. and Howard E. Smokler (editors), Studies in Subjective Probability. New York: Wiley.

SAVAGE, LEONARD J. 1971 Elicitation of Personal Probabilities and Expectations. Journal of the American Statistical Association 66:783–801.

SAVAGE, LEONARD J. 1973 Probability in Science: A Personalistic Account. Pages 417–428 in International Congress for Logic, Methodology and Philosophy of Science, Fourth, Bucharest, 1971, Logic, Methodology and Philosophy of Science: Proceedings. Edited by Patrick Suppes. Amsterdam: North-Holland.

SHUFORD, EMIR H. JR.; ARTHUR, ALBERT; and MASSENGILL,

EDWARD H. 1966 Admissible Probability Measurement Procedures. *Psychometrika* 31:125–145.

STAËL VON HOLSTEIN, CARL-AXEL S. 1970a *Assessment and Evaluation of Subjective Probability Distributions.* Stockholm School of Economics, Economic Research Institute.

STAËL VON HOLSTEIN, CARL-AXEL S. 1970b Measurement of Subjective Probability. Pages 146–159 in G. De Zeeuw, C. A. J. Vlek, and W. A. Wagenaar (editors), *Subjective Probability: Theory, Experiments, Applications.* Amsterdam: North-Holland. → Reprinted in *Acta Psychologica* 34 (1970):146–159.

STEGMÜLLER, WOLFGANG 1973 *Probleme und Resultate der Wissenschaftstheorie und analytischen Philosophie.* 4 vols. in 2. Volume 4: *Personelle und statistische Wahrscheinlichkeit.* Berlin and New York: Springer.

TVERSKY, AMOS; and KAHNEMAN, DANIEL 1974 Judgment Under Uncertainty: Heuristics and Biases. *Science* 185:1124–1131.

WINKLER, ROBERT L. 1967 The Assessment of Prior Distributions in Bayesian Analysis. *Journal of the American Statistical Association* 62:776–800.

WINKLER, ROBERT L.; and MURPHY, ALLEN H. 1968 "Good" Probability Assessors. *Journal of Applied Meteorology* 7:751–758.

PROBABILITY SAMPLING

See SAMPLE SURVEYS.

PROCESS CONTROL

See under QUALITY CONTROL, STATISTICAL.

PRODUCT–MOMENT CORRELATION

See MULTIVARIATE ANALYSIS, *article on* CORRELATION METHODS.

PROGRAMMING

"Programming"—more properly, "mathematical programming," to be distinguished from "computer programming"—is a term applied to certain mathematical techniques designed to solve the problem of finding the maximum or minimum of a function subject to several constraints that are usually expressed as inequalities. In itself, mathematical programming has no economic content, but mathematical programming problems arise frequently in the context of economics and operations research, where inequality restrictions are often encountered. A large variety of programming formulations exists, and mathematical programming has been applied in a considerable number of concrete cases. The existence of large electronic computers has made the actual computation of solutions to mathematical programming problems fairly routine.

Because of theoretical and computational differences among various programming formulations, it is convenient to classify programming problems according to whether they are linear or nonlinear and according to whether they are discrete or continuous. This article focuses mainly on linear continuous, nonlinear continuous, and linear discrete (integer) programming. Baumol (1958) provides an elementary introduction to these topics, and Dorfman, Samuelson, and Solow (1958) discuss them at an intermediate level.

Linear programming

Formulation. As an illustration of a linear programming problem, consider the case of an entrepreneur who may produce any or all of n commodities, the produced amounts of which are measured by $x_1, x_2, \cdots, x_n$. Assume that the profit per unit of the ith commodity is constant and is given by c_i. Assume, further, that the entrepreneur has m resources (land, labor of various kinds, raw materials, machines) available to him, their amounts being measured by $b_1, b_2, \cdots, b_m$. Assume, finally, that the activity of producing one unit of the ith commodity requires (or "uses up") a_{1i} units of resource 1, a_{2i} units of resource 2, and so on. The problem of choosing production levels for each commodity in such a manner that total profit is maximized can then be stated as follows: maximize

$$(1) \qquad z = c_1 x_1 + c_2 x_2 + \cdots + c_n x_n,$$

subject to

$$(2) \quad \begin{aligned} a_{11} x_1 + a_{12} x_2 + \cdots + a_{1n} x_n &\leqslant b_1, \\ a_{21} x_1 + a_{22} x_2 + \cdots + a_{2n} x_n &\leqslant b_2, \\ \vdots \qquad \vdots \qquad\qquad \vdots \qquad\quad \vdots \\ a_{m1} x_1 + a_{m2} x_2 + \cdots + a_{mn} x_n &\leqslant b_m; \end{aligned}$$

$$(3) \qquad x_1 \geqslant 0, x_2 \geqslant 0, \cdots, x_n \geqslant 0.$$

Equation (1), referred to as the objective function, represents total profit in this example. Inequalities (2) express the requirement that for each of the m resources the total amount of a resource used up in the production of all commodities must not exceed the available amount. Inequalities (3) express the requirement that production levels be nonnegative. Equation (1), together with inequalities (2) and (3), is called a *linear program.* The term "linear" refers to the fact that both the objective function and the inequalities (2) are linear. Linear programs in which the x_i measure the levels or intensities at which productive activities are operated are sometimes called *activity analysis problems.* In compact matrix notation the above linear program can be written as follows: maximize

$$c'x,$$

subject to

$$Ax \leqq b;$$

$$x \geqq 0.$$

(Here the symbol "$\leqq$" is used to indicate that one matrix is less than or equal to the other, element by element, without any requirement that strict inequality hold for any component. The definition of "$\geqq$" is analogous.)

A set of x-values $(x_1^0, x_2^0, \cdots, x_n^0)$ is called a *feasible solution* if it satisfies inequalities (2) and (3). If the set of all feasible solutions contains an element for which (1) attains a maximum, then that element is called the *optimal solution*—or simply the solution—of the linear program. The solution need not be unique.

The inequalities (2) may be rewritten as equalities by introducing m new nonnegative variables $v_1, \cdots, v_m$ and rewriting the constraints as

$$
\begin{aligned}
(2') \quad
& a_{11}x_1 + a_{12}x_2 + \cdots + a_{1n}x_n + v_1 \qquad\qquad = b_1, \\
& a_{21}x_1 + a_{22}x_2 + \cdots + a_{2n}x_n + \quad v_2 \qquad = b_2, \\
& \quad\vdots \qquad\quad \vdots \qquad\qquad\quad \vdots \qquad\qquad\quad \vdots \\
& a_{m1}x_1 + a_{m2}x_2 + \cdots + a_{mn}x_n + \qquad\quad v_m = b_m.
\end{aligned}
$$

The new variables, called *slack variables* or *disposal activities*, are also required to be nonnegative. Their introduction changes neither the set of feasible solutions for x nor the optimal solution(s). In a linear programming problem with m constraints of type $(2')$, a *basic feasible solution* is one in which at most m of the $n + m$ variables appear with nonzero values. A fundamental theorem of linear programming is that if a linear programming problem possesses a feasible solution with an associated profit level z_f, it also possesses a basic feasible solution with an associated profit level z_b such that $z_b \geqslant z_f$.

The mathematical meaning of this theorem is that if the linear programming problem has any solution, then at least one solution will always occur at an extreme point of the convex polyhedron defined by (2). The economic meaning, in terms of the previous example, is that the number of commodities produced will never exceed the number of resource limitations.

Duality. One of the fundamental facts of linear programming is that with every linear program one can associate another program, known as the dual of the first, that stands in a particular relationship to it. The relationship between a (primal) linear program and its dual program leads to a number of important mathematical theorems and meaningful economic interpretations.

Given the linear program defined by (1), (2), and (3), the corresponding dual program is to minimize

$$t = b_1 u_1 + b_2 u_2 + \cdots + b_m u_m,$$

subject to

$$
\begin{aligned}
& a_{11}u_1 + a_{21}u_2 + \cdots + a_{m1}u_m \geqslant c_1, \\
& a_{12}u_1 + a_{22}u_2 + \cdots + a_{m2}u_m \geqslant c_2, \\
& \quad\vdots \qquad\quad \vdots \qquad\qquad\quad \vdots \qquad\quad \vdots \\
& a_{1n}u_1 + a_{2n}u_2 + \cdots + a_{mn}u_m \geqslant c_n; \\
& u_1 \geqslant 0,\, u_2 \geqslant 0,\, \cdots,\, u_m \geqslant 0.
\end{aligned}
$$

It may be noted that the relationship between a primal problem and its dual is as follows: (a) If the primal problem is a maximization problem, then the dual problem is a minimization problem, and conversely. (b) The coefficients on the right-hand side of the inequalities (2) in the primal problem become the coefficients in the objective function in the dual, and conversely. (c) The coefficient on the left-hand side of the constraints (2) in the primal that is associated with the ith constraint and jth variable becomes associated with the ith variable in the jth constraint in the dual. (d) A new set of variables is introduced in the dual, and each dual variable is associated with a primal constraint. If a primal (dual) constraint is an equality, the corresponding dual (primal) variable is unrestricted as to sign; otherwise it is required to be nonnegative. (e) The symmetry of these relations implies that the dual of a dual is the primal.

The economic interpretation of the dual variables and the dual program is as follows: Let u_i be the imputed price of a unit of the ith resource. The dual program then minimizes the total imputed value of resources subject to the constraints that the imputed cost of producing a unit of any commodity not be less than the profit derived from producing a unit of that commodity. This interpretation becomes meaningful in the light of the following duality theorems: (1) If both the primal and the dual problems possess feasible solutions, then $z \leqslant t$ for any feasible solution; that is, the value of the (primal) objective function to be maximized does not exceed the value of the (dual) objective function to be minimized. (2) If both the primal and the dual problems possess feasible solutions, they both possess optimal solutions, and the maximum value of z is equal to the minimum value of t. (3) If in an optimal solution to the primal and dual problems a constraint in the primal (dual) problem is satisfied as a strict inequality, the corresponding dual (primal) variable has zero value; if a primal (dual) variable in the solution is positive, the corresponding dual (primal) constraint is satisfied as an equality. (4) In an optimal solution to the primal and dual problems, the value of the ith dual variable is equal to the improvement in the primal objective function resulting from an increase of one unit in the availability of the ith resource.

In terms of the earlier economic interpretation, the meaning of these theorems is as follows: If a resource is not fully used up (one of the primal constraints is satisfied with strict inequality), the corresponding imputed price of that resource is zero, from which we can also infer that an addition to the entrepreneur's supply of that resource will not contribute to profit. If the jth constraint of the dual program is satisfied as an inequality, diverting resources to production of the jth commodity will reduce profits resulting from production of all other commodities by an amount greater than the profits resulting from production of the jth commodity; hence, that commodity will not be produced. These relationships, expressed in theorem (4), are known as complementary slackness.

The discovery and development of the theory of linear programming and of the notions of duality can be credited to John von Neumann (see Dantzig 1963, p. 24), D. Gale, H. W. Kuhn, and A. W. Tucker (1951), L. V. Kantorovich (1960), G. B. Dantzig (1951), and others. A notable application of linear programming to the theory of the firm is due to R. Dorfman (1951).

Computational methods. Linear programs are normally solved with the aid of the *simplex method*, which consists of two phases: phase I provides a feasible solution to the linear program, and phase II, taking over at the point at which the task of phase I is completed, leads to the optimal solution (if one exists). In many problems a feasible solution can be found by inspection, and phase I can be by-passed.

Phase II of the simplex method is an iterative procedure by which a basic feasible solution to the linear program is carried into another basic feasible solution in such a manner that if the objective function is to be maximized, the objective function is monotone nondecreasing—that is, the value of the objective function at the new basic feasible solution is greater than or equal to the value of the objective function at the preceding basic feasible solu-

tion. It is a finite method (except in the case of degeneracy; see below) in that it terminates in a finite number of steps either with the optimal solution or with the indication that no maximum exists —that is, that the objective function is unbounded.

In order to solve the problem given by (1), (2), and (3), we first rewrite it in the form of the *condensed simplex tableau*:

$-x_1$	$\cdots$	$-x_n$	1	
a_{11}	$\cdots$	a_{1n}	b_1	$= v_1$
$\vdots$		$\vdots$	$\vdots$	$\vdots$
a_{m1}	$\cdots$	a_{mn}	b_m	$= v_m$
$-c_1$	$\cdots$	$-c_n$	z_0	$= z$

In this tableau the basic variables are $v_1, \cdots, v_m$, and the nonbasic variables are $x_1, \cdots, x_n$. The current solution is thus $v_1 = b_1, \cdots, v_m = b_m, x_1 = 0, \cdots, x_n = 0$. It is assumed in the current formulation that the b_i are nonnegative coefficients. An iteration consists of a pivot operation by which the roles of some v_i and x_j are exchanged. If the variable to enter the current solution is x_k and the variable to leave it is v_r, the exchange is effected by solving the constraint equation

$$v_r = -a_{r1}x_1 - \cdots - a_{rk}x_k - \cdots - a_{rn}x_n + b_r$$

for x_k and substituting the solution into the other equations. This substitution yields a new tableau that can be interpreted in the same fashion as the old one. The substitution also yields the rules by which the elements of the new tableau can be calculated from the old one. Specifically, denoting the elements of the new tableau by $a'_{ij}, b'_i, c'_j, z'_0$, we have

$$a'_{rk} = 1/a_{rk};$$

$$a'_{rj} = a_{rj}/a_{rk}, \qquad \text{for } j \neq k;$$
$$b'_r = b_r/a_{rk};$$

$$a'_{ik} = -a_{ik}/a_{rk}, \qquad \text{for } i \neq r;$$
$$c'_k = -c_k/a_{rk};$$

$$a'_{ij} = a_{ij} - a_{ik}a_{rj}/a_{rk}, \qquad \text{for } i \neq r, j \neq k;$$
$$c'_j = c_j - c_k a_{rj}/a_{rk}, \qquad \text{for } j \neq k;$$
$$b'_i = b_i - b_r a_{ik}/a_{rk}, \qquad \text{for } i \neq r;$$
$$z'_0 = z_0 + b_r c_k/a_{rk}.$$

The element a_{rk} is called the *pivot* of the iteration. It is chosen in the following manner: (a) Its column index can refer to any column k that has $c_k > 0$; this ensures that the objective function is nondecreasing. If there is no such c_k, the maximum has been reached. (b) After such a column k is chosen, if all $a_{ik} \leqslant 0$ in that column, the problem is unbounded. Otherwise we choose the $a_{rk} > 0$ for which b_i/a_{ik} is smallest. This ensures that the new basic solution is feasible.

It is standard procedure to choose for introduction into the solution that x_c from among all eligible ones for which c_k is largest. Extensive experimental evidence suggests that the number of iterations required to solve linear programs is highly sensitive to the particular pivot choice criterion employed and that many criteria exist which are more efficient than the standard one (Kuhn & Quandt 1963).

A characteristic of this form of the simplex method is that it is a *primal* algorithm—that is, the current solution to the primal problem is feasible throughout the computation. The dual simplex method is completely analogous and differs only in that dual feasibility is preserved throughout the computations.

The column vectors in a simplex tableau can be expressed in terms of the inverse (if it exists) of the matrix of coefficients corresponding to the basic variables. This inverse matrix is altered when a basic variable is replaced by a nonbasic variable. The change in the basic inverse can be expressed by multiplying the old inverse by a certain elementary matrix. From the computational point of view we can simplify the solution of a linear program by keeping track of the successive elementary matrices necessary to effect the requisite transformations. Using this procedure is called using the *product form of the inverse.*

If k denotes the pivot column and if $a_{ik} > 0$ and $b_i = 0$ at any stage in the solution of a linear program, the program is said to be degenerate. Degenerate linear programs can cycle—that is, it is possible that a particular sequence of simplex tableaus will repeat indefinitely without any further improvement in the objective function, even though the maximum has not been reached. Although artificial examples exhibiting the property of cycling have been constructed (Hoffmann 1953; Beale 1955), in practice, cycling does not seem to have occurred. Since the appearance of degeneracy in a problem is preceded by a nonunique choice of a pivot in a given column of a tableau, cycling could normally be avoided by making a random choice among potential pivots if more than one is available. Should this procedure fail to avoid degeneracy, there are various methods of perturbing the original problem that guarantee that cycling and degeneracy will be avoided. The development of the simplex method as a whole can be credited largely to G. B. Dantzig (Dantzig, Orden, & Wolfe 1954; Dantzig 1963).

Applications and special techniques

The diet problem. The diet problem is to find the cheapest diet that satisfies prescribed nutritional requirements. One of the first problems to be formulated as a linear programming problem (Stigler 1945), it is formally analogous to the activity analysis problem. Suppose there are n foods whose prices are $p_1, \cdots, p_n$ and m nutrients whose minimum requirements are $b_1, \cdots, b_m$. Let $x_1, \cdots, x_n$ be the amounts of each food included in the diet, and let a_{ij} be the amount of the ith nutrient in the jth food. Then the diet problem is to minimize

$$\sum_{j=1}^{n} p_j x_j,$$

subject to

$$\sum_{j=1}^{n} a_{ij} x_j \geqslant b_i, \qquad i = 1, \cdots, m;$$

$$x_j \geqslant 0, \qquad j = 1, \cdots, n.$$

Matrix games. Two-person zero-sum games can easily be formulated and solved as linear programs [see GAME THEORY]. Let a_{ij}, $i = 1, \cdots, m$, $j = 1, \cdots, n$, be the payoff from player II to player I if I employs his ith pure strategy and II his jth. Irrespective of the (pure) strategy employed by player II, player I, utilizing a maximin criterion, would wish to use the strategy that will ensure him an amount at least equal to some number V, called the value of the game. He will play his various pure strategies with probabilities $x_1, \cdots, x_m$. The requirement that his expected gain not be less than V (irrespective of the strategy employed by player II) can be expressed by

$$\sum_{i=1}^{m} a_{ij} x_i \geqslant V, \qquad j = 1, \cdots, n;$$

$$\sum_{i=1}^{m} x_i = 1;$$

$$x_i \geqslant 0, \qquad i = 1, \cdots, m.$$

We transform this statement of the problem into a linear program by defining $z_i = x_i/V$, $i = 1, \cdots, m$. Then, assuming that $V > 0$, the problem is to minimize

$$\frac{1}{V} = \sum_{i=1}^{m} z_i,$$

subject to

$$\sum_{i=1}^{m} a_{ij} z_i \geqslant 1, \qquad j = 1, \cdots, n;$$

$$z_i \geqslant 0, \qquad i = 1, \cdots, m.$$

If $V \leqslant 0$, we choose a λ large enough so that a solution can be found to $\sum_{i=1}^{m} a'_{ij} x_i \geqslant V$, $j = 1, \cdots, n$, where $a'_{ij} = a_{ij} + \lambda$. This transformation leaves the optimal mixed strategy unchanged. For further material on the application of linear programming to matrix games, see Tucker (1960).

Transportation and assignment problems. Assume that a commodity is available at m sources in amounts a_i, $i = 1, \cdots, m$, and is required at n destinations in amounts b_j, $j = 1, \cdots, n$. Assuming that the total supply $(\sum_{i=1}^{m} a_i)$ equals the total demand $(\sum_{j=1}^{n} b_j)$ and that the unit cost of shipping from source i to destination j is the constant c_{ij}, $i = 1, \cdots, m$, $j = 1, \cdots, n$, how can demands be satisfied so that no source is required to ship more than is available at that source and so that the total cost of shipping is minimized? The formulation of this problem, known as the transportation problem, is as follows: define x_{ij} as the amount shipped from source i to destination j; then minimize

$$\sum_{i=1}^{m} \sum_{j=1}^{n} c_{ij} x_{ij},$$

subject to

$$\sum_{i=1}^{m} x_{ij} = b_j, \qquad j = 1, \cdots, n;$$

$$\sum_{j=1}^{n} x_{ij} = a_i, \qquad i = 1, \cdots, m;$$

$$x_{ij} \geqslant 0, \qquad \text{for all } i, j.$$

The equations above express the requirements that all demands be satisfied and that no availabilities be exceeded.

One of the $m + n$ equalities above will always be linearly dependent on the others; thus, in the absence of degeneracy, a solution will consist of $m + n - 1$ nonzero shipments. An important property of the transportation problem is that if the a_i and b_j are integers, the solution of the transportation problem will also consist of integers. Variants of the transportation problem are obtained (a) by assigning capacity limits to the various routes from source to destination, expressed by constraints of the type $x_{ij} \leqslant k_{ij}$, and (b) by permitting a shipment from i to j to go via some intermediate destination(s), resulting in what is called the transshipment problem (Orden 1956).

The transportation problem is closely related to the assignment problem. Consider the problem of assigning n persons to n tasks in such a manner that exactly one person is assigned to each task. It is assumed that we can measure the effectiveness of each person in each task on some cardinal scale. If the effectiveness of the ith person in the jth task

is given by e_{ij}, we wish to maximize

$$\sum_{i=1}^{n} \sum_{j=1}^{n} e_{ij} x_{ij},$$

subject to

$$\sum_{i=1}^{n} x_{ij} = 1, \qquad j = 1, \cdots, n;$$

$$\sum_{j=1}^{n} x_{ij} = 1, \qquad i = 1, \cdots, n;$$

$$x_{ij} \geqslant 0.$$

The solution to the above mathematical problem *is* the solution to the assignment problem, since its formal analogy with the transportation problem guarantees that the solution values for the x_{ij} will be integers. Hence, each x_{ij} will be either 0 or 1; $x_{ij} = 0$ means that the ith person is not assigned to the jth task, and $x_{ij} = 1$ means that he is. The equality constraints ensure that one and only one man will be assigned to each task.

There are various methods for solving the transportation and the assignment problems. The most common are the simplex method—which is computationally easier in these cases than in a general linear programming problem—and variants of the Hungarian method due to H. W. Kuhn (1955). Some of the variants of the Hungarian method can also be used to solve the so-called network flow problem, the objective of which is to maximize the flow from a single source to a single destination (sink) through a network of intermediate nodes (Ford & Fulkerson 1962). The close relation between these network-oriented problems and the technique of graph theory has resulted in fruitful applications of graph theory to certain types of linear programming problems. The transportation problem was originated by F. L. Hitchcock (1941) and T. C. Koopmans (1951).

Decomposition in linear programming. Certain problems are characterized by the fact that they can be represented as two (or more) almost independent problems. A case in point is the production-scheduling problem of a corporation that has two divisions. Assume for argument's sake that each division faces production constraints that are independent of those faced by the other division. Thus, the amounts of labor, raw materials, and other material inputs used by one division in no way depend upon the amounts of these inputs used by the other division. There may, however, exist certain over-all corporate constraints, perhaps of a financial nature. In such circumstances the profit maximization problem of the corporation may be expressed as the problem of maximizing

$$\mathbf{c}_1'\mathbf{X}_1 + \mathbf{c}_2'\mathbf{X}_2,$$

subject to

$$\mathbf{A}_1\mathbf{X}_1 + \mathbf{A}_2\mathbf{X}_2 = \mathbf{b}_0,$$
$$\mathbf{B}_1\mathbf{X}_1 \qquad\quad = \mathbf{b}_1,$$
$$\qquad\quad \mathbf{B}_2\mathbf{X}_2 = \mathbf{b}_2;$$
$$\mathbf{X}_1 \geqq \mathbf{0},\ \mathbf{X}_2 \geqq \mathbf{0},$$

Here $\mathbf{X}_1$ and $\mathbf{X}_2$ are vectors of activity levels; $\mathbf{b}_0$, $\mathbf{b}_1$, and $\mathbf{b}_2$ are vectors of resource availabilities; and $\mathbf{A}_1$, $\mathbf{A}_2$, $\mathbf{B}_1$, and $\mathbf{B}_2$ are matrices of input coefficients specifying the amount of the ith resource necessary to sustain one unit of the jth activity. The first set of equalities represents the over-all corporate constraints; the second and third sets represent the two divisions' respective constraints.

Very large problems (involving tens of thousands of constraints) exhibiting this type of structure can be solved efficiently by the decomposition principle. The decomposition principle rests upon the following basic observations: (a) The solution of a linear program can always be expressed as a convex combination of the extreme points of the convex feasible set of solutions. This representation is called the executive or master program. (b) If the given problem is rewritten in terms of the extreme points, we obtain a new problem with substantially fewer constraints but with many more variables. (c) The solution method is not contingent on our finding all extreme points. On the contrary, columns of coefficients can be generated when needed (for introduction into the simplex tableau) by the solution of certain small subprograms, each consisting of the independent divisional constraints. (d) The solution process is an iterative one that consists of the following steps: (i) a feasible solution to the executive program is found; (ii) subprograms are solved to determine what new column is to be introduced into the executive program; (iii) the new (restricted) executive program is solved; (iv) the corresponding subprograms are solved again; and so on, until an optimal solution is reached in a finite number of steps.

The decomposition principle lends itself to applications of decentralized decision making. From the computational point of view, the decomposition principle extends the power of electronic computers (Dantzig 1963, chapter 23).

Nonlinear programming

Formulation. Cases of mathematical programming in which either the objective function or the constraints, or both, are nonlinear are referred to as instances of nonlinear programming. Examples of nonlinear programming arise frequently in economic contexts. The activity analysis example discussed in the section "Linear programming" contains two particularly severe restrictions: (a) the profit that can be obtained from producing an extra unit of a commodity is constant and is thus independent of the level of production, and (b) the amounts of additional inputs needed to produce an additional unit of a commodity are constant and do not depend on the level of production. Thus, the linear programming formulation accounts neither for the possibility of (eventually) declining average profit owing to negatively sloped demand functions nor for the possibility of increasing marginal amounts of inputs, which are required for the maintenance of constant unit additions to output because of the well-known phenomenon of diminishing returns. Nonlinear programming is well suited to dealing with the case of such realistic modifications of economic models. The general nonlinear programming problem can be formulated as follows: maximize

$$(4) \qquad\qquad f(x_1, \cdots, x_n),$$

subject to

$$(5) \qquad \begin{array}{ccc} g_1(x_1, \cdots, x_n) \geqslant 0, \\ \vdots & & \vdots \\ g_m(x_1, \cdots, x_n) \geqslant 0; \end{array}$$

$$(6) \qquad\qquad x_1 \geqslant 0, \cdots, x_n \geqslant 0.$$

Duality and the Kuhn–Tucker conditions. The necessary and sufficient conditions for a solution to the inequalities (5) and (6) to be an optimal solution are called the Kuhn–Tucker conditions (Kuhn & Tucker 1951, pp. 481–492). These conditions represent a powerful generalization of the duality theorems of linear programming. They are based on the assumption that the objective function f and the constraints $g_1, \cdots, g_m$ are differentiable, concave functions. (The function $h(x)$ is concave if, for any choice of two points x_1 and x_2 and $0 < \theta < 1$, the value of the function at any point between x_1 and x_2, $f[\theta x_1 + (1 - \theta)x_2]$, is not smaller than the value of the linear function between $f(x_1)$ and $f(x_2)$ given by $\theta f(x_1) + (1 - \theta)f(x_2)$.) The maximization problem can then be reformulated in terms of the differential calculus using Lagrangian multipliers. The new Lagrangian objective function is

$$\phi(\mathbf{x}, \mathbf{u}) = f(x_1, \cdots, x_n) + \sum_{i=1}^{m} u_i g_i(x_1, \cdots, x_n).$$

The so-called saddle value problem is to find

nonnegative vectors $x^0 = (x_1^0, \cdots, x_n^0)$ and $u^0 = (u_1^0, \cdots, u_m^0)$ such that

$$\phi(x, u^0) \leqslant \phi(x^0, u^0) \leqslant \phi(x^0, u).$$

(Here $\phi(x^0, u^0)$ is called the saddle value of ϕ, and the point (x^0, u^0) is called the saddle point of ϕ.) That is, the saddle value problem is to find x^0 and u^0 such that the Lagrangian function ϕ has a maximum with respect to x at x^0, a minimum with respect to u at u^0, and a saddle point at (x^0, u^0). The following are the crucial theorems: (1) A necessary and sufficient condition for x^0 to be a solution to the (primal) maximum problem is that there exist a u^0 such that x^0 and u^0 are a solution to the saddle value problem. (2) Necessary and sufficient conditions for x^0 and u^0 to be a solution to the saddle value problem are

$$(7) \quad \left.\frac{\partial\phi}{\partial x_j}\right|_{x_j=x_j^0} \leqslant 0; \quad \sum_{j=1}^{n} x_j^0 \left.\frac{\partial\phi}{\partial x_j}\right|_{x_j=x_j^0} = 0; \quad x_j^0 \geqslant 0,$$

for all j,

and

$$(8) \quad \left.\frac{\partial\phi}{\partial u_i}\right|_{u_i=u_i^0} \geqslant 0; \quad \sum_{i=1}^{m} u_i^0 \left.\frac{\partial\phi}{\partial u_i}\right|_{u_i=u_i^0} = 0; \quad u_i^0 \geqslant 0,$$

for all i.

If theorem (2) is applied to the linear case, it immediately reduces to the familiar duality theorems, with the middle conditions in (7) and (8) ensuring complementary slackness. The frequent case in which the constraints are linear but the objective function is not linear has an immediate economic interpretation: the first part of conditions (7) states, for example, that the imputed value of the resources necessary to produce a unit of the jth commodity must not be less than the *marginal* profit contribution of that commodity.

Recently, the conditions under which the Kuhn–Tucker conditions hold have been somewhat relaxed by H. Uzawa (1958).

Computational methods. A number of methods have been developed for solving nonlinear programs of various types. The success achieved by these methods depends in general on the configuration of the particular problem. Several methods require that the objective function and the constraints be concave; strict concavity is necessary for some methods. If the objective function is not concave, the maximum is generally not unique; thus, solution methods that are capable of solving nonlinear programs of this type usually obtain local maxima but not global maxima. Solution methods fall broadly into three classes: (*a*) gradient methods,

which, starting with some solution to the problem, evaluate the slope (gradient) of the nonlinear objective function and alter the current solution by moving in the direction of the steepest gradient; (*b*) the simplex method, applicable in the case of a quadratic objective function and linear constraints, which obtains its usefulness by virtue of the fact that in the case mentioned the Kuhn–Tucker conditions are expressible as linear equations; and (*c*) decomposition approaches, which arise from solving certain suitable linear programming problems based on evaluating the nonlinear objective function over a lattice of points in the space of feasible solutions (Wolfe 1963).

Discrete (integer) programming

Formulation. In a large variety of linear programming problems, one has to dispense with the usual assumption that the space of solutions is finely divisible. Cases in which the set of solutions consists of discrete points at which the components of the solution vector have integer values are called discrete or integer programming problems.

Integer programming problems arise essentially for one of two reasons that are perhaps conceptually distinct but are in practice not always distinguishable. One reason is that the values of some or all variables in a linear programming problem may have to be restricted to integers because fractional solutions may not make sense from the economic point of view. Certain economic activities (the building of a bridge, the dispatching of an aircraft) are by their nature indivisible, and any solution involving such variables must be integer. The other reason is that certain otherwise intractable mathematical problems, often of a combinatorial nature, have natural formulations in terms of integer-valued variables. In such instances fractional-valued variables in the solution must be ruled out for mathematical and logical reasons.

Applications. The following examples illustrate the types of problems that can be formulated as integer programs.

The fixed-charges transportation problem. Consider a standard transportation problem with m sources, n destinations, unit shipping costs c_{ij}, requirements b_j, and availabilities a_i. In addition, assume that a certain fixed charge e_{ij}, independent of the amount shipped, is incurred if x_{ij} is positive —that is, if the ith source ships to the jth destination—and is not incurred if x_{ij} is zero. An integer programming formulation is to minimize

$$\sum_i \sum_j c_{ij} x_{ij} + \sum_i \sum_j e_{ij} t_{ij},$$

subject to

$$\sum_i x_{ij} = b_j, \qquad\qquad j = 1, \cdots, n;$$

$$\sum_j x_{ij} = a_i, \qquad\qquad i = 1, \cdots, m;$$

$$x_{ij} \geqslant 0;$$

(9) $\qquad\qquad t_{ij} \leqslant x_{ij};$

(10) $\qquad\qquad \frac{1}{M} x_{ij} \leqslant t_{ij} \leqslant 1,$

where t_{ij} is an integer and M is any suitably large number, say $M = \max(a_i, b_j)$. Constraint (9) ensures that $t_{ij} = 0$ whenever $x_{ij} = 0$. Inequalities (10), together with the requirement that t_{ij} be an integer, ensure that $t_{ij} = 1$ if x_{ij} is positive. Hence, a cost e_{ij} will be incurred if and only if x_{ij} is positive.

The dispatching problem.   Let there be m customers whose orders must be delivered and n activities, each of which represents the act of delivering one or more customers' orders by truck. These activities $\mathbf{A}_j$ can be represented by column vectors of zeros and ones; the ith element of $\mathbf{A}_j$ is 0 if the jth activity does not deliver the ith customer's order and 1 if it does. These various activities arise from the fact that several orders can occasionally be combined in a single truck. If c_j represents the cost of the jth activity (per unit), an optimal dispatching pattern is found by minimizing

$$\sum_{j=1}^n c_j x_j,$$

subject to

$$\sum_{j=1}^n \mathbf{A}_j x_j = \mathbf{J};$$

$$x_j \geqslant 0,$$

where x_j is an integer and $\mathbf{J}$ is a column vector of ones. In the solution, $x_j = 1$ or $x_j = 0$, depending on whether the jth method of customer deliveries is or is not employed. This problem is closely related to the covering problem of graph theory.

The traveling salesman problem.   Given n cities and the distance between each two, what is the shortest tour that begins and ends in a given city and passes through every other city exactly once? This problem can be represented as an integer program in several ways.

Computational methods.   The computational methods employed in solving integer programs are due primarily to R. E. Gomory (1958). An outline of the working of these methods is as follows:

(a) The problem is solved by conventional methods as a linear program. If the solution consists of integers, that is the solution to the integer program. This is, in fact, always the case with the transportation and assignment problems.

(b) If the solution to the linear program is not integer, a new constraint is generated from the parameters of the previous solution with the property that the constraint excludes part of the feasible region (hence the name "Gomory-cut") without, however, excluding any integer points in the feasible region. The introduction of such a constraint adds a new, fictitious variable to the problem and renders the primal problem infeasible; hence, a pivot step by the dual simplex method can be undertaken.

(c) After a finite number of Gomory-cuts and dual simplex iterations, the integer solution to the problem can be achieved.

There exists some choice about how to generate new constraints. Several methods appear to be more successful with some problems than with others. The relative success of various computer codes in solving integer programs seems to depend in some sense on the structure of the problem. The relative computational efficiency of the various methods is only imperfectly understood.

It may be noted that the dual values in integer programs do not have all the properties and interpretations that dual values ordinarily have (Gomory & Baumol 1960). These problems do not possess the property of continuity, and the Kuhn–Tucker conditions are therefore not applicable. (For an excellent recent treatment of various aspects of integer programming, see Balinski 1965.)

Miscellaneous problems and methods

This final section is devoted to a brief description of some additional problems in programming.

Parametric programming.   Parametric programming deals with (linear) programming problems in which either the objective function or the constraining inequalities are parametrized as follows: maximize

$$\sum_{j=1}^n (c_j + k_1 e_j) x_j,$$

subject to

$$\sum_{j=1}^n a_{ij} x_j \leqslant b_i + k_2 d_i;$$

$$x_j \geqslant 0,$$

where k_1 and k_2 are parameters. Maximizing with respect to the x_j as before, one can deduce from such formulations the sensitivity of the solution to variations in the coefficients of the objective function and to variations in the right-hand sides of the first set of inequality constraints. By employing variants and extensions of the simplex method, problems of this type can be solved, and the range of k-values for which a solution exists can be determined.

Stochastic programming. If some of the coefficients in a programming problem are not known with certainty but can be regarded as depending upon the outcome of some chance event, the problem is one of stochastic programming. Such programming problems arise in a variety of situations in which it is desired to schedule production to meet an uncertain demand. Various formulations may differ in the following respects: (a) the coefficients in the objective function, in the left-hand sides of the inequality constraints, or in the right-hand sides of the inequality constraints may be random variables; (b) the decision problem may be nonsequential or may be sequential in that decision variables fall into two or more groups and decisions must be made with respect to each group at a different time; (c) the objective may be to maximize expected profit or to minimize the variance of profit, subject to a constraint on expected profit. Stochastic programming problems are often nonlinear, and solution methods may differ from case to case, involving ordinary linear programming, separable programming, and other techniques (Hadley 1964).

Dynamic programming. Dynamic programming, which was developed primarily by R. Bellman (1957), employs a set of powerful methods that can be applied to problems in which decisions can be represented as occurring sequentially. Mathematically it rests upon Bellman's principle of optimality, according to which an optimal set of n decisions is characterized by the fact that the last $n - 1$ decisions in the set are optimal with respect to the state resulting from the first decision, whatever that first decision may have been. Many problems in operations research are capable of being formulated as dynamic programs. These problems resemble the types of mathematical programming problems discussed above in that they characteristically involve maximization (or minimization) of some function subject to inequality constraints and to nonnegativity requirements. The computational difficulties involved in reaching solutions vary, depending on the problem.

RICHARD E. QUANDT

[*See also* OPERATIONS RESEARCH.]

BIBLIOGRAPHY

BALINSKI, M. L. 1965 Integer Programming: Methods, Uses, Computation. *Management Science* 12:253–313.

BAUMOL, WILLIAM J. 1958 Activity Analysis in One Lesson. *American Economic Review* 48:837–873.

BEALE, E. M. L. 1955 Cycling in the Dual Simplex Algorithm. *Naval Research Logistics Quarterly* 2:269–275.

BELLMAN, RICHARD 1957 *Dynamic Programming.* Princeton Univ. Press.

DANTZIG, GEORGE B. 1951 The Programming of Interdependent Activities: Mathematical Model. Pages 19–32 in Cowles Commission for Research in Economics, *Activity Analysis in Production and Allocation: Proceedings of a Conference.* Edited by Tjalling C. Koopmans. New York: Wiley.

DANTZIG, GEORGE B. 1963 *Linear Programming and Extensions.* Princeton Univ. Press.

DANTZIG, GEORGE B.; ORDEN, ALEX; and WOLFE, PHILIP 1954 *Notes on Linear Programming.* Part I: The Generalized Simplex Method for Minimizing a Linear Form Under Linear Inequality Restraints. RAND Corporation Memorandum RM-1264. Santa Monica, Calif.: The Corporation.

DORFMAN, ROBERT 1951 *Application of Linear Programming to the Theory of the Firm, Including an Analysis of Monopolistic Firms by Non-linear Programming.* Berkeley: Univ. of California Press.

DORFMAN, ROBERT; SAMUELSON, PAUL A.; and SOLOW, ROBERT M. 1958 *Linear Programming and Economic Analysis.* New York: McGraw-Hill.

FORD, LESTER R. JR.; and FULKERSON, D. R. 1962 *Flows in Networks.* Princeton Univ. Press.

GALE, DAVID; KUHN, HAROLD W.; and TUCKER, ALBERT W. 1951 Linear Programming and the Theory of Games. Pages 317–329 in Cowles Commission for Research in Economics, *Activity Analysis in Production and Allocation: Proceedings of a Conference.* Edited by Tjalling C. Koopmans. New York: Wiley.

GOMORY, RALPH E. (1958) 1963 An Algorithm for Integer Solutions to Linear Programs. Pages 269–302 in Robert L. Graves and Philip Wolfe (editors), *Recent Advances in Mathematical Programming.* New York: McGraw-Hill. → First appeared as Princeton–IBM Mathematics Research Project, Technical Report No. 1.

GOMORY, RALPH E.; and BAUMOL, WILLIAM J. 1960 Integer Programming and Pricing. *Econometrica* 28:521–550.

HADLEY, GEORGE 1964 *Nonlinear and Dynamic Programming.* Reading, Mass.: Addison-Wesley.

HITCHCOCK, FRANK L. 1941 The Distribution of a Product From Several Sources to Numerous Localities. *Journal of Mathematics and Physics* 20:224–230.

HOFFMAN, A. J. 1953 *Cycling in the Simplex Algorithm.* U.S. National Bureau of Standards, Report No. 2974. Washington: Government Printing Office

KANTOROVICH, L. V. 1960 Mathematical Methods of Organizing and Planning Production. *Management Science* 6:366–422.

KOOPMANS, TJALLING C. 1951 Optimum Utilization of the Transportation System. Volume 5, pages 136–145 in International Statistical Conference, Washington, D.C., 1947, *Proceedings.* Washington: The Conference.

KUHN, HAROLD W. 1955 The Hungarian Method for the Assignment Problem. *Naval Research Logistics Quarterly* 2:83–97.

KUHN, HAROLD W.; and QUANDT, RICHARD E. 1963 An Experimental Study of the Simplex Method. Volume 15, pages 107–124 in Symposium in Applied Mathematics, Fifteenth, Chicago and Atlantic City, 1962,

Experimental Arithmetic, High Speed Computing and Mathematics: Proceedings. Providence: American Mathematical Society.

KUHN, HAROLD; and TUCKER, A. W. 1951 Nonlinear Programming. Pages 481–492 in Berkeley Symposium on Mathematical Statistics and Probability, Second, *Proceedings.* Edited by Jerzy Neyman. Berkeley: Univ. of California Press.

ORDEN, ALEX 1956 The Transshipment Problem. *Management Science* 2:276–285.

STIGLER, GEORGE J. 1945 The Cost of Subsistence. *Journal of Farm Economics* 27:303–314.

TUCKER, A. W. 1960 Solving a Matrix Game by Linear Programming. *IBM Journal of Research and Development* 4:507–517.

UZAWA, HIROFUMI 1958 The Kuhn–Tucker Theorem in Concave Programming. Pages 32–37 in Kenneth J. Arrow, Leonid Hurwicz, and Hirofumi Uzawa, *Studies in Linear and Non-linear Programming.* Stanford Univ. Press.

WOLFE, PHILIP 1963 Methods of Nonlinear Programming. Pages 67–86 in Robert L. Graves and Philip Wolfe (editors), *Recent Advances in Mathematical Programming.* New York: McGraw-Hill.

Postscript

Among the most interesting developments in programming are a variety of computational methods. In the area of integer programming the most fruitful innovation is the branch-and-bound algorithm. It can be applied to all-integer or mixed-integer programming problems, and it is based on solving a sequence of artificial linear programming problems. These artificial problems are generated by first placing upper and lower bounds on the variables required to be integer and solving the resulting problem as an ordinary linear programming problem. Second, new constraints are generated for variables by choosing a variable x_j that was noninteger in the previous solution with value x_j^* and introducing two new problems with constraints $x_j \leqq [x_j^*]$ and $x_j \geqq [x_j^*] + 1$, where $[x_j^*]$ denotes the largest integer contained in x_j^*. Problems of this type either are infeasible or generate new problems until either an optimal integer solution is found or infeasibility is determined. For an elementary discussion the reader is referred to Wagner (1969) and for a detailed discussion of many variants of the algorithm to Garfinkel and Nemhauser (1972).

Another important area of advance is the computation of nonlinear programs. The number of methods available is too great to be enumerated, but two approaches deserve particular mention. They are the methods employing *barrier functions* and *penalty functions*. Let x be an n-vector, $f(x)$ a function to be minimized, and $g(x)$ a vector-valued function with m components; then consider the problem: minimize

$$f(x)$$

subject to

$$g(x) \geqq 0;$$
$$x \geqq 0.$$

If there exists an x in the interior of the set defined by the inequality constraints, one may construct $F(g(x))$ with the properties that F is bounded whenever $g \neq 0$ and that $F \to \infty$ as any component of $g \to 0$. An obvious example of such a barrier function is

$$F(g(x)) = i'\log g(x),$$

where $'$ denotes transposition, i is a column vector of 1's, and $\log g(x)$ denotes the vector whose components are the logarithms of the components of $g(x)$. Then consider the solutions to the sequence of the quasi-Lagrangean minimization problems: minimize

$$f(x) + \mu_j F(g(x))$$

over a sequence of positive $\mu_j \to 0$. It can be shown that, if certain conditions hold, the solution of the original problem is the same as the limit of the sequence of solutions to the transformed problem. The important observation is that the constrained minimization problem has been transformed into a sequence of unconstrained minimization problems and thus the vast literature of unconstrained optimization becomes applicable. If the constraints in the original problem are equalities, $g(x) = 0$, we can construct a penalty function $F(g(x))$ that is zero when the constraints are satisfied and is increasing in the components of $g(x)$; for example, $F = i'g(x)^2$, where $g(x)^2$ denotes the vector whose components are the squares of the components of $g(x)$. We then solve a sequence of quasi-Lagrangean problems over a sequence of $\mu_j \to \infty$; again the limit of the sequence of solutions x is equivalent to the solution of the optimal problem. (Osborne 1972; Anderssen et al. 1972). Thus here, too, techniques of unconstrained optimization can be employed for which there exists a vast and, by now, well-known literature (Bard 1974; Brent 1973; Conference on Numerical Methods for Nonlinear Optimization 1972; Murray 1972). Among the most noteworthy classes of algorithms are the quasi-Newton methods (Marquardt 1963; Goldfeld et al. 1966), the variable metric algorithms (Powell 1971), and algorithms requiring no evaluation of derivatives, such as conjugate gradient methods (Powell 1964; Brent 1973) and simplex methods— not to be confused with the simplex method for solving linear programs (Nelder & Mead 1964).

RICHARD E. QUANDT

ADDITIONAL BIBLIOGRAPHY

ANDERSSEN, R. S.; JENNINGS, L. S.; and RYAN, D. M. (editors) 1972 *Optimization*. St. Lucia (Australia): Univ. of Queensland Press. → Proceedings of a seminar held at the Australian National University, Dec. 8, 1971.

BARD, YONATHAN 1974 *Nonlinear Parameter Estimation*. New York: Academic Press.

BRENT, RICHARD P. 1973 *Algorithms for Minimization Without Derivatives*. Englewood Cliffs, N.J.: Prentice-Hall.

CONFERENCE ON NUMERICAL METHODS FOR NON-LINEAR OPTIMIZATION, UNIVERSITY OF DUNDEE, 1971 1972 *Numerical Methods for Non-linear Optimization*. Edited by F. A. Lootsma. London and New York: Academic Press.

GARFINKEL, ROBERT S.; and NEMHAUSER, GEORGE L. 1972 *Integer Programming*. New York: Wiley.

GOLDFELD, STEPHEN M.; QUANDT, RICHARD E.; and TROTTER, HALE F. 1966 Maximization by Quadratic Hill-climbing. *Econometrica* 34:541–551.

MARQUARDT, D. W. 1963 An Algorithm for Least-squares Estimation of Nonlinear Parameters. *Journal of the Society for Industrial and Applied Mathematics* 11:431–441.

MURRAY, WILLIAM A. (editor) 1972 *Numerical Methods for Unconstrained Optimization*. London and New York: Academic Press. → Based on a joint IMS/NPL conference at the National Physical Laboratory, Jan. 7–8, 1971.

NELDER, J. A.; and MEAD, R. 1964 A Simplex Method for Function Minimization. *Computer Journal* 7:308–313.

OSBORNE, M. R. 1972 On Penalty and Barrier Function Methods in Mathematical Programming. Pages 106–115 in R. S. Anderssen, L. S. Jennings, and D. M. Ryan (editors), *Optimization*. St. Lucia (Australia): Univ. of Queensland Press.

POWELL, M. J. D. 1964 An Efficient Method for Finding the Minimum of a Function of Several Variables Without Calculating Derivatives. *Computer Journal* 7:155–162.

POWELL, M. J. D. 1971 Recent Advances in Unconstrained Optimization. *Mathematical Programming* 1:26–57.

WAGNER, HARVEY M. (1969) 1975 *Principles of Operations Research*. 2d ed. Englewood Cliffs, N.J.: Prentice-Hall.

PROGRAMMING, COMPUTER

See COMPUTATION.

PROGRAMMING, LINEAR

See OPERATIONS RESEARCH; PROGRAMMING.

PROJECTIONS

See DEMOGRAPHY; PREDICTION; PREDICTION AND FORECASTING, ECONOMIC; TIME SERIES.

PROSPECTIVE STUDIES

See EXPERIMENTAL DESIGN; FALLACIES, STATISTICAL; PUBLIC POLICY AND STATISTICS.

PSYCHOMETRICS

Psychometrics, broadly defined, includes all aspects of the science of measurement of psychological variables and all research methodologies related to them. In addition to this article, the area of psychometrics is discussed in EXPERIMENTAL DESIGN; FACTOR ANALYSIS AND PRINCIPAL COMPONENTS; LATENT STRUCTURE; PSYCHOPHYSICS; QUANTAL RESPONSE; SOCIOMETRY; *and in numerous* IESS *articles less relevant to statistics.*

Measurement

Measurement is generally considered to be any procedure whereby numbers are assigned to individuals (used herein to mean persons, objects, or events) according to some rule. The rule usually specifies the categories of an attribute or some quantitative aspect of an observation, and hence defines a *scale*. A scale is possible whenever there exists a one-to-one relationship between some of the properties of a group of numbers and a set of operations (the measurement procedure) which can be performed on or observed in the individuals. Scales of measurement are commonly classified as nominal scales, ordinal scales, interval scales, and ratio scales; the variables they measure can be discrete (i.e., providing distinct categories that vary from each other in a perceptibly finite way) or continuous (i.e., not readily providing distinct categories; varying by virtually imperceptible degrees).

Nominal scales. In a nominal scale the numbers merely identify individuals or the categories of some attribute by which individuals can be classified. Letters or words or arbitrary symbols would do just as well. Simple identification is illustrated by the assignment of numbers to football players; classification by assigning numbers to such attributes as sex, occupation, national origin, or color of hair. We can cross-classify according to the categories of two or more attributes; e.g., sex by occupation, or sex by occupation by national origin. With nominal scales that classify, the variables are always treated as discrete. Sex, occupation, and national origin are genuine discrete variables. Color of hair, on the other hand, is a multidimensional continuous variable. If, for instance, we treat it as a discrete variable by establishing the categories blond, brunette, and redhead, the measure becomes unreliable to some degree, and some individuals will be misclassified. Where such misclassification can occur, the scale may be termed a *quasi-nominal* scale. Subject to the limitation of unreliability, it has the properties and uses of any other nominal scale.

The basic statistics used with nominal scales are the numbers, percentages, or proportions of individuals in the categories of an attribute or in the cells of a table of cross classification. Hypotheses about the distribution of individuals within the categories of one attribute, or about the association of attributes and categories in a table of cross classification, are usually tested with the chi-square test. Descriptive statistics used include the mode (the category which includes the largest number of individuals) and various measures of association, the commonest of which is the contingency coefficient. These statistics remain invariant when the order of the categories of each attribute is rearranged and when the numbers that identify the categories are changed. If the categories of an attribute have a natural order, this order is irrelevant to nominal scaling, and nominal-scale statistics do not use the information supplied by any such order. [See COUNTED DATA; STATISTICS, DE-SCRIPTIVE, article on ASSOCIATION; SURVEY ANALYSIS, article on THE ANALYSIS OF ATTRIBUTE DATA.]

Ordinal scales. An ordinal scale is defined by a set of postulates. We first introduce the symbol ">," and define it broadly as almost any asymmetrical relation; it may mean "greater than," "follows," "older than," "scratches," "pecks," "ancestor of," etc. We also define "$\neq$" to mean "is unequal to" or "is different from." Then, given any class of elements (say, $a, b, c, d, \cdots$) the relation $>$ must obey the following postulates:

If $a \neq b$, then $a > b$ or $b > a$.
If $a > b$, then $a \neq b$.
If $a > b$ and $b > c$, then $a > c$.

If $a, b, c, \cdots$, are the positive integers and $>$ means "greater than," these postulates define the ordinal numerals. If $a, b, c, \cdots$, are chicks, and $>$ means "pecks," the postulates define the behavioral conditions under which a "pecking order" exists. If $a, b, c, \cdots$, are minerals and $>$ means "scratches," conformity to the postulates determines the existence of a unique scratching order. If it is suggested that aggression implies pecking or that hardness implies scratching, conformity of the behavior of chicks or minerals to the postulates indicates whether aggression or hardness, each assumed to be a single variable, can be measured on an ordinal scale in terms of observations of pecking or by scratching experiments.

Where the variable underlying the presumed order is discrete or where it is possible for two or more individuals to have identical or indistinguishably different amounts of this variable, we must enlarge the concept of order to include equality.

To do so we define $\not\succ$ ("not greater than," "does not follow," "not older than," "does not scratch," "does not peck," "is not an ancestor of," etc.) and add the postulate:

If $a \not\succ b$ and $b \not\succ a$, then $a = b$.

For example, two chicks are equally aggressive if neither one pecks the other, and two minerals are equally hard if neither scratches the other.

For an ordinal scale, the relation between the ordinal numerals and the attribute they measure is monotonic. If one individual has more of the attribute than another does, he must have a higher rank in any group of which they are both members. But *differences* in this attribute are not necessarily associated with proportionately equal differences in ordinal numerals. The measurements may in fact be replaced by their squares or their logarithms (if the measurements are positive), their cube roots, or any one of many other monotonic functions without altering their ordinal positions in the series.

Medians and percentiles of ordinal distributions are themselves seldom of interest because each of them merely designates an individual, or two individuals adjacent in the order. Hypotheses involving ordinal scale data may be tested by the Wilcoxon–Mann–Whitney, Kruskal–Wallis, Siegel–Tukey, and other procedures. (Interval-scale data are often converted by ranking into ordinal scales to avoid the assumption of normality.) Both the Kendall and the Spearman rank correlation procedures apply to ordinal scales. [See NONPARAMETRIC STATISTICS, articles on ORDER STATISTICS and RANKING METHODS.]

Quasi-ordinal scales. Rankings made by judges are of course subject to errors. These are of two types: within-judge errors, where judges fail to discriminate consistently, and between-judge errors, where judges disagree in their rankings.

Within-judge discrimination can be evaluated, and the errors partially "averaged out," by the method of paired comparisons. A set of judgments such as $a > b$, $b > c$, $c > a$, which violates the third postulate, is termed a circular triad. If a judge's rankings provide some but not too many circular triads, a "best" single ranking is obtained by assigning a score of 1 to an individual each time he is judged better than another, summing the scores for each individual, and ranking the sums. This procedure yields a quasi-ordinal scale. A true ordinal scale exists only if there are no circular triads.

Between-judge agreement can be estimated by the coefficient of concordance or the average rank correlation, when several judges have ranked the same set of individuals on the same attribute. A true ordinal scale exists if all judges assign the same ranks to all individuals, so that the coefficient of concordance and the average rank correlation are both unity. If these coefficients are not unity but are still fairly high, the "best" single ranking of the individuals is obtained by summing all ranks assigned to each individual by the several judges, and then ranking these sums. This case is again a quasi-ordinal scale: an ordinal scale affected by some unreliability.

All statistical procedures which apply to ordinal scales apply also to quasi-ordinal scales, with the reservation that the results will be attenuated by scale unreliability.

Interval scales.   An interval scale has equal units, but its zero point is arbitrary. Two interval scales measuring the same variable may use different units as well as different arbitrary zero points, but within each scale the units are equal. The classic examples are the Fahrenheit and centigrade scales of temperature used in physics.

For an interval scale, the relation between the scale numbers and the magnitudes of the attribute measured is not only monotonic; it is linear. Hence, if two interval scales measure the same variable in different units and from different zero points, the relation between them must be linear also. The general linear equation, for variables X and Y, is of the form $Y = a + bX$. Thus, for the Fahrenheit and centigrade scales we have

$$°F = 32 + 1.8°C; \quad °C = .5556°F - 17.7778.$$

On interval scales, differences in actual magnitudes are reflected by *proportional* differences in scale units. Thus, if two temperatures are 18 units apart on the Fahrenheit scale, they will be 10 units apart on the centigrade scale, no matter where they are located on these scales (e.g., 0° to 18°F or 200° to 218°F).

Interval-scale units may be added and subtracted, but they may not be multiplied or divided. We *cannot* say that Fahrenheit temperature 64° means twice as hot as Fahrenheit 32°. Almost all ordinary statistical procedures may be applied to interval-scale measurements, with the reservation that measures of central tendency must be interpreted as depending upon the arbitrary zero points. Almost all other statistical procedures are functions of deviations of the measures from their respective means and, hence, involve only addition and subtraction of the scale units. [*See* MULTIVARIATE ANALYSIS, *article on* CORRELATION METHODS; STATISTICS, DESCRIPTIVE, *article on* LOCATION AND DISPERSION.]

Quasi-interval scales.   Suppose that to each successive unit of an interval scale we apply a *random* stretch or compression. In this context randomness means that if the actual length of each unit is plotted against its ordinal position, there would be no trend of any sort: the larger units would not occur more frequently at one end, at both ends, in the middle, or in either or both intermediate regions. If the largest unit is small compared with the range of the variable in the group measured, we have a quasi-interval scale. All ordinary statistical procedures apply to quasi-interval scales, with the reservation that they have reduced reliability: errors of measurement are built into the scale units.

Ratio scales.   A ratio scale has the properties of an interval scale and in addition a true zero point, the scale-value zero meaning absence of any amount of the variable measured. Classic examples from physics are length and weight. The relation between the actual quantities and the scale values is linear, and the equations, moreover, have no constants and are of the form $Y = bX$. If two ratio scales measure the same variable in different units, any measurement on one scale is the same multiple or fraction of the corresponding measurement on the other scale. Thus, if the length of an object is X inches, its length is also 2.54X centimeters. And if the length of another object is Y centimeters, its length is also .3937Y inches.

Ratio-scale units may be multiplied and divided as well as added and subtracted. A man who is six feet high *is* exactly twice as tall as is one who is three feet high. Statistics applicable to ratio scales include geometric means, harmonic means, coefficients of variation, standard scores, and most of the common transformations (square root, logarithmic, arcsine, etc.) used to achieve improved approximations to normality of data distributions, homogeneity of variances, and independence of sampling distributions from unknown parameters. [*See* STATISTICAL ANALYSIS, SPECIAL PROBLEMS OF, *article on* TRANSFORMATIONS OF DATA.]

Quasi-ratio scales.   A quasi-ratio scale is a ratio scale with random stretches and compressions applied to its units, in much the same way described for quasi-interval scales. All of the statistics appropriate to ratio scales apply also to quasi-ratio scales, with the reservation of reduced reliability.

Operational definitions.   For many of the variables of the physical sciences, and some of the variables of the social sciences, the variable itself is defined by the operations used in measuring it or in constructing the measuring instrument. Thus length can be defined as what is measured by a ruler or yardstick. If we have first a "standard inch," say two scratches on a piece of metal (originally, according to story, the length of the first joint of a king's index finger), we can lay off successive inches on a stick by using a pair of dividers. For smaller units we can subdivide the inch by successive halving, using the compass and straightedge of classical Euclidean geometry. Height, in turn, may be defined as length measured vertically, with vertical defined by a weight hanging motionless on a string.

Psychological and social variables, on the other hand, can less often be defined in such direct operational terms. For example, psychophysics and scaling have as major concerns the reduction of sensory, perceptual, or judgmental data to interval scale, quasi-interval scale, ratio scale, or quasi-ratio scale form. [See PSYCHOPHYSICS; see also Torgerson 1968.]

Test scores as measurements. In the context of this discussion, a test is usually simply a set of questions, often with alternative answers, printed in a paper booklet, together with the instructions given by an examiner; a test performance is whatever an examinee does in the test situation. The record of his test performance consists of his marks on the test booklet or on a separate answer sheet. If he is well motivated and has understood the examiner's instructions, we assume that the record of his test performance reflects his knowledge of and ability in the field covered by the test questions. For the simpler types of items, with a simple scoring procedure, we credit him with the score +1 for each item correctly marked and 0 for each item incorrectly marked. We know that the organization of knowledge of an area in a human mind is complex and that an examinee's answer to one question is, in consequence, not independent of his answers to other questions. His score on the test is supposed to represent his total knowledge of the field represented by all the items (questions and alternative answers). In order to justify using such a score, we must be able to make at least two assumptions: (1) that knowledge of the area tested is in some sense cumulative along a linear dimension and (2) that there is at least a rough one-to-one relation between the amount of knowledge that each individual possesses and his test score.

The scores on a test, then, should form a quasi-ordinal scale. But suppose we have a 100-item, five-alternative, multiple-choice test with items arranged in order of difficulty from very easy to extremely hard. Individual A gets the first 50 right and a random one-fifth of the remainder. Individual B gets all the odd-numbered items right and a random one-fifth of all the even-numbered items.

If a simple scoring formula that credits each correct response with a score of +1 is used, the score of each is 60. Yet most persons would tend to say that individual B has more of the ability measured by the test than has individual A. These cases are of course extreme, but in general we tend to attribute higher ability to an individual who gets more hard items right, even though he misses several easy items, than to one who gets very few hard items right but attains the same score by getting more easy items right. E. L. Thorndike (1926) distinguishes between *altitude of intellect* and *range of intellect* (at any given altitude) as two separate but correlated variables. He discusses ways of measuring them separately, but his suggestions have so far had little impact on the mainstream of psychometric practice. [*Additional discussion of test scores as measurements is provided in the section "Item analysis," subsection on "Indexes of difficulty"; see also* Joncich 1968.]

○Correction for guessing

When an objective test is given to an individual, the immediate aim is to assess his knowledge of the field represented by the test items or, more generally, his ability to perform operations of the types specified by these items. But with true–false, multiple-choice, matching, and other *recognition*-type items, it is also possible for the examinee to mark right answers by guessing. It is known that the guessing tendency is a variable on which large individual differences exist, and the logical purpose of the correction for guessing is to reduce or eliminate the expected advantage of the examinee who guesses blindly instead of omitting items about which he knows nothing. (See Messick 1968.)

○¹The earliest derivations were based on the *all-or-none assumption*, which holds that an examinee either knows the right answer to a given item with certainty and marks it correctly, or else he knows nothing whatever about it and the mark represents a blind guess, with probability $1/a$ of being right (where a is the number of alternatives, only one of which is correct). Under this assumption we infer that when the examinee has marked $a - 1$ items incorrectly, there were really a items whose answers he did not know, and that he guessed right on one of them and wrong on the other $a - 1$. Hence, for every $a - 1$ wrong answers we deduct one from the number of right answers for the one he presumably got right by guessing. The correction formula is then

$$S = R - W/(a - 1),$$

where S is the corrected score (the number of items whose answers the examinee presumably knew). R is the number right, and W is the number wrong. It is assumed that this formula will be correct *on the average*, although in any particular case an examinee may guess the correct answers to more or fewer than $W/(a-1)$ items.

There is empirical evidence (Lord 1964) that correction for guessing corrects fairly well for high guessing tendency, but not so well for extreme caution, since the examinee is credited with zero knowledge for every item he omits. If an examinee omits items about which he has some, but not complete, knowledge, he will still be penalized. Hence, instructions should emphasize the point that an item should be omitted *only* if an answer would be a *pure guess*. If an examinee has a "hunch," he should always play it; and if he can eliminate one alternative, he should always guess among the remainder. This is a matter of ethics applying to all tests whose items have "right" answers; an examinee should never be able to increase his most probable score by disobeying the examiner's instructions.

A timed power test should begin with easy items and continue to items too hard for the best examinee. The time limit should be so generous that every examinee can continue to the point where his knowledge becomes substantially 0 for every remaining item. In this case the correction formula will cancel, on the average, the advantage that would otherwise accrue to those examinees who, near the end of the test period, give random responses to all remaining items.

There are only three conditions under which the correction for guessing need not be used: (1) there is no time limit, and examinees are instructed to mark every item; (2) the time limit is generous, examinees are instructed to mark every item as they come to it, and a minute or two before the end of the session the examiner *instructs* them to record random responses to all remaining items; (3) the test is a pure speed test, with no item having any appreciable difficulty for any examinee. In this case, errors occur only when an examinee works too fast.

Item analysis

In the construction of standardized tests, item analysis consists of the set of procedures by which the items are pretested for difficulty and discrimination by giving them in an experimental edition to a group of examinees fairly representative of the target population for the test, computing an index of difficulty and an index of discrimination for each item, and retaining for the final test those items having the desired properties in greatest degree.

Difficulty refers to how hard an item is, to how readily it can be answered correctly. A test item possesses *discrimination* to the extent that "superior" examinees give the right answer to it oftener than do "inferior" examinees. "Superior" and "inferior" are usually defined by total scores on the experimental edition itself. This is termed *internal consistency* analysis. When the less discriminating items are eliminated, the test becomes more *homogeneous*. A perfectly homogeneous test is one in which the function or combination of related functions measured by the whole test is also measured by every item. A test may, however, be designed for a specific use—e.g., to predict college freshman grades—in which case "superior" and "inferior" may be defined externally by the freshman grade-point average.

Wherever possible, the experimental edition is administered to the item-analysis sample without time limit and with instructions to the examinees to mark every item. If the experimental session has a time limit, the subset of examinees who mark the last few items form a biased subsample, and there is no satisfactory way to correct for this bias (Wesman 1949).

The experimental group for an item analysis should be reasonably representative of the target population, and particularly representative with regard to age, school grade(s), sex, socioeconomic status, residence (city or country), and any other variables that might reasonably be expected to correlate substantially with total scores. Its range of ability should be as great as that of the target population. Beyond this, it does not have to be as precisely representative as does a norms sample. Item analyses based on a group from a single community (e.g., a city and the surrounding countryside) are often quite satisfactory if this community is representative of the target population on all of the associated variables.

There are two major experimental designs for item analysis. The first is called the upper-and-lower groups (ULG) design. On the basis of the total scores (or some external criterion scores) an upper group and a lower group are selected: usually the upper and lower 27 per cent of the whole sample, since this percentage is optimal. With the ULG design, the only information about the total scores (or the external criterion scores) that is used is the subgroup membership. Hence, this design calls for large experimental samples.

In the second design all the information in the data is used: for each item the distribution of total scores (or external criterion scores) of those who mark it correctly is compared with the distribution of those who mark it incorrectly. This is the item-total score (ITS) design. Here the sample size can be smaller.

Indexes of difficulty. With either design, a quasi-ordinal index of the difficulty of an item is provided by the per cent of the total sample who respond correctly. With the ULG design, a very slightly biased estimate is given by the average per cent correct in the upper and lower groups; a correction for this bias is found in the tables compiled by Fan (1952). For many purposes, however, an index of difficulty with units which in some sense form a quasi-interval scale is desired. With free-choice items, and the assumption that the underlying ability is normally distributed in the experimental sample, the normal deviate corresponding to the per cent correct yields a quasi-interval scale of difficulty. But under this assumption the distribution of difficulty of a recognition-type item will be skewed, with amount of skewness depending on the number of alternatives. The precise form of this distribution is not known. Common practice involves discarding items with difficulties not significantly higher than chance, even if they show high discrimination; redefining per cent correct as per cent correct above chance, $p' = (p - 1/a)/(1 - 1/a)$, where p and p' are now proportions rather than percentages; and treating as a very rough quasi-interval scale the normal deviates corresponding to these adjusted proportions.

Another method first replaces the raw total scores (or external criterion scores) with normalized standard scores to form a quasi-interval score scale. The distributions of these normalized standard scores for those who pass and fail the item are then formed and smoothed, and the difficulty of the item is taken as the score corresponding to the point of intersection of the two distributions. This is strictly an ITS procedure.

When item difficulties have a rectangular distribution ranging from easy for the least able examinee to hard for the most able, and when items are all equally discriminating on the average, the distribution of the test scores will be approximately the same as the distribution of the ability which underlies them; and these scores will form a quasi-interval scale. Almost the only tests which actually are so constructed are those for the measurement of general intelligence, such as the Stanford–Binet. Most tests have roughly normal, or at best mildly flat-topped, distributions of item difficulties. When applied to a group for which the mean item difficulty corresponds to the mean ability level and in which the ability is approximately normally distributed, the resulting score distribution tends to be flat-topped. Empirical data support this theoretical conclusion.

Tests constructed with all items of almost equal difficulty are useful for selection purposes; they have maximum reliability at the given ability level. With a rectangular distribution of item difficulties, a test is equally reliable at all scale levels, but its reliability at any one level is relatively low. With a normal or near-normal distribution of item difficulties, the reliability is at a maximum in the region of the modal difficulty and decreases toward the tails, but this decrease is less marked than it is in the case of a test whose items are all equally difficult.

Although scores on tests with near-normal distributions of item difficulties are frequently treated as forming quasi-interval scales, they should more properly be treated as forming only quasi-ordinal scales. All the strictures against treating percentile ranks as interval scales apply to such raw-score scales with only slightly diminished force.

Indexes of discrimination in ULG design. For some purposes we need only to eliminate items for which the number of right answers is not significantly greater in the upper group than in the lower group, using the chi-square test of association. This procedure is often used in the selection of items for an achievement test.

In other cases we may wish, say, to select the 100 "best" items from an experimental test of 150. Here "best" implies a quasi-ordinal index of discrimination for each item. Widespread-tails tetrachoric correlations are often employed (Fan 1952; Flanagan 1939). The correlation indexes are statistically independent of the item difficulties. Where we may need quasi-interval scales, the Fisher z'-transformation is commonly applied to the widespread-tails tetrachoric correlation, yielding at least a crude approximation to an interval scale.

A less common procedure is to use the simple difference between the per cents correct in upper and lower groups as the index of discrimination. This index is precisely the percentage of cases in which the item will discriminate correctly between a member of the upper group and a member of the lower group (Findley 1956).

Indexes of discrimination in ITS design. With the ITS design, a t-test may be used to test the hypothesis that the mean total (or external criterion) scores of those who do and do not mark the right answer to the item are equal. If we cannot assume normality of the score distribution, we can replace the raw scores with their ranks and use the two-group Wilcoxon–Mann–Whitney test with only slight loss of efficiency. [*See* LINEAR HYPOTHESES, *article on* ANALYSIS OF VARIANCE, *for a discussion of the t-test; see* NONPARAMETRIC STATISTICS, *article on* RANKING METHODS, *for a discussion of the Wilcoxon–Mann–Whitney test.*]

To obtain an ordinal index of discrimination, the biserial, point-biserial, or Brogden biserial correlation (1949) between the item and the total (or external criterion) scores may be used. A crude approximation to interval scaling is given by applying the Fisher z'-transformation to the biserial or Brogden biserial correlations. [*See* MULTIVARIATE ANALYSIS, *article on* CORRELATION METHODS.]

Item analysis with wide-range groups. Some tests are designed to be used over several consecutive ages or grades, and the mean growth of the underlying variable may be assumed to be roughly linear. In such cases an item may be very hard at the lowest level but very easy at the highest, and highly discriminating at one level but quite undiscriminating at another. In such cases we may plot for each item the per cent correct at successive ages or grades, or the intersections of the score distributions for those who pass and fail the item at each age or grade. Before using this latter procedure, the raw scores may be scaled by first assigning normalized standard scores at each age or grade, assuming that the underlying variable is normally distributed at each level, and then combining them into a single scale by adjusting for the mean differences from age to age or grade to grade. An item is then retained if it shows a regular increase from age to age or grade to grade, or if it shows a large increase from any one age or grade to the next and no significant decrease at any other level. The scale difficulty of each item is the score level at which the per cent correct is 50, or for recognition items, the score level at which p', defined as above, is .50.

Two-criterion item analysis. When a test is designed to predict a single external criterion, such as freshman grade-point average, success in a technical training course, or proficiency on a given job, we can do somewhat better than merely to select items on the basis of their correlations with the criterion measure. The original item pool for the experimental edition is deliberately made complex, in the hope that items of different types will assess different aspects of the criterion performance. The best subset of items will then be one in which the items have relatively high correlations with the criterion *and* relatively low correlations with the total score on the experimental edition. Methods of item selection based on this principle have been discussed by Gulliksen (1950, chapter 21) and by

Horst (1936), using in both cases the ITS design.

Inventory items. Aptitude, interest, attitude, and personality inventories usually measure several distinct traits or include items of considerable diversity, subsets of which are scored to indicate similarity between the examinee's answer pattern and those of a number of defined groups. The items are usually single words or, more commonly, short statements, and the examinee marks them as applicable or inapplicable to him, true or false, liked or disliked, or statements with which he agrees or disagrees. The scoring may be dichotomous (like or dislike), trichotomous (Yes, ?, No), or on a scale of more than three points (agree strongly, agree moderately, uncertain, disagree moderately, disagree strongly). Often the statements are presented in pairs or triplets and the examinee indicates which he likes most and which least, or which is most applicable to him and which least applicable. The distinction between inventories based on internal analysis and those based on external criteria is a major one.

For internal analysis, the items are first allocated by judgment to preliminary subscales, often on the basis of some particular theoretical formulation. Each item is then correlated with every subscale and reallocated to the subscale with which it correlates highest; items which have low correlations with all subscales are eliminated. If the subscales are theoretically determined, all items which do not correlate higher with the subscales to which they were assigned than to any other are eliminated. If the subscales are empirically determined, new subscale scores are computed after items are reallocated, and new item–subscale correlations are obtained; this process is repeated until the subscales "stabilize." Purely empirical subscales may also be constructed by rough factor analyses of the item data or by complete factor analyses of successive subsets of items. [*See* FACTOR ANALYSIS AND PRINCIPAL COMPONENTS.]

For a *normative* scale, the job is finished at this point. (All aptitude and achievement tests form normative scales.) But when the statements are presented in pairs or triplets, they form an *ipsative* or partly ipsative scale. For a perfectly ipsative scale, the items of each subscale must be paired in equal numbers with the items of every other subscale, and only *differences* among subscale scores are interpretable. The California Test of Personality and the Guilford and Guilford–Martin inventories are examples of normative scales. The Edwards Personal Preference Schedule is perfectly ipsative, and the Kuder Preference Record is partly ipsative.

In filling out inventories of these types, whose items do not have right or wrong answers, we want examinees to be honest and accurate. Normative inventories are easily fakeable; ipsative inventories somewhat less so. Response sets also affect inventory scores much more than they do aptitude and achievement test scores. Most of the better inventories therefore have special scales to detect faking and to correct for various types of response sets. In forming pairs and triplets, efforts are made to equalize within-set social desirability or general popularity, while each statement represents a different subscale. (See Clark 1968; Dahlstrom 1968; Ebel 1968; Fleishman 1968; Holtzman 1968; Messick 1968.)

Inventories constructed on the basis of external criteria use a base group, usually large ("normal" individuals, "normal" individuals of given sex, professional men in general, high school seniors in general, high school seniors of given sex, and the like), and a number of special groups (hospital or clinic patients with the same clear diagnosis, or men or women in a given occupation). An answer (alternative) is scored for a special group if the people in that group mark it significantly more often (scored positively) or significantly less often (scored negatively) than do the people in the base group. In some inventories the more highly significant or highly correlated answers are given greater positive or negative weights than are the less significant or less highly correlated answers. Inventories of this type are almost always normative, and new subscales can be developed whenever new special groups can be identified and tested. The outstanding examples are the Strong Vocational Interest Blank and the Minnesota Multiphasic Personality Inventory.

In inventories of this type, the same item may be scored for several subscales. In consequence there are inter-key correlations, and the reliabilities of differences between pairs of subscale scores vary with the numbers of common items. A further consequence is that general subscales based on factor analyses of individual subscale intercorrelations are difficult to evaluate, since the individual subscale scores are not experimentally independent. Similar difficulties arise in the interpretation of factor analyses of ipsative and partly ipsative scale scores.

Reliability

Reliability is commonly defined as the accuracy with which a test measures whatever it does measure. In terms of the previous discussion, it might be defined in some cases as a measure of how closely a quasi-ordinal or quasi-interval scale, based on summation of item scores, approximates a true ordinal or interval scale. The following treatment assumes quasi-interval scales, since reliability theories based entirely on the allowable operations of ordinal arithmetic, which do not define the concepts of variance and standard deviation, have not been worked out. However, definitions and results based on correlations probably apply to quasi-ordinal scales if the correlations are Spearman rank correlations. [See MULTIVARIATE ANALYSIS, *article on* CORRELATION METHODS.]

The raw score of an individual on a test may be thought of as consisting of the sum of two components: a true score representing his real ability or achievement or interest level or trait level, and an error of measurement. Errors of measurement are of two major types. One type reflects the limitation of a test having a finite number of items. Using ability and aptitude tests as the basis for discussion, the individual's true score would be his score on a test consisting of *all* items "such as" the items of the given test. On the finite test he may just happen to know the right answers to a greater or lesser proportion of the items than the proportion representing his true score. Errors of this type are termed *inconsistency* errors. A second type of error reflects the fact that the *working* ability of an individual fluctuates about his *true* ability. On some occasions he can "outdo himself": his working ability exceeds his true ability. On other occasions he cannot "do justice to himself": his working ability is below his true ability. Working ability fluctuates about true ability as a result of variations in such things as motivation, inhibitory processes, physical well-being, and external events that are reflected in variation in concentration, cogency of reasoning, access to memory, and the like. Such fluctuations occur in irregular cycles lasting from a second or two to several months. Errors of this type are termed *instability* errors.

If a second test samples the same universe of items as does the first and in the same manner (random sampling or stratified random sampling with the same strata), the two tests are termed *parallel forms* of the same test. Parallel forms measure the same true ability, but with different inconsistency errors.

The basic theorem which underlies all formulas of reliability, and of empirical validity as well, may be stated as follows: *In a population of individuals, the errors of measurement in different tests and in different forms of the same test are uncorrelated with one another and are uncorrelated with the true scores on all tests and forms.*

Coefficient and index of reliability. The *reliability coefficient*, R, may be defined as the ratio of the true score variance to the raw score variance; it is also the square of the correlation between the raw scores and the true scores. The *index of reliability* is the square root of the reliability coefficient; it is the ratio of the standard deviation of the true scores to the standard deviation of the raw scores, or the correlation between the true scores and the raw scores. These definitions are purely conceptual. They are of no computational value because the true scores cannot be measured directly.

Furthermore, where R_A and R_B are the reliability coefficients of the two parallel forms and ρ_{AB} is the correlation between them, it is implied in the basic theorem that $\sqrt{R_A R_B} = \rho_{AB}$. If, moreover, as is usually the case, the two forms are equally reliable, $R_A = R_B = \rho_{AB}$; i.e., the correlation between the two forms is the reliability coefficient of *each* of them. When we estimate ρ_{AB} by computing r_{AB}, the correlation in a sample, the estimate is not unbiased, but the bias is usually small if the sample is reasonably large.

Consistency coefficient. If two equally reliable parallel forms of a test are administered *simultaneously* (e.g., by merging them as odd and even items in the same test booklet), the reliability coefficient becomes a *consistency coefficient*, since instability errors affect both forms equally.

The split-half correlation (e.g., the correlation between odd and even items) provides the consistency coefficient of each of the half-tests. The consistency coefficient of the whole test, as estimated from the sample, is then derived from the Spearman–Brown formula:

$$R_{A+B} = \frac{2r_{AB}}{1 + r_{AB}},$$

where r_{AB} is the correlation between the half-tests. The more generalized Spearman–Brown formula is

$$R_n = \frac{nr_{AB}}{1 + (n - 1)r_{AB}},$$

where r_{AB} is again the correlation between the half-tests and R_n is the consistency coefficient of a parallel form n times as long as one half-test. In deriving the Spearman–Brown formula, we must assume that the half-tests are equally variable as well as equally reliable, but these requirements are not very stringent. Kuder and Richardson (1937) also present several formulas for the consistency of one form of a homogeneous test. Their most important formula (formula 20) was generalized and discussed at some length by Cronbach (1951).

Interform reliability coefficient. If two equally reliable parallel forms of a test are administered to the same group of examinees at two different times, the correlation between them is an interform reliability coefficient. The interform reliability is lower than the consistency because it is affected by instability errors, which increase with time. Reports of interform reliability should include the length of the time interval.

Stability coefficient. Instability errors are related to the interval between testings and are independent of the inconsistency errors. The stability coefficient may be defined as the interform reliability that would be found if both forms of the test were perfectly consistent. It may be estimated by the formula

$$s_{AB} = r_{AB} / \sqrt{C_A C_B},$$

where r_{AB} is the interform reliability coefficient and C_A and C_B are the consistency coefficients of the two forms, each computed from the split-half correlation and the Spearman–Brown formula, or from the Kuder–Richardson formula 20. Its value is independent of the lengths of the two forms of the test but dependent upon the time interval separating their administration.

The increase in interform reliability resulting from increase in test length may be estimated by the formula

$$R_n = \frac{nr_{AB}}{1 + (n - 1)C_{AB}},$$

where r_{AB} is the interform reliability and C_{AB} is the consistency of each form. The two forms must be assumed equally consistent, and C_{AB} is computed as $C_{AB} = \sqrt{C_A C_B}$. Then R_n is an estimate of the interform reliability of a parallel form n times as long as *one* of the two actual forms. If $n = 2$, the formula

$$R_{A+B} = \frac{2r_{AB}}{1 + C_{AB}}$$

gives the interform reliability of scores on the two forms combined.

Test–retest correlation. When the same form is given to the same examinees on two different occasions, the correlation is *not* a stability coefficient, and it would not be a stability coefficient even if every examinee had total amnesia of the first testing (and of nothing else) on the second occasion. In addition to the quantitative fluctuations in working ability which give rise to instability errors, there are qualitative fluctuations in perceptual organization, access to memory, and reasoning–procedure patterns. In consequence, the same set of items, administered on different occasions, gives rise to different reactions; and in consequence there are still some inconsistency errors. Perseveration effects, including but not limited to memory on the second occasion of some of the responses made to particular items on the first occasion, introduce artificial consistency, in varying amounts for different examinees. In consequence, test–retest coefficients cannot be clearly interpreted in terms of reliability theory.

○ **Standard error of measurement.** If several parallel forms, all equally reliable and with identical distributions of item difficulty and discrimination, could be given simultaneously to one examinee, the standard deviation of his scores would be the standard error of measurement of one form for him. The standard error of measurement is ordinarily reported as an estimate of the *average* standard error of measurement for all members of an examinee group. The formula is

$$SE_m = s\sqrt{1 - r_{AB}},$$

where s is the standard deviation of the total scores and r_{AB} is their consistency, computed by the split-half correlation and the Spearman–Brown formula or the Kuder–Richardson formula 20. The standard error of measurement may also be defined, for the whole sample or population of examinees, as the standard deviation of the inconsistency errors or the standard deviation of the differences between raw scores and the corresponding true scores. With this last definition we can also compute a standard error of measurement which includes instability errors over a given time period, by letting s represent a pooled estimate of the standard deviation of one form based on the data for both forms, and r_{AB} an interform reliability coefficient. In this case, SE_m is the standard error of measurement of one (cither) form.

Reliability and variability. The variance of true scores increases with the variability of ability in the group tested, while the variance of the errors of measurement remains constant or almost constant. By the variance-ratio definition of the reliability coefficient, it follows that this coefficient increases as the range of ability of the group measured increases. The reliability coefficient of a particular test is higher for a random sample of all children of a given age than for a sample of all children in a given grade, and lower for all children in a single class. When a reliability coefficient is reported, therefore, the sample on which it was computed should be described in terms which indicate as clearly as possible the range of ability of the subjects.

The formula relating variability to reliability is

$$R_{AB} = 1 - \frac{s^2}{S^2}(1 - r_{AB}),$$

where s^2 and r_{AB} are the variance and the reliability coefficient of the test for one group, and S^2 and R_{AB} are the variance and reliability coefficient for another group. The group means should be similar enough to warrant the assumption that the average standard error of measurement is the same in both groups. If a test author reports r_{AB} and s^2 in his manual, a user of the test need only compute S^2 for his group, and R_{AB} for that group can then be computed from the formula. Note that this formula applies exactly only if the test-score units form a quasi-interval scale over the whole score range of both groups.

Reliability at different score levels. If the test-score units do not form a quasi-interval scale, the standard error of measurement will be different at different score levels. If two forms of the test or two halves of one form are *equivalent,* and the experimental sample is large enough, the standard error of measurement may be computed for any given score level. Two parallel forms or half-tests are equivalent if their joint distributions of item difficulty and item discrimination are essentially identical. In this case, their score distributions will also be essentially identical.

To compute the standard error of measurement at a given score level, we select from a large experimental sample a subgroup whose total scores on the two forms or half-tests combined are equal within fairly narrow limits. The standard error of measurement of the total scores is then the standard deviation of the *differences* beween the half-scores. When, as is usually the case, the half-scores are based on splitting one form into equivalent halves administered simultaneously, the standard errors of measurement at different score levels are based only on inconsistency errors.

The reliability coefficient at a given score level, still referred to the variability of the whole group, is given by

$$r_{AB} = 1 - SE_m^2/s^2,$$

where SE_m is defined as above and s^2 is the total-score variance of the whole group.

Comparability. Two forms of a test, or two tests measuring different true-score functions, are termed comparable if and only if their units of measurement are equal. If the units do not form quasi-interval scales, they can be made comparable only if their score distributions are of the same shape and their standard errors of measurement are proportional at all score levels. Only equivalent forms have comparable raw-score units.

If two different tests have proportionally similar joint distributions of item difficulties and discriminations, they will meet these conditions. Meaningful interpretations of profiles of scores on different tests can be made only if the scores are comparable.

Validity

Test validity has to do with *what* a test measures and how well it measures what it is intended to measure, or what it is used to measure in any particular application if it is a multiple-use test.

Content validity. Content validity applies mainly to achievement tests, where the questions themselves define the function or combination of related functions measured and there is no external criterion of achievement with which the scores can be compared. The test developer should provide a detailed outline of both the topics and the processes —such as information, comprehension, application, analysis, synthesis, evaluation, etc.—that the test measures. A more detailed list of processes, with illustrative test questions from several fields, is given in *Taxonomy of Educational Objectives* (1963–1964). The item numbers of the test items are then entered in the cells of the outline, along with their indexes of difficulty and discrimination.

Evaluations of content validity are essentially subjective. The prospective user of the test may agree or disagree to a greater or lesser extent with the outline or the basis on which it was constructed, with the allocation of items to topics and processes, or with the author's classification of some of the items. If all such evaluations are positive, the test's validity is equal, for all practical purposes, to its interform reliability over some reasonable time period.

In constructing an achievement test, item analysis ordinarily consists only of the elimination of nondiscriminating items. If the test is to yield a single score, the various topics and processes must be sufficiently homogeneous to permit every item to correlate positively and significantly with the total score. All further elimination occurs in the balancing of the item contents against the requirements of the topic-by-process outline. In discussing *school* achievement tests, content validity is often termed curricular validity.

Empirical validity. Empirical validity is concerned with how well a test, either alone or in combination with others, measures what it is used to measure in some particular application. The empirical validity of a test is evaluated in terms of its correlation with an external *criterion* meas-

ure: an experimentally independent assessment of the trait or trait complex to be predicted. The term "prediction" is used here, without regard to time, to designate any estimate made from a regression equation, expectancy table, or successive-hurdles procedure. The term "forecast" will be used when we explicitly predict a *future* criterion. Empirical validity is also termed statistical validity and criterion validity.

There are two basically different types of criteria. The first may be termed *sui generis* criteria, criteria that exist without any special effort made to predict them. Examples include persistence in college, success or failure in a training course, dollar volume of sales, years of service in a company, and salary level. The unreliability of the criterion measure sets a natural upper limit for the validity of any predictor. The validity of a predictor or predictor battery is simply its correlation or multiple correlation with the criterion. We term such a correlation an index of *raw validity*.

The second type of criteria may be termed *constructed criteria,* and are developed upon the basis of a trait concept such as academic ability, job proficiency, or sales accomplishment. For academic achievement such a criterion might be grade-point average in academic subjects only. For job proficiency it might be based on quantity of output, quality of output, material spoilage, and an estimate of cooperation with other workers. For sales accomplishment it might be based on number of sales, dollar volume, new customers added, and an estimate of the difficulty of the territory. In any event, it must be accepted as essentially an *operational definition* of the trait concept and, hence, intrinsically content-valid. And since the error of measurement is no part of the operational definition of a trait concept, it is evident that we should predict *true* criterion scores rather than *raw* scores.

Concurrent and forecast validity. We can recognize two types of assessments of true validity: *concurrent* true validity and *forecast* true validity. In the first case, the criterion measure is usually one which is expensive or difficult to obtain, and the predictor is designed to be a *substitute measure* for it. In this case the predictor test or battery should be administered at the *middle* of the time interval over which the criterion behavior is observed; otherwise, instability errors will distort validity estimates.

For *forecast* true validity, the predictor test or battery is administered at some "natural" time: at or shortly before college entrance or admission to training or initial employment, and the criterion data should cover the later time period over which they will be most valid in terms of the criterion trait concept. In this case, the instability errors resulting from the earlier administration of the predictor test or battery are intrinsic to the prediction enterprise.

To obtain a quick rough estimate of forecast value, an investigator often tests present employees, students, or trainees at essentially the same time that the criterion data are obtained. This procedure should *not* be termed concurrent validation but, rather, something like *retroactive* validation.

Test selection and cross-validation. A common type of empirical validation study consists in administering to the experimental sample more predictors than are to be retained for the final battery. The latter then consists of that subset of predictors, of manageable length, whose multiple correlation most nearly approximates that of the whole experimental battery. Predictors commonly include scored biographical inventories, reference rating scales, and interview rating scales as well as tests.

When a subset of predictors is selected by regression procedures, its multiple correlation with the criterion is inflated: sampling error determines in part which predictors will be selected and what their weights will be. In *cross-validation,* the reduced predictor battery is applied to a second criterion group, using the weights developed from the first group. The aggregate correlation in the second group (now no longer a multiple correlation) is an unbiased estimate of the battery validity. In estimating forecast true validity, two criterion measures for each examinee are required only in the cross-validation sample. [*See* LINEAR HYPOTHESES, *article on* REGRESSION.]

The same situation arises in even more exaggerated form when predictor items are selected on the basis of their correlations with an external criterion measure. Each predictor requires a separate item-analysis sample. A different sample is required to determine predictor weights. And a still different sample is required to estimate the validity of the whole predictor battery.

Various split-sample methods have been devised to use limited data more effectively. Thus, test selection may be carried out on two parallel samples, keeping finally the subset of tests selected by both samples. The validity of the battery is estimated in each sample by using the weights from the other. The average of the two validity indexes is then a lower bound for the battery validity when the weights used are the averages of the weights from the two samples.

Validation procedures. The commonest methods of test selection and use are those described above, using multiple regression as the basic procedure. These procedures assume that the criterion elements are all positively and substantially correlated, and can be combined with suitable weights into a single criterion measure of at least moderate homogeneity. It is then further assumed that a low score on one predictor can be compensated for by high scores on others, a weighted total score on all predictors being accepted as a single predictor of criterion performance.

Some criteria, however, consist of elements which are virtually uncorrelated. In such cases the elements must be predicted separately. In practice there is usually one predictor for each element, although it is possible to predict an element by its multiple regression on two or more predictors. In this situation, the preferred procedure is the multiple cutoff procedure. Each criterion-element measure is dichotomized at a critical (pass–fail) level, the corresponding predictor level is determined, and a successful applicant must be above the critical levels on *all* predictors. A further refinement consists in rating the criterion elements on their importance to the total job and requiring a successful applicant to be above the critical levels on the predictors of all the more important elements, but permitting him to be a little (but not far) below this level on one or two of the predictors of the less important elements.

In predicting a dichotomous criterion, the most accurate predictions are made when the predictor cutoff score is at the point of intersection of the smoothed frequency curves of predictor scores for the upper and lower criterion groups. If the applicant group is large, however, the predictor cutoff score may be set one or two standard errors of measurement above this point.

Correction for attenuation. Correction for attenuation is a procedure for estimating what the correlation between two variables would be if both of them were perfectly reliable or consistent; i.e., if the correlation were based on the true scores of both variables. The unreliabilities of the variables attenuate (reduce) this correlation. To determine the proper correction, the experimental design must be such that the instability errors in the intercorrelation(s) are identical with those in the reliability or consistency coefficients.

Index and coefficient. In discussing formulas for validity, the term "index" rather than the term "coefficient" has been used, although the latter is the term commonly used. The square of each of these correlations, however, is a *coefficient of determination* (of raw criterion scores by raw predictor scores, true criterion scores by raw predictor scores, or true criterion scores by true predictor scores). The reliability coefficient is also a coefficient of determination (of true scores by raw scores on the same variable), and the *index* of reliability is its square root.

As the intrinsic validity (the true validity of a perfectly reliable predictor) approaches unity, the index of true validity approaches the index of reliability of the predictor, since in this case the predictor and criterion true scores are identical; and an unreliable predictor cannot predict anything better than it predicts its own true scores. The statement "The upper limit of a test's validity is the square root of its reliability" is erroneous: the upper limit of its validity is its reliability when both are expressed either as indexes or as coefficients. The error is due to the common practice of calling "indexes" of validity "coefficients" of validity.

Synthetic validity. Synthetic validation is test selection without a criterion variable; and synthetic validity is an estimate of empirical validity, also without a criterion variable. The procedure is based on job analysis and the accumulated experience from many previous empirical validation studies. The number of possible jobs, and the number of real jobs as well, greatly exceeds the number of distinct job elements and job-qualification traits. If previous studies have shown which qualification traits are required for each job element of a given job, and what predictors best predict each of these traits, predictors can be selected for the given job on the basis of a job analysis without a new empirical study, and rough estimates can even be made of the probable validity of the prediction battery. The procedures of job analysis are by now fairly well refined; there are substantial bodies of data on the qualification requirements for many job elements; and there are also fairly large amounts of data on the correlations between predictors and qualification traits and the intercorrelations among such predictors. Hence, synthetic validation is by now at least a practical art, and its methodology is approaching the status of an applied science. Synthetic validation is the only procedure which *can* be used when the number of positions in a given job category is too small to furnish an adequate sample for an empirical validation study.

○ **Factorial validity.** Factor analysis provides a way of answering the question "What does this test measure?" The simplest answer is merely a collection of correlations between the given test and a variety of other better-known tests. However, if the given test and several others have all been administered to the same large sample, the factor structure of the given test provides a much better answer.

When a large number of tests are administered serially in a factor analysis study, the instability errors are greater for tests far apart in the series than for tests administered consecutively. This leads to the generation of small instability factors and complicates making the decision about when to stop factoring. If there are two forms of each test, all the A forms might be given serially, and then after an interval of a week or more all the B forms might be given in a different serial order. Two parallel factor analyses could then be performed, one (for tests 1 and 2, say) using the correlation between 1A and 2B; the other, the correlation between 1B and 2A. The correlations between 1A and 2A and between 1B and 2B would not be used. Interform reliabilities consistent with the intercorrelations would be given by the correlations between 1A and 1B and between 2A and 2B. [*See* FACTOR ANALYSIS AND PRINCIPAL COMPONENTS.]

Construct validity. Construct validation is an attempt to answer the question "Does this test measure the trait it was designed to measure?" when no single criterion measure or combination of criterion measures can be considered a well-defined, agreed-upon, valid measure of the trait; and there is in fact no assurance that the postulated trait is sufficiently unitary to be measurable. We start with a trait *construct*: a hypothesis that a certain trait exists and is measurable. Then we build a test we hope will measure it. There are two questions, to be answered simultaneously: (1) Does the trait construct actually represent a measurable trait? (2) Does the test measure that trait rather than some other trait?

From the trait construct we draw conclusions about how the test should correlate with other variables. With some it ought to correlate fairly highly, and if in every such case it does, the resulting evidence is termed *convergent* validity. Variables of this type are in a sense *partial* criteria. With other variables it should have low correlations, and if in every such case it does, the resulting evidence is termed *discriminant* validity. Consider the trait construct "general intelligence" and the Stanford–Binet Scale.

(*a*) General intelligence should increase with age during the period of growth. Mean scores on the Stanford–Binet increase regularly throughout this period, but so do height and weight.

(*b*) General intelligence can be observed and rated with some validity by teachers. Children rated bright by their teachers make higher Stanford–

Binet scores than do children of the same age rated dull. There is some judgmental invalidity, however; docile and cooperative children are overrated, and classroom nuisances are underrated.

(c) The extremes of general intelligence are more certainly identifiable. Individuals judged mentally deficient make very low scores, but so do prepsychotics and children with intense negative attitudes toward school. Outstanding scholars, scientists, writers, and musical composers make high scores. Equally outstanding statesmen, executives, military leaders, and performing artists make somewhat lower but still quite high scores. This finding also agrees with the hypothesis, for people in the latter categories need high, but not so *very* high, general intelligence; and they also need special talents not so highly related to general intelligence.

(d) Items measuring diverse cognitive traits should correlate fairly highly with one another and should generate a large general factor if general intelligence is indeed a relatively unitary measurable trait. The items of the Stanford–Binet clearly do measure diverse cognitive traits, and their intercorrelations do generate a large general factor.

(e) Reliable homogeneous tests of clearly cognitive traits should correlate fairly highly with general intelligence. Tests of vocabulary, verbal and nonverbal reasoning, arithmetic problems, and visual–space manipulation correlate fairly highly with the Stanford–Binet, and not quite so highly with one another.

(f) Tests judged "borderline cognitive" should have positive but moderate correlations with general intelligence. Tests of rote memory, verbal fluency, mechanical knowledge, and the like do have positive but moderate correlations with the Stanford–Binet.

(g) Wholly noncognitive tests should have near-zero correlations with general intelligence. Tests of writing speed, visual acuity, physical strength, and the like do have near-zero correlations with the Stanford–Binet.

The full combination of these predictions and results, along with others not cited above, leads us to place considerable confidence in the trait status of the construct "general intelligence" and in the Stanford–Binet Scale (along with others) as a measure of this trait.

If there is even a single glaring discrepancy between theory and data, either the theory (the trait construct) must be revised, or the test must be considered invalid. A test of "social intelligence" was shown to correlate as highly with a test of general–verbal intelligence as other tests of general–verbal intelligence did. From this one finding

we must conclude either that social intelligence is not a trait distinct from general–verbal intelligence, or that the test in question is not a discriminatingly valid measure of social intelligence because it *is* a valid measure of general–verbal intelligence. (See R. L. Thorndike 1968.)

Construct validity is, in the end, a matter of judgment and confidence. The greater the quantity of supporting evidence, the greater the confidence we can place in both the trait construct and the test. But the possibility of a single glaring discrepancy is never wholly ruled out by any quantity of such evidence (see Cronbach & Meehl 1955).

Test norms

Since the raw scores on educational and psychological tests consist of arbitrary units with arbitrary zero points—units which in most cases vary systematically as well as randomly with score level—these individual scores can be interpreted intelligently only by comparing them with the *distributions* of scores of defined groups, or norms. The comparisons are facilitated by using various types of score transformations based on these distributions.

Norms may be local, regional, or national; they may refer to the whole population or to defined subgroups, such as sex and occupation. Local norms are usually determined by testing everyone in a particular group: all children in certain grades of a school system, all freshmen applying to or admitted to a college or university, all employees in specified jobs in a company, etc. Regional and national norms must be determined by sampling. Since random sampling of the whole defined population is never practical, much care is necessary in the design of the sampling procedure and in the statistical analysis of the data to assure the representativeness of the results. The principles and procedures of survey sampling are beyond the scope of this article, but one caution should be noted. Use of a "pick-up" sample, depending upon the vagaries of cooperation in various communities, however widespread over regions, rural and urban communities, etc., followed by weighting based on census data, is never wholly satisfactory, although it is the method commonly employed. When norms are based on a sample which omits some major regions or population subgroups entirely, no weighting system can correct the bias, and such norms must be used with extreme caution in these other regions and with omitted subgroups. [See SAMPLE SURVEYS.]

Grade norms. For elementary and junior high school achievement tests, grade norms are commonly employed. The data used are the mean or median scores of children in successive grades, and the unit is one-tenth of the difference between these averages for successive grades. There is probably some error in assuming one month of educational growth over the summer vacation, but this error in interpreting grade scores is small in comparison with the standard errors of measurement of even the best achievement tests.

Age norms. Age scores are used mainly with general intelligence tests, where they are termed mental ages. The data are the average scores of children within one or two months (plus or minus) of a birthday, or the average scores of all children between two given birthdays, and the unit is one-twelfth of the difference between the averages for successive ages; i.e., one month of mental age. For ages above 12 or 13, extrapolations and corrections of various sorts are made, so that mental ages above 12 or 13 no longer represent age averages but, rather, estimates of what these averages would be if mental growth did not slow down during the period from early adolescence to maturity. Thus, when we say that the mental age of the average adult is 15, we do *not* mean that for most people mental growth ceases at age 15. What we do mean is that the total mental growth from age 12 to maturity is about half the mental growth from age 6 to age 12; so that at age 15 the average person would have reached the mental level he actually reached at maturity (in the early or middle twenties) if mental growth from 12 to 15 had continued at the same average rate as from 6 to 12.

Age norms have been used in the past with elementary and junior high school achievement tests also, defining "arithmetic age," "reading age," "educational age," etc., in like manner; but they and the corresponding "arithmetic quotient," "reading quotient," and "educational quotient" are no longer used.

Quotient scores. The IQ as defined originally, for children up to age 12 or 13, is $100(MA/CA)$, the ratio of mental age to chronological age, multiplied by 100 to rid it of decimals. For older ages the divisor is modified so that the "equivalent chronological age" goes from 12 or 13 to 15 as the actual age goes from 12 or 13 to mental maturity. The *de facto* corrections have usually been crude.

The IQ would have the same standard deviation at all ages if the mental growth curve were a straight line, up to age 12 or 13, passing through true zero at birth, and if the standard deviation of mental ages were a constant fraction of the mean mental age at all chronological ages. These condi-

tions are met very roughly by age scales such as the Stanford–Binet and are not met at all well by any other tests. In consequence, and because of the troubles encountered in the extrapolation of mental age scales and the derivation of equivalent chronological ages beyond age 12 or 13, IQs are no longer computed for most tests by the formula $100(MA/CA)$. They have been replaced quite generally by "deviation IQs," which are standardized scores or normal-standardized scores for age groups; and even the names "intelligence" and "IQ" are tending to be replaced by other names because of their past widespread misinterpretation as measuring innate intellectual capacity. (See R. L. Thorndike 1968; Wright 1968.)

Modal-age grade scores. As we proceed from the first grade to the ninth, the age range within a grade increases, and the distribution of ages becomes skewed, with the longer tail in the direction of increased age. The reason is that retardation is considerably more common than acceleration, and the consequence is that in the upper grades the grade averages on tests are no longer equal to the averages for pupils making normal progress (one grade per year) through school. Modal-age grade scores are used to compare the level of an individual child's performance with the level representing normal school progress. The data are the means or medians, not of all children in a given grade but, rather, of those children whose ages are within six to nine months of the *modal* age (not the median or mean age) for the grade. The units are otherwise the same as those of total-group grade scores, with all the interpretive difficulties noted previously. Modal-age grade scores are recommended for judging the progress of an individual child; total-group grade scores for comparing classes, schools, and school systems.

When total-group grade scores are based on grade medians, they are about one-third closer to the corresponding modal-age grade scores than when they are based on grade means.

Standard scores and standardized scores. When individual scores are to be compared with those of a single distribution, rather than with the means or medians of successive groups (as is the case with grade and age scores), scores based on the mean and standard deviation of the distribution are frequently employed. Originally, standard scores were defined by the formula $Z = (x - \bar{x})/s$, where x is a raw score, $\bar{x}$ is the group mean, and s is the standard deviation. Thus Z-scores have a mean of zero and a standard deviation of unity; the variance is also unity, and the product–moment correlation between two sets of Z-scores is the same as the covariance: $r_{AB} = \sum Z_A Z_B / N$. For raw scores below the mean, Z-scores are negative. Thus the Z-score 1.2 corresponds to the raw score which is 1.2 standard deviations above the mean, and the Z-score −.6 corresponds to the raw score which is six-tenths of a standard deviation below the mean.

Because of the inconveniences in using negative scores and decimals, Z-scores are usually converted via a linear transformation into some other system having an arbitrary mean and an arbitrary standard deviation. These other systems are commonly termed standard-score systems, but the present writer prefers, like Ghiselli (1964), to reserve the term "standard score" for Z-scores and to call the other systems "standardized scores."

The units of a standardized-score system form a quasi-interval scale if and only if the raw scores form such a scale. When the item-difficulty distribution is roughly normal rather than rectangular, the units are smallest near the score corresponding to the modal difficulty, and they become progressively larger with distance in either direction from this level.

Normal-standardized scores. Normal-standardized scores are standard or standardized scores for a normal distribution having the same mean and standard deviation as the raw-score distribution. They are found by looking up in a table of the normal distribution the Z-scores corresponding to the percentile ranks of the raw scores in the actual distribution, and then subjecting these Z-scores to any desired arbitrary linear transformation. This procedure corrects for departure of the *score* distribution from normality, but it does not insure equality of units in any practical sense, even if the distribution of the underlying ability is also normal.

The phrase "normal-standardized scores" is to be preferred to the more common "normalized standard scores." To a mathematician, "standardizing" means reducing to Z-scores, and "normalizing" means producing scores each equal to $Z/\sqrt{N}$, with sum of squares (instead of standard deviation) unity, and has no reference to the normal distribution.

Percentiles and percentile ranks. A percentile is defined as that score (usually fractional) below which lies a given percentage of a sample or population. The median is the 50th percentile: half the group make scores lower than the median and half make scores higher. The lower quartile is the 25th percentile and the upper quartile is the 75th percentile. All score distributions are grouped distributions, even though the grouping interval may be only one score unit. Percentiles are computed by interpolation under the assumption that the abilities represented by the scores within an interval are evenly distributed across that interval.

Percentile ranks are the percentiles corresponding to given scores. Since a single score represents an *interval* on a continuous scale, its percentile rank should be the percentage of individuals who make lower scores plus half the percentage who make the given score. In practice they are frequently computed as simply the percentage who make lower scores, and occasionally as the percentage who make the same or lower scores. Neither of these errors is large in comparison with the error of measurement.

Percentiles and percentile ranks are sometimes given for grade groups, age groups, or normal-age grade groups with elementary and junior high school achievement tests and intelligence tests. They are used more commonly with high school and college tests, with the reference group all students in a given class (grade), or all college applicants, in the case of college entrance tests. For tests in particular subject areas, the reference groups are more commonly all students who have studied the subject for a given number of years in high school or college.

Strict warnings are commonly given against treating percentile ranks as though they form quasi-interval scales; but as noted above, raw scores, standardized scores, and normal-standardized scores may be little, if any, better in this respect when item-difficulty distributions are far from rectangular. It is quite possible, in fact, that for some not uncommon item-difficulty distributions, the percentile ranks may have more nearly the properties of an interval scale than the raw scores have.

Score regions. Centiles and deciles are the regions between adjacent percentiles and sets of ten percentiles. The first centile is the score region below the first percentile. The 100th centile is the region above the 99th percentile. The kth centile is the region between the $(k-1)$th and kth percentiles. The first decile is the region below the tenth percentile, sometimes termed the first decile point. The tenth decile is the region above the 90th percentile or ninth decile point. The kth decile is the region between the $10(k-1)$th and $10k$th percentiles, or the $(k-1)$th and kth decile points.

The term "quartile" is often used also to represent a region. The lower quartile, the median, and the upper quartile are the three quartile points. The first quartile is the region below the lower quartile, the second quartile the region between the lower quartile and the median, the third quartile the region between the median and the upper quartile, and the fourth quartile the region above the upper quartile.

Centiles, deciles, and quartiles are *equal-frequency* score regions. Stanines and stens define equal standard score or normal-standard score regions, with unequal frequencies.

Scaled scores. A few intelligence and achievement tests and test batteries are designed to cover wide ranges of ability, e.g., grades 3–9 inclusive. More commonly, however, they are issued for successive levels, such as primary, elementary, advanced (junior high school), and in some cases secondary (senior high school) and adult. In achievement test batteries, additional subject areas are usually included at the higher levels; and at the primary level, picture tests may replace tests which at other levels require reading. The successive levels usually have similar materials differing in average difficulty, but with the harder items at one level overlapping in difficulty the easier items at the next higher level.

With wide-range tests and tests issued at several levels, grade scores, age scores, modal-age grade scores, and grade, age, or modal-age grade percentile ranks or standardized scores may represent quite unequal units at different levels. Scaled score systems are designed to have units which are equal in some sense, at least on the average, from level to level throughout the range. They are based on assumptions about the shape of the underlying ability distribution within a grade or age group, and are derived by considering both the score distributions at successive grades or ages and the mean or median gains from grade to grade or age to age. None of these methods are wholly satisfactory.

Further problems arise when attempts are made to scale tests of different abilities in comparable units, since the relations between mean gains and within-grade or within-age variability are quite different for different functions. Thus, mean annual gain in reading is a much smaller fraction of within-group standard deviation than is mean annual gain in arithmetic; or, stated in terms of growth units, variability in reading is much greater than in arithmetic.

Equating. Before the scores on two tests, or even the scores on two forms of the same test, can be compared, the relations between their score scales must be established. The preferred experimental design is to give both tests to the same group of examinees: to half the group in the A–B order and to the other half in the B–A order.

The simplest method of establishing comparable scores is termed line-of-relation equating. Scores on test A and test B are considered comparable if they correspond to the same standard score. This method is satisfactory only if the item-difficulty distributions are of the same shape and are equally variable, which is seldom the case. The preferred method is termed equipercentile equating. Scores are considered comparable if they correspond to the same percentile. Selected percentiles are computed for each distribution, such as percentiles 1, 2, 3, 5, 10, 15, 20, 30, 40, 50, 60, 70, 80, 85, 90, 95, 97, 98, and 99. A two-way chart is prepared, with the scores on one form as ordinates and the scores on the other form as abscissas. For each of the selected percentiles a point is plotted representing the scores on the two tests corresponding to this percentile, and a smooth curve is drawn as nearly as possible through all these points. If, but only if, this curve turns out to be a straight line, will line-of-relation equating have been satisfactory.

If the two distributions are first smoothed, the equipercentile points are more likely to lie on a smooth curve, and the accuracy of the equating is improved.

This method of equating is satisfactory if the two tests are equally consistent. If they are not, the scores on each test should all be multiplied by the consistency coefficient of that test, and the resulting "estimated true scores" should be equated.

When new forms of a test are issued annually, a full norms study is usually conducted only once every five or ten years, and the norms for successive forms are prepared by equating them to the "anchor form" used in the last norms study.

Standards. In a few cases, standards of test performance can be established without reference to the performances of members of defined groups. Thus, in typing 120 words per minute with not more than one error per 100 words is a fairly high standard.

"Quality scales," in areas such as handwriting and English composition, are sets of specimens at equal intervals of excellence. A standard of handwriting legibility can be set by measuring the speed with which good readers read the various specimens. The standard would be the poorest specimen which is not read significantly slower than the best specimen. Units above this standard would then represent mainly increases in beauty; units below the standard, decreases in legibility. In English composition, the poorest specimen written in substantially correct grammar could be identified by a consensus of English teachers. Then units above the standard would represent mainly improvements in style; units below the standard, decreases in grammatical correctness.

Research is in progress to determine standards for multiple-choice tests of subject-matter achievement. When the items of such a test are arranged in order of actual difficulty, experienced teachers expert in the subject might be able to agree on the item which barely passing students should get right half the time. Given this item and the item analysis data, the passing score for the test can be determined fairly readily.

Expectancy tables. When the regression of a test or battery on a criterion has been determined from a representative sample of the same population, norms for the population can be expressed in terms of expected criterion scores. The predictor scores are usually expressed in fairly broad units. Then, if the criterion is dichotomous, the expectancy table gives for each score level the probability that a person at that score level will be in the upper criterion group. If the criterion is continuous, the criterion scores are grouped into quartiles, deciles, stanines, stens, or grade levels; and for each predictor score level the table gives probabilities for the several criterion levels. Thus if the criterion is a grade in a course, the expectancy table will show for each predictor score level the probabilities that the grade will be A, B, C, D, or F.

EDWARD E. CURETON

[*Directly related is the entry* MATHEMATICS. *Other relevant material may be found in* FACTOR ANALYSIS AND PRINCIPAL COMPONENTS; MULTIVARIATE ANALYSIS, *article on* CORRELATION METHODS; NONPARAMETRIC STATISTICS; STATISTICS, DESCRIPTIVE; *and in the biographies of* BIRNBAUM, PEARSON, *and* SPEARMAN. *See also* Adkins 1968; Beck 1968; Block 1968; Clark 1968; Dahlstrom 1968; Ebel 1968; Fleishman 1968; Henry 1968; Holtzman 1968; Joncich 1968; Lindzey & Thorpe 1968; Messick 1968; Pichot 1968; Rokeach 1968; Santostefano 1968; Smith 1968; R. L. Thorndike 1968; Torgerson 1968.]

BIBLIOGRAPHY

The standard work in psychometrics is Guilford 1936. It includes psychophysics and scaling, as well as the topics covered in this article. A somewhat more elementary treatment is given in Ghiselli 1964, and a somewhat more advanced treatment in Gulliksen 1950. Another general work is Stevens 1951, which treats with some care the foundations of measurement, a topic not covered by Guilford, Ghiselli, or Gulliksen.

►ADKINS, DOROTHY C. 1968 Thurstone, L. L. Volume 16, pages 22–25 in *International Encyclopedia of the Social Sciences*. Edited by David L. Sills. New York: Macmillan and Free Press.

►BECK, SAMUEL J. 1968 Projective Methods: II. The Rorschach Test. Volume 12, pages 568–573 in *International Encyclopedia of the Social Sciences*. Edited by David L. Sills. New York: Macmillan and Free Press.

►BLOCK, JACK 1968 Personality Measurement: I. Overview. Volume 12, pages 30–37 in *International Encyclopedia of the Social Sciences*. Edited by David L. Sills. New York: Macmillan and Free Press.

BROGDEN, HUBERT E. 1949 A New Coefficient: Application to Biserial Correlation and to Estimation of Selective Efficiency. *Psychometrika* 14:169–182.

►CLARK, KENNETH E. 1968 Vocational Interest Testing. Volume 16, pages 345–350 in *International Encyclopedia of the Social Sciences*. Edited by David L. Sills. New York: Macmillan and Free Press.

CRONBACH, LEE J. 1951 Coefficient Alpha and the Internal Structure of Tests. *Psychometrika* 16:297–334.

CRONBACH, LEE J.; and MEEHL, P. E. (1955) 1956 Construct Validity in Psychological Tests. Pages 174–204 in Herbert Feigl and Michael Scriven (editors), *The Foundations of Science and the Concepts of Psychology and Psychoanalysis*. Minneapolis: Univ. of Minnesota Press. → First published in Volume 52 of the *Psychological Bulletin*.

►DAHLSTROM, W. GRANT 1968 Personality Measurement: III. The Minnesota Multiphasic Personality Inventory. Volume 12, pages 43–48 in *International Encyclopedia of the Social Sciences*. Edited by David L. Sills. New York: Macmillan and Free Press.

►EBEL, ROBERT L. 1968 Achievement Testing. Volume 1, pages 33–39 in *International Encyclopedia of the Social Sciences*. Edited by David L. Sills. New York: Macmillan and Free Press.

FAN, CHUNG-TEH 1952 *Item Analysis Table*. Princeton, N.J.: Educational Testing Service.

FINDLEY, WARREN G. 1956 A Rationale for Evaluation of Item Discrimination Statistics. *Educational and Psychological Measurement* 16:175–180.

FLANAGAN, JOHN C. 1939 General Considerations in the Selection of Test Items and a Short Method of Estimating the Product–Moment Coefficient From the Data at the Tails of the Distribution. *Journal of Educational Psychology* 30:674–680.

►FLEISHMAN, EDWIN A. 1968 Aptitude Testing. Volume 1, pages 369–364 in *International Encyclopedia of the Social Sciences*. Edited by David L. Sills. New York: Macmillan and Free Press.

GHISELLI, EDWIN E. 1964 *Theory of Psychological Measurement*. New York: McGraw-Hill.

GUILFORD, JOY P. (1936) 1954 *Psychometric Methods*. 2d ed. New York: McGraw-Hill.

GULLIKSEN, HAROLD 1950 *Theory of Mental Tests*. New York: Wiley.

►HENRY, WILLIAM E. 1968 Projective Methods: III. The Thematic Apperception Test. Volume 12, pages 573–579 in *International Encyclopedia of the Social Sciences*. Edited by David L. Sills. New York: Macmillan and Free Press.

►HOLTZMAN, WAYNE H. 1968 Personality Measurement: II. Personality Inventories. Volume 12, pages 37–43 in *International Encyclopedia of the Social Sciences*. Edited by David L. Sills. New York: Macmillan and Free Press.

HORST, PAUL 1936 Item Selection by Means of a Maximizing Function. *Psychometrika* 1:229–244.

►JONCICH, GERALDINE 1968 Thorndike, Edward L. Volume 16, pages 8–14 in *International Encyclopedia of the Social Sciences*. Edited by David L. Sills. New York: Macmillan and Free Press.

KUDER, G. F.; and RICHARDSON, M. W. 1937 The Theory of the Estimation of Test Reliability. *Psychometrika* 2:151–160.

►LINDZEY, GARDNER; and THORPE, JOSEPH S. 1968 Projective Methods: I. Projective Techniques. Volume 12, pages 561–568 in *International Encyclopedia of the Social Sciences.* Edited by David L. Sills. New York: Macmillan and Free Press.

LORD, FREDERIC M. 1964 An Empirical Comparison of the Validity of Certain Formula-scores. *Journal of Educational Measurement* 1:29–30.

LYERLY, SAMUEL B. 1951 A Note on Correcting for Chance Success in Objective Tests. *Psychometrika* 16:21–30.

►MESSICK, SAMUEL 1968 Response Sets. Volume 13, pages 492–496 in *International Encyclopedia of the Social Sciences.* Edited by David L. Sills. New York: Macmillan and Free Press.

►PICHOT, PIERRE 1968 Binet, Alfred. Volume 2, pages 74–78 in *International Encyclopedia of the Social Sciences.* Edited by David L. Sills. New York: Macmillan and Free Press.

►ROKEACH, MILTON 1968 Attitudes: I. The Nature of Attitudes. Volume 1, pages 449–458 in *International Encyclopedia of the Social Sciences.* Edited by David L. Sills. New York: Macmillan and Free Press.

►SANTOSTEFANO, SEBASTIANO 1968 Personality Measurement: IV. Situational Tests. Volume 12, pages 48–55 in *International Encyclopedia of the Social Sciences.* Edited by David L. Sills. New York: Macmillan and Free Press.

►SMITH, M. BREWSTER 1968 Attitudes: II. Attitude Change. Volume 1, pages 458–467 in *International Encyclopedia of the Social Sciences.* Edited by David L. Sills. New York: Macmillan and Free Press.

STEVENS, S. S. (1951) 1958 Mathematics, Measurement, and Psychophysics. Pages 1–49 in S. S. Stevens (editor), *Handbook of Experimental Psychology.* New York: Wiley. → See especially pages 13–15 and 21–30.

Taxonomy of Educational Objectives. Edited by Benjamin S. Bloom. 2 vols. 1956–1964 New York: McKay. → Handbook 1: *The Cognitive Domain,* by B. S. Bloom and D. R. Kratwohl, 1956. Handbook 2: *The Affective Domain,* by D. R. Kratwohl, B. S. Bloom, and B. B. Masia, 1964.

THORNDIKE, EDWARD L. et al. 1926 *The Measurement of Intelligence.* New York: Columbia Univ., Teachers College.

THORNDIKE, ROBERT L. 1951 Reliability. Pages 560–620 in E. F. Lindquist (editor), *Educational Measurement.* Washington: American Council on Education.

►THORNDIKE, ROBERT L. 1968 Intelligence and Intelligence Testing. Volume 7, pages 421–429 in *International Encyclopedia of the Social Sciences.* Edited by David L. Sills. New York: Macmillan and Free Press.

►TORGERSON, WARREN S. 1968 Scaling. Volume 14, pages 25–39 in *International Encyclopedia of the Social Sciences.* Edited by David L. Sills. New York: Macmillan and Free Press.

WESMAN, ALEXANDER G. 1949 Effect of Speed on Item–Test Correlation Coefficients. *Educational and Psychological Measurement* 9:51–57.

►WRIGHT, JOHN C. 1968 Intellectual Development. Volume 7, pages 387–399 in *International Encyclopedia of the Social Sciences.* Edited by David L. Sills. New York: Macmillan and Free Press.

Postscript

[1]The correction for guessing can be derived without the all-or-none assumption by using instead a *linear assumption:* the amount of knowledge an examinee has of an item is related linearly to the probability that he will mark the right answer if he attempts the item. The amount of knowledge, k, is 0 if the probability is $1/a$ (for a alternatives), and 1 if the probability, p, is 1. Negative values of k (where $p < 1/a$) imply misinformation. This has the esthetic advantage that p can take any value from 0 to 1, rather than only the two values $1/a$ and 1. Since the same formula for the correction for guessing follows, however, there is no practical change in using the linear assumption (Cureton 1966). The lowest possible value for k is $-1/(a-1)$, and thus k can reach -1 (when $p = 0$) only for the two-choice or true–false item, where $a = 2$.

There is today a considerable and growing literature on *criterion-referenced testing:* the construction of tests designed to be interpreted in terms of *standards* rather than norms.

EDWARD E. CURETON

ADDITIONAL BIBLIOGRAPHY

CURETON, EDWARD E. 1966 The Correction for Guessing. *Journal of Experimental Education* 34:44–47.

KRISTOF, WALTER 1974 Estimation of Reliability and True Score Variance From a Split of a Test Into Three Arbitrary Parts. *Psychometrika* 39:491–499.

LORD, FREDERIC M.; and NOVICK, MELVIN R. 1968 *Statistical Theories of Mental Test Scores.* Reading, Mass.: Addison-Wesley.

THORNDIKE, ROBERT L. (editor) 1971 *Educational Measurement.* 2d ed. Washington: American Council on Education. → The first edition (1951) was edited by E. F. Lindquist.

PSYCHOPHYSICS

Psychophysics is the study of physical stimuli and their relation to sensory reactions. Of the energy that strikes the sensory surfaces of man and the animals, only a restricted fraction is capable of eliciting a reaction. Thus, visual responses in man are triggered by a narrow band of the vast electromagnetic spectrum (wavelengths between about 400 and 750 millimicrons); and auditory responses result from periodic displacements of the eardrum in the frequency range from about 20 to 20,000 cycles per second (cps). Over these stimulus ranges, neither the eye nor the ear is uniformly responsive: to produce a sensation may require thousands of times more energy at one wavelength or frequency than at another. It is the goal of psychophysics to map out the relations between the physical events and the psychological responses of organisms, and thus to provide a basic, over-all description of the function of the senses.

Major problems. The traditional questions posed by psychophysics fall into four groups. For a given sense modality we may ask about (1) the smallest detectable energy (the measurement of *sensitivity*); (2) the smallest detectable change in energy (the measurement of *resolving power*); (3) the configurations of energy that produce an invariant sensory effect, such as a constant loudness or color (the measurement of *static invariances*); (4) the way in which the magnitude of a sensory effect depends functionally on the stimulus (the measurement of *dynamic properties*).

Sensitivity

The problem of sensitivity involves the determination of the smallest detectable intensity of a stimulus (called the *absolute threshold*), often as a function of another stimulus dimension, such as wavelength, frequency, duration, or areal extent. The threshold of audibility, for example, depends on the frequency of the tone; sensitivity is greatest to frequencies between about 2,000 and 3,000 cps. In the vicinity of 20 and 20,000 cps (which are conveniently but arbitrarily called the "limits" of hearing), the threshold energy may rise to roughly 10^8 times the minimal value. A similar relation exists between the threshold of visibility and the wavelength of the stimulating light, except that the shape of the visibility curve depends on what part of the sensory surface is stimulated; foveal (cone) vision has maximum sensitivity at about 555 millimicrons; peripheral (rod) vision, at about 505 millimicrons; and the "limits" of visibility are also different for the two populations of receptors.

Variation in sensitivity. The absolute threshold is not a rigidly fixed value. Sensitivity fluctuates irregularly, so that a given stimulus level may trigger a response at one time but not at another. The threshold is usually defined statistically, e.g., as the energy level that is detected as often as not over a series of presentations.

Sensitivity is also subject to systematic variation, either of the permanent kind encountered in aging or in pathology of the sensory tissues, or of a temporary kind observed, for example, in the relatively rapid decline of visual sensitivity under exposure to light (light adaptation) and the subsequent gradual recovery of sensitivity in the dark (dark adaptation). These and many other systematic changes in sensitivity are frequently expressed as alterations of the absolute threshold.

The study of absolute thresholds reveals the exquisite sensitivity of the sense organs under optimal conditions. A periodic displacement of the eardrum through a distance equal to the diameter of a hydrogen molecule may suffice to produce an audible sound, and a couple of quanta of light absorbed at the retina may suffice to arouse a faint visual sensation.

Methods of measurement. Because it fluctuates, the threshold is difficult to measure, and the various methods that have been tried do not always yield the same value. The method of *adjustment* provides a rapid approximation; the observer is required to set the level of the stimulus so that it is just perceptible. The threshold may be defined as the average of several settings. In the method of limits, either the stimuli are presented in order of increasing magnitude until the observer reverses his response from "imperceptible" to "perceptible," or they are presented in order of decreasing magnitude until the observer reverses his response from "perceptible" to "imperceptible." The threshold may be defined as the average value that marks the reversal in response over several ascending and descending series. In the method of *constant stimuli*, fixed stimulus levels are presented several times, each in irregular order. The threshold may be defined as the stimulus value that is perceived on half the presentations. This value is interpolated from a plot relating the percentage of positive responses to the stimulus magnitude.

The methods of adjustment, limits, and constant stimuli are known as the *classical psychophysical methods* because they have continued in widespread use ever since G. T. Fechner described them in his *Elemente der Psychophysik* (1860), the monumental work that marks the establishment of psychophysics. (Reviews of the classical methods are given by Urban 1908; Titchener 1905; and Boring 1942.)

One of the difficulties inherent in these methods is the observer's awareness that a stimulus event actually takes place on each trial. When a "catch trial" (a feigned presentation of a stimulus) is given, observers will occasionally give an affirmative response (a "false alarm"). The knowledge that the observer's expectations and motivations come into play has stimulated the invention of new methods that offer the hope of better understanding and controlling the observer's response biases. An example is the *forced choice* method, in which at regular intervals a stimulus is presented or withheld and the observer must decide each time whether or not he detected it. Results obtained under this procedure reveal that detection may depend not only on the magnitude of the stimulus but also on the prearranged probability of a stimulus event. A high proportion of "no-stimulus" trials causes a relatively high incidence of "false alarms"; a low proportion of "no-stimulus" trials, on the other hand, causes a lower incidence of correct detections of actual stimulus events. The probability of a "Yes" or "No" response can also be systematically influenced by rewarding correct detections and punishing the false alarms.

The forced-choice experiments have done much to underscore and clarify the role of response variables in the measurement of thresholds. It is sometimes suggested that the "detection" model may actually do away with the conception of the threshold as a simple, determinable value marking the critical terminus of sensory experience. According to this view, the detection of a stimulus (the "signal") has much in common with the mathematical process of statistical decision. The observer is confronted with two distributions: that of the persistent background noise and that of the signal added to the noise. He decides from which of the two distributions a sample is taken in much the way that a statistician tests a statistical hypothesis. The decision will depend on the overlap of the distributions and also on the "pay-off matrix"—the consequences of false detections and failures of detection (see Swets 1964; Luce et al. 1963).

Interesting technological advances have recently been made in the field of threshold measurement. The Békésy audiometer, for example, uses the method of *tracking* for the efficient measurement of the just-audible intensity as a function of tonal frequency (Von Békésy 1928–1958). The observer "tracks" his threshold by pressing a key whenever the tone is audible and releasing it whenever the tone becomes inaudible. While the key is pressed, the level of the tone steadily decreases; while the key is not pressed, the level steadily increases. The observer may continue to track the threshold while the tonal frequency changes from one end of the audible spectrum to the other. On a moving paper chart, the stimulus level, which weaves back and forth across the threshold, is recorded continuously as a function of the frequency.

The tracking method has also been used to determine the visual thresholds of human observers and has been adapted to mapping the sensitivity functions of animals.

Resolving power

The second major concern of psychophysics is to measure the smallest detectable change in a stimulus (the so-called *difference threshold*). The problem may be to measure the just-noticeable differences in *intensity*, e.g., in the brightness of a light or in the concentration of a sweet solution, or in *quality*, e.g., in the hue of a colored light.

The capacity for resolving stimulus differences is expressed in terms of the Weber fraction, $\Delta I/I$, where ΔI stands for the increment that produces a just-noticeable change when added to the stimulus level I. The smaller the value of ΔI, the keener is the ability to discriminate.

Methods of measurement. Like the absolute threshold, the difference threshold is a fluctuating quantity, so that ΔI must be assessed by a statistical treatment of a series of measurements. Most of the methods used are versions of those used to measure absolute thresholds. In the method of *adjustment*, for example, the observer sets a comparison stimulus to match a standard fixed stimulus. The threshold, ΔI, may be defined as the average error or the standard deviation of several settings. The greater the variability of the settings, the grosser the discrimination and the larger the Weber fraction. A difference threshold may be regarded either as a measure of the precision or as a measure of the variability or "noisiness" of the sensory process.

In the method of *constant stimuli*, the observer judges whether each of a set of fixed discrete stimulus levels appears greater or smaller than a standard stimulus (a judgment of "equal" is also permitted by some experimenters). The difference threshold may be defined as the difference between a standard stimulus and a comparison stimulus that is perceived as being greater (or smaller) than the standard stimulus on a certain percentage of the trials. This value can be interpolated from a *poikilitic* (scatter) *function*.

In Figure 1 the ordinate represents the relative frequency with which the comparison stimulus is judged greater than the standard. The threshold, ΔI, is the difference between the stimulus magnitude that is perceived as being greater than the standard on 75 per cent of the trials (L) and the stimulus magnitude that is so perceived on 50 per cent of the trials (E), i.e., the stimulus that appears to match the standard stimulus. Often there is a small difference, called the time error, between the standard stimulus (S) and the stimulus value associated with the 50 per cent point. Urban (1908) provides a detailed discussion of poikilitic functions.

The nature of the difference threshold has often been studied by the method of *quantal increments*. From time to time, a small increment, ΔI, is added briefly to a steady stimulus, I. The task is to indicate whether the increment was detected. Of theo-

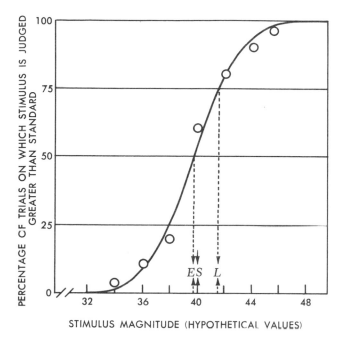

Figure 1 — A poikilitic (scatter) function determined by the method of constant stimuli

retical concern is the mathematical form of the poikilitic function that relates the proportion of detections to the size of the increment. If the precision were ultimately limited by nothing but the random "noisiness" of the sensory process, then the function would be expected to have the sigmoid shape (the integral of a bell-shaped distribution) that is predicted by the theory of random error. The usual result approximates this form. When pains are taken to aid the attention and to eliminate the extraneous sources of error, the obtained function may assume a linear rather than a sigmoid form and conform to a predictable slope. According to the *neural-quantum hypothesis*, a linear function of the appropriate slope demonstrates that discrimination is basically all-or-none and that sensation grows by the addition of minute but finite (or quantal) steps (see Von Békésy 1928–1958; S. S. Stevens 1961; Luce et al. 1963; Swets 1964).

Weber's law. A major aim in threshold measurement has been to test the famous generalization, credited to Ernst H. Weber but formalized and promoted by Fechner, that the Weber fraction is constant along a given sensory continuum, or that $\Delta I/I = k$. In other words, a just-noticeable change should occur when a constant fractional increment is added to a stimulus of any magnitude. Under good conditions the increment is about 1 per cent for brightness, 2 per cent for loudness, and 20 per cent for saltiness. The sensory systems

differ greatly in their resolving power, but for any one system it is the percentage change that matters most.

On most continua Weber's law holds over a substantial portion of the stimulus range (for loudness and brightness over at least 99.9 per cent of the range), but the law fails near the absolute threshold, where discrimination is relatively gross. Nevertheless, Weber's law stands as one of the oldest and broadest empirical generalizations of psychophysics—psychology's "law of relativity," as one writer put it (S. S. Stevens 1951).

Fechner's law. If discrimination has seemed to receive more than reasonable attention among students of the senses, the explanation is likely to be found in the significance that the founder of psychophysics attached to the subject. For Fechner, discrimination provided the key to the measurement of sensory magnitude. He began with the postulate that on a given sensory continuum all just-noticeable differences (*jnd*) represent subjectively equal units. Subjective equality of *jnd*s is a powerful (if questionable) assumption, because the integration of such units would provide a true scale of subjective magnitude. Since by Weber's law a *jnd* corresponds to a constant fractional increase in the stimulus, it follows that the number of *jnd*s grows in an arithmetic series when the stimulus intensity grows in a geometric series. Fechner concluded, therefore, that the magnitude of sensation is a logarithmic function of the stimulus. The logarithmic function implies that equal *ratios* of stimulus magnitude give rise to equal *differences* in subjective magnitude.

Plateau's power function. In contrast to the indirectness of the Fechnerian approach to the measurement of sensory magnitudes was an early experiment by the Belgian physicist Joseph A. F. Plateau, who asked a group of artists each to paint a gray that seemed to lie midway between a white sample and a gray sample (a version of a scaling method later termed *equisection*). Of historical interest is Plateau's conclusion that sensation grows as a power function rather than a logarithmic function of the stimulus. But the power function was subsequently given up by Plateau and virtually forgotten until the 1950s. With new techniques for the direct assessment of sensory magnitude, it was shown that Plateau's early conjecture about the form of the psychophysical function happened to be correct. The current approach to the problem is generally to regard as separate properties the resolving capacity of the sensory system and the functional dependence of sensory magnitude on the stimulus.

The static invariances of sensory systems

A third major problem of psychophysics is to determine those arrangements of stimuli that produce responses that are equivalent in some respect. The goal of this kind of measurement is to specify all the energy configurations in the environment that produce an invariant or equivalent sensory response.

For example, the goal may be to determine the combinations of intensity and duration of a flash target that produce the same apparent brightness. The level or the duration of a comparison flash is adjusted to match the brightness of a standard flash of fixed intensity and duration. The judgment requires a degree of abstraction because the task is to match for brightness without regard to a difference in apparent duration. A plot relating the duration and intensity that produce a constant (standard) brightness provides an example of an *equal sensation function*. Usually it is desirable to map the family of these equal sensation functions for a pair of parameters. In the present example this means that a function is obtained for each of a set of representative standard brightnesses along the brightness continuum. We learn from this family that, up to a critical duration (roughly 150 milliseconds), a decrease in the stimulus level can be offset by lengthening the flash. Moreover, the critical duration gets systematically shorter as the brightness is increased.

Measurement of the static invariances is common in psychophysics. Examples include the equal brightness functions relating energy and wavelength, the equal loudness functions relating sound pressure and tonal frequency, the equal pitch functions relating frequency and sound pressure (within limits, the apparent pitch of a tone can be altered by a change in sound pressure level), and the equal hue functions relating wavelength and light intensity. (The change in hue when intensity is altered has long been known as the Bezold–Brücke phenomenon.)

The measurement of invariance may call for complete equivalence. An example is the concept of metamerism in color vision. The measurement of metameric pairs (sample lights of identical appearance but different wavelength compositions) has made it possible to state the laws of color mixtures and to predict the color of a sample of any spectral composition.

Methods of measurement. Because of its speed and immediacy, the method of adjustment usually recommends itself for the mapping of equivalents. Other usable procedures, however, include constant stimuli, limits, and tracking. The measurement of equivalence is straightforward in principle, but any procedure is usually beset by constant errors, such as the "time error" (see Figure 1).

Dynamic properties of sensory systems

It is one thing to know the stimulus conditions that produce an invariant sensory effect and another thing to know how much larger one sensory effect is than another—e.g., how much brighter one luminance level appears than another or how much two tones seem to differ in pitch. A major problem of psychophysics is to learn how much the magnitude of the sensory response grows when the stimulus intensity increases.

The direct scaling methods. In the 1930s it became apparent to students of hearing that the logarithmic function fails to agree with the reports of observers who are asked to judge the relative loudness of stimuli. Attempts were made to measure the loudness function by a variety of direct methods. Subsequently, the direct methods were expanded and refined and finally applied to the study of all the major sensory continua.

The main feature of the direct methods is the attempt to match segments of the number continuum directly to segments of the sensory continuum. In the method of *magnitude estimation*, various fixed levels of the stimulus are presented one by one in irregular order, and the observer attempts to assign numbers to these levels in proportion to their subjective magnitude. The inverse of this procedure is *magnitude production*: a set of numbers is called out one by one to the observer, who adjusts the level of the stimulus so as to produce subjective magnitudes that are proportional to the numbers. In *ratio production*, a comparison stimulus is adjusted to appear in some fractional or multiplicative relation to a standard stimulus, and in *ratio estimation*, the observer estimates numerically the apparent ratio that corresponds to a pair of stimulus magnitudes. Variations on these procedures are numerous (S. S. Stevens 1958).

Two classes of continua. For a few continua, of which pitch is a noteworthy example, a scale of integrated *jnd*s turns out to agree well with direct judgment. S. S. Stevens (1957) called these continua *metathetic* and distinguished them from the large class of *prothetic* continua on all of which the *jnd* does not afford a constant unit of subjective magnitude. (Table 1 provides a partial list of prothetic continua.)

The psychophysical power functions. S. S. Stevens has also proposed a general psychophysical relation pertaining to all prothetic continua (1957). Equal stimulus ratios are held to correspond to

Table 1 — Representative exponents of the power functions relating psychological magnitude to stimulus magnitude on prothetic continua

Continuum	Exponent	Stimulus condition
Loudness	0.6	binaural
Loudness	0.54	monaural
Brightness	0.33	5° target—dark-adapted eye
Brightness	0.5	point source—dark-adapted eye
Lightness	1.2	reflectance of gray papers
Smell	0.55	coffee odor
Smell	0.6	heptane
Taste	0.8	saccharine
Taste	1.3	sucrose
Taste	1.3	salt
Temperature	1.0	cold—on arm
Temperature	1.5	warmth—on arm
Vibration	0.95	60 cps—on finger
Vibration	0.6	250 cps—on finger
Duration	1.1	white-noise stimulus
Repetition rate	1.0	light, sound, touch, and shocks
Finger span	1.3	thickness of wood blocks
Pressure on palm	1.1	static force on skin
Heaviness	1.45	lifted weights
Force of handgrip	1.7	precision hand dynamometer
Vocal effort	1.1	sound pressure of vocalization
Electric shock	3.5	60 cps—through fingers
Tactual roughness	1.5	felt diameter of emery grits
Tactual hardness	0.8	rubber squeezed between fingers
Viscosity	0.5	stirring silicone fluids
Visual velocity	1.2	moving spot of light
Visual length	1.0	projected line of light
Visual area	0.7	projected square of light

Source: Adapted from S. S. Stevens 1957.

equal sensation *ratios* (rather than to equal sensation *differences*, as Fechner had conjectured). In other words, the apparent magnitude ψ grows as a power function of the stimulus magnitude, or $\psi = k\phi^\beta$, where k is a constant of proportionality and β is the exponent. The size of β varies from one continuum to another. In Figure 2A are plotted the power functions in linear coordinates for three continua: apparent length ($\beta = 1$), brightness ($\beta = 0.33$), and the apparent intensity of an electric current passed through the fingers ($\beta = 3.5$). When plotted in log–log coordinates, as in Figure 2B, these same functions become straight lines whose slopes equal the values of the exponents. This is true because the logarithmic form of the power function is $\log \psi = \log k + \beta \log \phi$.

Table 1 shows that the size of the exponent may depend not only on the sense organ stimulated but also on the conditions of the stimulation. Note the difference between the monaural and the binaural loudness functions and the exponent's dependence on frequency for vibration magnitude.

Although the simple equation $\psi = k\phi^\beta$ holds for large stimulus values, in the neighborhood of the absolute threshold a more precise form is needed. The power equation can be written as $\psi = k(\phi - \phi_0)^\beta$,

where ϕ_0 approximates the absolute threshold. The correction for threshold brings into coincidence the zero of the stimulus scale and the zero of the sensation scale (Luce et al. 1963).

'The properties of sensory systems may reveal themselves as parametric shifts in the values of ϕ, k, and β. The changes in visual sensitivity that occur under light adaptation provide an example (Stevens & Stevens 1963). Light adaptation causes (1) an elevation in the absolute threshold (i.e., ϕ_0 increases), (2) an increase in the luminance necessary to produce a given subjective brightness (i.e., k decreases), and (3) a slight increase in the exponent β. The mapping of these parametric changes has made it possible to write the power

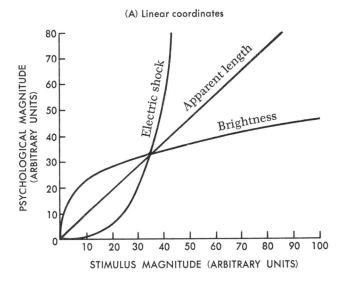

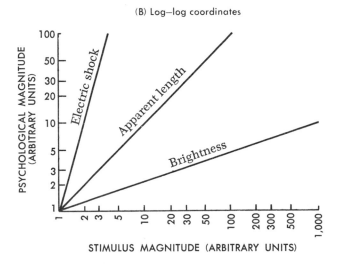

Figure 2 — The power function for three prothetic continua plotted in (A) linear coordinates and (B) log–log coordinates

Source: S. S. Stevens 1957.

function that pertains to any given level of adaptation and consequently to predict the subjective brightness produced by any luminance level when viewed by an eye adapted to any other luminance level.

Cross-modality validations. A method has been devised that circumvents the need for the observer to make numerical estimates of his sensation but leads to the same psychophysical power function (see S. S. Stevens 1961; and Luce et al. 1963). In *cross-modality matching* the task is to make the sensations in two different sense modalities appear equal in strength. The pairs of physical intensities that produce equal apparent intensities can be plotted as an equal sensation function. It turns out that the equal sensation function relating any two prothetic continua a and b is itself a power function of the form $\phi_a = k\phi_b^\gamma$, where ϕ_a and ϕ_b stand for physical intensity. The size of the exponent γ depends on which two continua are matched. Within the experimental error, γ is predictable from the psychophysical function governing the two continua. Given that $\psi_a = \phi_a^\alpha$ and $\psi_b = \phi_b^\beta$ (with suitable units of measurement), and given that $\psi_a = \psi_b$, the equation of the equal sensation function becomes $\phi_a = \phi_b^{\beta/\alpha}$. The exponent γ thus turns out to be the ratio of the exponents α and β.

Any continuum could be substituted for the number continuum and used as a "yardstick" to measure sensory magnitudes on all of the other sensory continua. In one set of experiments, for example, force of handgrip as registered on a dynamometer was used to assess subjective magnitudes on nine other prothetic continua (S. S. Stevens 1961). Many other examples could be cited to show that the psychophysical power law is able to predict both the form and the exponent of the equal sensation function obtained by cross-modality matching.

JOSEPH C. STEVENS

[*Statistical techniques applicable to psychophysical methods are described in* QUANTAL RESPONSE. *See also* Beck 1968; Boring 1968; Chapanis 1968; Dallenbach 1968; Granit 1968; Harris 1968; Jerison 1968; MacLeod 1968; Melzack 1968; Metzger 1968; Mote 1968; Mueller 1968; Pollack 1968; Ruch 1968; Small 1968; Torgerson 1968; Wenzel 1968.]

BIBLIOGRAPHY

►BECK, JACOB 1968 Helmholtz, Hermann von. Volume 6, pages 345–350 in *International Encyclopedia of the Social Sciences*. Edited by David L. Sills. New York: Macmillan and Free Press.

BORING, EDWIN G. 1942 *Sensation and Perception in the History of Experimental Psychology*. New York: Appleton.

►BORING, EDWIN G. 1968 Wundt, Wilhelm. Volume 16, pages 581–586 in *International Encyclopedia of the Social Sciences*. Edited by David L. Sills. New York: Macmillan and Free Press.

►CHAPANIS, A. 1968 Vision: III. Color Vision and Color Blindness. Volume 16, pages 329–336 in *International Encyclopedia of the Social Sciences*. Edited by David L. Sills. New York: Macmillan and Free Press.

►DALLENBACH, KARL M. 1968 Titchener, Edward B. Volume 16, pages 88–90 in *International Encyclopedia of the Social Sciences*. Edited by David L. Sills. New York: Macmillan and Free Press.

FECHNER, GUSTAV T. (1860) 1907 *Elemente der Psychophysik*. 3d ed. 2 vols. Leipzig: Breitkopf & Härtel.

►GRANIT, RAGNAR 1968 Senses: II. Central Mechanisms. Volume 14, pages 177–182 in *International Encyclopedia of the Social Sciences*. Edited by David L. Sills. New York: Macmillan and Free Press.

►HARRIS, ALBERT J. 1968 Vision: IV. Visual Defects. Volume 16, pages 336–340 in *International Encyclopedia of the Social Sciences*. Edited by David L. Sills. New York: Macmillan and Free Press.

►JERISON, HARRY J. 1968 Attention. Volume 1, pages 444–449 in *International Encyclopedia of the Social Sciences*. Edited by David L. Sills. New York: Macmillan and Free Press.

LUCE, R. DUNCAN; BUSH, ROBERT R.; and GALANTER, EUGENE (editors) 1963 *Handbook of Mathematical Psychology*. Volume 1. New York: Wiley.

►MACLEOD, ROBERT B. 1968 Weber, Ernst Heinrich. Volume 16, page 493 in *International Encyclopedia of the Social Sciences*. Edited by David L. Sills. New York: Macmillan and Free Press.

►MELZACK, RONALD 1968 Pain. Volume 11, pages 357–364 in *International Encyclopedia of the Social Sciences*. Edited by David L. Sills. New York: Macmillan and Free Press.

►METZGER, WOLFGANG 1968 Fechner, Gustav Theodor. Volume 5, pages 350–353 in *International Encyclopedia of the Social Sciences*. Edited by David L. Sills. New York: Macmillan and Free Press.

►MOTE, F. A. 1968 Senses: I. Overview. Volume 14, pages 172–177 in *International Encyclopedia of the Social Sciences*. Edited by David L. Sills. New York: Macmillan and Free Press.

►MUELLER, CONRAD G. 1968 Vision: I. Overview. Volume 16, pages 323–327 in *International Encyclopedia of the Social Sciences*. Edited by David L. Sills. New York: Macmillan and Free Press.

PIÉRON, HENRI (1945) 1952 *The Sensations: Their Functions, Processes, and Mechanisms*. New Haven: Yale Univ. Press; London: Müller. → First published as *Aux sources de la connaissance: La sensation, guide de vie*.

►POLLACK, IRWIN 1968 Information Theory. Volume 7, pages 331–337 in *International Encyclopedia of the Social Sciences*. Edited by David L. Sills. New York: Macmillan and Free Press.

►RUCH, THEODORE C. 1968 Skin Senses and Kinesthesis. Volume 14, pages 300–307 in *International Encyclopedia of the Social Sciences*. Edited by David L. Sills. New York: Macmillan and Free Press.

►SMALL, ARNOLD M. JR. 1968 Hearing. Volume 6, pages 336–339 in *International Encyclopedia of the Social Sciences*. Edited by David L. Sills. New York: Macmillan and Free Press.

STEVENS, JOSEPH C.; and STEVENS, S. S. 1963 Brightness Function: Effects of Adaptation. *Journal of the Optical Society of America* 53:375–385.

STEVENS, S. S. (editor) 1951 *Handbook of Experimental Psychology*. New York: Wiley.

STEVENS, S. S. 1957 On the Psychophysical Law. *Psychological Review* 64:153–181.

STEVENS, S. S. 1958 Problems and Methods of Psychophysics. *Psychological Bulletin* 55:177–196.

STEVENS, S. S. 1961 To Honor Fechner and Repeal His Law. *Science* 133:80–86.

STEVENS, S. S. 1966 A Metric for the Social Consensus. *Science* 151:530–541.

STEVENS, S. S.; and GALANTER, EUGENE 1957 Ratio Scales and Category Scales for a Dozen Perceptual Continua. *Journal of Experimental Psychology* 54:377–411.

SWETS, JOHN A. (editor) 1964 *Signal Detection and Recognition by Human Observers: Contemporary Readings*. New York: Wiley.

SYMPOSIUM ON PRINCIPLES OF SENSORY COMMUNICATION, ENDICOTT HOUSE, *1959* 1961 *Sensory Communication: Contributions*. Cambridge, Mass.: M.I.T. Press.

TITCHENER, EDWARD B. 1905 *Experimental Psychology: A Manual of Laboratory Practice*. Volume 2: *Quantitative Experiments*. London and New York: Macmillan.

►TORGERSON, WARREN S. 1968 Scaling. Volume 14, pages 25–39 in *International Encyclopedia of the Social Sciences*. Edited by David L. Sills. New York: Macmillan and Free Press.

URBAN, FRIEDRICH M. 1908 *The Application of Statistical Methods to the Problems of Psychophysics*. Philadelphia: Psychological Clinic Press.

VON BÉKÉSY, GEORG (1928–1958) 1960 *Experiments in Hearing*. New York: McGraw-Hill.

►WENZEL, BERNICE M. 1968 Taste and Smell. Volume 15, pages 514–516 in *International Encyclopedia of the Social Sciences*. Edited by David L. Sills. New York: Macmillan and Free Press.

Postscript

It has become universal practice to designate auditory frequency in terms of hertz (Hz) instead of cps; all references to cps in the main article can be read as Hz, since the numerical values are identical.

JOSEPH C. STEVENS

ADDITIONAL BIBLIOGRAPHY

¹MARKS, LAWRENCE E. 1974 *Sensory Processes: The New Psychophysics*. New York: Academic Press.

STEVENS, S. S. 1975 *Psychophysics: Introduction to Its Perceptual, Neural, and Social Prospects*. New York: Wiley.

PUBLIC POLICY AND STATISTICS

► *This article was specially written for this volume.*

Statistics takes its name from "state," for at its beginning statistics was regarded as information about political states, for example, about the sizes of populations and armed forces. Since those early days the importance of statistics, both as information and as methodology, has markedly increased for modern government and the forming of public policies.

To illustrate the breadth of applications of statistics in policy, the discussion is divided into the following three broad areas: information gathering for society, understanding policy effects, and making decisions.

Information gathering for society

Information gathering for society includes all those data-collecting and data-analyzing activities that inform us about the current state of society. (This contrasts with effecting changes, which is dealt with in the next section, "Understanding policy effects.") Selected types of information gathering discussed below are (1) opinion and preference surveys, (2) general surveys and indicators, and (3) special investigations.

Opinions, preferences, and elections. An election is a kind of large-scale opinion survey leading to results that directly influence public policy. Thus studies of the electoral process are studies of the process of social intelligence gathering. In one such study, Edward Tufte (1973) has discovered interesting differences in the relationships between votes won and seats won by opposing parties in elections to legislatures. The regression line relating the percentage of seats won to the percentage of votes won by one party in a series of elections describes important characteristics of the electoral districting during that period. For example, in elections to the New Jersey state legislature between 1926 and 1947, when the Democratic party polled 50 per cent of the votes, it took an estimated average of only 35 per cent of the seats; in the period 1947–1969 this had changed so that the Democrats took an estimated average of 48 per cent of the seats when polling 50 per cent of the votes. (When percentage of votes won is 50 per cent, a regression line should have an ordinate of 50 per cent of the seats won; if not, the system is biased against one of the parties.) [*See* LINEAR HYPOTHESES, *article on* REGRESSION.]

The slope of the regression line describes the responsiveness of the electoral districting scheme. The slope—called in these applications the "swing ratio"—estimates the increase in percentage of seats won by a party as its percentage of votes increases by 1 per cent. The larger the slope, the more responsive is legislative representation to voting shifts. From the end of World War II to about 1970 the swing ratio in Great Britain was 2.83, in New Zealand 2.27, and in the United States (for Congress) 1.93. Figure 1 graphs the data for Great Britain from which the swing ratio of 2.83 was estimated. Evidence from the last several congressional elections in the United States indicates a sharply declining swing ratio, probably dipping below 1.

This study of seats and votes shows how a standard statistical model can describe and measure important characteristics of electoral districting arrangements. Further, proposed schemes for redistricting of a political unit can be evaluated by simulating the seats–votes regression that would be generated by a series of elections to the legislature in that unit. So the model here performs a number of valuable functions for setting public policy: simplifying description, facilitating comparisons of electoral arrangements over time and between jurisdictions, providing a standard for evaluation of these arrangements, and allowing tests of hypothetical redistricting schemes.

Voting is a direct and institutionalized means by which the public in a democracy participates in determining objectives and priorities. Survey research provides ways to tap the public's opinions and preferences in greater detail. One unusual example is an interview survey study of happiness.

From Aristotle's *Nicomachean Ethics* to the American Declaration of Independence and beyond, the "pursuit of happiness" has been a primary and unchallenged objective of public policy, at least in the Western world. Aristotle wrote that "both the general run of men and people of superior refinement say that [the highest of all goods achievable by action] is happiness [endaimoria], . . . but with regard to what happiness is they differ, and the many do not give the same account as the wise" (quoted in Bradburn 1969, p. 6). Perhaps because of the formidable phenomenological difficulties referred to by Aristotle there have been few systematic studies of this central goal of policy.

One of these few is a series of interview surveys by Norman Bradburn (1969) in which self-reports on happiness were solicited by the question "Taken altogether, how would you say things are these days—would you say you are very happy, pretty happy, or not too happy?" Surveys taken at different times and in different cities in the United States show a high degree of consistency, with about a third of the people interviewed reporting "very happy," about half "pretty happy," and around 5 to 15 per cent reporting "not too happy." Additional information solicited in the interviews about the respondents' sex, age, education, income, race, marital status, job, and other personal characteristics provided the data for extensive tabular analyses of the relation of the happiness reports to these factors, both separately and jointly. Readers who have definite ideas about whether men report being

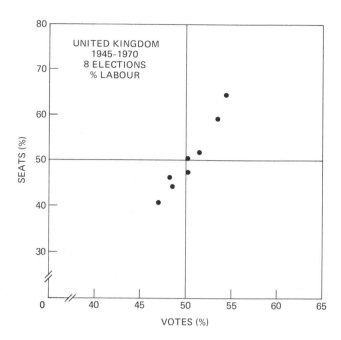

Figure 1 — The relationship between seats and votes

Source: Tufte 1973, p. 541.

happier than women, or the poor happier than the rich, or blacks happier than whites may find surprises both in the direction of the difference and in the magnitude. Among many other things, Bradburn finds men and women remarkably close in percentages for the three reported happiness categories, a steady decrease in percent reporting "not too happy" with increase in income, and whites clearly happier than blacks.

Richard Easterlin (1973) reviewed over thirty happiness surveys in different countries and at different times. Within countries a positive association between income and happiness was always found. Happiness did not, however, increase with income over time, nor were richer countries happier on the average than poorer. Easterlin theorizes that absolute increases in income are not important to happiness but that one's relative position within a culture is. Campbell et al. (1976) have broadened and extended the happiness studies to survey reports on satisfaction or quality of life in the United States in fifteen domains of living, including work, marriage, family life, health, neighborhood, friendship, housing, education, and savings.

Pre-election polls differ from opinion surveys in that their results can be checked for accuracy against the actual election results, and in pre-election polls accuracy is of the essence. So it was with some chagrin that four of five major pollsters found themselves predicting on the eve of the British general election in 1970 a Labour party victory with Harold Wilson, only to see the Tories under Edward Heath win by a margin of 3 per cent over Labour. The *Economist* (June 20, 1970) commented scathingly:

It was in character with this sunny June election that the opinion polls should have wound up suffering from sunstroke. . . . The polls carried out in early June show an astonishing spread of 10.4 per cent in their calculations of Labour lead. . . . The differences between them [the polls] are enough to rattle all but the most convinced sampling theorist (to say nothing of what they have been doing to the politicians throughout the campaign).

Yet a comprehensive post-election analysis of the polls' results by Richard Rose (1970) revealed that there was strong evidence that a late swing to the Conservatives was the overwhelming factor in the polls' apparent bias, and that even so, as indicated by computer simulations of the polls' sampling plans, sampling error alone could account for the kind of variation that so startled the *Economist*.

Polls—and the newspapers that publish them—have, however, usually been remiss in not stating estimates of error magnitudes. Rose (1974, p. 127) had the following comment on the accuracy of the polls, one that is pertinent to other uses of statistics in policy making:

Because these estimates cannot be unvaryingly and eternally accurate, this is no reason to avoid doctors or abandon the census and return to the 18th-century habit of settling by Parliamentary debate whether the population is increasing or decreasing. Nor is there any reason to avoid studying political behavior if accuracy is less than total.

In political surveys, as in many other forms of human endeavor, the choice is not between total knowledge and total ignorance, but rather between more or less knowledge or more or less accuracy. People who wish for a sure thing, whether on a general election, a stock market share, or a doctor's diagnosis, should not consult professionally qualified men but rather soothsayers, astrologers or others who make a living by taking money from those who believe that perfect foreknowledge is possible. Meanwhile, those who can conceive of degrees of accuracy can consider whether the degree of accuracy obtainable in pre-election polls is sufficient to the purpose at hand.

In the 1974 pre-election polls six of the seven major polls produced a correct forecast, in that they forecast correctly that the Conservatives would gain a larger share of the popular vote than Labour (38.8 per cent versus 38.0 per cent). None of these six, however, foresaw that Labour would win more seats in the House of Commons than the Conservatives, so they did not in the end predict the ultimate electoral victory of Labour that established Wilson again as prime minister. In predicting the vote shares, however, the polls increased their accuracy over the 1970 levels.

General surveys and indicators of conditions. Measurement defines a condition that can be, and inevitably will be, compared with standards, whether relative (other countries, or past years, or different parts of the country) or absolute. Furthermore, measurement throws a strong spotlight on conditions that might otherwise be hidden. Walter Lippmann (1922, p. 380) vividly described the impact of statistics in these terms: "The printing of comparative statistics of infant mortality is often followed by a reduction of the death rates of babies. Municipal officials and voters did not have, before publication, a place in their picture of the environment for those babies. The statistics made them visible, as visible as if the babies had elected an alderman to air their grievances." These themes about the importance of social statistics are also well expressed by Daniel Moynihan in his book *Maximum Feasible Misunderstanding* (1969, p. 30):

Statistics are used as mountains are climbed: because they are there. If one recalls that the nation went through the entire depression of the 1930's without ever really knowing what the unemployment rate was (the statistic was then gathered once each ten years by the Census Bureau), one gains a feeling for the great expansion of knowledge in this and related fields in the quarter century that followed. By the 1960's, the monthly employment data had become a vital, sensitive, and increasingly reliable source of information about American society, and that information increasingly insisted that although the majority of Americans were prosperous indeed, a significant minority were not.

Because such statistics have so much impact it is crucial that they be accurate. Accurate assessments of social and economic conditions require the carefully designed collection of data. Everyone's personal experience is limited and biased for these and other reasons:

(1) The stratification of society tends to mask a large part of social reality. Thus the unemployed may not be well known to the employed, or the undernourished to the adequately fed, and vice versa.

(2) Activities that are illegal or illicit may be hidden. For example, a survey in New York City revealed widespread sub rosa gambling to an extent surprising even to the citizens of that city. (See Fund for the City of New York 1972.)

(3) Many activities of a personal nature, such as sexual behavior, are not readily observed.

(4) One person's experience, direct or indirect, with unusual events like infant deaths, accidents, or crimes is usually too limited for good estimates of frequencies. Then, too, publicity accorded single incidents may badly bias perceptions of conditions. A celebrated example is Lincoln Steffens' description of the spurious "crime waves" generated by daily news-

792 PUBLIC POLICY AND STATISTICS

papers whenever reporters went on a binge of crime reporting. (See Steffens 1931, chapter 14.)

(5) Some conditions are defined theoretically and are not observable on an individual level. For example, the population of motor vehicle drivers can be imagined to be ranged in groups having different probabilities or rates of accidents per year. Since for any given person accidents are rare, however, these rates are not observable for particular individuals. Models for a distribution of rates for the entire population can nevertheless be defined. These are discussed further below.

Many of the regular statistical series of modern governments are based on sample surveys taken from specified relevant populations. Such sample surveys often yield better information than that made available as a by-product of regular administrative operations. An interesting example is the collection of data on crime in the United States. The traditional source of data on crime in the cities and states has been the Federal Bureau of Investigation, which regularly compiles totals of crimes reported to local police units throughout the country. Sample surveys, called "victimization" surveys, first done in the 1960s, established substantial differences between the numbers of crimes reported to the police and the actual numbers of crimes committed. (See Ennis 1967.) In these surveys, citizens were asked to recall instances within a specified time interval in which they were victims of crime. Comparisons with the usual FBI-reported crime totals showed, for example, some three times as many burglaries and 50 per cent more robberies than were reported to police. In the 1970s the U.S. Department of Justice began a major program of victimization surveys to measure crime rates in a number of large cities. For a comprehensive review of the purposes, scope, and methods, as well as the problems, of these surveys, see National Research Council (1976).

Many sources list the kinds of data that modern governments collect on social and economic conditions. For the United States the report of the President's Commission on Federal Statistics (1971) can be consulted as can other works available at library reference desks. Several countries now publish special compendia of social statistics or "social indicators" that trace the development over time of the state of health, employment, crime, and other major concerns of social policy. Examples are *Social Trends* for the United Kingdom and *Social Indicators* for the United States. These annual social reports complement the more prevalent economic reports issued by governments. Philip M. Hauser surveys the variety of uses, public and private, of official statistics in *Social Statistics in Use* (1975). [*See also* GOVERNMENT STATISTICS.]

Special investigations. The documentation of particular conditions (for example, alleged discrimination based on race, sex, age, and so on) often requires statistical analysis and inference. The history of legal challenges to racially discriminatory selection of jurors in certain American courts illustrates the usefulness of statistical methods. Michael O. Finkelstein (1966) has described the legal evolution of cases from nonrepresentation up through the so-called underrepresentation cases, like *Swain* v. *Alabama*. In these cases blacks were not actually barred by statute from jury service as had earlier been the case in some states, and indeed they did actually serve on juries, but their numbers on the panels from which jurors were chosen were less than proportionate to their numbers in the general population. Of the use of statistical analysis based on a binomial model for fair selection of jurors, Finkelstein writes (1966, p. 374):

A basic legal principle in the jury discrimination cases is that the selection of an improbably small number of Negroes is evidence of discrimination. This principle, which links a finding of discrimination to a determination of probabilities, opens the door to the use of statistical analysis in these cases. The mathematical methods described here have been used to calculate the probabilities which the law has established as relevant for determining the existence of discrimination.

In some significant cases the U.S. Supreme Court has made mention of probabilistic analyses (*Whitus* v. *Georgia*, 385 U.S. 545, 1967; *Alexander* v. *Louisiana*, 405 U.S. 625, 1972). [*See* STATISTICS AS LEGAL EVIDENCE.]

Following the 1970 draft lottery held in the United States to determine selectees for the army, questions were raised about the randomness of selection. It appeared that the lottery, in picking birth dates to determine the order of selection, had favored the early months of the year over the later months. Several analyses of the order of selection of the dates all showed a small probability of obtaining the outcome observed (or a more extreme one) if the lottery were truly random as intended; on the other hand, this outcome was much more probable if a biased order of selection by dates had been made, as other information about the conduct of the drawing in fact suggested. This was not the first time that draft lotteries in the United States had been found to be conducted in an amateurish fashion. After lotteries held during World War I and II, public outcries over nonrandom selections were heard. Stephen E. Fienberg (1971) describes the history of this experience and includes a description of the 1970 draft lottery, which was constructed with professional statistical advice to meet the announced standard of random selection. Fienberg also discusses the important social roles of injected randomness: equity, objectivity, and, more recently, confidentiality.

The subtle problem of estimating the probabilities of rare events, especially those having catastrophic effects, such as nuclear power plant accidents, has been frequently discussed, especially as environmental concern has mounted. Other examples are oil and natural gas spills, effects on the atmosphere of supersonic planes, the potential contamination of other planets by organisms of the earth or back contamination of the earth by returning spacecraft, and accidental nuclear war. Methods that have been employed or proposed for studying these probabilities include building probability models that use estimated probabilities for contributory events, using insurance rates as an indirect source of likelihood estimation, and bounding the chance of the event in question by the chance of events believed to be more likely. Dealing with dependent events in a system and with modes of occurrence of rare events that escape attention or are difficult to include in formal models are major hurdles. (See Fairley 1977; Mosteller 1977; Selvidge 1975; U.S. Nuclear Regulatory Commission 1975.)

Reporting.   The reporting of data and results of statistical analysis of data raises a number of issues.

How can numerical information be communicated to a wide audience in a way that both preserves accuracy and conveys appropriate information about errors? William H. Kruskal (1977) writes of a resurgence of interest in graphical presentation. He also notes how little explicit attention has been given to ways to achieve clarity and accuracy of prose in statistical commentary. These questions demand the joint attention of statisticians and psychologists.

It would be naive to forget that statistics may be used to mislead deliberately, especially in adversarial and public relations theaters of public policy making. Such uses may be discouraged or put in a critical spotlight if there are accepted standards of statistical presentation. An example of misdirection because of highly inaccurate raw data were so-called body counts of persons killed in the Vietnam war. A substantial percentage of U.S. Army officers close to the battlefields in that war later reported their belief that the body counts were often inflated. (See Kinnard 1975–1976.)

The responsibility of the statistician does not end with the collection and dissemination of data. Analysis and interpretation may be necessary to avoid misinterpretation and wrong inferences by a public audience. For a vigorous exposition of this view of the statistician's role, see Claus Moser (1976). Facts just do not speak for themselves and tables are rarely if ever self-explanatory. One defense against the misuse of raw data to "prove" anything is early, even preliminary and necessarily incomplete, discussion of valid and invalid uses.

Appropriate reporting of both nonsampling and sampling errors is a concern for statistics in the public policy domain as well as elsewhere. In reviewing the basic statistical data series maintained by the federal government, the President's Commission on Federal Statistics noted that the sources and the sizes of error were generally not well understood. Without information about possible errors, policy makers and others who use the data may be misled.

In the U.S. Bureau of the Census, interest in maintaining the reporting of errors has lead to the publication by Maria Gonzalez et al. (1975) of "Standards for Discussion and Presentation of Errors in Survey and Census Data." The introduction (p. 5) explains the purpose: "Each year the Bureau of the Census publishes a vast number of estimates on many subjects. It is the Bureau's responsibility to inform its data users of the important limitations of the estimates, both those due to sampling and those due to response and other nonsampling errors." The authors discuss, with the help of numerous examples, recommendations on the reporting of error information.

An early critical survey of sources and magnitudes of errors in economic data was given by Oskar Morgenstern (1950). Kruskal (1977) discusses a number of problems in official statistics, emphasizing the widespread—although not universal—inattention "to how the data were gathered, to their limitations, and in general to the error structure of sampling, selection, measurement, and subsequent handling" (p. 12). [See ERRORS, *article on* NONSAMPLING ERRORS.]

Ethical issues.   The report of the President's Commission on Federal Statistics (1971) presents a detailed discussion of the issues of privacy and confidentiality. Privacy as a privilege of not divulging certain kinds of personal information (such as religion or status as an unwed mother) has been an issue raised periodically in connection with the decennial census.

The statutory and constitutional requirements for maintaining confidentiality of some records after information has been obtained affects the manner of processing and storage of data. Since statistical summaries based on individual records are usually permitted, challenging problems have arisen concerned with ways to present summaries that do not reveal information about individuals. See, for example, Felligi and Phillips (1974).

With the increased interest in social experimentation, together with widening interest in randomized trials in medicine, has come considerable attention to the ethics of social and human experimentation. A Brookings conference explored these issues. [*See* Rivlin & Timpane 1975; *see also* ETHICAL ISSUES IN THE SOCIAL SCIENCES.]

Understanding policy effects

The presence of data and their analyses can help set a factual standard for public discussion of a policy issue. Erroneous claims that flourish in the absence of such factual background often receive short shrift in its presence. Examples are claims that higher-income people are unhappier (see Bradburn 1969) or that increasing police patrol will greatly decrease crime (see Kelling et al. 1974).

Problems in understanding policy effects. Yet in trying to understand some policy effects we often find ourselves without basic knowledge about the effects of either current practice or alternative policies. We can identify some of the reasons.

Changing population. Policies are very particular. They are, or should be, tailored to particular needs and expectations of particular groups at a particular time. What's more, we know that times and circumstances change: attitudes, leaders, technology, social arrangements. The particularity and the volatility of the factors that influence people's behavior pose severe problems for assessing policy effects and for the use of statistics in doing so. At the same time statistics is the science of variability and of uncertainty, so the analysis of the complex and fast-moving domain of the policy world poses a challenge to the subject. There is a paradox here: the more things change, the more we need rational planning, but the harder it is to plan because of the difficulties in generalizability.

Association versus causation. It is difficult to estimate changes in outcome variables following deliberate changes in instrumental variables. Studies using observational data are especially weak here, and properly randomized experiments may bring with them superior cause-and-effect inferences. [*See* CAUSATION; EXPERIMENTAL DESIGN, *article on* THE DESIGN OF EXPERIMENTS.]

In a designed experiment the investigator controls the assignment of the experimental treatments to the units. In a good experiment that assignment is randomized to avoid the bias of selection (like that of the nurse who helpfully assigns the healthier patients to the treatment known to be favored by the doctor). Randomization, together with the presence of at least one control group, also provides a fundamental yardstick for statistical inference.

Some of the advantages of designed experimentation flow from its prospective character. The investigator can specify the policies to be studied and can collect from the start the needed data. In contrast, the investigator who takes data from past observed experience may find it difficult to describe exactly what the policy was, and for data must make do with whatever management or other records are available.

The greatest advantage of the designed experiment, however, is that it allows better control of other variables that may be confounded with the effects of the policies under study. For example, if the new policy is an innovative educational program and parents are allowed to enroll their children or not as they choose, it may well be that children of more highly motivated parents will be enrolled and hence there will be no way of separating the effects of parental motivation from those of the educational program. In a designed experiment, however, the investigator's control over the assignment of treatments to units breaks the confounding of treatments with unit characteristics permitted by the self-selection of units into treatment categories. For example, the investigator might randomly assign half of the children volunteered for the innovative educational program to the program, and the other half to a control group.

The limitations of purely observational data and the parallel limitations on the capacity of statistical methods to control for important confounding effects have not been widely understood. It is rare to find a model estimated from observational data that claims to have the inclusiveness required for causal inference. Glen G. Cain (1975, p. 312) described the requirements for causal inference in these terms:

There is no question but that the theory must convincingly "close" the system defined by the process and environment of the program.[1] In other words, the model must be "complete" in the following sense: variables correlated with y must be 1) included in the model as "control variables"; or, if excluded, they must be 2) known to have a net or partial zero correlation with T; or 3) known to be *unvarying in the given environment or process;* or 4) known to be themselves completely *determined by* included variables (in which case we could say that the former have no "net" relation to y); or 5) known to be part of a set of omitted variables which tend to "offset" each other in their effects on y—i.e., where the expected value of y, given the omitted X's, is zero.

The informational requirements for such a complete model are formidable, of course. Incidentally, I would not object if someone claimed that the requirements amount to knowing the "selection" process in a more fundamental sense than that which is narrowly implied by the "direct selection" procedures which are determined by the administration of the program. If economists, for example, claim to estimate the net effect of labor unions on wages, we must have a model to "capture" (specify) the selection mechanism which distinguishes union member from nonunion member (or, perhaps, a worker covered by a union contract compared with one not covered). Here, the union status variable represents the T variable in our evaluation model, and all the caveats for estimating a union effect in a context of nonrandom assignments must apply. As I mentioned at the outset of this discussion, whether economics provides examples of credible empirical models—credible in the context in which they are intended to apply—is a large question.

---

[1] The term "closed system" was used by F. Mosteller, who raised the question at the conference and in subsequent correspondence of whether the theory provided a closed system (i.e., a complete model) for the purpose at hand. Mosteller quite properly remarked that the question of "bias" in the effects of a given variable, whether T or some other variable, has no meaning in the absence of such a complete model.

Cain and Watts (1973, p. 355) provide another illustration of the difficulties of causal inference from observational data in their discussion of regression studies of the effects of higher wages or income on the supply (hours worked) of labor. The effect of added income on hours worked is a question that arises in considering a federal income maintenance or negative income tax plan. In reviewing a large number of regression analyses using survey data on wages and on hours worked they write:

The first issue to be raised here is the potential bias in the measure of wage and income effects caused by omitted variables that are correlated both with these and with labor supply. The most likely candidates are (1) preferences for work relative to nonwork activities; (2) skills and/or productivity in relevant nonmarket work activities like home production; and (3) various unmeasured traits affecting wage, income, and labor supply such as the quality of education, training, work experience, and mental and physical health.

The general point about preferences is that personal traits—ambition, the "protestant ethic," a desire to retire in comfort or to leave abundant material goods to one's heirs, a dislike for spending time at home, or any number of other characteristics—could be "causal" to decisions to obtain high wages (or to accumulate nonhuman wealth) and to work many hours in the market. Clearly, because an income-maintenance program will change the effective wage rates and nonlabor income across all families in the eligible population,

the information we are looking for is the partial relationships between wage rates and income on labor supply, holding personal traits constant. Because the variables available in survey data offer at best meager control over such traits, the resulting estimates of wage and income effects on labor supply may well be biased.

The use of simultaneous equations models, including path analysis, although it represents a potentially superior form of specification because it permits reciprocal causation, does not remove the strictures about causal inference discussed above. (A review of these models can be found in Goldberger & Duncan 1973.) A simultaneous equations model for the deterrent effects of capital punishment on the homicide rate in the United States does lead to strong claims (Ehrlich 1975). The model was used in arguments on capital punishment before the U.S. Supreme Court. Claims for the model's validity are attacked by Bowers and Pierce (1975), who find the results highly sensitive to specification of form and to time period.

There are practical and ethical limits to the scope of social experimentation, although these have often been exaggerated. When experimentation is described as expensive and time consuming, we can sometimes ask whether the alternative, if it does not in fact produce the kinds of cause-and-effect inferences sought, is not in the long run even more expensive and less timely. When an experiment is described as unethical we can sometimes put the shoe on the other foot and ask whether the alternative, if it results in not really learning what the effects of policies are, is not less ethical. See, for example, the characterization in Gilbert et al. (1975, p. 182) of some observational studies as "fooling around with people." Yet we should avoid the untenable position that experiments alone can support cause-and-effect inference. Much scientific advance, not to mention practical knowledge, has been based on observational studies —and the "observational" sciences of astronomy and geology are not alone in this—even if, arguably, the social sciences have been especially hindered from lack of experimentation. A comprehensive discussion of the prospects for, and issues related to, more widespread use of the experimental method is in Riecken and Boruch (1974).

Theoretical thinness. In many studies of policy effects there is little theoretical guidance and therefore serious difficulty in identifying and controlling for important variables. This is a problem for both experiments and observational studies. An experiment that establishes an effect without theory always suffers from the possibility that the effect is peculiar to that population or that size.

The Kansas City Preventive Patrol Experiment tested alternative levels of police patrol (see Kelling et al. 1974). Fifteen patrol beats were divided into three groups of five each: one group was a control group with preventive patrol continued as usual; a second group were "reactive beats" into which police cars entered only in response to calls; and a third group were "proactive beats" in which police patrol visibility was increased two to three times. The findings of the experiment were primarily negative in that major measured quantities, such as crimes reported to the police, crimes reported by victims in surveys, and citizens' perceptions of police presence, did not vary in a statistically significant way by experimental area.

Richard C. Larson (1976) applies a simple theory for predicting the frequency of patrol-car passings to conclude that the range of preventive patrol coverages actually experienced in the experiment was smaller than expected and smaller in fact than the range experienced by other U.S. police departments. For this reason, though Larson finds valuable implications for police strategy in the findings, he concludes that the results should not be used to extrapolate the general value of a visible police presence to other cities. Hindsight suggests that even simple theory could have been helpful in designing the experiment, and that the absence of better theory even today limits the utility of a newly designed experiment. (See also Fienberg et al. 1976.)

Functions of statistics in understanding policy effects. Statistics is well suited to helping fill the gaps in needed information or basic knowledge because it is an all-purpose methodological discipline for empirical research, offering standard methods of data gathering and analysis with which to help answer questions. Thus, in considering a policy issue, one might first analyze the available relevant data. Second, one might gather additional data, perhaps by a specially designed survey or experiment. Third, one might make more fundamental investigations, testing basic theory.

Statistics and three groups of policy instruments. Three major groups of policy instruments whose effects we should strive to understand are (1) direct regulation of private actions, (2) economic incentives to affect private or local provision of goods and services, and (3) provision of social services by government. In investigations of policy effects, statistical methods are used in study designs to facilitate causal inferences and as methods of controlling for secondary variables. [See EXPERIMENTAL DESIGN, *articles on* THE DESIGN OF EXPERIMENTS *and* QUASI-EXPERIMENTAL DESIGN; ECONOMETRICS;

LINEAR HYPOTHESES; MULTIVARIATE ANALYSIS; SURVEY ANALYSIS; SIMULTANEOUS EQUATION ESTIMATION.]

Direct regulation. Direct regulation encompasses such policy options as prohibition, restricted use, registration, licensing, and setting standards. Scientific studies behind direct regulation frequently use statistical methods, for example, in connection with health and safety: smoking and health, air pollution and health, and the effectiveness of seat belts.

A study by Colton and Buxbaum (1968) of the effectiveness of state motor vehicle inspection in reducing motor vehicle accident fatalities illustrates some of the difficulties in investigating direct regulation. They found that states with inspection experienced lower fatality rates than states without: for white males the rates in 1960 were 30.8 fatalities per 100,000 for states with inspection and 44.6 for states without, a difference of 13.8. Did inspection itself reduce fatalities, or were there other differences between the groups of states that could account for the difference in rates? Why did some states choose to have inspection and some not? The first question was partially answered by making a covariance regression adjustment to the difference, adjusting for population density, per capita income, and the mortality rate from other kinds of accidents. The adjusted rate for inspection states was 35.5 and for noninspection states was 42.3, a difference of 6.8, about half the unadjusted difference. As the authors point out, however, the findings are inconclusive, for the remaining difference might perhaps be explained by such factors of adjustment as actual miles driven, safety of the highway network, and enforcement policies. In addition, there may be technical problems in doing the covariance adjustments.

Other study designs could be used, for example a study of rates in states before and after a change in the inspection law, or better yet, a study by randomized experiment of inspection versus noninspection in different localities. Or one might study in more detail the presumed causal links beween inspection (the policy) and fatality reduction (the policy objective). For example, to what extent is an official state policy of inspection actually enforced? What is the difference in mechanical reliability of the vehicles in inspection and noninspection states? Are there fewer accidents from mechanical sources in the inspection states? Does inspection have any perverse effects leading people to postpone repairs?

In some areas of policy, such as accident prevention, chance plays a large role that can be described by probability models to facilitate reasoning about policy effects. A well-known fallacy is the belief that because in a given year only a small percentage, say 5 per cent, of drivers, are involved in a large percentage, say 50 per cent, of accidents, then removing these 5 per cent from the road would reduce accidents by 50 per cent. The fallacy is most clearly revealed by noting that in a short interval of time even 100 per cent of the accidents will involve only a small percentage of the drivers, but removing these drivers from the road will not reduce accidents by 100 per cent. Accident occurrences might even be consistent with a completely random occurrence of accidents, happening with equal likelihood to every driver. For California drivers, Joseph Ferreira, Jr. (1970) fitted statistical models that assumed variable accident probabilities among individuals. He found most drivers to lie in the middle range of small-accident probability but also found evidence of smaller groups of very safe and very dangerous drivers. His model predicts expected savings of accidents from such long-advocated policies as stringent revocation of drivers' licenses after some kind of accident. Furthermore, the mistakes of such a policy can be studied quantitatively; for example, how many low-accident-probability drivers would through bad luck be caught in the same enforcement net that correctly removes the licenses of the truly dangerous drivers? Realistic modeling also considers the large numbers of persons with revoked licenses who continue to drive!

An advantage of a model is that policy effects can be studied on paper by manipulating the parameters of the model and observing the changes in the outcome variables of the model before expensive real-world experience with actual changes is attempted (Ferreira 1974). Of course, a disadvantage of a model is that effects of manipulations of the model and of the real world may be far apart because of limitations of the model. For example, Ferreira's models did not allow for learning as a result of an accident.

Economic incentives. Policy instruments that influence private action indirectly through economic incentives include regulation of market prices of private goods and services, government subsidies and transfers, wages paid and prices set by governmental organizations, government insurance, and taxation.

Large-scale experiments with the negative income tax or income supplements have been conducted during the 1970s in the United States. In the New Jersey experiment some 700 low-income families of the working poor were given income supplements of varying amounts, while some 600

families in a control group were not. The principal object was to see whether family members worked less, and if so how much less, when receiving income supplements of differing amounts and on differing terms. Different income supplements (or no income supplement) were assigned by a random scheme to the families chosen for the study. In this way the groups receiving various income supplements and the control group were made similar. Differences in reported work experience could then with more reason be attributed to income supplements rather than to anything else—such as aggressive application for a particular income-supplementation group. (See U.S. Department of Health, Education, and Welfare 1973.) As is always true of carefully designed experiments of this kind, analyses of the data continue for some time. The major initial finding in New Jersey, that there were no substantial decreases in work among groups who received income supplements, was corroborated upon publication of the full report, *The New Jersey Income-maintenance Experiment* (1976–1977).

Social services provision. Providers of such public services as education, police protection, or manpower training are called upon with increasing frequency to justify their progress in terms of demonstrable effects and not simply in terms of competent administration. Schools are asked to demonstrate effects on achievement scores over and above what can be explained in terms of the socioeconomic character of their student bodies (U.S. Department of Health, Education, and Welfare 1966). Preschool programs are asked to show how much they contribute to child development over and above maturation that occurs naturally (see, for example, Weisberg 1974). Police are asked for evidence that particular patrolling practices do minimize crime (Kelling et al. 1974; Fienberg et al. 1976). Manpower training programs are asked to show that trainees' job-market experience is more favorable than it would have been in the absence of the training.

Such programs have undergone many evaluations, and the predominant result has been negative in two senses. One is that social innovations to improve social services are at least as often failures as successes. In a review of 28 social and medical innovations that had been carefully evaluated, Gilbert, Light, and Mosteller (1975) judged only 6 clear substantial successes, 6 more marginal successes, 13 to offer no improvement, and 3 actually harmful.

The second sense is that no clearly successful strategy has been generally accepted for learning what works and how to improve programs over time. Both the themes of program inadequacy and methodological deficiency were underscored in a National Academy of Sciences final report (1974) on manpower training programs, which noted that after program expenditures of $6.8 billion between 1963 and 1971 and evaluation studies costing $180 million, no clear assessment of these programs' impacts could be made. It is of course often difficult to specify criteria of success. Ought we, for example, to be interested in short-term test scores or long-term social adjustment? Reliable and valid measures are often difficult to obtain.

The problems of evaluation are political and organizational as well as statistical. In an informative review of evaluative efforts Alice M. Rivlin (1971, p. 86) wrote:

One reason that not much has been learned from statistical analysis of the existing health, education, and social service systems . . . is the failure thus far to organize social service systems to facilitate investigation of their effectiveness. Major deviations from the established pattern are rare and their effects are hard to disentangle from the special circumstances that brought them about. Equally little has been learned from evaluation of federal government programs. The reasons are much the same. Headstart, Title I, model cities and other federal programs could have been designed to produce information on their effectiveness, but they were not.

Making decisions

A final group of statistical contributions relate to the making of decisions directly by providing (1) criteria for evidence, (2) methods for decisions about individuals, (3) normative theories of decision making, and (4) tools for reasoning about chance. Items (1) and (4) are relevant under the prior subheadings as well.

Criteria for evidence. In judicial or quasi-judicial proceedings, the issue of criteria for proof often arises. A preliminary question can concern admissibility: that is, what kind of evidence will be admitted into consideration? [*See* STATISTICS AS LEGAL EVIDENCE *for a discussion of how courts have treated this question for sample surveys, where the hearsay rule has sometimes been invoked to rule out the survey results based on personal interviews.*]

Specific numerical criteria for tests of significance have sometimes been established. In fair-employment proceedings in the United States, the Equal Employment Opportunity Commission requires employers to establish the validity of employment tests as predictors of on-the-job success. The commission's guidelines (1976) require significance tests of the null hypothesis of no validity

to meet the 5 per cent level customary in much published scientific literature. In considering evidence for jury discrimination, accompanied by significance tests, the U.S. Supreme Court seems to have regarded the figure of 1 in 20,000—on one occasion quoted as a descriptive significance level for a null hypothesis of no discrimination—as an actual standard for future evidence to meet. What standards are appropriate certainly depend upon, among other things, considerations of costs and benefits of mistakes for the different parties and on their relative burdens of proof. These are questions that need more attention. (See Finkelstein 1966; Kaplan 1968).

Regulatory commissions, which are often required to undertake extensive fact-finding investigations, have met problems in using the often complex statistical analyses put before them as evidence. Some of the issues that arise are how to encourage the parties to examine the relevant data; how to combine personal judgment with results of formal models in reaching a decision; and what rules to require of parties criticizing and objecting to the analyses. For econometric models of administrative hearings four protocols or rules are given by Finkelstein (1973). Constructive suggestions on the criticism of statistical studies are given by Bross (1960). He argues that it is not enough for a criticism of a model to show that some of its assumptions may not be fully supported. The criticism should show that the implications that have been drawn from the model will be importantly affected by the failure of these assumptions.

Individual decisions. In decision making that hinges on the prediction of an individual's success, such as admission to educational institutions or training programs, hiring for a job, or parole from a prison, it is sometimes possible to use a quantitative predictor of individual success, usually employing a regression analysis of available data on factors relevant to the prediction, that is superior to a purely subjective judgment based only on informal examination of the same data plus additional information such as personal interviews and recommendations. Dawes (1971) has written of the experience of the psychology department at the University of Oregon with a predictor of the academic success of applicants to the department. The regression uses as independent variables applicants' undergraduate grade-point average, graduate admissions test scores, and a quality index of undergraduate institutions. The admissions committee of the faculty also predicts the likely academic success of admitted students. Both the regression predictor and the committee rating were subsequently compared with the faculty's own rating of academic success after the students had completed a period of graduate study. The regression predictor predicted the faculty's rating better than the committee had. Thus the equation proved superior to the committee by using more consistently the same information available to the committee itself. More general discussions of statistical prediction in decision making are Meehl (1954) and Slovic and Lichtenstein (1971).

Decision theory. Statistical decision theory, including Bayesian decision theory, provides theory for incorporating the objectives, costs and benefits, and uncertainties of decision making in a formal scheme for choosing among alternative courses of action. Although this family of theories has not found wide application in actual decision contexts, there are examples: de Neufville and Keeney (1972), Howard et al. (1972), Kaplan (1968), and Zeckhauser et al. (1974). [*See also* BAYESIAN INFERENCE; DECISION THEORY.]

Reasoning about chance. Psychologists interested in human capabilities for reasoning about chance have developed evidence for characteristic errors or biases in such reasoning. Some of these errors are familiar to anyone acquainted with the counterintuitive outcomes of certain problems in probability theory. For a detailed review of the types of heuristics that people use in making assessments of probabilities, and of the biases that have been shown to be characteristically associated with these intuitions, see Tversky and Kahneman (1974). [*See also* PROBABILITY; FALLACIES, STATISTICAL.]

One example of elementary probability and statistics in reasoning about chance in a policy problem is furnished by Light's study (1973) of policies for dealing with abused children (physically injured, severely neglected, or sexually abused). Suppose that health examiners were able to detect an abused child 90 per cent of the time. And suppose they correctly detected a nonabused child 95 per cent of the time. Now if only 1 per cent of children nationally are abused, what percentage of children diagnosed as abused are actually abused? The percentage of 15.4, which can

be computed using Bayes' theorem or more directly by laying out a 2×2 table, surprises those not acquainted with the behavior of this kind of conditional probability. The reader can verify the number by using Table 1. Such a low value raises serious questions about a policy of nationwide health examinations conducted to identify abused children. Wallis and Roberts (1956, p. 328) discuss an example of cancer diagnosis with a similarly surprising conclusion.

The reputation of statistics in public affairs is too often an echo of Disraeli's reputed phrase "lies, damned lies, and statistics." Actually his reference was to the testimony of expert witnesses, not statisticians (see Cook 1913, pp. 433–434), but the "statistics can prove anything" notion has been long current. Darrell Huff traded on the idea in his amusing "how not to do it" book *How to Lie With Statistics* (1954). A response to this notion, only half facetious, is that a fool can prove anything with anything.

A serious response to the notion is to establish the more modest but important things that statisticians can do. These lie somewhere between "prove nothing" and "prove anything." An agreed-upon basis in fact, and often in quantitative fact, can be a big help in a contest, an argument, an adversary proceeding. Such a basis, however, is usually only a foundation for further analysis and debate about the relevance, assumptions, definitions, and so on. Using statistics doesn't settle the case, but it can elevate the discussion.

WILLIAM B. FAIRLEY

Table 1 – Abuse status of 10,000 children by diagnosis and by real condition (hypothetical data)

		REAL CONDITION		
		Abused	Nonabused	Row totals
DIAGNOSIS	Abused	90	495	585
	Nonabused	10	9405	9415
	Col. totals	100	9900	10000

BIBLIOGRAPHY

BOWERS, WILLIAM J.; and PIERCE, GLENN L. 1975 The Illusion of Deterrence in Isaac Ehrlich's Research on Capital Punishment. *Yale Law Journal* 85:187–208.

BRADBURN, NORMAN M. 1969 *The Structure of Psychological Well-being.* Chicago: Aldine.

BROSS, IRWIN D. J. 1960 Statistical Criticism. *Cancer* 13:394–400.

CAIN, GLEN G. 1975 Regression and Selection Models to Improve Nonexperimental Comparisons. Chapter 4 in Carl A. Bennett and Arthur A. Lumsdaine (editors), *Evaluation and Experiment: Some Critical Issues in Assessing Social Programs.* New York: Academic Press.

CAIN, GLEN G.; and WATTS, HAROLD W. (editors) 1973 *Income Maintenance and Labor Supply: Econometric Studies.* Chicago: Rand McNally.

CAMPBELL, ANGUS; CONVERSE, PHILIP E.; and RODGERS, WILLARD L. 1976 *The Quality of American Life: Perceptions, Evaluations, and Satisfactions.* New York: Russell Sage.

COLTON, THEODORE; and BUXBAUM, ROBERT C. 1968 Motor Vehicle Inspection and Motor Vehicle Accident Mortality. *American Journal of Public Health* 58:1090–1099.

Cook, Edward T. (1913) 1942 *The Life of Florence Nightingale.* 2 vols. in 1. New York: Macmillan.

Dawes, Robyn M. 1971 A Case Study of Graduate Admissions: Application of Three Principles of Human Decision Making. *American Psychologist* 26:180–188.

de Neufville, Richard; and Keeney, Ralph L. 1972 Use of Decision Analysis in Airport Development for Mexico City. Chapter 23 in Alvin W. Drake, Ralph L. Keeney, and Philip M. Morse (editors), *Analysis of Public Systems.* Cambridge, Mass.: M.I.T. Press.

Easterlin, Richard A. 1973 Does Money Buy Happiness? *Public Interest* [1973], no. 30:3–10.

Ehrlich, Isaac 1975 The Deterrent Effect of Capital Punishment: A Question of Life and Death. *American Economic Review* 65:397–417.

Elkana, Yehuda et al. (editors) 1977 *Toward a Metric of Science: The Advent of Science Indicators.* New York: Wiley.

Ennis, Philip. H. 1967 *Criminal Victimization in the United States: A Report of a National Survey.* U.S. President's Commission on Law Enforcement and Administration of Justice, Field Survey 2. National Opinion Research Center, Univ. of Chicago; Washington: Government Printing Office.

Equal Employment Opportunity Commission 1976 Guidelines on Employee Selection Procedures. Title 29, chapter 14, part 1607. *Federal Register* 41, part 228: S1983–S1986.

Fairley, William B. 1977 Evaluating the "Small" Probability of a Catastrophic Accident From the Marine Transportation of Liquefied Natural Gas. Pages 331–353 in William B. Fairley and Frederick Mosteller (editors), *Statistics and Public Policy.* Reading, Mass.: Addison-Wesley.

Fairley, William B.; and Mosteller, Frederick (editors) 1977 *Statistics and Public Policy.* Reading, Mass.: Addison-Wesley.

Fellegi, I. P.; and Phillips, J. L. 1974 Statistical Confidentiality: Some Theory and Applications to Data Dissemination. *Annals of Economic and Social Measurement* 3:399–409.

Ferreira, Joseph Jr. 1970 *Quantitative Models for Automobile Accidents and Insurance.* U.S. Department of Transportation, Automobile Insurance and Compensation Study. Washington: Government Printing Office.

Ferreira, Joseph Jr. 1974 The Long-term Effects of Merit-rating Plans on Individual Motorists. *Operations Research* 22, no. 5:954–978.

Fienberg, Stephen E. 1971 Randomization and Social Affairs: The 1970 Draft Lottery. *Science* 171:255–261.

Fienberg, Stephen E.; Larnitz, Kinley; and Reiss, Albert J. Jr. 1976 Redesigning the Kansas City Preventive Patrol Experiment. *Evaluation* 3:124–131.

Finkelstein, Michael O. 1966 The Application of Statistical Decision Theory to the Jury Discrimination Cases. *Harvard Law Review* 80:338–376.

Finkelstein, Michael O. 1973 Regression Models in Administrative Proceedings. *Harvard Law Review* 86: 1442–1475.

Fund for the City of New York 1972 *Legal Gambling in New York: A Discussion of Numbers and Sports Betting.* New York: The Fund.

Gilbert, John; Light, Richard; and Mosteller, Frederick 1975 Assessing Social Innovations: An Empirical Base for Policy. Chapter 2 in Carl A. Bennett and Arthur A. Lumsdaine (editors), *Evaluation and Experiment: Some Critical Issues in Assessing Social Programs.* New York: Academic Press.

Goldberger, Arthur S.; and Duncan, Otis Dudley (editors) 1973 *Structural Equation Models in the Social Sciences.* New York: Academic Press.

Gonzalez, Maria E. et al. 1975 Standards for Discussion and Presentation of Errors in Survey and Census Data. *Journal of the American Statistical Association* 70, part 2:5–23.

Hauser, Philip M. 1975 *Social Statistics in Use.* New York: Russell Sage.

Howard, R. A.; Matheson, J. E.; and North, D. W. 1972 The Decision to Seed Hurricanes. *Science* 176:1191–1202.

Huff, Darrell 1954 *How to Lie With Statistics.* New York: Norton.

Kaplan, John 1968 Decision Theory and the Factfinding Process. *Stanford Law Review* 20:1065–1092.

Kelling, George L. et al. 1974 *The Kansas City Preventive Patrol Experiment: A Summary Report.* Washington: Police Foundation.

Kinnard, Douglas 1975–1976 Vietnam Reconsidered: An Attitudinal Survey of U.S. Army General Officers. *Public Opinion Quarterly* 39:445–456.

Kruskal, William H. 1977 Taking Data Seriously. Chapter 6 in Yehuda Elkana et al. (editors), *Toward a Metric of Science: The Advent of Science Indicators.* New York: Wiley.

Larson, Richard C. 1976 What Happened to Patrol Operations in Kansas City? *Evaluation* 3:117–123.

Light, Richard J. 1973 Abused and Neglected Children in America: A Study of Alternative Policies. *Harvard Educational Review* 43:556–598.

Lippmann, Walter 1922 *Public Opinion.* New York: Macmillan. → A paperback edition was published by The Free Press in 1965.

Meehl, Paul E. 1954 *Clinical Versus Statistical Prediction: A Theoretical Analysis and a Review of the Evidence.* Minneapolis: Univ. of Minnesota Press.

Morgenstern, Oskar (1950) 1963 *On the Accuracy of Economic Observations.* 2d ed., completely rev. Princeton Univ. Press.

Moser, Claus 1976 The Role of the Central Statistical Office in Assisting Public Policy Makers. *American Statistician* 30, no. 2:59–67.

Mosteller, Frederick 1977 Assessing Unknown Numbers: Order of Magnitude Estimation. Pages 163–184 in William B. Fairley and Frederick Mosteller (editors), *Statistics and Public Policy.* Reading, Mass.: Addison-Wesley.

Moynihan, Daniel P. 1969 *Maximum Feasible Misunderstanding: Community Action in the War on Poverty.* New York: Free Press. → The Clarke A. Sanford Lectures on Local Government and Community Life, 1967. A paperback edition, with a new introduction by the author, was published in 1970.

National Academy of Sciences 1974 *Final Report of the Panel on Manpower Training Evaluation: The Use of Social Security Earnings Data for Assessing the Impact of Manpower Training Programs.* Washington: The Academy.

National Research Council, Committee on National Statistics 1976 *Surveying Crime.* Edited by Bettye K. Penick and Maurice E. B. Owens III. Report of the Panel for the Evaluation of Crime Surveys. Washington: National Academy of Sciences.

The New Jersey Income-maintenance Experiment. 3 vols. 1976–1977 New York: Academic Press. → Volume 1: *Operations, Surveys and Administration,* edited by D. Kershaw and I. Fair. Volume 2: *Labor–Supply Responses,* edited by Harold W. Watts and Albert Rees.

Volume 3: *Expenditures, Health and Social Behavior, and the Quality of the Evidence,* edited by Harold W. Watts and Albert Rees.

RIECKEN, HENRY W.; and BORUCH, ROBERT F. 1974 *Social Experimentation: A Method for Planning and Evaluating Social Intervention.* New York: Academic Press.

RIVLIN, ALICE M. 1971 *Systematic Thinking for Social Action.* Washington: Brookings. → The H. Rowan Gaither Lectures in Systems Science.

RIVLIN, ALICE M.; and TIMPANE, P. MICHAEL (editors) 1975 *Ethical and Legal Issues of Social Experimentation.* Washington: Brookings.

ROSE, RICHARD 1970 The Polls and the 1970 Election. Occasional Paper No. 7. Glasgow: Survey Research Centre, Univ. of Strathclyde.

ROSE, RICHARD (1974) 1975 The Polls and Election Forecasting in February 1974. Pages 109–130 in Howard R. Penniman (editor), *Britain at the Polls: The Parliamentary Elections of 1974.* Rev. & enl. ed. Foreign Affairs Study 14A. Washington: American Enterprise Institute for Public Policy Research.

SELVIDGE, JUDITH 1975 A Three-step Procedure for Assigning Probabilities to Rare Events. Pages 199–216 in Research Conference on Subjective Probability, Utility and Decision Making, Fourth, Rome, 1973, *Utility, Probability, and Human Decision Making.* Edited by Dirk Wendt and Charles Vlek. Dordrecht (Netherlands) and Boston: Reidel.

SLOVIC, PAUL; and LICHTENSTEIN, SARAH 1971 Comparison of Bayesian and Regression Approaches to the Study of Information Processing in Judgment. *Organizational Behavior in Human Performance* 6:649–744.

SOCIAL SCIENCE RESEARCH COUNCIL CONFERENCE ON SOCIAL EXPERIMENTS, BOULDER, *1974* 1976 *Experimental Testing of Public Policy: Proceedings.* Edited by Robert F. Boruch and Henry W. Riecken. Boulder, Colo.: Westview.

STEFFENS, LINCOLN 1931 *The Autobiography of Lincoln Steffens.* New York: Harcourt.

TUFTE, EDWARD R. 1973 The Relationship Between Seats and Votes in Two-party Systems. *American Political Science Review* 67:540–554.

TVERSKY, AMOS; and KAHNEMAN, DANIEL 1974 Judgment Under Uncertainty: Heuristics and Biases. *Science* 185:1124–1131.

U.S. DEPARTMENT OF HEALTH, EDUCATION, AND WELFARE 1973 *Summary Report: Graduated Work Incentive Experiment.* Washington: Government Printing Office.

U.S. DEPARTMENT OF HEALTH, EDUCATION, AND WELFARE, OFFICE OF EDUCATION 1966 *Equality of Educational Opportunity.* By James S. Coleman et al. Washington: Government Printing Office.

U.S. NUCLEAR REGULATORY COMMISSION 1975 *Reactor Safety Study.* Washington: Government Printing Office. → A draft summary. The final report was published as *Assessment of Accident Risks in U.S. Commercial Nuclear Power Plants,* NUREG 75/014.

U.S. OFFICE OF MANAGEMENT AND BUDGET, STATISTICAL POLICY DIVISION 1973 *Social Indicators, 1973: Selected Statistics on Social Conditions and Trends in the United States.* Washington: Government Printing Office.

U.S. PRESIDENT'S COMMISSION ON FEDERAL STATISTICS 1971 *Federal Statistics: Report.* 2 vols. Washington: Government Printing Office.

WALLIS, W. ALLEN; and ROBERTS, HARRY V. 1956 *Statistics: A New Approach.* Glencoe, Ill.: Free Press.

WEISBERG, HERBERT I. 1974 *Short Term Cognitive Effects of Head Start Programs: A Report on the Third Year of Planned Variation, 1971–72.* Cambridge, Mass.: Huron Institute.

ZECKHAUSER, RICHARD; SHEARER, GAIL; and MEMISHIAN, PAMELA 1974 Decision Analysis for Flight in the Stratosphere. Teaching and Research Materials, No. 18T. Public Policy Program, Harvard Univ.

QUALITATIVE ANALYSIS

For methods of obtaining, analyzing, and describing qualitative data, see COUNTED DATA; EVALUATION RESEARCH; GRAPHIC PRESENTATION; INTERVIEWING IN SOCIAL RESEARCH; SURVEY ANALYSIS; TABULAR PRESENTATION.

QUALITY CONTROL, STATISTICAL

"Quality control," in its broadest sense, refers to a spectrum of managerial methods for attempting to maintain the quality of manufactured articles at a desired level. "Statistical quality control" can refer to all those methods that use statistical principles and techniques for the control of quality. In this broad sense, statistical quality control might be regarded as embracing in principle all of statistical methodology. Some areas of statistics, however, have, both historically and as currently used, a special relationship to quality control; it is these areas that are discussed in the articles under this heading. The methods of quality control described in these articles have applications in fields other than industrial manufacturing. For example, process control has been used successfully as an administrative technique in large-scale data-handling organizations, such as census bureaus. Reliability concepts are important in engineering design and seem potentially useful in small-group research.

I. ACCEPTANCE SAMPLING   *H. C. Hamaker*
II. PROCESS CONTROL   *E. S. Page*
III. RELIABILITY AND LIFE TESTING   *Marvin Zelen*

I
ACCEPTANCE SAMPLING

It is generally recognized that mass-production processes inevitably turn out a small amount of product that does not satisfy the specification requirements. Certainly, however, the fraction of defective product should be kept under control, and *acceptance sampling* is one of the methods used for this purpose. Although primarily used in manufacturing situations, the techniques of acceptance sampling can also be applied to nonmanufacturing operations, such as interviewing or editing.

Mass products are usually handled in discrete lots, and these are the units to which acceptance sampling applies. A sample from each lot is inspected, and on the basis of the data provided by the sample it is decided whether the lot as a whole shall be *accepted* or *rejected*.

The actual procedures differ according to the nature of the product and the method of inspection. A distinction must be made between discrete products (nuts and bolts, lamps, radios) and bulk products (coal, fertilizer, liquids). Another important distinction is that between inspection by the method of attributes, where each item inspected is simply classified as defective or nondefective, and inspection by the method of variables, where meaningful numerical measurements on the sample are used to sentence a lot.

The following discussion will mainly concern attribute inspection, which is the technique most widely applied. Variables inspection will be mentioned only briefly.

Systematic investigations of acceptance sampling problems were initiated by two now classic papers of Dodge and Romig (1929–1941), which were later published in book form. Most of the basic concepts discussed here, such as the AOQL, the LTPD, and single and double sampling, go back to their work.

Further important developments took place during World War II, when acceptance sampling was extensively applied to military supplies. Research was carried out by the Statistical Research Group, Columbia University, and published in two books (1947; 1948).

Since then there has been a steady flow of publications concerned with a variety of both theoretical and practical aspects, such as modifications of sampling procedures, economic principles, the development of sampling standards, or applications in particular situations.

Method of attributes

Basic concepts. Attribute sampling applies to discrete products that are classified as defective or nondefective, regardless of the nature and seriousness of the defects observed. The simplest procedure consists in a *single sampling plan:*

From each lot submitted, a random sample of size n is inspected. The lot is accepted when the number of defectives, x, found in the sample is less than or equal to the acceptance number, c. When x is greater than or equal to the rejection number, $c + 1$, the lot is rejected.

Suppose $n = 100$ and $c = 2$; that is, 2 per cent defectives are permitted in the sample. This does not mean that a lot containing 2 per cent defectives will always be accepted. Chance fluctuations come into play; sometimes the sample will be better and sometimes worse than the lot as a whole. Statistical theory teaches that if the size of the sample is much smaller than the size of the lot, then a lot with 2 per cent defectives has a *probability of acceptance* $P_A = 0.68 = 68$ per cent. On the average, from 100 such lots, 68 will be accepted and 32 rejected.

It is clear that the probability of acceptance is a function of the *per cent defective*, p, in the lot. A plot of P_A as a function of p is called the *operating characteristic curve*, or the O.C. curve. Every sampling plan has a corresponding O.C. curve, which conveniently portrays its practical performance. The O.C. curves for three single sampling plans are presented in Figure 1.

O.C. curves always have similar shapes, and for practical purposes they can be sufficiently specified by two parameters only. The most important parameters that have been proposed are the following:

(1) *Producer's risk point* (PRP) $= p_{95} =$ that per cent defective for which the probability of acceptance, P_A, is 95 per cent.

(2) *Consumer's risk point* (CRP), more often called the lot tolerance percentage defective (LTPD) $= p_{10} =$ that per cent defective for which $P_A = 10$ per cent.

(3) *Indifference quality* (IQ) or *point of control* $= p_{50} =$ that per cent defective for which $P_A = 50$ per cent.

(4) *Average outgoing quality limit* (AOQL), which, however, applies only to rectifying inspection, that is, inspection whereby rejected lots are completely screened (100 per cent inspection) and accepted after the removal of all defective items. It can be shown that under rectifying inspection the average outgoing quality (AOQ), that is, the average per cent defective in the accepted lots, can never surpass a certain upper limit, even under the most unfavorable circumstances. This upper limit is the AOQL.

These four parameters are essentially characteristics of the O.C. curves or of the corresponding sampling plans. But the choice of a sampling plan cannot be determined by looking at the O.C. curves alone; in some way the choice must be in keeping with the capabilities of the production process under consideration. Realizing this, Dodge and Romig introduced the *process average*, that is, the average per cent defective found in the samples, as one of the quantities to be used in deciding on a sampling plan. Their sampling plans are such that for lots with a per cent defective equal to the process average the probability of acceptance is always fairly high.

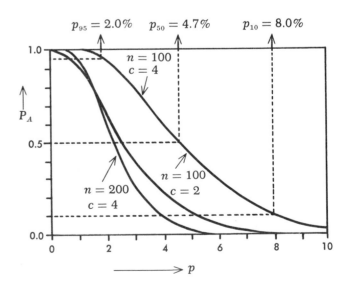

Figure 1 — O.C. curves for three single sampling plans

In practice this is not quite satisfactory, because it means that the acceptance sampling system is more lenient toward a poor producer with a high process average than toward a good one with a low average.

Hence the process average has now been superseded by the *acceptable quality level* (AQL). This is again a per cent defective and may be considered as an upper limit of acceptability for the process average. Definitions have varied in the course of time. In the earliest version of the *Military Standard 105*, the *105A*, which was developed for acceptance sampling of military supplies during World War II and published in 1950, the AQL was defined as a "nominal value to be specified by the U.S. Government." The sampling plans in this standard are so constructed that the probability of acceptance for lots that meet the AQL standard is higher than 90 per cent. This has led other authors to define the AQL instead as the per cent defective with a probability of acceptance of 95 per cent, thus identifying the AQL with the PRP. Thereby, however, the AQL would be essentially a characteristic of a sampling plan and no longer a tolerance requirement of production. The uncertainty seems to have been finally settled by defining the AQL as the maximum per cent defective that, for purposes of acceptance sampling, can be considered satisfactory as a process average. This definition has been adopted in the latest version of the *Military Standard 105*, the *105D*, and in a standard on terminology established by the American Society for Quality Control.

Strictly, the O.C. curve of a single sampling plan also depends on the size of the lot, N, but if, as in most cases, the sample is only a small fraction of the lot, the O.C. curve is almost completely determined by the sample size, n, and the acceptance number, c; the influence of the lot size is unimportant and can be ignored. Since, moreover, the per cent defective is generally very low, a few per cent only, the Poisson formula can be used for computing the O.C. curves [see DISTRIBUTIONS, STATISTICAL, *article on* SPECIAL DISCRETE DISTRIBUTIONS].

An important logical distinction between contexts must be made: does the O.C. curve relate to the specific lot at hand or to the underlying process that produced the lot? Definitions of consumer's risk, etc., also may be made in either of these contexts. In practice, however, the numerical results are nearly identical, whether one regards as basic a particular lot or the underlying process.

It will be noticed that the O.C. curve of a sampling plan has a close resemblance to the power curve of a one-sided test of a hypothesis [see HYPOTHESIS TESTING]. Indeed, acceptance sampling can be considered as a practical application of hypothesis testing. If on the basis of the findings in the sample the hypothesis $H_0 : p \leqslant$ PRP is tested with a significance level of 5 per cent, the lot is rejected if the hypothesis is rejected and the lot is accepted if the hypothesis is not rejected. Likewise $H_0 : p =$ PRP; $H_1 : p =$ LTPD can be considered as two alternative hypotheses tested against each other with errors of the first and second kind, of 5 per cent and 10 per cent respectively.

Curtailed, double, multiple, and sequential sampling. Curtailed sampling means that inspection is stopped as soon as the decision to accept or reject is evident. For example, with a single sampling plan with $n = 100$, $c = 2$, inspection can be stopped as soon as a third defective item is observed. This reduces the number of observations without any change in the O.C. curve. The gain is small, however, because bad lots, which are rejected, occur infrequently in normal situations.

Better efficiency is achieved by double sampling plans, which proceed in two stages. First, a sample of size n_1 is inspected, and the lot is accepted if $x_1 \leqslant c_1$ and rejected if $x_1 \geqslant c_2$, x_1 being the number of defectives observed. But when $c_1 < x_1 < c_2$, a second sample, of size n_2, is inspected; and the lot is then finally accepted if the total number of defectives, $x_1 + x_2$, is $\leqslant c_3$ and is rejected otherwise. The basic idea is that clearly good or bad lots are sentenced by the first sample and only in doubtful cases is a second sample required.

Multiple sampling operates on the same principle but in more than two successive steps, often six or eight. After each step it is decided again whether to accept, reject, or proceed to inspection of another sample. In (fully) sequential sampling, inspection is carried out item by item and a three-way decision taken after each item is inspected.

The basic parameters discussed above apply also to double, multiple, and sequential sampling plans. These procedures can be so constructed that they possess O.C. curves almost identical with that of a given single sampling plan; on the average, however, they require fewer observations. The amount of saving is on the order of 25 per cent for double sampling and up to 50 per cent for multiple and sequential sampling.

Disadvantages of these plans as compared with single sampling plans are more complicated administration and a variable inspection load. Double sampling is fairly often adopted, but multiple and sequential sampling are used only in cases where the cost of inspection is very high, so that a high economy is essential. In many situations single sampling is preferred for the sake of simplicity.

The choice of a sampling plan. There are many considerations involved in choosing a sampling plan.

Sampling standards. If numerical values are specified for two of the four parameters listed above, the O.C. curve is practically fixed and the corresponding sampling plan may be derived by computation or from suitable tables. In practice, however, the use of standard sampling tables is much more common.

The best known of these tables is the *Military Standard 105A* (U.S. Department of Defense, Standardization Division 1950), which prescribes a sampling plan in relation to an AQL and the size of the lot; the sample size is made to increase with the lot size so as to reduce the risk of wrong decisions. Three separate tables give single, double, and multiple plans. The user can also choose between different inspection levels, corresponding to systematic changes in all sample sizes; a sample size $n = 50$ at level I is increased to $n = 150$ at level III for example.

The *Military Standard 105* also contains rules for a transition to *tightened* or *reduced inspection*. The idea is that, when the sample data indicate that the average per cent defective in the lots received is significantly higher than the required AQL, the consumer should tighten his inspection in order to prevent the acceptance of too many bad lots and to stimulate the producer to improve his production. If, on the other hand, the average per cent defective is significantly lower than the AQL, this indicates that the production process is tightly controlled; inspection is then less essential and the amount of inspection can be reduced.

The *Military Standard 105* has several times been revised; the successive versions are known as the *Military Standard 105A, 105B, 105C,* and *105D* (U.S. Department of Defense, Standardization Division 1963). The differences between *A, B,* and *C* are relatively unimportant; but in *105D,* issued in 1963, more drastic changes were introduced. These changes were proposed by a special committee of experts jointly appointed by the departments of defense in the United States, Great Britain, and Canada. The basic principles underlying the *Military Standard 105,* however, have not been altered.

The *Military Standard 105* was originally intended for acceptance sampling of military supplies, but it has also been widely applied in industry for other purposes. Other standards have been established by large industrial firms or by governmental offices in other countries. Sometimes these are only modifications of the *Military Standard 105A;* sometimes they have been developed independently and from different points of view. The merits and demerits of a number of these standards are discussed by Hamaker (1960).

Acceptance sampling as an economic problem. Many authors have attempted to deduce an optimum sampling plan from economic considerations. Given (*a*) the cost of inspection of an item, (*b*) the a priori distribution, that is, the distribution of the per cent defective among the lots submitted for inspection, (*c*) the loss caused by accepted defectives, and (*d*) the loss due to rejected nondefectives, one can, in principle, derive an optimum sampling plan, for which the total cost (cost of inspection plus losses) is a minimum.

These economic theories have had little practical success except in isolated cases, mainly, I believe, because their basis is too restricted and because they require detailed information not readily available in industry. The a priori distribution is usually not known and cannot easily be obtained; the consequences of rejecting a lot depend to a very high degree on the stock available and may consequently vary from day to day. Rejection often does not mean that a lot is actually refused; it only means that the inspector cannot accept without consultation with a higher authority or some other authorization.

Since even the most simple products can show a variety of defects, some much more serious than others, it is not easy to see how the loss due to accepted defectives should be estimated. In the *Military Standard 105* this problem has been solved by a classification into critical, major, and minor defects, with a separate AQL for each class. An alternative method, *demerit rating,* consists in scoring points, say 10, 3, and 1, for critical, major, and minor defects and sentencing the lot by the total score resulting from a sample. The advantage is a single judgment instead of three separate judgments, but the theory of O.C. curves no longer applies in a simple and straightforward manner. There are practical problems as to how different defects should be classified and what scores should be assigned to the different classes. These should be solved by discussions between parties interested in the situation envisaged. Some case studies have been described in the literature. No attempts have so far been made to incorporate the classification of defects into economic theories; that would make them yet more complicated and unworkable. This is a decided drawback because in actual practice the degree of seriousness of the defects observed always has a considerable influence on the final decisions concerning the lots inspected. Besides, the economic theories assume that the prevention of accepted defectives is the only purpose of acceptance sampling and this is not correct. It also serves to show the consumer's interest in good quality and to stimulate the producer to take good care of his production processes. Some large firms have developed vendor rating systems for this very purpose. The information supplied by the samples is systematically collected for each supplier separately and is used in deciding where to place a new contract.

It should not be concluded that no attention is devoted to the economic aspects of the problem of choosing a sampling plan. In acceptance sampling there is always a risk of taking a wrong decision by accepting a bad lot or rejecting a good lot. The larger the lot, the more serious are the economic consequences, but the risk of wrong decisions can be reduced by using larger samples, which give steeper O.C. curves. All existing sampling tables prescribe increased sample sizes with increasing lot sizes, and this practice is derived from economic considerations.

Also, before installing sampling inspection it is always necessary to make some inquiries about the situation to be dealt with. Do bad lots occur, and if so how frequently? How bad are those bad lots? Do accepted defectives cause a lot of trouble? The answers to such questions as these provide crude information about the a priori distribution and the economic aspects, and in choosing an AQL and an inspection level this information is duly taken into account.

Looking at the problem from this point of view, industrial statistics always uses a rough Bayesian approach [see BAYESIAN INFERENCE]. A statistician will never be successful in industry if he does not properly combine his statistical practices with existing technical knowledge and experience and with cost considerations. It is only when an attempt is made to apply the Bayesian principles in a more precise way that problems arise, because the basic parameters required for that purpose cannot easily be estimated with sufficient accuracy. [It is interesting in this connection to compare the empirical Bayesian approach discussed in DECISION THEORY.]

Method of variables

In sampling by the method of variables a numerical quality characteristic, x, is measured on each item in the sample. It is usually supposed that x has a normal distribution [see DISTRIBUTIONS, STATISTICAL] and that a product is defective when x falls beyond a single specification limit or outside a specification interval: that is, when $x < L$ and/or $x > U$. The basic idea is that from the mean, $\bar{x}$, and standard deviation, s, computed from the sample, an estimate of the per cent defective in the lot can be derived, and this estimate can be compared with the AQL required. For a single specification limit the criterion operates as follows: If x is a quality such as length, and the specification requires that not more than a small percentage, p, of the units in the lot shall have a length greater than U (upper limit), then the lot is accepted when $U - \bar{x} \geqslant ks$ and rejected when $U - \bar{x} < ks$, where k is a constant that depends on p and the sample size n and is derived from the theory. For double limits the technique is somewhat more complicated. When the standard deviation is known, s is replaced by σ and a different value of k has to be used.

On the basis of earlier suggestions, the theory of sampling by variables was worked out in detail by the Statistical Research Group, Columbia University (1947), and by Bowker and Goode (1952). This theory led to the establishment of the *Military Standard 414* in 1957. This standard has a structure similar to the *Military Standard 105A*; lot size and AQL are the main parameters that determine a sampling plan. Tables are given for one and two specification limits, and for σ both known and unknown.

The advantage of sampling by variables is a more effective use of the information provided by the sample and consequently fewer observations. Where the *Military Standard 105A* prescribes a sample size $n = 150$, the *Military Standard 414* uses $n = 50$ when σ is unknown and n ranging from 8 to 30 when σ is known.

Disadvantages are that the performance and handling of measurements require more highly trained personnel and that the assumption of a normal distribution is a risky one when it is not known under what circumstances a lot has been produced. For these reasons sampling by variables has found only limited application; sampling by attributes is often preferred.

Present-day reliability requirements have led to an increased interest in life testing procedures and hence to the development of acceptance sampling techniques based on the exponential and the Weibull distributions [see QUALITY CONTROL, STATISTICAL, *article on* RELIABILITY AND LIFE TESTING].

Acceptance sampling of bulk materials. Acceptance sampling of bulk material constitutes a separate problem. A liquid can be homogenized through stirring, and then the analysis of a single specimen will suffice. Solid material, such as fertilizer or coal, is handled in bales, barrels, wagons, etc. Then, there may be a variability within bales and additional variability among bales. Extensive research is often needed for each product separately before an adequate acceptance sampling procedure can be developed. A good example is Duncan's investigation of fertilizer (1960).

The theory of sampling by attributes is fairly complete. Research into the economics of acceptance sampling will probably continue for quite a while, but this seems to be an interesting academic exercise that will not lead to drastic changes in industrial practices. A common international sampling standard would be of great practical value, but since in different countries different standards are already established, this is a goal that will not be easily reached.

The theory of sampling by variables may perhaps require some further development. In the *Military Standard 414* the sample sizes prescribed when σ is unknown are three times the size of those for σ known, even for samples as large as 100 or more items. Further research may make possible the reduction of this ratio.

Perhaps the most important new developments will come from new fields of application. It is, for example, recognized that using accountancy to check a financial administration can also be considered as an acceptance sampling procedure and should be dealt with as such. However, the nature of the material and the requirements to be satisfied are entirely different from those in the technological sector. Industrial techniques cannot be taken over without considerable modification, and suitable methods have to be developed afresh. (See Vance & Neter 1956; Trueblood & Cyert 1957.)

H. C. HAMAKER

BIBLIOGRAPHY

BOWKER, ALBERT H.; and GOODE, HENRY P. 1952 *Sampling Inspection by Variables*. New York: McGraw-Hill.

COLUMBIA UNIVERSITY, STATISTICAL RESEARCH GROUP 1947 *Selected Techniques of Statistical Analysis for Scientific and Industrial Research, and Production and Management Engineering*. New York: McGraw-Hill. → See especially Chapter 1 on the use of variables in acceptance inspection for per cent defective.

COLUMBIA UNIVERSITY, STATISTICAL RESEARCH GROUP 1948 *Sampling Inspection*. New York: McGraw-Hill. → Describes theory and methods of attribute inspection developed during World War II.

DEMING, W. EDWARDS 1960 *Sample Design in Business Research*. New York: Wiley. → Contains practical examples of the use of samples in business administration; some of these applications can be considered as problems in acceptance sampling.

DODGE, HAROLD F.; and ROMIG, HARRY G. (1929–1941) 1959 *Sampling Inspection Tables: Single and Double Sampling*. 2d ed., rev. & enl. New York: Wiley; London: Chapman. → This book is a republication of fundamental papers published by the authors in 1929 and 1941 in the *Bell System Technical Journal*.

DUNCAN, ACHESON J. (1952) 1965 *Quality Control and Industrial Statistics*. 3d ed. Homewood, Ill.: Irwin. → Contains five chapters on acceptance sampling.

DUNCAN, ACHESON J. 1960 An Experiment in the Sampling and Analysis of Bagged Fertilizer. *Journal of the Association of Official Agricultural Chemists* 43: 831–904. → A good example of the research needed to establish acceptance-sampling procedures of bulk material.

GRANT, EUGENE L. (1946) 1964 *Statistical Quality Control*. 3d ed. New York: McGraw-Hill. → The merits and demerits of various acceptance-sampling procedures and standards as well as *Military Standard 105D* are discussed in detail.

HALD, H. A. 1960 The Compound Hypergeometric Distribution and a System of Single Sampling Inspection Plans Based on Prior Distributions and Costs. *Technometrics* 2:275–340. → Considers sampling from an economic point of view, and contains a fairly complete list of references to earlier literature on this aspect of the sampling problem.

HAMAKER, HUGO C. 1958 Some Basic Principles of Acceptance Sampling by Attributes. *Applied Statistics* 1:149–159. → Contains a discussion of the difficulties hampering the practical application of economic theories.

HAMAKER, HUGO C. 1960 Attribute Sampling in Operation. International Statistical Institute, *Bulletin* 37, no. 2:265–281. → A discussion of the merits and demerits of a number of existing sampling standards.

Quality Control and Applied Statistics: Abstract Service. → Published since 1956. A useful source of information giving fairly complete abstracts of papers published in statistical and technical journals. Of special importance are the classification numbers 200–299: *Sampling Principles and Plans*, and number 823: *Sampling for Reliability*.

TRUEBLOOD, ROBERT M.; and CYERT, RICHARD M. 1957 *Sampling Techniques in Accounting*. Englewood Cliffs, N.J.: Prentice-Hall.

U.S. DEPARTMENT OF DEFENSE, STANDARDIZATION DIVISION 1950 *Military Standard 105A*. Washington: Government Printing Office. → A sampling standard widely applied. In 1959 this standard was slightly revised in *Military Standard 105B*. Recently, a further, more drastic revision has been effected jointly by the departments of defense in Canada, Great Britain, and the United States.

U.S. DEPARTMENT OF DEFENSE, STANDARDIZATION DIVISION 1957 *Military Standard 414*. Washington: Government Printing Office. → A standard for variables inspection corresponding to *Military Standard 105A*.

U.S. DEPARTMENT OF DEFENSE, STANDARDIZATION DIVISION 1963 *Military Standard 105D*. Washington: Government Printing Office. → Earlier versions of this standard are known as *Military Standard 105A, 105B*, and *105C*.

VANCE, LAWRENCE L.; and NETER, JOHN 1956 *Statistical Sampling for Auditors and Accountants*. New York: Wiley.

Postscript

As stated in the main article, a common international standard on acceptance sampling would be of great practical importance. It is therefore to be considered a great step forward that the International Standardisation Organisation (ISO) adopted, in 1973, the *Military Standard 105D* as one of its official standards, indicated as *ISO/DIS 2859*. Moreover, a second standard, *ISO/DIS 3319*, has been added, explaining the basic principles underlying attribute sampling and the practical use of the *ISO 2859* tables. Both these standards are available in English and in French. The standard *105D* is likewise recommended by the International Electrical Commission for the sampling of lots of electrical components. It is to be hoped that, in consequence of these developments, this standard will soon replace the variety of national standards that have been used in the past.

Furthermore, a greatly simplified and abbreviated version of the *Military Standard 414* has been worked out by the Department of Defence in Great Britain and will presumably also be brought out as an ISO standard in the near future.

Practical considerations. Sampling standards do not provide ready-made solutions for all possible situations. In the *Military Standard 105D*, one of the main parameters used in selecting a sampling plan is the AQL, which implies that this standard is primarily designed for acceptance sampling of a continuing series of lots where the concept of a process average makes sense. However, the standard also provides full information on the O.C. curves of all its sampling plans, both in the form of graphs and of numerical tables. These may be consulted in situations in which the main tables cannot be applied directly; for example, in sampling inspection of a single or isolated lot.

Many other practical questions should be considered in organizing an acceptance sampling operation. A precise definition of a defect should be established. The size and composition of the lots to be inspected may require careful attention. On the average, large lots require less inspection than a set of smaller lots of the same total size, but large lots are more difficult to sample and can generally be expected to be less homogeneous in quality. It should be clearly stated how rejected lots are to be disposed of; they can, for instance, be destined for some other purpose, be sold at a reduced price, be subjected to 100 per cent inspection with removal or reworking of the defective pieces, or they can be scrapped as waste. The choice among such alternatives will depend on the nature of the defects observed, and someone must be given the authority to decide.

Consequently, a sampling standard must be considered only as a tool, the effective use of which requires a certain amount of common sense and practical experience.

H. C. Hamaker

II
PROCESS CONTROL

The present article discusses that aspect of statistical quality control relating to the control of a routinely operating process. The traditional and most common field of use is in controlling the quality of manufactured products, but applications are possible in fields as diverse as learning experiments, stock exchange prices, and error control in the preparation of data for automatic computers. Process control of this kind is usually effected by means of charts that exhibit graphically the temporal behavior of the process; hence, the subject is sometimes called, somewhat superficially, "control charts."

Inspection for control. The concept of quality control in its industrial context and the first widely used methods were introduced by W. A. Shewhart (1931). One of the prime concerns was to detect whether the items of output studied had characteristics that behaved like independent observations from a common statistical distribution, that is, whether groups of such items had characteristics behaving like random samples. If the procedure suggested acceptance of the hypothesis of a single distribution, with independence between observations, the production process was said to be "in control" (Shewhart 1931, p. 3), although at this stage the quality of the output, determined by the parameters of the distribution, might not be acceptable. A major contribution was the explicit recognition that such a state of control is necessary before any continuous control of the process can succeed. When the initial examination of the sample data shows the process to be out of control in the above sense, reasons connected with the operation of the process are sought (the search for "assignable causes"; Shewhart 1931, p. 13), improved production methods are introduced, and a further examination is made to see if control has been achieved. Thus, this initial phase of study of a process employs a significance test of a hypothesis in a way encountered in other applications of statistics. It should be noted, however, that often no alternative hypotheses are specified and that, indeed, they are frequently only vaguely realized at the beginning of the investigation.

Inspection of a process in control. When the process is in a state of control and its output has the relevant quality characteristics following a distribution with the values of the parameters at their targets (that is, those values of the parameters chosen to cause the output to meet the design specifications with as much tolerance as possible), one task is to determine when the process departs from this state, so that prompt restoring action may be taken. A quality control scheme for this purpose needs, therefore, to give a signal when action is demanded. One of the features of importance for a process inspection scheme is thus the speed at which it detects a change from target. In conflict with the desire for rapid detection of a change is the necessity for infrequent signals demanding action when no change from target has occurred. The "errors" of signaling a change when none has occurred and of failing to give the signal immediately after a change are similar to the two types of error familiar in hypothesis testing. Whereas in the latter case the probability of such errors is a suitable measure of their occurrence, in quality control the repetition of sampling as the process continues operation and the possibility of combining several samples to decide about a signal make probabilities less convenient measures of error behavior. Instead, the average run length (A.R.L.), defined as the average number of samples taken up to the appearance of a signal, gives a convenient means of comparison for different process inspection schemes (Barnard 1959, p. 240); the A.R.L. is easily related both to the amount of substandard output produced between any change and the signal and to the frequency of unnecessary interference with a controlled process.

Ancillary tasks of process inspection. In addition to providing a signal after a change, a process inspection scheme may be required to yield other information. For example, the magnitude of change may need to be estimated so that a dial may be adjusted. Again, coupled with the provision of a signal to take remedial action may be a rule about the destination of any recent production for which there is evidence of a fall in standard, and so the position of any change may need estimation; in this case the scheme is partly one for the deferred sentencing of output, that is, for retrospective acceptance or rejection. In other cases a satisfactory record of a process inspection scheme may guarantee the acceptance of the output by a consumer.

In these and other manufacturing applications where the aim is financial gain, a comparison of alternative schemes in monetary terms should be attempted and not, as so frequently has happened in practice, ignored. Although quantification of many aspects of real situations is difficult or impossible, a monetary comparison can avoid the confusion of using in one situation a measure appropriate for comparing schemes in a totally different situation.

Types of inspection schemes. Schemes for the various applications base their rules for the appropriate action to be taken on the results either of single samples of observations or of sequences of such samples; the sequences may be of fixed or variable length. All such schemes are closely related, for those employing a single sample or a fixed number of samples can be exhibited as special cases of schemes using a variable number of samples. However, the appearance of the graphical records is quite different. In the first two cases the individual sample results are marked on the chart (Figure 1), while in the last case the sums of all the previous observations, corrected for any expected trend, are plotted (Figure 2). The application of such techniques to different distributions changes the schemes only slightly; the constants involved differ, and sometimes attention is concentrated upon evidence of changes in the distribution parameter in one direction only—for example, increases in the fraction of defective articles produced or in the variance of a measured characteristic are often particularly interesting.

Besides the normal distribution model to be discussed below, another useful model is the Poisson for applications concerning the number of occur-

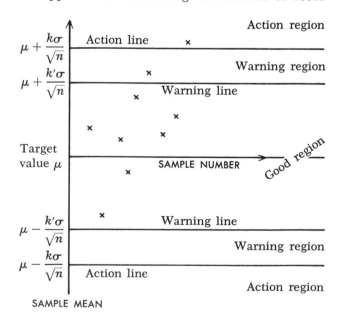

Figure 1 — Chart with warning lines

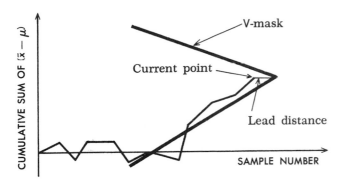

Figure 2 — Cumulative sum chart

rences of a particular type in a given item or length of time, for example, the number of blemishes in sheets of glass of fixed size. The most commonly used schemes are those for controlling the mean of a normal distribution; they are typical of those for other distributions, both continuous and discrete, although schemes for each distribution need separate consideration for calculation of their properties.

The samples used in this sort of activity are usually small, of perhaps four to six observations.

The Shewhart procedure. The classical Shewhart procedure for providing a signal when a process in control departed from its target mean μ used two "action lines" drawn at $\mu \pm k\sigma/\sqrt{n}$, where σ is the process standard deviation estimated from many earlier observations, n is the size of sample, and k is a constant chosen pragmatically to be 3 (Shewhart 1931, p. 277) or 3.09 (Dudding & Jennett 1942, p. 64). The signal of a change was received when any sample point, the mean of a sample, fell outside the action lines (Shewhart 1931, p. 290). These rules are such that only about 1 in 500 samples would yield a point outside the action lines if the output were in control with the assumed parameter values; the 1-in-500 choice rests on experience and lacks any foundation from consideration of costs, but the success of these schemes has, in many applications, afforded abundant justification of their use and represented a major advance in process control. For processes where information is required about small changes in the mean, the single small-sample schemes were insufficiently sensitive—small changes need large samples for quick detection—but limitations in sampling effort and a reluctance to sacrifice the possibility of rapid detection of large changes with small samples have caused additions to the original Shewhart scheme.

Warning line schemes. Charts of warning line schemes have drawn on them the action lines and two other lines—warning lines—at $\mu \pm k'\sigma/\sqrt{n}$, where k' is a constant less than k (Dudding & Jennett 1942, p. 14; Grant 1946, art. 159). Accordingly, the chart is divided into action, warning, and good regions (Figure 1). A change in parameter is signaled by rules such as the following: Take action if m out of the last n sample points fell in one of the warning regions or if the last point falls in an action region (Page 1955).

Schemes using runs of points are special cases (Grant 1946, art. 88); some that are popular base their action rule on runs of sample points on one side of the target mean (case $k' = 0$, $k = \infty$, $m = n$, that is, the action regions disappear and the two warning regions are separated only by a line at the target value: action is taken when a long enough sequence of consecutive points falls in one region).

These schemes retain some of the advantages of small samples and seek to combine the results of a fixed number of samples in a simple way to increase the sensitivity for sustained small changes in parameter.

● *Cumulative sum schemes.* An extension of this idea is provided by the cumulative sum schemes (or cusum schemes), which enjoy the advantages of both large and small samples by combining the relevant information from all recent samples (Barnard 1959, p. 270). Instead of plotting individual sample means, $\bar{x}$, on the chart, the differences of these means from the target value, $\bar{x} - \mu$, are cumulated and the running total plotted after each sample is taken (Figure 2). A change in process mean causes a change in the direction of the trend of plotted points. One method of defining the conditions for a signal is to place a V-mask on the chart and take action if the arms of the V obscure any of the sample points. The angle of the V and the position of the vertex relative to the last plotted point (the lead distance) can be chosen to achieve the required A.R.L.s. Tables of these constants exist for several important distributions (Ewan & Kemp 1960; Goldsmith & Whitfield 1961; Kemp 1961; Page 1962; 1963). Alternative methods of recording may be adopted for schemes to detect one-sided or two-sided deviations in the parameter. These schemes are based upon the corresponding methods for recording the one-sided or two-sided repeated sequential tests of Wald-type, to which the cusum schemes are equivalent [Woodward & Goldsmith 1966; *see also* SEQUENTIAL ANALYSIS]. Gauging devices may be used instead of measuring in order to speed the manual operation of such a scheme or to make automatic performance simpler (Page 1962).

● **Related developments.** An increasing use of cusum schemes has prompted theoretical studies of the properties of estimators derived from them; for example, the point of apparent significant change in the direction of the path as an estimate of the point at which the change in population parameter occurred (Hinkley 1971; 1972). Attention has been given to schemes for the complete control of the process, that is, procedures for automatically detecting departures from target and making adjustments of appropriate sizes to the control variables (Box & Jenkins 1962; 1964; 1970). Naturally, such methods of control are applicable only to those processes for which both measurements of the quality characteristics and the adjustments can be made automatically. Serial correlation, of course, affects process inspection schemes, and in those (perhaps comparatively rare) cases where the characteristics of the run length distribution are critical in the application, both the process itself and the scheme need special study. Broadly speaking, positive serial correlations decrease the A.R.L. of cusum schemes from the independent case (Johnson & Bagshaw 1974). Other work has examined different stochastic models of process behavior (Barnard 1959; Bather 1963) and has attempted a comprehensive study of all the costs and savings of a process inspection scheme (Duncan 1956; Schmidt & Taylor 1973). In industrial applications the financial benefits accruing from the operation of such a scheme are usually of paramount importance, and, however difficult it may be to assess them quantitatively, they deserve careful consideration at all stages of the selection and operation of the scheme.

E. S. PAGE

[See also the biography of SHEWHART.]

BIBLIOGRAPHY

BARNARD, G. A. 1959 Control Charts and Stochastic Processes. *Journal of the Royal Statistical Society* Series B 21:239–271. → Introduces the V-mask cusum chart and estimation methods.

BATHER, J. A. 1963 Control Charts and the Minimization of Cost. *Journal of the Royal Statistical Society* Series B 25:49–80.

BOX, G. E. P., and JENKINS, G. M. 1962 Some Statistical Aspects of Adaptive Optimization and Control. *Journal of the Royal Statistical Society* Series B 24:297–343.

BOX, G. E. P.; and JENKINS, G. M. 1964 Further Contributions to Adaptive Quality Control; Simultaneous Estimation of Dynamics: Non-zero Costs. International Statistical Institute, *Bulletin* 40:943–974.

DUDDING, B. P.; and JENNETT, W. J. 1942 *Quality Control Charts*. London: British Standards Institution.

DUNCAN, ACHESON J. (1952) 1959 *Quality Control and Industrial Statistics*. Rev. ed. Homewood, Ill.: Irwin.

DUNCAN, ACHESON J. 1956 The Economic Design of $\bar{x}$ Charts Used to Maintain Current Control of a Process. *Journal of the American Statistical Association* 51:228–242.

EWAN, W. D.; and KEMP, K. W. 1960 Sampling Inspection of Continuous Processes With No Autocorrelation Between Successive Results. *Biometrika* 47:363–380. → Gives tables and a nomogram for one-sided cusum charts on a normal mean and fraction defective.

GOLDSMITH, P. L.; and WHITFIELD, H. 1961 Average Run Lengths in Cumulative Chart Quality Control Schemes. *Technometrics* 3:11–20. → Graphs of V-mask schemes for normal means.

GRANT, EUGENE L. (1946) 1964 *Statistical Quality Control*. 3d ed. New York: McGraw-Hill. → Many examples of the Shewhart chart for different distributions.

KEMP, K. W. 1961 The Average Run Length of the Cumulative Sum Chart When a V-mask Is Used. *Journal of the Royal Statistical Society* Series B 23:149–153. → Gives tables of cusum schemes for a normal mean.

PAGE, E. S. 1954 Continuous Inspection Schemes. *Biometrika* 41:100–115. → Introduces cusum schemes.

PAGE, E. S. 1955 Control Charts With Warning Lines. *Biometrika* 42:242–257.

PAGE, E. S. 1962 Cumulative Sum Schemes Using Gauging. *Technometrics* 4:97–109.

PAGE, E. S. 1963 Controlling the Standard Deviation by Cusums and Warning Lines. *Technometrics* 5:307–315. → Gives tables of cusum schemes for a normal range.

SHEWHART, WALTER A. 1931 *Economic Control of Quality of Manufactured Product*. Princeton, N.J.: Van Nostrand. → The classic volume introducing control chart methods.

ADDITIONAL BIBLIOGRAPHY

BOX, G. E. P.; and JENKINS, G. M. (1970) 1976 *Time Series Analysis: Forecasting and Control*. Rev. ed. San Francisco: Holden-Day.

HINKLEY, D. V. 1971 Inference About the Change-point From Cumulative Sum Tests. *Biometrika* 58:509–523.

HINKLEY, D. V. 1972 Time-ordered Classification. *Biometrika* 59:509–523.

JOHNSON, R. A.; and BAGSHAW, M. 1974 The Effect of Serial Correlation on the Performance of Cusum Tests. *Technometrics* 16:103–112.

SCHMIDT, J. W.; and TAYLOR, R. E. 1973 A Dual Purpose Cost Based Quality Control System. *Technometrics* 15:151–166.

WOODWARD, ROBERT H.; and GOLDSMITH, P. L. 1966 *Cumulative Sum Techniques*. London: Oliver & Boyd; Princeton, N.J.: Van Nostrand.

III

RELIABILITY AND LIFE TESTING

Technology has been characterized since the end of World War II by the development of complex systems containing large numbers of subsystems, components, and parts. This trend to even larger and more complex systems is accelerating with the development of space vehicles, electronic computers, and communications and weapons systems. Many of these systems may fail or may operate inefficiently if a single part or component fails. Hence, there is a high premium on having the components operate efficiently so that the system operates in a reliable, trustworthy manner.

In order to have reliable systems, it is not only necessary initially to design the system to be reliable, but also, once the system is in operation, to have appropriate maintenance and check-out schedules. This requires quantitative estimates of the reliability of the entire system, as well as reliability estimates for the major components, parts, and circuits that make up the system. While formalized definitions of reliability vary, there is general agreement that it refers to the probability of satisfactory performance under clearly specified conditions.

Reliability of components and the system. An important problem is to predict the reliability of a system from knowledge of the reliability of the components that make it up. For example, a satellite system may be regarded as composed of subsystems that perform the propulsion, guidance, communication, and instrument functions. All subsystems must function in order for the satellite to function. It is desired to predict the reliability of the satellite from knowledge of the reliability of the basic components or subsystems.

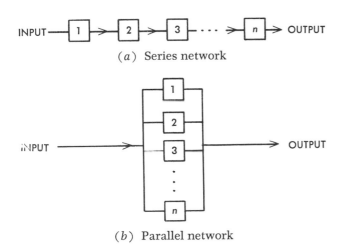

(a) Series network

(b) Parallel network

Figure 1

More abstractly, consider a system made up of n major components, each of which must function in order that the whole system function. An idealization of this system is to regard the components as connected in a series network (Figure 1a), although the actual connections may be much more complex. The input signal to a component is passed on to a connecting component only if the component is functioning. Let p_i refer to the probability that the ith component functions, and assume that

the components operate *independently* of one another. Then the probability of the entire system functioning correctly is equal to the product of the probabilities for each major component; that is,

Reliability of series system: $R = p_1 p_2 \cdots p_n$.

It is clear that the reliability of the system is no greater (and generally rather less) than the reliability of any single component. For example, if a system has three major components with reliabilities of .95, .90, and .90, the reliability of the entire system would be $R = (.95)(.90)(.90) = .7695$.

One way of improving the reliability of a system is to introduce redundancy. Redundancy, as usually employed, refers to replacing a low-reliability component by several components having identical functions. These are connected so that it is necessary only that *at least one* of these components function in order for the system to function. Auxiliary power systems in hospital operating rooms and two sets of brakes on automobiles are common examples of redundancy. Redundant components may be idealized as making up a parallel series (cf. Figure 1b). When there are n components in parallel, such that each component operates independently of the others, the reliability of the system is given by

Reliability of parallel system: $R = 1 - \prod_{i=1}^{n}(1 - p_i)$.

Since $1 - R = \prod_i (1 - p_i) \leqslant 1 - p_i$ for any i, it can be seen that $R \geqslant p_i$ for any i. Consequently using redundancy always increases the reliability of a system (assuming $0 < p_i < 1$). The most frequent use of redundancy is when a single component is replaced by n identical components in parallel. Then the reliability of this parallel system is $R = 1 - (1 - p)^n$, where p refers to the probability of a single component functioning.

The formulas for the reliability of series and parallel systems are strictly valid only if the components function independently of one another. In some applications this may not be true, as failure of one component may throw added stress on other components. In practice, also, an entire system will usually be made up of both series and parallel systems.

Note that either in the series or parallel network, one can regard the system as functioning if one can take a "path" from the input to the output via functioning components. There is only one such path for components in series whereas there are n possible paths for the parallel system. The concept of finding paths of components for which the system functions is the basis for predicting the re-

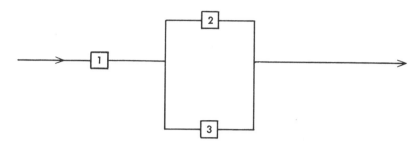

(*a*) Two-terminal network: $\phi(\mathbf{X}) = X_1[1 - (1 - X_2)(1 - X_3)]$

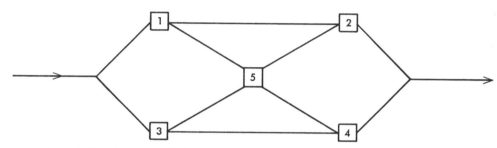

(*b*) Bridge network: $\phi(\mathbf{X}) = 1 - (1 - X_1X_2)(1 - X_3X_4)(1 - X_1X_4X_5)(1 - X_2X_3X_5)(1 - X_1X_2X_5)(1 - X_3X_4X_5)$

Figure 2

liability of more complicated systems. Consider n components having only two states of performance; that is, a component is either in a functioning or failing state. Let X_i take the value 1 if component i is functioning and 0 otherwise. Also define a function $\phi(\mathbf{X})$, which depends on the state of the n components through the vector $\mathbf{X} = (X_1, X_2, \cdots, X_n)$, such that $\phi(\mathbf{X}) = 1$ if the system functions and $\phi(\mathbf{X}) = 0$ if the system is not functioning. This function is termed the *structure function* of the system. The structure functions for series and parallel systems are

series system:

$$\phi(\mathbf{X}) = \prod_{i=1}^{n} X_i = \min(X_1, X_2, \cdots, X_n),$$

parallel system:

$$\phi(\mathbf{X}) = 1 - \prod_{i=1}^{n}(1 - X_i) = \max(X_1, X_2, \cdots, X_n).$$

Some examples of other structure functions are shown in Figure 2.

The reliability of the system with structure function $\phi(\mathbf{X})$ is obtained by taking the expectation of $\phi(\mathbf{X})$. When the components function independently of one another, the reliability of the system is a function only of the vector $\mathbf{p} = (p_1, p_2, \cdots, p_n)$ and can be obtained by replacing each X_i by p_i:

(1) $\qquad R(\mathbf{p}) = E[\phi(\mathbf{X})] = Pr\{\phi(\mathbf{X}) = 1\}.$

Barlow and Proschan (1965) have an excellent exposition of the properties of structure functions,

as well as a presentation of optimum redundancy techniques.

Predicting reliability from sample data. The development in the preceding section assumed that the reliability of the individual components was known. In the practical situation, this is not the case. Although experiments may be conducted on each component in order to obtain an estimate of its reliability, it is often more feasible to test subsystems or major components [*for a general discussion of these methods, see* ESTIMATION].

Suppose that n_i independent components of type i are tested, with the result that s_i components function and $n_i - s_i$ components fail. Then $\hat{p}_i = s_i/n_i$ is an estimate of p_i, the reliability of the component. When such information is available on all components in a system, then an estimate of the reliability of the system, $R(\hat{\mathbf{p}})$, is obtained by replacing p_i by $\hat{p}_i$ in (1). However, one often requires an interval estimate which may take the form of a lower confidence interval; that is, one may wish to find a number $R_\alpha(\mathbf{s})$, $\mathbf{s} = (s_1, s_2, \cdots, s_n)$, which is a function of the sample information, such that

$$Pr\{R(\mathbf{p}) \geqslant R_\alpha(\mathbf{s})\} = 1 - \alpha, \qquad 0 < \alpha < 1,$$

for all $\mathbf{p}$. This general problem has not yet been solved satisfactorily, although progress has been made in a few special cases for components in series. These special methods are reviewed in Lloyd and Lipow (1962).

A general method of computing confidence intervals is to simulate the operation of the system using the sample information. The simulation consists of "building" a set of systems out of the tested components, using the data from each component only once. The proportion of times $\phi(\mathbf{X}) = 1$ is an estimate of $R(\mathbf{p})$, and one can use the theory associated with the binomial distribution to calculate a confidence interval for $R(\mathbf{p})$. A good discussion and summary of such simulation techniques is given by Rosenblatt (1963). These same ideas and techniques may be useful for the study of human organizations, particularly in the transmission of information. [*See* SIMULATION.]

Time-dependent reliability. Often the reliability of a system or component is defined in terms of the equipment functioning for a given period of time. When a complex system is capable of repair, the system is usually repaired after a failure and the failure characteristics of the system are described by the *time between failures*. On the other hand, if a part such as a vacuum tube fails, it is replaced (not repaired) and one refers to the *time to failure* to describe the failure characteristics of a population of nominally identical tubes. Let T be a random variable denoting the time to failure of a component or times between failure of a system, and define $f(t)$ to be its probability density function. Then the probability of the component functioning to time t (or the time between failures of a system being greater than t) is the survivorship function

$$S(t) = Pr\{T > t\} = \int_t^\infty f(x)dx.$$

A useful quantity associated with the failure distribution is the hazard function defined by

$$h(t) = f(t)/S(t).$$

For small positive Δt, $h(t)\Delta t$ is approximately the probability that a component will fail during the time interval $(t, t + \Delta t)$, given that the component has been in satisfactory use up to time t; that is,

$$Pr\{t < T \leqslant t + \Delta t | T > t\} = \frac{Pr\{t < T \leqslant t + \Delta t\}}{Pr\{T > t\}}$$

$$\cong \frac{f(t)}{S(t)} \Delta t = h(t)\Delta t.$$

Sometimes $h(t)$ is called the instantaneous failure rate or force of mortality [*see* LIFE TABLES].

When $h(t)$ is an increasing (decreasing) function of t, then the longer the component has been in use, the greater (smaller) the probability of immediate failure. These failure laws are referred to as positive (negative) aging. If $h(t) = $ constant, independent of t, the conditional probability of failure does not depend on the length of time the components have been in use. Such failures are called random failures.

Knowledge of the hazard function enables one to calculate the survivorship function from the relation

$$S(t) = \exp\left[-\int_0^t h(x)dx\right] = e^{-H(t)},$$

where $H(t) = \int_0^t h(x)dx$. A particularly important class of hazard functions is given by

$$h(t) = pt^{p-1}/\theta^p.$$

Since $H(t) = \int_0^t h(x)dx = t^p/\theta^p$, the survivorship function is

$$S(t) = \exp[-(t/\theta)^p].$$

This distribution is called the Weibull distribution. The parameter p is termed the shape parameter, and θ is the scale parameter. If $p > 1$ ($p < 1$) there is positive (negative) aging. The importance of the Weibull distribution arises from the fact that positive, negative, and no aging depend only on the shape parameter p. When $p = 1$, $h(t) = 1/\theta$ (which is independent of t) and the distribution reduces to the simple exponential distribution

$$S(t) = e^{-t/\theta}.$$

[*See* DISTRIBUTIONS, STATISTICAL, *article on* SPECIAL CONTINUOUS DISTRIBUTIONS, *for more information on these distributions and those discussed below.*]

Two other distributions that are useful in describing the failure time of components are the gamma and log-normal distributions. With respect to the gamma distribution, negative or positive aging occurs for $0 < p < 1$ and $p > 1$ respectively. The hazard function for the log-normal distribution increases to a maximum and then goes to zero as $t \to \infty$. Buckland (1964), Epstein (1962), Govindarajulu (1964), and Mendenhall (1958) have compiled extensive bibliographies on the topics discussed in this section.

The exponential distribution. The exponential distribution (sometimes called the negative exponential distribution) has been widely used in applications to describe the failure of components and systems. One reason for its popularity is that the mathematical properties of the distribution are very tractable. The above-mentioned bibliographies cite large numbers of papers dealing with statistical techniques based on the exponential failure law.

One of the properties of the exponential distribution is that the conditional probability of failure does not depend on how long the component has been in use. This is not true for most applications to components and parts; however, the exponential distribution may be appropriate for some complex systems. If a complex system is composed of a large number of components such that the failure of any component of the system will cause the system to fail, then under general mathematical conditions it has been proved that the distribution of times between failures tends to an exponential distribution when the number of components becomes large (cf. Khintchine [1955] 1960, chapter 5). One of these conditions is that the times between failures of the system be relatively short compared to the failure time of each component.

Using the exponential distribution when it is not appropriate may result in estimates, decisions, and conclusions that are seriously in error. This has motivated research into methods that assume only an increasing failure rate (IFR) or decreasing failure rate (DFR). Barlow and Proschan (1965) present a very complete summary of results dealing with IFR and DFR hazard functions.

MARVIN ZELEN

BIBLIOGRAPHY

BARLOW, RICHARD E.; and PROSCHAN, FRANK 1965 *Mathematical Theory of Reliability.* New York: Wiley. → A good development of special aspects of theory.

BUCKLAND, WILLIAM R. 1964 *Statistical Assessment of the Life Characteristic: A Bibliographic Guide.* New York: Hafner. → A large bibliography on life and fatigue testing.

EPSTEIN, BENJAMIN 1962 Recent Developments in Life Testing. International Statistical Institute *Bulletin* 39, no. 3:67–72.

GOVINDARAJULU, ZAKKULA 1964 A Supplement to Mendenhall's "Bibliography on Life Testing and Related Topics." *Journal of the American Statistical Association* 59:1231–1291. → An excellent bibliography covering the period 1958–1962.

KHINTCHIN, ALEKSANDR IA. (1955) 1960 *Mathematical Methods in the Theory of Queueing.* London: Griffin. → First published in Russian.

LLOYD, DAVID K.; and LIPOW, MYRON 1962 *Reliability: Management, Methods, and Mathematics.* Englewood Cliffs, N.J.: Prentice-Hall. → Discusses theory and applications.

MENDENHALL, WILLIAM 1958 A Bibliography on Life Testing and Related Topics. *Biometrika* 45:521–543. → Includes important works up to 1957.

ROSENBLATT, JOAN R. 1963 Confidence Limits for the Reliability of Complex Systems. Pages 115–137 in Marvin Zelen (editor), *Statistical Theory of Reliability.* Madison: Univ. of Wisconsin Press.

ZELEN, MARVIN (editor) 1963 *Statistical Theory of Reliability.* Madison: Univ. of Wisconsin Press. → Contains survey and expository articles on reliability theory.

Postscript

Developments in reliability have been motivated by important applications to safety problems of nuclear power reactors and computer software. These have led to a new technique called *fault tree analysis.* The technique involves a schematic model representing relations between events. The graphical model serves two purposes. (1) It displays all logical interconnections between events in a complex system. It aids in determining possible causes of an accident and can lead to discoveries of failure combinations that may not have been recognized. (2) The fault tree provides a convenient and efficient format to compute the probability of a system failure. Helpful references on this subject are the papers of the 1974 Conference on Reliability and Fault Tree Analysis (1975).

The area of life testing has been much influenced by applications in the biomedical sciences with respect to the analysis of the survival of individuals having chronic diseases. The major problem is the development of models for survival distributions that take into account known concomitant variables affecting survival. For example, in analyzing the survival of cancer patients, one should take into account demographic variables (age, sex), disease-related variables (history of tumor, anatomic staging), and method of treatment (primary treatment, secondary treatments). The simplest situation is one in which the survival time (T) follows an exponential distribution having a mean that depends on a known variable x; that is, $E(T|x) = \theta(x)$. Possible models for $\theta(x)$ are $\theta(x) = ae^{bx}$, $a + bx$, and $(a + bx)^{-1}$. A large number of techniques have been developed, for which the most important reference is Cox (1972).

MARVIN ZELEN

ADDITIONAL BIBLIOGRAPHY

BARLOW, RICHARD E.; and PROSCHAN, FRANK 1974 *Statistical Theory of Reliability and Life Testing: Probability Models.* New York: Holt.

CONFERENCE ON RELIABILITY AND BIOMETRY, FLORIDA STATE UNIVERSITY, 1973 1974 *Reliability and Biometry: Statistical Analysis of Lifelength; Papers.* Edited by Frank Proschan and R. J. Serfling. Philadelphia: Society for Industrial and Applied Mathematics.

CONFERENCE ON RELIABILITY AND FAULT TREE ANALYSIS, UNIVERSITY OF CALIFORNIA, BERKELEY, 1974 1975 *Reliability and Fault Tree Analysis: Theoretical and Applied Aspects of System Reliability and Safety Assessment; Papers.* Edited by Richard E. Barlow, Jerry B. Fussell, and Nozer D. Singpurwalla. Philadelphia: Society for Industrial and Applied Mathematics.

COX, D. R. 1972 Regression Models and Life-tables. *Journal of the Royal Statistical Society* Series B 34: 187–220.

LEVENBACH, G. J. 1965 Systems Reliability and Engineering: Statistical Aspects. *American Scientist* 53: 375–384.

MANN, NANCY R.; SCHAFER, RAY E.; and SINGPURWALLA, NOZER D. 1974 *Methods for Statistical Analysis of Reliability and Life Data.* New York: Wiley.

QUANTAL RESPONSE

The response of an experimental or survey subject is called "quantal" if it is a dichotomous response—for example, dead–alive or success–failure. Such dichotomous variables are common throughout the social sciences, but the term "quantal response" and the statistical analyses associated with the term refer specifically to situations in which dichotomous observations are taken in a series of groups that are ordered on some underlying metric. The probability of a specific response —for example, dead—is then taken to be a function of this underlying variable, and it is this function (or some characteristic of it) that is of interest. The first developments of techniques for the analyses of quantal response were in psychophysics [*see* PSYCHOPHYSICS], where the functional relationship between the probability of a psychological response (for example, affirmation of a sensation) and the amount of, or change in, stimulus was of interest.

The purpose of this article is to point out the areas in which the techniques for quantal response find application, to describe the techniques, and to discuss in detail a few of the applications in terms of methods and areas of investigation and of fulfillment of the statistical assumptions.

Quantal response data arise when subjects (more generally, experimental units) are exposed to varying levels of some treatment or stimulus and when it is noted for each subject whether or not a specific response is exhibited. In psychophysical experiments the stimulus might be a sound (with varying levels of intensity measured in decibels) or a light (with variations in color measured in wave lengths). The subject states whether he has heard the sound or can discern that the light is different in color from some standard. In the assay of a tranquilizing drug for toxicity, the treatment variable is the dose of the drug and the response variable might be the presence or absence of nausea or some other dichotomous indication of toxicity.

Descriptive statistics for quantal data

Table 1 contains a typical set of quantal response data. The data were treated by Spearman (1908) and represent the proportion (p_i) of times

Table 1 — Proportion of times a sound of varying intensity was judged higher than a standard sound of intensity 1,772 decibels

Intensity (decibels)	Logarithm of intensity (x_i)	Proportion of high judgments (p_i)
1,078	3.03	.00
1,234	3.09	.00
1,402	3.15	.11
1,577	3.20	.28
1,772	3.25	.48
1,972	3.30	.71
2,169	3.34	.83
2,375	3.37	.91
2,579	3.41	.95
2,793	3.45	.98
3,011	3.48	1.00

Source: Data from Spearman 1908.

a subject could distinguish a sound of varying intensity (x_i) from a given standard sound. As Figure 1 shows, the proportion of positive responses tends to increase from 0 to 1, in an S-shaped, or sigmoid, curve, as the level of x increases. It is often assumed that points like those plotted in Figure 1 approximate an underlying smooth sigmoid curve that represents the "true" circumstances.

The value of x for which the sigmoid curve has value $\frac{1}{2}$ is taken to be the most important descriptive aspect of the curve by most writers. This halfway value of x, denoted by Med, corresponds to the median if the sigmoid curve is regarded as a cumulative frequency curve. For values of x below Med, the probability of response is less than $\frac{1}{2}$; for

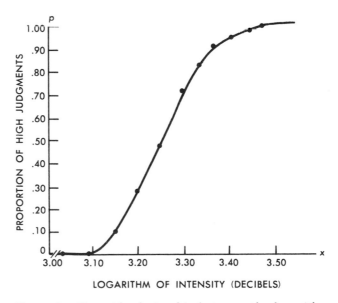

Figure 1 — *Sigmoid relationship between the logarithm of the intensity of a sound and the proportion of times it is judged higher than a standard sound (intensity 1,772, log intensity 3.25)*

values of x above Med, the probability of response is greater than $\frac{1}{2}$. Of course, in describing a set of data a distinction must be made between the true Med for the underlying smooth curve and an estimate of Med from the data. The value Med is often called the threshold value in research in sensory perception (see Guilford 1936). This value is not to be confused with the conceptual stimulus value (or increment) that induces *no* response, that is, the 0 per cent response value, sometimes called the absolute threshold. (See Corso 1963 for a review of these concepts.) In pharmacological research the dose at which 50 per cent of the subjects respond is called the E.D. 50 (effective dose fifty), or L.D. 50 (lethal dose fifty) if the response is death.

A natural method of estimating Med would be to graph the p_i against x_i (as in Figure 1), graduate the data with a smooth line by eye, and estimate Med from the graph. In many cases this method is sufficient. The method lacks objectivity, however, and provides no measure of reliability, so that a number of arithmetic procedures have been developed.

Unweighted least squares. One early method of estimating Med was to fit a straight line to the p_i by unweighted least squares [see LINEAR HYPOTHESES, *article on* REGRESSION]. This standard procedure leads easily to an estimator of Med,

$$\hat{M}_1 = \frac{\frac{1}{2} - \bar{p}}{\hat{B}_1} + \bar{x},$$

where $\bar{p}$ and $\bar{x}$ are averages over the levels of x, and $\hat{B}_1$, the estimated slope of the straight line, is

$$\hat{B}_1 = \frac{\sum p_i(x_i - \bar{x})}{\sum(x_i - \bar{x})^2}.$$

Minimum normit least squares. The method of unweighted least squares is simple and objective, but it applies simple linear regression to data that usually show a sigmoid trend. Furthermore, the method does not allow for differing amounts of random dispersion in the p_i.

Empirical and theoretical considerations have suggested the assumption that the expected value of p_i is related to the x_i by the sigmoidal function, $\Phi[(x_i - \text{Med})/\sigma]$, where σ is the standard deviation of the distribution of the x_i, and Φ denotes the standard normal cumulative distribution. [See DISTRIBUTIONS, STATISTICAL, *article on* SPECIAL CONTINUOUS DISTRIBUTIONS.] Correspondingly, the probability of response at any x is assumed to be $\Phi[(x - \text{Med})/\sigma]$. Notice that $p = \frac{1}{2}$ at $x = \text{Med}$.

The inverse function of Φ, $Z(p)$, is called the standard normal deviate of p, or the normit. To eliminate negative numbers, $Z(p) + 5$ is used by some and is called the probit. The normit, $Z(p)$, is linearly related to x:

$$(1) \qquad Z(p) = -\frac{\text{Med}}{\sigma} + \frac{1}{\sigma}x.$$

If x_i is plotted against $Z(p_i)$ from Table 1, a sensibly linear relationship is indeed obtained (see Figure 2), so that the normal assumption seems reasonable for these data.

The minimum normit least squares method (Berkson 1955) is to fit a straight line to the $Z(p_i)$ by *weighted* least squares [see LINEAR HYPOTHESES, *article on* REGRESSION]. The weights are chosen to approximate the reciprocal variances of the $Z(p_i)$. Specifically, letting $y_i = Z(p_i)$ and letting z_i be the *ordinate* of the standard normal distribution with cumulative probability p_i, the weight, w_i, for y_i will be

$$w_i = \frac{n_i z_i^2}{p_i(1 - p_i)}.$$

The resultant estimator of Med is

$$\hat{M}_2 = -\frac{\tilde{y}}{\hat{B}_2} + \tilde{x},$$

where $\tilde{x} = \sum w_i x_i / \sum w_i$ and $\tilde{y} = \sum w_i y_i / \sum w_i$ are weighted means and $\hat{B}_2$ is the estimator of slope,

$$(2) \qquad \hat{B}_2 = \frac{\sum w_i y_i(x_i - \tilde{x})}{\sum w_i(x_i - \tilde{x})^2}.$$

This method is called the minimum normit chi-squared method by Berkson (1955).

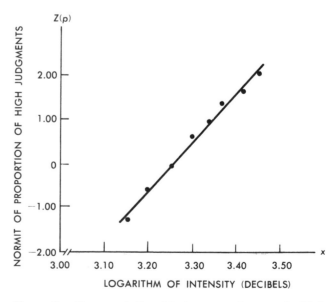

Figure 2 — *Linear relationship between the normit, Z(p), and the logarithm of sound intensity for Spearman's data*

Maximum likelihood estimators. The method of maximum likelihood, one of general application in statistics, may be applied to quantal response analysis [see ESTIMATION, *article on* POINT ESTIMATION]. A detailed discussion is given by Finney (1947; 1952). In essence, the technique may be regarded as a weighted least squares procedure, similar to the minimum normit procedure, using y_i and w_i, slightly different from the normits and weights of the preceding section. The values are obtained from special tables. The modified weighted least squares procedure is iterated, using new weights stemming from the prior computations, until stability in the estimates is achieved.

Estimation using the logistic function. Although the cumulative normal distribution has been most heavily used in quantal response analysis, the logistic function was proposed early and has been advocated for quantal assay in several fields (Baker 1961; Emmens 1940). The function is

$$p = \frac{1}{1 + e^{-(A+Bx)}}.$$

Notice that for $p = \frac{1}{2}$, $x = \text{Med} = -A/B$. This function is of sigmoid form and is practically indistinguishable from a cumulative normal, with appropriate choice of A and B.

One advantage of the logistic function is that x can be expressed as a simple function of p, that is,

$$Y = \ln \frac{p}{1-p} = A + Bx.$$

Thus, $y_i = \ln p_i/(1 - p_i)$ will be approximately linearly related to x_i. The method of approximate weighted least squares can be used to fit a straight line to the data. The resulting estimator for Med is

$$\hat{M}_3 = -\tilde{y}/\hat{B}_3 + \tilde{x},$$

where $\tilde{y}$ and $\tilde{x}$ represent weighted means of the $y_i = \ln [p_i/(1 - p_i)]$ and x_i, with weights $w_i = n_i p_i (1 - p_i)$, and the estimated slope is

$$\hat{B}_3 = \frac{\sum w_i y_i (x_i - \tilde{x})}{\sum w_i (x_i - \tilde{x})^2}.$$

The method of maximum likelihood could be applied with the logistic function, but again it proves to be an iterative regression method. However, some simplifications in this case allow the maximum likelihood solution to be graphed; this has been done for certain configurations of x-levels common in drug assay (Berkson 1960).

Spearman–Karber estimation. Finally, an important method of estimating Med was described by Spearman (1908) and attributed by him to

German psychophysicists. The method is most convenient when the difference between x-levels is constant, for then the estimator is

$$\hat{M}_4 = x_k + \tfrac{1}{2}d - d\sum_{i=1}^{k} p_i,$$ (3)

where x_k is the highest level of x and d is the constant difference between levels of x. Note that an additional lower level with $p_0 = 0$ would not change the value of $\hat{M}_4$. Neither would an additional higher level of x with $p_{k+1} = 1$, since x_k would be increased by d, and an additional d would be subtracted. A more general formula applicable in the case of unequal spacing of the x_i is easily available. The expression for $\hat{M}_4$ above does not require that the group sizes (n_i) be equal. The calculation of $\hat{M}_4$ can be intuitively justified by noting that it is equivalent to reconstructing a histogram from the cumulative distribution formed by graphing the p_i against the x_i as a broken line increasing from 0 to 1. The proportion in the histogram between x_i and x_{i+1} would be $p_{i+1} - p_i$. The mean of the histogram ($\hat{M}_4'$) would be the sum of the midpoints of the intervals, $x_i + \tfrac{1}{2}d$, multiplied by their respective relative frequencies, $p_{i+1} - p_i$:

$$\hat{M}_4' = \sum_{i=0}^{k} (x_i + \tfrac{1}{2}d)(p_{i+1} - p_i),$$

where $p_0 = 0$, $p_{k+1} = 1$, by definition. It is easily shown that when both are applicable, $\hat{M}_4 = \hat{M}_4'$, thus demonstrating that the Spearman estimator is simply the mean of a histogram constructed from the quantal response data. Although $\hat{M}_4$ resembles a mean, it can be regarded as an estimator of the median, Med, for the usual symmetrical sigmoidal curve, since mean and median will be equal in these cases.

The Spearman–Karber estimator has been criticized because it seems to depend on the possibly fallacious assumption that further x-levels on either end of the series actually used would have resulted in no responses and all responses, respectively. It should be emphasized here that the heuristic justification above makes use of the unobserved values, $p_0 = 0$ and $p_{k+1} = 1$, but the Spearman–Karber estimator must be judged on the basis of theory and performance, not on the basis of the heuristic reasoning that suggested the method. Theoretical and practical results suggest that the Spearman–Karber method has no superior as a general method for estimating Med from quantal data. (See Brown 1961 for a review and for additional results on this point.) Furthermore, the Spearman–Karber estimator is nonparametric in that no function relating p and x appears in the definition of the estimator.

Other methods. Other methods of analyzing quantal data are described by Finney (1952). None seems as desirable for general use as the methods described above. The more frequently mentioned procedures are (*a*) methods based on moving average interpolation of the p_i, (*b*) methods based on the angular function as an alternative to the normal or logistic, and (*c*) the Reed–Muench and Dragstedt–Behrens method used primarily in biology.

Sequential estimation of Med. Occasionally, data on quantal response can be collected most economically in a sequential way and analyzed as they are gathered. As an example, in a clinical trial of a psychotherapeutic drug, individual patients may be allocated to various dose-levels of drug and placebo as they are admitted to, and diagnosed at, a mental hospital. Since the data on treatment effectiveness will become available sequentially and not too rapidly, the evaluation might well be done sequentially. Sequential collection and analysis of quantal response data will yield values of Med (the L.D. 50, or threshold value, etc.) that are more reliable than the fixed-sample-size investigation for comparable numbers of observations. The added precision of the sequential procedure is attained by choosing levels for further observations in the light of the data observed. The result is a concentration of *x*-levels in the range where the most information on Med is gained—namely, at *x* levels in the vicinity of Med. Several sources (Cochran & Davis 1964; Wetherill 1963) give appropriate methods of carrying out sequential experimentation and analysis specifically for quantal data. [*See* SEQUENTIAL ANALYSIS *for a general discussion.*]

Estimation of slope of curve. All of the preceding discussion of computational procedures has been concerned with computation of an estimator for Med. Another characteristic of the data shown in Figure 1 is the steepness with which the p_i rise from 0 to 1 as the level of *x* is increased. It has been noted that if the function relating *p* to *x* is regarded as a cumulative frequency function, then Med is the median of the distribution. Similarly, the steepness is related (inversely) to the variability of the frequency function. In particular, if the normal frequency function is used, the slope of the straight line (eq. 1) relating the normal deviate of *p* to *x* is simply the inverse of the standard deviation of the normal function. Therefore, if the regression methods that stem from least squares or maximum likelihood are used, the resulting value for the slope, for example, $\hat{B}_2$ in expression (2), could be used to estimate σ. If the logistic functional form is used, the slope can be converted to an estimate

of the standard deviation by multiplying the inverse of the slope by a constant, $\pi/\sqrt{3}$. The nonparametric Spearman–Karber estimation of Med does not provide as a side product an estimator of slope or standard deviation. If a value for the standard deviation is desired, a corresponding Spearman procedure for estimating the standard deviation is available (Cornfield & Mantel 1950).

Occasionally, an estimator of a quantile other than the median is of interest. If the tolerance distribution is of normal form, the *p*th quantile, x_p, can be estimated by

$$\hat{x}_p = \hat{M} + z_p\hat{\sigma},$$

where $\hat{M}$ and $\hat{\sigma}$ are estimators of the median and standard deviation of the tolerance distribution and z_p is the *p*th quantile of the standard normal distribution.

Reliability of estimators

The computational procedures described above can be carried out on any set of quantal response data for the purpose of summary or concise description. However, it is apparent that the data are subject to random variation, and this variation, in turn, implies that estimators of Med or σ computed from a given set of quantal response data should be accompanied by standard errors to facilitate proper evaluation. Valid standard errors can be computed only on the basis of a careful examination of the sources and nature of the variation in each specific application, but some widely applicable procedures will be discussed in this section.

Measuring reliability of the estimators. The estimators of Med discussed in previous paragraphs will be unbiased, for practical purposes, unless the number of x_i levels used is small (say, two or three) or the x_i levels are so widely spaced or poorly chosen that the probability of response is either 0 or 1 at each x_i. The ideal experiment will have several x_i levels, with probabilities of response ranging from 5 per cent to 95 per cent. [*See* ESTIMATION, *article on* POINT ESTIMATION, *for a discussion of unbiasedness.*]

A simple method for measuring the reliability of an estimator is to carry out the experiment in several independent replications or complete repetitions. This series of experiments will provide a sequence of estimates for the desired parameter (Med, for example). The mean of the estimates can be taken as *the* estimator of the parameter, and the standard error of this mean can be computed as the standard deviation of the estimates divided by the square root of the number of estimates. It will be a valid measure of the reliability of the mean.

Statistical model. The disadvantage of the procedure described above is that it may not yield error limits as narrow as a method that is tailor-made to the known characteristics of the particular investigation. Furthermore, the statistical properties of the procedure may not be as easily ascertained as for a method based on a more specific mathematical model. The following model seems to describe quite a number of situations involving quantal response; the estimation procedures discussed above are appropriate to this model.

Take P_i to be the expected value of p_i at the level x_i, with n_i subjects at this ith level of x_i and with r_i ($= p_i n_i$) of the n_i subjects responding. Assume that the x_i are fixed and known without error and that each of the n_i subjects at x_i has a probability, P_i, of responding, independent of all other subjects, whatever their x_i level. This implies that the observed p_i are independent, binomially distributed proportions. In particular, each p_i is an unbiased estimator of its P_i, with variance $[P_i(1 - P_i)]/n_i$.

The assumption of complete independence among all subjects in the investigation must be carefully checked. For example, if the animals at a given dose-level (x_i) are littermates, or if the persons at each level of ability (x_i) are tested as a group, there may be a serious violation of this assumption, and any assessment of standard error must acknowledge the dependence. (See Finney [1952] 1964, pp. 136–138, for a discussion of procedures appropriate for this type of dependence.) In what follows, complete independence is assumed.

Standard errors and confidence limits. In the parametric regression methods, either maximum likelihood or minimum normit chi-squared, estimated standard errors for the estimators of Med and σ are

$$SE_{\hat{M}} = \frac{1}{\hat{B}} \left[\frac{1}{\sum w_i} + \frac{\hat{M}^2}{\sum w_i (x_i - \bar{x})^2} \right]^{\frac{1}{2}}$$

$$SE_{\hat{B}} = \frac{1}{\hat{B}^2 \sqrt{\sum w_i (x_i - \bar{x})^2}},$$

where $\hat{B}$, w_i, and $\bar{x}$ are as defined in previous sections. The same formulas apply for logistic applications. If 95 per cent confidence limits are desired, it is usually satisfactory to take the estimate plus and minus two standard errors. A more exact procedure, developed by Fieller, is discussed in detail by Finney (1952).

The standard error for the nonparametric Spearman procedure is simple and rapidly computed. From expression (3) it can be seen that the Spearman estimator involves only the sum of independent binomially distributed random variables. The usual estimator of standard error is therefore

$$SE_{\hat{M}_4} = d \sqrt{\sum_{i=1}^{k} \frac{p_i(1 - p_i)}{n_i}},$$

although there is evidence that it is slightly better to replace n_i with $n_i - 1$. It is useful to note that the Spearman estimator has a standard deviation that depends on σ, the standard deviation of the curve relating P to x. If the curve is normal or logistic, the relationship (Brown 1961) is approximately

$$SD_{\hat{M}_4} \cong \sqrt{\frac{d\sigma}{2n}}.$$

Thus, if σ can be approximated from past experience, this last expression can be used to plan an investigation, with respect to the choice of d and n.

Choice of computational procedure. In many applications a graphical estimate of Med (and, perhaps, σ) seems to be sufficient. However, if a more objective estimator is desired, with a measure of reliability, some one of the computational techniques must be chosen. Finney ([1952] 1964, p. 540) presents a review of work comparing the several computational techniques. The results of this work can be described as follows:

(*a*) The estimators discussed above give quite comparable values in practice, with no estimator clearly more reasonable than another as a descriptive measure.

(*b*) The reliabilities of the weighted normit or logit least squares, the maximum likelihood, and the Spearman estimator are the same for practical purposes.

(*c*) The choice of specific functional form—that is, normal, logistic, or some other—does not affect the estimators much and should be made on the basis of custom in the field of application and mathematical convenience.

In summary, the Spearman–Karber estimator, $\hat{M}_4$, or a noniterative, least-squares procedure, $\hat{M}_1$, with standard error given by $SE_{\hat{M}_4}$ or $SE_{\hat{M}}$, respectively, seems acceptable for many applications. If a parametric estimator is used, the choice of functional form can usually be made on the basis of availability of tables and computational ease.

Some quantal response investigations

Presented below in some detail are several examples of the application of the quantal response techniques discussed above.

Quantal response in sensory perception. A typical experiment in sensory perception will call for the presentation of a continuous sound at 1,000 cycles per second, interrupted regularly by sounds with small increment in cycles per second (see, for example, Stevens et al. 1941). The subject tells whether or not he detected the increment. A sequence of increments of the same size may be followed by a sequence of increments of increased size, ending with a sequence of quite easily detectable, large increments. Modifications of this procedure call for random ordering of the magnitudes of increment, varying the background signal, or varying the conditions of the observer—for example, his motivation, fatigue, or training. Recent modifications involve random time of presentation of the increment and variations in the length of the signal increment. Other types of stimulus present new possibilities for experimental procedure.

In many experimental situations the proportions of positive responses to the sequence of stimuli at varying levels resemble independent binomially distributed proportions, with expected values ranging from 0 for small stimulus (or stimulus increment) to 1 for large stimulus. The proportion of responses as a function of the stimulus level (or increment) often has a sigmoidal shape. Then the methods discussed above for estimating the stimulus (Med, or threshold) that gives 50 per cent positive response, with standard error, are applicable. [See PSYCHOPHYSICS.]

The present research emphasis is not on establishing the existence of thresholds and measuring their values, but on determining the factors that cause them to vary. At present, evidence points to great sensitivity of the threshold to the way in which the sequence of stimuli is given, the experimental environment (for example, noise), and the psychological and physiological state of the subject.

The work of Stevens, Morgan, and Volkmann (1941) and that of Miller and Garner (1944), among others, indicate that well-trained observers, tested under ideal conditions for discrimination, can produce linear (rather than sigmoid) functions relating probability of response to stimulus (increment) level. These results have been used as a basis for a new theory of discrimination called the neural quantum theory. The statistical methods that depend on a sigmoidal function do not apply to this type of data but could be adapted so that they would apply.

Quantal data in pharmacological research. The strength of drugs is often measured in terms of the amount of a drug necessary to induce a well-defined quantal response in some biological subject. (Often, dosage is measured on a logarithmic scale, in order to attain an approximately sigmoid curve.) Since biological subjects vary in their responses to the same dose of a drug, the strength must be carefully defined.

If a large number of subjects are divided into subgroups and if these groups are given varying doses, ranging from a relatively small dose for one group to a large dose for another group, the proportion of subjects responding per group will increase gradually from 0 for the smallest-dose group to 1 for the group receiving the highest dose. This curve is called the dose–response curve in biological assay. The E.D. 50, the dose at which 50 per cent of the subjects respond, is used to characterize the strength of a preparation.

If the subjects are randomly allocated to the dose groups and if the dose–response curve is sigmoidal in shape, the proportions of responses in the groups often may be taken to be binomially distributed proportions, and the estimation methods discussed above will be appropriate (Finney 1952).

It is convenient to regard each subject as having an individual absolute threshold for response to the drug. A dose lower than the threshold of a subject would fail to induce a response; a dose higher than the threshold would induce a response. The thresholds or tolerances of the subjects will form a distribution. The E.D. 50 is the median of this tolerance distribution, and the dose–response function is simply the cumulative distribution of the tolerances or absolute thresholds. It should be emphasized that the dose–response curve is directly estimated by the observed proportions, whereas the underlying tolerance distribution is not ordinarily directly observed. In fact, postulation of the tolerance distribution is not essential to the validity of the assay model or estimation procedure.

Experience has shown that characterization of the strength of a drug by its E.D. 50 is unreliable, since the subjects themselves may change in sensitivity over a period of time. It is common practice, at present, to assay a drug by comparing its effect on the subjects to the effect of a standard preparation. The result is expressed as the relative potency of the test preparation compared to the standard preparation. If z_T units of the test drug perform like ρz_T units of the standard for any z_T (that is, if the test drug acts like a dilution of the standard drug) this relation will hold regardless of the fluctuation in the sensitivity of the subjects. This implies that the E.D. 50 for the test preparation will be ρ^{-1} times the E.D. 50 of the standard. On the log dose scale the log(E.D. 50)'s will differ by $\log\rho$, and the estimation is usually carried out by estimating the E.D. 50's for test and standard preparations on the log scale, taking the difference and then taking the antilog of the difference. The estimation methods above can be used for each preparation separately, although some refinements are useful in combining the two analyses.

The use of a standard preparation and the potency concept are especially necessary in clinical trials of tranquilizers or analgesics. In these cases, random variation is large, and stability from one trial to the next is difficult to achieve. Inclusion of doses of standard preparation for comparison of dose–response functions seems to be essential for each trial. [See SCREENING AND SELECTION.]

Quantal response in mental testing. The theory of mental tests developed by Lawley (1943) and elaborated by Lord (1952) and by others assumes a probability of correct response to each item that depends on the ability level of the responder. This functional relationship between probability of correct response and ability level was taken by Lawley to be a normal function, but Maxwell (1959) proposed the logistic function as an alternative and Baker (1961) presented empirical evidence that the logistic is more economical to use and fits mental-test data as well as the integrated normal.

In Lawley's theory of mental testing, each item is characterized by a measure of difficulty, the ability level corresponding to 50 per cent probability of correct response. This is the mean (Med) of the normal cumulative distribution and is called the limen value for the item. The standard deviation of the normal distribution is a measure of the variation in the probability of response over ability levels. The inverse of this standard deviation is defined as the discriminating power of the item.

Although Ferguson (1942) and Baker (1961) have described in detail the methods of item analysis for estimating the limen and the discriminating power, with particular reference to the data typical of mental-test investigations, this method is rarely used at present and is not described in the introductory manuals on test construction.

There is a noteworthy difference in the typical mental data and the analogous data of psychophysics or drug assay. Although a detailed discussion of the theory of testing is not in order here, it should be noted that the x variable is ability. There would be difficulties in obtaining groups of subjects at specified ability levels. The usual procedure is to administer the test to a large group and to stratify them with respect to their total scores on the test after it is given. If the number of items in the test is large and if the number of ability groups is not too small, the usual assumptions for quantal assay will be well approximated. [See PSYCHOMETRICS.]

Other uses in the social sciences. There are other areas of the social sciences in which the quantal response model may prove useful. In market analysis, for example, the probability of purchasing specific items will depend on some underlying, continuous variable, such as income, amount of education, or exposure to advertising. Data on this dependence might well be available in the form of proportions of purchasers or of intending purchasers at increasing levels of the continuous variable. Such data could be summarized by estimating the level at which the probability attains some specified value, such as 50 per cent, and the rate at which the probability increases with the continuous variable. This technique could be adapted to the summary of many sample survey collections of data.

The concept of quantal response can be useful in estimating or verifying the validity of an average that might ordinarily be obtained by depending on the memory of persons interviewed. For example, the average age at which a child is able to take several steps without aid might be ascertained by questioning a large group of mothers whose children have been walking for some time. A less obvious, but probably more valid, method of obtaining the estimate would be to interview groups of mothers with children of various ages, ranging from six months to thirty months. The proportion of children walking at each age could be recorded, and this set of quantal response data could be analyzed by the methods discussed above to obtain an estimate of the age at which 50 per cent of the children walk, with a standard error for the estimate. This method has been used to estimate the age of menarche through interviews with adolescent schoolgirls.

BYRON W. BROWN, JR.

[See also COUNTED DATA; PSYCHOPHYSICS.]

BIBLIOGRAPHY

BAKER, FRANK B. 1961 Empirical Comparison of Item Parameters Based on the Logistic and Normal Functions. *Psychometrika* 26:239–246.

BERKSON, JOSEPH 1955 Estimate of the Integrated Normal Curve by Minimum Normit Chi-square With Particular Reference to Bio-assay. *Journal of the American Statistical Association* 50:529–549.

BERKSON, JOSEPH 1960 Nomograms for Fitting the Logistic Function by Maximum Likelihood. *Biometrika* 47:121–141.

BROWN, BYRON W. JR. 1961 Some Properties of the Spearman Estimator in Bioassay. *Biometrika* 48:293–302.

COCHRAN, WILLIAM G.; and DAVIS, MILES 1964 Stochastic Approximation to the Median Effective Dose in Bioassay. Pages 281–297 in *Stochastic Models in Medicine and Biology: Proceedings of a Symposium . . . 1963.* Edited by John Gurland. Madison: Univ. of Wisconsin Press.

CORNFIELD, JEROME; and MANTEL, NATHAN 1950 Some New Aspects of the Application of Maximum Likelihood to the Calculation of the Dosage Response Curve. *Journal of the American Statistical Association* 45: 181–210.

CORSO, JOHN F. 1963 A Theoretico–Historical Review of the Threshold Concept. *Psychological Bulletin* 60: 356–370.

EMMENS, C. W. 1940 The Dose/Response Relation for Certain Principles of the Pituitary Gland, and of the Serum and Urine of Pregnancy. *Journal of Endocrinology* 2:194–225.

FERGUSON, GEORGE A. 1942 Item Selection by the Constant Process. *Psychometrika* 7:19–29.

FINNEY, DAVID J. (1947) 1962 *Probit Analysis: A Statistical Treatment of the Sigmoid Response Curve.* 2d ed. Cambridge Univ. Press.

FINNEY, DAVID J. (1952) 1964 *Statistical Method in Biological Assay.* 2d ed. New York: Hafner.

GUILFORD, JOY P. (1936) 1954 *Psychometric Methods.* 2d ed. New York: McGraw-Hill.

LAWLEY, D. N. 1943 On Problems Connected With Item Selection and Test Construction. Royal Society of Edinburgh, *Proceedings* 61A:273–287.

LORD, F. 1952 *A Theory of Test Scores.* Psychometric Monograph No. 7. New York: Psychometric Society.

MAXWELL, A. E. 1959 Maximum Likelihood Estimates of Item Parameters Using the Logistic Function. *Psychometrika* 24:221–227.

MILLER, G. A.; and GARNER, W. R. 1944 Effect of Random Presentation on the Psychometric Function: Implications for a Quantal Theory of Discrimination. *American Journal of Psychology* 57:451–467.

SPEARMAN, C. 1908 The Method of "Right and Wrong Cases" ("Constant Stimuli") Without Gauss's Formulae. *British Journal of Psychology* 2:227–242.

STEVENS, S. S.; MORGAN, C. T.; and VOLKMANN, J. 1941 Theory of the Neural Quantum in the Discrimination of Loudness and Pitch. *American Journal of Psychology* 54:315–335.

WETHERILL, G. B. 1963 Sequential Estimation of Quantal Response Curves. *Journal of the Royal Statistical Society* Series B 25:1–48. → Contains 10 pages of discussion by P. Armitage et al.

Postscript

Chmiel (1976) reviews work on Spearman procedures in bioassay and presents a thorough investigation of the statistical characteristics of the Spearman estimator of the standard deviation. Wesley (1976) reviews Bayesian techniques in bioassay, proposes some new Bayesian methods, and applies various of the new adaptive techniques to the bioassay problem. [*See* BAYESIAN INFERENCE.]

Corso (1973) discusses the continuing controversy regarding quantal assay models and statistical techniques in psychophysics. Lord (1974) presents quantal response models and techniques that have been developed for psychological testing, especially Bayesian methods, and a variant of the up-and-down sequential methods.

For applications of quantal methods in general, the 1970 monograph by D. R. Cox is extremely useful.

BYRON W. BROWN, JR.

ADDITIONAL BIBLIOGRAPHY

CHMIEL, JOHN 1976 Some Properties of Spearman-type Estimators of the Variance and Percentiles in Bioassay. *Biometrika* 63:621–626.

CORSO, JOHN F. 1973 Neural Quantum Controversy in Sensory Psychology. *Science* 181:467–468.

COX, D. R. 1970 *The Analysis of Binary Data.* London: Methuen; New York: Halsted.

LORD, FREDERIC M. 1974 Individualized Testing and Item Characteristic Curve Theory. Volume 2, pages 106–126 in David H. Krantz et al. (editors), *Contemporary Developments in Mathematical Psychology.* San Francisco: Freeman.

WESLEY, MARGARET NAKAMURA 1976 Estimating the Mean of the Tolerance Distribution. Stanford Univ., Div. of Biostatistics, Technical Report, No. 17.

QUASI-EXPERIMENTAL DESIGN
See under EXPERIMENTAL DESIGN.

QUETELET, ADOLPHE

Lambert Adolphe Jacques Quetelet (1796–1874), best known for his contributions to statistics, was born in the Belgian city of Ghent. When he was seven his father died, and Quetelet, on finishing secondary school at the age of 17, was forced to earn his own living. He accepted a post as teacher of mathematics in a secondary school at Ghent, but his true inclination at the time was toward the arts, not the sciences. For a time he was an apprentice in a painter's studio, and he later produced several canvases of his own, which were well received. He wrote poetry of some distinction and collaborated on an opera with his old school friend Germinal Dandelin. Only through the influence of Jean Guillaume Garnier, a professor of mathematics at the newly created University of Ghent, was Quetelet finally persuaded to turn from his artistic endeavors to the full-time study of mathematics (although he continued to dabble in poetry until about the age of thirty). His doctoral dissertation, in which he announced the discovery of a new curve, *la focale*, was the first to be presented at the university (on July 24, 1819), and was widely acclaimed as an original contribution to analytic geometry. As a result, at age 23 Quetelet found himself called to Brussels to occupy the chair of elementary mathematics at the Athenaeum. Only a few months later, early in 1820, he was further honored by being elected to membership in the Académie Royale des Sciences et des Belles-Lettres de Bruxelles.

The rapid pace set in the first part of Quetelet's scientific career never slackened; his productivity over the next fifty years was phenomenal. The record of his activities during the ten or so years after his arrival at Brussels well illustrates his prodigious capacity for work: Immediately after he assumed the position at the Athenaeum, he began publishing a vast array of essays, mostly in mathematics and in physics; they appeared initially in the *Nouveaux mémoires* of the academy and later in *Correspondances mathématiques et physiques*, a journal he founded in 1825 and edited (for the first two years together with Garnier) until its dissolution in 1839. For a time this was the foremost journal of its kind in Europe, attracting contributions from the most eminent scientists on the Continent. In 1824 Quetelet added to his duties the task of delivering a series of public lectures at the Brussels Museum, first in geometry, probabilities, physics, and astronomy and later in the history of the sciences. There, as at the Athenaeum, he quickly became renowned as a great teacher, and his lectures were always crowded, by regular students and auditors and by eminent scientists who came from all over Europe to hear him. He continued these lectures until 1834, when the museum was absorbed by the University of Brussels. Offered a chair in mathematics at the new university, he declined in order to devote himself to his many researches. However, the public lectures were resumed in 1836, at the Military School at Brussels, founded two years earlier.

During this same period Quetelet published several elementary works in natural science and in mathematics, designed to expose these fields to a wide popular audience. *Astronomie élémentaire*, published in 1826, was soon followed by *Astronomie populaire* (1827a). (Several biographers—notably Reichesberg [1896] and Hankins [1908]—asserted that the latter work immediately achieved the "distinction," accorded some earlier publications in astronomy, of being placed on the *Index librorum prohibitorum*. Lottin [1912, pp. 34–37] proved this to be a myth.) In 1827 Quetelet also published a summary of his course in physics at the museum, entitled *Positions de physique, ou résumé d'un cours de physique générale* (1827b). Robert Wallace, in the preface to his English translation of the work in 1835, signaled its importance as follows: "No other work in the English language contains such an extensive and succinct account of the different

branches of physics, or exhibits such a general knowledge of the whole field in so small a compass."

The following year, 1828, saw the publication of *Instructions populaires sur le calcul des probabilités*, which Quetelet identified as a résumé of the introductory lectures to his courses in physics and astronomy at the museum. This work marked Quetelet's shift from exclusive concentration on mathematics and the natural sciences to the study of statistics and, eventually, to the investigation of social phenomena.

Interestingly enough, Quetelet considered none of these works his main preoccupation or his main accomplishment in the ten years from 1823 to 1832. What concerned him most during this time was the project of establishing an observatory at Brussels. It is still unknown how Quetelet came to adopt this as a prime objective. What is certain, however, is that his activities directed toward establishing the observatory brought about, quite fortuitously, the major change of orientation in his scientific career. Upon accepting Quetelet's proposal for an observatory, the minister of education promptly sent him to Paris to acquaint himself with the latest astronomical techniques and instruments. There he was warmly received by the astronomers François Arago and Alexis Bouvard and was introduced by Bouvard to the coterie of French intellectuals gathered around the illustrious mathematicians Poisson, Laplace, and Jean Baptiste Fourier. These men had for some time been engaged in laying the foundations of modern probability theory, and several of them had analyzed empirical social data in their work.

It was instruction from these mathematicians, particularly Laplace, together with the stimulation of continual informal contact with their group at the École Polytechnique, that aroused in Quetelet the keen interest in statistical research and theory, based on the theory of probabilities, that was to become the focus of all his scientific work. In later reminiscences he said that after he had become acquainted with the statistical ideas of his French masters, he immediately thought of applying them to the measurement of the human body, a topic he had become curious about when he was a painter. One direct effect of learning the theory of probabilities was to make Quetelet realize "the need to join to the study of celestial phenomena the study of terrestrial phenomena, which had not been possible until now. . . ." The crucial impact of the Paris experience on his thinking is evident a few sentences later, where he said, "Thus, it was among the learned statisticians and economists of that time that I began my labors . . ." (1870).

After his return from Paris in 1823, the project of the observatory moved along by fits and starts, at first held up by difficulties over financing and by some disagreements between Quetelet and the architect and later interrupted by the Belgian revolution of 1830 (during which the half-finished observatory was used as a makeshift fortress and suffered some structural damage). Quetelet finally took up residence in the nearly completed observatory in 1832.

Organization of data collection

While the observatory was under construction, Quetelet's interest in statistics, which had crystallized during his visit to Paris, coupled with his manifest abilities as an organizer led him to become more and more active in projects requiring the collection of empirical social data. When the Royal Statistics Commission was formed in 1826, he became correspondent for Brabant. (From 1814 to 1830 Belgium was under Dutch rule; thus, "Royal" in this case refers to the House of Orange.) Quetelet's first publications covered quantitative information about Belgium which could be used for practical purposes, including mortality tables with special reference to actuarial problems of insurance. In 1827 he analyzed crime statistics, again with a practical eye to improving the administration of justice. In 1828 he edited a general statistical handbook on Belgium, which included a great deal of comparative material obtained from colleagues he had come to know during his stays in France and also in England. At his urging, a census of the population was taken in 1829, the results of which were published separately for Holland and Belgium after the revolution of 1830.

In 1841, largely through Quetelet's efforts, the Commission Centrale de Statistique was organized, and this soon became the central agency for the collection of statistics in Belgium. Quetelet served as its president until his death, and under his direction it performed its functions with remarkable thoroughness and efficiency, setting a standard for similar organizations throughout Europe. In 1833 he was delegated official representative to the meeting of the British Association for the Advancement of Science and there played a key role in the formation of a statistical section. Dissatisfied with the narrow scope of the section, he urged its chairman, Babbage, to organize the Statistical Society of London. This was accomplished in 1834, and the society survives today as the Royal Statistical Society (having been renamed in 1877).

In his work in statistics, as in the natural sciences, Quetelet placed great emphasis on the need for uniformity in methods of data collection and tabulation and in the presentation of results. His principal goal, in all his organizational endeavors, was to see this realized in practice. In 1851 Quetelet proposed to a group of scientists gathered at the Universal Exposition in London a plan for international cooperation in the collection of statistical information. The idea was heartily approved, and progress on it was so rapid that in 1853 the first International Statistical Congress was held at Brussels. At the initial session of the congress Quetelet was chosen president, and he naturally devoted his opening address to the importance of uniform procedures and terminology in official statistical publications. During the next twenty-five years the congress was enormously effective in spurring the development of official statistics in Europe, establishing permanent lines of communication between statisticians, and improving comparability. Internal dissension and controversy weakened the congress during the 1870s and eventually led to its collapse in 1880. By that time, however, Quetelet's original proposition that there must be some international organization to maintain uniformity and promote cooperation in the collection and analysis of official statistics had been so fully accepted that it was only a matter of five years before a new organization, the International Statistical Institute, was established to continue the work of the congress.

To round out this picture of Quetelet's successes as an organizer, we need only mention the preeminent role he played in the Académie Royale des Sciences et des Belles-Lettres. When he was chosen a member in 1820, the academy was near to closing down, with only about half a dozen superannuated members attending its sessions and virtually no publications to its name. Quetelet brought new life and vigor into the association, quickly assuming the major responsibility for its activities, recruiting into its ranks many of his young scientific colleagues, and fortifying its publications with his own numerous scientific writings. He was made director for the years 1832 and 1833, and in 1834 he was elected perpetual secretary, an office he occupied for the next forty years, during which time he was considered "the guiding spirit of the academy."

Social research and theory

The two memoirs which form the basis for all of Quetelet's subsequent investigations of social phenomena appeared in 1831. By then he had decided that he wanted to isolate, from the general pool of statistical data, a special set dealing with human beings. He first published a memoir entitled *Recherches sur la loi de la croissance de l'homme* (1831a), which utilized a large number of measurements of people's physical dimensions. A few months later he published statistics on crime, under the title *Recherches sur le penchant au crime aux différens ages* (1831b). While the emphasis in these publications is on what we would call the life cycle, both of them also include many multivariate tabulations, such as differences in the age-specific crime rates for men and women separately, for various countries, and for different social groups. (As noted by Hankins [1908, p. 55] and by Lottin [1912, pp. 128–138], the core idea contained in the second memoir—the constancy in the "budget" of crimes from year to year in each age group—can be traced back to a memoir read to the academy on December 6, 1828, and published early in 1829. Thus, Quetelet was probably right in claiming priority for the idea [1835; see p. 96 in 1842 edition] over A. M. Guerry, who published it under the title "Statistique comparée de l'état de l'instruction et du nombre des crimes" [1832].)

In 1833 Quetelet published a third memoir giving developmental data on weight (1833a). By this time he had formed the idea of a social physics, and in 1835 he combined his earlier memoirs into a book entitled *Sur l'homme et le développement de ses facultés*, with the subtitle *Physique sociale* (see 1835). Quetelet republished this work in an augmented version, with the titles reversed, in 1869. To prevent confusion, the first edition is usually referred to as *Sur l'homme*, the second as *Physique sociale*. Included in the later edition, and one of its highlights, is a long essay by the English astronomer John Herschel: it had first been published in the *Edinburgh Review* (1850) as a review article on Quetelet's *Letters . . . on the Theory of Probabilities* and other work by Quetelet, and since it was highly favorable, Quetelet made it the introduction to his *Physique sociale*. To the constancy in crime rates noted earlier, this work added demonstrations of regularities in the number of suicides from year to year and in the rate of marriage for each sex and age cohort. Although Quetelet was convinced that many other regularities existed, these three—in crimes, suicides, and rates of marriage—were the only regularities in man's "moral" characteristics (i.e., those involving a choice of action) actually demonstrated in his writings. Thus, with the publication of *Sur l'homme*, all of Quetelet's basic ideas became available to a broader public.

Basic principles. It was in writings published in the 1830s that Quetelet established the theoretical foundations of his work in moral statistics or, to use the modern term, sociology. First there was the idea that social phenomena in general are extremely regular and that the empirical regularities can be discovered through the application of statistical techniques. Furthermore, these regularities have causes: Quetelet considered his averages to be "of the order of physical facts," thus establishing the link between physical laws and social laws. But rather than attach a theological interpretation to these regularities—as Süssmilch and others had done a century earlier, finding in them evidence of a divine order—Quetelet attributed them to social conditions at different times and in different places [see the biography of SÜSSMILCH]. This conclusion had two consequences: It gave rise to a large number of ethical problems, casting doubt on man's free will and thus, for example, on individual responsibility for crime; and in practical terms it provided a basis for arguing that meliorative legislation can alter social conditions so as to lower crime rates or rates of suicide.

On the methodological side, two key principles were set forth very early in Quetelet's work. The first states that "Causes are proportional to the effects produced by them" (1831b, p. 7). This is easy to accept when it comes to man's physical characteristics; it is the assumption that allows us to conclude, for example, that one man is "twice as strong" as another (the cause) simply because we *observe* that he can lift an object that is twice as heavy (the effect). Quetelet proposed that a scientific study of man's moral and intellectual qualities is possible only if this principle can be applied to them as well. (The role this principle played in Quetelet's theories is discussed below.) The second key principle advanced by Quetelet is that large numbers are necessary in order to reach any reliable conclusions—an idea that can be traced to the influence of Laplace (1812), Fourier (1826), and Poisson (1837). The interweaving of these principles with the theoretical ideas summarized above is illustrated in the following:

It seems to me that *that which relates to the human species, considered en masse, is of the order of physical facts;* the greater the number of individuals, the more the influence of the individual will is effaced, being replaced by the series of general facts that depend on the general causes according to which society exists and maintains itself. These are the causes we seek to grasp, and when we do know them, we shall be able to ascertain their effects in social matters, just as we ascertain effects from causes in the physical sciences. (1831b, pp. 80–81)

Quetelet was greatly concerned that the methods he adopted for studying man in all his aspects be as "scientific" as those used in any of the physical sciences. His solution to this problem was to develop a methodology that would allow full application of the theory of probabilities. For in striking contrast to his contemporary Auguste Comte, Quetelet believed that the use of mathematics is not only the *sine qua non* of any exact science but the measure of its worth. "The more advanced the sciences have become," he said, "the more they have tended to enter the domain of mathematics, which is a sort of center toward which they converge. We can judge of the perfection to which a science has come by the facility, more or less great, with which it may be approached by calculation" (1828, p. 230).

Pattern of work. Before proceeding to a more detailed exposition of Quetelet's work in moral statistics, we should note his method of publication. Quetelet's literary background and the fact that his humanist friends remained an important reference group for him help to explain the manner in which he published his works. When he had new data or had developed a new technique or idea, he first announced his discovery in brief notes, usually in the reports of the academy or in *Correspondances* and sometimes in French or English journals. Once such notes had appeared, he would elaborate the same material into longer articles and give his data social and philosophical interpretations. He would finally combine these articles into books which he hoped would have a general appeal. He obviously felt very strongly that empirical findings should be interpreted as much as possible and made interesting to readers with broad social and humanistic concerns.

Quetelet further extended his influence through the voluminous correspondence he maintained with scientists, statesmen, and men of letters throughout Europe and America. Liliane Wellens-De Donder has identified approximately 2,500 correspondents, including such names as Gauss; Ampère; Faraday; Alexander von Humboldt; James A. Garfield, then a U.S. congressman, who solicited Quetelet's advice on means of improving the census; Joseph Henry; Lemuel Shattuck; Charles Wheatstone; Louis René Villermé; and Goethe, who befriended Quetelet when the latter visited Germany in 1829 (see Wellens-De Donder 1964). Probably Quetelet's most famous correspondence was with the princes Ernest and Albert of Saxe-Coburg and Gotha, whom he tutored in mathematics beginning in 1836, at the request of their uncle, Leopold I, king of the Belgians. Although the princes left Belgium to attend school in Germany shortly after they began studying under Quetelet, the lessons continued for many years by

correspondence. Quetelet's second major work on moral statistics, *Letters Addressed to H.R.H. the Grand Duke of Saxe-Coburg and Gotha, on the Theory of Probabilities, as Applied to the Moral and Political Sciences* (1846), shows his side of the correspondence. (The title refers only to Ernest, who as reigning duke was head of the house of Coburg.) *Du système social et des lois qui le régissent* (1848), was dedicated to Albert, with whom Quetelet had established an especially close friendship. Quetelet's profound influence on Albert's thinking is clearly shown in the keynote address Albert delivered to the fourth meeting of the International Statistical Congress in London, on July 16, 1860 (see Schoen 1938).

The average man. Quetelet's conceptualization of social reality is dominated by his notion of the average man, or *homme moyen*. In his preface to *Du système social*, he himself identified this as his central concept and traced its development through his writings (1848, pp. vii–ix). In *Sur l'homme*, he said, he had developed the idea that the characteristics of the average man can be presented only by giving the mean *and* the upper and lower limits of variation from that mean. In the *Lettres* he had shown that "regarding the height of men of one nation, the individual values group themselves symmetrically around the mean according to . . . the *law of accidental causes*" (p. viii); and further, that for a nation the average man "is actually the *type* or the standard and that other men differ from him, by more or by less, only through the influence of accidental causes, whose effects become calculable when the number of trials is sufficiently large . . . . In this new work," Quetelet continued, "I show that the law of accidental causes is a general law, which applies to individuals as well as to peoples and which governs our moral and intellectual qualities just as it does our physical qualities. Thus, what is regarded as accidental ceases to be so when observations are extended to a considerable number of cases" (p. ix). It is no wonder, then, that discussions of Quetelet's theories and researches on society invariably take as a starting point his concept of the *homme moyen*.

His first approach to the concept was through the measurement of physical characteristics of man, in particular height and weight (1831a). He conceived of the average height of a group of individuals of like age as the mean around which the heights of all persons of that age "oscillate," although just how this oscillation takes place Quetelet could not say. He did suggest, even in this first exposition of the concept, that similar means and oscillations might be observed if moral and intellectual, not just physical, qualities of men were studied.

In his essay *Recherches sur le penchant au crime*, the term *homme moyen* appears for the first time. There, also, we find the first statement of the idea that if one were to determine the *homme moyen* for a nation, he would represent the *type* for that nation; and if he were determined for all mankind, he would represent the *type* for the entire human species.

The next advance in the development of the concept came in a memoir published in 1844, in which Quetelet first took note of the fact that his observations were symmetrically distributed about the mean—in almost exactly the pattern to be anticipated (*prévu*) from the binomial and normal distributions—and went on to speculate about the likelihood that all physical characteristics might be distributed in the same way. By applying the theory of probabilities, he was then able to derive a theoretical frequency distribution for height, weight, or chest circumference that coincided remarkably with the empirical distributions in his data for various groups.

An interesting result obtained by applying this method, also first published in the memoir of 1844, was Quetelet's discovery of draft evasion in the French army. By noting the discrepancy between the distribution of height of 100,000 French conscripts and his prediction (i.e., the theoretical distribution calculated by assuming a probable error of 49 millimeters), he came to the conclusion that some 2,000 men had escaped service by somehow shortening themselves to just below the minimum height. Thus, quite by accident, Quetelet emerged with the first practical, although perhaps somewhat trivial, application of his statistical techniques.

In his discussions of the average man, Quetelet had up to this time limited himself to calculating the means and distributions of only a few physical characteristics. The task he now set for himself was to extend the concept to all of man's physical traits (thus forming the basis for what he called "social physics") and, thence, to all moral and intellectual qualities as well ("moral statistics"). Furthermore, he planned to apply the concept to collectivities of all sizes, ranging from the small group to the whole of mankind, and expected that it would hold equally well for any time in human history. Quetelet had suggested these extensions in earlier works, but the grand generalization did not emerge in its final form until the publication of *Du système social*. In the first pages Quetelet announced his theme: "There is a general law which governs our universe . . . ; it gives to every-thing that breathes an infinite variety. . . . That law, which science has long misunderstood and which has until now remained useless in practice, I shall call the *law of accidental causes*" (1848, p. 16). A few lines later he elaborated his over-all viewpoint:

. . . among organized beings all elements vary around a mean state, and . . . variations, which arise from accidental causes, are regulated with such harmony and precision that we can classify them in advance numerically and by order of magnitude, within their limits.

One part of the present work is devoted to demonstrating the law of accidental causes, both for physical man and for moral and intellectual man, considering him individually, as well as in the aggregate. . . . (p. 17)

Concept of causality. It is obvious that an explanation of what Quetelet meant by "accidental causes" and by "law" is critical for an understanding of his conception of the average man. Quetelet hypothesized that every mean he presented resulted from the operation of constant causes, while the variations about the mean were due to "perturbative" or "accidental" causes. "Constant causes," he explained, "are those which act in a continuous manner, with the same intensity and in the same direction" ([1846] 1849, p. 107). Among the constant causes he named are sex, age, profession, geographical latitude, and economic and religious institutions. (As a category parallel to constant causes, Quetelet sometimes mentioned "variable causes," which are those that "act in a continuous manner, with energies and intensities that change" [*ibid.*]. The seasons are cited as the type case, although Quetelet meant to include as variable causes all periodical phenomena.) "Accidental causes only manifest themselves fortuitously, and act indifferently in any direction" (*ibid.*). Quetelet frequently classed man's free will as an accidental cause (although occasionally he claimed that it played no role at all), but insisted that its operation is constrained within very narrow limits. The essence of Quetelet's theory is that, given sufficient data over time, the shape and extent of variations about the mean state which result from accidental causes can be "classified in advance" with a high degree of accuracy, through the application of the theory of probabilities of independent events.

Quetelet's conception of "law" depended on whether he was talking about man's physical attributes, his moral traits, or all human characteristics. Thus, in *Du système social* we find these three distinct uses of the term. In the early part of the book, he referred to a trend in a series of averages over time as a law: "If we knew what [man's mean] height had been from one century to another, we would have a series of sizes which would express the law of development of humanity as regards height" (1848, p. 11). Later, in presenting the law of propensity to crime, he used the term to denote a regular pattern of correlations: *The propensity to crime increases quite rapidly toward adulthood; it reaches a maximum and then decreases until the very end of life.* This law appears to be constant, and varies only with respect to the magnitude of the maximum and the time of life when it occurs" (p. 86). (By way of contrast, his law of propensity to suicide posits a direct variation with age "until the most advanced age" [p. 88].)

It is important to note in this connection that Quetelet went far beyond such simple two-variable correlations in his studies of man. Numerous three-variable and four-variable tables appear throughout his work (see esp. 1835, vol. 2). In one case, for example, he presented a table that shows the relationship of mean weight to age, sex, and occupation (*ibid.*, p. 91); similar tables show the breakdown of crimes in various groups by sex and level of education (p. 297), by age and sex (p. 302), and by age, sex, and the type of court in which the crime was tried (p. 308). These remarkable anticipations of modern techniques went largely unnoticed by Quetelet's contemporaries, and only in recent times have social scientists rediscovered and fully explored the possibilities of multivariate analysis—a striking discontinuity in the history of empirical social research that surely deserves further study and explanation.

The two uses of the term "law" illustrated above are similar in that they both refer in some way to a correlation—in one case between height and century, and in the other between age and the incidence of some social act. Quetelet's third type of law, the "law of accidental causes," is quite another thing; it is simply the assertion that every human trait is normally distributed about a mean and that the larger the number of observations, the more closely the empirical distribution will coincide with the theoretical probability distribution. In sum, the word "law" is used alternatively to refer to a trend in a series of specific empirical findings, an empirical generalization, and an assertion (or, in effect, a theory) that a certain type of regularity exists in all human phenomena.

Measurement. Perhaps one reason Quetelet had trouble maintaining a single conception of "law" is that the types of measures he used to substantiate laws were few in number and inadequate to his purpose. Limiting himself to the manipulation of data gathered from available official statistical publications, he was forced to improvise new and different techniques for physical and moral qualities and so emerged with different kinds of laws, according to the type of phenomenon in question. It comes as something of a surprise to realize that in all Quetelet's research on man, for example, he actually used only three kinds of measurements. (He did suggest some others but never applied them to his data or attempted to collect the data that would make them applicable. Lazarsfeld's 1961 essay analyzes the kinds of measures that are mentioned in *Physique sociale* in the light of modern ideas on quantification.)

First, he determined the empirical distribution of some human trait in a group and computed the mean—which he then identified as a characteristic of the average man for the group. He repeatedly asserted that similar distributions and means could be found for moral and intellectual characteristics, presented some rough hypothetical curves, but never performed any such calculations using a set of empirical observations. A second type of measurement involved counting the number of certain social events, such as crimes or marriages, that occurred during a series of years among particular groups; the average of these yearly counts was taken to be the probable number of such events that would occur in each group during the next year. As a third type of measurement Quetelet used rates—the number of crimes or other events in each age group, divided by the number of persons in that group. Quetelet regarded the results as the respective probabilities of committing a crime at various ages; he called these probabilities the propensity to crime—the *penchant au crime* or, alternatively, the *tendance au crime*—at each age. (At one point in *Du système social* [1848, p. 93] he proposed substituting the word *possibilité* for *penchant*, but he reverted to the use of *penchant* throughout the later text.) Although Quetelet said again and again that his concern was with groups, not with individuals, there are many instances where he clearly uses the *penchant au crime* derived by this method to refer to a characteristic of *each* member of a given age group.

Apparent and real propensities. Quetelet developed the concept of *penchant* in order to overcome the methodological and theoretical problems he encountered in trying to found a science that would deal with all aspects of man. As long as he restricted himself to physical characteristics there was no problem; the technique for obtaining individual measurements was obvious, and moreover, plentiful data of this sort had already been collected by many agencies and was available for analysis. Once he moved on to moral and intellectual traits, however, the only data available were rates (of crime, or suicide, or marriage) for different populations. To parallel his analysis of physical characteristics, Quetelet would have needed measurements taken on each individual in a group over a period of time, and so far no one had collected such data. Occasionally, in the more speculative portions of his writings, Quetelet was able to suggest how such individual measurements might be made: to measure a scholar's productivity, for example, he thought of counting the number of publications the man produced.

Unfortunately, he never applied such ideas in his statistical work. Instead, he made do with the data at his disposal by establishing the critical distinction between "apparent propensities" and "real propensities." Apparent propensities are those that can be calculated as outlined above, using the population rates found in official statistical publications. The information needed is only the number of acts (crimes, suicides, marriages), the age of each actor, and the total population, distributed according to age. The real propensity is what causes the observable regularity to appear; and this propensity, Quetelet claimed, cannot be ascertained from direct observation. It can be known only by its effects.

The following passage illustrates how Quetelet related the two types of propensities. Commenting on computations of the probability of marriage for certain city dwellers, he said:

This probability may be considered as giving, in cities, the measure of the *apparent tendency* to marriage of a Belgian aged 25 to 30. I say *apparent tendency* intentionally, to avoid confusion with the *real tendency*, which may be quite different. One man may have, throughout his life, a real tendency to marry without ever marrying; another, on the contrary, carried along by fortuitous circumstances, may marry without having the least propensity to marriage. The distinction is essential.   (1848, p. 77)

The cue to the way Quetelet visualized the relation between apparent tendencies and real tendencies lies in his repeated statement that "causes are proportional to effects." Thus, if one thinks of the apparent tendencies as being caused by the underlying real tendencies, "the error that may result from substituting the value of the one for the other can be calculated directly by the theory of probabilities" (*ibid.*, p. 78).

Rather than solve the problem, this does no more than identify it. Quetelet's difficulty arose from the fact that he never clearly separated the problem into its two components. One is the question that continues to be of interest today: how may manifest data be related to latent dispositions? The other is the methodological problem of whether measurement techniques analogous to those applied to individual physical traits could be developed for moral and intellectual characteristics as well.

On the theoretical side, Quetelet failed to recognize that his dispositional concept of *penchant* could just as reasonably be applied in the study of physical attributes as in the study of moral or intellectual traits. One can, for example, conceive of studying the "tendency to obesity" in a population, which would parallel, in the physical, Quetelet's notion of a *penchant au crime*. On the methodological side, he failed to realize that some of the techniques he himself suggested for measuring individual personality or intelligence were exact counterparts to his "direct" measurements of physical traits. Especially surprising—in view of his extensive use of crime statistics—is the fact that the idea of analyzing, for example, "repeated offenders," completely eluded him. This would have provided him with quantitative individual "measures" of criminal behavior, corresponding to measures of size or weight. It seems likely that these shortcomings in Quetelet's work were due, not to sheer lack of insight, but, at least in part, to the inadequacies of the data available at the time. It is only benefit of hindsight that allows us to identify such gaps, and one cannot gainsay Quetelet's merit in having made the first attempt to handle what remain to this day crucial problems in the analysis of empirical social phenomena.

After Quetelet

In 1855 Quetelet suffered a stroke, from which he never totally recovered. He resumed his work very soon afterward but never again produced any new ideas. His publications from then on, although numerous, were largely compendia of prior essays or summaries of new researches which supported his earlier ideas. His son Ernest virtually took over the running of the observatory after 1855. Quetelet died on February 17, 1874, and as Hankins put it, "was buried with honors befitting one of the earth's nobility." A statue of him, funded by popular subscription, was unveiled at Brussels in 1880.

Quetelet's concern with the distribution of human characteristics was destined for an interesting future. His basic idea was that certain social processes (corresponding to his interplay of causes) would explain the final distribution of certain observable data. This notion has been amply justified by modern mathematical models regarding the distribution of, for example, income, words, or city sizes. But Quetelet concentrated exclusively on the binomial and normal (Gaussian) distributions, which presuppose the independence of the events studied. Today many other distributions are known, based on more complicated processes; especially in the social sciences, "contagious events," which depend upon each other, are in the center of interest. Quetelet remained unaware of alternative mathematical possibilities. Nonetheless, his basic idea was not only correct but probably influenced directly writers who had begun to broaden this whole field, such as Poisson and Lexis.

In general, however, Quetelet's contemporaries focused their attention primarily on his concept of the average man and his proposition that social phenomena reproduce themselves with extreme regularity. For different reasons these ideas quickly became the subject of the most vigorous and widespread debate among nineteenth-century statisticians, philosophers, and social scientists, while most of Quetelet's other ideas remained largely unnoticed.

The debate over the *homme moyen* was set off by Quetelet's suggestion that the means of various traits could be combined to form one paradigmatic human being, who would represent the "type" for a group, a city, a nation, or even for all of mankind. Typical of the early criticism was that of Cournot (1843), who reasoned from a mathematical analogy: just as the averages of the sides of many right triangles do not form a right triangle, so the averages of physical traits would certainly not be compatible. Combining them would not, as Quetelet claimed, produce a "type for human beauty" or a "type for physical perfection," but a monstrosity. Quetelet's insistence that the average man be considered no more than a "fictitious being" was taken as simply an evasion of the issue; his attempts to reply directly to Cournot proved unconvincing. (A recent essay by Guilbaud discusses the Cournot problem as it relates to current statistical concepts like that of "aggregation"; Guilbaud 1952.)

Quetelet's inability to refute Cournot's criticism encouraged others to publish similar attacks until, in 1876, Bertillon issued what is usually considered the definitive statement, which pretty well put an end to the debate. Here the criticism was applied not only to Quetelet's notion of combining average physical characteristics but also to his ideas about moral and intellectual traits. Surely,

Bertillon said, Quetelet was mistaken in believing that his average man would represent the ideal of moral virtue or intellectual perfection. Such a man would, on the contrary, be the personification of mediocrity or, to use Bertillon's apt phrase, the *type de la vulgarité* (p. 311). Thereafter the average man was generally viewed as a concept not worth taking seriously, although sporadic attempts have been made to revive it. (The latest is that of Maurice Fréchet, who suggests that Quetelet's *homme moyen* could be "rehabilitated" as the concept of the *homme typique;* by defining the *homme typique* as a *particular* individual in the group—whose traits taken as a whole come *closest* to the average—Fréchet manages to avoid most of the criticisms leveled against Quetelet's concept; Fréchet 1955.)

A second and even more widespread controversy centered on the question of what implications ought to be drawn from the startling regularities Quetelet had demonstrated in his studies of social phenomena. Did Quetelet's proposition that "society prepares the crime and the guilty person is only the instrument by which it is executed" (1835, p. 108 in 1842 edition) imply that human beings have no free will at all? Quetelet's philosophical speculations on the subject certainly left room for this interpretation. The result was a heated and long-lasting debate between those who supported Quetelet's "deterministic" explanation of social regularities and those who argued that only by taking the individual as the starting point for analysis could one arrive at an explanation of human behavior. Free will, the latter group contended, must be considered a prime determinant of action, not classed as a practically negligible "accidental cause." (One by-product of the debate was the formation of a "German school" of moral statisticians, headed by Moritz Wilhelm Drobisch, one of Quetelet's most vehement opponents; see Drobisch 1867.) The over-all result of the controversy was not so much to refute Quetelet's ideas as to argue them into oblivion. In P. E. Fahlbeck's opinion (1900), its most important effect was that until the end of the nineteenth century statisticians were so involved in discussing the implications of Quetelet's propositions that they made little effort to confirm empirically the nature and extent of the regularities Quetelet had discovered. (A detailed discussion of the controversy over Quetelet's determinism and over the concept of the average man appears in Lottin 1912, pp. 413–458.)

Only in recent years has Quetelet's sociological work begun to receive due recognition. His conviction that a scientific study of social life must be based on the application of quantitative methods and mathematical techniques anticipated what has become the guiding principle of modern social research. Some of the specific methods he employed and advocated—e.g., the substitution of one-time observation of a population for repeated observations of the individual, and his early attempts at multivariate analysis—are as important today as they were new in his time. The same may be said of his efforts to transform statistics from the mere clerical task of collecting important facts about the state (hence the term "statistics") to an exact method of observation, measurement, tabulation, and comparison of results, which would serve as the scaffolding upon which he could erect his science of moral statistics. On these grounds alone it is difficult to dispute Sarton's description of *Sur l'homme* as "one of the greatest books of the nineteenth century" ([1935] 1962, p. 229); or, for that matter, his choice of Quetelet over Comte as the "founder of sociology."

DAVID LANDAU AND PAUL F. LAZARSFELD

[*For the historical context of Quetelet's work, see* SOCIAL RESEARCH, THE EARLY HISTORY OF, *and the biographies of* GAUSS; LAPLACE; POISSON. *For discussion of the subsequent development of Quetelet's ideas, see* GOVERNMENT STATISTICS; STATISTICS, DESCRIPTIVE, *article on* LOCATION AND DISPERSION; *and the biographies of* GALTON; GINI; PEARSON. *See also* Cerase 1968; Clark 1968; Reiss 1968.]

WORKS BY QUETELET

(1826) 1834 *Astronomie élémentaire.* 3d ed., rev. & corrected. Brussels: Tircher.

(1827*a*) 1832 *Astronomie populaire.* 2d ed., rev. Brussels: Remy.

(1827*b*) 1834 *Positions de physique, ou résumé d'un cours de physique générale.* 2d ed. Brussels: Tircher. → An English translation was published by Sinclair in 1835, as *Facts, Laws and Phenomena of Natural Philosophy: Or, Summary of a Course of General Physics.*

1828 *Instructions populaires sur le calcul des probabilités.* Brussels: Tarlier. → Translation of extract in the text was provided by David Landau. An English translation of the entire work was published by Weale in 1849, as *Popular Instructions on the Calculation of Probabilities*, with appended notes by Richard Beamish.

1831*a* *Recherches sur la loi de la croissance de l'homme.* Brussels: Hayez. → A 32-page pamphlet. Also published in Volume 7 of the *Nouveaux mémoires* of the Académie Royale des Sciences, des Lettres et des Beaux-Arts de Belgique.

(1831*b*) 1833 *Recherches sur le penchant au crime aux différens âges.* 2d ed. Brussels: Hayez. → An 87-page pamphlet. Translations of extracts in the text were provided by David Landau. Also published in Volume

7 of the *Nouveaux mémoires* of the Académie Royale des Sciences, des Lettres et des Beaux-Arts de Belgique.

1833*a* *Recherches sur le poids de l'homme aux différens âges.* Brussels: Hayez. → A 44-page pamphlet. Also published as part of Volume 7 of the *Nouveaux mémoires* of the Académie Royale des Sciences, des Lettres et des Beaux-Arts de Belgique.

1833*b* *Lettre à M. Villermé, sur la possibilité de mesurer l'influence des causes qui modifient les éléments sociaux. Annales d'hygiène publique et de médecine légale* 1st Series 9:309 only.

(1835) 1869 *Physique sociale: Ou, essai sur le développement des facultés de l'homme.* 2 vols. Brussels: Muquardt. → First published as *Sur l'homme et le développement de ses facultés: Physique sociale.* An English translation was published by Chambers in 1842 as *A Treatise on Man and the Development of His Faculties.*

1844 *Sur l'appréciation des documents statistiques, et en particulier sur l'appréciation des moyennes.* Belgium, Commission Centrale de Statistique, *Bulletin* 2:205–286.

(1846) 1849 *Letters Addressed to H.R.H. the Grand Duke of Saxe-Coburg and Gotha, on the Theory of Probabilities, as Applied to the Moral and Political Sciences.* London: Layton. → First published in French.

1848 *Du système social et des lois qui le régissent.* Paris: Guillaumin. → Translations of extracts in the text were provided by David Landau.

1870 *Des lois concernant le développement de l'homme.* Académie Royale des Sciences, des Lettres et des Beaux-Arts de Belgique, *Bulletin* 2d Series 29:669–680. → Translations of extracts in the text were provided by David Landau.

SUPPLEMENTARY BIBLIOGRAPHY

BERTILLON, ADOLPHE 1876 *La théorie des moyennes en statistique. Journal de la Société de Statistique de Paris* 17:265–271, 286–308.

BUCKLE, HENRY THOMAS (1857–1861) 1913 *The History of Civilization in England.* 2d ed. 2 vols. New York: Hearst.

►CERASE, FRANCESCO P. 1968 Niceforo, Alfredo. Volume 11, pages 174–175 in *International Encyclopedia of the Social Sciences.* Edited by David L. Sills. New York: Macmillan and Free Press.

►CLARK, TERRY N. 1968 Bertillon, Jacques. Volume 2, pages 69–71 in *International Encyclopedia of the Social Sciences.* Edited by David L. Sills. New York: Macmillan and Free Press.

Correspondances mathématiques et physiques. → Published from 1825 to 1839.

COURNOT, ANTOINE AUGUSTIN 1843 *Exposition de la théorie des chances et des probabilités.* Paris: Hachette. → See especially pages 213–214.

Discours [prononcés aux funérailles de Quetelet], by N. de Keyser et al. 1874 Académie Royale des Sciences, des Lettres et des Beaux-Arts de Belgique, *Bulletin* 2d Series 37:248–266.

DROBISCH, MORITZ W. 1867 *Die moralische Statistik und die menschliche Willensfreiheit.* Leipzig: Voss.

DURKHEIM, ÉMILE (1897) 1951 *Suicide: A Study in Sociology.* Glencoe, Ill.: Free Press. → First published in French. See pages 300–306 for a discussion of Quetelet's concept of the "average man."

FAHLBECK, PONTUS E. 1900 La régularité dans les

choses humaines ou les types statistiques et leurs variations. *Journal de la Société de Statistique de Paris* 41:188–201.

[FOURIER, JEAN BAPTISTE] 1826 Mémoire sur les résultats moyens déduits d'un grand nombre d'observations. Volume 3, pages ix–xxxi in Seine (Dept.), *Recherches statistiques sur la ville de Paris et le département de la Seine.* Paris: Imprimerie Royale.

FRÉCHET, MAURICE 1955 Rehabilitation de la notion statistique de l'homme moyen. Pages 310–341 in Maurice Fréchet, *Les mathématiques et le concret.* Paris: Presses Universitaires de France.

GALTON, FRANCIS (1869) 1952 *Hereditary Genius: An Inquiry Into Its Laws and Consequences.* New York: Horizon Press. → See the Appendix for a discussion of some of the mathematical aspects of the "average man." A paperback edition was published in 1962 by World.

GILLISPIE, C. C. 1963 Intellectual Factors in the Background of Analysis by Probabilities. Pages 431–453 in Symposium on the History of Science, Oxford, 1961, *Scientific Change: Historical Studies in the Intellectual, Social, and Technical Conditions for Scientific Discovery and Technical Invention, From Antiquity to the Present.* New York: Basic Books.

GINI, CORRADO 1914a L'uomo medio. *Giornale degli economisti e rivista di statistica* 3d Series 48:1–24.

GINI, CORRADO 1914b Sull'utilità delle rappresentazioni grafiche. *Giornale degli economisti e rivista di statistica* 3d Series 48:148–155.

GUERRY, ANDRÉ MICHEL 1832 Statistique comparée de l'état de l'instruction et du nombre des crimes. *Revue encyclopédique* 55:414–424.

GUILBAUD, GEORGES TH. (1952) 1966 Theories of the General Interest and the Logical Problem of Aggregation. Pages 262–307 in Paul F. Lazarsfeld and Neil W. Henry (editors), *Readings in Mathematical Social Science.* Chicago: Science Research Associates. → See especially pages 271–292. First published in French in Volume 5 of *Économie appliquée.*

HALBWACHS, MAURICE 1912 *La théorie de l'homme moyen: Essai sur Quetelet et la statistique morale.* Paris: Alcan.

HANKINS, FRANK H. 1908 *Adolphe Quetelet as Statistician.* New York: Longmans.

[HERSCHEL, JOHN F. W.] 1850 [Review of] *Letters.* . . . *Edinburgh Review* 92:1–57.

KNAPP, GEORG F. 1871 Bericht über die Schriften Quetelet's zur Socialstatistik und Anthropologie. *Jahrbücher für Nationalökonomie und Statistik* 17:167–174, 342–358, 427–445.

KNAPP, GEORG F. 1872 A. Quetelet als Theoretiker. *Jahrbücher für Nationalökonomie und Statistik* 18:89–124.

LAPLACE, PIERRE SIMON DE (1812) 1820 *Théorie analytique des probabilités.* 3d ed., rev. Paris: Courcier.

LAZARSFELD, PAUL F. 1961 Notes on the History of Quantification in Sociology: Trends, Sources and Problems. *Isis* 52:277–333.

LOTTIN, JOSEPH 1912 *Quetelet, statisticien et sociologue.* Paris: Alcan; Louvain: Institut Supérieur de Philosophie.

MAILLY, NICOLAS É. 1875a Essai sur la vie et les ouvrages de Lambert-Adolphe-Jacques Quetelet. Académie Royale des Sciences, des Lettres et des Beaux-Arts de Belgique, *Annuaire* 41:109–297.

MAILLY, NICOLAS É. 1875b Eulogy on Quetelet. Smithsonian Institution, *Annual Report* [1874]:169–183. →

Abstract of "Notice sur Adolphe Quetelet," first published in Series 2, Volume 38 of the *Bulletin* of the Académie Royale des Sciences, des Lettres et des Beaux-Arts de Belgique.

POISSON, SIMÉON DENIS 1837 *Recherches sur la probabilité des jugements en matière criminelle et en matière civile, précédées des règles générales du calcul des probabilités.* Paris: Bachelier.

REICHESBERG, NAUM 1896 Der berühmte Statistiker, Adolf Quetelet, sein Leben und sein Wirken: Eine biographische Skizze. *Zeitschrift für schweizerische Statistik* 32:418–460.

►REISS, ALBERT J. JR. 1968 Sociology: I. The Field. Volume 15, pages 1–23 in *International Encyclopedia of the Social Sciences.* Edited by David L. Sills. New York: Macmillan and Free Press.

SARTON, GEORGE (1935) 1962 Quetelet (1796–1874). Pages 229–242 in George Sarton, *Sarton on the History of Science.* Edited by Dorothy Stimson. Cambridge, Mass.: Harvard Univ. Press.

SCHOEN, HARRIET H. 1938 Prince Albert and the Application of Statistics to the Problems of Government. *Osiris* 5:276–318.

WELLENS-DE DONDER, LILIANE 1964 La correspondance d'Adolphe Quetelet. *Archives et bibliothèques de Belgique* 35:49–66.

WOLOWSKI, L. 1874 Éloge de Quetelet. *Journal de la Société de Statistique de Paris* 15:118–126.

QUEUES

There are many situations in industry and everyday life in which *customers* require *service* from limited service facilities. Much work has been done on techniques for predicting the amount of congestion in such systems. Applications include the following: telephone calls requiring service at an exchange; aircraft requiring an opportunity to take off or land; cars requiring a changed traffic light, or an opportunity to turn from a minor road into a major road; products at one stage of an industrial process requiring entrance to the next stage of processing; machines requiring attention from an operator; people requiring transport by bus, train, or airplane; and patients requiring attention by a doctor or admission to a hospital.

The objective of most investigations of queueing is to modify the system to make it more efficient. For example, the object of a study of queueing at traffic lights might be to formulate rules for adjusting the timing of the lights to the rates of flow of traffic. Such matters fall within the general subject of operations research. There seems to be as yet little work on queueing of a fundamental sociological nature.

Methods of investigation. There are, essentially, four methods of investigating queueing problems: (1) direct observation of practical situations, (2) planned experiments under artificial conditions, (3) simulation, and (4) mathematical analysis. Published work is largely on mathematical analysis, but this does not properly reflect the relative importance of the methods. The last sections of this article deal with simulation and mathematical analysis.

Some elementary results. Some important theoretical ideas and results can be illustrated with a simple example. Consider customers arriving at a shop having one server who serves only one customer at a time. Unserved customers queue up waiting their turn for service. Suppose that the instants at which customers arrive are distributed in a stable statistical pattern in which the mean interval between the arrival of successive customers is a minutes. Let the time taken to serve a customer have a stable frequency distribution with mean s min.

In a very long time, T min., about T/a customers will arrive, whereas, even if the server works continuously, the number of customers served in that time is about T/s. Hence, if $T/a > T/s$, that is, if $s/a > 1$, customers arrive more rapidly than they can be served and the queue will grow until the system changes; for example, a second server may be obtained or customers may be deterred from joining the queue.

The first step in many queueing investigations is to see whether the average service capability is enough to meet average demands. Suppose that this is so, that is, $s/a < 1$. It is reasonable to expect a state of statistical equilibrium in which the number of customers queueing fluctuates with a stable frequency distribution. Also, in a long time T, the server will need to work for a total time of about sT/a, in order to serve T/a customers. Hence, there will be no customers in the system for a proportion $1 - (s/a)$ of the time. The dimensionless quantity s/a, called the *traffic intensity*, measures the average demand on the system relative to the average service capability. The qualitative behavior of the system is determined in the first place by whether the traffic intensity exceeds, or is less than, one. (Nearly always, the case of exact balance is unstable, leading eventually to very long queues.)

The next general point is that when $s/a < 1$ the amount of congestion depends on the random fluctuations present. Thus, suppose that $a = 1$, $s = 0.9$. If arrivals are perfectly regular every minute and if service always takes 0.9 min., an exact cycle of working is set up and no congestion arises. But if there is appreciable random variation in the arrivals or in the service times, there will, from time to time, be several customers at the service point simultaneously, and congestion will result. It can

be shown that if arrivals are completely random (defined later) and the service time is constant, then the mean time spent queueing before the start of service is 4.05 min. and the server is idle 10 per cent of the time. Further, if both arrivals and service times are random and if the standard deviation of service time is equal to the mean, the mean time spent queueing rises to 8.1 min. With regular arrivals and a particular form of distribution of service time with standard deviation equal to the mean, the mean queueing time is 3.76 min. These represent substantial congestion.

The important general points are (1) if there is appreciable random variation, substantial congestion may result if the system is loaded near to its limiting average capacity, and (2) the amount of congestion depends in an essential way on the statistical variability in the system.

The amount of congestion can be reduced in various ways, for example, by having more servers, by reducing the variability in the system, or by reducing the traffic intensity. Suppose, however, that the traffic intensity is reduced to 0.8, by reduction of the service time to a constant, $s = 0.8$ min. Then the mean queueing time drops from 4.05 min. to 1.6 min. The proportion of time that the server is idle rises from 0.1 to 0.2. An analysis of the costs associated with congestion and with the service mechanism is usually required to find the best way to run the system.

General description of queueing systems. The main features that determine a queueing system are the input or arrival system, the service mechanism, and the queue discipline. Many possibilities arise, only a few being described here.

The main features of the input are the average rate of arrival (including its dependence, if any, on time) and the local statistical character of the arrival pattern. The two simplest types of arrival used in mathematical work are *regular* arrivals and *random* arrivals. The latter, often called a Poisson process, is a very special form defined as follows. In any small time interval $(t, t + \Delta t)$ there is a chance $\Delta t/a$ of a customer arriving, quite independently of arrivals in other time periods. The average interval between arrivals is a; the distribution of x, the interval between successive arrivals, is negative exponential, having density function $a^{-1} \exp(-x/a), x > 0$ [*see* DISTRIBUTIONS, STATISTICAL].

The service mechanism is described by the service capacity, the service availability, and the properties of service time. The service capacity is the maximum number of customers that can be served at a time, one in the single-server queue discussed above. In some applications—for example, many transport problems—service is available only at certain times, and this must be specified; or service may not be available until a group of customers of specified size has collected (batch service). Finally, one must specify the statistical properties of the service time, usually by giving a probability distribution. Perhaps the most common type has a unimodal positively skewed density function with a standard deviation of 20 to 50 per cent of the mean. In mathematical work, a particularly important density is the negative exponential, with equation $s^{-1}e^{-x/s}$, where s is the mean service time.

The queue discipline specifies what happens to customers who cannot be served immediately. Sometimes there may be *losses*, for example, customers may have to be rejected because of lack of waiting room. Then the probability of loss should be calculated. If there are no losses, customers queue awaiting their turn for service. One then needs to specify how customers are selected for service from the queue. Even in a single-server system there are many possibilities: for example, first-come, first-served, that is, customers queue in order of arrival; random selection from a queue of customers; non-pre-emptive priority, in which the customers are divided into two or more priority classes so that when a customer is to be selected for service the one with the highest priority is chosen; and pre-emptive priority, in which a high-priority customer is served immediately upon arrival.

This classification applies to a single point of congestion. Often there will be several such interlinked points.

Arrival pattern, service mechanism, and queue discipline determine the system. The resulting amount of congestion can be described by the distributions of three quantities: the number of customers in the system at an arbitrary time, the time a customer spends queueing before being served, and the server's busy and idle periods. Which quantity or combination of quantities should be considered depends on the costs associated with congestion.

Investigating the behavior of queues. Several of the usual methods used in investigating queues are described below.

Simulation. The structure of the system is specified statistically, for example, by giving numerically the frequency distribution of service time. The behavior of the system is then *simulated*, that is, reconstructed empirically using pencil and paper (in simple cases), an electrical or hydraulic analogue machine, or, most commonly, digital computation on an electronic computer. The behavior over a long period is found and relevant properties, such as queueing times, measured. The procedure is normally repeated for a range of values of the important parameters. This approach is empirical observation of an idealized statistical model. It is sometimes called the Monte Carlo method (Florida . . . 1956). [*See* COMPUTATION.]

Simulation is not necessary for the one-server queueing model for which numerical results were given above, because simple mathematical results can be found. If used, however, simulation would proceed roughly as follows. From tables of random numbers, or from a computer program for generating pseudo-random numbers, values are formed representing the intervals between the arrivals of customers and the service times of customers. The arrival instants and the instants at which service is started and completed can be built up and anything of interest, such as the frequency distribution of queueing time, measured [*see* SIMULATION].

An important advantage of this method is its extreme flexibility. Systems far too complex for mathematical analysis can be investigated. A disadvantage is that many simulations under a wide range of conditions may be necessary to understand fully the system's behavior. Simulation is, however, the most generally applicable technique for investigating specific practical problems.

Mathematical analysis. The theory of stochastic processes deals with systems that change in time probabilistically. To get working results about queues from the theory, it is usually necessary to consider very simplified models of the real problem. The usefulness of the theory is partly that skillful simplification may leave the essential features preserved and, perhaps more importantly, that intensive study of simplified models can lead to valuable qualitative understanding. A good example of qualitative conclusions arising from a simplified mathematical model comes from an investigation by Tanner (1961) of a model of overtaking on a two-lane road. Tanner's calculations show that under certain circumstances there is a density of slow traffic, well below the capacity of the road, above which overtaking is effectively impossible. In many circumstances, high acceleration is more important than high speed, and a cutting of safety margins makes little difference to the average speed of a fast car over its whole journey. Mathematical investigation of simplified models combined with simulation of more complex models is a powerful method of investigation.

Essentially three types of mathematical investigation can be made: (1) to find rigorously the conditions for the existence of a unique equilibrium statistical distribution, (2) to find the equilibrium distribution assuming that it exists, and (3) to examine the transient behavior when, for example, the parameters change in time. The answer to (1) is usually qualitatively obvious, although the rigorous proofs often call for delicate arguments. Answers to (2) form the bulk of the literature and are relevant when long-run behavior is investigated. When behavior over a short period or under changing conditions is required, the transient behavior needs consideration, but not many usable results are yet available.

Application of Markov processes. The general idea behind the mathematical theory is to relate the probability distributions of the states of the system at different times. Consider the transition probabilities determining the probability distribution of the system at time $t + \Delta t$, given the state at time t. If these transition probabilities depend only on the state at t and not also on what happens before t, the process is a Markov process [see MARKOV CHAINS]. It is in principle then easy to find simultaneous differential equations describing the transient behavior of the system and ordinary linear equations for the equilibrium behavior.

The Markov property holds in simple form when arrivals are random and the distribution of service time is negative exponential, with density $s^{-1}e^{-x/s}$. Simplicity is gained by having random arrivals because the probability of an arrival in the small time interval $(t, t + \Delta t)$ is $\Delta t/a$, *independently of all other occurrences.* The analogous property of the exponential distribution of service time is that if service of a customer is in progress at time t, the probability that service is completed in $(t, t + \Delta t)$ is $\Delta t/s$, where s is the mean service time, *independently of the length of time for which the particular service operation has been going on.*

For example, consider the single-server queue with random arrivals and exponentially distributed service time. Let $p_n(t)$ be the probability at time t that there are n customers in the system, including the one, if any, being served. Consider $p_0(t + \Delta t)$, where $\Delta t > 0$ is very small. The ways in which the queue may be empty at $t + \Delta t$ are that (1) the queue is empty at t and no customer arrives in $(t, t + \Delta t)$, (2) there is one customer present at t whose service is completed in $(t, t + \Delta t)$, and (3) two or more events occur in $(t, t + \Delta t)$. The probability of (3) is negligible. Thus $p_0(t, t + \Delta t)$ is the sum of the probabilities of (1) and (2). That is,

$$p_0(t + \Delta t) = p_0(t)\{1 - \Delta t/a\} + p_1(t)\Delta t/s,$$

from which

(1) $$\frac{dp_0(t)}{dt} = -\frac{1}{a}p_0(t) + \frac{1}{s}p_1(t).$$

Similarly, if $n \geqslant 1$, $p_n(t + \Delta t)$ is the sum of the probabilities that (a) there are n customers at t and there is no arrival or completion of service in $(t, t + \Delta t)$, (b) there are $n - 1$ customers at t and one customer arrives in $(t, t + \Delta t)$, and (c) there are $n + 1$ customers at t and service is completed in $(t, t + \Delta t)$. Therefore

(2) $$\frac{dp_n(t)}{dt} = -\left(\frac{1}{a} + \frac{1}{s}\right)p_n(t) + \frac{1}{a}p_{n-1}(t) + \frac{1}{s}p_{n+1}(t),$$
$$n = 1, 2, \cdots.$$

This set of simultaneous linear differential equations can be solved to give the transient behavior of the system. The answer is rather complicated. If it is assumed that an equilibrium distribution exists, the equations must have a solution, $\{p_n\}$, that is independent of time. Thus, directly from (1) and (2),

$$0 = -\frac{p_0}{a} + \frac{p_1}{s},$$

$$0 = -\left(\frac{1}{a} + \frac{1}{s}\right)p_n + \frac{1}{a}p_{n-1} + \frac{1}{s}p_{n+1},$$
$$n = 1, 2, \cdots.$$

Also, since the $\{p_n\}$ form a probability distribution, $1 = p_0 + p_1 + p_2 + \cdots$; the solution is

$$p_0 = 1 - r,$$
$$p_1 = r(1 - r),$$
$$\vdots \qquad \vdots$$
$$p_n = r^n(1 - r),$$
$$\vdots \qquad \vdots$$

where $r = s/a$ is the traffic intensity. The mean of the distribution is $r/(1 - r)$. When $r = 0.9$, the case examined previously, the mean number in the system is 9; this is the average over a long time of the number of customers present. Because arrivals are random, an arriving customer also finds on the average 9 customers ahead, each taking on the average 0.9 min. to be served. Thus the mean time spent queueing, for the queue discipline first-come, first-served, is 9×0.9 min. = 8.1 min. The general formula is that in a single-server queue, with random arrivals, exponential service time, and the queue discipline first-come, first-served,

$$(3) \qquad \frac{\text{mean time spent queueing}}{\text{mean service time}} = \frac{1}{1 - r},$$

where r is the traffic intensity.

The method used above to get linear equations can be used for more complex systems, provided that arrivals are independent and random and that service time distributions are exponential. However, consider even the simple single-server queueing system with random arrivals but with a non-exponential distribution of service time; here the argument fails, because the probability that service of a customer is completed in $(t, t + \Delta t)$ is not constant but depends on how long service has been in progress. There are various ways round this difficulty, of which, one—the method of the imbedded Markov process—depends on considering the system only at instants at which service of a customer is completed. The generalization of (3) is the Pollaczek–Khinchin formula, $\frac{1}{2}r(1 + c^2)/(1 - r)$, where c is the ratio of the standard deviation of service time to the mean.

Future developments. The number of new queueing situations is unlimited; almost each new application shows a fresh combination of input, service mechanism, and queue discipline. Simulation is usually the best method for tackling the more complex situations, and there is scope for further work on increasing the efficiency of simulation techniques. Two fields where further mathematical work is likely are those of networks of queues and of the automatic control of queueing systems.

History. The first major theoretical investigation was made in 1909 by A. K. Erlang, Copenhagen Telephone Co. Between the two world wars, important mathematical work on queueing was done by T. C. Fry (United States), A. Ia. Khinchin (Soviet Union), and Felix Pollaczek (France). After 1945, interest in the subject grew rapidly, both because of the much increased attention paid to operations research, and because the appropriate mathematical tool, the theory of stochastic processes, was being widely studied in other contexts. There is now an extensive literature on the mathematical problems connected with queueing. Except for the special fields of telephone engineering and traffic studies, relatively little has been written about applications.

D. R. Cox

[*See also* MODELS, MATHEMATICAL.]

BIBLIOGRAPHY

Introductory theoretical books in English are Khinchin 1955, Morse 1958, Cox & Smith 1961, *and* Riordan 1962. Saaty 1961 *gives a more extensive review of the subject and has a long bibliography.* Takács 1962 *and* Beneš 1963 *are more specialized and mathematical.* Syski 1960 *deals particularly with applications to telephone engineering. Empirical and theoretical work on road traffic problems is best approached through the publications of the Road Research Laboratory (Great Britain) and the Highway Research Board (United States).* White 1962 *contains a critical discussion of work on the somewhat related problem of the sizes of casual groups of people.*

BENEŠ, VÁCLAV E. 1963 *General Stochastic Processes in the Theory of Queues.* Reading, Mass.: Addison-Wesley.

COX, DAVID R.; and SMITH, W. L. 1961 *Queues.* London: Methuen; New York: Wiley.

FLORIDA, UNIVERSITY OF, GAINESVILLE, STATISTICAL LABORATORY 1956 *Symposium on Monte Carlo Methods, Held at the University of Florida . . . March 16 and 17, 1954.* Edited by Herbert A. Meyer. New York: Wiley.

KHINCHIN, ALEXANDR IA. (1955) 1960 *Mathematical Methods in the Theory of Queueing.* London: Griffin; New York: Hafner. → First published in Russian.

MORSE, PHILIP M. 1958 *Queues, Inventories and Maintenance: The Analysis of Operational Systems With Variable Demand and Supply.* New York: Wiley.

RIORDAN, JOHN 1962 *Stochastic Service Systems.* New York: Wiley.

SAATY, THOMAS L. 1961 *Elements of Queueing Theory, With Applications.* New York: McGraw-Hill.

SYSKI, RYSZARD 1960 *Introduction to Congestion Theory in Telephone Systems.* Edinburgh: Oliver & Boyd.

TAKÁCS, LAJOS 1962 *Introduction to the Theory of Queues.* New York: Oxford Univ. Press.

TANNER, J. C. 1961 Delays on a Two-lane Road. *Journal of the Royal Statistical Society* Series B 23:38–63.

WHITE, H. 1962 Chance Models of Systems of Casual Groups. *Sociometry* 25:153–172.

Postscript

A large number of books and articles on queues have appeared since the main article was published; particularly interesting is an elementary introduction by Leibowitz (1968) with some emphasis on traffic problems.

D. R. Cox

ADDITIONAL BIBLIOGRAPHY

LEIBOWITZ, MARTIN A. 1968 Queues. *Scientific American* 219, Aug.:96–103.

QUOTA SAMPLING

See SAMPLE SURVEYS, *article on* NONPROBABILITY SAMPLING.

R

RANDOM NUMBERS

Random numbers are numbers generated by a mechanism that produces irregularity, in a sense that will be made precise. There are four major uses of random numbers: (1) To protect against selective bias in the acquisition of information from sample surveys and experiments. In these same contexts, random numbers provide a known probability structure for statistical calculations. (2) To gain insight, by simulation, into the behavior of complex mechanisms or models. (3) To study theoretical properties of statistical procedures, such as efficiency of estimation and power of statistical tests. (4) To obtain approximate solutions to other mathematical problems. In each of these applications, random numbers are used to simulate observations from an arbitrary probability distribution. It is often useful to think of random numbers as an idealization of the successive outcomes of a carefully made gambling device, say a roulette wheel.

Definitions. A definition of random numbers is a specification of the probabilistic properties of the process that generates them, rather than a specification of the properties of the numbers themselves. One can only partially test and verify the numbers themselves; they are only specific outcomes of the process. To clarify the distinction between a process and an outcome of the process, consider the example of coin tossing. In stating that the probability of getting a head in flipping an honest coin is $\frac{1}{2}$, one implies that before the coin is flipped the process of flipping gives probability $\frac{1}{2}$ of observing a head. Once the coin is flipped, however, either a head or a tail shows; the situation is determined and not probabilistic.

Two concepts are involved in specifying the process generating random numbers: independence and a specific type of nondeterminism. In terms of the coin-flipping process, independence implies that the process that flips the coin is not influenced by prior flips and nondeterminism implies that one cannot state with certainty whether a head or a tail will occur on a flip. In the case of an honest coin one can only say that the probability is $\frac{1}{2}$ that a head will occur.

A process is said to generate random numbers in a strict, or theoretical, sense when the process provides random variables that are (1) independent in the sense of probability and (2) governed by the same continuous uniform distribution. Independence implies that knowledge of some numbers generated by the process cannot help in predicting other numbers. The uniform distribution requirement ensures that the numbers generated lie within an interval; the probability that a random number lies in a subinterval is given by the length of the subinterval divided by the length of the entire interval [see DISTRIBUTIONS, STATISTICAL; PROBABILITY].

A process generates random (decimal) digits when it provides random variables that are independent and take the values 0, 1, $\cdots$, 9 with equal probability $\frac{1}{10}$, that is, the random variables follow a discrete uniform distribution. Similar definitions apply to random digits for bases other than 10, for any set of contiguous integers, or, in general, for any set following a discrete uniform distribution.

No real process can truly generate random numbers in the strict sense, since real processes must always generate rational numbers. The term "random numbers" is, however, often applied to processes that generate independent, equispaced, equally probable rational numbers when those numbers are so numerous that the discrete process may be considered an approximation to a continuous one. It is necessary, nonetheless, to keep this distinction between theory and practice in mind. Similarly the terms "random digits" and "random numbers" are in practice applied to real processes that presumably can never provide values that are *exactly* independent or uniform. Further, the term "random numbers" is often loosely used to apply to any of the above possibilities. Finally, these terms are even sometimes used to mean any number-producing process—probabilistic or not—where the produced values appear approximately equally often and where there is no evident pattern in the series of outcomes. See Hammersley and Handscomb (1964), Shreider (1962), or Young (1962) for additional discussion. This terminological thicket is not so tangled as might appear at first, since the meaning intended is usually clear from context.

Uses of random numbers. Random numbers, digits, etc. permit the construction of probability samples, which in turn permit the use of present-day statistical inference. In sample surveys, units must be selected from the sampled population by probability sampling if statistical analysis, in the ordinary sense, is properly to be done. The simplest case is that in which each unit of the population has an identification number (from 1 to N, say). Random integers (in the interval 1 to N) are chosen and the corresponding units form the sample. In practice, sampling without replacement is usually used, and here random permutations are needed; these are discussed below. Also in practice, sampling designs of various degrees of complexity are common, using devices of stratification, clustering, etc. [see SAMPLE SURVEYS].

In a similar way, random integers or permutations are used in experimental design for deciding which experimental treatments are to be applied to the various physical units: mice, men, fields, etc. In both these cases, the use of a random device serves to protect against selective bias, for example, assigning the healthiest subjects to one experimental therapy. Of equal importance, however, is the fact that the use of a random device provides a firm foundation for statistical inference [see EXPERIMENTAL DESIGN].

Random numbers can also be very useful in studying models of complex physical mechanisms or abstract theories, whenever a model involves variables with probabilistic (stochastic) elements from arbitrary probability distributions. For example, in models of fatigue or of learning one can simulate the performance of many individuals by generating observations from a hypothetical probability distribution depicting performance and then using the generated results to estimate actual performance. In essence, one is sampling from the hypothetical population to estimate how the "real" population can be expected to perform. Random numbers are used to obtain observations from an arbitrary probability distribution function. If observations $\{x\}$ are desired from a probability distribution $F(x)$, and if $\{u\}$ denotes values of random numbers in the unit interval $(0,1)$, that is, observations from the unit uniform distribution, then an observation value x from $F(x)$ is obtained by generating a value for u and solving for x in the relationship $x = F^{-1}(u)$. For the derivation and conditions necessary for this result, see, for example, Mood (1950, p. 107). More economical ways of obtaining a conversion of random numbers to observational values of $F(x)$ can often be achieved For example, several methods exist for obtaining observations from the normal distribution function (see Box & Muller 1958; Muller 1958; Teichroew 1965; Hull & Dobell 1962).

The use of random numbers has also been considered for solving physical or mathematical problems that appear to be completely nonprobabilistic. For example, assume that a mathematical function is given and it is desired to find (by integration) the area under its curve between specified limits of the independent variable. The solution to this problem can be approximated by use of random numbers. Imagine, for convenience, that the area to be estimated lies within the unit square; then by generating pairs of random numbers that lie in the unit square and comparing the number that lie under the curve to the total number of pairs generated, one obtains an estimate of the area under the curve. The general basis for this approach is considered, for example, by George W. Brown (1956, p. 283).

With the advent of modern digital computers this approach to mathematical problems has been given considerable attention and is referred to as the Monte Carlo (sampling) approach (see Florida, Univ. of 1956; Hammersley & Handscomb 1964; Shreider 1962). Indeed, the Monte Carlo approach can be used to explore theoretical questions of statistics that are not readily solved by direct analysis (see Teichroew 1965). Without going into detail, it can be stated that random numbers can also be used to mix strategies in complex decision processes, an application of growing usefulness [*see* DECISION THEORY].

Sources of random numbers. Physical and mathematical processes have both been employed to generate random numbers. It is very difficult to determine whether these sources provide satisfactory numbers; the closeness of agreement between the real-world process and the hypothetical model has been measured by rather general purpose statistical tests, which at best can give only partial answers.

Until recently, the major sources of random numbers have been published tables derived primarily by using physical devices, some of which have been improved with the aid of mathematical processing. A list of prepared tables of random numbers is presented below.

The physical processes used include (1) drawing capsules from a bowl, with replacement and remixing after each drawing; (2) flashing a beam of light at irregular time intervals on a sectioned rotating disk, so as to obtain section numbers; (3) recording digitized electronic pulses whose time distribution of occurrences is supposedly random; and (4) counting the number of output pulses generated by the radioactive decay of cobalt 60 during constant time intervals.

Physical processes have several disadvantages as sources of random numbers. Unless one stores the numbers generated by a physical process one cannot have them available and cannot regenerate the specific sequence used if it becomes necessary to check or repeat a computation. Furthermore, the early types of physical processes were difficult to maintain and use. Interest has shifted to the use of mathematical processes that can be programmed for digital computers, since large tables cannot as yet be economically entered and stored in the high-speed memories of computers. Most such mathematical processes center about a recurrence relation, where the next number of a sequence is derived from one or more prior numbers of the sequence.

These mathematical processes are deterministic, that is, any number of a sequence is a specified function of prior numbers of the sequence. For this reason, one could easily raise objections to the use of these processes and prefer the use of physical processes. However, as will be noted below, there is a strong pragmatic justification for these mathematical procedures.

Several methods use the output of a physical process as initial "random" numbers, which are improved by use of mathematical relationships such as those provided by Horton and Smith (1949). These techniques have been called compound randomization techniques, and they appear to hold much promise.

History. Tables of random numbers and random digits are a relatively recent development, having been available only since 1927. This is surprising, considering that, almost a century earlier, mechanical methods and shuffling devices for random numbers were already employed to examine probability problems. Samples for studying theoretical properties of various statistics had been reported in 1908 by "Student" (W. S. Gosset), who investigated the distribution of the standard deviation and the ratio of the mean to the standard deviation of samples by shuffling and drawing at "random" from a population of 3,000 cards (see Gosset 1907–1937). This study and earlier sampling studies using dice to estimate correlations among events—for example, Darbishire in 1907 and Weldon in 1906—are reviewed in Teichroew (1965).

Even after it was recognized that random numbers would provide a reliable and economical tool in selecting samples, a satisfactory method to generate them still remained to be found. Karl Pearson was instrumental in encouraging L. H. C. Tippett to develop a table of random digits. In the introduction to that table, Pearson mentioned that Tippett found that drawing numbered cards from a bag was unsatisfactory and that better results were obtained by using digits selected from a 1925 census report. With this approach, Tippett (1927) provided a table of 41,600 digits. Subsequently several analyses of this table were made and larger tables using other methods have since appeared. With the introduction of punched-card equipment, calculators, and digital computers, other methods of creating random numbers have been considered. Interest has focused on additional ways to test for the adequacy of numbers produced by various methods. Recognizing that in long sequences of random numbers, one might observe patterns of numbers that might be objectionable for a specific use, Kendall and Smith (1938, p. 153) provided criteria of local randomness by which to judge whether a set of numbers is acceptable for a specific application. One could raise objections to the use of the concept of local randomness to include or exclude specific sequences of numbers because each has one or more sets of subsequences that do or do not occur in some desired order. The practical and philosophical questions here are not solved and they raise doubt in some minds about the very foundations of probability theory. For example, G. Spencer Brown (1957) has given considerable discussion to the apparent paradox resulting from the exclusion of specific patterns of numbers.

The first of the mathematical generation processes, which is discussed below, appears to have been suggested by John von Neumann, who proposed the middle square method; this was followed by a multiplicative congruence method by Lehmer, which appeared in 1951.

In passing it should be noted that although Fisher did not directly contribute to the development of techniques for random numbers until 1938, when he and Yates provided their tables, his influence in suggesting the use of randomization in experimental design undoubtedly stimulated much interest in the subject.

Tables of random digits. Since the publication of the table of 41,600 digits by Tippett in 1927, several major tables have appeared. In 1938–1939 two were published: the first, by Fisher and Yates (1938), contains 15,000 digits based on entries in a table of logarithms; the second, by Kendall and Smith (1938; 1939), contains 100,000 digits based on readings from a rotating disk. In conjunction with the publication of their table, Kendall and Smith introduced their four tests of randomness and the concept of local randomness. Several workers have examined these three tables. In 1953, Good raised the question of the need for a modification of one of the tests by Kendall and Smith. From 1939 through the middle of the 1950s several other tables were developed. The most ambitious and largest was the one developed by workers at the RAND Corporation, using the output of an electronic pulse generator and a compound randomization technique to "improve the numbers" (see Brown 1951). These tables originally had limited distribution and were also available on punched cards. In 1955, a table appeared as a book under the title *A Million Random Digits With 100,000 Normal Deviates* (RAND Corp. 1955). This book is now more or less the standard source. The table of normal deviates was derived by converting random numbers to observations from the unit normal distribution by the method of inversion mentioned earlier. Several earlier workers had also published tables of normal deviates based upon, for example, Tippett's table. See Section 13 of Greenwood and Hartley (1962) for an extensive list of tables on random digits and random observations from other distributions.

Each publication of random digits contains, in addition, instructions for the use of the tables. There are unresolved questions about the consequences of the repeated use of the same table.

As previously mentioned, mathematical methods for producing approximately random numbers were introduced for reasons of convenience and efficiency. When a sequence of random numbers is used in a computation, it is often desirable to be able to reproduce the sequence exactly so that comparisons can be made among related computations that use the same sequence. The mathematical generation processes that have been provided to date all satisfy the requirement of reproducibility; that is, given a definite starting point in the process, the same sequence of subsequent numbers can be obtained in repeated applications of the process. The mathematical processes that have been introduced reflect the fact that the electromechanical calculators and early digital computers had very limited memories and generally could perform only one arithmetical or logical operation at a time. It was, therefore, not economical to store large volumes of random numbers, and the generation processes, to conserve memory and computer time, usually used at each step only one or two prior numbers of the process, which were combined by simple computer operations. It remains to be seen whether more involved generation processes will be introduced to take fuller advantage of advances in the speed and capacity of computers.

Mathematical generation processes. A mathematical process for obtaining sequences of numbers in a deterministic manner cannot generate sequences of truly random numbers. Despite this limitation, however, deterministic processes have a practical advantage over random processes, in that a random process can generate very undesirable, although unlikely, sequences of numbers, which would provide questionable samples or experimental designs. One can, on the other hand, sometimes design a deterministic process that generates numbers in a sufficiently haphazard arrangement to satisfy criteria of approximate randomness essential to an application. Processes can be designed so as to give special importance or weight to specific sequences of numbers. Such weighting techniques are similar to stratification techniques used in sampling applications; these ideas are considered, for example, by George W. Brown (1956), Daniel Teichroew (1965), and H. A. Meyer (Florida, Univ. of 1956). Hammersley and Handscomb (1964, p. 27) use the term "quasi random numbers" when it is known that a sequence of numbers is nonrandom but possesses particular desirable statistical properties.

To prevent misunderstanding about the interpretation of numbers generated by a deterministic mathematical process, the process will be referred to as one that generates pseudo random numbers when the numbers are produced with the intent to use them as if they were random numbers. Use of the term "pseudo random numbers" relates to the intent rather than the performance of the process, and the suitability of a process that generates pseudo random numbers must be judged relative to each specific application. For example, assume that in a specific application it is necessary to be able to generate even and odd numbers with the same probability. Further, assume there exists some method for generating statistically independent numbers, with r as the probability that a number generated will be odd and $q = 1 - r$ as the probability that it will be even. It is desired that r equal q. The following device can be employed to satisfy the requirement: Generate and use the numbers in pairs. If the pair is odd–odd (with probability r^2) or even–even (with probability q^2), ignore the pair. If the pair is odd–even, call it an odd occurrence; if the pair is even–odd, call it an even occurrence. Note that the probabilities of an odd occurrence or an even occurrence are now equal; namely, $r(1 - r) = (1 - r)r$ so that the conditional probabilities are both $\frac{1}{2}$.

If the generation process is being performed to obtain p-position numbers in a given number system (the number 725 is a three-position decimal number), then the largest number of distinct numbers possible is the base of the number system raised to the pth power. For example, in the decimal base the limit is 10^p; thus the number of distinct numbers that can be specified using three positions is limited to 1,000. Although it follows that the number of possible distinct pseudo random numbers is limited by the number system and the number of positions used, this limitation is not necessarily serious. Some of the numbers will be repeated whenever a sufficiently long sequence of p-position numbers is generated. If the sequence being generated begins to consist of numbers in a cycle pattern that will be repeated without interruption, then the length of this cycle is called the period of the process. The early generation methods were designed to provide sequences with a period as large as possible for a given p. This was partially motivated by the belief that it might help ensure adequate haphazardness of the numbers. The period of a specific process can be either smaller or larger than the number of distinct p-position numbers that can be generated. The determination of the period, if any, of a process can be very difficult and can depend on how many

prior numbers are used in determining the next number to be generated. It is often felt that it is important for the period to be longer than the number of pseudo random numbers needed for a specific application or experiment.

In addition to seeking a long period, it is customary to be sure that a process does not contain the same number too often. Some processes also need to be checked to ensure that they do not degenerate, that is, begin to repeat a single number pattern with a short period. In spite of the deterministic nature of a process, it may be very difficult to analyze its behavior except by subjecting actual outcomes to statistical tests. For example, the behavior of the middle square method, which was proposed by von Neumann, has been difficult to analyze. The method proposed by Lehmer, however, is much more tractable to analysis by number theory, and it has features in common with many other methods proposed by subsequent workers, such as the multiplicative congruence method described below. Hull and Dobell (1962) have compiled a very comprehensive survey of all mathematical methods and studies made through the middle of 1962.

Multiplicative congruence. The term "congruence" means the following: two integers A and B are said to be congruent modulo M, written $A \equiv B$ (mod M), if A and B have the same remainder when divided by M. A sequence $x_1, x_2, \cdots$ can be generated by the multiplicative congruence relation, such as $x_{n+1} \equiv kx_n \pmod{M}$, as follows: Assume that k is a given constant and it is multiplied by the first number of the sequence. The result of the multiplication is divided by M and the remainder of this division process is specified to be the next number of the sequence, x_2. Thereafter x_3 is obtained by repeating the multiplication and division process using x_2 in place of x_1. Studies of this process have been made concerning the selection of k and x_1 relative to M to ensure a large period. As mentioned by Lehmer, the particular relation given here has the undesirable feature that the right-hand digit positions of the generated numbers are not satisfactory individually as random digits. To overcome this limitation and to try to get a period as large as M, Rotenberg suggested modifying the relation by adding a constant, that is, $x_{n+1} \equiv kx_n + c \pmod{M}$. However, the choice of k and c relative to M not only influences the period but also influences the extent of correlation between successive numbers of the sequence.

Other recurrence relations. Other types of recurrence relations, such as using two prior numbers instead of one, have been suggested; and these are reviewed in the paper by Hull and Dobell (1962), as is the method suggested by Rotenberg. However, much more work remains to be done before the behavior of the methods that have been proposed is fully understood. Although pseudo random numbers that pass the statistical tests applied to them have been produced, specific digit positions of these numbers have failed to be satisfactory for use as pseudo random digits. One can, however, get pseudo random digits by judicious use of pseudo random numbers or by putting pseudo random numbers through a compound randomization procedure of the type suggested by Horton and Smith (1949) or Walsh (1949). Studies have been made of the digit positions of irrational numbers such as π and e as pseudo random digits since they have infinite periods. For example, Metropolis, Reitwiesner, and von Neumann (1950) studied some of the properties of the first 2,000 digits of π and e, and Stoneham (1965) studied the first 60,000 digits of e. It is important to keep in mind that although one can pose and apply a large number of meaningful tests for randomness, any given sequence of numbers or digits cannot be expected to meet all such tests with an arbitrary level of statistical significance.

Random permutations. Random permutations represent a special application of random digits in the design of experiments. Here "permutation" is used to indicate that a set of objects has been assigned a specific order. Random permutations are used when it is desired, for example, to assign (order) a set of n different objects to n positions on a line at random. The number of different arrangements possible for n different objects is $n!$. For example, three objects labeled 1, 2, 3 can be arranged in 3! (3! = 3·2·1 = 6) different ways: 123, 132, 213, 231, 312, 321.

A process is said to generate random permutations of size n if each of the $n!$ different possible permutations is independently generated with equal probability. The probability of observing a specific permutation will be $1/n!$. Random permutations of size n may be based on random digits as follows. If n is a p-position number, then p-position integers will be selected from a table of random digits with the following restrictions: If the number selected is equal to or less than n and if it has not already been selected for the permutation, then it should be included as part of the permutation; otherwise it should be rejected. Continue this process until all n places in the permutation have been filled. Although this method is practicable when n is quite small, there are better methods that do not reject so many random digits

(see Cochran & Cox 1950, p. 569; Fisher & Yates 1938, p. 34; Walsh 1957, p. 355).

The idea of having random permutations available as a table is due to George W. Snedecor. Tables of random permutations of size 9 and 16 can be found in Cochran and Cox (1950, p. 577) and Moses and Oakford (1963). For other sources, see Greenwood and Hartley (1962, p. 460).

Sampling without replacement. Sampling without replacement is a type of probability sampling of a finite population where each item of the population has an equal and independent probability of being selected but each item is allowed to be included in a sample only once. A random permutation of size n represents the extreme situation of sampling without replacement where the entire population of size n is included in the sample in a specified order. If one is interested in selecting, without replacement, random samples of size r from a population of size n, then random permutations of size n can be used to accomplish the selection by using the first r numbers of each permutation. There are other methods of using random numbers to select samples without replacement. For example, a recent set of techniques that can be used on digital computers is due to Fan, Muller, and Rezucha (1962, p. 387). With respect to the direct use of random digits to select samples without replacement, Jones (1959) has studied how many numbers must be used to obtain a sample of given size, taking into account the need to reject duplicate and undesired numbers.

MERVIN E. MULLER

[*Other relevant material may be found in* ERRORS, *article on* NONSAMPLING ERRORS, *and in the biographies of* FISHER *and* GOSSET.]

BIBLIOGRAPHY

Box, G. E. P.; and MULLER, MERVIN E. 1958 A Note on the Generation of Random Normal Deviates. *Annals of Mathematical Statistics* 29:610–611.

○BROWN, G. SPENCER. → See under Spencer Brown.

BROWN, GEORGE W. 1951 History of RAND's Random Digits: Summary. Pages 31–32 in U.S. National Bureau of Standards, Applied Mathematical Series, *Monte Carlo Method*. Edited by Alston S. Householder et al. Washington: Government Printing Office.

BROWN, GEORGE W. 1956 Monte Carlo Methods. Pages 279–303 in Edwin F. Beckenbach (editor), *Modern Mathematics for the Engineer*. New York: McGraw-Hill.

COCHRAN, WILLIAM G. (1953) 1963 *Sampling Techniques*. 2d ed. New York: Wiley.

COCHRAN, WILLIAM G.; and COX, GERTRUDE M. (1950) 1957 *Experimental Designs*. 2d ed. New York: Wiley.

DARBISHIRE, A. D. 1907 Some Tables for Illustrating Statistical Correlation. Manchester Literary and Philosophical Society, *Memoirs and Proceedings* 51, no. 16.

FAN, C. T.; MULLER, M. E.; and REZUCHA, I. 1962 Development of Sampling Plans by Using Sequential (Item by Item) Selection Techniques and Digital Computers. *Journal of the American Statistical Association* 57:387–402.

○FISHER, R. A.; and YATES, FRANK (1938) 1963 *Statistical Tables for Biological, Agricultural and Medical Research.* 6th ed., rev. & enl. New York: Hafner; Edinburgh: Oliver & Boyd.

FLORIDA, UNIVERSITY OF, GAINESVILLE, STATISTICAL LABORATORY 1956 *Symposium on Monte Carlo Methods, Held at the University of Florida . . . March 16 and 17, 1954.* Edited by Herbert A. Meyer. New York: Wiley.

GOOD, I. J. 1953 The Serial Test for Sampling Numbers and Other Tests for Randomness. Cambridge Philosophical Society, *Proceedings* 49:276–284.

[GOSSET, WILLIAM S.] (1907–1937) 1943 *"Student's" Collected Papers.* Edited by E. S. Pearson and John Wishart. London: Biometrika Office, University College. → Gosset wrote under the pseudonym "Student."

GREENWOOD, JOSEPH A.; and HARTLEY, H. O. 1962 *Guide to Tables in Mathematical Statistics.* Princeton Univ. Press. → See especially pages 454–468, "Tables of Random Samples."

HAMMERSLEY, JOHN M.; and HANDSCOMB, DAVID C. 1964 *Monte Carlo Methods.* New York: Wiley.

HORTON, H. BURKE; and SMITH, R. TYNES 1949 A Direct Method for Producing Random Digits in Any Number System. *Annals of Mathematical Statistics* 20:82–90.

HULL, T. E.; and DOBELL, A. B. 1962 Random Number Generators. *SIAM Review* 4:230–254.

JONES, HOWARD L. 1959 How Many of a Group of Random Numbers Will Be Usable in Selecting a Particular Sample? *Journal of the American Statistical Association* 54:102–122.

KENDALL, M. G.; and SMITH, B. BABINGTON 1938 Randomness and Random Sampling Numbers. *Journal of the Royal Statistical Society* Series A 101:147–166.

KENDALL, M. G.; and SMITH, B. BABINGTON 1939 *Tables of Random Sampling Numbers.* Cambridge Univ. Press.

LEHMER, D. H. 1951 Mathematical Methods in Large-scale Computing Units. Pages 141–146 in Symposium on Large-scale Digital Calculating Machinery, Second, 1949, *Proceedings.* Cambridge, Mass.: Harvard Univ. Press.

METROPOLIS, N. C.; REITWIESNER, G.; and VON NEUMANN, J. 1950 Statistical Treatment of Values of First 2,000 Decimal Digits of *e* and *π* Calculated on the ENIAC. *Mathematical Tables and Other Aids to Computation* 4:109–111.

MOOD, ALEXANDER M. 1950 *Introduction to the Theory of Statistics.* New York: McGraw-Hill.

MOSES, LINCOLN E.; and OAKFORD, ROBERT V. 1963 *Tables of Random Permutations.* Stanford (Calif.) Univ. Press.

MULLER, MERVIN E. 1958 An Inverse Method for the Generation of Random Normal Deviates on Large-scale Computers. *Mathematical Tables and Other Aids to Computation* 12:167–174.

RAND CORPORATION 1955 *A Million Random Digits With 100,000 Normal Deviates.* Glencoe, Ill.: Free Press.

SHREIDER, IULII A. (editor) (1962) 1964 *Method of Statistical Testing: Monte Carlo Method.* Amsterdam and New York: Elsevier. → First published in Russian.

○SPENCER BROWN, G. 1957 *Probability and Scientific Inference.* London: Longmans. → The author's surname is Spencer Brown, but common library practice is to alphabetize his works under Brown.

STONEHAM, R. G. 1965 A Study of 60,000 Digits of Transcendental *e.* *American Mathematical Monthly* 72:483–500.

TEICHROEW, DANIEL 1965 A History of Distribution Sampling Prior to the Era of the Computer and Its Relevance to Simulation. *Journal of the American Statistical Association* 60:27–49.

TIPPETT, LEONARD H. C. (1927) 1952 *Random Sampling Numbers.* Cambridge Univ. Press.

VON NEUMANN, JOHN 1951 Various Techniques Used in Connection With Random Digits. Pages 36–38 in U.S. National Bureau of Standards, Applied Mathematics Series, *Monte Carlo Method.* Edited by Alston S. Householder. Washington: Government Printing Office.

WALSH, JOHN E. 1949 Concerning Compound Randomization in the Binary System. *Annals of Mathematical Statistics* 20:580–589.

WALSH, JOHN E. 1957 An Experimental Method for Obtaining Random Digits and Permutations. *Sankhyā: The Indian Journal of Statistics* 17:355–360.

WELDON, W. F. R. 1906 Inheritance in Animals and Plants. Pages 81–109 in T. B. Strong (editor), *Lectures on the Method of Science.* Oxford: Clarendon.

YOUNG, FREDERICK H. 1962 *Random Numbers, Mathematical Induction, Geometric Numbers.* Boston: Ginn.

Postscript

There has been continuing extensive interest in both the theoretical foundations and the applications of random number generation techniques. The theoretical efforts have been related to the questions of what is randomness, what are random numbers, what are good computational procedures, and what are good programs for testing pseudo random number generators.

Knuth (1969) and Jansson (1966) provide excellent summaries of the subject as well as original research contributions. Halton (1970) provides a review with emphasis on the Monte Carlo method. Sowey (1972) provides an excellent chronological and classified bibliography on random number generation and testing from 1927 to 1971. On the theoretical side one sees a very interesting integration of disciplines involving the foundations of probability (based upon the work of Kolmogorov, Popper, von Mises, and Wald) and the foundations of logic concerning questions of consistency, complexity, and computability, and information theory (based upon the work of Church, Gödel, Shannon, and Turing). Much of this activity is summarized in Knuth (1969).

Some of these contributions relate the definition of randomness to the complexity of a digital computer procedure (algorithm). These approaches include the proposition that a sequence of p numbers is random, if given the first $p - 1$ numbers, the complexity (haphazardness) of the procedure to specify the *p*th number is maximal. Papers of special interest here are Chaitin (1966), Chaitin's nontechnical summary (1975), the nontechnical comments by Gardner (1968), and Good (1969), Martin-Löf (1966; 1969), and Stentiford and Lewin (1973).

Interest has continued in the development and evaluation of generation procedures for pseudo random numbers. An important development has been that of a new class of tests, now identified as spectral and lattice tests, to evaluate the distribution of *n* points $(X_{i+1}, X_{i+2}, \cdots, X_{i+n})$, $i = 0, 1, \cdots$. These efforts were stimulated by Coveyou and MacPherson (1967), who introduced a spectral test based upon Fourier analysis to study the distribution of the observations in *n*-space, where $2 \leqslant n \leqslant 10$. It has now been realized that performing a direct lattice analysis instead of a spectral test based upon Fourier analysis is more directly understandable. See in particular Coveyou (1970), Beyer (1972), and Beyer et al. (1971); see also Good and Gover (1967). Theoretical interest has continued in congruential recurrence relationships for generating pseudo random numbers. For example, Jansson (1964) has provided results for the correlations of the generated series; see also Dieter and Aherns (1971). Other papers of interest include Ahrens et al., (1970), Dieter (1971; 1972), Gustavson and Liniger (1970), Lewis et al. (1969), Marsaglia (1970; 1972), and Miller and Prentice (1968). The spectral and lattice tests are complex and involve considerable computing effort. The use of these tests has been reassuring, in that computational methods already in serious question, such as the use of multiplicative relations with a multiplier consisting of sums of powers of the number two, $\sum_{s=0}^{m} 2^i s$ have been rejected by these tests. Furthermore, those congruential recurrence relationships that appeared to be satisfactory continue to be considered acceptable, but there is much more insight into their properties. Other tests have indicated that the middle square method may no longer be considered a justifiable procedure.

The problems of testing for randomness clearly deserve and will receive more attention. As a nonspecialist begins to study in depth questions of randomness or pseudo random number generators, one easily gets the impression that the subject is complex and confusing. Indeed, confusion can readily enter at various levels of understanding. Popper (1959) helps one to appreciate some of the depths of the difficulties as they relate to an acceptable foundation for probability theory. For a specific application, a computational method for generating pseudo random numbers is adequate (regardless of its values elsewhere) only insofar as it meets the needs of that application with respect to that application's requirements for approximate statistical independence of observations or approximate uniformity of observations.

The generation of random permutations and order statistics continues to evoke interest and attention; see in particular Durstenfeld (1964) and Pike (1965) for a permutation algorithm, Knuth (1969) for other algorithms and applications, and Rabinowitz and Berenson (1974) for a review of methods for order statistics. Ord-Smith (1970) discusses generation of sets of permutations, but not necessarily in a random order.

The use of pseudo random numbers to generate directly or indirectly observations from distributions other than the uniform, for example, the normal distribution, has also continued to receive attention. See Jansson (1966), Knuth (1969), and Sowey (1972) for references. See also Bell (1968), Brent (1974), Chay et al., (1975), and Forsythe (1972) concerning recent methods for generating normal deviates.

Practical advice on using a table of random numbers appears in Wallis and Roberts (1956, pp. 334–337, 631).

For an interesting commentary on the role of chance and randomness in social affairs dating back to biblical usage, see Fienberg (1971). He summarizes in particular the use of randomization in the 1970 U.S. Selective Service Draft. He points out that such a draft selection technique appears to have been used in Austria–Hungary between 1889 and the start of World War I, and he also discusses why a technical definition as given in this paper (independence and uniformity) is necessary in applications requiring a probability structure, and is different from a commonplace dictionary definition, for example, "proceeding without aim."

MERVIN E. MULLER

ADDITIONAL BIBLIOGRAPHY

AHRENS, J. H.; DIETER, U.; and GRUBE, A. 1970 Pseudorandom Numbers: A New Proposal for the Choice of Multiplicators. *Computing* 6:121–138.

BELL, J. R. 1968 Algorithm 334, Normal Random Deviates. Association for Computing Machinery, *Communications of the ACM* 11:498 only.

BEYER, W. A. 1972 Lattice Structure and Reduced Bases of Random Vectors Generated by Linear Recurrences. Pages 361–370 in S. K. Zaremba (editor), *Applications of Number Theory to Numerical Analysis*. New York: Academic Press.

BEYER, W. A.; ROOF, R. B.; and WILLIAMSON, D. 1971 The Lattice Structure of Multiplicative Congruential Pseudo-random Vectors. *Mathematics of Computation* 25:345–363.

BRENT, RICHARD P. 1974 Algorithm 488, a Gaussian

Pseudo-random Number Generator. *Association for Computing Machinery, Communications of the ACM* 17:704–706.

CHAITIN, GREGORY J. 1966 On the Length of Programs for Computing Finite Binary Sequences. *Journal of the Association for Computing Machinery* 13:547–569.

CHAITIN, GREGORY J. 1975 Randomness and Mathematical Proof. *Scientific American* 226, May:47–52.

CHAY, S. C.; FARDO, R. D.; and MAZUMDAR, M. 1975 On Using the Box–Muller Transformation With Multiplicative Congruential Pseudo-random Number Generators. *Applied Statistics* 24:132–135.

COVEYOU, R. R. 1970 Random Number Generation Is Too Important to Be Left to Chance. Society for Industrial and Applied Mathematics, *Studies in Applied Mathematics* 3:70–111.

COVEYOU, R. R.; and MACPHERSON, R. D. 1967 Fourier Analysis of Uniform Random Number Generators. *Journal of the Association for Computing Machinery* 14:100–119.

DIETER, U. 1971 Pseudo-random Numbers: The Exact Distribution of Pairs. *Mathematics of Computation* 25:855–883.

DIETER, U. 1972 Statistical Interdependence of Pseudo Random Numbers Generated by the Linear Congruential Method. Pages 287–317 in S. K. Zaremba (editor), *Applications of Number Theory to Numerical Analysis.* New York: Academic Press.

DIETER, U.; and AHRENS, J. H. 1971 An Exact Determination of Serial Correlations of Pseudo-random Numbers. *Numerische Mathematik* 17:101–123.

DURSTENFELD, RICHARD 1964 Algorithm 235, Random Permutation. Association for Computing Machinery, *Communications of the ACM* 7:420 only.

FIENBERG, STEPHEN E. 1971 Randomization and Social Affairs: The 1970 Draft Lottery. *Science* 171:255–261.

FORSYTHE, GEORGE E. 1972 Von Neumann's Comparison Method for Random Sampling From the Normal and Other Distributions. *Mathematics of Computation* 26:817–826.

GARDNER, MARTIN 1968 On the Meaning of Randomness and Some Ways of Achieving It. *Scientific American* 219, July:116–121.

GOOD, I. J. 1969 How Random Are Random Numbers? *American Statistician* 23, no. 4:42–45.

GOOD, I. J.; and GOVER, T. N. 1967 The Generalized Serial Test and the Binary Expansion of $\sqrt{2}$. *Journal of the Royal Statistical Society* Series A 130:102–107. → See also the corrigendum and addition in Volume 131 (1968), page 434.

GUSTAVSON, FRED G.; and LINIGER, WERNER 1970 A Fast Random Number Generator With Good Statistical Properties. *Computing* 6:221–226.

HALTON, JOHN H. 1970 A Retrospective and Prospective Survey of the Monte Carlo Method. *SIAM Review* 12:1–63.

JANSSON, BIRGER 1964 Autocorrelations Between Pseudo-random Numbers. *Nordisk tidskrift for informationsbehandling* 4:6–27.

JANSSON, BIRGER 1966 *Random Number Generators.* Stockholm: Almquist & Wiksell.

KNUTH, DONALD E. 1969 *The Art of Computer Programming.* Volume 2: *Seminumerical Algorithms.* Reading, Mass.: Addison-Wesley.

LEWIS, P. A. W.; GOODMAN, A. S.; and MILLER, J. M. 1969 A Pseudo-random Number Generator for the System/360. *IBM Systems Journal* 8:136–146.

MARSAGLIA, GEORGE 1970 Regularities in Congruential Random Number Generators. *Numerische Mathematik* 16:8–10.

MARSAGLIA, GEORGE 1972 The Structure of Linear Congruential Sequences. Pages 249–285 in S. K. Zaremba (editor), *Applications of Number Theory to Numerical Analysis.* New York: Academic Press.

MARTIN-LÖF, PER 1966 The Definition of Random Sequences. *Information and Control* 9:602–619.

MARTIN-LÖF, PER 1969 Algorithms and Randomness. *Review of the International Statistical Institute* 37:265–272.

MILLER, J. C. P.; and PRENTICE, M. J. 1968 Additive Congruential Pseudo-random Number Generators. *Computer Journal* 11:341–346.

ORD-SMITH, R. J. 1970 Generation of Permutation Sequences: Part 1. *Computer Journal* 13:152–155.

PIKE, M. C. 1965 Remark on Algorithm 235, Random Permutation. Association for Computing Machinery, *Communications of the ACM* 8:445 only.

POPPER, KARL R. (1935) 1959 *The Logic of Scientific Discovery.* New York: Basic Books; London: Hutchinson. → First published as *Logik der Forschung.*

RABINOWITZ, M.; and BERENSON, M. L. 1974 A Comparison of Various Methods of Obtaining Random Order Statistics for Monte Carlo Computations. *American Statistician* 28, no. 1:27–29.

SOWEY, E. R. 1972 A Chronological and Classified Bibliography on Random Number Generation and Testing. *International Statistical Review* 40:355–371.

STENTIFORD, F. W. M.; and LEWIN, W. W. 1973 The Evolutionary Approach to the Concept of Randomness. *Computer Journal* 16:148–151.

WALLIS, W. ALLEN; and ROBERTS, HARRY V. 1956 *Statistics: A New Approach.* Glencoe, Ill.: Free Press.

RANDOM WALK

See MARKOV CHAINS; QUEUES.

RANDOMIZATION

See EXPERIMENTAL DESIGN; RANDOM NUMBERS; *and the biographies of* FISHER *and* PEIRCE.

RANDOMNESS

See PROBABILITY; RANDOM NUMBERS; SAMPLE SURVEYS; STATISTICS.

RANK CORRELATION

See NONPARAMETRIC STATISTICS, *article on* RANKING METHODS; STATISTICS, DESCRIPTIVE, *article on* ASSOCIATION; *and the biography of* SPEARMAN.

RANK–SIZE RELATIONS

If each object in some collection is characterized by its size, it is possible to rank-order the objects from the largest to the smallest. One may then view the rank as the horizontal axis of a coordinate system on which the objects are arranged, so that the largest is assigned the horizontal coordinate $x = 1$, the next largest $x = 2$, etc., assuming no ties. The vertical coordinate, y, can be viewed as the size of each object. For example, if the objects are cities of the United States rank-ordered according to population (as reported in U.S. Bureau of the Census 1964), New York will have coordinates $x = 1$, $y = 7{,}781{,}984$, Chicago will have coordinates $x = 2$, $y = 3{,}550{,}404$, San Francisco will have $x = 12$, $y = 740{,}316$, etc. Such data can be represented on a bar graph. If the number of objects is very large, the bar graph can be approximated by a continuous curve through the pairs of coordinates so defined, as in Figure 1. When ties occur the tied observations are given consecutive ranks.

Zipf's law. A question investigated at great length by George K. Zipf (1949) concerned the mathematical properties of rank–size curves obtained from collections of many different sorts of objects. In particular, Zipf examined the rank–size curves of cities (by population), biological genera (by number of species), books (by number of pages), and many other collections.

Zipf found that in most instances the rank–size curves were very nearly segments of rectangular hyperbolas, that is, curves whose equations are of the form $xy = \text{constant}$, or, as expressed in logarithmic coordinates, $\log x + \log y = \text{constant}$. Therefore, when such rank–size curves are plotted on log–log paper they are very nearly straight lines with slopes close to -1. At least, this was the case with the collections that Zipf singled out for attention. The relation *rank* $\times$ *size = constant* is sometimes referred to as Zipf's law.

The principle of least effort. Zipf attempted to derive his law from theoretical considerations, which he summarized in the so-called principle of least effort. The connection between this principle and the rank–size law is by no means clear, and Zipf's theoretical arguments now have at most only historical interest. However, his work attracted wide attention and spurred investigations more rigorous and theoretically more suggestive than his own.

The following discussion will concern Zipf's law in its statistical–linguistic context. Let the collection be the words used in some large verbal output, say in a book or in a number of issues of a news-

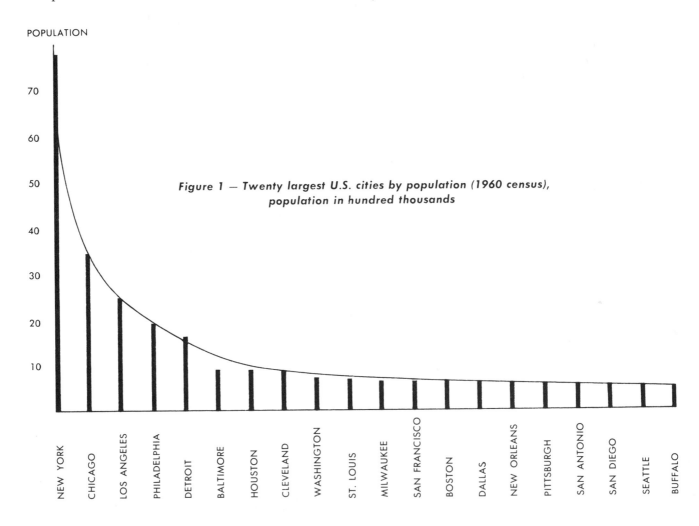

POPULATION

Figure 1 — Twenty largest U.S. cities by population (1960 census), population in hundred thousands

NEW YORK · CHICAGO · LOS ANGELES · PHILADELPHIA · DETROIT · BALTIMORE · HOUSTON · CLEVELAND · WASHINGTON · ST. LOUIS · MILWAUKEE · SAN FRANCISCO · BOSTON · DALLAS · NEW ORLEANS · PITTSBURGH · SAN ANTONIO · SAN DIEGO · SEATTLE · BUFFALO

paper. Such a collection is called a "corpus." Let the size assigned to each word be the number of times the word appears in the corpus. Then in most cases the "largest" (that is, the most frequently occurring) word will be *the*, the next "largest" will be *and*, etc. (Thorndike & Lorge 1944). By the nature of the ordering, the curve will be monotonically decreasing (a J curve)—at first steeply falling, then gradually flattening out—for by the time the low frequency words are reached, there will be many having the same (small) number of occurrences. Their ranks, however, will keep increasing, because in a rank–size graph objects having the same size are nevertheless assigned consecutive ranks. Therefore, larger and larger blocks of the bar graph will have the same height, which means that the continuous curve through the bar graph will become increasingly flatter. The rectangular hyperbola, being asymptotic to the horizontal axis, also has this property, which in part accounts for the good agreement between the hyperbola and the rank–size graph.

It is known that the most frequently occurring words in any language are usually the shorter ones. In English, for example, these are articles, prepositions, and conjunctions. If the use of a word represents effort by the speaker, and if the speaker tries to minimize effort, he can be expected to use the shortest words most frequently. But there are comparatively few short words, because there are comparatively few combinations to be formed from few letters (or phonemes). Consequently, the number of different high-ranking words (those with low rank numbers) will be small. By a similar argument, the words well along in the ranking are, on the average, the longer ones, and so their numbers will be large; thus, many will have equal frequencies. It is, in fact, observed that the largest number of different words in a typical corpus of some thousands of words are those used only once.

In developing his argument from the principle of least effort in the context of language statistics, Zipf postulated opposing tendencies on the part of the speaker and on the part of the hearer. From the point of view of the speaker, so his argument goes, the language most economical of effort would be one with very few words, each word having many meanings. From the point of view of the hearer, on the contrary, the ideal language would be one in which each word has a unique meaning, since in that case the labor of matching meaning with context would be saved. A balance is struck between the effort-economizing tendency of the speaker and that of the hearer by a certain distribution of *ranges* of meaning associated with the distribution of frequencies with which words are used.

Whatever merit Zipf's principle of least effort has in the context of language statistics, his arguments to the effect that the same principle is responsible for the rank–size distributions of the great variety of collections from cities (ranked by populations) to applicants for marriage licenses (ranked by distances between the homes of the bride and the groom) seem of questionable relevance. It is not clear how the principle of least effort operates in each instance to produce the observed rank–size curve. Similar distributions may be traceable to mathematically isomorphic processes, but Zipf, in his principle of least effort, emphasizes not the possible mathematical genesis of his law but its alleged origins in the nature of human behavior.

That this can be misleading can be seen in a hypothetical case of a scholar who, having noted that the weights of beans, of rabbits, and of people are normally distributed, concludes that the normal distribution is a manifestation of a "life force" (because beans, rabbits, and people are biological objects) and seeks the manifestation of this "force" in all instances where the normal distribution is observed. The normal distribution may arise, and presumably often does, from a certain kind of interplay of chance events (roughly, addition of many nearly independent and not wildly dissimilar random variables). Thus, insofar as the normal distribution arises in many contexts, its genesis may be the common statistical structure of the contexts, not their contents—that is, not whether they involve beans or people, animate or inanimate objects. [*See* PROBABILITY, *article on* FORMAL PROBABILITY, *for a discussion of how the normal distribution arises from the interplay of chance events; this is called the central limit theorem.*]

Statistical rationales for Zipf's law

Explanations of Zipf's law and of rank–size relations in general, like the explanations of the normal distribution discussed above, are to be sought in the statistical structure of the events that might generate these relations instead of in the nature of the objects to which the relations apply.

Consider a frequency density in which the horizontal axis represents nonnegative size while the vertical axis represents the relative frequency with which objects of a given size are encountered in some large collection. Call this frequency density $f(x)$, so that $\int_0^\infty Nf(x)\,dx = N$, where N is the number of objects in the collection. Consider now

$G(x) = \int_x^\infty Nf(t)\, dt$, which is the number of objects having sizes greater than x. But this is exactly the *rank* (according to size) of the object whose size is x, assuming that the ranks of objects of equal size have been assigned arbitrarily among them. Therefore the rank–size curve is essentially the integral of the size–frequency curve. Any mathematical theory that applies to the one will apply to the other after the transformation just described has been performed. In particular, Zipf's law holds $(G(x) = K/x)$ if and only if the frequency density, $f(x)$, is of form K/x^2, since $\int_x^\infty (K/t^2)\, dt = K/x$.

Investigators seeking a statistical rationale for Zipf's rank–size curves studied the associated frequency density curves. Probabilistic models underlying some common size–frequency distributions are well known. For example, if there is a collection of objects characterized by sizes, and the size of each object is considered as the sum of many random (at least approximately independent) variables, none of which has a dispersion dominating the dispersions of others, then the resulting distribution of sizes will be approximately a normal distribution. This probabilistic process adequately accounts for the normal distributions so frequently observed in nature. On the other hand, suppose that each object suffers repeated increments or decrements that are proportional to the sizes of the objects on which they impinge. Then the ultimate equilibrium distribution will be a so-called logarithmic normal one, that is, a frequency distribution of a random variable whose logarithm is normally distributed. Such frequency distributions are also commonly observed [*see* DISTRIBUTIONS, STATISTICAL, *article on* SPECIAL CONTINUOUS DISTRIBUTIONS].

The problem of finding a statistical rationale for Zipf's law, therefore, is that of finding a probabilistic process that would result in an equilibrium distribution identical with the derivative of Zipf's rank–size curve—that is, one where the frequency would be inversely proportional to the square of the size.

Models yielding Yule distributions. Herbert Simon (1955) proposed such a model for the size–frequency distribution of verbal outputs. The essential assumption of the model is that as the corpus is being created (in speaking or writing) the probability of a particular word being added to the already existing list is proportional to the total number of occurrences of words in that frequency class and, moreover, that there is also a nonnegative probability that a new word will be added to the list. In some variants of Simon's

model, the latter probability is a constant; in others it decreases as the corpus grows in size (to reflect the depletion of the vocabulary of the speaker or writer). The resulting equilibrium frequency distribution coincides with Zipf's, at least in the high-frequency range.

Moreover, it appears that similar models are plausible rationales for many other distributions. Essentially it is assumed that the increments impinging on objects are proportional to the sizes of the objects and also that new objects are added to the population according to a certain probability law. The first assumption leads to a logarithmic normal distribution. Combined with the second assumption (the "birth process") it leads to so-called Yule distributions, which greatly resemble Zipf's.

If the derivation indicated by Simon is accepted, the principle of least effort becomes entirely superfluous, for clearly it is the probabilistic structure of events rather than their content that explains the frequency density and hence the rank–size distributions that Zipf considered to be prima facie evidence for the principle of least effort.

Information-theoretical models. The principle of least effort was not entirely abandoned by those who sought rationales for Zipf's law, at least for its manifestation in language statistics. Benoît Mandelbrot (1953), for example, restated the principle of least effort as follows. Assume, as Zipf did, that there is an effort or cost associated with each word. Then if the speaker is to economize effort, clearly he should select the cheapest word and speak only that word. However, discourse of this sort would not convey any information, since if the same word is spoken on all occasions, the hearers know in advance what is going to be said and get no new information from the message. The problem is, according to Mandelbrot, not to minimize effort (or cost, as he calls it) *unconditionally* but rather to minimize it *under the constraint* that a certain average of information per word must be conveyed. Equivalently, the problem is to maximize the information per word to be conveyed under the constraint that a certain average cost of a word is fixed.

Here Mandelbrot was able to utilize the precise definition of the *amount of information* conveyed by a message, as formulated in the mathematical theory of communication (Shannon & Weaver 1949). Having cast the problem into mathematical form, Mandelbrot was able to derive Zipf's rank–size curve as a consequence of an assumption related to the principle of least effort, namely, the minimization of cost, given the amount of information to be conveyed (see Pollack 1968).

Actually, Mandelbrot's derived formula was more general than Zipf's. While Zipf used rank × size = constant, Mandelbrot obtained from his model the formula

$$P_r = P(r + m)^{-B},$$

in which P_r, being the frequency of occurrence, represents size, r is rank, and P, m, and B are constants. If $B = 1$ and $m = 0$, Mandelbrot's formula reduces to Zipf's rank–size law.

As would be expected, the generalized formula fits most rank–size verbal-output curves better than Zipf's; in addition, it is derived rigorously from plausible assumptions.

Investigations of rank–size relations. Zipf himself suggested a generalization of his rank–size law, namely,

$$(\text{rank})^q \times \text{size} = \text{constant}, \qquad q > 0.$$

With q as an extra free parameter, clearly more of the observed rank–size curves could be fitted than without it. Of greater importance to a rank–size theory is a rationale for introducing the exponent q. However, Zipf's arguments on this score are as vague as those related to the originally postulated law.

Frank A. Haight (1966) pointed out the dependence of the observed size–frequency relationship on the way the data are grouped. Suppose, for example, the size–frequency relationship of cities is examined. Clearly, in order to obtain several cities in each population class the populations must be rounded off—say, to the nearest thousand or ten thousand. Haight has shown that if Z is the number of digits rounded off in the grouping and if Zipf's generalized rank–size law, given above, holds for cities, then the size–frequency distribution will be given by

$$p_n(Z) = (2Z)^{-1/q} \left\{ \left[\frac{2Z}{2n - 1} \right]^{1/q} - \left[\frac{2Z}{2n + 1} \right]^{1/q} \right\},$$

where $[x]$ is the integral part of x. Here $p_n(Z)$ is the fraction of cities with population near n (rounded off by Z digits).

As Z becomes large, this distribution tends to the zeta distribution, namely,

$$p_n = (2n - 1)^{-1/q} - (2n + 1)^{-1/q}.$$

The zeta distribution gives a fairly good fit to the populations of the world metropolitan areas rounded off to the nearest million. (There are 141 such areas with populations close to one million, 46 with populations close to two million, etc.) The number of accredited colleges and universities in the United States with student populations rounded

off to the nearest thousand is also well fitted by the zeta distribution.

Robert H. MacArthur (1957) treated the problem of the relative abundance of species in a natural organic population by means of a model isomorphic to a random distribution of $n - 1$ points on a line segment. The distances between the points then represent the "sizes of biological niches" available to the several species and therefore the abundance of the species. Thus the size of the rth rarest species among n species turns out to be proportional to

$$\sum_{i=1}^{r} (n - i + 1)^{-1}.$$

It appears, then, that the original rank–size law proposed by Zipf is only one of many equally plausible rank–size laws. Clearly, if objects can be arranged according to size, beginning with the largest, *some* monotonically decreasing curve will describe the data. The fact that many of these curves are fairly well approximated by hyperbolas proves nothing, since an infinitely large number of curves resemble hyperbolas sufficiently closely to be identified as hyperbolas. No theoretical conclusion can be drawn from the fact that many J curves look alike. Theoretical conclusions can be drawn only if a rationale can be proposed that implies that the curves *must* belong to a certain class. The content of the rationales becomes, then, the content-bound theory. Specifically, the constants contained in the proposed mathematical model can receive a content interpretation.

For example, in Mandelbrot's model $1/B$ is interpreted as the "temperature" of the verbal output. Taken out of context the "temperature of a language sample" seems like an absurd notion. But the term is understandable in the meaning of the exact mathematical analogue of temperature, as the concept is derived in statistical thermodynamics. In this way, formal structural connections are established between widely different phenomena, which on a priori grounds would hardly have been suspected to be in any way related.

Such discoveries are quite common. A strict mathematical analogy was found between the distribution of the number of bombs falling on districts of equal areas in London during the World War II bombing and the distribution of the numbers of particles emitted per unit time by a radioactive substance. The unifying principle is to be found neither by examining bombs nor by examining radioactive substances but rather by inquiring into the probabilistic structure of the events in question.

Having noted that the rank–size relation is simply another way of viewing the size–frequency relation, it can be seen that all the studies of the latter are relevant to the former. Lewis F. Richardson (1960) gathered extensive data on the incidence of "deadly quarrels"—that is, wars, riots, and other encounters resulting in fatalities [see the biography of RICHARDSON]. Designating the size of a deadly quarrel by the logarithm of the number of dead, he studied the associated size–frequency relation, seeking to derive a law of "organization for aggression." He believed he had found evidence for such a law in the circumstance that the size–frequency relation governing Manchurian bandit raids was very similar to the one governing Chicago gangs in the prohibition era. Here one might also interpose the objection that the similarity may have nothing to do with aggression as such, being simply a reflection of the probabilities governing the formation, growth, and dissolution of human groups. Curiously, a comparison of the distribution of sizes of casual groups such as people gathered around swimming pools (Coleman & James 1961) turned out to be different from that of gangs. Moreover, the former type is derived from a stochastic process in which single individuals can join or leave a group, while the latter derive from a process in which no individual can leave unless the whole group disintegrates (Horvath & Foster 1963). These results are intriguing, because the difficulty with which an individual may leave a gang is well known, and so Richardson's conjecture may have been not without foundation.

It appears, therefore, that the search for the stochastic processes underlying observed rank–size or size–frequency relations can result in important theoretical contributions.

ANATOL RAPOPORT

BIBLIOGRAPHY

COLEMAN, JAMES S.; and JAMES, JOHN 1961 The Equilibrium Size Distribution of Freely-forming Groups. Sociometry 24:36–45.

HAIGHT, FRANK A. 1966 Some Statistical Problems in Connection With Word Association Data. Journal of Mathematical Psychology 3:217–233.

HORVATH, WILLIAM J.; and FOSTER, CAXTON C. 1963 Stochastic Models of War Alliances. General Systems 8:77–81.

MACARTHUR, ROBERT H. 1957 On the Relative Abundance of Bird Species. National Academy of Sciences, Proceedings 43:293–295.

MANDELBROT, BENOÎT 1953 An Informational Theory of the Statistical Structure of Language. Pages 486–502 in Willis Jackson (editor), Communication Theory. New York: Academic Press; London: Butterworth. → Contains three pages of discussion of Mandelbrot's article.

PIELOU, E. C.; and ARNASON, A. NEIL 1966 Correction

to One of MacArthur's Species-abundance Formulas. Science 151:592 only. → Refers to MacArthur 1957.

►POLLACK, IRWIN 1968 Information Theory. Volume 7, pages 331–337 in International Encyclopedia of the Social Sciences. Edited by David L. Sills. New York: Macmillan and Free Press.

RAPOPORT, ANATOL 1957 Comment: The Stochastic and the "Teleological" Rationales of Certain Distributions and the So-called Principle of Least Effort. Behavioral Science 2:147–161.

RICHARDSON, LEWIS F. 1960 Statistics of Deadly Quarrels. Pittsburgh: Boxwood.

SHANNON, CLAUDE E.; and WEAVER, WARREN (1948–1949) 1959 The Mathematical Theory of Communication. Urbana: Univ. of Illinois.

SIMON, HERBERT A. 1955 On a Class of Skew Distribution Functions. Biometrika 42:426–439. → Reprinted in Simon's Models of Man.

THORNDIKE, EDWARD L.; and LORGE, IRVING 1944 The Teacher's Word Book of 30,000 Words. New York: Columbia Univ., Teachers College.

U.S. BUREAU OF THE CENSUS 1964 Population and Land Area of Urbanized Areas: 1960 and 1950. Table 22 in U.S. Bureau of the Census, Census of Population: 1960. Volume 1: Characteristics of the Population. Part 1. Washington: Government Printing Office.

ZIPF, GEORGE K. 1949 Human Behavior and the Principle of Least Effort: An Introduction to Human Ecology. Reading, Mass.: Addison-Wesley.

Postscript

The rank–size relation was investigated with respect to rivers by Luna B. Leopold (1962). Rivers or channels are classified by "orders." A channel of order 1 is a channel without tributaries. A channel of order 2 is one with tributaries only of order 1, and so on recursively. In terms of size, the river of order 10 (in the United States, only the Mississippi) has rank 1. There are eight rivers of order 9 (the Columbia is one), and so on. The number of rivers of each order and their average lengths are shown in Table 1.

Table 1 — Rivers classified by order and length

Order*	Number	Average length (miles)	Average drainage area (sq. mi.)
10	1	1,800	1,250,000
9	8	777	264,000
8	41	338	55,600
7	200	147	11,700
6	950	64	2,460
5	4,200	28	518
4	18,000	12	109
3	80,000	5.3	23
2	350,000	2.3	4.7
1	1,570,000	1	1

* The order numbers are based on the determination of the smallest order using maps of scale 1:62,500. That is, maps of larger scales would reveal smaller streams and so a larger number of streams and orders.

Source: Adapted from Leopold 1962, table 1, p. 512.

If the frequency, that is, the number of streams of a given average length, is plotted against the average lengths, we obtain the size–frequency form of Zipf's law, $f(x) = K/x^2$, where f is frequency, x is the corresponding average length, and K is a constant. In fact the data of Table 1 are fitted very well by the equation $\log f + 2 \log x = 6.5$.

Note, however, that this "corroboration" of Zipf's law depends crucially on the definition of size. If we define the "size" of a stream by its drainage area, y, the data of Table 1 are fitted very well by $\log f + \log y = 6.3$. This is hardly surprising, since we would expect the drainage area to be approximately proportional to the squared length of a stream (with tributaries). The latter equation, however, is *not* in accord with the size–frequency form of Zipf's law.

Clearly there is no a priori reason to choose either the length of a stream or its drainage area to represent "size" in preference to the other. Hence the size–frequency relation (or, equivalently, the rank–size relation) cannot be conceived as representing a "law of nature," as Zipf conceived it, not even a law reflecting a particular material context. As pointed out above, the basis of any noted regularity in rank–size relations is to be sought in the statistical structure of events that can be supposed to generate this regularity rather than in the nature of the events in question. On the other hand, the statistical approach does not preclude the drawing of parallels between processes with similar dynamics even though of widely dissimilar content, for example (to venture a guess), between drainage systems of rivers and vascular systems. It may not be coincidental that the picture of a typical drainage region looks strikingly like a leaf.

Bruce M. Hill (1970) derived Zipf's law for the number of species in genera from a stochastic process in which species are allocated to genera in accordance with so-called Bose–Einstein statistics. That is, if $\mathbf{L} = (L_1, L_2, \cdots, L_M)$ is a vector with L_i the number of species in the ith genus, then all such vectors (with no empty genera) are equi-probable:

$$Pr\left[\mathbf{L}|M, N\right] = \binom{N-1}{M-1}^{-1},$$

where $N = \sum_{i=1}^{m} L_i$.

On the other hand, if the allocation were in accordance with so-called Maxwell–Boltzman statistics, that is,

$$Pr\left[\mathbf{L}|M, N\right] = (N-M)! \left(\prod_{i=1}^{m}(L_i - 1)!\right)^{-1} M^{-(N-M)},$$

a very different rank–size relation would result. In this way, a manifestation of Zipf's law in a biological context (for instance, in the rank–size rela-

tion observed in lizard genera) suggests a Bose–Einstein type of allocation of species to genera rather than, say, a Maxwell–Boltzman type of allocation.

The question arises of how such an allocation is to be interpreted. Hill offers an interesting approach to this question from the point of view of the classificatory procedure (rather than a relatively objective stochastic process), in which the prior expectations of the classifier play a part.

Early history of rank–size laws. E. U. Condon (1928) refers to word counts by Leonard P. Ayres (1915) and by Godfrey Dewey (1923) and points out that the rank–frequency plots of these counts are very well fitted by what would later be called Zipf's law. It is interesting to note that Condon's paper antedates Zipf's early studies (1932) by four years. It is also interesting to note that Condon ventures an "explanation" of the rank–size relation in terms of "Weber's law of psychology," another example of interpreting statistical regularities as laws of nature.

Alfred J. Lotka (1925), in mentioning the rank–size relations observed in the populations of American cities, refers to "Auerbach's law" and gives the date of publication of the relevant paper erroneously as 1923. Actually, the paper appeared ten years earlier (Auerbach 1913). In it the approximate constancy of rank × size is established for the 47 largest German cities (census of 1910). As in many instances of this "law," the magnitude of the product is smaller for the largest elements of the set examined. On the whole, rank × size rises steadily until about rank 15 and remains fairly constant thereafter. As we have seen, this discrepancy can be accounted for by Mandelbrot's formula

$$P_r = P(r + m)^{-B},$$

where P_r is size, r is rank, and P, m, and B are positive constants ($B \cong 1$), whatever this may mean in the present instance.

Zipf discussed the constancy of rank × size in 1935 (Zipf 1935). The germ of the idea of the "principle of least effort" seems to have occurred to him earlier, as can be seen from the following passage: "Observing the speech of many hundreds of millions of people, we have demonstrated . . . that the conspicuousness or intensity of any element of a language is inversely proportional to its frequency. Using X for frequency and Y for conspicuousness, we can express our thesis thus: $XY = n$" (Zipf 1929).

This statement can be taken to mean "The greater the effort (intensity), the less frequently it is made." Inherent in this statement is, first, a failure to realize that it has no theoretical content unless intensity is mathematically defined, and, second, that it contradicts Zipf's law. For the relationship expressed is not *rank × size = constant* but rather *size × frequency = constant*, whereas if rank × size = constant, then $(size)^2 × frequency$ must be constant. We can, of course, assume that somehow intensity is to be taken as proportional to $(size)^2$, but Zipf says nothing about it. Apparently, to him any inverse relation meant inverse proportionality.

It is interesting to note that Auerbach made a similar mistake. He calls the regularity of the rank–size relation an empirical one "like so many laws of exact science." Nevertheless, he does question the possible theoretical underpinnings of the so-called law. The first task, he suggests, is to ascertain whether the rank–size relation holds in many different contexts. As we have seen, Zipf undertook this task energetically and, perhaps, thereby earned the honor of having the law named after him. Not infrequently, discoveries are named after their publicizers, as in the case of Amerigo Vespucci. In this connection, Auerbach notes that if people are ranked by their fortunes, it turns out that "there are four times as many, not twice as many half-million-aires than millionaires." He goes on to say, "Where these differences come from is precisely the subject of a theoretical investigation" (Auerbach 1913, p. 76). However, the relation in question is a size–frequency relation, and insofar as millionaires and half-millionaires are concerned, the relationship $(size)^2 × frequency = constant$ is satisfied, or at least so Auerbach states; he gives no documentation. To test the rank–size relation, Auerbach should have compared not the numbers of millionaires and half-millionaires but rather the fortunes of the richest person, the second richest, and so on.

ANATOL RAPOPORT

ADDITIONAL BIBLIOGRAPHY

AUERBACH, FELIX 1913 Das Gesetz der Bevölkerungs-konzentration. *Petermanns geographische Mitteilungen* 59:74–76.

AYRES, LEONARD P. 1915 *A Measuring Scale for Ability in Spelling.* New York: Russell Sage.

CONDON, E. U. 1928 Statistics of Vocabulary. *Science* 67:300 only.

DEWEY, GODFREY (1923) 1950 *Relativ Frequency of English Speech Sounds.* Rev. ed. Cambridge, Mass.: Harvard Univ. Press.

HILL, BRUCE M. 1970 Zipf's Law and Prior Distributions for the Composition of a Population. *Journal of the American Statistical Association* 65:1220–1232.

HILL, BRUCE M. 1974 The Rank–Frequency Form of Zipf's Law. *Journal of the American Statistical Association* 69:1017–1026.

LEOPOLD, LUNA B. 1962 Rivers. *American Scientist* 50:511–537.

LOTKA, ALFRED J. (1925) 1957 *Elements of Mathe-matical Biology.* New York: Dover. → First published as *Elements of Physical Biology.*

ZIPF, GEORGE K. 1929 Relative Frequency as a Determinant of Phonetic Change. *Harvard Studies in Classical Philology* 40:1–95.

ZIPF, GEORGE K. 1932 *Selected Studies of the Principle of Relative Frequency in Language.* Cambridge, Mass.: Harvard Univ. Press.

ZIPF, GEORGE K. (1935) 1965 *The Psycho-biology of Language: An Introduction to Dynamic Philology.* Cambridge, Mass.: M.I.T. Press.

RANKING METHODS
See under NONPARAMETRIC STATISTICS.

REGRESSION
See under LINEAR HYPOTHESES; *see also the biography of* GALTON.

RELIABILITY
See ERRORS; EXPERIMENTAL DESIGN; PSYCHOMETRICS; QUALITY CONTROL, STATISTICAL.

RÉNYI, ALFRÉD

▶ *This article was specially written for this volume.*

Alfréd Rényi (1921–1970), distinguished mathematician, was an innovator in probability theory, mathematical statistics, information theory, and other branches of mathematics.

Rényi was born in Budapest on March 20, 1921. Growing up in a family of scientific workers—his maternal grandfather was the well-known Hungarian philosopher B. Alexander—he developed many early interests, but soon turned to mathematics. After some years in an English-language high school, he took his final examination at the secondary school in 1939. Because of racial laws, he could not immediately attend the university, so he became an apprentice in a shipyard for half a year. As a second winner of the yearly mathematical competition in 1940, however, he was admitted to the University of Budapest to study mathematics and physics under L. Fejér, among others. When he finished his studies in 1944 he was imprisoned by the Nazis in a labor camp, but after a short time he was able to escape. In the occupied city, Rényi, wearing a military uniform, helped many fugitives from German oppression.

After the 1945 liberation of Budapest, Rényi went to the University of Szeged to study for his PH.D. with F. Riesz. Awarded the degree—his thesis was on the theory of Fourier–Stieltjes series—he returned to Budapest, worked for a year as statistician, and in 1946 married the mathematician Catherine Schulhof. That same year they obtained scholarships to the University of Leningrad, where Rényi studied under the guidance of Yu. V. Linnik. There he wrote his famous paper on the Goldbach conjecture (1947), completing it as his candidate dissertation in one year instead of the usual three. It was his connection with Linnik that led Rényi to the field of probability theory.

Returning home, Rényi published a sequence of papers, and simultaneously played a leading role in the reorganization of Hungarian research and education in mathematics. In 1947 he became an assistant lecturer at the University of Budapest and in the next year a *Privatdozent*. In the years 1948–1950, while he was a professor at the University of Debrecen, a school of probability theory grew up around him. This was a field where—apart from the significant, but isolated, productivity of Charles Jordan—hardly anything had been done in Hungary.

At this same time, Rényi was director of the Mathematical Institute of the Hungarian Academy of Sciences, head of the Department of the Theory of Probability in the institute, chairman of the Department of Probability Theory in the University of Budapest, acting secretary general of the János Bolyai Mathematical Society, secretary of the Department of Mathematics and Physics in the Hungarian Academy of Sciences, and secretary of the newly organized National Postgraduate Degree Granting Board. He was then 29 years old. His initiative and working capacity, surpassing all belief, continued unabated during the next 20 years.

In 1950 Rényi began writing his monumental textbook on probability theory and began working intensively on the applications of statistical methods to such fields as industry, economics, and agriculture. He played a leading role in the organization of the First Hungarian Mathematical Congress in 1950 and was chairman of the organizing committee of the second congress in 1960. In the intervening ten years he organized yearly conferences in theoretical and applied probability theory, mathematical statistics, and information theory. These soon became international events. During this productive decade he was also one of the initiators of various mathematical competitions for students at all levels, from elementary school through university.

In appreciation of his scientific work, Rényi was elected in 1949 as a corresponding member of the Hungarian Academy of Sciences and in 1956 as a full member. In 1949 he won the silver grade of the Kossuth Prize, in 1954 the gold grade. He was elected to membership in the International Statistical Institute, and in 1965 he became a vice-chairman. He was a fellow of the Institute of Mathematical Statistics. He was a member of the editorial board of ten mathematical periodicals in different countries; among them he was founder of the Publications of the Mathematical Institute of the Hungarian Academy of Sciences (now Studia Scientiarum Mathematicarum Hungaricae). He was visiting professor at Michigan State University (1961), University of Michigan (1964), Stanford University (1966), Cambridge University (1968), and the University of North Carolina (1968).

With his quickly receptive mind Rényi was able to work simultaneously in very different fields, and he became a scientific partner of P. Erdős, P. Turán, and other mathematicians in Hungary and elsewhere. He also wrote several papers jointly with his wife Catherine, a talented mathematician. Catherine Rényi died unexpectedly in August 1969, and shortly thereafter the first signs of a lung cancer appeared in her husband. He died on February 1, 1970.

Probability theory and mathematical statistics. Rényi's first achievement in probability theory was a purely probabilistic formulation of the large-sieve method (1947; 1958c). He used mixing sequences to study central limit theorems and ergodic theory. He also investigated the algebra of distributions, composed Poisson distributions, stochastic dependence and its measurement, the law of the iterated logarithm, projections of probability distributions, and the inequality of Kolmogorov. He made major contributions to the theory of order statistics and set up an axiomatic foundation of probability theory based on conditional probability. He was interested in applications in the theory of breaking, storage problems, determination of electric energy needs, rational dimensioning of compressors and air tanks, chemical reactions (he accomplished a probabilistic derivation and extension of the law of mass action of Gouldberg–Waage), the chemical countercurrent distribution in the case of noncomplete diffusion, construction of an interneuronal synapse model, and regulation of prices.

Later, Rényi considered information theory, especially the mathematical concept and interpretation of entropy, and statistical physics. Still later, information-theoretical foundations of mathematical statistics claimed his attention. Besides these, Rényi derived results about ergodic properties of representation of real numbers, mixing sequences of random variables, random graphs, and sampling from finite populations. In this last case, and in the case of the sum of a random number of independent random variables, he proved the central limit theorem under general conditions. A new, independent field of his research work was in the theory of search, which attracted wide interest.

In the last years of his life his textbook (1954b) was published successively in German, French, English, and Czech—each edition being different from the previous one, as he improved and rewrote. His highly original book on the foundations of probability theory appeared posthumously (1970) and was based on his concept of conditional probability. He dealt with the applicability of probability theory in other branches of mathematics, especially in number theory and analysis, and he turned more and more to basic questions of mathematics.

Other branches of mathematics. Rényi's scientific papers always contained either a clever innovation or a solution of a difficult problem. Equally at home working in combinatorics, measure theory, complex or real analysis, number theory, algebra, and geometry, he obtained results in all these fields. His most famous contribution to the other parts of mathematics consisted, no doubt, in proving first the quasi-Goldbach conjecture according to which every sufficiently large integer can be represented as a sum of a prime and of an almost prime (that is, an integer containing a restricted number of primes).

Sometimes Rényi humorously remarked that the whole of mathematics is probability theory, but he considered in a very serious way the applications of probabilistic methods in other fields of mathematics.

Popularization and teaching of mathematics. Rényi's expository and pedagogical ability is exhibited by his successful textbook, by his so-called *Dialogues* (1967), and by his *Letters on Probability* (1972). A sparkling wit, a feeling for philosophy and knowledge of it, and a keen didactic disposition show in these works. After writing his Socratic dialogues, Rényi extended his considerations to the applications of mathematics in the same style, and then defined his position on the foundation and role of probability theory in an imaginary correspondence between Fermat and Pascal. He wrote often on the teaching of probability theory and mathematical statistics. In a series of television broadcasts in Budapest under the title "Games and Mathematics" he popularized not only probability theory but also mathematics itself. His ideas concerning "reality and mathematics" appear in his essay "Die Sprache des Buches der Natur" (1968) and are restated in "Ars mathematica" (1970), which also contains his ideas on information theory.

Rényi established a group in the Mathematical Institute of the Hungarian Academy of Sciences to work on the problems of teaching mathematics in public high schools; he did a great deal toward the introduction and expansion of the specialization in applied mathematics in the University of Budapest; and he often lectured to or consulted with technical, agricultural, and economic societies or institutions. Through the Society for the Popularization of Scientific Knowledge, he exerted great influence not only on the improvement of teaching mathematics in high schools but also on an extended series of public lectures on the applications of mathematics.

Scientific influence. A large number of papers followed Rényi's initiatives. The method given in his work on order statistics was widely applied and his new test was considered by several subsequent authors. His work on mixing, on series of random numbers, on random graphs, and on information theoretical statistics also had wide influence.

ISTVÁN VINCZE

[*See also* PROBABILITY; NONPARAMETRIC STATISTICS, *article on* ORDER STATISTICS.]

WORKS BY RÉNYI

(1947) 1962 On the Representation of an Even Number as a Sum of a Prime and an Almost Prime Number. *Transactions of the American Mathematical Society* Series 2 19:299–321. → First published in Russian.

1948 Simple Proof of a Theorem of Borel and of the Law of the Iterated Logarithm. *Mathematisk tidsskrift* Series B [1948]:41–48.

1949 Un nouveau théorème concernant les fonctions indépendantes et ses applications à la théorie des nombres. *Journal de mathématiques pures et appliquées* 28:137–149.

1950 JÁNOSSY, LAJOS; RÉNYI, ALFRÉD; and ACZEL, JÁNOS On Composed Poisson Distributions. Part 1. *Acta mathematica Academiae Scientiarum Hungaricae* 1:209–224.

1950 RÉNYI, ALFRÉD; and ERDŐS, PÁL Some Problems and Results on Consecutive Primes. *Simon stevin* 27:115–125.

1951 PUKÁNSZKY, L.; and RÉNYI, ALFRÉD On the Approximation of Measurable Functions. Tudományegyetem Debrecen, Matematikai Intézet, *Publicationes mathematicae* 2:146–149.

1952 Sztochasztikus függetlenség és függvények teljes rendszere (Stochastic Independence and Complete System of Functions). Pages 299–316 in Hungarian Mathematical Congress, First, 1950, *Comptes rendus*. Budapest: Akadémia Kiadó. → With English and Russian summaries.

1952 RÉNYI, ALFRÉD; and TURÁN, PÁL On the Zeros of Polynomials. *Acta mathematica Academiae Scientiarum Hungaricae* 3:275–284.

1953a On the Theory of Order Statistics. *Acta mathematica Academiae Scientiarum Hungaricae* 4:191–231.

1953b Kémiai reakciók tárgyalása a sztochasztikus folyamatok elmélete segítségével (Treatment of the Theory of Chemical Reactions With the Aid of the Theory of Stochastic Processes). Magyar Tudományos

Akadémia, Alkalmazott Matematikai Intézet, *Közleményei* 2:85–101.

1953   RÉNYI, ALFRÉD et al.   A raktárkészlet pótlásáról: I. A törzskészlet (On the Replacement Policy in Stocks). Magyar Tudományos Akadémia, Alkalmazott Matematikai Intézet, *Közleményei* 2:187–201.

1954a   A valószinűségszámitás új axiomatikus felépitése (A New Axiomatic Foundation of the Theory of Probability). Magyar Tudományos Akadémia, Matematika–Fizika Osztály, *Közlemények* 4:369–427.

1954b   *Valószinűségszámitás* (Theory of Probability). Budapest: Tankönyvkiadó. → Later published in German, French, English, and Czech, with revisions in each successive edition; see below (1962b; 1966b; 1970a; 1972).

1954   HAJÓS, GYÖRGY; and RÉNYI, ALFRÉD   Elementary Proofs of Some Basic Facts in the Theory of Order Statistics. *Acta mathematica Academiae Scientiarum Hungaricae* 5:1–6.

1955   On a Combinatorial Problem in Connection With the Improving of the Lucerne. *Matematikai lapok* 6:151–164.

1955   HÁJEK, JAROSLAV; and RÉNYI, ALFRÉD   Generalization of an Inequality of Kolmogorov. *Acta matematica Academiae Scientiarum Hungaricae* 6:281–283.

1956   BALATONI, JÁNOS; and RÉNYI, ALFRÉD   Az entrópia fogalmáról (On the Notion of Entropy). Magyar Tudományos Akadémia, Matematikai Kutató Intézet, *Közleményei* 1:9–40.

1956   PALÁSTI, ILONA; and RÉNYI, ALFRÉD   A Monte-Carlo módszer mint minimax stratégia (Monte Carlo Methods as Minimax Strategies). Magyar Tudományos Akadémia, Matematikai Kutató Intézet, *Közleményei* 1:529–545.

1956   PRÉKOPA, ANDRÁS; RÉNYI, ALFRÉD; and URBANIK, KÁROLY   O predel'nom raspredelenii dliia summ nezavisymykh sluchainykh velichen na bikompaktnykh kommutativnykh topologicheskikh gruppakh (On the Limit Distribution of Sums of Independent Random Variables on Bicompact Commutative Topological Groups). *Acta mathematica Academiae Scientiarum Hungaricae* 7:11–16.

1956   RÉNYI, ALFRÉD; and SZENTÁGOTHAI, JÁNOS   Az ingerületatvitel valószínűsége egy egyszerű konvergens kapcsolású interneuronális synapsis-modelben (The Probability of Synaptic Transmissions in Simple Models of Interneuronal Synapses With Convergent Coupling). Magyar Tudományos Akadémia, Matematikai Kutató Intézet, *Közleményei* 1:83–91. → With Russian and English summaries.

1957a   On the Asymptotic Distribution of the Sum of a Random Number of Independent Random Variables. *Acta mathematica Academiae Scientiarum Hungaricae* 8:193–199.

1957b   Representations for Real Numbers and Their Ergodic Properties. *Acta mathematica Academiae Scientiarum Hungaricae* 8:477–493.

1957   ERDŐS, PÁL; and RÉNYI, ALFRÉD   A Probabilistic Approach to Problems of Diophantine Approximation. *Illinois Journal of Mathematics* 1:303–315.

1958a   On Mixing Sequences of Sets. *Acta mathematica Academiae Scientiarum Hungaricae* 9:215–228.

1958b   (Probabilistic Methods in Number Theory.) *Acta sinica* 4:465–510. → In Chinese.

1958c   On the Probabilistic Generalization of the Large Sieve of Linnik. Magyar Tudományos Akadémia, Matematikai Kutató Intézet, *Közleményei* 3:199–206.

1958   RÉNYI, ALFRÉD; and TURÁN, PÁL   On a Theorem of Erdős and Kac. *Acta arithmetica* 4:71–84.

1959a   ERDŐS, PÁL; and RÉNYI, ALFRÉD   On Random Graphs. Part 1. Tudományegyetem Debrecen, Matematikai Intézet, *Publicationes mathematicae* 6:290–297.

1959b   ERDŐS, PÁL; and RÉNYI, ALFRÉD   On the Central Limit Theorem for Samples From a Finite Population. Magyar Tudományos Akadémia, Matematikai Kutató Intézet, *Közleményei* 4:49–61.

1961   On Random Generating Elements of a Finite Boolean Algebra. Tudományegyetem Szeged, *Acta scientiarum mathematicarum* 22:75–81.

1962a   Théorie des éléments saillants d'une suite d'observations. Université de Clermont-Ferrand, Faculté de Sciences, *Annales* 2, no. 8:1–12.

1962b   *Wahrscheinlichkeitsrechnung, mit einem Anhang über Informationstheorie.* Berlin: Deutscher Verlag der Wissenschaften. → The German edition of Rényi's textbook; see above (1954b).

1963a   On Stable Sequences of Events. *Sankhyā* Series A 25:293–302.

1963b   Un dialogue. *Cahiers rationalistes* 33:4–32.

1963   ERDŐS, PÁL; NEVEU, J.; and RÉNYI, ALFRÉD   An Elementary Inequality Between the Probability of Events. *Mathematica scandinavia* 13:99–104.

1963   ERDŐS, PÁL; and RÉNYI, ALFRÉD   On Random Matrices. Magyar Tudományos Akadémia, Matematikai Kutató Intézet, *Közleményei* 8A:455–461.

1963   RÉNYI, ALFRÉD; and SULANKE, R.   Über die konvexe Hülle von *n* zufällig gewählten Punkten. *Zeitschrift für Wahrscheinlichkeitstheorie und Verwandte Gebiete* 2:75–84.

1964a   On an Extremal Property of the Poisson Process. Institute of Statistical Mathematics, Tokyo, *Annals* 16:129–133.

1964b   On Two Mathematical Models of the Traffic on a Divided Highway. *Journal of Applied Probability* 1:311–320.

1964c   A Socratic Dialogue on Mathematics. *Canadian Mathematical Bulletin* 7:441–462.

(1965) 1967   *Dialogues on Mathematics.* San Francisco: Holden-Day. → First published in Hungarian.

1966a   On the Amount of Missing Information and the Neyman–Pearson Lemma. Pages 281–288 in F. N. David (editor), *Research Papers in Statistics: Festschrift for J. Neyman.* London: Wiley.

1966b   *Calcul des probabilités, avec un appendice sur la théorie d'information.* Paris: Dunod. → The French edition of Rényi's textbook; see above (1954b).

(1967) 1972   *Letters on Probability.* Detroit: Wayne State Univ. Press. → First published in Hungarian.

1968   Die Sprache des Buches der Natur. *Neue Sammlung* 8:117–123.

1969   *On the Mathematical Theory of Trees.* Amsterdam: North-Holland.

1969   ERDŐS, PÁL; and RÉNYI, ALFRÉD   On Random Entire Functions. *Zastosowania matematyki* 10:47–55.

1970a   *Probability Theory.* Edited by H. A. Lauwerier and W. T. Koiter. Amsterdam: North-Holland; New York: American Elsevier. → The English edition of Rényi's textbook; see above (1954b).

1970b   *Foundations of Probability.* San Francisco: Holden-Day.

1970c   Ars mathematica. Pages 201–204 in Tore Dalenius et al. (editors), *Scientists at Work: Festschrift in Honour of Herman Wold.* Uppsala (Sweden): Almqvist & Wiksell.

1970   Rényi, Alfréd; and Erdős, Pál   On a New Law of Large Numbers. *Journal d'analyse mathématique* 23:103–111.

1972   *Teorie pravděpodobnosti.* Prague: Academia. → The Czech edition of Rényi's textbook; see above (1954*b*).

Selected Papers. 3 vols. Edited and annotated by Pál Turán. Budapest: Akadémia Kiadó, 1977. → 158 articles written during 1947–1970 and offering a general view of Rényi's mathematical activities and their effects on mathematical development.

SUPPLEMENTARY BIBLIOGRAPHY

Kendall, David G. 1970 Obituary: Alfréd Rényi. *Journal of Applied Probability* 7:509–522.

Révész, P.; and Vincze, István 1972 Alfréd Rényi, 1921–1970. *Annals of Mathematical Statistics* 43, no. 6:i–xvi.

Schmetterer, Leopold 1972 Alfréd Rényi, in Memoriam. Volume 2, pages xxv–l in Berkeley Symposium on Mathematical Statistics and Probability, Sixth, 1970, *Proceedings.* Berkeley: Univ. of California Press.

Vincze, István 1970 Öt éve halt meg Rényi Alfréd (Alfréd Rényi Died Five Years Ago). *Természet világa* (The World of Nature) [1970]:86 only.

RESEARCH DESIGN

See Experimental design; Epidemiology; Evaluation research; Organizations: methods of research; Public policy and statistics; Statistics as legal evidence; *and the biographies of* Fisher *and* Youden.

RESIDUALS

See Data analysis, exploratory.

RESPONSE SURFACES

See under Experimental design.

RETROSPECTIVE STUDIES

See Experimental design, *article on* quasi-experimental design; Fallacies, statistical; Public policy and statistics.

RICE, STUART

Stuart Arthur Rice is distinguished for his contributions to the growth of behavioral approaches in social science, to the progress of the United States government's statistical activities, and to bettering the relations between government and business. In particular, he improved statistical reporting, the organization of international statistics, and the collection of statistics in various foreign countries.

Rice received his PH.D. in 1924 from Columbia University, where he studied primarily under Franklin H. Giddings. Sociological training at Columbia emphasized quantitative research, and Rice produced in the 1920s a series of pioneer studies (e.g., 1924; 1928) which used quantitative methods to test a number of hypotheses regarding political behavior. He hoped his studies would be models for an objective, empirical, value-free science of politics. One hypothesis he investigated was that political behavior may be primarily the resultant of attitudes. He assumed that attitudes are normally distributed: abnormal distributions result from disturbing factors. He implied that political attitudes lie, for the most part, along one basic continuum of radicalism–conservatism. Along with William F. Ogburn and others, Rice used election statistics to analyze differences between various subgroups, thus anticipating later ecological voting studies. Following the leadership of A. Lawrence Lowell, he also used roll calls to detect differences between particular legislators and legislative blocs.

Another topic which Rice investigated in his early studies was the change over time in such political behavior as voter turnout and party preferences. He pioneered in the use of panel techniques to show the impact of observed stimuli upon the political attitudes of students, as measured by simple questionnaires.

At the time that Rice made his studies of political behavior, empirical research in social psychology was just beginning. L. L. Thurstone had not yet perfected his attitude-measurement techniques. Rice's use of the normal curve as a model of political behavior was soon outdated. More recently, social scientists commanding new data have elaborated his analysis of the areal distribution of political preferences. Again, the sample survey of the general population has replaced Rice's indirect inferences with the direct measurement of both actual voting and the individual characteristics of citizens. Yet, considering the tools available in the 1920s, Rice made a notable attempt to show the possibilities of an empirical approach to the study of political behavior.

Rice compiled, under the auspices of the Social Science Research Council, a critical review of the methods employed in the social sciences (1931). Case studies of the methods employed in outstanding contributions to social science were arranged in a theoretical setting designed to portray different methodological approaches, without regard to traditional boundaries in subject matter. The basic classification of methods was devised by Rice, as chief investigator and editor, with the assistance of Harold D. Lasswell.

Rice then turned from research and teaching to administrative work in the federal government. Here he did much to professionalize the activities of the Bureau of the Census. He also made notable contributions to the development of social statistics in the United States, the study of the effects of unemployment, the analysis of world standards of living, the establishment of a rationalized system of federal statistics, and the projecting of the statistical needs of the United Nations.

Rice was born in northern Minnesota in 1889. He attended the University of Washington, obtaining his B.A. there in 1912 and his M.A. in 1915, and was employed for some years in welfare administration at the local, state, and regional level. After an unsuccessful venture into minor party politics in 1920, he became interested in exploring the rational character of social movements and pursued the objective analysis of sociological data at Columbia University. He taught at Dartmouth College from 1923 to 1926 and at the Wharton School of Finance and Commerce of the University of Pennsylvania from 1926 to 1933.

His first appointment in Washington was as assistant director of the Bureau of the Census, a position he held until 1936. As president of the American Statistical Association in 1933, he was influential in creating the advisory services which in turn produced the central administrative structure for the development and coordination of the federal statistical system. For nearly two decades he was director of the Office of Statistical Standards of the Bureau of the Budget, charged by law with control of statistical work within the government. He won wide praise from the business community for his successful endeavors to reduce the reporting burdens imposed by federal questionnaires. At the beginning of World War II, Rice instigated the creation of the Inter-American Statistical Institute, which supplies the statistical staff of the Pan American Union. At the end of the war, he was active in the establishment of the Statistical Office of the United Nations and served as the first chairman of the UN Statistical Commission, where he displayed skill and ingenuity in working toward greater adequacy and comparability of world statistics.

As president of the International Statistical Institute from 1947 to 1953, he promoted collaboration among the world's statisticians and at the same time pointed out that the conceptions of statistics in the Soviet Union were an impediment to the growth of international statistics. As head of the Statistical Mission to Japan, he helped develop the statistical services of that country. Upon retiring from federal service in 1954, Rice organized and became president of the Surveys & Research Corporation, a consulting firm offering economic, statistical, and management services to United States and foreign governmental agencies, as well as to private industry and nonprofit organizations.

HAROLD F. GOSNELL

[*For discussion of the subsequent development of Rice's ideas, see* GOVERNMENT STATISTICS; Eulau 1968. *For the historical context of Rice's work, see* Berman 1968; Hankins 1968; Jaffe 1968.]

WORKS BY RICE

1923   The Effect of Unemployment Upon the Worker and His Family. Pages 99–109 in *Business Cycles and Unemployment: Report and Recommendations of a Committee of the President's Conference on Unemployment.* New York: McGraw-Hill.

1924   *Farmers and Workers in American Politics.* Columbia University Studies in History, Economics and Public Law, Vol. 113, No. 2. New York: Columbia Univ. Press.

1928   *Quantitative Methods in Politics.* New York: Knopf.

1930   AMERICAN STATISTICAL ASSOCIATION, COMMITTEE ON SOCIAL STATISTICS *Statistics in Social Studies.* Edited by Stuart A. Rice. Philadelphia: Univ. of Pennsylvania Press.

1931   SOCIAL SCIENCE RESEARCH COUNCIL, COMMITTEE ON SCIENTIFIC METHOD IN THE SOCIAL SCIENCES *Methods in Social Science: A Case Book . . . .* Edited by Stuart A. Rice. Univ. of Chicago Press.

1932   The Field of the Social Sciences. Pages 613–632 in James H. S. Bossard (editor), *Man and His World.* New York: Harper. → Bossard's book includes three other articles by Stuart Rice, on pages 633–669.

1933   RICE, STUART et al. *Next Steps in the Development of Social Statistics.* Ann Arbor, Mich.: Edwards.

1933   WILLEY, MALCOLM M.; and RICE, STUART *Communication Agencies and Social Life.* New York: McGraw-Hill.

SUPPLEMENTARY BIBLIOGRAPHY

►BERMAN, MILTON 1968 Lowell, A. Lawrence. Volume 9, pages 479–480 in *International Encyclopedia of the Social Sciences.* Edited by David L. Sills. New York: Macmillan and Free Press.

►EULAU, HEINZ 1968 Political Behavior. Volume 12, pages 203–214 in *International Encyclopedia of the Social Sciences.* Edited by David L. Sills. New York: Macmillan and Free Press.

►HANKINS, FRANK H. 1968 Giddings, Franklin H. Volume 6, pages 175–177 in *International Encyclopedia of the Social Sciences.* Edited by David L. Sills. New York: Macmillan and Free Press.

►JAFFE, A. J. 1968 Ogburn, William Fielding. Volume 11, pages 277–281 in *International Encyclopedia of the Social Sciences.* Edited by David L. Sills. New York: Macmillan and Free Press.

Postscript

Rice's distinguished career ended with his death on June 4, 1969, in the midst of his continuing activities in the field of international statistics. Dunn (1969) and Sibley (1969) are among the many articles of obituary and appreciation that appeared.

HAROLD F. GOSNELL

ADDITIONAL BIBLIOGRAPHY

DUNN, H. L. 1969 Stuart A. Rice, 1889–1969. *International Statistical Review* 37:332–334.

SIBLEY, ELBRIDGE 1969 Stuart Arthur Rice, 1889–1969. *American Statistician* 23, no. 4:47–48.

RICHARDSON, LEWIS FRY

Lewis Fry Richardson (1881–1953), British meteorologist and student of the causes of war, was born at Newcastle-on-Tyne, the youngest of seven children in a Quaker family. He attended the Newcastle Preparatory School, then Bootham School at York. His inclination toward science seems to have been inspired by J. Edmund Clark, a master at Bootham School, who was a meteorologist. However, while still in his teens, Richardson was convinced that "science ought to be subordinate to morals." After leaving Bootham in 1898, he attended Durham College of Science at Newcastle, then King's College at Cambridge.

Several appointments followed, each of short duration. Richardson worked as an assistant in the National Physics Laboratory, as chemist for a peat company, and as director of the physical and chemical laboratory of a lamp company. At the outbreak of World War I he was superintendent of Eskdalemuir Observatory of the Meteorology Office. Thenceforth, meteorology became one of his abiding scientific concerns, and he contributed some thirty papers and a book (1922) to that field. In 1926 his scientific achievements were recognized by his election as fellow of the Royal Society.

For Richardson, Quakerism is firmly identified with pacifism. Accordingly he declared himself a conscientious objector, a stand that subsequently barred him from university appointments. Nevertheless he participated directly in the war in a noncombatant capacity, as a member of the Friends' Ambulance Unit, attached to the 16th French Infantry Division. After the armistice he returned to the Meteorology Office, but in 1920, when that office became part of the Air Ministry, Richardson resigned his position. He next took charge of the physics department at Westminster Training College and in 1929 became principal of Paisley Technical College and School of Art, his last post. Richardson retired in 1940 in order to devote all of his time to the study of war, searching both for its causes and for means to prevent it (see Wright 1968).

It is with his work on war that Richardson's name is most frequently associated. Although his pacifist convictions were surely a source of his dedication to this study, on which he spent at least 35 years, it is likely that his deep involvement with meteorology influenced his method of research. The prediction of weather is notoriously difficult, even though the determinants of weather are entirely understood. The principles governing the several variables—the motion of air masses, the changes of pressures and temperatures, the onset of precipitation—are all known, but the interactions among all these factors are so complex that even if the atmospheric conditions over the globe were precisely known at a given moment, the calculation of future states for even a short period would be a superhuman task. At one time Richardson estimated (1922) that it would take 60,000 human computers working at high speed to compute tomorrow's world weather charts before tomorrow's weather arrived. The vital lesson that Richardson therefore learned from the problems of meteorological prediction—one that has been confirmed by modern computing technology—is that events which seem to be governed by chance (as does the weather to one ignorant of the dynamics of air masses) are in fact governed by laws and can be predicted *if* enough information can be processed.

The link between this insight and Richardson's approach to the phenomenon of war is in his rejection of the idea that war is a rational, or at least a purposeful, form of behavior, as is often assumed in the conventional political conception of international relations. Richardson viewed war instead in Tolstoyan fashion, as a massive phenomenon governed by forces akin to the forces of nature, over which individuals have little or no control. Accordingly, he ignored all those intricacies of diplomatic–strategic analysis usually pursued by political historians and turned his attention to quasi-mechanical and quantifiable processes which, he assumed, govern the dynamics of the international system of sovereign states. Neither the contents of memoranda and ultimatums nor dynastic claims, territorial ambitions, and networks of alliances play an explicit role in Richardson's theory of war. Instead, one finds differential equations purporting to represent the interactions among states or the spread of attitudes and moods (like the spread of communicable diseases) among the populations of those states. The equations are quite similar to those representing physical or chemical systems, at times tending toward equilibria, at times moving away from equilibria at accelerated rates and culminating in explosions.

Richardson harbored no illusions concerning the adequacy of his mathematical theory in providing an "explanation" of war, much less a means of preventing it. In omitting strategic calculations, considerations of prudence, and other "rational" factors as determinants of war or peace, he was not asserting the irrelevance of these factors. Rather, he sought to build a viable theory of war *in vacuo*, as it were, an admittedly crude but tractable model upon which a more sophisticated theory could be developed. "The equations are merely a description of what people would do if they did not stop to think," he wrote (1960*a*, p. 12). Some of his equations did fit what actually happened in the years preceding World War I, and Richardson concluded that the Great Powers in that instance were acting *as if* they were driven by mechanical forces.

In spite of the formal, mechanistic character of the equations that Richardson proposed as a model of international relations, he thought of the causes of war as primarily psychological. The underlying psychology was that of a mass, much simplified by the averaging out of the many opposing pressures, devoid of self-insight or foresight; it was not the psychology of an individual, with a large range of choices, moral convictions, and idiosyncratic preferences. Richardson's first paper in this vein was entitled "Mathematical Psychology of War." Written in 1919 and privately printed, it was not published until 1935. In the intervening years Richardson studied psychology as an external student of University College in London, receiving the special B.SC. degree in 1929. His work in psychology is strongly oriented toward the quantitative approach, and some of it shows the same influence of the "meteorological orientation" as do his studies on war (1937).

Richardson published the complete version of his mathematical theory of war, *Generalized Foreign Politics* (1939), on the very eve of World War II. Its point of departure is a pair of differential equations representing a hypothetical interaction between two rival states. The components of interaction are (1) mutual stimulation of armaments buildup, assuming that each nation's rate of increase of armaments expenditures is positively proportional to the other nation's current expenditures; (2) self-inhibition of armaments, assuming that the rate of change of armaments expenditures is negatively proportional to the already existing armaments burden; and (3) constant stimulants to armaments buildup in the form of grievances or ambitions of states. These latter may also be negative, in which case they are interpreted as reservoirs of good will. Cooperation between the rival states, for example, in the form of trade, is interpreted as negative armaments expenditures.

Richardson then examined the dynamics of the postulated system. The solution of the equations (given the initial conditions and the parameters of interaction) determines the time course of the armaments expenditures. By substituting selected values for the parameters, Richardson was able to obtain a good fit of the predicted time course for the armaments buildup of the rival European blocs (the Central Powers and the Entente) in the years preceding World War I. However, the large number of free parameters and the small number of points representing the time course make this "corroboration" of the theory less than impressive.

Of considerably greater interest is the theoretical result deduced from the model, namely that depending on the relative magnitudes of the stimulation and inhibition parameters (but not of the grievance–good-will parameters), the system may be either stable or unstable. If the parameters are such that the system is stable, then an armaments balance is possible (this might also be interpreted as arms control). However, if the parameters are such that the system is unstable, then such a balance is not possible. The system must move one way or the other, depending on the initial conditions, either toward total disarmament and beyond to ever-increasing cooperation or into a runaway race, presumably followed by war.

By noting the rate of disarmament of Great Britain following World War I and the rate of rearmament of Germany prior to World War II, Richardson was able to get rough estimates of the parameters in question. He concluded that the parameters were well within the region of instability and, moreover, that the initial conditions prior to World War I made it touch and go whether the system would move toward peace or war. Possibly just a slightly lower armaments level or just a little more interbloc trade would have pushed the system toward a united Europe instead of toward world war (see Singer 1968).

Following the analysis of the arms race of 1908–1914, Richardson attempted to analyze the similar process that started shortly after Hitler's rise to power. The disappearance of the gold standard as a basis for the measure of expenditures and the scantiness of data from the U.S.S.R. made the analysis of the arms race preceding World War II extremely difficult, beyond anything that a lone investigator could accomplish in a lifetime. Still, Richardson's theory of the mutually stimulating arms race did point to a second world war.

Whether or not these conclusions ought to be taken seriously is a difficult question. Certainly, controlled experiments on the scale of international relations cannot be brought to bear on the critical aspects of Richardson's theory. Yet whatever the explanatory or predictive merits of the theory, one cannot deny that it invites us to see the phenomenon of war from an unusual point of view. This point of view may have been stated earlier (Richardson cites Thucydides as a proponent of the mutual stimulation theory of arms races and wars), but the quantitative implications of rigorously formulated models based on this view seem never to have been worked out.

Besides the two-nation problem of mutual stimulation in an arms race, Richardson also posed the N-nation problem. Again, the relevance of his results to real international dynamics is an open question because of the vastly simplified assumptions on which his models are based. Nevertheless, the results are interesting, not because of the answers they provide but because of the questions they raise. Thus, Richardson found, for example, that in an arms race involving three nations, the situation can be stable for each of the three pairs separately but unstable for all three taken together. This result may be relevant to the currently acute N-nation nuclear force problem (see Kahn 1968).

Following his retirement in 1940, Richardson started extensive empirical investigations. He performed the monumental labor of gathering a vast variety of data related to all "deadly quarrels" that have been known to occur since the end of the Napoleonic Wars. Richardson's treatment of the data again reveals his view of war and of violence in general as something with which the whole human race is afflicted. This view is diametrically opposed to that represented by strategic thinkers and most clearly expressed by Karl von Clausewitz, who saw war as an instrument of national policy and a normal form of intercourse among civilized states (see Paret 1968).

To Richardson, war is a particular case of a "deadly quarrel," defined as a violent encounter among human beings resulting in one or more deaths. He placed all such encounters on a scale of magnitude (defined by the logarithm of the number of dead). Thus, single murders appear on this scale as deadly quarrels of magnitude 0 ($\log_{10} 1$

$= 0$), small riots with some ten victims as deadly quarrels of magnitude 1 ($\log_{10} 10 = 1$), and so on. The two world wars appear on this scale as deadly quarrels of magnitude 7.

Richardson sought to establish a relation between the magnitudes of deadly quarrels and the relative frequency of their occurrence, analogous to the relations established by George K. Zipf (1949) between ranks and sizes of a great variety of objects [see RANK–SIZE RELATIONS]. But unlike Zipf, who singled out the rank–size relation as a unifying principle of fundamental importance, Richardson treated this relation as one of many to be examined in the search for regularities from which, he hoped, the laws governing human violence would emerge.

Richardson sought to relate the frequencies of wars not only to their magnitudes but to every other conceivable factor that could be extracted from the data. He examined the effect of the existence of common frontiers between the combatants and of the existence of a common language, of a common religion, and of a common government (as in civil wars). He made a list of all the "pacifiers" that have been supposed at one time or another to counteract the tendency to violent outbreaks, such as distraction by sports, diversion of hatred to other groups, the direction of hatred inward, armed strength as a deterrent, collective security, or intermarriage among potentially hostile groups. None of the indices related to these pacifiers shows up as a significant contributing factor either to the likelihood of a "deadly quarrel" or to its prevention. Possible exceptions are international trade and allegiance to a common government. Thus, hardly anything in the way of new knowledge as to the "causes" of wars has emerged from this monumental analysis, unless one views as new the refutation of established notions by negative results. In particular, neither armed might nor collective security measures (contrary to widespread opinion) emerge as significant war-preventing influences.

The failure of the "statistics of deadly quarrels" to generate a theory on the causes of wars was to be expected in view of the magnitude of the problem. Because Richardson worked alone and had no access to modern computing machinery, the bulk of his effort was absorbed in tedious data gathering and routine calculations. It is likely, of course, that any crudely empirical brute-force attack on the causes of wars is inherently doomed to failure. The nature of the primary contributing causes may be shifting rapidly and may be quite different in different cultural milieus, so that lumping together all the deadly quarrels in the world for a period of some 120 years may be statistically meaningless. Still, the pioneering significance of this work and Richardson's gallantry in attacking the formidable problem singlehanded should not be underestimated. Whatever conclusions he drew or failed to draw, his work remains a rich collection of data.

Two of Richardson's books were published in 1960, under the titles of *Arms and Insecurity* and *Statistics of Deadly Quarrels* (1960a; 1960b). The former, edited by Nicolas Rashevsky and Ernesto Trucco, contains Richardson's mathematical theories of arms races, worked out in great detail, and the data relevant to those theories. The latter, edited by Quincy Wright and C. C. Lienau, contains the analysis of the voluminous data on violence ranging from murders to world wars. The two books were widely reviewed and helped establish Richardson's reputation as the pioneer of research into the causes of war.

Following the publication of these two books, a number of investigators in England and the United States undertook to combine Richardson's methods of mathematical model construction and data analysis in the field of international relations. In particular, content analysis methods have been used in attempts to obtain indices of internation hostility; various attempts have been made to extend Richardson's theory of arms races to cover the present period; and his methods of correlation analysis, designed as a search for the causative factors of wars, have been modified and refined. In Volume 8 of *General Systems* some of these investigations were published or reprinted under the heading "After Richardson" (Rummel 1963; Smoker 1963).

The principal value of Richardson's contribution lies in the idea of bringing quantitative analysis to bear upon the possible massive system-dynamic determinants of war. His work stands at one end of a spectrum, the other end of which is represented by game-theoretic analysis, where it is supposed that international relations can be viewed as interplays of strategies calculated by "rational players." This latter view is by far the more prevalent among political scientists. Richardson's quasi-deterministic view of international relations is complementary to the strategic view, which assumes rationality in the pursuit of "interests" but leaves unanalyzed the genesis of the interests. The strategic view may inquire how nations conduct (or would conduct, if they were rational) a diplomatic–military game but says nothing about how the game got started, why enmities are built up between some states and not between others, or, of course, why states behave so frequently and so

clearly against their own interests. Although Richardson did not shed much direct light on these matters, his approach raises important questions that are too often ignored in the purely diplomatic–military approach to international relations.

ANATOL RAPOPORT

[*See also* Alger 1968; Galtung 1968; Kaplan 1968; Kelman 1968; Rapoport 1968; Singer 1968; Wright 1968.]

WORKS BY RICHARDSON

1922 *Weather Prediction by Numerical Process.* Cambridge Univ. Press.

1933 The Measurability of Sensations of Hue, Brightness or Saturation. Pages 112–114 in Physical Society, London, *Report of a Joint Discussion on Vision.* London: The Society.

1935 Mathematical Psychology of War. *Nature* 135: 830–831; 136:1025 only.

1937 Hints From Physics and Meteorology as to Mental Periodicities. *British Journal of Psychology* 28:213–215.

1939 *Generalized Foreign Politics: A Study in Group Psychology.* Cambridge Univ. Press.

1960a *Arms and Insecurity: A Mathematical Study of the Causes and Origins of War.* Edited by Nicolas Rashevsky and Ernesto Trucco. Pittsburgh: Boxwood; Chicago: Quadrangle.

1960b *Statistics of Deadly Quarrels.* Edited by Quincy Wright and C. C. Lienau. Pittsburgh: Boxwood; Chicago: Quadrangle.

WORKS ABOUT RICHARDSON

BOYCE, ANNE O. 1889 *Records of a Quaker Family: The Richardsons of Cleveland.* London: Harris.

GOLD, E. 1954 Lewis Fry Richardson 1881–1953. Royal Society of London, *Obituary Notices of Fellows* 9:216–235. → Includes 4 pages of bibliography.

RAPOPORT, ANATOL 1957 Lewis F. Richardson's Mathematical Theory of War. *Journal of Conflict Resolution* 1:249–299.

SHEPPARD, P. A. 1953 Dr. L. F. Richardson, F.R.S. *Nature* 172:1127–1128.

SUPPLEMENTARY BIBLIOGRAPHY

▶ALGER, CHADWICK F. 1968 International Relations: I. The Field. Volume 8, pages 61–69 in *International Encyclopedia of the Social Sciences.* Edited by David L. Sills. New York: Macmillan and Free Press.

▶GALTUNG, JOHAN 1968 Peace. Volume 11, pages 487–496 in *International Encyclopedia of the Social Sciences.* Edited by David L. Sills. New York: Macmillan and Free Press.

HORVATH, WILLIAM J.; and FOSTER, CAXTON C. 1963 Stochastic Models of War Alliances. *General Systems* 8:77–81.

▶KAHN, HERMAN 1968 Nuclear War. Volume 11, pages 214–223 in *International Encyclopedia of the Social Sciences.* Edited by David L. Sills. New York: Macmillan and Free Press.

▶KAPLAN, MORTON A. 1968 Systems Analysis: IV. International Systems. Volume 15, pages 479–486 in *International Encyclopedia of the Social Sciences.* Edited by David L. Sills. New York: Macmillan and Free Press.

▶KELMAN, HERBERT C. 1968 International Relations:

III. Psychological Aspects. Volume 8, pages 75–83 in *International Encyclopedia of the Social Sciences*. Edited by David L. Sills. New York: Macmillan and Free Press.

▶PARET, PETER 1968 Clausewitz, Karl von. Volume 2, pages 511–513 in *International Encyclopedia of the Social Sciences*. Edited by David L. Sills. New York: Macmillan and Free Press.

▶RAPOPORT, ANATOL 1968 Systems Analysis: I. General Systems Theory. Volume 15, pages 452–458 in *International Encyclopedia of the Social Sciences*. Edited by David L. Sills. New York: Macmillan and Free Press.

RUMMEL, R. J. 1963 Dimensions of Conflict Behavior Within and Between Nations. *General Systems* 8:1–50.

▶SINGER, J. DAVID 1968 Disarmament. Volume 4, pages 192–202 in *International Encyclopedia of the Social Sciences*. Edited by David L. Sills. New York: Macmillan and Free Press.

¹SMOKER, PAUL 1963 A Mathematical Study of the Present Arms Race. *General Systems* 8:51–59.

▶WRIGHT, QUINCY 1968 War: I. The Study of War. Volume 16, pages 453–468 in *International Encyclopedia of the Social Sciences*. Edited by David L. Sills. New York: Macmillan and Free Press.

ZIPF, GEORGE K. 1949 *Human Behavior and the Principle of Least Effort: An Introduction to Human Ecology*. Reading, Mass.: Addison-Wesley.

Postscript

The extent of Richardson's influence on subsequent research is not easy to evaluate. His work, summarized in two posthumous books, is duly cited by workers concerned with statistical and "systemic" approaches to the phenomenon of war; but whether this work actually provided the impetus for subsequent ramifications or was itself an early manifestation of what developed into a style is difficult to say. In some circles, quantitative social science enjoys prestige, because it is associated with masses of "hard" data and with occasional resort to mathematical deductive techniques that guarantee the validity of conclusions derived from precisely stated initial assumptions. At the same time, the accuracy or the relevance of the assumptions are always open to question. Hence, controversies are generated concerning the appropriateness of *models* of social phenomena, and the identification of theories with models is very much in the spirit of modern social science. Because mathematical models make the underlying assumptions explicit, these controversies appear to be more fruitful and constructive than those that have characterized the humanistic approach in the social sciences. In an intellectual climate where the "hardening" of a field of inquiry is identified with its growing sophistication or maturation, the pioneers of this hardening process assume the status of mentors. Richardson achieved this status, even though none of his works can be considered as a substantive concrete contribution to a theory of war, as a "theory" is understood in the context of a "hard" science.

R. J. Rummel (1965; 1966; 1969) was the leading investigator in a project called Dimensionality of Nations undertaken at the University of Hawaii. A principal aim of the project was to discern "profiles" of nations in terms of a number of quantitative indices, singled out by factor analysis performed on large quantities of raw data and relating to incidence of internal and external conflict. In this connection, a question naturally arises concerning the relative contributions to the dynamics of international systems of the attributes of individual nations on the one hand and, on the other, of the "field," in which the nations are immersed, that is the "forces" generated by the interrelations among the participants in the international arena. Some of Rummel's work concerns this question.

Structural aspects of the international system occupies a central position in the work of J. David Singer and Melvin Small (1966a; 1966b), who were particularly interested in the role of the network of alliances as an enhancing or an inhibiting factor in the incidence and severity of wars.

The predominantly empirical approach to the correlates or the genesis of war, represented by the authors mentioned, involves four phases: (1) decisions concerning what data to gather; (2) the gathering of the data; (3) data processing, that is, the derivation of indices singled out for attention in step (1); (4) the interpretation of the relations revealed thereby. Phases (2) and (3) are technical in the sense that carrying out the corresponding tasks can be progressively improved by discovery of new sources, by more efficient methods of information retrieval, by more powerful computing technology, and so on, in short by techniques not available to Richardson. Phases (1) and (4), however, in the absence of a good theory, depend on luck, or intuition, or trial and error. In this respect, the "state of the art" cannot, perhaps, be said to have advanced significantly beyond what it was in Richardson's time. Interesting results are obtained, but what they mean is still a matter of conjecture. To make an example, Richardson discovered a similarity between the distribution of sizes of Chicago gangs and that of the severity of bandit raids in Manchuria, and he wondered whether both might be manifestations of "organization for aggression." In their search for the correlates of war in alliance structures, Singer and Small found that in the nineteenth century strong bipolarity of alliance structure was negatively correlated with some indices of war but that in the twentieth century bipoliarity was positively correlated with the same indices. It is intriguing to conjecture that the "balance of power" principle was a pacifying influence in the nineteenth century but a destabilizing one in the twentieth. In assessing the value of this and similar approaches, of which Richardson was a pioneer, it is important to keep in mind that it is not so much the validity of the conjectures that is the principal issue but rather the possibility of generating such conjectures by mechanical data processing, that is, without preconceived ideas.

Further developments of the Richardsonian theory of arms races have been on lines of more or less ad hoc modifications of the differential equation models and utilization of the much more abundant data available since World War II. Paul Smoker (1963) analyzed the U.S.–U.S.S.R. arms race, drawing conclusions concerning its various phases on the basis of estimated parameters in the differential equations. John C. Lambelet (1971) analyzed the arms race in the Middle East, using a model considerably more elaborate than Richardson's, on the basis of which he was able to differentiate the reactions of the various participants of that race (which Richardson did not do) and to single out the role of separate factors, for instance, the elasticity of defense spending, the political influence of the military, and so on.

Lambelet concluded, incidentally, that the Middle East arms race was not of the "explosive" variety and predicted an "equilibrium" to be reached around 1975. However, it should be noted that a steady (nonexplosive) growth of armament expenditures is also included as an instance of a "dynamic equilibrium" in Lambelet's formulation. And, of course, all such prognoses are predicated on the legitimacy of extrapolations of current trends, barring unforeseen "extraneous" factors.

Investigations aimed specifically at uncovering the causes or, as it is sometimes said more cautiously, the correlates of war have often been presented as contributions to knowledge that may make it possible to avoid war by controlling the relevant conditions. P. E. Chase (1969), for example, suggested the use of control theory for preventing runaway arms races. Possibly Richardson had something of the sort in mind when he embarked on his work. This conception of peace research (as work of this kind came to be called) must be tempered by the realization that even a well-founded theory with an excellent prediction record (which is at present nowhere on the horizon) would not necessarily confer the power of control (tacitly expected of a mature science), if

only because presently no institutions exist designed and empowered to use scientific knowledge for the purpose that "knowledge about the causes of wars" is sought. The significance of peace research, a field said to have been founded by Richardson, is to be sought elsewhere, namely in its influence on changing the conception of war, particularly of making obsolete the notion that war is a normal instrument to be used judiciously in the pursuit of national interests, and that the robustness or the safety of a nation is meaningfully measured by its war-making potential. The systemic conception of war with its own built-in dynamic precludes such a conception.

ANATOL RAPOPORT

ADDITIONAL BIBLIOGRAPHY

CHASE, P. E. 1969 Feedback Control Theory and Arms Races. *General Systems* 14:137–149.

LAMBELET, JOHN C. 1971 A Dynamic Model of the Arms Race in the Middle East, 1953–1965. *General Systems* 16:145–167.

RUMMEL, R. J. 1965 A Field Theory of Social Action With Application to Conflict Within Nations. *General Systems* 10:183–211.

RUMMEL, R. J. 1966 The Dimensionality of Nations Project. Pages 109–129 in Richard L. Merritt and Stein Rokkan (editors), *Comparing Nations: The Use of Quantitative Data in Cross-national Research.* New Haven: Yale Univ. Press.

RUMMEL, R. J. 1969 Indicators of Cross-national and International Patterns. *American Political Science Review* 63:127–147.

SINGER, J. DAVID; and SMALL, MELVIN 1966a Formal Alliances, 1815–1939. *Journal of Peace Research* 3:1–32.

SINGER, J. DAVID; and SMALL, MELVIN 1966a Formal Alliance Commitments and War Involvement, 1815–1945. Peace Research Society (International), *Papers* 5:109–140.

¹SMOKER, PAUL 1963 A Pilot Study of the Present Arms Race. *General Systems* 8:61–76.

RIDGE REGRESSION

See LINEAR HYPOTHESES, *article on* REGRESSION.

RISK

See DECISION MAKING, *article on* ECONOMIC ASPECTS; DECISION THEORY.

ROBUSTNESS

See DATA ANALYSIS, EXPLORATORY; ERRORS, *article on* EFFECTS OF ERRORS IN STATISTICAL ASSUMPTIONS; NONPARAMETRIC STATISTICS, *article on* ORDER STATISTICS; STATISTICAL ANALYSIS, SPECIAL PROBLEMS OF, *article on* OUTLIERS.

RUNS

See under NONPARAMETRIC STATISTICS.

S

SAMPLE SURVEYS

There is hardly any part of statistics that does not interact in some way with the theory or the practice of sample surveys. The differences between the study of sample surveys and the study of other statistical topics are primarily matters of emphasis.

The field of survey research is closely related to the statistical study of sample surveys [see SURVEY ANALYSIS]. Survey research is more concerned with highly multivariate data and complex measures of relationship; the study of sample surveys has emphasized sampling distributions and efficient design of surveys.

I
THE FIELD

The theory of sample surveys is mathematical and constitutes a part of theoretical statistics. The practice of sample surveys, however, involves an intimate mixture of subject matter (such as demography, psychology, consumer research, medicine, engineering) with theory. The germ of a study lies in the subject matter. Translation of a substantive question into a stimulus (question or test) enables man to inquire of nature and to quantify the result in terms of estimates of what the same inquiry would produce were it to cover every unit of the population.

Sampling, properly applied, does more. It furnishes, along with an estimate, an index of the precision thereof—that is, a margin of the uncertainty, for a stated probability, that one may reasonably ascribe to accidental variations of all kinds, such as variability between units (that is, between households, blocks, patients), variability of the interviewer or test from day to day or hour to hour, variations in coding, and small, independent, accidental errors in transcription and card punching.

The techniques of sampling also enable one to test the performance of the questionnaire and of the investigators and to test for differences between alternative forms of the questionnaire. They enable one to measure the extent of under-coverage or over-coverage of the prescribed units selected and also to measure the possible effects of differences between investigators and of departures from prescribed rules of interviewing and coding.

This article describes *probability sampling*, with special reference to studies of human populations, although the same theory and methods apply to studies of physical materials, to accounting, and to a variety of other fields. The main characteristic of probability sampling is its use of the theory of probability to maximize the yield of information for an allowable expenditure of skills and funds. Moreover, as noted above, the same theory enables one to estimate, from the results themselves, margins of uncertainty that may reasonably be attributed to small, accidental, independent sources of variation. The theory and practice of probability sampling are closely allied to the design of experiments.

The principal alternatives to probability sampling are judgment sampling and convenience sampling [see SAMPLE SURVEYS, *article on* NONPROBABILITY SAMPLING].

Uses of sampling. Probability sampling is used in a wide variety of studies of many different kinds

867

of populations. Governments collect and publish monthly or quarterly current information in such areas as employment, unemployment, expenditures and prices paid by families for food and other necessaries, and condition and yield of crops.

In modern censuses only basic questions are asked of every person, and most census information is elicited for only a sample of the people, such as every fourth household or every twentieth. Moreover, a large part of the tabulation program is carried out only on a sample of the information elicited from everyone.

Sampling is the chief tool in consumer research. Samples of records, often supplemented by other information, furnish a basis on which to predict the impact that changes in economic conditions and changes in competitive products will have on a business.

Sampling is an important tool in supervision and is helpful in many other administrative areas, such as studies of use of books in a library to improve service and to make the best use of facilities.

Sampling—what is it? Everyone acquires information almost daily from incomplete evidence. One decides on the basis of the top layer of apples in a container at the fruit vendor's whether to buy the whole container. The top layer is a good sample of the whole if the apples are pretty well mixed; it is a bad sample and may lead to a regrettable purchase if the grocer has put the best ones on top.

The statistician engaged in probability sampling takes no chances on inferences drawn exclusively from the top layer or from any other single layer. He uses random numbers to achieve a standard degree of mixing, thereby dispersing the sample throughout the container and giving to every sampling unit in the frame an ascertainable probability of selection [see RANDOM NUMBERS]. He may use powerful techniques of stratification, ratio estimation, etc., to increase accuracy and to decrease costs. For instance, in one type of stratified sampling he in effect divides the container of apples into layers, mixes the apples in each layer, and then takes a sample from each layer.

Some history of sampling. Sir Frederick Morton Eden estimated the number of inhabitants of Great Britain in 1800 at nine million, using data on the average number of people per house in selected districts and on the total number of houses on the tax-rolls, with an allowance for houses not reported for taxation. The first census of Great Britain, in 1801, confirmed his estimate. Messance in 1765 and Moheau in 1778 obtained estimates of the population of France by multiplying the ratio of the number of inhabitants in a sample of districts to the number of births and deaths therein by the number of births and deaths reported for the whole country. Laplace introduced refinements in 1786 and calculated that 500,000 was the outside margin of error in his estimate of the population of France, with odds of 1,161 : 1. His estimate and its precision were more successful than those of the complete census of France that was attempted at the same time. [See the biography of LAPLACE.]

A. N. Kiaer used systematic selection in a survey of Norwegian workers in 1895, as well as in special tabulations from the census of Norway in 1900 and from the census of Denmark in 1901 and in a study of housing in Oslo in 1913.

Bowley in 1913 used a systematic selection of every twentieth household of working-class people in Reading (England) and computed standard errors of the results.

Tabulation of the census of Japan in 1921, brought to a halt by the earthquake of 1923, went forward with a sample consisting of the records of every thousandth household. The results agreed with the full tabulation, which was carried out much later. The Swedish extraordinary census of 1935 provides a good example of the use of sampling in connection with total registrations.

One strong influence on American practice came in the 1930s from Margaret H. Hogg, who had worked under Bowley. Another came when controversies over the amount of unemployment during the depressions of 1921 and 1929 called for improved methods of study—Hansen's sample of postal routes for estimates of the amount of unemployment in 1936 gained recognition for improved methods; without it the attempt at complete registration of unemployed in the United States at the same time would have been useless.

Mahalanobis commenced in 1932 to measure the yield of jute in Bengal and soon extended his surveys to yields of rice and of other crops. In 1952 all of India came under the national surveys, the scope of which included social studies and studies of family budgets, sickness, births, and deaths. Meanwhile, the efforts of statisticians, mainly in India and England, had brought advances in methodology for estimation of yield per acre by random selection of small plots to be cut and harvested.

A quarterly survey of unemployment in the United States, conducted through interviews in a sample of households within a sample of counties, was begun in 1937. It was soon made monthly, and in 1942 it was remodeled much along its present lines (Hansen et al. 1953, vol. 1, chapter 9).

Sampling was used in the census of the United

States in 1940 to extend coverage and to broaden the program of tabulation and publication. Tabulation of the census of India in 1941 was carried out by a 2 per cent sample. Subsequent censuses in various parts of the world have placed even greater dependence on sampling, not only for speed and economy in collection and tabulation but also for improved reliability. The census of France used sampling as a control to determine whether the complete Census of Commerce of 1946 was sufficiently reliable to warrant publication; the decision was negative (Chevry 1949). [*For further history, see* Stephan 1948. *Some special references to history are contained in* Deming (1943) 1964, p. 142. *See also* STATISTICS, *article on* THE HISTORY OF STATISTICAL METHOD.]

Misconceptions about sampling. Sampling, of course, possesses some disadvantages: it does not furnish detailed information concerning every individual person, account, or firm; furthermore, error of sampling in very small areas and subclasses may be large. Many doubts about the value of sampling, however, are based on misconceptions. Some of the more common misconceptions will now be listed and their fallacies pointed out.

It is ridiculous to think that one can determine anything about a population of 180 million people, or even 1 million people, from a sample of a few thousand. The number of people in the country bears almost no relation to the size of the sample required to reach a prescribed precision. As an analogy (suggested by Tukey), consider a basket of black and white beans. If the beans are really mixed, a cupful would determine pretty accurately the proportion of beans that are black. The cupful would still suffice and would give the same precision for a whole carload of beans, provided the beans in the carload were thoroughly mixed. The problem lies in mixing the beans. As has already been noted, the statistician accomplishes mixing by the use of random numbers.

Errors of sampling are a hazard because they are ungovernable and unknown. Reliability of a sample is a matter of luck. Quality and reliability of data are built in through proper design and supervision, with aid from the theory of probability. Uncertainty resulting from small, independent, accidental errors of a canceling nature and variation resulting from the use of sampling are in any case determinable afterward from the results themselves.

Errors of sampling are the only danger that one has to worry about in data. Uncertainty in statistical studies may arise from many sources. Sampling is but one source of error. [*See below, and see also* ERRORS, *article on* NONSAMPLING ERRORS].

Electronic data-processing machines, able to store and retrieve information on millions of items with great speed, eliminate any need of sampling. This is a fanciful hope. The inherent accuracy of original records as edited and coded is the limitation to the accuracy that a machine can turn out. Often, complete records are flagrantly in error or fail to contain the information that is needed. Moreover, machine-time is expensive; sampling reduces cost by reducing machine-time.

A "complete" study is more reliable than a sample. Data are the end product of preparation and of a long series of procedures—interviewing, coding, editing, punching, tabulation. Thus, error of sampling is but one source of uncertainty. Poor workmanship and structural limitations in the method of test or in the questionnaire affect a complete count as much as they do a sample. It is often preferable to use funds for improving the questionnaire and tests rather than for increasing the size of the sample.

Statistical parts of sampling procedure. A sampling procedure consists of ten parts. In the following list, M will denote those parts that are the responsibility of the expert on the subject matter, and S will denote those that are the responsibility of the statistician. (The technical terms used will be defined below.)

(*a*) Formulation of the problem in statistical terms (probability model) so that data will be meaningful (*M, S*). A problem is generated by the subject matter, not by statistical theory.

(*b*) Decision on the universe (*M*). The universe follows at once from a careful statement of the problem.

(*c*) Decision on the frame (*M, S*). Decision on the type and size of sampling units that constitute the frame (*S*).

(*d*) Procedure for the selection of the sample (*S*).

(*e*) Procedure for the calculation of estimates of the characteristics desired (averages, totals, proportions, etc.) (*S*).

(*f*) Procedure for the calculation of standard errors (*S*).

(*g*) Design of statistical controls, to permit detection of the existence and extent of various nonsampling errors (*S*).

(*h*) Editing, coding, tabulation (*M, S*).

(*i*) Evaluation of the statistical reliability of the results (*S*).

(*j*) Uses of the data (*M*).

Definitions of terms

The technical terms that have been used above and that will be needed for further discussion will now be defined.

Universe of study. The universe consists of all the people, firms, material, conditions, units, etc., that one wishes to study, whether accessible or not. The universe for any study becomes clear from a careful statement of the problem and of the uses intended for the data. Tabulation plans disclose the content of the universe and of the information desired. Examples of universes are (*i*) the housewives aged 20–29 that will live in the Metropolitan Area of Detroit next year, (*ii*) all the school children in a defined area, (*iii*) all the pigs in a country, both in rural areas and in towns.

Frame. The frame is a means of access to the universe (Stephan 1936) or to enough of the universe to make a study worthwhile. A frame is composed of sampling units. A sampling unit commonly used in house-to-house interviewing is a compact group or segment of perhaps five consecutive housing units. A frame is often a map, divided up—either explicitly or tacitly—into labeled areas. In a study concerned with professional men, for example, the frame might be the roster of membership of a professional society, with pages and lines numbered. The sampling unit might be one line on the roster or five consecutive lines.

Without a frame probability sampling encounters numerous operational difficulties and inflated variances (see, for example, the section "Sampling moving populations," below).

In the types of problems to be considered here (with the exception of those treated in the section "Sampling moving populations," below) there will be a frame, and every person, or every housing unit, will belong to one sampling unit, or will have an ascertainable probability of belonging to it. In the sampling of stationary populations, a sampling procedure prescribes rules by which it is possible to give a serial number to any sampling unit, such as a small area. A random number will then select a definite sampling unit from the frame and will lead to investigation of all or a subsample of the material therein that belongs to the universe.

Selection of persons within a dwelling unit. Some surveys require information concerning individuals, and in such cases it may be desirable, for various reasons (contagion, fatigue, and so on), to interview only one eligible person in a dwelling unit that lies in a selected segment. In such surveys, the interviewer may make a list of the eligible people in each dwelling unit that falls in the sample and may select therefrom, on the spot, by a scheme based on random numbers, one person to interview. Appropriate weights are applied in tabulation (Deming 1960, p. 240).

Nominal frame and actual frame. One must often work with a frame that fails to include certain areas or classes that belong to the universe. A list of areas that contain normal families may not lead to all the consumers of a product, as some consumers may live in quasi-normal quarters, such as trailers and dormitories. Extension of the sampling procedure into these quarters may present problems. Fortunately, the proportion of people in quasi-normal households is usually small (mostly 1 per cent to 3 per cent in American cities), and one may therefore elect to omit them.

A frame may be seriously impaired if it omits too much of certain important classes that by definition belong to the nominal frame. It is substantive judgment, aided by calculation, that must decide whether a proposed frame is satisfactory.

Sampling from an incomplete frame. Almost every frame is in some respects out of date at the time of use. It is often possible, however, to use an obsolete or incomplete frame in a way that will erase the defects in the areas that fall into the sample. One may, for example, construct rules by which to select large sampling units from an incomplete frame and then to amend those units, by local inquiry, in order to bring them up to date. Selection of small areas within the larger area, with the appropriate probability, will maintain the prescribed over-all probability of selection.

Sampling for rare characteristics. One sometimes wishes to study a rare class of people when there is no reliable list of that class. One way to accomplish this is to carry out a cheap, rapid test in order to separate a sample of households into two groups (strata)—one group almost free of the rare characteristic, the other heavily populated with it—and then to investigate a sample drawn from each group. Optimum sampling fractions and weights for consolidation may be calculated by the theory of stratified sampling (discussed below; see also Kish in Symposium . . ., 1965).

Equal complete coverage of a frame. The equal complete coverage of a frame is by definition the result that would be obtained from an investigation of all sampling units in a given frame, carried out by the same field workers or inspectors, using the same definitions and procedures, and exercising the same care as they exercised on the sample, and at about the same period of time. The adjective "equal" signifies that the same methods must be used for the equal complete coverage as for the sample.

Some operational definitions. *Sampling error.* Suppose that for a given frame, sampling units bear the serial numbers 1, 2, 3, and on to N. However it be carried out, and whatever be the rules for coding and for adjustment for nonresponse, a complete coverage of the N sampling units of the frame would yield the numerical values

$$a_1, a_2, a_3, \cdots, a_N \text{ for } x,$$
$$b_1, b_2, b_3, \cdots, b_N \text{ for } y.$$

In a survey of unemployment, for example, the x-characteristic of a person might be the property of being unemployed and his y-characteristic the property of belonging to the labor force. Then a_1, the x-population of sampling unit No. 1 (which might consist of five successive households), would be the count of people that have the x-characteristic in that sampling unit. That is, a_1 would be the count of unemployed persons in the five households. Similarly, b_1, the y-population, would be the count of people in the labor force in those same households. Then a_1/b_1 would be the proportion unemployed in the sampling unit of five households.

Again, x might refer to expenditure for bread and y to expenditure for all food. Then a_1/b_1 would be the proportion of money that goes for bread in sampling unit No. 1, expenditure for all food being the base.

Here, the people with the x-characteristic form a subclass of those with the y-characteristic, but this may not be so in other surveys. Thus, the x-characteristic and the y-characteristic might form a dichotomy, such as passed and rejected or male and female. One often deals with multiple characteristics, but two will suffice here.

Denote the sum of the x-values and of the y-values in the N sampling units by

$$A = a_1 + a_2 + a_3 + \cdots + a_N = Na = x\text{-total},$$
$$B = b_1 + b_2 + b_3 + \cdots + b_N = Nb = y\text{-total},$$

which makes a and b the average x-value and the average y-value per sampling unit in the frame, as in Table 1. For example, A might be the total number unemployed in the whole frame and B the total number of people in the labor force. Then $\phi = A/B$ would be the proportion of people in the labor force that are unemployed.

An operational definition of the sampling process and of the consequent error of sampling is contained in the following experiment.

(*a*) Take for the frame N cards, numbered serially 1 to N. Card i shows a_i and b_i for the values of the x-characteristic and y-characteristic.

(*b*) Draw a sample of n cards, following the specified sampling procedure (which will invariably require selection by random numbers).

Table 1 illustrates the notation for the frame and for the results of a sample. The serial numbers on the cards in the sample are not their serial numbers in the frame but denote instead the ordinal number as drawn by random numbers. Sample card No. 1 could be any card from 1 to N in the frame. In general, another sample would be composed of different cards, as the drawings are random.

(*c*) Form estimates by the formulas specified in the sampling plan. For illustration, one may

Table 1 — Some notation for frame and sample

	FRAME			SAMPLE	
Serial number of sampling unit	x-value	y-value	Serial number in order drawn in sample	x-value	y-value
1	a_1	b_1	1	x_1	y_1
2	a_2	b_2	2	x_2	y_2
$\vdots$	$\vdots$	$\vdots$	$\vdots$	$\vdots$	$\vdots$
N	a_N	b_N	n	x_n	y_n
Total	A	B		x	y
Average per sampling unit	$a = A/N$	$b = B/N$		$\bar{x} = x/n$	$\bar{y} = y/n$
Variance*	$\sigma_a^2 = \dfrac{1}{N}\sum_{i=1}^{N}(a_i - a)^2$	$\sigma_b^2 = \dfrac{1}{N}\sum_{i=1}^{N}(b_i - b)^2$		$s_x^2 = \dfrac{1}{n}\sum_{i=1}^{n}(x_i - \bar{x})^2$	$s_y^2 = \dfrac{1}{n}\sum_{i=1}^{n}(y_i - \bar{y})^2$
Standard deviation	σ_a	σ_b		s_x	s_y

* Some authors define variances by means of N–1 and n–1 rather than N and n.

form, from the sample, estimators like

$$(1) \qquad \bar{x} = \frac{1}{n} \sum_{i=1}^{n} x_i,$$

$$(2) \qquad \bar{y} = \frac{1}{n} \sum_{i=1}^{n} y_i,$$

$$(3) \qquad X = N\bar{x},$$

$$(4) \qquad Y = N\bar{y},$$

$$(5) \qquad f = \bar{x}/\bar{y}.$$

If (1), (2), (3), (4), and (5) are used as estimators of a, b, A, B, and ϕ, respectively, and if the results of the complete coverage were known, then one could, for any experiment, compute errors of sampling, such as

$$(6) \qquad \Delta\bar{x} = \bar{x} - a,$$

$$(7) \qquad \Delta\bar{y} = \bar{y} - b,$$

$$(8) \qquad \Delta f = f - \phi.$$

It is an exciting fact that a single sample—provided that it is big enough (usually 25, 30, or more sampling units), and provided that it is designed properly and skillfully in view of possible statistical peculiarities of the frame and is carried out in reasonable conformance with the specifications—will make possible an estimate, based on theory to follow later, of the important characteristics of the distribution of sampling variation of all possible samples that could be drawn from the given equal complete coverage and processed by the specified sampling procedure.

Standard error and mathematical bias. We continue our conceptual experiment.

(*d*) Return the sample of n cards to the frame, and repeat steps (*b*) and (*c*) by the same sampling procedure, to form a new sample and new estimates $\bar{x}$, $\bar{y}$, f. Repeat these steps again and again, 10,000 times or more.

Explicit statements will now be confined to $\bar{x}$. The 10,000 experiments give an empirical distribution for $\bar{x}$, by which one may count the number of samples for which $\bar{x}$ lies between, for example, 100 and 109. We visualize an underlying theoretical distribution of $\bar{x}$, which the empirical distribution approaches closer and closer as the number of repetitions increases.

We are typically concerned with relationships between (*i*) the empirical distribution of $\bar{x}$ and (*ii*) the theoretical distribution of $\bar{x}$, for the given sampling procedure. Study of these relationships helps in the use of sampling for purposes of making estimates of characteristics of the frame.

Let a be the characteristic of the complete coverage that the generic symbol $\bar{x}$ estimates. Then if

$$(9) \qquad E\bar{x} = a,$$

the sampling procedure is said to be unbiased. (The symbol E denotes *expectation*, the mean of the theoretical distribution of $\bar{x}$.) But if

$$(10) \qquad E\bar{x} = a + C, \qquad C \neq 0,$$

the sampling procedure has the mathematical bias C. In any case, the variance of the distribution of $\bar{x}$ is

$$(11) \qquad \sigma_{\bar{x}}^2 = E(\bar{x} - E\bar{x})^2,$$

and its square root, $\sigma_{\bar{x}}$, is the standard error of the sampling procedure for the estimator $\bar{x}$. Thus, a sampling procedure has, for any estimator, an expected value, a standard error, and possibly a mathematical bias (see the section "Possible bias in ratio estimators," below).

Uncertainty from accidental variation. Under the conditions stated above, the margin of uncertainty in the estimator $\bar{x}$ that is attributable to sampling and to small, independent, accidental variations, including random error of measurement (Type III in the next section), may be estimated, for a specified probability, as $t\hat{\sigma}_{\bar{x}}$, where $\hat{\sigma}_{\bar{x}}$ is an estimator of $\sigma_{\bar{x}}$. The factor t depends on the probability level selected for the margin of uncertainty (which will in turn depend on the risks involved) and also on the number of degrees of freedom in the estimator $\hat{\sigma}_{\bar{x}}$. In large samples the distributions of most estimators are nearly normal, except for frames that exhibit very unusual statistical characteristics. The standard deviation, $\sigma_{\bar{x}}$, then contains nearly all the information regarding the margin of uncertainty of $\bar{x}$ that is attributable to accidental variation. Presentation of the results of a survey requires careful consideration when there is reason to question the approximate normality of estimators (Fisher 1956, p. 152; Shewhart 1939, p. 106).

Random selection. It is never safe to assume, in statistical work, that the sampling units in a frame are already mixed. A frame comes in layers that are different, owing to geographic origin or to order of production. Even blood, for example, has different properties in different parts of the body.

A random variable is the result of a random operation. A system of selection that depends on the use, in a standard manner, of an acceptable table of random numbers is acceptable as a random selection. Methods of selection that depend on physical mixing, shuffling, drawing numbers out of a hat, throwing dice, are not acceptable as random, because they have no predictable behavior. Neither are schemes that merely remove the choice of sampling units from the judgment of the interviewer. Pseudo-random numbers, generated under competent skill, are well suited to certain types of statistical investigation [*see* RANDOM NUMBERS].

Types of uncertainty in statistical data

All data, whether obtained by a complete census or by a sample, are subject to various types of uncertainty. One may reduce uncertainties in data by recognizing their existence and taking steps for improvement in future surveys. Sample design is an attempt to strike an economic balance between the different kinds of uncertainty. There is no point, for example, in reducing sampling error far below the level of other uncertainties.

Three types of uncertainty. The following discussion will differentiate three main types of uncertainty.

Type I. Uncertainty of Type I comprises built-in deficiencies, or structural limitations, of the frame, questionnaire, or method of test.

Any reply to a question, or any record made by an instrument, is only a response to a stimulus. What stimulus to apply is a matter of judgment. Deficiencies in the questionnaire or in the method of test may therefore arise from incomplete understanding of the problem or from unsuitable methods of investigation. Structural limitations are independent of the size or kind of sample. They are built in: a recanvass will not discover them, nor will calculation of standard errors or other statistical calculations detect them.

Some illustrations of uncertainty of Type I are the following:

(*a*) The frame may omit certain important segments of the universe.

(*b*) The questionnaire or method of test may fail to elicit certain information that is later found to be needed. The questionnaire may contain inept definitions, questions, and sequences. Detailed accounting will give results different from those given by mere inquiry about total expenditure of a family for some commodity; date of birth gives a different age from that given in answer to the simple question, How old are you? There may be differential effects of interviews depending on such variables as sex and race of the interviewer.

(*c*) Use of telephone or mail may yield results different from those obtained by personal interview.

(*d*) Judgments of coders or of experts in the subject matter may differ.

(*e*) The date of the survey has an important effect on some answers.

Type II. Uncertainty of Type II includes operational blemishes and blunders—for example:

(*f*) One must presume the existence of errors of a noncanceling nature (persistent omission of sampling units designated, persistent inclusion of sampling units not designated, persistent favor in recording results).

(*g*) One must presume the existence of bias from nonresponse.

(*h*) Information supplied by coders for missing or illegible entries may favor high or low values.

(*i*) There may be a large error, such as a unique blunder.

Type III. Uncertainty of Type III is caused by random variation. Repeated random samples drawn from the same frame will give different results. Besides, there are inherent uncorrelated, nonpersistent, accidental variations of a canceling nature that arise from inherent variability of investigators, supervisors, editors, coders, punchers, and other workers and from random error of measurement.

Standard error of an estimator. The standard error of a result includes the combined effects of all kinds of random variation, including differences within and between investigators, supervisors, coders, etc. By proper design, however, it is possible to get separate estimates of some of these differences.

A small standard error of a result signifies (*i*) that the variation between repeated samples will be small and (*ii*) that the result of the sample agrees well with the equal complete coverage of the same frame. It usually tells little about uncertainties of Type II and never anything about uncertainties of Type I.

Limitations of statistical inference. Statistical inference (estimates, standard errors, statistical tests) refers only to the frame that was sampled and investigated. No statistical calculation can, by itself, detect or measure nonsampling errors, although side experiments or surveys may be helpful. No statistical calculation can detect defects in the frame. No statistical calculation can bridge the gap between the frame covered and the universe. This is as true of probability sampling as it is of judgment sampling, and it is true for a complete census of the frame as well.

Comparison of surveys. Substantial differences in results may come from what appear to be inconsequential differences in questionnaires or in methods of hiring, training, and supervision of interviewers and coders or in dates of interviewing. The sampling error in a sample is thus not established by comparison against a complete census unless the complete census is the equal complete coverage for the sample.

Recalls on people not at home. Many characteristics of people that are not at home at first call, or that are reluctant to respond, may be very different from the average. What is needed is response from everyone selected, including those that are hard to get. To increase the initial size of the sample is no solution. Calculations that cover a wide variety of circumstances show that the amount of information per dollar expended on a survey increases with the number of recalls, the only practicable limit being the time for the completion of the survey. Good sample design therefore specifies that four to six well-timed recalls be made or specifies that recalls continue until the level of response reaches a prescribed proportion. Special procedures, such as intensive subsampling of those not at home on the first or second call, have been proposed (see Leven 1932; Hansen & Hurwitz 1946; Deming 1960).

Surveys by post. One can often effect important economies by starting with a mail survey of a fairly large sample properly drawn from a given frame, then finishing with a final determined effort in the form of personal interviews on all or a fraction (one in two or one in three) of the people that failed to reply (Leven 1932). Mail surveys require a frame, in the form of a list of names with reasonably accurate addresses, and provision for keeping records of mailings and of returns. They are therefore especially adaptable to surveys of members of a professional society, subscribers to a journal, or subscribers to a service. [*For further discussion of mail surveys, see* ERRORS, *article on* NONSAMPLING ERRORS.]

Simple designs for enumerative purposes

The aim in an *enumerative* study is to count the number of people in an area that have certain characteristics or to estimate a quantity, perhaps their annual income, regardless of how they acquired these characteristics. The aim in an *analytic* study is to detect differences between classes or to measure the effects of different treatments.

For illustration consider a study of schizophrenics. One enumerative aim might be to estimate the number of children born to schizophrenic parents before onset of the disease or before the first admission of one of the parents to a hospital for mental diseases. Further aims of the same study might be analytic, such as to discover differences in fertility or in duration of hospitalization caused by different treatments, differences between communities, or differences between time periods.

The finite multiplier typified by $1/n - 1/N$ (to be seen later) appears in estimators for enumerative purposes. It has no place in estimators for analytic purposes.

Optimum allocation of effort for an enumerative aim may not be optimum for an analytic aim.

Moreover, what is optimum for one enumerative characteristic may not be optimum for another. Hence, it will usually be necessary to compromise between competitive aims.

Enumerative aims will occupy most of the remaining space in this article.

The theory presented in this section is for the design commonly called *simple random sampling*. This is often a practicable design, and the theory forms a base for more complex designs.

A simple procedure of selection and some simple estimators. Definitions of "frame," "sample," and other terms were introduced above. In addition, it will be convenient to define the *coefficient of variation*. For the x-population and y-population of the frame, the coefficients of variation are defined as

$$(12) \qquad C_a = \frac{\sigma_a}{a}, \qquad C_b = \frac{\sigma_b}{b}, \qquad a > 0, b > 0.$$

In like manner, the symbol C_x denotes the coefficient of variation of the empirical or theoretical distribution of the random variable x. The square, C_x^2, of the coefficient of variation C_x is called the *rel-variance* of x. The coefficient of variation is especially useful for characteristics (such as height) that are positive. It is often helpful to remember, for example, that $C_{\bar{x}} = C_x = C_X$ because x, $\bar{x}$, and X are constant multiples of each other.

The procedure of selection specified earlier gives every member of the frame the same probability of selection as every other member, wherefore

$$(13) \qquad \begin{aligned} E\bar{x} &= a, \\ E\bar{y} &= b. \end{aligned}$$

That is, $\bar{x}$ and $\bar{y}$ are unbiased estimators of a and b, respectively. Moreover,

$$(14) \qquad \begin{aligned} X &= N\bar{x}, \\ Y &= N\bar{y} \end{aligned}$$

are unbiased estimators of A and B.

Often, a ratio such as

$$(15) \qquad \phi = \frac{A}{B} = \frac{a}{b}$$

is of special interest. The sample gives

$$(16) \qquad f = \frac{X}{Y} = \frac{x}{y} = \frac{\bar{x}}{\bar{y}}$$

as an estimator of ϕ. If the total y-population, B, is known from another source, such as a census, A may be estimated by the formula

$$(17) \qquad X' = Bf.$$

This estimator X' is called a *ratio estimator*. It will be more precise than the estimator $X = N\bar{x}$ in (14) if the correlation between x_i and y_i is high. Other estimators will be discussed later (for example, regression estimators). Theory provides a basis for the choice of estimator.

Possible bias in ratio estimators.   Necessary and sufficient conditions for there to be no bias in f as an estimator of ϕ are that $Ey \neq 0$ and that x_i/y_i and y_i be uncorrelated—that is, that $E[(x/y)y] = E(x/y)Ey$. In practice, if bias exists at all, it is usually negligible when the sample contains more than three or four sampling units.

Sampling with and without replacement.   Usually, in the sampling of finite populations, one permits a sampling unit to come into the sample only once. In statistical language, this is sampling *without replacement*. Tests of physical materials are sometimes destructive, and a second test would be impossible. To draw without replacement, one simply disregards a random number that appears a second time (or uses tables of so-called random permutations).

There are circumstances, however, in which one accepts the random numbers as they come and permits a sampling unit to come into the sample more than once. This is sampling *with replacement.*

Hereafter, most equations will be written for sampling without replacement. It is a simple matter to drop the fraction $1/N$ from any formula to get the corresponding formula for sampling with replacement. Actually, in practice, samples are usually such a small part of the frame that the fraction $1/N$ is ignored, even though the sampling be done without replacement.

Variances.   The variances of the estimator $\bar{x}$ derived from the sampling procedure described earlier are

$$(18) \quad \begin{cases} \text{without} & \sigma_{\bar{x}}^2 = \dfrac{N-n}{N-1}\dfrac{\sigma^2}{n} \\ \text{replacement} & \cong \left(\dfrac{1}{n} - \dfrac{1}{N}\right)\sigma^2, \\[2mm] \text{with} & \sigma_{\bar{x}}^2 = \dfrac{\sigma^2}{n}. \\ \text{replacement} & \end{cases}$$

(The sign $\cong$ indicates an approximation that is sufficiently close in most practice.)

Similar expressions hold for $\bar{y}$. For the ratio $f = \bar{x}/\bar{y}$, the approximation

$$(19) \quad C_f^2 \cong \left(\frac{1}{n} - \frac{1}{N}\right)(C_a^2 + C_b^2 - 2abC_{ab})$$

is useful if n be not too small. Here

$$(20) \quad C_{ab} = \frac{1}{Nab}\sum (a_i - a)(b_i - b)$$

is the rel-covariance of the x-population and y-population per sampling unit in the frame.

When the ratio estimator of the total x-population is derived as in eq. (17), eq. (19) gives the same approximation for $C_{x'}^2$.

Estimate of aggregate characteristic—number of units in class unknown.   It often happens in practice that one wishes to estimate the aggregate value of some characteristic of a subclass of a group when the total number of units in the subclass is unknown. For example, one might wish to estimate the aggregate income of women aged 15 or over that live in a certain district, are gainfully employed, and have at least one child under 12 years old at home (this specification defines the universe). The number of women that meet this specification is not known. An estimate of the average income per woman of this specification, prepared from a sample, suffers very little from this gap in available knowledge, but an estimate of the total income of all such women is not so fortunate.

As an illustration, suppose that the frame is a serialized list of N women aged 15 or over and that the sample is a simple random sample of n of these women, drawn with replacement by reading out n random numbers between 1 and N. Information on the n women is collected, and it is noted which ones belong to the specified subclass—that is, which ones live in a certain district, are gainfully employed, and have at least one child under 12 years old at home. Suppose that this number is n_s and that the average income of the n_s women is $\bar{x}_s$. Of course, n_s is a random variable with a binomial distribution.

What is the rel-variance of $\bar{x}_s$? Let C_s^2 be the rel-variance between incomes of the women in the frame that belong to the subclass. It is a fact that the conditional rel-variance of $\bar{x}_s$, for samples of size n_s of the specified subclass, will be C_s^2/n_s, just as if the women of this subclass had been set off beforehand into a separate stratum and a sample of size n_s had been drawn from it.

The conditional expected value of $\bar{x}_s$ over all samples of fixed size n_s in the subclass has moreover the convenient property of being the average income of all the women in the frame that belong to this subclass. It is for this reason that the conditional rel-variance of $\bar{x}_s$ is useful for assessing the precision of a sample at hand. For purposes of design, one uses the rel-variance of $\bar{x}_s$ over all samples of size n, which is $C_s^2 E(1/n_s)$, or very nearly $C_s^2[1 + Q/nP]/nP$, where P is the proportion of all women 15 or over that meet the specification of the subclass, and $P + Q = 1$.

In contrast, any estimator, X_s, of the aggregate income of all the women in the specified subclass will not have such convenient properties as $\bar{x}_s$. The conditional expectation of X_s, for samples of size n_s, is not equal to the aggregate income of all the women in the frame that belong to the subclass. The conditional rel-variance of X_s for a sample of size n_s at hand, although equal to the conditional rel-variance of $\bar{x}_s$, therefore requires careful interpretation. Instead of attempting to interpret the conditional rel-variance of X_s, one may elect to deal with the variance of X_s in all possible samples of size n. Thus, if X_s is set equal to $(N/n)n_s\bar{x}_s$ (here N/n is used as an expansion factor equal to the reciprocal of the probability of selection), it is a fact that the rel-variance of X_s over all samples of size n will be $(C_s^2 + Q)/nP$ (see Deming 1960, p. 129).

The problem with X_s arises from the assumption that N_s, the number of women in the frame that meet the specification of the subclass, is unknown. If N_s were known, one could form the estimator $X_s = N_s\bar{x}_s$, which would have all the desirable properties of $\bar{x}_s$.

One way to reduce the variance of the total income, X_s, of the specified class is (1) to select from the frame a large preliminary sample, (2) by an inexpensive investigation to classify the units of the preliminary sample into two classes, those that belong to the specified class and those that do not, (3) to investigate a sample of the units that fell into the specified class, to acquire information on income. The preliminary sample provides an estimate of N_s, and the final sample provides an estimate of $\bar{x}_s$. The product gives the estimate $X_s = N_s\bar{x}_s$ for the total income in the specified class. (For the variance of X_s and for optimum sizes of samples, see Hansen, Hurwitz, & Madow, 1953, vol. 1, pp. 65 and 259.)

If, further, N were not known and only the probability, π, of selection, to be applied to every sampling unit in the frame, were known, both n and n_s will be random variables, and there will be a further inflation of the rel-variance of any estimator of the aggregate income of all the women in the specified subclass. Thus, if X_s be set equal to $n_s\bar{x}_s/\pi$ for such an estimator, then the unconditional rel-variance of X_s will be $(C_s^2 + 1)/nP$. The conditional rel-variance of $\bar{x}_s$, however, is still C_s^2/n_s.

It may be noted that for a small subclass there is little difference between $C_s^2 + Q$ and $C_s^2 + 1$.

Examples are common. Thus, one might read out a two-digit random number for each line of a register, following the rule that the item listed on a line will be drawn into the sample if the random number is 01. If counts from outside sources are not at hand or are not used, then the rel-variance of an estimator, X_s, of the total number or total value of any subclass of items on the register contains the factor $C_s^2 + 1$.

Use of thinning digits. Reduction of the probability of selection of units of specified characteristics (such as items of low value) through the use of thinning digits may produce either the factor $C_s^2 + Q$ or the factor $C_s^2 + 1$ in the rel-variance of an estimator of an aggregate, depending on the mode of selecting the units.

Estimates of variances. Estimates of variances are supplied by the sample itself, under proper conditions, as was discussed above. Some of the more important estimators follow, denoted by a circumflex (ˆ). For the variance of $\bar{x}$,

$$(21) \qquad \hat{\sigma}_{\bar{x}}^2 = \left(\frac{1}{n} - \frac{1}{N}\right)\frac{1}{n-1}\sum(x_i - \bar{x})^2,$$

with a similar expression for $\hat{\sigma}_{\bar{y}}^2$. For the covariance,

$$(22)$$
$$\hat{\sigma}_{\bar{x}\bar{y}} = \left(\frac{1}{n} - \frac{1}{N}\right)\frac{1}{n-1}\sum(x_i - \bar{x})(y_i - \bar{y}).$$

Eqs. (21) and (22), with N infinite, were developed by Gauss (1823). These estimators are unbiased; $\sqrt{\hat{\sigma}_{\bar{x}}^2}$ is a slightly biased estimator of $\sigma_{\bar{x}}$, but the bias is negligible for n moderate or large.

Under almost all conditions met in practice, one may set

$$(23) \qquad t = \frac{\bar{x} - E\bar{x}}{\hat{\sigma}_{\bar{x}}}$$

and compare this quantity with tabulated values of t to find the margin of uncertainty in $\bar{x}$ for any specified probability. Such calculations give excellent approximations unless the distribution of sampling units in the frame is highly skewed. Extreme skewness may often be avoided by stratification (discussed below).

A useful approximate estimator for the rel-variance of $f = X/Y = \bar{x}/\bar{y}$ is

$$(24) \quad \hat{C}_f^2 \cong \left(\frac{1}{n} - \frac{1}{N}\right)\frac{1}{(n-1)\bar{x}^2}\sum(x_i - fy_i)^2.$$

This formula is derived by combination of eqs. (19), (21), and (22). In accordance with a previous remark, one may take $\hat{C}_{X'} = \hat{C}_f$, where X' is the ratio estimator of A as given by (17).

Size of frame usually not important. Because of the way in which N enters the variances, the size of the frame has little influence on the size of sample required for a prescribed precision, unless the sample is 20 per cent or more of the total frame.

For instance, the sample required to reach a specified precision would be the same for the continental United States as for the Boston Metropolitan Area, on the assumption that the underlying variances encountered are about the same for the entire United States as for Boston.

Special form for attributes (0,1 variate). In many studies a sampling unit gives only one of two possible observations, such as *yes* or *no*, *male* or *female*, *heads* or *tails*. The above equations then assume a simple form.

If each person in a frame is a sampling unit, and if $a_i = 1$ for *yes*, $a_i = 0$ for *no*, then the total *x*-population, *A*, in the frame is the total number of *yes* observations that would be recorded in the equal complete coverage, and *a* is the proportion *yes*, commonly denoted by *p*. The variance between the a_i in the frame is

$$(25) \qquad \sigma_a^2 = pq,$$

where $p + q = 1$.

The random variate, x_i, will take the value 0 or 1;

$$(26) \qquad x = \sum x_i$$

will be the number of *yes* observations in the sample, and

$$(27) \qquad \hat{p} = x/n$$

will be the proportion *yes* in the sample. Replacement of $\bar{x}$ by $\hat{p}$ in previous equations shows that $\hat{p}$ is an unbiased estimator of the proportion *yes* in the frame and that

$$(28) \qquad \sigma_{\hat{p}}^2 = \left(\frac{1}{n} - \frac{1}{N}\right) pq.$$

It is important to note that this variance is valid only if each sampling unit produces the value 0 or 1. It is not valid, for instance, for a sample of segments of area if there is more than one person per segment, or if the segments are clustered (as discussed below).

For an estimate of the variance of $\hat{p}$ (provided the sampling procedure meets the conditions stated) one may use

$$(29) \qquad \hat{\sigma}_{\hat{p}}^2 = \left(\frac{1}{n} - \frac{1}{N}\right) \hat{p}\hat{q},$$

where $\hat{p} + \hat{q} = 1$.

How good is an estimator of a variance? The variance of the estimator $\hat{\sigma}_{\bar{x}}^2$ in eq. (21) depends on the standardized fourth moment, β_2, of the frame and on the number of degrees of freedom for the estimator. Thus, if one defines

$$(30) \qquad \beta_2 = \frac{1}{N} \sum_{i=1}^{N} \left(\frac{a_i - a}{\sigma}\right)^4,$$

then the rel-variance of the estimator $\hat{\sigma}_{\bar{x}}^2$ of eq. (21) will be $(\beta_2 - 1)/n$, which diminishes with *n*.

Systematic selection. A simple and popular way to spread the sample over the frame is to select every *Z*th unit, with a random start between 1 and Z, where $Z = N/n$. This is called systematic sampling with a single random start, and it is one form of patterned sampling. In certain kinds of materials, specifically those in which nearby sampling units are, on the average, more similar than units separated by a longer interval, systematic sampling will be slightly more efficient than stratified random sampling (Cochran 1946).

A disadvantage of systematic sampling with a single random start is that there is no strictly valid way to make a statistical estimate of the standard error of a result so obtained. This is because the single start is equivalent to the selection of only one sampling unit from the Z possible sampling units that could be formed. One may nevertheless, under proper conditions, get a useful approximation to the rel-variance by using the sum of squares of successive pairs. Eq. (21) with $n = 2$ and $N = Z$ gives the estimator

$$(31) \qquad \hat{C}_{\bar{x}}^2 = \left(1 - \frac{2}{Z}\right) \frac{\sum(x_{i1} - x_{i2})^2}{[\sum(x_{i1} + x_{i2})]^2},$$

where the summation runs over all pairs.

Hidden and unsuspected periodicities often turn up, and in such cases the above formula may give a severe underestimate or overestimate of the variance. For example, every *n*th household might be nearly in phase with the natural periodicity of income, rent, size of family, and other characteristics associated with corners and with the configuration of dwelling units within areas and within apartment houses. Systematic sampling of physical elements or of time intervals can lead to disaster.

A statistician will therefore justify a single random start and use of eq. (31) only if he has had long experience with a body of material.

Instead of a single random start between 1 and N/n, one may take two random starts between 1 and $2N/n$ and every $(2N/n)$th sampling unit thereafter. Extension to multiple random starts is obvious. Two or more random starts give a valid estimate of the variance. Fresh random starts every six or eight zones will usually reap any possible advantage of systematic sampling and will avoid uncertainty in estimation of the variance.

Efficiency of design. The relative efficiency of two sampling procedures, I and II, that give normally distributed estimators of some characteristic are by definition the ratio of the inverses of the variances of these estimators for the same size, n, of sample. In symbols (E denotes efficiency),

$$(32) \qquad \frac{E_I}{E_{II}} = \frac{\sigma_{II}^2}{\sigma_I^2}, \qquad n_I = n_{II}.$$

This concept of efficiency is due to Fisher (1922).

Comparison of costs is usually more important than comparison of numbers of cases. Let the costs be c_I and c_{II} for equal variances. Then

$$(33) \qquad \frac{E_I}{E_{II}} = \frac{c_{II}}{c_I}, \qquad \sigma_I = \sigma_{II}.$$

Comparison of efficiencies of estimators whose distributions depart appreciably from normality require special consideration.

Sampling moving populations. A possible procedure in sampling moving populations is to count and tag all the people visible from a number of enumerators' posts through a period of a day or a week (the first round) and then to repeat the count from the same or different posts some time later (the second round). The n_1 people counted and tagged in the first round constitute a mobile frame for the second round. If the number of people counted in the first round is n_1, and if the number counted in the second round is n_2, with an intersection of n_{12} for people counted in both rounds, then an estimator of the total number of mobile inhabitants in the whole area is $\hat{N} = n_1 n_2 / n_{12}$ (Yates 1949, p. 43; Deming & Keyfitz 1967).

More complex designs

Considerations of cost—clustering. The total cost of a survey includes cost of preparing the frames and cost of travel to the units selected. In some surveys it may be possible to get more information per unit cost by enlarging the sampling unit, a procedure commonly called *clustering*. One may, for example, define a sampling unit as comprising all the dwellings in a compact segment of area. Further, one may, with experience and care, subsample dwelling units from a selected cluster or select one member of a family where two or more members qualify for the universe. Again, in a national survey, one may restrict the sample to a certain number of counties that will come into the sample by a random process. Or, in a survey of a city, one may restrict the sample to a random selection of blocks.

Any such plan reduces the interviewer's expenses for travel and reduces the cost of preparing the frame. However, restriction of the sample usually also increases variances, unless the total number of households in the sample be increased as compensation. It should be remembered, though, that the actual precision obtained by the use of cluster sampling may be nearly as good as that obtained by an unrestricted random selection of the same number of dwelling units with no clustering.

Theory indicates the optimum balance between enlargement of the sampling unit and the number of sampling units to include in the sample. Obviously, the theory is more complex than that discussed in the last section. Stratification, ratio estimators, and regression estimators are additional techniques that, under certain conditions, yield further increases in efficiency (see below).

An example. The following illustration refers to a sample of a city: (i) Suppose that it has been determined in advance that for the main purposes of a survey the optimum size of areal unit is a compact group of five dwelling units, called a segment. (ii) A sampling unit within the city will consist of $\bar{n}$ segments from a larger number of segments contained in a block. The $\bar{n}$ segments of a sampling unit (if $\bar{n} > 1$) should be scattered over the block. A good way to effect this scatter is by a systematic selection. (iii) The m sampling units in the city will be selected by random numbers. For simplicity, assume that all blocks in the city contain an equal number, $\bar{N}$, of segments. Suppose that there are

> M blocks in the whole city,
> $\bar{N}$ segments in a block,
> $N = M\bar{N}$ segments in the whole city,
> $\bar{n}$ segments in a sampling unit,
> $\bar{N}/\bar{n}$ sampling units in a block,
> $M\bar{N}/\bar{n}$ or $N/\bar{n}$ sampling units in the whole city,
> m sampling units in the sample.

Then if

$$(34) \qquad \bar{x} = x/m\bar{n},$$

one may take

$$(35) \qquad X = N\bar{x}$$

for an estimator of the x-population in the whole city. For this estimator,

$$(36) \qquad C_X = C_{\bar{x}},$$

and

$$(37) \qquad \operatorname{var} \bar{x} = \frac{N - m\bar{n}}{N - \bar{n}} \left(\frac{\sigma_b^2}{m} + \frac{\bar{N} - \bar{n}}{\bar{N} - 1} \frac{\sigma_w^2}{m\bar{n}} \right).$$

If m is small compared with M,

$$(38) \qquad \operatorname{var} \bar{x} \cong \frac{\sigma_b^2}{m} + \frac{\bar{N} - \bar{n}}{\bar{N} - 1} \frac{\sigma_w^2}{m\bar{n}}.$$

If, also, $\bar{n}$ is small compared with $\bar{N}$,

$$(39) \qquad \operatorname{var} \bar{x} \cong \frac{\sigma_b^2}{m} + \frac{\sigma_w^2}{m\bar{n}}.$$

Here σ_b^2 is the variance between blocks of the mean x-population per sampling unit, and σ_w^2 is the average variance between sampling units within blocks.

Important principle in size of secondary unit. Suppose that the cost of adding one more block to the sample is c_1 (cost of maps, preparation, delineation of segments, travel) and that the cost of an interview in an additional sampling unit is c_2. Then the total cost of the survey will be

$$(40) \qquad K = mc_1 + m\bar{n}c_2.$$

In eq. (37) var $\bar{x}$ will be at its minimum for a fixed cost K if

$$(41) \qquad \bar{n} = \frac{\sigma_w}{\sigma_b} \sqrt{\frac{c_1}{c_2}}, \qquad \text{optimum } \bar{n}.$$

This equation was derived by both L. H. C. Tippett and Shewhart, independently, in 1931.

Note that m does not appear in this equation. That is, the optimum value of $\bar{n}$ on the basis of the cost function (40) is independent of m, the number of sampling units in the sample (and very nearly independent of the number of blocks in the sample).

The optimum m is found by substituting the optimum $\bar{n}$ from eq. (41) into eq. (40) and solving for m. Of course, it is necessary to assume values for σ_w/σ_b and for $\sqrt{c_1/c_2}$ to do this. (Because each sampling unit will usually fall in a different block, m will usually be exactly or nearly as large as the number of blocks in the sample.) Usual numerical values of $\sigma_w : \sigma_b$ and of $c_1 : c_2$ lead to small values of $\bar{n}$ and to large values of m. Efficient design therefore usually requires a small sample from a block and dispersion of the sample into a large number of blocks.

Extension of this theory to a national sample, and to stratified designs and ratio estimators, leads to the same principle.

Variation in size of segment will increase var $\bar{x}$ by the factor $1 + C_u^2/n$, where C_u^2 is the rel-variance of the distribution of the number of dwelling units per segment. A similar factor, $1 + C_{N_i}^2/m$, measures the increase in var $\bar{x}$ from variation in the number of segments per block.

Replicated designs for ease in estimation of variance. Replication of a sample in two or more interpenetrating networks of samples will provide a basis for rapid calculation of a valid estimate of the standard error of any result, regardless of the complexity of the procedure of selection and of the formulas for the formation of estimates [Mahalanobis 1944; Deming 1950; 1960; *see also* INDEX NUMBERS, *article on* SAMPLING].

Stratified sampling

The primary aim of stratified sampling is to increase the amount of information per unit of cost. A further aim may be to obtain adequate information about certain strata of special interest.

One way to carry out stratification is to rearrange the sampling units in the frame so as to separate them into classes, or strata, and then to draw sampling units from each class. The goal should be to make each stratum as homogeneous as possible, within limitations of time and cost. Stratification is equivalent to blocking in the design of an experiment. It is often a good plan (*i*) to draw a preliminary sample from the frame without stratification; (*ii*) to classify into strata the units in the preliminary sample; and (*iii*) to draw, for the final sample, a prescribed number of sampling units from each stratum so formed. Step (*i*) will sometimes require an inexpensive investigation or test of every sampling unit in the preliminary sample to determine which stratum it belongs to.

Stratification is one way to make use of existing information concerning the frame other than the information obtained from investigating the sampling units in the final sample itself. Other ways to use existing information are through ratio estimators and regression estimators (see below).

In practice a frame is to some extent naturally stratified to begin with. Thus, areas in geographic order usually are already pretty well stratified in respect to income, occupation, density of population, tastes of the consumer, and other characteristics. No frame arrives thoroughly mixed, and any plan of sampling should be applied by zones, so as to capture the natural stratification. Theory serves as a guide to determine whether further stratification would be profitable.

Plans of stratification for enumerative studies. Several plans of stratified sampling for enumerative studies will now be described.

The notation and definitions to be used in this discussion are given in tables 2 and 3. (Note that $\bar{N}$ and $\bar{n}$ are defined differently here than they were earlier.) These tables are presented in terms of

Table 2 — Notation and definitions for the frame (M = 2 strata)

STRATUM	NUMBER OF SAMPLING UNITS In the frame	NUMBER OF SAMPLING UNITS In the sample	STRATUM'S PROPORTION OF SAMPLING UNITS IN THE FRAME	POPULATION Average per sampling unit in the stratum	POPULATION Total in the stratum	BETWEEN THE POPULATIONS OF THE SAMPLING UNITS WITHIN THE STRATUM Standard deviation	BETWEEN THE POPULATIONS OF THE SAMPLING UNITS WITHIN THE STRATUM Variance
1	N_1	n_1	$P_1 = \dfrac{N_1}{N}$	a_1	$A_1 = N_1 a_1$	σ_1	σ_1^2
2	N_2	n_2	$P_2 = \dfrac{N_2}{N}$	a_2	$A_2 = N_2 a_2$	σ_2	σ_2^2
Total for the frame	N	n	1	—	A	—	—
Unweighted average per stratum	$\bar{N} = \dfrac{N}{M}$	$\bar{n} = \dfrac{n}{M}$	$\dfrac{1}{M}$	—	$\bar{A} = \dfrac{A}{M}$	—	—
Weighted average per sampling unit	—	—	—	$a = \dfrac{A}{N}$	—	$\bar{\sigma}_w$	σ_w^2

Source: Deming 1960, p. 286.

two strata ($M = 2$), but extension to a greater number of strata follows obviously. The following additional definitions are needed:

$$(42) \qquad \sigma_R^2 = Q_1 \sigma_1^2 + Q_2 \sigma_2^2 + Q_3 \sigma_3^2,$$

the average reverse variance between sampling units within strata, and

$$(43) \qquad \bar{\sigma}_R = Q_1 \sigma_1 + Q_2 \sigma_2 + Q_3 \sigma_3,$$

the average reverse standard deviation between sampling units within strata, where $Q_i + P_i = 1$.

Plan A (no stratification): The scheme of sampling described above will be designated plan A. It is needed here for comparison, and also because it constitutes the basis for selection from any stratum.

Note that in plan A, as in plans B, D, F, and H, below, all the sampling units in the frame have equal probability of selection, namely n/N, wherefore $E\bar{x} = a$ and $EX = A$.

P_i known—whole frame classified. Two sampling plans for which the proportions in each stratum are known (or ascertainable) and the the whole frame is classified will now be described.

Plan B (proportionate sampling): Decide with the help of eq. (47) the size, n, of the sample required.

Compute next

$$(44) \qquad n_i = nN_i/N = nP_i.$$

Draw by random numbers, as in plan A, a sample of size n_i from stratum i. Investigate every member of the sample, and calculate

$$(45) \qquad X_1 = N_1 x_1/n_1, \qquad X_2 = N_2 x_2/n_2, \qquad \text{etc.}$$

(For simplicity, most formulas will henceforth be written for two strata, in conformance with tables 2 and 3. Extension to more strata is obvious.) Here, $n_i/N_i = n/N$, wherefore

$$(46) \qquad \begin{aligned} X &= X_1 + X_2 \\ &= \frac{N}{n}(x_1 + x_2) \\ &= N\frac{x}{n} = N\bar{x} \end{aligned}$$

and

$$(47) \qquad \operatorname{var} \bar{x} = \left(\frac{1}{n} - \frac{1}{N}\right)\sigma_w^2, \qquad \text{plan B.}$$

The n_i of eq. (44) and later expressions will not in general be integers. In practice one uses the closest integer; the effects on variance formulas are usually completely negligible.

Table 3 — Notation and definitions for the sample

Stratum	Population in the sample	Mean population per sampling unit	Estimated total population	Variance of this estimator*
1	x_1 x-population in stratum 1	$\bar{x}_1 = \dfrac{x_1}{n_1}$	$X_1 = N_1 \dfrac{x_1}{n_1}$	$\operatorname{var} X_1$
2	x_2 x-population in stratum 2	$\bar{x}_2 = \dfrac{x_2}{n_2}$	$X_2 = N_2 \dfrac{x_2}{n_2}$	$\operatorname{var} X_2$
Sum	x	—	X	$\operatorname{var} X$

* The variances are additive only if the N_i (or P_i) are known and used in the estimator X.

Source: Deming 1960, p. 287.

Plan C (Neyman sampling): Decide with the help of eq. (49) the size, n, of the sample required. Compute next the Neyman allocation (Neyman 1934),

$$(48) \qquad n_i = nP_i\sigma_i/\bar{\sigma}_w.$$

Draw by random numbers, as in plan A, a sample of size n_i from stratum i. Investigate every member of the sample. Form estimators X_1, X_2, and $X = X_1 + X_2$. Form $\bar{x} = X/N$ for an unbiased estimator of a. Here

$$(49) \qquad \text{var } \bar{x} = \frac{(\bar{\sigma}_w)^2}{n} - \frac{\sigma_w^2}{N}, \qquad \text{plan C.}$$

The Neyman allocations are the optimal n_i for minimizing var $\bar{x}$ when the P_i are known.

P_i known—only a sample classified. One may, in appropriate circumstances, require only the classification of a preliminary sample drawn from the frame. The decision hinges on the costs of classification and the expected variances of the plans under consideration.

Plan D: Decide with the help of eq. (50) the size, n, of the sample required. Draw the sample as in plan A. Classify the sampling units into strata. The number, n_i, of sampling units drawn from stratum i will be a random variable. Carry out the investigation of every unit of the sample. Form X_1, X_2, X, and $\bar{x}$ as in plan B. Then

$$
\begin{array}{c}
\text{plan B} \\
\overbrace{\phantom{\left(\frac{1}{n} - \frac{1}{N}\right)}}
\end{array}
$$
$$(50) \quad \text{var } \bar{x} = \left(\frac{1}{n} - \frac{1}{N}\right)\left(\sigma_w^2 + \frac{1}{n}\sigma_R^2\right), \quad \text{plan D.}$$

Plan E: Decide with the help of eq. (52) the size, n, of the final sample. Draw by random numbers a preliminary sample of size n'. Thin (reduce) by random numbers the strata of the preliminary sample to reach the Neyman ratios

$$(51) \qquad \frac{n_1}{n_1'} : \frac{n_2}{n_2'} : \cdots = \sigma_1 : \sigma_2 : \cdots$$

and simultaneously the total sample, n. Here n_1', n_2', etc., are the sizes of the preliminary sample in the several strata, and n_1, n_2, etc., are the sizes of the final sample. For greatest economy, choose n' so that one stratum will require no thinning. Carry out the investigation of every unit of the final sample. Form the estimators X_1, X_2, and $X = X_1 + X_2$. Then $\bar{x} = X/N$ will again be an unbiased estimator

of a, but now

$$
\begin{array}{c}
\text{plan C} \\
\end{array}
$$
$$(52) \quad \text{var } \bar{x} = \overbrace{\frac{(\bar{\sigma}_w)^2}{n} - \frac{\sigma_w^2}{N}}^{} + \frac{1}{n}\left(\frac{1}{n'} - \frac{1}{N}\right)\bar{\sigma}_w\bar{\sigma}_R
$$
$$
\hspace{4cm} \text{plan E}
$$
$$
\cong \frac{1}{n}\left[(\bar{\sigma}_w)^2 + \frac{1}{n'}\bar{\sigma}_w\bar{\sigma}_R\right],
$$

the latter form useful if N is large relative to n'.

Sequential classification of units into strata. We now describe two plans in which the sample-sizes, n_i, are reached sequentially, with considerable saving under appropriate conditions.

Plan F: Determine the desired sample-sizes, n_i, as in plan B. Draw by random numbers one unit at a time from the frame, and classify it into its proper stratum. Continue until the quotas, n_i, are all filled. Form X as in plan B; var $\bar{x}$ will be the same as for plan B.

Plan G: This is the same as plan F except that the sample sizes, n_i, are fixed as in plan C. Form X as in plan C; var $\bar{x}$ will be the same as for plan C.

P_i not known in advance. When the proportions, P_i, in the frame are unknown, estimates thereof must come from a sample, usually a preliminary sample of size $N' > n$, where n is the size of the final sample.

Plan H: Decide with the help of eq. (55) the size, n, for the final sample. Compute the optimum size, N', of the preliminary sample by the formula

$$(53) \qquad \frac{n}{N'} = \frac{\sigma_w}{\sigma_b}\sqrt{\frac{c_1}{c_2}},$$

where c_1 is the average cost of classifying a sampling unit in the preliminary sample, and c_2 is the average cost of the final investigation of one sampling unit.

The procedure is to draw as in plan A a preliminary sample of size N' and to classify it into strata. Treat the preliminary sample as a frame of size N'. Then thin all strata of the preliminary sample proportionately to reach the final total size, n. Carry out the investigation of every sampling unit in the final sample. An unbiased estimator of a is

$$(54) \qquad \bar{x} = \frac{x}{n},$$

where x is the total x-population in the sample. Then

$$(55) \qquad \text{var } \bar{x} \cong \frac{\sigma_w^2}{n} + \frac{\sigma_b^2}{N'} = \frac{\sigma^2}{N'} + \left(\frac{1}{n} - \frac{1}{N'}\right)\sigma_w^2,$$
$$
\hspace{5cm} \text{plan H,}
$$

is an excellent approximation if N be large relative to N'.

Plan I: Decide with the help of eq. (59) the size, n, for the final sample. Compute the optimum size, N', of the preliminary sample, using the equation (Neyman 1938)

$$(56) \qquad \frac{n}{N'} = \frac{\bar{\sigma}_w}{\sigma_b}\sqrt{\frac{c_1}{c_2}}.$$

Draw as in plan A a preliminary sample of size N'. Classify it as in plan H. Thin the strata differentially to satisfy the Neyman ratios

$$(57) \qquad \frac{n_1}{N_1'} : \frac{n_2}{N_2'} : \cdots = \sigma_1 : \sigma_2 : \cdots$$

and to reach the desired final total sample-size, n. Carry out the investigation of every sampling unit in the final sample. An unbiased estimator of a is

$$(58) \qquad \bar{x} = \frac{1}{N'}\left(\frac{N_1'}{n_1}\,x_1 + \frac{N_2'}{n_2}\,x_2\right),$$

for which

$$(59) \qquad \operatorname{var}\bar{x} \cong \frac{(\bar{\sigma}_w)^2}{n} + \frac{\sigma_b^2}{N'}, \qquad \text{plan I,}$$

is an excellent approximation if N be large relative to N' and to n.

One may use plan F or plan G in combination with plan H or plan I to reap the benefit of many strata without actually classifying the entire preliminary sample, N' (Koller 1960).

Gains of stratified sampling. Gains of stratified sampling can be evaluated by comparing variances. Denote by A, B, and C the variances of the estimators of a calculated by the plans A, B, and C. Then

$$(60)$$
$$\frac{A-B}{A} = \frac{\sigma^2 - \sigma_w^2}{\sigma^2} = \frac{\sigma_b^2}{\sigma^2} = \sum_{i<j} P_iP_j\left(\frac{a_j - a_i}{\sigma}\right)^2,$$

$$(61)$$
$$\frac{B-C}{B} \cong \frac{\sigma_w^2 - (\bar{\sigma}_w)^2}{\sigma_w^2} = \sum_{i<j} P_iP_j\left(\frac{\sigma_j - \sigma_i}{\sigma_w}\right)^2.$$

For example, if $P_1 = .6$, $P_2 = .4$, and $\sigma_w^2 = .8\sigma^2$, $(A-B)/A$ would be $(1-.8)/1 = .2$, meaning that 100 interviews selected according to plan B would give rise to the same variance as 125 selected according to plan A.

The gains of plans F and G over plan A are the same as the gains of plans B and C over plan A. The average gains in repeated trials of plans D and E are less. If σ_R^2 and $\bar{\sigma}_w\bar{\sigma}_R$ are large, plans D and E will usually not be good choices. For large samples, however, in circumstances where σ_R^2 and $\bar{\sigma}_w\bar{\sigma}_R$ are not large, the gains of plans D and E may be almost equal to the gains of plans B and C, at considerably less cost.

Eqs. (60) and (61) show that the gain to be expected from the proposed formation of a new stratum, i, will not be impressive unless its proportion, P_i, be appreciable, or unless its σ_i or its a_i be widely divergent from the average.

Stratification to estimate over-all ratio. The case to be used for illustrating stratified sampling to estimate an over-all ratio consists of three strata: stratum 1 for large units (for example, high incomes or large farms), stratum 2 for medium-sized units, and stratum 3 for small units. Here stratum 1 is to be covered 100 per cent; obvious modifications take care of the case in which stratum 1 is not sampled completely.

First take as an estimator of ϕ

$$(62) \qquad f = \frac{X}{Y} = \frac{A_1 + N_2\bar{x}_2 + N_3\bar{x}_3}{B_1 + N_2\bar{y}_2 + N_3\bar{y}_3},$$

in the notation of tables 1 and 2, with B_i as the value of the y-characteristic in stratum i of the frame. Optimum allocation to strata 2 and 3 is very nearly reached if both

$$(63a) \qquad n_2 = \frac{B_2s_2}{B_2s_2 + B_3s_3}$$

and

$$(63b) \qquad n_3 = \frac{B_3s_3}{B_2s_2 + B_3s_3},$$

wherein s_2 and s_3 are the standard deviations of the ratio of x to y in strata 2 and 3.

If, as is often the case, s_2 and s_3 do not differ much, or if little is known about them in advance, one can still make an important gain in efficiency by setting $n_2 : n_3 = B_2 : B_3$ or $n_2 : n_3 = A_2 : A_3$.

Another estimator of the ratio ϕ is

$$(64) \qquad f = P_1f_1 + P_2f_2 + P_3f_3,$$

wherein $P_i = B_i/B$ and $f_i = \bar{x}_i/\bar{y}_i$. This estimator is sometimes preferred when f_i varies greatly from stratum to stratum, and when there can be no trouble with small denominators. The allocation of sample for this estimator is, for practical purposes, the same as in eq. (63) (Cochran [1953] 1963, p. 175; Hansen et al. 1953, vol. 1, p. 209).

Sequential adjustment of size of sample. It is sometimes possible, when decision on the size of sample is difficult, or when time is short, to break the sample in advance into two portions, 1 and 2, each being a valid sample of the whole frame. Portion 1 is definitely to be carried through to completion, but portion 2 will be used only if required. This may be called a two-stage sequential method. It is practicable where the investigation is to be carried out by a small number of experts that will stay on the job as long as necessary but not where a field force must be engaged in advance for a definite period.

Modifications for differing costs. If investigating a sampling unit in a particular stratum is three or more times as costly as the average investigation, it may be wise to decrease the sample in the costly stratum and to build up the sample in other strata (Deming 1960, p. 303).

Considerations for planning. In order to plan a stratified sample, certain assumptions are necessary. Fair approximations to the relevant ratios, such as $\sigma_w : \sigma$, $\bar{\sigma}_w : \sigma$, $\bar{\sigma}_w : \sigma_b$, $\sqrt{c_1 : c_2}$, will provide excellent allocation. On the other hand, bad approximations to these ratios, or failure to use theory at all, can lead to serious losses.

The required good approximations to these ratios may come from prior experience, or from probing the knowledge of experts in the subject matter. For example, the distribution of intelligence quotients in the stratum between 90 and 110, if rectangular, would provide $\sigma^2 = (110 - 90)^2/12$, or 33, whence $\sigma = 5.7$. Other shapes have other variances, but shape is fortunately not critical (Deming 1950, p. 262; 1960, p. 260). A stratum with very high values should be set off for special treatment and possibly sampled 100 per cent.

Stratification for analytic studies. As mentioned earlier, the aim in an analytic study is to detect differences between classes or to measure the effects of different treatments.

The general formula for the variance of the difference between two means, $\bar{x}_A$ and $\bar{x}_B$, derived from independent samples of sizes n_A and n_B drawn by random numbers singly and without stratification from, for example, two groups of patients, A and B, is

$$(65) \qquad \text{var} \, (\bar{x}_A - \bar{x}_B) = \sigma_A^2/n_A + \sigma_B^2/n_B,$$

wherein σ_A^2 and σ_B^2 are the respective variances between the patients within the two groups.

For such analytic studies the optimum allocation of skill and effort is found by setting

$$(66) \qquad \frac{n_A}{n_B} = \frac{\sigma_A}{\sigma_B} \sqrt{\frac{c_B}{c_A}},$$

wherein c_A and c_B are the costs per case. Note that the sizes of the groups do not enter into this formula and that it is different from the optimum allocation in enumerative problems.

In many analytic studies σ_A and σ_B will be about equal, and so will the costs c_A and c_B. In such circumstances, the best allocation is

$$(67) \qquad n_A = n_B.$$

Regression estimators

We have already seen reduction in variance resulting from use of prior or supplementary knowledge concerning the frame. Use of prior knowledge of N to form the estimator $X = N\bar{x}$ is an instance. Prior knowledge of B to form the ratio estimator is another instance. This section, on *regression estimators*, describes other ways to use prior or supplementary knowledge concerning the frame. Regression estimators include the simple estimator, $\bar{x}$, and the ratio estimator, fb, as special cases, but they also include many other estimators, some of them highly useful. Like the ratio estimator of a total, these additional estimators are applicable only if independent and fairly reliable information is available about the y-population in the frame. Any estimator that takes advantage of supplementary information may have considerable advantage over the simple estimator, $\bar{x}$, if the correlation, ρ, between x_i and y_i is high, but this condition is not in itself sufficient.

Specific forms of regression estimators. Assume simple random selection and write the regression estimator in the form

$$(68) \qquad \bar{x}_i = \bar{x} + m_i(b - \bar{y}),$$

wherein b is known independently from some source such as a census. The subscript i on $\bar{x}_i$ here differentiates the several specific forms of regression estimators.

Regression estimators are closely allied with the analysis of covariance [*see* LINEAR HYPOTHESES, *article on* ANALYSIS OF VARIANCE]. The four cases to be considered here are taken largely from Hansen, Hurwitz, and Madow (1953).

Simple estimator. If m_i is taken as zero, the regression estimator obtained is $\bar{x}_1 = \bar{x}$, seen earlier. This procedure makes no use of supplemental information. Under the assumption that N is large relative to n, the variance of this estimator is

$$(69a) \qquad \text{var} \, \bar{x}_1 = \sigma_{\bar{x}}^2;$$

likewise,

$$(69b) \qquad \text{var} \, \bar{y}_1 = \sigma_{\bar{y}}^2.$$

Difference estimator. The estimator $\bar{x}_2$, often called the difference estimator, is practicable if prior knowledge (such as prior surveys of a related type) provides a rough approximation to the regression coefficient $\beta = \rho\sigma_{\bar{x}}/\sigma_{\bar{y}}$. This estimator is

$$(70) \qquad \bar{x}_2 = \bar{x} + m_2(b - \bar{y}),$$

where m_2 is any approximate slope not derived from the sample under consideration. The variance of $\bar{x}_2$ is

$$
\begin{aligned}
(71) \quad \operatorname{var} \bar{x}_2 &= \sigma_{\bar{x}}^2(1-\rho^2) + \sigma_{\bar{y}}^2(m_2-\beta)^2 \\
&= \sigma_{\bar{x}}^2(1-\rho^2+\rho^2 e^2),
\end{aligned}
$$

where $\beta = \rho\sigma_{\bar{x}}/\sigma_{\bar{y}}$ and $e = (m_2-\beta)/\beta$. (Note that $\rho e = (\sigma_{\bar{y}}/\sigma_{\bar{x}})(m_2-\beta)$ even if $\rho = 0$.)

Least squares regression estimator. If m_i is chosen as

$$
(72) \qquad m_3 = \frac{\sigma_{\bar{x}\bar{y}}}{\sigma_{\bar{y}}^2} = \rho\,\frac{\sigma_{\bar{x}}}{\sigma_{\bar{y}}},
$$

then the equation

$$
(73) \qquad \bar{x}_3 = \bar{x} + m_3(b-\bar{y})
$$

gives the so-called least squares regression estimator. The variance of this estimator is

$$
(74) \qquad \operatorname{var} \bar{x}_3 = \sigma_{\bar{x}}^2(1-\rho^2) + R.
$$

Here R is a remainder in the Taylor series involving $1/n^2$ and higher powers; this remainder will be negligible if n is large.

Ratio estimator. If m_i is chosen as

$$
(75) \qquad m_4 = \bar{x}/\bar{y} = f,
$$

the ratio estimator is

$$
(76) \qquad \bar{x}_4 = fb.
$$

The variance of $\bar{x}_4$ is

$$
(77) \qquad \operatorname{var} \bar{x}_4 = \sigma_{\bar{x}}^2(1 + C_{\bar{y}}^2/C_{\bar{x}}^2 - 2\rho C_{\bar{y}}/C_{\bar{x}}) + R',
$$

R' being another remainder. For large n and for $C_{\bar{x}} \cong C_{\bar{y}}$,

$$
(78) \qquad \operatorname{var} \bar{x}_4 \cong 2\sigma_{\bar{x}}^2(1-\rho).
$$

It follows that for large n and for $m_2 \cong \beta$ and $C_{\bar{x}} \cong C_{\bar{y}}$,

$$
(79) \qquad \frac{\operatorname{var} \bar{x}_4}{\operatorname{var} \bar{x}_2} \cong \frac{2}{1+\rho}.
$$

Comparison of regression estimators. If the correlation, ρ, between x_i and y_i is moderate or high, but the line of regression of x on y misses the origin by a wide margin, then the estimator $\bar{x}_3$ will show substantial advantages over $\bar{x}_4$ and $\bar{x}_1$. If the y-variate shows relatively wide spread (that is, if $C_{\bar{y}}$ is much greater than $C_{\bar{x}}$), the ratio estimator $\bar{x}_4$ may be far less precise than the simple estimator

$\bar{x}_1 = \bar{x}$, even when ρ is high, especially if the line of regression misses the origin by a wide margin. On the other hand, if the line of regression passes through the origin ($\rho C_{\bar{x}} = C_{\bar{y}}$), or nearly through it, $\bar{x}_3$ and $\bar{x}_4$ will have about the same variance, but $\bar{x}_4$ may be much easier to compute.

Estimator of b subject to sampling error. It often happens that the y-population per sampling unit is not known with the reliability of a census but comes instead from another and bigger sample. This circumstance introduces additional terms into the variances. Let n be the size of the present sample and n' the size of the sample that provides the estimate of b. We suppose that the variance of this estimate of b is $n\sigma_{\bar{y}}^2/n'$. The resulting variances of the regression estimators are shown in Table 4.

W. EDWARDS DEMING

BIBLIOGRAPHY

BOWLEY, ARTHUR L. (1901) 1937 *Elements of Statistics.* 6th ed. New York: Scribner; London: King.

CHEVRY, GABRIEL 1949 Control of a General Census by Means of an Area Sampling Method. *Journal of the American Statistical Association* 44:373–379.

COCHRAN, WILLIAM G. 1946 Relative Accuracy of Systematic and Stratified Random Samples for a Certain Class of Populations. *Annals of Mathematical Statistics* 17:164–177.

COCHRAN, WILLIAM G. (1953) 1963 *Sampling Techniques.* 2d ed. New York: Wiley.

DALENIUS, TORE 1962 Recent Advances in Sample 'Survey Theory and Methods. *Annals of Mathematical Statistics* 33:325–349.

DEMING, W. EDWARDS (1943) 1964 *Statistical Adjustment of Data.* New York: Dover.

DEMING, W. EDWARDS 1950 *Some Theory of Sampling.* New York: Wiley.

DEMING, W. EDWARDS 1960 *Sample Design in Business Research.* New York: Wiley.

DEMING, W. EDWARDS; and KEYFITZ, NATHAN 1965 Theory of Surveys to Estimate Total Population. Volume 3, pages 141–144 in World Population Conference, Belgrade, August 30–September 10, 1965, *Proceedings.* New York: United Nations.

FISHER, R. A. (1922) 1950 On the Mathematical Foundations of Theoretical Statistics. Pages 10.307a–10.368 in R. A. Fisher, *Contributions to Mathematical Statistics.* New York: Wiley. → First published in Volume 222 of the *Philosophical Transactions,* Series A, of the Royal Society of London.

Table 4 — Rel-variances when estimator of b is subject to sampling error

Estimator	Case I: sample of size n is drawn as a subsample of n'	Case II: samples of size n and n' are independent
$\bar{x}_1$	$C_{\bar{x}}^2$	Same as in Case I
$\bar{x}_2$	$C_{\bar{x}}^2[1 - \rho^2(1-e^2)(1-n/n')]$	$C_{\bar{x}}^2[1 - \rho^2(1-e^2) + \rho^2(1+e)^2 n/n']$
$\bar{x}_3$	$C_{\bar{x}}^2[1 - \rho^2(1-n/n')]$	Same as in Case I
$\bar{x}_4$	$C_{\bar{x}}^2 - (2\rho C_{\bar{x}}C_{\bar{y}} - C_{\bar{y}}^2)(1-n/n')$	$C_{\bar{x}}^2 - (2\rho C_{\bar{x}}C_{\bar{y}} - C_{\bar{y}}^2)(1-n/n') + (2C_{\bar{y}}n/n')(C_{\bar{y}} - \rho C_{\bar{x}})$

○FISHER, R. A. (1956) 1973 *Statistical Methods and Scientific Inference.* 3d ed., rev. & enl. New York: Hafner; Edinburgh: Oliver & Boyd.

GAUSS, CARL FRIEDRICH 1823 *Theoria combinationis observationum erroribus minimis obnoxiae.* Göttingen (Germany): Dieterich. → A French translation was published in Gauss' *Méthode des moindres carrés* (1855). An English translation of the French was prepared as *Gauss's Work (1803–1826) on the Theory of Least Squares,* by Hale F. Trotter; Statistical Techniques Research Group, Technical Report, No. 5, Princeton Univ., 1957.

HANSEN, MORRIS H.; and HURWITZ, WILLIAM N. 1943 On the Theory of Sampling From Finite Populations. *Annals of Mathematical Statistics* 14:333–362.

HANSEN, MORRIS H.; and HURWITZ, WILLIAM N. 1946 The Problem of Non-response in Sample-surveys. *Journal of the American Statistical Association* 41: 517–529.

HANSEN, MORRIS H.; HURWITZ, WILLIAM N.; and MADOW, WILLIAM G. 1953 *Sample Survey Methods and Theory.* 2 vols. New York: Wiley.

KISH, LESLIE 1965 *Survey Sampling.* New York: Wiley. → A list of errata is available from the author.

KOLLER, SIEGFRIED 1960 *Aussenhandelsstatistik: Untersuchungen zur Anwendung des Stichprobenverfahrens.* Pages 361–370 in Germany (Federal Republic), Statistisches Bundesamt, *Stichproben in der amtlichen Statistik.* Stuttgart (Germany): Kohlhammer.

LEVEN, MAURICE 1932 *The Income of Physicians: An Economic and Statistical Analysis.* Univ. of Chicago Press.

MAHALANOBIS, P. C. 1944 On Large-scale Sample Surveys. Royal Society of London, *Philosophical Transactions* Series B 231:329–451.

MAHALANOBIS, P. C. 1946 Recent Experiments in Statistical Sampling in the Indian Statistical Institute. *Journal of the Royal Statistical Society* Series A 109: 326–378. → Contains eight pages of discussion.

MOSER, C. A. 1949 The Use of Sampling in Great Britain. *Journal of the American Statistical Association* 44:231–259.

NEYMAN, JERZY 1934 On the Two Different Aspects of the Representative Method: The Method of Stratified Sampling and the Method of Purposive Selection. *Journal of the Royal Statistical Society* Series A 97:558–606.

NEYMAN, JERZY 1938 Contribution to the Theory of Sampling Human Populations. *Journal of the American Statistical Association* 33:101–116. → See especially equation 40 on page 110.

QUENOUILLE, M. H. 1959 *Rapid Statistical Calculations.* London: Griffin; New York: Hafner. → Pages 5–7 show estimators of the standard deviation by use of the range.

SATTERTHWAITE, F. E. 1946 An Approximate Distribution of Estimates of Variance Components. *Biometrics* 2:110–114.

SHEWHART, WALTER A. 1939 *Statistical Method From the Viewpoint of Quality Control.* Washington: U.S. Department of Agriculture, Graduate School.

STEPHAN, FREDERICK F. 1936 Practical Problems of Sampling Procedures. *American Sociological Review* 1:569–580.

STEPHAN, FREDERICK F. 1948 History of the Uses of Modern Sampling Procedures. *Journal of the American Statistical Association* 43:12–39.

STEPHAN, FREDERICK F.; and MCCARTHY, PHILIP J. 1958 *Sampling Opinions: An Analysis of Survey Procedure.* New York: Wiley.

STUART, ALAN 1962 *Basic Ideas of Scientific Sampling.* London: Griffin; New York: Hafner.

SYMPOSIUM ON CONTRIBUTIONS OF GENETICS TO EPIDEMIOLOGIC STUDIES OF CHRONIC DISEASES, ANN ARBOR, MICHIGAN, 1963 1965 *Genetics and the Epidemiology of Chronic Diseases.* U.S. Public Health Service, Publication No. 1163. Washington: Government Printing Office. → See especially "Selection techniques for rare traits," by Leslie Kish, pages 165–176.

YATES, FRANK (1949) 1960 *Sampling Methods for Censuses and Surveys.* 3d ed., rev. & enl. New York: Hafner. → Earlier editions were also published by Griffin.

ADDITIONAL BIBLIOGRAPHY

DEMING, W. EDWARDS; and HANSEN, MORRIS H. 1967 Review of Regression Estimators. Pages 147–153 in Heinrich von Strecker and Willi R. Bihn (editors), *Die Statistik in der Wirtschaftsforschung: Festgabe für Rolf Wagenführ zum 60. Geburtstag.* Berlin: Drucker & Humblot.

GILBERT, RICHARD O. 1973 Approximations of the Bias in the Jolly–Seber Capture–Recapture Model. *Biometrics* 29:501–526.

KOLLER, SIEGFRIED 1969 Use of Nonrepresentative Surveys for Etiological Problems. Pages 235–246 in Symposium on the Foundations of Survey Sampling, University of North Carolina, 1968, *New Developments in Survey Sampling.* Edited by Norman L. Johnson and Harry Smith, Jr. New York: Wiley.

II

NONPROBABILITY SAMPLING

Nonprobability sampling refers to the selection of sampling units for a statistical study according to some criterion other than a probability mechanism. In contrast to probability sampling, nonprobability sampling has two major disadvantages: (1) the possibility of bias in the selection of sampling units and (2) the impossibility of calculating sampling error from the sample.

The attitude toward sampling held by most social scientists has come full circle: whereas it was once necessary to argue strenuously the case for deliberately using a probability mechanism in the selection of material for study, that case is now so widely accepted that any nonprobabilistic sampling procedure is regarded with suspicion. It will be argued here that such suspicion is generally well founded but that it is not to be equated with condemnation, since there are occasions when nonprobability sampling is the only procedure open to the investigator.

The issues involved in nonprobability sampling may be sharply focused by considering first the status of the inferences made by an archeologist. He can usually expect to find only a fraction of the material that he would like in order to examine the causes and motives of past events, and that fraction will generally be small for remote periods. Every object that he would judge relevant is a potential witness in his court of inquiry, but many of them have long since been destroyed or lost. Yet there is no contradicting the fact that some strongly convincing arguments can be made, even about the remote past. Does it follow, then, that probability sampling is inessential to any social scientist's inferences about the present?

As soon as it is drawn, this analogy is seen to be false. The archeologist usually has no choice; he must, in the court metaphor, accept what witnesses offer themselves. The situation of another social scientist is different: he is rather in the position of having too many potential witnesses, among whom he must choose on grounds of cost and time. How is he to choose? He cannot fairly exclude those potential witnesses whose evidence may not be to his taste, for if he did, he would implicitly abandon his judicial position and become merely an advocate. (The fact that some have apparently undergone this transformation is not good reason for others to follow them.) If some sources of evidence must be chosen in preference to others, they can only conscientiously be chosen in some way unrelated to the nature of their evidence: they must be selected by a probability mechanism.

The value of probability sampling is, therefore, not that it ensures employment for statisticians, but that it provides a guarantee of freedom from selection bias on the part of the investigator, bias that may be strong even if quite unconsciously produced (Yates [1949] 1960, chapter 2). The absence of selection bias is the first advantage of probability sampling.

Nonprobability sampling always has the characteristic that the sampling procedure is ill-defined. One literally cannot know what chance of selection any individual in the population has had. It is this lack of precise definition that leads directly to the second disadvantage of nonprobability sampling: there is no way of estimating the sampling error of a nonprobability sample from the sample itself, while such estimation is typically possible when probability sampling is used.

Despite what has just been said, nonprobability samples will always be important in the social sciences. Confronted with a remote and hostile tribe, the anthropologist will not be inclined to imperil his foothold of hospitality and cooperation by selecting informants at random. In advanced societies, selected individuals or institu-

tions may refuse to cooperate in an inquiry and usually cannot be compelled to do so. Even in such circumstances, however, at least an attempt at a probability sample should always be the investigator's object. He may fail to achieve it but cannot be worse off than if he had not tried at all. The unshakable optimism of all candidates for political office must be partly attributed to reports from supporters who have largely been sampling the faithful.

Nonprobability samples can arise in several ways: through the investigator's ignoring or denying the force of the above considerations; through his recognizing their force but claiming that the objectives of probability sampling can just as well be achieved by substitute methods; or finally, through his lack of success in attempts to achieve probability samples. About the first category nothing more will be said here, but a more detailed examination of the other two will be made.

Representative methods. Most of the substitutes suggested for probability sampling have in common some attempt to make the sample "representative" of the population from which it was selected, for example, by arranging that the proportions of men and women be the same in both or (more rarely) that the average income be the same in both. In practice, the most frequent substitute is quota sampling (Moser 1952), in which distributions by sex, age, "class," and sometimes additional characteristics are equalized between sample and population and the interviewers are otherwise free to select cases. Such representative methods are to be sharply distinguished from stratified probability sampling, where the population is divided into groups prior to sampling and each group is sampled separately by probability methods.

There are several distinct criticisms to be made of representative procedures. In the first place, the agreement of even quite a large number of averages or percentages between sample and population in no way guarantees high accuracy in other respects, and very large biases are still possible (Neyman 1934). Second, the lack of definition in the sampling procedure is not corrected by the imposition of such agreements; sampling error is still not calculable.

The most important criticism, however, is a paradoxical one. There is no such thing as a "representative," "unbiased," "fair," or otherwise "acceptable" sample: such adjectives are strictly applicable to the sampling process that produces the sample, and their application to the sample itself is at best a piece of verbal shorthand and at worst a sign of muddled thinking. A sample can be judged only in relation to the process that produced it. The central concepts of selection bias and sampling error have no meaning except in this context (Stuart 1962). Thus, an ill-defined sampling process, such as is involved in nonprobability samples, can only produce a sample with ill-defined properties of bias and sampling error.

Embedding in a probability framework. The statement that there is no way of calculating sampling error for nonprobability samples is always true but can be made irrelevant by the device of embedding a nonprobability sampling procedure within a higher-order probability framework. Suppose that a nonprobability sampling procedure is proposed in which 1,000 families are to be interviewed in a certain area and that ten interviewers are to be used. If each interviewer is independently given an assignment to interview 100 families by identical methods, there will be not one, but ten nonprobability samples, and the variation among the results of these ten samples may be used to estimate the sampling error attached to a sample of 100 families and thence, by a natural extension, to estimate the error for a sample of 1,000 families. Since the sample is designed as ten equivalent independent subsamples, in effect the sampling is of ten members from the population of all possible samples obtainable by that method with those interviewers.

In practice, it may be difficult or practically impossible to achieve independent subsamples in some contexts. For example, in the construction of index numbers the choice of commodity items is usually made by a panel of experts, so that the items chosen constitute a nonprobability sample. If an attempt is made to measure the variation between the judgments of several experts at the same level of expertise, interaction phenomena that make the measurement difficult are immediately encountered. For example, the experts are likely to have consulted one another; they are likely to know the criteria of selection used by their peers; they are bound to meet in the course of examining files containing background information, and so on. Thus, although independent judgments are in principle possible, the practical problem of arranging for independence is very difficult. Now, almost certainly, in the simple interviewing example, differences between the interviewers will affect the results of the survey (through personal characteristics, differences in thoroughness, etc.), and such differences will inflate the sampling error when it is estimated as above. If it were important to eliminate differences between interviewers from the sampling error, a slight modification in sampling design would suffice. Each interviewer must be asked to carry out at least two independent sample assignments, say four assignments of 25 families each. It will then be possible, by exactly the same argument as before, to estimate variability "within interviewers" only, thus excluding variability "among interviewers" from sampling error estimates. Further improvements in design are also possible. [See INDEX NUMBERS, *article on* SAMPLING.]

It will be seen that nonprobability sampling treated in this way utilizes the theory of experiment design, and, as always in that subject, care is necessary to ensure that assignments are allocated at random to interviewers. An incidental advantage of such experiment designs, which is often of greater practical importance than the original purpose of estimating sampling error, is that they make it possible to see whether interviewers are varying so much in performance that further training or other action is called for.

Such evidence as is available from designs of this sort indicates (Moser & Stuart 1953; Stephan & McCarthy 1958) that the sampling errors of quota sampling are considerably larger than those of comparable probability sampling. (If sampling is multistage, the inflation will of course only apply to the stage or stages at which quota sampling is used—this is commonly the final stage only.) Quota sampling can therefore only be justified, if at all, on grounds of reduction of costs, but it is doubtful whether, in the light of current cost structures, this factor is large enough to offset the excess sampling error of quota sampling. Some crude numerical guide is contained in the introduction to the tables by Stuart (1963).

Of course, experiment designs of this sort may be used with any ordinary probability sampling scheme (Mahalanobis 1946). The point here is that experiment designs rescue nonprobability sampling schemes from one of their worst deficiencies, the impossibility of calculating sampling error estimates, which does not hold for probability sampling. Nothing, however, can rescue nonprobability sampling from the ever-present danger of selection biases, a danger particularly acute for those forms (like quota sampling) that allow considerable freedom to interviewers to select cases.

Taken in conjunction with the sampling error and cost considerations, the bias danger suggests that, whatever may have been the case when it originated, quota sampling is now, in terms of reliable information delivered per unit cost, uneconomical as well as anachronistic.

Incompletely achieved probability samples. It has been noted that the major unremovable drawback of nonprobability samples is the danger of bias in the selection procedure. Inevitably, this danger must also arise when an intended probability sample is, through lack of cooperation, through inaccurate fieldwork, or through accidentally missing records, reduced to a fraction of the number of cases originally selected. The selected sample is, for one or another of these reasons, incompletely achieved, and the sampler's duty is to survey the damage done to his original plan and try to assess the importance of the danger to his purposes. Strictly speaking, an incompletely achieved probability sample ceases to be a probability sample, although it usually continues to be called one. Intentions are not the same as achievement, and the sampler must see whether the courtesy title is deserved.

Before discussing what can be done in such situations, it is useful to compare the reaction of the probability sampler confronted by an incompletely achieved sample with that of the nonprobability sampler in the same position. It is an empirical fact that the probability sampler is usually much more worried about the incomplete sample than is the nonprobability sampler, because the stringency of a probability sampling scheme draws much greater attention to any weaknesses in the achieved sample. By contrast, the nonprobability sampler usually finds it easier to bring his sample numbers up to the required level. His sampling scheme is less stringent and can more easily be patched up. Advocates of nonprobability sampling (principally quota sampling) have been known to categorize this as an advantage for their method, but we shall take the contrary view. It is because nonprobability sampling is so ill-defined that its definition can be extended to cover incomplete sample fulfillment. In the kingdom of the one-eyed, the blind man is less easily seen. Perhaps the least appreciated, because it is the most troublesome, feature of probability sampling is the urge for sample completeness that it imposes upon its practitioners.

Faced with a seriously incomplete sample, the probability sampler will usually redouble his efforts, but however hard he tries, he will ultimately have to face the fact that some part of his sample is not to be completed. The characteristic rates of noncompletion vary greatly with types of sample. For probability samples of individuals in the United Kingdom, it is possible to achieve completion rates of 90 per cent, although 80 per cent is nearer the general average. For probability samples of households in United Kingdom government household-budget inquiries, the completion rates range around 70 per cent. Few surveys involve a more onerous burden upon the respondent than the keeping of a detailed household budget, so it is probably reasonable to state that completion rates in the United Kingdom should never, with careful fieldwork, fall below about 70 per cent. The range of noncompletion rates likely to be encountered is thus of the order of 10 to 30 per cent. For many purposes, the sample results will not be seriously biased by even a noncompletion rate of one-quarter. For example, if 60 per cent of all households in the completed three-quarters of the sample display a certain characteristic, the noncompleted quarter of the sample would need to have under 40 per cent or over 80 per cent with that characteristic for the bias in the completed sample to exceed 5 per cent.

The crucial question to be examined in assessing the likely magnitude of bias due to incompleteness is whether the causes of incompleteness are related to the questions of interest. In a study of women's cosmetics usage, it would be running a serious risk to ignore a very low completion rate for teen-agers, who are particularly heavy users of cosmetics; if the completed sample showed a very low percentage of teen-agers compared with the selected sample (or known recent population data), special efforts would have to be made to increase that percentage. But if the completed sample showed an excess of women aged 45–55 and a shortage of women over 65, this would be unlikely to exert a significant influence on cosmetics-usage findings. However, if the subject studied were demand for domestic help, it would be the shortage of the elderly that would be the threat. Data from published population figures or from specially undertaken supplementary probability samples can often be used to check and, with appropriate methods of statistical analysis, to improve estimates suspected of bias, whether it is due to sample incompleteness or not (Moser & Stuart 1953; United Nations 1960, sec. 13). The simple examples just given show that such check data can only be used in conjunction with the experienced judgment of the investigator. If extreme imbalances of the completed sample are revealed, they are just cause for suspicion of the sample results, but the converse does not hold: no finite amount of external checking of this kind can ever fully validate a nonprobability sample, since the crucial variable to check may be overlooked or not checkable. There is no completely satisfactory substitute for a fully achieved prob-

.ability sample. Any shortfall from this status is a potential threat to the inferences drawn from the sample. The investigator may be able to judge in some cases that the threat is not likely to be serious, but he can do no more than this. Insofar as a probability sample is fully achieved, it obviates the need to make such judgments.

To return to the first analogy, it can now be seen that the archeologist is in a situation closely resembling that of the investigator with an incomplete nonprobability sample. Another social scientist can usually improve on this situation at least to the extent of starting with a probability sample of his material. The incompleteness of his sample may on occasion compel him to make judgments of an inconclusive kind about the quality of his sample, but this is not a valid reason for extending the scope of the inconclusive judgment to the whole sampling procedure.

ALAN STUART

[*For methods of accomplishing randomization, see* EXPERIMENTAL DESIGN, *article on* THE DESIGN OF EXPERIMENTS; RANDOM NUMBERS. *Also related are* ERRORS, *article on* NONSAMPLING ERRORS; EXPERIMENTAL DESIGN, *article on* QUASI-EXPERIMENTAL DESIGN; INDEX NUMBERS, *article on* SAMPLING.]

BIBLIOGRAPHY

MAHALANOBIS, P. C. 1946 Recent Experiments in Statistical Sampling in the Indian Statistical Institute. *Journal of the Royal Statistical Society* Series A 109: 326–378. → Contains eight pages of discussion.

MOSER, CLAUS A. 1952 Quota Sampling. *Journal of the Royal Statistical Society* Series A 115:411–423.

MOSER, CLAUS A. 1958 *Survey Methods in Social Investigation.* New York: Macmillan.

MOSER, CLAUS A.; and STUART, ALAN 1953 An Experimental Study of Quota Sampling. *Journal of the Royal Statistical Society* Series A 116:349–405. → Contains 11 pages of discussion.

NEYMAN, JERZY 1934 On the Two Different Aspects of the Representative Method: The Method of Stratified Sampling and the Method of Purposive Selection. *Journal of the Royal Statistical Society* Series A 97: 558–606.

STEPHAN, FREDERICK F.; and McCARTHY, PHILIP J. 1958 *Sampling Opinions: An Analysis of Survey Procedure.* New York: Wiley.

STUART, ALAN 1962 *Basic Ideas of Scientific Sampling.* London: Griffin; New York: Hafner.

STUART, ALAN 1963 Standard Errors for Percentages. *Applied Statistics* 12:87–101.

UNITED NATIONS, DEPARTMENT OF ECONOMIC AND SOCIAL AFFAIRS 1960 *A Short Manual on Sampling.* Volume 1: Elements of Sample Survey Theory. United Nations, Statistical Office, Studies in Methods, Series F, No. 9. New York: United Nations.

YATES, FRANK (1949) 1960 *Sampling Methods for Censuses and Surveys.* 3d ed., rev. & enl. London: Griffin; New York: Hafner.

SAMPLING

See INDEX NUMBERS; SAMPLE SURVEYS.

SAVAGE, LEONARD JIMMIE

▶ *This article was specially written for this volume.*

Leonard Jimmie Savage (1917–1971) was among the keenest, most imaginative, and most influential statistical thinkers of our time; his premature death just before his 54th birthday was an incalculable loss to statistics in particular and to science in general.

Jimmie Savage (he hardly ever used the name Leonard) contributed to many areas of statistics and mathematics, as well as to economics and other social and natural sciences. His major work, in terms of influence, widespread publication, and controversy, was undoubtedly his development of personal probability, especially in its relations to statistical inference. By the 1940s, the subjective, or personal Bayesian, approach to statistics had for years been under a cloud of contumely. Statistical training in the United States, and no doubt elsewhere, scarcely mentioned the personal Bayesian approach, passed it off as a rarely applicable consequence of the elementary theorem of Bayes, or actively scorned the possibility that personal views might reasonably enter into scientific inference via a properly considered subjective prior distribution and the application of Bayes' theorem. [*See* BAYESIAN INFERENCE; PROBABILITY, *article on* INTERPRETATIONS.]

There were, of course, exceptions. For example, Frank P. Ramsey (1931) had worked out an axiomatics for personal probability and utility in the 1920s, but it was not widely influential. B. O. Koopman (see, for example, 1941) and I. J. Good (1950) may also be mentioned; I do not try for completeness. Among such few earlier thinkers actively pressing the relevance of personal probability to inference, the steadiest, deepest, and most prolific was Bruno de Finetti. When Jimmie Savage first read de Finetti's works, he was "like stout Cortez . . . upon a peak in Darien"; more accurately, and with apologies to Keats, Savage viewed with new eyes what most had called a small puddle and saw it as a Pacific, an ocean worthy of the most vigorous exploration. Interaction between de Finetti and Savage continued actively throughout the rest of Savage's life in a relationship of great worth to both.

Savage's work in personal probability and statistics was, in the first instance, a careful axiomatization of personal probability from first principles of consistency of opinion. This found expression in *The Foundations of Statistics* ([1954] 1972), a volume that contains a great deal more than formal axiomatics: in particular, informal justification and commentary, together with highly useful criticisms of conventional statistical teaching and practice— areas that certainly contained their quotas of philosophical sitting ducks. *Foundations* also includes important contributions to the theory of utility.

The 1954 book was enormously influential. It persuaded a number of statisticians and others that the personal probability approach to statistics was respectable, and perhaps the single correct approach. At the same time, it hardened the opinions of many anti-Bayesians, who might grant the excellent mathematical developments in Savage's axiomatization, but who felt more skeptical than before about its applications to statistics in science and in daily life. As so often happens, those of a more eclectic or neutral bent might be scorned from both sides.

Yet of course the debates were, and are, more complicated; any one-dimensional description is bound to mislead. For there are many variants of personal, or subjective, probability and of the Bayesian approach to statistics. [*A taxonomy of those variants is given in* PROBABILITY, *article on* INTERPRETATIONS, *by de Finetti.*]

Savage's concern with subjective probability and the foundations of statistics continued to be central in his intellectual development. Perhaps particularly useful to the student are the 1962 colloquium proceedings *The Foundations of Statistical Inference* and the two 1967 papers by Savage published in philosophical journals. His 1971 paper on elicitation of personal probabilities is an excellent discussion of that partly psychological portion of the field. Chapter 8 in de Finetti (1972) is Savage's English summary of a joint paper in Italian by Savage and de Finetti. The brief, nontechnical summary is about choice of initial probabilities.

We must not neglect Savage's other, diverse contributions to probability and statistics. His paper with Paul R. Halmos (1949) on sufficiency [*see* SUFFICIENCY] first put the famous Neyman factorization condition for sufficiency on a firm, broad mathematical basis that set the stage for important later developments. His paper with R. J. Nunke (1952) on finitely additive measures, and his paper with Edwin Hewitt (1955) on symmetric measures, again represented major steps forward— in these cases, steps of mathematical formalization and development starting from initial work by de Finetti. Savage and R. R. Bahadur (1956) brilliantly settled a nagging question about the impossibility of a particular kind of nonparametric procedure. His paper on stopping times, with his brother, I. Richard Savage, himself a distinguished statistician, dealt carefully with a question of interest to both Bayesians and non-Bayesians (1965).

With Lester E. Dubins, Jimmie Savage worked out a major influential theory of gambling for circumstances in which the relevant loss function presses one to participate in a game known to have negative expectation. An illustrative paradigm is that of a man in Monte Carlo with only a small amount of money, no other resources, and a life-or-death need to be in New York within two days. He has perhaps no choice but to gamble and hope to increase his small initial stake to cover the all-important cost of transportation.

Savage also worked in the foundations of economics; see in particular two papers with Milton Friedman on utility (1948; 1952), and a paper with James H. Lorie on rationing capital (1955). Finally, and along a very different line, Savage's paper on rereading R. A. Fisher (1976) brings to the reader a store of information about statistics, Fisher, Savage, and the reader himself.

I forgo detailed discussions of these works, and of many others not cited here; it is expected that a full scholarly appreciation of Savage's work will be published before long.

Yet one important aspect of his scientific career will not be easy to include in a conventional appreciation. He lives on in his contributions to scores—perhaps hundreds—of publications in which Jimmie Savage played an important role, but was not a joint author. He was, when he wished to be, the model applied statistician, the beau ideal of helpfulness to a scientist planning an investigation or struggling to make sense of data. His deep background in mathematics and statistics, his broad scientific knowledge, his concern with first principles and digging under mere traditional convention, and his lambent intelligence made him simultaneously the best and the most awesome of scientific colleagues.

That combination is not a paradox. Savage usually suffered gladly those less gifted than he, but he suffered not at all those who wanted only routine turning of handles or incantation of standard phrases. I remember a consultation between Jimmie Savage and two research physicians. He began to ask, courteously but persistently, about the reasons for the particular therapeutic regimen under trial. The questions were clear, pertinent, but not statistical in the narrow sense of statistics apparently held by the physicians. They became uncomfortable, almost as if one of their patients had begun to auscultate *their* chests, and the discussion broke up in a cloud of misunderstanding.

I have also been present at dozens of conversations, some brief and informal, at which Savage magically would get at the true heart of a problem or its solution and wholly change the course of a scientist's research. A medical example may serve here as well. A novel cure for ulcers had been in tentative use for years at a university hospital, but a control group had not been formed for comparison. The responsible physicians, realizing this lack, came to Savage for help, and he provided intellectual and moral aid toward the best solution under the circumstances: specifically, (1) trying for comparisons with population statistics and (2) forming a retroactive quasi-control group from other patients at the same hospital.

Jimmie Savage was born to a family of comfortable means in Detroit. His eyesight was congenitally poor and his early education uneven. Not until college were his high mathematical abilities fully recognized, either by himself or others, but once they were, progress was rapid. He received his bachelor of science degree from the University of Michigan in 1938 and his doctor of philosophy degree there in 1941. Following some postdoctoral years as a mathematician at the Institute for Advanced Study, at Cornell, and at Brown, Jimmie Savage spent a brief yet crucial period at Columbia University's wartime Statistical Research Group. There he interacted with a group of statisticians that included Churchill Eisenhart, Milton Friedman, Abraham Girshick, Harold Hotelling, Frederick Mosteller, Abraham Wald, W. Allen Wallis, and Jacob Wolfowitz. [*See the biographies of* GIRSHICK; HOTELLING; WALD.] Small wonder that Savage became fascinated by problems of statistics, absorbed by conversational osmosis and by reading a tremendous amount of statistical background, and began to make his own fresh contributions.

In 1947 Savage went to the University of Chicago, where at first he worked on applications of mathematics and statistics to biology. In 1949 he was a joint founder with W. Allen Wallis of a new statistics department at Chicago, which grew and prospered. Jimmie Savage was its highly effective chairman during the years 1956–1960.

Personal circumstances led (despite the appeals of his Chicago colleagues that he stay) to a move to the University of Michigan in 1960. In 1964 he went to Yale University, where he became Eugene Higgins professor of statistics. It is tragic that he died so soon after the achievement of personal tranquility and his installation and leadership in a vigorous new statistics group at a great private university. I mourn the insights we would have had during Savage's later life.

Encyclopedia biographies are usually written in an impersonal style, as if there were positive merit in the outward clothing of impossible objectivity. That would be especially inappropriate in the case of Jimmie Savage, who lived his life whole, and who stressed so eloquently the importance of taking personal opinion into explicit account. This would be an incomplete picture were I not to present my own perception of an important interaction between Savage's personal and scientific lives, interwined as they were. In his development of personal probability, Savage moved more and more to a proselytizing position. Personal probability was not only useful and interesting to study; it became for him the only sensible approach to probability and statistics. Thus, orthodoxy of neoradicalism developed: if one were not in substantial agreement with him, one was inimical, or stupid, or at the least inattentive to an important scientific development.

This attitude, no doubt sharpened by personal difficulties and by the mordant rhetoric of some anti-Bayesians, exacerbated relationships between Jimmie Savage and many old professional friends. The problem had a special poignancy for those who, like myself, took an eclectic point of view. These difficulties eased greatly during his last years at Yale; perhaps it is wrong to write of them now, but one must know something of them to appreciate the life and work of a great scientist.

I have earlier pointed out that Jimmie Savage's brother, I. Richard Savage, is also a major figure in the statistical world. Jimmie Savage's son, Sam L. Savage, is a computer scientist.

Among Savage's scientific honors, one might especially list his John Simon Guggenheim Memorial Foundation fellowship in 1951–1952, his 1963–1964 fellowship at the Center for Advanced Study in the Behavioral Sciences, his 1958 presidency of the Institute of Mathematical Statistics, and his 1963 honorary D.SC. from the University of Rochester.

The Sterling Memorial Library at Yale University maintains the Leonard J. Savage archives, a collection of manuscripts, publications, correspondence, and other materials from Savage's files or donated by colleagues. He was an indefatigable correspondent, and the archives will become an important source for the history of statistical development in our times.

WILLIAM H. KRUSKAL

WORKS BY SAVAGE

Savage's collected works, edited by William A. Ericson, are planned for publication under the joint sponsorship of the American Statistical Association and the Institute of Mathematical Statistics.

(1948) 1976 FRIEDMAN, MILTON; and SAVAGE, LEONARD J. The Utility Analysis of Choices Involving Risk. Pages 20–50 in Stephen H. Archer and Charles A. D'Ambrosio (editors), *Theory of Business Finance: A Book of Readings.* 2d ed. New York: Macmillan. → First published in *Journal of Political Economy* 56:279–304 and often reprinted.

1949 HALMOS, PAUL R.; and SAVAGE, LEONARD J. Application of the Radon–Nikodym Theorem to the Theory of Sufficient Statistics. *Annals of Mathematical Statistics* 15:225–241.

1952 FRIEDMAN, MILTON; and SAVAGE, LEONARD J. The Expected Utility Hypothesis and the Measurability of Utility. *Journal of Political Economy* 60:463–474.

1952 NUNKE, R. J.; and SAVAGE, LEONARD J. On the Set of Values of a Nonatomic, Finitely Additive, Finite Measure. *American Mathematical Society, Proceedings* 3:217–218.

(1954) 1972 *The Foundations of Statistics.* Rev. ed. New York: Dover.

1955 HEWITT, EDWIN; and SAVAGE, LEONARD J. Symmetric Measure on Cartesian Products. *American Mathematical Society, Transactions* 80:470–501.

1955 LORIE, JAMES H.; and SAVAGE, LEONARD J. Three Problems in Rationing Capital. *Journal of Business* 28:229–239.

1956 BAHADUR, R. R.; and SAVAGE, LEONARD J. The Nonexistence of Certain Statistical Procedures in Nonparametric Problems. *Annals of Mathematical Statistics* 27:1115–1122.

1962 SAVAGE, LEONARD J. et al. *The Foundations of Statistical Inference: A Discussion.* London: Methuen; New York: Wiley. → Joint Statistics Seminar, University of London, 1959.

(1963) 1965 EDWARDS, WARD; LINDMAN, HAROLD; and SAVAGE, LEONARD J. Bayesian Statistical Inference for Psychological Research. Volume 2, pages 519–568 in R. Duncan Luce, Robert R. Bush, and Eugene Galanter (editors), *Readings in Mathematical Psychology.* New York: Wiley. → First published in *Psychological Review* 70:193–242.

1965 SAVAGE, LEONARD J.; and DUBINS, LESTER E. *How to Gamble If You Must: Inequalities for Stochastic Processes.* New York: McGraw-Hill.

1965 SAVAGE, LEONARD J.; and SAVAGE, I. RICHARD Finite Stopping Time and Finite Expected Stopping Time. *Journal of the Royal Statistical Society* Series B 27:284–289.

1967a Difficulties in the Theory of Personal Probability. *Philosophy of Science* 34:305–310.

1967b Implications of Personal Probability for Induction. *Journal of Philosophy* 64:593–607.

1971 Elicitation of Personal Probabilities and Expectations. *Journal of the American Statistical Association* 66:783–801.

1976 On Rereading R. A. Fisher. *Annals of Statistics* 4:441–483. → Discussion on pages 483–500.

SUPPLEMENTARY BIBLIOGRAPHY

DE FINETTI, BRUNO 1972 *Probability, Induction and Statistics.* New York: Wiley. → See especially pages v–vi, "In Memory of Leonard Jimmie Savage."

FIENBERG, STEPHEN E.; and ZELLNER, ARNOLD (editors) 1975 *Studies in Bayesian Econometrics and Statistics: In Honor of Leonard J. Savage.* Amsterdam: North-Holland; New York: American Elsevier.

GOOD, I. J. 1950 *Probability and the Weighing of Evidence.* London: Griffin; New York: Hafner.

KOOPMAN, B. O. 1941 Intuitive Probabilities and Sequences. *Annals of Mathematics* Series 2 42:169–187.

LINDLEY, DENNIS V. 1972 Obituary: L. J. Savage. *Journal of the Royal Statistical Society* Series A 135:462–463.

RAMSEY, FRANK P. 1931 *The Foundations of Mathematics and Other Logical Essays.* London: Kegan Paul; New York: Harcourt.

SCALES OF MEASUREMENT

See PSYCHOMETRICS; STATISTICS, DESCRIPTIVE.

SCALING, MULTIDIMENSIONAL

▶ *This article was specially written for this volume. The concepts of bilinear methods, often referred to herein, are discussed in detail in* FACTOR ANALYSIS AND PRINCIPAL COMPONENTS, *article on* BILINEAR METHODS. *Further discussion of the Boynton–Gordon color-naming data also appears there.*

Multidimensional scaling (MDS), a measurement methodology that emerged in psychology and the behavioral sciences, has come to be of general interest as a descriptive or data-analytic statistical approach. Originally, two-way MDS was aimed at accounting for subjective judgments of similarity or dissimilarity among a set of objects—usually called stimuli by psychologists—in terms of a geometric model assuming that similarity (dissimilarity) varies inversely (directly) with distance in an underlying metric space. (Three-way methods were originally conceived as ways to handle the *individual* differences among human subjects in such judgments of similarities or dissimilarities.)

It has become apparent, however, that the objects may be almost any entities whatever, and that measures of similarity or dissimilarity may be *derived* measures (derived, say, from traditional multivariate data), or they may be various kinds of more general (subjective or objective) measures of "proximity," not necessarily limited to direct judgments of similarity or dissimilarity. It has also become apparent that data sources other than human subjects may define the third way in a three-way data array. Therefore, both two- and three-way multidimensional scaling can be viewed as general data-analytic tools. For a survey of other senses in which the word "scaling" is used in psychology, see Cliff (1973).

1. Dyadic two-way data

The most common forms of multidimensional scaling deal with what we call *dyadic two-way data,* for which there are I objects called U_i and a variable or measurement that involves *two* of the objects at a time. For example, the objects might be countries, and the variable might be the volume of trade *from* country i to country i'. These data may be represented as an $I \times I$ matrix, where row i corresponds to U_i and column i' corresponds to $U_{i'}$. Since the rows and columns correspond to the same objects, the matrix is square. Correlation and covariance matrices, probably the most familiar examples of dyadic two-way data, will be discussed in detail below.

Another widely used kind of dyadic data comes from direct judgments by human subjects of how similar two objects appear. Dyadic data based on perceived similarity can be collected by many different means, sometimes much more efficiently than by direct pairwise judgments. Thus, it is sometimes useful to ask subjects to form groups of the objects that are similar in some respect. By counting how often two objects appear together in the same group, we derive a measure of similarity. It is generally best to permit each subject to form groups that overlap (that is, permit him to use an object several times in forming several different groups). By asking the subject to describe with a word or brief phrase what the common element is for each group he forms, and using these labels as part of the analysis, Wish has been able to obtain rich results (see Wish & Carroll 1974). Other examples of dyadic data include the mail volume, traffic volume, telephone call volume, and so on, between cities, states, or regions, volume of citations *in* one journal or field *to* another journal or field, and the number of times two articles are cited together in the same list of references. Two unusual examples of dyadic data are estimates of physical distance among different parts of a protein macromolecule, and the serological immunological reaction between a serum from one strain or species and the antiserum from another.

Multidimensional scaling of this type can be applied directly only to dyadic data, but it can be applied to other two-way data indirectly. One method is to scale dyadic data *derived* from other two-way data, by methods to be discussed. On the other hand, factor analysis and principal components can be viewed as applying either directly to two-way data or directly to dyadic data, according to the nature of the data involved, and according to the personal opinion of the investigator.

Perhaps the most widely familiar examples of dyadic data are correlation and covariance matrices. Note that these are derived from underlying two-way data arrays. Many other methods have been used to derive dyadic data from multivariate data. If we consider each row of a multivariate data matrix as a profile (or vector), then the matrix of squared distances (or of distances) among the profiles is an important way of deriving a dyadic matrix. Any measure of profile similarity, or dissimilarity, can be used in forming a dyadic matrix.

We can form a dyadic matrix from a two-way $I \times J$ data matrix by working either with *rows* or with *columns*. For example, if we take squared Euclidean distances between rows, the result will be $I \times I$; if we take correlations between the columns, the result will be $J \times J$. It is often sensible and useful to derive and analyze dyadic matrices in both directions (that is, both by comparing rows and by comparing columns). There is, however, a certain deceptiveness in the formal duality between rows and columns that permits us to compute covariance, correlation, Euclidean distance, or any other index between two rows as easily as between two columns. This duality suggests that any such index is equally sensible to compute in either direction. This conclusion need not be correct, however, because changes to, say, the columns that are not seen as significantly altering the data would be viewed quite differently if applied to the rows. For example, suppose the (i,j)th element of the matrix is the score of individual i on ability test j. For such a data matrix, we may well believe a priori that adding a constant to everybody's score on one particular test, or multiplying everybody's score on one particular test by a positive constant, does not really change the data in any significant way. In other words, a linear transformation (with positive multiplier) of a column is not important. The same idea most emphatically would not apply to a row. The implications of this distinction between rows and columns are complex and not fully understood, but it clearly suggests that a different treatment of rows and columns may be appropriate. As a practical matter in cases like that described, we favor the indices like covariance and correlation among columns, and indices like squared Euclidean distance among rows, although our arguments (not presented here) are only heuristic and suggestive.

2. Two-way multidimensional scaling

The phrase "multidimensional scaling" (introduced by Torgerson 1958) has two meanings in general use today. The narrower definition, adopted in this article, is rather precise. The broader definition refers rather more vaguely to a class of methods that includes also the bilinear methods and other approaches useful in studying perception and other subjective phenomena.

The input to multidimensional scaling usually consists of a matrix of two-way dyadic data, or some part of such a matrix. The entries in the matrix are dissimilarities, similarities, or other proximities of the most diverse sort among a single class of I objects U_i. The objective of multidimensional scaling is to reveal or discover relationships among the objects by making a "map" of the objects in which distances among the objects reflect their relative similarities: two objects should be close together in the map if they are similar, far apart if they are dissimilar, and more generally, there should be a *systematic relationship between the interpoint distances and the proximities*. In many applications, it is not clear in advance whether such a map is possible, or what dimensionality is required to portray it. There are ways of using multidimensional scaling indirectly that start with a two-way data matrix $\mathbf{X}$ like that used in the bilinear methods, although these ways have not emerged as major applications. Used in one of these special ways, multidimensional scaling is an alternative to the bilinear methods, which may be advantageous when the data points lie near a mildly curved subspace rather than a flat subspace.

In its narrower definition, multidimensional scaling is a method for constructing a configuration of points (usually in low-dimensional Euclidean space) from the distances between them, perhaps corrupted by error, or from some sort of information about these corrupted distances, such as their rank order. Most commonly the input consists of a single matrix (often symmetric) of the two-way dyadic data denoted by the $I \times I$ matrix $\mathbf{\Delta}$ with entries $\delta_{ii'}$. Frequently only part of this matrix is available, such as the part below or the part above the main diagonal, often excluding the diagonal itself, or other systematic portions. In addition, individual entries may be missing in a nonsystematic way. Technically the $\delta_{ii'}$ are called "dissimilarities" if a larger value indicates the objects are less alike, and "similarities" in the opposite case. Nevertheless, we shall often refer to the values as dissimilarities except where it is necessary to make a distinction. Replicated measurements of each $\delta_{ii'}$ may also be used, but we shall not discuss this possibility.

The space involved is generally low-dimensional Euclidean space, let us say of dimension R. The data analyst must choose the dimensionality R. Often, this choice is made by trying several values, and comparing the results that multidimensional scaling produces for each of them. (How to per-

form the comparison, and how to arrive at a final choice of R, is a large topic in itself, which is thoroughly discussed in Kruskal & Wish 1977.) In R-dimensional space each point can be represented by R coordinates, for which we use the notation (discussed below) $a_{i\rightarrow} = (a_{i1}, \cdots, a_{iR})$ because there is a useful analogy between these points and the factor scores used in the bilinear methods. It also means that distances $d_{ii'}$ between these points are measured by the familiar formula

$$d_{ii'}^2 = \sum_{r=1}^{R} (a_{ir} - a_{i'r})^2.$$

We shall denote the entire configuration of points $a_{i\rightarrow}$ by the matrix $\mathbf{A}$, which is $I \times R$.

Two key questions concern what is meant by "a systematic relationship" between the distances $d_{ii'}$ and the dissimilarities $\delta_{ii'}$, and how we calculate the configuration of points so as to achieve it. (Discussion of these points is deferred.)

Although formal statistical models have not received much attention in multidimensional scaling, it is helpful to think of the data values as being generated in accordance with one of two equations:

$$(1) \quad \text{or} \quad \begin{matrix} \delta_{ii'} = g\,(d_{ii'} + \text{random error}) \\ \delta_{ii'} = g\,(d_{ii'}) + \text{random error,} \end{matrix}$$

where the $d_{ii'}$ are distances among an unknown true underlying configuration of points, and g is an unknown monotonic function that belongs to some assumed family of functions. The functional relationship g gives explicit meaning to the phrase "a systematic relationship" between the distances and the dissimilarities. In discussing methods actually used in multidimensional scaling, it is often helpful to work with the inverse function g^{-1}, which we denote by f, so that the first equation becomes

$$f(\delta_{ii'}) = d_{ii'} + \text{random error.}$$

If nothing is assumed about g except that it belongs to the family of all monotonic increasing functions when the data are dissimilarities (or instead that it belongs to the family of all monotonic decreasing functions when the data are similarities), then it is only the *rank order* of the dissimilarities that conveys useful information. This very popular approach is referred to as *nonmetric* multidimensional scaling (Shepard 1962; Kruskal 1964a; 1964b; Guttman 1968) since the metric information in the dissimilarities is not used.

An illustrative example: Boynton and Gordon color-naming data. Before proceeding to more detailed discussion, we illustrate many of these concepts by using some data due to Boynton and Gordon (1965), which are discussed in detail in Shepard and Carroll (1966). For this purpose, we form dyadic data from these nondyadic data by a preliminary step prior to applying multidimensional scaling. The Boynton–Gordon data comprise a 23 × 4 matrix **X** whose general entry, x_{ij}, is the frequency with which three subjects chose the jth color name (red, yellow, blue, or green) as appropriate for stimulus i, which was one of 23 pure spectral colors. We use the "city-block" distance to define

$$\delta_{ii'} = \sum_{j=1}^{4} |x_{ij} - x_{i'j}|.$$

(It is more common to use the Euclidean distance

$$\left[\sum_{j=1}^{4} (x_{ij} - x_{i'j})^2 \right]^{1/2}$$

or its square for $\delta_{ii'}$. For these data, dissimilarities defined by the squared Euclidean distance formula were also analyzed, and gave very similar results to the city-block dissimilarities.) Since $\delta_{ii'} = \delta_{i'i}$ and $\delta_{ii} = 0$, only the lower half of the matrix ($\delta_{ii'}$ for which $i > i'$) was used, because the other values convey no additional information.

In this case, the dissimilarities $\delta_{ii'}$ were analyzed by an unusual variant of multidimensional scaling, which assumes that g is a second-order polynomial $g(d) = c_0 + c_1 d + c_2 d^2$ and minimizes the sum of the squared residual errors from the fitting equation based directly on the equation above. (This procedure is unusual in that most multidimensional scaling with polynomial fitting fits the polynomial to f rather than to g, because this is what some generally available computer programs now permit.) The results, which consist in part of an estimated configuration **A** and an estimated function g, are displayed in Figures 1 and 2.

The configuration clearly shows a distorted color circle like that seen by using a bilinear analysis. [See FACTOR ANALYSIS AND PRINCIPAL COMPONENTS, article on BILINEAR METHODS, for this analysis.] However, there is one obvious difference, and that is the rather definite bend at yellow (one of the color names used in collecting the data) where the earlier figure showed only a gentle curve. Since this bend also shows up in the bilinear analysis if we use three dimensions instead of two, it is presumably real, and is associated with the color name used in the subjective task. Thus, in this particular case, multidimensional scaling has produced a slightly more revealing result. We believe that such improvements may occur moderately often with real data, although this "dimension-reduction application" of scaling has not been used frequently enough to result in a clear conclusion.

Figure 2 shows the scatter diagram of $\delta_{ii'}$ versus $d_{ii'}$. This consists of $253 = (23 \times 22)/2$ points $(d_{ii'}, \delta_{ii'})$, one for each available dissimilarity. The

estimated function g is not plotted in this diagram; however, the relationship between the distances and the dissimilarities is very clearly displayed. It is evident from the straightness of this relationship that the term $c_2 d^2$ of second degree is small compared to the linear part of g for distances in the range of interest.

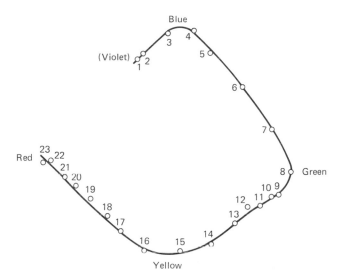

Figure 1 – **Two-dimensional representation of the 23 color profiles based on a proximity analysis of the d measures**

Source: Shepard & Carroll 1966, p. 575.

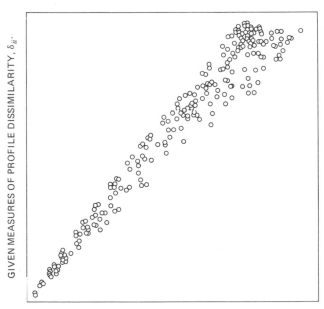

Figure 2 – **Relation between the given d measures and the interpoint distances in the two-dimensional solution of Figure 1**

Source: Shepard & Carroll 1966, p. 575.

3. Metric and nonmetric methods for two-way scaling

Indeterminacy. Like other kinds of solutions, two-way multidimensional scaling solutions are subject to some indeterminacies. Since we are concerned only with distances between the points $a_{i \to}$, the configuration may be shifted by a constant shift. This is the *minor indeterminacy*, and is exactly the same as that in the bilinear methods. The configuration may also be subjected to an orthogonal rotation (which includes the possibility of reflection). This is the *fundamental indeterminancy* of two-way scaling, and differs from that of bilinear methods, where oblique rotations are also permitted. In most forms of multidimensional scaling, there is a third indeterminacy: it makes no difference if we expand or contract the configuration uniformly in all directions. This is called the *indeterminacy of scale*.

Arrow notation. To aid our frequent use of rows and columns of matrices, we use a nonstandard notation to indicate them: $x_{i \to}$ (pronounced "x sub i row" or "x i row") indicates the ith row of **X**; $x_{\downarrow j}$ (pronounced "x sub column j" or "x column j") indicates the jth column of **X**. The horizontal arrow in $x_{i \to}$ reminds us that a row runs horizontally from left to right, and the vertical arrow in $x_{\downarrow j}$ reminds us that a column runs vertically from top to bottom. In each case, the arrow is written in place of the missing subscript.

Objective functions. We can now give a more formal and precise account of two-way multidimensional scaling. The data consist of an $I \times I$ matrix **Δ** of dissimilarities, or perhaps many such matrices. Each matrix may be incomplete, in the sense that not all entries are present. For some R, chosen by the data analyst as mentioned above, the method estimates an $I \times R$ matrix **A** whose rows $a_{i \to}$ form the points (or vectors) of a configuration in R-dimensional space. Only the distances $d_{ii'}$ between these points are fitted to the data. The dominant approach to fitting is to introduce some objective function that measures or defines how well any given configuration fits the data. This function compares the distances $d_{ii'}$ with the dissimilarities $\delta_{ii'}$ in some way (quite a few alternatives are in use), and then tries to choose points $a_{i \to}$ that optimize the objective function.

Regression. One way to compare distances with the dissimilarities is to use regression. Geometrically, using regression means that we fit some curve or function to the points with coordinates $(\delta_{ii'}, d_{ii'})$ that appear on the scatter diagram (like the one in Figure 2). This fitting can be done in two ways. For the Boynton–Gordon data, the dissimilarities $\delta_{ii'}$ (as "dependent variable") were regressed on the distances $d_{ii'}$ (as the "independent variable"). This means that the function being fitted is g, that the deviations in the scatter diagram (between the points and the curve) are measured along the dissimilarity axis, and that the sum of the squared deviations (or some other total deviation) is minimized. This way of using regression seems appropriate if the random error occurs after the function g is applied to the distances, as in the second of equations (1). It is, however, far more common (for reasons mentioned below) to regress the distances $d_{ii'}$ (as the "dependent variable") on the dissimilarities $\delta_{ii'}$ (as the "independent variable"). This means that the function being fitted is f and that deviations are measured along the distance axis, which seems appropriate if the random errors occur before the function g is applied, as in the first of equations (1). For nonmetric scaling, where the function being fitted is merely assumed to be increasing or to be decreasing, there are very great computational advantages to taking the latter approach. These advantages have been influential, both directly and indirectly, in making the latter approach more common.

In regression methods, a typical objective function is

$$\sqrt{\frac{\sum(d_{ii'} - \hat{d}_{ii'})^2}{\sum d_{ii}^2}},$$

which is called *stress* (Kruskal 1964*a*; 1964*b*). Here $\hat{d}_{ii'}$ is the value assumed by the fitted curve at the place where the "independent" variable δ takes on the value $\delta_{ii'}$. A denominator something like the one shown is necessary if the solution is subject to the indeterminacy of scale, but could be omitted in other cases (and typically would be omitted when the distances are used as the independent variable).

Classical MDS. The earliest method of MDS, now sometimes called the *classical* method (Torgerson 1958), is still of considerable importance largely because it is computationally efficient. It is based on the following theorem, which is due (in a slightly different form) to Young and Householder (1938). Suppose there are I points, $a_{i \to}$, which are centered at the origin (that is, $\sum a_{i \to} = 0_{\to}$). Suppose **D** is the matrix whose (i,j)th element d_{ij}^2 is the squared Euclidean distance from $a_{i \to}$ to $a_{j \to}$. Suppose we find the grand mean of the matrix **D** (the mean of all I^2 entries), and subtract it from each entry. Suppose then we find the mean of each row and each column of the modified matrix (the same for the ith row as for the ith column of course), and subtract from each entry the mean of its row and the mean of its column. (This entire process is called double-centering by psychometricians, and "taking out the grand mean and the row and column effects" by statisticians.) Suppose we now multiply this "double centered" matrix by $-\frac{1}{2}$. Then the (i,i') element of the resulting matrix $\mathbf{C}$ is

$$c_{ii'} = (a_{i\rightarrow} - \bar{a}_{\rightarrow}) \cdot (a_{i'\rightarrow} - \bar{a}_{\rightarrow})$$

$$= \sum_j (a_{ij} - \bar{a}_j)(a_{i'j} - \bar{a}_j)$$

$$= \sum_j a_{ij} a_{i'j},$$

where $\bar{a}_{\rightarrow} = (\bar{a}_1, \cdots, \bar{a}_R)$ is the mean of the $a_{i\rightarrow}$, which, by assumption equals $0_{\rightarrow}$. Since the entries of $\mathbf{D}$ do not change if we add a constant vector to every $a_{i\rightarrow}$, we can require that $\bar{a}_{\rightarrow} = 0_{\rightarrow}$ without any loss of generality. This convention (placing the origin of the coordinate system at the centroid of all the points) is, therefore, almost universally used in MDS work. This has the additional advantage that the expression for $\mathbf{C}$ simplifies, because the $\bar{a}_{\rightarrow}$ terms drop out, to a form expressed more compactly in matrix notation, as $\mathbf{C} = \mathbf{AA'}$. The steps of forming $\mathbf{C}$ from $\mathbf{D}$ will be called the *Young–Householder process*. Recovery of $\mathbf{A}$ from $\mathbf{C}$ is sometimes called matrix factoring, and is a central step in many methods of factor analysis. One widely used method for matrix factoring (described below) is based on finding eigenvectors and eigenvalues of $\mathbf{C}$. Notice that $\mathbf{C}/(I-1)$ is the covariance matrix of the points $a_{i\rightarrow}$, which is the reason for using the letter $\mathbf{C}$ in this context. It is often called $\mathbf{B}$ or the $\mathbf{B}$ matrix in older writings on multidimensional scaling. Also notice that the elements of $\mathbf{C}$ are scalar products of the points $a_{i\rightarrow}$.

The classical method of multidimensional scaling consists of three stages. The first stage, which is sometimes considered a preliminary step, not part of the scaling process proper, often fails to receive enough attention. In this stage, the matrix of dissimilarities is processed in a simple way to make it suitable as input to the Young–Householder process, that is, to make it resemble a matrix of squared Euclidean distances. The concept of a matrix "resembling" a matrix of squared distances is only intuitive, and has no precise meaning. In practice, a minimum requirement is that the entries in the matrix should be nonnegative, they should be zero along the diagonal, and that they should be true dissimilarities, as opposed to similarities. Sometimes other characteristics are required, for example, that the triangle inequality should hold for the square roots of the entries, or that the dispersion

of the entries should be large enough relative to the mean of the entries. One useful method of processing is simply to transform the dissimilarities to rank numbers, possibly followed by some of the other steps described next. Often the processing to put dissimilarities into squared distance form is nothing more than a transformation, possibly linear, applied to each entry separately. In many contexts, it seems appropriate to assume that the original dissimilarities are distances (with random error), plus some additive constant. In this case, processing consists of estimating the additive constant involved (a much discussed topic), subtracting the estimated constant from each dissimilarity, and then squaring the result to obtain an estimate of the squared distance. Sometimes other steps are helpful, for example, a matrix of exports among countries might be standardized by dividing each entry by the total exports from the source and by the total imports to the destination. Fortunately, a rough and ready choice of first stage is often adequate, because the subsequent stages are quite robust against deviations from the ideal input.

The second stage of classical scaling is simply to apply the Young–Householder process to the result of the first stage, to yield a matrix $\mathbf{C}$. In view of the Young–Householder theorem, the matrix $\mathbf{C}$ should resemble a matrix of scalar products, whether or not it actually has that form. To indicate that $\mathbf{C}$ is thought of as a matrix of scalar products, it is common to say that $\mathbf{C}$ "has scalar product form," and to refer to the Young–Householder process as converting a matrix from "squared distance form" to "scalar product form."

The third stage of classical scaling is to find the $I \times R$ matrix $\mathbf{A}$ such that $\mathbf{AA'}$ gives least-squares fit to $\mathbf{C}$. This least-squares fitting can be done efficiently by use of methods for calculating eigenvectors and eigenvalues. [See FACTOR ANALYSIS AND PRINCIPAL COMPONENTS, *article on* BILINEAR METHODS.] In particular, each column $a_{\downarrow r} = (a_{1r}, \cdots, a_{Ir})'$ of $\mathbf{A}$ can be taken to be an eigenvector of $\mathbf{C}$ multiplied by the square root of the corresponding eigenvalue, where the R largest eigenvalues are used to get the least-squares fit in R dimensions.

The classical method of scaling is computationally very efficient, and the configurations it yields are in practice very similar, in many cases, to those provided by later methods. Its objective function (the sum of squared residual differences between $\mathbf{C}$ and $\mathbf{AA'}$) is, however, a less reliable guide to how well the configuration fits the data than some other objective functions, since this objective function is inflated by a nonlinear relationship between the dissimilarities and the distances, where some other methods are not. Also, this objective function is a measure of fit for the scalar product form of the data and not for original dissimilarities; this makes it a less direct criterion of goodness of fit. Furthermore, classical scaling has the important practical limitation that it does not permit the use of incomplete data: all the dissimilarities must be observed. Although there are adaptations and modifications of classical scaling to overcome these two limitations, classical scaling seems to lose its special computational efficiency in proportion to how well the limitations are overcome.

Comparison of different approaches to multidimensional scaling. We have discussed the classical method of multidimensional scaling, which is inherently metric, and the regression approach, which includes both metric and nonmetric varieties. There are also other approaches, which we cannot go into here. Which approach should one use? Unfortunately, no one approach is uniformly better than the others. We shall briefly indicate some of the advantages and disadvantages of different approaches. A much fuller discussion of this kind can be found in Kruskal and Wish (1977, chapter 5).

There is a fortunate empirical observation of overriding importance. For a given choice of dimensionality R, and for most interesting sets of data, all methods of multidimensional scaling tend to give configurations that are very similar to one another. In other words, the configuration obtained is very robust under variation of method. This robustness is closely associated with the robustness that classical scaling has under different choices for the first stage. (Use of rank numbers as the first stage makes classical scaling particularly robust.) It is frequently true for good data from a well-designed experiment containing a large enough number of objects that the method of scaling will not affect the configuration enough to influence the valid conclusions one can draw. Some important exceptions will be discussed below.

On the other hand, it is important to remember that multidimensional scaling is *not* robust in some other ways. Most notably, the results depend very strongly on the domain from which the objects are chosen, and in particular on how broad the domain is. For example, even if one uses the selfsame matrix of dissimilarities, scaling a systematic subset of the stimuli (that is, one part of the matrix) may reveal interesting dimensions that do not appear when the whole matrix is scaled at once. A famous example of this occurs in Shepard (1963), which analyzes several sets of similarities among

Morse code stimuli. One data set includes 36 symbols (26 letters and 10 digits), and the most notable subset consists of the 10 digits.

Also, the goodness-of-fit value, such as stress, is definitely *not* robust under choice of methods. For example, even though nonmetric and various metric uses of the Kruskal–Shepard method typically yield extremely similar configurations in some given dimensionality R, they may yield very different values of stress. This is important for several reasons. First, the comparison of stress in different dimensionalities can be very helpful in selecting the most appropriate dimensionality, as discussed in detail in Kruskal and Wish (1977, chapter 3). Second, the goodness-of-fit value is helpful in deciding whether the configuration is worth serious examination, or whether it should be discarded.

Furthermore, the meaning of a certain numerical value for goodness of fit depends very sensitively on just which measure was used, and on almost everything else. The interpretation of these values is one of the most treacherous areas of multidimensional scaling. In this connection, it is perhaps worth mentioning that the least-squares goodness-of-fit measure, which is naturally associated with classical scaling, becomes more reliable if the data are transformed to rank numbers in the first stage.

There are exceptions to the robustness of configuration under variations in the method and the detail of approach. Most notably, if the number of objects is just barely large enough to make scaling in a certain dimensionality sensible, then the configuration becomes quite sensitive. Basically, the difficulty in this case is that the data are not redundant enough to overwhelm the assumptions built into the method. Such situations are extremely tricky, and there are no easy answers apart from using more objects.

Another nonrobustness of configuration is that many nonmetric methods sometimes yield what are called "degenerate" configurations. Although there is no precise definition of this term, such configurations are frequently characterized by having most or all the points very close to a small number of locations—usually no more than three. Such a solution (from a nonmetric method) is not very informative, although it may reveal a clustering of the objects or a few outliers. Another nonrobustness is that a few points in the configuration may shift position wildly under change of method, while most of the configuration stays fairly constant. This phenomenon suggests that these objects are peculiar, and may require another dimension, or may not fit with the space in a natural way at all.

In view of the varying advantages and disadvantages of the different methods, it can be useful to scale the same data twice, once by some nonmetric method and once by some metric method. If the configurations are substantially different, it would be important to understand why, before proceeding to interpret either one.

Dimension reduction. The use of MDS for dimension reduction (illustrated above with the Boynton–Gordon data) starts with a two-way array of data **X** that is not dyadic, and forms a dyadic matrix from it by some preliminary process, such as taking squared Euclidean distances among the rows. Multidimensional scaling is then applied to the dyadic matrix. This will yield a single configuration, corresponding to the objects U_i or the variables V_j depending on which way the dyadic matrix was formed. Suppose that squared Euclidean distances between the rows are used to form the dyadic matrix and that classical scaling is applied (with no preliminary processing before applying the modified Young–Householder process, of course). Then there is a theorem that the resulting configuration **A** is the same as that formed by modern principal components or Eckart–Young factor analysis applied to **X** (up to the indeterminacies, of course, and assuming **X** is at least *single*-centered, by subtracting column—or variable—means).

This theorem (which is explained more fully in the next section) shows the strong parallelism between the bilinear methods and the dimension-reduction application of multidimensional scaling. However, it also shows that, if one is not careful about how the latter is done, there is no hope of doing better than the bilinear methods. The use of *nonmetric* scaling (Shepard 1962; Kruskal 1964*a*; 1964*b*; Guttman 1968) is believed to offer the largest profit in this direction, since it can accommodate a configuration of points $x_{i\rightarrow}$ that lie on a mildly curved surface better than do the bilinear methods. It is possible that dyadic matrices other than squared Euclidean distance might offer advantages (with classical scaling) over the bilinear methods, but evidence is lacking.

The dimension-reduction application of multidimensional scaling shows relationships only among the objects (or only among the variables). It provides, however, an easy basis for subsequent analysis of the variables (or of the objects). The most common approach to this is to perform linear or quadratic regression of the original variables over the configuration of points $a_{i\rightarrow}$. If the variables are categorical with a small number of categories, another common technique is to examine (for each variable) the regions of R-space that contain the

points from each category. If $R = 2$, this can be done most easily by drawing curves to separate the object points from different categories. Sometimes it is possible to draw smooth noncrossing curves for this purpose. Sometimes there are extra indications of meaningfulness, such as all or most of the regions being convex, or the separating curves being drawable as straight lines. For further discussion, including a general nontechnical overview of multidimensional scaling, especially in the case of two-way data, see Carroll and Wish (1974*a*), and Kruskal and Wish (1977).

Clustering. Other techniques providing useful adjuncts to MDS are methods of clustering [*see, for example,* Johnson 1967; Hartigan 1975; *see also* CLUSTERING]. A clustering solution will often provide complementary information to a scaling solution, and can sometimes be quite helpful as an aid in interpreting scaling configurations (either in a dimensional or in a nondimensional "configurational" way). It is often useful, in fact, to try to draw "contour plots" on MDS solutions, where the contours are defined on the basis of a hierarchical clustering solution (see Shepard 1972 for some good examples). The use of contour plots is necessarily limited to two dimensions, of course, although it can often be helpfully applied to two-dimensional subspaces of higher dimensional MDS solutions. For a general discussion of relations between MDS and hierarchial clustering, hierarchical tree structure models, and other "discrete" and "hybrid" models, see Carroll (1976).

4. Some relationships between the bilinear methods and multidimensional scaling

Some important relationships between the bilinear methods and multidimensional scaling can best be understood through the use of Figure 3 and Table 1. Basically, the diagram shows *types of matrices* and *operations* that lead from one type of matrix to another (or back to the same type, in which case the operation arrow loops back on itself). Each arrow indicates an operation that can be applied to any suitable input matrix. Incidentally, **D** and **S** matrices often arise directly, from sources outside the diagram. An arrow is drawn as a double line if it is conceived of as a substantial data analysis technique or with a single line if it is conceived of as a relatively simple preprocessing step. One double arrow splits and leads both to **A** and to **B** because the procedure starts with one matrix and yields two.

In Figure 3, **X** is any two-way $I \times J$ data matrix. **D** is any dyadic matrix of dissimilarities. If **D** is

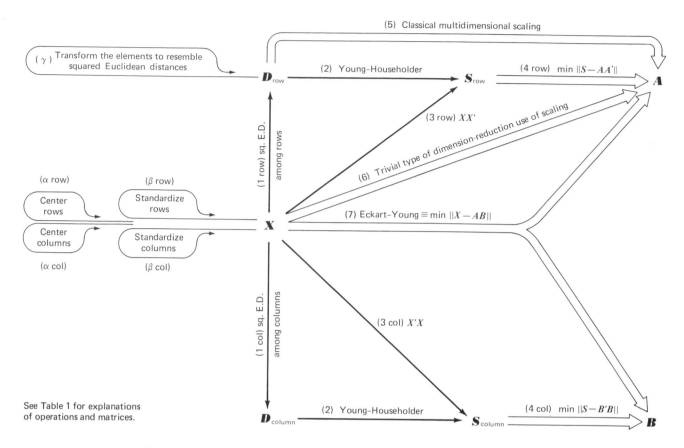

Figure 3 – Matrices and operations in MDS and the bilinear methods

derived from $\mathbf{X}$, then the subscript "row" or "column" indicates whether the dissimilarities are between rows or columns, but if $\mathbf{D}$ has some other source, then the subscript merely indicates whether $\mathbf{D}$ is $I \times I$ or $J \times J$. $\mathbf{S}$ is what we call a "scalar-product-type" matrix, which means a correlation matrix, a covariance matrix, a cross-products matrix, or any other matrix that we find useful to treat in the same way as one of the preceding. Finally, $\mathbf{A}$ and $\mathbf{B}$ indicate reduced forms of the input data, to be examined and used by the data analyst. [*The conceptual meaning of $\mathbf{A}$ and $\mathbf{B}$ is the same here as it is in* FACTOR ANALYSIS AND PRINCIPAL COMPONENTS, *article on* BILINEAR METHODS.] This classification of types of matrices is merely a convenience for present purposes.

Many definitions and theorems are easy to visualize in terms of different paths through the network. These are listed at the end of this section, starting with simpler facts, and slowly proceeding to the more difficult and interesting ones. For example, the first entry in this listing is

(1 row) = (α col) followed by (1 row).

This entry states the elementary fact that calculating the matrix of squared (Euclidean) distances

among the rows of $\mathbf{X}$, which is symbolized by (1 row), yields the same result whether or not $\mathbf{X}$ has been column-centered first, which is symbolized by (α col) and means subtracting the column mean of each column from all the entries in that column. This entry can be visualized as asserting that the two paths on the network shown in Figure 4 yield the same result.

Two subsequent entries indicate that the full process of classical multidimensional scaling consists of (γ) followed by operation (5), which is often described elsewhere as being classical multidimensional scaling; and that operation (5) consists of operation (2), the Young–Householder process, followed by (4 row), which is least-squares matrix factoring. Thus we see the full process of classical multidimensional scaling decomposed into three successive stages, (γ), (2), and (4 row), which are the three stages used to describe it earlier.

One of the entries is the Young–Householder theorem, which states that the two dotted paths shown in Figure 5 yield the same result (and similarly for their column counterparts in the lower half of Figure 3).

The listing also describes several methods from outside multidimensional scaling, such as principal

Table 1 — Operations used in Figure 3

Operation	Explanation
(α)	Center every row (or center every column): calculate the mean value of the row and subtract it from each element.
(β)	Standardize every row (or standardize every column): calculate the root-mean-square of the row elements and divide every element by it.
(γ)	Transform the elements to resemble squared Euclidean distances, as described in section 2.
(1)	Calculate the matrix of squared Euclidean distances among the rows or columns of X.
(2)	Apply the modified Young–Householder procedure to D: subtract the grand mean from every element, then center rows and columns, and finally multiply by $-\frac{1}{2}$.
(3)	Form matrix XX' of scalar products among the rows (or matrix $X'X$ of scalar products among the columns).
(4 row)	Find the matrix A that minimizes the sum of the squares of $S-AA'$.
(4 col)	Find the matrix B that minimizes the sum of the squares of $S-B'B$.
(5)	The classical method of scaling, but excluding the preliminary step (γ).
(6)	A trivial application of multidimensional scaling for dimension reduction on the objects (see text).
(7)	Eckart–Young factor analysis, which is essentially finding A and B to minimize the sum of squares of $X-AB$.

components, Eckart–Young factor analysis, eigenvectors, and singular value decomposition, to help clarify the relationship between these methods and multidimensional scaling. [*Figure 3 is also useful in connection with* FACTOR ANALYSIS AND PRINCIPAL COMPONENTS, *article on* BILINEAR METHODS.]

Figure 3 heavily emphasizes least-squares methods, since it is for these that most of the nice theorems can be proved. However, since other methods are often approximated by corresponding least-squares methods, the diagram has conceptual value beyond the methods actually shown on it. For example, consider operation (6). This is one kind of dimension reduction by the use of multidimensional scaling. Precisely because this is the same as (7) (in the sense described below), which is much more efficient computationally, this method is not interesting. However, when we modify this method by substituting other types of multidimensional scaling for operation (5), or perhaps by modifying step (1 row), the modification does become of value, as described previously.

More specifically, if we use (1) followed by metric scaling of a restricted type, such as one in which linear regression is used to relate the dissimilarities to distance, the change in the objective function permits different and possibly better results. This is, in all likelihood, the reason for mild improvement in the configuration for the

two-dimensional Boynton–Gordon configuration between the principal components solution and the multidimensional scaling solution. Even though the procedure allowed for a curved relationship between the dissimilarities and the distances, it turned out to be linear anyhow (as is clear from Figure 2). In case a curved relationship really exists, the improvements can be greater. This possibility is illustrated by a *one*-dimensional scaling solution to the Boynton–Gordon data, given in Shepard and Carroll (1966), which the bilinear methods cannot come close to matching. This results because the one-dimensional structure is drastically nonlinear in a way that multidimensional scaling can accommodate only with some effort (see the analysis in Shepard & Carroll 1966), but that the bilinear methods cannot accommodate at all. Only in special circumstances, however, is the dimension-reduction application of multidimensional scaling better than the bilinear methods, and this method is not recommended as a routine replacement for them.

We see from the diagram that the step of "factoring" S (either into $B'B$ or into AA'), which is often presented as the essence of factor analysis, is one step of classical multidimensional scaling. Thus we see that the bilinear methods can be used to perform one kind of multidimensional scaling, and that multidimensional scaling, in its dimen-

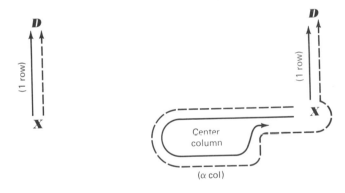

Figure 4 – Two paths from Figure 3 giving the same result

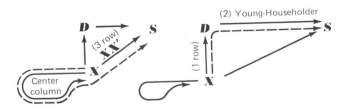

Figure 5 – Parts of Figure 3 illustrating the Young–Householder theorem

sional-reduction application, can be used to do something similar to the bilinear methods. The relationship between the two areas is very complex.

Some facts that can be stated in terms of Figure 3:

$$(1\text{ row}) = (\alpha\text{ col})\text{ followed by }(1\text{ row}),$$
$$(1\text{ col}) = (\alpha\text{ row})\text{ followed by }(1\text{ col}).$$

$(\alpha\text{ row})$ followed by $(\alpha\text{ col})$ gives the same result as $(\alpha\text{ col})$ followed by $(\alpha\text{ row})$.

Sample covariance matrix of **X**: the operation of forming this consists of $(\alpha\text{ col})$, then (3 col), then division by I or $I - 1$.

Sample correlation matrix of **X**: the operation of forming this consists of $(\alpha\text{ col})$, then $(\beta\text{ col})$, then (3 col).

Operation (5), the classical method of multi-dimensional scaling, but excluding the preliminary transformation of the dissimilarities: this is defined as (2) followed by (4 row).

Full process of classical multidimensional scaling: this is defined as (γ) followed by (5).

Young–Householder theorem: $(\alpha\text{ col})$ followed by (3 row) = (1 row) followed by (2), according to the (modified) Young–Householder theorem. Similarly, $(\alpha\text{ row})$ followed by (3 col) = (1 col) followed by (2).

Modern principal components: this is often described as consisting of $(\alpha\text{ col})$, then optional use of $(\beta\text{ col})$, then (3 col), then (4 col).

Principal factor analysis and Hotelling principal components: same description as the preceding, except that the diagonal entries of **S** are modified between (3 col) and (4 col).

Operation (6), an uninteresting type of dimension reduction by use of multidimensional scaling: this is defined as (1 row) followed by (5).

Summary of a few of the preceding facts:

$$(6) = (1\text{ row}) + (5) = (1\text{ row}) + (2) + (4\text{ row})$$
$$= (\alpha\text{ col}) + (3\text{ row}) + (4\text{ row}).$$

Operation (7), Eckart–Young factor analysis: if we resolve the fundamental indeterminacy by requiring **BB'** to be the identity matrix, and if we resolve the minor indeterminacy in the customary manner, then the matrix **A** we get this way is the same as the matrix **A** we get from (3 row) followed by (4 row). (This is one of the subtler theorems of this field.) Thus, in a sense, $(\alpha\text{ col})$ followed by (7) is the same as (6).

Operation (7), Eckart–Young factor analysis: if we resolve the fundamental indeterminacy by requiring **A'A** to be the identity matrix and if we resolve the minor indeterminacy in the customary manner, then the matrix **B** we get this way is the same as the matrix **B** we get from (3 col) followed

by (4 col). (This paragraph is the mirror image of the previous one).

Eigenvectors: the solution to (4 row) and (4 col) can be accomplished by the use of the R largest eigenvalues of **S** and their corresponding eigenvectors. (The eigenvectors are scaled by the square roots of the corresponding eigenvalues.) This is one of the most basic theorems of this subject.

Singular-value decomposition: the solution to the minimization involved in (7) can be accomplished by use of the R largest singular values of **X** and their corresponding singular vectors. The work of Eckart and Young constitutes the first modern exposition of this mathematics, although the phrase "singular-value decomposition" was introduced later. Conceptually, it is interesting to note that because the singular-value decomposition generalizes and subsumes the eigenvectors of symmetric matrices, the solution to (4 row) and (4 col) can also be accomplished by use of the singular-value decomposition.

5. Trilinear methods for three-way data

The bilinear methods and multidimensional scaling deal essentially only with two-way data (either general or dyadic), despite various minor generalizations that make use of richer data in a limited way. In this section we introduce individual differences scaling and other trilinear methods for dealing with three-way data: some of these methods use the extra information contained in a three-way array, as opposed to a two-way array, much more fully to achieve more penetrating insight. The contributions made by the trilinear methods are similar to those for the bilinear methods: they reveal relationships among the objects, variables, or other entities by *dimension reduction*, and they suggest what the *basic underlying variables* may be. In addition they may reveal relationships among the subjects or other data sources comprising the "third way" in the three-way data. The greater power of individual differences scaling and some other methods for three-way data seems to result in large part from the following mathematical and empirical facts: (1) the rotational position of the configurations they provide is uniquely determined, unlike the situation in two-way multidimensional scaling, and (2) this uniqueness results directly from the requirement that the model fit the data well, unlike the situation for the bilinear methods, which need supplementary requirements like simple structure. (Some qualifications are discussed below.) It is an empirical fact that the dimensions that these methods provide directly (without subsequent rotation of any kind) turn out to be convincing candidates for basic variables substantially more often than those provided by any of the bilinear methods. Although this is not unreasonable, in view of the much richer data these methods utilize, learning how to use the richer data is a substantial step forward in data analysis.

The three-way methods of greatest demonstrated value are individual differences scaling, or INDSCAL for short (from *IN*dividual *Differences SCAL*ing) and the dimension-reduction application of IND-SCAL. The input to this method is an array of three-way dyadic data, $\delta_{ii'k}$, where each value represents the dissimilarity between two objects U_i and $U_{i'}$ for one subject or condition W_k. Often it is assumed that $\delta_{ii'k}$ and $\delta_{i'ik}$ would be equal except for disturbances, and only one or the other may actually be measured. (Existing computer programs accept only data forming half the full three-way array, namely, terms for which $i > i'$. Where separate measurements are taken for $\delta_{ii'k}$ and $\delta_{i'ik}$, they must be averaged or combined in some way before the application of INDSCAL.) The input may be thought of as a series of k separate (two-way) matrices, $\delta_{\downarrow \to k'}$, one for each subject or condition W_k. Not only does INDSCAL accommodate differences among these matrices (of a prescribed but apparently very common type), it actually *relies* on these differences to identify basic underlying variables. If the matrices are too much alike (fortunately not too common an occurrence), the special strength INDSCAL has in identifying basic variables (as compared with the bilinear methods and two-way multidimensional scaling) can and does fail.

The mathematical model for INDSCAL. In mathematical terms, the INDSCAL model (Horan 1969; Carroll & Chang 1970) assumes

$$g_k(\delta_{ii'k}) \cong d_{ii'k}$$

(where $\cong$ means "equals, except for error terms," whose nature will not be further specified at this time) where

$$d_{ii'k} = \sqrt{\sum_{t=1}^{T} w_{kt}\,(x_{it} - x_{i't})^2},$$

which can readily be seen to be a simple generalization of the Euclidean metric assumed in two-way multidimensional scaling. The generalization, of course, involves introducing the new parameters, w_{kt}, which can be viewed as a set of dimension weights for different subjects (w_{kt} being the weight the kth subject applies to the tth dimension). Geometrically, the square roots of the w_{kt}'s define the factors by which the dimensions are stretched or compressed for different subjects. The x's, of course, define the coordinates of the stimuli or other objects in what, in INDSCAL, is called the "group stimulus space." The configuration of subject points defined by the w's is what is called the subject space.

The INDSCAL method of analysis. The INDSCAL method of analysis will not be described in detail here. The original, and still most widely used algorithm, is the metric procedure described by Carroll and Chang (1970). This procedure can be viewed as a direct three-way generalization of the classical metric method of two-way MDS described earlier. The basic input consists of a three-way ($n \times n \times m$) array consisting of m square ($n \times n$) symmetric matrices of similarities or dissimilarities data (one matrix for each subject or other data source). In the first stage, each matrix is processed by procedures like those described for the two-way case, to generate estimated squared Euclidean distances. In the second stage, each of these matrices is converted to scalar product form by the modified Young–Householder procedure. In the third stage, a trilinear decomposition method is applied. This least-squares trilinear decomposition, now called CANDECOMP (for *CAN*onical *DE*-*COMP*osition) is a direct generalization of the least-squares matrix factoring applied in the two-way case. Because each of the original matrices is assumed symmetric in the INDSCAL procedure, each of the scalar-product-form matrices is also symmetric. However, CANDECOMP can also be applied to an array without this symmetry property, and thereby provides a multiway generalization of the Eckart–Young method of factor analysis.

Although INDSCAL was developed with direct measures of similarity or dissimilarity in mind, it can of course be used much as two-way scaling with various derived measures of dissimilarity. Any method for deriving dyadic data from two-way data can be used to derive dyadic three-way data from plain three-way data, simply by applying the method to each "slice" in some direction. It turns out that the use of squared Euclidean distances for this purpose is valuable in connection with the three-way methods we shall discuss. For example, suppose that the three-way data x_{ijk} are based on I countries by J subjective rating scales by K subjects. As illustrated in Wish and Carroll (1974), good use can be made of the squared distances among countries across subjects for each rating scale, that is,

$$\delta_{ii'j}^{2} = \sum_{k} (x_{ijk} - x_{i'jk})^2.$$

Also useful in the same study were the squared distances among countries across rating scales for each subject, although in this case the squared distances were pooled across predetermined groups of subjects before further analysis, largely to reduce the size of the data for computational reasons:

$$\delta_{ii'\tilde{k}} = \sum_{\substack{k \text{ in} \\ \text{group } \tilde{k}}} \sum_j (x_{ijk} - x_{i'jk})^2.$$

It is not fully clear yet which ways to derive dyadic three-way data from plain three-way data are sensible and useful.

An illustrative example of INDSCAL: Helm's color–distance data. In the data on color perception collected by Helm (1964), the U_i are 10 Munsell color chips varying in hue (with saturation and brightness maintained fairly constant), and there are 16 "subjects" W_k (representing only 14 distinct subjects, two of whom each carried out the experimental task on two separate occasions). The value $\delta_{ii'k}$ is perceived psychological distance (measured on a ratio scale) of color chips i and i' for subject W_k, as determined by a preliminary stage of data processing from an interesting experimental task, in which subjects arranged three color chips at a time in a plane so as to reflect the relative subjective distances among the three colors. In contrast to many other experiments in color perception, subjects with deficient color vision were deliberately included in the experiment, which turned out to be quite helpful.

The most important output from INDSCAL is shown in Figure 6, part (a): the configuration of points (the "group stimulus space") that correspond to the colors U_i. This configuration, which conforms quite well with the standard representation of the color circle, plays exactly the same role as does the corresponding configuration in two-way MDS, and is analogous to the configuration of points (as opposed to vectors) in the reduced diagram of the bilinear methods. As in the configurations for the Boynton–Gordon data presented earlier, the dimensions shown here correspond to the red-versus-green and blue-versus-yellow dimensions, which are known to play a central role in human color vision. However, there is an important difference. The configuration produced by two-way multidimensional scaling of the Boynton–Gordon data is not provided with a meaningful rotational position by the method, and was deliberately rotated by Shepard into the position shown so as to make the already known red–green and blue–yellow oppositions in reading the diagram. Neither two-way scaling nor the bilinear methods produce a meaningfully unique orientation of the coordinate system.

INDSCAL also produces a configuration of points (the "subject space") that correspond to the subjects W_k; see Figure 6, part (b). These are roughly analogous to the vectors in the reduced diagram of the bilinear methods but have a very different geometric meaning. (There is nothing to correspond to the subject space in two-way multidimensional scaling as such.) This configuration also has a well-defined rotational position, and (unlike the stimulus configuration) a well-defined origin. Four subjects (one of whom is represented in two replications) stand out as making very little use of the red–green dimension (that is, as having small coordinates on dimension 2). These subjects (and none of the others) have a red–green color deficiency as measured by separate standard tests for this purpose. Thus, this analysis is also helpful in revealing relationships among the subjects, which is a common experience with INDSCAL.

INDSCAL can also provide a *separate* configuration of points corresponding to the colors U_i for each subject W_k. Four of these configurations are shown in Figure 6, parts (c)–(f). Each can be derived from the group configuration of points mentioned above in a very special way, namely, by shrinking or stretching in the group configuration by a constant multiplier along each coordinate axis, with different multipliers used for different axes. Each of these individual configurations corresponds to a different pattern of multipliers or weights. The multipliers for subject W_k are the square roots of the coordinates of the point corresponding to W_k in the configuration of subjects. As we see in the figure, subject CD1 (with color-deficient vision) has a very small coordinate on dimension 2, so that his private configuration of colors varies very little along that dimension. Informally, we may say that this subject "makes very little use of dimension 2," or that this dimension is not very "salient" to this subject. Although, in this particular application, we may presume that this reflects this subject's inability to perceive the dimension, nonuse of a dimension in other applications may occur for quite different reasons. By contrast, subject N7, who is located right in the middle of the normal subjects, has a private configuration of colors that is almost identical to the group configuration. For other applications of INDSCAL, see Carroll and Chang (1970), Carroll (1972), or Wish and Carroll (1974). General discussions of multidimensional scaling are provided by Carroll and Wish (1974a; 1974b) and by Kruskal and Wish (1977).

As we have seen, each private configuration is obtained mathematically from the group configuration and the subject configuration. INDSCAL com-

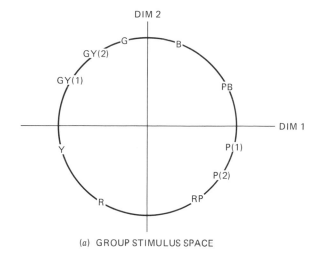

(a) GROUP STIMULUS SPACE

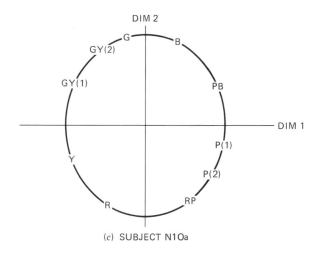

(c) SUBJECT N10a

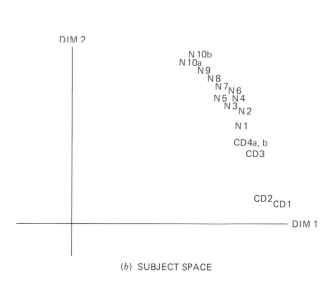

(b) SUBJECT SPACE

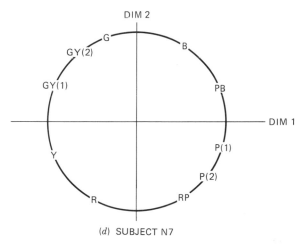

(d) SUBJECT N7

Figure 6 — Two-dimensional INDSCAL analysis of Helm's color–distance data

The coding of the stimuli is as follows: R = red; Y = yellow; GY(1) = green–yellow; GY(2) = a green–yellow containing more green than GY(1); G = green; B = blue; PB = purple–blue; P(1) = purple; P(2) = a purple containing more red than P(1); RP = red–purple. In the subject space, part (b), CD1 through CD4 are four color-deficient subjects (CD4a and CD4b are two replications of one subject) and N1 through N10 are ten "normal" subjects (N10a and N10b are replications of the same subject). At right are shown the "private" perceptual spaces for two "normal" subjects, parts (c) and (d), and two red–green color-deficient subjects, parts (e) and (f).

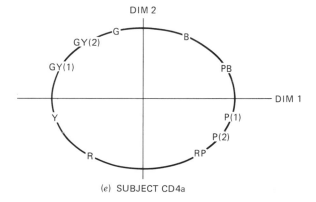

(e) SUBJECT CD4a

Source: The full figure, slightly modified, first appeared in Bell Laboratories Record, May 1971, p. 150, and has since been often reprinted.

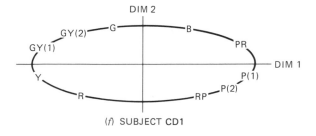

(f) SUBJECT CD1

pares the *k*th private configuration to the matrix $\delta_{\downarrow \to k}$, and adjusts both the group configuration and the subject configuration at the same time, so as to minimize the total deviation for all the comparisons.

Because the unique determination of rotational position for individual differences scaling is so important and is so different from the more familiar bilinear methods, we give a hypothetical example here, which provides some geometrical insight into this phenomenon, as well as a precise statement later about the relevant mathematical facts. Suppose that the group configuration consists of four points at the corners of a square, oriented in the usual way (parallel to the coordinate axes). Stretching or shrinking along the coordinate axes (as INDSCAL permits the subjects to distort the configuration) will always turn this square into a rectangle, although one of any conceivable size and shape. On the other hand, suppose that the group configuration is a square oriented at 45 degrees to the coordinate axes. Then, stretching or shrinking along the coordinate axes will always turn the square into a *rhombus*, although one of any conceivable size and shape. Even if we discard the given rotational position for the separate configurations, and rotate them each freely and arbitrarily, we can still tell the difference between a family of rectangles (oriented at odd angles) from a family of rhombuses (oriented at odd angles). Furthermore, the only rotational positions in which all the rectangles can be derived from a single figure by the permitted stretching or shrinking is the one in which their sides are all parallel to the coordinate axes. (Of course, the side connecting the points for two particular objects U_1 and U_2 in one rectangle must be parallel to the same coordinate axis as the corresponding side of every other rectangle.) Similar reasoning applies to the rhombuses. In fact, this reasoning works quite generally for other configurations and for spaces of dimensionality higher than 2.

Other models and methods for three-way or multiway data. Space does not permit a detailed discussion of other approaches to analysis of three- or multiway data, so only a mention of the most important ones, with references, will be given. The oldest approach to individual difference multidimensional scaling is the Tucker–Messick (1963) points-of-view analysis. Tucker's (1964) three-mode factor analysis was the first general model and method for analysis of general three-way multivariate data. This model has been extended by Tucker (1972) to include as a special case a three-mode multidimensional scaling model. Harshman's (1970) PARAFAC model, first proposed as a three-way factor analytic approach, turns out to be mathematically equivalent to the three-way CANDECOMP model underlying INDSCAL. PARAFAC 2 is an interesting generalization of the INDSCAL model proposed by Harshman (1972), for which, however, there is as yet no fully satisfactory numerical method for fitting to data. IDIOSCAL (Carroll & Chang 1970; 1972; Carroll & Wish 1974*a*; 1974*b*) is a general three-way multidimensional scaling model that includes INDSCAL, Harshman's PARAFAC-2, and Tucker's three-mode scaling all as special cases.

Other approaches attempt to synthesize various aspects of different models. For example, the TRAIS approach of Cohen (1974) and a rather similar approach being taken by McCallum (1976) both involve synthesis of aspects of three-mode scaling and INDSCAL. Many more such eclectic approaches will undoubtedly emerge in the future, as well as entirely new models for individual differences multidimensional scaling and for multilinear data analysis.

A final new method that bears mentioning is one implemented in a computer program called ALSCAL, devised by Takane, Young, and de Leeuw (1977). ALSCAL incorporates both two-way scaling and three-way scaling in terms of the INDSCAL model, allowing either metric or nonmetric analyses, and allows an option for analysis of unordered categorical data (often called "nominal scale" data), thus allowing MDS analyses under the weakest possible assumptions. ALSCAL uses a different fit measure than any of those described above. This is called SStress (for "squared-distance stress") by Takane, Young, and de Leeuw. ALSCAL also has options for dealing with missing data and various kinds of conditionality in the data that may be useful. Experience with this new computer program is not extensive enough to compare it adequately with alternative procedures, but ALSCAL appears at the moment to be a very promising new development.

J. DOUGLAS CARROLL AND
JOSEPH B. KRUSKAL

BIBLIOGRAPHY

BOYNTON, ROBERT M.; and GORDON, JAMES 1965 Bezold–Brücke Hue Shift Measured by Color-naming Technique. *Journal of the Optical Society of America* 55:78–86.

CARROLL, J. DOUGLAS 1972 Individual Differences and Multidimensional Scaling. Volume 1, pages 105–155 in Roger N. Shepard, A. K. Romney, and S. Nerlove (editors), *Multidimensional Scaling: Theory and Applications in the Behavioral Sciences.* New York: Academic Press.

CARROLL, J. DOUGLAS 1976 Spatial, Non-spatial and Hybrid Models for Scaling. *Psychometrika* 41:439–463.

CARROLL, J. DOUGLAS; and CHANG, JIH-JIE 1970 Analysis of Individual Differences in Multidimensional Scaling via an N-way Generalization of "Eckart–Young" Decomposition. *Psychometrika* 35:283–319.

CARROLL, J. DOUGLAS; and CHANG, JIH-JIE 1972 *IDIOSCAL* (Individual Differences In Orientation SCALing): A Generalization of INDSCAL Allowing IDIOsyncratic Reference Systems as Well as an Analytic Approximation to INDSCAL. Paper presented at meetings of the Psychometric Society, Princeton, N.J.

CARROLL, J. DOUGLAS; and WISH, MYRON 1974a Multidimensional Perceptual Models and Measurement Methods. Volume 2, pages 391–447 in Edward C. Carterette and M. P. Friedman (editors), *Handbook of Perception*. Volume 2: *Psychophysical Judgment and Measurement*. New York: Academic Press.

CARROLL, J. DOUGLAS; and WISH, MYRON 1974b Models and Methods for Three-way Multidimensional Scaling. Volume 2, pages 57–105 in David H. Krantz et al. (editors), *Contemporary Developments in Mathematical Psychology*. Volume 2: *Measurement, Psychophysics, and Neural Information Processing*. San Francisco: Freeman.

CLIFF, NORMAN 1973 Scaling. *Annual Review of Psychology* 24:473–506.

COHEN, H. S. 1974 Three-mode Rotation to Approximate INDSCAL Structure. Paper presented at the annual meeting of the Psychometric Society, Palo Alto, Calif.

GUTTMAN, LOUIS 1968 A General Nonmetric Technique for Finding the Smallest Coordinate Space for a Configuration of Points. *Psychometrika* 33:469–506.

HARSHMAN, R. A. 1970 Foundations of the PARAFAC Procedure: Models and Conditions for an Explanatory Multi-modal Factor Analysis. Thesis, University of California, Los Angeles.

HARSHMAN, R. A. 1972 PARAFAC 2: Mathematical and Technical Notes. Pages 30–44 in University of California, Los Angeles, Working Papers in Phonetics, No. 22.

HARTIGAN, JOHN A. 1975 *Clustering Algorithms*. New York: Wiley.

HELM, CARL E. 1964 Multidimensional Ratio Scaling Analysis of Perceived Color Relations. *Journal of the Optical Society of America* 54:256–262.

HORAN, C. B. 1969 Multidimensional Scaling: Combining Observations When Individuals Have Different Perceptual Structures. *Psychometrika* 32:241–254.

JOHNSON, S. C. 1967 Hierarchical Clustering Schemes. *Psychometrika* 32:241–254.

KRUSKAL, JOSEPH B. 1964a Multidimensional Scaling by Optimizing Goodness of Fit to a Nonmetric Hypothesis. *Psychometrika* 29:1–27.

KRUSKAL, JOSEPH B. 1964b Nonmetric Multidimensional Scaling: A Numerical Method. *Psychometrika* 29:115–129.

KRUSKAL, JOSEPH B.; and WISH, MYRON 1977 *Multidimensional Scaling*. Beverly Hills, Calif.: Sage.

McCALLUM, ROBERT C. 1976 Transformation of a Three-mode Multidimensional Scaling Solution to INDSCAL Form. *Psychometrika* 41:385–400.

SHEPARD, ROGER N. 1962 The Analysis of Proximities: Multidimensional Scaling With an Unknown Distance Function. *Psychometrika* 27:125–140, 219–246.

SHEPARD, ROGER N. 1963 Analysis of Proximities as a Technique for the Study of Information Processing in Man. *Human Factors* 5:33–48.

SHEPARD, ROGER N. 1972 Psychological Representation of Speech Sounds. Pages 67–113 in Edward E. David, Jr. and Peter B. Denes (editors), *Human Communication: A Unified View*. New York: McGraw-Hill.

SHEPARD, ROGER N.; and CARROLL, J. DOUGLAS 1966 Parametric Representation of Nonlinear Data Structures. Pages 561–592 in International Symposium on Multivariate Analysis, Dayton, Ohio, 1965, *Multivariate Analysis: Proceedings*. Edited by Paruchuri R. Krishnaiah. New York: Academic Press.

TAKANE, Y.; YOUNG, F. W.; and DE LEEUW, J. 1977 Nonmetric Individual Differences Multidimensional Scaling: An Alternating Least Squares Method With Optimal Scaling Features. *Psychometrika* 42:7–67.

TORGERSON, WARREN S. 1958 *Theory and Methods of Scaling*. New York: Wiley.

TUCKER, L. R. 1964 The Extension of Factor Analysis to Three-dimensional Matrices. Pages 109–127 in Norman Fredriksen and H. Gulliksen (editors), *Contributions to Mathematical Psychology*. New York: Holt.

TUCKER, L. R. 1972 Relations Between Multidimensional Scaling and Three-mode Factor Analysis. *Psychometrika* 37:3–27.

TUCKER, L. R.; and MESSICK, S. J. 1963 An Individual Differences Model for Multidimensional Scaling. *Psychometrika* 28:333–367.

WISH, MYRON; and CARROLL, J. DOUGLAS 1974 Applications of Individual Differences Scaling to Studies of Human Perception and Judgments. Volume 2, pages 449–491 in Edward C. Carterette and M. P. Friedman (editors), *Handbook of Perception*. Volume 2: *Psychophysical Judgment and Measurement*. New York: Academic Press.

WOLD, HERMAN 1966 Estimation of Principal Components and Related Models by Iterative Least Squares. Pages 391–420 in International Symposium on Multivariate Analysis, Dayton, Ohio, 1965, *Multivariate Analysis: Proceedings*. Edited by Paruchuri R. Krishnaiah. New York: Academic Press.

YOUNG, G.; and HOUSEHOLDER, A. S. 1938 Discussion of a Set of Points in Terms of Their Mutual Distances. *Psychometrika* 3:19–22.

SCIENCE

The articles under this heading were first published in IESS *with four companion articles less relevant to statistics.*

I

THE HISTORY OF SCIENCE

As an independent professional discipline, the history of science is a new field still emerging from a long and varied prehistory. Only since 1950, and initially only in the United States, has the majority of even its youngest practitioners been trained for, or committed to, a full-time scholarly career in the field. From their predecessors, most of whom were historians only by avocation and thus derived their goals and values principally from some other field, this younger generation inherits a constellation of sometimes irreconcilable objectives. The resulting tensions, though they have relaxed with increasing maturation of the profession, are still perceptible, particularly in the varied primary audiences to which the literature of the history of science continues to be addressed. Under the circumstances any brief report on development and current state is inevitably more personal and prognostic than for a longer-established profession.

Development of the field. Until very recently most of those who wrote the history of science were practicing scientists, sometimes eminent ones. Usually history was for them a by-product of pedagogy. They saw in it, besides intrinsic appeal, a means to elucidate the concepts of their specialty, to establish its tradition, and to attract students. The historical section with which so many technical treatises and monographs still open is contemporary illustration of what was for many centuries the primary form and exclusive source for the history of science. That traditional genre appeared in classical antiquity both in historical sections of technical treatises and in a few independent histories of the most developed ancient sciences, astronomy and mathematics. Similar works—together with a growing body of heroic biography—had a continuous history from the Renaissance through the eighteenth century, when their production was much stimulated by the Enlightenment's vision of science as at once the source and the exemplar of progress. From the last fifty years of that period come the earliest historical studies that are sometimes still used as such, among them the historical narratives embedded in the technical works of Lagrange (mathematics) as well as the imposing separate treatises by Montucla (mathematics and physical science), Priestley (electricity and optics), and Delambre (astronomy). In the nineteenth and early twentieth centuries, though alternative approaches had begun to develop, scientists continued to produce both occasional biographies and magistral histories of their own specialties, for example, Kopp (chemistry), Poggendorff (physics), Sachs (botany), Zittel and Geikie (geology), and Klein (mathematics).

A second main historiographic tradition, occasionally indistinguishable from the first, was more explicitly philosophical in its objectives. Early in the seventeenth century Francis Bacon proclaimed the utility of histories of learning to those who would discover the nature and proper use of human reason. Condorcet and Comte are only the most famous of the philosophically inclined writers who, following Bacon's lead, attempted to base normative descriptions of true rationality on historical surveys of Western scientific thought. Before the nineteenth century this tradition remained predominantly programmatic, producing little significant historical research. But then, particularly in the writings of Whewell, Mach, and Duhem, philosophical concerns became a primary motive for creative activity in the history of science, and they have remained important since.

Both of these historiographic traditions, particularly when controlled by the textual–critical techniques of nineteenth-century German political history, produced occasional monuments of scholarship, which the contemporary historian ignores at his peril. But they simultaneously reinforced a concept of the field that has today been largely rejected by the nascent profession. The objective of these older histories of science was to clarify and deepen an understanding of *contemporary* scientific methods or concepts by displaying their evolution. Committed to such goals, the historian characteristically chose a single established science or branch of science—one whose status as sound knowledge could scarcely be doubted—and described when, where, and how the elements that in his day constituted its subject matter and presumptive method had come into being. Observations, laws, or theories which contemporary science had set aside as error or irrelevancy were seldom considered unless they pointed a methodological moral or explained a prolonged period of apparent sterility. Similar selective principles governed discussion of factors external to science. Religion, seen as a hindrance, and technology, seen as an occasional prerequisite to advance in instrumentation, were almost the only such factors which received attention. The outcome of this approach has recently been brilliantly parodied by the philosopher Joseph Agassi.

Until the early nineteenth century, of course, characteristics very much like these typified most historical writing. The romantics' passion for distant times and places had to combine with the scholarly standards of Biblical criticism before even general historians could be brought to recognize the interest and integrity of value systems other than their own. (The nineteenth century is, for example, the period when the Middle Ages were first observed to have a history.) That transformation of sensibility which most contemporary historians would suppose essential to their field was not, however, at once reflected in the history of science. Though they agreed about nothing else, both the romantic and the scientist–historian continued to view the development of science as a quasi-mechanical march of the intellect, the successive surrender of nature's secrets to sound methods skillfully deployed. Only in this century have historians of science gradually learned to see their subject matter as something different from a chronology of accumulating positive achievement in a technical specialty defined by hindsight. A number of factors contributed to this change.

Probably the most important was the influence, beginning in the late nineteenth century, of the history of philosophy. In that field only the most partisan could feel confident of his ability to distinguish positive knowledge from error and superstition. Dealing with ideas that had since lost their appeal, the historian could scarcely escape the force of an injunction which Bertrand Russell later phrased succinctly: "In studying a philosopher, the right attitude is neither reverence nor contempt, but first a kind of hypothetical sympathy, until it is possible to know what it feels like to believe in his theories." That attitude toward past thinkers came to the history of science from philosophy. Partly it was learned from men like Lange and Cassirer who dealt historically with people or ideas that were also important for scientific development. (Burtt's *Metaphysical Foundations of Modern Physical Science* and Lovejoy's *Great Chain of Being* were, in this respect, especially influential.) And partly it was learned from a small group of Neo-Kantian epistemologists, particularly Brunschvicg and Meyerson, whose search for quasi-absolute categories of thought in older scientific ideas produced brilliant genetic analyses of concepts which the main tradition in the history of science had misunderstood or dismissed.

These lessons were reinforced by another decisive event in the emergence of the contemporary profession. Almost a century after the Middle Ages had become important to the general historian, Pierre Duhem's search for the sources of modern science disclosed a tradition of medieval physical thought which, in contrast to Aristotle's physics, could not be denied an essential role in the transformation of physical theory that occurred in the seventeenth century. Too many of the elements of Galileo's physics and method were to be found there. But it was not possible, either, to assimilate it quite to Galileo's physics and to that of Newton, leaving the structure of the so-called Scientific Revolution unchanged but extending it greatly in time. The essential novelties of seventeenth-century science would be understood only if medieval science were explored first on its own terms and then as the base from which the "New Science" sprang. More than any other, that challenge has shaped the modern historiography of science. The writings which it has evoked since 1920, particularly those of E. J. Dijksterhuis, Anneliese Maier, and especially Alexandre Koyré, are the models which many contemporaries aim to emulate. In addition, the discovery of medieval science and its Renaissance role has disclosed an area in which the history of science can and must be integrated with more traditional types of history. That task has barely begun, but the pioneering synthesis by Butterfield and the special studies by Panofsky and Frances Yates mark a path which will surely be broadened and followed.

A third factor in the formation of the modern historiography of science has been a repeated insistence that the student of scientific development concern himself with positive knowledge as a whole and that general histories of science replace histories of special sciences. Traceable as a program to Bacon, and more particularly to Comte, that demand scarcely influenced scholarly performance before the beginning of this century, when it was forcefully reiterated by the universally venerated Paul Tannery and then put to practice in the monumental researches of George Sarton. Subsequent experience has suggested that the sciences are not, in fact, all of a piece and that even the superhuman erudition required for a general history of science could scarcely tailor their joint evolution to a coherent narrative. But the attempt has been crucial, for it has highlighted the impossibility of attributing to the past the divisions of knowledge embodied in contemporary science curricula. Today, as historians increasingly turn back to the detailed investigation of individual branches of science, they study fields which actually existed in the periods that concern them, and they do so with an awareness of the state of other sciences at the time.

Still more recently, one other set of influences has begun to shape contemporary work in the history of science. Its result is an increased concern, deriving partly from general history and partly from German sociology and Marxist historiography, with the role of nonintellectual, particularly institutional and socioeconomic, factors in scientific development. Unlike the ones discussed above, however, these influences and the works responsive to them have to date scarcely been assimilated by the emerging profession. For all its novelties, the new historiography is still directed predominantly to the evolution of scientific ideas and of the tools (mathematical, observational, and experimental) through which these interact with each other and with nature. Its best practitioners have, like Koyré, usually minimized the importance of non-intellectual aspects of culture to the historical developments they consider. A few have acted as though the obtrusion of economic or institutional considerations into the history of science would be a denial of the integrity of science itself. As a result, there seem at times to be two distinct sorts of history of science, occasionally appearing between the same covers but rarely making firm or fruitful contact. The still dominant form, often called the "internal approach," is concerned with the substance of science as knowledge. Its newer rival, often called the "external approach," is concerned with the activity of scientists as a social group within a larger culture. Putting the two together is perhaps the greatest challenge now faced by the profession, and there are increasing signs of a response. Nevertheless, any survey of the field's present state must unfortunately still treat the two as virtually separate enterprises.

Internal history. What are the maxims of the new internal historiography? Insofar as possible (it is never entirely so, nor could history be written if it were), the historian should set aside the science that he knows. His science should be learned from the textbooks and journals of the period he studies, and he should master these and the indigenous traditions they display before grappling with innovators whose discoveries or inventions changed the direction of scientific advance. Dealing with innovators, the historian should try to think as they did. Recognizing that scientists are often famous for results they did not intend, he should ask what problems his subject worked at and how these became problems for him. Recognizing that a historic discovery is rarely quite the one attributed to its author in later textbooks (pedagogic goals inevitably transform a narrative), the historian should ask what his subject thought he had discovered and what he took the basis of that discovery to be. And in this process of reconstruction the historian should pay particular attention to his subject's apparent errors, not for their own sake but because they reveal far more of the mind at work than do the passages in which a scientist seems to record a result or an argument that modern science still retains.

For at least thirty years the attitudes which these maxims are designed to display have increasingly guided the best interpretive scholarship in the history of science, and it is with scholarship of that sort that this article is predominantly concerned. (There are other types, of course, though

the distinction is not sharp, and much of the most worthwhile effort of historians of science is devoted to them. But this is not the place to consider work like that of, say, Needham, Neugebauer, and Thorndike, whose indispensable contribution has been to establish and make accessible texts and traditions previously known only through myth.) Nevertheless, the subject matter is immense; there have been few professional historians of science (in 1950 scarcely more than half a dozen in the United States); and their choice of topic has been far from random. There remain vast areas for which not even the basic developmental lines are clear.

Probably because of their special prestige, physics, chemistry, and astronomy dominate the historical literature of science. But even in these fields effort has been unevenly distributed, particularly in this century. Because they sought contemporary knowledge in the past, the nineteenth-century scientist–historians compiled surveys which often ranged from antiquity to their own day or close to it. In the twentieth century a few scientists, like Dugas, Jammer, Partington, Truesdell, and Whittaker, have written from a similar viewpoint, and some of their surveys carry the history of special fields close to the present. But few practitioners of the most developed sciences still write histories, and the members of the emerging profession have up to this time been far more systematically and narrowly selective, with a number of unfortunate consequences. The deep and sympathetic immersion in the sources which their work demands virtually prohibits wide-ranging surveys, at least until more of the field has been examined in depth. Starting with a clean slate, as they at least feel they are, this group naturally tries first to establish the early phases of a science's development, and few get beyond that point. Besides, until the last few years almost no member of the new group has had sufficient command of the science (particularly mathematics, usually the decisive hurdle) to become a vicarious participant in the more recent research of the technically most developed disciplines.

As a result, though the situation is now changing rapidly with the entry both of more and of better-prepared people into the field, the recent literature of the history of science tends to end at the point where the technical source materials cease to be accessible to a man with elementary college scientific training. There are fine studies of mathematics to Leibniz (Boyer, Michel); of astronomy and mechanics to Newton (Clagett, Costabel, Dijksterhuis, Koyré, and Maier), of electricity to Coulomb (Cohen), and of chemistry to Dalton

(Boas, Crosland, Daumas, Guerlac, Metzger). But almost no work within the new tradition has as yet been published on the mathematical physical science of the eighteenth century or on any physical science in the nineteenth.

For the biological and earth sciences, the literature is even less well developed, partly because only those subspecialties which, like physiology, relate closely to medicine had achieved professional status before the late nineteenth century. There are few of the older surveys by scientists, and the members of the new profession are only now beginning in any number to explore these fields. In biology at least there is prospect of rapid change, but up to this point the only areas much studied are nineteenth-century Darwinism and the anatomy and physiology of the sixteenth and seventeenth centuries. On the second of these topics, however, the best of the book-length studies (e.g., O'Malley and Singer) deal usually with special problems and persons and thus scarcely display an evolving scientific tradition. The literature on evolution, in the absence of adequate histories of the technical specialties which provided Darwin with both data and problems, is written at a level of philosophical generality which makes it hard to see how his *Origin of Species* could have been a major achievement, much less an achievement in the sciences. Dupree's model study of the botanist Asa Gray is among the few noteworthy exceptions.

As yet the new historiography has not touched the social sciences. In these fields the historical literature, where it exists, has been produced entirely by practitioners of the science concerned, Boring's *History of Experimental Psychology* being perhaps the outstanding example. Like the older histories of the physical sciences, this literature is often indispensable, but as history it shares their limitations. (The situation is typical for relatively new sciences: practitioners in these fields are ordinarily expected to know about the development of their specialties, which thus regularly acquire a quasi-official history; thereafter something very like Gresham's law applies.) This area therefore offers particular opportunities both to the historian of science and, even more, to the general intellectual or social historian, whose background is often especially appropriate to the demands of these fields. The preliminary publications of Stocking on the history of American anthropology provide a particularly fruitful example of the perspective which the general historian can apply to a scientific field whose concepts and vocabulary have only very recently become esoteric.

External history.   Attempts to set science in a cultural context which might enhance understanding both of its development and of its effects have taken three characteristic forms, of which the oldest is the study of scientific institutions. Bishop Sprat prepared his pioneering history of the Royal Society of London almost before that organization had received its first charter, and there have since been innumerable in-house histories of individual scientific societies. These books are, however, useful principally as source materials for the historian, and only in this century have students of scientific development started to make use of them. Simultaneously they have begun seriously to examine the other types of institutions, particularly educational, which may promote or inhibit scientific advance. As elsewhere in the history of science, most of the literature on institutions deals with the seventeenth century. The best of it is scattered through periodicals (the once standard book-length accounts are regrettably out-of-date) from which it can be retrieved, together with much else concerning the history of science, through the annual "Critical Bibliography" of the journal *Isis* and through the quarterly *Bulletin signalétique* of the Centre National de la Recherche Scientifique, Paris. Guerlac's classic study on the professionalization of French chemistry, Schofield's history of the Lunar Society, and a recent collaborative volume (Taton) on scientific education in France are among the very few works on eighteenth-century scientific institutions. For the' nineteenth, only Cardwell's study of England, Dupree's of the United States, and Vucinich's of Russia begin to replace the fragmentary but immensely suggestive remarks scattered, often in footnotes, through the first volume of Merz's *History of European Thought in the Nineteenth Century*.

Intellectual historians have frequently considered the impact of science on various aspects of Western thought, particularly during the seventeenth and eighteenth centuries. For the period since 1700, however, these studies are peculiarly unsatisfying insofar as they aim to demonstrate the influence, and not merely the prestige, of science. The name of a Bacon, a Newton, or a Darwin is a potent symbol: there are many reasons to invoke it besides recording a substantive debt. And the recognition of isolated conceptual parallels, e.g., between the forces that keep a planet in its orbit and the system of checks and balances in the U.S. constitution, more often demonstrates interpretive ingenuity than the influence of science on other areas of life. No doubt scientific concepts, particularly those of broad scope, do help to change extrascientific ideas. But the analysis of their role in producing this kind of change demands immersion in the literature of science. The older historiography of science does not, by its nature, supply what is needed, and the new historiography is too recent and its products too fragmentary to have had much effect. Though the gap seems small, there is no chasm that more needs bridging than that between the historian of ideas and the historian of science. Fortunately there are a few works to point the way. Among the more recent are Nicolson's pioneering studies of science in seventeenth- and eighteenth-century literature, Westfall's discussion of natural religion, Gillispie's chapter on science in the Enlightenment, and Roger's monumental survey of the role of the life sciences in eighteenth-century French thought.

The concern with institutions and that with ideas merge naturally in a third approach to scientific development. This is the study of science in a geographical area too small to permit concentration on the evolution of any particular technical specialty but sufficiently homogeneous to enhance an understanding of science's social role and setting. Of all the types of external history, this is the newest and most revealing, for it calls forth the widest range of historical and sociological experience and skill. The small but rapidly growing literature on science in America (Dupree, Hindle, Shryock) is a prominent example of this approach, and there is promise that current studies of science in the French Revolution may yield similar illumination. Merz, Lilley, and Ben-David point to aspects of the nineteenth century on which much similar effort must be expended. The topic which has, however, evoked the greatest activity and attention is the development of science in seventeenth-century England. Because it has become the center of vociferous debate both about the origin of modern science and about the nature of the history of science, this literature is an appropriate focus for separate discussion. Here it stands for a type of research: the problems it presents will provide perspective on the relations between the internal and external approaches to the history of science.

The Merton thesis. The most visible issue in the debate about seventeenth-century science has been the so-called Merton thesis, really two overlapping theses with distinguishable sources. Both aim ultimately to account for the special productiveness of seventeenth-century science by correlating its novel goals and values—summarized in the program of Bacon and his followers—with other aspects of contemporary society. The first, which owes something to Marxist historiography, emphasizes the extent to which the Baconians hoped to learn from the practical arts and in turn to make

science useful. Repeatedly they studied the techniques of contemporary craftsmen—glassmakers, metallurgists, mariners, and the like—and many also devoted at least a portion of their attention to pressing practical problems of the day, e.g., those of navigation, land drainage, and deforestation. The new problems, data, and methods fostered by these novel concerns are, Merton supposes, a principal reason for the substantive transformation experienced by a number of sciences during the seventeenth century. The second thesis points to the same novelties of the period but looks to Puritanism as their primary stimulant. (There need be no conflict. Max Weber, whose pioneering suggestion Merton was investigating, had argued that Puritanism helped to legitimize a concern with technology and the useful arts.) The values of settled Puritan communities—for example, an emphasis upon justification through works and on direct communion with God through nature—are said to have fostered both the concern with science and the empirical, instrumental, and utilitarian tone which characterized it during the seventeenth century.

Both of these theses have since been extended and also attacked with vehemence, but no consensus has emerged. (An important confrontation, centering on papers by Hall and de Santillana, appears in the symposium of the Institute for the History of Science edited by Clagett; Zilsel's pioneering paper on William Gilbert can be found in the collection of relevant articles from the *Journal of the History of Ideas* edited by Wiener and Noland. Most of the rest of the literature, which is voluminous, can be traced through the footnotes in a recently published controversy over the work of Christopher Hill.) In this literature the most persistent criticisms are those directed to Merton's definition and application of the label "Puritan," and it now seems clear that no term so narrowly doctrinal in its implications will serve. Difficulties of this sort can surely be eliminated, however, for the Baconian ideology was neither restricted to scientists nor uniformly spread through all classes and areas of Europe. Merton's label may be inadequate, but there is no doubt that the phenomenon he describes did exist. The more significant arguments against his position are the residual ones which derive from the recent transformation in the history of science. Merton's image of the Scientific Revolution, though long-standing, was rapidly being discredited as he wrote, particularly in the role it attributed to the Baconian movement.

Participants in the older historiographic tradition did sometimes declare that science as they conceived it owed nothing to economic values or religious doctrine. Nevertheless, Merton's emphases on the importance of manual work, experimentation, and the direct confrontation with nature were familiar and congenial to them. The new generation of historians, in contrast, claims to have shown that the radical sixteenth- and seventeenth-century revisions of astronomy, mathematics, mechanics, and even optics owed very little to new instruments, experiments, or observations. Galileo's primary method, they argue, was the traditional thought experiment of scholastic science brought to a new perfection. Bacon's naive and ambitious program was an impotent delusion from the start. The attempts to be useful failed consistently; the mountains of data provided by new instruments were of little assistance in the transformation of existing scientific theory. If cultural novelties are required to explain why men like Galileo, Descartes, and Newton were suddenly able to see well-known phenomena in a new way, those novelties are predominantly intellectual and include Renaissance Neoplatonism, the revival of ancient atomism, and the rediscovery of Archimedes. Such intellectual currents were, however, at least as prevalent and productive in Roman Catholic Italy and France as in Puritan circles in Britain or Holland. And nowhere in Europe, where these currents were stronger among courtiers than among craftsmen, do they display a significant debt to technology. If Merton were right, the new image of the Scientific Revolution would apparently be wrong.

In their more detailed and careful versions, which include essential qualification, these arguments are entirely convincing, up to a point. The men who transformed scientific theory during the seventeenth century sometimes talked like Baconians, but it has yet to be shown that the ideology which a number of them embraced had a major effect, substantive or methodological, on their central contributions to science. Those contributions are best understood as the result of the internal evolution of a cluster of fields which, during the sixteenth and seventeenth centuries, were pursued with renewed vigor and in a new intellectual milieu. That point, however, can be relevant only to the revision of the Merton thesis, not to its rejection. One aspect of the ferment which historians have regularly labeled "the Scientific Revolution" was a radical programmatic movement centering in England and the Low Countries, though it was also visible for a time in Italy and France. That movement, which even the present form of Merton's argument does make more comprehensible, drastically altered the appeal, the locus, and the nature of much scientific research during the seventeenth century, and the changes have been permanent. Very likely, as contemporary historians argue, none of these novel features played a large role in transforming scientific concepts during the seventeenth century, but historians must learn to deal with them nonetheless. Perhaps the following suggestions, whose more general import will be considered in the next section, may prove helpful.

Omitting the biological sciences, for which close ties to medical crafts and institutions dictate a more complex developmental pattern, the main branches of science transformed during the sixteenth and seventeenth centuries were astronomy, mathematics, mechanics, and optics. It is their development which makes the Scientific Revolution seem a revolution in concepts. Significantly, however, this cluster of fields consists exclusively of classical sciences. Highly developed in antiquity, they found a place in the medieval university curriculum where several of them were significantly further developed. Their seventeenth-century metamorphosis, in which university-based men continued to play a significant role, can reasonably be portrayed as primarily an extension of an ancient and medieval tradition developing in a new conceptual environment. Only occasionally need one have recourse to the Baconian programmatic movement when explaining the transformation of these fields.

By the seventeenth century, however, these were not the only areas of intense scientific activity, and the others—among them the study of electricity and magnetism, of chemistry, and of thermal phenomena—display a different pattern. As sciences, as fields to be scrutinized systematically for an increased understanding of nature, they were all novelties during the Scientific Revolution. Their main roots were not in the learned university tradition but often in the established crafts, and they were all critically dependent both on the new program of experimentation and on the new instrumentation which craftsmen often helped to introduce. Except occasionally in medical schools, they rarely found a place in universities before the nineteenth century, and they were meanwhile pursued by amateurs loosely clustered around the new scientific societies that were the institutional manifestation of the Scientific Revolution. Obviously these are the fields, together with the new mode of practice they represent, which a revised Merton thesis may help us understand. Unlike that in the classical sciences, research in these fields added little to man's understanding of nature during the seventeenth century, a fact which has made them easy to ignore when evaluating Merton's viewpoint. But the achievements of the late eighteenth and of the nineteenth centuries will not be comprehensible until they are taken fully into account. The Baconian program, if initially barren of conceptual fruits, nevertheless inaugurated a number of the major modern sciences.

Internal and external history. Because they underscore distinctions between the earlier and later stages of a science's evolution, these remarks about the Merton thesis illustrate aspects of scientific development recently discussed in a more general way by Kuhn. Early in the development of a new field, he suggests, social needs and values are a major determinant of the problems on which its practitioners concentrate. Also during this period, the concepts they deploy in solving problems are extensively conditioned by contemporary common sense, by a prevailing philosophical tradition, or by the most prestigious contemporary sciences. The new fields which emerged in the seventeenth century and a number of the modern social sciences provide examples. Kuhn argues, however, that the later evolution of a technical specialty is significantly different in ways at least foreshadowed by the development of the classical sciences during the Scientific Revolution. The practitioners of a mature science are men trained in a sophisticated body of traditional theory and of instrumental, mathematical, and verbal technique. As a result, they constitute a special subculture, one whose members are the exclusive audience for, and judges of, each other's work. The problems on which such specialists work are no longer presented by the external society but by an internal challenge to increase the scope and precision of the fit between existing theory and nature. And the concepts used to resolve these problems are normally close relatives of those supplied by prior training for the specialty. In short, compared with other professional and creative pursuits, the practitioners of a mature science are effectively insulated from the cultural milieu in which they live their extra-professional lives.

That quite special, though still incomplete, insulation is the presumptive reason why the internal approach to the history of science, conceived as autonomous and self-contained, has seemed so nearly successful. To an extent unparalleled in other fields, the development of an individual technical specialty can be understood without going beyond the literature of that specialty and a few of its near neighbors. Only occasionally need the historian take note of a particular concept, problem, or technique which entered the field from outside. Nevertheless, the apparent autonomy of the internal approach is misleading in essentials, and the passion sometimes expended in its defense has obscured important problems. The insulation of a mature scientific community suggested by Kuhn's analysis is an insulation primarily with respect to concepts and secondarily with respect to problem structure. There are, however, other aspects of scientific advance, such as its timing. These do depend critically on the factors emphasized by the external approach to scientific development. Particularly when the sciences are viewed as an interacting group rather than as a collection of specialties, the cumulative effects of external factors can be decisive.

Both the attraction of science as a career and the differential appeal of different fields are, for example, significantly conditioned by factors external to science. Furthermore, since progress in one field is sometimes dependent on the prior development of another, differential growth rates may affect an entire evolutionary pattern. Similar considerations, as noted above, play a major role in the inauguration and initial form of new sciences. In addition, a new technology or some other change in the conditions of society may selectively alter the felt importance of a specialty's problems or even create new ones for it. By doing so they may sometimes accelerate the discovery of areas in which an established theory ought to work but does not, thereby hastening its rejection and replacement by a new one. Occasionally, they may even shape the substance of that new theory by ensuring that the crisis to which it responds occurs in one problem area rather than another. Or again, through the crucial intermediary of institutional reform, external conditions may create new channels of communication between previously disparate specialties, thus fostering cross-fertilization which would otherwise have been absent or long delayed.

There are numerous other ways, including direct subsidy, in which the larger culture impinges on scientific development, but the preceding sketch should sufficiently display a direction in which the history of science must now develop. Though the internal and external approaches to the history of science have a sort of natural autonomy, they are, in fact, complementary concerns. Until they are practiced as such, each drawing from the other, important aspects of scientific development are unlikely to be understood. That mode of practice has hardly yet begun, as the response to the Merton thesis indicates, but perhaps the analytic categories it demands are becoming clear.

The relevance of the history of science. Turning in conclusion to the question about which judgments must be the most personal of all, one may ask about the potential harvest to be reaped from the work of this new profession. First and foremost will be more and better histories of science. Like any other scholarly discipline, the field's primary responsibility must be to itself. Increasing signs of its selective impact on other enterprises may, however, justify brief analysis.

Among the areas to which the history of science relates, the one least likely to be significantly affected is scientific research itself. Advocates of the history of science have occasionally described their field as a rich repository of forgotten ideas and methods, a few of which might well dissolve contemporary scientific dilemmas. When a new concept or theory is successfully deployed in a science, some previously ignored precedent is usually discovered in the earlier literature of the field. It is natural to wonder whether attention to history might not have accelerated the innovation. Almost certainly, however, the answer is no. The quantity of material to be searched, the absence of appropriate indexing categories, and the subtle but usually vast differences between the anticipation and the effective innovation, all combine to suggest that reinvention rather than rediscovery will remain the most efficient source of scientific novelty.

The more likely effects of the history of science on the fields it chronicles are indirect, providing increased understanding of the scientific enterprise itself. Though a clearer grasp of the nature of scientific development is unlikely to resolve particular puzzles of research, it may well stimulate reconsideration of such matters as science education, administration, and policy. Probably, however, the implicit insights which historical study can produce will first need to be made explicit by the intervention of other disciplines, of which three now seem particularly likely to be effective.

Though the intrusion still evokes more heat than light, the philosophy of science is today the field in which the impact of the history of science is most apparent. Feyerabend, Hanson, Hesse, and Kuhn have all recently insisted on the inappropriateness of the traditional philosopher's ideal image of science, and in search of an alternative they have all drawn heavily from history. Following directions pointed by the classic statements of Norman Campbell and Karl Popper (and sometimes also significantly influenced by Ludwig Wittgenstein), they have at least raised problems that

the philosophy of science is no longer likely to ignore. The resolution of those problems is for the future, perhaps for the indefinitely distant future. There is as yet no developed and matured "new philosophy" of science. But already the questioning of older stereotypes, mostly positivistic, is proving a stimulus and release to some practitioners of those newer sciences which have most depended upon explicit canons of scientific method in their search for professional identity.

A second field in which the history of science is likely to have increasing effect is the sociology of science. Ultimately neither the concerns nor the techniques of that field need be historical. But in the present underdeveloped state of their specialty, sociologists of science can well learn from history something about the shape of the enterprise they investigate. The recent writings of Ben-David, Hagstrom, Merton, and others give evidence that they are doing so. Very likely it will be through sociology that the history of science has its primary impact on science policy and administration.

Closely related to the sociology of science (perhaps equivalent to it if the two are properly construed) is a field that, though it scarcely yet exists, is widely described as "the science of science." Its goal, in the words of its leading exponent, Derek Price, is nothing less than "the theoretic analysis of the structure and behavior of science itself," and its techniques are an eclectic combination of the historian's, the sociologist's, and the econometrician's. No one can yet guess to what extent that goal is attainable, but any progress toward it will inevitably and immediately enhance the significance both to social scientists and to society of continuing scholarship in the history of science.

THOMAS S. KUHN

[*See also* Belaval 1968; Hellman 1968.]

BIBLIOGRAPHY

Agassi, Joseph 1963 *Towards an Historiography of Science.* History and Theory, Beiheft 2. The Hague: Mouton.

►Belaval, Yvon 1968 Koyré, Alexandre. Volume 8, pages 447–449 in *International Encyclopedia of the Social Sciences.* Edited by David L. Sills. New York: Macmillan and Free Press.

Ben-David, Joseph 1960 Scientific Productivity and Academic Organization in Nineteenth-century Medicine. *American Sociological Review* 25:828–843.

Boas, Marie 1958 *Robert Boyle and Seventeenth-century Chemistry.* Cambridge Univ. Press.

Boyer, Carl B. (1939) 1949 *The Concepts of the Calculus: A Critical and Historical Discussion of the Derivative and the Integral.* New York: Hafner. → A paperback edition was published in 1959 by Dover as *The History of the Calculus and Its Conceptual Development*.

BUTTERFIELD, HERBERT (1950) 1957 *The Origins of Modern Science, 1300–1800*. 2d ed., rev. New York: Macmillan. → A paperback edition was published in 1962 by Collier.

CARDWELL, DONALD S. L. 1957 *The Organisation of Science in England: A Retrospect*. Melbourne and London: Heinemann.

CLAGETT, MARSHALL 1959 *The Science of Mechanics in the Middle Ages*. Madison: Univ. of Wisconsin Press.

COHEN, I. BERNARD 1956 *Franklin and Newton: An Inquiry Into Speculative Newtonian Experimental Science and Franklin's Work in Electricity as an Example Thereof*. American Philosophical Society, Memoirs, Vol. 43.. Philadelphia: The Society.

COSTABEL, PIERRE 1960 *Leibniz et la dynamique: Les textes de 1692*. Paris: Hermann.

CROSLAND, MAURICE 1963 The Development of Chemistry in the Eighteenth Century. *Studies on Voltaire and the Eighteenth Century* 24:369–441.

DAUMAS, MAURICE 1955 *Lavoisier: Théoricien et expérimentateur*. Paris: Presses Universitaires de France.

DIJKSTERHUIS, EDWARD J. (1950) 1961 *The Mechanization of the World Picture*. Oxford: Clarendon. → First published in Dutch.

DUGAS, RENÉ (1950) 1955 *A History of Mechanics*. Neuchâtel (Switzerland): Éditions du Griffon; New York: Central Book. → First published in French.

DUHEM, PIERRE 1906–1913 *Études sur Léonard de Vinci*. 3 vols. Paris: Hermann.

DUPREE, A. HUNTER 1957 *Science in the Federal Government: A History of Policies and Activities to 1940*. Cambridge, Mass.: Belknap.

DUPREE, A. HUNTER 1959 *Asa Gray: 1810–1888*. Cambridge, Mass.: Harvard Univ. Press.

FEYERABEND, P. K. 1962 Explanation, Reduction and Empiricism. Pages 28–97 in Herbert Feigl and Grover Maxwell (editors), *Scientific Explanation, Space, and Time*. Minnesota Studies in the Philosophy of Science, Vol. 3. Minneapolis: Univ. of Minnesota Press.

GILLISPIE, CHARLES C. 1960 *The Edge of Objectivity: An Essay in the History of Scientific Ideas*. Princeton Univ. Press.

GUERLAC, HENRY 1959 Some French Antecedents of the Chemical Revolution. *Chymia* 5:73–112.

GUERLAC, HENRY 1961 *Lavoisier; the Crucial Year: The Background and Origin of His First Experiments on Combustion in 1772*. Ithaca, N.Y.: Cornell Univ. Press.

HAGSTROM, WARREN O. 1965 *The Scientific Community*. New York: Basic Books.

HANSON, NORWOOD R. (1958) 1961 *Patterns of Discovery: An Inquiry Into the Conceptual Foundations of Science*. Cambridge Univ. Press.

►HELLMAN, C. DORIS 1968 Sarton, George. Volume 14, pages 15–17 in *International Encyclopedia of the Social Sciences*. Edited by David L. Sills. New York: Macmillan and Free Press.

HESSE, MARY B. 1963 *Models and Analogies in Science*. London: Sheed & Ward.

HILL, CHRISTOPHER 1965 Debate: Puritanism, Capitalism and the Scientific Revolution. *Past and Present* No. 29:88–97. → Articles relevant to the debate may also be found in numbers 28, 31, 32, and 33.

HINDLE, BROOKE 1956 *The Pursuit of Science in Revolutionary America, 1735–1789*. Chapel Hill: Univ. of North Carolina Press.

INSTITUTE FOR THE HISTORY OF SCIENCE, UNIVERSITY OF WISCONSIN, *1957* 1959 *Critical Problems in the History of Science: Proceedings*. Edited by Marshall Clagett. Madison: Univ. of Wisconsin Press.

JAMMER, MAX 1961 *Concepts of Mass in Classical and Modern Physics*. Cambridge, Mass.: Harvard Univ. Press.

JOURNAL OF THE HISTORY OF IDEAS 1957 *Roots of Scientific Thought: A Cultural Perspective*. Edited by Philip P. Wiener and Aaron Noland. New York: Basic Books. → Selections from the first 18 volumes of the *Journal*.

KOYRÉ, ALEXANDRE 1939 *Études galiléennes*. 3 vols. Actualités scientifiques et industrielles, Nos. 852, 853, and 854. Paris: Hermann. → Volume 1: *À l'aube de la science classique*. Volume 2: *La loi de la chute des corps: Descartes et Galilée*. Volume 3: *Galilée et la loi d'inertie*.

KOYRÉ, ALEXANDRE 1961 *La révolution astronomique: Copernic, Kepler, Borelli*. Paris: Hermann.

KUHN, THOMAS S. 1962 *The Structure of Scientific Revolutions*. Univ. of Chicago Press. → A paperback edition was published in 1964.

LILLEY, S. 1949 Social Aspects of the History of Science. *Archives internationales d'histoire des sciences* 2:376–443.

MAIER, ANNELIESE 1949–1958 *Studien zur Naturphilosophie der Spätscholastik*. 5 vols. Rome: Edizioni di "Storia e Letteratura."

MERTON, ROBERT K. (1938) 1967 *Science, Technology and Society in Seventeenth-century England*. New York: Fertig.

MERTON, ROBERT K. 1957 Priorities in Scientific Discovery: A Chapter in the Sociology of Science. *American Sociological Review* 22:635–659.

METZGER, HÉLÈNE 1930 *Newton, Stahl, Boerhaave et la doctrine chimique*. Paris: Alcan.

MEYERSON, ÉMILE (1908) 1964 *Identity and Reality*. London: Allen & Unwin. → First published in French.

MICHEL, PAUL-HENRI 1950 *De Pythagore à Euclide*. Paris: Édition "Les Belles Lettres."

NEEDHAM, JOSEPH 1954–1965 *Science and Civilisation in China*. 4 vols. Cambridge Univ. Press.

NEUGEBAUER, OTTO (1951) 1957 *The Exact Sciences in Antiquity*. 2d ed. Providence, R.I.: Brown Univ. Press. → A paperback edition was published in 1962 by Harper.

NICOLSON, MARJORIE H. (1950) 1960 *The Breaking of the Circle: Studies in the Effect of the "New Science" Upon Seventeenth-century Poetry*. Rev. ed. New York: Columbia Univ. Press. → A paperback edition was published in 1962.

O'MALLEY, CHARLES D. 1964 *Andreas Vesalius of Brussels, 1514–1564*. Berkeley and Los Angeles: Univ. of California Press.

PANOFSKY, ERWIN 1954 *Galileo as a Critic of the Arts*. The Hague: Nijhoff.

PARTINGTON, JAMES R. 1962– *A History of Chemistry*. New York: St. Martins. → Volumes 2–4 were published from 1962 to 1964; Volume 1 is in preparation.

PRICE, DEREK J. DE SOLLA 1966 The Science of Scientists. *Medical Opinion and Review* 1:81–97.

ROGER, JACQUES 1963 *Les sciences de la vie dans la pensée française du XVIIIᵉ siècle: La génération des animaux de Descartes à l'Encyclopédie*. Paris: Colin.

SARTON, GEORGE 1927–1948 *Introduction to the History of Science*. 3 vols. Baltimore: Williams & Wilkins.

SCHOFIELD, ROBERT E. 1963 *The Lunar Society of Birmingham: A Social History of Provincial Science and Industry in Eighteenth-century England.* Oxford: Clarendon.

SHRYOCK, RICHARD H. (1936) 1947 *The Development of Modern Medicine.* 2d ed. New York: Knopf.

SINGER, CHARLES J. 1922 *The Discovery of the Circulation of the Blood.* London: Bell.

STOCKING, GEORGE W. JR. 1966 Franz Boas and the Culture Concept in Historical Perspective. *American Anthropologist* New Series 68:867–882.

TATON, RENÉ (editor) 1964 *Enseignement et diffusion des sciences en France au XVIIIᵉ siècle.* Paris: Hermann.

THORNDIKE, LYNN (1923–1958) 1959–1964 *A History of Magic and Experimental Science.* 8 vols. New York: Columbia Univ. Press.

TRUESDELL, CLIFFORD A. 1960 *The Rational Mechanics of Flexible or Elastic Bodies 1638–1788: Introduction to Leonhardi Euleri* Opera omnia Vol. X et XI seriei secundae. Leonhardi Euleri Opera omnia, Ser. 2, Vol. 11, part 2. Turin (Italy): Fussli.

VUCINICH, ALEXANDER S. 1963 *Science in Russian Culture.* Volume 1: A History to 1860. Stanford Univ. Press.

WESTFALL, RICHARD S. 1958 *Science and Religion in Seventeenth-century England.* New Haven: Yale Univ. Press.

WHITTAKER, EDMUND 1951–1953 *A History of the Theories of Aether and Electricity.* 2 vols. London: Nelson. → Volume 1: *The Classical Theories.* Volume 2: *The Modern Theories, 1900–1926.* Volume 1 is a revised edition of *A History of the Theories of Aether and Electricity From the Age of Descartes to the Close of the Nineteenth Century,* published in 1910. A paperback edition was published in 1960 by Harper.

YATES, FRANCES A. 1964 *Giordano Bruno and the Hermetic Tradition.* Univ. of Chicago Press.

Postscript

Some of the issues discussed in the main article are further developed in Kuhn (1971). Kuhn's discussion of the literature on the role of Hermeticism in the Scientific Revolution repairs an unfortunate omission from the main article. Further indication of trends in the study of the history of science is provided by Teich and Young (1973).

THOMAS S. KUHN

ADDITIONAL BIBLIOGRAPHY

KUHN, THOMAS S. 1971 The Relations Between History and History of Science. *Daedalus* 100:271–304.

TEICH, MIKULÁŠ; and YOUNG, ROBERT M. (editors) 1973 *Changing Perspectives in the History of Science.* London: Heinemann. → Essays in honor of Joseph Needham.

II

THE PHILOSOPHY OF SCIENCE

The general topic of the philosophy of science can be divided into subareas by subject matter: the philosophy of physics, the philosophy of biology, the philosophy of the social sciences, and so on. But it may also be divided into discussion of *structural* problems in science, on the one hand, and *substantive* problems within specified sciences, on the other. Structural problems are those traditionally associated with such topics as scientific inference, classification, explanation, prediction, measurement, probability, and determinism. Substantive problems arise when a question is asked such as "How can learning theory be axiomatized?" or "How should anxiety be defined?" Substantive problems, as a class, merge gradually into abstract scientific problems, which are properly handled by scientists themselves.

The substantive philosophy of science usually requires some professional training in the relevant field, whereas the structural philosophy of science may require only a general scientific education. Both require professional training in philosophy or, to be more specific, in logic, broadly conceived (that is, not only in classical or in symbolic logic). The history of science is also of very great importance to the philosopher of science, for it provides him with more numerous and more diverse examples of conceptual schemes and theories than he could ever find in contemporary science, and such schemes and theories are his raw data.

It will be seen, then, that a large part of the philosophy of science stands to scientific theorizing, classifying, and so on, as experimental design stands to scientific experimenting. For the philosophy of science is simply the discussion of general criteria for theories, classifications, and the rest, and is necessary as long as the intuitions or assertions of scientists about what constitutes good procedure are in conflict or might be improved by analysis. Curiously enough, however, many scientists reject the philosophy of science as irrelevant to their own activities, although they constantly talk it and teach it and illustrate its relevance in their work, sometimes under the title "methodology" and sometimes just as advice without a title. It is only the hyperspecialized, rock-ribbed empiricist who can entirely avoid discussing the subject in his scientific work and teaching, and even he usually contradicts himself by espousing operationism as the epitome of empiricism—a rather elementary mistake which would have grave consequences for the validity and utility of his results if he acted on it. Often the antagonism of science to philosophy seems due to a low tolerance of ambiguity, and particularly to insecurity about involvement with the more debatable and less decidable issues of philosophy. This is surely an immature reaction, unless it can be shown that these issues need not be raised in the teaching and practice of science.

This article will provide (a) a brief discussion of those aspects of the structural philosophy of science which have special relevance for the social sciences and (b) some reference to substantive issues in the philosophy of particular social sciences. In general, these aspects will be intertwined, in order to illustrate the immediate relevance of the structural topics to substantive issues.

Main concerns of the philosophy of science. Science has been said to be concerned with observation, description, definition, classification, measurement, experimentation, generalization, explanation, prediction, evaluation, and control of the world. This list is of course much too comprehensive; to be at all useful it has to be narrowed down in the course of examining individual scientific activities (for instance, instead of analyzing description in general, we might ask what *kind* of describing is scientific, as opposed to poetic). In any case, there are as many (though probably no more) points of view about main topics in the philosophy of science as about main topics in any of the sciences. If the area of dispute sometimes appears greater than it is, this is because when a subtopic in philosophy is settled, it is given a new name and called a science. Philosophy is the mother of sciences and has spawned physics, astronomy, symbolic logic, economics, and psychology, among other subjects. The domain of fundamental and still unresolved conceptual disputes at any time is called philosophy and, of course, bears the cross of conflict. But, perhaps for that very reason, it is a residue that eternally regenerates, is perpetually fertile. The social sciences began with a similar handicap; what was well known about human behavior was called common sense, and only the muddy residue was left for psychology and the other social sciences to clarify. It is no wonder that a subject is difficult and debatable when it is defined to exclude the easy and certain.

Because "science" is a term for an activity as well as for a body of knowledge, we ought not to be surprised if it is continuous with the prescientific activities directed to the same ends, and if these, in turn, are an extension of prelinguistic adaptive behavior. This synoptic view of man's cognitive development and current equipment should—and, I believe, does—provide some criteria for good scientific practice. We can begin by applying it to the various procedures connected with scientific concept formation: observation, description, definition, and classification, with measurement and generalization as closely related activities.

Concept clarification. It is now widely accepted in the philosophy of science that we must qualify or reject the traditional idea that the basic data from which the scientist forms his hypotheses and classifications are provided by a faculty of observation which is somehow independent of science. Not only has the attempt by sense-data theorists and phenomenologists to identify these basic data been a total failure, but the failure begins to appear explicable. We have come to see the "observation language"—the nonscientific language in which such basic data were supposed to be reported—as really just a "theory language," with somewhat higher interjudge reliability than more abstract language, yet still subject to perceptual errors in its application and even to pervasive errors caused by infiltration by unnoticed theory. When errors of the latter kind are recognized, we may redescribe our observations or we may retain the old language and simply reinterpret it, perhaps metaphysically. Thus Feyerabend has pointed out (in conversation with the author) that the expression "The sun is rising," although it originally incorporated a false astronomical theory, has been retained as a statement of pure observation. So there is no absolutely basic observation level on which all theory rests and, even more clearly, no sharp distinction between observation and theories. We must think instead of a constant interaction and exchange between observation, which seems immune from correction, and speculative theory. Indeed, the development of new instruments may simply render observable the hitherto unobservable (as when the electron microscope was developed) or the observed mistaken (as when it became possible to measure the curvature of the seemingly flat earth). Does this relativization of the distinction between observation and theory destroy the distinction's utility? Not at all. Indeed, it strengthens our understanding of the hierarchical structure of science as a whole when we realize that observables of the nth level (for instance, in neurophysiology) may include some hypothetical constructs from the (n + 1)th level (in this case, individual psychology).

What leads us, then, to introduce concepts at a given level? Not just the pressure of their existence on our blank brain. Rather, the preconscious selective effect of certain environmental and internal pressures began the process of concept formation in our ancestors, as in other organisms—a sequence of events that can be summarized as "discrimination under drive conditions." The preverbal organism comes to react to certain stimulus configurations that represent ingestibles or predators, and it is simply an extension of this process when an interpersonal vocabulary is introduced to refer to these configurations. The psychologist refers to both the preverbal and the verbal stages of this process as concept formation, and although many philosophers prefer to restrict the term to the verbal stage, there is no doubt that the verbal stage develops from the preverbal one, whether or not it becomes qualitatively different in the course of development. (See Kendler & Kendler 1968.)

Definition. Given the pragmatic view of concept formation outlined above, it is very natural to abandon classical definition theory (in which a definition is a set of logically necessary and sufficient conditions for the use of a term) and to adopt instead an approach based on the notions of indicator and of indicator clusters. The rationale for such an approach may be stated as follows. Within a formal system, such as mathematics or an invented language, we can find examples of classical definitions, but seldom, if ever, can we find them within a natural language or the language of a developing science. The reason for this is simply that natural phenomena are extremely untidy, and a language for referring to them must be flexible enough to accommodate this untidiness if it is to be scientifically useful. For instance, the concepts "subject," "length," "response," "group," "aggression," and "temperature" cannot be given classical definitions (other than trivially circular ones). But we have no serious difficulty in learning or teaching these concepts, which we do by giving paradigm examples, contrasts, and explicitly approximate or conditional definitions. Any paradigm provides us, if it can be successfully analyzed, with a set of sufficient conditions for the application of the term, but it is not possible to insist that paradigms, let alone analysis of them, can never turn out to be erroneous. So the sense in which these conditions are sufficient is still not that of logical sufficiency. It is still more obvious that we often cannot specify any properties that are logically necessary for the application of the concept. We therefore adopt the notion of "weighted indicators," or criteria—that is, properties which are relevant to the application of the term, but in varying degrees. Some of the properties may, in some cases, be virtually necessary; some small groups of them may be sufficient; but these are special cases. The clustering of these properties in property space indicates an entity for which it is useful to have a name, but there is typically considerable scatter and no sharp boundary to the cluster (the so-called open texture of concepts). The charm of a concise translation (that is, of a classical definition) is not often provided by the cluster approach, but it does preserve that most valuable feature of classi-

cal theory—a reliable verbal response that can be successfully taught. Contrasting and approximating, which are the other procedures involved in specification of meaning, depend on the psychological superiority of discrimination skills over absolute recognition skills and involve no new logical points.

At this point, it should be stressed that there are two elements in the theory of definition: the pragmatic and the literal. We have been talking about the limitations of the literal theory of definition as a kind of translation. But there is the independent question of how one can best—that is, most usefully—select the terms that one will mention in the definition, whether it is a translation or a cluster of indicators. A famous answer to this has often been taken as both literal and pragmatic, which makes it nonsensical instead of good pragmatic advice of a limited range of applicability. I refer to operationism (for which see especially Bridgman 1927). There are, of course, some areas where it is to be taken very seriously as advice, and psychology is one of them. (In theoretical physics, contrary to the common impression, it is extremely rare to seek or find operational definitions.) But if operationism is taken as requiring that definitions actually contain nothing but operations, then it is absurd, a category error —and, in addition, it is self-refuting, since the concept of an operation cannot be defined operationally. Definitions should therefore be of such a kind that we can apply some independently determinable criteria to decide when the defined term should be used (we can, if we wish, call the process of determining whether these criteria apply "operations"). But we cannot require that the only content of definitions should be operations, or we find ourselves caught in the dilemma of deciding when we have only one operation and, hence, only one concept . . . and so on.

Classification. The traditional requirements for classification also need to be relaxed. Instead of the Aristotelian requirement of exclusive and exhaustive subcategories, we must settle for minimized overlap and maximized coverage, balancing these desiderata against considerations of simplicity and applicability. The development of a scientific classification scheme is essentially a remapping of property space, using a new system of coordinates or a new projection. Hence, assessment of the scheme's value very much depends, as in the map analogy, on the purposes for which it is intended (which may be as diverse as fast labeling, representation of a new theoretical insight, and easy recall) and very little on the idealized and context-free notion of "cutting Nature at the joints," which was the goal of earlier taxonomists (and which still admirably expresses the feeling inspired by a successful classification scheme).

Measurement. The urge to measure is a kind of conditioned reflex among empirically trained scientists, and with some reason. Measurement requires—or, when developed, provides—an inter-subjectively reliable procedure for applying the potentially infinite descriptive vocabulary of number. For describing any continuous variable we need such a vocabulary, and for describing many discrete variables it is exceedingly useful because it creates the possibility of applying some mathematical apparatus. But the social scientist all too often assumes that scaling and measuring will make it likely that we can discover useful laws. In fact, many scales disguise the existence of regularities, and many others fail to reveal them. The descriptive utility of a scale must be considered independently of its theoretical fertility, for the two considerations may lead in different directions. The proliferation of scales, like the proliferation of statistical analyses, often merely clogs the channels of communication. There is no magic about numbers per se, and only hard thinking and good fortune, in combination, achieve a worthwhile new scale. The same is true of classification. The cluster concept is peculiarly suited to the needs of statistical scaling and to those of statistical inference in general. [See CLUSTERING; see also Torgerson 1968.]

Evaluation. The notion of measurement phases quite gradually into that of evaluation. "Pecking order," "peer-group ranking," and "intelligence quotient" can all, in certain contexts, appear as value-impregnated scales. This is often disputed by those who think "value" is a dirty word. Social scientists are peculiarly susceptible to the charms of the Weberian myth of a value-free social science, which rests on a one-sided interpretation of the distinction between facts and values. In its more defensible form this distinction amounts to the assertion that (a) evaluation depends on establishing certain facts; and (b) evaluation goes beyond the facts on which it depends. To this it can be objected that evaluation may well *result* in facts, albeit facts different from those with which it begins.

Thus *Consumer Reports* can often substantiate evaluations such as "Brand X is the best dishwasher on the market today" on the basis of performance data plus data about the consumers' wants. And that evaluation, when suitably supported, simply states a fact about Brand X. It is one thing to be cautious about importing *dubious* value premises into an argument; it is another to suppose that all value premises or conclusions are dubious, "mere matters of opinion." It is not at all dubious that dishwashers should get the dishes clean, that clocks should keep good time, and so on. It may simply be a fact that Brand X of dishwasher excels at its task. Since the evaluation is nothing more than a combination of these facts, it is itself a fact, although it is also an evaluation.

Precisely this analysis applies to myriad issues in the social sciences—evaluation of the relative merits of protective tariff barriers and direct subsidies or of different forms of treatment for dysfunctional neurosis, to name only two examples. Of course, there are many other issues where only a relativized answer is possible, since legitimate differences of taste exist (that is, differences where no argument can establish the superiority of any particular taste). But this in no way distinguishes evaluation from theoretical interpretation. It will not do to say that "in principle" these differences can be resolved in the theoretical area when more facts are available, but not in the evaluation area. Not only are there very fundamental disagreements about the function of theory that make this improbable, but additional facts often reduce value disagreement. In any case, *relativized* answers, which are still value judgments, admit of no disagreement, and therefore do not involve subjectivity. It is simply naive to suppose that fields must, if they are to be regarded as objective, always contain answers to questions of the form "Which is the best X?" Neither mathematics nor engineering has this property, or any corresponding property with respect to "Which is *the* true solution to X?," where the conditions of X are incompletely specified.

As to the final defense of the distinction between fact and value—that a particular choice of ultimate values, or ends, cannot be proved true—three answers apply: (1) No such choice has to be proved true, only relevant. (2) No one has yet produced a demonstrably ultimate value, and it seems clear that the hierarchical model is as inappropriate here as it is in epistemology. (There are no ultimate, or basic, facts, only some that are more reliable than others; and all can—under pressure—be given support by appeal to others.) (3) In the case of moral values, where some need for a single answer can be demonstrated, there is a perfectly good way of handling any moral disputes that are likely to arise in the social sciences. This is by appealing to the equality of prima-facie rights—the fundamental axiom of democracy. Justification of this is itself one of the most important tasks of the social sciences and requires contributions from game theory, political science, economics, sociology, and psychology, as well as from the obvious candidate, anthropology. It is the problem of deciding on the best system of sublegal mores—insofar, of course, as any answer is possible. But this *type* of problem is already familiar: it is an efficiency problem.

Indeed, the social sciences have for too long been reticent about pointing out the extent to which it is possible both to give answers to moral problems and to refute some of the current answers to them. For some reason, the sound point that there are many areas of morality where there cannot yet be any proof of the superiority of a particular answer is taken to imply that acting as if all answers were correct is defensible. But of course the only defensible position in such a situation is a compromise or, if possible, a suspension of action. In my opinion, the axiom of equal prima-facic rights can be shown to be superior to any alternative, and also to be capable of generating a complete moral system. It achieves both of these ends by being the only such axiom that can be satisfactorily supported, and of course it roughly coincides with the Golden Rule and other basic maxims (details of this line of argument will be found in Scriven 1966). At any rate, it is time that the possibility of this kind of "strategic ethics" was examined by all the strategists who have sprung up to look at other kinds of social interaction in terms of the language of games.

In summary, the scientific approach to evaluation necessarily occurs in both applied science (finding the best solution to a practical problem) and pure science (finding the best estimate, hypothesis, or experimental design). Moreover, the element of evaluation enters into much "straight" measurement—that of intelligence, creativity, and social acceptability, for instance. The reason for this is as simple as the reason for introducing new classifications: we have to order our knowledge if we are to use it efficiently. It follows that the most appropriate kind of order is determined by the use to which that knowledge is to be put. One might put the point paradoxically and say that all science is applied science. The application may be to efficient explanation, accurate prediction, or improvement of the world, but it is still—in one important sense—an application. Nor can we assess (that is, evaluate) theories or instruments unless we have some analysis of their proposed use.

Data analysis. There are obviously limitations on the memory storage capacity of the human brain that would make it impotent in the face of even a modest amount of data about the path of a projectile or a planet, if these data were presented in discrete form. But if we can find some simple pattern that the data follow, especially if it is a pattern that we are familiar with from some other, preferably visual, experience, then we can handle it quite easily. (The preference for visual models is due to the vastly greater channel capacity of the visual modality; the notion of "handling" involves more than simple recovery, and even that involves more than storage capacity, since it includes look-up and read-out time.) This simple point can be taken as the start for a complete data-processing approach to the aspects of structural philosophy of science that remain to be discussed. It also pervades the parts already discussed, for the introduction of concepts is itself a device to cope with the "buzzing, blooming confusion" of sense experience.

Generalizations and laws. The simplest examples of the cognitive condensation described in the preceding section are laws and generalizations. They are so necessary that we are prepared to forgive them a very high degree of inaccuracy if only they will simplify the range of possibilities for us. Contrary to Karl Popper's view, which stresses the equivalence of a law to the denial of counterinstances, the present view suggests that we are rightly very unimpressed by counterinstances in this irregular world, and extremely thankful if we can find a regularity that provides a reasonable approximation over a good range of instances. This is as true in the physical sciences as in the social, for the laws of gases, the laws of motion, gravitation, radiation, and so on, are all merely approximations at best. There are occasional exceptions (for example, the third law of thermodynamics), but the important feature of a law is that it introduces a pattern where none was before. In this sense its truth is its utility, not its lack of error. A great deal has been made in the literature of the fact that a law will support counterfactual inferences while an accidental generalization will not (Nagel 1961, pp. 68–73; compare Strawson 1952, pp. 85, 197, 200). On the present view, all this means is that we think natural laws are true of continuing or pervasive phenomena, i.e., of the nature of things. All other attempts to give logical analyses of the concept of law—for instance, the view that laws cannot contain reference to particular entities, or must support predictions—seem to me simply arbitrary.

Explanation. The task of explanation is the integration of "new" phenomena (whether subjectively or objectively new makes no difference) into the structure of knowledge. Typically, this consists in fitting these phenomena into a pattern with which we are already familiar. But explanation is not simply reduction to the familiar. Indeed, it may involve reducing the familiar to the unfamiliar, as Popper (1963, p. 63) has pointed out. But it must involve reducing the uncomprehended phenomenon to phenomena that are comprehended, and it is hardly surprising if this sometimes increases the sense of familiarity.

It has long been argued that the way in which explanation is performed in science always involves subsuming the phenomenon to be explained under a known law [see SCIENTIFIC EXPLANATION]. In view of the known deficiencies in our scientific laws, explanation, if this theory were acceptable, would in any case be a rather imprecise affair. But the restriction to laws is too confining, unless one treats the term "law" so loosely that it covers almost any statement except a pure particular. The real danger in this model of explanation—the "deductive" model, as it is usually called—is that it seems to support the view that explanations are closely related to predictions. In contrast, the "pattern" model of explanation, which I prefer, allows for retrospectively applicable patterns as well as for those which can be grasped in advance of their completion. It happens that this is very important in the behavioral sciences, since there is, and always will be, a real shortage of "two-way" laws (that is, laws that both predict and explain). This has typically been treated by social scientists as a sign of the immaturity of their subject; but in fact it is simply a sign of its nature and is very like the situation in the "messier" areas of the physical sciences, especially engineering. When a bridge fails, or a mob riots, or a patient commits suicide, there are often a number of possible explanations, of which only one "fits the facts." We thus have no difficulty in saying that this one is *the* explanation, although the facts available before the event were not enough to permit us to tell what would happen. Indeed, it may be that the chances in advance were very strongly against the actual outcome, and that we would therefore have been entitled to make a well-founded scientific prediction of the event's nonoccurrence. It is not helpful to suggest in such cases that we are explaining by deduction from a law. What sort of law? That some factors *can* cause others? That *only* certain factors cause a certain effect? These are indeed general propositions, but they do not look much like laws of the usual kind, and they do not support any predictions at all. It is not in the least important whether or not we call them

laws, but it is important to understand that we often support explanations by appealing to general claims that will not support predictions.

Prediction. On the other hand, it is often possible to predict on the basis of laws that do not explain. Any reliable correlation, however little we understand it, will give us a basis for predicting (that is, of course, provided there is temporal separation between the two correlated factors). To be able to predict, we need only to be able to infer that an event must occur at some conveniently remote time in the future. To explain, on the other hand, we have only to show *why* something occurred, not necessarily that it *had* to occur.

For a long time it was thought that the only way we could be sure to avoid *ad hoc* explanations was to require that (*a*) the grounds for the explanation be known to be true and (*b*) these same grounds imply only one possible outcome, the event or phenomenon to be explained. But there are many ways in which we can show that a particular explanation is correct in a particular case without having to show that the factors mentioned in the explanation would enable one to infer the explained event.

For example, we may demonstrate the presence and operation of a particular cause by exhibiting some clues that uniquely identify its *modus operandi*. Of course, we may still have reason to believe that there are certain other factors present which, in conjunction with the factors we have identified, would enable us to make a formal inference that the event we have explained was bound to occur. But in many cases we do not know what those other factors are, and we do not know the laws that would make the inference possible. In fact, our belief that we may still be able to make this inference merely indicates that we believe in determinism. It certainly does not prove that a good scientific explanation demonstrates the necessity of what it explains.

No doubt it is tempting to remark that a *complete* explanation would have this property, but even this approach is wrong. In the first place, there are many so-called complete scientific explanations that do not have it. Second, this approach recommends a use of the term "complete explanation" that has the unpromising property of identifying mere subsumption under a generalization as complete explanation. But the genius of Freud, for instance, was not manifested by proving that the "explanation" of obsessive behavior in a particular neurotic was that this behavior was common in neurotics. Rather, he claimed to show a causal connection between certain other factors and obsessive behavior; and a causal connection is neither as strong nor as weak as a high correlation, it is simply quite different.

Causes. We are often able to identify causes in the social sciences without being able to give the laws and other factors in conjunction with which they operate to bring about their effects. The basic control group study is a case in point. It may give us excellent grounds for supposing, let us say, that a particular drug accelerates the subjects' rate of response, but it does not tell us what other factors are necessary conditions for this effect. Of course, we know that it is some set of the factors which are present, but we do not know which set. In more complex cases, even though a particular factor does not reliably lead to this effect, we may establish it as the cause simply by elimination of the alternative possible causes, either through their absence or through the presence of clues showing that, although present, they could not have been active this time. Of course, causal explanations are incomplete in the sense that there are always further interesting questions about the phenomenon that we would like answered. But this is also true even when we can subsume the explanation under a law: for instance, we can ask why the law is true. But causal explanations are often complete in the sense that they tell us exactly what we need to know, for our interest in a phenomenon is often very specific. Thus explanation is a context-dependent notion; explanations are devices for filling in our understanding, and the notion of *the* explanation makes sense only when there is a standard background of knowledge and understanding, as in the normal pedagogical development of an academic subject.

Understanding. The key notion behind that of explanation, and hence that of cause, is understanding. It must not be thought of purely as a subjective feeling; the feeling is only something associated with it. The condition of understanding itself is an objectively testable one; in fact, it regularly *is* tested by examinations, which are supposed to stress the detection of "real understanding," not just rote knowledge. Understanding is integrated, related knowledge; more generally—so that the definition will apply to the understanding possessed by machines, animals, and children—it is the capacity to produce the appropriate response to novel stimuli within a certain range or field. It is a characteristic of tests for understanding that they present the student with new problems, and the possession of understanding is valuable just because it does embody this capacity to handle novelty.

Approaching it in this way, we see immediately that the question of how the brain can provide this capacity is of interest not only from a neurophysiological point of view but possibly also as a way of obtaining better insight into the nature of understanding itself. In fact, a remarkable overlap emerges between the requirements of efficient storage and understanding, as it is commonly conceived. Efficient storage, both to economize on storage capacity and to facilitate fast recovery and input, must use mapping, modeling, and simplifying procedures. But these are exactly what we use in the process of trying to understand phenomena, even in the more pragmatic process of trying to improve our classifications, descriptions, and predictions. Thus, to mention one example of how fruitful this conception is, the idea of simplicity, which has often been recognized as a criterion for satisfactory scientific theories, can be seen as equivalent to the familiar notion of economy, not as some mysterious aesthetic requirement. Nor is the whole aim of understanding to be seen as a concession to human weakness; for the most sophisticated and capacious computers we shall ever design will be hard pressed to store even part of the brain's data and will therefore have to employ procedures that economize on storage space. Putting the matter more strongly, a simple map stores an infinity of facts with very little complication in the read-out procedures, in comparison with item storage, and an enormous improvement in speed of readout (that is, of recall); it also handles a quantity of information that would be beyond the capacity of ordinary item storage. (The difference between the use of models and the use of mnemonics is that the model stores an indefinite number of facts, which the mnemonic does not, and the model stores them in a way that is more directly related to truth.) Models, from analogies to axiomatizations, are the key to understanding. (See Martindale 1968.)

Experimentation. The use of applied science for the purpose of controlling reality has its counterpart within the methodology of science itself in the area of experimentation. Thus applied science is to pure science as experimentation is to pure observation. In each case the armory of relevant methods is importantly different: the "purer" subjects are more concerned with understanding and description; the more applied ones, with causation and manipulation. Indeed, it does not appear possible to give an analysis of "cause" without some reference to manipulation, though a great deal can be said before coming to the irreducible element of intervention or action. [*See* CAUSATION.]

A number of topics have not been discussed here, although they are closely related. For example, the problem of the reduction of one science or theory to another is a combination of a problem about explanation with one about definition. I have also said very little, even implicitly, about the issue of free will versus determinism, and wish only to state that it now seems possible to reconcile the two positions, provided we first make a distinction between predictive and explanatory determinism. In conclusion, I should stress my conviction that there is very little in the social sciences that does not have a parallel in the physical sciences, but it has not been to these parallels that social scientists have turned for paradigms. They have turned instead to the absurdly oversimple paradigms of Newtonian mechanics and astronomy. Even there, the significant fact is that the problem of predicting the motion of pure point-masses, moving under the sole influence of the very simple force of gravity, passes from the realm where solutions and predictions are possible to the realm where solutions give way to (at best) explications as soon as the number of bodies is increased from two to three. It is surely absurd to imagine that the forces acting on human beings are simpler than those involved in the "three-body problem." And if this is so, then we need a new "methodology of the complex domain," perhaps along the lines indicated above.

MICHAEL SCRIVEN

[*Directly related are the entries* CAUSATION; EXPERIMENTAL DESIGN; MODELS, MATHEMATICAL; MULTIVARIATE ANALYSIS; PREDICTION; PROBABILITY; SCIENCE, *article on* THE HISTORY OF SCIENCE; SCIENTIFIC EXPLANATION; STATISTICS, *article on* THE FIELD; *and the biographies of* GALTON; PEARSON; PEIRCE; YULE. *See also* Albert 1968; Belaval 1968; Bendix 1968; Kaplan 1968; Martindale 1968; Rapoport 1968; Rosenfield 1968; Rotwein 1968; Rynin 1968; Torgerson 1968; Williams 1968.]

BIBLIOGRAPHY

Contemporary philosophers of science can be roughly classified as either post-positivists (who form the majority) or linguistic analysts. For examples of work by leading post-positivists, see Popper 1934; 1957; Carnap 1950; Reichenbach 1951; Hempel 1952; 1965; Braithwaite 1953; Nagel 1961; Feyerabend 1962; Kemeny & Snell 1962; Suppes & Zinnes 1963. *There are many important topics on which these authors disagree, but it seems fair to say that among the predecessors they would be prepared to acknowledge are such founders of modern positivism as Karl Pearson, Moritz Schlick, and Ernst Mach. Other outstanding earlier writers who, although somewhat untypical of the field, have contributed to the philosophy of science in general, not just the post-positivist branch of it, include C. S. Peirce, Max Weber, N. R. Campbell, P. W. Bridgman, and Harold Jeffreys. Works by philosophers of science who use at least some of the methods of the linguistic analysts, even*

if they would not all classify themselves as such, include, in the study of the physical sciences, Toulmin 1953; Polanyi 1958; *in the study of the social sciences,* Dray 1957; Winch 1958; Kaplan 1964; Scriven 1966; *and, in the field of measurement and probability,* Ellis 1966; Hacking 1965. *Certain trends in the history of science, particularly in the emphasis, developed in* Butterfield 1950 *and* Kuhn 1962, *and by Alexandre Koyré, on autonomous conceptual schemes as units of progress, have influenced the philosophy of science. These trends have in turn been influenced by philosophical works such as* Feyerabend 1962; Hesse 1963. *Among the extremely valuable contributions to the philosophy of science made by practicing social scientists, special note should be taken of* Simon 1947–1956; Skinner 1953; Meehl 1954; Stevens 1958; Chomsky 1966. *Of the general treatises available in 1966, only* McEwen 1963 *and* Kaplan 1964 *are concerned principally with the social sciences, although* Nagel 1961 *and* Hempel 1965 *deal with them to some extent.* Lazarsfeld 1954; Natanson 1963; Braybrooke 1966 *are interesting collections that focus exclusively on the social sciences. As an introduction to the philosophical problems of the social sciences,* Kaplan 1964 *is to be recommended to layman and expert alike, although the work takes an approach somewhat different from that presented in this article.*

►ALBERT, ETHEL M. 1968 Values: II. Value Systems. Volume 16, pages 287–291 in *International Encyclopedia of the Social Sciences.* Edited by David L. Sills. New York: Macmillan and Free Press.

►BELAVAL, YVON 1968 Koyré, Alexandre. Volume 8, pages 447–449 in *International Encyclopedia of the Social Sciences.* Edited by David L. Sills. New York: Macmillan and Free Press.

►BENDIX, REINHARD 1968 Weber, Max. Volume 16, pages 493–502 in *International Encyclopedia of the Social Sciences.* Edited by David L. Sills. New York: Macmillan and Free Press.

BRAITHWAITE, R. B. 1953 *Scientific Explanation: A Study of the Function of Theory, Probability and Law in Science.* Cambridge Univ. Press. → A paperback edition was published in 1960 by Harper.

BRAYBROOKE, DAVID (editor) 1966 *Philosophical Problems of the Social Sciences.* New York: Macmillan.

BRIDGMAN, P. W. (1927) 1946 *The Logic of Modern Physics.* New York: Macmillan.

BUTTERFIELD, HERBERT (1950) 1957 *The Origins of Modern Science, 1300–1800.* 2d ed., rev. New York: Macmillan. → A paperback edition was published in 1962 by Collier.

CAMPBELL, NORMAN R. (1920) 1957 *Foundations of Science: The Philosophy of Theory and Experiment.* New York: Dover. → First published as *Physics: The Elements.*

CARNAP, RUDOLF (1950) 1962 *The Logical Foundations of Probability.* 2d ed. Univ. of Chicago Press.

CHOMSKY, NOAM 1966 *Cartesian Linguistics: A Chapter in the History of Rationalist Thought.* New York: Harper.

DRAY, WILLIAM (1957) 1964 *Laws and Explanation in History.* Oxford Univ. Press.

ELLIS, BRIAN 1966 *Basic Concepts of Measurement.* Cambridge Univ. Press.

FEIGL, HERBERT; and SCRIVEN, MICHAEL (editors) 1956 *The Foundations of Science and the Concepts of Psychology and Psychoanalysis.* Minnesota Studies in the Philosophy of Science, Vol. 1. Minneapolis: Univ. of Minnesota Press.

FEYERABEND, P. K. 1962 Explanation, Reduction and Empiricism. Pages 28–97 in Herbert Feigl and Grover Maxwell (editors), *Scientific Explanation, Space, and*

Time. Minnesota Studies in the Philosophy of Science, Vol. 3. Minneapolis: Univ. of Minnesota Press.

HACKING, IAN 1965 *Logic of Statistical Inference*. Cambridge Univ. Press.

HEMPEL, CARL G. 1952 Fundamentals of Concept Formation in Empirical Science. Volume 2, number 7, in *International Encyclopedia of Unified Science*. Univ. of Chicago Press.

HEMPEL, CARL G. 1965 *Aspects of Scientific Explanation, and Other Essays in the Philosophy of Science*. New York: Free Press.

HESSE, MARY B. 1963 *Models and Analogies in Science*. London: Sheed & Ward.

JEFFREYS, HAROLD (1931) 1957 *Scientific Inference*. 2d ed. Cambridge Univ. Press.

KAPLAN, ABRAHAM 1964 *The Conduct of Inquiry: Methodology for Behavioral Science*. San Francisco: Chandler.

►KAPLAN, ABRAHAM 1968 Positivism. Volume 12, pages 389–395 in *International Encyclopedia of the Social Sciences*. Edited by David L. Sills. New York: Macmillan and Free Press.

KEMENY, JOHN G.; and SNELL, J. LAURIE 1962 *Mathematical Models in the Social Sciences*. Boston: Ginn.

►KENDLER, HOWARD H.; and KENDLER, TRACY S. 1968 Concept Formation. Volume 3, pages 206–211 in *International Encyclopedia of the Social Sciences*. Edited by David L. Sills. New York: Macmillan and Free Press.

KUHN, THOMAS S. 1962 *The Structure of Scientific Revolutions*. Univ. of Chicago Press. → A paperback edition was published in 1964.

LAZARSFELD, PAUL F. (editor) (1954) 1955 *Mathematical Thinking in the Social Sciences*. 2d ed., rev. Glencoe, Ill.: Free Press.

McEWEN, WILLIAM P. 1963 *The Problem of Social-scientific Knowledge*. Totowa, N.J.: Bedminster.

►MARTINDALE, DON 1968 Verstehen. Volume 16, pages 308–313 in *International Encyclopedia of the Social Sciences*. Edited by David L. Sills. New York: Macmillan and Free Press.

MEEHL, PAUL E. (1954) 1956 *Clinical Versus Statistical Prediction: A Theoretical Analysis and a Review of the Evidence*. Minneapolis: Univ. of Minnesota Press.

NAGEL, ERNEST 1961 *The Structure of Science: Problems in the Logic of Scientific Explanation*. New York: Harcourt.

NATANSON, MAURICE (editor) 1963 *Philosophy of the Social Sciences: A Reader*. New York: Random House.

POLANYI, MICHAEL 1958 *Personal Knowledge: Towards a Post-critical Philosophy*. Univ. of Chicago Press.

POPPER, KARL R. (1935) 1959 *The Logic of Scientific Discovery*. New York: Basic Books. → First published as *Logik der Forschung*.

POPPER, KARL R. 1957 *The Poverty of Historicism*. Boston: Beacon.

POPPER, KARL R. 1963 *Conjectures and Refutations: The Growth of Scientific Knowledge*. New York: Basic Books; London: Routledge.

►RAPOPORT, ANATOL 1968 Systems Analysis: I. General Systems Theory. Volume 15, pages 452–458 in *International Encyclopedia of the Social Sciences*. Edited by David L. Sills. New York: Macmillan and Free Press.

REICHENBACH, HANS (1951) 1959 Probability Methods in Social Science. Pages 121–128 in Daniel Lerner and Harold D. Lasswell (editors), *The Policy Sciences:*

Recent Developments in Scope and Method. Stanford Univ. Press.

►ROSENFIELD, LEONORA COHEN 1968 Cohen, Morris R. Volume 2, pages 540–542 in *International Encyclopedia of the Social Sciences*. Edited by David L. Sills. New York: Macmillan and Free Press.

►ROTWEIN, EUGENE 1968 Hume, David. Volume 6, pages 546–550 in *International Encyclopedia of the Social Sciences*. Edited by David L. Sills. New York: Macmillan and Free Press.

►RYNIN, DAVID 1968 Schlick, Moritz. Volume 14, pages 52–56 in *International Encyclopedia of the Social Sciences*. Edited by David L. Sills. New York: Macmillan and Free Press.

SCRIVEN, MICHAEL 1956 A Possible Distinction Between Traditional Scientific Disciplines and the Study of Human Behavior. Pages 88–130 in Herbert Feigl and Michael Scriven (editors), *The Foundations of Science and the Concepts of Psychology and Psychoanalysis*. Minnesota Studies in the Philosophy of Science, Vol. 1. Minneapolis: Univ. of Minnesota Press.

SCRIVEN, MICHAEL 1959 Truisms as the Grounds for Historical Explanations. Pages 443–475 in Patrick L. Gardiner (editor), *Theories of History*. Glencoe, Ill.: Free Press.

SCRIVEN, MICHAEL 1962 Explanations, Predictions and Laws. Pages 170–230 in Herbert Feigl and Grover Maxwell (editors), *Scientific Explanation, Space, and Time*. Minnesota Studies in the Philosophy of Science, Vol. 3. Minneapolis: Univ. of Minnesota Press.

SCRIVEN, MICHAEL 1965 An Essential Unpredictability in Human Behavior. Pages 411–425 in Benjamin B. Wolman and Ernest Nagel (editors), *Scientific Psychology: Principles and Approaches*. New York: Basic Books.

SCRIVEN, MICHAEL 1966 *Primary Philosophy*. New York: McGraw-Hill.

SIMON, HERBERT A. (1947–1956) 1957 *Models of Man; Social and Rational: Mathematical Essays on Rational Human Behavior in a Social Setting*. New York: Wiley.

SKINNER, B. F. (1953) 1964 *Science and Human Behavior*. New York: Macmillan.

STEVENS, S. S. 1958 Problems and Methods of Psychophysics. *Psychological Bulletin* 55:177–196.

STRAWSON, P. F. 1952 *Introduction to Logical Theory*. London: Methuen; New York: Wiley.

SUPPES, PATRICK; and ZINNES, JOSEPH L. 1963 Basic Measurement Theory. Volume 1, pages 1–76 in R. Duncan Luce, Robert R. Bush, and Eugene Galanter (editors), *Handbook of Mathematical Psychology*. New York: Wiley.

►TORGERSON, WARREN S. 1968 Scaling. Volume 14, pages 25–39 in *International Encyclopedia of the Social Sciences*. Edited by David L. Sills. New York: Macmillan and Free Press.

TOULMIN, STEPHEN 1953 *The Philosophy of Science: An Introduction*. London: Hutchinson's University Library. → A paperback edition was published in 1960 by Harper.

►WILLIAMS, ROBIN M. JR. 1968 Values: I. The Concept of Values. Volume 16, pages 283–287 in *International Encyclopedia of the Social Sciences*. Edited by David L. Sills. New York: Macmillan and Free Press.

WINCH, PETER (1958) 1963 *The Idea of a Social Science and Its Relation to Philosophy*. London: Routledge; New York: Humanities.

SCIENTIFIC EXPLANATION

There is little disagreement that explanation is a major aim of science; there is much about the conditions that a proposed explanation must satisfy. Most of this disagreement, especially in history and in the social sciences, concerns the so-called nomological and deductive models of satisfactory explanation. There is also a more general controversy: many—perhaps most—scientists and philosophers maintain that these models clearly apply to the natural sciences and should apply to the social sciences as well, while others insist that they do not apply to any science, natural or social. For the most part, however, I shall bypass this controversy and discuss the applicability of the two models to the social sciences.

But first let us raise the logically prior question "What is an explanation?" There are three traditional answers to this: (1) to explain is to remove perplexity; (2) to explain is to change the unknown to the known; (3) to explain an event or type of event is to give its causes.

The first two answers have the virtue of reminding us of the psychological conditions that frequently obtain when we request an explanation. But this virtue is, in a way, the defect of these two approaches: they are too psychological, in the sense that they do not stipulate the logical conditions that must obtain before we have a satisfactory explanation. Clearly, not every way of removing perplexity can count as an explanation. Again, it should be noted that both of these approaches fail to point out that when we explain an event or type of event, we attempt to relate it to others like it and to give a systematic account of phenomena which will show how they are interrelated. In short, these answers fail to emphasize the systematic interconnection between explanation and theory construction.

The third answer does not share in these defects. The thesis that to explain an event is to give its cause is one that goes back to Aristotle and has been defended at length by Mill (1843) and other distinguished logicians. But this approach, whatever its virtues, cannot count as an adequate theory of explanation. At best it is a partial account of the explanation of events, but not of regularities, dispositions, and other types of phenomena which we try to explain. Moreover, since the term "cause" is not in itself a very clear one, a theory which builds on it without attempting any explication of it is not a reliable guide. However, many philosophers and social scientists have been convinced by the arguments offered by Hume and his followers that a causal statement entails a lawlike one, i.e., that a statement of the form "Event A caused event B" entails a statement of the form "Events like A regularly precede and are contiguous with events like B." If this analysis be granted, then the approach to explanation which emphasizes causation is compatible with the deductive and nomological models.

Nomological and deductive models. Supporters of the nomological model of scientific explanation insist that no event or regularity is explained by a proposed explanation unless it contains a scientific law (*nomos* is Greek for law). Supporters of the deductive model accept this but go further; they maintain that an event or regularity is explained if a sentence describing that event or regularity is deduced from a set of premises containing (*a*) a scientific law and (*b*) any other premises that may be required to make the deduction logically correct (provided, of course, that these premises are true).

Thus, given a sentence P that can be used to describe an event, the nomological model states that the event is not explained by a series of sentences $S_1, S_2, \cdots, S_n$ unless this series contains a scientific law. To this the deductive model adds that if $S_1, S_2, \cdots, S_n$ contains one or more laws, and if P can be deduced from $S_1, S_2, \cdots, S_n$, then $S_1, S_2, \cdots, S_n$ is a satisfactory conjunction of explaining sentences. This assumes that P stands for a sentence describing a particular event. When we want to explain not an event, but a trend or a regularity, all the sentences needed, both descriptive and explanatory, would of course be far more complex, but the over-all form of the explanation would stay the same.

On occasion some defenders of the deductive or the nomological models add that the explaining premises $S_1, S_2, \cdots, S_n$ must contain or constitute a theory. But this addition does not clarify matters at all, because the meaning of the term "theory" is itself unclear. Sometimes the term is applied to any systematically interconnected set of sentences; sometimes it appears to mean the same as "hypothesis"; sometimes it is used interchangeably with "law." Clearly, if "theory" and "law" are to be used interchangeably, acceptance of either the deductive or the nomological model requires us to believe that when we explain, we appeal to or use a theory. Since the term "theory" has so many uses, it is perhaps best to avoid using it in this context and to say simply that the premises of nomological and deductive explanations contain laws.

Candor demands, however, the admission that we are far from a completely satisfactory statement of either model. Thus, in the nomological model the meaning of "law" is just as much in dispute as that of "theory," while in the deductive model the kinds of sentences that may appear in the series $S_1, S_2, \cdots, S_n$ need to be specified in greater detail. On the other hand, there is little point in being concerned with such details if this whole approach to explanation is not valid. To see whether it is, I shall try to remove some misunderstandings about the nature of the nomological and deductive models, and in so doing to counter some of the arguments against them and admit the truth of others, at least in qualified form.

Do the models involve determinism? Both the deductive and the nomological models have been criticized on the grounds that they involve commitment to determinism. But this criticism is irrelevant. One may accept either model and still be agnostic about the thesis that every event is covered by laws—a thesis frequently given as one specification of determinism. Thus, philosophers who accept this thesis frequently argue that any sentence truly descriptive of an event is in principle deducible from some series of sentences containing one or more laws. But clearly this is not the same as claiming, with the supporters of the nomological and deductive models, that if a sentence truly descriptive of an event is deduced from a series of sentences containing a law, then the event is explained.

Of course, many defenders of the nomological and deductive models do accept an even stronger version of determinism than the thesis that every event is covered by laws. But they often accept it as a "guiding principle of inquiry," to use C. S. Peirce's phrase. Thus, they may insist that the scientist should not merely seek a variety of general laws for explaining various events and regularities but should construct and confirm one general theory that would explain all events and regularities, or at least all those within the purview of a particular science. But obviously a guiding principle is not the same thing as an article of faith, and many supporters of the nomological and deductive models agree that there may very well be no such general theory or theories for scientists to discover.

Can social scientists use the models? The deductive and nomological models are sometimes criticized on the grounds that social scientists have not succeeded in confirming any laws and therefore cannot use either model. To this it can be objected that some social scientists at least claim to have discovered laws, and that in any case only true or completely acceptable explanations, of which social scientists admit to having provided very few, are supposed to conform to the models' requirements.

But though the criticism as it stands is irrelevant, it does indicate that it is not enough to consider only true explanations or fully acceptable ones; we must consider not only the truth of proposed explanations but also their degrees of acceptability. Clearly, when a scientist proposes an explanation and employs a generalization, he obviously does not know that the generalization is true and, hence, that he is in possession of a law. The nomological and deductive models must therefore be supplemented, as follows. Let us use the term "lawful statement" to denote any highly confirmed lawlike statement—that is, any statement that (a) would be a law if it were true and (b) has in fact been found to hold true on numerous occasions. Explanations containing such statements have, other things being equal, a high degree of acceptability, and many of them also turn out to be true explanations—that is, the lawful sentences that they contain are true. But, of course, explanations that at one time were highly acceptable because the lawlike sentences they contained were at that time highly confirmed may turn out to be false (a classic case of this is Newtonian mechanics). This distinction between explanations that are acceptable at a given time and true explanations that simply conform to the deductive model is not *ad hoc*. Rather, it corresponds to the distinction, usually applied to statements, between "true" and "well confirmed."

Here it may be objected that social scientists do not refer to laws and, hence, that it is misleading to appeal to laws when discussing the logic of social scientific explanation. But this objection is a weak one. Social scientists tacitly employ many laws that are so well known that no special reference to them is needed. Furthermore, it should be emphasized that many social scientists do in point of fact refer to laws—economists, for instance, to Say's law of markets, and sociologists to Michels' "iron law of oligarchy." It may also be objected that the distinction between a highly confirmed generalization and a lawlike one is artificial. But once again the answer is readily available. Thus, in everyday life we would not treat the generalization "All American presidents are Christian" as lawlike or as a candidate for a law, even though all the evidence to date for it is positive and, in that sense, it is well confirmed. Observe further that the distinction we are discussing here is parallel to, although admittedly identifiable with, the frequently made distinction between causal statements, on the one hand, and statements of mere association, or statements of correlations which may be spurious, on the other.

What, then, are we to make of proposed explanations that do not contain lawful statements? We certainly do not always know on what law, if any, an investigator has based his explanation of certain data, even if we are inclined to accept what he says. As an argument against the nomological and deductive models this is not a very strong one, since most social scientists would admit that such proposed explanations should be replaced by more carefully stated ones. Besides, if the acceptability of such an explanation depends, as it usually does, on the degree to which it approximates a nomological or a deductive explanation, then it cannot be called an alternative to these latter, but merely an inferior substitute for them.

Motivational explanations. It is a commonplace that social scientists offer motivational explanations, but it is not at all obvious how such explanations should be interpreted from a logical point of view. Note first that when a motivational explanation is offered, there is no necessary commitment to the thesis that there are private internal entities called motives which are responsible for behavior or action; all that is involved is some reference to the wants, preferences, or aims of the individual or group whose action is being described or explained. In this sense all Freudian theory, for instance, makes tacit or explicit reference to motives, and so do all explanations by economists that refer to the utility of the agent or group whose behavior is being discussed. Similarly, tacit reference is made to the aims of a person when sociologists or social psychologists appeal, in order to make sense of his behavior, to the reference group with which he identifies himself. What then are we attributing to a person when we attribute a motive to him? The answer can readily be suggested: we are attributing to him a disposition to behave in a certain way or a disposition to prize certain outcomes over others. But this constitutes another possible objection to the nomological and deductive models. When a motivational explanation is offered, no appeal is made to a general law about the way human beings behave but, rather, to a dispositional statement about an individual. But it is very difficult to give the exact grounds upon which we can distinguish between laws and dispositional statements. Admittedly, it seems possible to do without general

laws if we restrict our attention to those cases in which the action seems to have been inspired by only one motive and in which there is no need, for explanatory purposes, to consider what the agent thought about the alternative ways available to him for the satisfaction of his motive. But when we consider the more usual cases of action—action performed out of many motives and action after deliberation about alternatives—it will not do to consider only the agent's disposition. Can we handle both these latter cases by reference to law? The answer is not clear. Some social scientists have thought it plausible to assume, as at least a first approximation, that insofar as people are rational, they try to maximize expected utility. But many suspect that the generalization tacitly appealed to here is a disguised tautology. Nevertheless, the defense of this generalization, or of alternatives to it, indicates that in order to explain actions for which there are many motives, some social scientists do look for laws and, hence, act in accord with the deductive model.

Others suggest that we should not seek general laws but that we should instead construct ideal types. To construct an ideal type explanation for an action A by a person P is (1) to impute at least one motive, T, to P; (2) to list the alternatives confronted by P; (3) to show that an action of the same type as A is a rational one—perhaps the only rational one—for P to perform, given T; (4) to show that performing A, given alternatives A_1, A_2, $\cdots$, A_n, is an efficient way of satisfying T. If, given the same motive and set of alternatives, P does not perform A, defenders of the ideal type approach would suggest that we must institute a search for the factors making for irrationality on the part of P.

One argument against this version of the ideal type model of explanation is that the motive imputed to the agent may in fact be the wrong one; another is that the model is incomplete unless the terms "rational" and "efficient" are explicated. Moreover, ideal type explanation sometimes appears to depend on a general law about the behavior of rational agents and, hence, to approximate either the deductive model or, at least, the nomological one.

Functional explanations. Conclusions similar to those drawn above can, I think, be applied to so-called functional explanations, for these either lack prima-facie acceptability or, when they have it, are not evidently at variance with the deductive or nomological models. Thus, a common type of functional "explanation" may state merely that a certain institution plays an indispensable role in a society. Such a statement provides us with nothing more than a necessary condition and, hence, lacks prima-facie acceptability as an explanation. However, the stronger types of functional explanation do show that, holding certain things constant, the presence of a certain mechanism is a sufficient condition for the existence of a certain state of affairs. In my opinion, explanations of this type conform to the deductive model, for to say that S is a sufficient explanatory condition for T is, I think, equivalent to saying that the statement "If S, then T" is not only true but also lawful.

Of course, we need more complex laws than the ones thus far discussed if we are to provide descriptions and explanations of events in systems that are purposive and self-corrective. But it is far from obvious that this requires a special kind of explanation called teleological, or functional, explanation, and even less obvious that societies—at least complex ones—can be analyzed in teleological terms. Certainly, societies are not totally purposive, nor do such purposes as they may sometimes be said to have (making war, for example) always remain the same. At most, societies may tend to reorganize themselves in order to keep certain properties constant. But it is hard to say which properties these are (see Parsons 1968).

Causal explanation. It has been maintained at least since Aristotle that to explain an event is to give its causes. But this does not necessarily involve controverting the nomological or the deductive models, since it can be argued that to give the causes for an event is to present a deductive explanation containing a causal law. Given this approach, we may then try to explicate the uses of the term "causal law."

A statement is occasionally called a causal law if it specifies either a sufficient condition or necessary and sufficient conditions for a certain type of occurrence. In logic, such a statement may take any of these forms: (1) if A then B; (2) A if and only if B; (3) A is a function of B (when A and B stand in the place of terms designating measurable properties). Deterministic laws, which describe how a system, with states that are described mathematically, evolves over time, are also called causal laws. Finally, the term "causal" is applied to laws of the form "if A then B" when (1) A denotes a type of event that comes just before an event of type B; (2) A and B are events or episodes in bodies or agents that are spatially contiguous; (3) the occurrence of A is to be considered a sufficient condition for the occurrence of B even though an event of type B might occur without one of type A preceding it. Stimulus–response generalizations

frequently meet these three criteria—criteria that represent perhaps the most legitimate use of the term "causal law."

Not all explanations that appeal, whether overtly or tacitly, to the notion of cause make use of causal laws in any of the senses discussed above. Accordingly, it is not evident that all causal explanations should conform to the nomological model, and still less to the deductive. Nevertheless, the view I am advocating has good philosophical precedent. At least since Hume many philosophers have argued that to assert of an event, A, that it caused another event, B, is equivalent to appealing tacitly to a law of the form "Events of type A are sufficient conditions of events of type B and precede them."

This view is not completely persuasive. Often we say that an event A caused an event B when we mean only that the event A was a necessary condition for the event B. Also, we sometimes refer to an event A as the cause or a cause of an event B when we mean to assert that A was either a sufficient or a necessary condition for the occurrence of B only when conjoined with other such conditions. Moreover, when we say that A caused B, we are often unprepared to specify a general law relating events of type A to events of type B. Thus, we might be convinced that a given remark had caused the hearer to blush, but we might not be ready at the same time to cite a general law about the relation between remarks and blushing. Nevertheless, it does seem that when we specify a cause for an event and believe that this specification provides us with an explanation of the event, we tacitly commit ourselves to *some* generalization, however vague. Thus, to revert to the illustration given above, a person who insists that X blushed because Y said something might be challenged by Z to the effect that X didn't blush the last time he heard a similar remark, and that therefore the present remark couldn't have been responsible for his blushes. But this observation, though it helps to diminish the distance between explanations that contain causal laws and explanations that merely appeal to the notion of cause, does not abolish that distance completely. Whether it can ever be abolished is still under debate.

Statistical explanation. The status of statistical explanation, and its relation to nomological and deductive explanation, are topics subject to much philosophical controversy.

Statistical explanations are explanations containing lawlike statements based on observation of statistical regularities and/or on the statistical theory of probability. It is generally accepted that probabilistic statements, as interpreted in the sciences, cannot be finally confirmed or even disconfirmed by observational evidence. It is also accepted that we cannot deduce from any statistical generalization a statement to the effect that any particular event must occur. [*See* PROBABILITY.]

It is not so widely recognized, however, that the problem of statistical explanation is much harder to deal with in the case of statistical generalizations that are either spatially or temporally restricted. Thus, if we knew that 90 per cent of the people in Milwaukee are Democrats and that 85 per cent of all living graduates of Yale University are Republicans, we could not use these generalizations as they stand to explain the voting habits of Yale graduates living in Milwaukee. It is therefore most important, when discussing statistical explanations, to appeal to the nomological, and not the deductive, model, for clearly we cannot expect to construct deductive explanations on every occasion when we use statistical generalizations for explanatory purposes. However, it would be a mistake to think that we can never have a deductive explanation under these circumstances, for we may deduce a statistical generalization from a conjunction of two or more other statistical generalizations.

It is important to note that we have not discussed the very complex but relatively typical situation that arises when social scientists who are using statistical data in an attempt to distinguish between the effects of several factors note that one factor is more relevant than the others for the purpose of explaining the phenomenon. Discussion of this and other such situations has been omitted because their logic is under much dispute. Many of the issues involved have been considered by Nagel (1952), though for some of the problems that are raised by statistical data and their analysis, the interested reader should consult works by Blalock (1964) and Boudon (1965; see also Ando et al. 1963).

Amplifying the models. We have discussed some objections to the nomological and deductive models; it is now time to consider amplifying them in order to meet these objections. Let us begin by emphasizing the trivial point that we never explain an event as such, but only selected aspects of it. In other words, it is not the event itself that is explained but the event under a given description. Thus, if a person sits down on a tack and yells, we may say that he yelled because he sat on the tack. But we obviously are not ready to claim that we can explain the specific pitch or duration of the yell in the same terms.

Observe, however, that an explanation may be totally satisfactory once we specify what aspect of the event we are trying to explain. This is of special relevance to the discussion of historical explanation. Historians have often noted that any event can be described in many ways and that some explanations of events are satisfactory only if the events are described schematically. Thus, we might be able to explain that a group migrated because it hoped to improve its lot but not be able to explain why it migrated on a specific date and to a specific country. Some historians, noting that there seems to be no end to the detail in which a given event can be described, have concluded that, for this reason, no event can ever be explained, nor can the causes of an event be fully specified. Against this view, I would like to urge that we discuss not the explanation of an event but the explanation of an event under a given description. This would allow us to accept some explanations as totally satisfactory even if we wanted to replace them with explanations of the same event under a more refined description.

Finally, it should be noted that in order to give an explanation of an event of a certain type, it is not necessary to cite laws or theories about events of that type. This is often overlooked in discussion of how to explain certain social phenomena; for instance, many social scientists seem to believe that in order to explain instances of crime or divorce, they need a general theory of crime or divorce. But this is dubious. To explain why a man slipped on a banana peel, we do not need a general theory of slipping. Rather, the laws that we need to cite are general laws in which terms like "slipping" do not occur. The essential point is that in order to explain an event, we often must redescribe it in terms of a given theory; moreover, two events that in ordinary life are both classified under the same general term may, for purposes of explanation, have to be described differently and explained by different general theories. To take an everyday example: A man who hurries to meet a friend and a man who flees to avoid disaster may both be running, yet the explanations for their behavior are different. More complicated examples, requiring explanations based on abstract theories, would require that the events needing explanation be redescribed in terms drawn from those theories. Although this type of redescription was not required in our example, the point remains that two events ordinarily classified as instances of the same type may have different explanations. The same applies to explanations involving not events but groups, epochs, social systems, and societies; we must specify what aspects of these entities we want to have explained.

SIDNEY MORGENBESSER

[*Directly related are the entries* CAUSATION; EXPERIMENTAL DESIGN; PREDICTION; SCIENCE, *article on* THE PHILOSOPHY OF SCIENCE; *see also* Cancian 1968. *Other relevant material may be found in* MULTIVARIATE ANALYSIS; STATISTICS, DESCRIPTIVE; *and the biography of* PEIRCE; *see also* Birney 1968; Gardiner 1968; Kaplan 1968; O'Kelly 1968; Rosenfield 1968.]

BIBLIOGRAPHY

ANDO, ALBERT; FISHER, F. M.; and SIMON, HERBERT A. 1963 *Essays on the Structure of Social Science Models.* Cambridge, Mass.: M.I.T. Press. → See especially pages 5–31 and 107–112.

ARGYLE, MICHAEL 1957 *The Scientific Study of Social Behaviour.* London: Methuen.

►BIRNEY, ROBERT C. 1968 Motivation: II. Human Motivation. Volume 10, pages 514–522 in *International Encyclopedia of the Social Sciences.* Edited by David L. Sills. New York: Macmillan and Free Press.

BLALOCK, HUBERT M. JR. 1964 *Causal Inferences in Nonexperimental Research.* Chapel Hill: Univ. of North Carolina Press.

BOUDON, RAYMOND 1965 A Method of Linear Causal Analysis: Dependence Analysis. *American Sociological Review* 30:365–374.

BRAITHWAITE, R. B. 1953 *Scientific Explanation.* Cambridge Univ. Press.

BROWN, ROBERT R. 1963 *Explanation in Social Science.* Chicago: Aldine.

BUNGE, MARIO 1959 *Causality: The Place of the Causal Principle in Modern Science.* Cambridge, Mass.: Harvard Univ. Press. → A paperback edition was published in 1963 by World.

►CANCIAN, FRANCESCA M. 1968 Functional Analysis: II. Varieties of Functional Analysis. Volume 6, pages 29–43 in *International Encyclopedia of the Social Sciences.* Edited by David L. Sills. New York: Macmillan and Free Press.

CHRIST, CARL F. 1966 *Econometric Models and Methods.* New York: Wiley. → See especially the Foreword by Jacob Marshak.

COHEN, MORRIS R.; and NAGEL, ERNEST 1934 *An Introduction to Logic and Scientific Method.* New York: Harcourt. → The first part of the original text was published in a paperback edition in 1962 by Harcourt, under the title *An Introduction to Logic.*

DRAY, WILLIAM 1957 *Laws and Explanation in History.* Oxford Univ. Press.

FEIGL, HERBERT; and MAXWELL, GROVER E. (editors) 1962 *Scientific Explanation, Space, and Time.* Minnesota Studies in the Philosophy of Science, Vol. 3. Minneapolis: Univ. of Minnesota Press. → See especially pages 231–273, "Explanation, Prediction and 'Imperfect Knowledge,'" by May Brodbeck.

FODOR, JERRY A.; and KATZ, JERROLD J. (editors) 1964 *The Structure of Language: Readings in the Philosophy of Language.* Englewood Cliffs, N.J.: Prentice-Hall.

►GARDINER, PATRICK 1968 History: I. The Philosophy of History. Volume 6, pages 428–434 in *International Encyclopedia of the Social Sciences.* Edited by David L. Sills. New York: Macmillan and Free Press.

GIBSON, QUENTIN B. (1960) 1963 *The Logic of Social Inquiry.* New York: Humanities.

GREENWOOD, ERNEST 1945 *Experimental Sociology: A Study in Method.* New York: Columbia Univ. Press.

HEMPEL, CARL G. (1952) 1963 Typological Methods in the Social Sciences. Pages 210–230 in Maurice A. Natanson (editor), *Philosophy of the Social Sciences: A Reader.* New York: Random House.

HEMPEL, CARL G. 1965 *Aspects of Scientific Explanation, and Other Essays in the Philosophy of Science.* New York: Free Press.

KAPLAN, ABRAHAM 1964 *The Conduct of Inquiry: Methodology for Behavioral Science.* San Francisco: Chandler.

►KAPLAN, ABRAHAM 1968 Positivism. Volume 12, pages 389–395 in *International Encyclopedia of the Social Sciences.* Edited by David L. Sills. New York: Macmillan and Free Press.

LOWE, ADOLPH 1965 *On Economic Knowledge.* New York: Harper.

MADDEN, EDWARD H. 1962 *Philosophical Problems of Psychology.* New York: Odyssey. → See especially Chapter 3.

MILL, JOHN STUART (1843) 1961 *A System of Logic, Ratiocinative and Inductive: Being a Connected View of the Principles of Evidence and the Methods of Scientific Investigation.* London: Longmans. → See especially Book 6.

MORGENBESSER, SIDNEY 1966 Is It a Science? *Social Research* 33:255–271.

NAGEL, ERNEST (1952) 1953 The Logic of Historical Analysis. Pages 688–700 in Herbert Feigl and May Brodbeck (editors), *Readings in the Philosophy of Science.* New York: Appleton. → First published in *Scientific Monthly.*

NAGEL, ERNEST 1961 *The Structure of Science: Problems in the Logic of Scientific Explanation.* New York: Harcourt. → See especially pages 582–588.

►O'KELLY, LAWRENCE I. 1968 Motivation: I. The Concept. Volume 10, pages 507–514 in *International Encyclopedia of the Social Sciences.* Edited by David L. Sills. New York: Macmillan and Free Press.

►PARSONS, TALCOTT 1968 Systems Analysis: II. Social Systems. Volume 15, pages 458–473 in *International Encyclopedia of the Social Sciences.* Edited by David L. Sills. New York: Macmillan and Free Press.

POINCARÉ, HENRI (1908) 1952 *Science and Method.* New York: Dover. → First published in French.

POPPER, KARL R. (1935) 1959 *The Logic of Scientific Discovery.* New York: Basic Books. → First published as *Logik der Forschung.*

REICHENBACH, HANS 1951 Probability Methods in Social Science. Pages 121–128 in Daniel Lerner and Harold D. Lasswell (editors), *The Policy Sciences: Recent Developments in Scope and Method.* Stanford Univ. Press.

►ROSENFIELD, LEONORA COHEN 1968 Cohen, Morris R. Volume 2, pages 540–542 in *International Encyclopedia of the Social Sciences.* Edited by David L. Sills. New York: Macmillan and Free Press.

SCHOEFFLER, SIDNEY 1955 *The Failures of Economics: A Diagnostic Study.* Cambridge, Mass.: Harvard Univ. Press.

SEGERSTEDT, TORGNY T. 1966 *The Nature of Social Reality.* Copenhagen: Munksgaard.

SKINNER, B. F. 1953 *Science and Human Behavior.* New York: Macmillan.

TOULMIN, STEPHEN E. 1953 *The Philosophy of Science: An Introduction.* London: Longmans. → A paperback edition was published in 1960 by Harper.

WINCH, PETER 1958 *The Idea of a Social Science and Its Relation to Philosophy.* London: Routledge; New York: Humanities.

WOLMAN, BENJAMIN B.; and NAGEL, ERNEST (editors) 1965 *Scientific Psychology.* New York: Basic Books. → See especially "Concerning an Incurable Vagueness in Psychological Theories," by Benbow F. Ritchie, and "Mathematical Learning Theory," by Richard C. Atkinson and Robert C. Calfe.

Postscript

Philosophers and social scientists continue to debate the thesis that explanation in the social sciences is different in nature from explanation in the natural ones. Some philosophers argue that in the social sciences explanation is by appeal to reasons and in the natural sciences by appeal to causes. A review of these issues is to be found in Ryan (1973). Davidson (1963), Pears (1975), and others question the dualism between reasons and causes. Davidson (1967) also questions some aspects of the Humean approach to causation.

Philosophers continue to debate the nomological model of social-scientific explanation (see Körner 1975). Articles on emergence and reduction in the social sciences are collected in O'Neill (1973). There has been some work on teleological explanation, especially in biology, that is of interest to the philosophy of the social sciences (see Woodfield 1976; Wright 1976).

For a review of some issues in the logic of explanation in psychology, see Fodor (1975), and in sociology see Blalock (1971). An important case study in explanation is David et al. (1976).

SIDNEY MORGENBESSER

ADDITIONAL BIBLIOGRAPHY

BLALOCK, HUBERT M. JR. (editor) 1971 *Causal Models in the Social Sciences.* Chicago: Aldine.

DAVID, PAUL A. et al. 1976 *Reckoning With Slavery: Critical Study in the Quantitative History of American Negro Slavery.* Oxford Univ. Press.

DAVIDSON, DONALD 1963 Actions, Reasons, and Causes. *Journal of Philosophy* 60:685–700. → Reprinted in Ryan (1973).

DAVIDSON, DONALD 1967 Causal Relations. *Journal of Philosophy* 4:691–703.

FODOR, JERRY A. 1975 *The Language of Thought.* New York: Crowell.

KÖRNER, STEPHAN (editor) 1975 *Explanation.* New Haven: Yale Univ. Press.

O'NEILL, JOHN (editor) 1973 *Modes of Individualism and Collectivism.* London: Heinemann; New York: St. Martin's.

PEARS, DAVID 1975 *Questions in the Philosophy of Mind.* New York: Barnes & Noble.

RYAN, ALAN (editor) 1973 *The Philosophy of Social Explanation.* Oxford Univ. Press.

WOODFIELD, ANDREW 1976 *Teleology.* Cambridge Univ. Press.

WRIGHT, LARRY 1976 *Teleological Explanation: An Etiological Analysis of Goals and Functions.* Berkeley: Univ. of California Press.

SCREENING AND SELECTION

Screening and selection procedures are statistical methods for assigning individuals to two or more categories on the basis of certain tests or measurements that can be made upon them. Of concern usually is some desired trait or characteristic of the individuals that cannot be measured directly. All that can be done is to obtain an estimate for each individual from the results of the available tests and then to make the assignment on the basis of these estimates. The central statistical problem is to evaluate the properties of alternative schemes for utilizing the available data to make the assignments in order to choose the scheme that best achieves whatever objectives are considered to be most relevant for the particular application.

For example, some educational selection schemes may be regarded in this light. The individuals might be high school students and the categories "admit to college" and "do not admit to college." The desired trait is future success in college, but only tests at the high school level are available. (As will be seen, many examples of selection and screening are somewhat more complex than this, particularly in their use of more than one level of screening. In this educational context one might instead use three categories: "admit," "put on waiting list," and "do not admit.")

Denote by N the number of individuals to be assigned and let c be the number of categories. For any individual, let Y denote the unknown value of the desired trait and let $X_1, \cdots, X_p$ denote the measurements or scores that can be obtained and used as predictors of Y. The screening or selection procedure is a scheme that specifies in terms of $X_1, \cdots, X_p$ how each individual is to be assigned to one of the c categories.

The screening may be done at one stage—that is, all the measurements $X_1, \cdots, X_p$ become available before an individual is assigned—or it may be multistage. The advantage of a multistage procedure is that it may allow some individuals to be assigned at an early stage or, at least, to be eliminated from contention for the categories of interest, thus permitting the resources available for performing the tests to be concentrated on fewer individuals in the later stages.

The terms *screening* and *selection* are largely synonymous, although in particular applications one or the other may be preferred. Sometimes, in order to avoid the possible connotation that certain categories may be more desirable than others, a neutral term such as *allocation* is used. The term *classification* has a different shade of meaning, referring to the identification of which of several distinct distributions each individual belongs to (in taxonomy, for example, classification involves the assignment of an organism to its proper species) [see MULTIVARIATE ANALYSIS, *article on* CLASSIFICATION AND DISCRIMINATION].

Various formulations of screening and selection problems have been proposed and investigated. A bibliography containing more than five hundred references has been given by Federer (1963). Of special interest is the case $c = 2$, in which the individuals are separated into two categories, a selected group and the remainder; the success of the screening procedure is judged by the values of Y for the individuals in the selected group. This article will be primarily devoted to this case of two categories.

The case where N is large

This section deals with the case of separation into two categories where either the number, N, of individuals is large enough for their Y-values to be considered as forming a continuous distribution or, alternatively, the N individuals themselves are considered as a random selection from a conceptually infinite population. The object of the screening procedure is to produce a distribution of Y-values in the selected group that is, in an appropriate sense, an improvement on the original distribution. For example, one might want the selected group to have as high a median Y as possible.

In many applications, the feature of the distribution of Y that is considered most important is its mean. For example, in a plant-breeding program, Y might stand for the crop yield of the individual varieties, and the purpose of the program might be to select a set of varieties whose mean Y is as high as possible. The difference in the mean Y for the selected varieties from that of the original group is referred to as the "advance" or the "gain due to selection."

Cochran (1951) summarized the mathematical basis for selection procedures designed to maximize the mean Y in the selected set. He showed that the optimum selection rule to use at each stage should be based on the regression of Y on the X's that are known at that stage. In the case where the joint distribution of these variates is multivariate normal, this regression is the linear combination of the X's that has maximum correlation with Y. [See MULTIVARIATE ANALYSIS, *article on* CORRELATION METHODS.]

Plant selection. In a plant-breeding program for improving the yield of a particular crop, a large number of potential new varieties become available in any year. These are tested in successive plantings, and the better-yielding types in each planting are selected to produce the seed for the next sowing, until finally a small fraction of the original number remain as possible replacements for the standard varieties in commercial use.

Finney (1966) and Curnow (1961) have carried out an extensive theoretical and numerical investigation of a fairly general type of selection procedure that is particularly applicable to plant selection. The problem is to reduce in k stages an initial set of N candidates to a predetermined fraction, π, called the "selection intensity," using a fixed total expenditure of resources, A. At stage r $(1 \leqslant r \leqslant k)$, the candidates selected at stage $r - 1$ are tested, using resources A_r, to obtain for each a score, X_r, which estimates Y with a precision dependent upon A_r; the fraction, P_r $(0 < P_r \leqslant 1)$, having the highest scores are selected, and the remainder are discarded. Stage 0 consists in selecting at random a fraction, P_0, of the initial set. The problem is to choose P_r and A_r to maximize the expected mean Y-value in the selected group, subject to $A_1 + \cdots + A_k = A$ and $P_0 P_1 \cdots P_k = \pi$, where A and π are given. The authors found that approximately optimum results were obtained with $A_r = A/k$, $P_r = \sqrt[k]{\pi}$, for $r = 1, \cdots, k$, called the "symmetric" scheme, usually with $P_0 = 1$, although in some circumstances a value $P_0 < 1$ effected further improvement. Three or four stages at most were sufficient.

In the context of plant selection, the N candidates are the new crop varieties produced in a particular year, and the resource expenditure, A, is the area of land available for testing, which must be divided into separate portions for varieties being tested for the first time, varieties selected on the basis of last year's tests to be tested in stage 2, and so on. For example, suppose $N = 200$ varieties are started in a two-year program to select 8 to compare with the standard commercial types (thus, $\pi = .04$). Then at the end of each year, $\sqrt{.04} = \frac{1}{5}$ of the varieties should be selected—that is, 40 at the end of the year 1 and 8 at the end of year 2—with equal areas of land to be divided among the 200 varieties in year 1 and the 40 varieties in year 2.

Drug screening. In drug screening, the problem is to screen a large supply of chemical compounds by means of a biological test, usually in laboratory animals, in order to select for further testing the few that may possess the biological activity desired. Here Y stands for the unknown activity level of a compound (averaged over a conceptual population of animals), estimates being provided by the test results, $X_1, X_2, \cdots$. The distribution of Y in the population of compounds available for screening will usually have a large peak at $Y = 0$, since most of the compounds do not possess the activity being sought unless a specific class of compounds chemically related to known active compounds is being screened. The number of compounds available for testing usually exceeds the capacity of the testing facilities; therefore, part of the problem in drug screening is to determine the optimum number, N, of compounds to screen in a given period of time.

The mean Y in the selected group does not have as much relevance in drug screening as it does in plant selection. Instead, a value, a, is usually specified such that a drug is of interest if its activity equals or exceeds level a. The screening procedure is then designed to maximize the number of compounds in the selected group having $Y \geqslant a$, usually subject to the requirement that the total number of compounds selected over a certain period of time is fixed. Davies (1958) and King (1963) have considered in detail this approach to the statistical design of drug-screening tests.

Educational selection. The consideration of selection procedures to allocate school children to different "streams" is necessarily much more complex than in the applications considered above. For one thing, there can be no question of rejection; the object, at least in principle, is to provide the education most suitable for each child. Furthermore, it can be expected that the characteristic Y of each child will be altered by the particular stream in which he may be placed. An admirable discussion of the problems was given by Finney (1962), who described, as an illustration of the methodology, a simplified mathematical model of the educational selection process then in operation in the British school system.

Finney considered university entrance as a two-stage selection process: the first stage is the separation of students at the age of 11+ into those who will receive a grammar school education and those who will go instead to a secondary school, and the second stage is university entrance. Denoting by X_1 the composite score of all test results available at the first stage, by X_2 the composite score at university entrance, and by Y the "suitability" of a student for university study as determined by his subsequent university grades, Finney considered Y, X_1, and X_2 to have a multivariate normal distribution with correlation coefficients estimated from available data. He studied the effects that varying the proportion of students admitted to grammar school, as well as the relative proportions admitted to universities from the two types

of school, had on the average value of Y in the selected group and on the proportion of university entrants having $Y \geqslant a$. Interesting numerical results are presented, but the main feature of the paper is its demonstration of how the approach can bring about a clearer insight into the issues involved in a selection process.

The case where N is small

Rather different methods of approach have been developed for selection when the number N is small enough so that the individual values of Y, rather than their distribution, may be considered. The object of the selection procedure is expressed in terms of the Y's; for example, the object may be to select the individual with the largest value of Y, to select the individuals with the t largest values, or to rank the N individuals in order according to their values of Y. These are special cases of the general goal of dividing the N individuals into c categories, containing respectively the n_c individuals with the highest values of Y, the n_{c-1} individuals with the next highest values, and so on, down to the n_1 individuals with the lowest values of Y. Bechhofer (1954) developed expressions for the probability of a correct assignment of the N individuals to the c categories.

Selecting the "best" of N candidates. Of most frequent interest in practical applications is the selection of the best of several candidates, "best" being interpreted to mean the one with the largest Y. Estimates of the unknown Y's are obtained from experimentation, and the problem usually is to decide how much experimentation needs to be done.

In a single-stage selection procedure, an experiment consisting of taking n observations for each candidate is performed, and the candidate with the highest observed mean is selected. The probability of a correct selection is the probability that the candidate with the highest Y will also have the highest observed mean value; this is a function not only of the number n of observations but also of the unknown configuration of values of Y.

Bechhofer (1954) recommended that the experimenter choose n so that the probability of a correct selection would exceed a specified value, P, whenever the best value of Y exceeded the others by at least a specified amount, d. His paper contains tables of the required value of n, calculated on the assumption that the Y's are in the "least favorable" configuration, which in this case is the configuration where all the Y's except the best one are equal and are less than the best one by the amount d.

Another approach to determining the value of n is based on striking an optimum balance between the cost of taking observations, which is a function of n, and the economic loss incurred if some candidate other than the best one is selected. The loss due to an incorrect selection is assumed to be a function, usually linear, of the difference between the largest Y and the selected Y. With the probability of selecting any particular candidate taken into account, an expected loss or risk function, which is a function of n and the unknown Y's, is determined. Somerville (1954) showed how the minimax principle can be used to determine the optimum sample size; this procedure is appropriate if no prior information is available about the unknown Y-values [see DECISION THEORY]. In the case where prior information is available, Dunnett (1960) showed how such information can be utilized in determining the sample size; he also made numerical comparisons between alternative procedures.

Gupta (1965) considered the situation in which the experimenter is willing to select a larger group than is actually needed and is concerned with guaranteeing a specified probability, P, that the best candidate is included in the selected group. In this way, the need to specify a minimum difference, d, as in Bechhofer's method, is avoided, but there is the drawback of not necessarily having a unique selection. Gupta investigated the effect of the sample size, n, and the configuration of the Y's on the expected size of the selected group.

Sequential procedures have also been investigated; for example, Paulson (1964) considered a sequential method for dropping candidates from contention at each stage until only one remains, so as to achieve a specified probability, P, that the best one is selected whenever its value Y exceeds the others by at least d. [See SEQUENTIAL ANALYSIS.]

A medical selection problem. An interesting method (Colton 1963) for selecting the better of two medical treatments uses some of the principles discussed above but also contains ingenious points of difference. In the problem considered, there is a fixed number of patients to be treated. A clinical trial is performed on a portion of them, with equal numbers being given each treatment. On the basis of the trial, one of the two treatments is selected to treat the remainder of the patients. The problem is to determine how many patients to include in the trial in order to maximize the expected total number receiving the better drug. Sequential procedures for making the selection are also discussed.

Tournaments

A tournament is a series of contests between pairs of N players (or teams) with the object either of selecting the best player or of ranking the players in order. It may be regarded as a selection procedure in which the experiments consist of paired comparisons between candidates. David (1959) studied some properties of two types of tournaments, the knockout and the round robin. Glenn (1960) compared the round robin and several variations of the knockout tournament for the case of four contestants; he found that a single knockout tournament with each contest being on a "best two out of three" basis achieved the highest probability that the best player will win, but at the "expense" of requiring a higher average number of games.

Other problems

There are many other interesting topics in screening and selection. One is *group screening*, in which a single test is performed on several candidates as a group to determine whether any of them possess the characteristic of interest. When an affirmative answer is obtained, further tests are performed to determine which ones possess the characteristic. One application of this procedure is in blood testing for the presence of some disease; a great saving in the number of tests necessary is accomplished by physically pooling several samples and making a single test. Another application is in factor screening in industrial research (see Sobel & Groll 1959; Watson 1961).

In practice, many screening problems are multivariate—that is, there is more than one trait, Y, of interest, and the traits are likely to be correlated. Sometimes the measurements on the several traits are reduced to a single variate by combination into a suitable index, perhaps with weights determined by the economic worth of each trait. Sometimes only one trait is dealt with at a time, the candidates considered for selection on the basis of trait Y_r being those who have previously been selected on the basis of $Y_1, \cdots, Y_{r-1}$ in turn. Much work remains to be done to determine the best procedures for use in multivariate screening. (See Rao 1965 for a treatment of some of the mathematical problems.)

CHARLES W. DUNNETT

[See also CLUSTERING; MULTIVARIATE ANALYSIS, article on CLASSIFICATION AND DISCRIMINATION; STATISTICAL ANALYSIS, SPECIAL PROBLEMS OF, article on OUTLIERS.]

BIBLIOGRAPHY

BECHHOFER, R. E. 1954 A Single-sample Multiple Decision Procedure for Ranking Means of Normal Populations With Known Variances. *Annals of Mathematical Statistics* 25:16–39.

COCHRAN, WILLIAM G. 1951 Improvement by Means of Selection. Pages 449–470 in Berkeley Symposium on Mathematical Statistics and Probability, Second, *Proceedings*. Berkeley: Univ. of California Press.

COLTON, THEODORE 1963 A Model for Selecting One of Two Medical Treatments. *Journal of the American Statistical Association* 58:388–400.

CURNOW, R. N. 1961 Optimal Programmes for Varietal Selection. *Journal of the Royal Statistical Society* Series B 23:282–318. → Contains eight pages of discussion.

DAVID, H. A. 1959 Tournaments and Paired Comparisons *Biometrika* 46:139–149.

DAVIES, O. L. 1958 The Design of Screening Tests in the Pharmaceutical Industry. International Statistical Institute, *Bulletin* 36, no. 3:226–241.

DUNNETT, CHARLES W. 1960 On Selecting the Largest of *k* Normal Population Means. *Journal of the Royal Statistical Society* Series B 22:1–40. → Includes 10 pages of discussion.

FEDERER, WALTER T. 1963 Procedures and Designs Useful for Screening Material in Selection and Allocation, With a Bibliography. *Biometrics* 19:553–587.

FINNEY, D. J. 1962 The Statistical Evaluation of Educational Allocation and Selection. *Journal of the Royal Statistical Society* Series A 125:525–549. → Includes "Discussion on Dr. Finney's Paper," by T. Lewis et al.

FINNEY, D. J. 1966 An Experimental Study of Certain Screening Processes. *Journal of the Royal Statistical Society* Series B 28:88–109.

GLENN, W. A. 1960 A Comparison of the Effectiveness of Tournaments. *Biometrika* 47:253–262.

GUPTA, SHANTI S. 1965 On Some Multiple Decision (Selection and Ranking) Rules. *Technometrics* 7:225–245.

KING, E. P. 1963 A Statistical Design for Drug Screening. *Biometrics* 19:429–440.

PAULSON, EDWARD 1964 A Sequential Procedure for Selecting the Population With the Largest Mean from *k* Normal Populations. *Annals of Mathematical Statistics* 35:174–180.

RAO, C. RADHAKRISHNA 1965 Problems of Selection Involving Programming Techniques. Pages 29–51 in IBM Scientific Computing Symposium on Statistics, Yorktown Heights, N.Y., 1963 *Proceedings*. White Plains, N.Y.: IBM Data Processing Division.

SOBEL, MILTON; and GROLL, PHYLLIS A. 1959 Group Testing to Eliminate Efficiently All Defectives in a Binomial Sample. *Bell System Technical Journal* 38:1179–1252.

SOMERVILLE, PAUL N. 1954 Some Problems of Optimum Sampling. *Biometrika* 41:420–429.

WATSON, G. S. 1961 A Study of the Group Screening Method. *Technometrics* 3:371–388.

Postscript

Marshall and Olkin (1968) consider the problem of screening and allocation from a decision–theoretic point of view. The conditional distribution of Y, the variable that would determine to which category each item belonged if its value could be observed, given the values of the screening variables $X_1, \cdots, X_k$, is assumed to be known. A loss function must be specified to measure the loss associated with assigning an item to each category as a function of the value y of the variable Y. A particular procedure for classifying items will be optimal if it minimizes the expected loss or risk. Marshall and Olkin show how optimal procedures can be devised in a wide range of situations and discuss some interesting potential applications, such as screening for the presence of a disease and the use of the "breathalyser" test as a quick screening procedure for motorists suspected of driving with elevated levels of alcohol in their bloodstream. The main problem in applying this approach is to obtain a realistic assessment of the loss functions.

Drug screening continues to be a major area of application. Mantel (1960) in addition to the previously cited authors, has discussed the basic principles. An article in *Science* (T.H.M. 1974) described the large-scale screening program instituted by the National Cancer Institute to uncover new compounds with anti-cancer activity. The compounds are tested on mice in which human tumors have been induced to grow. The initial screening test is carried out in mice with leukemia, and a positive result is defined as an average 25 per cent increase in the survival time of the mice. If a chemical passes this test twice in succession, it is subjected to two further screening stages, one in mouse lung tumors and one in skin tumors. A composite index derived from scores obtained on all the tests is used to decide whether the compound is worth trying out in humans. However, before this can be done, other tests must be carried out to study the compound's pharmacological properties and to determine what potential forms of toxicity it might possess.

In addition to the search for compounds to help in the fight against cancer, there is also the problem of detecting compounds that are carcinogens, that is, cancer-causing. Any new chemical food additive or insecticide has to be tested in animals to determine whether it enhances the growth of tumors. This is a much more difficult problem than screening for anti-cancer agents, because tumors induced by a carcinogen may take the entire lifespan of the animal to develop and be observed. Elashoff, Sobel, and Schneiderman (1974) have studied the application of group screening principles to obtain information regarding tumorigenicity on several compounds simultaneously, in order to be able to test more compounds and also determine whether certain combinations of chemicals might be tumorigenic, even though individually they are not.

CHARLES W. DUNNETT

ADDITIONAL BIBLIOGRAPHY

ELASHOFF, ROBERT M.; SOBEL, MILTON; and SCHNEIDERMAN, M. A. 1974 A Proposal for Economical First-stage Screening for Tumorigens With a Possible "Joint Action" Bonus. Pages 365–374 in Satellite Symposium on Statistical Aspects of Pollution Problems, Harvard Business School, 1971, *Statistical and Mathematical Aspects of Pollution Problems: Papers.* Edited by John W. Pratt. New York: Dekker.

MANTEL, NATHAN 1960 Principles in Chemotherapeutic Screening. Volume 4, pages 293–306 in Berkeley Symposium on Mathematical Statistics and Probability, Fourth, *Proceedings.* Berkeley: Univ. of California Press.

MARSHALL, ALBERT W.; and OLKIN, INGRAM 1968 A General Approach to Some Screening and Classification Problems. *Journal of the Royal Statistical Society* Series B 30:407–435. → Discussion on pages 435–443.

T. H. M. 1974 Screening for Drugs: A Massive Undertaking. *Science* 184:971 only.

SEASONAL ADJUSTMENT OF TIME SERIES
See under TIME SERIES.

SELECTION
See SCREENING AND SELECTION.

SEQUENTIAL ANALYSIS

Sequential analysis is the branch of statistics concerned with investigations in which the decision whether or not to stop at any stage depends on the observations previously made. The motivation for most sequential investigations is that when the ends achieved are measured against the costs incurred (including the cost of making observations), sequential designs are typically more efficient than nonsequential designs; some disadvantages of the sequential approach are discussed later.

The term "sequential" is occasionally extended to cover also investigations in which various aspects of the design may be changed according to the observations made. For example, preliminary experience in an experiment may suggest changes in the treatments being compared; in a social survey a small pilot survey may lead to modifications in the design of the main investigation. In this article attention will be restricted mainly to the usual situation in which termination of a single investigation is the point at issue.

In a sequential investigation observations must be examined either one by one as they are collected or at certain stages during collection. A sequential procedure might be desirable for various reasons. The investigator might wish to have an up-to-date record at any stage, either for general information or because the appropriate sample size depends on quantities that he can estimate only from the data themselves. Alternatively, he may have no intrinsic interest in the intermediate results but may be able

to achieve economy in sample size by taking them into account. Three examples will illustrate these points:

(1) An investigator may wish to estimate to within 10 per cent the mean weekly expenditure on tobacco per household. In order to determine the sample size he would need an estimate of the variability of the expenditure from household to household, and this might be obtainable only from the survey itself.

(2) A physician wishing to compare the effects of two drugs in the treatment of some disease may wish to stop the investigation if at some stage a convincing difference can already be demonstrated using the available data.

(3) A manufacturer carrying out inspection of batches of some product may be able to pass most of his batches with little inspection but may carry out further inspection of batches of doubtful quality. A given degree of discrimination between good and bad batches could be achieved in various ways, but a sequential scheme will often be more economical than one in which a sample of constant size is taken from each batch [*see* QUALITY CONTROL, STATISTICAL *for further discussion of such applications*].

The most appropriate design and method of analysis of a sequential investigation depend on the purpose of the investigation. The statistical formulation of that purpose may take one of a number of forms, usually either estimation of some quantity to a given degree of precision or testing a hypothesis with given size and given power against a given alternative hypothesis. Economy in number of observations is typically important for sequential design. Some details of particular methods are given in later sections.

Sometimes a sequential investigation, although desirable, may not be practicable. To make effective sequential use of observations they must become available without too great a delay. It would not be possible, for example, to do a sequential analysis of the effect of some social or medical policy if this effect could not be assessed until five years had elapsed. In other situations it may be possible to scrutinize the results as they are obtained, but only at very great cost. An example might be a social survey in which data could be collected rather quickly but in which a full analysis would be long and costly.

History. An important precursor of the modern theory of sequential analysis was the work done in 1929 by Dodge and Romig (1929–1941) on double sampling schemes. Their problem was to specify sampling inspection schemes that discriminated between batches of good and bad quality. The first stage of sampling would always be used, but the second stage would be used only if the results of the first were equivocal; furthermore, the size of the second sample and the acceptance criteria might depend on the first stage results. Bartky (1943) generalized this idea in his "multiple sampling," which allowed many stages, and his procedure was very closely related to a particular case of the general theory of sequential analysis that Wald was developing simultaneously.

This theory, developed for the testing of military equipment during World War II, is summarized in Wald's book *Sequential Analysis* (1947). It represents a powerful exploitation of a single concept, the "sequential probability ratio test," which has provided the basis of most subsequent work. Related work proceeding simultaneously in Great Britain is summarized by Barnard (1946). Whereas Wald's theory· provided the specification of a sampling scheme satisfying given requirements, Barnard's work was devoted to the converse problem of examining the properties of a given sequential scheme. Barnard drew attention to the close analogy between sequential inspection schemes and games of chance. Indeed, some of the solutions to gaming problems provided by seventeenth-century and eighteenth-century mathematicians are directly applicable to modern sequential schemes.

Postwar theoretical development, stimulated primarily by Wald's work, has perhaps outrun practical applications. Many recent workers have apparently felt that the standard sequential theory does not provide answers to the right questions, and a number of new lines of approach have been attempted.

Sequential estimation. Suppose that in a large population a proportion, p, of individuals show some characteristic (or are "marked") and that in a random sample of size n the number of marked individuals is X. Then the proportion of marked individuals is X/n. By standard binomial distribution theory, the standard error of X/n is $\sqrt{p(1-p)/n}$. The standard error expressed as a proportion of the true value is therefore $\sqrt{(1-p)/np}$, and when p is small this will be approximately $(np)^{-\frac{1}{2}}$. Now np is the mean value of $X = n(X/n)$. Intuitively, therefore, one could achieve an approximately constant proportional standard error by choosing a fixed value of X. That is, sampling would be continued until a predetermined number of marked individuals had been found. This is called a "stopping rule." The sample size, n, would be a random variable. If p happened to be very small, n would tend to be very large; an increase in p would tend

to cause a decrease in n. This procedure is called "inverse sampling," and its properties were first examined by Haldane (1945).

At first sight it seems natural to estimate p by X/n. This estimator is slightly biased under inverse sampling, and some statisticians would use the modified estimator $(X-1)/(n-1)$, which is unbiased. Others feel that X/n is preferable despite the bias. In most practical work the difference is negligible.

Inverse sampling is one of the simplest methods of sequential estimation. One could define different stopping rules and for any particular stopping rule examine the way in which the precision of X/n varied with p; conversely, one could specify this relationship and ask what stopping rule would satisfy the requirement.

If a random variable is normally distributed with mean μ and variance σ^2, a natural requirement might be to estimate μ with a confidence interval of not greater than a given length at a certain probability level. With samples of fixed size, the length of the confidence interval depends on σ, which is typically unknown. The usual Student t procedure provides intervals of random and unbounded length. Stein (1945) describes a two-sample procedure (with preassigned confidence-interval length) in which the first sample provides an estimate of σ. This estimate then determines the size of the second sample; occasionally a second sample is not needed. This scheme leads naturally to a general approach to sequential estimation. Suppose that, as in inverse sampling, one proceeds in a fully sequential manner, taking one observation at a time, and stops when the desired level of precision is reached. This precision may be determined by customary standard error formulas. Anscombe (1953) showed that in large samples this procedure will indeed yield estimates of the required level of precision. Thus, suppose an investigator wished to estimate the mean number of persons per household in a certain area, with a standard error of 0.1 person, and he had little initial evidence about the variance of household size. He could sample the households until the standard error of the mean, given by the usual formula $s/\sqrt{n}$, fell as low as 0.1. A practical difficulty might be that of ensuring that the sampling was random.

Sequential hypothesis testing. Suppose that one wishes to test a specific hypothesis, H_0, in such a way that if H_0 is indeed true it will usually be accepted and that if an alternative hypothesis, H_1, is true H_0 will usually be rejected. In the most elementary case H_0 and H_1 are *simple* hypotheses; that is, each specifies completely the probability distribution of the generic random variable, X. Suppose $f_0(x)$ and $f_1(x)$ are the probabilities (or probability densities) that X takes the value x when H_0 and H_1 are true, respectively.

Sequential probability ratio test (SPRT). Wald proposed the following method of sequential hypothesis testing. Independent observations, X_i, are taken sequentially and result in values, x_i. Define two positive constants, A_0 and A_1. At the nth stage, calculate

$$\frac{f_{1n}}{f_{0n}} = \frac{f_1(x_1) \cdots f_1(x_n)}{f_0(x_1) \cdots f_0(x_n)}.$$

If $f_{1n}/f_{0n} \leq A_0$, accept H_0; if $f_{1n}/f_{0n} \geq A_1$, reject H_0. If $A_0 < f_{1n}/f_{0n} < A_1$, take the next observation and repeat the procedure.

Wald called this procedure the "sequential probability ratio test" (SPRT). The ratio f_{1n}/f_{0n}, normally called the "likelihood ratio," plays an important part in the Neyman–Pearson theory of hypothesis testing, a fact that probably largely explains Wald's motivation. The likelihood ratio also occurs naturally in Bayesian inference [*see* BAYESIAN INFERENCE].

Let α be the probability of rejecting H_0 when it is true, and let β be the probability of accepting H_0 when H_1 is true. It can be shown that $A_0 \geq \beta/(1-\alpha)$ and $A_1 \leq (1-\beta)/\alpha$ and that the SPRT with $A_0 = \beta/(1-\alpha)$ and $A_1 = (1-\beta)/\alpha$ will usually have error probabilities α' and β' rather close to α and β. (The inequalities arise because sampling usually stops when the bounds A_0 and A_1 are slightly exceeded rather than equaled.)

The number of observations, n, required before a decision is reached is a random variable. Wald gave approximate formulas for $E_0(n)$ and $E_1(n)$, the mean number of observations when H_0 or H_1 is true. Wald conjectured, and Wald and Wolfowitz (1948) proved, that no other test (sequential or not) having error probabilities equal to α' and β' can have lower values for either $E_0(n)$ or $E_1(n)$ than those of the SPRT.

As an example, suppose that H_0 specifies that the proportion, p, of "marked" individuals in a large population is p_0 and that H_1 states that p is p_1 where $p_1 > p_0$. At the nth stage, if x marked individuals have been found, the likelihood ratio is

$$\frac{f_{1n}}{f_{0n}} = \frac{p_1^x(1-p_1)^{n-x}}{p_0^x(1-p_0)^{n-x}}.$$

Sampling will continue as long as

$$\frac{\beta}{1-\alpha} < \frac{p_1^x(1-p_1)^{n-x}}{p_0^x(1-p_0)^{n-x}} < \frac{1-\beta}{\alpha},$$

using the appropriate formulas for the bounds. Taking logarithms, this inequality is

$$\log\left(\frac{\beta}{1-\alpha}\right)$$
$$< x\log\left(\frac{p_1}{p_0}\right) + (n-x)\log\left(\frac{1-p_1}{1-p_0}\right)$$
$$< \log\left(\frac{1-\beta}{\alpha}\right)$$

or

$$a_0 + bn < x < a_1 + bn,$$

where a_0, a_1, and b are functions of p_0, p_1, α, and β. The SPRT can thus be performed as a simple graphical procedure, with coordinate axes for x and n and two parallel boundary lines, $x = a_0 + bn$ and $x = a_1 + bn$ (see Figure 1). The successive values of X are plotted to form a "sample path," and sampling stops when the sample path crosses either of the boundaries.

Graphic solutions with parallel lines are also obtained for the test of the mean of a normal distribution with known variance and for the parameter of a Poisson distribution.

The SPRT is clearly a powerful and satisfying procedure in situations where one of two simple hypotheses is true and where the *mean* number of observations is an appropriate measure of the sampling effort. Unfortunately, some or all of these conditions may not hold. A continuous range of hypotheses must usually be considered; the hypotheses may not be simple; and the variability of the number of observations, as well as its mean, may be important.

Suppose that there is a single parameter, θ, describing the distribution of interest, and further suppose that H_0 and H_1 specify two particular values of θ: θ_0 and θ_1. For every value of θ, including θ_0 and θ_1, quantities of interest are the proba-

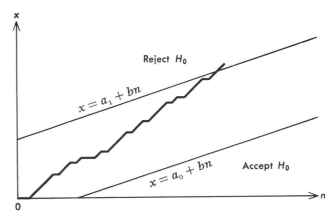

Figure 1 — Sequential probability ratio test for binomial sampling

bility, $L(\theta)$, of accepting H_0 (called the "operating characteristic" or O.C.) and the average number of observations, $E_\theta(n)$ (called the "average sample number," ASN, rather than the "average sample size," for obvious reasons). Approximate formulas for both these quantities are found in Wald's book (1947). The O.C. is an approximately smooth function between 0 and 1, taking the values $L(\theta_0) = 1 - \alpha$, $L(\theta_1) = \beta$ (see Figure 2). The ASN, that is, $E_\theta(n)$, normally has a maximum for a value of θ between θ_0 and θ_1 (see Figure 3). In the binomial problem discussed above, for example, the maximum ASN occurs close to the value $p = b$, for which, on the average, the sample path tends to move parallel to the boundaries. It is remarkable, though, that in many situations of practical interest this maximum value of the ASN is less than the size of the nonsequential procedure that tests H_0 and H_1 with the same error probabilities as the SPRT.

Closed procedures. It can be shown in most cases that an SPRT, although defined for indefinitely large n, must stop some time. Individual sample numbers may, however, be very large and the great variability in sample number from one sample to another may be a serious drawback. Wald suggested that the schemes should be "truncated" by taking the most appropriate decision if a boundary had not been reached after some arbitrary large number of readings. The properties of the SPRT are, however, affected unless the trun-

cation sample size is very high, and a number of authors (Bross 1952; Armitage 1957; Anderson 1960; Schneiderman & Armitage 1962) have recently examined closed procedures (that is, procedures with an upper bound to the number of observations) of a radically different type. In general, it seems possible to find closed procedures which are only slightly less efficient than the corresponding SPRT at H_0 and H_1 but which are more efficient in intermediate situations. (For an account of the use of closed procedures in trials of medical treatments, see Armitage 1960.) Frequently the hypotheses to be tested will be composite rather than simple, in that the probability distribution of the observations is not completely specified. For example, in testing the mean of a normal distribution the variance may remain unspecified. Wald's approach to this problem was not altogether satisfactory and recent work (for example, following Cox 1952) has tended to develop analogues of the SPRT using sufficient statistics where possible [*see* SUFFICIENCY]. A sequential *t*-test of this type was tabulated by Arnold (see U.S. National Bureau of Standards 1951). The usual approximations to the O.C. and ASN do not apply directly in these situations. Cox (1963) has described a large-sample approach based on maximum likelihood estimates of the parameters.

Two-sided tests of hypotheses. In many statistical problems two-sided tests of hypotheses are more appropriate than one-sided tests. In an experi-

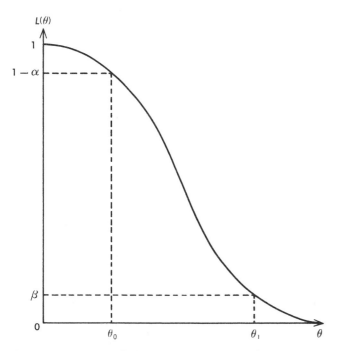

Figure 2 — Typical form of operating characteristic (O.C.)

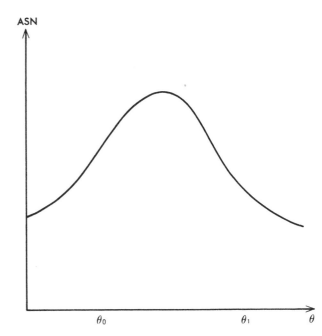

Figure 3 — Typical form of average sample number curve

ment to compare two treatments, for example, it is customary to specify a null hypothesis that no effective difference exists and to be prepared to reject the hypothesis if differences in either direction are demonstrable. One approach to two-sided sequential tests (used, for example, in the standard sequential *t*-test) is to allow an alternative composite hypothesis to embrace simple hypotheses on both sides of the null hypothesis. It may be more appropriate to recognize here a three-decision problem, the decisions being to accept the null hypothesis (that is, to assert no demonstrable difference), to reject it in favor of an alternative in one direction, or to reject it in the other direction. A useful device then is to run simultaneously two separate two-decision procedures, one to test H_0 against H_1 and the other to test H_0 against H_1' where H_1 and H_1' are alternatives on different sides of H_0 (see Sobel & Wald 1949).

Other approaches. Wald's theory and the sort of developments described above are in the tradition of the Neyman–Pearson theory of hypothesis testing, with its emphasis on risks of accepting or rejecting certain hypotheses when these or other hypotheses are true. This approach is arbitrary in many respects; for example, in the SPRT there is no clear way of choosing values of α and β or of specifying an alternative to a null hypothesis. Much theoretical work is now based on statistical decision theory [*see* DECISION THEORY; *see also* Wald 1950]. The aim here is to regard the end product of a statistical analysis as a decision of some sort, to measure the gains or losses that accrue, under various circumstances, when certain decisions are taken, to measure in the same units the cost of making observations, and to choose a rule of procedure that in some sense leads to the highest expectation of gain. Prior probabilities may or may not be attached to various hypotheses. With certain assumptions, the SPRT emerges as an optimal solution for the comparison of two simple hypotheses, but there is no reason to accept it as a general method of sequential analysis. Chernoff (1959) has developed a large-sample theory of the sequential design of experiments for testing composite hypotheses. The aim is to minimize cost in the limiting situation in which costs of wrong decisions far outweigh costs of experimentation. Account is taken of the choice between different types of observation (for example, the use of either of two treatments). A somewhat different approach (for instance, Wetherill 1961) is to stop an investigation as soon as the expected gain achieved by taking a further observation is outweighed by the cost of the observation. The formulation of the

problem requires the specification of prior probabilities and its solution involves dynamic programming [*see* PROGRAMMING].

The recent interest shown in statistical inference by likelihood, with or without prior probabilities, has revealed a conflict between this approach and the more traditional methodology of statistics, involving significance tests and confidence intervals [*see* LIKELIHOOD]. In the likelihood approach, inferences do not depend on stopping rules. There is, on this view, no need to have a separate theory of estimation for inverse sequential sampling or to require that investigations should follow a clearly defined stopping rule before their results can be rigorously interpreted.

The effect of all this work is at present hard to assess. The attraction of a global view is undeniable. On the other hand, the specification of prior distributions and losses may be prohibitively difficult in most scientific investigations.

A number of workers have studied problems involving progressive changes in experimental conditions. Robbins and Monro (1951) give a method for successive approximation to the value of an independent variable, in a regression equation, corresponding to a specified mean value of the dependent variable. Similar methods for use in stimulus–response experiments are reviewed by Wetherill (1963). In industrial statistics much attention has been given to the problem of estimating, by a sequence of experiments, the set of operating conditions giving optimal response. This work has been stimulated mainly by G. E. P. Box, whose *evolutionary operation* is a method by which continuous adjustments to operating conditions can be made [*see* EXPERIMENTAL DESIGN, *article on* RESPONSE SURFACES].

P. ARMITAGE

[*Directly related are the entries* ESTIMATION; HYPOTHESIS TESTING; SCREENING AND SELECTION.]

BIBLIOGRAPHY

Useful bibliographies and surveys of sequential analysis are given in Jackson 1960; Johnson 1961; *and* Wetherill 1966. Wald 1947 *remains an important source book; an elementary exposition of the SPRT, with examples, is contained in* Columbia University, Statistical Research Group 1945. Armitage 1960 *is concerned with medical applications.* Federer 1963 *discusses sequential procedures in screening and selection problems.*

ANDERSON, T. W. 1960 A Modification of the Sequential Probability Ratio Test to Reduce the Sample Size. *Annals of Mathematical Statistics* 31:165–197.

ANSCOMBE, F. J. 1953 Sequential Estimation. *Journal of the Royal Statistical Society* Series B 15:1–29. → Contains eight pages of discussion.

ARMITAGE, P. 1957 Restricted Sequential Procedures. *Biometrika* 44:9–26.

[1]ARMITAGE, P. 1960 *Sequential Medical Trials.* Oxford: Blackwell; Springfield, Ill.: Thomas.

BARNARD, G. A. 1946 Sequential Tests in Industrial Statistics. *Journal of the Royal Statistical Society* Supplement 8:1–26. → Contains four pages of discussion.

BARTKY, WALTER 1943 Multiple Sampling With Constant Probability. *Annals of Mathematical Statistics* 14:363–377.

BROSS, I. 1952 Sequential Medical Plans. *Biometrics* 8:188–205.

CHERNOFF, HERMAN 1959 Sequential Design of Experiments. *Annals of Mathematical Statistics* 30:755–770.

COLUMBIA UNIVERSITY, STATISTICAL RESEARCH GROUP 1945 *Sequential Analysis of Statistical Data: Applications.* New York: Columbia Univ. Press.

COX, D. R. 1952 Sequential Tests for Composite Hypotheses. Cambridge Philosophical Society, *Proceedings* 48:290–299.

COX, D. R. 1963 Large Sample Sequential Tests for Composite Hypotheses. *Sankhyā: The Indian Journal of Statistics* 25:5–12.

DODGE, HAROLD F.; and ROMIG, HARRY G. (1929–1941) 1959 *Sampling Inspection Tables: Single and Double Sampling.* 2d ed., rev. & enl. New York: Wiley; London: Chapman. → This book is a republication of fundamental papers published by the authors in 1929 and in 1941 in the *Bell System Technical Journal.*

FEDERER, WALTER T. 1963 Procedures and Designs Useful for Screening Material in Selection and Allocation, With a Bibliography. *Biometrics* 19:553–587.

HALDANE, J. B. S. 1945 On a Method of Estimating Frequencies. *Biometrika* 33:222–225.

JACKSON, J. EDWARD 1960 Bibliography on Sequential Analysis. *Journal of the American Statistical Association* 55:561–580.

JOHNSON, N. L. 1961 Sequential Analysis: A Survey. *Journal of the Royal Statistical Society* Series A 124:372–411.

ROBBINS, HERBERT; and MONRO, SUTTON 1951 A Stochastic Approximation Method. *Annals of Mathematical Statistics* 22:400–407.

SCHNEIDERMAN, M. A.; and ARMITAGE, P. 1962 A Family of Closed Sequential Procedures. *Biometrika* 49:41–56.

SOBEL, MILTON; and WALD, ABRAHAM 1949 A Sequential Decision Procedure for Choosing One of Three Hypotheses Concerning the Unknown Mean of a Normal Distribution. *Annals of Mathematical Statistics* 20:502–522.

STEIN, CHARLES M. 1945 A Two-sample Test for a Linear Hypothesis Whose Power is Independent of the Variance. *Annals of Mathematical Statistics* 16:243–258.

U.S. NATIONAL BUREAU OF STANDARDS 1951 *Tables to Facilitate Sequential t-Tests.* Edited by K. J. Arnold. National Bureau of Standards, Applied Mathematics Series, No. 7. Washington: U.S. Department of Commerce.

WALD, ABRAHAM 1947 *Sequential Analysis.* New York: Wiley.

WALD, ABRAHAM (1950) 1964 *Statistical Decision Functions.* New York: Wiley.

WALD, ABRAHAM; and WOLFOWITZ, J. (1948) 1955 Optimum Character of the Sequential Probability Ratio Test. Pages 521–534 in Abraham Wald, *Selected Papers in Statistics and Probability.* New York: McGraw-Hill. → First published in Volume 19 of the *Annals of Mathematical Statistics.*

WETHERILL, G. BARRIE 1961 Bayesian Sequential Analysis. *Biometrika* 48:281–292.

WETHERILL, G. BARRIE 1963 Sequential Estimation of Quantal Response Curves. *Journal of the Royal Statistical Society* Series B 25:1–48. → Includes nine pages of discussion.

[2]WETHERILL, G. BARRIE 1966 *Sequential Methods in Statistics.* New York: Wiley.

Postscript

Ghosh (1971) has written a useful book concentrating mainly on Wald's theory. Second editions of Armitage (1960) and Wetherill (1966) have been published. Interesting proposals for sequential design in clinical trials, with data-dependent treatment allocation leading to preponderant use of the better treatment, have been made by Zelen (1969), Flehinger and Louis (1971; 1973), and Hoel et al. (1975), among others.

P. ARMITAGE

ADDITIONAL BIBLIOGRAPHY

[1]ARMITAGE, P. (1960) 1975 *Sequential Medical Trials.* 2d ed. New York: Wiley.

FLEHINGER, B. J.; and LOUIS, T. A. 1971 Sequential Treatment Allocation in Clinical Trials. *Biometrika* 58:419–426.

FLEHINGER, B. J.; and LOUIS, T. A. 1973 Sequential Medical Trials With Data-dependent Treatment Allocation. Volume 4, pages 43–51 in Berkeley Symposium on Mathematical Statistics and Probability, Sixth, *Proceedings.* Berkeley: Univ. of California Press.

GHOSH, BHASKAR KUMAR 1971 *Sequential Tests of Statistical Hypotheses.* Reading, Mass.: Addison-Wesley.

HOEL, DAVID G.; SOBEL, MILTON; and WEISS, GEORGE H. 1975 A Survey of Adaptive Sampling for Clinical Trials. Volume 1, pages 29–61 in Robert M. Elashoff (editor), *Perspectives in Biometrics.* New York: Academic Press.

[2]WETHERILL, G. BARRIE (1966) 1975 *Sequential Methods in Statistics.* 2d ed. New York: Wiley.

ZELEN, MARVIN 1969 Play the Winner Rule and the Controlled Clinical Trial. *Journal of the American Statistical Association* 64:131–146.

SERIAL CORRELATION

See TIME SERIES.

SHEWHART, WALTER A.

▶ *This article was specially written for this volume.*

Walter A. Shewhart was born on March 18, 1891, in New Canton, Illinois; he died on March 13, 1967. After receiving his PH.D. in physics from the University of California in 1917, he began his professional life as an engineer with the Western Electric Company. In 1925 he joined the Bell Laboratories and in 1940 became a member of the technical staff.

Statistical control.   Shewhart is best known for his invention of the statistical control of quality, together with methods by which to reach the state of statistical control, and methods to judge whether the state has been achieved (1925; 1926b; 1931a; 1934). Statistical control corresponds to the abstract idea of a sequence of independent identically distributed random variables [for a detailed discussion, see QUALITY CONTROL, article on PROCESS CONTROL].

Statistical control applies to the output of human effort, as well as to the output of a machine. It includes quantity of output, as well as quality. Quality does not necessarily mean high quality: it means quality suited to the market.

Statistical control achieves economies in several ways: (1) knowledge of what specifications the process can meet economically, and (2) minimization of loss from waste of materials, idle machinetime, incoming materials and assemblies unsuited to the process, machines or workers unsuited to the process, meaningless specifications of incoming or outgoing materials (1931a).

Shewhart's operational definition of statistical control is not restricted in application to manufacture, although it originated in that setting. It is also relevant to any repetitive situation in which stability and predictability are important. Examples include methods of measurement in the laboratory, errors made by keypunch operators, and times of arrival of trains.

Statistical control can be achieved and monitored by statistical methods, chiefly by use of the control chart or by the run-chart. Statistical methods detect the existence of special causes, which, when eliminated, leave the process in control and reduce the spread and erratic behavior of output. The same methods measure the amount of dispersion chargeable to the system itself, dispersion that the production worker cannot in ordinary circumstances reduce. Alteration of the system, to reduce trouble, is the responsibility of management. [These methods are discussed in QUALITY CONTROL, article on PROCESS CONTROL.]

Other contributions.   Shewhart made other contributions to knowledge and methodology. In particular, he wrote at length about the inherent need for operational definitions and specifications. Characteristics (round, red, old, unemployed, lateness of a train, etc.) can be communicated only in terms

that are operational, which necessarily means statistical terms. In order that a sentence have scientific meaning, it must be testable on the basis of measurements obtained by carrying out previously specified operations. The meaning of the sentence is the method of its testing. (Of course in many cases the measurements may be available only in principle, or theoretically.) Further, an operational definition inevitably varies with need and use, and with the state of knowledge; hence there is no underlying true value of any measured characteristic. [See SCIENCE, article on PHILOSOPHY OF SCIENCE for further discussion of the problems of definition in science.]

Original scientific writings, Shewhart believed, should preserve all the evidence, including, beyond the data themselves, descriptions of the process of measurement, comparison of observers, and records of materials used and of any environmental condition that could affect the results. By the same standards, he argued that summaries of data via such statistics as the average and variance should be used only if they lead to substantially the same conclusions as the full evidence would lead to (1939).

Shewhart is also known for his insistence that physical laws that are taught as deterministic can be understood only in a statistical sense. For example, the Newtonian law that distance traveled by a body falling freely in vacuum for time t is $\frac{1}{2}gt^2$, where g is the acceleration of gravity, is actually a statement of expectation. Any set of observations will show departure from the law. In the ideal state, these observations will show statistical control.

The precision of a result, Shewhart pointed out, stands throughout time but the accuracy of the same result changes with changes in the operational definition of the thing being measured.

Shewhart stressed the importance of subject matter. The results of a statistical investigation must be useful to the expert on the subject studied. The statistician in practice must accordingly learn something of the substantive field he works in, for choices of sampling unit, of method of selection, and of estimators will depend on the material (people, patients in hospitals, soil, potatoes, industrial product) and must be efficient for the problem at hand, but not beyond the capability of the people who will carry out the work.

He always asked the questions "What will the results refer to?" "How will you use the results?" Generalization to the future, to people, patients, and soils beyond the range of those sampled and tested, is the responsibility of the substantive expert. Statistical theory cannot help in such generalization.

Shewhart made other original contributions. He published in 1931 the basic formula for the optimal number of units to measure in all strata of investigation as a function of costs and of variances within and between strata (1931b). This result, basic in surveys and experiments, was published the same year by Tippett (1931) and later by Neyman (1934; 1938). Shewhart worked on the theory of sampling from nonnormal frames, and illustrated theory with examples. [See SAMPLE SURVEYS.]

W. EDWARDS DEMING

WORKS BY SHEWHART

1924 Some Applications of Statistical Methods to the Analysis of Physical and Engineering Data. *Bell System Technical Journal* 3:43–87.

1925 The Application of Statistics as an Aid in Maintaining Quality of a Manufactured Product. *Journal of the American Statistical Association* 20:546–548.

1926a Correction of Data for Errors of Measurement. *Bell System Technical Journal* 5:11–26.

1926b Quality Control Charts. *Bell System Technical Journal* 5:593–603.

1927 Quality Control. *Bell System Technical Journal* 6:722–735.

1928a Economic Aspects of Engineering Applications of Statistical Methods. Franklin Institute, Philadelphia, *Journal* 205:395–405.

1928b Note on the Probability Associated With the Error of a Single Observation. *Journal of Forestry* 26:601–607.

1928 SHEWHART, WALTER A.; and WINTERS, F. W. Small Samples: New Experimental Results. *Journal of the American Statistical Association* 23:144–153.

1931a *Economic Control of Quality of Manufactured Product*. New York: Van Nostrand.

1931b Random Sampling. *American Mathematical Monthly* 38:245–270.

1931c Application of Statistical Method in Engineering. *Journal of the American Statistical Association* 26:214–221.

1933 Annual Survey of Statistical Technique: Developments in Sampling Theory. *Econometrica* 1:225–237.

1934 Some Aspects of Quality Control. *Mechanical Engineering* 56:725–730.

1938 Application of Statistical Methods to Manufacturing Problems. Franklin Institute, Philadelphia, *Journal* 226:163–186.

1939 *Statistical Method From the Viewpoint of Quality Control*. Washington: U.S. Department of Agriculture, Graduate School.

1941 Contribution of Statistics to the Science of Engineering. Pages 97–124 in University of Pennsylvania, Bicentennial Conference, Philadelphia, 1941, *Fluid Mechanics and Statistical Methods in Engineering*. By Hugh L. Dryden et al. Philadelphia: Univ. of Pennsylvania Press.

1946 The Advancing Statistical Front. *Journal of the American Statistical Association* 41:1–15.

1958 Nature and Origin of Standards of Quality. *Bell System Technical Journal* 37:1–22.

SUPPLEMENTARY BIBLIOGRAPHY

DEMING, W. EDWARDS 1968 Walter A. Shewhart, 1891–1967. *Review of the International Statistical Institute* 36:372–375.

NEYMAN, JERZY 1934 On the Two Different Aspects of the Representative Method: The Method of Stratified Sampling and the Method of Purposive Selection. *Journal of the Royal Statistical Society* 97:558–606. → Discussion on pages 607–625.

NEYMAN, JERZY 1938 Contribution to the Theory of Sampling Human Populations. *Journal of the American Statistical Association* 33:101–116. → See especially page 110.

TIPPETT, L. H. C. (1931) 1952 *The Methods of Statistics*. 4th ed., rev. London: Williams & Norgate. → See especially page 177.

SIGNAL DETECTION

See MODELS, MATHEMATICAL; MULTIVARIATE ANALYSIS, *article on* CLASSIFICATION AND DISCRIMINATION.

SIGNIFICANCE, TESTS OF

The topic of significance testing is treated in detail in HYPOTHESIS TESTING, *a term that is usually regarded as synonymous with "tests of significance," although some statisticians feel that there are important distinctions. The present article describes the basic ideas of significance testing, outlines the most important elementary tests, and reviews problems related to the philosophy and application of significance testing.*

Significance tests are statistical procedures for examining a hypothesis in the light of observations. Many significance tests are simple and widely used, and have a long history; the idea of significance testing goes back at least to the eighteenth century. There has been much confusion about significance testing, and consequently it has been much misused. Despite their apparent simplicity (and sometimes because of it), significance tests have generated many controversies about meaning and utility.

A significance test starts with observations and with a hypothesis about the chance mechanism generating the observations. From the observations a *test statistic* is formed. Large values of the test statistic (or small values, or both, depending on circumstances) lead to strong skepticism about the hypothesis, whereas other values of the test statistic are held to be in conformance with the hypothesis. Choice of the test statistic usually depends upon the alternatives to the hypothesis under test.

Basic ideas

As an example of significance testing, suppose that the hypothesis under test, often called the *null hypothesis*, or H_0, is that sleep deprivation has no effect on a certain kind of human skill, say arithmetic ability as measured by a particular type of test. Some population must be specified; suppose that it consists of all students at a particular college who are willing to participate in a sleep-deprivation experiment. One hundred students are chosen at random, tested for arithmetic skill, deprived of a night's sleep, and then tested again. For each student the initial test result minus the second test result is regarded as the basic observation, datum, or score. Suppose, for the present, that if the null hypothesis is false, the effect of sleep deprivation is either to increase average score or to decrease it; either direction might hold, so the *alternative hypotheses* are two-sided.

The analyst would, of course, be concerned about possible practice and motivational effects; a helpful device might be the use of a control group that would also be tested twice, but with a normal night's sleep. The effect of practice might be lessened by the presentation of a sequence of tests before the experiment proper begins. Another matter of concern is whether the basic score used, the *difference* between test results, is more appropriate than the ratio of test results or some other combination of them. For present purposes such issues are not discussed.

From the 100 scores a test statistic is then formed. The choice might well be the Student *t*-statistic, which is the observed average score divided by an estimate of its own variability. Details of the *t*-statistic are given later.

The nub of the significance-testing concept is that if the null hypothesis really is true, then it is unlikely that the Student *t*-statistic would be greater than, say, 2.6 in unsigned numerical value. The probability of that event, under the usual—but possibly dangerous—assumptions that the scores behave like statistically independent random variables with the same normal distribution, is about .01, or 1 out of 100. If, on the other hand, the null hypothesis is false, the probability of the event is greater than .01, much greater if sleep deprivation has a large effect.

Hence, if a Student *t*-statistic greater than 2.6 or less than −2.6 is observed, *either* the null hypothesis is false, *or* the null hypothesis is true and an event of small probability has occurred, *or* something is wrong with the underlying assumptions. If those assumptions seem satisfactory, and if the Student *t*-statistic is, say, 2.7, most people would prefer to act as if the null hypothesis were false. The cut-off number, 2.6, and the associated probability, .01, are cited for the sake of specific illustration. If 2.6 is replaced by a larger number,

the associated probability is smaller; if it is replaced by a smaller number, the probability is larger.

In formal significance testing one decides beforehand on the cut-off number. If 2.6 is chosen, then there is a .01 probability of erroneously doubting the null hypothesis when it is in fact true.

In the example, 2.6 is called the *critical value* of the (unsigned) Student *t*-statistic, and .01 is called the *level of significance* or (usually synonymously) the *size* of the test. The level of significance, frequently denoted by α, is also called the probability of Type I error, or of error of the first kind; this is the error of falsely doubting the null hypothesis when it is in fact true. In the present context, if the unsigned Student *t*-statistic turns out to be greater than the critical value, it is often said that the statistic (or the test, or the sample) is *statistically significant*; this terminology can lead to confusion, as will be described below.

From this approach, then, a significance test is a procedure, agreed upon beforehand, that will result in one of two conclusions (or actions, or viewpoints): if the test statistic—here the unsigned Student *t*-statistic—is greater than the critical value, one tends to act as if the null hypothesis is false; otherwise one tends to act in some different way, not necessarily as if the null hypothesis is true but at least as if its falsity is not proved. These two kinds of conclusions or actions are conventionally called, respectively, *rejection* and *acceptance* of the null hypothesis. Such nomenclature, with its emotional overtones and its connotations in everyday speech, has led to confusion and needless controversy. The terms "rejection" and "acceptance" should be thought of simply as labels for two kinds of actions or conclusions, the first more appropriate if the null hypothesis is false and the second more appropriate if it is true or at least not known to be false. In the example, rejection and acceptance might correspond to different attitudes toward a theory of sleep or to different recommendations for desirable amounts of sleep. (In some contexts, such as acceptance sampling in industry, the terms "acceptance" and "rejection" may be used literally.)

There is further room for confusion in the varying ways that significance testing is described. In the example it would be accurate to say that the hypothesis of no sleep-deprivation effect (on average score) is being tested. But it might also be said, more loosely, that one is testing the effect of sleep deprivation, and this might be misinterpreted to mean that the null hypothesis asserts a positive mean score under sleep deprivation. One reason for this ambiguity is that the null hypothesis is often set up in the hope that it will be rejected. In the comparison of a new medical treatment with an old one, equality of effectiveness might be uninteresting, so only rejection of the null hypothesis of equality would excite the experimenter, especially if he had invented the new treatment. On the other hand, accepting the null hypothesis may itself be very important—for example, in an experiment to test the theory of relativity. Usually, however, there is a basic asymmetry between the null hypothesis and the other hypotheses.

¹The two-decision procedure discussed above is a highly simplified framework for many situations in which significance tests are used. Although there are cases in which the above viewpoint makes direct sense (for example, industrial acceptance sampling, crucial experiments for a scientific theory), most uses of significance tests are different. In these more common uses the analyst computes that level of significance at which the observed test statistic would just lead to rejection of the null hypothesis. In the example, if the observed Student *t*-statistic were 1.8, the resulting *sample level of significance*, or *observed significance level*, would be .075, again under the conventional assumptions listed earlier. From this second viewpoint the observed (or sample) significance level is a measure of rarity or surprise; the smaller the level, the more surprising the result if the null hypothesis is true. Naturally, if the result is very surprising under the null hypothesis, one is likely to conclude that the null hypothesis is false, but it is important to know quantitatively the magnitude of the surprise. The sample level of significance is often called the *P*-value, and the notation $P = .075$ (or $p = .075$) may be used. (For other views on measuring surprise, see Weaver 1948; Good 1956.)

Usually a very rough knowledge of the degree of surprise suffices, and various simplifying conventions have come into use. For example, if the observed significance level lies between .05 and .01, some authors say that the sample is *statistically significant*, while if it is less than .01, they call it *highly statistically significant*. The adverb "statistically" is often omitted, and this is unfortunate, since statistical significance of a sample bears no necessary relationship to possible subject-matter significance of whatever true departure from the null hypothesis might obtain (see Boring 1919).

From one point of view, significance testing is a device for curbing overenthusiasm about apparent effects in the data. For example, if the sleep-deprivation experiment showed a positive effect, an experimenter enthusiastic about a theory of sleep psychology might cry, "Eureka! Sleep deprivation really does have the predicted consequence."

After he himself has had a good night's sleep, however, he might compute the appropriate test statistic and see, perhaps, that the observed effect—or a greater one—could occur with probability .2 under the null hypothesis of no real underlying effect. This would probably dampen his initial enthusiasm and might prevent one of the many premature announcements of discoveries that clutter the literature of some sciences. (The sleep example is solely for illustration; I have no reason to think that experimenters on sleep are anything but cautious in their statistical analyses.)

On the other hand, it is easy to be overconservative and throw out an interesting baby with the nonsignificant bath water. Lack of statistical significance at a conventional level does not mean that no real effect is present; it means only that no real effect is clearly seen from the data. That is why it is of the highest importance to look at power and to compute confidence intervals, procedures that will be discussed later.

The null hypothesis, despite its name, need not be that the average effect of sleep deprivation is zero. One might test the null hypothesis that the average effect is 10 score units. Or one might test such a null hypothesis as: the average effect of sleep deprivation is not positive; or, again: the average effect lies between −10 and 10 score units. The last two null hypotheses, unlike the previous ones, do not fix the hypothesized average effect at a single value.

Standard tests

There are a number of popular standard significance tests in the statistical armamentarium. The tests described here include those on means, variances, correlations, proportions, and goodness of fit.

Mean. The null hypothesis in a test on means, as in the original example, is that the expected value (mean) of a distribution is some given number, μ_0. This hypothesis is to be tested on the basis of a random sample of n observations on the distribution; that is, one considers n independent random variables, each with that same distribution whose mean is of interest. If the variance, σ^2, of the distribution is known, and if $\bar{X}$ denotes the sample mean (arithmetic average), a widely useful test statistic is

$$(1) \qquad \frac{\bar{X} - \mu_0}{\sigma/\sqrt{n}},$$

which has mean zero and standard deviation one under the null hypothesis. Often the absolute (unsigned numerical) value of (1) is the test statistic.

If the underlying distribution of interest is normal, then (1) is itself normally distributed, with variance one and mean zero (that is, it has a unit normal distribution) under the null hypothesis. Even if the underlying distribution is not normal, (1) has approximately the unit normal distribution under the null hypothesis, provided that n is not too small.

In practice σ^2 is usually unknown, and an estimator of it, s^2, is used instead. The most usual estimator is based on the sum of squared deviations of the observations from $\bar{X}$,

$$(2) \quad s^2 = \frac{1}{n-1}[(X_1 - \bar{X})^2 + \cdots + (X_n - \bar{X})^2],$$

where the X_i are the observations. (The usual name for s^2 is "sample variance," and that for s is "sample standard deviation.") Then (1) becomes the Student (or t-) statistic

$$(3) \quad \frac{\bar{X} - \mu_0}{s/\sqrt{n}}.$$

If the X_i are normal, then (3) has as its distribution under the null hypothesis the Student (or t-) distribution with $n - 1$ degrees of freedom (df). If n is large (and often it need not be very large), then in quite general circumstances the distribution of (3) under the null hypothesis is approximately the unit normal distribution. [See DISTRIBUTIONS, STATISTICAL, *articles on* SPECIAL CONTINUOUS DISTRIBUTIONS *and* APPROXIMATIONS TO DISTRIBUTIONS.]

Two samples. Suppose that the observations form *two* independent random samples of sizes n_1 and n_2: $X_{11}, \cdots, X_{1n_1}$ and $X_{21}, \cdots, X_{2n_2}$. Here the first subscript denotes the sample (1 or 2), and the second subscript denotes the observation within the sample (1 through n_1 or n_2). For example, in the sleep-deprivation experiment the first sample might be of male students and the second of female students. The null hypothesis might be that the average effect of sleep deprivation on arithmetic skill is the same for male and female students. Note that this null hypothesis says nothing about the magnitude of the average effect; it says only that the average effect is the same for the two groups.

More generally, suppose the null hypothesis is that the two underlying means are equal: $\mu_1 = \mu_2$, where μ_1 is the expected value of any of the X_{1i} and μ_2 that of any of the X_{2i}. If the variances are known, say σ_1^2 and σ_2^2, the usual test statistic is

$$(4) \quad \frac{\bar{X}_1 - \bar{X}_2}{\sqrt{\dfrac{\sigma_1^2}{n_1} + \dfrac{\sigma_2^2}{n_2}}},$$

where $\bar{X}_1$ and $\bar{X}_2$ are the two sample means. If the X's are normally distributed, (4) has the unit normal as null distribution—that is, as its distribution when the null hypothesis holds. Even if the X's are not normally distributed, the unit normal distribution is a good approximation to the null distribution of (4) for n_1, n_2 large. (If, instead of the null hypothesis $\mu_1 = \mu_2$, one wishes to test $\mu_1 = \mu_2 + \delta$, where δ is some given number, one need only replace $\bar{X}_1 - \bar{X}_2$ in the numerator of (4) with $\bar{X}_1 - \bar{X}_2 - \delta$.)

If the variances are not known, and if s_1^2 and s_2^2 are estimators of them, a common test statistic is

$$(5) \quad \frac{\bar{X}_1 - \bar{X}_2}{\sqrt{\dfrac{s_1^2}{n_1} + \dfrac{s_2^2}{n_2}}},$$

which, for n_1, n_2 large, has the unit normal distribution as an approximation to the null distribution. The null distribution is better approximated by a t-distribution [*details can be found in* ERRORS, *article on* EFFECTS OF ERRORS IN STATISTICAL ASSUMPTIONS].

If the variances are unknown but may reasonably be assumed to be equal ($\sigma_1^2 = \sigma_2^2$), then a common test statistic is the two-sample Student statistic,

$$(6) \quad \frac{\bar{X}_1 - \bar{X}_2}{\sqrt{\dfrac{(n_1 - 1)s_1^2 + (n_2 - 1)s_2^2}{n_1 + n_2 - 2}\left(\dfrac{1}{n_1} + \dfrac{1}{n_2}\right)}},$$

where s_1^2, s_2^2 are defined in terms of (2). If the observations are normally distributed, (6) has as null distribution the t-distribution with $n_1 + n_2 - 2$ df. This holds approximately, in general, if the X's are mildly nonnormally distributed and if σ_1^2 and σ_2^2 differ somewhat but n_1 and n_2 are not very different.

Paired samples. The two-sample procedure described above should be distinguished from the *paired* two-sample procedure. The latter might be used if the subjects are fraternal twins of opposite sex, with both members of a sampled pair of twins observed. Then the male and female samples would no longer be independent, since presumably the members of a pair of twins are more nearly alike in most respects than are two random subjects. In this case one would still use $\bar{X}_1 - \bar{X}_2$ in the numerator of the test statistic, but the denominator would be different from those given above. Although it is not always possible, pairing (more generally, blocking) is often used to increase the power of a test. The underlying idea is to make comparisons within a relatively homogeneous set of experimental units. [See EXPERIMENTAL DESIGN.]

[*These significance tests, and generalizations of them, are further discussed in* HYPOTHESIS TESTING; LINEAR HYPOTHESES, *article on* ANALYSIS OF VARIANCE; *and* ERRORS, *article on* EFFECTS OF ERRORS IN STATISTICAL ASSUMPTIONS.]

Variances. The null hypothesis in a test on variances may be that σ^2 has the value σ_0^2; then the usual test statistic is s^2 as defined in (2), or, more conveniently,

$$(7) \qquad (n-1)s^2/\sigma_0^2.$$

Under the null hypothesis, (7) has the chi-square distribution with $n-1$ degrees of freedom, provided that the observations are normal, independent, and identically distributed. Unlike the null distributions for the tests on means, the null distribution of (7) is highly sensitive to deviations from normality.

In the two-sample case a common null hypothesis is $\sigma_1^2 = \sigma_2^2$. The usual test statistic is s_1^2/s_2^2, which, under normality, has as null distribution the F-distribution with $n_1 - 1$ and $n_2 - 1$ df.

[*These tests are discussed further in* LINEAR HYPOTHESES, *article on* ANALYSIS OF VARIANCE; VARIANCES, STATISTICAL STUDY OF.]

Correlation. The most common procedure in simple correlation analysis is to test the null hypothesis that the population correlation coefficient is zero, using as test statistic the sample correlation coefficient. Under the assumptions of bivariate normality and a random sample, the null distribution is closely related to a t-distribution. One can also test the null hypothesis that the population correlation coefficient has some nonzero value, say .55. Special tables or approximate methods are required here [*see* MULTIVARIATE ANALYSIS, *article on* CORRELATION METHODS].

Proportions. *Single sample.* The simplest case in the analysis of proportions is that of a sample proportion and the null hypothesis that the corresponding population probability is p_0, some given number between 0 and 1. The sample proportion and population probability correspond to some dichotomous property: alive–dead, heads–tails, success–failure. It is convenient to summarize the sample in the form of a simple table

$$\boxed{\begin{array}{c|c|c} N & n-N & n \end{array}},$$

where n is the sample size, N the number of sample observations having a stated property, $n-N$ the number not having this property, and N/n the sample proportion. The usual sampling assumptions, which need examination in each application, are that the observations are statistically independent and that each has the same probability of

having the stated property. As a result of these assumptions, N has a binomial distribution.

The usual test statistic is

$$(8) \qquad \frac{N - np_0}{\sqrt{np_0(1 - p_0)}},$$

which, for n not too small, has approximately the unit normal distribution under the null hypothesis. The square of this test statistic, again for n not too small and under the null hypothesis, has approximately a chi-square distribution with one df.

Notice that (8) is really a special case of (1) if N is regarded as the sum of n independent observations taking values 1 and 0 with probabilities p_0 and $1 - p_0$.

Two samples. The data may consist of two sample proportions, and the null hypothesis may be that the corresponding population proportions are equal. It is convenient to express such data in terms of a 2×2 table like Table 1, where the two

Table 1 — Summary of two samples for analysis of proportions

	Number having property	Number not having property	Totals
Sample 1	N_{11}	N_{12}	n_{1+}
Sample 2	N_{21}	N_{22}	n_{2+}
Totals	N_{+1}	N_{+2}	$n_{++} = n$

samples (of sizes n_{1+}, n_{2+}) correspond to the upper two lines and the bottom and right rims are marginal totals—for example, $N_{11} + N_{12} = n_{1+}$. Capital letters denote random variables and lower-case letters nonrandom quantities. The usual assumptions, which must be examined in each application, are that N_{11} and N_{21} are independently and binomially distributed. A common test statistic is

$$(9) \qquad \frac{(N_{11}/n_{1+}) - (N_{21}/n_{2+})}{\sqrt{\left(\dfrac{1}{n_{1+}} + \dfrac{1}{n_{2+}}\right)\dfrac{N_{+1}N_{+2}}{n^2}}},$$

which, for n's not too small, has approximately the unit normal distribution under the null hypothesis. (Note the great similarity of test statistics (6) and (9), which are related in much the same way as (1) and (8).)

If the unsigned value of (9) is the test statistic, one may equivalently consider its square, which is expressible in the simpler form

$$(10) \qquad n \frac{(N_{11}N_{22} - N_{12}N_{21})^2}{n_{1+}n_{2+}N_{+1}N_{+2}}.$$

Under the null hypothesis, and for n's not too small, (10) has approximately a chi-square distribution with one *df*. Large values of (10) are statistically significant.

Chi-square statistics. The above test statistics, in squared form, are special cases of the chi-square test statistic; their null distributions are approximately chi-square ones. Such test statistics can generally be thought of in terms of data tabulations of the kind shown above; the statistic is the sum of terms given by the mnemonic expression

$$\frac{(\text{observed} - \text{"expected"})^2}{\text{"expected"}},$$

summed over the cells in the body of the table. For example, in the first, very simple table for a sample proportion, there are two cells in the body of the table, with N and $n - N$ observed frequencies in them. The corresponding frequencies expected under the null hypothesis are np_0 and $n(1 - p_0)$. Applying the mnemonic expression gives as the chi-square statistic

$$\frac{(N - np_0)^2}{np_0} + \frac{[n - N - n(1 - p_0)]^2}{n(1 - p_0)},$$

which can readily be shown to equal the square of (8). The one *df* for the approximate chi-square distribution is sometimes described thus: There are *two* observed frequencies, but they are restricted in that their sum must be n. Hence, there is *one df*.

Similarly, (10) can be obtained from the table for two proportions by means of the mnemonic expression. In this case the null hypothesis does not completely specify the expected frequencies, so they must be estimated from the data. That is why quotation marks appear around "expected" in the mnemonic. [*Further details, and related procedures, including the useful continuity correction, are discussed in* COUNTED DATA.]

Association for counted data. Suppose that the individuals in a sample of people are cross-classified by hair color and eye color, with perhaps five categories for each attribute. The null hypothesis is that the two attributes are statistically independent. A standard procedure to test for association in this *contingency table* is another chi-square test, which is based essentially on (10) above. Although the test statistics are the same and have the same approximate null distribution, the power functions are different. [*This test and related ones are discussed in* COUNTED DATA.]

Nonparametric tests. The tests for means, variances, and correlation, discussed above, are closely tied to normality assumptions, or at least to approximate normality. Analogous tests of significance, without the normality restriction, have been devised [*these are discussed in* NONPARAMETRIC STATISTICS; HYPOTHESIS TESTING].

General approximation. In many cases a test statistic, T, is at hand, but its null distribution is difficult to express in usable form. It is often useful to approximate this null distribution in terms of μ_0, the expected value of T under the null hypothesis, and S, an estimator of the standard deviation of T under the null hypothesis. The approximation takes $(T - \mu_0)/S$ as roughly unit normal under the null hypothesis. This approximation is, intuitively and historically, very important. The first tests of proportions discussed above are special cases of the approximation.

Goodness of fit. A number of significance tests are directed to the problem of goodness of fit. In these cases the null hypothesis is that the parent distribution is some specific one, or a member of some specific family. Another problem often classed as one of goodness of fit is that of testing, on the basis of two samples, the hypothesis that two parent populations are the same. [*See* GOODNESS OF FIT.]

Other aspects of significance testing

Alternative hypotheses and power. A significance test of a null hypothesis is usually held to make sense only when one has at least a rough idea of the hypotheses that hold if the null hypothesis does not; these are the *alternative hypotheses* (or alternate hypotheses). In testing that a population mean, μ, is zero, as in the example at the start of the article, the test is different if one is interested only in positive alternatives ($\mu > 0$), if one is interested only in negative alternatives ($\mu < 0$), or if interest extends to both positive and negative alternatives.

When using such a test statistic as the Student *t*-statistic, it is important to keep the alternatives in mind. In the sleep-deprivation example the original discussion is appropriate when both positive and negative alternatives are relevant. The null hypothesis is rejected if the test statistic has either a surprisingly large positive value or a surprisingly large negative value. On the other hand, it might be known that sleep deprivation, if it affects arithmetic skill at all, can only make it poorer. This means that the expected average score (where the score is the initial test result minus the result after sleep deprivation) cannot possibly be negative but must be zero or positive. One would then use the so-called right-tail test, rejecting the null hypothesis only if the Student t-statistic is large. For example, deciding to reject H_0 when the statistic is greater than 2.6 leads to a level of significance of .005. Similarly, right-tail P-values would be computed. If the Student statistic observed is 1.8, the right-tail P-value is .037; that is, the probability, under the null hypothesis, of observing a Student statistic 1.8 or larger is .037. There are also left-tail tests and left-tail P-values.

For a test considered as a two-action procedure, the *power* is the probability of (properly) rejecting the null hypothesis when it is false. Power, of course, is a function of the particular alternative hypothesis in question, and *power functions* have been much studied. The probability of error of Type II is one minus power; a Type II error is acceptance of the null hypothesis when it is false. By using formulas, approximations, tables, or graphs for power, one can determine sample sizes so as to control both size and power. [*See* EXPERIMENTAL DESIGN, *article on* THE DESIGN OF EXPERIMENTS.]

A common error in using significance tests is to neglect power considerations and to conclude from a sample leading to "acceptance" of the null hypothesis that the null hypothesis holds. If the power of the test is low for relevant alternative hypotheses, then a sample leading to acceptance of H_0 is also very likely if those alternatives hold, and the conclusion is therefore unwarranted. Conversely, if the sample is very large, power may be high for alternatives "close to" the null hypothesis; it is important that this be recognized and possibly reacted to by decreasing the level of significance. These points are discussed again later.

It is most important to consider the power function of a significance test, even if only crudely and approximately. As Jerzy Neyman wrote, perhaps with a bit of exaggeration, ". . . if experimenters realized how little is the chance of their experiments discovering what they are intended to discover, then a very substantial proportion of the experiments now in progress would have been abandoned in favour of an increase in size of the remaining experiments, judged more important" (1958, p. 15). [*References to Neyman's fundamental and pathbreaking contributions to the theory of testing are given in* HYPOTHESIS TESTING.]

There is, however, another point of view for which significance tests of a null hypothesis may be relevant without consideration of alternative hypotheses and power. An illuminating discussion of this viewpoint is given by Anscombe (1963).

Combining significance tests. Sometimes it is desirable to combine two or more significance tests into a single one without reanalyzing the detailed data. For example, one may have from published materials only the sample significance levels of two tests on the same hypothesis or closely related ones. (Discussions of how to do this are given in Mosteller & Bush [1954] 1959, pp. 328–331; and Moses 1956. See also Good 1958.) Of course, a combined analysis of all the data is usually desirable, when that is possible.

²Preliminary tests of significance. A desired test procedure is often based on an assumption that may be questionable—for example, equality of the two variances in the two-sample t-test. It has frequently been suggested that a preliminary test of the assumption (as null hypothesis) be made and that the test of primary interest then be carried out only if the assumption is accepted by the preliminary test. If the assumption is rejected, the test of primary interest requires modification.

Such a two-step procedure is difficult to analyze and must be carried out with caution. One relevant issue is the effect on the primary test of a given error in the assumption. Another is that the preliminary test may be much more sensitive to errors in other underlying assumptions than is the main significance test. [*See* ERRORS, *article on* EFFECTS OF ERRORS IN STATISTICAL ASSUMPTIONS. *A discussion of preliminary significance tests, with references to prior literature, is given in* Bancroft 1964. *Related material is given in* Kitagawa 1963.]

Relation to confidence sets and estimation. It often makes sense, although the procedure is not usually described in these terms, to compute appropriate sample significance levels not only for the null hypothesis but also for alternative hypotheses as if—for the moment—each were a null hypothesis. In this way one obtains a measure of how surprising the sample is for both the null and the alternative hypotheses. If parametric values corresponding to those hypotheses not surprising (at a specified level) for the sample are considered together, a *confidence region* is obtained. [*See* ESTIMATION, *article on* CONFIDENCE INTERVALS AND REGIONS.]

In any case, a significance test is typically only one step in a statistical analysis. A test asks, in effect, Is anything other than random variation appearing beyond what is specified by the null hypothesis? Whatever the answer to that question is—but especially if the answer is Yes—it is almost always important to estimate the magnitudes of underlying effects. [*See* ESTIMATION.]

Relation to discriminant analysis. Significance testing, historically and in most presentations, is asymmetrical: control of significance level is more important than control of power. The null hypothesis has a privileged position. This is sometimes reasonable—for example, when the alternative hypotheses are diffuse while the null hypothesis is sharp. In other cases there is no particular reason to call one hypothesis "null" and the other "alternative" and hence no reason for asymmetry in the treatment of the two kinds of error. This symmetric treatment then is much the same as certain parts of the field called discriminant analysis. [See MULTIVARIATE ANALYSIS, *article on* CLASSIFICATION AND DISCRIMINATION.]

Dangers, problems, and criticisms

Some dangers and problems of significance testing have already been touched on: failure to consider power, rigid misinterpretation of "accept" and "reject," serious invalidity of assumptions. Further dangers and problems are now discussed, along with related criticisms of significance testing.

[3] Nonsignificance is often nonpublic. Negative results are not so likely to reach publication as are positive ones. In most significance-testing situations a negative result is a result that is not statistically significant, and hence one sees in published papers and books many more statistically significant results than might be expected. Many—perhaps most—statistically nonsignificant results never see publication.

The effect of this is to change the interpretation of published significance tests in a way that is hard to analyze quantitatively. Suppose, to take a simple case, that some null hypothesis is investigated independently by a number of experimenters, all testing at the .05 level of significance. Suppose, further, that the null hypothesis is true. Then any one experimenter will have only a 5/100 chance of (misleadingly) finding statistical significance, but the chance that *at least one* experimenter will find statistical significance is appreciably higher. If, for example, there are six experimenters, a rejection of the null hypothesis by at least one of them will take place with probability .265, that is, more than one time out of four. If papers about experiments are much more likely to be published when a significance test shows a level of .05 (or less) than otherwise, then the nonpublication of nonsignificant results can lead to apparent contradictions and substantive controversy. If the null hypothesis is false, a similar analysis shows that the "power" of published significance tests may be appreciably higher than their nominal power. (Discussions of this problem are given in Sterling 1959; and Tullock 1959.)

[4] Complete populations. Another difficulty arises in the use of significance tests (or any other procedures of probabilistic inference) when the data

consist of a complete census for the relevant population. For example, suppose that per capita income and per capita dollars spent on new automobiles are examined for the 50 states of the United States in 1964. The formal correlation coefficient may readily be computed and may have utility as a descriptive summary, but it would be highly questionable to use sampling theory of the kind discussed in this article to test the null hypothesis that the population correlation coefficient is zero, or is some other value. The difficulty is much more fundamental than that of nonnormality; it is hard to see how the 50 pairs of numbers can reasonably be regarded as a sample of any kind. Some statisticians believe that permutation tests may often be used meaningfully in such a context, but there is no consensus. [*For a definition of permutation tests, see* NONPARAMETRIC STATISTICS. *Further discussion of this problem and additional references are given in* Hirschi & Selvin 1967, chapter 13. *An early article is* Woofter 1933.]

Target versus sampled populations. Significance tests also share with all other kinds of inference from samples the difficulty that the population sampled from is usually more limited than the broader population for which an inference is desired. In the sleep-deprivation example the sampled population consists of students at a particular college who are willing to be experimental subjects. Presumably one wants to make inferences about a wider population: all students at the college, willing or not; all people of college age; perhaps all people. [*See* STATISTICS; ERRORS, *article on* NONSAMPLING ERRORS.]

Neglect of power by word play. A fallacious argument is that power and error of the second kind (accepting the null hypothesis when it is false) need not be of concern, since the null hypothesis is never really accepted but is just not rejected. This is arrant playing with words, since a significance test is fatuous unless there is a question with at least two possible answers in the background. Hence, both kinds of probabilities of wrong answers are important to consider. Recall that "accept" and "reject" are token words, each corresponding to a conclusion that is relatively more desirable when one or another true state of affairs obtains.

To see in another way why more than Type I error alone must be kept in mind, notice that one can, without any experiment or expense, achieve zero level of significance (no error of the first kind) by never rejecting the null hypothesis. Or one can achieve any desired significance level by using a random device (like dice or a roulette wheel) to decide the issue, without any substantive experiment. Of course, such a procedure is absurd because, in the terminology used here, its power equals its level of significance, whatever alternative hypothesis may hold.

Difficulties with significance level. If significance tests are regarded as decision procedures, one criticism of them is based on the arbitrariness of level of significance and the lack of guidance in choosing that level. The usual advice is to examine level of significance and power for various possible sample sizes and experimental designs and then to choose the least expensive design with satisfactory characteristics. But how is one to know what is satisfactory? Some say that if satisfaction cannot be described, then the experimenter has not thought deeply enough about his materials. Others oppose such a viewpoint as overmechanical and inappropriate in scientific inference; they might add that the arbitrariness of size, or something analogous to it, is intrinsic in any inferential process. One cannot make an omelet without deciding how many eggs to break.

Unconventional significance levels. Probably the most common significance levels are .05 and .01, and tables of critical values are generally for these levels. But special circumstances may dictate tighter or looser levels. In evaluating the safety of a drug to be used on human beings, one might impose a significance level of .001. In exploratory work, it might be quite reasonable to use levels of .10 or .15 in order to increase power. What is of central importance is to know what one is doing and, in particular, to know the properties of the test that is used.

Nonconstant significance levels. For many test situations it is impossible to obtain a sensible test with the same level of significance for all distributions described by a so-called composite null hypothesis. For example, in testing the null hypothesis that a population mean is not positive, the level of significance depends generally on the value of the nonpositive mean; the more it departs from zero, the smaller the probability of Type I error. In such cases the usual approach is to think in terms of the *maximum* probability of Type I error over all distributions described by the null hypothesis. In the nonpositive mean case, the maximum is typically attained when the mean is zero.

5 Necessarily false null hypotheses. Another criticism of standard significance tests is that in most applications it is known beforehand that the null hypothesis cannot be *exactly* true. For example, it seems most implausible that sleep deprivation should have literally no effect at all on arithmetic

ability. Hence, why bother to test a null hypothesis known at the start to be false?

One answer to this criticism can be outlined as follows: A test of the hypothesis that a population mean has a specified value, μ_0, is a simplification. What one really wants to test is whether the mean is *near* μ_0, as near as makes no substantive difference. For example, if a new psychological therapy raises the cure rate from 51 per cent to 51.1 per cent, then even if one could discover such a small difference, it might be substantively uninteresting and unimportant. For "reasonable" sample sizes and "reasonable" significance levels, most standard tests have power quite close to the level of significance for alternative hypotheses close to the null hypothesis. When this is so and when, in addition, power is at least moderately large for alternatives interestingly different from the null hypothesis, one is in a satisfactory position, and the criticism of this section is not applicable. The word "reasonable" is in quotation marks above because what is in fact reasonable depends strongly on context. To examine reasonableness it is necessary to inspect, at least roughly, the entire power function. Many misuses of significance testing spring from complete disregard of power.

A few authors, notably Hodges and Lehmann (1954), have formalized the argument outlined above by investigating tests of null hypotheses such as the following: The population mean is in the interval $(\mu_0 - \Delta, \mu_0 + \Delta)$, where μ_0 is given and Δ is a given positive number.

There are, to be sure, null hypotheses that are not regarded beforehand as surely false. One well-known example is the null hypothesis that there is no extrasensory effect in a parapsychological experiment. Other examples are found in crucial tests of well-formulated physical theories. Even in these instances the presence of small measurement biases may make interpretation difficult. A minuscule measurement bias, perhaps stemming from the scoring method, in an extensive parapsychological experiment may give a statistically significant result although no real parapsychological effect is present. In such a case the statistical significance may reflect only the measurement bias, and thus much controversy about parapsychology centers about the nature and magnitude of possible measurement biases, including chicanery, unconscious cues, and biases of scoring.

A sharp attack on significance testing, along the lines of this section and others, is given by L. J. Savage (1954, chapter 16).

Several tests on the same data. Frequently, two or more hypothesis tests are carried out on the same data. Although each test may be at, say, the .05 level of significance, one may ask about the *joint* behavior of the tests. For example, it may be important to know the probability of Type I error for both of two tests together. When the test statistics are statistically independent, there is no problem; the probability, for example, that at least one of two tests at level .05 will give rise to Type I error is $1 - (.95)^2 = .0975$. But in general the tests are statistically dependent, and analogous computations may be difficult. Yet it is important to know, for example, when two tests are positively associated, in the sense that given a Type I error by one, the other is highly likely to have a Type I error.

When a moderate to large number of tests are carried out on the same data, there is generally a high probability that a few will show statistical significance even when all the null hypotheses are true. The reason is that although the tests are dependent, they are not completely so, and by making many tests one increases the probability that at least some show statistical significance. [*Some ways of mitigating these problems for special cases are described in* LINEAR HYPOTHESES, *article on* MULTIPLE COMPARISONS.]

One-sided versus two-sided tests. There has been much discussion in the social science literature (especially the psychological literature) of when one-sided or two-sided tests should be used. If a one-sided test is performed, with the choice of sidedness made tendentiously *after* inspection of the data, then the nominal significance level is grossly distorted; in simple cases the actual significance level is twice the nominal one.

Suppose that in the sleep-deprivation example the data show an average score that is positive. "Aha," says the investigator, "I will test the null hypothesis with a right-tail test, against positive means as alternatives." Clearly, if he pursues that policy for observed positive average scores and the opposite policy for observed negative average scores, the investigator is unwittingly doing a *two*-tail test with double the stated significance level. This is an insidious problem because it is usually easy to rationalize a choice of sidedness *post hoc*, so that the investigator may be fooling both himself and his audience.

The same problem occurs in the choice of test statistic in general. If six drugs are compared in their effects on arithmetic skill and only the observations on those two drugs with least and greatest observed effects are chosen for a test statistic, with no account taken of the choice in computing critical values, an apparently statistically significant result may well just reflect random error, in the

sense that the true significance level is much higher than the nominal one. [*Some methods of dealing with this are described in* LINEAR HYPOTHESES, *article on* MULTIPLE COMPARISONS.]

Three-decision procedures. Many writers have worried about what to do after a two-tail test shows statistical significance. One might conclude, for example, that sleep deprivation has an effect on average score, but one wishes to go further and say that it has a positive or a negative effect, depending on the sign of the sample average. Yet significance testing, regarded stringently as a two-decision procedure, makes no provision for such conclusions about sign of effect when a two-sided test rejects the null hypothesis.

What is really wanted in such a case may well be a *three*-decision procedure: one either accepts H_0, rejects H_0 in favor of alternatives on one side, or rejects H_0 in favor of alternatives on the other side. There is no reason why such a procedure should not be used, and it is, in effect, often used when the user says that he is carrying out a significance test. A major new consideration with this kind of three-decision procedure is that one now has *six*, rather than *two*, kinds of error whose probabilities are relevant; for each possible hypothesis there are two erroneous decisions. In some simple cases symmetry reduces the number of different probabilities from six to three; further, some of the probabilities may be negligibly small, although these are probabilities of particularly serious errors (for a discussion, see Kaiser 1960). A three-decision procedure may often be usefully regarded as a composition of two one-sided tests. [*See* HYPOTHESIS TESTING.]

A variety of other multiple-decision procedures have been considered, generally with the aim of mitigating oversimplification in the significance-testing approach. A common case is that in which one wants to choose the best of several populations, where "best" refers to largest expected value. [*A pioneering investigation along this line is* Mosteller 1948. *See also* DECISION THEORY; SCREENING AND SELECTION.]

Hypotheses suggested by data. In the course of examining a body of data, the analyst may find that the data themselves suggest one or more kinds of structure, and he may decide to carry out tests of corresponding null hypotheses from the very data that suggested these hypotheses. One difficulty here is statistical dependence (described above) between the tests, but there is a second difficulty, one that appears even if there is only a single significance test. In a sense, this difficulty is just a general form of the *post hoc* one-sided choice problem.

Almost any set of data, of even moderate size and complexity, will show anomalies of some kind when examined carefully, even if the underlying probabilistic structure is wholly random—that is, even if the observations stem from random variables that are independent and identically distributed. By looking carefully enough at random data, one can generally find some anomaly—for example, clustering, runs, cycles—that gives statistical significance at customary levels although no real effect is present. The explanation is that although any particular kind of anomaly will occur in random data with .05 statistical significance just 5 times out of 100, so many kinds of anomalies are possible that at least one will very frequently appear. Thus, use of the same data both to suggest and to test hypotheses is likely to generate misleading statistical significance.

Most sets of real data, however, are not completely random, and one does want to explore them, to form hypotheses and to test these same hypotheses. One can sometimes use part of the data for forming hypotheses and the remainder of the data to examine the hypotheses. This is, however, not always feasible, especially if the data are sparse. Alternatively, if further data of the same kind are available, then one can use the earlier data to generate hypotheses and the later data to examine them. Again, this approach is not always possible.

As an example, consider one of the earliest instances of significance testing, by Daniel Bernoulli and his son John in 1734, as described by Todhunter ([1865] 1949, secs. 394–397). Astronomers had noticed that the orbital planes of the planets are all close together. Is this closeness more than one might expect if the orbital planes were determined randomly? If so, then presumably there is a physical reason for the closeness. The Bernoullis first attempted to make precise what is meant by randomness; in modern terminology this would be specification of randomness as a null hypothesis. They then computed the probability under randomness that the orbital planes would be as close together as, or closer than, the observed planes. This corresponds to deciding on a test statistic and computing a one-tail *P*-value. The resulting probability was very small and the existence of a physical cause strongly suggested. [*For biographical material on the Bernoullis, see* BERNOULLI FAMILY.]

Todhunter, in his *History*, described other early instances of significance testing, the earliest in 1710 by John Arbuthnot on the human sex ratio at birth (the relevant sections in Todhunter [1865] 1949 are 343–348, 617–622, 888, 915, and 987). Significance testing very much like that of the Bernoullis continues; a geophysical example, relating to the surprising degree of land–water antipodality on the earth's surface, has appeared recently (Harrison 1966). The whole topic received a detailed discussion by Keynes ([1921] 1952, chapter 25). Up-to-date descriptions of astronomical theories about the orbital planes of the planets have been given by Struve and Zebergs (1962, chapter 9).

Quite aside from the issue of whether it is reasonable to apply probabilistic models to such unique objects as planetary orbits (or to the 50 states, as in an earlier example), there remains the difficulty that a surprising concatenation was noted in the data, and a null hypothesis and test statistic fashioned around that concatenation were considered. Here there is little or no opportunity to obtain further data (although more planets were discovered after the Bernoullis and, in principle, planetary systems other than our sun's may someday be observable). Yet, in a way, the procedure seems quite reasonable as a means of measuring the surprisingness of the observed closeness of orbital planes. I know of no satisfactory resolution of the methodological difficulties here, except for the banal moral that when testing hypotheses suggested during the exploration of data, one should be particularly cautious about coming to conclusions. (In the Bayesian approach to statistics, the problem described here does not arise, although many statisticians feel that fresh problems are introduced [*see* BAYESIAN INFERENCE].)

An honest attempt at answering the questions, What else would have surprised me? Was I bound to be surprised? is well worthwhile. For example, to have observed planetary orbits symmetrically arranged in such a way that their perpendiculars nearly formed some rays of a three-dimensional asterisk would also have been surprising and might well be allowed for.

Difficulties in determining reference set. Some statisticians have been concerned about difficulties in determining the proper underlying reference probabilities for sensibly calculating a *P*-value. For example, if sample size is random, even with a known distribution, should one compute the *P*-value as if the realized sample size were known all along, or should one do something else? Sample size can indeed be quite random in some cases; for example, an experimenter may decide to deal with all cases of a rare mental disorder that come to a particular hospital during a two-month period. [*Three papers dealing with this kind of problem are* Barnard 1947; Cox 1958; *and* Anscombe 1963. *See also* LIKELIHOOD.]

Random sample sizes occur naturally in that

part of statistics called sequential analysis. Many sequential significance testing procedures have been carefully analyzed, although problems of determining reference sets nonetheless continue to exist. [See SEQUENTIAL ANALYSIS.]

Optional stopping. Closely related to the discussion of the preceding section is the problem of optional stopping. Suppose that an experimenter with extensive resources and a tendentious cast of mind undertakes a sequence of observations and from time to time carries out a significance test based on the observations at hand. For the usual models and tests, he will, sooner or later, reach statistical significance at any preassigned level, even if the null hypothesis is true. He might then stop taking observations and proclaim the statistical significance as if he had decided the sample size in advance. (The mathematical background of optional stopping is described in Robbins 1952.)

For the standard approach to significance testing, such optional stopping is as misleading and reprehensible as the suppression of unwanted observations. Even for an honest experimenter, if the sampling procedure is not firmly established in advance, a desire to have things turn out one way or another may unconsciously influence decisions about when to stop sampling. If the sampling procedure is firmly established in advance, then, at least in principle, characteristics of the significance test can be computed in advance; this is an important part of sequential analysis.

Optional stopping is, of course, relevant to modes of statistical analysis other than significance testing. It poses no problem for approaches to statistics that turn only on the observed likelihood, but many statisticians feel that these approaches are subject to other difficulties that are at least equally serious. [See BAYESIAN INFERENCE; LIKELIHOOD.]

Simplicity and utility of hypotheses. It is usually the case that a set of data will be more nearly in accord with a complicated hypothesis than with a simpler hypothesis that is a special case of the complicated one. For example, if the complicated hypothesis has several unspecified parameters whereas the simpler one specializes by taking some of the parameters at fixed values, a set of data will nearly always be better fit by the more complicated hypothesis than by the simpler one just because there are more parameters available for fitting: a point is usually farther away from a given line than from a given plane that includes the line; in the polynomial regression context, a linear regression function will almost never fit as well as a quadratic, a quadratic as well as a cubic, and so on.

Yet one often prefers a simpler hypothesis to a better-fitting more complicated one. This preference, which undoubtedly has deep psychological roots, poses a perennial problem for the philosophy of science. One way in which the problem is reflected in significance testing is in the traditional use of small significance levels. The null hypothesis is usually simpler than the alternatives, but one may be unwilling to abandon the null hypothesis unless the evidence against it is strong.

Hypotheses may be intrinsically comparable in ways other than simplicity. For example, one hypothesis may be more useful than another because it is more closely related to accepted hypotheses for related, but different, kinds of observations.

The theory of significance testing, however, takes no explicit account of the simplicity of hypotheses or of other aspects of their utility. A few steps have been made toward incorporating such considerations into statistical theory (see Anderson 1962), but the problem remains open.

Importance of significance testing. Significance testing is an important part of statistical theory and practice, but it is only one part, and there are other important ones. Because of the relative simplicity of its structure, significance testing has been overemphasized in some presentations of statistics, and as a result some students come mistakenly to feel that statistics is little else than significance testing.

Other approaches to significance testing. This article has been limited to the customary approach to significance testing based on the frequency concept of probability. For other concepts of probability, procedures analogous to significance testing have been considered. [See BAYESIAN INFERENCE. *An extensive discussion is given in* Edwards et al. 1963.] Anscombe (1963) has argued for a concept of significance testing in which only the null hypothesis, not the alternatives, plays a role.

WILLIAM H. KRUSKAL

[See also HYPOTHESIS TESTING.]

BIBLIOGRAPHY

ANDERSON, T. W. 1962 The Choice of the Degree of a Polynomial Regression as a Multiple Decision Problem. *Annals of Mathematical Statistics* 33:255–265.

ANSCOMBE, F. J. 1963 Tests of Goodness of Fit. *Journal of the Royal Statistical Society* Series B 25:81–94.

BAKAN, DAVID 1966 The Test of Significance in Psychological Research. *Psychological Bulletin* 66:423–437.

BANCROFT, T. A. 1964 Analysis and Inference for Incompletely Specified Models Involving the Use of Preliminary Test(s) of Significance. *Biometrics* 20:427–442.

BARNARD, G. A. 1947 The Meaning of a Significance Level. *Biometrika* 34:179–182.

BORING, EDWIN G. 1919 Mathematical vs. Scientific Significance. *Psychological Bulletin* 15:335–338.

COX, DAVID R. 1958 Some Problems Connected With Statistical Inference. *Annals of Mathematical Statistics* 29:357–372.

EDWARDS, WARD; LINDMAN, HAROLD; and SAVAGE, LEONARD J. 1963 Bayesian Statistical Inference for Psychological Research. *Psychological Review* 70:193–242.

GOOD, I. J. 1956 The Surprise Index for the Multivariate Normal Distribution. *Annals of Mathematical Statistics* 27:1130–1135.

GOOD, I. J. 1958 Significance Tests in Parallel and in Series. *Journal of the American Statistical Association* 53:799–813.

HARRISON, CHRISTOPHER G. A. 1966 Antipodal Location of Continents and Oceans. *Science* 153:1246–1248.

HIRSCHI, TRAVIS; and SELVIN, HANAN C. 1967 *Methods in Delinquency Research.* New York: Free Press. → See especially Chapter 13, "Statistical Inference."

HODGES, J. L. JR.; and LEHMANN, E. L. 1954 Testing the Approximate Validity of Statistical Hypotheses. *Journal of the Royal Statistical Society* Series B 16:261–268.

KAISER, HENRY F. 1960 Directional Statistical Decisions. *Psychological Review* 67:160–167.

KEYNES, JOHN MAYNARD (1921) 1952 *A Treatise on Probability.* London: Macmillan.

KITAGAWA, TOSIO 1963 Estimation After Preliminary Tests of Significance. *University of California Publications in Statistics* 3:147–186.

MOSES, LINCOLN E. 1956 Statistical Theory and Research Design. *Annual Review of Psychology* 7:233–258.

MOSTELLER, FREDERICK 1948 A *k*-sample Slippage Test for an Extreme Population. *Annals of Mathematical Statistics* 19:58–65.

MOSTELLER, FREDERICK; and BUSH, ROBERT R. (1954) 1959 Selected Quantitative Techniques. Volume 1, pages 289–334 in Gardner Lindzey (editor), *Handbook of Social Psychology.* Cambridge, Mass.: Addison-Wesley.

NEYMAN, JERZY 1958 The Use of the Concept of Power in Agricultural Experimentation. *Journal of the Indian Society of Agricultural Statistics* 9, no. 1:9–17.

ROBBINS, HERBERT 1952 Some Aspects of the Sequential Design of Experiments. American Mathematical Society, *Bulletin* 58:527–535.

○SAVAGE, LEONARD J. (1954) 1972 *The Foundations of Statistics.* Rev. ed. New York: Dover.

STERLING, THEODORE D. 1959 Publication Decisions and Their Possible Effects on Inferences Drawn From Tests of Significance—or Vice Versa. *Journal of the American Statistical Association* 54:30–34.

STRUVE, OTTO; and ZEBERGS, VELTA 1962 *Astronomy of the 20th Century.* New York: Macmillan.

TODHUNTER, ISAAC (1865) 1949 *A History of the Mathematical Theory of Probability From the Time of Pascal to That of Laplace.* New York: Chelsea.

TULLOCK, GORDON 1959 Publication Decisions and Tests of Significance: A Comment. *Journal of the American Statistical Association* 54:593 only.

WEAVER, WARREN 1948 Probability, Rarity, Interest, and Surprise. *Scientific Monthly* 67:390–392.

WITTENBORN, J. R. 1952 Critique of Small Sample Statistical Methods in Clinical Psychology. *Journal of Clinical Psychology* 8:34–37.

WOOFTER, T. J. JR. 1933 Common Errors in Sampling. *Social Forces* 11:521–525.

Postscript

[1] A good discussion of significance tests from this viewpoint of evaluating the evidence against a hypothesis is given by Kalbfleisch and Sprott (1976).

[2] The literature on preliminary tests of significance continues to grow slowly. See, for example, Ahsanullah and Ehsanes Saleh (1972).

[3] Greenwald (1975) is an interesting paper related to the problem that nonsignificance is often nonpublic. It carries much further the discussions of Sterling (1959) and Tullock (1959) to give a detailed model for the research cum publication process, with special attention to the distribution of published significance levels. Relevant empirical material is also included.

[4] Further discussion of complete populations is given by Morrison and Henkel (1969; 1970), Winch and Campbell (1969), Finifter (1972), and Selvin (1968). Winch and Campbell defend the permutation approach.

[5] Greenwald (1975), mentioned above, is also relevant to the criticism of necessarily false null hypotheses.

[6] The question of simplicity and utility of hypotheses has, of course, received attention from philosophers, especially logicians, in their discussions of evidence and confirmation. An interesting paper in that literature is Hullett and Schwartz (1967).

Loose use of "statistical significance." The term "statistically significant," and its slight grammatical variations, have a reasonably precise meaning save for ambiguity in how small the P-value is to be before statistical significance is claimed.

Sometimes, however, "statistical significance" is used simply to mean "based on many observations," or perhaps "impressive." In some instances one is tempted to speculate that the usage is that of a convenient technical-sounding expression to generate spurious confidence.

Three examples of such misleading or erroneous usage follow:

The vastness of this country, the high mobility rate of many of its inhabitants and its statistically significant immigrant population all contribute to the need for an efficient postal service.   (O'Meara 1976)

It is difficult to test our results . . . against observations because no statistically significant *global* record of temperature back to 1600 has been constructed. (Scheider & Mass 1975, p. 745)

. . . statistically significant numbers of faunal remains and human skeletons have been added over the last 4000 years.   (Laughlin 1975, p. 510)

Confused pseudo-Bayesian interchange in prob-

abilistic interpretation. One sometimes meets a statement of significance testing that confusedly presents a *P*-value as if it were the probability of the null hypothesis. A typical example appears in Vanderpol et al. (1975): "The correlation coefficient between the 20 pairs of data points is 0.72 with a probability of less than 0.1 percent that this effect is due to random occurrence. . . ." Another example appears in a supposed exposition of the correct meaning of significance test (Schatz 1976, p. 6): ". . . a sampling . . . or experimental result is deemed 'statistically significant' when the calculated probability of its being solely an artifact of chance is below a specified low value." A third example appears in a paper that also is troubled— without notice—by the complete population problem (Hoch 1976, p. 858): "Statistical significance here means that there is only one chance in 20 that the variable really has no effect."

Such statements are misleading because the standard significance testing approach computes sample tail probabilities under one or more hypotheses, *not* probabilities of one or more hypotheses given the sample. Thus the first of the above quotations, for example, should have correctly said that the probability is less than 0.1 per cent that a correlation coefficient of 0.72 or more occurs under the hypothesis of zero correlation in the (normal) parent population.

In a Bayesian approach [*see* BAYESIAN INFERENCE] one would, to be sure, compute probabilities of hypotheses, but the computations might be rather different from those of standard significance testing. Even if the numbers turn out to be the same— as they may—their interpretations are completely different.

An excellent brief discussion of this expository problem is given by Mosteller and Tukey (1968, p. 183). See also Elashoff and Snow (1971).

(In addition to the confusion of interchange found in quotations like the above, a companion confusion is the implication that the null hypothesis must be one of "random occurrence" or of chance only.)

Remarks on references. Among the treatments of significance testing in recent textbooks of statistics, I note especially that of Cox and Hinkley (1974).

Several statistical anthologies have appeared with particular emphasis on reprinting articles dealing with significance testing, including some of the articles cited in the main bibliography. Among these anthologies are Morrison and Henkel (1970), Steger (1971), and Lieberman (1971).

WILLIAM H. KRUSKAL

ADDITIONAL BIBLIOGRAPHY

AHSANULLAH, M.; and EHSANES SALEH, A. K. 1972 Estimation of Intercept in a Linear Regression Model With One Dependent Variable After a Preliminary Test on the Regression Coefficient. *International Statistical Review* 40:139–145.

COX, DAVID R. 1977 The Role of Significance Tests. *Scandinavian Journal of Statistics* 4:49–63. → Discussion on pages 63–70.

COX, DAVID R.; and HINKLEY, D. V. 1974 *Theoretical Statistics*. London: Chapman & Hall.

ELASHOFF, JANET D.; and SNOW, RICHARD E. 1971 *"Pygmalion" Reconsidered: A Case Study in Statistical Inference; Reconsideration of the Rosenthal–Jacobson Data on Teacher Expectancy*. Worthington, Ohio: Charles A. Jones. → Published under the auspices of the National Society for the Study of Education. Pages 16–17 relate to the pseudo-Bayesian interchange discussed in the postscript.

FINIFTER, BERNARD M. 1972 The Generation of Confidence: Evaluating Research Findings by Random Subsample Replication. Pages 112–175 in Herbert L. Costner (editor), *Sociological Methodology, 1972*. San Francisco: Jossey-Bass.

FREEDMAN, DAVID; and LANE, DAVID 1978 *Significance Testing in a Nonstochastic Setting*. Technical Report No. 317, School of Statistics, University of Minnesota. → Relevant to the complete population question and to similar questions.

GREENWALD, ANTHONY G. 1975 Consequences of Prejudice Against the Null Hypothesis. *Psychological Bulletin* 82:1–20.

HOCH, IRVING 1976 City Size Effects, Trends, and Policies. *Science* 193:856–863.

HULLETT, JAMES; and SCHWARTZ, ROBERT 1967 Grue: Some Remarks. *Journal of Philosophy* 64:259–271.

KALBFLEISCH, J. G.; and SPROTT, D. A. 1976 On Tests of Significance. Volume 2, pages 259–272 in W. L. Harper and C. A. Hooker (editors), *Foundations of Probability Theory, Statistical Inference and Statistical Theories of Science*. Volume 2: *Foundations and Philosophy of Statistical Inference*. Higham, Mass.: Reidel. → Proceedings of an international colloquium held at the University of Western Ontario, May 1973.

LAUGHLIN, WILLIAM S. 1975 Aleuts: Ecosystem, Holocence History, and Siberian Origin. *Science* 189:507–515.

LIEBERMAN, BERNHARDT (editor) 1971 *Contemporary Problems in Statistics: A Book of Readings for the Behavioral Sciences*. New York: Oxford Univ. Press.

MORRISON, DENTON E.; and HENKEL, RAMON E. 1969 Significance Tests Reconsidered. *American Sociologist* 4:131–140.

MORRISON, DENTON E.; and HENKEL, RAMON E. (editors) 1970 *The Significance Test Controversy: A Reader*. Chicago: Aldine.

MOSTELLER, FREDERICK; and TUKEY, JOHN W. 1968 Data Analysis, Including Statistics. Volume 2, pages 80–203 in Gardner Lindzey and Elliot Aronson (editors), *Handbook of Social Psychology*, 2d ed. Volume 2: *Research Methods*. Reading, Mass.: Addison-Wesley.

O'MEARA, MARY 1976 The More Important Letters. *New York Times* March 10, p. 38, col. 3.

SANATHANAN, LALITHA 1974 Critical Power Function and Decision Making. *Journal of the American Statistical Association* 69:398–402. → This article relates to the discussion of power and to the distinction between statistical significance and subject-matter significance.

Schatz, Gerald S. 1976 "Statistically Significant"? National Academy of Sciences, *News Report* 26, no. 8:6 only.

Scheider, Stephen H.; and Mass, Clifford 1975 Volcanic Dust, Sunspots, and Temperature Trends. *Science* 190:741–746.

Selvin, Hanan C. 1968 The Computer Analysis of Observational Data. Pages 227–238 in Journées Internationales d'Études sur les Méthodes de Calcul Dans les Sciences de l'Homme, Rome, 1966, *Calcul et formalisation dans les sciences de l'homme.* Paris: Editions du Centre National de la Recherche Scientifique.

Steger, Joseph A. (editor) 1971 *Readings in Statistics for the Behavioral Scientist.* New York: Holt.

Vanderpol, A. H. et al. 1975 Aerosol Chemical Parameters and Air Mass Character in the St. Louis Region. *Science* 190:570 only.

Winch, Robert J.; and Campbell, Donald T. 1969 Proof? No. Evidence? Yes.: The Significance of Tests of Significance. *American Sociologist* 4:140–143.

SIMULATION

I. Individual Behavior	*Allen Newell and Herbert A. Simon*
II. Economic Processes	*Irma Adelman*
III. Political Processes	*Charles F. Hermann*

I
INDIVIDUAL BEHAVIOR

"Simulation" is a term now generally employed to denote an approach to the construction of theories that makes essential use of computers and computer-programming languages. In some applications simulation techniques are used in investigating formal mathematical theories—for example, stochastic learning theories. In other applications, the theoretical models of the systems that are simulated are essentially nonquantitative and nonnumerical.

We may define simulation more specifically as a method for analyzing the behavior of a system by computing its time path for given initial conditions, and given parameter values. The computation is said to "simulate" the system, because it moves forward in time step by step with the movement of the system it is describing. For the purposes of simulation, the system must be specified by laws that define its behavior, during any time interval, in terms of the state it was in' at the beginning of that interval.

Simulation has always been an important part of the armory of applied mathematics. The introduction of modern computers, however, has so reduced the cost of obtaining numerical solutions to systems of equations that simulation has taken on vastly increased importance as an analytic tool. As a consequence, the relative balance of advantage has shifted from simplifying a theory in order to make it solvable toward retaining complexity in order to increase its accuracy and realism. With simulation and other techniques of numerical analysis, computers permit the study of far more complicated systems than could have been investigated at an earlier time.

Simulation with mathematical theories. Consider first the simulation of systems that are described by mathematical models (Newell & Simon 1963*a*, pp. 368–373). A model consists of a system of equations containing variables, literal constants (often called parameters), and numerical constants. The system of equations may be solved symbolically, by expressing the several variables explicitly in terms of the literal and numerical constants; or the system may be solved numerically, by substituting particular numerical values for the literal constants and then solving the equations for these special cases. For many complex systems of equations, symbolic solutions cannot be obtained, only numerical solutions. The former are clearly preferred when obtainable, for once a symbolic solution has been found, the special cases are readily evaluated by substituting numerical values for the parameters in the solutions instead of in the original equations.

To take a highly simplified example, a learning theory might postulate that the rate at which learning occurs is proportional to the amount of material remaining to be learned. If, for simplicity of exposition, we divide time into discrete intervals and define the variable $x(t)$ as the amount of material that has been learned up to the end of the tth time interval, then the theory can be expressed by a simple difference equation:

$$(1) \qquad x(t+1) - x(t) = a[b - x(t)],$$

where a is a literal constant that we can name intelligence for this learning task, and b is a literal constant that represents the total amount to be learned. The symbolic solution to this simple system is well known to be:

$$(2) \qquad x(t) = b - [b - \bar{x}(0)](1-a)^t,$$

where $\bar{x}(0)$ is a literal constant denoting the amount that had already been learned up to the time $t = 0$. Particular numerical solutions can now be found by assigning numerical values a, b, and $\bar{x}(0)$ and by solving (2) for $x(t)$ for any desired values of t. Alternatively, numerical solutions—learning curves—for the system can as readily be computed for given a, b, and $\bar{x}(0)$ directly from equation (1), transferring the $x(t)$ term to the right-hand side and solving successively for $x(1)$,

$x(2)$, $x(3)$, $\cdots$. This latter procedure—solving (1) numerically—would be called simulation.

The advantages of simulation as a technique for analyzing dynamic systems are not evident from the excessively simple equation (1). Suppose, however, that a, instead of being a literal constant, designated some function, $a[b,x(t)]$, of b and $x(t)$. Only for very special cases could (1) then be solved symbolically to obtain an explicit expression, like (2), for $x(t)$. In general, when we wish to know the time paths of systems expressed by such equations, numerical solution for special cases—that is, simulation—is our main recourse.

A great many of the mathematical theories currently used for the study of learning and other individual behavior are probabilistic or stochastic in character. Such theories pose formidable mathematical problems, and explicit symbolic solutions for their equations are seldom obtainable. Unlike a deterministic system, the solution of the equations of a stochastic system does not specify a single time path for the system but assigns a probability distribution to all possible paths of the system. In some cases, certain properties of this probability distribution—certain means and variances—can be obtained symbolically, but the specification of the entire probability distribution can seldom be obtained symbolically. In order to estimate the parameters of the distribution numerically, the system can be simulated by Monte Carlo techniques. That is, random variables can be introduced into the model, and the equations analogous to (1) solved a number of times, each time with different values of the random variables assigned and using the appropriate probabilities. The numerical solutions to (1) will trace out *possible* paths of the system, and the probability that a particular path will be followed in the simulation will be proportional to its probability under the assumptions of the model. Then, from the numerical results of the Monte Carlo simulation, numerical estimates can be made of the parameters of the probability distribution. [*See* RANDOM NUMBERS.]

Information-processing theories. The greatest use of simulation for the study of individual behavior has involved so-called information-processing theories, which are different in important respects from classical mathematical theories. Simulation has exactly the same relation to information-processing theories as to other formal theories—it is a method for discovering how the system described by the theory will behave, in particular circumstances, over a period of time. However, almost no methods other than simulation exist for investigating information-processing theories. For this reason, a phrase like "simulation of cognitive processes" often refers both to an information-processing theory of cognitive processes and to the investigation of the theory by simulation, in the narrower sense of the term. (See Pollack 1968.)

Information-processing theories of individual behavior take the form of computer programs, usually written in programming languages called list-processing languages, especially devised for this purpose. The information-processing theories undertake to explain how a system like the human central nervous system carries out such complex cognitive processes as solving problems, forming concepts, or memorizing. The explanation is sufficiently detailed to predict—by computer simulation—behavior in specific problem situations. That is to say, the theory aims at predicting, not merely some quantitative aspects of behavior (number of errors in a learning situation, for example), but the actual concrete behaviors and verbal outputs of subjects placed in the very same situation and confronted with the identical task. (See Kendler & Kendler 1968; Taylor 1968.)

A specific example will help make clear exactly what is meant by this. Clarkson (1963) has constructed an information-processing theory to explain how a bank officer selects portfolios of securities for trust funds. The trust officer, who has been told the purpose of the trust and the amount of money available for investment, prepares a list of stocks and bonds that can be bought with this sum. If the computer program is provided (as input) the same information about the purpose of the trust and its size, it will also produce (as output) a list of stocks and bonds. A first test of the program, as a theory, is whether it will pick the same list of companies and the same number of shares in each as the trust officer. In an actual test, for example, the program predicted the trust officer would buy 60 shares of General American Transportation, 50 Dow Chemical, 10 IBM, 60 Merck and Company, and 45 Owens-Corning Fiberglas. The trust officer in fact bought 30 Corning Glass, 50 Dow Chemical, 10 IBM, 60 Merck and Company, and 50 Owens-Corning Fiberglas.

A second and more severe test of the theory is whether, in reaching its final choices, it goes through the same stages of analysis, weighs the same factors, considers the same alternatives, as the human it is simulating. As one technique for making this kind of test, the human subject is asked to perform the task while thinking aloud, and his stream of verbalizations (protocol) is recorded. The human protocols are then compared with the *trace* produced by the computer program while it is performing the same task (for an example of this technique, which Clarkson also used, see Newell & Simon 1963*b*).

Omar K. Moore and Scarvia B. Anderson (1954) devised a task that required the subject to "recode" certain symbolic expressions. One problem was to recode $L1$: $(r \supset -p) \cdot (-r \supset q)$ into $L0$: $-(-q \cdot p)$. (For purposes of the experiment and of this example, these strings of symbols can be treated as "code messages," whose meaning need not be known. The expression $L1$ may be read "parenthesis r horseshoe minus p parenthesis dot parenthesis minus r horseshoe q parenthesis.") The recoding had to be carried out according to certain rules, which were numbered from 1 to 12. At one point in his thinking-aloud protocol, a subject said, "Now I'm looking for a way to get rid of the horseshoe inside the two brackets that appear on the left and right side of the equation. And I don't see it. Yeah, if you apply rule 6 to both sides of the equation, from there I'm going to see if I can apply rule 7."

The same recoding task, recoding $L1$ into $L0$ according to the rules specified by Moore and Anderson, was given to a program, the General Problem Solver (GPS), expressing an information-processing theory of human problem solving. One portion of the trace produced by the program read as follows:

Goal: Apply rule 7 to $L1$.
　　Subgoal: Change "horseshoe" to "wedge" in left side of $L1$.
　　Subgoal: Apply rule 6 to the left side of $L1$.
　　Result: $L4$: $(-r \vee -p) \cdot (-r \supset q)$.
　　Subgoal: Apply rule 7 to $L4$.
　　Subgoal: Change "horseshoe" to "wedge" in right of $L4$.
　　Subgoal: Apply rule 6 to right side of $L4$.

Part of the test of the adequacy of GPS as a theory of human problem solving would be to decide how closely the segment of the human protocol reproduced above corresponded to the segment of the computer trace. For example, in the illustrative fragment cited here, both the human subject and GPS went down a blind alley, for rule 7 turned out to be not applicable to the expression recoded by rule 6. A theory of human problem solving must predict and explain the mistakes people make, as well as their ability sometimes to achieve solutions.

Information-processing theories can also be tested in more orthodox ways than by comparing

protocols with traces. A program can be simulated in an experimental design identical with one that has been employed for human subjects. Statistics from the computer runs can be compared with the statistics of the human performance. This is the principal means that has been used to test the Elementary Perceiver and Memorizer (EPAM), an information-processing theory of human rote learning that will be described briefly later (Feigenbaum 1963). By having EPAM learn nonsense syllables by the standard serial anticipation method, at various simulated memory drum speeds and for lists of different lengths, quantitative predictions were made of the shape of the so-called serial position curve (relative numbers of errors during the learning process for different parts of the list of syllables). These predictions showed excellent agreement with the published data from human experiments. In other experiments in the learning of paired nonsense syllables, EPAM predicted quantitatively the effects upon learning rate of such variables as degree of familiarity with the syllables and degree of similarity between syllables. These predictions also corresponded closely with the published data from several experiments. (See Postman 1968; Underwood 1968.)

In summary, an information-processing theory is expressed as a computer program which, exactly like a system of difference or differential equations, predicts the time path of a system from given initial conditions for particular values of the system parameters. The theory predicts not merely gross quantitative features of behavior but the actual stream of symbolic outputs from the subject. Such theories can be subjected to test in numerous ways; among others, by comparing the behaviors, including the thinking-aloud protocols, of subjects with the computer traces produced by the simulating programs.

Evaluation. The examples reveal both some of the strengths and some of the difficulties in the information-processing approach to theory construction and theory testing. On the positive side, the theories are written in languages that are capable of representing stimuli and responses directly and in detail, without an intermediate stage of translation into mathematical form. This has the further advantage of avoiding virtually all problems of stimulus scaling. The stimuli themselves, and not scales representing some of their gross characteristics, serve as inputs to the theory. Although the computer traces produced by the simulations are not in idiomatic English, the comparison of trace with protocol can be handled by relatively unambiguous coding techniques.

One of the main difficulties of the approach is

closely related to one of its strengths—the detail of its predictions. Since different human subjects do not behave in exactly the same way in identical task situations, a simulation that predicts correctly the detail of behavior of one individual will predict the detail of behavior of others incorrectly. Presumably, a theory of individual behavior should consist of general statements about how human beings behave, rather than statements about how a particular human being behaves. At a minimum we would want to require of a theory that modifications to fit the behaviors of different subjects should change only relatively superficial features of the theory—values of particular parameters, say —and that they not necessitate fundamental reconstruction of the whole simulation program. It remains to be seen how fully this requirement will be met by information-processing theories. Up to the present, only a modest amount of investigation has been made of the possibilities of creating variants of simulation programs to fit different subjects.

Generality is needed, too, in another direction. Most of the early simulation programs were constructed to explain human performance in a single task—discovering proofs for theorems in symbolic logic, say, or making moves in chess. It seems unreasonable to suppose that the programs a human uses to solve problems in one task environment are entirely specific to that environment and separate and independent from those he uses for different tasks. The fact that people can be scaled, even roughly, by general intelligence argues against such total specialization of skills and abilities. Hence, an important direction of research is to construct theories that can be applied over a wide range of tasks.

An example of such a theory is GPS, already mentioned. GPS can simulate behavior in any problem environment where the task can be symbolized as getting from a given state of affairs to a desired state of affairs. A wide range of problem-solving tasks can be expressed in this format. (To say that GPS can simulate behavior in any such environments means, not that the program will predict human behavior correctly and in detail, but that the program can at least be made to operate and produce a prediction, a trace, in that environment, for comparison with the human behavior.)

Thus, GPS has actually been given tasks from about a dozen task environments—for example, discovering proofs for theorems in logic, solving the missionaries-and-cannibals puzzle, and solving trigonometric identities—and there is every reason to suppose it can handle a wide range of others.

Levels of explanation. There has been great diversity of opinion in individual psychology as to the appropriate relation between psychological theory and neurophysiology. One extreme of the behaviorist position holds that the laws of psychology should state functional relations between the characteristics of stimulus situations, as the independent variables, and response situations, as the dependent variables, with no reference to intervening variables. A quite different position holds that explanation in psychology should relate behavior to biological mechanisms and that the laws of psychology are essentially reducible to laws of neurophysiology.

Information-processing theories represent a position distinct from either of these (Simon & Newell 1964). Unlike the extreme behaviorist theories, they make strong assumptions about the processes in the central nervous system (CNS) that intervene between stimulus and response and endeavor to explain how the stimulus produces the response. On the other hand, the information-processing theories do not describe these intervening processes in terms of neurophysiological mechanisms but postulate much grosser elements: symbol structures and elementary information processes that create and modify such structures and are, in turn, controlled by the structures.

Behaviorist theories may be called one-level theories, because they refer only to directly observable phenomena. Neurophysiological theories are two-level theories, for they refer both to behavioral events and underlying chemical and biological mechanisms. Information-processing theories are at least three-level theories, for they postulate elementary information processes to explain behavior, with the suggestion that the elementary information processes can subsequently be explained (in one or more stages of reduction) in chemical and biological terms. (See Hodos & Brady 1968; King 1968; Lindsley 1968; McCleary 1968; Riss 1968.)

Elementary information processes and symbol structures play the same role, at this intermediate level of explanation, as that played by chemical reactions, atoms, and molecules in nineteenth-century chemistry. In neither case is it claimed that the phenomena cannot be reduced, at least in principle, to a more microscopic level (neurophysiology in the one case, atomic physics in the other); it is simply more convenient to have a division of labor between psychologists and neurophysiologists (as between chemists and atomic physicists) and to use aggregative theories at the information-processing level as the bridge between them.

Some specific theories. The remainder of this article will be devoted to an account of the psychological substance of some current information-processing theories of cognition. Because space prohibits a complete survey of such theories, the examples will be restricted to problem solving, serial pattern recognition, and rote memory processes. Theories of other aspects of pattern recognition and concept formation will be omitted (Hunt 1962; Uhr 1966), as will be theories proposing a parallel, rather than serial, organization of cognitive processes (Reitman 1965).

The theories to be described are incorporated in a number of separate programs, which have not yet been combined into a single information-processing theory of "the whole cognitive man." Nevertheless, the programs have many similar components, and all incorporate the same basic assumptions about the organization and functioning of the CNS. Hence, the theories are complementary and will be discussed here, not as separate entities, but as components of a theory of cognition (Simon & Newell 1964).

The basic assumptions about the organization and functioning of the CNS are these:

(1) The CNS contains a *memory*, which stores symbols (discriminable patterns) and composite structures made up of symbols. These composite structures include *lists* (ordered sets of symbols, e.g., the English alphabet) and *descriptions* (associations between triads of symbols). The relation "black, opposite, white," to be translated "white is the opposite of black," is an example of a description.

(2) The CNS performs certain *elementary processes* on symbols: storing symbols, copying symbols, associating symbols in lists and descriptions, finding symbols on lists and in descriptions, comparing symbols to determine whether they are identical or different.

(3) The elementary processes are organized hierarchically into *programs*, and the CNS incorporates an *interpretive process* capable of executing such programs—determining at each stage what elementary process will be carried out next. The interpretation is serial (not all information-processing theories share this assumption, however), one process being executed at a time.

The memory, symbol structures, elementary processes, and interpreters are the information-processing mechanisms in terms of which observed human behavior is to be explained. Programs constructed from these mechanisms are used to simulate problem solving, memorizing, serial pattern recognizing, and other performances.

Problem-solving processes. A central task of any theory of human problem solving is to explain how people can find their way to solutions while exploring enormous "mazes" of possible alternative paths. It has been calculated that if a chess player were to consider all the possible outcomes of all possible moves, he would have to examine some 10^{120} paths. In fact he does not do this—obviously he could not, in any event—but conducts a highly selective search among a very much smaller number of possibilities. Available empirical data indicate that a problem solver usually explores well under a hundred paths in this and other problem-solving "mazes."

Experiments based on information-processing theories have shown that such selective searches can be accomplished successfully by programs incorporating rules of thumb, or heuristics, for determining which paths in the maze are likely to lead to solutions (Feigenbaum & Feldman 1963, part 1, secs. 2, 3). Some of the heuristics are very specific to a particular task environment. As a person becomes skilled in such an environment, he learns to discriminate features of the situation that have diagnostic value, and he associates with those features responses that may be appropriate in situations of the specified kind. Thus, an important component in specific skills is a "table of connections," stored in memory, associating discriminable features with possibly relevant actions. For example, a chess player learns to notice when the enemy's king is exposed, and he learns to consider certain types of moves (e.g., "checking" moves) whenever he notices this condition. (See Jerison 1968.)

Much of the behavior of problem solvers in chess-playing and theorem-proving tasks can be explained in terms of simple "feature-noticing" programs and corresponding tables of connections, or associations. But the programs must explain also how these components in memory are organized into effective programs for selective search. These organizing programs are relatively independent of the specific kind of problem to be solved but are applicable instead (in conjunction with the noticing processes and associations) to large classes of task environments. They are general heuristics for problem solving.

The central GPS heuristic is an organization of elementary processes for carrying out *means–end analysis*. These processes operate as follows: the symbolized representations of two situations are compared; one or more differences are noted between them; an operator is selected among those associated with the difference by the table of connections; and the operator is applied to the situation that is to be changed. After the change has been made, a new situation has been created, which can be compared with the goal situation, and the process can then be repeated.

Consider, for example, a subject trying to solve one of the Moore–Anderson "recoding" problems. In this case, the problem is to change $L1$: $r \cdot (-p \supset q)$ to $L0$: $(q \vee p) \cdot r$. The subject begins by saying, "You've got to change the places of the r to the other side and you have to change the ... minus and horseshoe to wedge, and you've got to reverse the places from p and q to q and p, so let's see.... Rule 1 (i.e., $A \cdot B \to B \cdot A$) is similar to it because you want to change places with the second part...."

The means–end chain in this example includes the following: notice the difference in order between the terms of $L1$ and $L0$; find an operator (rule 1) that changes the order.

Another kind of organization, which may be called *planning and abstraction*, is also evident in problem solving. In planning, some of the detail of the original problem is eliminated (abstracted). A solution is then sought for the simplified problem, using means–end analysis or other processes. If the solution is found for the abstract problem, it provides an outline or plan for the solution of the original problem with the detail reinserted.

The programs for means–end analysis and for planning, each time they are activated, produce a "burst" of problem-solving activity directed at a particular subgoal or organization of subgoals. The problem-solving program also contains processes that control and organize these bursts of activity as parts of the total exploratory effort. (See Taylor 1968.)

Rote memory processes. The noticing processes incorporated in the problem-solving theory assume that humans have subprograms for discriminating among symbol structures (e.g., for noticing differences between structures) and programs for elaborating the fineness of these discriminations. They also assume that associations between symbol structures can be stored (e.g., the table of connections between differences and operators). Discrimination and association processes and the structures they employ constitute an important part of the detail of problem-solving activity but are obscured in complex human performances by the higher-level programs that organize and direct search.

In simpler human tasks—memorizing materials by rote—the organizing strategies are simpler, and the underlying processes account for most of the behavioral phenomena. It has been possible to provide a fairly simple information-processing explanation for the acquisition of discriminations and associations (Feigenbaum 1963). (Our description follows EPAM, the program for rote learning mentioned above.) In the learning situation, there develops in memory (a) a "sorting net," for discriminating among stimuli and among responses, and (b) compound symbol structures, stored in the net, that contain a partial representation, or "image," of the stimulus as one component and a partial image of the response as another.

When stimuli or responses are highly similar to each other, greater elaboration of the sorting net is required to discriminate among them, and more detail must be stored in the images. A stimulus becomes "familiar" as a result of repeated exposure and consequent elaboration of the sorting net and its stored image. We have already noted that these simple mechanisms have been shown to account, quantitatively as well as qualitatively, for some of the main features of learning rates that have been observed in the laboratory. (See Lawrence 1968.)

Recognition of periodic patterns. As an example of an information-processing explanation of some of the phenomena of concept formation, we consider how, according to such a theory, human subjects detect and extrapolate periodic patterns —e.g., the letter pattern $A\,B\,M\,C\,D\,M \ldots$ (Simon & Newell 1964, pp. 293–294).

At the core of the explanation is the hypothesis that in performing these tasks humans make use of their ability to detect a small number of basic relations and operations: for example, the relations "same" and "different" between two symbols or symbol structures; the relation "next" on a familiar alphabet; and the operations of finding the difference between and the quotient of a pair of numbers. (These discriminations are not unlike those required as components in problem-solving programs to detect features of symbol structures and differences between them.) With a relatively simple program of these processes, the pattern in a sequence can be detected, the pattern can be described by a symbol structure, and the pattern description can be used by the interpretive process to extrapolate. In the example $A\,B\,M\,C\,D\,M$, the relation "next" is detected in the substring $A\,B\,C\,D$, and the relation "same" in the M's. These relations are used to describe a three-letter pattern, whose next member is the "next" letter after D, i.e., E.

The information-processing theory of serial pattern detection has proved applicable to a wide range of cognitive tasks—including the Thurstone Letter Series Completion Test, symbol analogies tests, and partial reinforcement experiments. (See Feldman 1963; Kendler & Kendler 1968.)

This article has described the use of simulation in constructing and testing theories of individual behavior, with particular reference to nonnumerical information-processing theories that take the form of computer programs. Examples of some of the current theories of problem solving, memorizing, and pattern recognition were described briefly, and the underlying assumptions about the information-processing characteristics of the central nervous system were outlined. Various techniques were illustrated for subjecting information-processing theories to empirical test.

ALLEN NEWELL AND HERBERT A. SIMON

[*Directly related are the entries* DECISION MAKING, *especially the article on* PSYCHOLOGICAL ASPECTS; MODELS, MATHEMATICAL; *see also* Pollack 1968; Taylor 1968. *Other relevant material may be found in* MATHEMATICS; PROGRAMMING; *see also* Hodos & Brady 1968; Johnson 1968; King 1968; Lindsley 1968; Riss 1968; Ross 1968; Vinacke 1968; Zajonc 1968.]

BIBLIOGRAPHY

CLARKSON, GEOFFREY P. E. 1963 A Model of the Trust Investment Process. Pages 347–371 in Edward A. Feigenbaum and Julian Feldman (editors), *Computers and Thought.* New York: McGraw-Hill.

FEIGENBAUM, EDWARD A. 1963 The Simulation of Verbal Learning Behavior. Pages 297–309 in Edward A. Feigenbaum and Julian Feldman (editors), *Computers and Thought.* New York: McGraw-Hill.

FEIGENBAUM, EDWARD A.; and FELDMAN, JULIAN (editors) 1963 *Computers and Thought.* New York: McGraw-Hill.

FELDMAN, JULIAN 1963 Simulation of Behavior in the Binary Choice Experiment. Pages 329–346 in Edward A. Feigenbaum and Julian Feldman (editors), *Computers and Thought.* New York: McGraw-Hill.

GREEN, BERT F. JR. 1963 *Digital Computers in Research: An Introduction for Behavioral and Social Scientists.* New York: McGraw-Hill.

►HODOS, WILLIAM; and BRADY, JOSEPH V. 1968 Nervous System: III. Brain Stimulation. Volume 11, pages 150–157 in *International Encyclopedia of the Social Sciences.* Edited by David L. Sills. New York: Macmillan and Free Press.

HUNT, EARL B. 1962 *Concept Learning: An Information Processing Problem.* New York: Wiley.

►JERISON, HARRY J. 1968 Attention. Volume 1, pages 444–449 in *International Encyclopedia of the Social Sciences.* Edited by David L. Sills. New York: Macmillan and Free Press.

►JOHNSON, DONALD M. 1968 Reasoning and Logic. Volume 13, pages 344–350 in *International Encyclopedia of the Social Sciences.* Edited by David L. Sills. New York: Macmillan and Free Press.

►KENDLER, HOWARD H.; and KENDLER, TRACY S. 1968 Concept Formation. Volume 3, pages 206–211 in *International Encyclopedia of the Social Sciences.* Edited by David L. Sills. New York: Macmillan and Free Press.

►KING, FREDERICK A. 1968 Nervous System: II. Structure and Function of the Brain. Volume 11, pages 130–150 in *International Encyclopedia of the Social Sciences.* Edited by David L. Sills. New York: Macmillan and Free Press.

►LAWRENCE, DOUGLAS H. 1968 Learning: V. Discrimination Learning. Volume 9, pages 143–148 in *International Encyclopedia of the Social Sciences.* Edited by David L. Sills. New York: Macmillan and Free Press.

►LINDSLEY, DONALD 1968 Nervous System: IV. Electroencephalography. Volume 11, pages 157–161 in *International Encyclopedia of the Social Sciences.* Edited by David L. Sills. New York: Macmillan and Free Press.

►MCCLEARY, ROBERT A. 1968 Learning: VII. Neurophysiological Aspects. Volume 9, pages 154–160 in *International Encyclopedia of the Social Sciences.* Edited by David L. Sills. New York: Macmillan and Free Press.

MILLER, GEORGE A.; GALANTER, EUGENE; and PRIBRAM, K. H. 1960 *Plans and the Structure of Behavior.* New York: Holt.

MOORE, OMAR K.; and ANDERSON, SCARVIA B. 1954 Modern Logic and Tasks for Experiments on Problem Solving Behavior. *Journal of Psychology* 38:151–160.

NEWELL, ALLEN; and SIMON, HERBERT A. 1963a Computers in Psychology. Volume 1, pages 361–428 in R. Duncan Luce, Robert R. Bush, and Eugene Galanter (editors), *Handbook of Mathematical Psychology.* New York: Wiley.

¹NEWELL, ALLEN; and SIMON, HERBERT A. 1963b GPS, a Program That Simulates Human Thought. Pages 279–293 in Edward A. Feigenbaum and Julian Feldman (editors), *Computers and Thought.* New York: McGraw-Hill.

►POLLACK, IRWIN 1968 Information Theory. Volume 7, pages 331–337 in *International Encyclopedia of the Social Sciences.* Edited by David L. Sills. New York: Macmillan and Free Press.

►POSTMAN, LEO 1968 Learning: VIII. Verbal Learning. Volume 9, pages 160–168 in *International Encyclopedia of the Social Sciences.* Edited by David L. Sills. New York: Macmillan and Free Press.

REITMAN, WALTER R. 1965 *Cognition and Thought: An Information-processing Approach.* New York: Wiley.

►RISS, WALTER 1968 Nervous System: I. Evolutionary and Behavioral Aspects of the Brain. Volume 11, pages 126–130 in *International Encyclopedia of the Social Sciences.* Edited by David L. Sills. New York: Macmillan and Free Press.

►ROSS, LEONARD E. 1968 Learning Theory. Volume 9, pages 189–197 in *International Encyclopedia of the Social Sciences.* Edited by David L. Sills. New York: Macmillan and Free Press.

¹SIMON, HERBERT A.; and NEWELL, ALLEN 1964 Information Processing in Computer and Man. *American Scientist* 52:281–300.

►TAYLOR, DONALD W. 1968 Problem Solving. Volume 12, pages 505–511 in *International Encyclopedia of the Social Sciences.* Edited by David L. Sills. New York: Macmillan and Free Press.

UHR, LEONARD (editor) 1966 *Pattern Recognition.* New York: Wiley.

►UNDERWOOD, BENTON J. 1968 Forgetting. Volume 5, pages 536–542 in *International Encyclopedia of the Social Sciences.* Edited by David L. Sills. New York: Macmillan and Free Press.

►Vinacke, W. Edgar 1968 Thinking: I. The Field. Volume 15, pages 608–615 in *International Encyclopedia of the Social Sciences*. Edited by David L. Sills. New York: Macmillan and Free Press.

►Zajonc, Robert B. 1968 Thinking: II. Cognitive Organization and Processes. Volume 15, pages 615–622 in *International Encyclopedia of the Social Sciences*. Edited by David L. Sills. New York: Macmillan and Free Press.

Postscript

Information-processing theories have now attained wide use and acceptance in psychology. There has been a corresponding burgeoning in the use of simulation as a tool for formulating and testing such theories, along with a growing eclecticism in which simulation methods and protocol analysis are employed side by side with more conventional experimental techniques. Anderson and Bower (1973) illustrates these trends. Simulation of process detail has, in turn, exerted pressure for a higher density, per unit of time, of observations of subjects' behavior, with a resulting growing interest and activity in chronometric and eye-movement studies.

Problem-solving theory is surveyed in Newell and Simon (1972), and models of memory are surveyed in Anderson and Bower (1973). Simulation has been employed in many new areas, as, for example, the modeling of children's performance of Piagetian tasks (Klahr 1973). Important progress has been made in simulating natural language and semantic processes (for example, Winograd 1972; Ninth Symposium on Cognition 1974), and this recent focus on "the meaning of meaning" is bringing about a reintegration of linguistics and psycholinguistics with general cognitive theory, as the notion of linguistic deep structure is being gradually absorbed by the concept of internal semantic representations in long-term memory.

The methodology of simulation has been undergoing continual evaluation (Hunt 1968) and improvement. A detailed comparison between information-processing theories and stochastic theories of concept formation has shown that in many cases the latter can be derived formally from the former by aggregation (Gregg & Simon 1967). Semiformal methods for encoding protocols are described by Newell and Simon (1972), and Waterman and Newell (1971) have constructed a system, PAS-II, for automatic protocol analysis by computer.

Cognitive simulation continues to influence, and to be influenced by, developments in the theory of programming languages. Newell and Simon (1972) have hypothesized that human information-processing programs have the form of *production systems*, programs composed of independent instructions called productions. Each production consists of a condition part and an action part. Whenever the conditions of a production are satisfied (by symbols in short-term memory or perceptual symbols), the actions are executed, thus obviating the need for a conventional program control structure. Several cognitive performances have now been simulated with production systems (Newell 1973; Klahr 1973).

Allen Newell and Herbert A. Simon

ADDITIONAL BIBLIOGRAPHY

Anderson, John R.; and Bower, Gordon H. 1973 *Human Associative Memory*. Washington: Winston.

Gregg, Lee W.; and Simon, Herbert A. 1967 Process Models and Stochastic Theories of Simple Concept Formation. *Journal of Mathematical Psychology* 4: 246–276.

Hunt, Earl B. 1968 Computer Simulation: Artificial Intelligence Studies and Their Relevance to Psychology. *Annual Review of Psychology* 19:135–168.

Klahr, David 1973 A Production System for Counting, Subitizing, and Adding. Pages 527–546 in Symposium on Cognition, Eighth, Carnegie–Mellon University, 1972, *Visual Information Processing: Proceedings*. Edited by William G. Chase. New York: Academic Press.

Newell, Allen 1973 Production Systems: Models of Control Structures. Pages 463–526 in Symposium on Cognition, Eighth, Carnegie–Mellon University, 1972, *Visual Information Processing: Proceedings*. Edited by William G. Chase. New York: Academic Press.

¹Newell, Allen; and Simon, Herbert A. 1972 *Human Problem Solving*. Englewood Cliffs, N.J.: Prentice-Hall. → Contains much of the material of Newell and Simon (1963b) and Simon and Newell (1964).

Symposium on Cognition, Ninth, Carnegie–Mellon University, 1973 1974 *Knowledge and Cognition*. Edited by Lee W. Gregg. Potomac, Md.: Lawrence Erlbaum.

Waterman, Donald; and Newell, Allen 1971 Protocol Analysis as a Task for Artificial Intelligence. *Artificial Intelligence* 2:285–318.

Winograd, Terry 1972 *Understanding Natural Language*. New York: Academic Press.

II

ECONOMIC PROCESSES

Simulation is at once one of the most powerful and one of the most misapplied tools of modern economic analysis. "Simulation" of an economic system means the performance of experiments upon an analogue of the economic system and the drawing of inferences concerning the properties of the economic system from the behavior of its analogue. The analogue is an idealization of a generally more complex real system, the essential properties of which are retained in the analogue.

While this definition is consistent with the denotation of the word, the connotation of simulation among economists active in the field today is much more restricted. The term "simulation" has been generally reserved for processes using a physical or mathematical analogue and requiring a modern high-speed digital or analogue computer for the execution of the experiments. In addition, most economic simulations have involved some stochastic elements, either as an intrinsic feature of the model or as part of the simulation experiment field. Thus, while a pencil-and-paper calculation on a two-person Walrasian economy would, according to the more general definition, constitute a simulation experiment, common usage of the term would require that, to qualify as a simulation, the size of the model must be large enough, or the relationships complex enough, to necessitate the use of a modern computing machine. Even the inclusion of probabilistic elements in the pencil-and-paper game would not suffice, in and of itself, to transform the calculations into a bona fide common-usage simulation. In order to avoid confusion, the term "simulation" will be used hereafter in its more restrictive sense.

To clarify the concept of simulation, it might be useful to indicate how the simulation process is related, on the one hand, to the analytic solution of mathematical models of an economy, and on the other, to the construction of mathematical and econometric models of economic systems.

The relationship between simulation and econometric and mathematical formulations of economic theories is quite intimate. It is these descriptions of economic processes that constitute the basic inputs into a simulation. After a mathematical or econometric model is translated into language a computing machine can understand, the behavior of the model, as described by the machine output, represents the behavior of the economic system being simulated. The solutions obtained by means of simulation techniques are quite specific. Given a particular set of initial conditions, a particular set of parameters, and the time period over which the model is to be simulated, a single simulation experiment yields a particular numerically specified set of time paths for the endogenous variables (the variables whose values are determined or explained by the model). A variation in one or more of the initial conditions or parameters requires a separate simulation experiment which provides a different set of time paths. Comparisons between the original solution and other solutions obtained under specific variations in assumptions can then be used to infer some of the properties of the relationships between input and output quantities in the system under investigation. In general, only very partial inferences concerning these relationships can be drawn by means of simulation experiments. In addition, the results obtained can be assumed to

be valid only for values of the parameters and initial conditions close to those used in the simulation experiments. By contrast, traditional mathematical approaches for studying the implications of an economic model produce general solutions by deductive methods. These general solutions describe, in functional form, the manner in which the model relates the endogenous variables to any set of initial conditions and parameters and to time.

The models formulated for a simulation experiment must, of course, represent a compromise between "realism" and tractability. Since modern computers enable very large numbers of computations to be performed rapidly, they permit the step-by-step solution of systems that are several orders of magnitude larger and more complicated than those that can be handled by the more conventional techniques. The representation of economic systems to be investigated with the aid of simulation techniques can therefore be much more complex; there are considerably fewer restrictions on the number of equations and on the mathematical forms of the functions that can be utilized. Simulation, therefore, permits greater realism in the extent and nature of the feed-back mechanisms in the stylized representation of the economy contained in the econometric or mathematical model.

The impetus for the use of simulation techniques in economics arises from three major sources. First, both theory and casual observation suggest that an adequate description of the dynamic behavior of an economy must involve complex patterns of time dependencies, nonlinearities, and intricate interrelationships among the many variables governing the evolution of economic activity through time. In addition, a realistic economic model will almost certainly require a high degree of disaggregation. Since analytic solutions can be obtained for only very special types of appreciably simplified economic models, simulation techniques, with their vastly greater capacity for complexity, permit the use of more realistic analogues to describe real economic systems.

The second driving force behind the use of simulation, one that is not unrelated to the first, arises from the need of social scientists in general and of economists in particular to find morally acceptable and scientifically adequate substitutes for the physical scientist's controlled experiments. To the extent that the analogue used in the simulation represents the relevant properties of the economic system under study, experimentation with the analogue can be used to infer the results of analogous experiments with the real economy. The effects in the model of specific changes in the values of particular policy instruments (e.g., taxes, interest rates, price level) can be used to draw at least qualitatively valid inferences concerning the probable effects of analogous changes in the real economic system. Much theoretical analysis in economics is aimed at the study of the probable reactions of an economy to specified exogenous changes, but any economic model that can be studied by analytic techniques must of necessity omit so many obviously relevant considerations that little confidence can be placed in the practical value of the results. Since a simulation study can approximate the economy's behavior and structure considerably more closely, simulation experiments can, at least in principle, lead to conditional predictions of much greater operational significance.

Finally, the mathematical flexibility of simulation permits the use of this tool to gain insights into many phenomena whose intrinsic nature is not at all obvious. It is often possible, for example, to formulate a very detailed quantitative description of a particular process before its essential nature is sufficiently well understood to permit the degree of stylization required for a useful theoretical analysis. Studies of the sensitivity of the results to various changes in assumptions can then be used to disentangle the important from the unimportant features of the problem.

Past uses of simulation. The earliest applications of simulation approaches to economics employed physical analogues of a hydraulic or electrical variety. Analogue computers permit the solution of more or less complex linear or nonlinear dynamic systems in which time is treated as a continuous variable. They also enable a visual picture to be gained of adjustment processes. On the other hand, they are much slower than digital computers and can introduce distortions into the results because of physical effects, such as "noise" and friction, which have no conscious economic analogue.

Subsequent economic simulations have tended to rely primarily on digital computers. As the speed and memory capacity of the computers have improved, the economic systems simulated have become increasingly more complex and more elaborate, and the emphasis in simulation has shifted from use as a mathematical tool for solving systems of equations and understanding economic models to use as a device for forecasting and controlling real economies. In addition, these improvements in computer design have permitted the construction of microanalytic models in which aggregate relationships are built up from specifica-

tions concerning the behavioral patterns of a large sample of microeconomic units.

As early as 1892, Irving Fisher recommended the use of hydraulic analogies "not only to obtain a clear and analytical *picture* of the interdependence of the many elements in the causation of prices, but also to employ the mechanism as an instrument of investigation and by it, study some complicated variations which could scarcely be successfully followed without its aid" (1892, p. 44). It was not until 1950, however, that the first hydraulic analogues of economic systems were constructed. Phillips (1950) used machines made of transparent plastic tanks and tubes through which colored water was pumped to depict the Keynesian mechanism of income determination and to indicate how the production, consumption, stocks, and prices of a commodity interact. Electrical analogues were used to study models of inventory determination (Strotz et al. 1957) and to study the business cycle models of Kalecki (Smith & Erdley 1952), Goodwin (Strotz et al. 1953), and others.

The shift to the use of digital computers began in the late 1950s. In a simulation study on an IBM 650, Adelman and Adelman (1959) investigated the dynamic properties of the Klein–Goldberger model of the U.S. economy by extrapolating the exogenous variables and solving the system of 25 nonlinear difference equations for a period of one hundred years. In this process no indications were found of oscillatory behavior in the model. The introduction of random disturbances of a reasonable order of magnitude, however, generated cycles that were remarkably similar to those characterizing the U.S. economy. On the basis of their study, they concluded that (1) the Klein–Goldberger equations may represent good approximations to the behavioral relationships in the real economy; and (2) their results are consistent with the hypothesis that the fluctuations experienced in modern highly developed societies are due to random perturbations.

Duesenberry, Eckstein, and Fromm (1960) constructed a 14-equation quarterly aggregative econometric model of the U.S. economy. Simulation experiments with this model were used to test the vulnerability of the U.S. economy to depressions and to assess the effectiveness of various automatic stabilizers, such as tax declines, increases in transfer payments, and changes in business savings.

A far more detailed and, indeed, the most ambitious macroeconometric simulation effort to date, at least on this side of the iron curtain, is being carried out at the Brookings Institution in Washington, D.C. The simulation is based on a 400-equation quarterly econometric model of the U.S. economy constructed by various experts under the auspices of the Social Science Research Council (Duesenberry et al. 1965). The Brookings model has eight production sectors. For each of the non-government sectors there are equations describing the determinants of fixed investment intentions, fixed investment realizations, new orders, unfilled orders, inventory investment, production worker employment, nonproduction worker employment, production worker average weekly hours, labor compensation rates per hour, real depreciation, capital consumption allowances, indirect business taxes, rental income, interest income, prices, corporate profits, entrepreneurial income, dividends, retained earnings, and inventory valuation adjustment. The remaining expenditure components of the national product are estimated in 5 consumption equations, 11 equations for nonbusiness construction, and several import and export equations. For government, certain nondefense expenditures, especially at the state and local levels, are treated as endogenous variables, while the rest are taken as exogenous. On the income side of the accounts there is a vast array of additional equations for transfer payments, personal income taxes, and other minor items. The model also includes a demographic sector containing labor force, marriage rate, and household formation variables and a financial sector that analyzes the demand for money, the interest rate term structure, and other monetary variables. Finally, aside from a battery of identities, the model also incorporates two matrices: an input–output matrix that translates constant dollar GNP (gross national product) component demands into constant dollar gross product originating by industry, and a matrix that translates industry prices into implicit deflators for GNP components.

The solution to the system is achieved by making use of the model's block recursive structure. First, a block of variables wholly dependent on lagged endogenous variables is solved. Second, an initial simultaneous equation solution is obtained for variables in the "quantity" block (consumption, imports, industry outputs, etc.) using predetermined variables and using prices from the previous period in place of current prices. Third, a simultaneous equation solution is obtained for variables in the "price" block using the predetermined variables and the initial solution from the quantity block. Fourth, the price block solution is used as a new input to the quantity block, and the second and third steps are iterated until each variable changes (from iteration to iteration) by no more than 0.1 per cent. Fifth, the three blocks are solved recursively.

Simulation experiments show that the system yields predictions both within and beyond the sample period that lie well within the range of accuracy of other models. Simulation studies have also been undertaken to determine the potential impact of personal-income and excise-tax reductions, of government expenditure increases, and of monetary tightness and ease. Stochastic simulations are also being carried out to ascertain the stability properties of the system.

A rather different approach to the estimation of behavioral relationships for the entire socioeconomic system has been developed by Orcutt and his associates (Orcutt et al. 1961). Starting with the observed behavior of microeconomic decision units rather than macroeconomic aggregates, they estimate behavioral relationships for classes of decision units and use these to predict the behavior of the entire system. For this purpose, functions known as operating characteristics are estimated from sample surveys and census data. These functions specify the probabilities of various outcomes, given the values of a set of status variables which indicate the relevant characteristics of the microeconomic units at the start of a period.

Their initial simulation experiments were aimed at forecasting demographic behavior. For example, to estimate the number of deaths occurring during a given month, the following procedure was applied: an operating characteristic,

$$\log P_i(m) = F_1[A_i(m - \tfrac{1}{2}), R_i, S_i] + (m - m_0)F_2[A_i(m - \tfrac{1}{2}), R_i, S_i] + F_3(m),$$

was specified. Here, $P_i(m)$ indicates the probability of the death of individual i in month m, $A_i(m - \tfrac{1}{2})$ is the age of individual i at the start of month m, R_i denotes the race of individual i, and S_i denotes the sex of individual i. The function F_1 is a set of four age-specific mortality tables, one for each race and sex combination, which describe mortality conditions prevailing during the base month m_0. The function F_2 is used to update each of these mortality tables to month m. The function F_3 is a cyclic function which accounts for seasonal variations in death rates.

This and other operating characteristics were used in a simulation of the evolution of the U.S. population month by month between 1950 and 1960. A representative sample of the U.S. population consisting of approximately 10,000 individuals was constructed. In each month of the calculation a random drawing determined, in accordance with the probabilities specified by the relevant operating characteristics, whether each individual in the sample would die, marry, get divorced, or give birth. A regression analysis was then used to compare the actual and predicted demographic changes during the sample period. Close agreement was obtained. This approach is currently being extended to permit analysis of the consumption, saving, and borrowing patterns of U.S. families, the behavior of firms, banks, insurance companies, and labor unions, etc. When all portions of the microanalytic model have been fitted together, simulation runs will be made to explore the consequences of various changes in monetary and fiscal policies.

So far, we have discussed simulation experiments for prediction purposes only. However, the potential of simulation as an aid to policy formulation has not been overlooked by development planners in either free-enterprise economies or socialist countries. A microanalytic simulation model has been formulated at the U.S.S.R. Academy of Sciences in order to guide the management of the production system of the Soviet economy (Cherniak 1963). The model is quite detailed and elaborate. It starts with individual plants at a specific location and ends up with interregional and interindustry tables for the Soviet Union as a whole. In addition, the U.S.S.R. is contemplating the introduction of a 10-step joint man–computer program. The program is to be cyclic, with odd steps being man-operated and even steps being computer-operated. The functions fulfilled by the man-operated steps include the elaboration of the initial basis of the plan, the establishment of the criteria and constraints of the plan, and the evaluation of results. The computer, on the other hand, will perform such tasks as summarizing and balancing the plan, determining optimal solutions to individual models, and simulating the results of planning decisions. This procedure will be tested at a level of aggregation corresponding to an industrial sector of an economic council and will ultimately be extended to the economy as a whole. I have been unable to find any information on the progress of this work more recent than that in Cherniak (1963).

To take an example from the nonsocialist countries, a large-scale interdisciplinary simulation effort was undertaken at Harvard University, at the request of President John F. Kennedy, in order to study the engineering and economic development of a river basin in Pakistan. The major recommendations of the report (White House 1964) are currently being implemented by the Pakistani government, with funds supplied by the U.S. Agency for International Development.

In Venezuela a 250-equation dynamic simulation model has been constructed by Holland, in conjunction with the Venezuelan Planning Agency and the University of Venezuela. This model, which is based on experience gained in an earlier simulation of a stylized underdeveloped economy (Holland & Gillespie 1963), is being used in Venezuela to compare, by means of sensitivity studies, the repercussions of alternative government expenditure programs on government revenues and on induced imports, as well as to check the consistency of the projected rate of growth with the rate of investment and the rate of savings. An interesting macroeconomic simulation of the Ecuadorian economy, in which the society has been disaggregated into four social classes, is currently being carried out by Shubik (1966).

Monte Carlo studies. The small sample properties of certain statistical estimators cannot be determined using currently available mathematical analysis alone. In such cases simulation methods are extremely useful. By simulating an economic structure (including stochastic elements) whose parameters are known, one generates samples of "observations" of a given sample size. Each sample is then used to estimate the parameters by several estimation methods. For each method, the distribution of the estimates is compared with the true values of the parameters to determine the properties of the estimator for the given sample size. This approach, known as the Monte Carlo sampling method, has frequently been applied by econometricians in studies of the small sample properties of alternative estimators of simultaneous equation models [*see* SIMULTANEOUS EQUATION ESTIMATION].

Limitations and potential. To a large extent, the very strength of simulation is also its major weakness. As pointed out earlier, any simulation experiment produces no more than a specific numerical case history of the system whose properties are to be investigated. To understand the manner of operation of the system and to disentangle the important effects from the unimportant ones, simulation with different initial conditions and parameters must be undertaken. The results of these sensitivity studies must then be analyzed by the investigator and generalized appropriately. However, if the system is very complex, these tasks may be very difficult indeed. To enable interpretation of results, it is crucial to keep the structure of the simulation model simple and to recognize that, as pointed out in the definition, simulation by no means implies a blind imitation of the actual situation. The simulation model should express only the logic of the simulated system together with the elements needed for a fruitful synthesis.

Another major problem in the use of simulation is the interaction between theory construction and simulation experiments. In many cases simulation has been used as an alternative to analysis, rather than as a supplementary tool for enriching the realm of what can be investigated by other, more conventional techniques. The inclination to compute rather than think tends to permeate a large number of simulation experiments, in which the investigators tend to be drowned in a mass of output data whose general implications they are unable to analyze. There are, of course, notable exceptions to this phenomenon. In the water-resource project (Dorfman 1964), for example, crude analytic solutions to simplified formulations of a given problem were used to pinpoint the neighborhoods of the solution space in which sensitivity studies to variations in initial conditions and parameters should be undertaken in the more complex over-all system. In another instance, insights gained from a set of simulation experiments were used to formulate theorems which, once the investigator's intuition had been educated by means of the simulation, were proved analytically. Examples of such constructive uses of simulation are unfortunately all too few.

Finally, in many practical applications of simulation to policy and prediction problems, insufficient attention is paid to the correspondence between the system simulated and its analogue. As long as the description incorporated in the simulation model appears to be realistic, the equivalence between the real system and its analogue is often taken for granted, and inferences are drawn from the simulation that supposedly apply to the real economy. Clearly, however, the quality of the input data, the correspondence between the behavior of the outputs of the simulation and the behavior of the analogous variables in the real system, and the sensitivity of results to various features of the stylization should all be investigated before inferences concerning the real world are drawn from simulation experiments.

In summary, simulation techniques have a tremendous potential for both theoretical analysis and policy-oriented investigations. If a model is chosen that constitutes a reasonable representation of the economic interactions of the real world and that is sufficiently simple in its structure to permit intelligent interpretation of the results at the present state of the art, the simulations can be used to acquire a basic understanding of, and a qualitative feeling for, the reactions of a real economy to various types of stimuli. The usefulness of the technique will depend crucially, however, upon the validity of the representation of the system to be simulated and upon the quality of the compromise between realism and tractability. Presumably, as the capabilities of both high-speed computers and economists improve, the limitations of simulation will be decreased, and its usefulness for more and more complex problems will be increased.

IRMA ADELMAN

[*For additional treatment of the terminology in this article and other relevant material, see* ECONOMETRIC MODELS, AGGREGATE.]

BIBLIOGRAPHY

General discussions of the simulation of economic processes can be found in Conference on Computer Simulation 1963; Holland & Gillespie 1963; Orcutt et al. 1960; 1961. *A comprehensive general bibliography is* Shubik 1960. *On the application of analogue computers to problems in economic dynamics, see* Phillips 1950; 1957; Smith & Erdley 1952; Strotz et al. 1953; 1957. *Examples of studies using digital computers are* Adelman & Adelman 1959; Cherniak 1963; Dorfman 1964; Duesenberry et al. 1960; 1965; Fromm & Taubman 1966; White House 1964.

ADELMAN, IRMA; and ADELMAN, FRANK L. (1959) 1965 The Dynamic Properties of the Klein–Goldberger Model. Pages 278–306 in American Economic Association, *Readings in Business Cycles.* Homewood, Ill.: Irwin. → First published in Volume 27 of *Econometrica.*

CHERNIAK, IU. I. 1963 The Electronic Simulation of Information Systems for Central Planning. *Economics of Planning* 3:23–40.

CONFERENCE ON COMPUTER SIMULATION, UNIVERSITY OF CALIFORNIA, LOS ANGELES, *1961* 1963 *Symposium on Simulation Models: Methodology and Applications to the Behavioral Sciences.* Edited by Austin C. Hoggatt and Frederick E. Balderston. Cincinnati, Ohio: South-Western Publishing.

DORFMAN, ROBERT 1964 Formal Models in the Design of Water Resource Systems. Unpublished manuscript.

DUESENBERRY, JAMES S.; ECKSTEIN, OTTO; and FROMM, GARY 1960 A Simulation of the United States Economy in Recession. *Econometrica* 28:749–809.

DUESENBERRY, JAMES S. et al. (editors) 1965 *The Brookings Quarterly Econometric Model of the United States.* Chicago: Rand McNally.

FISHER, IRVING (1892) 1961 *Mathematical Investigations in the Theory of Value and Prices.* New Haven: Yale Univ. Press.

FROMM, GARY; and TAUBMAN, PAUL 1966 Policy Simulations With an Econometric Model. Unpublished manuscript.

HOLLAND, EDWARD P.; and GILLESPIE, ROBERT W. 1963 *Experiments on a Simulated Under-developed Economy: Development Plans and Balance-of-payments Policies.* Cambridge, Mass.: M.I.T. Press.

ORCUTT, GUY H. et al. 1960 Simulation: A Symposium. *American Economic Review* 50:893–932.

ORCUTT, GUY H. et al. 1961 *Microanalysis of Socioeconomic Systems: A Simulation Study.* New York: Harper.

PHILLIPS, A. W. 1950 Mechanical Models in Economic Dynamics. *Economica* New Series 17:283–305.

PHILLIPS, A. W. 1957 Stabilisation Policy and the Time-forms of Lagged Responses. *Economic Journal* 67:265–277.

SHUBIK, MARTIN 1960 Bibliography on Simulation, Gaming, Artificial Intelligence and Allied Topics. *Journal of the American Statistical Association* 55:736–751.

Shubik, Martin 1966 Simulation of Socio-economic Systems. Part 2: An Aggregative Socio-economic Simulation of a Latin American Country. Cowles Foundation for Research in Economics, *Discussion Paper No. 203.*

Smith, O. J. M.; and Erdley, H. F. 1952 An Electronic Analogue for an Economic System. *Electrical Engineering* 71:362–366.

Strotz, R. H.; Calvert, J. F.; and Morehouse, N. F. 1957 Analogue Computing Techniques Applied to Economics. American Institute of Electrical Engineers, *Transactions* 70, part 1:557–563.

Strotz, R. H.; McAnulty, J. C.; and Naines, J. B. Jr. 1953 Goodwin's Nonlinear Theory of the Business Cycle: An Electro-analog Solution. *Econometrica* 21:390–411.

White House—[U.S.] Department of Interior, Panel on Waterlogging and Salinity in West Pakistan 1964 *Report on Land and Water Development in the Indus Plain.* Washington: Government Printing Office.

Postscript

The technique of simulation appears to be gaining acceptance as a tool of economic analysis, forecasting, and planning. It is applied increasingly to gain better understanding of the implied dynamic behavior of more realistically specified descriptions of various aspects of an economy, as well as to interdisciplinary problems. Large-scale applications have included descriptions of the labor market and discrimination (Holt 1977; Bergmann 1974), the urban housing market (Engle et al. 1972), the tax and social security systems (Pechman & Okner 1974), the derivation of the macro-behavior implications of micro-system specification (Bergmann 1974; Fair 1974–1976; Adelman & Robinson 1978), forecasting to the year 2000 (Forrester 1971; Meadows et al. 1972), development planning of antipoverty policy (Adelman & Robinson 1978), modeling of urban political systems (Adelman & Adelman 1974). Much of relevance for economic theory and public policy is being learned from these models. They demonstrate the potential of simulation as a technique. The caveats mentioned in the main article retain their validity, however, despite increased user caution.

Irma Adelman

ADDITIONAL BIBLIOGRAPHY

Adelman, Frank L.; and Adelman, Irma 1974 Simulation of City Politics. Volume 1, pages 89–112 in Windsor Conference on Comparative Urban Economics and Development, 1972, *Urban and Social Economics in Market and Planned Economies.* 2 vols. Edited by Alan A. Brown, Joseph A. Licari, and Egon Neuberger. New York: Praeger.

Adelman, Irma; and Robinson, Sherman 1978 *Income Distribution Policy in Developing Countries: A Case Study of Korea.* Stanford Univ. Press.

Bergmann, Barbara R. 1974 A Microsimulation of the Macroeconomy With Explicitly Represented Money Flows. *Annals of Economic and Social Measurement* 3:475–489.

Engle, Robert F. III et al. 1972 An Econometric Simulation Model of Intra-metropolitan Housing Location: Housing, Business, Transportation and Local Government. *American Economic Review* 62, no. 2:87–97.

Fair, Ray C. 1974–1976 *A Model of Macroeconomic Activity.* 2 vols. Cambridge, Mass.: Ballinger. → Volume 1: *The Theoretical Model.* Volume 2: *The Empirical Model.*

Forrester, Jay W. (1971) 1973 *World Dynamics.* 2d ed. Cambridge, Mass.: Wright-Allen.

Holt, Charles C. 1977 Modeling a Segmented Labor Market. Pages 83–119 in Phyllis A. Wallace (editor), *Women, Minorities, and Employment Discrimination.* Lexington, Mass.: Heath.

Meadows, Donella H. et al. (1972) 1974 *The Limits to Growth: A Report of the Club of Rome's Project on the Predicament of Mankind.* 2d ed. New York: Universe.

Pechman, Joseph A.; and Okner, Benjamin A. 1974 *Who Bears the Tax Burden? Studies in Government Finance.* Washington: Brookings.

III
POLITICAL PROCESSES

A political game or political simulation is a type of model that represents some aspect of politics. The referent, or "reality," represented by a simulation-gaming technique may be some existing, past, or hypothetical system or process. Regardless of the reference system or process depicted in a game or simulation, the model is always a simplification of the total reality. Some political features will be excluded. Those elements of political phenomena incorporated in the model are reduced in complexity. The simplification and selective incorporation of a reference system or process produce the assets of parsimony and manageability as well as the liability of possible distortion. These attributes are, of course, equally applicable to other kinds of models, whether they be verbal, pictorial, or mathematical.

Simulations and games differ from other types of models in that their interrelated elements are capable of assuming different values as the simulation or game operates or unfolds. In other words, they contain rules for transforming some of the symbols in the model. For example, a game might begin with the description of a particular situation circulated to players who then are instructed to make responses appropriate for their roles. Initial reactions of some players lead to action by others; they in turn provoke new responses. In this manner not only do situations evolve, but basic changes also occur in the relationships between the players.

When the value of one element in the simulation is changed, related properties can be adjusted accordingly. Because of this ability to handle time and change, Brody has described simulations and games as "operating models" (1963, p. 670).

Games versus simulations. To the users of both techniques, the distinctions between games and simulations are still ambiguous, as some current definitions will illustrate.

Brody (1963) and Guetzkow and his associates (1963) distinguish between (1) machine, or computer, simulations, (2) man, or manual, games, and (3) mixed, or man–computer, simulations. Generally the "machine" referred to is a digital computer, although some simulations involve other types of calculating equipment. The extension of the term "simulation" to cover those models which are a mixture of man and machine does not occur in other definitions. For example, the concept "simulation" is confined by Pool, Abelson, and Popkin (1965) to models that are completely operated on a computer. Rapoport (1964) defines a simulation as a technique in which both the assessment of the situation and the subsequent decisions are made in accordance with explicit and formal rules. When either assessment or decisions are made by human beings without formal rules, the technique is described as a game. When both assessment and decision are made by men, Rapoport stipulates that the device is a scenario. A different interpretation is given by Shubik, who suggests that a game "is invariably concerned with studying human behavior or teaching individuals" (1964, p. 71). On the other hand, a simulation is designated as the reproduction of a system or organism in which the actual presence of humans is not essential because their behavior is one of the givens of the simulated environment. The necessary involvement of humans in games also appears in the definitions of Thorelli and Graves (1964) and Dawson (1962). In Dawson's conceptualization all operating models are simulations, but only those which introduce human players are games.

In these definitions of games and simulations, two distinguishing criteria are recurrent. The techniques are differentiated either by the role specified for human participants or by the role assigned formal rules of change or transformation. When human participation is the distinguishing criterion, a game is described as an operating model in which participants are present and a simulation as a model without players. With rules of transformation as the criterion, the type of model without extensive use of formal rules is a game and the technique with formal rules for handling change is a simulation.

The two distinguishing criteria can be interpreted as opposite poles of a single differentiating property. A model designed with only a limited number of its operating rules explicitly stated requires human players and administrators to define rules as the game proceeds. Conversely, a model designed with the interrelationships between its units specified in formal rules has less use for human decision makers during its operation. Thus, formal rules and human participants are alternative means for establishing the dynamic relationships between a model's units. Given this interpretation, two operational characteristics can be used for separating games and simulations. One definitional approach is to reserve the term "simulation" for models that are completely programmed (i.e., all operations are specified in advance) and that confine the role of humans to the specification of inputs. Alternatively, simulations can be defined as operating models that, although not necessarily excluding human decision makers, involve such complex programmed rules that either computer assistance or a special staff using smaller calculating equipment is required to determine the consequences of the rules. The latter definition of simulation is used in the remainder of this essay. Accordingly, a game is an operating model whose dynamics are primarily determined by participants and game administrators with a minimum of formal rules (i.e., neither computer nor calculation staff).

The use of "game" to identify one type of dynamic model necessitates distinguishing political gaming from the theory of games and from "parlor games." Occasionally, game theory and political gaming are applicable to similar political problems. The theory of games, however, is a mathematical approach to conflict situations in which the outcome of play is dependent on the choices of the conflicting parties. To employ game theory, specified conditions must be fulfilled; for example, the players must have knowledge of all their alternatives and be able to assign utilities to each one. Political gaming does not require those conditions, nor does it contain formal solutions to conflict situations. Political gaming also may be confused with "parlor games" designed for entertainment. The two activities can be distinguished by political gaming's instructional or research focus, its greater complexity, and its more elaborate effort to reproduce a political reference system or process.

To summarize, political games and simulations are operating models whose properties are substituted for selected aspects of past, present, or hypothetical political phenomena. Although consensus is lacking, many definitions imply that games are operating models in which the pattern of interaction and change between the represented units is made by human participants and administrators as the model is played. In simulations, the interaction and change among elements represented in the model are specified in formal rules of transformation which are frequently programmed on a computer. A proposed operational distinction is that a game becomes a simulation when the specified rules become so detailed as to require a separate calculating staff or a computer.

Operating models relevant to politics

Development. One origin of political gaming and simulation is the war game used to explore military strategy and tactics. The *Kriegsspiel*, or war game, was introduced in the German general staff late in the first quarter of the nineteenth century. Before the beginning of World War II both the Germans and the Japanese—and perhaps, the Russians—are reported (Goldhamer & Speier [1959] 1961, p. 71) to have extended some war games from strictly military exercises to operations which included the representation of political features.

The current development of simulations and games for the study of politics also has been influenced by the use of operating models in the physical sciences (e.g., the study of aerodynamics in wind tunnels), the availability of electronic computers, and the interest in experimental research throughout the social sciences. The extension of the small-group laboratory studies to include the examination of larger social systems has been a vital contribution.

RAND international politics game. The all-man game created at the RAND Corporation in the 1950s was one of the earliest post-World War II games conducted in the United States. As the game is described by Goldhamer and Speier (1959), it involves a scenario, a control team, and a group of players who are divided into teams. To establish the setting for the game, the scenario describes the relevant world features that exist at the outset of play. Each team of national policy makers decides on appropriate policy in response to the circumstances described in the scenario. Action taken by a nation is submitted to the control team, which judges its feasibility and plausibility. If the control team rules that a given policy is realistic, the action is assumed to have occurred. Appropriate teams are notified of the new development. The game evolves as nations respond to the moves of other teams and to the events introduced by the control team (e.g., technological innovations, assassinations, etc.).

In addition to the exercises conducted at RAND, this international politics game has provided the basic format for games at various institutions, including the Center for International Studies at the Massachusetts Institute of Technology (Bloomfield & Whaley 1965). In contrast to the RAND games, most of the university exercises have been conducted with students for instructional purposes. The problems presented by the game scenarios have ranged from historical events to future hypothetical developments. In some games, teams have represented various branches or departments within the same government rather than different nations.

Legislative game. The legislative game was developed by Coleman (1964) to examine collective decision making and bargaining. Although it is an all-man game like the RAND model, the initial conditions of the legislative game are established by a deck of especially designed cards rather than a scenario. Moreover, the game is structured exclusively by rules and the behaviors of players. It does not involve a control team. Six to eleven players assume the role of legislators representing the hypothetical constituencies or regions indicated on randomly assigned cards. These legislators attempt to remain in office by passing issues or bills the voters in their regions favor and by defeating items the voters oppose. In the critical bargaining period which precedes the legislators' vote on each issue, any player may agree to support issues which are not relevant to his constituents in exchange for a vote from another legislator on an issue crucial for his re-election. After the game's eight issues have been passed, defeated, or tabled, the legislators are able to determine whether they will be re-elected for another session.

Public opinion game. The public opinion game was designed by Davison (1961) as an instructional device to demonstrate such factors as the development of cross-pressures, the difference between private and public opinion, and the role of both psychological and social forces. Each participant is given the description of an individual whose role the player assumes. The final public opinion of every hypothetical individual is a weighted composite of (1) his private opinion before the issue is discussed, (2) his religious organization's position, (3) his secondary organization's position, and (4) his primary group's position. Although the public opinion game is an all-man exercise like both the RAND and legislative games, it contains to a greater degree than the other two models quantitative rules that determine the outcome. The players, in various combinations, are able to choose the positions assumed by each of the elements which influence an individual's final public opinion. Once the positions are taken, however, specific rules indicate what values shall be assigned to each opinion component and how they shall be compiled. If the public opinion game incorporated more determinants of opinion (the author acknowledges many simplifications), the resulting complexity of the rules probably would require a calculation staff or computer, thus transforming the game into a simulation according to the present definition.

Inter-Nation Simulation. The man–machine simulation developed by Guetzkow and his associates (1963), Inter-Nation Simulation (INS), is an example of an operating model whose dynamics are so complex that a calculation staff is required. In each of a series of 50-minute to 70-minute periods at least one exchange occurs between the structured calculations and the participants. Participants assume one of several functionally defined roles that are specified for each government. Unlike the RAND game, neither nations nor the positions within them correspond to specific counterparts in world politics.

In each simulated nation the decision makers allocate military, consumer, and "basic" resources in pursuit of objectives of their own choosing. After negotiating with other nations or alliances, the decision makers allocate their resources for trade, aid, blockades, war, research and development, defense preparations, and the generation of new resources. Programmed rules are used by the calculation staff to determine the net gain or loss in various types of resources that have resulted from the decisions. The programmed rules also determine whether the actions of the government have maintained the support of the politically relevant parts of their nation. If these elites, symbolically represented in the calculations, are dissatisfied, a revolution or election may establish a new government. The calculated results are returned to each nation, and a new cycle of interactions and decisions is begun. Versions of the Inter-Nation Simulation have been used in various institutions in the United States as well as by the Institute for Behavioral Science in Tokyo, Japan, and University College of London, England.

Benson and TEMPER simulations. The dynamics of INS are created by a combination of participant activity and quantitative rules. By comparison, the operations of the international simulation constructed by Benson (1961) are completely programmed on a computer. Nine major nations and nine tension areas in contemporary world politics are contained in the program. To begin a simulation exercise, the operator supplies the description of the nine major nations along such dimensions as aggressiveness, alliance membership, foreign bases, trade, and geographical location. From these initial inputs the computer determines the nature of the international system, the extent of each nation's interest in various areas, and other necessary indices.

The operator then announces an action by one of the major nations against one of the tension areas. Actions are described in terms of the degree of national effort required—from diplomatic protest to total war. From this information the computer program determines the response of each of the other nations. The effects of the initial action and resulting counteractions are then computed for each nation and the international system. With these new characteristics, the operator may specify another action.

A much more complex computer simulation of international relations, TEMPER (Technological, Economic, Military and Political Evaluation Routine), is currently being developed for the Joint War Games Agency of the U.S. Joint Chiefs of Staff (1965). Represented in TEMPER is the interaction between political, economic, cultural, and military features of as many as 39 nations, which are selected from 117 nations included in the program.

Crisiscom. Whereas the Benson simulation and TEMPER are computer representations of international systems, Crisiscom (Pool & Kessler 1965) depicts the psychological mechanisms in individuals that affect their information processing. In its present form, the decision makers represented in Crisiscom deal with foreign policy, and the computer inputs are messages that concern the interaction between international actors. The policy makers and the material they handle, however, might be adapted for other levels of political decision making.

The messages fed into the computer are assessed for affect (attitude or feeling of one actor toward another) and salience (the importance attached to an actor or interaction). Each message is given attention, set aside, or forgotten by the simulated decision maker, according to programmed instructions that reflect psychological observations about the rejection and distortion of information. For example, more attention is given a message that does not contradict previous interpretations. This filtering of communications determines what issues are selected by the policy makers for attention and decision. The filtering program not only influences the process for the selection of issues but also gradually alters the decision maker's affect matrix. The affect matrix represents the decision maker's perception of how each actor feels toward all the other actors in the world. In a test of Crisiscom the kaiser and the tsar were simulated. Both were given identical messages about events in the week prior to the beginning of World War I. At the end of the simulated week the differences in their affect matrices and the events to which they were attending reflected quite plausibly differences between the actual historical figures.

Simulmatics election simulation. The Simulmatics Corporation has applied an all-computer simulation to the 1960 and 1964 American presidential elections (Pool et al. 1965). The electorate is represented in the computer by 480 types of voters. Every voter type is identified by a combination of traits that include geographical region, city size, sex, race, socioeconomic status, political party, and religion. For each voter type information is stored on 52 political characteristics, for example, intention to vote, past voting history, and opinions on various political issues. The empirical data for the 52 "issue-clusters" are drawn from national opinion polls conducted during campaigns since 1952. The aggregation of surveys, made possible by the stability of political attitudes in the United States, permits a more detailed representation of voters than is obtained in any single poll. Before the simulation can be applied to a particular election, the operators must decide what characteristics and issues are most salient. Equations or rules of transformation are written to express the impact of the issues on different kinds of voters.

In the 1960 campaign, a simulation whose equations applied cross-pressure principles to the voters' religion and party ranked the nonsouthern states according to the size of the Kennedy vote. The simulation's ordering of the states correlated .82 with the actual Kennedy vote. In 1964, the equations introduced into the computer program primarily represented the effects of three "issue-clusters" (civil rights, nuclear responsibility, and social welfare). This time the percentage of the vote in each state obtained by Johnson was estimated, and the correlation with the actual election was .90.

Other operating models. The described operating models illustrate various constructions and procedures but do not form an exhaustive list of those techniques currently used in political inquiry. For example, the Simulmatics project is not the only computer simulation of political elections. McPhee (1961), Coleman (1965), and Abelson and Bernstein (1963) have constructed simulations representing not only the election outcome but also the process by which voters form opinions during the campaign. The last model differs from the others by representing a local community referendum.

Other games and simulations range from local politics games (Wood 1964) to games of international balance of power (Kaplan et al. 1960); from simulations of disarmament inspection (Singer & Hinomoto 1965) to those of a developing nation (Scott & Lucas 1966). In addition, other operating models not directly concerned with the study of political phenomena are relevant. Abstract games of logic developed by Layman Allen are used in the study of judicial processes at the Yale Law School. Bargaining games, such as those advanced by Thomas Schelling (1961), and the substantial number of business and management simulations (K. Cohen et al. 1964; American Management Association 1961; Thorelli & Graves 1964) have political applications, as do some of the numerous war games (Giffin 1965).

Uses of political gaming and simulation

The development and operation of a game or simulation demand time, effort, and finances. Moreover, operating models are beset by such difficulties as the question of their political validity. Despite these problems, games and simulations have been considered to be useful techniques in research, instruction, and policy formation.

Research. Many of the research attractions of operating models can be summarized in their ability to permit controlled experimentation in politics. In political reality, determination of the effects of a phenomenon can be severely hindered by confounding influences of potentially related events. Through a carefully designed model, an experimenter can isolate a property and its effects by either deleting the competing mechanisms from the model or holding them constant. Not only can a game or simulation be manipulated, but it can also be subjected to situations without the permanent and possibly harmful consequences that a comparable event might create in the real world.

This ability of the researcher to control the model permits replications and increased access. To establish a generalized pattern, repeated observations are required. A simulation or game can be assigned the same set of initial conditions and played over and over again, whereas in its reference system the natural occurrence of a situation might appear infrequently and perhaps only once. Moreover, replications of an operating model may reveal a greater range of alternative responses to a situation than could otherwise be identified.

Increased access to the objects of their research

also leads students of politics to consider operating models. Players in a political game can be continuously observed; a detailed set of their written and verbal communications can be secured; and they can be asked to respond to interviews, questionnaires, or other test instruments at points selected by the researcher. Computer simulations can be instructed to provide a "print-out" at any specified stage of their operation, thus providing a record for analysis. The diversity of means by which data are readily obtained from such models contrasts with the limited access to the crowded lives of actual policy makers and their often sensitive written materials.

In addition to control, replication, and access, gaming and simulation contribute in at least two ways to theory building. First, the construction of a game or a simulation requires the developer to be explicit about the units and relationships that are to exist in his model and presumably in the political reference system it represents. An essential component of the political process which has been ignored in the design of a game is dramatized for the discerning observer. A computer simulation necessitates even more explication of the relationships among the model's components. An incomplete program may terminate abruptly or result in unintelligible output. Thus, in constructing an operating model a relationship between previously unconnected findings may be discovered. Alternatively, a specific gap in knowledge about a political process may be pinpointed, and hypotheses, required by the model, may be advanced to provide an explanation.

A second value for theory building results from the operating, or dynamic, quality of games and simulations. Static statements of relationships can be transformed into processes which respond to change without the restrictions imposed by such alternatives as linear regression models. According to Abelson and Bernstein (1963, p. 94), ". . . the static character of statistical models and their reliance on linear assumptions seem to place an upper limit on their potential usefulness. . . . Computer simulation offers a technique for formulating systemic models, thus promising to meet this need."

It should be recognized that these research assets are not shared equally by all political games and simulations. For example, a computer simulation may not generate the range of potential alternatives to a situation provided by a series of games played with policy experts. On the other hand, control and replication are more easily achieved in a programmed computer simulation than in a political game in which players are given considerable latitude in forming their own patterns of interaction.

Instruction. Simulations involving human participants and games are being used in graduate and undergraduate instruction as well as in secondary school teaching. Several evaluations of these techniques for college and university teaching have been reported in political science (e.g., Alger 1963; Cohen 1962) and in other fields (e.g., Thorelli & Graves 1964, especially pp. 25–31; Conference on Business Games 1963). One of these reports (Alger 1963, pp. 185–186) summarizes the frequently cited values of a simulation or game as a teaching aid: "(1) It increases student interest and motivation; (2) It serves as a laboratory for student application and testing of knowledge; (3) It provides insight into the decision-maker's predicament; (4) It offers a miniature but rich model that facilitates comprehension of world realities."

In contrast to the positive evaluations of users like Alger, a critical assessment of the instructional merits of political gaming and simulation has been offered by Cohen (1962). On the basis of a game he conducted, Cohen questioned (1) whether the game increased motivation among students not already challenged by the course; (2) whether the game stimulated interest in more than a narrow segment of the entire subject matter; (3) whether the game misrepresented or neglected critical features of political reality, thereby distorting the image of political phenomena; and (4) whether a comparable investment by both instructor and students in more conventional modes of learning might have offered a larger increment in education.

Systematic educational research has yet to clarify the differing evaluations of games and simulations for instruction. In one educational experiment with a political simulation (Robinson et al. 1966), undergraduate students in each of three courses were divided into two groups controlled for intelligence, grade-point average, and certain personality characteristics. One group augmented its regular course activities by a weekly discussion of relevant case studies. The other group in each course participated in a continuing simulation for an equal amount of time. Although most measures of student interest did not indicate significant differences between the groups, students evinced more interest in the case studies, while attendance was higher in the simulation group. Similarly, no direct and unmediated difference was found between the two groups on either mastery of facts or principles. To date, the limited educational research on gaming in other fields (e.g., Conference on Business Games 1963) also has produced ambiguous results.

It may be that the more knowledge students have in the subject area of a model or game, the more complex and detailed it must be to provide a satisfactory learning experience. One instructional use of games and simulations that may prove particularly useful with sophisticated students is to involve them in the construction of a model rather than to simply have them assume the role of players.

Policy formation.   In some respects the application of games and simulations as adjuncts to policy formation is only an extension of their use for research and instruction. Policy makers have employed games or man–machine simulations as training aids for governmental officials in the same manner as management-training programs have incorporated business games. In the Center for International Studies game at the Massachusetts Institute of Technology, "questionnaires returned by participants have revealed that responsible officials . . . place a uniformly high value on the special benefits the games provide, particularly in sharpening their perspectives of alternatives that could arise in crisis situations" (Bloomfield & Whaley 1965, p. 866).

Simulation and gaming research performed under government contract or with more or less direct policy implications has concerned such issues as the systemic consequences of the proliferation of nuclear weapons (Brody 1963), the impact on political decision making of situational characteristics associated with crisis (Hermann 1965), and the potential role of an international police force at various stages of disarmament (Bloomfield & Whaley 1965). An application of simulation research for political campaigns is reflected in the election simulations conducted under contract to the National Committee of the Democratic Party by the Simulmatics Corporation (Pool et al. 1965). One of the major areas of U.S. government-related activity involving operating models is war gaming and its extension to cover relevant political aspects of national security. Public material on these activities, however, is limited.

Validity—evaluating correspondence

How are the elements of political reality estimated in a way to make their transmission to an operating model accurate? What are the implications of representing political properties by deterministic or stochastic processes? What are the relative advantages of games compared to simula-

tions? If humans act as participants, how can the motivations of actual political leaders be reproduced? Underlying most of these questions is the problem of validity. Operating models are by definition representations of an existing or potential system or process. They are constructed for the information they can provide about a selected reference system. If in comparison to the performance of the reference system the simulation or game produces spurious results, the model is of little worth. An estimation of the "goodness of fit" or extent of similarity between an operating model and external criteria is, therefore, the central problem of validity.

Different types of criteria or standards can be used for evaluating the correspondence between operating models and their reference system (Hermann 1967). To date, however, the only effort at validity that has been applied to most simulations and games uses no specific criterion of comparison. In this approach, called face validity, the model's realism is based on impressionistic judgments of observers or participants. If participants are used, face validity may concern estimates of their motivation or involvement. Unfortunately, participants can be motivated in a game that incorrectly represents the designated political environment. Furthermore, if the operating model involves the substitution of one property for another, some feature may give the appearance of being quite unreal even though it replicates the performance of the real-world property it replaces.

When specific criteria for correspondence are established, model validity may be determined by comparing events, variables and parameters, or hypotheses. In the first case, the product or outcome of the operating model is compared to the actual events it was intended to replicate (e.g., the correlation between the state-by-state election returns in the 1960 U.S. presidential campaign and those projected by the Simulmatics Corporation's model; Pool et al. 1965). The second approach compares the variables and parameters that constitute a model with the real-world properties they are intended to represent (e.g., a current study uses factor analysis of the core variables of a simulation and the quantifiable real-world indices for which the simulation variables have been substituted; validity judgments are made from comparisons of the factors and of the variables that load on each factor; see Chadwick 1966). The third approach to validity compares the statistical results of a number of hypotheses tested in an operating model with comparable tests of the same hypotheses conducted with data drawn from the reference system (e.g., Zinnes 1966). Confirmation of a variety of hypotheses in both the operating model and in the actual political system increases confidence in the model's validity.

With the exception of face validity, each of the described validity approaches requires that values for specified properties be determined with precision for both the operating model and its reference system. Not only is this procedure often difficult, but in some instances it may not be appropriate. Imagine a political game designed as an aid for policy making whose purpose is to display as many different alternative outcomes to an initial situation as possible. The fact that the subsequent course of actual events leads to only one of those outcomes —or perhaps, none of them—may not reduce the utility of the exercise for increasing the number of alternatives considered by the policy makers. In other instances, an operating model may be consistent with one of the described validity criteria but remain unsatisfactory for some purposes. One illustration is an election simulation in which the winning political party is decided by a stochastic process. With correct probability settings and frequent elections in the model, the simulation's distribution of party victories would closely approximate election outcomes in the reference system. Despite the apparent event validity, the simulation would be undesirable for instruction in election politics. The naive participant might conclude from his experience with the collapsed campaign process that party victory is determined exclusively by chance.

In summary, a variety of validity strategies exist. The appropriate strategy varies from model to model and with the purpose for which the game or simulation was designed. Each operating model must be independently validated. At any given time the confidence in the accuracy with which a game or simulation represents its intended political reference system will be a matter of degree. Evaluation of these operating models for the study of politics will depend upon the development and application of procedures for measuring the extent to which each model's purposes are fulfilled.

CHARLES F. HERMANN

[See also Alger 1968; Brodie 1968.]

BIBLIOGRAPHY

ABELSON, ROBERT P.; and BERNSTEIN, ALEX 1963 A Computer Simulation Model of Community Referendum Controversies. *Public Opinion Quarterly* 27:93–122.

ALGER, CHADWICK F. 1963 Use of the Inter-Nation Simulation in Undergraduate Teaching. Pages 150–189 in Harold Guetzkow et al., *Simulation in International Relations: Developments for Research and Teaching*. Englewood Cliffs, N.J.: Prentice-Hall.

►ALGER, CHADWICK F. 1968 International Relations: I. The Field. Volume 8, pages 61–69 in *International Encyclopedia of the Social Sciences*. Edited by David L. Sills. New York: Macmillan and Free Press.

AMERICAN MANAGEMENT ASSOCIATION 1961 *Simulation and Gaming: A Symposium.* Report No. 55. New York: The Association.

BENSON, OLIVER 1961 A Simple Diplomatic Game. Pages 504–511 in James N. Rosenau (editor), *International Politics and Foreign Policy: A Reader in Research and Theory.* New York: Free Press.

BLOOMFIELD, L. P.; and WHALEY, B. 1965 The Political–Military Exercise: A Progress Report. *Orbis* 8:854–870.

►BRODIE, BERNARD 1968 Strategy. Volume 15, pages 281–288 in *International Encyclopedia of the Social Sciences*. Edited by David L. Sills. New York: Macmillan and Free Press.

BRODY, RICHARD A. 1963 Some Systemic Effects of the Spread of Nuclear Weapons Technology: A Study Through Simulation of a Multi-nuclear Future. *Journal of Conflict Resolution* 7:663–753.

CHADWICK, R. W. 1966 Developments in a Partial Theory of International Behavior: A Test and Extension of Inter-Nation Simulation Theory. Ph.D. dissertation, Northwestern Univ.

COHEN, BERNARD C. 1962 Political Gaming in the Classroom. *Journal of Politics* 24:367–381.

COHEN, KALMAN J. et al. 1964 *The Carnegie Tech Management Game: An Experiment in Business Education.* Homewood, Ill.: Irwin.

COLEMAN, JAMES S. 1964 Collective Decisions. *Sociological Inquiry* 34:166–181.

COLEMAN, JAMES S. 1965 The Use of Electronic Computers in the Study of Social Organization. *Archives européennes de sociologie* 6:89–107.

CONFERENCE ON BUSINESS GAMES, TULANE UNIVERSITY, *1961* 1963 *Proceedings.* Edited by William R. Dill et al. New Orleans, La.: Tulane Univ., School of Business Administration.

DAVISON, W. P. 1961 A Public Opinion Game. *Public Opinion Quarterly* 25:210–220.

DAWSON, RICHARD E. 1962 Simulation in the Social Sciences. Pages 1–15 in Harold Guetzkow (editor), *Simulation in Social Science: Readings.* Englewood Cliffs, N.J.: Prentice-Hall.

GIFFIN, SIDNEY F. 1965 *The Crisis Game: Simulating International Conflict.* Garden City, N.Y.: Doubleday.

GOLDHAMER, HERBERT; and SPEIER, HANS (1959) 1961 Some Observations on Political Gaming. Pages 498–503 in James N. Rosenau (editor), *International Politics and Foreign Policy: A Reader in Research and Theory.* New York: Free Press.

GUETZKOW, HAROLD et al. 1963 *Simulation in International Relations: Developments for Research and Teaching.* Englewood Cliffs, N.J.: Prentice-Hall.

HERMANN, C. F. 1965 *Crises in Foreign Policy Making: A Simulation of International Politics.* China Lake, Calif.: U.S. Naval Ordnance Test Station, Contract N123 (60S30) 32779A.

HERMANN, C. F. 1967 Validation Problems in Games and Simulations With Special Reference to Models of International Politics. *Behavioral Science* 12:216–233.

KAPLAN, MORTON A.; BURNS, ARTHUR L.; and QUANDT, RICHARD E. 1960 Theoretical Analysis of the "Balance of Power." *Behavioral Science* 5:240–252.

McPHEE, WILLIAM N. (1961) 1963 Note on a Campaign Simulator. Pages 169–183 in William N. McPhee, *Formal Theories of Mass Behavior.* New York: Free Press.

POOL, ITHIEL DE SOLA; and KESSLER, A. 1965 The Kaiser, the Tsar, and the Computer: Information Processing in a Crisis. *American Behavioral Scientist* 8, no. 9:31–38.

POOL, ITHIEL DE SOLA; ABELSON, ROBERT P.; and POPKIN, SAMUEL L. 1965 *Candidates, Issues, and Strategies: A Computer Simulation of the 1960 and 1964 Presidential Elections.* Rev. ed. Cambridge, Mass.: M.I.T. Press.

RAPOPORT, ANATOL 1964 *Strategy and Conscience.* New York: Harper.

ROBINSON, JAMES A. et al. 1966 Teaching With Inter-Nation Simulation and Case Studies. *American Political Science Review* 60:53–65.

SCHELLING, T. C. 1961 Experimental Games and Bargaining Theory. *World Politics* 14:47–68.

SCOTT, ANDREW M.; and LUCAS, WILLIAM A. 1966 *Simulation and National Development.* New York: Wiley.

SHUBIK, MARTIN 1964 Game Theory and the Study of Social Behavior: An Introductory Exposition. Pages 3–77 in Martin Shubik (editor), *Game Theory and Related Approaches to Social Behavior: Selections.* New York: Wiley.

SINGER, J. DAVID; and HINOMOTO, HIROHIDE 1965 Inspecting for Weapons Production: A Modest Computer Simulation. *Journal of Peace Research* 1:18–38.

THORELLI, HANS B.; and GRAVES, ROBERT L. 1964 *International Operations Simulation: With Comments on Design and Use of Management Games.* New York: Free Press.

WOOD, R. C. 1964 Smith–Massachusetts Institute of Technology Political Game, Documents 1–5. Unpublished manuscript, Massachusetts Institute of Technology, Department of Political Science.

U.S. JOINT CHIEFS OF STAFF, JOINT WAR GAMES AGENCY 1965 *TEMPER.* Volume 1: Orientation Manual. DDC AD 470 375L. Washington: The Agency.

ZINNES, DINA A. 1966 A Comparison of Hostile Behavior of Decision-makers in Simulate and Historical Data. *World Politics* 18:474–502.

SIMULTANEOUS EQUATION ESTIMATION

The distinction between partial and general equilibrium analysis in economic theory is well grounded (see Arrow 1968). Early work in econometrics paid inadequate attention to this distinction and overlooked for many years the possibilities of improving statistical estimates of individual economic relationships by embedding them in models of the economy as a whole [*see* ECONOMETRIC MODELS, AGGREGATE]. The earliest studies in econometrics were concerned with estimating parameters of demand functions, supply functions, production functions, cost functions, and similar tools of economic

analysis. The principal statistical procedure used was to estimate the α's in the relation

$$y_t = \sum_{i=1}^{n} \alpha_i x_{it} + u_t, \qquad t = 1, 2, \cdots, T,$$

using the criterion that $\sum_{t=1}^{T} u_t^2$ be minimized. This is the principle of "least squares" applied to a single equation in which y_t is chosen as the dependent variable and $x_{1t}, \cdots, x_{nt}$ are chosen as the independent variables. The criterion is the minimization of the sum of squared "disturbances" (u_t) which are assumed to be unobserved random errors. The estimation of the unknown parameters α_i is based on the sample of T observations of y_t and $x_{1t}, \cdots, x_{nt}$. This is the usual statistical model and estimation procedure that is used in controlled experimental situations where the set of independent variables consists of selected, fixed variates for the experimental readings on y_t, the dependent variable. [*See* LINEAR HYPOTHESES, *article on* REGRESSION.]

However, economics, like most other social sciences, is largely a nonexperimental science, and it is generally not possible to control the values of $x_{1t}, \cdots, x_{nt}$. The values of the independent variables, like those of the dependent variable, are produced from the general outcome of economic life, and the econometrician is faced with the problem of making statistical inferences from nonexperimental data. This is the basic reason for the use of simultaneous equation methods of estimation in econometrics. In some situations $x_{1t}, \cdots, x_{nt}$ may not be controlled variates, but they may have a one-way causal influence on y_t. The main point is that least squares yields desirable results only if u_t is independent of $x_{1t}, \cdots, x_{nt}$, that is, if $E(u_t x_{it}) = 0$ for all i and t.

Properties of estimators. If the x_{it} are fixed variates, estimators of α_i obtained by minimizing $\sum u_t^2$ are *best linear unbiased* estimators. They are linear estimators because, as shown below, they are linear functions of y_t. An estimator, $\hat{\alpha}_i$, of α_i is called unbiased if

$$E\hat{\alpha}_i = \alpha_i,$$

i.e., if the expected value of the estimator equals the true value. An estimator is best if among all unbiased estimators it has the least variance, i.e., if

$$E(\hat{\alpha}_i - \alpha_i)^2 \leqslant E(\tilde{\alpha}_i - \alpha_i)^2,$$

where $\tilde{\alpha}_i$ is any other unbiased estimator. Clearly, the properties of being unbiased and best are desirable ones. These properties are defined without reference to sample size. Two related but weaker properties, which are defined for large samples, are *consistency* and *efficiency.*

An estimator, $\hat{\alpha}_i$, is consistent if $\text{plim } \hat{\alpha}_i = \alpha_i$, that is, if

$$\lim_{T \to \infty} P(|\hat{\alpha}_i - \alpha_i| < \epsilon) = 1, \qquad \text{for any } \epsilon > 0.$$

This states that the probability that $\hat{\alpha}_i$ deviates from α_i by an amount less than any arbitrarily small ϵ tends to unity as the sample size T tends to infinity.

Consider now the class of all consistent estimators that are normally distributed as $T \to \infty$. An *efficient* estimator of α_i is a consistent estimator whose asymptotic normal distribution has a smaller variance than any other member of this class. [*See* ESTIMATION.]

Inconsistency of least squares. The choice of estimators, $\hat{\alpha}_i$, such that $\sum_{t=1}^{T}(y_t - \sum_{i=1}^{n} \alpha_i x_{it})^2$ is minimized is formally equivalent to the empirical implementation of the condition that $E(u_t x_{it}) = 0$, since the first-order condition for a minimum is

$$\sum_{t=1}^{T}[(y_t - \sum_{i=1}^{n} \alpha_i x_{it})x_{it}] = 0.$$

On the one hand, the α_i are estimated so as to minimize the residual sum of squares. On the other hand, they are estimated so that the residuals are uncorrelated with $x_{1t}, \cdots, x_{nt}$. The possible inconsistency of this method is clearly revealed by the latter criterion, for if it is assumed that the u_t are independent of $x_{1t}, \cdots, x_{nt}$ when they actually are not, the estimators will be inconsistent. This is shown by the formula

$$\text{plim } \hat{\alpha}_i = \alpha_i + \text{plim } \frac{1}{|M|} \sum_{j=1}^{n} m_{uj} M_{ji},$$

where M is the moment matrix whose typical element is $\sum_t x_{it} x_{jt}$; $|M|$ is the determinant of M; m_{uj} is $\sum_t u_t x_{jt}$; and M_{ji} is the j, i cofactor of M. The inconsistency in the estimator is due to the nonvanishing probability limit of m_{uj}. In a nonexperimental sample of data, such as that observed as the joint outcome of the uncontrolled simultaneous economic process, we would expect many or all the x_{it} in a problem to be dependent on u_t.

Identifying restrictions. Since economic models consist of a set of simultaneous equations generating nonexperimental data, the equations of the model must be *identified* prior to statistical estimation. Unless some restrictions are imposed on specific relationships in a linear system of simultaneous equations, every equation may look alike to the statistician faced with the job of estimating the unknown coefficients. The economist must place a priori restrictions, in advance of statistical

estimation, on each of the equations in order to identify them. These restrictions may specify that certain coefficients are known in advance—especially that they are zero, for this is equivalent to excluding the associated variable from an economic relation. Other restrictions may specify linear relationships between the different coefficients.

Consider the generalization of a single equation,

$$y_t = \sum_{i=1}^{m} \alpha_i z_{it} + u_t,$$

where $E(z_{it} u_t) = 0$ for all i and t, to a whole system,

$$\sum_{j=1}^{n} \beta_{ij} y_{jt} + \sum_{k=1}^{m} \gamma_{ik} z_{kt} = u_{it}, \qquad i = 1, 2, \cdots, n,$$

where $E(z_{kt} u_{it}) = 0$ for all k, i, and t. Every variable enters every equation linearly without restriction, and the statistician has no way of distinguishing one relation from another. Zero restrictions, if imposed, would have the form $\beta_{rs} = 0$ or $\gamma_{pq} = 0$, for some r, s, p, or q. In many equations, we may be interested in specifying that sums or differences of variables are economically relevant combinations, i.e., that $\beta_{rs} = \beta_{ru}$ or that $\beta_{rv} = -\gamma_{rw}$ or, more generally, that

$$\sum_{s=1}^{n_r} w_s \beta_{rs} + \sum_{s=1}^{m_r} v_s \gamma_{rs} = 0.$$

The last restriction implies that a homogeneous linear combination of parameters in the rth equation is specified to hold on a priori grounds. The weights w_s and v_s are known in advance.

If general linear restrictions are imposed on the equations of a linear system, we may state the following rule: an equation in a linear system is identified if it is not possible to reproduce by linear combination of some or all of the equations in the system an equation having the same statistical form as the equation being estimated.

If the restrictions are of the zero type, a necessary condition for identification of an equation in a linear system of n equations is that the number of variables excluded from that equation be greater than or equal to $n - 1$. A necessary and sufficient condition is that it is possible to form at least one nonvanishing determinant of order $n - 1$ out of those coefficients, properly arranged, with which the excluded variables appear in the $n - 1$ other equations (Koopmans et al. 1950).

Criteria for identifiability are stated here for linear equation systems. A more general treatment in nonlinear systems is given by Fisher (1966). [*See* STATISTICAL IDENTIFIABILITY.]

Alternative estimation methods

Assuming that we are dealing with an identified system, let us turn to the problems of estimation. In the system of equations above, the y_{jt} are *endogenous* or *dependent* variables and are equal in number to the number of equations in the system, n. The z_{kt} are *exogenous* variables and are assumed to be *independent* of the disturbances, u_{it}.

In one of the basic early papers in simultaneous equation estimation (Mann & Wald 1943), it was shown that large-sample theory would, under fairly general conditions, permit lagged values of endogenous variables to be treated like purely exogenous variables as far as consistency in estimation is concerned. Exogenous and lagged endogenous variables are called *predetermined* variables.

Early econometric studies, for example, that of Tinbergen (1939), were concerned with the estimation of a number of individual relationships in which the possible dependence between variables and disturbances was ignored. These studies stimulated Haavelmo (1943) to analyze the consistency problem, for he noted that the Tinbergen model contained many single-equation least squares estimates of equations that were interrelated in the system Tinbergen was constructing, which was intended to be a theoretical framework describing the economy that generated the observations used.

The lack of independence between disturbances and variables can readily be demonstrated. Consider the two-equation system

$$\beta_{11}y_{1t} + \beta_{12}y_{2t} + \sum_{k=1}^{m}\gamma_{1k}z_{kt} = u_{1t}$$

$$\beta_{21}y_{1t} + \beta_{22}y_{2t} + \sum_{k=1}^{m}\gamma_{2k}z_{kt} = u_{2t}.$$

The z_{kt} are by assumption independent of u_{1t} and u_{2t}. Some of the γ's are specified to be zero or are otherwise restricted so that the two equations are identified. Suppose we wish to estimate the first equation. To apply least squares to this equation, we would have to select either y_1 or y_2 as the dependent variable. Suppose we select y_1 and set β_{11} equal to unity. We would then compute the least squares regression of y_1 on y_2 and the z_k according to the relation

$$y_{1t} = -\beta_{12}y_{2t} - \sum_{k=1}^{m}\gamma_{1k}z_{kt} + u_{1t},$$

which incorporates all the identifying restrictions on the γ's.

For this procedure to yield consistent estimators, y_{2t} must be independent of u_{1t}. The question is whether the existence of the second equation has any bearing on the independence of y_{2t} and u_{1t}. Multiplying the second equation by u_{1t} and forming expectations, we have

$$E(y_{2t}u_{1t}) = -(\beta_{21}/\beta_{22})E(y_{1t}u_{1t}) + (1/\beta_{22})E(u_{1t}u_{2t}).$$

From the first equation (with $\beta_{11} = 1$), we have

$$E(y_{1t}u_{1t}) = -\beta_{12}E(y_{2t}u_{1t}) + E(u_{1t}^2).$$

Combining these two expressions, we obtain

$$E(y_{2t}u_{1t}) = \frac{E(u_{1t}u_{2t}) - \beta_{21}E(u_{1t}^2)}{\beta_{22} - \beta_{21}\beta_{12}}.$$

In general, this expression does not vanish, and we find that y_{2t} and u_{1t} are not independent.

The maximum likelihood method. The maximum likelihood method plays a normative role in the estimation of economic relationships, much like that played by perfect competition in economic theory. This method provides consistent and efficient estimators under fairly general conditions. It rests on specific assumptions, and it may be hard to realize all these assumptions in practice or, indeed, to make all the difficult calculations required for solution of the estimation equations.

For the single-equation model, the maximum likelihood method is immediately seen to be equivalent to ordinary least squares estimation for normally distributed disturbances. Let us suppose that $u_1, \cdots, u_T$ are T independent, normally distributed variables. The T-element sample has the probability density function

$$p(u_1, \cdots, u_T) = \left(\frac{1}{\sqrt{2\pi}\sigma}\right)^T \exp\left(-\frac{1}{2\sigma^2}\sum_{t=1}^{T}u_t^2\right).$$

By substitution we can transform this joint density of $u_1, \cdots, u_T$ into a joint density of $y_1, \cdots, y_T$, given $x_{11}, \cdots, x_{n1}, x_{12}, \cdots, x_{n2}, \cdots, x_{nT}$, namely,

$$p(y_1, \cdots, y_T | x_{11}, \cdots, x_{nT})$$
$$= \left(\frac{1}{\sqrt{2\pi}\sigma}\right)^T \exp\left[-\frac{1}{2\sigma^2}\sum_{t=1}^{T}\left(y_t - \sum_{i=1}^{n}\alpha_i x_{it}\right)^2\right].$$

This function will be denoted as L, the likelihood function of the sample, and is seen to depend on the unknown parameters $\alpha_1, \cdots, \alpha_n$ and σ. We maximize this function by imposing the following conditions:

$$\frac{\partial \log L}{\partial \alpha_i} = -\frac{1}{\sigma^2}\sum_{t=1}^{T}\left[\left(y_t - \sum_{j=1}^{n}\alpha_j x_{jt}\right)x_{it}\right] = 0,$$
$$i = 1, \cdots, n,$$

$$\frac{\partial \log L}{\partial \sigma} = -\frac{T}{\sigma} + \frac{1}{\sigma^3}\sum_{t=1}^{T}\left(y_t - \sum_{j=1}^{n}\alpha_j x_{jt}\right)^2 = 0.$$

These are recognized as the "normal" equations of single-equation least squares theory and the estimation equation for the residual variance—apart from adjustment for degrees of freedom used in estimating σ^2.

In a system of simultaneous equations, we wish to estimate the parameters in

$$\sum_{j=1}^{n}\beta_{ij}y_{jt} + \sum_{k=1}^{m}\gamma_{ik}z_{kt} = u_{it}, \qquad i = 1, \cdots, n.$$

Here we have n linear simultaneous equations in n endogenous and m exogenous variables. The parameters to be estimated are the elements of the $n \times n$ coefficient matrix

$$\mathbf{B} = (\beta_{ij}),$$

the $n \times m$ coefficient matrix

$$\mathbf{\Gamma} = (\gamma_{ik}),$$

and the $n \times n$ variance–covariance matrix

$$\mathbf{\Sigma} = (\sigma_{ij}).$$

The variances and covariances are defined by

$$\sigma_{ij} = E(u_i u_j).$$

A rule of normalization is applied for each equation,

$$\boldsymbol{\beta}_i \mathbf{\Sigma} \boldsymbol{\beta}_i' = 1.$$

In practice, one element of $\boldsymbol{\beta}_i = (\beta_{i1}, \cdots, \beta_{in})$ in each equation is singled out and assigned a value of unity.

The likelihood function for the whole system is

$$L = \mathrm{mod}\,|\mathbf{B}|^T\left(\frac{1}{\sqrt{2\pi}}\right)^{Tn}|\mathbf{\Sigma}|^{-T/2} \times$$

$$\exp\left\{-\tfrac{1}{2}\sum_{t=1}^{T}[(\mathbf{B}\mathbf{y}_t' + \mathbf{\Gamma}\mathbf{z}_t')'\mathbf{\Sigma}^{-1}(\mathbf{B}\mathbf{y}_t' + \mathbf{\Gamma}\mathbf{z}_t')]\right\},$$

where $\mathbf{y}_t = (y_{1t}, \cdots, y_{nt})$, $\mathbf{z}_t = (z_{1t}, \cdots, z_{mt})$, $|\mathbf{\Sigma}|$ is the determinant of $\mathbf{\Sigma}$, and $\mathrm{mod}\,|\mathbf{B}|$ is the absolute value of the determinant of $\mathbf{B}$ (Koopmans et al. 1950). The matrix $\mathbf{B}$ enters this expression as the Jacobian of the transformation from the variables $u_{1t}, \cdots, u_{nt}$ to $y_{1t}, \cdots, y_{nt}$. The problem of maximum likelihood estimation is to maximize L or $\log L$ with respect to the elements of $\mathbf{B}$, $\mathbf{\Gamma}$, and $\mathbf{\Sigma}$. This is especially difficult compared with the similar problem for single equations shown above, because $\log L$ is highly nonlinear in the unknown parameters, a difficult source of nonlinearity coming from the Jacobian expression $\mathrm{mod}\,|\mathbf{B}|$.

Maximizing $\log L$ with respect to $\mathbf{\Sigma}^{-1}$, we obtain the maximum likelihood estimator of $\mathbf{\Sigma}$, which is

$$\hat{\mathbf{\Sigma}} = (\mathbf{B}\ \mathbf{\Gamma})\mathbf{M}(\mathbf{B}\ \mathbf{\Gamma})',$$

where $\mathbf{M}$ is the moment matrix of the observations, i.e.,

$$\mathbf{M} = \frac{1}{T}\begin{bmatrix} y_{11} & \cdots & y_{1T} \\ \vdots & & \vdots \\ y_{n1} & \cdots & y_{nT} \\ \hline z_{11} & \cdots & z_{1T} \\ \vdots & & \vdots \\ z_{m1} & \cdots & z_{mT} \end{bmatrix}\begin{bmatrix} y_{11} & \cdots & y_{n1} & z_{11} & \cdots & z_{m1} \\ \vdots & & \vdots & \vdots & & \vdots \\ y_{1T} & \cdots & y_{nT} & z_{1T} & \cdots & z_{mT} \end{bmatrix}$$

$$= \begin{bmatrix} \mathbf{M}_{yy} & \mathbf{M}_{yz} \\ \mathbf{M}_{zy} & \mathbf{M}_{zz} \end{bmatrix}.$$

Substitution of $\hat{\mathbf{\Sigma}}$ into the likelihood function yields the *concentrated form* of the likelihood function

$$\log L = \mathrm{Const.} + T\log\mathrm{mod}\,|\mathbf{B}| - (T/2)\log|\hat{\mathbf{\Sigma}}|,$$

where Const. is a constant. Hence, we seek estimators of $\mathbf{B}$ and $\mathbf{\Gamma}$ that maximize

$$\log\mathrm{mod}\,|\mathbf{B}| - \tfrac{1}{2}\log|\hat{\mathbf{\Sigma}}|$$
$$= \log\mathrm{mod}\,|\mathbf{B}| - \tfrac{1}{2}\log|(\mathbf{B}\ \mathbf{\Gamma})\mathbf{M}(\mathbf{B}\ \mathbf{\Gamma})'|.$$

In the single-equation case we minimize the one-element variance expression, written as a function of the α_i. In the simultaneous equation case, we maximize $\log\mathrm{mod}\,|\mathbf{B}| - \tfrac{1}{2}\log|\hat{\mathbf{\Sigma}}|$, but this can be shown to be equivalent (Chow 1964) to minimization of $|\hat{\mathbf{\Sigma}}|$, subject to the normalization rule

$$|\mathbf{B}\mathbf{M}_{yy}\mathbf{B}'| = C,$$

where C is a constant. This normalization is direction normalization, and as long as it is taken into account, scale normalization (such as $\beta_{ii} = 1$, cited previously) is arbitrary. Viewed in this way, the method of maximum likelihood applied to a system of equations appears to be a natural generalization of the method of maximum likelihood applied to a single equation, in which case we minimize σ^2 subject to a direction-normalization rule.

Recursive systems. The concentrated form of the likelihood function shows clearly that a new element is introduced into the estimation process, through the presence of the Jacobian determinant, which makes calculations of the maximizing values of $\mathbf{B}$ and $\mathbf{\Gamma}$ highly nonlinear. It is therefore worthwhile to search for special situations in which estimation methods simplify at least to the point of being based on linear calculations.

It is evident that the concentrated form of the likelihood function would lend itself to simpler methods of estimating $\mathbf{B}$ and $\mathbf{\Gamma}$ if $|\mathbf{B}|$ were a known constant. This would be the case if $\mathbf{B}$ were triangular, for then, by a scale normalization, we would have $\beta_{ii} = 1$ and $|\mathbf{B}| = 1$. If $\mathbf{B}$ is triangular, the system of equations is called a recursive system. We then simply minimize $|\hat{\mathbf{\Sigma}}|$ with respect to the unknown coefficients; this can be looked upon as a generalized variance minimization, an obvious analogue of least squares applied to a single equation.

If, in addition, it can be assumed that $\mathbf{\Sigma}$ is diagonal, maximum likelihood estimators become a series of successive single-equation least squares estimators. Since the matrix $\mathbf{B}$ is assumed to be triangular, there must be an equation with only one unlagged endogenous variable. This variable (with unit coefficient) is to be regressed on all the predetermined variables in that equation. Next, there will be an equation with one new endogenous variable. This variable is regressed on the preceding endogenous variable and all the predetermined variables in that equation. In the third equation, another new endogenous variable is introduced. It is regressed on the two preceding endogenous variables and all the predetermined variables in that equation, and so on.

If $\mathbf{\Sigma}$ is not diagonal, a statistically consistent procedure would be to use values of the endogenous variables computed from preceding equations in the triangular array instead of using their actual values. Suppose one equation in the system specifies y_1 as a function of certain z's. We would regress y_1 on these z's and then compute values of y_1 from the relation

$$\hat{y}_{1t} = \sum_{k=1}^{m} \hat{\gamma}_{1k} z_{kt},$$

where the $\hat{\gamma}_{1k}$ are the least squares regression estimators of the γ_{1k}. (Some of the γ_{1k} are zero, as a result of the identifying restrictions imposed prior to computing the regression.) Suppose a second equation in the system specifies y_2 as a function of y_1 and certain z's. Our next step would be to regress y_2 on $\hat{y}_1$ and the included z's and then compute values of y_2 from the relation

$$\hat{y}_{2t} = \hat{\beta}_{21} \hat{y}_{1t} + \sum_{k=1}^{m} \hat{\gamma}_{2k} z_{kt}.$$

The procedure would be continued until all n equations are estimated.

Methods of dealing with recursive systems have been studied extensively by Wold, and a summary appears in Strotz and Wold (1960). A recursive system without a diagonal $\mathbf{\Sigma}$-matrix is found in Barger and Klein (1954). One of the most familiar types of recursive systems studied in econometrics is the cobweb model of demand and supply for agricultural products [see BUSINESS CYCLES: MATHEMATICAL MODELS].

Limited-information maximum likelihood. Another maximum likelihood approach that is widely used is the limited-information maximum likelihood method. It does not hinge on a specific formulation of the model, as do methods for recursive systems; it is a simplified method because it neglects information. As we have seen, identifying restrictions for an equation takes the form of specifying zero values for some parameters or of imposing certain linear relations on some parameters. The term "limited information" refers to the fact that only the restrictions relating to the particular equation (or subset of equations) being estimated are used. Restrictions on other equations in the system are ignored when a particular equation is being estimated.

Let us again consider the linear system

$$\mathbf{B}\mathbf{y}_t' + \mathbf{\Gamma}\mathbf{z}_t' = \mathbf{u}_t'.$$

These equations make up the *structural form* of the system and are referred to as structural equations. We denote the *reduced form* of this system by

$$\mathbf{y}_t' = -\,\mathbf{B}^{-1}\mathbf{\Gamma}\mathbf{z}_t' + \mathbf{B}^{-1}\mathbf{u}_t'$$

or

$$\mathbf{y}_t' = \mathbf{\Pi}\mathbf{z}_t' + \mathbf{v}_t'.$$

From the reduced form equations select a subset corresponding to the n_1 endogenous variables in a particular structural equation, say equation i, which is

$$\sum_{j=1}^{n_1} \beta_{ij} y_{jt} + \sum_{k=1}^{m_1} \gamma_{ik} z_{kt} = u_{it}.$$

The summation limit m_1 indicates the number of predetermined variables included in this equation; we have excluded all zero elements in γ_i and indexed the z's accordingly. Form the joint distribution of $v_{1t}, \cdots, v_{n_1 t}$ over the sample observations and maximize it with respect to the unknown parameters in the ith structural equation, subject to the restrictions on this equation alone. The restrictions usually take the form

$$0 = \sum_{j=1}^{n_1} \beta_{ij} \pi_{jk}, \qquad k = m_1 + 1, \cdots, m,$$

where there are m predetermined variables in the whole system; that is, the γ_{ik}, $k = m_1 + 1, \cdots, m$, are specified to be zero. The estimated coefficients, $\hat{\beta}_{ij}$, $\hat{\gamma}_{ik}$, and $\hat{\sigma}_1^2$, obtained from this restricted likelihood maximization are the limited-information estimators. Methods of obtaining these estimators and a study of their properties are given in Anderson and Rubin (1949).

Linear regression calculations are all that are needed in this type of estimation, save for the extraction of a characteristic root of a matrix with dimensionality $n_1 \times n_1$. A quickly convergent series of iterations involving matrix multiplication leads to the computation of this root and associated vector. The vector obtained, properly normalized by making one coefficient unity, provides estimates of the β_{ij}. The estimates of the γ_{ik} are obtained from

$$\hat{\gamma}_{ik} = -\sum_{j=1}^{n_1} \hat{\beta}_{ij} \hat{\pi}_{jk}, \qquad k = 1, \cdots, m_1,$$

where the $\hat{\pi}_{jk}$ are least squares regression coefficients from the reduced form equations.

It is significant that both full-information and limited-information maximum likelihood estimators are *essentially* unchanged no matter which variable is selected to have a unit coefficient in each equation. That is to say, if we divide through an estimated equation by the coefficient of any endogenous variable, we get a set of coefficients that would have been obtained by applying the estimation methods under the specification that the same variable have the unit coefficient. Full-information and limited-information maximum likelihood estimators are invariant under this type of scale normalization. Other estimators are not.

Two-stage least squares. The classical method of least squares multiple regression applied to a single equation that is part of a larger simultaneous system is inconsistent by virtue of the fact that some of the "explanatory" variables in the regression (the variables with unknown coefficients) may not be independent of the error variable. If we can "purify" such variables to make them independent of the error terms, we can apply ordinary least squares methods to the transformed variables. The method of two-stage least squares does this for us.

Let us return to the equation estimated above by limited information. Choose y_1, say, as the dependent variable, that is, set β_{i1} equal to unity. In place of $y_{2t}, \cdots, y_{n_1 t}$, we shall use

$$\hat{y}_{jt} = \sum_{k=1}^{m} \hat{\pi}_{jk} z_{kt}, \qquad j = 2, \cdots, n_1,$$

as explanatory variables. The $\hat{y}_{jt}$ are *computed* values from the least squares regressions of y_j on

all the z_k in the system ($k = 1, \cdots, m$). The coefficients $\hat{\pi}_{jk}$ are the computed regression coefficients. The regression of y_1 on $\hat{y}_2, \cdots, \hat{y}_{n_1}, z_1, \cdots, z_{m_1}$ provides a two-stage least squares estimator of the single equation. All the equations of a system may be estimated in this way. This can be seen to be a generalization, to systems with nontriangular Jacobians, of the method suggested previously for recursive models in which the variance–covariance matrix of disturbances is not diagonal.

We may write the "normal" equations for these least squares estimators as

$$\begin{bmatrix} \sum\limits_{t=1}^{T} y_{1t}\hat{\boldsymbol{y}}_t' \\[1em] \sum\limits_{t=1}^{T} y_{1t}\boldsymbol{z}_t^{*'} \end{bmatrix} = \begin{bmatrix} \boldsymbol{M}_{\hat{y}\hat{y}} & \boldsymbol{M}_{\hat{y}z^*} \\[0.5em] \boldsymbol{M}_{z^*\hat{y}} & \boldsymbol{M}_{z^*z^*} \end{bmatrix} \begin{bmatrix} -\boldsymbol{b}' \\[0.5em] -\boldsymbol{c}' \end{bmatrix}.$$

In this notation $\hat{\boldsymbol{y}}_t$ is the vector of computed values $(\hat{y}_{2t}, \cdots, \hat{y}_{n_1 t})$; $\boldsymbol{z}_t^*$ is the vector $(z_{1t}, \cdots, z_{m_1 t})$;

$$\begin{bmatrix} \boldsymbol{M}_{\hat{y}\hat{y}} & \boldsymbol{M}_{\hat{y}z^*} \\[0.5em] \boldsymbol{M}_{z^*\hat{y}} & \boldsymbol{M}_{z^*z^*} \end{bmatrix}$$

$$= \begin{bmatrix} \hat{y}_{21} & \cdots & \hat{y}_{2T} \\ \vdots & & \vdots \\ \hat{y}_{n_1 1} & \cdots & \hat{y}_{n_1 T} \\ \hline z_{11} & \cdots & z_{1T} \\ \vdots & & \vdots \\ z_{m_1 1} & \cdots & z_{m_1 T} \end{bmatrix} \begin{bmatrix} \hat{y}_{21} & \cdots & \hat{y}_{n_1 1} & z_{11} & \cdots & z_{m_1 1} \\ \vdots & & \vdots & \vdots & & \vdots \\ \hat{y}_{2T} & \cdots & \hat{y}_{n_1 T} & z_{1T} & \cdots & z_{m_1 T} \end{bmatrix};$$

$\boldsymbol{b}$ is the estimator of the vector $(\beta_{i2}, \cdots, \beta_{in_1})$; and $\boldsymbol{c}$ is the estimator of the vector $(\gamma_{i1}, \cdots, \gamma_{im_1})$. It should be noted that $\boldsymbol{M}_{\hat{y}z^*} = \boldsymbol{M}_{yz^*} = \boldsymbol{M}_{z^*y}'$. It should be further observed that

$$\boldsymbol{M}_{\hat{y}\hat{y}} = \boldsymbol{M}_{yz} \boldsymbol{M}_{zz}^{-1} \boldsymbol{M}_{zy}.$$

In this expression the whole vector $\boldsymbol{z}_t = (z_{1t}, \cdots, z_{mt})$, which includes all the predetermined variables in the system, is used for the evaluation of the relevant moment matrices.

k-Class estimators. Theil (1958) and Basmann (1957), independently, were the first to advocate the method of two-stage least squares. Theil suggested a whole system of estimators, called the *k*-class. He defined these as the solutions to

$$\begin{bmatrix} \sum_{t=1}^{T} y_{1t} \boldsymbol{y}'_t - k \sum_{t=1}^{T} y_{1t} \hat{\boldsymbol{v}}'_t \\[2ex] \sum_{t=1}^{T} y_{1t} \boldsymbol{z}^{*\prime}_t \end{bmatrix}$$

$$= \begin{bmatrix} \boldsymbol{M}_{vv} - k \boldsymbol{M}_{yz} \boldsymbol{M}_{zz}^{-1} \boldsymbol{M}_{zy} & \boldsymbol{M}_{yz^*} \\[2ex] \boldsymbol{M}_{z^*y} & \boldsymbol{M}_{z^*z^*} \end{bmatrix} \begin{bmatrix} -\boldsymbol{b}' \\[2ex] -\boldsymbol{c}' \end{bmatrix} .$$

In this expression $\hat{\boldsymbol{v}}_t$ is the vector of residuals computed from the reduced form regressions of $y_{2t}, \cdots, y_{n_1 t}$ on all the z_{kt}. If $k = 0$, we have ordinary least squares estimators. If $k = 1$, we have two-stage least squares estimators. If $k = 1 + \lambda$ and λ is the smallest root of the determinantal equation

$$|\boldsymbol{M}_{yz} \boldsymbol{M}_{zz}^{-1} \boldsymbol{M}_{zy} - \boldsymbol{M}_{yz^*} \boldsymbol{M}_{z^*z^*}^{-1} \boldsymbol{M}_{z^*y} - \lambda(\boldsymbol{M}_{vv} - \boldsymbol{M}_{yz} \boldsymbol{M}_{zz}^{-1} \boldsymbol{M}_{zy})| = 0,$$

we have limited-information maximum likelihood estimators. This is a succinct way of showing the relationships between various single-equation methods of estimation. Of these three members of the *k*-class, ordinary least squares is not consistent; the other two are.

● **Three-stage least squares.** The original derivation of two-stage least squares estimates was obtained by an application of Aitken's generalized method of least squares. The equation to be estimated will be written as

$$\boldsymbol{y}_1 = -\boldsymbol{Y}_i \boldsymbol{\beta}_i - \boldsymbol{Z}_i \boldsymbol{\gamma}_i + \boldsymbol{u}_i ,$$

where

$$\boldsymbol{y}_1 = \begin{bmatrix} y_{11} \\ y_{12} \\ \cdot \\ \cdot \\ \cdot \\ y_{1T} \end{bmatrix}; \ \boldsymbol{Y}_i = \begin{bmatrix} y_{21} & y_{31} & \cdots & y_{n_1 1} \\ y_{22} & y_{32} & \cdots & y_{n_1 2} \\ \cdot & \cdot & & \cdot \\ \cdot & \cdot & & \cdot \\ \cdot & \cdot & & \cdot \\ y_{2T} & y_{3T} & \cdots & y_{n_1 T} \end{bmatrix}; \ \boldsymbol{\beta}_i = \begin{bmatrix} \beta_{i2} \\ \beta_{i3} \\ \cdot \\ \cdot \\ \cdot \\ \beta_{in_1} \end{bmatrix}$$

$$\boldsymbol{\gamma}_i = \begin{bmatrix} \gamma_{i1} \\ \gamma_{i2} \\ \cdot \\ \cdot \\ \cdot \\ \gamma_{im_1} \end{bmatrix}; \ \boldsymbol{Z}_i = \begin{bmatrix} z_{11} & z_{21} & \cdots & z_{m_1 1} \\ z_{12} & z_{22} & \cdots & z_{m_1 2} \\ \cdot & \cdot & & \cdot \\ \cdot & \cdot & & \cdot \\ \cdot & \cdot & & \cdot \\ z_{1T} & z_{2T} & \cdots & z_{m_1 T} \end{bmatrix}; \ \boldsymbol{u}_i = \begin{bmatrix} u_{i1} \\ u_{i2} \\ \cdot \\ \cdot \\ \cdot \\ u_{iT} \end{bmatrix} .$$

Form the product

$$\boldsymbol{Z}' \boldsymbol{y}_1 = \boldsymbol{Z}' \boldsymbol{Y}_i \boldsymbol{\beta}_i - \boldsymbol{Z}' \boldsymbol{Z}_i \boldsymbol{\gamma}_i + \boldsymbol{Z}' \boldsymbol{u}_i ,$$

where $\boldsymbol{Z}_i$ is a submatrix of $\boldsymbol{Z}$, which is the data matrix for the whole set of exogenous variables. As long as $m > n_1 - 1$, the overidentified case (necessary condition), we can regard this setup as implying the regression of the moment quantity $\sum_{t=1}^{T} z_{it} y_{1t}$ on $\sum_{t=1}^{T} z_{it} y_{jt}$ and $\sum_{t=1}^{T} z_{it} z_{kt}$ ($j = 2, \cdots, n_1$; $k = 1, 2, \cdots, m_1$). The data set for the regression runs from 1 to m. The disturbance has variance $\sigma^2 \boldsymbol{X}' \boldsymbol{X}$. The formulas for the generalized least squares regression with the indicated variance reduce to those for two-stage least squares estimation of $\boldsymbol{\beta}_i$ and $\boldsymbol{\gamma}_i$.

$$\text{est} \begin{bmatrix} -\boldsymbol{\beta}_i \\ -\boldsymbol{\gamma}_i \end{bmatrix}$$

$$= [(\boldsymbol{Z}'\boldsymbol{Y}_i, \boldsymbol{Z}'\boldsymbol{Z}_i)' \ (\boldsymbol{Z}'\boldsymbol{Z})^{-1} \ (\boldsymbol{Z}'\boldsymbol{Y}_i, \boldsymbol{Z}'\boldsymbol{Z}_i)]^{-1} \times$$
$$(\boldsymbol{Z}'\boldsymbol{Y}_i, \boldsymbol{Z}'\boldsymbol{Z}_i)' \ (\boldsymbol{Z}'\boldsymbol{Z})^{-1} \boldsymbol{Z}'\boldsymbol{y}_1$$

$$= \begin{bmatrix} \boldsymbol{Y}'_i \boldsymbol{Z} (\boldsymbol{Z}'\boldsymbol{Z})^{-1} \boldsymbol{Z}' \boldsymbol{Y}_i & \boldsymbol{Y}'_i \boldsymbol{Z}_i \\ \boldsymbol{Z}'_i \boldsymbol{Y}_i & \boldsymbol{Z}'_i \boldsymbol{Z}_i \end{bmatrix}^{-1} \begin{bmatrix} \boldsymbol{Y}'_i \boldsymbol{Z} (\boldsymbol{Z}'\boldsymbol{Z})^{-1} \boldsymbol{Z}' \boldsymbol{y}_1 \\ \boldsymbol{Z}'_i \boldsymbol{y}_1 \end{bmatrix} .$$

These are standard formulas for two-stage least squares estimators, using data matrix notation.

▶ The method of three-stage least squares is a natural extension of the method of generalized least squares as applied to the previous case. The complete system of equations can be written as

$$\begin{bmatrix} \boldsymbol{y}_1 \\ \boldsymbol{y}_2 \\ \cdot \\ \cdot \\ \cdot \\ \boldsymbol{y}_n \end{bmatrix} = \begin{bmatrix} \boldsymbol{X}_1 & 0 & \cdots & 0 \\ 0 & \boldsymbol{X}_2 & \cdots & 0 \\ \cdot & & & \cdot \\ \cdot & & & \cdot \\ \cdot & & & \cdot \\ 0 & 0 & \cdots & \boldsymbol{X}_n \end{bmatrix} \begin{bmatrix} \delta_1 \\ \delta_2 \\ \cdot \\ \cdot \\ \cdot \\ \delta_n \end{bmatrix} + \begin{bmatrix} \boldsymbol{u}_1 \\ \boldsymbol{u}_2 \\ \cdot \\ \cdot \\ \cdot \\ \boldsymbol{u}_n \end{bmatrix} ,$$

where

$$\boldsymbol{X}_i = (\boldsymbol{Y}_i, \boldsymbol{Z}_i),$$

$$\delta_i = - \begin{bmatrix} \boldsymbol{\beta}_i \\ \boldsymbol{\gamma}_i \end{bmatrix} .$$

Now form the equations

$$\begin{bmatrix} \boldsymbol{Z}'\boldsymbol{y}_1 \\ \boldsymbol{Z}'\boldsymbol{y}_2 \\ \cdot \\ \cdot \\ \cdot \\ \boldsymbol{Z}'\boldsymbol{y}_n \end{bmatrix} = \begin{bmatrix} \boldsymbol{Z}'\boldsymbol{X}_1 & 0 & \cdots & 0 \\ 0 & \boldsymbol{Z}'\boldsymbol{X}_2 & \cdots & 0 \\ \cdot & & & \cdot \\ \cdot & & & \cdot \\ \cdot & & & \cdot \\ 0 & 0 & \cdots & \boldsymbol{Z}'\boldsymbol{X}_n \end{bmatrix} \begin{bmatrix} \delta_1 \\ \delta_2 \\ \cdot \\ \cdot \\ \cdot \\ \delta_n \end{bmatrix} + \begin{bmatrix} \boldsymbol{Z}'\boldsymbol{u}_1 \\ \boldsymbol{Z}'\boldsymbol{u}_2 \\ \cdot \\ \cdot \\ \cdot \\ \boldsymbol{Z}'\boldsymbol{u}_n \end{bmatrix} .$$

The three-stage least squares estimator of δ is the

Aitken generalized least squares estimator of this system.

est $\boldsymbol{\delta} =$
$[\mathbf{X}' \ (\mathbf{S}^{-1} \otimes \mathbf{Z}(\mathbf{Z}'\mathbf{Z})^{-1}\mathbf{Z}')\mathbf{X}]^{-1} \ \mathbf{X}' \ (\mathbf{S}^{-1} \otimes \mathbf{Z}(\mathbf{Z}'\mathbf{Z})^{-1}\mathbf{Z}')\boldsymbol{y},$

where $\mathbf{S}$ is the estimated covariance matrix of residuals from the two-stage least squares estimate of each single equation. We can write this out as

$$
\text{est} - \begin{bmatrix} \beta \\ \gamma \end{bmatrix} = \begin{bmatrix} S^{11}\mathbf{X}'_1\mathbf{Z}(\mathbf{Z}'\mathbf{Z})^{-1}\mathbf{Z}'\mathbf{X}_1 & \cdots & S^{1n}\mathbf{X}'_1\mathbf{Z}(\mathbf{Z}'\mathbf{Z})^{-1}\mathbf{Z}'\mathbf{X}_n \\ & \cdot & \\ & \cdot & \\ & \cdot & \\ S^{n1}\mathbf{X}'_n\mathbf{Z}(\mathbf{Z}'\mathbf{Z})^{-1}\mathbf{Z}'\mathbf{X}_1 & \cdots & S^{nn}\mathbf{X}'_n\mathbf{Z}(\mathbf{Z}'\mathbf{Z})^{-1}\mathbf{Z}'\mathbf{X}_n \end{bmatrix} \times \begin{bmatrix} \sum S^{1i}\mathbf{X}'\mathbf{Z}(\mathbf{Z}'\mathbf{Z})^{-1}\mathbf{Z}'\boldsymbol{y}_i \\ \cdot \\ \cdot \\ \cdot \\ \sum S^{ni}\mathbf{X}'\mathbf{Z}(\mathbf{Z}'\mathbf{Z})^{-1}\mathbf{Z}'\boldsymbol{y}_i \end{bmatrix},
$$

where $(S^{ij}) = \mathbf{S}^{-1}$.

▶ Both two- and three-stage least squares estimators can thus be interpreted as Aitken estimators, one for single equations in a system and one for whole systems of equations. This class of estimators was first derived by Theil and Zellner (see Zellner & Theil 1962).

Theil and Zellner termed their method three-stage least squares because they first derived two-stage least squares estimators for each single equation in the system. They computed the residual variance for each equation and used these as estimators of the variances of the true (unobserved) random disturbances. They then used Aitken's generalized method of least squares (1935) to estimate all the equations in the system simultaneously. Aitken's method applies to systems of equations in which the variance–covariance matrix for disturbances is a general known positive definite matrix. Theil and Zellner used the two-stage estimator of this variance–covariance matrix as though it were known. The advantage of this method is that it is of the full-information variety, making use of restrictions on all the equations of the system.

Other methods. If the conditions for identification of a single equation are such that there are just enough restrictions to transform linearly and uniquely the reduced form coefficients into the structural coefficients, an indirect least squares method of estimation can be used. Exact identification under zero-type restrictions would enable one to solve

$$
0 = \sum_{j=1}^{n} \beta_{ij}\pi_{jk}, \qquad k = m_1 + 1, \cdots, m,
$$

for a unique set of estimated β_{ij}, apart from scale normalization, given a set of estimated π_{jk}. The latter would be determined from least squares estimators of the reduced forms. Since there are $n_1 - 1$ of the β_{ij} to be determined, the necessary condition for exact identification here is that $n_1 - 1 = m - m_1$.

If there is *underidentification*, i.e., too few a priori restrictions, structural estimation cannot be completed but unrestricted reduced forms can be estimated by the method of least squares. This is the most information that the econometrician can extract when there is lack of identification. Least squares estimators of the reduced form equations are consistent in the underidentified case, but estimates of the structural parameters cannot be made.

Instrumental variables. The early discussion of estimation problems in simultaneous equation models contained, on many occasions, applications of a method known as the method of instrumental variables. In estimating the ith equation of a linear system, i.e.,

$$
\sum_{j=1}^{n_1} \beta_{ij}y_{jt} + \sum_{k=1}^{m_1} \gamma_{ik}z_{kt} = u_{it},
$$

we may choose $(n_1 - 1) + m_1$ variables that are independent of u_{it}. These are known as the instrumental set. Naturally, the exogenous variables in the equation $(z_{1t}, \cdots, z_{m_1 t})$ are possible members of this set. In addition, we need $n_1 - 1$ more instruments from the list of exogenous variables in the system but not in the ith equation. For this problem let these be denoted as $x_{2t}, \cdots, x_{n_1 t}$. Since $E(z_{st}u_{it}) = 0$, $s = 1, \cdots, m_1$, and $E(x_{rt}u_{it}) = 0$, $r = 2, \cdots, n_1$, we can estimate the unknown parameters from

$$
\sum_{j=1}^{n_1} \beta_{ij}\sum_{t=1}^{T} y_{jt}x_{rt} + \sum_{k=1}^{m_1} \gamma_{ik}\sum_{t=1}^{T} z_{kt}x_{rt} = 0, \qquad r = 2, \cdots, n_1,
$$

$$
\sum_{j=1}^{n_1} \beta_{ij}\sum_{t=1}^{T} y_{jt}z_{st} + \sum_{k=1}^{m_1} \gamma_{ik}\sum_{t=1}^{T} z_{kt}z_{st} = 0, \qquad s = 1, \cdots, m_1.
$$

With a scale-normalization rule, such as $\beta_{i1} = 1$, we have $(n_1 - 1) + m_1$ linear equations in the same number of unknown coefficients. In exactly identified models there is no problem in picking the x_{rt}, for there will always be exactly $n_1 - 1$ z's excluded from the ith equation. The method is then identical with indirect least squares. If $m - m_1 > n_1 - 1$, i.e., if there are more exogenous variables outside the ith equation than there are endogenous variables minus one, we have overidentification, and the number of possible instrumental variables exceeds the minimum needed. In order to avoid the problem of subjective or arbitrary choice among instruments, we turn to the methods of limited information or two-stage least squares. In fact, it is instructive to consider how the method of two-stage least squares resolves this matter. In place of single variables as instruments, it uses linear combinations of them. The computed values

$$\hat{y}_{jt} = \sum_{k=1}^{m} \hat{\pi}_{jk} z_{kt}, \qquad j = 2, \cdots, n_1,$$

are the new instruments. We can view the method either as the regression of y_1 on $\hat{y}_2, \cdots, \hat{y}_{n_1}$, $z_1, \cdots, z_{m_1}$ or as instrumental-variable estimators with $\hat{y}_{2t}, \cdots, \hat{y}_{n_1 t}, z_{1t}, \cdots, z_{m_1 t}$ as the instruments. Both come to the same thing. The method of instrumental variables yields consistent estimators.

Subgroup averages. The instrumental-variables method can be applied in different forms. One form was used by Wald (1940) to obtain consistent estimators of a linear relationship between two variables each of which is subject to error. This gives rise to a method that can be used in estimating econometric systems. Wald proposed that the estimator of β in

$$y_t = \alpha + \beta x_t,$$

where y_t and x_t are both measured with error, be computed from

$$\beta = \frac{\sum_{t=\frac{T}{2}+1}^{T} y_t - \sum_{t=1}^{\frac{T}{2}} y_t}{\sum_{t=\frac{T}{2}+1}^{T} x_t - \sum_{t=1}^{\frac{T}{2}} x_t}.$$

He proposed ordering the sample in ascending magnitudes of the variable x. From two halves of the sample, we determine two sets of mean values of y and x. The line joining these means will have a slope given by $\hat{\beta}$. Wald showed the conditions under which these estimates are consistent.

This may be called the method of subgroup averages. It is a very simple method, which may readily be applied to equations with more than two parameters. The sample is split into as many groups as there are unknown parameters to be determined in the equation under consideration. If there are three parameters, for example, the sample may be split into thirds and the parameters

estimated from

$$\bar{y}_1 = \alpha + \beta \bar{x}_1 + \gamma \bar{z}_1,$$
$$\bar{y}_2 = \alpha + \beta \bar{x}_2 + \gamma \bar{z}_2,$$
$$\bar{y}_3 = \alpha + \beta \bar{x}_3 + \gamma \bar{z}_3.$$

The extension to more parameters is obvious. The method of subgroup averages can be shown to be a form of the instrumental-variables method by an appropriate assignment of values to "dummy" instrumental variables.

Subgroup averages is a very simple method, and it is consistent, but it is not very efficient.

Simultaneous least squares. The simultaneous least squares method, suggested by Brown (1960), minimizes the sum of squares of all reduced form disturbances, subject to the parameter restrictions imposed on the system, i.e., it minimizes

$$\sum_{t=1}^{T} \sum_{i=1}^{n} v_{it}^2,$$

subject to restrictions. Suppose that the v_{it} are expressed as functions of the observables and parameters, with all restrictions included; then Brown's method minimizes the sum of the elements on the main diagonal of Σ_v, where Σ_v is the variance–covariance matrix of reduced form disturbances, whereas full-information maximum likelihood minimizes $|\Sigma_v|$.

Brown's method has the desirable property of being a full-information method; it is distribution free; it is consistent; but it has the drawback that its results are not invariant under linear transformations of the variables. This drawback can be removed by expressing the reduced form disturbance in standard units

$$v'_{it} = v_{it}/\sigma_{y_i}$$

and minimizing

$$\sum_{t=1}^{T} \sum_{i=1}^{n} (v'_{it})^2.$$

Evaluation of alternative methods

The various approaches to estimation of whole systems of simultaneous equations or individual relationships within such systems are *consistent* except for the single-equation least squares method. If the system is recursive and disturbances are independent between equations, least squares estimators are also consistent. In fact, they are maximum likelihood estimators for normally distributed disturbances. But generally, ordinary least squares estimators are not consistent. They are included in the group of alternatives considered here because they have a time-honored status and because they have minimum variance. In large-sample theory, maximum likelihood estimators of parameters are generally efficient compared with all other estimators. That is why we choose full-information maximum likelihood estimators as norms. They are consistent and efficient. Least squares estimators are minimum-variance estimators if their variances are estimated about estimated (inconsistent) sample means. If their variances are measured about the *true*, or population, values, it is not certain that they are efficient.

Limited-information estimators are less efficient than full-information maximum likelihood estimators. This should be intuitively obvious, since full-information estimators make use of more a priori information; it is proved in Klein (1960). Two-stage least squares estimators have asymptotically the same variance–covariance matrix as limited-information estimators, and three-stage (or simultaneous two-stage) least squares estimators have the same variance–covariance matrix as full-information maximum likelihood estimators. Thus, asymptotically the two kinds of limited-information estimators have the same efficiency, and the two kinds of full-information estimators have the same efficiency. The instrumental-variables or subgroup-averages methods are generally inefficient. Of course, the instrumental-variables method can be pushed to the point where it is the same as two-stage least squares estimation and can thereby gain efficiency.

A desirable aspect of the method of maximum likelihood is that its properties are preserved under a single-valued transformation. Thus, efficient estimators of structural parameters by this method transform into efficient estimators of reduced form parameters. The apparently efficient method of least squares may lose its efficiency under this kind of transformation. In applications of models, we use the reduced form in most cases, not the individual structural equations; therefore the properties under conditions of transformation from structural to reduced form equations are of extreme importance. Limited-information methods are a form of maximum likelihood methods. Therefore the properties of limited information are preserved under transformation.

To obtain limited-information estimators of the single equation

$$\sum_{j=1}^{n_1}\beta_{ij}y_{jt} + \sum_{k=1}^{m_1}\gamma_{ik}z_{kt} = u_{it},$$

we maximize the joint likelihood of $v_{1t}, \cdots, v_{n_1 t}$ in

$$y_{jt} = \sum_{l=1}^{m}\pi_{jl}z_{lt} + v_{jt}, \qquad j = 1, \cdots, n_1,$$

subject to the restrictions on the ith equation. In this case only the n_1 reduced forms corresponding to $y_{1t}, \cdots, y_{n_1 t}$ are used. It is also possible to simplify calculations, and yet preserve consistency (although at the expense of efficiency), by using fewer than all m predetermined variables in the reduced forms. In this sense the reduced forms of limited-information estimation are not necessarily unique, and the same endogenous variable appearing in different structural equations of a system may not have the same reduced form expression for each equation estimator. There is yet another sense in which we may derive reduced forms for the method of limited information. After each equation of a complete system has been estimated by the method of limited information, we can derive algebraically a set of reduced forms for the whole system. These would, in fact, be the reduced forms used in forecasting, multiplier analysis, and similar applications of systems. The efficiency property noted above for limited and full information has not been proved for systems of this type of reduced forms, but this has been studied in numerical analysis (see below).

Ease of computation. Finally we come to an important practical matter in the comparison of the different methods of estimation—relative ease of computation. Naturally, calculations are simpler and smaller in magnitude for single-equation least squares than for any of the other methods except that of subgroup averages. The method of instrumental variables is of similar computational complexity, but for equations with four or more variables it pays to have the advantage of symmetry in the moment matrices, as is the case with single-equation least squares. This is hardly a consideration with modern electronic computing machines, but it is worth consideration if electric desk machines are being used.

The next-simplest calculations are those for two-stage least squares. These consist of a repeated application of least squares regression techniques of calculation, but the first regressions computed are of substantial size. There are as many independent variables in the regression as there are predetermined variables in the system, provided there are enough degrees of freedom. Essentially, the method amounts to the calculation of parameters and computed dependent variables in

$$\hat{y}_{jt} = \sum_{k=1}^{m}\hat{\pi}_{jk}z_{kt}, \qquad j = 2, \cdots, n_1.$$

Only the "forward" part of this calculation by the standard Gauss–Doolittle method need be made in order to obtain the moment matrix of the y_{jt}. In the next stage we compute the regression

$$y_{1t} = -\sum_{j=2}^{n_1} b_{ij}\hat{y}_{jt} - \sum_{k=1}^{m_1} c_{ik}z_{kt}.$$

Two important computing problems arise in the first stage. In many systems $m > T$; i.e., there are insufficient degrees of freedom in the sample for evaluation of the reduced forms. We may choose a subset of the z_{kt}, or we may use principal components of the z_{kt} (Kloek & Mennes 1960). Systematic and efficient ways of choosing subsets of the z_{kt} have been developed by taking account of the recursive structure of the model (Fisher 1965). In many economic models m has been as large as 30 or more, and it is often difficult to make sufficiently accurate evaluation of the reduced form regression equations of this size, given the amount of multicollinearity found in economic data with common trends and cycles. The same procedures used in handling the degrees-of-freedom problem are recommended for getting round the difficulties of multicollinearity. Klein and Nakamura (1962) have shown that multicollinearity problems are less serious in ordinary than in two-stage least squares. They have also shown that these problems increase as we move on to the methods of limited-information and then full-information maximum likelihood.

Limited-information methods require all the computations of two-stage least squares and, in addition, the extraction of a root of an $n_1 \times n_1$ determinantal equation. The latter calculation can be done in a straightforward manner by iterative matrix multiplication, usually involving fewer than ten iterations.

Both limited information and two-stage least squares are extremely well adapted to modern computers and can be managed without much trouble on electric desk machines.

Three-stage least squares estimators involve the computation of two-stage estimators for each equation of a system, estimation of a variance–covariance matrix of structural disturbances, and simultaneous solution of a linear equation system of the order of all coefficients in the system. This last step may involve a large number of estimating equations for a model of 30 or more structural equations.

All the previous methods consist of standard linear matrix operations. The extraction of a characteristic root is the only operation that involves nonlinearities, and the desired root can quickly be found by an iterative process of matrix multiplication. Full-information maximum likelihood methods, however, are quite different. The estima-

tion equations are highly nonlinear. For small systems of two, three, or four equations, estimates have been made without much trouble on large computers (Eisenpress 1962) and on desk machines (Chernoff & Divinsky 1953). The problem of finding the maximum of a function as complicated as the joint likelihood function of a system of 15 to 20 or more equations is, however, formidable. Electronic machine programs have been developed for this purpose. The most standardized sets of full-information maximum likelihood calculations are for systems that are fully linear in both parameters and variables. Single-equation methods require linearity only in unknown parameters, and this is a much weaker restriction. Much progress in computation has been made since the first discussion of these econometric methods of estimation, in 1943, but the problem is far from solved, and there is no simple, push-button computation. This is especially true of full-information maximum likelihood.

Efficient programs have recently been developed for calculating full-information maximum likelihood estimates in either linear or nonlinear systems, and these have been applied to models of as many as 15 structural equations, involving more than 60 unknown parameters.

Generalization of assumptions. The basis for comparing different estimation methods or for preferring one method over another rests on *asymptotic* theory. The property of consistency is a large-sample property, and the sampling errors used to evaluate efficiency measures are asymptotic formulas. Unfortunately, samples of economic data are frequently not large, especially time series data. The amount of small-sample bias or the small-sample confidence intervals for parameter estimators are not generally known in specific formulas. Constructed numerical experiments, designed according to Monte Carlo methods, have thrown some light on the small-sample properties. These are reported below.

Another assumption sometimes made for the basic model is that the error terms are mutually independent. We noted above that successive least squares treatment of equations in recursive systems is identical with maximum likelihood estimation when the variance–covariance matrix of structural disturbances is diagonal. This implies mutual independence among contemporaneous disturbances. In a time series model we usually make another assumption, namely, that

$$E(u_{it}u_{jt'}) = 0, \qquad t \neq t', \text{ for all } i, j.$$

The simplest way in which this assumption can be

modified is to allow the errors to be related in some linear autoregressive process, such as

$$u_{it} = \sum_{j=1}^{n} \sum_{k=1}^{p} \rho_{ijk} u_{j,t-k} + e_{it}, \qquad i = 1, 2, \cdots, n,$$

where $E(e_{it} e_{jt'}) = 0$ ($t \neq t'$, for all i, j). In a formal sense joint maximum likelihood estimation of structural parameters and autoregressive coefficients, ρ_{ijk}, can be laid out in estimation equations, but there are no known instances where these have been solved on a large scale, for the estimation equations are very complicated. For single-equation models or for recursive systems which split into a series of single-equation regressions, the autoregressive parameters of first order have been jointly estimated with structural parameters (Barger & Klein 1954). The principal extensions to larger systems have been in cases where the autoregressive parameters are known a priori. Then it is easy to make known autoregressive transformations of the variables and proceed as in the case of independent disturbances. [See TIME SERIES.]

Related to the above two points is the treatment of lagged values of endogenous variables as predetermined variables. The presence of lagged endogenous variables reflects serial correlation among endogenous variables rather than among disturbances. In large samples it can be shown that for purposes of estimation we are justified in treating lagged variables as predetermined, but in small samples we incur bias on this account.

Another assumption regarding the disturbances in simultaneous equation systems is that they are mainly due to neglected or unmeasurable variables that affect or disturb each equation of the model. They are regarded as errors in behavior or technology. From a formal mathematical point of view, they could equally well be regarded as a direct error in observation of the normalized dependent variable in each equation, assuming that the system is written so that there is a different normalized dependent variable in each equation. There is an implicit assumption that the exogenous variables are measured without error. If we change the model to one in which random errors enter through disturbances to each relation and also through inaccurate observation of each individual variable, we have a more complicated probability scheme, whose estimation properties have not been developed in full generality. This again has been a case for numerical treatment by Monte Carlo methods.

The procedures of estimating simultaneous equation models as though errors are mutually independent when they really are not and as though variables are accurately measured when they really

are not are *specification* errors. Other misspecifications of models can occur. For simplicity we assume linearity or, at least, linearity in unknown parameters, but the true model may have a different functional form. Errors may not follow the normal distribution, as we usually assume. [See ERRORS, *article on* EFFECTS OF ERRORS IN STATISTICAL ASSUMPTIONS.]

Full-information methods are sensitive to specification error because they depend on restrictions imposed throughout an entire system. Single-equation methods depend on a smaller set of restrictions. If an investigator has particular interest in just one equation or in a small sector of the economy, he may incur large specification error by making too superficial a study of the parts of the economy that do not particularly interest him. There is much to be said for using single-equation methods (limited information or two-stage least squares) in situations where one does not have the resources to specify the whole economy adequately.

There are numerous possibilities for specifying models incorrectly. These probably introduce substantial errors in applied work, but they cannot be studied in full generality for there is no particular way of showing all the misspecifications that can occur. We can, however, construct artificial numerical examples of what we believe to be the major specification errors. These are discussed below.

Sampling experiments. The effect on estimation methods of using simplified assumptions that are not fully met in real life often cannot be determined by general mathematical analysis. Econometricians have therefore turned to constructing sampling experiments with large-scale computers to test proposed methods of estimation where (1) the sample is small; and (2) there is specification error in the statement of the model, such as (*a*) nonzero parameters assumed to be zero, (*b*) dependent exogenous variables and errors assumed to be independent, (*c*) imperfectly measured exogenous variables assumed to be perfectly measured, or (*d*) serially correlated errors assumed to be not serially correlated.

So-called Monte Carlo methods are used to perform the sampling experiments that conceptually underlie sampling error calculations. These sampling experiments are never, in fact, carried out with nonexperimental sources of data, for we cannot relive economic life over and over again; but we can instruct a machine to simulate such an experiment.

Consider a single equation to be estimated by

different methods, for example,

$$y_t = \alpha + \beta x_t + u_t, \qquad t = 1, 2, \cdots, T.$$

Fix α and β at, say, 3.0 and 0.5, respectively, and set $T = 30$. This would correspond to the process

$$y_1 = 3.0 + 0.5x_1 + u_1$$
$$\vdots \quad \vdots \quad \vdots \quad \vdots$$
$$y_{30} = 3.0 + 0.5x_{30} + u_{30}.$$

We also fix the values of the predetermined variables $x_1, x_2, \cdots, x_{30}$ once and for all. We set $T = 30$ to indicate that we are dealing with a 30-element small sample. A sample of 30 annual observations would be the prototype.

Employing a source of random numbers scaled to have a realistic standard deviation and a zero mean, we draw a set of random numbers $u_1, \cdots, u_{30}$. We then instruct a machine to use $u_1, \cdots, u_{30}$ and $x_1, \cdots, x_{30}$ to compute $y_1, \cdots, y_{30}$ from the above formulas. From the samples of data, $y_1, \cdots, y_{30}$ and $x_1, \cdots, x_{30}$, we estimate α and β by the methods being studied. Let $\hat{\alpha}$ and $\hat{\beta}$ be the estimated values. We then draw a new set of random numbers, $u_1, \cdots, u_{30}$, and repeat the process, using the same values of $x_1, \cdots, x_{30}$. From many such repetitions, say 100, we have sampling distributions of $\hat{\alpha}$ and $\hat{\beta}$. Means of these distributions, when compared with α ($= 3.0$) and β ($= 0.5$), indicate bias, if any, and standard deviations or root-mean-square values about 3.0 or 0.5 indicate efficiency. From these sampling distributions we may compare different estimators of α and β.

What we have said about this simple type of experiment for a single equation can readily be extended to an entire system:

$$\mathbf{B}\boldsymbol{y}_t' + \boldsymbol{\Gamma}\boldsymbol{z}_t' = \boldsymbol{u}_t', \qquad t = 1, 2, \cdots, T.$$

In this case we must start with assumed values of $\mathbf{B}$ and $\boldsymbol{\Gamma}$. We choose a T-element vector of values for each element of $\boldsymbol{z}_t$, the predetermined variables, and *repeated* T-element vectors of values for each element of $\boldsymbol{u}_t$. The random variables are chosen so that their variance–covariance matrix equals some specified set of values. As in the single-equation case, $T = 30$ or some likely small-sample value. The $\boldsymbol{z}_t$ are often chosen in accordance with the values of predetermined variables used in actual models. In practice, Monte Carlo studies of simultaneous equation models have dealt with small systems having only two, three, or four equations.

Two sets of results are of interest from these studies. Estimates of individual elements in $\mathbf{B}$ and $\boldsymbol{\Gamma}$ can be studied and compared for different esti-

mators; estimates of $\mathbf{B}^{-1}\boldsymbol{\Gamma}$, the reduced-form coefficients, can be similarly investigated. In addition, we could form some over-all summary statistic, such as standard error of forecast, for different estimators.

The simplest Monte Carlo experiments have been made to test for small-sample properties alone; they have not introduced measurement errors, serial correlation of disturbances, or other specification errors. Generally speaking, these studies clearly show the bias in single-equation least squares estimates where some of the "independent" variables in the regression calculation are not independent of the random disturbances. Maximum likelihood estimators (full or limited information) show comparatively small bias. The standard deviations of individual parameter estimators are usually smallest for the single-equation least squares method, but this standard deviation is computed about the biased sample mean. If estimated about the true mean, least squares sometimes does not show up well, indicating that bias outweighs efficiency. Full-information maximum likelihood shows up as an efficient method, whether judged in terms of variation about the sample or the true mean. Two-stage least squares estimators appear to have somewhat smaller variance about the true values than do limited-information estimators, and both methods measure up to the efficiency of single-equation least squares methods when variability is measured about the true mean.

Asymptotically, limited-information and two-stage estimators have the same variance–covariance matrices, and they are both inefficient compared with full-information estimators. The Monte Carlo results for small samples are not surprising, although the particular experiments studied give a slight edge to two-stage estimators.

When specification error is introduced, in the form of making an element of $\boldsymbol{\Gamma}$ zero in the estimation process when it is actually nonzero in the population, we find that full-information methods are very sensitive. Both limited-information and two-stage estimators perform better than full-information maximum likelihood. Two-stage estimators are the best among all methods examined in this situation. Limited-information estimators are very sensitive to intercorrelation among predetermined variables.

The principal result for Monte Carlo estimators of reduced form parameters is that transformed single-equation least squares values lose their efficiency properties. Being seriously biased as well, these estimates show a poor over-all rating when used for estimating reduced forms for a system as a whole. Full-information estimators, which are shown in these experiments to be sensitive to specification error, do better in estimating reduced form coefficients than in estimating structural coefficients. Their gain in making use of all the a priori information outweighs the losses due to the misspecification introduced and, in the end, gives them a favorable comparison with ordinary least squares estimators of the reduced form equations that make no use of the a priori information and have no specification error.

If a form of specification error is introduced in a Monte Carlo experiment by having common time trends in elements of z_t and u_t, so that they are not independent as hypothesized, we find that limited-information estimators are as strongly biased as are ordinary least squares values. If time trend is introduced as an additional variable, however, the limited-information method has small bias.

When observation errors are imposed on the z_t, both least squares and limited-information estimators show little change in bias but increases in sampling errors. In this model, it turns out as before that the superior efficiency of least squares estimators of individual structural parameters does not carry over to the estimators of reduced form parameters.

A comprehensive sampling-experiment study of alternative estimators under correctly specified and under misspecified conditions is given in Summers (1965), and Johnston (1963) compares results from several completed Monte Carlo studies. This approach is in its infancy, and further investigations will surely throw new light on the relative merits of different estimation methods.

For some years economists were digesting the modern approach to simultaneous equation estimation introduced by Haavelmo, Mann and Wald, Anderson and Rubin, and Koopmans, Rubin, and Leipnik, and there was a period of little change in this field. Since the development of the two-stage least squares method by Theil, there have been a number of developments. The methods are undergoing interpretation and revision. New estimators are being suggested, and it is likely that many new results will be forthcoming in the next few decades. Wold (1965) has proposed a method based on iterative least squares that recommends itself by its adaptability to modern computers, its consistency, and its capacity to make use of a priori information on all equations simultaneously and to treat some types of nonlinearity with ease. Also, excellent recent books, by Christ (1966), Gold-

berger (1964), and Malinvaud (1964), greatly aid instruction in this subject.

LAWRENCE R. KLEIN

[See also LINEAR HYPOTHESES, article on REGRESSION.]

BIBLIOGRAPHY

AITKEN, A. C. 1935 On Least Squares and Linear Combination of Observations. Royal Society of Edinburgh, *Proceedings* 55:42–48.

ANDERSON, T. W.; and RUBIN, HERMAN 1949 Estimation of the Parameters of a Single Equation in a Complete System of Stochastic Equations. *Annals of Mathematical Statistics* 20:46–63.

►ARROW, KENNETH J. 1968 Economic Equilibrium. Volume 4, pages 376–389 in *International Encyclopedia of the Social Sciences*. Edited by David L. Sills. New York: Macmillan and Free Press.

BARGER, HAROLD; and KLEIN, LAWRENCE R. 1954 A Quarterly Model for the United States Economy. *Journal of the American Statistical Association* 49:413–437.

BASMANN, R. L. 1957 A Generalized Classical Method of Linear Estimation of Coefficients in a Structural Equation. *Econometrica* 25:77–83.

BROWN, T. M. 1960 Simultaneous Least Squares: A Distribution Free Method of Equation System Structure Estimation. *International Economic Review* 1:173–191.

CHERNOFF, HERMAN; and DIVINSKY, NATHAN 1953 The Computation of Maximum-likelihood Estimates of Linear Structural Equations. Pages 236–269 in Cowles Commission for Research in Economics, *Studies in Econometric Method*. Edited by William C. Hood and Tjalling C. Koopmans. New York: Wiley.

CHOW, GREGORY C. 1964 A Comparison of Alternative Estimators for Simultaneous Equations. *Econometrica* 32:532–553.

CHRIST, C. F. 1966 *Econometric Models and Methods*. New York: Wiley.

EISENPRESS, HARRY 1962 Note on the Computation of Full-information Maximum-likelihood Estimates of Coefficients of a Simultaneous System. *Econometrica* 30:343–348.

FISHER, FRANKLIN M. 1965 Dynamic Structure and Estimation in Economy-wide Econometric Models. Pages 589–635 in James S. Duesenberry et al., *The Brookings Quarterly Econometric Model of the United States*. Chicago: Rand McNally.

FISHER, FRANKLIN M. 1966 *The Identification Problem in Econometrics*. New York: McGraw-Hill.

GOLDBERGER, ARTHUR S. 1964 *Econometric Theory*. New York: Wiley.

HAAVELMO, TRYGVE 1943 The Statistical Implications of a System of Simultaneous Equations. *Econometrica* 11:1–12.

JOHNSTON, JOHN 1963 *Econometric Methods*. New York: McGraw-Hill.

KLEIN, LAWRENCE R. 1960 The Efficiency of Estimation in Econometric Models. Pages 216–232 in Ralph W. Pfouts, *Essays in Economics and Econometrics: A Volume in Honor of Harold Hotelling*. Chapel Hill: Univ. of North Carolina Press.

KLEIN, LAWRENCE R.; and NAKAMURA, MITSUGU 1962 Singularity in the Equation Systems of Econometrics: Some Aspects of the Problem of Multicollinearity. *International Economic Review* 3:274–299.

KLOEK, T.; and MENNES, L. B. M. 1960 Simultaneous Equations Estimation Based on Principal Components of Predetermined Variables. *Econometrica* 28:45–61.

KOOPMANS, TJALLING C.; RUBIN, HERMAN; and LEIPNIK, R. B. (1950) 1958 Measuring the Equation Systems of Dynamic Economics. Pages 53–237 in Tjalling C. Koopmans (editor), *Statistical Inference in Dynamic Economic Models*. Cowles Commission for Research in Economics, Monograph No. 10. New York: Wiley.

MALINVAUD, EDMOND (1964) 1966 *Statistical Methods of Econometrics*. Chicago: Rand McNally. → First published in French.

MANN, H. B.; and WALD, ABRAHAM 1943 On the Statistical Treatment of Linear Stochastic Difference Equations. *Econometrica* 11:173–220.

STROTZ, ROBERT H.; and WOLD, HERMAN 1960 A Triptych on Causal Chain Systems. *Econometrica* 28:417–463.

SUMMERS, ROBERT 1965 A Capital Intensive Approach to the Small Sample Properties of Various Simultaneous Equation Estimators. *Econometrica* 33:1–41.

THEIL, HENRI (1958) 1961 *Economic Forecasts and Policy*. 2d ed., rev. Amsterdam: North-Holland Publishing.

TINBERGEN, JAN 1939 *Statistical Testing of Business-cycle Theories*. Volume 2: Business Cycles in the United States of America: 1919–1932. Geneva: League of Nations, Economic Intelligence Service.

WALD, ABRAHAM 1940 The Fitting of Straight Lines if Both Variables Are Subject to Error. *Annals of Mathematical Statistics* 11:284–300.

WOLD, HERMAN 1965 A Fix-point Theorem With Econometric Background. *Arkiv för Matematik* 6:209–240.

ZELLNER, ARNOLD; and THEIL, HENRI 1962 Three-stage Least Squares: Simultaneous Estimation of Simultaneous Equations. *Econometrica* 30:54–78.

Postscript

Work has continued in several fruitful directions for estimation of systems of simultaneous equations. The principal directions concern generalization to deal with serially correlated errors, dynamic structures, and iterated instrumental variables estimators. The method of two-stage least squares has been applied to estimation of single equations in complete systems with errors satisfying

$$u_{it} = \sum_{j=1}^{n} \sum_{k=1}^{p} p_{ijk} u_{j,\,t-k} + e_{it}\,,$$

$$E(e_{it} e_{jt'}) = 0, \qquad\qquad t \neq t', \text{ all } i, j.$$

Usually these methods are restricted to the case where $p_{ij} = 0$ if $i \neq j$. Most work has been done with two-stage least squares estimators for the serially correlated case, but results have been developed for other methods as well.

The case of serially correlated errors has been treated, where lagged endogenous variables are present in the equation. Additionally, lag distributions have been considered for individual equations, with or without the presence of serially correlated errors.

A natural outgrowth of Wold's fixed point method, in which ordinary least squares estimators are iterated, is one in which instrumental variables are determined from estimated reduced forms, with all restrictions on parameters imposed, and used iteratively in the estimation of each single equation. Such methods provide consistent estimates and have the desirable properties of using up degrees of freedom quite economically and also being easy to use in nonlinear systems. Given an initial estimate of the individual equations of a system with coefficient matrices $\mathbf{B}^{(0)}$ $\mathbf{\Gamma}^{(0)}$, instruments are determined as

$$\mathbf{y}_t^{(1)} = -(\mathbf{B}^{(0)})^{-1} \mathbf{\Gamma}^{(0)} \mathbf{Z}_t\,.$$

The included predetermined variables and components of $\mathbf{y}_t^{(1)}$ (corresponding to included endogenous variables) are used as instruments, equation by equation. New estimated equations are solved in reduced form to generate new instruments and the process goes on iteratively in an obvious way. Some investigators are iterating these instrumental variable estimators until convergence is obtained, while others are stopping after one or two iterations. The method is being investigated for the case of dynamically generated instruments from system simulation over time. These methods appear to be quite promising, but no conclusive verdict is yet available.

LAWRENCE R. KLEIN

ADDITIONAL BIBLIOGRAPHY

BRUNDY, J. M.; and JORGENSON, D. W. 1971 Efficient Estimation of Simultaneous Equations by Instrumental Variables. *Review of Economics and Statistics* 53:207–224.

CHOW, GREGORY C. 1968 Two Methods of Computing Full-information Maximum Likelihood Estimates in Simultaneous Stochastic Equations. *International Economic Review* 9:100–112.

DHRYMES, PHOEBUS J. 1970 *Econometrics: Statistical Foundations and Applications*. New York: Harper.

DHRYMES, PHOEBUS J. 1971 *Distributed Lags: Problems of Estimation and Formulation*. San Francisco: Holden-Day.

DHRYMES, PHOEBUS J. 1972 Simultaneous Equations Inference in Econometrics. Institute of Electrical and Electronics Engineers, *Transactions on Automatic Control* 17:427–438.

DUTTA, M.; and LYTTKENS, E. 1974 Iterative Instrumental Variables Method and Estimation of a Large

Simultaneous System. *Journal of the American Statistical Association* 69:967–986.

EISENPRESS, HARRY; and GREENSTADT, JOHN 1966 The Estimation of Nonlinear Econometric Systems. *Econometrica* 34:851–861.

FAIR, RAY C. 1970 The Estimation of Simultaneous Equation Models With Lagged Endogenous Variables and First Order Serially Correlated Errors. *Econometrica* 38:507–516.

HENDRY, D. F. 1971 Maximum Likelihood Estimation of Systems of Simultaneous Regression Equations With Errors Generated by a Vector Autoregressive Process. *International Economic Review* 12:257–272.

KLEIN, LAWRENCE R. (1973) 1974 *A Textbook of Econometrics.* 2d ed. Englewood Cliffs, N.J.: Prentice-Hall.

THEIL, HENRI 1971 *Principles of Econometrics.* New York: Wiley.

SIZE DISTRIBUTIONS IN ECONOMICS

The size distributions of certain economic and socioeconomic variables—incomes, wealth, firms, plants, cities, etc.—display remarkably regular patterns. These patterns, or distribution laws, are usually skew, the most important being the Pareto law (see Allais 1968) and the log-normal, or Gibrat, law (below). Some disagreement about the patterns actually observed still exists. The empirical distributions often approximate the Gibrat law in the middle ranges of the variables and the Pareto law in the upper ranges. The study of size distributions is concerned with explaining why the observed patterns exist and persist. The answer may be found in the conception of the distribution laws as the steady state equilibria of stochastic processes that describe the underlying economic or demographic forces. A steady state equilibrium is a macroscopic condition that results from the balance of a great number of random microscopic movements proceeding in opposite directions. Thus, in a stationary population a constant age structure is maintained by the annual occurrence of approximately constant numbers of births and deaths—the random events par excellence of human life.

The steady state explanation is evidently inspired by the example of statistical mechanics in which the macroscopic conditions are heat and pressure and the microscopic random movements are performed by the molecules. Characteristically, the steady state is independent of initial conditions, i.e., the initial size distribution. In economic applications this is important because it means that the pattern determined by certain structural constants tends to be re-established after a disturbance is imposed on the process. This will only be the case,

however, if the process leading to the steady state is really ergodic, that is, if the influence of initial conditions on the state of the system becomes negligible after a certain time; and it will be relevant in practice only if this time interval is sufficiently short.

The idea that the stable pattern of a distribution might be explained by the interplay of a multitude of small random events was first demonstrated in the case of the normal distribution. The central limit theorem shows that the addition of a great number of small independent random variables yields a variable that is normally distributed, when properly centered and scaled. A stochastic process that leads to a normal distribution is the random walk on a straight line with, for example, a 50 per cent probability each of a step in one direction and a step in the opposite direction. It is only natural that attempts to explain other distribution patterns should have started from this idea. The first extension was to allow the random walk to proceed on a logarithmic scale. The resulting distribution is log-normal on the natural scale and is known as the log-normal or Gibrat distribution. The basic assumption, in economic terms, is that the chance of a certain proportionate growth or shrinkage is independent of the size already reached—the law of proportionate effect. This law was proposed by J. C. Kapteyn, by Francis Galton, and, later, by Gibrat (1931).

○ Let size (of towns, firms, incomes) at time t be denoted by $Y(t)$, and let $\epsilon(t)$ represent a random variable with a certain distribution. We have

$$Y(t) - Y(t-1) = \epsilon(t)Y(t-1),$$
$$Y(t) = (1 + \epsilon(t))Y(t-1)$$
$$= Y(0)(1 + \epsilon(1)) \cdots (1 + \epsilon(t)),$$

where $Y(0)$ is size at time 0, the initial period. For small time intervals the logarithm of size can be represented as the sum of independent random variables and an initial size which will become negligible as t grows:

$$\log Y(t) = \log Y(0) + \epsilon(1) + \epsilon(2) \cdots + \epsilon(t).$$

If the random variables ϵ are identically distributed with mean m and variance σ^2, the distribution of $\log Y(t)$ will be approximately normal with mean mt and variance $\sigma^2 t$.

This random walk corresponds to the process of diffusion in physics which is illustrated by the so-called Brownian movement of particles of dust put into a drop of liquid. Since it implies an ever growing variance, the idea of Gibrat is not itself enough to provide an explanation for a stable distribution. There must be a stabilizing influence to offset the tendency of the variance to increase; indeed, a distinguishing feature of the various theories presently to be reviewed lies in the kind of stabilizer they introduce to offset the diffusion. Two interesting cases may be noted here. One possibility is to modify the law of proportionate effect and assume that the chances of growth decline as size increases. This approach has been taken by Kalecki (1945), who assumes a negative correlation between the size and the jump and obtains a Gibrat law with constant variance. Another possibility is to combine the diffusion process of the random walk with a steady inflow of new, small units (firms, cities, incomes). Some units may continue indefinitely to increase in size, but their weight will be offset by that of a continuous stream of many new, small entrants, so that both the mean and the variance of the distribution will remain constant. This approach, which leads to the Pareto law, has been taken by Simon (1955).

Review of various models. Descriptions of various models will illustrate the methods employed. Models differ with regard to the distribution law explained, the field of application (towns, incomes, etc.), and the type of stochastic process used.

Champernowne's model. Champernowne (1953) presents a model that explains the Pareto law for the size distribution of incomes. The stochastic process employed is the so-called Markov chain [*see* MARKOV CHAINS]. The model is based on a matrix of probabilities of transition from one income class to another in a certain interval of time, say a year. The rows are the income classes of one year, the columns the income classes of the next year. The income classes are chosen in such a way that they are equal on the logarithmic scale (for example, incomes from 1 to 10, from 10 to 100, etc.). The probability of a jump from one income class to the next income class in the course of a year is assumed to be independent of the income from which the jump is made (the law of proportionate effect). The number of income earners is constant.

The number of income earners in income class s is then determined as follows. The number of incomes in class s at time $t + 1$ is

$$f(s, t+1) = \sum_{u=-n}^{1} f(s-u,,t)p(u),$$

where s, u, and t take on integer values, $p(u)$ is the probability of a jump over u intervals (i.e., the transition probability), and the size of the jump is constrained to the range $+1$, $-n$. In the steady state equilibrium reached after a sufficiently long time has passed, the action of the transition matrix leaves the distribution unchanged. We then have

$$f(s) = \sum_{u=-n}^{1} f(s-u)p(u), \qquad s > 0,$$

as $t \to \infty$. This difference equation is solved by putting $f(s) = z^s$. The characteristic equation

$$g(z) = \sum_{u=-n}^{1} z^{1-u}p(u) - z = 0$$

has two positive real roots, one of which is unity. To assure that the other root will be between 0 and 1, Champernowne introduces the following stability condition:

$$g'(1) = -\sum_{u=-n}^{1} up(u) > 0.$$

The relevant solution is $f(s) = b^s$, $0 < b < 1$, which gives the number of incomes in income class s. If the lower bound of this class is the log of the income Y_s, then the probability of an income exceeding Y_s is given by

$$\log P(Y_s) = s \log b.$$

Since s is determined by

$$\log Y_s = sh + \log Y_{\min},$$

where h is the class interval and $Y_{\min}$ is the lower boundary of the lowest income class, it follows that

$$\log P(Y_s) = \gamma - \alpha \log Y_s,$$

where the parameters γ and α are determined by b, h, and $Y_{\min}$. This is the Pareto law with Pareto coefficient α.

Champernowne's stability condition implies that the mathematical expectation of a change in income is negative. This counteracts the diffusion. How can the stability condition be justified on economic grounds? It may be connected with the fact that in this model every income earner who drops out is replaced by a new income earner. Since, in practice, young people have on the average lower and more uniform incomes than old people, the replacement of old income earners by young ones usually means a drop in income. Thus, Champernowne's stability condition, as far as its economic basis is concerned, is very similar to the entry of new, small units that act as a stabilizer in Simon's model.

Rutherford's model. Rutherford's model (1955) leads, in his opinion, to the Gibrat law for the size distribution of incomes. Newly entering income earners, assumed to be log-normally distributed at the start, are subject to a random walk and thus to increasing variance during their lifetimes. The process of birth and death of income earners, which is explicitly introduced into the model, acts as the stabilizer.

The distribution of total income is obtained by summing the distributions for all age cohorts that contribute survivors. Rutherford's method is to derive the moments of the distribution by integration over time of the moments for the entrance groups. The distribution is built up "synthetically" from the moments, as it were. In the absence of an analytical solution with a definite distribution law, some disagreement remains about the result.

Simon's model.   In Simon's model (1955), which leads to what he calls the Yule distribution, the aggregate growth of firms, cities, or incomes is given a priori. The stochastic process apportions this given increment to various units according to certain rules, which are weakened forms of the law of proportionate effect and rules of new entry. As a consequence of this procedure, there is no possibility of shrinkage of individual units. The given aggregate emphasizes the interdependence of fortunes of different firms (the gain of one is the loss of another)—a point that is neglected in other models, such as that of Steindl (1965). On the other hand, the aggregate is, in reality, not given; it is not independent of the action of the firms, which may increase their total market by advertising, product innovation, and so on.

The process of apportionment may be described as follows. We may conveniently think of populations of cities, so that $f(n, N)$ is the frequency of cities with n inhabitants in a total urban population of N; to be realistic, we shall assume that a city exceeds a certain minimum number of inhabitants; n will measure the excess over this minimum, and N will correspondingly be the sum of these excess populations. An additional urban inhabitant is allocated to a new city with a probability α and to an existing city, of any size class, with a probability proportionate to the number of (excess) inhabitants in that size class. Then,

$$f(n, N + 1) - f(n, N)$$
$$= \frac{1 - \alpha}{N} [(n - 1)f(n - 1, N) - nf(n, N)],$$

$$f(1, N + 1) - f(1, N)$$
$$= \alpha - \frac{1 - \alpha}{N} f(1, N).$$

We assume that there is a steady state solution in which the frequencies of all classes of cities change in the same proportion, that is, in which

$$\frac{f(n, N + 1)}{f(n, N)} = \frac{N + 1}{N}, \qquad \text{for all } n.$$

Using this relation and defining a relative frequency of cities as $f(n) = f(n, N)/(\alpha N)$, we obtain from the above equations

$$f(n) = f(n - 1) \frac{(1 - \alpha)(n - 1)}{1 + (1 - \alpha)n}$$
$$f(1) = \frac{1}{2 - \alpha};$$

or, setting $1/(1 - \alpha) = \rho$,

$$f(n) = \frac{(n - 1)(n - 2) \cdots 2 \cdot 1}{(n + \rho)(n + \rho - 1) \cdots (2 + \rho)} f(1)$$
$$= \frac{\Gamma(n)\Gamma(\rho + 2)}{\Gamma(n + \rho + 1)} f(1).$$

This expression is the Yule distribution. Using a property of the Γ-function, it can be shown that the Yule distribution asymptotically approaches the Pareto law for large values of n, that is, $f(n) \to n^{-\rho-1} f(1) \Gamma(\rho + 2)$ as $n \to \infty$.

This model is applicable to cases in which size is measured by a stock, for example, number of employees of a firm. Simon provides an alternative interpretation of it that applies to flows, such as income and turnover of firms. For example, the total flow of income is given, and each dollar is apportioned to existing and new income earners according to the rules given above.

Using simulation techniques, Ijiri and Simon (1964) show that the pattern of the Yule distribution persists if serial correlation of the growth of individual firms in different periods is assumed. This finding is important because, in reality, growth is often affected by "constitutional" factors, such as financial resources and research done in the past.

The model of Wold and Whittle.   Wold and Whittle (1957) present a model of the size distribution of wealth in which stability is provided by the turnover of generations, as in Rutherford's model. On the death of a wealth owner, his fortune is divided among his heirs (in equal parts, as a simplification). The diffusion effect is provided by the growth of wealth of living proprietors, which proceeds deterministically at compound interest. The model is shown to lead to a Pareto distribution, the Pareto coefficient depending on the number of heirs to an estate and the ratio of the growth rate of capital to the mortality rate of the wealth owners.

Steindl's models.   Steindl's models (1965, chapters 2, 3) are designed to explain the size distribution of firms, but they can equally well be applied to the size distribution of cities. The distribution laws obtained are, for large sizes, identical with the Pareto law. Like Rutherford's model, Steindl's models rest on a combination of two stochastic processes. One is a birth-and-death process of the population of cities or firms; the other is a stochastic process of the growth of the city or firm itself.

The way in which the interplay of these two processes brings about the Pareto law can be explained in elementary terms. We start with the size distribution of cities. The number of cities can be explained by a birth process, if we assume that cities do not die. Let us assume that new cities are appearing at a constant rate, ϵ, the birth rate of cities. The number of cities increases exponentially, and the age distribution of cities at a given moment of time is

$$(1) \qquad R(t) = R(0)e^{-\epsilon t}, \qquad \epsilon > 0,$$

where t is age and $R(t)$ is the number of cities with age in excess of t; in other words, $R(t)$ is the rank of the town aged $t + dt$, and $R(0)$ is the total number of towns existing at the moment of time considered. The size of the city—its number of inhabitants—increases, on the average, with age. If the rate of births plus immigration, λ, and of deaths plus emigration, μ, are constant, we obtain an exponential growth function for the size of the city:

$$(2) \qquad n(t) = e^{(\lambda-\mu)t}, \qquad \lambda > \mu.$$

Eliminating t between eqs. (1) and (2), we get

$$(3) \qquad \ln R = - \frac{\epsilon}{\lambda - \mu} \ln n + \ln R(0).$$

This is the Pareto law (if $\epsilon/(\lambda - \mu) > 1$), and the Pareto coefficient is seen to be the ratio of the growth rate of the number of cities to the growth rate of a city.

This demonstration, which on the face of it is deterministic in character, can be supplemented by a graphical illustration in which the stochastic features are included. In Figure 1 the distribution of cities according to age is plotted in the vertical $(\ln R, t)$ plane. The abscissa shows the age of the city, and the ordinate shows the log of the rank of the city. Each city is thus represented by a dot, and the regression line fitted to these points represents relation (1). In the horizontal $(t, \ln n)$ plane, we show the exponential growth of cities with age, as in relation (2). Again each city may be represented by a dot showing age and size. The scatter diagram in the horizontal plane may be regarded as a stochastic transformation of the time variable into the size variable. If the size of each city has been found on the scatter diagram, the cities can be reordered according to size; we then obtain, in the third $(\ln R, \ln n)$ plane, the transformed

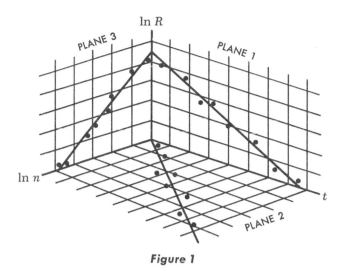

Figure 1

relation (3) between the number of cities (rank) and the size of a city.

If firms are studied, we must take into account the death of firms. We might assume that a firm dies when it ceases to have customers. We can imagine that the age distribution in plane 1 of Figure 1 includes the dead firms; they are automatically eliminated in the transformation to size, being transferred to the size class below one. In the exponential relation (1), ϵ must now represent the net rate of growth of the number of firms if the birth of firms is assumed to be a constant ratio of the population.

Figure 1 illustrates how the evolution in time of the number of firms (cities) is mapped onto the cross section of sizes. This process may be compared to sedimentation in geology, where a historical development is revealed in a cross section of the layers. We can also see how irregularities in the evolution over time will affect the size distribution. If an exceptional spurt of births of new firms occurs at one point of time (after a war, for example), the regression line in plane 1 will be broken and its upper part shifted upward in a parallel fashion. The same thing will happen to the transformed distribution in plane 3.

The complete model for firms may be described as follows. The size of a firm is measured by the number of customers attached to it. This is governed by a birth-and-death process. Let us denote by $o(\Delta t)$ a magnitude that is small in comparison with Δt. There is a chance $\lambda \Delta t + o(\Delta t)$ of a customer's being acquired and a chance $\mu \Delta t + o(\Delta t)$ of a customer's being lost in a short period of time, Δt; multiple births and deaths have a chance of $o(\Delta t)$. The probability that a firm has n or more than n customers is given by

$$P(n) = \int_0^\infty P(n, t) r(t) \, dt,$$

where $P(n, t)$ is the probability that a firm of age t has n or more than n customers. The term $r(t)$ is the density of the age distribution of firms, including dead firms; for large t it is the steady state of a renewal process and is given by $r(t) = ce^{-\epsilon t}$, where ϵ is the net rate of growth of the firm population and c is a constant. The number of firms with less than one customer, $P(0, t) - P(1, t)$, equals the dead firms. The value of $P(n, t)$ is obtained as the solution of a birth-and-death process for the customers of a firm:

$$P(n, t) = \left(1 - \frac{\mu}{\lambda}\right)\frac{(1 - \alpha)^{n-1}}{(1 - \alpha')^n}, \quad n = 1, 2, \cdots,$$

where $\alpha = e^{-(\lambda-\mu)t}$, $\alpha' = (\mu/\lambda)e^{-(\lambda-\mu)t}$, and $\lambda > \mu$. This expression can be expanded in series and inserted in the above integral; this yields, integrating term by term,

$$P(n) = C\sum_{\nu=0}^{\infty}\sum_{k=0}^{n-1}\frac{\left(\frac{-\mu}{\lambda}\right)^{\nu+k}\binom{-n}{\nu}\binom{n-1}{k}}{\nu + k + \omega}$$

$$= C\sum_{\nu=0}^{\infty}(-\mu/\lambda)^{\nu}\binom{-n}{\nu}\mathrm{B}(n, \omega + \nu),$$

where $\omega = \epsilon/(\lambda - \mu) > 0$ and $\mathrm{B}(n, \omega + \nu)$ is the Beta integral. Hence,

$$P(n) = C\sum_{\nu=0}^{\infty}\left(\frac{\mu}{\lambda}\right)^{\nu}\frac{1}{\nu\mathrm{B}(n, \nu)}\mathrm{B}(n, \omega + \nu).$$

If $\mu/\lambda < 1$, we can neglect the terms with ν above a certain value. Thus, if $n \to \infty$ and ν has a moderate value, we can use the approximations

$$\mathrm{B}(n, \omega + \nu) \simeq n^{-\omega-\nu}\Gamma(\omega + \nu),$$
$$\mathrm{B}(n, \nu) \simeq n^{-\nu}\Gamma(\nu);$$

therefore, as $n \to \infty$,

$$P(n) \to C'n^{-\omega}.$$

The following features of the solution may be remarked: Since the approximation depends on the value of μ/λ, which is the mortality of firms of high age, the smaller the mortality of firms, the greater will be the proportion of the distribution that conforms to Pareto's law. The mean of the distribution will be finite if $\omega > 1$. This is important in connection with disequilibria, which can arise through changes in λ, μ, and ϵ. It can be shown that the Pareto solution applies to the growing firm ($\lambda > \mu$, the above case) and, in a modified form, to the shrinking firm ($\lambda < \mu$); but it does not obtain for the stationary firm ($\lambda = \mu$).

The above solution for the distribution according to customers can be shown to be valid also for the distribution according to sales, if firms grow mainly by acquiring more customers and not by getting bigger customers. This is often true in retail trade but not in manufacturing. An alternative model assumes the other extreme—that firms grow only by getting bigger orders. This model is based on the theory of collective risk. The capital of the firm, a continuous variable, is subject to sudden jumps at the instant when orders are executed and to a continuing drain of costs, which is represented deterministically by an exponential decline. The steady state solution obtained from this process is, for large values of capital, identical with the Pareto law; for moderate values, the distribution has a mode and represents, albeit with some complications, a modification of the "first law of Laplace," which was proposed by Fréchet (1939) for income distributions.

Size as a vector. It would be natural to measure the size of a firm by a vector, including employment, output, capital, etc., and apply the steady state concept to the joint distribution of several variables. Regression and correlation coefficients obtained in a cross section could then be regarded, like the Pareto coefficient, as characteristics of the steady state. It may be guessed that the growth of the number of firms will have an influence on these parameters as well.

Practically no work has been done in this direction, but it is the only way to clear up the meaning of cross-section data and their relation to time series data and to the theoretical parameters of the underlying stochastic process. The situation in economics is totally unlike that in physics, where the processes are stationary and the ergodic law establishes the identity of time and phase averages. (Only the cosmogony of F. Hoyle, in which the continuous creation of matter offsets the expansion of the universe to establish a steady state of the cosmos, offers a parallel to the growth processes considered above.) The surprise expressed at one time at the difference in estimates of income elasticities from cross-section data and from time series data appears naive in this light because we could only expect them to be equal if the processes generating households, incomes, and consumption were stationary.

But the population of households or the population of firms is not stationary. A cross section of firms shows the growth path of the firm through its different stages of evolution; but the number of firms of a given age depends on the past growth of the total number of firms, and this may influence the regression coefficient. Moreover, the growth path is not unique, because there are several processes superimposed upon one another (growth paths depending on age of firm, age of equipment, age of the management, etc.). For example, the short-run and long-run cost curves are inevitably mixed up in a cross section of firms. [See Cross-section analysis.]

How much "stability" and why. The starting point of the theories here reviewed is the stability of distributions, but stability must not be taken literally. The distributions do change in time, but the change is usually slow. The tail of the distribution of firms or, to a lesser extent, of wealth is composed of very old units, and time must pass before it can be affected by, for example, a change in new entry rates or in growth rates of firms. Thus, the reason for the quasi stability of distributions is that the stock of firms, etc., revolves only slowly. Indirectly this also accounts for the quasi stability of the distribution of incomes, because income is largely determined by wealth or its equivalent in the form of education. An even more enduring influence on the income distribution is the differentiation of skills and professions, which evolves slowly, as a secular process.

The explanations advanced in this article do not exclude the possibility that distribution patterns may change abruptly—for example, as a consequence of taxation, in the case of net incomes; or as a consequence of a big merger movement, in the case of firms.

Josef Steindl

[*Directly related is the entry* Rank–size relations. *See also* Lebergott 1968.]

BIBLIOGRAPHY

▶Allais, Maurice 1968 Pareto, Vilfredo: I. Contributions to Economics. Volume 11, pages 399–411 in *International Encyclopedia of the Social Sciences.* Edited by David L. Sills. New York: Macmillan and Free Press.

Champernowne, D. G. 1953 A Model of Income Distribution. *Economic Journal* 63:318–351.

Frechet, Maurice 1939 Sur les formules de répartition des revenus. International Statistical Institute, *Revue* 7:32–38.

Gibrat, Robert 1931 *Les inégalités économiques.* Paris: Sirey.

Ijiri, Yuji; and Simon, Herbert A. 1964 Business Firm Growth and Size. *American Economic Review* 54:77–89.

Kalecki, Michael 1945 On the Gibrat Distribution. *Econometrica* 13:161–170.

▶Lebergott, Stanley 1968 Income Distribution: II. Size. Volume 7, pages 145–154 in *International Encyclopedia of the Social Sciences.* Edited by David L. Sills. New York: Macmillan and Free Press.

Mansfield, Edwin 1962 Entry, Gibrat's Law, Innovation, and the Growth of Firms. *American Economic Review* 52:1023–1051.

Rutherford, R. S. G. 1955 Income Distributions: A New Model. *Econometrica* 23:277–294.

Simon, Herbert A. (1955) 1957 On a Class of Skew Distribution Functions. Pages 145–164 in Herbert A. Simon, *Models of Man: Social and Rational.* New York: Wiley. → First published in Volume 42 of *Biometrika.*

Simon, Herbert A.; and Bonini, Charles P. 1958 The Size Distribution of Business Firms. *American Economic Review* 48:607–617.

Steindl, Josef 1965 *Random Processes and the Growth of Firms: A Study of the Pareto Law.* London: Griffin; New York: Hafner.

Wold, H. O. A.; and Whittle, P. 1957 A Model Exploring the Pareto Distribution of Wealth. *Econometrica* 25:591–595.

Zipf, George K. 1949 *Human Behavior and the Principle of Least Effort: An Introduction to Human Ecology.* Reading, Mass.: Addison-Wesley.

Postscript

The diffusion process assumed in some of the above models has been studied directly on the basis of individualized data for German retail firms (Steindl 1965) but more recently also for Austrian manufacturing firms (Steindl 1972*a*). The variance of the logarithm of sales is shown to increase with time at a rate that is different in different industries. The question arises whether this diffusion constant has an economic meaning; it is tentatively suggested that it might be regarded, in some sense, as a measure of the "dynamics" of an industry (technological change in the widest sense, with resulting competition).

Wold and Whittle's model of wealth distribution has been reformulated by Steindl (1972*b*) using an age-dependent branching process. The Pareto coefficient of wealth distribution is seen to depend on the speed of accumulation over the generations within a wealth dynasty, and on the rate at which new wealth dynasties appear.

It is noted that a constant Pareto coefficient is compatible with growing concentration of wealth in a few hands, if the sample of wealth holders grows in time, and wealth sizes, which before were mere theoretical possibilities, become actualized.

Josef Steindl

ADDITIONAL BIBLIOGRAPHY

Champernowne, D. G. 1973 *The Distribution of Income Between Persons.* New York: Cambridge Univ. Press.

Ijiri, Yuji; and Simon, Herbert A. 1967 A Model of Business Firm Growth. *Econometrica* 35:348–355.

Ijiri, Yuji; and Simon, Herbert A. 1971 Effects of Mergers and Acquisitions on Business Firm Concentration. *Journal of Political Economy* 74:314–322.

IJIRI, YUJI; and SIMON, HERBERT A. 1974 Interpretations of Departures From the Pareto Curve Firm-size Distributions. *Journal of Political Economy* 82:315–331.

SINGH, AJITH; and WHITTINGTON, GEOFFREY 1975 The Size and Growth of Firms. *Review of Economic Studies* 42:15–26.

STEINDL, JOSEF 1972*a* Diffusion der Umsätze als Maßstab für die Dynamik einer Branche. Österreichisches Institut für Wirtschaftsforschung, *Monatsberichte* 45:56–63.

STEINDL, JOSEF 1972*b* The Distribution of Wealth After a Model of Wold and Whittle. *Review of Economic Studies* 39:263–279.

SKEWNESS

See ERRORS, *article on* EFFECTS OF ERRORS IN STATISTICAL ASSUMPTIONS; STATISTICS, DESCRIPTIVE, *article on* LOCATION AND DISPERSION.

SLUTSKY, EUGEN

Eugen Slutsky, or Evgenii Evgenievich Slutskii (1880–1948), Russian economist, statistician, and mathematician, was born in Yaroslavl province, the son of a schoolteacher. He entered the University of Kiev as a student of mathematics in 1899 but was expelled three years later for revolutionary activities. From 1903 to 1905 he studied engineering at the Institute of Technology in Munich. After the revolution of 1905 he returned to Russia, obtaining a degree in law from the University of Kiev in 1911. For a while he taught jurisprudence in a technical college, but he became interested in political economy and in 1918 he received a degree in that subject and became a professor at the Kiev Institute of Commerce. In 1926 he left Kiev for Moscow to join the Kon'iunkturnyi Institut (i.e., the Institute for the Study of Business Cycles). From 1931 to 1934 he was on the staff of the Central Institute of Meteorology. Moscow State University conferred an honorary degree in mathematics on him in 1934. From that time until his death he held an appointment at the Mathematical Institute of the Academy of Sciences of the Soviet Union.

One of Slutsky's first published works was his famous article on the theory of consumer behavior (1915). Slutsky here developed some ideas of F. Y. Edgeworth's and Vilfredo Pareto's on the relationship between the utility function and prices, income, and consumption. His main achievement was to show that, with money income fixed, any change in the price of a commodity can be divided into two parts. The first part is the change in relative prices, with real (not money) income fixed. This is called the substitution effect; the consumer maintains approximately a given indifference level. The second part is the balance of the price change (a proportional shift in all prices), which can be translated into an equivalent change in money income, with prices constant, causing a variation in real income. This is called the income effect; the consumer shifts from one indifference level to another. The two effects turn out to be independent and additive ("Slutsky's relation").

The distinction drawn between the substitution effect and the income effect is often left ambiguous. There is a choice as to whether to define the holding of real income fixed as (*a*) holding the consumer's utility fixed, or (*b*) holding a Lespeyre measure of the purchasing power of his money income fixed. Slutsky clearly uses (*a*) in his mathematical analysis and (*b*) in his verbal explanation. The difference is, however, of the second order of smalls and unimportant when considering small changes in price. (For a further discussion of this point, see Mosak 1942.)

Although in R. D. G. Allen's opinion (1950) the present theory of consumer behavior is as much a development of Slutsky's work as of Pareto's, Slutsky's paper attracted no attention until Allen discovered it in the mid-1930s (1936).

In his later years Slutsky did very little work in pure economics but made a considerable contribution to mathematical statistics and the theory of probability. His first paper on this subject, "On the Criterion of Goodness of Fit of the Regression Lines and on the Best Method of Fitting Them to the Data" (1913), was written eight years before the work of R. A. Fisher on the same subject. Slutsky was interested in the problem of formalizing the theory of probability. In the article "Über stochastische Asymptoten und Grenzwerte" (1925) he examined the accurate definition of the concept of asymptotic convergence in probability, and so made an important contribution to the theory of probability (see also 1928; 1938).

Slutsky was one of the originators of the theory of stochastic processes. In his work "The Summation of Random Causes as the Source of Cyclic Processes" (1927) he proved that the "periodical" oscillations in economic, meteorological, and other time series do not necessarily show the presence of any underlying periodic cause; such oscillations are typical of all serially correlated random sequences, including those which result from taking a moving sum in a sequence of mutually independent, purely random quantities. By repeatedly smoothing such sequences, one can obtain a new sequence, which will in any limited period of time be closely approximated (with probability one) by a sine curve. This effect of averaging random series was studied by Slutsky independently of G. U. Yule.

Slutsky's study of stochastic processes centered on serial, or lag, correlation, that is, correlation between two members of a time series separated by a fixed time interval. In many important applications serial correlation depends only on interval length, not on position along the time series. The concept of serial correlation has broad applications in many branches of science and technology, including radio engineering. Slutsky's theory was generalized by A. Khinchin (1934).

Slutsky wrote several articles on the estimation of parameters of stochastic processes. His complete works on the theory of probability and mathematical statistics have been published (see *Izbrannye trudy*).

In the last years of his life Slutsky studied the problem of computing tables for functions of several variables. This work resulted in the posthumous publication of *Tablitsy dlia vychisleniia nepolnoi Γ-funktsii i funktsii veroiatnosti χ²* (Tables for the Computation of the Incomplete Γ-function and the χ^2 Probability Distribution; 1950).

A. A. KONÜS

[*For the historical context of Slutsky's work, see the biography of* EDGEWORTH; *see also* Allais 1968. *For discussion of the subsequent development of Slutsky's ideas, see* TIME SERIES; *see also* Georgescu-Roegen 1968.]

WORKS BY SLUTSKY

1913   On the Criterion of Goodness of Fit of the Regression Lines and on the Best Method of Fitting Them to the Data. *Journal of the Royal Statistical Society* 77:78–84. → Published in Russian in 1960 in *Izbrannye trudy* (Selected Works).

1915   Sulla teoria del bilancio del consumatore. *Giornale degli economisti* 3d Series 51:1–26. → Published in Russian in 1963 by the Akademiia Nauk SSSR, in *Economiko-matematicheskie metody*, Volume 1.

1925   Über stochastische Asymptoten und Grenzwerte. *Metron* 5, no. 3:3–89. → Published in Russian in 1960 in *Izbrannye trudy*.

(1927) 1937   The Summation of Random Causes as the Source of Cyclic Processes. *Econometrica* 5:105–146. → First published in Russian. Reprinted in 1960 in *Izbrannye trudy*.

1928   Sur les fonctions éventuelles continuées, intégrables et dérivables dans le sens stochastique. Académie des Sciences, Paris, *Comptes rendus hebdomadaires des séances* 187:878–880. → Published in Russian in 1960 in *Izbrannye trudy*.

1938   Sur les fonctions aléatoires presque périodiques et sur la décomposition des fonctions aléatoires stationnaires en composantes. Pages 33–55 in *Colloque consacré à la théorie des probabilités*. Part 5: Les fonctions aléatoires, by S. Bernstein, E. Slutsky, and H. Steinhaus. Actualités scientifiques et industrielles, No. 738. Paris: Hermann. → Published in Russian in 1960 in *Izbrannye trudy*.

1950   *Tablitsy dlia vychisleniia nepolnoi Γ-funktsii i funktsii veroiatnosti χ²*. Edited by A. N. Kolmorgorov. Moscow: Akademiia Nauk SSSR. → Published posthumously.

Izbrannye trudy: Teoriia veroiatnostei, matematicheskaia statistika. Moscow: Akademiia Nauk SSSR, 1960. → Contains papers originally published between 1913 and 1938.

SUPPLEMENTARY BIBLIOGRAPHY

►ALLAIS, MAURICE 1968 Pareto, Vilfredo: I. Contributions to Economics. Volume 11, pages 399–411 in *International Encyclopedia of the Social Sciences*. Edited by David L. Sills. New York: Macmillan and Free Press.

ALLEN, R. D. G. 1936 Professor Slutsky's Theory of Consumers' Choice. *Review of Economic Studies* 3: 120–129. → A summary of Slutsky's 1915 article on consumer behavior.

ALLEN, R. D. G. 1950 The Work of Eugen Slutsky. *Econometrica* 18:209–216. → Includes a bibliography on pages 214–216.

►GEORGESCU-ROEGEN, NICHOLAS 1968 Utility. Volume 16, pages 236–267 in *International Encyclopedia of the Social Sciences*. Edited by David L. Sills. New York: Macmillan and Free Press.

KHINCHIN, A. 1934 Korrelationstheorie der stationären stochastischen Prozesse. *Mathematische Annalen* 109: 604–615.

MOSAK, JACOB 1942 On the Interpretation of the Fundamental Equation of Value Theory. Pages 69–74 in University of Chicago, Department of Economics, *Studies in Mathematical Economics and Econometrics in Memory of Henry Schultz*. Edited by Oskar Lange, Francis McIntyre, and Theodore O. Yntema. Univ. of Chicago Press.

SOCIAL EXPERIMENTATION

See EXPERIMENTAL DESIGN; PUBLIC POLICY AND STATISTICS.

SOCIAL INDICATORS

See PREDICTION; PUBLIC POLICY AND STATISTICS.

SOCIAL MOBILITY

Social mobility is the movement of individuals, families, and groups from one social position to another. The *theory of social mobility* attempts to account for the frequencies with which these movements occur.

The study of social mobility relates a present to

a past social position. It thus forms part of the more general study of *social selection*, i.e., of how people get distributed into different social positions. It is, however, hardly possible to study effectively the influence of past social position except in the context of other influences that determine the individual's present social status. In practice, it has become increasingly difficult to separate the fields of social mobility and social selection. No rigorous separation has been attempted in the following discussion.

Studies of *intergenerational mobility* compare the social positions of parent and offspring; studies of *career mobility* compare the social positions of the same individual at different times. *Group mobility* is concerned with changes in the social position of groups possessing a relatively homogeneous status (for example, castes, intellectuals, artisans).

In the context of mobility studies, *social position*, or *social status*, signifies a certain rank with respect to the possession of goods (values) esteemed and desired by most members of a society. The changes in social position that interest the theory of social mobility are primarily variations in occupation, prestige, income, wealth, power, and social class. A high or low rank in one of these values is often associated with a roughly corresponding rank in most of the other values; consequently, position with respect to one of these values, and more especially a constellation of them, provides a measure of what in many societies is viewed as success in life. Studies in social mobility do not usually concern themselves with the possession of aesthetic, moral, and spiritual values. This is presumably due to the supposition, correct for most societies, that these goods do not measure "success in life." Nor does their possession seem to lead, except in a limited number of societies, to the attainment of those material goods whose pursuit is more evident in human behavior and whose possession tends to limit the amount possessed by others and to provide opportunities for—or at least the illusion of—control over one's own and others' destinies.

Mobility and moral critiques of society

Political and moral critiques of society have both inspired and been inspired by studies in social and, more particularly, intergenerational mobility. This is understandable in view of the concern of intergenerational studies to show the relation of an individual's life chances (income, occupation, prestige, etc.) to the social circumstances (parental status) in which he is brought up. Casual observation, historical studies, and quantitative inquiry have long made it evident that despite considerable intergen-

erational movement up and down the social ladder, many children of high-status and low-status parents retain, when adults, approximately the same status levels as their parents. Egalitarian sentiments are affronted by a social process that appears to condemn many children to the inferior life circumstances of their families or to guarantee to other children, by virtue of the more favored position of their families, a high degree of fortune. Observation of these continuities in family social position has led to the characterization of societies as "open" or "closed" according to the degree to which the adult status of offspring is independent of (*open society*), or dependent on (*closed society*), the social status of the parents.

Independence is sometimes interpreted, in this context, in its strictly statistical sense: societies are egalitarian (open, democratic) to the extent that children coming from very different family backgrounds have the same probability of achieving a specified status level. This criterion, however, could be met without any commonly accepted view of social justice being fulfilled. This distribution of social positions might equally well be achieved were children assigned to occupations or income levels by a state lottery or by a purely whimsical procedure of an absolute ruler.

Societies are also sometimes viewed as being more or less open according to the frequency with which social mobility occurs. This criterion is not consistent with the criterion of statistical independence. The maximum frequency of social mobility occurs (for example, in a two-status system) when all the children of high-status parents fall to low status and all children of low-status parents achieve high status. This entails certain consequences of dubious accord with commonly accepted notions of social equity: offspring of low-status parents must resign themselves to seeing *their* children excluded from high status; and the future social position of children is as fully fixed by that of their parents as in the most rigid caste society.

Evidently, moral and political critiques of society must rest not on distributions of children's status by parents' status (see Tables 1–4) but on the nature of the processes that produce these distributions. Late eighteenth-century and nineteenth-century writers on these problems were concerned in their critiques of society with the institutional and personal factors involved in mobility; they aspired toward a society in which merit and talent were rewarded and opportunities for their development and exercise were freely available. Contemporary writers have by no means abandoned this theme; but with the increased availability of quantitative data they have been more disposed to substitute summary statistical indices of the frequency of mobility for critical analysis of the total process of social selection. This seems to assume either that certain frequencies are in themselves desirable without regard to the institutional and personal factors involved in social selection or that the frequency of mobility in itself provides secure knowledge concerning the nature and moral legitimacy of the selection system. Neither of these assumptions appears justified.

The mobility table

A central feature of most quantitative studies of intergenerational mobility is the mobility table (see Tables 1–4). An understanding of the technical problems associated with it is essential to the interpretation of the large body of data such tables have made available.

The mobility scale. If no additional assumptions are made, the social positions (e.g., occupations) between which movement may occur remain a series of discrete, qualitatively different categories. The study of social mobility is, however, closely related to questions concerning social achievement and, more particularly, concerning the role of the parental social position in determining filial status. Interest attaches, then, to the amount of *upward* and *downward* mobility (or conversely, the amount of nonmovement, or *inheritance*). The terms "upward" and "downward" imply an *ordering* of the categories along some quantitatively defined axis. When the social positions are defined in terms of (occupational) prestige or income level, an ordering is more readily made, although in the first case not without difficulty. Often, however, the social position categories represent occupational classes (professional, entrepreneurial, clerical, skilled, etc.) or social classes (upper, middle, lower) arrived at, not by one clear-cut criterion (e.g., income, education), but by a mixture of criteria that often leaves the meaning of ordering on a *single* axis, and hence the interpretation of the data, in doubt. The difficulty is not entirely resolved by reducing the multiple criteria to a single score, since a considerable measure of arbitrariness enters into the weight assigned to each criterion in the total score.

Even if a satisfactory ordering is achieved, the categories may not mean—and in the empirical literature generally have not meant—that the social positions represent *scale* positions. It is, therefore, not always possible to say whether movement ("distance") from social position A to social position B is as great as, greater than, or less than, movement from B to C. Recourse is sometimes had to speaking of the number of "steps" (ordered classes) through which movement has occurred. Clearly, however, these steps do not necessarily have the same significance, and this creates difficulties in relating the probability of movement between two positions to the "distance" separating them.

Mobility and number of social categories. A fundamental datum of the mobility table, the proportion of sons who manifest upward or downward mobility (or conversely, inheritance of the parental position), can be made larger or smaller by a simple change in the number of occupational or other classes representing the range of social positions. Thus, the amount of mobility (or inheritance) that a mobility table reveals is dependent on how fine (or broad) the social position categories are—a point that requires particular attention when intersociety comparisons are being made. The tendency toward arbitrariness in the number of occupational or other classes employed is increased both by the practical necessities of research (data availability, manageability) and by the use of multiple criteria in constructing the social-position classes.

Mobility within the life-span. Social position often varies during the life-span of the individual. It would, therefore, be desirable to make comparisons of the social position of father and son at several time points during their careers. This would also throw light on career mobility and bring its investigation into closer relation to intergenerational studies. Practical difficulties, especially in the specification of the father's status at several time points, have generally discouraged the use of this procedure. If only a single time point in each career is used to fix the social position of the father and his son, a choice must be made appropriate to the objectives of the study.

When the aim is simply to measure the frequency with which sons attain a higher, lower, or similar status to that of their fathers, it is reasonable to match their occupations at corresponding ages, preferably a fairly mature age, when the occupational career has become relatively stable. Other choices are to compare the highest status levels achieved in each generation and to compare the occupations pursued over the longest number of years.

On the other hand, when analysis is directed toward understanding why the son attains the particular status that distinguishes his position, quite different considerations enter. The question is then which point in the parental occupational career best aids in the prediction of the filial status. Consequently, the investigator must choose a hypothesis to be tested; the question of comparability in the ages of father and son is not involved.

Sampling problems of father–son studies. Intergenerational mobility tables are usually constructed by obtaining from a sample of subjects (sons) their occupations and those of their fathers. But the probability of major interest is the probability that a son will attain a certain status given that the parent has a certain status. From this standpoint, and some others, it would be preferable to select a sample of fathers and trace the occupations of *all* of their sons, since the desired probability (when calculated from a table based on a sample of sons) is subject to error. This procedure is generally avoided because of the considerable age the fathers would have had to attain in order to ensure that the sons have likewise attained a fairly mature age and stable occupation. In addition, the fathers who have survived to that age would be a biased sample of the fathers of the current generation.

Adjusting for size of occupation. The mobility table provides an estimate of the relative frequency with which the sons of any particular class of fathers will be found in their fathers' occupation (inheritance) or in an occupation of higher or lower status (mobility). However, the entries in the mobility table reflect not only the effect of the father's status on the son's occupational locus but also the effect of the size of each occupational group. The probability that the son of a cabinetmaker will also become a cabinetmaker is a function not only of the special advantages and motivations that may accrue to him from the nature of the paternal occupation (and family circumstances correlated with it) but also of the number of cabinetmakers that the society requires or supports. It is often desirable, then, to separate the component of mobility that is due to the current occupational distribution from the component that reflects the influence of the parental status. One way of effecting this separation is to relate the probability of movement from position E_j to position E_k to the number of positions at E_k currently available in the society.

The original entries, in percentage form, are the probability that a son will be in class E_k given that the father is in class E_j, written

$$P(E_k|E_j) = \frac{\text{sons, with fathers in } E_j, \text{ who enter } E_k}{\text{all sons with fathers in } E_j}.$$

This expression is then divided by the proportion, $P(E_k)$, of all sons who have entered occupa-

tion E_k, which gives

$$M_{jk} = \frac{P(E_k|E_j)}{P(E_k)}$$

where M_{jk}, the mobility ratio, is the transformed entry that expresses the amount of movement (or in the diagonal cells, inheritance) from E_j to E_k relative to the number of "openings" at E_k in the society (see Tables 1 and 2).

Adjustment for occupational birth rate. The chances of movement into a particular occupation are also affected by differences in the birth rate of the various occupational classes. If, for instance, doctors had very few children, then, assuming a constant size of the medical profession, the medical replacements of the next generation will tend to come more largely from the children of other occupational groups.

The effects on mobility of the occupational distribution and of differential occupational birth rates are sometimes termed *structural components* of intergenerational mobility. These are to be contrasted with the effect of parental status (and of factors correlated with it).

Interpreting mobility ratios. The calculation of M_{jk} provides a standard in terms of which mobility may be viewed as high or low. If sons were distributed in occupations on a purely random (chance) basis, then the sons of any given parental class would enter the various occupations simply in proportion to the size of that occupation in the society. In this case M_{jk} has the value of 1.0. A value of M_{jk} greater (or less) than 1.0 signifies that sons from a particular class of fathers are entering an occupation more (or less) frequently than would be expected on a purely chance basis. Thus, deviations of M_{jk} from 1.0 indicate the operation of factors associated with "father's status."

Findings of intergenerational studies

Tables 1–4 provide illustrative intergenerational mobility findings. In Tables 1 and 2 the upper entries in each cell are percentages, and the lower, parenthetic entries are the cell values for M_{jk} (see above). These are summary tables and do not permit analysis in terms of particular age groups or other demographic subdivisions. These and numerous other tables to be found in the literature suggest the following statements:

(1) Mobility tables uniformly show deviation from random distribution, that is, they show that filial status is statistically (and positively) dependent on parental status in varying degrees (see Tables 3 and 4). Impressive as this relationship

Table 1 — Sons' occupation by fathers' occupation, Indianapolis, 1940, percentage distribution and mobility ratios

FATHERS' OCCUPATION

SONS' OCCUPATION	Professional	Semiprofessional	Proprietors, managers, etc.*	Clerical and sales	Skilled	Semiskilled	Unskilled	Protective service.	Personal service	Farming	Total %	Total N
Professional	28.3% (5.1)	15.8% (2.8)	7.7% (1.4)	7.7% (1.4)	3.3% (0.6)	2.5% (0.4)	2.4% (0.4)	2.5% (0.4)	4.9% (0.9)	3.8% (0.7)	5.5	548
Semiprofessional	6.3 (2.0)	19.3 (6.2)	3.4 (1.1)	5.2 (1.7)	2.9 (0.9)	2.1 (0.7)	1.5 (0.5)	0.8 (0.3)	4.3 (1.4)	1.7 (0.5)	3.1	307
Proprietors, managers, etc.*	7.6 (1.1)	3.5 (0.5)	17.6 (2.7)	7.6 (1.2)	4.3 (0.6)	4.1 (0.6)	2.8 (0.4)	6.6 (1.0)	5.5 (0.8)	5.9 (0.9)	6.6	656
Clerical and sales	27.9 (1.3)	17.5 (0.8)	30.6 (1.4)	42.2 (1.9)	19.1 (0.9)	17.3 (0.8)	13.1 (0.6)	22.8 (1.0)	17.1 (0.8)	15.2 (0.7)	22.1	2,188
Skilled	15.4 (0.7)	23.7 (1.1)	14.3 (0.6)	15.1 (0.7)	32.3 (1.5)	18.4 (0.8)	15.4 (0.7)	17.0 (0.8)	22.6 (1.0)	23.1 (1.1)	21.9	2,163
Semiskilled	9.5 (0.4)	12.3 (0.4)	19.8 (0.7)	16.4 (0.6)	26.9 (1.0)	43.2 (1.6)	30.0 (1.1)	31.5 (1.2)	29.9 (1.1)	28.8 (1.1)	27.1	2,678
Unskilled	2.5 (0.4)	2.6 (0.4)	2.5 (0.4)	2.4 (0.3)	5.6 (0.8)	5.3 (0.8)	28.6 (4.1)	8.7 (1.3)	3.7 (0.5)	8.9 (1.3)	6.9	684
Protective service	0.8 (0.4)	1.8 (0.8)	1.6 (0.7)	1.3 (0.6)	2.1 (0.9)	2.2 (1.0)	2.4 (1.0)	8.3 (3.6)	1.8 (0.8)	3.6 (1.5)	2.3	229
Personal service	1.5 (0.4)	3.5 (1.0)	2.1 (0.6)	1.9 (0.6)	3.0 (0.9)	4.3 (1.3)	3.6 (1.1)	1.2 (0.4)	10.4 (3.1)	5.1 (1.5)	3.4	334
Farming	0.2 (0.2)	0.0 (0.0)	0.5 (0.5)	0.2 (0.2)	0.6 (0.6)	0.6 (0.6)	0.3 (0.3)	0.4 (0.4)	0.0 (0.0)	4.2 (3.9)	1.1	105
Total %	100.0	100.0	100.1	100.1	100.1	100.0	100.1	99.8	100.2	100.3	100.0	
Total N	474	114	1,203	1,092	2,729	1,520	720	241	164	1,635		9,892

* Includes officials.

Source: Adapted from Rogoff 1953, tables 4 and 53, pp. 48, 118.

Table 2 — Sons' occupation by fathers' occupation, U.S. national sample, 1957, percentage distribution and mobility ratios

FATHERS' OCCUPATION

SONS' OCCUPATION	Professional	Business	White collar	Skilled manual	Semiskilled	Unskilled	Farmer	Total %	Total N
Professional	40.4% (4.8)	18.3% (2.2)	20.3% (2.4)	8.5% (1.0)	2.3% (0.3)	1.5% (0.2)	2.5% (0.3)	8.4	86
Business	19.1 (1.4)	25.8 (2.0)	17.4 (1.3)	13.6 (1.0)	6.3 (0.5)	6.1 (0.5)	11.2 (0.8)	13.2	135
White collar	12.8 (0.9)	22.5 (1.6)	24.6 (1.8)	15.6 (1.1)	17.2 (1.2)	10.6 (0.8)	8.4 (0.6)	14.0	143
Skilled manual	19.1 (0.7)	15.0 (0.6)	20.3 (0.8)	42.2 (1.6)	28.9 (1.1)	36.4 (1.4)	21.6 (0.8)	26.5	271
Semiskilled	2.1 (0.1)	12.5 (0.7)	10.1 (0.6)	14.6 (0.8)	32.8 (1.9)	27.3 (1.6)	16.5 (1.0)	17.3	177
Unskilled	4.3 (0.5)	1.7 (0.2)	5.8 (0.6)	4.5 (0.5)	10.2 (1.1)	15.2 (1.7)	13.5 (1.5)	9.1	93
Farmer	2.1 (0.2)	4.2 (0.4)	1.4 (0.1)	1.0 (0.1)	2.3 (0.2)	3.0 (0.3)	26.4 (2.3)	11.5	118
Total %	99.9	100.0	99.9	100.0	100.0	100.1	100.1	100.0	
Total N	47	120	69	199	128	66	394		1,023

Source: Adapted from Jackson & Crockett 1964, p. 7.

Table 3 — Sons' occupational status by fathers' occupational status, Great Britain, 1949, percentage distribution

SONS' PRESENT STATUS CATEGORY	FATHERS' STATUS CATEGORY (1)	(2)	(3)	(4)	(5)	(6)	(7)	Total %	N
(1) Professional; high administrative	38.8%	10.7%	3.5%	2.1%	0.9%	0.0%	0.0%	2.9	103
(2) Managerial; executive	14.6	26.7	10.1	3.9	2.4	1.3	0.8	4.6	159
(3) Inspectional; supervisory; other nonmanual (higher grade)	20.2	22.7	18.8	11.2	7.5	4.1	3.6	9.4	330
(4) Inspectional; supervisory; other nonmanual (lower grade)	6.2	12.0	19.1	21.2	12.3	8.8	8.3	13.1	459
(5) Skilled manual; routine grades nonmanual	14.0	20.6	35.7	43.0	47.3	39.1	36.4	40.9	1,429
(6) Semiskilled manual	4.7	5.3	6.7	12.4	17.1	31.2	23.5	17.0	593
(7) Unskilled manual	1.5	2.0	6.1	6.2	12.5	15.5	27.4	12.1	424
Total %	100.0	100.0	100.0	100.0	100.0	100.0	100.0	100.0	
N	129	150	345	518	1,510	458	387		3,497

Source: Adapted from Glass 1954, p. 183.

may appear to casual inspection, it is equally apparent that the sons of most classes of fathers are distributed in substantial numbers throughout most of the status classes. Evidently, then, the status of the father permits considerable variation in the status of the son. A more precise summary statement of the over-all relationship suggested by the available studies is that probably not more than one-quarter of the variance in filial status is accounted for by parental status; and this includes the effect of some factors correlated with, but not properly included in, parental status (e.g., race).

(2) The sons are most heavily overrepresented (as compared with random expectation) in the diagonal cells, that is, in those cells representing inheritance or a continuity by the son of the parental status. This necessarily implies underrepresentation in some other cells. This underrepresentation is generally spread over a larger number of cells and is, therefore, less striking, except at times when the parental and filial statuses are in very marked contrast. It follows, then, that for those sons who enter an occupational class different from that of their fathers, *which* particular other occupational class will be entered is, generally, less dependent upon the paternal status.

(3) The above findings lend themselves to two rather different emphases. On the one hand, it is probably correct to say that only one-quarter or less of the over-all variance in filial status is accounted for by parental status and that consequently other factors, taken collectively, play a more important role in determining the status of the son than does parental status. At the same time, it is possible to select particular cells and quite correctly emphasize the large deviations from random expectation in these cases: for example, the

considerable excess representation in professional occupations of sons of professional fathers; or the considerable deficiency among professional workers of sons of unskilled workers.

(4) The probability that the sons of a particular class of fathers, E_j, will achieve a given status level, E_k, is inversely proportional to the status "distance" between social positions E_j and E_k. Since the status "distance" is itself a function of several variables, such as education and income, this summary formulation cloaks a number of more specific relations of interest (see below).

(5) Much discussion has been devoted in recent years to two questions: (*a*) whether the rate of mobility has changed during the last generation or two; and (*b*) whether European societies show a lower rate of mobility (a greater continuity of family status from generation to generation) than the United States. Despite the number of mobility studies now available, numerous difficulties with respect to their design and comparability preclude confident answers to the foregoing questions. The following statements are therefore *tentative*.

Studies of the United States and most Western industrial societies with increasing urban sectors and considerable provision for education show in the last two generations no substantial over-all change in the tendency of sons to inherit the father's occupational class, at least when these are rather broadly defined. The available data do not suffice unequivocally to detect smaller changes that may in fact have occurred. It is evident that the decline in agricultural employment and some skilled crafts, and the emergence of new occupations, operate to reduce occupational inheritance. But such changes have not been confined to any one generation and therefore do not necessarily

**Table 4 — Sons' social status at age 30 by fathers' social status at age 30,
Denmark, 1954**

FATHERS' SOCIAL STATUS

SONS' SOCIAL STATUS	(1) (Highest)	(2)	(3)	(4)	(5) (Lowest)	Total %	Total N
(1) (Highest)	41.9%	5.1%	2.1%	.9%	.6%	2.3	45
(2)	25.8	30.8	12.8	3.9	2.2	9.4	185
(3)	16.1	30.3	36.0	20.7	13.4	24.1	474
(4)	12.9	23.9	33.3	45.2	36.9	36.9	726
(5) (Lowest)	3.2	9.8	15.8	29.3	46.9	27.4	540
Total %	99.9	99.9	100.0	100.0	100.0	100.1	
Total N	31	234	531	675	499		1,970

Source: Adapted from Svalastoga 1959, p. 324.

produce drastic alterations in the mobility pattern. This is particularly true when social position is defined in terms of occupational prestige. The new occupations may leave unaltered the relative frequency of different status levels in the society and the distribution of the labor force among them. The increased access to educational facilities and the growth of large-scale enterprises also suggest that the data should show a decreasing dependence of filial status on parental status; but it may well be that more limited access to education in earlier years was offset by the correspondingly weaker emphasis on formal educational requirements and a greater reliance on apprenticeship and learning "on the job."

The United States has often been viewed as a society in which individual effort and merit are rewarded more substantially than in European societies and where family background counts for less than elsewhere in the distribution of status positions. Available studies, on the other hand, show no striking differences between the amount of mobility in the United States and western European societies (compare Tables 1 and 2 with Tables 3 and 4). However, most international comparisons have, in the search for comparability, reduced the data of individual studies to a least common denominator which denudes them of much of their value, and it is dubious whether, even so, comparability has been achieved. Nonetheless, there is one characteristic of U.S. society, absent in any similar degree from European societies, that may account for the difference between common assumptions and research findings; namely, a large Negro population subject to severe handicaps in the selection process. The position of the Negro in U.S. society adds to the correlation between parental and filial status in the lower strata of the status hierarchy. When the upward mobility rate of unskilled white workers is considered separately,

it is appreciably higher than the upward mobility rate for the total unskilled group. The assumption that the United States has higher mobility rates than European countries may, then, rest in part on a disregard of a sizable sector of the society.

Finally, if, as suggested above, parental status accounts at most for only one-quarter of the variance in the distribution of filial status, considerable differences could exist in the operation of different social-selection systems without necessitating correspondingly great differences in the specific effect of parental status.

Determinants of social mobility

Although the mobility table has been the principal product of many mobility studies, it leaves unanswered many questions central to the theory of social selection and social mobility. The "variable" parental occupation or status embraces, or rather conceals, a host of more specific influences. Research on social mobility is now increasingly directed toward untangling the roles of these more specific variables and shows a corresponding decrease of interest in simply adding new mobility tables to those already available. Further, with respect to the question of how people get sorted into different occupations or status levels, the mobility table can provide at best only a very partial answer, that is, an answer in terms of the statistical dependence of filial status on parental status. But there are clearly many other factors that determine occupational and status selection.

Father's occupation. Parental occupation or status is related to the probability of filial entry into an occupation in two principal ways: (*a*) the father's occupational status may be correlated with a variety of filial attributes, such as education, intelligence, and race, that affect the son's occupational locus; (*b*) the parental occupational status may affect filial occupational locus more directly:

the father's occupational experience may influence his son's occupational interests and may provide him with special knowledge, experience, incentives, and opportunities for access to it or other occupations.

Educational level. A substantial portion—substantial, relative to other variables—of the variation in status is accounted for by variations in educational level. Educational level is, of course, in considerable measure dependent upon the status level of parents. This dependence is lessened by increases in the society's investment in educational facilities and the degree to which these make educational opportunities available without respect to social origin. To the extent that this occurs, the relation of education to status achievement is freed from an intermediate dependence on parental status. Sons of similar parental status show variations in educational levels, and the effect of these variations on status certainly cannot be ascribed to parental status. However, the extension of educational opportunities may at the same time reduce the correlation between status achievement and education. The more widespread a certain level of education becomes (for example, primary or secondary education), the less will variations in status, especially in the lower reaches of the status hierarchy, depend on variations in educational level—provided, of course, that increased access to education does not change the status significance of the occupational groups.

Intelligence and mobility. Part of the effect of education on status achievement is due to the correlation of education with intelligence. In the process of status selection, variations in intelligence operate to influence the achieved level of social status, both by leading to variations in educational level and (for persons of similar educational level) by facilitating the advancement of those of greater intelligence. Certain educational attainments have, however, become such decisive prerequisites for entry into many occupational positions that high intelligence without the added attainment of a corresponding education is unable to produce its full potential effect. Even were educational opportunities commensurate with intellectual capabilities, the correlation of intelligence with status achievement would be limited by the dependence of achievement on motivational and other personality characteristics. An equally important limitation is the fact that, except in a certain gross sense, intelligence is not an overriding criterion in occupational selection and advancement.

Intelligence as customarily measured depends in part, in its turn, on environmental circumstances. But the present state of investigation also leads one to conclude that, even as currently measured, intelligence has a major genetic component. Consequently, part of the dependence of filial on parental status is due to two sources of correlation between parental and filial intelligence—social and genetic.

Recent studies have been effective in demonstrating the existence of complex interactions between parental status, education, and intelligence and in establishing that each of these variables plays a significant role in determining filial status, both independently and by mediating the influence of the others. But tested models do not now exist that enable one to state quantitatively the probable change in status ensuing from an increment in one of the variables (occurring at a specified stage of an individual's life), given the values of the other variables. The introduction of a genetic component of intelligence complicates the task of building such a model but at the same time appears indispensable if such models are to be used to derive a picture of the evolution of the mobility or social-selection system over time.

Discrimination. In some societies the existence of sizable racial or other groups subject to various modes of discrimination increases the dependence of filial status on parental status. The common obstacle shared by father and son tends to show up as a correlation between their status positions. Since status position as defined by mobility studies does not usually include the criterion of race, this necessarily inflates the degree of dependence of filial status on paternal status.

Other social handicaps. Relatively little attention has been paid in mobility studies to the role of special deficits that are not severe enough to exclude persons from the labor force and yet act as powerful handicaps to occupational achievement. High-grade mental deficiency, physical disabilities, chronic disease, mental disorders, alcoholism, etc., taken together, have a sufficiently high incidence and a sufficiently decided effect on occupational achievement to influence mobility tables. It is possible that a substantial part of the cases of extreme downward mobility can be accounted for in this way.

A neglected source of downward mobility. A further source of downward mobility may be viewed either as the result of a bias in the design of mobility studies or as an intrinsic feature of the parent–son status relationship. Mobility studies generally draw a sample of the gainfully occupied within the age range chosen for the investigation. The sample usually includes unmarried subjects and married subjects with and without children. The father sample, arrived at through the son sample, will, however, necessarily include only persons of the preceding generation who have at least one son and are therefore (in most cases) married. Since marriage and sometimes fertility are associated with greater occupational stability, mental and physical health, and general achievement, the fathers of the subjects will represent a special sample of their generation. If this is viewed as a bias in comparing the status of fathers and sons, it could be overcome by choosing the subject (son) sample only from those in the labor force who are married and have at least one child. Since, however, all sons *must* come from the special (father) sample of the preceding generation, it is more useful to view the differential character of the father and son groups as one source of downward mobility. Given the biological and social significance of marriage and, perhaps, fertility, fathers are, other things being equal, superior as a group to an *unselected* sample of their sons. Consequently, one should expect that for this reason a number of the sons will arrive at status positions inferior to those of their fathers.

How important is parental status?

The variables cited above account for a large measure of the filial status distribution. Does this mean then that the emphasis on the derivation of filial status from parental status has been misplaced? In part, the answer is certainly, Yes. But there is, nonetheless, a danger that the reduction of parental status to a series of more specific and often independently operating factors may lead to a neglect of those sources of influence in the parental occupational situation that exercise a direct influence on the son's occupational destiny by giving him special knowledge, incentives, and opportunities with respect to particular occupations. In dealing with sons who are professionals (or even more so, who are white-collar workers) it is fairly easy to account for their occupational status without making an appeal to the *specific* occupational locus of their parents. But if we are required to predict, not which sons will become professional or white–collar, but rather which sons will become doctors or cabinetmakers, then whether the parent is or is not a doctor or a cabinetmaker is still of considerable importance.

Behind the propensity of some classes of fathers to produce sons who follow in their occupational footsteps appear to lie certain relationships that, however, can be stated only very tentatively. Sons seem to be more likely to pursue their fathers' occupations under certain conditions: (1) if the fathers are self-employed; (2) if the self-employed fathers utilize a substantial capital in the pursuit of their self-employed occupations; (3) if entry into the father's occupation is regulated by licensing, examinations, union control, apprenticeship, or other obstacles that the parental status may aid the son to overcome; (4) if the parental occupation requires special training or education. Naturally, these relationships operate more effectively if the occupation involved provides satisfactory rewards relative to alternatives open to the sons.

There are numerous other individual and family attributes that affect the probability that a child will attain a given status position. The number of children in the family and the birth order of a child may in some institutional settings be particularly important. The motivations and aspirations of young people in different sectors of society obviously play an important role in determining the manner in which various individual and family assets and handicaps exercise their influence.

Social mobility and social structure

Finally, at least a brief word must be said about the relation of social selection and social mobility to the principal institutional structures of society. The research to which we owe the many mobility tables now available has been mostly pursued in Western industrialized societies with large urban sectors. This has made it easier to disregard the role of major institutional differences in the formation of social-selection systems. Current studies have tended to confine their attention more particularly to demographic and technological changes and to the role of educational institutions and the practices which affect access to them. Although there is a considerable literature on employment procedures, the study of social mobility has not adequately taken account of the fact that the occupational distribution process is a dual process, in which two sets of preferences and decisions, those of the employee and those of the employer or manager, confront each other, and that the distributive outcome is affected by the supply and demand of various qualities.

There are other major institutional features that, because of their relative stability, best reveal their relation to social mobility when it is studied over quite long time spans. Even the most stable political, juridical, and economic institutions of a society are of capital importance for the mobility process. Thus, the occupational selection process in Western society is decisively influenced by the nature of the labor contract, by laws relating to freedom of movement, and, ultimately, by the distribution of political power, together with the political and social sentiments associated with this distribution. As increasingly complex models of the social-selection process are developed, it will become necessary to specify more explicitly and exactly the institutional environment to which the model has application. Perhaps, too, it will become possible to relate parameter values to changes in the institutional environment and thus to unify the interests of quantitative research and comparative historical inquiry.

HERBERT GOLDHAMER

[*Other relevant material may be found in* MARKOV CHAINS. *See also* Barber 1968; Berreman 1968; Form 1968; Hodge & Siegel 1968; Lipset 1968; Stinchcombe 1968.]

BIBLIOGRAPHY

Bibliographical resources for the study of social mobility are excellent. For the earlier literature Sorokin 1927–1941 *is still valuable.* Lipset & Bendix 1959 *provides an extensive listing of relevant literature in the course of a broad survey of occupational mobility, as well as a presentation of the authors' own research. Consult also* Mack et al. 1957 *and* Miller 1960. *These sources can be brought up to date by consulting* Sociological Abstracts.

For general surveys of social mobility, see Sorokin 1927–1941; Lipset & Bendix 1959; Barber 1957; Bendix & Lipset 1953. *Studies of social mobility with national, regional, and special occupational samples are too numerous to list in detail. Representative studies for the United States are* Taussig & Joslyn 1932; Warner & Abegglen 1955; Rogoff 1953; Jackson & Crockett 1964; Jaffe & Carleton 1954. *For Great Britain,* Ginsberg 1932; Glass 1954. *For France,* Bresard 1950; Desabie 1956. *For the Scandinavian countries,* Geiger 1951; Carlsson 1958; Svalastoga 1959. *For Germany,* Bavaria, Statistisches Landesamt 1930; Janowitz 1958; Bolte 1959. *For Japan,* Nishira 1957. *For Italy,* Chessa 1912; Livi 1950. *For the USSR,* Inkeles 1950. *For other countries, and for additional material on the foregoing countries, consult* World Congress of Sociology. *On the relation of education and intelligence to social selection and social mobility, see, in addition to the major sources already cited,* Anderson et al. 1952; Conway 1958; Halsey et al. 1961; Duncan & Hodge 1963.

On the technical and methodological problems of studying social mobility, see especially Carlsson 1958; Svalastoga 1959; Duncan & Hodge 1963. *A developing literature on mathematical models of the mobility process includes* Prais 1955; Blumen et al. 1955; Beshers & Reiter 1963; White 1963. *A number of topics in social mobility (career mobility, three-generational mobility, social mobility of women, etc.) receive scant or no attention in this article. However, the reader will have no difficulty in finding material on these topics if he consults the more general works in the literature cited here.*

ANDERSON, C. ARNOLD; BROWN, J. C.; and BOWMAN, M. J. 1952 Intelligence and Occupational Mobility. *Journal of Political Economy* 60:218–239.

BARBER, BERNARD 1957 *Social Stratification: A Comparative Analysis of Structure and Process.* New York: Harcourt.

►BARBER, BERNARD 1968 Stratification, Social: I. Introduction. Volume 15, pages 288 296 in *International Encyclopedia of the Social Sciences.* Edited by David L. Sills. New York: Macmillan and Free Press.

BAVARIA, STATISTISCHES LANDESAMT 1930 *Sozialer Auf und Abstieg im deutschen Volk.* Beiträge zur Statistik Bayerns, Vol. 117. Munich: Lindauer.

BENDIX, REINHARD; and LIPSET, SEYMOUR M. (editors) (1953) 1966 *Class, Status, and Power: Social Stratification in Comparative Perspective.* 2d ed. New York: Free Press.

►BERREMAN, GERALD D. 1968 Caste: I. The Concept of Caste. Volume 2, pages 333–339 in *International Encyclopedia of the Social Sciences.* Edited by David L. Sills. New York: Macmillan and Free Press.

BESHERS, JAMES M.; and REITER, STANLEY 1963 Social Status and Social Change. *Behavioral Science* 8:1–13.

BLUMEN, ISADORE; KOGAN, M.; and McCARTHY, P. J. 1955 *The Industrial Mobility of Labor as a Probability Process.* Ithaca, N.Y.: Cornell Univ. Press.

BOLTE, KARL M. 1959 *Sozialer Aufstieg und Abstieg: Eine Untersuchung über Berufsprestige und Berufsmobilität.* Stuttgart (Germany): Enke.

BRESARD, MARCEL 1950 Mobilité sociale et dimension de la famille. *Population* 5:533–566.

CARLSSON, GÖSTA 1958 *Social Mobility and Class Structure.* Lund (Sweden): Gleerup.

CHESSA, FEDERICO 1912 *La trasmissione ereditaria delle professioni.* Turin (Italy): Bocca.

CONWAY, J. 1958 The Inheritance of Intelligence and Its Social Implications. *British Journal of Statistical Psychology* 11:171–190.

DESABIE, J. 1956 La mobilité sociale en France. *Bulletin d'information* 1:25–63.

DUNCAN, OTIS D.; and HODGE, R. W. 1963 Education and Occupational Mobility: A Regression Analysis. *American Journal of Sociology* 68:629–644.

►FORM, WILLIAM H. 1968 Occupations and Careers. Volume 11, pages 245–254 in *International Encyclopedia of the Social Sciences.* Edited by David L. Sills. New York: Macmillan and Free Press.

GEIGER, THEODOR 1951 *Soziale Umschichtungen in einer dänischen Mittelstadt.* Copenhagen: Munksgaard.

GINSBERG, MORRIS 1932 *Studies in Sociology.* London: Methuen. → See especially pages 160–174, "Interchange Between Social Classes."

GLASS, DAVID V. (editor) 1954 *Social Mobility in Britain.* London: Routledge.

HALSEY, A. H.; FLOUD, JEAN; and ANDERSON, C. ARNOLD (editors) 1961 *Education, Economy, and Society: A Reader in the Sociology of Education.* New York: Free Press.

►HODGE, RORERT W.; and SIEGEL, PAUL M. 1968 Stratification, Social: III. The Measurement of Social Class. Volume 15, pages 316–325 in *International Encyclopedia of the Social Sciences.* Edited by David L. Sills. New York: Macmillan and Free Press.

INKELES, ALEX 1950 Social Stratification and Mobility in the Soviet Union: 1940–1950. *American Sociological Review* 15:465–479.

JACKSON, ELTON F.; and CROCKETT, HARRY J. JR. 1964 Occupational Mobility in the United States: A Point Estimate and Trend Comparison. *American Sociological Review* 29:5–15. → Provides a table based on a U.S. national sample, and reviews earlier national sample studies.

JAFFE, ABRAM J.; and CARLETON, R. O. 1954 *Occupational Mobility in the United States: 1930–1960.* New York: King's Crown Press. → A cohort analysis of the American working force, with projections of mobility trends.

JANOWITZ, MORRIS 1958 Social Stratification and Mobility in West Germany. *American Journal of Sociology* 64:6–24.

►LIPSET, SEYMOUR M. 1968 Stratification, Social: II. Social Class. Volume 15, pages 296–316 in *International Encyclopedia of the Social Sciences.* Edited by David L. Sills. New York: Macmillan and Free Press.

LIPSET, SEYMOUR M.; and BENDIX, REINHARD 1959 *Social Mobility in Industrial Society.* Berkeley: Univ. of California Press.

LIVI, LIVIO 1950 Sur la mesure de la mobilité sociale: Résultats d'un sondage effectué sur la population italienne. *Population* 5:65–76.

MACK, RAYMOND W.; FREEMAN, L.; and YELLIN, S. 1957 *Social Mobility; Thirty Years of Research and Theory: An Annotated Bibliography.* Syracuse Univ. Press.

MILLER, S. M. 1960 Comparative Social Mobility: A Trend Report and Bibliography. *Current Sociology* 9, no. 1:1–89.

NISHIRA, SIGEKI 1957 Cross-national Comparative Study on Social Stratification and Social Mobility. Institute of Statistical Mathematics, Tokyo, *Annals* 8:181–191.

PRAIS, S. J. 1955 Measuring Social Mobility. *Journal of the Royal Statistical Society* Series A 118:56–66.

ROGOFF, NATALIE 1953 *Recent Trends in Occupational Mobility.* Glencoe, Ill.: Free Press.

Sociological Abstracts. → Published since 1952. Consult recent entries under "Social Stratification"; "Sociology of Occupations and Professions"; and "Sociology of Education."

SOROKIN, PITIRIM A. (1927–1941) 1959 *Social and Cultural Mobility.* Glencoe, Ill.: Free Press. → Reprints *Social Mobility* and Chapter 5 from Volume 4 of *Social and Cultural Dynamics.*

►STINCHCOMBE, ARTHUR L. 1968 Stratification, Social: IV. The Structure of Stratification Systems. Volume 15, pages 325–332 in *International Encyclopedia of the Social Sciences.* Edited by David L. Sills. New York: Macmillan and Free Press.

SVALASTOGA, KAARE 1959 *Prestige, Class, and Mobility.* Copenhagen: Gyldendal.

TAUSSIG, FRANK W.; and JOSLYN, CARL S. 1932 *American Business Leaders: A Study in Social Origins and Social Stratification.* New York: Macmillan.

WARNER, W. LLOYD; and ABEGGLEN, JAMES C. 1955 *Occupational Mobility in American Business and Industry, 1928–1952.* Minneapolis: Univ. of Minnesota Press.

WHITE, HARRISON C. 1963 Cause and Effect in Social Mobility Tables. *Behavioral Science* 8:14–27.

WILENSKY, HAROLD L. 1966 Measures and Effects of Social Mobility. Pages 98–140 in Neil J. Smelser and Seymour M. Lipset (editors), *Social Structure and Mobility in Economic Development.* Chicago: Aldine.

WORLD CONGRESS OF SOCIOLOGY *Transactions.* → Published since 1954. Transactions of the first congress, held in 1950, were not published.

Postscript

The study of social mobility has progressed largely along two major fronts. In the first of these, the son's achievement is measured by a numerical prestige score for his occupation. Systems of equations are set up to relate this score to background characteristics of the son, and to relate these background characteristics to one another. Thus the son's prestige score is seen as a function of several interrelated variables, one of which is the father's prestige score, together with statistical error. If we focus on how father's prestige statistically affects son's prestige, then mobility can be seen as, in part, the intergenerational transmission of occupational prestige. Such prestige studies (for example Blau & Duncan 1967; Duncan et al. 1972) have used statistical techniques like multiple classification analysis (Andrews, Morgan, & Sonquist [1967] 1973), path analysis, and structural equations. [*For discussion of some of these techniques, see* SURVEY ANALYSIS; SIMULTANEOUS EQUATION ESTIMATION.]

In the main article, it is mentioned that father's occupation explains about a quarter of the variance in son's occupation in the United States, although it is also mentioned that, as the number of categories increases, the relationship weakens. Blau and Duncan (1967, pp. 169–174) utilized continuous scales of prestige, as described in the previous paragraph, with data collected in the United States in 1962. For sons aged 20–64 at that time, the product–moment correlation between the son's prestige and that of the father (when the son was about 16 years old) was .405. If path analysis is used, we further find that most of the effect of father's prestige is transmitted via its effects upon son's education and son's first job; in a regression equation including all of these variables, the (standardized) coefficient for the direct effect of father's prestige on son's prestige is only .115.

The second major front of development has been along lines laid out in the main article. The present social position of the son and the earlier (or father's) social position are measured categorically, using such categories as occupational groups. Use of the mobility ratio M_{jk} has declined since it is now known that it does not adequately control for the marginal distributions of the mobility table (see Duncan 1966). In more recent work, the impact of the origin and destination distributions upon the interior of the table is represented by multiplicative row-and-column-effect parameters. The population may also be assumed to be subject to other structural effects pertaining to the mobility process itself. Thus, the population may be divided into stayers, along the main diagonal of the mobility table, and movers, off the main diagonal (see Blumen, Kogan, & McCarthy 1955; Goodman 1965; Pullum 1975). Movers may be further classified, for example, according to the direction of their movement (if the occupational groups can be ordered by prestige) and their distance from the main diagonal. The probability that an individual will have a particular combination of current and earlier social positions is then expressed as a product of effect parameters for these various characteristics. (Equivalently, the logarithm of the probability is assumed to be the sum of effect parameters.)

Most of the development of these models is due to Goodman (1971). He was able to describe British data from Glass (1954) remarkably well with such a model. One of the best-fitting models used four types of structural parameters: (1) origin effects, (2) destination effects, (3) effects for the absolute distance from the main diagonal, and (4) effects for the barriers between strata that have been passed over by mobile persons. The parameter estimates are themselves relevant to the grouping of occupations, to an assessment of a group's inertia, or to the permeability of the barriers between adjacent groups. For further discussion of these techniques, with examples, see Bishop, Fienberg, and Holland (1975). [*See also* SURVEY ANALYSIS, *article on* METHODS OF SURVEY ANALYSIS.]

Two other formal developments based on categorical measurement should be mentioned. Spilerman (1972) and others have proposed a dynamic process in which the probability of changing category depends (discretely or continuously) upon time, or age, or duration. In this view, it is explicitly recognized that the observed mobility table is simply a snapshot of the unfolding history of the sampled people and families. Matras (1967) and Pullum (1975) have proposed frameworks in which inter- and intragenerational mobility are combined, with the explicit incorporation of differential fertility.

The main article points out the lack of an adequate articulation between social mobility and social structure. Harrison White (1970) has since developed the notion of vacancy chains, with dual, interlocked sets of jobs and of men. Except for a few bureaucratic fields, however, it is not possible to delineate the set of jobs as distinct from the set of occupants.

Both of the major developments in quantitative sociological methodology have become established principally in the substantive context of social mobility. Specifically, structural equations have been used when social position is measured by a continuous, interval-level variable, and log-linear models have been used when social position is measured by a categorical variable. (An open research question concerns the relationships between the two approaches when the categories may be considered as rounding intervals for an underlying continuous scale.) To an extent, the development of these techniques in conjunction with the analysis of social mobility may be a historical accident, but it reflects the centrality of social mobility to sociological theory and research. We can expect it to continue to be an entry point for techniques with implications for other sociological topics.

THOMAS W. PULLUM

ADDITIONAL BIBLIOGRAPHY

ANDREWS, FRANK M.; MORGAN, JAMES N.; and SONQUIST, JOHN A. (1967) 1973 *Multiple Classification Analysis: A Report on a Computer Program for Multiple Regression Using Categorical Predictors.* Rev. ed. Ann Arbor: Survey Research Center, Univ. of Michigan.

BARTHOLOMEW, DAVID J. (1967) 1973 *Stochastic Models for Social Processes.* 2d ed. New York: Wiley.

BISHOP, YVONNE M. M.; FIENBERG, STEPHEN E.; and HOLLAND, PAUL W. 1975 *Discrete Multivariate Analysis.* Cambridge, Mass.: M.I.T. Press.

BLAU, PETER M.; and DUNCAN, OTIS DUDLEY 1967 *The American Occupational Structure.* New York: Wiley.

BOUDON, RAYMOND 1973 *Mathematical Structures of Social Mobility.* Amsterdam: Elsevier; San Francisco: Jossey-Bass.

DUNCAN, OTIS DUDLEY 1966 Methodological Issues in the Analysis of Social Mobility. Pages 51–97 in Neil J. Smelser and Seymour M. Lipset (editors), *Social Structure and Mobility in Economic Development.* Chicago: Aldine.

DUNCAN, OTIS DUDLEY; FEATHERMAN, DAVID L.; and DUNCAN, BEVERLY 1972 *Socioeconomic Background and Achievement.* New York: Seminar Press.

GOODMAN, LEO A. 1965 On the Statistical Analysis of Mobility Tables. *American Journal of Sociology* 70: 564–585.

GOODMAN, LEO A. 1969 How to Ransack Social Mobility Tables and Other Kinds of Cross-classification Tables. *American Journal of Sociology* 75:1–40.

GOODMAN, LEO A. 1971 Some Multiplicative Models for the Analysis of Cross-classified Data. Volume 1, pages 649–696 in Berkeley Symposium on Mathematical Statistics and Probability, Sixth, *Proceedings.* Volume 1: *Theory of Statistics.* Berkeley: Univ. of California Press.

HAUSER, ROBERT M. et al. 1975 Structural Changes in Occupational Mobility Among Men in the United States. *American Sociological Review* 40:585–598.

HOPE, KEITH (editor) 1972 *The Analysis of Social Mobility: Methods and Approaches.* Oxford: Clarendon.

MATRAS, JUDAH 1967 Social Mobility and Social Structure: Some Insights From the Linear Model. *American Sociological Review* 32:608–614.

PULLUM, THOMAS W. 1975 *Measuring Occupational Inheritance.* Amsterdam: Elsevier; New York: American Elsevier.

SPILERMAN, SEYMOUR 1972 Extensions of the Mover–Stayer Model. *American Journal of Sociology* 78:599–626.

WHITE, HARRISON C. 1970 *Chains of Opportunity: System Models of Mobility in Organizations.* Cambridge, Mass.: Harvard Univ. Press.

SOCIAL RESEARCH, THE EARLY HISTORY OF

This article was first published in IESS *as "Sociology: III. The Early History of Social Research" with two companion articles less relevant to statistics.*

Beginings of research in England

The idea that social topics could be subjected to quantitative analysis first acquired prominence in England in the latter half of the seventeenth century. The influence of Francis Bacon had already created a favorable intellectual climate; now a political motive was added by the growing currency of the notions that good government should be based on precise information and that population was a primary source of national power and wealth. Moreover, the rise of the life insurance business and the general expansion of commerce and trade called for a rational and calculable foundation in statistical fact. Curiosity and fear also played a part in the interest in vital statistics: people wanted to know how many had died in the Great Plague of 1665 and other epidemics and what effect this would have on population growth. Finally, the size of London, compared with that of Paris and Amsterdam, became a matter for English national pride, and there was a demand for statistics illustrating the city's growth.

Political arithmetic. The English political arithmeticians, as they called themselves, never formed an organized school. Single individuals of varied backgrounds pursued political arithmetic as an avocation. They were compelled to work with existing administrative records, which were incomplete, of varying quality, and scattered in parishes all over the country. John Graunt, a London draper who lived from 1620 to 1674, was the first, in his *Natural and Political Observations Made Upon the Bills of Mortality* (1662), to make a systematic analysis of the London parish records on christenings and deaths. In order to assess the validity and completeness of the bills of mortality, he described in great detail how they were recorded and assembled. Graunt drew attention to the fact that the mortality rate was higher in London than in the rest of the country; he estimated, by several independent methods, the total population of the city; he noted the rapid recovery of the London population after the Great Plague; and he estimated the extent of migration into the city. Graunt's friend William Petty, who lived from 1623 to 1687, coined the term "political arithmetic"; he was in turn seaman, physician, professor of anatomy, inventor, and land surveyor in Ireland, as well as being one of the founders of the Royal Society. During the period 1671–1676 he wrote *The Political Anatomy of Ireland*, which was based on his personal observations and experiences and in which he formulated a general theory of government founded on concrete empirical knowledge. Edmund Halley, the astronomer, published the first life tables (1693), based on mortality records of the city of Breslau in Silesia, which had been forwarded to the Royal Society by a number of intermediaries (including Leibniz). Their advantage over Graunt's London bills was that the age of death was recorded. Assuming a stable population and a constant rate of birth, Halley calculated the chances of surviving to any given age. The actuarial techniques needed for life insurance were thereafter slowly perfected, calculations being based on the information accumulated by the insurance companies.

Political arithmetic also had a hand in reshaping the descriptive "state-of-the-kingdom" literature: numerical data were added to geographical, biographical, and historical descriptions. Gregory King, who lived from 1648 to 1712, in his "Natural and Political Observations and Conclusions Upon the State and Condition of England" (1696), divided the entire population into 24 strata, from lords spiritual and temporal to vagrants, and estimated not only the total number of families but also the average family income and the share of the total national wealth that each family enjoyed.

The rise of demography. Demography as a science grew out of political arithmetic. The notion became accepted that regularities in human events, similar to the laws of natural science, existed; initially, these regularities were taken to be a demonstration of divine order and benevolence, but progressively the concept of regularity was secularized. In his "Argument for Divine Providence" (1710), based on the observed number of christenings of infants of each sex in London from 1629 to

1710, John Arbuthnott argued, from the exact balance that he found maintained between the numbers of males and females, that polygamy was contrary to the law of nature and justice. Noting that in every year the number of male births slightly exceeded that of female births, he showed, by analogy with a fair-coin-tossing experiment, that in a large number of binomial trials both an extreme unbalance in the sex ratio and an even split between the sexes were very improbable events. The Reverend William Derham's *Physico-theology: Or, a Demonstration of the Being and Attributes of God From His Works of Creation* (1713), another demographic classic, ran through 13 English editions and one each in French, German, and Swedish.

Interaction of English and Continental work. The works of Derham and the other English political arithmeticians were a major influence upon the physician–pastor Johann Peter Süssmilch, whose *Die göttliche Ordnung in den Veränderungen des menschlichen Geschlechts* (1741) was the most complete demographic compendium of the time. In Süssmilch's work, in particular in the second, enlarged edition of *Die göttliche Ordnung,* published in 1761, the field of demography was clearly defined for the first time. Süssmilch systematically introduced the concepts of fertility, mortality, and nuptiality (marriage rates) and assessed their effects upon population size.

Süssmilch in turn provided much information and inspiration for Thomas Malthus when the latter wrote his influential *Essay on Population* (1798), in which the notion of an inexorable law of population growth was linked with the problems of pauperism and food supply. But in the absence of censuses, demography and vital statistics in England remained on a shaky empirical foundation. At the end of the eighteenth century the most sophisticated analytic and empirical work on population was being carried on in France. Indeed, by that date the promise of political arithmetic as a major instrument of government and as a foundation for an empirical social science had not been fulfilled. The new school of political economists, in particular Adam Smith, was highly skeptical of the methods and data of the political arithmeticians, as well as of their protectionist and state-interventionist economic policies. In over-all terms, political arithmetic split into a number of specialized branches and did not join the mainstream of thought on social theory and on social problems.

Beginnings of social research in France

The social research conducted in France at the time when political arithmetic was flourishing in

England contrasted with it in three respects: because of monarchical centralization, most inquiries were conceived and carried out by the administration; the results were kept secret; and the data related to the kingdom as a whole.

Colbert's inquiries. The search for qualitative or numerical information on French society far antedates Colbert (who was controller general of finances from 1661 until his death in 1683), but it was he who was largely responsible for systematizing the previously scattered efforts in this field. The general inquiries that he instituted consisted of descriptions of the territorial units governed by intendants. Two trial runs (one in Alsace in 1657, the other throughout Alsace, Lorraine, and Trois-Évêchés in 1663) preceded the 1664 inquiry. A uniform circular (Esmonin 1956) asked the intendants for information about the existing maps of the district, ecclesiastical matters (especially the "credit and influence" of the bishops), the military government and the nobility, the administration of justice, and the condition of the district's finances and economic life.

The first special inquiry ordered by Colbert was one into manufactures, in 1665. He also concerned himself with inquiries into the state of the population: in 1667 he issued an ordinance on the keeping of parish registers, and three years later, under the influence of Graunt, he instituted publication of yearly data on births, marriages, and deaths.

Vauban's inquiries. Sébastien le Prestre, marquis de Vauban, undertook several far-reaching inquiries, with the aid of the military authorities. Vauban was commissioner general of fortifications from 1677 until he was forced to retire in disgrace thirty years later. He set out to describe territorial units: his papers contain the *Agenda pour faire l'instruction du dénombrement des peuples et la description des provinces* (possibly dating from 1685), as well as 24 memoirs describing provinces, *élections*, or cities, including the celebrated "Description de l'élection de Vezelay," written in 1686. In his *Méthode générale et facile pour faire le dénombrement des peuples*, written in 1686 (see 1707), Vauban recommended taking the census by counting individuals rather than households. And indeed, between 1682 and 1701 a number of censuses were taken in the manner that Vauban recommended. In subsequent censuses, however, the administration returned to the old method. (For Vauban's other inquiries, see Dumazedier 1968.)

The Grande Enquête. An impressive series of 32 memoirs, describing each of the administrative districts of France, was begun in 1697. The survey was made by the intendants, at the instigation of the duke of Beauvilliers, governor of Louis XIV's grandson, the duke of Burgundy. Known as the Grande Enquête, it was compiled for the purpose of demonstrating to the duke of Burgundy the undesirable consequences of Louis XIV's policy of war and excessive taxation (Esmonin 1954; 1956). Each memoir was based on a questionnaire sent to one of the district intendants and included a description of the territory and of "the nature of the peoples," a census of the population, the number and reputation of the clergy and nobles, and the answers to 15 questions on the district's economic life. Many copies were made for high dignitaries; the intendants also made use of the original memoir for various publications (Gille 1964). It is estimated that about 900 manuscripts relate to this vast inquiry, which served as a guide for the intendants until a new one was directed by Bertin in 1762. L'état de la France, by Henri de Boulainvilliers (1727), sheds light on the contents of the Grande Enquête and contains some savage criticisms.

Research in France in the later eighteenth century

Administrative inquiries were resumed in 1724, shortly after the death of the regent and the end of the financial crisis into which the collapse of "Law's system" in 1720 had plunged the country. With a few exceptions these inquiries were nationwide. However, the general descriptions favored by Colbert and his followers now tended to be replaced by the study of specific problems, such as manufactures, public administration, beggary, and wages. In 1730 Orry, a worthy successor to Colbert, instituted a general economic inquiry, and in 1745 another of his inquiries, on the "resources of the people" and on militia recruiting, led to a census (authentic according to Dainville 1952; fictitious according to Gille 1964). In addition, the intendants were instructed to spread rumors concerning increases in town dues and the raising of a militia and then to make conscientious reports (which have survived) of the citizens' reactions. Great pains were taken to make the answers comparable and to involve in the inquiry scholars from outside the government. A member of the French Academy was entrusted with drawing up the final document.

Scholars and learned societies. About 1750 the French government yielded its place as the leading exponent of social research to learned societies and private individuals, who were to dominate the field until 1804, when Napoleon, then first consul, had himself proclaimed emperor.

At the same time that political arithmetic was losing its impetus in England, it began to take hold in France, where its exponents confined themselves to population studies, thereby avoiding criticisms such as Diderot made of Petty (see "Arithmétique politique," 1751). The advances made in the calculus of probabilities during this period led Deparcieux to draw up a mortality table, published in 1746 in his Essai sur les probabilités de la durée de la vie humaine. Buffon made use of these data in the second volume of his Histoire naturelle, published in 1749.

More original studies tried to measure the population on the basis of the number of births given by the parish registers. Between 1762 and 1770 the abbé Expilly, with the help of a huge number of subscribers and correspondents who answered his questionnaires, published his Dictionnaire géographique, historique et politique des Gaules et de la France (see Esmonin 1957). From 1764 on, he ceased basing his estimate of the total population on the number of hearths. Instead, he had the parish registers examined for the years 1690–1701 and 1752–1763, drawing up lists of names of the inhabitants of each parish in order to establish the population size on the basis of the number of births. The results for 9,000 parishes, published in 1766, showed that contrary to current opinion the population had increased. The population was estimated by Expilly at 22 million for the entire kingdom. The same method was employed in the Recherches sur la population des généralités d'Auvergne, de Lyon, de Rouen et de quelques provinces et villes du royaume, a work published in 1766 under the name of Messance, receiver of taxes for the district of Saint-Étienne (the actual author may have been La Michodière, intendant of Lyons). The author of this work made use of variable coefficients in order to estimate the number of inhabitant–birth ratio in each area (the highest such coefficient used by Expilly was 25); he, too, concluded that the population had increased and estimated it at 23,909,400.

These efforts were crowned by the Recherches et considérations sur la population de la France written in 1774 and published in 1778 under the name of Moheau, secretary of Montyon, the former intendant of Auvergne, who was undoubtedly the real author (Chevalier 1948; Esmonin 1958). Montyon compared the various research methods and came to the conclusion that the firmest foundation was that offered by study of the parish registers and of the number of births. By these means he estimated the population to be 23,687,409. His great originality lies in his analysis of the distribution of the population (by age groups and sex, by civil status, and by marital condition) and of the natural and social factors affecting fertility.

Research undertaken by the Academy. During the latter half of the eighteenth century the French Academy of Sciences was responsible for the development of social research in two areas. The first was the application of the calculus of probabilities to quantitative social data. In this connection, there was a controversy in 1760/1761 between d'Alembert and Bernoulli on the statistical measurement of vaccination, and Laplace wrote a memorandum in 1778 on the ratio in Paris of the sexes at birth, and another in 1786 on births, marriages, and deaths; Condorcet (1785) used the calculus of probabilities to study jury verdicts and election results (see Rosèn 1955; Westergaard 1932).

The second area was that of the technical problems on which the government frequently consulted the Academy. Thus, in 1762 Deparcieux prepared a report for the government on the supply of drinking water to Paris, and in 1764 one on floods. In 1785 the new statutes of the Academy, thanks to Lavoisier and Condorcet, included the specification that an agricultural section should be set up. In the same year Calonne, the controller general, formed an agricultural commission, of which Lavoisier was a member; and the Academy appointed a committee for reforming the Hôtel-Dieu.

Lavoisier's labors on behalf of Calonne's commission supplied material for the publication, in 1791, of *Résultats d'un ouvrage intitulé: De la richesse territoriale du royaume de France* (this work actually appeared under the auspices of the Taxation Committee of the Constituent Assembly). Lavoisier obtained an estimate of cereal production by combining data on the size of the population (derived from Messance and Montyon) with data on consumption. From the population, he passed to the number of plows, to a "hypothetical census" of livestock, and to the area of land under cultivation. He advocated the centralization of official statistics and their publication.

The committee for reforming the Hôtel-Dieu (comprising Bailly, Lavoisier, Laplace, and Tenon) engaged in a vast inquiry, from 1785 to 1789, on the organization of hospitals in France and in Europe. The official conclusions, based on documents, questionnaires, and direct observation, were presented in three reports by Bailly (1790). In addition, Tenon (1788) published a detailed description of how Parisian hospitals were organized and went on to examine their deficiencies and to propose steps toward a more rational organization. Early in 1790 Cabanis published his *Observations sur les hôpitaux,* in which he proposed more radical

measures than those proposed by the commissioners of the Academy. A new generation had appeared, the generation of the ideologues.

The ideologues and the Institut. Social research began again after the fall of Robespierre. In 1795 the Convention set up the Institut National des Sciences et des Arts, intended to replace the academies, which had been suppressed in 1793. The Institut included a "second class," devoted to the moral and political sciences, the official doctrine of which was the ideology laid down by Destutt de Tracy and Cabanis. This doctrine, with its stress on the analysis of language and of signs and its notion of a perceptible relationship between the moral and the physical, influenced empirical social research in ethnography and hygiene and also affected government administration.

In ethnography the outstanding figure was Volney, first a physician and then an Orientalist and traveler, who gave an exact description and systematic analysis of Middle Eastern society in his *Voyage en Égypte et en Syrie,* published in 1787. Taking his inspiration from a questionnaire compiled in 1762 by the German Orientalist Michaelis and from the instructions to representatives of the Ministry of the Interior, he addressed his *Questions de statistique à l'usage des voyageurs* (1795) to diplomatic agents (Gaulmier 1951). Other ethnographic activity included the foundation, in 1800, of the Société des Observateurs de l'Homme in Paris. Two years later the Institut organized a scientific expedition to New Holland, for which Gérando wrote *Considérations sur les diverses méthodes à suivre dans l'observation des peuples sauvages,* published in 1801.

The reform of medical education ordered by the Convention in 1794, under the influence of Cabanis, Pinel, and Bichat—ideologues all three—put great emphasis on hygiene and forensic medicine (Rosen 1946; 1958; Foucault 1963). Hygiene was seen as linked up with welfare work and philanthropy (Cabanis 1803). The minister of the interior under the Directory, François de Neufchâteau, a physiocrat and ideologue, had translations made of a collection of English and German works on "humanitarian establishments." In 1802 the Conseil de Salubrité de la Seine was established. It was used as a model by industrial cities after 1815. In it the physician appeared as a social inquirer and reformer. The hygienist movement continued to develop in the period of industrialization, when it attained its apogee (the complete works of Cabanis were published in the early 1820s). Gérando's career is an illustration of the movement's continuity; he was general secretary of the Ministry of the Interior under the Empire and published *Le visiteur du pauvre* in 1820 and *De la bienfaisance publique* in 1839.

Other research in continental Europe

In other continental European states in the eighteenth century, census-type information was occasionally assembled for the government by individuals appointed specifically for that purpose. No permanent machinery for collecting and tabulating the information existed, and no standardized methods were used. It is therefore not surprising that the results were generally incomplete and unreliable.

The aim of these surveys was, for the most part, to obtain information useful for taxation and military planning—information that was naturally intended to be kept secret. In Austria, Belgium, and several other countries, occasional enumerations of the population, dwellings, livestock and other aspects of agriculture, commerce, industry, and the army were undertaken, usually only for a given region or city. In Denmark population enumerations took place in 1769 and in 1787. In Sweden, by the law of 1686, parish registers had to be kept of the number of births and deaths and of migration, as well as a list of parish members. The size of the population was of particular concern because of a suspected population decline in the early eighteenth century. In 1748 provision was made for the regular deposit and analysis of these records at a central location. The parish clergy completed the local enumeration for their parishes on standardized forms and forwarded them through the church bureaucracy. The summary for 1749, prepared by Per Wargentin, is probably the oldest national census report. Additional data were published in 1761 by the Swedish Academy of Science. Much of this information, however, remained unanalyzed, even though, unlike French work in this area, the Swedish census was made public.

In the numerous German kingdoms, principalities, and free cities of the seventeenth century, civic reconstruction became the major concern after the devastation of the Thirty Years' War. As a result, a need was felt for systematic information about countries and states. The term "statistics" derives from the activities designed to fulfill this need. Originally, "statistics" meant a mixture of geography, history, law, political science, and public administration. Hermann Conring, who died in 1681, was a professor at Brunswick who developed a set of categories for the purpose of characterizing the state. He was especially concerned with interstate comparisons. He was explicit about his method, classified his sources, and gave criteria for evaluating their reliability. By the early eighteenth century his system was being widely taught at German universities to future civil servants. The later academic school of "statistics," whose outstanding representatives were Achenwall, Schlözer, and Nieman, centered in Göttingen and further perfected Conring's system. In the early nineteenth century these descriptive statisticians were challenged by the "table statisticians," who used the increasingly available quantitative data to make numerous cross-classifications.

At this time, too, a number of statistical associations were formed in Germany, and several states created statistical agencies. After lengthy and bitter polemics, the older statistical tradition slowly underwent a three-way division of labor. The academic discipline of political science and public administration (*Staatswissenschaften*) continued the descriptive tradition. Political economy (*Volkswirtschaftslehre*) became established in the universities and combined the historical and descriptive with the newer, quantitative methods. Finally, statisticians monopolized the statistical agencies and census bureaus.

Research in nineteenth-century Britain

British decennial censuses were started in 1801, under the direction of John Rickman. The first three concentrated on the enumeration of inhabitants, families, and dwellings. The clergy filled returns for each parish, and the quality of the returns left much to be desired. The 1831 census was the first to probe for occupation. It gained added importance because precise demographic data were needed for parliamentary reform. This new information was quickly incorporated into such works as Patrick Colquhoun's *Treatise on the Wealth, Power, and Resources of the British Empire* (1814), William Playfair's *Statistical Breviary* (1801), and John R. McCulloch's *Descriptive and Statistical Account of the British Empire* (1837). They were the equivalents of today's statistical abstracts, yearbooks, who's whos, and information almanacs.

Rural surveys. The latter part of the eighteenth century was the period of the agricultural revolution in Britain. Novel techniques of social research were brought to bear upon the pressing rural problems of the day. Arthur Young, who lived until 1820, carried out social investigations of rural areas. Unlike the extensive travel literature that focused merely on the peculiar customs of rural folk, Young's works described the actual way of life of the rural areas and assessed their agricultural resources with a view to improving cultivation and husbandry. Young's extensive travel accounts covered the rural scene in England, France, and Ireland (1771a; 1771b; 1780; 1793).

It remained for the Scotsman John Sinclair to introduce quantitative techniques into rural surveys. Sinclair was a wealthy landowner, scientific farmer, traveler, member of Parliament, and writer on many topics. He was familiar with the German statistical tradition, and when in 1755 a private population census of Scotland was carried out through the agency of the Scottish clergy, he conceived work along the same lines but far more ambitious. This work eventually became a monumental social statistical inquiry so broad that it took over seven years to complete. It was published in 21 volumes, between 1791 and 1799, as *The Statistical Account of Scotland*. Sinclair defined statistics as "an inquiry into the state of a country for the purpose of ascertaining the quantum of happiness enjoyed by its inhabitants and the means of its future improvement" (1791–1799, vol. 20, p. xiv). In several appendixes, contained in the last volume, he gave a clear account of his methodology and techniques. He enlisted the cooperation of the Scottish clergy, from whom he eventually obtained parish accounts for each of the 881 parishes of Scotland. The heart of the inquiry was a questionnaire schedule with over a hundred separate questions. In later stages he also used a shorter form, which merely required the respondents to fill numbers into tables. In the appendixes, he reprinted the 23 follow-up letters which over the years he had sent to nonrespondents, in which he alternately begged, cajoled, argued, and threatened. In the end he was forced to send several "statistical missionaries" to complete certain parish returns. He presented a table of the number of returned schedules by date. He urged the ministers to send him all manner of information and the results of their own studies, in addition to replying to his inquiry. The questionnaire itself is divided into several parts. The first 40 questions deal with the geography, geology, and natural history of the parish. Questions 41 to 100 deal with the population: age, sex, occupation, religion, estate and profession (nobility, gentry, clergy, attorneys, etc.), births, deaths, suicides, murders, and number of unemployed, paupers, habitual drunkards, etc. Questions 101 to 116 deal with agricultural produce, husbandry, and minerals; a series of miscellaneous items at the end inquires into wages, prices, history of the parish, character of the people, patterns of land tenure, and comparisons of present conditions in the parish with earlier periods. Sinclair had a thoroughly modern attitude toward quantification. When he made inquiries about the character of the parish population, he asked whether the people were fond of military life and wanted to know the number of enlistments in recent years. When he asked, "Are people disposed to humane and generous actions," he wanted to know whether they "protect and relieve the shipwrecked, etc." The quality of the returns varied widely; understandably, there were major difficulties in analyzing, summarizing, and publishing them. Sinclair employed several paid assistants to compile county tabulations from the parish returns, but a definitive quantitative exploitation of the data was not published until 1825, after Sinclair had retired from public life. In the end, *The Statistical Account of Scotland* was a collection of the parish accounts, with a few county summary tabulations by Sinclair and his assistants.

Sinclair's *Statistical Account*, however, was a useful precedent for a census and demonstrated that one could be made. He tried to induce European governments to establish a decennial census, and his efforts certainly contributed to the adoption of the census by many nations in the first half of the nineteenth century. His two-volume summary was translated into French. However, his method of choosing the parish as the reporting unit and the clergy as reporting agents was even then outdated, since the parish, as a unit of local government and administration, had already largely broken down under the impact of the industrial revolution and the growth of cities. Later social investigators of industrial and urban problems had to devise a different approach, the house-to-house survey, which Charles Booth was still using at the end of the century.

Research on problems of industrialization. In order to understand the extraordinary outpouring of social research in the period from 1780 to 1840 in Britain, it must be remembered that this was a time of great efforts aimed at reforming outdated social institutions, including the poor laws, the educational system, local government, public health institutions, and Parliament itself. Independent authorities, commissions, and improvement associations were set up and were staffed by lawyers, businessmen, ministers, educators, physicians, and other civic leaders. Many of these reformers doubled as social researchers. Thus, social policy, social research, reform, and legislation formed a part of a single, broad effort. Many physicians were engaged in reform activities and social investigations, for two reasons at least. First, they were daily reminded, through their contacts with working-class patients and the poor, of the magnitude of the problem, as far as health, diet, poverty, and unsanitary living conditions were concerned. Second, the miasmatic theory of the origin and spread of diseases, which was then prevalent, gave their reforming outlook a justification from the point of view of medical science.

The most frequent pattern of action, involving social policy, social research, reform, and legislation, was as follows. First, a social evil was recognized by an individual or a small group, who often initiated research into the topic. Second, as a result of this initiative, other studies and local experiments and improvements were undertaken by larger, more organized groups. Third, these efforts stirred up and molded public opinion, attracted government attention, and finally led to government action in the form of boards of inquiry and royal commissions, which, when successful, led in turn to legislation attempting to correct the evil. Finally, legislation provided for inspection systems and other institutionalized means of controlling the implementation of social change.

Of the numerous reformer–investigators, perhap the most outstanding were Howard, Eden, Kay-Shuttleworth, and Chadwick. John Howard, a country squire, became a tireless researcher into prison conditions and an advocate of prison reform. By his own account, he traveled over 42,000 miles, all over Great Britain and Europe, during his investigations. In *State of the Prisons* (1777) and *An Account of the Principal Lazarettos in Europe* (1789), he described in detail the conditions within hundreds of prisons and prison hospitals: how the prisoners passed the time; what they were given for food; what illnesses and discomforts they suffered; the manner of prison administration; the number of prisoners, by sex and crime category, in every English prison; and so on. His books were filled with comparisons of the treatment of criminals in different countries and with suggestions for improvements. Thanks to his efforts and those of his backers, more humane treatment was introduced in many prisons.

Sir Frederick Morton Eden was a businessman. The inflation of 1794–1795 was the immediate cause of his empirical investigation into the number and conditions of the poor. He visited many parishes and carried on a voluminous correspondence with the local clergy. The fruit of his efforts, the three-volume *State of the Poor*, was published in 1797. Most of the work is taken up by parochial reports, which contain information, some of it in numerical detail, on the size of the population, the number of houses that paid taxes, the principal manufactures, the typical wages in the principal

occupations, the rent of farms, the prices of foods, the friendly societies and their membership, the number of poor, and detailed description of conditions of the parish workhouses, among other things. He also presented 43 detailed family budgets of laborers, weavers, miners, masons, and other workers. He used a paid investigator for much of this detailed work, an innovation later adopted by the statistical societies. He based his recommendation for reform of the poor-law system on his findings. His budgetary studies and detailed empirical method represent an innovation in social research which was later developed further by Le Play and his school.

The physician James Phillips Kay-Shuttleworth was an active sanitary reformer, one of the founders of the Manchester Statistical Society, and, later, assistant poor-law commissioner. His early surveys were published as *The Moral and Physical Condition of the Working Classes Employed in the Cotton Manufacture in Manchester* (1832). These set a model for the more comprehensive surveys undertaken by the statistical societies. After 1840 he devoted his energies to introducing and developing a national system of education.

Edwin Chadwick was a civil servant who was active on parliamentary and other commissions throughout his life. More than any other person he was responsible for the *Report on the Sanitary Condition of the Labouring Population of Gt. Britain*, published in 1842, which eventually led to the Public Health Act of 1848 and the establishment of the Central Board of Health. The report set a precedent for subsequent administrative and parliamentary investigations into social problems.

In the 1830s a large number of local statistical societies were founded by citizens active in social reform. The two oldest and most active, those of Manchester and London, have survived to the present day. In addition, a statistical section of the British Association for the Advancement of Science was founded in 1833, at the insistence of Babbage, Malthus, and Quetelet.

The aim of these statistical societies was social improvement based on matter-of-fact, quantitative inquiries into problems of society. The societies organized committees of inquiry to carry out research into the health, living conditions, education, religious practices, and working conditions of the lower classes. The surveys were often large undertakings that took several months and many hundreds of pounds to complete. Paid agents were sent to hold door-to-door interviews based on a prepared schedule of questions. Results were tabulated centrally and presented at the annual meetings of the

British Association or of the societies themselves and often appeared in such publications as the *Journal of the Royal Statistical Society.* Dozens of such surveys were completed, some making use of quite sophisticated multivariate cross-tabulations. Other studies were based on existing institutional records.

Decline of research related to social problems. However, toward the end of the 1840s a decline in social research set in. The major aims of the researcher-reformers were increasingly fulfilled as Parliament passed the Factory Acts and numerous other bills and measures; the economic conditions of the working people noticeably improved, and their political activities collapsed with the defeat of Chartism. Few original surveys were undertaken, and the methodology of those that were undertaken was inferior. Many of the local statistical societies themselves passed out of existence. In Manchester a large number of the original founding members became increasingly absorbed in civic and political activity. In the London Statistical Society, interest shifted to public health, vital statistics, the health of troops, mortality in the colonies, the duration of life in different occupations, etc., for which secondary analysis of published statistics sufficed. There was generally a lack of interest in methodology, whether in questions of study design, techniques of data collection, or analysis. For a time the National Association for the Promotion of Social Science, founded in 1857, rallied individuals concerned with the empirical investigation of labor relations, education, and social problems. But increasingly it was social Darwinism which shaped the intellectual life of Britain. The separation of the social research organizations from any academic context accounts in part for the lack of institutionalization of social research in Britain in the mid-century period. Absence of continuity in recruiting interested researchers, of continuous improvement in the methods of research, of a tradition oriented beyond the short-range goals of social betterment, and of regular financial backing—these were some of the disadvantages that this lack of institutionalization entailed.

The discovery of evolution, the advances in the biological sciences, and the increasing acceptance of race and heredity as fundamental categories in social analysis produced a shift in the intellectual climate. Whereas earlier researchers linked crime with indigence and lack of education, the newer outlook searched for evidence of hereditary degeneracy and other physical and psychological impairments. The lower classes, the destitute, criminals, and other unfortunate groups were often considered an inferior species of humanity. Social policy to ease their condition was viewed by many with alarm because it would prevent the fittest from being the only ones to survive and would thus slowly degenerate the stock of the entire nation. The same sort of educated, middle-class or upper-middle-class individual with a scientific turn of mind who in the 1830s and 1840s might have joined the statistical societies and conducted door-to-door parish surveys of the working people later in the century became a member of anthropological, ethnographic, and eugenics societies and spent his time in the study of primitive cultures or modern genealogies, making hundreds of cranial and other physical measurements in endless efforts at classification and typology. The burden of proof devolved upon the environmentalists.

Booth and the study of the poor. Starting in the 1880s there took place a revival in the scientific study of the poor, which culminated in the monumental social investigations of Charles Booth. Booth came from a wealthy family of Liverpool shipowners. His object was to show one-half of London how the other half lived, more particularly, "the numerical relation which poverty, misery, and depravity bear to regular earnings, and to describe the general condition under which each class lives" (1889–1891, vol. 1, p. 6). Gathering around him such researchers as Beatrice Potter Webb and Octavia Hill and other social workers and social economists, Booth started work in 1886 on what was to become the *Life and Labour of the People in London,* published from 1889 to 1891 in many volumes and several editions. London at the time already had four million inhabitants. Booth at first collected available information from the census and from the four hundred school-attendance officers, who kept records on every poor family. These records were cross-checked with information available to the police, sanitary inspectors, friendly societies, and the numerous charitable organizations and agencies dealing with a wage-earning clientele. Later, Booth followed up these data with participant observation of particular streets and households, and for a time he was himself a lodger with various workingmen's families.

Booth divided the families of London into eight classes, according to the amount and regularity of their earnings and, for the well-to-do, the number of their domestic servants. In his analysis he proceeded to characterize each city block according to the predominant class of the families in it. Assigning a color to each of the eight classes, he prepared colored maps of the entire city. He proceeded to describe, district by district, street by street, the style of life, the problems, and the prospects of the families living in them, including their religious practices, their recreation, and the use made by them of public houses and of the voluntary organizations to be found in their district. He devoted several volumes to description of the wages and working conditions in the trades of the city. The end result was the most detailed and large-scale social description ever achieved, which stirred up the contemporary social conscience and eventually led to the Old Age Pension Act of 1908, a legal minimum wage in the "sweated" trades, state provision for the sick and the disabled, and the start of unemployment insurance.

Booth's work inspired urban surveys by other investigators, who perfected his techniques. The most notable of these was B. Seebohm Rowntree, whose *Poverty: A Study of Town Life*, published in 1901, dealt with one city, and Arthur L. Bowley and A. R. Burnett-Hurst, whose *Livelihood and Poverty*, published in 1915, was a multicity study, the first in which sampling was used systematically in place of a complete enumeration. Booth's analysis of the causes of poverty left much to be desired, as contemporary statistical critics pointed out. One of these, G. Udny Yule, using Booth's data, subjected social data for the first time to multiple-regression and correlation techniques (1899). However, the systematic application of the statistical techniques developed by Galton, Edgeworth, Pearson, and Yule had to wait until the twentieth century, when these techniques entered social research by way of the biological, agricultural, and psychological sciences [see STATISTICS, *article on* THE HISTORY OF STATISTICAL METHOD].

Research in nineteenth-century France

The creation in 1800/1801 of the *Statistique de la République* (Gille 1964, pp. 121–147) and the suppression in 1803 of the second class of the Institut gave the lead once more to governmental inquiries. Until 1806 there prevailed great enthusiasm for statistics, sustained by the *Annales de statistique*, published from 1802 to 1804, and the Société de Statistique, which lasted from 1803 to 1806. The Bureau de Statistique published statistical memoirs by the various *départements*; these memoirs were based on the general inquiry ordered in 1801 by Chaptal, then minister of the interior (Pigeire 1932), and by Duquesnoy. The questions asked of the prefects relate to the location, condition, and movement of the population in 1789 and 1801; the "state of the citizens" and the changes therein from 1789 to 1801 for six categories; the religious and lay ways of life, habits,

and customs; and finally, the changes of agriculture and industry since 1789. Many errors were discovered, leading to cessation of publication and reorganization of the Bureau de Statistique in 1806. However, works based on the data collected continued to appear until about 1810. The Bureau also received regular reports from the prefects and undertook such special inquiries as the so-called census of 1801, which was conducted by the *maires* (Reinhard 1961), and an estimate of the population carried out in 1802 with the collaboration of Laplace and described in his *Théorie analytique des probabilités* (published in 1812). After 1806 the Bureau was in charge of Coquebert de Monbret, who was interested in inquiries on special subjects: an industrial and an agricultural inquiry in 1806, and the next year an inquiry into what was described as the means of support, kind of occupation, and various religions of more than sixteen social categories. A second census of the population was carried out in 1806.

The idea of a general inquiry was picked up again in 1810. A questionnaire containing 334 items was sent to the prefects, and a general estimate of the population was made in the following year. But the failure of the general inquiry led to a return to special inquiries (for instance, into industry and communications), which taken together form the *Exposé de la situation de l'Empire, présenté au corps législatif dans sa séance du 25 février 1813* by Jean-Pierre Montalivet, minister of the interior. The Bureau de Statistique had already been abolished, in September 1812, and from that time on the collection of industrial and agricultural statistics was a government prerogative.

Research on problems of industrialization. The importance given to social research by the various revolutionary governments explains the initial distrust of such studies on the part of the monarchy, which had been restored in 1815. However, urbanization and industrialization were accompanied by social problems, which were played up in parliamentary debates and in the press, and a number of institutions were created in order to study these problems.

The philanthropic movement continued under the Restoration, with the Société Royale pour l'Amélioration des Prisons and, outside Paris, the Société Industrielle de Mulhouse. Villermé's early classic of penology, *Des prisons, telles qu'elles sont, et telles qu'elles devraient être*, appeared in 1820. Public health research founded on observation and quantification increased (see Ackerknecht 1948). The force of the movement resulted in the establishment of the Académie Royale de Médecine (1820), the Conseil Supérieur de Santé (1822), the creation of health councils (*conseils de salubrité*) in the provinces between 1822 and 1830, and the founding in 1829 of the *Annales d'hygiène publique et de médecine légale*, which affirmed the role of the physician as investigator and social reformer.

Statistical research had been resumed with the census of Paris ordered in 1817 by the prefect Chabrol and published in the six volumes of *Recherches statistiques sur la ville de Paris et le département de la Seine* (see Seine 1821–1860). These volumes contained data not only on the population but also on goods consumed, levels of wealth, causes of death, suicide, etc. There also appeared the *Compte général de l'administration de la justice criminelle* (see France, Ministère de la Justice), which classified the types of crimes, as well as the criminals, and the *Comptes présentés au roi sur le recrutement de l'armée*, in which the conscripts' degree of education was made public (see France, Ministère de la Guerre). Publication of this information made possible the rise of "moral statistics," a field in which one of the earliest achievements was a study of crime and education by Balbi and Guerry (1829).

The July revolution of 1830 promoted social research in two ways. The reinstatement in 1832 of the second class of the Institut, under the title Académie des Sciences Morales et Politiques, enabled public health workers and statisticians to undertake their own research, since the new academy sponsored prize competitions. Two winners of these competitions were Frégier's *Des classes dangereuses de la population dans les grandes villes*, published in 1840, and Buret's *De la misère des classes laborieuses en Angleterre et en France*, published the same year. In 1832 there was created the Statistique Générale de la France, which began its publications in 1834 and in 1836 carried out the first trustworthy census of the whole of France in 15 years and which provided scholars with extensive and solid data.

Social research was also stimulated by the labor question, to which Villeneuve-Bargemont had drawn attention in 1828 with his unpublished *Rapport sur le département du Nord* (see Beautot 1939–1943) and which could no longer be ignored after the insurrections of 1831 and 1834. There were an increasing number of inquiries and books devoted to the working classes (Rigaudias 1936). In 1834 the Académie des Sciences Morales directed Benoiston de Chateauneuf and Villermé to find out as precisely as possible the physical and moral condition of the working classes. Villermé's researches, conducted in the mill towns from 1835 to 1840,

were published in 1840 as *Tableau de l'état physique et moral des ouvriers employés dans les manufactures de coton, de laine et de soie*.

Guerry's "statistique morale." Guerry, an attorney born in Tours, took an interest in the official statistics on crime. In his *Essai sur la statistique morale de la France* (1833) he studied the relationship of two social variables, the crime rate and the level of education. He came up against two methodological problems: the absence of any measure of statistical correlation, and the difficulty of using collective measures, such as the crime rates or average level of education for a whole county, to explore questions of individual behavior, such as whether the better educated commit more or fewer crimes than the less well educated. Having established, over a six-year period, that the crime rate tended to remain stable—a fact he attributed to the systematic and constant nature of the causal forces at work—Guerry tried to compare the ecological distribution of criminality with that of education, ranking the 85 *départements* of France by their crime rate, on the one hand, and their level of literacy, on the other, and then looking to see whether a county at the extreme of one of these two distributions had the same position on the other. Since the counties ranked at the extremes of the distribution for education were not the same counties that were at the extremes of the distribution for crimes against the *person* but were the same counties as those at the extremes of the distribution for crimes against *property*, Guerry was led to conclude that there was no negative correlation between education and criminality and that the intervening variable might be level of industrialization. (For other research by Guerry, see Douglas 1968.)

Parent-Duchâtelet and prostitution. Parent-Duchâtelet, a member of the group that had founded the *Annales d'hygiène*, was responsible for two important works: *Hygiène publique* (1836), which brought together the 30 reports drawn up since 1825 for the Conseil de Salubrité de la Seine, and a two-volume work entitled *On Prostitution in the City of Paris* (1834), one of the best inquiries of the period in this field. The origins of this inquiry were both philanthropic and administrative. A philanthropist friend of the author's, eager to help prostitutes by publishing the truth about them, discovered that they lived in a world apart (what would nowadays be called a subculture) and that the first thing that had to be done was to get to know them. The municipality, which had succeeded in curbing prostitution and as a result was receiving requests for information from abroad, wanted an "evaluation" of what it had done [*see* EVALUATION RESEARCH]. In his research Parent-Duchâtelet combined the use of documents, such as police files; personal observation in the field and interviews (something unusual in the France of that period, when researchers, including Villermé, obtained their information indirectly, through observer reports); and statistical method (he compiled about 150 tables). He tried to determine the number of prostitutes and how the total varied over time; the prostitutes' regional and social origins; their physiological and social characteristics; their attitudes toward institutions such as marriage and religion; the reasons that had led them to a life of prostitution; and ways in which they left it. The study ended with an acknowledgment of the inevitability of prostitution and set forth arguments for the moral and material necessity of caring for prostitutes (a matter debated at the time) and of providing institutions designed to receive repentant prostitutes. Parent-Duchâtelet appended a draft law for checking the offenses against public decency that were caused by prostitution. In its scientific neutrality on a problem laden with moral taboos, as well as the use of direct observation and interviews, his whole approach was astonishingly modern.

Villermé's study of textile workers. Born in Paris in 1782, Villermé was 52 years old when he began his great inquiry into the conditions of textile workers. He spent six years observing workers in the most important centers of textile manufacturing. His two-volume report is the crowning work of a career that began in 1819 and was devoted to both hygiene and statistics. Between 1819 and 1834 Villermé published more than forty articles, in seven different journals (Guérard 1864). Some of these articles are simply collections of observations on problems of hygiene or other social problems; they reveal Villermé's skill in observation, which he had developed while serving as a surgeon in the Napoleonic armies from 1802 to 1814 and while studying to become a physician (he received his medical degree in 1818). Beginning in 1822 he became interested chiefly in the mortality statistics of Paris and of France, especially as they were related to income: he wrote several papers showing that the mortality rate of the poor was much higher than that of the well-to-do (Vedrenne-Villeneuve 1961).

In his study of the textile workers, Villermé used both statistical data and his own qualitative observations. He found some statistics in the annual reports of the *départements*, but for the most part he collected them himself. The statistics dealt with the number of workers (difficult to establish in the absence of a census by occupation, which did not exist in France until 1851); the average rate of pay for different kinds of workers (the data were provided by owners and foremen); the length of the working day; demographic information (births, marriages, number of children, number of illegitimate births); and the budgets of working families. Villermé's own observations concerned the cleanliness of the workshops and of the workers' dwellings, and he took note of the workers' clothing and diet and various aspects of their behavior (for example, the amount of alcoholism and prostitution).

Both when he used statistics and when he used qualitative observations, Villermé made use of indicators, without designating them as such. Thus, he took a high number of illegitimate births to be a reliable index of the disruption of customs, and he interpreted qualitative indices such as being paid monthly (rather than by the day or week), having wine with Sunday dinner, using window curtains, or owning an umbrella as signs of affluence.

Like Guerry in his *Statistique morale*, Villermé ran into two difficulties in his interpretation of statistical data: first, the lack of a measure of correlation; second and more important, the ecological character of the data, which described predominantly working-class neighborhoods but not the workers themselves. These handicaps made it necessary for him to perform statistical calculations that were often very interesting, for example, when he demonstrated that at Amiens more than 70 per cent of working-class conscripts were rejected by the army for reasons of health, as against 50 per cent of nonworking-class conscripts (1840, vol. 1, pp. 311–317).

In the first volume of his work, Villermé brought together all his facts, arranging them in almost the same way for every industrial area he studied, a procedure that enabled him to present a systematic description. The second volume, which he presented in 1837 to the Académie des Sciences Morales, covers most of the same categories in an analytical way.

Villermé's inquiry is methodologically less highly developed than that of Parent-Duchâtelet. He did not interview workers, with the exception of the silk workers of Lyon; he used only informants. But Villermé's subject was far more controversial. At the end of his careful description, he maintained that the lot of the workers had slightly improved, a conclusion that was attacked in socialist circles, especially by Buret. At the same time, Villermé's facts about wages and budgets dramatically revealed the inadequacy of the workers' resources and their miserable living conditions, and his denunciation of the employing of very young children in factories made him the target of attacks by orthodox liberals and supporters of the established order. The controversy about Villermé has continued into recent times: Rigaudias (1936) accepted without question the position taken by Buret, while Fourastié (1951) applauded the precision of Villermé's descriptive and statistical information and the solidity of his conclusions.

The facts about child labor collected by Villermé were discussed and criticized in England, in the debates of May and June 1839 in both the House of Lords and the House of Commons, and they played a part in the enactment of the law regulating child labor that was passed on March 22, 1841.

Inquiry into agricultural and industrial labor. The National Assembly that was elected after the fall of Louis XVIII decided in May 1848 to undertake an inquiry into the state of agricultural and industrial labor. The inquiry took up the same subjects that Guerry, Villermé, and Buret had dealt with. Completed by the end of 1850, it was the last large-scale official inquiry that sought to study major problems in France as a whole. After 1852 the government authorized either statistical surveys of the whole country or detailed studies of limited problems. As for studies by individuals, these were hereafter generally monographs; the work of Le Play illustrates this development (see Pitts 1968).

The inquiry by the National Assembly precipitated a lively ideological debate concerning social research (Rigaudias 1936). Between 1840 and 1848 the socialists had repeatedly and vainly demanded an official inquiry into working-class conditions (Ledru-Rollin's Workers' Petition was rejected by the Chamber of Deputies in 1845); but now, with Louis Blanc as their spokesman, the socialists demanded the establishment of a ministry of labor. Moderates and conservatives, on the other hand, opposed such a ministry, favoring instead an inquiry of the kind they had earlier rejected. Social research based on the observation of facts became identified with moderate and conservative bourgeois ideology and was therefore rejected by the various strains of socialist thought. Open socialist opposition to such research continued until the time of Durkheim and his school.

The questionnaire used by the inquiry contained 29 questions about major problems in each district, the general state of industry, the economic and social condition of the industrial workers, and the general state of agriculture. Answers were to be supplied by district commissions, presided over by the justice of the peace of the district and composed of one employer and one worker for each industrial specialty. The carrying out of the inquiry was hampered by the imprecision of the questionnaire (stressed in Gille 1964), and by doubts, shared by the government and the workers, as to its usefulness. (In the opinion of Rigaudias, the workers' lack of confidence constituted the essential obstacle.) Although responses were received from 76 per cent of the districts—a high rate of response—the Assembly ordered, after a brief debate, that the questionnaires be placed in the files of the Ministry of Industry and Commerce, with no provision for publication of the results.

Even to the present day, only a small part of the results of the inquiry of 1848 has been published by historians (Gossez 1904; Kahan-Rabecq 1934/1935; 1939; Vincienne & Courtois 1958). There are gaps in the documents in the Archives Nationales: missing are the materials on 22 *départements,* on almost all the large industrial centers, and on all the large industrial cities except Marseilles. Investigations, made in the archives of the *départements,* of the working of the commissions (Vidalenc 1948; Agulhon 1958; Guillaume 1962; all of these are cited in Gille 1964, p. 216) show that the answers to the questionnaire reflected the point of view of the leading citizens of the towns—sometimes, indeed, simply those of the presiding justice of the peace.

The failure of the great inquiry of 1848 served as a justification for the abandonment of national social inquiries conducted by representatives of the government. It also led to the breach between socialism and empirical social research. The publication in 1855 of Frédéric Le Play's *Ouvriers européens* marked the beginning of a new kind of research and was the most important event in the history of social research during the Second Empire. Like the hygienists earlier, especially Villermé, Le Play used direct observation and the monographic method. He was more like the traditionalists and the socialists (and unlike the hygienists) in basing a scheme for the global reorganization of society on his "method of observation," yet he differed from the socialists in striving for the restoration of traditional principles, rather than for a new social order.

Finally, it may be noted that Le Play, like all social philosophers and researchers before him, remained outside the university system. Only with the generation of Tarde and Durkheim and the establishment of the Office (later Ministry) of Labor, endorsed by the parliamentary socialists, was there a reconciliation of empirical social research with socialist thought and with the university system.

Research in nineteenth-century Germany

Unlike social research in Britain, which was mainly the work of private individuals and voluntary associations, the bulk of German social research in the nineteenth century was carried out by academic scholars and professional organizations. At that time the German universities constituted the most advanced system of higher education in the world. The historical school of economics, which rejected the views of the British political economists, contributed most to social research and to the gradual emergence of sociology as a distinct discipline. These economists sympathized with the underprivileged and firmly believed that research would contribute to a progressive social policy and the solution of social problems.

In the 1860s and 1870s, as a result of Quetelet's influence, moral statistics and demography became important areas of research in Germany. Ernst Engel, who was to become the head of the Prussian and German statistical bureaus, made use of both Le Play's and Quetelet's budget data and methodology to demonstrate his "law." Engel's law stated that regardless of the total size of the family budget, there exists the same priority of needs, as measured by the total amounts spent on certain types of expenditures and that, furthermore, the poorer a family, the larger will be the proportion of its budget spent on food alone. M. W. Drobisch, G. F. Knapp, Adolf Wagner, Johannes Conrad, Wilhelm Lexis, Alexander von Oettingen and others subjected statistical data on crime, suicide, marriage, jury convictions, conscripts' characteristics, and other economic and social data to analytic scrutiny in their varied attempts to establish empirical constancies in social life. Unfortunately, these attempts were involved with the prolonged and sterile debate on free will and determinism that Quetelet's ideas had provoked. However, more and more areas of social life were gradually made the object of quantitative study: occupations and social mobility, higher education, voting, the circulation figures of newspapers, the clientele of public libraries, the composition of military and political elites, and many others. At best, important monographs on single topics were completed. At worst, voluminous compilations of statistics with no theoretical underpinning were endlessly gathered and offered as proof that a new social science discipline had come into existence.

Tönnies and his sociographic method. Ferdinand Tönnies throughout his life fought against the narrow conception of empirical sociology as the mere compilation of facts. To this he opposed his own notion of "sociography," in which systematic observation, case studies, and other qualitative methods were included, together with statistical description. The goal of sociography was to arrive at empirical laws by the method of induction, applied to systematically collected information. Sociography was to be one of the three branches of sociology, equal in importance to theory. Tönnies himself had studied statistics with Engel, who impressed upon him the importance of Quetelet's accomplishments. Starting in 1895, and continuing for the rest of his life, he published intensive statistical monographs on land tenure and agrarian social structure, demography, crime, suicide, and voting. Most of the studies dealt with his home state of Schleswig-Holstein. Tönnies developed a measure of association and a method for the analysis of correlation between time series. In later life he repeatedly called for the creation of sociographic observatories, where specialists of many disciplines, together with members of the liberal professions and educated laymen, would collaborate in studying the facts of social life, especially those considered of moral significance. However, many of Tönnies' methods were not free from error, and he was unable to arouse the enthusiasm of his colleagues for sociographic observatories or sociography.

Research concerning the agrarian problem. After the unification of Germany in 1871, the agrarian problem received a great deal of attention; rural poverty and ignorance were prevalent, and there was a gradual displacement of native German peasants by Polish wage workers, especially in East Prussia. Many peasants emigrated; under the impact of the capitalist methods of agricultural production, introduced by the *Junker* with the help of protective tariffs, others were gradually reduced to the level of a rural proletariat. Everywhere the earlier, paternalistic type of labor relations was disappearing. The first large-scale survey of agricultural laborers to concern itself with these problems was undertaken in 1874–1875, on behalf of the Congress of German Landowners, by Theodor von der Goltz.

The Verein für Socialpolitik, founded in 1872, conducted, among other nationwide surveys, two on agricultural laborers, one on usury in rural areas, and one on cottage industry. The Verein was part professional association, part pressure group, and part research organization. At its conventions the implications of survey findings were debated with an eye to influencing social policy and legislation; further surveys were planned by an executive committee that enlisted the cooperation of key professors, who in turn brought their students into the survey. Max Weber's first empirical work was done under the auspices of the Verein (Weber 1892).

In the Verein surveys a schedule of questions was usually drawn up by topic. Questionnaires on such topics as land tenure, production, wages, living conditions, composition of the work force, and the extent of theft and drunkenness were then answered by landowners, ministers, doctors, notaries, teachers, members of agricultural societies, and other informed persons. The weaknesses of this methodology were that it assumed accurate knowledge on the part of the informants, that the questions were imprecisely put and were grouped in a haphazard fashion, that low response rates were achieved, and that the returns were only imperfectly exploited. In short, perfection of the techniques of survey research did not concern the Verein members; only Gottlieb Schnapper-Arndt (1888) subjected their methodology to a sharp critique. He had previously published *Fünf Dorfgemeinden auf dem Hohen Taunus* (1883), a very detailed field study of rural life, based on several months of painstaking participant observation and explicitly indebted to Le Play. Unfortunately, Schnapper-Arndt's impact upon his contemporaries remained minimal.

Religious organizations, such as the Evangelical-Social Congress, also conducted rural surveys. Their main interest was understandably in the morals, religion, and literacy of the rural population, and they sought the information primarily from ministers.

Studies of industrial workers. With the retirement of Bismarck and the repeal in 1890 of the laws banning socialist political activity, the working-class question began to receive much attention. Repeated socialist successes at the polls alarmed the German middle and upper classes. Increasing international competition and trade union demands for shorter working hours brought up the issue of productivity and the ability of German industry to compete against Britain in world markets.

In 1890 a young theology student named Paul Göhre, later a Social Democratic deputy in the Reichstag, decided to find out what he hoped would be the whole truth about the working classes. He spent three months working in a factory, pretending to be an apprentice and sharing in every way, both on and off the job, the daily life of the work force. Every night he wrote down his experiences of the day in the form of field notes, which he later published (Göhre 1891). His book was a remarkable document of the social structure on the factory floor and of the life styles, aspirations, and religious conceptions of the workers, and it received widespread recognition from the academic and general public alike. Göhre and Max Weber teamed up a year later as research directors for a study of agricultural laborers under the auspices of the aforementioned Evangelical-Social Congress. Göhre later edited a series of workers' autobiographies.

Adolf Levenstein, a self-educated worker, undertook, from 1907 to 1911, what is probably the first large-scale attitude and opinion survey on record. He sent out 8,000 questionnaires to miners, steelworkers, and textile workers, using a snowball procedure which started with his many friends. He achieved a 63 per cent rate of return, which is remarkably high. The questionnaire itself, despite technical shortcomings, explored the workers' attitudes on many of the important issues of the day —their material and political hopes and wishes, and their aspirations, religious beliefs, political activities, recreational and cultural pursuits, satisfaction or boredom with their work, and drinking habits, in addition to the standard information on social background and wages. Levenstein at first refused to make the findings public, but was persuaded by Weber and others to code and tabulate the answers, which eventually appeared in book form (Levenstein 1912). In many ways Levenstein followed Weber's advice (1909) on how to analyze an attitude survey, which foreshadowed some present-day procedures.

Weber himself at this time was the principal moving force behind the Verein für Socialpolitik survey of industrial workers. This was planned as a large-scale attempt to determine the occupational careers, social origins, and style of life of the workers; some of the data were to be obtained directly from the workers themselves. Weber also intended to test a number of hypotheses about worker productivity. In particular he wanted to find out to what extent the laboratory methods of experimental psychophysics might be adapted to field experiments and surveys in a factory setting. In preparation for these tasks, he spent a whole summer observing the workers in a textile mill and analyzing their production figures. At the same time, he drew up a plan of procedure for the Verein researchers and wrote an explanation of the theoretical and methodological underpinning of the entire undertaking. The survey came to an unhappy end when the vast majority of the workers refused to cooperate in the study. Nevertheless, Weber's preparatory conceptual and statistical studies were acclaimed at the time as pathbreaking; they were indeed highly sophisticated, although they reflected the then current psychophysical approach to worker productivity. Weber's own attitude toward the often tedious tasks involved in social research is evident in his famous address, "Science as a Vocation," in which he said, "No sociologist . . . should think himself too good, even in his old age, to make tens of thousands of quite trivial computations in his head and perhaps for months at a time" ([1919] 1946, p. 135). Weber's own plans for encouraging social research in Germany remained unfulfilled. However, it is clear from the works reviewed in this article that there was no lack of historical precedents for his stress on the collection and analysis of empirical social data. Rewriting of the history of sociology to take full account of these precedents is a task that is long overdue.

BERNARD LÉCUYER AND
ANTHONY R. OBERSCHALL

[See also the biographies of CONDORCET; GRAUNT; KING; LAPLACE; PETTY; QUETELET; SÜSSMILCH. See also Bendix 1968; Cole 1968; Heberle 1968; Parsons 1968; Pfautz 1968; Pitts 1968.]

BIBLIOGRAPHY

ACKERKNECHT, E. H. 1948 Hygiene in France, 1815–1848. *Bulletin of the History of Medicine* 22:117–155.

AGULHON, MAURICE 1958 L'enquête du Comité du Travail de l'Assemblée Constituante (1848): Étude critique de son exécution dans deux départements du Midi. *Annales du Midi* 70:73–85.

Annales d'hygiène publique et de médecine légale. → Published from 1829 to 1950. From 1923 called *Annales d'hygiène publique, industrielle et sociale.*

Annales de statistique. → Published from 1802 to 1804. Superseded by *Archives statistiques de la France.*

ARBUTHNOTT, JOHN 1710 An Argument for Divine Providence, Taken From the Constant Regularity Observ'd in the Births of Both Sexes. Royal Society of London, *Philosophical Transactions* 27:186–190.

Arithmétique politique. 1751 Volume 1, pages 678–680 in *Encyclopédie, ou, Dictionnaire raisonné des sciences, des arts et des métiers.* Edited by Denis Diderot. Paris: Briasson.

BAILLY, JEAN S. 1790 *Discours et mémoires.* Volume 2. Paris: de Bure.

BALBI, ADRIANO; and GUERRY, ANDRÉ M. 1829 *Statistique comparée de l'état de l'instruction et du nombre des crimes dans les divers arrondissements des académies et des cours royales de France.* Paris: Renouard.

BEAUTOT, ÉMILE 1939–1943 Le département du Nord sous la Restauration: Rapport du préfet Villeneuve-Bargemont en 1828. *Revue du Nord* 25:243–277; 26:21–45.

▶BENDIX, REINHARD 1968 Weber, Max. Volume 16, pages 495–502 in *International Encyclopedia of the Social Sciences.* Edited by David L. Sills. New York: Macmillan and Free Press.

BOOTH, CHARLES et al. (1889–1891) 1902–1903 *Life and Labour of the People in London.* 17 vols. London: Macmillan.

BOULAINVILLIERS, HENRI DE (1727) 1752 *L'état de la France.* 2 vols. London: Wood & Palmer. → Written in 1711; published posthumously.

BOWLEY, ARTHUR L.; and BURNETT-HURST, A. R. 1915 *Livelihood and Poverty: A Study in the Economic Conditions of Working-class Households in Northampton, Warrington, Stanley and Reading.* London: Bell.

BUFFON, GEORGES-L. 1749 *Histoire naturelle générale et particulière avec la description du cabinet du roy.* Volume 2: Histoire naturelle de l'homme. Paris: Imprimerie Royale.

BURET, EUGÈNE 1840 *De la misère des classes laborieuses en Angleterre et en France.* 2 vols. Paris: Paulin.

CABANIS, PIERRE JEAN G. 1790 *Observations sur les hôpitaux.* Paris: Imprimerie Nationale.

CABANIS, PIERRE JEAN G. (1803) 1956 Quelques principes et quelques vues sur les secours publics. Volume 2, pages 1–63 in Pierre Jean G. Cabanis, *Oeuvres philosophiques.* Paris: Presses Universitaires de France.

CHEVALIER, LOUIS 1948 Préface à Moheau. *Population* 3:211–232.

▶COLE, MARGARET 1968 Webb, Sidney and Beatrice. Volume 16, pages 487–491 in *International Encyclopedia of the Social Sciences.* Edited by David L. Sills. New York: Macmillan and Free Press.

COLQUHOUN, PATRICK 1814 *A Treatise on the Wealth, Power, and Resources of the British Empire.* London: Mawman.

CONDORCET 1785 *Essai sur l'application de l'analyse à la probabilité des décisions rendues à la pluralité des voix.* Paris: Imprimerie Royale.

DAINVILLE, FRANÇOIS DE 1952 Un dénombrement inédit au XVIIIe siècle: L'enquête du contrôleur général Orry—1745. *Population* 7:49–68.

DEPARCIEUX, ANTOINE 1746 *Essai sur les probabilités de la durée de la vie humaine.* Paris: Guérin & Delatour. → A supplement entitled *Addition à l'essai . . .* was published in 1760.

DERHAM, WILLIAM (1713) 1742 *Physico-theology: Or, a Demonstration of the Being and Attributes of God, From His Works of Creation.* 10th ed. London: Innys.

Dictionnaire géographique, historique et politique des Gaules et de la France. Edited by Jean-J. Expilly. 6 vols. 1762–1770 Paris: Desaint & Saillant.

▶DOUGLAS, JACK D. 1968 Suicide: I. Social Aspects. Volume 15, pages 375–385 in *International Encyclopedia of the Social Sciences.* Edited by David L. Sills. New York: Macmillan and Free Press.

▶DUMAZEDIER, JOFFRE 1968 Leisure. Volume 9, pages 248–254 in *International Encyclopedia of the Social Sciences.* Edited by David L. Sills. New York: Macmillan and Free Press.

EDEN, FREDERICK M. (1797) 1928 *The State of the Poor.* 3 vols. London: Routledge.

ESMONIN, EDMOND 1954 Quelques données inédites sur Vauban et les premiers recensements de population. *Population* 9:507–512.

ESMONIN, EDMOND 1956 Les mémoires des intendants pour l'instruction du duc de Bourgogne (étude critique). Société d'Histoire Moderne, Paris, *Bulletin* 55:12–21.

ESMONIN, EDMOND 1957 L'abbé Expilly et ses travaux de statistique. *Revue d'histoire moderne et contemporaine* 4:241–280.

ESMONIN, EDMOND 1958 Montyon, véritable auteur des *Recherches et considérations sur la population* de Moheau. *Population* 13:269–283.

FOUCAULT, MICHEL 1963 *Naissance de la clinique: Une archéologie du regard médical.* Paris: Presses Universitaires de France.

FOURASTIÉ, JEAN (1951) 1960 *The Causes of Wealth.* Glencoe, Ill.: Free Press. → First published in French.

FRANCE, MINISTÈRE DE LA GUERRE *Comptes présentés au roi sur le recrutement de l'armée.* → Published from 1819 to 1908. Until 1837 published as *Comptes présentés en exécution de la loi du 10 mars 1818 sur le recrutement de l'armée.*

FRANCE, MINISTÈRE DE LA JUSTICE *Compte général de l'administration de la justice criminelle en France et en Algérie.* → Published from 1827 to 1907; includes Algeria from 1853.

FRÉGIER, H. A. 1840 *Des classes dangereuses de la population dans les grandes villes, et des moyens de les rendre meilleures.* 2 vols. Paris: Baillière.

GAULMIER, JEAN 1951 *L'idéologue Volney (1757–1820): Contribution à l'histoire de l'orientalisme en France.* Beirut: Presses de l'Imprimerie Catholique.

GÉRANDO, JOSEPH-MARIE DE 1801 *Considérations sur les diverses méthodes à suivre dans l'observation des peuples sauvages.* Paris: Société des Observateurs de l'Homme.

GÉRANDO, JOSEPH-MARIE DE (1820) 1837 *Le visiteur du pauvre.* 4th ed. Paris: Colas.

GÉRANDO, JOSEPH-MARIE DE 1839 *De la bienfaisance publique.* 4 vols. Paris: Renouard.

GILLE, BERTRAND 1964 *Les sources statistiques de l'histoire de France, des enquêtes du XVIIᵉ siècle à 1870.* Paris, École Pratique des Hautes Études, Centre de Recherches d'Histoire et de Philologie, Publications, Section IV, Série V: Hautes études médiévales et modernes, No. 1. Geneva: Droz; Paris: Minard.

GÖHRE, PAUL 1891 *Drei Monate Fabrikarbeiter und Handwerksbursche.* Leipzig: Grünow.

GOSSEZ, A. M. 1904 *Le département du Nord sous la Deuxième République (1848–1852): Étude économique et politique.* Lille (France): Leleu.

GRAUNT, JOHN (1662) 1939 *Natural and Political Observations Made Upon the Bills of Mortality.* Edited and with an introduction by Walter F. Willcox. Baltimore: Johns Hopkins Press.

GREAT BRITAIN, POOR LAW COMMISSIONERS (1842) 1965 *Report on the Sanitary Condition of the Labouring Population of Gt. Britain,* by Edwin Chadwick. Edinburgh Univ. Press.

GUÉRARD, ALPHONSE 1864 Notice sur M. Villermé. *Annales d'hygiène publique et de médecine légale* Second Series 21:162–177.

GUERRY, ANDRÉ M. 1833 *Essai sur la statistique morale de la France.* Paris: Crochard.

GUILLAUME, P. 1962 Département de la Loire. Pages 429–450 in Congrès des Sociétés Savantes de Paris et des Départements, 86ᵉ, *Actes.* Paris: Imprimerie Nationale.

HALLEY, EDMUND 1693 An Estimate of the Degrees of the Mortality of Mankind, Drawn From Curious Tables of the Births and Funerals at the City of Breslaw. . . . Royal Society of London, *Philosophical Transactions* 17:596–610.

►HEBERLE, RUDOLF 1968 Tönnies, Ferdinand. Volume 16, pages 98–103 in *International Encyclopedia of the Social Sciences.* Edited by David L. Sills. New York: Macmillan and Free Press.

HOWARD, JOHN (1777) 1929 *State of the Prisons.* New York: Dutton. → First published as *State of Prisons in England and Wales.*

HOWARD, JOHN (1789) 1791 *An Account of the Principal Lazarettos in Europe.* . . . 2d ed. London: Johnson.

KAHAN-RABECQ, MARIE M. 1934/1935 Les réponses hâvraises à l'enquête sur le travail industriel et agricole en 1848. *La Révolution de 1848* 31:95–113.

KAHAN-RABECQ, MARIE M. 1939 *L'Alsace économique et sociale sous le règne de Louis-Philippe.* Volume 2: Réponses du département du Haut-Rhin à l'enquête faite en 1848. Paris: Presses Modernes.

KAY-SHUTTLEWORTH, JAMES PHILLIPS 1832 *The Moral and Physical Condition of the Working Classes Employed in the Cotton Manufacture in Manchester.* London: Ridgway.

KING, GREGORY (1696) 1936 Natural and Political Observations and Conclusions Upon the State and Condition of England. Pages 12–56 in *Two Tracts by Gregory King.* Edited by George E. Barnett. Baltimore: Johns Hopkins Press. → The manuscript of 1696 was first published in 1802, in George Chalmers' *Estimate of the Comparative Strength of Great-Britain.*

LAPLACE, PIERRE SIMON DE (1812) 1820 *Théorie analytique des probabilités.* 3d ed., rev. Paris: Courcier. → This work also published as Volume 7 of *Oeuvres de Laplace.*

LAVOISIER, ANTOINE L. (1791) 1819 *Résultats d'un ouvrage intitulé: De la richesse territoriale du royaume de France.* Paris: Huzard. → Published in 1791 by order of the National Assembly.

LEVENSTEIN, ADOLF 1912 *Die Arbeiterfrage mit besonderer Berücksichtigung der sozialpsychologischen Seite des modernen Grossbetriebes und der psychophysischen Einwirkungen auf die Arbeiter.* Munich: Reinhardt.

McCULLOCH, JOHN R. (1837) 1854 *A Descriptive and Statistical Account of the British Empire.* 4th ed., rev. 2 vols. London: Longmans. → First published as *A Statistical Account of the British Empire.*

MALTHUS, THOMAS R. (1798) 1958 *An Essay on Population.* 2 vols. New York: Dutton. → First published as *An Essay on the Principle of Population.* A paperback edition was published in 1963 by Irwin.

MESSANCE, RECEVEUR DES TAILLES DE L'ÉLECTION DE ST-ÉTIENNE 1766 *Recherches sur la population des généralités d'Auvergne, de Lyon, de Rouen et de quelques provinces et villes du royaume* . . . . Paris: Durand.

MOHEAU, M. (1778) 1912 *Recherches et considérations sur la population de la France, 1778.* Paris: Geuthner. → Written in 1774.

MONTALIVET, JEAN-PIERRE 1813 *Exposé de la situation de l'Empire, présenté au corps législatif dans sa séance du 25 février 1813.* Paris: Imprimerie Impériale.

PARENT-DUCHÂTELET, ALEXANDRE J. B. (1834) 1857 *On Prostitution in the City of Paris.* 3d ed. 2 vols. London: Burgess. → First published in French.

PARENT-DUCHÂTELET, ALEXANDRE J. B. 1836 *Hygiène publique, ou Mémoires sur les questions les plus importantes de l'hygiène appliquée aux professions et aux travaux d'utilité publique.* 2 vols. Paris: Baillière.

►PARSONS, TALCOTT 1968 Durkheim, Émile. Volume 4, pages 311–320 in *International Encyclopedia of the Social Sciences.* Edited by David L. Sills. New York: Macmillan and Free Press.

PETTY, WILLIAM (1671–1676) 1963 The Political Anatomy of Ireland. Volume I, pages 121–231 in William Petty, *The Economic Writings. . . .* New York: Kelley. → Written between 1671 and 1676; first published posthumously, in 1691.

►PFAUTZ, HAROLD W. 1968 Booth, Charles. Volume 2, pages 124–126 in *International Encyclopedia of the Social Sciences.* Edited by David L. Sills. New York: Macmillan and Free Press.

PIGEIRE, JEAN 1932 *La vie et l'oeuvre de Chaptal, 1756–1832.* Paris: Domat-Montchrestien.

►PITTS, JESSE R. 1968 Le Play, Frédéric. Volume 9, pages 84–91 in *International Encyclopedia of the Social Sciences.* Edited by David L. Sills. New York: Macmillan and Free Press.

PLAYFAIR, WILLIAM 1801 *The Statistical Breviary.* London: Wallis.

REINHARD, MARCEL 1961 *Étude de la population pendant la Révolution et l'Empire.* Gap (France): Jean.

RIGAUDIAS, HILDE (WEISS) 1936 *Les enquêtes ouvrières en France entre 1830 et 1848.* Paris: Alcan.

ROSEN, GEORGE 1946 The Philosophy of Ideology and the Emergence of Modern Medicine in France. *Bulletin of the History of Medicine* 20:328–339.

ROSEN, GEORGE 1955 Problems in the Application of Statistical Analysis to Questions of Health: 1700–1800. *Bulletin of the History of Medicine* 29:27–45.

ROSEN, GEORGE 1958 *A History of Public Health.* New York: MD Publications.

ROWNTREE, B. SEEBOHM (1901) 1922 *Poverty: A Study of Town Life.* New ed. London and New York: Longmans.

SCHNAPPER-ARNDT, GOTTLIEB 1883 *Fünf Dorfgemeinden auf dem Hohen Taunus.* Leipzig: Duncker & Humblot.

SCHNAPPER-ARNDT, GOTTLIEB 1888 *Zur Methodologie sozialer Enquêten.* Frankfurt am Main (Germany): Auffarth.

SEINE (DEPT.) 1821–1860 *Recherches statistiques sur la ville de Paris et le département de la Seine.* 6 vols. Paris: Imprimerie Royale.

SINCLAIR, JOHN 1791–1799 *The Statistical Account of Scotland: Drawn Up From the Communications of the Ministers of the Different Parishes.* 21 vols. Edinburgh: Creech.

SÜSSMILCH, JOHANN PETER (1741) 1788 *Die göttliche Ordnung in den Veränderungen des menschlichen Geschlechts, aus der Geburt, dem Tode und der Fortpflanzung.* 3 vols. Berlin: Verlag der Buchhandlung der Realschule.

TENON, JACQUES RENÉ 1788 *Mémoires sur les hôpitaux de Paris.* Paris: Royez.

VAUBAN, SÉBASTIEN LE PRESTRE DE (1707) 1943 *Projet d'une dîme royale.* Paris: Guillaumin. → The papers "Méthode générale . . . ," and "Description de l'élection de Vezelay," were written in 1686 and later incorporated as part 2, chapter 10 of the above volume. An English translation was published in 1708 as *A Project for a Royal Tythe, or General Tax.*

VEDRENNE-VILLENEUVE, EDMONDE 1961 L'inégalité sociale devant la mort dans la première moitié du XIXe siècle. *Population* 16:665–699.

VIDALENC, JEAN 1948 Les résultats de l'enquête sur le travail prescrite par l'Assemblée Constituante dans le département de l'Eure. Pages 325–341 in Congrès Historique du Centenaire de la Révolution de 1848, *Actes.* Paris: Presses Universitaires de France.

VILLENEUVE-BARGEMONT, ALBAN 1837 Études spéciales sur le département du Nord. Pages 216–227 in Alban Villeneuve-Bargemont, *Économie politique chrétienne: Ou recherches sur la nature et les causes de paupérisme en France et en Europe.* Brussels: Meline, Cans. → Written in 1828.

VILLERMÉ, LOUIS R. 1820 *Des prisons, telles qu'elles sont, et telles qu'elles devraient être.* Paris: Méquignon-Marvis.

VILLERMÉ, LOUIS R. 1840 *Tableau de l'état physique et moral des ouvriers employés dans les manufactures de coton, de laine et de soie.* 2 vols. Paris: Renouard.

VINCIENNE, MONIQUE; and COURTOIS, HÉLÈNE 1958 Notes sur la situation religieuse en France en 1848, d'après l'enquête cantonale ordonnée par le Comité du Travail. *Archives de sociologie des religions* 6:104–118.

VOLNEY, CONSTANTIN (1787) 1959 *Voyage en Égypte et en Syrie.* Paris: Mouton. → An English edition was published in 1798 by Tiebout.

VOLNEY, CONSTANTIN 1795 *Questions de statistique à l'usage des voyageurs.* Paris: Courcier.

WEBER, MAX 1892 *Die Verhältnisse der Landarbeiter im ostelbischen Deutschland.* Verein für Socialpolitik, Schriften, Vol. 55. Leipzig: Duncker & Humblot.

WEBER, MAX 1909 Zur Methodik sozialpsychologischer Enquêten und ihrer Bearbeitung. *Archiv für Sozialwissenschaft und Sozialpolitik* 29:949–958.

WEBER, MAX (1919) 1946 Science as a Vocation. Pages 129–156 in Max Weber, *From Max Weber: Essays in Sociology.* Translated and edited by H. H. Gerth and C. W. Mills. New York: Oxford Univ. Press. → First published as "Wissenschaft als Beruf."

WESTERGAARD, HARALD 1932 *Contributions to the History of Statistics.* London: King.

YOUNG, ARTHUR 1771a *The Farmer's Tour Through the East of England, Being the Register of a Journey Through Various Counties of This Kingdom, to Enquire Into the State of Agriculture, &c.* 4 vols. London: Printed for W. Strahan.

YOUNG, ARTHUR 1771b *A Six Months Tour Through the North of England.* 4 vols. London: Printed for W. Strahan.

YOUNG, ARTHUR 1780 *A Tour in Ireland: With General Observations on the Present State of That Kingdom.* 2 vols. Dublin: Printed for G. Bonham. → Volume 1 was printed by Whitestone, Sleater [etc.]. Volume 2 was printed by J. Williams.

YOUNG, ARTHUR (1793) 1950 *Travels in France During the Years 1787, 1788 & 1789.* Edited by Constantia Maxwell. Cambridge Univ. Press. → First published in two volumes, as *Travels During the Years 1787, 1788, 1789, Undertaken More Particularly With a View of Ascertaining the Cultivation, Wealth, Resources and National Prosperity of the Kingdom of France.*

YULE, G. UDNY 1899 An Investigation Into the Causes of Changes in Pauperism in England, Chiefly During the Last Two Intercensal Decades. Part 1. *Journal of the Royal Statistical Society* 62:249–286.

Postscript: Research in the United States at the turn of the century

Early academic sociologists in the United States tended to be textbook writers, not social researchers. The bulk of empirical social research up to World

War I was conducted outside the universities by social workers, philanthropists, public health and charity workers, journalists and reformers, and some academic social pathologists, all of them loosely allied in a broad reform movement that included settlement houses and social surveys. Their models were the British researcher–reformers, especially Booth.

First started in Britain, settlements were houses in the slums, staffed by college graduates. One of their activities was the collection and publication of social data on slum conditions with the purpose of pressuring city government into making improvements. Jane Addams's *Hull House Maps and Papers* (1895), subtitled *A Presentation of Nationalities and Wages in a Congested District of Chicago*, grew out of these concerns. It includes a notable essay by Florence Kelley on sweat shops. Robert Woods and his collaborators worked out of Andover House in Boston. He had the notion that university settlements were "laboratories of social science." The fruits of his investigations were published in *The City Wilderness* (1898), *Americans in Process* (1902), and other books. These were the first systematic, empirical studies of the adaptation and assimilation of immigrants in the United States, topics that became a central concern of sociologists.

The most scholarly early city investigation was *The Philadelphia Negro* (1899) by W. E. B. Du Bois. Hired by the University of Pennsylvania for the study, Du Bois spent 15 months of field work among Philadelphia blacks. He attended religious, business, political, and social meetings and conducted house-to-house visitations in families, during which six schedules were filled out, including information on job discrimination. Results were analyzed in tables and charts, and backed up with systematic white–black comparisons. Du Bois later organized a sociology department and "sociological laboratory" at Atlanta University. The series of studies of the condition of different black social strata at the turn of the century conducted by Du Bois and his students represents a lasting contribution to American social history.

The social survey movement flowered in the second decade of the twentieth century. The origin of the social survey idea was in the Charity Organization Society's effort to arouse public opinion for social reform. The movement was backed financially by the then recently founded Russell Sage Foundation. The moving force in the movement was Paul Kellogg, editor of *Charities and Commons* (later *Survey*) magazine. Findings of social surveys were publicized at huge exhibits. The largest survey effort took place in Pittsburgh. *The Pittsburgh Survey* was published in six volumes (1909–1914) filled with social data and photographs, and led to a civic exhibit in Pittsburgh itself that included lectures, conventions, and meetings of civic associations. A series of smaller surveys followed, the most notable of which, directed by Shelby M. Harrison, took place in Springfield. *The Springfield Survey* was published in three volumes (1918–1920). By the late 1920s, a bibliography listed 2,775 titles and projects in which the social survey technique had been used in some manner (Eaton & Harrison 1930).

Not until the late 1930s, and the years of World War II, however, was the survey technique methodologically perfected and systematically applied in professionally directed social scientific research projects. The transition from armchair sociology and from applied sociology concerned with social pathology to social research in the contemporary sense was made at Columbia University and the University of Chicago, the two leading graduate centers of sociological study at the time.

Columbia faculty and graduates pioneered the use of statistical techniques and research methods. A unique feature of the Columbia department was a statistical laboratory equipped with computational facilities. Students were required to collect reports of the charitable organizations in New York and put them into "scientific shape" by means of tabulations and statistical analysis. Under Richmond Mayo-Smith and Franklin H. Giddings the level of sophistication remained low. But a younger generation of social scientists, notably William F. Ogburn and F. Stuart Chapin, spearheaded the application of statistical techniques developed by British mathematical statisticians in U.S. social research. Ogburn (1918–1919) used multiple regression techniques in his analysis of the standard of living in Washington, D.C. Chapin's *Field Work and Social Research* (1920) was the first modern textbook on the topic and included expositions of random sampling, the drafting of interview schedules, coding, and data tabulation by means of statistical machines and punch cards. Stuart Rice, another Columbia graduate, was a pioneer in the use of statistical techniques in political science.

At Chicago, W. I. Thomas led the shift from social problems to systematic field work and theoretical concerns. His most important work expressing this transition was *The Polish Peasant in Europe and America* (1918–1920), written in collaboration with Florian Znaniecki. During eight periods of field work, Thomas collected altogether 8,000 documents for analysis. *The Polish Peasant* captured the totality of social change for an immigrant group in the process of adjusting to American society. Many of the subsequent concerns of Chicago sociology were present in that work: immigration, assimilation, family disorganization in an urban setting, culture change and nationalities, and an emphasis on process rather than on structure. Methodologically it exemplified the Chicago style of research, stronger on the side of discovery than of proof.

Later in the 1920s, a series of organizational innovations at the University of Chicago facilitated the institutionalization of social research in an academic setting. These innovations included among others the appointment of research professors, faculty promotion based on research output, student dissertations under close faculty guidance concentrating on related topics, research funding by foundations and public agencies, and interdisciplinary collaboration. Thus arose the singularly concentrated and coordinated research effort by Chicago sociology in urban sociology and ecology, which resulted in its preeminent position in the United States during the 1920s and 1930s.

BERNARD LÉCUYER AND
ANTHONY R. OBERSCHALL

[See also SURVEY ANALYSIS, *article on* METHODS OF SURVEY ANALYSIS.]

ADDITIONAL BIBLIOGRAPHY

ADDAMS, JANE 1895 *Hull House Maps and Papers: A Presentation of Nationalities and Wages in a Congested District of Chicago; Together With Comments and Essays on Problems Growing out of the Social Conditions, by Residents of Hull House, a Social Settlement.* New York: Crowell.

CHAPIN, F. STUART 1920 *Field Work and Social Research.* New York: Century.

DU BOIS, W. E. B. (1899) 1973 *The Philadelphia Negro: A Social Study; Together With a Special Report on Domestic Service by Isabel Eaton.* Millwood, N.Y.: Kraus Reprint. → Also published in paperback and hardbound editions in 1967 by Schocken.

EATON, ALLEN H.; and HARRISON, SHELBY M. 1930 *A Bibliography of Social Surveys: Reports on Fact-finding Studies Made as a Basis for Social Action.* New York: Russell Sage.

OGBURN, WILLIAM F. 1918–1919 Analysis of the Standard of Living in the District of Columbia in 1916. *Journal of the American Statistical Association* 16: 374–389.

The Pittsburgh Survey: Findings in Six Volumes. 1909–1914 New York: Charities Publication Committee; Survey Associates. → Volume 1: Elizabeth Beardsley Butler, *Women and the Trades: Pittsburgh, 1907–1908,* 1909. Volume 2: Crystal Eastman, *Work Accidents and the Law,* 1910. Volume 3: John A. Fitch, *The Steel Workers,* 1910. Volume 4: Margaret F. Byington, *Homestead: The Households of a Mill Town,* 1910. Volume 5: Paul Underwood Kellogg (editor), *The Pittsburgh District: Civic Frontage,* 1914. Volume 6: Paul Underwood Kellogg (editor), *Wage-earning Pittsburgh,* 1914. Reprints of all six volumes are published by Arno.

The Springfield Survey: A Study of Social Conditions in an American City. 3 vols. 1918–1920 New York: Russell Sage. → The survey was directed by Shelby M. Harrison.

THOMAS, W. I.; and ZNANIECKI, FLORIAN (1918–1920) 1958 *The Polish Peasant in Europe and America.* 2d ed. 2 vols. New York: Dover. → Reprinted in 1971 by Octagon.

WOODS, ROBERT ARCHEY (editor) 1898 *The City Wilderness: A Settlement Study by the Residents and Associates of the South End House.* Boston: Houghton Mifflin. → Reprinted by several publishers.

WOODS, ROBERT ARCHEY (editor) 1902 *Americans in Process: A Settlement Study; North and West Ends, Boston.* Boston: Houghton Mifflin. → The 1902 and the 1903 editions are reprinted by Patterson Smith and Arno, respectively.

SOCIAL STATISTICS

See CENSUS; GOVERNMENT STATISTICS; PUBLIC POLICY AND STATISTICS; SAMPLE SURVEYS; STATISTICS; VITAL STATISTICS.

SOCIAL SURVEYS

See EVALUATION RESEARCH; PANEL STUDIES; SAMPLE SURVEYS; SURVEY ANALYSIS.

SOCIOMETRY

The term "sociometry" has several meanings, but historically the closest association is with the work of J. L. Moreno, particularly his analysis of interpersonal relations in *Who Shall Survive?* (1934). Sociometry is traditionally identified with the analysis of data collected by means of the *sociometric test*—a type of questionnaire in which, roughly speaking, each member of a group is asked with which members he would most like to carry out some activity. The sociometric test was developed by Moreno and his associates, who made brilliant use of it in their own research.

In the early 1940s a number of pronouncements were made on what the meaning of the word "sociometry" should be, and Moreno himself (1943) urged that the meaning should not be restricted to the instruments he had developed. Although his sentiment and that of others was that "sociometry" should refer generally to the measurement of social phenomena, the fact is that the instruments developed by Moreno are still the almost universal reference point for the term.

Data collection. The form of the sociometric question and the setting in which data are collected permit a great variety of alternatives. The question must indicate to the subjects the setting or scope of choice. Thus, if the setting is a classroom it is appropriate to phrase the question accordingly ("With which students in the classroom would you like to discuss this problem?"). Otherwise, the subjects might choose such persons as the teacher, friends or relatives outside the classroom, or even experts whom they do not know personally. The planned activity (in this case, a classroom discussion) should also be clearly defined, so that the subjects know for what purpose the choice is made. This procedure may be contrasted with *ratings*, by means of which a person attributes characteristics to others but does not have to decide whether he wishes to associate with any of them.

In the "applied" use of the sociometric technique, say by the classroom teacher, the choice of criteria may be related to practical objectives. On the one hand, the teacher may wish to restructure the group subsequently on the assumption that more effective work can be carried out by children when they themselves have selected their co-workers. On the other hand, the purpose of the teacher may run just counter to this type of restructuring, and the objective might be to force more interaction with persons one might ordinarily avoid. In the early days of sociometry, the experimenters, possibly because of the therapeutic concerns of Moreno and his associates, often felt obliged to restructure the group regardless of the circumstances. But with the more general use of the technique, this implication has disappeared. In fact, sociometric questions are often phrased hypothetically ("Assuming there could be another meeting of this group, with which persons would you most want to participate?"). There is no indication that a hypothetical phrasing of the question is less useful than one that implies actual restructuring of the group, but there is a continued mandate for relevance in the phrasing of the question.

As in any research procedure, the use of the sociometric question requires attention to the general abilities of the subjects; to take an obvious example, children who cannot yet read or write will have to be interviewed. Under such circumstances the social setting for asking the questions must ensure privacy and confidentiality, and the interviewer should make sure that the child is not intimidated by, or made fearful of, the situation; otherwise, the validity of the responses may be affected. Collecting data through the sociometric question, however, is intrinsically a simple procedure and should be adaptable to most situations.

Variations range from the use of a single question asking for a simple listing of choices on one criterion to a battery of questions in which ordered choices and rejections are requested on many criteria.

Questionnaire construction. In utilizing sociometric questions, the form in which the data are collected determines the types of results that can be obtained. For example, the number of choices requested may be limited or unlimited. If the choices are unlimited, the total number of choices made in the group may be compared for different groups of the same size or for the same group on different occasions; if the choices are limited, however, comparisons based on the total number of choices are meaningless, since the total number of choices is determined by the instructions. But the total number of choices in a group has the advantage of being a relatively simple score to understand, and it has associations with *group cohesiveness* and *morale*. For instance, Goodacre (1951) found that a high rate of choosing within the group was associated with a high standard of group effectiveness; reduced to the simplest interpretation, it appears that if members of the group consider each other good for the operation of the group, the group is likely to be successful. While this may not seem a profound finding on the surface, it is at present virtually the only dependable association with group effectiveness beyond the predictions that successful groups will continue to be successful and that groups composed of persons with high demonstrated ability will be successful.

Permitting an unlimited number of choices is required if *networks* of social relationships are to be traced. Obviously, a clique structure involving ten people cannot effectively be found in a larger group if only three choices are permitted to each person. Since Moreno was originally concerned with the analysis of group structure at this level, the procedure traditionally recommended has been the use of unlimited choices. However, many alternative procedures have been suggested, including the following: a limited specific number of choices; a limited number of ordered choices; an ordered ranking of the entire group; paired comparisons within the group; estimates of the amount of time one wants to spend with others; guessing who has a particular characteristic or reputation; and rating each person within the group for particular characteristics. Of course, the possibility of negative choices (rejections) and the use of multiple criteria increase the number of alternative choices.

Sociometric description

The patterns of expressed choices can be represented graphically in the *sociogram*, which involves the use of some geometric figure to indicate each person (for example, a circle with a name in it) and connecting lines or arrows showing the direction of the choices. Although the use of sociograms in early studies was haphazard, a number of empirically based and theoretically important concepts were derived from them by Moreno and others.

The simplest concept is that of the *unchosen*, who may be viewed as the person socially isolated by others. In the early tradition, the *isolate* is the person who makes no choices and receives none; in this sense he is totally apart from the group. However, in common use, "isolate" has had the same meaning as "unchosen." The term "underchosen" is also found in the literature, but it tends to be less desirable, as it implies some expected level of being chosen. The *rejected* person, of course, can be distinguished only when the sociometric test has requested rejections ("With whom would you like to do this activity *least*?") as well as choices. Rejection of one person by another implicitly involves active dislike, while ignoring a person or not choosing him could indicate merely a lack of sufficient contact for the development of a crystallized attitude. When both positive and negative forms of the question are used in larger groups, however, persons who are unchosen when the positive-choice form of the sociometric question is used tend to be the ones who are rejected in the negative form of the question, and vice versa.

The highly chosen person has been viewed as being in a desirable position. The term "overchosen" was commonly encountered in the early sociometric literature but is no longer widely used. The concept of the *sociometric star* has had some appeal; the image evoked by the concept is one of the highly chosen persons surrounded by persons who are less chosen than the "star." The *popular leader* is a similar concept, and being highly chosen is most frequently associated with some notion of leadership. Popularity (being highly chosen) and leadership are not synonymous, however, and the distinction has been clearly indicated in the literature (Criswell & Jennings 1951). While the popular person may be the leader, there may be other persons with whom power resides, and the *power figure*—that is, the person chosen by others who are in key positions—may not be a popular person at all. The distinction between popularity and leadership arises most clearly in the sociometric literature in consideration of the content of sociometric questions. Helen H. Jennings (1947) has discussed the difference between *sociogroup* and *psychegroup*, distinguished according to whether the basis of choice lies in the task area or the social area. When the task-oriented (sociogroup) question is used, the highly chosen person is likely to be the leader; on the other hand, when the socially oriented (psychegroup) question is used, the highly chosen person is likely to be the popular or personally attractive person.

The concepts mentioned thus far, while related to the structure of relationships, refer to particular persons or types of persons within the structure. But even in the earliest sociometric studies, a great deal of attention was paid to networks of relationships. Part of the attraction was that arrangements or relationships between persons were easily named; for example, *mutual pair* and *mutual rejection* are obvious concepts. The description of relationships between persons becomes most intricate in the area that has come to be called *relational analysis*, which commonly involves both the calculation of all possible choices and rejections in a given situation and some attempt at predicting how the actual choices and rejections will be distributed (see, for instance, Tagiuri 1952). The study of even more complex arrangements of relationships has led to the use of more complex names for them. Geometric names, such as "triangle" and "quadrangle," have proved to have only limited applicability; more important theoretically have been looser configurations such as "chains" and "rings," which enter into the analysis of clique structure.

The sociogram—problems and proposals. The sociogram was ubiquitous in the early development of sociometric techniques, but although important concepts were associated with such diagrams, they were frequently used for display rather than for analytic purposes. When the person constructing the diagram had a specific implicit hypothesis, the diagrams could be dramatic. For example, the absence of choices across sex lines in elementary classroom groups would appear dramatic if the symbols for all the boys were located on one end of the page and those for the girls on the other. Similarly, if racial or ethnic groups were segregated in sections of the paper, the relative absence of connecting lines between these groups could be seen easily, in contrast to the many connecting lines *within* them. Analysis of clique structure, on the other hand, or the location of complex choice networks, can be almost impossible to represent in the form of a sociogram.

Alternative proposals for construction of sociograms have been numerous. Some techniques, such as the "target" sociogram (Northway 1940), emphasize *choice status*, indicated by concentric circles with the most chosen person as the centermost circle and patterns of relationships shown in the usual way with arrows. This alternative has not been used extensively, nor have the other techniques that emphasize the choice status of the person. Another such method (Powell 1951) suggests the use of symbols of different sizes, with large symbols, for example, meaning a large number of choices received by the person; yet another method (Proctor & Loomis 1951) makes use of physical distance between points on the sociogram to represent the "choice distance" between persons (mutual pairs, very close; mutual rejections, very distant). Still more complex descriptive devices, such as multidimensional diagrams (Chapin 1950), are also used, but the fact is that the analytic utility of sociograms appears to be small. Their descriptive utility, however, has remained, and the trend in this direction has been toward the simplification of diagrams, the procedure of minimizing crossing lines being common (Borgatta 1951).

Sociometric analysis

Analytic techniques, as contrasted with descriptive techniques, have stressed both the development of meaningful indexes of choice and the need for systematic analysis of the total choice matrix. Indexes are usually developed with a view to applying particular concepts, and it should be recognized that even such arbitrary classifications as "unchosen" and "highly chosen" are already indexes of the simple concept of sociometric choice. But indexes in sociometry usually represent more complex classifications and are often directed toward making different sets of data comparable. For example, an index may take into account the number of persons choosing, so that groups of different sizes are made comparable. Many problems arise in the construction of indexes, however, and the literature abounds with cautions that the attempt to "take something into account" in an index may not only fail but may also involve even more serious problems than those the researcher is trying to alleviate.

Matrix techniques. Beginning with the work of Moreno and Jennings (1938), considerable attention has been given to the question of the statistical significance of findings. Earlier approaches to this problem (see, for instance, Bronfenbrenner 1943)

are now generally regarded as impractical, but the discussion they provoked has resulted in emphasis on the *models* underlying choices in a group. In order to make sociometric data more amenable to statistical manipulation, Forsyth and Katz (1946) proposed that the cumbersome device of the sociogram be replaced by a matrix of $N \times N$ dimension (where N is the number of people in the group); choices or rejections could then be indicated clearly by marking the appropriate cell in the matrix with a plus or minus sign—so that, for instance, a plus in the tenth column of the fifth row would record that the fifth person had chosen the tenth.

In their original study, Forsyth and Katz attached special importance to choices recorded near the main diagonal of the matrix (the diagonal itself, of course, indicated self-choices); they also paid some attention to clusters of mutual choices and to adjacent clusters that had some members in common. A sophisticated variation of this procedure by Beum and Brundage (1950) was subsequently generalized for efficient computer analysis (Borgatta & Stolz 1963).

Both of the techniques just described depend essentially on the notion of rearranging the data as already given in the $N \times N$ matrix. In contrast, the matrix multiplication approach suggested by Leon Festinger (1949) has emphasized the identification of more formally defined structures. Subsequent work of this kind has been particularly directed toward naming and detecting ever more complex patterns of relationships (see, for instance, Luce 1950; Katz 1953; Harary & Ross 1957). This is a definite advance from the earlier sociometric studies, which were concerned mainly with patterns of mutual choice.

Other approaches to analysis of the matrix of choices have made use of graph theory (Harary & Norman 1953; Ramanujacharyulu 1964) and factor analysis (MacRae 1960); factor analysis is related to the rearrangement techniques noted above, so that the two approaches complement each other. There is also a technique based on cluster analysis (Bock & Husain 1950) that has aroused favorable comment (Ragsdale 1965).

Interest in the development of these analytic procedures reached a peak in the early 1950s; since then, the number of studies published has fallen off but seems likely to maintain a steady level. Nevertheless, there have been few applications of these procedures, possibly because they call for types of data that are not readily accessible in many adult social situations.

Reliability and validity. Although some research

has been done on the question of the reliability and validity of sociometric procedures, it has received little attention in recent years. One early review (Mouton et al. 1955) indicated some of the limitations of sociometric procedures from the point of view of the stability of measures. Among other problems, the stability of the measuring instrument is confounded with the stability of persons and social structures. Validity is especially difficult to assess in sociometry, since the sociometric indexes are so often seen as the criteria to be predicted. Intrinsically, sociometric information represents the objective depicting of the situation on the basis of the most relevant judges—those with whom one participates. Thus, there has been some tendency to emphasize the prediction of sociometric status on the basis of other characteristics rather than to use sociometric status to predict other variables.

Applications of sociometry

Sociometric procedures have been incorporated into many different types of studies. For example, in small group research one of the common types of information collected in post-meeting questionnaires is the set of sociometric ratings on criteria relevant to the group participation. On this score, it should be emphasized that sociometric procedures as classically defined have tended to merge with more general procedures for obtaining peer ratings and rankings. The structure of self rankings and peer rankings has been systematically explored by various researchers, with some convergence on the types of content involved and some crystallization of information about the stability of measures (Borgatta 1964). Content corresponding to that initially identified by Jennings (1947) with task and with social concerns has continued to be central, but other concepts have also been found to recur in analyses.

Sociometric procedures have also been important to the development of several other research areas. For example, study of the impact of group structure on the characteristics of its members or on group consequences, such as efficiency of task completion or morale of the group, has made necessary a more formal development of notions of communication networks. An extensive review of this research literature (Glanzer & Glaser 1961) has suggested the limitations of such approaches and has placed them in their historical context.

Sociometric techniques remain pervasive in the social sciences, having relevance to personality research, small group research, analysis of networks

of communication and group structures, and to special topics such as the reputational study of social status in the community and the study of segregation patterns.

EDGAR F. BORGATTA

[*Other relevant material may be found in* CLUSTERING; FACTOR ANALYSIS AND PRINCIPAL COMPONENTS; GRAPHIC PRESENTATION. *See also* Hare 1968; Homans 1968; Kuper 1968; Schachter 1968; Sherif & Sherif 1968.]

BIBLIOGRAPHY

BEUM, CORLIN O.; and BRUNDAGE, EVERETT G. 1950 A Method for Analyzing the Sociomatrix. *Sociometry* 13:141–145.

BOCK, R. DARRELL; and HUSAIN, SURAYA Z. 1950 An Adaptation of Holzinger's B-coefficients for the Analysis of Sociometric Data. *Sociometry* 13:146–153.

BORGATTA, EDGAR F. 1951 A Diagnostic Note on the Construction of Sociograms and Action Diagrams. *Group Psychotherapy* 3:300–308.

BORGATTA, EDGAR F. 1964 The Structure of Personality Characteristics. *Behavioral Science* 9:8–17.

BORGATTA, EDGAR F.; and STOLZ, WALTER 1963 A Note on a Computer Program for Rearrangement of Matrices. *Sociometry* 26:391–392.

BRONFENBRENNER, URIE 1943 A Constant Frame of Reference for Sociometric Research. *Sociometry* 6:363–397.

CHAPIN, F. STUART 1950 Sociometric Stars as Isolates. *American Journal of Sociology* 56:263–267.

CRISWELL, JOAN H.; and JENNINGS, HELEN H. 1951 A Critique of Chapin's "Sociometric Stars as Isolates." *American Journal of Sociology* 57:260–264.

FESTINGER, LEON 1949 The Analysis of Sociograms Using Matrix Algebra. *Human Relations* 2:153–158.

FORSYTH, ELAINE; and KATZ, LEO 1946 A Matrix Approach to the Analysis of Sociometric Data: Preliminary Report. *Sociometry* 9:340–347.

GLANZER, MURRAY; and GLASER, ROBERT 1961 Techniques for the Study of Group Structure and Behavior: 2. Empirical Studies of the Effects of Structure in Small Groups. *Psychological Bulletin* 58:1–27.

GOODACRE, DANIEL M. 1951 The Use of a Sociometric Test as a Predictor of Combat Unit Effectiveness. *Sociometry* 14:148–152.

HARARY, FRANK; and NORMAN, ROBERT Z. 1953 *Graph Theory as a Mathematical Model in Social Science.* Research Center for Group Dynamics, Publication No. 2. Ann Arbor: Univ. of Michigan, Institute for Social Research.

HARARY, FRANK; and ROSS, IAN C. 1957 A Procedure for Clique Detection Using the Group Matrix. *Sociometry* 20:205–215.

►HARE, A. PAUL 1968 Groups: IV. Role Structure. Volume 6, pages 283–288 in *International Encyclopedia of the Social Sciences.* Edited by David L. Sills. New York: Macmillan and Free Press.

►HOMANS, GEORGE CASPAR 1968 Groups: I. The Study of Groups. Volume 6, pages 259–265 in *International Encyclopedia of the Social Sciences.* Edited by David L. Sills. New York: Macmillan and Free Press.

JENNINGS, HELEN H. 1947 Sociometric Differentiation of the Psychegroup and the Sociogroup. *Sociometry* 10:71–79.

KATZ, LEO 1953 A New Status Index Derived From Sociometric Analysis. *Psychometrika* 18:39–43.

►KUPER, LEO 1968 Segregation. Volume 14, pages 144–150 in *International Encyclopedia of the Social Sciences.* Edited by David L. Sills. New York: Macmillan and Free Press.

LINDZEY, GARDNER; and BORGATTA, EDGAR F. 1954 Sociometric Measurement. Volume 1, pages 405–448 in Gardner Lindzey (editor), *Handbook of Social Psychology.* Cambridge, Mass.: Addison-Wesley.

LUCE, R. DUNCAN 1950 Connectivity and Generalized Cliques in Sociometric Group Structure. *Psychometrika* 15:169–190.

LUCE, R. DUNCAN; and PERRY, ALBERT D. 1949 A Method of Matrix Analysis of Group Structure. *Psychometrika* 14:95–116.

MACRAE, DUNCAN JR. 1960 Direct Factor Analysis of Sociometric Data. *Sociometry* 23:360–371.

MORENO, JACOB L. (1934) 1953 *Who Shall Survive? Foundations of Sociometry, Group Psychotherapy and Sociodrama.* Rev. & enl. ed. Beacon, N.Y.: Beacon House.

MORENO, JACOB L. 1943 Sociometry and the Cultural Order. *Sociometry* 6:299–344.

MORENO, JACOB L.; and JENNINGS, HELEN H. 1938 Statistics of Social Configurations. *Sociometry* 1:342–374.

MORENO, JACOB L. et al. (editors) 1960 *The Sociometry Reader.* Glencoe, Ill.: Free Press.

MOUTON, JANE S.; BLAKE, ROBERT R.; and FRUCHTER, BENJAMIN 1955 The Reliability of Sociometric Measures. *Sociometry* 18:7–48.

NORTHWAY, MARY L. 1940 A Method for Depicting Social Relationships Obtained by Sociometric Testing. *Sociometry* 3:144–150.

POWELL, REED M. 1951 A Comparative Social Class Analysis of San Juan Sur, and Attiro, Costa Rica. *Sociometry* 14:182–202.

PROCTOR, CHARLES H.; and LOOMIS, CHARLES P. 1951 Analysis of Sociometric Data. Part 2, pages 561–585 in *Research Methods in Social Relations, With Especial Reference to Prejudice,* by Marie Jahoda, Morton Deutsch, and Stuart W. Cook. New York: Dryden.

RAGSDALE, R. G. 1965 Evaluation of Sociometric Measures Using Stochastically Generated Data. Ph.D. dissertation, Univ. of Wisconsin.

RAMANUJACHARYULU, C. 1964 Analysis of Preferential Experiments. *Psychometrika* 29:257–261.

ROSS, IAN C.; and HARARY, FRANK 1952 On the Determination of Redundancies in Sociometric Chains. *Psychometrika* 17:195–208.

►SCHACHTER, STANLEY 1968 Cohesion, Social. Volume 2, pages 542–546 in *International Encyclopedia of the Social Sciences.* Edited by David L. Sills. New York: Macmillan and Free Press.

►SHERIF, MUZAFER; and SHERIF, CAROLYN W. 1968 Groups: III. Group Formation. Volume 6, pages 276–283 in *International Encyclopedia of the Social Sciences.* Edited by David L. Sills. New York: Macmillan and Free Press.

TAGIURI, RENATO 1952 Relational Analysis: An Extension of Sociometric Method With Emphasis Upon Social Perception. *Sociometry* 15:91–104.

SPEARMAN, C. E.

Charles Edward Spearman (1863–1945) is known for two major contributions to behavioral science: a methodological one—what we now call factor analysis; and a substantive one—the development of a rational basis for determining the concept of general intelligence and for validating intelligence testing. In addition, his name is associated eponymously with the Spearman rank-order correlation coefficient. But few psychologists today would agree with his judgment that his most important work was the enunciation of noegenetic cognitive laws.

Spearman's work on factor analysis and on intelligence are historically intertwined, and it would be difficult to say whether his philosophical interest in the notion of a single general ability forced him to study statistical correlational methods more creatively or whether his intrinsic love of clear and ingenious methods generated his two-factor theory of intelligence. The former is more likely. In any case, his 1904 article, "'General Intelligence' Objectively Determined and Measured," is a landmark in psychological thought and involves both of his major interests.

Factor analysis

Methodologically, Spearman began by recognizing that E. L. Thorndike, Clark Wissler, and James McKeen Cattell had failed to discover the structure of abilities through correlational methods; more particularly, they had been unable to find a general factor, because they had not allowed for the systematic influence of random error of measurement. Spearman demonstrated the *attenuating* effect of error on the correlation coefficient. Furthermore, he realized that the attenuation correction formula made it possible to discover what any two intercorrelated variables, X and Y, have in common with any other two intercorrelated variables, W and Z. This insight, backed by an evaluation of the standard error of the tetrad difference, led to the invention of factor analysis capable of demonstrating a single common factor plus specific factors. This made it possible to explain the individual differences in test scores as due primarily to differences in a *single general ability* as well as to something quite specific to each test. As Cyril Burt has pointed out, the beginning of the concept of factor analysis may be found in Karl Pearson's work, but Spearman's development of the concept cannot be explicitly traced to Pearson.

In the first quarter of this century the study, by

tests, of individual differences grew apace, and Spearman's discovery of factor analysis, as well as his statistical contributions in such formulae as the rank correlation coefficient and the Spearman–Brown prophecy formula, prevented the work in this field from becoming completely chaotic. (Since the more active and less scholarly failed to understand what Spearman was saying, the field became a sorry mess notwithstanding, especially in intelligence testing.) Spearman's two-factor theory of intelligence—or of "g" as he preferred to symbolize the discovered general factor—states that any cognitive performance is a function of two "factors"— the general ability common to most cognitive performances and an ability specific to a given test. Since it is possible to determine this general factor objectively, disputes about the validity of intelligence tests can be settled by assessing the loadings of the tests on the general factor. High loadings have been found particularly for analogies, classifications, and series—either in words or in perception material—and for problem solving.

Two major developments followed in psychology. In the first place, Spearman's important example taught psychologists to look beyond particular concrete criteria and test scores to underlying factors. Not all learned the lesson; some self-styled practical psychologists called factor analysis "mysticism"—a curious name for a basically scientific procedure. Second, it gave to intelligence testing a more positive theoretical basis than had the basically atheoretical empirical approaches of men like Binet and Wechsler, whom Spearman severely criticized, and whose work ultimately led, as Spearman and others clearly foresaw, to the scientifically cynical view that "intelligence is what intelligence tests measure." If the methodological elegance of Spearman's contributions had been more generally appreciated, many pointless experiments based on arbitrary "intelligence tests" would never have been conducted and the generations of psychologists who were working between 1900 and 1925 would have been spared many wrong leads and outright misconceptions.

Even among those who took up factor analysis, Spearman's theories soon ran into difficulties (see, e.g., Thurstone 1947). Godfrey Thompson pointed out that an alternative model could fit the same statistics (1939). Experiments also appeared showing that the correlational hierarchy requiring the postulation of a single general ability was found only with *certain* choice of variables. Spearman has been reproached by some for arbitrarily removing variables that produced "group factors," and his

first major book, *The Nature of "Intelligence" and the Principle of Cognition* (1923), certainly shows how impatient he was to establish a general factor. At that time he was concerned only with purifying the concept of intelligence. By the time his second book, *The Abilities of Man*, appeared in 1927, he had accepted the reality of several group factors— perseveration, oscillation, persistence, and fluency —and had, with his students and associates, done more than anyone else to define them.

It was at this point that Hotelling, Truman Kelley, and Thurstone generalized Spearman's factor analysis into what we now know as multiple factor analysis, and in so doing they incorporated purely mathematical notions that lay ready for such integration with Spearman's methodological ones. As a result, all broad factors found then assumed equal status. Psychologists, unfortunately, have been slower than physicists to perceive the importance of mathematical models to the scientific growth of their ideas, and around 1930, when Thurstone began to publish, only a minority— though an impressive minority—reacted favorably or even intelligently to these radical ideas. Among clinicians, for example, a common reaction was that factors were "unreal abstractions" unrelated to their problems. Or, again, psychologists teaching the history or methodology of their discipline often mistook multiple factor analysis for a revival of faculty psychology, oblivious to the vast difference between creating a faculty by giving it a name and discovering a functional unity by correlation. The manipulative experimenters, in the classical tradition of Wundt, were puzzled by the absence of manipulation in factor analytic investigations, for Spearman, although he was trained by Wundt (he obtained his PH.D. at Leipzig in 1908), rejected the fine atomism of experiments on perception and sensation and sought instead to "connect the psychics of the laboratory with those of real life" (quoted in Burt & Myers 1946, p. 68). Incidentally, gestalt psychology, which was moving in the same direction at the same time, never recognized that what we now call the multivariate experimental method, built on procedures implicit in and developed on the basis of Spearman's work, contains an effective holistic and "real life" treatment of social and general behavior.

Both Spearman's method and his specific views on ability and other structures have necessarily been further developed, as are all fertile contributions to science. Factor analysis is now multiple factor analysis; and multivariate experimental design, assessing the simultaneous effects of many variables, is recognized as a new principle of research. The general factor that Spearman sought is now regarded as being a second order rather than a primary factor, and it is thought to consist perhaps of two factors—fluid and crystallized general intelligence.

Noegenetic laws

Spearman's cognitive laws have not had the important impact on the development of psychology that his contributions to methodology and his work on intelligence have had. His interest in establishing these laws was rooted in his profound sense of the history and philosophy of science, which made him keenly aware of the absence of any adequate general laws in psychology. He believed that English associationism (of Locke, Hume, Bain, et al.) was the only existent systematic attempt to formulate such laws (apart from a few theories limited, for example, to perception, such as the Weber–Fechner law or the merely descriptive reflexological laws of conditioning), but he regarded the laws of association as only anoegenetic explanations of the *reproduction* of mental content, not as explanatory of the genesis of *new* mental content. The noegenetic laws, in contrast, assert that the perception of two fundaments tends to evoke a relation between them, and that the presentation of a fundament and a relation will tend to educe a new fundament. This applies in principle to even the simplest cognitive activity, as well as to the processes determined by the general intelligence factor. Whether the "tendency" to perceive a relation between, say, π and e, eventuates in a perception depends on the intelligence of the perceiver. An analogies test, for example, immediately illustrates in its two parts both of these noegenetic laws. In his penultimate (and slender) book, *Creative Mind* (1930), Spearman developed further the implications of his noegenetic laws, aware that this aspect of his work had received little recognition.

Spearman's last major work, *Psychology Down the Ages* (1937), was an ambitious attempt to describe and interpret the development of psychology over two thousand years. The book has many powerful ideas and insights, even if Spearman's earnest commitment to the validity of his own theories prevented it from being an ideally detached history; Cyril Burt has described it as an attempt to show "how all the acceptable formulations were really dim foreshadowings of the fundamental noegenetic laws" (see Burt & Myers 1946, p. 71). The book permitted Spearman, in his early seventies, a leisurely return to the contemplative philosophical interests of his youth.

In Spearman's case, it seems particularly necessary to relate his scientific creativity to his life and his personality. It would be hard to imagine a life pattern less similar to the academic norm. Coming from an English family of established status and some eminence, Spearman became an officer in the regular army because, he said, this offered him more leisure and freedom to study than did other professions. He served in the Burmese war and held the rank of major. Resigning after the Boer War, at 40, he was recommended by McDougall to a newly created position at University College, London. His first book, *The Nature of "Intelligence,"* appeared in his sixtieth year. Although a person of very definite opinions, whose students worked on thesis topics that fitted into his own monumental work (and enjoyed it), he possessed remarkable charm and a capacity to stimulate and reassure. As Burt, who worked with and succeeded him, remarked, "Few have possessed his gift of coördinating the research interests of pupils . . . on one single dominating and fertile theme" (see Burt & Myers 1946, p. 71). On retiring he went to America and with a former student, Karl Holzinger, worked on a unitary-traits project.

In addition to making a major contribution to the theory of human abilities in the first quarter of the twentieth century, Spearman gave great impetus to those multivariate experimental methods that have since revolutionized other areas, and thus he takes his place with the few great names in psychology during that period.

RAYMOND B. CATTELL

[*For discussion of the subsequent development of Spearman's ideas, see* EXPERIMENTAL DESIGN; FACTOR ANALYSIS AND PRINCIPAL COMPONENTS; NONPARAMETRIC STATISTICS, *article on* RANKING METHODS. *See also* Adkins 1968; Gates 1968; Joncich 1968; Thorndike 1968; Tiedeman 1968.]

WORKS BY SPEARMAN

1904 "General Intelligence" Objectively Determined and Measured. *American Journal of Psychology* 15:201–293.

(1923) 1927 *The Nature of "Intelligence" and the Principle of Cognition.* 2d ed. London: Macmillan.

1927 *The Abilities of Man: Their Nature and Measurement.* London: Macmillan.

(1930) 1931 *Creative Mind.* New York: Appleton.

1937 *Psychology Down the Ages.* 2 vols. London: Macmillan.

1950 SPEARMAN, C. E.; and JONES, LLEWELLYN W. *Human Ability.* London: Macmillan. → A continuation of Spearman's *The Abilities of Man* (1927).

SUPPLEMENTARY BIBLIOGRAPHY

►ADKINS, DOROTHY C. 1968 Thurstone, L. L. Volume 16, pages 22–25 in *International Encyclopedia of the Social Sciences*. Edited by David L. Sills. New York: Macmillan and Free Press.

BURT, CYRIL; and MYERS, C. S. 1946 Charles Edward Spearman, 1863–1945. *Psychological Review* 53:67–71.

►GATES, ARTHUR I. 1968 Cattell, James McKeen. Volume 2, pages 344–346 in *International Encyclopedia of the Social Sciences*. Edited by David L. Sills. New York: Macmillan and Free Press.

►JONCICH, GERALDINE 1968 Thorndike, Edward L. Volume 16, pages 8–14 in *International Encyclopedia of the Social Sciences*. Edited by David L. Sills. New York: Macmillan and Free Press.

THOMAS, FRANK C. 1935 *Ability and Knowledge: The Standpoint of the London School*. London: Macmillan.

THOMPSON, GODFREY (1939) 1951 *The Factorial Analysis of Human Ability*. 5th ed. Boston: Houghton Mifflin.

THOMPSON, GODFREY 1947 Charles Spearman: 1863–1945. Royal Society of London, *Obituary Notices of Fellows* 5:373–385.

►THORNDIKE, ROBERT L. 1968 Intelligence and Intelligence Testing. Volume 7, pages 421–429 in *International Encyclopedia of the Social Sciences*. Edited by David L. Sills. New York: Macmillan and Free Press.

THURSTONE, LOUIS L. 1947 *Multiple-factor Analysis: A Development and Expansion of* The Vectors of Mind. Univ. of Chicago Press.

►TIEDEMAN, DAVID V. 1968 Kelley, Truman L. Volume 8, pages 358–360 in *International Encyclopedia of the Social Sciences*. Edited by David L. Sills. New York: Macmillan and Free Press.

SPECTRAL ANALYSIS
See TIME SERIES.

STATIONARITY
See TIME SERIES.

STATISTICAL ANALYSIS, SPECIAL PROBLEMS OF

I. OUTLIERS	*F. J. Anscombe*
II. TRANSFORMATIONS OF DATA	*Joseph B. Kruskal*
III. GROUPED OBSERVATIONS	*N. F. Gjeddebæk*
IV. TRUNCATION AND CENSORSHIP	*Lincoln E. Moses*

I
OUTLIERS

In a series of observations or readings, an *outlier* is a reading that stands unexpectedly far from most of the other readings in the series. More technically, an outlier may be defined to be a reading whose *residual* (explained below) is excessively large. In statistical analysis it is common practice to treat outliers differently from the other readings; for example, outliers are often omitted altogether from the analysis.

The name "outlier" is perhaps the most frequently used in this connection, but there are other common terms with the same meaning, such as "wild shot," "straggler," "sport," "maverick," "aberrant reading," and "discordant value."

Perspective. It is often found that sets of parallel or similar numerical readings exhibit something close to a "normal" pattern of variation [*see* DISTRIBUTIONS, STATISTICAL, *article on* SPECIAL CONTINUOUS DISTRIBUTIONS]. From this finding of normal variation follows the interest of statisticians in simple means (that is, averages) of homogeneous sets of readings, and in the method of least squares for more complicated bodies of readings, as seen in regression analysis and in the standard methods of analysis for factorial experiments [*see* LINEAR HYPOTHESES].

Sometimes a set of readings does not conform to the expected pattern but appears anomalous in some way. The most striking kind of anomaly—and the one that has attracted most attention in the literature of the past hundred years—is the phenomenon of outliers. One or more readings are seen to lie so far from the values to be expected from the other readings as to suggest that some special factor has affected them and that therefore they should be treated differently. Outliers have often been thought of simply as "bad" readings, and the problem of treating them as one of separating good from bad readings, so that conclusions can be based only on good readings and the bad observations can be ignored. Numerous rules have been given for making the separation, the earliest being that proposed by B. Peirce (1852). None of these rules has seemed entirely satisfactory, nor has any met with universal acceptance.

Present-day thinking favors a more flexible approach. Outliers are but one of many types of anomaly that can be present in a set of readings; other kinds of anomalies are, for example, heteroscedasticity (nonconstant variability), nonadditivity, and temporal drift. The question of what to do about such anomalies is concerned with finding a satisfactory specification (or model) of the statistical problem at hand. Since no method of statistical analysis of the readings is uniquely best, tolerable compromises must be sought, so that as many as possible of the interesting features of the data can be brought out fairly and clearly. It is important that outliers be noticed, but the problem of dealing with them is not isolated from other problems of statistical analysis.

An illustration. Suppose that a psychologist arranges to have a stimulus administered to a group of 50 subjects and that the time elapsing before each subject gives a certain response is observed. From the resulting set of times he wishes to calculate some sort of mean value or measure of "central tendency," for eventual comparison with similar values obtained under different conditions. [*See* STATISTICS, DESCRIPTIVE, *article on* LOCATION AND DISPERSION.]

Just as he is about to calculate a simple arithmetic mean of the 50 readings, the psychologist notices that 3 of the readings are considerably larger than all the others. Should he include these outliers in his calculation, discard them, or what? Several different answers seem reasonable, according to the circumstances.

It may occur to the psychologist that the outliers have been produced by some abnormal condition, and he therefore inquires into the conduct of the trial. Perhaps he discovers that, indeed, three of the subjects behaved abnormally by going to sleep during the test, these being the ones yielding the longest response times. Now the psychologist needs to consider carefully what he wants to investigate. If he decides that he is interested only in the response times of subjects who do not go to sleep and explicitly defines his objective accordingly, he may feel justified in discarding the outlying readings and averaging the rest. But, on the other hand, he may decide that the test has been incorrectly administered, because no subject should have been allowed to go to sleep; then he may discard not only the outliers but all the rest of the readings and order a repetition to be carried out correctly.

In experimental work it is often not possible to verify directly whether some abnormal condition was associated with an outlier; nothing special is known about the reading except that it is an outlier. In that case, the whole distributional pattern of the readings should be examined. Response times, in particular, are often found to have a skew distribution, with a long upper tail and short lower tail, and this skewness may be nearly removed by taking logarithms of the readings or by making some other simple rescaling transformation [*see* STATISTICAL ANALYSIS, SPECIAL PROBLEMS OF, *article on* TRANSFORMATIONS OF DATA]. The psychologist may find that the logarithms of his readings have a satisfactorily normal pattern without noticeable outliers. If he then judges that the arithmetic mean of the log-times, equivalent to the geometric mean of the original times, is a satisfactory measure of central tendency, his outlier problem will have been solved without the rejection or special treatment of any reading.

Finally, it may happen that even after an appropriate transformation of the readings or some other modification in the specification, there still remain one or more noticeable outliers. In that case, the psychologist may be well advised to assign reduced weight to the outliers when he calculates the sample mean, according to some procedure designed to mitigate the effect of a long-tailed error distribution or of possible gross mistakes in the readings.

Thus, several actions are open to the investigator, and there is no single, simple rule that will always lead him to a good choice.

Terminology. An outlier is an observation that seems discordant with some type of pattern. Although in principle any sort of pattern might be under consideration, as a matter of fact the notion of outlier seems rarely to be invoked, except when the expected pattern of the readings is of the kind to which the method of least squares is applicable and fully justifiable [*see* LINEAR HYPOTHESES, *article on* REGRESSION]. That is, we have a series of readings $y_1, y_2, \cdots, y_n$, and we postulate, in specifying the statistical problem, that

$$(1) \qquad y_i = \mu_i + e_i, \qquad i = 1, 2, \cdots, n,$$

where the e_i are "errors" drawn independently from a common normal distribution having zero mean and where the expected values, μ_i, are specified in terms of one or several unknowns. For a homogeneous sample of readings (a single, simple sample) it is postulated that all the μ_i are equal to the common mean, μ; for more complicated bodies of readings a "linear hypothesis" is usually postulated, of form

$$(2) \qquad \mu_i = \mu + x_{i1}\beta_1 + x_{i2}\beta_2 + \cdots + x_{ir}\beta_r,$$

where the x_{ij} are given and μ and the β_j are parameters. The object is to estimate (or otherwise discuss the value of) some or all of the parameters, $\mu, \beta_1, \beta_2, \cdots$, and the variance of the error distribution.

Let Y_i denote the estimate of μ_i obtained by substituting in the right side of (2) the least squares estimates of the parameters—that is, the parameter values that minimize the expression $\sum_i (y_i - \mu_i)^2$, given the assumed relationship among the μ_i. The residuals, z_i, are defined by

$$z_i = y_i - Y_i.$$

Relative to the specification (1) and (2), an outlier is a reading such that the corresponding absolute value of z is judged to be excessively large.

Causes of outliers. It is convenient to distinguish three ways in which an outlier can occur:

(*a*) a mistake has been made in the reading, (*b*) no mistake has been made, but the specification is wrong, (*c*) no mistake has been made, the specification is correct, and a rare deviation from expectation has been observed.

In regard to (*a*), a mistake may be made in reading a scale, in copying an entry, or in some arithmetic calculation by which the original measurements are converted to the reported observations. Apparatus used for making measurements may fail to function as intended—for example, by developing a chemical or electrical leak. A more subtle kind of mistake occurs when the intended plan of the investigation is not carried out correctly even though the act of observation itself is performed perfectly. For example, in a study of ten-year-old children it would be a mistake to include, by accident, as though relating to those children, material that was in fact obtained from other persons, such as teachers, parents, or children of a different age.

In regard to (*b*), the specification can be wrong in a variety of ways [*see* ERRORS, *article on* EFFECTS OF ERRORS IN STATISTICAL ASSUMPTIONS]. The errors may have a nonnormal distribution or may not be drawn from a common distribution at all. The expression (2) for the expected values may be incorrect, containing too few terms or terms of the wrong sort.

In regard to (*c*), any value whatsoever is theoretically observable and is consistent with any given normal distribution of errors. But readings differing from the mean of a normal distribution by more than some three or four standard deviations are so exceedingly rare as to be a virtual impossibility. Usually the normal-law linear-hypothesis specification cannot be regarded as more than a rough approximation to the truth. Moreover, one can never entirely rule out the possibility that a mistake of some sort has been made. Thus, when a reading is seen to have a large residual, explanation (*a*) or (*b*) usually seems more plausible than (*c*). However, if one wishes to reach a verdict on the matter, one will do well to examine, not just the outliers, but all the readings for evidence of mistakes and of an incorrect specification.

Preferred treatment. Suppose that it could be known for sure whether the cause of a particular outlier was of type (*a*), (*b*), or (*c*) above. What action would then be preferred?

(*a*) If it were known that the outlying reading had resulted from a mistake, the observer would usually choose, if he could, either to correct the mistake or to discard the reading. One sufficiently gross error in a reading can wreck the whole of a statistical analysis. The danger is particularly great

when the data of an investigation are processed by an automatic computer, since it is possible that no one examines the individual readings. It is of great importance that such machine processing yield a display of the residuals in some convenient form, so that gross mistakes will not pass unnoticed and the conformity of the data to the specification can be checked [*see* COMPUTATION].

However, it is not invariably true that if the observer becomes aware that a reading was mistaken, and if he cannot correct the mistake, he will be wise to discard the reading, as though it had never been. For example, suppose that a new educational test is tried on a representative group of students in order to establish norms—that is, to determine the distribution of crude scores to be expected. When the trial is finished and the crude scores have been obtained, some circumstance (perhaps the occurrence of outliers) prompts an investigation of the trial, and then evidence comes to hand that about a quarter of the students cheated, a listing of those involved being available. Should the scores obtained by the cheaters be discarded and norms for the test be based on the scores of the noncheaters? Surely that would be misleading, for if the possibility of cheating will persist in the future, as in the trial, the scores of the cheaters should obviously not be excluded. If, on the other hand, a change in the administration of the test will prevent such cheating in the future, then for a correct norming the scores that the cheaters would have obtained, had they not been allowed to cheat, must be known. It would be rash to assume that these scores would be similar to the scores of the actual noncheaters. But that is what is implied by merely rejecting the cheaters' scores. Conceivably, a change in the administration of the test might even have affected the scores of those who did not cheat. Thus, the trial would have to be run afresh under the new system of administration in order to yield a fair distribution of scores. (For further discussion of this sort of situation, see Kruskal 1960.)

(*b*) Consider now the next imagined case, where it is known that no mistakes in observation were made but that the specification was wrong. What to do with any particular reading would then be a secondary question; attention should first be directed toward improving the specification. The appropriateness of the expression (2) for the expected values, μ_i, may possibly be improved by transformation of the observations, y_i, or the associated values, x_{ij}, or by a change in the form of the right side so that further terms are introduced or a nonlinear function of the parameters is postulated. Sometimes it is appropriate to postulate unequal variances for the errors, e_i, depending perhaps on the expectations, μ_i, or on some associated x-variable. A consequence of making such changes in the specification will usually be that the standard least squares method of analysis will be applied to modified data.

As for the assumption that the distribution of the errors, e_i, is normal, there are many situations where this does seem to agree roughly with the facts. In no field of observation, however, are there any grounds for thinking that the normality assumption is accurately true, and in cases where it is roughly true, extensive investigation has sometimes revealed a frequency pattern having somewhat longer tails than those of a normal distribution. Jeffreys (1939), Tukey (1962), Huber (1964), and others have considered various systems of assigning reduced weight to readings having large residuals, in order to make the least squares method less sensitive to a possibly long-tailed distribution of errors, while preserving as nearly as possible its effectiveness when the error distribution is normal and not greatly changing the computational procedure. One such modified version of the least squares method is described below.

(*c*) Finally, suppose that it were known that an outlier was caused neither by a mistake in observation nor by an incorrect specification, but that it was simply one of the rare deviations that must occasionally occur. Then one would wish to perform the usual statistical analysis appropriate to the specification; the outlier would be included with full weight and treated just like the other observations. This is so because the ordinary least squares estimates of the parameters (together with the usual estimate of common variance) constitute a set of sufficient statistics, and no additional information can be extracted from the configuration of the readings [*see* SUFFICIENCY].

Conclusions. This exposition has considered what action would be preferred if it were known for sure whether an outlier had arisen (*a*) from a mistake, (*b*) from an incorrect specification, or (*c*) by mere chance, the specification and the technique of observation both being correct. It has been shown that a different action would be preferred in each case. Usually, no such knowledge of the cause of an outlier is available in practice. A compromise is therefore necessary.

Obviously, all reasonable efforts should be made to prevent mistakes in observation and to find a specification and a method of statistical analysis consonant with the data. That is so in any case, although outliers may naturally stimulate a closer scrutiny of the specification and a more vigorous search for mistakes. But however careful one is, one can never be certain that undetected mistakes have not occurred and that the specification and plan of statistical analysis are completely appropriate. For the purpose of estimating parameters in a linear hypothesis like (2) above, some modification of the customary least squares procedure giving reduced weight to outliers seems to be advisable. The harmful effects of mistakes in observation and of a long-tailed distribution of errors are thereby mitigated, while negligible damage is done if no mistakes or specification errors have been made. Suitable computational procedures have not yet been fully explored, but it seems likely that considerable attention will be directed to this topic in the near future.

A modified least squares method. One type of modified least squares method for estimating the parameters for the statistical specification indicated at (1) and (2) above is as follows:

Positive numbers, K_1 and K_2, are chosen, with $K_2 > K_1$. Instead of taking as estimates of the parameters the values that minimize the sum of squares $\sum_i (y_i - \mu_i)^2$, one takes the values that minimize the sum of values of a function,

$$\sum_i \psi(y_i - \mu_i),$$

where $\psi(\cdot)$ is the square function for small values of the argument but increases less rapidly for larger values and is constant for very large values. Specifically, one minimizes the following composite sum:

$$(3) \quad \sum_{(1)} (y_i - \mu_i)^2 + \sum_{(2)} (2K_1 |y_i - \mu_i| - K_1^2) + \sum_{(3)} (2K_1 K_2 - K_1^2),$$

where $\sum_{(1)}$ denotes the sum over all values of i such that $|y_i - \mu_i| \leqslant K_1$, $\sum_{(2)}$ denotes the sum over all values of i such that $K_1 < |y_i - \mu_i| \leqslant K_2$, and $\sum_{(3)}$ denotes the sum over all remaining values of i. Minimizing (3) is, roughly speaking, equivalent to the ordinary least squares method modified by giving equal weight to all those readings whose residual does not exceed K_1 in magnitude, reduced weight (inversely proportional to the magnitude of the residual) to those readings whose residual exceeds K_1 but not K_2 in magnitude, and zero weight to those readings whose residual exceeds K_2 in magnitude. The minimization is a problem in quadratic programming [*see* PROGRAMMING]. If K_1 is large enough for most readings to come in the sum $\sum_{(1)}$, there are rapidly converging iterative procedures. The necessity of iterating comes from the fact that until the parameters have been estimated and the residuals calculated, it is not possible to say with certainty which readings contribute to the sum $\sum_{(1)}$, which to $\sum_{(2)}$, and which to $\sum_{(3)}$. As soon as all the readings have been correctly assigned, a single set of linear equations determines the values of the parameters, as in the ordinary least squares method. The ordinary method results when K_1 is set so large that all readings come in $\sum_{(1)}$. At another extreme, if K_1 is chosen to be infinitesimally small but K_2 is chosen to be very large, all readings come in $\sum_{(2)}$, and the result is the method of least absolute deviations. For general use, as a slight modification of the ordinary least squares method, K_1 and K_2 should be chosen so that only a small proportion of the readings come in $\sum_{(2)}$ and scarcely any in $\sum_{(3)}$. For example, K_1 might be roughly equal to twice the estimated standard deviation of the error distribution, and K_2 might be three or four times as large as K_1.

Such a procedure leads to the exclusion of any very wild reading from the estimation of the parameters. Less wild readings are retained with less than full weight.

F. J. ANSCOMBE

[*See also* DISTRIBUTIONS, STATISTICAL, *article on* MIXTURES OF DISTRIBUTIONS; NONPARAMETRIC STATISTICS, *article on* ORDER STATISTICS.]

BIBLIOGRAPHY

ANSCOMBE, F. J. 1967 Topics in the Investigation of Linear Relations Fitted by the Method of Least Squares. *Journal of the Royal Statistical Society* Series B 29:1–52. → Includes 23 pages of discussion and recent references.

ANSCOMBE, F. J.; and TUKEY, JOHN W. 1963 The Examination and Analysis of Residuals. *Technometrics* 5:141–160.

CHAUVENET, WILLIAM 1863 *Manual of Spherical and Practical Astronomy.* Philadelphia: Lippincott. → See especially the appendix on the method of least squares, sections 57–60.

DANIEL, CUTHBERT 1960 Locating Outliers in Factorial Experiments. *Technometrics* 2:149–156.

[GOSSET, W. S.] 1927 Errors of Routine Analysis, by Student [pseud.]. *Biometrika* 19:151–164.

HUBER, PETER J. 1964 Robust Estimation of a Location Parameter. *Annals of Mathematical Statistics* 35:73–101.

JEFFREYS, HAROLD (1939) 1961 *Theory of Probability.* 3d ed. Oxford: Clarendon.

KRUSKAL, WILLIAM H. 1960 Some Remarks on Wild Observations. *Technometrics* 2:1–3.

PEIRCE, BENJAMIN 1852 Criterion for the Rejection of Doubtful Observations. *Astronomical Journal* 2:161–163.

TUKEY, JOHN W. 1962 The Future of Data Analysis. *Annals of Mathematical Statistics* 33:1–67.

WRIGHT, THOMAS W. 1884 *A Treatise on the Adjustment of Observations, With Applications to Geodetic Work and Other Measures of Precision.* New York: Van Nostrand. → See especially sections 69–73.

Postscript

There has been much study of modifications and alternatives to the method of least squares, more or less in the spirit of the last section of the main article. (See Huber 1972; Andrews et al. 1972; Tukey 1977.)

This upsurge of interest in "robust" methods of fitting a linear hypothesis (equation (2) in the main article) when it is expected that the errors, e_i (in equation (1) of the main article), are distributed with longer tails than the normal, comes from a belief that longer tails often need to be reckoned with, and from the discovery of effective methods costing little more than least squares to carry out. There is not, however, at present much solid information concerning the statistical properties of errors actually encountered. Some data sets indicate clearly, through evident outliers, that the errors have a longer-tailed distribution (or at least not a homogeneous normal distribution). But other data sets do not so indicate, and indeed may even suggest a distribution with shorter tails than the normal. There is no general practice of accumulating evidence concerning distribution shape, either by appropriate graphical displays, or by calculating estimates of skewness and kurtosis from the residuals. Only long-tailed distributions with outliers catch the eye.

Whether robust methods should generally be used, in preference to the method of least squares, depends on whether thought and imagination will be brought to bear on the results. Some kinds of treatment of readings, or data processing, are routine or automatic, as much so as possible—for example, chemical or other determinations made in an automated laboratory or an automatic navigation system. Procedures are required for combining individual readings and arriving at a definitive report or action. For the sake of standardization and reproducibility, an analytical chemist making laboratory measurements should preferably know nothing of the provenance of the samples he works on, and should confine his judgments to the performance of the technique. In all automatic data processing, where raw readings are combined and summarized without human consideration or intervention, robust methods of combining the readings are advisable. These methods may range from a simple rule for rejecting outliers from readings made in triplicate, before averaging what is left, to more sophisticated procedures requiring substantial computation. It is more important to avoid grossly wrong results than to treat the raw readings with the utmost statistical efficiency. Only the designer of the method is interested in the statistical properties of the original readings.

At the other extreme is intelligent consideration of a body of statistical data, for the purpose of reaching the best understanding of the phenomena represented. A good theoretical description or model for the data is sought, based on careful and open-minded examination of the data in the light of available experience and theory. Determination of a definitive value is not the prime objective; and insofar as values are determined, robust methods for doing so are not particularly desirable, because all features of the theoretical description under consideration are of interest. Not only are parameters estimated, but goodness of fit is assessed by graphical displays and other means. [*See* GOODNESS OF FIT.] Alternative theoretical descriptions may be entertained and compared. The method of least squares applied to fit a linear structure of the type of equations (1) and (2) is well supported by theory: judgments concerning adequacy of the structure may be aided by various kinds of significance tests. Because of this good theoretical support, and not merely because of ease of execution, the method of least squares, followed by due examination of residuals, is often the most attractive first step in intelligent analysis of statistical data. Only if the computations are done blindly—as they are described in most textbooks, are implemented in many computer programs, and must necessarily be in automatic data processing—is there frequent reason to fear that least squares may be misleading (Anscombe 1973).

F. J. ANSCOMBE

[*See also* ERRORS, *article on* EFFECTS OF ERRORS IN STATISTICAL ASSUMPTIONS; DATA ANALYSIS, EXPLORATORY.]

ADDITIONAL BIBLIOGRAPHY

ANDREWS, DAVID F. et al. 1972 *Robust Estimates of Location: Survey and Advances.* Princeton Univ. Press.

ANSCOMBE, F. J. 1973 Graphs in Statistical Analysis. *American Statistician* 27, no. 1:17–21.

[1][GOSSET, W. S.] 1958 *"Student's" Collected Papers.* Edited by E. S. Pearson and John Wishart. Cambridge Univ. Press. → Includes Gosset (1927), cited in the bibliography to the main article. Gosset wrote under the pseudonym "Student."

HUBER, PETER J. 1972 Robust Statistics: A Review. *Annals of Mathematical Statistics* 43:1041–1067. → The 1972 Wald lecture.

MORGENSTERN, OSKAR (1950) 1970 *On the Accuracy of Economic Observations.* 2d ed., completely rev. Princeton Univ. Press. → See especially pages 30–34.

TUKEY, JOHN W. (1970) 1977 *Exploratory Data Analysis.* Reading, Mass.: Addison-Wesley. → First published in a mimeographed "limited preliminary edition."

II
TRANSFORMATIONS OF DATA

It is often useful to apply a transformation, such as $y = \log x$ or $y = 1/x$, to data values, x. This change can simplify relationships among the data and improve subsequent analysis.

Many transformations have been used, including $y = \sqrt{x}$, $y = x^c$, where c is a constant, and $y = \frac{1}{2} \log [x/(1 - x)]$. Sometimes a beneficial transformation is constructed empirically from the data themselves, rather than given as a mathematical expression.

The general effect of a transformation depends on the shape of its plotted curve on a graph. It is this curve, rather than the mathematical formula, that has central interest. Transformations with similar curves will have similar effects, even though the formulas look quite different. Graphical similarity, however, must be judged cautiously, as the eye is easily fooled.

The benefits of transforming

The relationship of y to other variables may be simpler than that of x. For example, y may have a straight-line relationship to a variable u although x does not. As another example, y may depend "additively" on u and v even though x does not.

Suppose the variance of x is not constant but changes as other variables change. In many cases it is possible to arrange for the variance of y to be nearly constant.

In some cases the distribution of y may be much more like the normal (Gaussian) distribution or some other desired distribution than is that of x.

Thus, the benefits of transforming are usually said to be (1) simpler relationships, (2) more stable variance, (3) improved normality (or closeness to another standard distribution). Where it is necessary to choose between these, (1) is usually more important than (2), and (2) is usually more important than (3). However, many authors have remarked that frequently (although not invariably) a single transformation achieves two or all three at once.

In some cases analysis of y, however illuminating, is not to the point, because some fact about x itself is needed (usually the expected value) and the corresponding fact about y is not an acceptable substitute. If so, it is often better not to transform, although sometimes it is desirable and feasible to obtain the necessary fact about x from information about y.

The ultimate profit. The benefits listed above are not ultimate profit but merely a means to achieve it. The ultimate profit, however, is difficult to describe, for it occurs during the creative process of interpreting data. A transformation may directly aid interpretation by allowing the central information in the data to be expressed more succinctly. It may permit a subsequent stage of analysis to be simpler, more accurate, or more revealing.

Later in this article an attempt will be made to illustrate these elusive ideas. However, the first examples primarily illustrate the immediate benefits, rather than the ultimate profit.

Some simple examples

Some of the most profitable transformations seem almost as basic as the laws of nature. Three such transformations in psychology are all logarithmic: from sound *pressure* to the decibel scale of sound *volume* in the study of hearing; from light intensity to its logarithm in the study of vision; and from tone *frequency* in cycles per second to tone *pitch* on the musical scale. In each case many benefits are obtained.

To simplify curves. Transformations may be used to display the relationship between two variables in a simple form—for example, as a straight line or as a family of straight lines or of parallel curves. One or both variables may be transformed. Figure 1 is a simple illustration, using hypothetical data.

One easy way to transform while plotting is to use graph paper with a special scale. Two widely used special scales are the logarithmic scale and the normal probability scale, which correspond, respectively, to the logarithmic and "probit" transformations. [*See* GRAPHIC PRESENTATION.]

To stabilize variance. The analysis of many experiments is simpler if the variance of response is approximately constant for different conditions. Crespi (1942, pp. 483–485) described very clearly how he first used the *time* required by a rat to run down a 20-foot runway as a measure of its eagerness to obtain food but found that the response variance differed greatly for experimental conditions with different average responses. This hampered the intended analyses of variance, which require approximate constancy of variance. However, the transformation to *speed* = (20 feet)/(time) removed this difficulty entirely. In short, the reciprocal transformation helpfully stabilized the variance.

Furthermore, Crespi indicated reasons why in this context speed should be better than time as a measure of eagerness. It happens quite often that performance times, latency times, reaction times, and so forth can benefit from a reciprocal or a logarithmic transformation.

To improve normality. Many statistical techniques are valid only if applied to data whose distribution is approximately normal. When the data are far from normal, a transformation can often be used to improve normality. For example, the distribution of personal income usually is skewed strongly to the right and is very nonnormal. Applying the logarithmic transformation to income data often yields a distribution which is quite close to

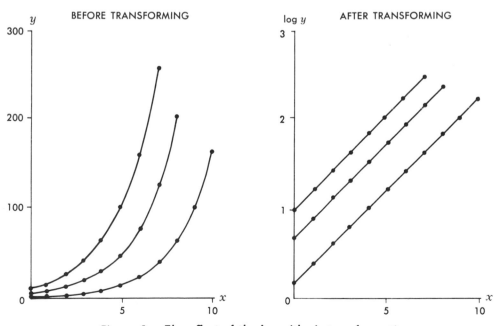

Figure 1 — The effect of the logarithmic transformation

normal. Thus, to test for difference of income level between two groups of people on the basis of two samples, it may well be wiser to apply the usual tests to logarithm of income than to income itself.

To aid interpretation. In a letter to *Science*, Wald (1965) made a strong plea for plotting spectra (of electromagnetic radiation) as a function of frequency rather than of wavelength; frequency plots are now much commoner. Frequency and wavelength are connected by a reciprocal transformation, and most of the reasons cited for preferring the frequency transformation, both in Wald's letter and in later, supporting letters, have to do with ease of interpretation. For example,

frequency is proportional to the very important variable, energy. On a frequency scale, but not on a wavelength scale, the area under an absorption band is proportional to the "transition probability," the half-width of the band is proportional to the "oscillator strength," the shape of the band is symmetrical, and the relationship between a frequency and its harmonics is easier to see.

To quantify qualitative data. Qualitative but ordered data are often made quantitative in a rather arbitrary way. For example, if a person orders n items by preference, each item may be given its numerical rank from 1 to n [see NON-PARAMETRIC STATISTICS, *article on* RANKING METH-

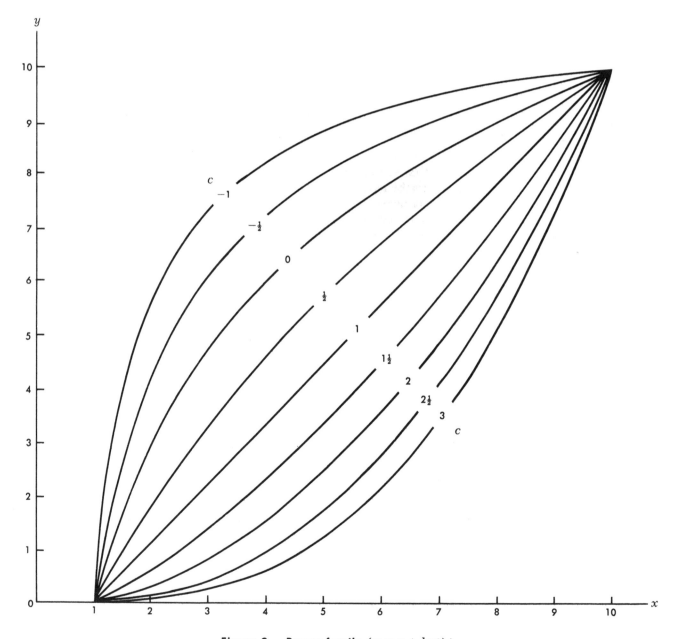

Figure 2 — Power family $(y = a + bx^c)$*

* For $c = 0$, $y = a + b \log x$.

ODS]. In such cases the possibility of transforming the ranks is especially relevant, as the original space between the ranks deserves no special consideration. Quite often the ranks are transformed into "normal scores," so as to be approximately normally distributed [see PSYCHOMETRICS].

To deal with counted data. Sometimes the number of people, objects, or events is of interest. For example, if qualified applicants to a medical school are classified by age, ethnic group, and other characteristics, the number in each group might be under study. [See COUNTED DATA.]

If the observed values have a wide range (say, the ratio of largest to smallest is at least 2 or 3), then a transformation is especially likely to be beneficial. Figure 2 shows a whole family of transformations which are often used with counted data and in many other situations as well. Each of these is essentially the same as $y = x^c$ for some constant c. (For greater ease of comparison, however, the plotted curves show $y = a + bx^c$. By a standard convention, $y = \log x$ substitutes for $y = x^0$, in order to have the curves change smoothly as c passes through the value 0. No other function would serve this purpose.)

For example, suppose the observed values x come from Poisson distributions, which are quite common for counted data. Then the variance of x equals its expected value, so different x's may have very different variances. However, the variance of $y = \sqrt{x} = x^{\frac{1}{2}}$ is almost constant, with value $\frac{1}{4}$ (unless the expected value of x is very close to 0). For various purposes, such as testing for equality, y is better than x. Often, simpler relationships also result from transforming.

Other values of c are also common. In only four pages Taylor (1961) has displayed 24 sets of counted biological data to which values of c over the very wide range from 0.65 to -0.54 are appropriate!

If the observations include 0 or cover a very wide range, it is common to use various modifications, such as $y = (x + k)^c$, with k a small constant, often $\frac{1}{2}$ or 1. There has been considerable investigation of how well various modifications stabilize the variance under certain assumptions, but most of these modifications differ so slightly that interest in them is largely theoretical.

To deal with fractions. A very common form of data is the fraction or percentage, p, of a group who have some characteristic (such as being smokers). If the observed percentages include some extreme values (say, much smaller than 10 per cent or much larger than 90 per cent), a transformation is *usually* beneficial. Figure 3 shows the

three transformations most frequently used with fractions: angular, probit, and logistic. Their formulas are given in Table 8, below. Of course, nothing prevents the use of transformations in the absence of extreme values, or the use of other transformations.

The upper part of Figure 3 displays the three transformations in a different way, showing how they "stretch the ends" of the unit interval relative to the middle.

Suppose a study is designed to compare different groups of men for propensity to smoke. The variance of p (the observed proportion of smokers in a sample from one group) depends on the true proportion, p^*, of that group. For extreme values of p^*, the variance of p gets very small. The nonconstant variance of p hinders many comparisons, such as tests for equality.

One possible remedy is to transform. Each of the three transformations mentioned is likely to make the variance more nearly stable. Each can be justified theoretically in certain circumstances. (For example, if p has the binomial distribution, the angular transformation is indicated.) However, transforming fractions often has practical value, even in the absence of such theory, and this value may include benefits other than variance stabilization.

There is an important caution to keep in mind when using the angular transformation for proportions, the square root transformation for Poisson data, or other transformations leading to theoretically known variances under ideal conditions. These transformations may achieve stabilization of variance even where ideal conditions do not hold. The stabilized variance, however, is often much larger than the theoretically indicated one. Thus, when using such transformations it is almost always advisable to use a variance estimated from the transformed values themselves rather than the variance expected under theoretical conditions.

To deal with correlation coefficients. Fisher's z-transformation is most commonly used on correlation coefficients, r, and occasionally on other variables which go from -1 to $+1$ [see MULTIVARIATE ANALYSIS, *article on* CORRELATION METHODS]. In Figure 3 the curve of Fisher's z-transformation coincides with the curve of the logistic transformation, because the two are algebraically identical when $r = 2p - 1$. Generally speaking, remarks similar to those made about fractions apply to correlation coefficients.

To improve additivity. Suppose x is influenced by two other variables, u and v; for example, suppose x has the values shown in Table 1, for two unspecified values of u and three unspecified values of v. Examine the values of x. You will note that the difference between corresponding entries in the two rows is always 20, *whichever column they*

are in. Likewise, the difference between corresponding entries in any two columns is independent of the particular row they are in. (The difference is 1 for the first two columns, 7 for the last two

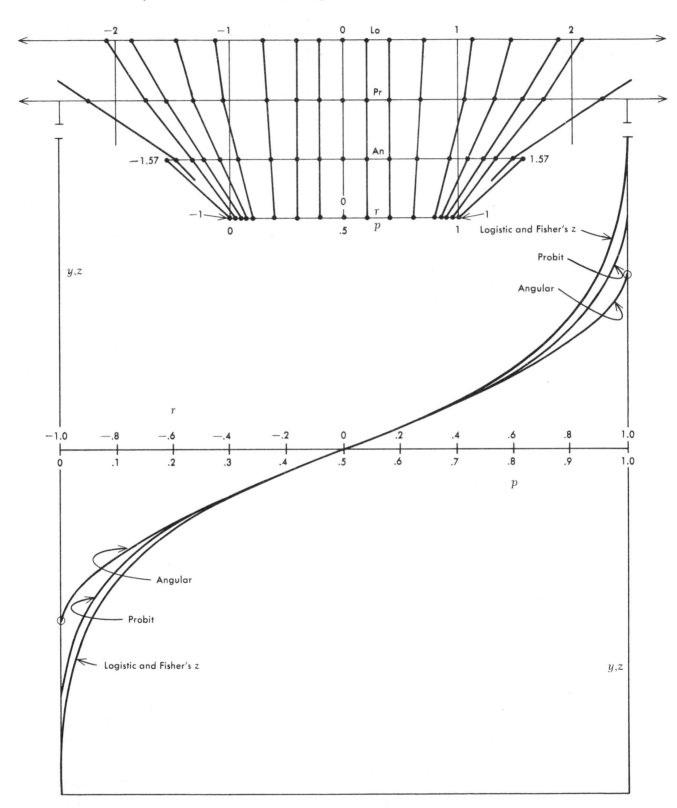

Figure 3 — Two-bend transformations*

* In the top part of the figure, the scale shown on the Lo axis is used for the Pr, An, and r axes as well.

Table 1 — Values of x

	v_1	v_2	v_3	
u_1	27	28	35	$a_1 = -10$
u_2	47	48	55	$a_2 = +10$

$$b_1 = -3 \quad b_2 = -2 \quad b_3 = +5 \qquad m = 40$$

columns, and 8 for the first and third columns.) These relations, which greatly simplify the study of how x depends on u and v, are referred to as *additivity*. (For a discussion of additivity in another sense, see Scheffé 1959, pp. 129–133.)

An alternative definition of additivity, whose equivalence is established by simple algebra, is phrased in terms of addition and accounts for the name. Call x additive in u and v if its values can be reconstructed by adding a row number (shown here as a_i) and a column number (shown here as b_j), perhaps plus an extra constant (shown here as m). For example, considering x_{11} in the first row and column,

$$27 = (-10) + (-3) + 40,$$

while, in general,

$$x_{ij} = a_i + b_j + m.$$

The extra constant (which is not essential, because it could be absorbed into the row numbers or the column numbers) is commonly taken to be the "grand mean" of all the entries, as it is here, and the row and column numbers are commonly taken, as here, to be the row and column means less the grand mean. [See LINEAR HYPOTHESES, *article on* ANALYSIS OF VARIANCE.]

To understand how additivity can be improved by a transformation, consider tables 2 and 3. Neither is additive. Suppose Table 2 is transformed by $y = \sqrt{x}$. This yields Table 4, which is

Table 2 — Values of x

	v_1	v_2	v_3
u_1	1	4	9
u_2	4	9	16
u_3	9	16	25

Table 3 — Values of x

	v_1	v_2	v_3
u_1	1.2	4.1	9.2
u_2	4.3	9.4	16.2
u_3	9.1	16.5	25.4

Table 4 — Values of y

	v_1	v_2	v_3	a_i
u_1	1	2	3	-1
u_2	2	3	4	0
u_3	3	4	5	$+1$
b_j	-1	0	$+1$	$3 = m$

Table 5 — Values of x

	v_1	v_2
u_1	1	0
u_2	0	1

Table 6 — Values of x

	v_1	v_2	v_3
u_1	0	1	2
u_2	1	2	0
u_3	2	0	1

clearly additive. The same transformation applied to Table 3 would yield values which are additive to a good approximation. Usually, approximate additivity is the best one can hope for.

Whether or not a transformation can improve additivity depends on the data. Thus, for tables 5 and 6, which are clearly nonadditive, no one-to-one transformation can produce even approximate additivity.

The concept of additivity is also meaningful and important when x depends on three variables, u, v, and w, or even more. Transformations are just as relevant to improve additivity in these cases.

Empirical transformations. Sometimes transformations are constructed as tables of numerical values and are not conveniently described by mathematical formulas. In this article such transformations are called empirical.

One use of empirical transformations is to improve additivity in u and v. J. B. Kruskal (1965) has described a method for calculating a monotonic transformation of x, carefully adapted to the given data, that improves additivity as much as possible according to a particular criterion.

Sometimes it is worthwhile to transform quantitative data into numerical ranks $1, 2, \cdots$ in order of size [*for example, this is a preliminary step to many nonparametric, or distribution-free, statistical procedures; for discussion, see* NONPARAMETRIC STATISTICS, *article on* RANKING METHODS]. This assignment of ranks can be thought of as an empirical transformation which leaves the data uniformly distributed. If normal scores are used instead of ranks, the empirical transformation leaves the distribution of the transformed data nearly normal.

Basic concepts

Linear transformations. A linear transformation $y = a + bx$, with b not 0, is often convenient, to shift the decimal point (for example, $y = 1,000x$), or to avoid negative values (for example, $y = 5 + x$) or for other reasons. Such transformations (often called coding) have no effect whatsoever on the properties of interest here, such as additivity, linearity, variance stability, and normality. Thus, linearly related variables are often considered equivalent or even the "same" in the context of transformations. To study the form of a transformation, a linearly related variable may be plotted instead. Thus, the comparison of power transformations $y = x^c$ in Figure 2 has been simplified by plotting $y = a + bx^c$, with a and b chosen for each c to make the curve go through two fixed points.

Monotonic transformations. A transformation is called (monotonic) increasing if, as in $y = \log x$, y gets larger as x gets larger. On a graph its curve always goes up as it goes to the right. A decreasing transformation, like $y = 1/x$ (where x is positive), goes the other way.

Data transformations of practical interest (in the sense of this article) are almost always either increasing or decreasing, but not mixed. The term "monotonic" (or, more precisely, "strictly monotonic") covers both cases. Sometimes the word "monotone" is used for "monotonic."

Region of interest. The region of interest in using a transformation is the region on the x-axis in which observed data values might reasonably be found or in which they actually lie. The characteristics of a transformation are relevant only in its region of interest. In particular, it need be monotonic only there.

Even though $y = x^2$ is not monotonic (it increases for x positive and decreases for x negative), it can be sensible to use $y = x^2$ for observations, x, that are necessarily positive.

Mild and strong transformations. If the graph of a transformation is almost a straight line in the region of interest, it is described as mild, or almost linear. If, on the contrary, it is strongly curved, it is called strong. Note, however, that visual impressions are sensitive to the sizes of the relative scales of the x-axis and y-axis.

A very mild transformation is useful, as a preliminary step, only when the subsequent analysis seeks maximum precision. On the other hand, if a strong transformation is appropriate, it may provide major benefits even for very approximate methods, such as visual inspection of graphical display (and for precise analysis also, of course).

The strength of a transformation depends critically on the region of interest. For x from 1 to 10, $y = \log x$ is fairly strong, as Figure 2 shows. From 5 to 6 it is quite mild, and from 5 to 5.05 it is virtually straight. From 1 to 1,000,000 it is very strong indeed.

Effect of transforming on relations among averages. Transforming can change the relationship among average values quite drastically. For example, consider the hypothetical data given in Table 7 for two rats running through a 20-foot channel. Which rat is faster? According to average time, rat 2 is slightly faster, but according to av-

Table 7—Hypothetical times and speeds

	Time (seconds)	Corresponding speed (feet per second)
Rat 1		
Trial 1	50.0	0.4
Trial 2	10.0	2.0
Trial 3	5.6	3.6
Average	21.9	2.0
Rat 2		
Trial 1	20.0	1.0
Trial 2	20.0	1.0
Trial 3	20.0	1.0
Average	20.0	1.0

erage speed, it is only half as fast! Thus, even the answer to this simple question may be altered by transformation. More delicate questions are naturally much more sensitive to transformation. This shows that use of the correct transformation may be quite important in revealing structure.

The effect of transformations on expected values and variances. The transform of the expected value does *not* equal the expected value of the transform, although they are crude approximations of each other. This is clearer in symbols. Because interest usually lies in estimating the expected value, $E(x)$, from information about $y = f(x)$, rather than vice versa, it is convenient to invert the transformation and write $x = g(y)$, where g is f^{-1}. Then, to a crude approximation,

$$E(x) = E[g(y)] \cong g[E(y)].$$

The milder the transformation is, the better this approximation is likely to be, and for linear transformations it is precisely correct.

In practice one usually uses an estimate for $E(y)$, often the average value, $\bar{y}$. Substituting the estimate for the true value is a second step of approximation: $E(x)$ is crudely approximated by $g(\bar{y})$.

One simple improvement, selected from many which have been used, is

$$E(x) \cong g[E(y)] + \tfrac{1}{2}g''[E(y)] \cdot \mathrm{var}\,(y),$$

where g'' denotes the second derivative of g and $\mathrm{var}\,(y)$ denotes the variance of y. (This approximation, since it is based on a Taylor expansion through the quadratic term, is exact if $g(y)$ is a quadratic polynomial, a situation that occurs for $y = \sqrt{x + k}$.) Substituting estimates, such as $\bar{y}$ for $E(y)$ and s_y^2, the sample variance, for $\mathrm{var}\,(y)$ is a second step of approximation:

$$E(x) \cong g(\bar{y}) + \tfrac{1}{2}g''(\bar{y}) \cdot s_y^2.$$

Advanced work along these lines is well represented by Neyman and Scott (1960).

For variances, the simplest approximation is

$$\text{var}(x) \cong \text{var}(y) \cdot \{g'[E(y)]\}^2,$$

where g' is the first derivative of g. This is fairly good for mild transformations and perfect for linear ones. After estimates are substituted, the estimate becomes

$$\text{var}(x) \cong s_y^2 \cdot [g'(\bar{y})]^2.$$

One-bend transformations. Most transformations of practical use have only one bend (as in Figure 2) or two bends (as in Figure 3). A curious fact is that the increasing one-bend transformations most often used bend downward, like $y = \log x$, rather than upward, like $y = x^2$. For decreasing transformations, turn the graph upside down and the same thing holds true.

Some important families of one-bend transformations are

$y = \log(x + k),$	logarithmic family;
$y = \sqrt{x + k},$	square-root family;
$y = x^c,$	power family;
$y = e^{cx}$ (or 10^{cx} or k^x),	exponential family;
$y = (x + k)^c,$	"simple" family.

Here k and c are constants, and $e \cong 2.718$ is a familiar constant. The region of interest is generally restricted either to $x \geqslant 0$ or to $x \geqslant -k$. Of course, each family also includes linearly related transformations (for example, $y = a + bx^c$ is in the power family).

The "simple" family (named and discussed in Tukey 1957) obviously includes two of the other families and, by a natural extension to mathematical limits, includes the remaining two as well.

Two-bend transformations. Only those two-bend transformations mentioned above (angular, probit, logistic, and z) are in general use, although the log–log transformation, given by $y = \log(-\log x)$,

is sometimes applied. A whole family of varying strengths would be desirable (such as $p^c - (1 - p)^c$), but no such family appears to have received more than passing mention in the literature.

Several formulas and names for two-bend transformations are presented in Table 8. Tables of these transformations are generally available. (For references, consult Fletcher et al. 1946 and the index of Greenwood & Hartley 1962.)

When to use and how to choose a transformation

Many clues suggest the possible value of transforming. Sometimes the same clue not only gives this general indication but also points to the necessary transformation or to its general shape.

When using a transformation, it is *essential* to visualize its plotted curve over the region of interest. If necessary, actually plot a few points on graph paper. (In several published examples the authors appear to be unaware that the transformations are so mild as to be useless in their context.)

If a quantity, such as the expected value, is needed for x itself, consider whether this quantity is best found by working directly with x or indirectly through the use of some transform. However, need should be judged cautiously; although it is often real, in many cases it may vanish on closer examination.

A word of caution: "outliers" can simulate some of the clues below and falsely suggest the need for transforming when, in fact, techniques for dealing with outliers should be used [*see* STATISTICAL ANALYSIS, SPECIAL PROBLEMS OF, *article on* OUTLIERS].

Some simple clues. Very simple and yet very strong clues include counted data covering a wide range, fraction data with some observations near 0 or 1, and correlation coefficient data with some

Table 8 — Two-bend transformations

Forward forms	Backward forms	Names
$y = \arcsin \sqrt{p}$ $= \sin^{-1} \sqrt{p}$	$p = \sin^2 y$	angular arcsine inverse sine
$y = \text{Erf}^{-1}(p)$ $= \Phi^{-1}(p)$ $y = \Phi^{-1}(p) + 5$ (often)	$p = \text{Erf}(y) = \Phi(y)$ $=$ normal probability up to y	probit[a] normit phi–gamma
$y = \frac{1}{2} \log \dfrac{p}{1 - p}$	$p = \dfrac{e^{2y}}{e^{2y} + 1}$	logistic[b] logit
$z = \frac{1}{2} \log \dfrac{1 + r}{1 - r}$ $= \text{arctanh } r$	$r = \dfrac{e^{2z} - 1}{e^{2z} + 1}$	Fisher's z[b] z hyperbolic arctangent

a. Under the name "probit" the addition of 5 is usual, to avoid negative values.

b. Natural logarithms are generally used here.

values near -1 or $+1$. Generally, observations which closely approach an *intrinsic boundary* may benefit by a transformation which expands the end region, perhaps to infinity.

If several related curves present a complex but systematic appearance, it is often possible to simplify them, as in Figure 1. For example, they might all become straight lines, or the vertical or horizontal spacing between curves might become constant along the curves.

A very nonnormal distribution of data is a clue, although, by itself, a weak one. For this purpose, nonnormality is best judged by plotting the data on "probability paper" [*see* GRAPHIC PRESENTATION]. The general shape of a normalizing transformation can be read directly from the plot.

Nonconstant variance. Suppose there are many categories (such as the cells in a two-way table) and each contains several observations. Calculate the sample variance, s^2_{ij}, and the average value, a_{ij}, in each category. If the s^2_{ij} vary substantially, make a scatter plot of the s^2_{ij} against the a_{ij}. If the s^2_{ij} tend to change systematically along the a-axis, then a variance-stabilizing transformation is possible and often worthwhile. Usually the scatter plot is more revealing if plotted on paper with both scales logarithmic, so that $\log s^2_{ij}$ is plotted against $\log a_{ij}$.

To choose the transformation, the relationship between the s^2_{ij} and the a_{ij} must be estimated. Using even a crude estimate may be better than not transforming at all. Suppose the estimated relationship is $s^2_{ij} = g(a_{ij})$. (Commonly, this is $s^2_{ij} = ka^c_{ij}$. This is a straight line on log–log paper. Taylor [1961] contains many such examples.) It can be shown that $y = \int [\sqrt{g(x)}]^{-1} dx$ stabilizes the variance (approximately). If $s^2_{ij} = ka_{ij}$ (for Poisson distributions, $k = 1$), this leads essentially to $y = \sqrt{x}$. For $s^2_{ij} = ka^c_{ij}$, it leads essentially to $y = x^d$, with $d = 1 - c/2$ (or $y = \log x$ if $d = 0$).

Removable nonadditivity. When nonadditivity can be removed by a transformation, it is almost always worthwhile to do so. To recognize nonadditivity is often easy, either by direct examination or by the size of the interaction terms in an analysis of variance [*see* LINEAR HYPOTHESES, *article on* ANALYSIS OF VARIANCE]. To decide how much of it is removable may be harder.

With experience, in simple cases one can often recognize removable nonadditivity by direct examination and discover roughly the shape of the required transformation. If Tukey's "one degree of freedom for non-additivity" (see Moore & Tukey 1954; or Scheffé 1959, pp. 129–133) yields a large value of F, some nonadditivity is removable. Closely

related is the scatter-plot analysis of residuals given by Anscombe and Tukey (1963, sec. 10). Kruskal's method (1965) directly seeks the monotone transformation leaving the data most additive.

How to choose the transformation. In addition to the methods mentioned above, one or more transformations may be tried quite arbitrarily. If significant benefits result, then the possibility of greater benefits from a stronger, weaker, or modified transformation may be investigated.

[1] A whole family of transformations can, in effect, be tried all at once, with the parameter values chosen to optimize some criterion. The important paper by Box and Cox (1964) gives a good discussion of this method. Another useful approach is provided in Kruskal (1965).

An illustration—galvanic skin response

It has long been known that electrical resistance through the skin changes rapidly in response to psychological stimuli—a phenomenon known as galvanic skin response (GSR) or electrodermal response. Originally, the only scale used for analyzing GSR was that of electrical resistance, R, measured in ohms (or in kilohms). As early as 1934, Darrow (see 1934; 1937) suggested the use of electrical conductance, $C = 1/R$, measured in mhos (or in micromhos), and later $\log C$, as well as various modifications. More recently, other scales, such as $\sqrt{C}$, have also been used, and the topic has continued to receive attention up to the present.

Although agreement on "the best scale" has not been reached, many authors who treat this question agree that R is a very poor scale and that both C and $\log C$ provide substantial improvement. Most experimenters now use either C or $\log C$, but a few still use R (and some fail to specify which scale they are using), more than thirty years after Darrow's original paper!

Lacey and Siegel (1949) have discussed various transformations, using their own experimental results. Their final recommendation is C. Using 92 subjects, they measured the resistance of each one twice, first while the subject was sitting quietly, then just after the subject had received an unexpected electric shock. Call the two values R_0 (before shock) and R_1 (after shock). Each subject received only one shock, and presumably all the shocks were of the same strength.

For any scale y (whether R, $\log C$, C, or another), let y_0 and y_1 be the two values (before and after) and let $\mathrm{GSR} = y_1 - y_0$. (A separate question, omitted here, is whether $y_1 - y_0$ itself should be transformed before use as the GSR.)

The major use of GSR is to measure the strength

of a subject's response to a stimulus. For this use it is desirable that the size of the GSR not depend on extraneous variables, such as the subject's basal resistance. Thus, in the equation $y_1 = y_0 + $ GSR the value of GSR should be independent of y_0. This is a form of additivity.

Figure 4 shows GSR plotted against y_0 for the two scales R and C. On the R scale it is obvious that the

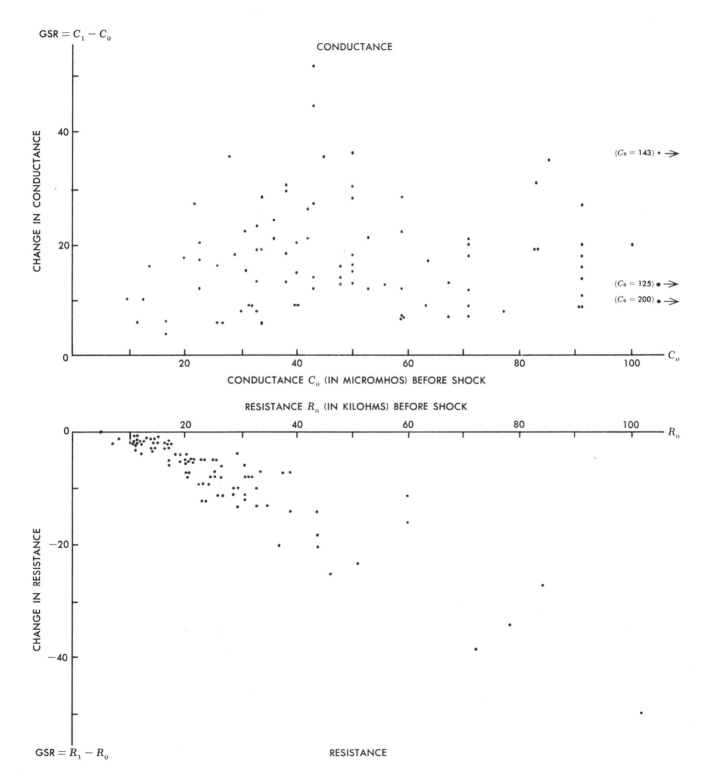

Figure 4 — *Galvanic skin response (GSR) plotted against initial level on two scales**

* GSR represents postshock level of resistance or conductance minus preshock level. The same 92 observations are plotted in both parts of the figure.

Source: Data from Lacey & Siegel 1949.

GSR has a strong systematic dependence on R_0 (in addition to its random fluctuation). The C scale is in strong contrast, for there the GSR displays relatively little systematic dependence on C_0. The corresponding plot for the intermediate $\log C$ scale (not shown here) displays a distinct but intermediate degree of dependence.

Another desirable property, often specified in papers on this topic, is that y_0 should be approximately normal. Plots on "probability paper" (not shown here) display $\log C_0$ as quite nicely straight but $\sqrt{C_0}$ and $\sqrt{R_0}$ as definitely curved (skew), and C_0 and R_0 as even more curved. Thus, a conflict appears: the requirement of normality points to the $\log C$ scale, while additivity points to the C scale. For the major use of GSR, additivity is more important and must dominate unless some resolution can be found.

An intermediate scale might provide a resolution. Schlosberg and Stanley (1953) chose $\sqrt{C}$. A significantly different intermediate scale, with a plausible rationale, is $\log (R - R^*)$, where R^* depends only on the electrodes, electrode paste, and so forth. For the data of the Lacey and Siegel study, the value of R^* should be a little more than 4 kilohms.

Some other topics

[2]Multivariate observations can be transformed. Even linear transformations are significant in this case, and these have been the main focus of interest so far. [See FACTOR ANALYSIS AND PRINCIPAL COMPONENTS; MULTIVARIATE ANALYSIS.]

Where detailed mathematical assumptions can safely be made, transformations can sometimes be used in a more complex way (based on the maximum likelihood principle) to obtain greater precision. This is often called the Bliss–Fisher method. How widely this method should be used has been the subject of much controversy. An article by Fisher (1954, with discussion) gives a good statement of that author's views, together with concise statements by five other eminent statisticians.

Similarity and dissimilarity measures of many kinds are sometimes transformed into spatial distances, by the technique of multidimensional scaling. Briefly, if δ_{ij} is a measure of dissimilarity between objects i and j, multidimensional scaling seeks points (in r-dimensional space) whose interpoint distances, d_{ij}, are systematically related to the dissimilarities. The relationship between δ_{ij} and d_{ij} may usefully be considered a transformation (for further information on multidimensional scaling, see Kruskal 1964).

An ingenious and appealing application of transformations (to three sets of data) appears in Shepard (1965). In each case the data can be represented by several plotted curves on a single graph. Each curve is essentially unimodal, but the peaks occur at different places along the x-axis. The x-variable is monotonically transformed so as to give the different curves the same shape, that is, to make the curves the same except for location along the x-axis. The transformed variables appear to have subject-matter significance.

JOSEPH B. KRUSKAL

[*See also* ERRORS; GRAPHIC PRESENTATION.]

BIBLIOGRAPHY

General discussion of transformations accompanied by worthwhile applications to data of intrinsic interest may be found in Bartlett 1947, Box & Cox 1964, Kruskal 1965, Moore & Tukey 1954, *and* Snedecor 1937. *Actual or potential applications of great interest may be found in* Lacey & Siegel 1949, Shepard 1965, Taylor 1961, *and* Wald 1965. *For tables, consult* Fletcher, Miller & Rosenhead 1946, *and* Greenwood & Hartley 1962. *Large and useful bibliographies may be found in* Grimm 1960 *and* Lienert 1962.

ANSCOMBE, F. J.; and TUKEY, JOHN W. 1963 The Examination and Analysis of Residuals. *Technometrics* 5: 141–160.

BARTLETT, M. S. 1947 The Use of Transformations. *Biometrics* 3:39–52. → Down-to-earth, practical advice. Widely read and still very useful.

BOX, G. E. P.; and COX, D. R. 1964 An Analysis of Transformations. *Journal of the Royal Statistical Society* Series B 26:211–252. → Starts with a very useful general review. The body of the paper, although important and equally useful, requires some mathematical sophistication.

CRESPI, LEO P. 1942 Quantitative Variation of Incentive and Performance in the White Rat. *American Journal of Psychology* 55:467–517.

DARROW, CHESTER W. 1934 The Significance of Skin Resistance in the Light of Its Relation to the Amount of Perspiration (Preliminary Note). *Journal of General Psychology* 11:451–452.

DARROW, CHESTER W. 1937 The Equation of the Galvanic Skin Reflex Curve: I. The Dynamics of Reaction in Relation to Excitation-background. *Journal of General Psychology* 16:285–309.

FISHER, R. A. 1954 The Analysis of Variance With Various Binomial Transformations. *Biometrics* 10: 130–151. → Contains a statement of the Bliss–Fisher method and a controversy over its scope. Do not overlook the important 11-page discussion, especially the remarks by Cochran and Anscombe, whose views are recommended.

FLETCHER, ALAN; MILLER, JEFFERY C. P.; and ROSENHEAD, LOUIS (1946) 1962 *An Index of Mathematical Tables.* 2d ed. 2 vols. Reading, Mass.: Addison-Wesley.

GREENWOOD, J. ARTHUR; and HARTLEY, H. O. 1962 *Guide to Tables in Mathematical Statistics.* Princeton Univ. Press.

GRIMM, H. 1960 Transformation von Zufallsvariablen. *Biometrische Zeitschrift* 2:164–182.

KRUSKAL, JOSEPH B. 1964 Multidimensional Scaling by Optimizing Goodness of Fit to a Nonmetric Hypothesis. *Psychometrika* 29:1–27.

KRUSKAL, JOSEPH B. 1965 Analysis of Factorial Experiments by Estimating Monotone Transformations of the Data. *Journal of the Royal Statistical Society* Series B 27:251–263.

LACEY, OLIVER L.; and SIEGEL, PAUL S. 1949 An Analysis of the Unit of Measurement of the Galvanic Skin Response. *Journal of Experimental Psychology* 39: 122–127.

LIENERT, G. A. 1962 Über die Anwendung von Variablen-Transformationen in der Psychologie. *Biometrische Zeitschrift* 4:145–181.

MOORE, PETER G.; and TUKEY, JOHN W. 1954 Answer to Query 112. *Biometrics* 10:562–568. → Includes Tukey's "One Degree of Freedom for Non-additivity" and an interesting example of how to choose a transformation, explained clearly and with brevity.

MOSTELLER, FREDERICK; and BUSH, ROBERT R. (1954) 1959 Selected Quantitative Techniques. Volume 1, pages 289–334 in Gardner Lindzey (editor), *Handbook of Social Psychology.* Cambridge, Mass.: Addison-Wesley.

NEYMAN, J.; and SCOTT, E. L. 1960 Correction for Bias Introduced by a Transformation of Variables. *Annals of Mathematical Statistics* 31:643–655.

SCHEFFÉ, HENRY 1959 *The Analysis of Variance.* New York: Wiley. → See especially pages 129–133.

SCHLOSBERG, HAROLD S.; and STANLEY, WALTER C. S. 1953 A Simple Test of the Normality of Twenty-four Distributions of Electrical Skin Conductance. *Science* 117:35–37.

SHEPARD, ROGER N. 1965 Approximation to Uniform Gradients of Generalization by Monotone Transformations of Scale. Pages 94–110 in David I. Mostofsky (editor), *Stimulus Generalization.* Stanford Univ. Press.

SNEDECOR, GEORGE W. S. (1937) 1957 *Statistical Methods: Applied to Experiments in Agriculture and Biology.* 5th ed. Ames: Iowa State Univ. Press. → See especially pages 314–327.

TAYLOR, L. R. 1961 Aggregation, Variance and the Mean. *Nature* 189:732–735.

TUKEY, JOHN W. 1957 On the Comparative Anatomy of Transformations. *Annals of Mathematical Statistics* 28:602–632.

WALD, GEORGE 1965 Frequency or Wave Length? *Science* 150:1239–1240. → See also several follow-up letters, under the heading "Frequency Scale for Spectra," *Science* 151:400–404 (1966).

Postscript

[1]In the last several years the approach described by Box and Cox (1964) for choosing transformations has received considerable further development. An easy-to-read paper by Draper and Hunter (1969) presents a practical description of some simple but useful methods. Some other useful papers are Andrews (1971), Schlesselman (1973), Atkinson (1973), and Hinkley (1975).

An approach like that of Box and Cox to choosing a nonlinear transformation of multivariate data is explored (for the bivariate case) in Andrews, Gnanadesikan, and Warner (1971).

[2]Even though linear transformations are not interesting for univariate observations, they are for multivariate observations. It is often useful to apply a linear transformation to the coordinates so that the transformed coordinates are uncorrelated and of equal variance, in other words, so that the sample covariance matrix of the transformed coordinates has zeroes off the main diagonal and a constant value on the main diagonal. Applying such a transformation is sometimes called "spherizing" or "sphericizing" the data. (This name is based on the shape of the ellipsoid that is the geometric equivalent of the covariance matrix. For the transformed coordinates the ellipsoid is a sphere.) A linear transformation having this property can be found by a classical factorization of the sample covariance matrix.

Of even greater value is the linear transformation that spherizes the pooled covariance matrix of several subsamples. Such transformations are closely related to discriminant analysis, Hotelling's T^2, and other multivariate techniques. [See MULTIVARIATE ANALYSIS, *articles on* THE FIELD *and on* CLASSIFICATION AND DISCRIMINATION.] With the recent development of robust methods for estimating covariance matrices, such as Gnanadesikan and Kettenring (1972, p. 87 ff.), it is an attractive idea to use a linear transformation that spherizes the robustly estimated covariance matrix, prior to examining the data by other means.

In the field of test theory and psychophysical measurement, Levine has been developing some very novel methods of genuine practical value for transforming data that constitute several monotonic curves, as shown in Figure 5 (1970; 1975, and other papers). A typical application would be to the item–test regression curves from a single test. (When a group of examinees is given a multiple choice test, there is one such curve for each item. To construct the curves, the examinees are grouped together by their total score, for example, there may be one group for each possible score, or

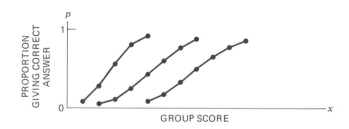

Figure 5 – Item–test regression curves*

* Each curve represents a single item.

there might be coarser groups. For a single item, and for every score-group of examinees, the fraction of examinees in the group who gave the correct answer on that item is plotted against the score of the group.) Levine's methods permit estimation of location and scale parameters for the item difficulty curves under extremely weak assumptions by making use of such advanced mathematical concepts as the derived subgroup (in group theory) and solutions of Abel's equation (in functional equations). The item difficulty curves are conceptual analogues to the item-test regression curves, and show the expected fraction correct for each level of ability of the examinee.

A transformational approach is now being frequently used to deal with observed fractions, contingency tables, binary data, and so on. To illustrate the approach, let p_{ij} be the observed fraction of all n_{ij} cases in cell (i,j) that have some property (so that each p_{ij} lies between 0 and 1), and suppose that the different p_{ij} are statistically independent. The transformational approach is to assume that $E(p_{ij})$ is some function of a simple linear model, for example,

$$E(p_{ij}) = f(m + a_i + b_j).$$

An interesting class of applications occurs in linguistics; see for example Cedergren and Sankoff (1974), where $f(x)$ is e^{-x} or $1 - e^{-x}$ but we don't know which, and where m, a_i, and b_j are all $\geqslant 0$. These authors use the direct approach of transforming by f^{-1} (for each f) and then using standard methods of analysis. There is a more structured approach to situations of this kind, which does not involve direct transformation. One version of this approach is presented by Cox (1970) and called "logistic–linear models" since he assumes f to be the logistic function, $f(x) = e^x/(1 + e^x)$. An advantage of this approach is that it can handle situations where the p_{ij} are always or frequently 0 or 1. Another version called "log-linear models" assumes $f(x) = e^x$, and is extensively developed in two books, Bishop et al. (1975) and Haberman (1974), as well as many articles. [See also COUNTED DATA; SURVEY ANALYSIS, *article on* METHODS OF SURVEY ANALYSIS.]

A new general discussion of transformations and a bibliography will be found in Hoyle (1973).

An unusual application of transformations occurs in international bridge tournaments, where ordinary rubber bridge scores are converted to International Match Points by use of an S-shaped transformation.

JOSEPH B. KRUSKAL

ADDITIONAL BIBLIOGRAPHY

ANDREWS, DAVID F. 1971 A Note on the Selection of Data Transformations. *Biometrika* 58:249–254.

ANDREWS, DAVID F.; GNANADESIKAN, RAMANATHAN; and WARNER, J. L. 1971 Transformations of Multivariate Data. *Biometrics* 27:825–840.

ATKINSON, A. C. 1973 Testing Transformations to Normality. *Journal of the Royal Statistical Society* Series B 35:473–479.

BISHOP, YVONNE M. M.; FIENBERG, STEPHEN E.; and HOLLAND, PAUL W. 1975 *Discrete Multivariate Analysis.* Cambridge, Mass.: M.I.T. Press.

CEDERGREN, H. A.; and SANKOFF, D. 1974 Variable Rules: Performance as a Statistical Reflection of Competence. *Language* 50:333–355.

COX, D. R. 1970 *Analysis of Binary Data.* London: Methuen.

DRAPER, NORMAN R.; and HUNTER, WILLIAM G. 1969 Transformations: Some Examples Revisited. *Technometrics* 11:23–40.

GNANADESIKAN, RAMANATHAN; and KETTENRING, J. J. 1972 Robust Estimates, Residuals, and Outlier Detection With Multiresponse Data. *Biometrics* 28:81–124.

HABERMAN, SHELBY J. 1974 *The Analysis of Frequency Data.* Univ. of Chicago Press.

HINKLEY, D. V. 1975 On Power Transformations to Symmetry. *Biometrika* 62:101–111.

HOYLE, M. H. 1973 Transformations: An Introduction and a Bibliography. *International Statistical Review* 41:203–223.

LEVINE, MICHAEL V. 1970 Transformations That Render Curves Parallel. *Journal of Mathematical Psychology* 7:410–443.

LEVINE, MICHAEL V. 1975 Additive Measurement With Short Segments of Curves. *Journal of Mathematical Psychology* 12:212–224.

SCHLESSELMAN, JAMES J. 1973 Data Transformation in Two-way Analysis of Variance. *Journal of the American Statistical Association* 68:369–378.

III
GROUPED OBSERVATIONS

When repeated empirical measurements are made of the same quantity, the results typically vary; on the other hand, ties, or repetitions of measurement values, often occur, because measurements are never made with perfect fineness. Length, for example, although theoretically a continuous quantity, must be measured to the closest inch, centimeter, or some such; psychological aptitudes that are conceptually continuous are measured to a degree of fineness consonant with the test or instrument used. In many cases it is desirable and efficient to measure less finely than current techniques permit, because the expense of refined measurement may outweigh the value of the added information. Also, grouping of measurements is often carried out after they are made, in order to enhance convenience of computation or presentation.

Table 1 — Thirty-two persons grouped according to their body weight (in pounds)

Weight	Number of persons	Mid-value of class
137.5–147.5	2	142.5
147.5–157.5	5	152.5
157.5–167.5	4	162.5
167.5–177.5	5	172.5
177.5–187.5	7	182.5
187.5–197.5	5	192.5
197.5–207.5	3	202.5
207.5–217.5	0	212.5
217.5–227.5	0	222.5
227.5–237.5	0	232.5
237.5–247.5	1	242.5

Source: Adapted with permission of The Macmillan Company from STATISTICS: A NEW APPROACH by W. Allen Wallis and Harry V. Roberts. Copyright 1956 by The Free Press of Glencoe, A Corporation.

No matter how the data are obtained, unless there are only a few it is usually convenient to present them in groups or intervals; that is, the data are organized in a table that shows how many observations are in each of a relatively small number of intervals. The motivations are clarity of description and simplification of subsequent manipulation. Table 1 presents an example of such grouped data. (The data are taken from Wallis & Roberts 1956; this work should be consulted for details of computation and for the useful discussion in secs. 6.2.1, 7.4.5, and 8.5.2.) If it is assumed that the weights were originally measured only to the nearest pound, then there is no ambiguity at the class boundaries about which group is appropriate for each measurement. Some information is, of course, lost when the data are recorded in these coarser groups. Such loss of information can, however, be compensated for by making some more observations, and so a purely economic problem results: which is cheaper—to make fewer but finer measurements or to make more but coarser measurements?

A second problem is that most mathematical and statistical tools developed for the treatment of data of this kind presuppose exact measurements, theoretically on an infinitely fine scale. To what extent can use be made of these tools when the data are grouped? How must the theory be modified so that it may legitimately be applied to grouped observations?

Another problem is that bias may arise in cases where an observation is more likely to fall in some part of a relatively wide group than in other parts of that group. Theoretically this holds, of course, for any size of group interval, but when group intervals are less than one-quarter of the standard deviation of the distribution, there will, in practice, be little trouble.

Usually one refers all the observations falling in a group to the midpoint of that group and then works with these new "observations" as if they were exact. What does such a procedure imply? Is it justified? And what should be done if the grouping is very coarse or if there are open intervals to the right and/or left? What should be done in the case where the groups can only be ranked, because their limits have no numerical values?

Estimation of μ and σ in the normal case. Most work on problems of grouped observations has centered about simple normal samples, where one is interested in inferences about mean and variance. Suppose N observations are regarded as a random sample from a normal population with mean μ and standard deviation σ. The observations have fallen in, say, k groups. The maximum likelihood (ML) estimators of μ and σ can be found whether or not the grouping is into equal intervals and whether or not the end intervals are open. The method is a bit troublesome because of its iterative character, but tables have been constructed for facilitation of the work (see Gjeddebæk 1949; advice on finer details is given in Kulldorff 1958a; 1958b). The ML method can be used with assurance in cases where there is some doubt about the admissibility of using the so-called simple estimators of μ and σ.

The simple estimators are obtained if all observations are referred to the midpoints of their respective groups and standard estimators m and s are calculated for the mean and standard deviation, as outlined in Wallis and Roberts (1956, secs. 7.4.5, 8.5.2). These estimators of μ and σ have almost the same variances and expected mean squares as the ML estimators when the grouping intervals are not wider than 2σ, and N, the sample size, is no more than 100. The simple estimators have biases that do not go to zero as N increases. If, however, interval width is no more than 2σ and $N \leqslant 100$, the bias is negligible.

If the grouping is equidistant with groups of size $h\sigma$, and $N < 100$, the efficiencies, E, of the ML mean (or of the simple mean, m) relative to the ungrouped mean are as presented in Table 2. Here E indicates how many ungrouped observations are equivalent to 100 grouped observations made with the group size in question. The above figures can be obtained from a formula given by R. A. Fisher (1922), $E = (1 + h^2/12)^{-1}$. If the phase relationship between true mean and group limits is taken into account, there are only very small changes (see Gjeddebæk 1956).

Table 2 — *Efficiency of the ML and simple estimators of the mean from grouped data, relative to the ungrouped mean*

h	E (in per cent)
0.2	99.7
0.4	98.7
0.6	97.1
0.8	94.9
1.0	92.3
1.2	89.3
1.4	86.0
1.6	82.4
1.8	78.7
2.0	75.0

ML grouped and simple estimators of σ have an efficiency of about 58 per cent for $h = 2$. For estimating σ, groups should not be wider than 1.6σ, so that "phase relationship" will not play a role. For equidistant grouping with intervals of length 1.6σ, all efficiencies for estimating σ will be about 70 per cent.

Practical conclusions. Because σ depends on both natural variation and measurement variation, it may be seen, for example, that measurements of height in man lose little efficiency when recorded in centimeters rather than in millimeters. It will also be seen that if a considerable reduction of cost per observation can be obtained by measuring with a coarse scale, the information obtained per unit of cost may be increased by such coarse measuring. Even with such large group intervals as 2σ, it is necessary to make only four grouped observations for each three ungrouped observations to obtain the same amount of information about μ.

An instance can be quoted of the advantage of refraining from using a scale to its utmost capacity. Consider the routine weighing of such things as tablets or other doses of substances weighing between 200 and 600 milligrams. Sometimes, in practice, even tenths of milligrams are taken into account here. The last significant figure obtained by such a weighing will, as a rule, require preposterous labor and will divert attention from the foregoing, more important figures. Thus, carrying out the weighing to this point will do more harm than good in the long run. Also, the accuracy of reading ought to be kept in reasonable proportion to the accuracy inherent in the measured pieces; in other words, observations ought to be grouped with due regard to the size of their natural variation. Here the table of efficiencies will be useful, as it immediately reveals at what point the coarseness of a scale essentially influences the accuracy. In this connection it is worthwhile to stress that an impression of great accuracy from a result with many decimal places is misleading if the natural variation of the results makes illusory most of the decimal places. For example, it was an exaggeration of scientific accuracy when it was said that every cigarette smoked by a person cuts off 14 minutes and 24 seconds from his lifetime. Of course, there was some publicity value in the statement. If, however, you are told that whenever you have smoked 100 cigarettes you have lost a day and a night of your lifetime, you will conclude that some scientist has given this opinion with due regard to the obvious uncertainty of such an estimate. From the viewpoint of a statistician, this statement seems a bit more honest, despite the fact that exactly the same thing is expressed as before. (Example 82C of section 3.10 of Wallis & Roberts 1956 gives a similar provocative use of figures.)

Sheppard correction. When the variance, σ^2, is estimated from data grouped into intervals of width $h\sigma$, it has been proposed that the best estimator of σ^2 is $s^2 - (h^2/12)\sigma^2$, where $-(h^2/12)\sigma^2$ is the so-called Sheppard correction. There has been persistent confusion about a statement by R. A. Fisher that the Sheppard correction should be avoided in hypothesis testing. Given N grouped observations, calculate the simple estimators m of the mean and s^2 of the variance, with the size of group intervals, $h\sigma$, the same for the two calculations. From the efficiency, E (see discussion above), it follows that the variance of m is $(\sigma^2/N)(1 + h^2/12)$. According to Sheppard, $s^2 - (h^2/12)\sigma^2$ is the best estimator of σ^2; thus, it must follow that the best estimator of the quantity $\sigma^2(1 + h^2/12)$ is simply s^2, and hence the best estimator of the variance of m is s^2/N. That is the same expression used for ungrouped observations, and so s^2 should be used without Sheppard's correction when m and s^2 are brought together in a testing procedure or in a statement of confidence limits for μ. Obvious modifications may be made when h is different for the grouping used to calculate m and s^2. Sheppard's correction should also be avoided in analysis of variance, as the same group intervals are used for "within" and "between" estimators of the variance. In practice, Sheppard's correction is not very useful, because an isolated estimate of variance is seldom required. For that case, however, the correction is well justified. A serious drawback is that it cannot be used when grouping is not equidistant. Here the maximum likelihood method seems to be the only reasonable precise one.

When the Sheppard-corrected estimator of σ^2 is available, it has the same efficiency as the ML estimator, at least for N less than 100 (see Gjeddebæk 1959a and the above discussion of the inconsistency of the simple estimators for large N).

Very coarse groups. When a grouping must be very coarse, the problem arises of where to place the group limits most efficiently. The problem is compounded because the efficiency depends on the site of the unknown true mean. Sometimes—for example, for quality-control purposes—it is known where the true mean ought to be, so the theory is often useful in these instances (see Ogawa 1962).

To investigate the consequences of very wide grouping, consider the weights example once again. Let the groups be very wide—40 pounds. Depending on which group limits are deleted, one of the four situations illustrated in Table 3 results. The ML method presupposes an underlying normal distribution, and therefore it gives rather small *s*-values in the last two extreme situations, but all in all, Table 3 demonstrates an astonishingly small effect of such coarse grouping. [*For a discussion of optimum grouping, see* NONPARAMETRIC STATISTICS, *article on* ORDER STATISTICS.]

Grouping ordinal data. If the group limits have no numerical values by nature but are ranked, a different set of problems arises. By use of probits the group limits may be given numerical values, and then the situation may be treated according to the principle of maximum likelihood. In reality a two-step use of that principle is involved, and this gives rise to distribution problems. (The method is discussed in Gjeddebæk 1961; 1963.) From a hypothesis-testing point of view, the grouping of ordinal data may be regarded as the introduction of ties [*see* NONPARAMETRIC STATISTICS, *article on* RANKING METHODS, *for a discussion of such ties*].

Other aspects. Extensions of techniques using grouped data to distributions other than the normal, to multivariate data, and to procedures other than estimation of parameters are possible and have been investigated. For example, Tallis and Young (1962) have discussed the multivariate case and have given hints on hypothesis testing. Kulldorff (1961) has investigated the case of the exponential distribution, and Gjeddebæk (1949; 1956; 1957; 1959a; 1959b; 1961; 1963) has worked on methods parallel to *t*-testing and *F*-testing. In addition, P. S. Swamy (1960) has done considerable work on these extensions. A comprehensive treatment of rounding errors is given by Eisenhart (1947).

N. F. GJEDDEBÆK

[*See also* STATISTICS, DESCRIPTIVE.]

BIBLIOGRAPHY

EISENHART, CHURCHILL 1947 Effects of Rounding or Grouping Data. Pages 185–233 in Columbia University, Statistical Research Group, *Selected Techniques of Statistical Analysis for Scientific and Industrial Research, and Production and Management Engineering,* by Churchill Eisenhart, Millard W. Hastay, and W. Allen Wallis. New York: McGraw-Hill.

FISHER, R. A. (1922) 1950 On the Mathematical Foundations of Theoretical Statistics. Pages 10.308a–10.368 in R. A. Fisher, *Contributions to Mathematical Statistics.* New York: Wiley. → First published in the *Philosophical Transactions,* Series A, Volume 222, of the Royal Society of London.

FISHER, R. A. (1925) 1958 *Statistical Methods for Research Workers.* 13th ed. New York: Hafner. → Previous editions were also published by Oliver & Boyd. See especially section 19, Appendix D, "Adjustment for Grouping."

GJEDDEBÆK, N. F. 1949 Contribution to the Study of Grouped Observations: I. Application of the Method of Maximum Likelihood in Case of Normally Distributed Observations. *Skandinavisk aktuarietidskrift* 32:135–159.

GJEDDEBÆK, N. F. 1956 Contribution to the Study of Grouped Observations: II. Loss of Information Caused by Groupings of Normally Distributed Observations. *Skandinavisk aktuarietidskrift* 39:154–159.

GJEDDEBÆK, N. F. 1957 Contribution to the Study of Grouped Observations: III. The Distribution of Estimates of the Mean. *Skandinavisk aktuarietidskrift* 40:20–25.

GJEDDEBÆK, N. F. 1959a Contribution to the Study of Grouped Observations: IV. Some Comments on Simple Estimates. *Biometrics* 15:433–439.

GJEDDEBÆK, N. F. 1959b Contribution to the Study of Grouped Observations: V. Three-class Grouping of Normal Observations. *Skandinavisk aktuarietidskrift* 42:194–207.

GJEDDEBÆK, N. F. 1961 Contribution to the Study of Grouped Observations: VI. *Skandinavisk aktuarietidskrift* 44:55–73.

Table 3 — *Alternate groupings of 32 persons according to their body weight (in pounds)*

WEIGHT	NUMBER OF PERSONS	MEAN Simple	MEAN ML	STANDARD DEVIATION Simple	STANDARD DEVIATION ML
Situation 1					
107.5–147.5	2				
147.5–187.5	21	177.5	177.7	22.0	21.9
187.5–227.5	8				
227.5–267.5	1				
Situation 2					
117.5–157.5	7				
157.5–197.5	21	175.0	175.1	24.2	24.1
197.5–237.5	3				
237.5–277.5	1				
Situation 3					
127.5–167.5	11				
167.5–207.5	20	175.0	174.7	18.0	17.1
207.5–247.5	1				
Situation 4					
137.5–177.5	16				
177.5–217.5	15	178.8	178.9	19.6	18.5
217.5–257.5	1				
Results of 10-pound grouping for comparison		176.6	176.6	21.0	21.0

GJEDDEBÆK, N. F. 1963 On Grouped Observations and Adjacent Aspects of Statistical Theory. *Methods of Information in Medicine* 2:116–121.

KULLDORFF, GUNNAR 1958a Maximum Likelihood Estimation of the Mean of a Normal Random Variable When the Sample Is Grouped. *Skandinavisk aktuarietidskrift* 41:1–17.

KULLDORFF, GUNNAR 1958b Maximum Likelihood Estimation of the Standard Deviation of a Normal Random Variable When the Sample Is Grouped. *Skandinavisk aktuarietidskrift* 41:18–36.

KULLDORFF, GUNNAR 1961 *Contributions to the Theory of Estimation From Grouped and Partially Grouped Samples.* Uppsala (Sweden): Almqvist & Wiksell.

OGAWA, JUNJIRO 1962 Determinations of Optimum Spacings in the Case of Normal Distribution. Pages 272–283 in Ahmed E. Sarhan and Bernard G. Greenberg (editors), *Contributions to Order Statistics.* New York: Wiley.

STEVENS, W. L. 1948 Control by Gauging. *Journal of the Royal Statistical Society* Series B 10:54–98. → A discussion of Stevens' paper is presented on pages 98–108.

SWAMY, P. S. 1960 Estimating the Mean and Variance of a Normal Distribution From Singly and Doubly Truncated Samples of Grouped Observations. Calcutta Statistical Association *Bulletin* 9, no. 36.

TALLIS, G. M.; and YOUNG, S. S. Y. 1962 Maximum Likelihood Estimation of Parameters of the Normal, Log-normal, Truncated Normal and Bivariate Normal Distributions From Grouped Data. *Australian Journal of Statistics* 4, no. 2:49–54.

WALLIS, W. ALLEN; and ROBERTS, HARRY V. 1956 *Statistics: A New Approach.* Glencoe, Ill.: Free Press.

Postscript

Some further references that will be useful to the reader include Bross (1958), Fisher (1955), Haitovsky (1973), and Mantel (1963). Gjeddebæk (1970) contains a rather thorough treatment of problems arising with grouping of observations that are by nature more or less rankable. It would be useful reading to those doing practical work in that domain.

N. F. GJEDDEBÆK

ADDITIONAL BIBLIOGRAPHY

BROSS, IRWIN D. J. 1958 How to Use Ridit Analysis. *Biometrics* 14:18–38.

FISHER, R. A. 1955 Answer [to a query concerning grouping]. *Biometrics* 11:237–238.

GJEDDEBÆK, N. F. 1970 *On Grouped Observations.* Søborg (Denmark): Egetforlag/Ferrosan.

HAITOVSKY, YOEL 1973 *Regression Estimation From Grouped Observations.* London: Griffin; New York: Hafner.

MANTEL, NATHAN 1963 Chi-square Tests With One Degree of Freedom: Extentions of the Mantel–Haenszel Procedure. *Journal of the American Statistical Association* 58:690–700.

IV

TRUNCATION AND CENSORSHIP

Statistical problems of truncation and censorship arise when a standard statistical model is appropriate for analysis except that values of the random variable falling below—or above—some value are not measured at all (truncation) or are only counted (censorship). For example, in a study of particle size, particles below the resolving power of observational equipment will not be seen at all (truncation), or perhaps small particles will be seen and counted, but will not be measurable because of equipment limitations (censorship). Most of the existing theory for problems of this sort takes the limits at which truncation or censorship occurs to be known constants. There are practical situations in which these limits are not exactly known (indeed, the particle-size censorship example above might involve an inexactly ascertainable limit), but little theory exists for them. Truncation is sometimes usefully regarded as a special case of selection: if the probability that a possible observation having value x will actually be observed depends upon x, and is, say, $p(x)$ (between 0 and 1), selection is occurring. If $p(x) = 1$ between certain limits and 0 outside them, the selection is of the type called (two-sided) truncation.

More particularly, if values below a certain lower limit, a, are not observed at all, the distribution is said to be truncated on the left. If values larger than an upper limit, b, are not observed, the distribution is said to be truncated on the right. If only values lying between a and b are observed, the distribution is said to be doubly truncated. One also uses the terms "truncated sampling" and "truncated samples" to refer to sampling from a truncated distribution. (This terminology should not be confused with the wholly different concept of truncation in sequential analysis.)

In *censored sampling*, observations are measured only above a, only below b, or only between a and b; but in addition it is known how many unmeasured observations occur below a and above b. Censorship on the left corresponds to measuring observations only above a; censorship on the right corresponds to measuring only below b; and double censorship corresponds to measuring only between a and b. In the case of double censorship, the total sample consists of l, the number of observations to the left of the lower limit a; r, the number of observations to the right of the upper limit b; and $X_1, \cdots, X_n$, the values of the observations occurring between the limits. In one-sided censorship, either l or r is 0. A second kind of censorship arises when, without regard to given limits a or b, the l smallest and/or r largest observations are identified but not measured. This is type II censorship. In consequence, "type I censoring" is a name applied to the case described earlier. (In older literature the word "truncation" may be used for any of the foregoing.)

Some of the definitions mentioned above may be illustrated by the following examples:

Suppose X is the most advanced year of school attained for people born in 1930 where the information is obtained by following up records of those who entered high school; then the distribution is truncated on the left, because every possible observation is forced to be larger than 8. In particular, every observation of a sample from this distribution must be larger than 8. Censorship on the left, in this case, would occur if the sample were drawn from the population of people born in 1930 and if, in addition to noting the most advanced year of school attained by those who entered high school, the number of sample members who did not enter high school were ascertained. Similarly, if two hours is allowed for an examination and the time of submission is recorded only for late papers, the distribution of time taken to write the examination is censored on the left at two hours.

Truncation on the right would apply to a 1967 study of longevity of people born in 1910, as inferred from a comprehensive survey of death certificates. Censorship on the right would occur in this case if the sample were based on birth certificates, rather than death certificates, since then the number of individuals with longevity exceeding the upper observable limit would be known.

The distribution of height of U.S. Navy enlistees in records of naval personnel is a doubly truncated distribution because of minimum and maximum height requirements for enlistment. An example of type II censorship on the right would be given by the dates of receipt of the first 70 responses to a questionnaire that had been sent out to 100 people.

Goals.   In dealing with truncated distributions, a key issue is whether the conclusions that are sought should be applicable to the entire population or only to the truncated population itself. For instance, since the navy, in purchasing uniforms, need consider only those it enlists, the truncated, not the untruncated, population is the one of interest. On the other hand, if an anthropologist wished to use extensive navy records for estimating the height distribution of the entire population or its mean height, his inferences would be directed, not to the truncated population itself, but to the untruncated population. (Perhaps this anthropologist

should not use naval enlistees at all, since the sample may not be representative for other reasons, such as educational requirements and socioeconomic factors.) In a study in which treated cancer patients are followed up for five years (or until death, if that comes sooner), the observed survival times would be from a distribution truncated on the right. This is also a natural example of censored sampling. The censored sample would provide the information relevant to setting actuarial rates for five-year (or shorter) term insurance policies. For assessing the value of treatment the censored sample is less adequate, since the remainder of the survival-time distribution is also important.

In cases in which the truncated population itself is of interest, few essentially new problems are posed by the truncation; for example, the sample mean and variance remain unbiased estimators of these parameters of the truncated population, and (at least with large samples) statistical methods that are generally robust may ordinarily be confidently applied to a truncated distribution. On the other hand, if data from the truncated sample are to be used for reaching conclusions about the untruncated distribution, special problems do arise. For instance, the sample mean and variance are not reasonable estimators of those parameters in the untruncated distribution, nor are medians or other percentiles directly interpreted in terms of the untruncated distribution. The situation for means and variances in censored samples, although not identical, is similar.

Estimation and testing under censorship.   Suppose, now, that a sample of observations is censored and that the purpose is to make estimates of parameters in the population. In this case the data consist of $l, X_1, \cdots, X_n$, and r. Let $N = l + n + r$. Usually N is regarded as fixed. For type II censorship, if both l and r are less than $N/2$, then the sample median has the same distributional properties as it would if no censorship had been imposed. Similarly, such statistics as the interquartile range, certain linear combinations of order statistics, and estimates of particular percentiles may be more or less usable, just as if censoring had not occurred (depending upon the values of $l, n,$ and r). This fact allows censoring to be deliberately imposed with advantage where the investigator knows enough to be sure that the sample median or other interesting quantiles will be among $X_1, \cdots, X_n$ and where the cost of taking the sample is greatly reduced by avoiding exact measurement of a substantial part of the sample. Generally, the precision obtainable from a sample of size N where censorship has been imposed can never be greater than the precision obtainable from a sample of size N from the same distribution without censoring (Raja Rao 1958), but the censored sample may be cheaper to observe. Sometimes censorship is deliberately imposed for another reason. If the investigator fears that some of the observations in samples are actually errors (coming from a "contaminating" distribution), he may deliberately choose to censor the smallest one or two (or more) and the largest one or two (or more) and use only the intermediate values in the statistical analysis. Censorship of this form is called "trimming"; it is akin to a related technique called Winsorizing. [*See* STATISTICAL ANALYSIS, SPECIAL PROBLEMS OF, *article on* OUTLIERS; NONPARAMETRIC STATISTICS, *article on* ORDER STATISTICS.]

Since censorship will generally cause some off-center part of the distribution to be the one furnishing $X_1, \cdots, X_n$, it is clear that the sample mean, $\bar{X}$, based only on those observations will generally be a seriously biased estimator of the population mean, μ; similarly, s^2, the sample variance of $X_1, \cdots, X_n$, will tend to be too small to be a good estimator of the population variance, σ^2. Thus, in using a censored sample to reach conclusions about the parameters (other than quantiles), such as the mean and standard deviation of the population, it is necessary to make some assumptions about the underlying distribution in order to arrive at estimators with known properties.

Even with strong assumptions, it is difficult to obtain procedures such as confidence intervals of exact confidence coefficient. (Halperin 1960 gives a method for interval estimation of location and scale parameters with bounded confidence coefficient.) If samples are large, then more satisfactory results are available through asymptotic theory.

With respect to testing hypotheses where two samples singly censored at the same point are available, it has been shown by Halperin (1960) that Wilcoxon's two-sample test can be applied in an adapted form and that for samples of more than eight observations with less than 75 per cent censoring, the normal approximation to the distribution of the (suitably modified) Wilcoxon statistic holds well.

Normal distributions and censorship. The case which has been most studied is, naturally, that of the normal distribution. For type I censorship, by use of a and b, together with l, r, and $X_1, \cdots, X_n$, it is possible to calculate the maximum likelihood estimators in the normal case. The calculation is difficult and requires special tables. Among other methods for the normal distribution are those based on linear combinations of order statistics.

[*See* NONPARAMETRIC STATISTICS, *article on* ORDER STATISTICS.]

Cohen (1959) gives a useful treatment of these problems, and a rather uncomplicated method with good properties is offered by Saw (1961), who also presents a survey of the more standard estimation methods. It is interesting to observe that a sample of size N which is censored on one side and contains n measured values is more informative about μ, the population mean, than would be an uncensored sample of n observations (Doss 1962). The additional information is obviously furnished by l (or r). The same thing is true in the estimation of σ, providing that censored observations lie in a part of the distribution with probability less than one-half (Doss 1962). Although in uncensored samples from the normal distribution the sample mean, $\bar{X}$, and standard deviation, s, are statistically independent, they are not so with one-sided censoring; indeed, the correlation between $\bar{X}$ and s grows as the fraction censored increases (Sampford 1954).

Estimation and testing under truncation. Samples from distributions truncated at points a and b generally give less information relating to that part of the distribution which has been sampled than do samples from the same distribution with censoring at those points. (In the case of type I censorship, l and r do afford some idea of whether the center or the left-hand side or the right-hand side is furnishing the observations $X_1, \cdots, X_n$.) It follows that no distribution-free relationships between percentiles, location parameters, or dispersion parameters of the untruncated distribution and the truncated one can hold. To reach any conclusions about the untruncated population, it is essential to have some assumptions about the underlying probability law.

If certain statistics are jointly sufficient for a random sample from the untruncated distribution, then those same statistics remain sufficient for a sample from the truncated distribution (Tukey 1949; Smith 1957).

The amount of information in the truncated sample may be greater than, less than, or equal to that afforded by an untruncated sample of the same size from the same distribution. Which of these alternatives applies depends upon what the underlying distribution is and how the truncation is done (Raja Rao 1958).

For the normal distribution, a truncated sample is always less informative about both μ and σ than an untruncated sample having the same number of observations (Swamy 1962). (However, an inner truncated sample from a normal distribution, that is, one in which only observations *outside* an interval are observed, may be more informative about σ^2 than an untruncated sample with the same number of observations.)

Testing two-sample hypotheses from truncated samples can be done on the basis of distributional assumptions. In addition, a little can be said about distribution-free procedures. Lehmann showed that if two continuous distributions, F and G, are being compared by Wilcoxon's test (or any rank test), where $G = F^k$, then whatever F is, the distribution of the test statistic depends only on k and the two sample sizes. Truncation on the right at point b gives the truncated cumulative distribution functions $F_b(x) = F(x)/F(b)$ and $G_b(x) = F^k(x)/F^k(b)$, so it is still true that $G_b(x) = [F_b(x)]^k$. Thus, truncation on the right does not affect the properties of Wilcoxon's test against Lehmann (1953) alternatives. On the other hand, truncation on the left at point a leads to the relations $G_a(x) = F^k(x)/[1 - F^k(a)]$ and $F_a(x) = F(x)/[1 - F(a)]$, and it is not true that $G_a(x) = [F_a(x)]^k$. Further, it can be shown that the noncentrality parameter of the test, $P(X < Y)$, declines as a grows (that is, as more and more of the distribution is truncated). Thus, truncation on the left does affect the test against Lehmann alternatives. By a similar argument, if F and G are related by $1 - G = (1 - F)^k$, then truncation on the left leaves the relationship unaltered, while truncation on the right does not.

In comparing two treatments of some disease where time to recurrence of the disease is of interest, *random* censorship is sometimes encountered. For example, death by accidental injury may intervene before the disease has recurred. Such an observation has been subjected to censorship by a random event. Problems of this sort are treated by Gehan (1965).

Bivariate cases. The bivariate case occurs when truncation or censorship is imposed on each member of a sample, or possible sample, in terms of one variable, say, X, while another variable, Y, is the one of principal interest. For example, in studying income data, an investigator might take as his sample all tax returns submitted before the delinquency date. He would then have censored the sample on the date of submission of the tax return, but his interest would apply to some other properties of these data, such as the taxable income reported. This kind of bivariate truncation or censoring is common in social science. Estimation of the parameters of the multivariate normal distribution when there is truncation or censorship has been treated by Singh (1960) in the case of mutually independent variables. The estimation equations in the case of truncation are the usual univariate

equations, which may be separately solved. But in the presence of censorship, when only the number of unobservable vectors is known and there is no information as to which components led to unobservability, the estimating equations require simultaneous solution.

If X and Y are independent, the truncation on X does not affect the distribution of Y. Otherwise, in general, it will; and it may do so very strongly. When dependence of X and Y exists, then censorship or truncation affords opportunities for large and subtle bias, on the one hand, and for experimental strategies, on the other. For example, the selection of pilots, students, domestic breeding stock, all represent choosing a sample truncated in terms of one variable (an admission score or preliminary record of performance), with a view to ensuring large values of a different variable in the truncated (retained) portion of the population. Generally, the larger the correlation between X and Y, the greater the improvement obtainable in Y by truncation on X. [*See* SCREENING AND SELECTION.]

A second example of truncation of one variable with the eye of purpose fixed on a second is afforded by "increased severity testing." This engineering method may be illustrated by taking a lot of resistors designed to tolerate a low voltage over a long service life and exposing them to a short pulse of very high voltage. Those not failing are assumed satisfactory for their intended use, provided that service life is not shortened as a result of the test. This is seen as an example of bivariate truncation if one attributes to each resistor two values: X, its service life at the high voltage, and Y, its service life at the low voltage. Truncation on X presumably increases the mean value of Y in the retained, or truncated, population. Analogues of this may have relevance for psychology in such areas as stress interviews or endurance under especially difficult experimental tasks.

▶ Indeed, Cochran (1951) has shown that in the bivariate normal case the improvement in the mean of Y that can be obtained through left truncation on X is exactly proportional to ρ_{XY} (of the untruncated population). So, if $\rho_{XY} = .6$ then selection on X will yield 60 per cent of the increase in the average of Y that could be obtained by truncation on Y *itself*. Thus, as an index of the efficacy of a predictor used for selection, ρ is more natural than ρ^2 (which indexes "explained variation").

In the bivariate case, truncation of one variable can greatly affect the correlation between the two variables. For example, although height and weight exhibit a fairly strong correlation in adult males, this correlation virtually disappears if we consider only males with height between 5 feet 6 inches and 5 feet 9 inches. Although there is considerable variation in weight among men of nearly the same height, little of this variation is associated with variation in height. On the other hand, *inner* truncation—omitting cases with intermediate values of one variable—will produce spuriously high correlation coefficients. Thus, in a sample of males of heights less than 5 feet 4 inches or greater than 6 feet 6 inches, virtually all the variation in weight will be associated with variation in height. In a linear regression situation, where there is inner or outer truncation, the slope of the regression line continues to be unbiasedly estimated. But the correlation coefficient has a value that may depend so strongly on the truncation (or, more generally, selection) that there may be little if any relationship between the correlations in the truncated and untruncated populations. [*See* ERRORS, *article on* NONSAMPLING ERRORS.]

To show how truncation can enormously affect the correlation coefficient, consider X and Y, two random variables with a joint distribution such that

$$Y = \alpha + \beta X + e,$$

where α and β are constants and where e is a random variable uncorrelated with X. This kind of simple linear structure frequently arises as a reasonable assumption. Assume that β is not 0.

Let σ_e^2 and σ_X^2 be the variances of e and X respectively; immediate computation then shows that the covariance between X and Y is $\beta\sigma_X^2$, while the variance of Y is $\beta^2\sigma_X^2 + \sigma_e^2$. It follows that ρ^2, the square of the correlation coefficient between X and Y, is

$$\rho^2 = \frac{\beta^2}{\beta^2 + (\sigma_e^2/\sigma_X^2)}.$$

Hence, if the structure stays otherwise the same but the marginal distribution of X is changed so that σ_X^2 becomes very large, ρ^2 becomes nearly unity. In particular, if the marginal distribution of X is changed by truncating *inside* the interval $(-d, d)$, it is readily shown that as d becomes indefinitely large, so will σ_X^2.

Similarly, if σ_X^2 becomes nearly 0, so will ρ^2. In particular, if X is truncated outside a small enough interval, σ_X^2 will indeed become nearly 0.

▶ When X and Y are positively correlated, truncation on the left (choosing high values of X as in selection aimed at getting high values of Y) alters all the parameters. The means of X and Y are increased, their standard deviations are reduced, and the correlation between X and Y is reduced. This last fact can lead to an interesting outcome.

It may be that U and V are each positively correlated with Y and that $\rho_{UY} > \rho_{VY}$. Now if selection occurs by truncation on U the value of ρ_{UY} is decreased, while that of ρ_{VY} is also decreased, but less sharply, So in the selected group U may have a lower correlation with Y than does V, and thus seem to be the poorer selector, judging from the cases that have been selected.

▶ Calculations related to questions of truncation of multivariate normal distributions are facilitated by the moment-generating function for the truncated multivariate normal given by Tallis (1961).

Still more difficulties arise if truncation in a bivariate (or multivariate) population is accomplished not by truncation on X or Y alone but on a function of them. If X and Y were utterly independent and a sample were drawn subject to the restraint $|X - Y| \leq a$, then all the points (X, Y) in the sample would be required to lie in a diagonal strip of slope 1 and vertical (or horizontal) width $2a$ units. Obviously, very high "correlation" might be observed! Some follow-up studies embody a bias of just this form. Suppose that survival of husband and wife is to be studied by following up all couples married during a period of 40 years. Suppose further that A_H, the age of the husband at marriage, and A_W, the age of the wife at marriage, are highly correlated (as they are) and that L_H and L_W, their lifetimes, are completely independent of each other statistically. Then the correlation between L_H and L_W observed in a follow-up study may be very high or very low, depending upon how the sample is truncated. Consider several methods:

(1) All couples married in the 40-year period are followed up until all have died—making a study of about a hundred years' duration; then there is no truncation of L_H or L_W, and since (by assumption) they are statistically independent, the observed correlation will, except for sampling error, be zero.

(2) All couples whose members have both died during the 41st year of the study furnish the sample values of L_H and L_W; now there will be a very high correlation. Write $L_H = A_H + T_H$ and $L_W = A_W + T_W$, where T represents life length after marriage. The curious sampling scheme just proposed ensures that $|T_H - T_W| \leq 1$, so T_H and T_W are highly correlated and A_H and A_W are also correlated; hence L_H and L_W in such a truncated sample will be highly correlated.

(3) All couples whose members have both died by the end of the 40th year (the cases "complete" by then) furnish the data. In this not infrequently used design, a fictitious correlation will be found. Those couples married during the last year and

with both members dead will have values of T_H and T_W which are nearly equal, and they will have correlated values of A_H and A_W; such couples will contribute strongly to a positive correlation. Those couples married two years before the end of the study and with both members dead will have values of T_H and T_W differing at most by 2, and correlated values of A_H and A_W; they will also contribute—not quite so strongly—to a positive correlation. By continuation of this reasoning, it is seen that the same kind of bias (diminishing with progress toward the earliest marriage) affects the entire sample. A detailed numerical example of this problem is given by Myers (1963).

It is probably wise to view with great caution studies that are multivariate in character (involve several observable random aspects) and at the same time use samples heavily truncated or censored on one or more of the variables or—especially—on combinations of them.

LINCOLN E. MOSES

BIBLIOGRAPHY

COHEN, A. CLIFFORD JR. 1959 Simplified Estimators for the Normal Distribution When Samples Are Singly Censored or Truncated. *Technometrics* 2:217–237.

DOSS, S. A. D. C. 1962 On the Efficiency of BAN Estimates of the Parameters of Normal Populations Based on Singly Censored Samples. *Biometrika* 49:570–573.

GEHAN, EDMUND A. 1965 A Generalized Wilcoxon Test for Comparing Arbitrarily Singly-censored Samples. *Biometrika* 52:203–223.

HALPERIN, MAX 1960 Extension of the Wilcoxon–Mann–Whitney Test to Samples Censored at the Same Fixed Point. *Journal of the American Statistical Association* 55:125–138.

LEHMANN, E. L. 1953 The Power of Rank Tests. *Annals of Mathematical Statistics* 24:23–43.

MYERS, ROBERT J. 1963 An Instance of the Pitfalls Prevalent in Graveyard Research. *Biometrics* 19:638–650.

RAJA RAO, B. 1958 On the Relative Efficiencies of BAN Estimates Based on Doubly Truncated and Censored Samples. National Institute of Science, India, *Proceedings* 24:366–376.

SAMPFORD, M. R. 1954 The Estimation of Response-time Distributions: III. Truncation and Survival. *Biometrics* 10:531–561.

SAW, J. G. 1961 Estimation of the Normal Population Parameters Given a Type I Censored Sample. *Biometrika* 48:367–377.

SINGH, NAUNIHAL 1960 Estimation of Parameters of a Multivariate Normal Population From Truncated and Censored Samples. *Journal of the Royal Statistical Society* Series B 22:307–311.

SMITH, WALTER L. 1957 A Note on Truncation and Sufficient Statistics. *Annals of Mathematical Statistics* 28:247–252.

SWAMY, P. S. 1962 On the Joint Efficiency of the Estimates of the Parameters of Normal Populations Based on Singly and Doubly Truncated Samples. *Journal of the American Statistical Association* 57:46–53.

TUKEY, JOHN W. 1949 Sufficiency, Truncation, and Selection. *Annals of Mathematical Statistics* 20:309–311.

Postscript

Random censorship. Studies conducted over a period of time frequently entail random censorship (usually on the right because of loss to follow-up, sometimes on the left where a late entrant to the study has already experienced the event of interest —at an earlier, but unknown, time). Methodology for analyzing such data has developed greatly.

A basic paper is by Kaplan and Meier (1958). The problem they treat can be thought of as follows. Each subject is observed until a certain event occurs (for example, death, or changing jobs, or marriage). The elapsed time until that event occurs for subject i is X_i. It is desired to estimate the c.d.f. $F_X(t) = Pr\{X \leqslant t\}$, or the survival function $P_X(t) = 1 - F_X(t)$. Unfortunately, some values of X are censored (on the right) by loss from follow-up or by the end of the study intervening before the event occurs.

Kaplan and Meier propose a nonparametric estimator that they call the product–limit estimator (now widely known as the Kaplan–Meier estimator). They show that if the factors causing censorship are independent of the variable X then their estimator is consistent, with negligible bias. They also give its asymptotic variance. Turnbull (1974) extends this work to estimating $F_X(t)$ when some observations are left censored and some are right censored (and time is counted discretely—as in months—rather than continuously).

Sometimes a c.d.f. gives more information than we wish to use, and interest attaches to some special property of it, such as its mean or median. Sander (1975) finds the median of the Kaplan–Meier c.d.f. to be much more stable than the mean. This is a natural result because the mean is strongly affected by the extremely large values, and it is in that part of the range that the probabilities are poorly determined because of right censoring.

Nonparametric comparison of two survival functions, each estimated in the presence of right censoring, is treated by Efron (1967) and also by Cox (1972), as a special case of a theory he proposes for handling general regression problems where the dependent variable is censored on the right.

An interesting analysis of survival after heart transplants is given by Turnbull et al. (1974); they apply three nonparametric and three different parametric analyses to a set of data (given in the article itself) involving right censorship.

LINCOLN E. MOSES

ADDITIONAL BIBLIOGRAPHY

COCHRAN, WILLIAM G. 1951 Improvement by Means of Selection. Volume 2, pages 449–470 in Berkeley Symposium on Mathematical Statistics and Probability, Second, *Proceedings*. Volume 2: *Theory of Statistics*. Berkeley: Univ. of California Press.

COX, D. R. 1972 Regression Models and Life Tables. *Journal of the Royal Statistical Society* Series B 34: 187–220. → Includes discussion.

EFRON, BRADLEY 1967 The Two Sample Problem With Censored Data. Volume 4, pages 831–853 in Berkeley Symposium on Mathematical Statistics and Probability, Fifth, *Proceedings*. Volume 4: *Biology and Problems of Health*. Berkeley: Univ. of California Press.

KAPLAN, E. L.; and MEIER, PAUL 1958 Nonparametric Estimation From Incomplete Observations. *Journal of the American Statistical Association* 53:457–481.

SANDER, JOAN M. 1975 A Comparison of Three Estimators Derived From the Kaplan–Meier Estimator. Technical Report No. 11. Division of Biostatistics, Stanford University.

TALLIS, G. M. 1961 The Moment Generating Function of the Truncated Multi-normal Distribution. *Journal of the Royal Statistical Society* Series B 23:223–229.

TURNBULL, BRUCE W. 1974 Nonparametric Estimation of a Survivorship Function With Doubly Censored Data. *Journal of the American Statistical Association* 69:169–173.

TURNBULL, BRUCE W.; BROWN, BYRON W. JR.; and HU, MARIE 1974 Survivorship Analysis of Heart Transplant Data. *Journal of the American Statistical Association*. 69:74–80.

STATISTICAL IDENTIFIABILITY

Identifiability is a statistical concept referring to the difficulty of distinguishing among two or more explanations of the same empirical phenomena. Unlike traditional statistical problems (for example, estimation and hypothesis testing), identifiability does not refer to sampling fluctuations stemming from limited data; rather, nonidentifiability, or the inability to distinguish among explanations, would exist even if the statistical distribution of the observables were fully known.

A model represents an attempt to describe, explain, or predict the values of certain variables as the outputs of a formally described mechanism. Yet it is evident that given any specified set of facts or observations to be explained, an infinite number of models are capable of doing so. One way of describing all scientific work is as the task of distinguishing among such eligible models by the introduction of further information.

The problem of identification, as usually encountered, is essentially the same phenomenon in a more restricted context. Suppose that the *form* of the explanatory model is regarded as specified, but that it involves unknown parameters. Suppose further that the observational material to be explained is so abundant that the basic statistical distributions may be regarded as known. (In practice this will rarely be the case, but identifiability considerations require thinking in these terms.) An important task then is to select from all possible *structures* (sets of values for the unknown parameters) contained in the model the particular one that, according to some criterion, best fits the observations. It may happen, however, that there are two, several, or even an infinite number of structures generating precisely the same distribution for the observations. In this case no amount of observation consistent with the model can distinguish among such structures. The structures in question are thus *observationally equivalent*.

It may, however, be the case that some specific parameter or set of parameters is the same in all observationally equivalent structures. In such a case, that set is said to be *identifiable*. Parameters whose values are not the same for all observationally equivalent structures are not identifiable; their values can never be recovered solely by use of observations generated by the model.

In its simplest form lack of identifiability is easy to recognize. Suppose, for example, that a random variable, X, is specified by the model to be distributed normally, with expectation or mean the difference between two unknown parameters, $EX = \theta_1 - \theta_2$. It is evident that observations on X can be used to estimate $\theta_1 - \theta_2$, which is identifiable, but that the individual parameters, θ_1 and θ_2, are not identifiable. The θ_i can be recovered only by combining outside information with observations on X or by changing the whole observational scheme. In cases such as this, the θ_i are sometimes said not to be *estimable*. Observations on X do restrict the θ_i, since their difference can be consistently estimated, but can never distinguish the true θ_i generating the observations from among all θ_i with the same difference. Although one way of describing the situation is to note that the likelihood function for a random sample has no unique maximum but has a ridge along the line $\theta_1 - \theta_2 = \bar{x}$, the sample average, it is instructive to note further that the problem persists even if the model is nonstochastic and X is a constant.

(In the context of the general linear hypothesis model, the concept of *estimability* has been developed by R. C. Bose, C. R. Rao, and others, apparently independently of the more general identifiability concept. [*See* LINEAR HYPOTHESES, *article on* REGRESSION; *for history, references, and discussion, see* Reiersøl 1964.] Estimability of a linear parameter, in the linear hypothesis context, means that an unbiased estimator of the parameter exists; within its domain of discussion, estimability is equivalent to identifiability.)

In more complicated cases, lack of identifiability may be less easy to recognize. Because of rounding errors or sample properties, numerical "estimates" of unidentifiable parameters may be obtained, although such estimates are meaningless. As a fanciful, although pertinent, example, suppose that in the situation of the previous paragraph there were two independent observations on X, say, X_1 and X_2, and that, by rounding or other error, these observations were regarded as having expectations not quite equal to $\theta_1 - \theta_2$, say,

$$EX_1 = .99\theta_1 - 1.01\theta_2,$$
$$EX_2 = 1.01\theta_1 - .99\theta_2.$$

It is then easy to see that the least squares (and maximum likelihood) estimators would appear to exist and would be

$$\hat{\theta}_1 = -\frac{99}{4}X_1 + \frac{101}{4}X_2,$$

$$\hat{\theta}_2 = -\frac{101}{4}X_1 + \frac{99}{4}X_2,$$

so that $\hat{\theta}_1 - \hat{\theta}_2 = \frac{1}{2}(X_1 + X_2)$, which last does make good sense. The effect of underlying nonidentifiability, with the coefficients slightly altered from unity, is that the variance of $\hat{\theta}_1$ (or $\hat{\theta}_2$) is very large, about 2,500 times the variance of $\frac{1}{2}(X_1 + X_2)$.

In other cases numerical estimates may be obtained in finite samples when in fact no consistent estimator exists and, for example, the matrix inverted in obtaining the numbers is guaranteed to be singular in the probability limit. In such cases it is of considerable interest to know what restrictions on the form or parameters of the model are necessary or sufficient for the identification of subsets of parameters. Analyses of identifiability are typically devoted to this question.

The identification problem can arise in many contexts. Wherever a reasonably complicated underlying mechanism generates the observations, and the parameters of that mechanism are to be estimated, the identification problem may be encountered. Examples are factor analysis and the analysis of latent structures. [See FACTOR ANALYSIS AND PRINCIPAL COMPONENTS; LATENT STRUCTURE; *for the analysis of identifiability in factor analysis, see* Reiersøl 1950a.] A further example occurs in the analysis of accident statistics, where the occurrence of approximately negative binomial counts led to the concept of accident proneness on the false assumption that a negative binominal can *only* be generated as a mixture of Poisson distributions. [See DISTRIBUTIONS, STATISTICAL, *articles on* SPECIAL DISCRETE DISTRIBUTIONS *and* MIXTURES OF DISTRIBUTIONS; FALLACIES, STATISTICAL.]

An important case, and one in which the analysis is rich, is that of a system of simultaneous equations such as those frequently encountered in econometrics. The remainder of this article is accordingly devoted to a discussion of identifiability in that context.

Identifiability of a structural equation

Suppose the model to be investigated is given by

$$(1) \qquad \boldsymbol{u}_t = \boldsymbol{A}\boldsymbol{x}_t = [\mathbf{B} \quad \boldsymbol{\Gamma}]\begin{bmatrix} \boldsymbol{y}_t \\ \boldsymbol{z}_t \end{bmatrix},$$

where $\boldsymbol{u}_t$ is an M-component column vector of random disturbances (with properties to be specified below) and $\boldsymbol{x}_t$ is an N-component column vector of variables, partitioned into $\boldsymbol{y}_t$, the M-component vector of *endogenous* variables to be explained by the model, and $\boldsymbol{z}_t$, the $\Lambda = (N - M)$-component vector of *predetermined* variables determined outside the current working of the model (one element of $\boldsymbol{z}_t$ can be taken to be identically unity).

The elements of $\boldsymbol{z}_t$ can thus either be determined entirely outside the model (for example, they can be treated as fixed) or represent lagged values of the elements of $\boldsymbol{y}_t$. The assumption that $\boldsymbol{z}_t$ is determined outside the current working of the model requires, in any case, that movements in the elements of $\boldsymbol{u}_t$ not produce movements in those of $\boldsymbol{z}_t$. In its weakest form this becomes the assumption that the elements of $\boldsymbol{z}_t$ are asymptotically uncorrelated with those of $\boldsymbol{u}_t$, in the sense that

$$\operatorname*{plim}_{T\to\infty}\left\{\frac{1}{T}\sum_{t=1}^{T}\boldsymbol{z}_t'\boldsymbol{u}_t\right\} = 0,$$

where the prime denotes transposition and plim denotes stochastic convergence (convergence in probability). As with all prior assumptions required for identification, this assumption is quite untestable within the framework of the model.

The t subscript denotes the number of the observation and will be omitted henceforth. In (1), $\boldsymbol{A}$ is an $M \times N$ matrix of parameters to be estimated and is partitioned into $\mathbf{B}$ and $\boldsymbol{\Gamma}$, corresponding to the partitioning of $\boldsymbol{x}$. As the endogenous variables are to be explained by the model, $\mathbf{B}$ (an $M \times M$ square matrix) is assumed nonsingular. Finally, the $\boldsymbol{u}$ vectors for different values of t are usually assumed (serially) uncorrelated, with common mean $\mathbf{0}$ and with common covariance matrix $\boldsymbol{\Sigma}$. Thus, $\boldsymbol{\Sigma}$ is $M \times M$ and is in general unknown and not diagonal, as its typical element is the covariance between contemporaneous disturbances from different equations of the model. (Normality of u is also generally assumed for estimation purposes but has so far been of little relevance for identifiability discussions in the present context; for its importance in another context, see Reiersøl 1950*b*.)

Such models occur frequently in econometrics, in contexts ranging from the analysis of particular markets to studies of entire economies. The study of identification and, especially, estimation in such models has occupied much of the econometric literature since Haavelmo's pathbreaking article (1943).

For definiteness, this article will concentrate on the identifiability of the first equation of (1), that is, on the identifiability of the elements of the first row of A, denoted by A_1. In general, one is content if A_1 is identifiable after the imposition of a normalization rule, since the units in which the variables are measured are arbitrary. This will be understood when A_1 is spoken of as identifiable.

It is not hard to show that the joint distribution of the elements of y, given z, depends only on the parameters Π and Ω of the *reduced form* of the model,

$$
\text{(2)} \quad
\begin{aligned}
y &= \Pi z + v, \\
\Pi &= -B^{-1}\Gamma, \\
v &= B^{-1}u, \\
\Omega &= B^{-1}\Sigma B^{-1\prime},
\end{aligned}
$$

so that observations generated by the model can at most be used to estimate Π and Ω (the variance–covariance matrix of the elements of v). Since the elements of z are assumed asymptotically uncorrelated with those of u, such estimation can be consistently done by ordinary least squares regression, provided that the asymptotic variance–covariance matrix of the elements of z exists. On the other hand, if (1) is premultiplied by any nonsingular $M \times M$ matrix, the resulting structure has the same reduced form as the original one, so that all such structures are observationally equivalent. Unless outside information is available restricting the class of such transformations to those preserving A_1 (up to a scalar multiple), A_1 is not identifiable.

Examination of the nonstochastic case, in which $u = 0$, provides another way to describe the phenomenon. Here the investigator can, at most, obtain values of the Λ predetermined variables and observe the consequences for the endogenous variables. This can be done in, at most, Λ independent ways. Let the N-rowed matrix whose tth column consists of the tth observation on x so generated be denoted by X. Then X has rank Λ in the most favorable circumstances (which are assumed to hold here),

and $AX = 0$. It is easy to see that Π can be recovered from this.

On the other hand, $A_1X = 0$ expresses all that observational evidence can tell about A_1. Since X has rank Λ and has $M = N - \Lambda$ rows and since A has rank M, the rows of A are a basis for the row null space of X, whence the true A_1 can be distinguished by observational information from all vectors which are not linear combinations of the rows of A but not from vectors which are. The second part of this corresponds to the obvious fact that, without further information, there is nothing to distinguish the first equation of (1) from any linear combination of those equations. If, returning to the stochastic case, one replaces X by $\begin{bmatrix} \Pi \\ I \end{bmatrix}$, the same analysis remains valid. The condition that $A_1 \begin{bmatrix} \Pi \\ I \end{bmatrix} = 0$ embodies all the information about A_1 which can be gleaned from the reduced form, and the reduced form, as has been noted, is all that can be recovered from observational evidence, even in indefinitely large samples.

This is the general form of the classic example (Working 1927) of a supply and a demand curve, both of which are straight lines (see Boulding 1968; Fox 1968). Only the intersection of the two curves is observable, and the demand curve cannot be distinguished from the supply curve or any other straight line through that intersection. Even if the example is made stochastic, the problem clearly remains unless the stochastic or other specification provides further (nonobservational) information enabling the demand curve to be identified.

Thus, the identification problem in this context is as follows: what necessary or sufficient conditions on prior information can be stated so that A_1 can be distinguished (up to scalar multiplication) from all other vectors in the row space of A? Equivalently, what are necessary or sufficient conditions on prior information that permit the recovery of A_1, given Π and Ω?

If A_1 cannot be so recovered, the first equation of (1) is called *underidentified*; if, given any Π and Ω, there is a unique way of recovering A_1, that equation is called *just identified*; if the prior information is so rich as to enable the recovery of A_1 in two or more different and nonequivalent ways, that equation is called *overidentified*. In the last case, while the true reduced form yields the same A_1 whichever way of recovery is followed, this is in general not true of sample estimates obtained without imposing restrictions on the reduced form. The problem of using overidentifying information to avoid this difficulty and to secure greater efficiency is the central problem of simultaneous equation estimation but is different in kind from the identification problem discussed here (although the two overlap, as seen below).

Homogeneous linear restrictions on a single equation

The most common type of prior identifying information is the specification that certain variables in the system do not in fact appear in the first equation of (1), that is, that certain elements of $\mathbf{A}_1$ are zero. Such exclusion restrictions form the leading special case of homogeneous linear restrictions on the elements of $\mathbf{A}_1$.

Thus, suppose φ to be an $N \times K$ matrix of known elements, such that $\mathbf{A}_1 \varphi = \mathbf{0}$. Since $\mathbf{A}_1$ can be distinguished by observational information from any vector not in the row space of $\mathbf{A}$, it is obviously sufficient, for the identification of $\mathbf{A}_1$, that the equation $(\eta' \mathbf{A})\varphi = \mathbf{0}$ be satisfied only for η' a scalar multiple of $(1 \quad 0 \quad \cdots \quad 0)$. If there is no further information on $\mathbf{A}_1$, this is also necessary. This condition is clearly equivalent to the requirement that the rank of $\mathbf{A}\varphi$ be $M - 1$, a condition known as the *rank condition*, which is due to Koopmans, Rubin, and Leipnik (1950, pp. 81–82), as is much of the basic work in this area.

Since the rank of $\mathbf{A}\varphi$ cannot exceed that of φ, a necessary condition for the identifiability of $\mathbf{A}_1$ under the stated restrictions is that the number of those restrictions which are independent be at least $M - 1$, a requirement known as the *order condition*. In the case of exclusion restrictions, this becomes the condition that the number of predetermined variables excluded from the first equation of (1) must be at least as great as the number of included endogenous variables.

While the order condition does not depend on unknown parameters, the rank condition does. However, if the order condition holds and the rank condition does not fail identically (because of restrictions on the other rows of $\mathbf{A}$), then the rank condition holds almost everywhere in the space of the elements of $\mathbf{A}$. This has led to a neglect of the rank condition in particular problems, a neglect that can be dangerous, since the rank condition may fail identically even if the order condition holds. Asymptotic tests of identifiability (and of *over*identifying restrictions) are known for the linear restriction case and should be used in doubtful cases.

The difference between the rank of $\mathbf{A}\varphi$ (appearing in the rank condition) and the rank of φ (ap-

pearing in the order condition) is the number of restrictions on the reduced form involving φ (see Fisher 1966, pp. 45–51).

Other restrictions

The case of restrictions other than those just discussed was chiefly investigated by Fisher, in a series of articles leading to a book (Fisher 1966), which in part examined questions opened by Koopmans, Rubin, and Leipnik (1950, pp. 93–110; Wald 1950, on identification of individual parameters, should also be mentioned). While generalizations of the rank and order conditions tend to have a prominent place in the discussion, other results are also available. The restrictions considered fall into two categories: first, restrictions on the elements of Σ; second, more general restrictions on the elements of $\mathbf{A}_1$.

Working (1927) had observed, for example, that if the supply curve is known to shift greatly relative to the demand curve, the latter is traced out by the intersection points. When such shifting is due to a variable present in one equation but not in the other, the identifiability of the latter is due to an exclusion restriction. On the other hand, such shifting may be due to a greater disturbance variance, which suggests that restrictions on the relative magnitude of the diagonal elements of Σ can be used for identification. This is indeed the case, provided those restrictions are carefully stated. The results are related to the conditions for the *proximity theorem* of Wold (1953, p. 189) as to the negligible bias (or inconsistency) of least squares when the disturbance variance or the correlations between disturbance and regressors are small.

Wold's work (1953, pp. 14, 49–53, and elsewhere) on recursive systems, which showed least squares to be an appropriate estimator if $\mathbf{B}$ is triangular and Σ diagonal, and Fisher's matrix generalization (1961) to block-recursive systems, suggested the study of identifiability in such cases and the extension to other cases in which particular off-diagonal elements of Σ are known to be zero (disturbances from particular pairs of equations uncorrelated). Special cases had been considered by Koopmans, Rubin, and Leipnik (1950, pp. 103–105) and by Koopmans in his classic expository article (1953, p. 34). Aside from the generalization of the rank and order conditions, the results show clearly the way in which such restrictions interact with those on $\mathbf{A}$ to make the identifiability of one equation depend on that of others.

Finally, certain special cases and the fact that equations nonlinear in the parameters can frequently (by Taylor series expansion) be made linear in the parameters but nonlinear in the variables, with nonlinear constraints on the parameters, led to the study of identification with nonlinear (or nonhomogeneous) constraints on A_1. This is a much more difficult problem than those already discussed, as it may easily happen that the true A_1 can be distinguished from any other vector in some neighborhood of A_1 without over-all identifiability holding. As might be expected, local results (based on the rank and order conditions) are fairly easy to obtain, but useful global results are rather meager.

Other specifications of the model

The Taylor series argument just mentioned, as well as the frequent occurrence of models differing from (1) in that they are linear in the parameters and disturbances but not in the variables, also led Fisher to consider identifiability for such models. In these models, it may turn out that nonlinear transformations of the structure lead to equations in the same form as the first one, so that the result that A_1 can be observationally distinguished from vectors not in the row space of A can fail to hold (although such cases seem fairly special). Provided a systematic procedure is followed for expanding the model to include all linearly independent equations resulting from such transformations, the rank and order conditions can be applied directly to the expanded model. In such application, linearly independent functions of the same variable are counted as separate variables. It is clear also that such nonlinear transformations are restricted if there is information on the distribution of the disturbances, but the implications of this remain to be worked out.

A somewhat similar situation arises when the assumption that the elements of u are not serially correlated is dropped and the elements of z include lagged values of the endogenous variables. In this case it is possible that the lagging of an equation can be used together with linear transformation of the model to destroy identification. In such cases there may be underidentified equations of the reduced form as well as of the structure. This possibility was pointed out in an example by Koopmans, Rubin, and Leipnik (1950, pp. 109–110) but was shown to be of somewhat limited significance by Fisher (1966, pp. 168–175). He showed that the problem cannot arise if there is sufficient independent movement among the present and lagged values of the truly exogenous variables, a result connected to one of those for models nonlinear in the variables. In such cases the rank condition remains necessary and sufficient. The problem in nearly or

completely self-contained models awaits further analysis.

Is identifiability discrete or continuous?

Identifiability is apparently a discrete phenomenon. A set of parameters apparently either is or is not identifiable. This was emphasized by Liu (1960, for example), who pointed out that the prior restrictions used to achieve identification, like the very specification of the model itself, are invariably only approximations. Liu argued strongly that if the true, exact specification and prior restrictions were written down, the interrelatedness of economic phenomena would generally make structural equations underidentified.

Liu's argument that badly misspecified structures and restrictions that are used to lead to identification in fact only lead to trouble is clearly true; true, also, is his contention that econometric models and restrictions are only approximations and that those approximations may not be good ones. More troublesome than this, however, are the apparent implications of his argument as to the possibility of having a "good" approximation. If identifiability disappears as soon as any approximation enters in certain ways, no matter how close that approximation might be to the truth, then structural estimation ceases altogether to be possible.

This issue of principle was settled by Fisher (1961), who showed that identifiability can be considered continuous, in the sense that the probability limits of estimators known to be consistent under correct specification approach the true parameters as the specification errors approach zero, a generalization of Wold's proximity theorem. If the equation to be estimated is identifiable under correct specification, the commission of minor specification errors leads to only minor inconsistency.

A number of other questions are then raised, however. Among them are the following: How good does an approximation have to be to lead to only minor inconsistency? To what extent should only approximate restrictions be imposed to achieve identification? What about overidentification, where the trade-off may be between consistency and minor gains in variance reduction? What can be said about the relative robustness of the different simultaneous equation estimators to the sorts of minor specification error discussed by Liu?

Clearly, once identifiability is considered continuous, the identification problem tends to merge with the estimation problem, rather than be logically prior to it. It seems likely that both can best be approached through an explicit recognition of the approximate nature of specification, for example, by a Bayesian analysis with exact prior restrictions replaced by prior distributions on the functions of the parameters to be restricted [see BAYESIAN INFERENCE]. Work on this formidable problem is just beginning (see, for example, Drèze 1962; Reiersøl 1964).

FRANKLIN M. FISHER

[See also DISTRIBUTIONS, STATISTICAL, *article on* MIXTURES OF DISTRIBUTIONS; LATENT STRUCTURE; SIMULTANEOUS EQUATION ESTIMATION.]

BIBLIOGRAPHY

►BOULDING, KENNETH E. 1968 Demand and Supply: I. General. Volume 4, pages 96–104 in *International Encyclopedia of the Social Sciences*. Edited by David L. Sills. New York: Macmillan and Free Press.

ᴵDRÈZE, JACQUES 1962 The Bayesian Approach to Simultaneous Equations Estimation. O.N.R. Research Memorandum No. 67. Unpublished manuscript, Northwestern University.

FISHER, FRANKLIN M. 1961 On the Cost of Approximate Specification in Simultaneous Equation Estimation. *Econometrica* 29:139–170.

FISHER, FRANKLIN M. 1966 *The Identification Problem in Econometrics*. New York: McGraw-Hill.

►FOX, KARL A. 1968 Demand and Supply: II. Econometric Studies. Volume 4, pages 104–111 in *International Encyclopedia of the Social Sciences*. Edited by David L. Sills. New York: Macmillan and Free Press.

HAAVELMO, TRYGVE 1943 The Statistical Implications of a System of Simultaneous Equations. *Econometrica* 11:1–12.

KOOPMANS, TJALLING C. 1953 Identification Problems in Economic Model Construction. Pages 27–48 in William C. Hood and Tjalling C. Koopmans (editors), *Studies in Econometric Method*. Cowles Commission for Research in Economics, Monograph No. 14. New York: Wiley.

KOOPMANS, TJALLING C.; RUBIN, H.; and LEIPNIK, R. B. (1950) 1958 Measuring the Equation Systems of Dynamic Economics. Pages 53–237 in Tjalling C. Koopmans (editor), *Statistical Inference in Dynamic Economic Models*. Cowles Commission for Research in Economics, Monograph No. 10. New York: Wiley.

LIU, TA-CHUNG 1960 Underidentification, Structural Estimation, and Forecasting. *Econometrica* 28:855–865.

REIERSØL, OLAV 1950a On the Identifiability of Parameters in Thurstone's Multiple Factor Analysis. *Psychometrika* 15:121–149.

REIERSØL, OLAV 1950b Identifiability of a Linear Relation Between Variables Which Are Subject to Error. *Econometrica* 18:375–389.

REIERSØL, OLAV 1964 Identifiability, Estimability, Phenorestricting Specifications, and Zero Lagrange Multipliers in the Analysis of Variance. *Skandinavisk aktuarietidskrift* 46:131–142.

WALD, A. (1950) 1958 Note on the Identification of Economic Relations. Pages 238–244 in Tjalling C. Koopmans (editor), *Statistical Inference in Dynamic Economic Models*. Cowles Commission for Research in Economics, Monograph No. 10. New York: Wiley.

WOLD, HERMAN 1953 *Demand Analysis: A Study in Econometrics*. New York: Wiley.

WORKING, E. J. (1927) 1952 What Do Statistical "Demand Curves" Show? Pages 97–115 in American Economic Association, *Readings in Price Theory*. Edited by G. J. Stigler and K. E. Boulding. Homewood, Ill.: Irwin. → First published in Volume 41 of the *Quarterly Journal of Economics*.

Postscript

Probably the most important development in the decade since the main article was published is the work of Hannan (1969; 1971) on identification in systems with autoregressive–moving average disturbances (ARMA).

FRANKLIN M. FISHER

ADDITIONAL BIBLIOGRAPHY

BOWDEN, ROGER 1973 The Theory of Parametric Identification. *Econometrica* 41:1064–1074.

ᴵDRÈZE, JACQUES 1976 Bayesian Limited Information Analysis of the Simultaneous Equations Model. *Econometrica* 44:1045–1075.

HANNAN, EDWARD J. 1969 The Identification of Vector Mixed Autoregressive–Moving Average Systems. *Biometrika* 56:223–225.

HANNAN, EDWARD J. 1971 The Identification Problem for Multiple Equation Systems With Moving Average Errors. *Econometrica* 39:751–765.

ROTHENBERG, THOMAS J. 1971 Identification in Parametric Models. *Econometrica* 39:577–592.

WEGGE, LEON L. 1965 Identifiability Criteria for a System of Equations as a Whole. *Australian Journal of Statistics* 7, no. 3:67–77.

STATISTICAL INFERENCE
See STATISTICS, *article on* THE FIELD.

STATISTICS

The articles under this heading provide an introduction to the field of statistics and to its history. The first article also includes a survey of the statistical articles in the encyclopedia. At the end of the second article there is a list of the biographical articles that are of relevance to statistics.

I. THE FIELD	*William H. Kruskal*
II. THE HISTORY OF STATISTICAL METHOD	*Maurice G. Kendall*

I
THE FIELD

A scientist confronted with empirical observations goes from them to some sort of inference, decision, action, or conclusion. The end point of this process may be the confirmation or denial of some complicated theory; it may be a decision about the next experiment to carry out; or it may simply be a narrowing of the presumed range for some constant of nature. (The end point may even be the conclusion that the observations are worthless.) An end point is typically accompanied by a statement, or at least by a feeling, of how sure the scientist is of his new ground.

These inferential leaps are, of course, never made only in the light of the immediate observations. There is always a body of background knowledge and intuition, in part explicit and in part tacit. It is the essence of science that a leap to a false position—whether because of poor observational data, misleading background, or bad leaping form—is sooner or later corrected by future research.

Often the leaps are made without introspection or analysis of the inferential process itself, as a skilled climber might step from one boulder to another on easy ground. On the other hand, the slope may be steep and with few handholds; before moving, one wants to reflect on direction, where one's feet will be, and the consequences of a slip.

Statistics is concerned with the inferential process, in particular with the planning and analysis of experiments or surveys, with the nature of observational errors and sources of variability that obscure underlying patterns, and with the efficient summarizing of sets of data. There is a fuzzy boundary, to be discussed below, between statistics and other parts of the philosophy of science.

Problems of inference from empirical data arise, not only in scientific activity, but also in everyday life and in areas of public policy. For example, the design and analysis of the 1954 Salk vaccine tests in the United States were based on statistical concepts of randomization and control. Both private and public economic decisions sometimes turn on the meaning and accuracy of summary figures from complex measurement programs: the unemployment rate, the rate of economic growth, a consumer price index. Sometimes a lack of statistical background leads to misinterpretations of accident and crime statistics. Misinterpretations arising from insufficient statistical knowledge may also occur in the fields of military and diplomatic intelligence.

There is busy two-way intellectual traffic between statisticians and other scientists. Psychologists and physical anthropologists have instigated and deeply influenced developments in that branch of statistics called multivariate analysis; sociologists sometimes scold statisticians for not paying more attention to the inferential problems arising in surveys of human populations; some economists are at once consumers and producers of statistical methods.

Theoretical and applied statistics. Theoretical statistics is the formal study of the process leading from observations to inference, decision, or whatever be the end point, insofar as the process can be abstracted from special empirical contexts. This study is not the psychological one of how scientists actually make inferences or decisions; rather, it deals with the consequences of particular modes of inference or decision, and seeks normatively to find good modes in the light of explicit criteria.

Theoretical statistics must proceed in terms of a more or less formal language, usually mathematical, and in any specific area must make assumptions—weak or strong—on which to base the formal analysis. Far and away the most important mathematical language in statistics is that of probability, because most statistical thinking is in terms of randomness, populations, masses, the single event embedded in a large class of events. Even approaches like that of personal probability, in which single events are basic, use a highly probabilistic language. [*See* PROBABILITY.]

But theoretical statistics is not, strictly speaking, a branch of mathematics, although mathematical concepts and tools are of central importance in much of statistics. Some important areas of theoretical statistics may be discussed and advanced without recondite mathematics, and much notable work in statistics has been done by men with modest mathematical training. [*For discussion of nonstatistical applications of mathematics in the social sciences, see, for example,* MATHEMATICS; MODELS, MATHEMATICAL; *and the material on mathematical economics in* ECONOMETRICS.]

Applied statistics, at least in principle, is the informed application of methods that have been theoretically investigated, the actual leap after the study of leaping theory. In fact, matters are not so simple. First, theoretical study of a statistical procedure often comes after its intuitive proposal and use. Second, there is almost no end to the possible theoretical study of even the simplest procedure. Practice and theory interact and weave together, so that many statisticians are practitioners one day (or hour) and theoreticians the next.

The art of applied statistics requires sensitivity to the ways in which theoretical assumptions may fail to hold, and to the effects that such failure may have, as well as agility in modifying and extending already studied methods. Thus, applied statistics in the study of public opinion is concerned with the design and analysis of opinion surveys. The main branch of theoretical statistics used here is that of sample surveys, although other kinds of theory may also be relevant—for example, the theory of Markov chains may be useful for panel studies, where the *same* respondents are asked their opinions at successive times. Again, applied statistics in the study of learning includes careful design and analysis of controlled laboratory experiments, whether with worms, rats, or humans. The statistical theories that enter might be those of experimental design, of analysis of variance, or of quantal response. Of course, nonstatistical, substantive knowledge about the empirical field— public opinion, learning, or whatever—is essential for good applied statistics.

Statistics is a young discipline, and the number of carefully studied methods, although steadily growing, is still relatively small. In the applications of statistics, therefore, one usually reaches a point of balance between thinking of a specific problem in formal terms, which are rarely fully adequate (few problems are standard), and using methods that are not as well understood as one might hope. (For a stimulating, detailed discussion of this theme, see Tukey 1962, where the term "data analysis" is used to mean something like applied statistics.)

The word "statistics" is sometimes used to mean, not a general approach like the one I have outlined, but—more narrowly—the body of specific statistical methods, with associated formulas, tables, and traditions, that are currently understood and used. Other uses of the word are common, but they are not likely to cause confusion. In particular, "statistics" often refers to a set of numbers describing some empirical field, as when one speaks of the mortality statistics of France in 1966. Again, "a statistic" often means some numerical quantity computed from basic observations.

Variability and error; patterns. If life were stable, simple, and routinely repetitious, there would be little need for statistical thinking. But there would probably be no human beings to do statistical thinking, because sufficient stability and simplicity would not allow the genetic randomness that is a central mechanism of evolution. Life is not, in fact, stable or simple, but there are stable and simple aspects to it. From one point of view, the goal of science is the discovery and elucidation of these aspects, and statistics deals with some general methods of finding patterns that are hidden in a cloud of irrelevancies, of natural variability, and of error-prone observations or measurements.

Most statistical thinking is in terms of variability and errors in observed data, with the aim of reaching conclusions about obscured underlying patterns. What is meant by natural variability and by errors of measurement? First, distinct experimental and observational units generally have different characteristics and behave in different ways: people vary in their aptitudes and skills; some mice learn more quickly than others. Second, when a quantity or quality is measured, there is usually an error of measurement, and this introduces a second kind of dispersion with which statistics deals: not only will students taught by a new teaching method react in different ways, but also the test that determines how much they learn cannot be a perfect measuring instrument; medical blood-cell counts made independently by two observers from the same slide will not generally be the same.

In any particular experiment or survey, some sources of variability may usefully be treated as constants; for example, the students in the teaching experiment might all be chosen from one geographical area. Other sources of variability might be regarded as random—for example, fluctuations of test scores among students in an apparently homogeneous group. More complex intermediate forms of variability are often present. The students might be subdivided into classes taught by different teachers. Insofar as common membership in a class with the same teacher has an effect, a simple but important pattern of dependence is present.

The variability concept is mirrored in the basic notion of a *population* from which one samples. The population may correspond to an actual population of men, mice, or machines; or it may be conceptual, as is a population of measurement errors. A population of numerical values defines a distribution, roughly speaking, and the notion of a *random variable*, fluctuating in its value according to this distribution, is basic. For example, if a student is chosen at random from a school and given a reading-comprehension test, the score on the test— considered in advance of student choice and test administration—is a random variable. Its distribution is an idealization of the totality of such scores if student choice and testing could be carried out a very large number of times without any changes because of the passage of time or because of interactions among students. [*For a more precise formulation, see* PROBABILITY.]

Although much statistical methodology may be regarded as an attempt to understand regularity through a cloud of obscuring variability, there are many situations in which the variability itself is the object of major interest. Some of these will be discussed below.

Planning. An important topic in statistics is that of sensible planning, or design, of empirical studies. In the above teaching example, some of the more formal aspects of design are the following: How many classes to each teaching method? How many students per class to be tested? Should variables other than test scores be used as well—for example, intelligence scores or personality ratings?

The spectrum of design considerations ranges from these to such subject-matter questions as the following: How should the teachers be trained in a new teaching method? Should teachers be chosen so that there are some who are enthusiastic and some who are skeptical of the new method? What test should be used to measure results?

No general theory of design exists to cover all, or even most, such questions. But there do exist many pieces of theory, and—more important—a valuable statistical point of view toward the planning of experiments.

History. The history of the development of statistics is described in the next article [see Statistics, *article on* the history of statistical method]. It stresses the growth of method and theory; the history of statistics in the senses of vital statistics, government statistics, censuses, economic statistics, and the like, is described in relevant separate articles [see Census; Demography; Government statistics; Life tables; Public policy and statistics; Social research, the early history of; Vital statistics]. Two treatments of the history of statistics with special reference to the social sciences are by Lundberg (1940) and Lazarsfeld (1961).

It is important to distinguish between the history of the word "statistics" and the history of statistics in the sense of this article. The word "statistics" is related to the word "state," and originally the activity called statistics was a systematic kind of comparative political science. This activity gradually centered on numerical tables of economic, demographic, and political facts, and thus "statistics" came to mean the assembly and analysis of numerical tables. It is easy to see how the more philosophical meaning of the word, used in this article, gradually arose. Of course, the abstract study of inference from observations has a long history under various names—such as the theory of errors and probability calculus—and only comparatively recently has the word "statistics" come to have its present meaning. Even now, grotesque misunderstandings abound—for example, thinking of statistics as the routine compilation of uninteresting sets of numbers, or thinking of statistics as mainly a collection of mathematical expressions.

Functions. My description of statistics is, of course, a personal one, but one that many statisticians would generally agree with. Almost any characterization of statistics would include the following general functions:

(1) to help in summarizing and extracting relevant information from data, that is, from observed measurements, whether numerical, classificatory, ordinal, or whatever;

(2) to help in finding and evaluating patterns shown by the data, but obscured by inherent random variability;

(3) to help in the efficient design of experiments and surveys;

(4) to help communication between scientists (if a standard procedure is cited, many readers will understand without need of detail).

There are some other roles that activities called "statistical" may, unfortunately, play. Two such misguided roles are

(1) to sanctify or provide seals of approval (one hears, for example, of thesis advisers or journal editors who insist on certain formal statistical procedures, whether or not they are appropriate);

(2) to impress, obfuscate, or mystify (for example, some social science research papers contain masses of undigested formulas that serve no purpose except that of indicating what a bright fellow the author is).

Some consulting statisticians use more or less explicit declarations of responsibility, or codes, in their relationships with "clients," to protect themselves from being placed in the role of sanctifier. It is a good general rule that the empirical scientist use only statistical methods whose rationale is clear to him, even though he may not wish or be able to follow all details of mathematical derivation.

A general discussion, with an extensive bibliography, of the relationship between statistician and client is given by Deming (1965). In most applied statistics, of course, the statistician and the client are the same person.

An example. To illustrate these introductory comments, consider the following hypothetical experiment to study the effects of propaganda. Suppose that during a national political campaign in the United States, 100 college students are exposed to a motion picture film extolling the Democratic candidate, and 100 other students (the so-called control group) are not exposed to the film. Then all the students are asked to name their preferred candidate. Suppose that 95 of the first group prefer the Democratic candidate, while only 80 of the second group have that preference. What kinds of conclusions might one want about the effectiveness of the propaganda?

(There are, of course, serious questions about how the students are chosen, about the details of film and questionnaire administration, about possible interaction between students, about the artificiality of the experimental arrangement, and so on. For the moment, these questions are not discussed, although some will be touched on below.)

If the numbers preferring the Democratic candidate had been 95 and 5, a conclusion that a real effect was present would probably be reached without much concern about inferential methodology (although methodological questions would enter any attempt to estimate the magnitude of the effect). If, in contrast, the numbers had both been 95, the conclusion "no effect observed" would be immediate, although one might wonder about the possibility of observing the tie by chance even if an underlying effect were present. But by and large it is the middle ground that is of greatest statistical interest: for example, do 95 and 80 differ enough in the above context to suggest a real effect?

The simplest probability model for discussing the experiment is that of analogy with two weighted coins, each tossed 100 times. A toss of the coin corresponding to the propaganda is analogous to selecting a student at random, showing him the motion picture, and then asking him which candidate he prefers. A toss of the other coin corresponds to observing the preference of a student in the control group. "Heads" for a coin is analogous, say, to preference for the Democratic candidate. The hypothetical coins are weighted so that their probabilities of showing heads are unknown (and in general not one-half), and interest lies in the difference between these two unknown heads probabilities.

Suppose that the students are regarded as chosen randomly from some large population of students, and that for a random propagandized student there is a probability p_A of Democratic preference, whereas a random nonpropagandized student has probability p_B of Democratic preference. Suppose further that the individual observed expressions of political preference are statistically independent; roughly speaking, this means that, even if p_A and p_B were known, and it were also known which groups the students are in, prediction of one student's response from another's would be no better than prediction without knowing the other's response. (*Lack* of independence might arise in various ways, for example, if the students were able to discuss politics among themselves during the interval between the motion picture and the questionnaire.) Under the above conditions, the probabilities of various outcomes of the experiment, for any hypothetical values of p_A and p_B, may be computed in standard ways.

In fact, the underlying quantities of interest, the so-called parameters, p_A and p_B, are not known; if they were, there would be little or no reason to do the experiment. Nonetheless, it is of fundamental importance to think about possible values of the parameters and to decide what aspects are of primary importance. For example, is $p_A - p_B$ basic? or perhaps p_A/p_B? or perhaps $(1 - p_B)/(1 - p_A)$, the ratio of the probabilities of an expressed Republican preference (assuming that preference is between Democratic and Republican candidates only)? The choice makes a difference: if $p_A = .99$ and $p_B = .95$, use of a statistical procedure sensitive to $p_A - p_B$ ($= .04$ in this example) might suggest that there is little difference between the parameters, whereas a procedure sensitive to the ratio $(1 - p_B)/(1 - p_A)$ (here $.05/.01 = 5$) might show a very large effect. These considerations are, unhappily, often neglected, and such neglect may result in a misdirected or distorted analysis. In recent discussions of possible relationships between cigarette smoking and lung cancer, controversy arose over whether ratios or differences of mortality rates were of central importance. The choice may lead to quite different conclusions.

¹Even apparently minor changes in graphical presentation may be highly important in the course of research. B. F. Skinner wrote of the importance to his own work of shifting from a graphical record that simply shows the times at which events occur (motion of a rat in a runway) to the logically equivalent *cumulative* record that shows the number of events up to each point of time. In the latter form, the *rate* at which events take place often becomes visually clear (see Skinner 1956, p. 225). This general area is called descriptive statistics, perhaps with the prefix "neo." [*See* STATISTICS, DESCRIPTIVE; GRAPHIC PRESENTATION; TABULAR PRESENTATION.]

As suggested above, the assumption of statistical independence might well be wrong for various reasons. One is that the 100 students in each group might be made up of five classroom groups that hold political discussions. Other errors in the assumptions are quite possible. For example, the sampling of students might not be at random from the same population: there might be self-selection, perhaps with the more enterprising students attending the motion picture. Another kind of deviation from the original simple assumptions (in this case planned) might come from balancing such factors as sex and age by stratifying according to these factors and then selecting at random within strata.

When assumptions are in doubt, one has a choice of easing them (sometimes bringing about a more complex, but a more refined, analysis) or of studying the effects of errors in the assumptions on the analysis based on them. When these effects are small, the errors may be neglected. This topic, sometimes called *robustness* against erroneous assumptions of independence, distributional form, and so on, is difficult and important. [*See* ERRORS, *article on* EFFECTS OF ERRORS IN STATISTICAL ASSUMPTIONS.]

Another general kind of question relates to the design of the experiment. Here, for example, it may be asked in advance of the experiment whether groups of 100 students are large enough (or perhaps unnecessarily large); whether there is merit in equal group sizes; whether more elaborate structures—perhaps allowing explicitly for sex and age —are desirable; and so on. Questions of this kind may call for formal statistical reasoning, but answers must depend in large part on substantive knowledge. [*See* EXPERIMENTAL DESIGN.]

It is important to recognize that using better measurement methods or recasting the framework of the experiment may be far more important aspects of design than just increasing sample size. As B. F. Skinner said,

. . . we may reduce the troublesome variability by changing the condition of the experiment. By discovering, elaborating, and fully exploiting every relevant variable, we may eliminate *in advance of measurement* the individual differences which obscure the difference under analysis.   (1956, p. 229)

In the propaganda experiment at hand, several such approaches come to mind. Restricting oneself to subjects of a given sex, age, kind of background, and so on, might bring out the effects of propaganda more clearly, perhaps at the cost of reduced generality for the results. Rather than by asking directly for political preference, the effects might be better measured by observing physiological reactions to the names or pictures of the candidates, or by asking questions about major political issues. It would probably be useful to try to follow the general principle of having each subject serve as his own control: to observe preference both before and after the propaganda and compare the numbers of switches in the two possible directions. (Even then, it would be desirable to keep the control group—possibly showing it a presumably neutral film—in order to find, and try to correct for, artificial effects of the experimental situation.)

Such questions are often investigated in side studies, ancillary or prior to the central one, and these pilot or instrumental studies are very important.

For the specific simple design with two groups, and making the simple assumptions, consider (conceptually in advance of the experiment) the two observed proportions of students expressing preference for the Democratic candidate, P_A and P_B, corresponding respectively to the propagandized and the control groups. These two *random variables,* together with the known group sizes, contain all relevant information from the experiment itself, in the sense that only the proportions, not the particular students who express one preference or another, are relevant. The argument here is one of sufficiency [*see* SUFFICIENCY, *where the argument and its limitations are discussed*]. In practice the analysis might well be refined by looking at sex of student and other characteristics, but for the moment only the simple structure is considered.

In the notational convention to be followed here, random variables (here P_A and P_B) are denoted by capital letters, and the corresponding parameters (here p_A and p_B) by parallel lower-case letters.

Estimation.   The random variables $100P_A$ and $100P_B$ have binomial probability distributions depending on p_A, p_B, and sample sizes, in this case 100 for each sample [*see* DISTRIBUTIONS, STATISTICAL, *article on* SPECIAL DISCRETE DISTRIBUTIONS]. The fundamental premise of most statistical methods is that p_A and p_B should be assessed on the basis of P_A and P_B in the light of their possible probability distributions. One of the simplest modes of assessment is that of point estimation, in which the result of the analysis for the example consists of two numbers (depending on the observations) that are regarded as reasonable estimates of p_A and p_B [*see* ESTIMATION, *article on* POINT ESTIMATION]. In the case at hand, the usual (not the only) estimators are just P_A and P_B themselves, but even slight changes in viewpoint can make matters less clear. For example, suppose that a point estimator were wanted for p_A/p_B, the ratio of the two underlying probability parameters. It is by no means clear that P_A/P_B would be a good point estimator for this ratio.

[2] Point estimators by themselves are usually inadequate in scientific practice, for some indication of *precision* is nearly always wanted. (There are, however, problems in which point estimators are, in effect, of primary interest: for example, in a handbook table of natural constants, or in some aspects of buying and selling.) An old tradition is to follow a point estimate by a "±" (plus-or-minus sign) and

a number derived from background experience or from the data. The intent is thus to give an idea of how precise the point estimate is, of the spread or dispersion in its distribution. For the case at hand, one convention would lead to stating, as a modified estimator for p_A,

$$P_A \pm \sqrt{\frac{P_A(1 - P_A)}{100}},$$

that is, the point estimator plus or minus an estimator of its standard deviation, a useful measure of dispersion. (The divisor, 100, is the sample size.) Such a device has the danger that there may be misunderstanding about the convention for the number following "$\pm$"; in addition, interpretation of the measure of dispersion may not be direct unless the distribution of the point estimator is fairly simple; the usual presumption is that the distribution is approximately of a form called normal [see DISTRIBUTIONS, STATISTICAL, *article on* SPECIAL CONTINUOUS DISTRIBUTIONS].

To circumvent these problems, a confidence interval is often used, rather than a point estimator [see ESTIMATION, *article on* CONFIDENCE INTERVALS AND REGIONS]. The interval is random (before the experiment), and it is so constructed that it covers the unknown true value of the parameter to be estimated with a preassigned probability, usually near 1. The confidence interval idea is very useful, although its subtlety has often led to misunderstandings in which the interpretation is wrongly given in terms of a probability distribution for the parameter.

There are, however, viewpoints in which this last sort of interpretation is valid, that is, in which the parameters of interest are themselves taken as random. The two most important of these viewpoints are Bayesian inference and fiducial inference [see BAYESIAN INFERENCE; FIDUCIAL INFERENCE; PROBABILITY, *article on* INTERPRETATIONS]. Many variants exist, and controversy continues as the philosophical and practical aspects of these approaches are debated [see LIKELIHOOD *for a discussion of related issues*].

Hypothesis testing. In the more usual viewpoint another general approach is that of hypothesis (or significance) testing [see HYPOTHESIS TESTING; SIGNIFICANCE, TESTS OF]. This kind of procedure might be used if it is important to ascertain whether p_A and p_B are the same or not. Hypothesis testing has two aspects: one is that of a two-decision procedure leading to one of two actions with known controlled chances of error. This first approach generalizes to that of decision theory and has generated a great deal of literature in theoretical statistics [see DECISION THEORY]. In this theory

of decision functions, costs of wrong decisions, as well as costs of observation, are explicitly considered. Decision theory is related closely to game theory, and less closely to empirical studies of decision making [see GAME THEORY; DECISION MAKING].

The second aspect of hypothesis testing—and the commoner—is more descriptive. From its viewpoint a hypothesis test tells how surprising a set of observations is under some *null hypothesis* at test. In the example, one would compute how probable it is under the null hypothesis $p_A = p_B$ that the actual results should differ by as much as or more than the observed 95 per cent and 80 per cent. (Only recently has it been stressed that one would also do well to examine such probabilities under a variety of hypotheses other than a traditional null one.) Sometimes, as in the propaganda example, it is rather clear at the start that some effect must exist. In other cases, for example, in the study of parapsychology, there may be serious question of any effect whatever.

There are other modes of statistical analysis, for example, classification, selection, and screening [see MULTIVARIATE ANALYSIS, *article on* CLASSIFICATION AND DISCRIMINATION; SCREENING AND SELECTION]. In the future there is likely to be investigation of a much wider variety of modes of analysis than now exists. Such investigation will mitigate the difficulty that standard modes of analysis, like hypothesis testing, often do not exactly fit the inferential needs of specific real problems. The standard modes must usually be regarded as approximate, and used with caution.

One pervasive difficulty of this kind surrounds what might be called exploration of data, or data-dredging. It arises when a (usually sizable) body of data from a survey or experiment is at hand but either the analyst has no specific hypotheses about kinds of orderliness in the data or he has a great many. He will naturally wish to explore the body of data in a variety of ways with the hope of finding orderliness: he will try various graphical presentations, functional transformations, perhaps factor analysis, regression analysis, and other devices; in the course of this, he will doubtless carry out a number of estimations, hypothesis tests, confidence interval computations, and so on. A basic difficulty is that any finite body of data, even if wholly generated at random, will show orderliness of some kind if studied long and hard enough. Parallel to this, one must remember that most theoretical work on hypothesis tests, confidence intervals, and other inferential procedures looks at their behavior in isolation, and supposes that the procedures are selected in advance of data inspection. For example, if a hypothesis test is to be made of the null hypothesis that mean scores of men and women on an intelligence test are equal, and if a one-sided alternative is chosen after the fact in the same direction as that shown by the data, it is easy to see that the test will falsely show statistical significance, when the null hypothesis is true, twice as often as the analyst might expect.

On the other hand, it would be ridiculously rigid to refuse to use inferential tools in the exploration of data. Two general mitigating approaches are (1) the use of techniques (for example, multiple comparisons) that include explicit elements of exploration in their formulation [*see* LINEAR HYPOTHESES, *article on* MULTIPLE COMPARISONS], and (2) the splitting of the data into two parts at random, using one part for exploration with no holds barred and then carrying out formal tests or other inferential procedures on the second part.

This area deserves much more research. Selvin and Stuart have given a statement of present opinions, and of practical advice. [*See* Selvin & Stuart 1966; *see also* SURVEY ANALYSIS. STATISTICAL ANALYSIS, SPECIAL PROBLEMS OF, *article on* TRANSFORMATIONS OF DATA, *and* Torgerson 1968 *are also relevant.*]

Breadth of inference. Whatever the mode of analysis, it is important to remember that the inference to which a statistical method directly relates is limited to the population actually experimented upon or surveyed. In the propaganda example, if the students are sampled from a single university, then the immediate inference is to that university only. Wider inferences—and these are usually wanted—presumably depend on subject-matter background and on intuition. Of course, the breadth of direct inference may be widened, for example, by repeating the study at different times, in different universities, in different areas, and so on. But, except in unusual cases, a limit is reached, if only the temporal one that experiments cannot be done now on future students.

Thus, in most cases, a scientific inference has two stages: the direct inference from the sample to the sampled population, and the indirect inference from the sampled population to a much wider, and usually rather vague, realm. That is why it is so important to try to check findings in a variety of contexts, for example, to test psychological generalizations obtained from experiments within one culture in some very different culture.

Formalization and precise theoretical treatment of the second stage represent a gap in present-day statistics (except perhaps for adherents of Bayesian

methodology), although many say that the second step is intrinsically outside statistics. The general question of indirect inference is often mentioned and often forgotten; an early explicit treatment is by von Bortkiewicz (1909); a modern discussion in the context of research in sexual behavior is given by Cochran, Mosteller, and Tukey (1954, pp. 18–19, 21–22, 30–31).

[3] An extreme case of the breadth-of-inference problem is represented by the case study, for example, an intensive study of the history of a single psychologically disturbed person. Indeed, some authors try to set up a sharp distinction between the method of case studies and what they call statistical methods. I do not feel that the distinction is very sharp. For one thing, statistical questions of measurement reliability arise even in the study of a single person. Further, some case studies, for example, in anthropology, are of a tribe or some other group of individuals, so that traditional sampling questions might well arise in drawing inferences about the single (collective) case.

Proponents of the case study approach emphasize its flexibility, its importance in attaining subjective insight, and its utility as a means of conjecturing interesting theoretical structures. If there is good reason to believe in small relevant intercase variability, then, of course, a single case does tell much about a larger population. The investigator, however, has responsibility for defending an assumption about small intercase variability. [*Further discussion will be found in* INTERVIEWING IN SOCIAL RESEARCH; *see also* Becker 1968.]

Other topics

Linear hypotheses. One way of classifying statistical topics is in terms of the kind of assumptions made, that is—looking toward applications—in terms of the structure of anticipated experiments or surveys for which the statistical methods will be used. The propaganda example, in which the central quantities are two proportions with integral numerators and denominators, falls under the general topic of the analysis of counted or qualitative data; this topic includes the treatment of so-called chi-square tests. Such an analysis would also be applicable if there were more than two groups, and it can be extended in other directions. [*See* COUNTED DATA.]

If, in the propaganda experiment, instead of proportions expressing one preference or the other, numerical scores on a multiquestion test were used to indicate quantitatively the leaning toward a candidate or political party, then the situation might come under the general rubric of linear hypotheses. To illustrate the ideas, suppose that there were more than two groups, say, four, of which the first was exposed to no propaganda, the second saw a motion picture, the third was given material to read, and the fourth heard a speaker, and that the scores of students under the four conditions are to be compared. Analysis-of-variance methods (many of which may be regarded as special cases of regression methods) are of central importance for such a study [see LINEAR HYPOTHESES, *articles on* ANALYSIS OF VARIANCE *and* REGRESSION]. Multiple comparison methods are often used here, although —strictly speaking—they are not restricted to the analysis-of-variance context [see LINEAR HYPOTHESES, *article on* MULTIPLE COMPARISONS].

If the four groups differed primarily in some quantitative way, for example, in the number of sessions spent watching propaganda motion pictures, then regression methods in a narrower sense might come into play. One might, for example, suppose that average test score is roughly a linear function of number of motion picture sessions, and then center statistical attention on the constants (slope and intercept) of the linear function.

Multivariate statistics. "Regression" is a word with at least two meanings. A meaning somewhat different from, and historically earlier than, that described just above appears in statistical theory for multivariate analysis, that is, for situations in which more than one kind of observation is made on each individual or unit that is measured [see MULTIVARIATE ANALYSIS]. For example, in an educational experiment on teaching methods, one might look at scores not only on a spelling examination, but on a grammar examination and on a reading-comprehension examination as well. Or in a physical anthropology study, one might measure several dimensions of each individual.

The simplest part of multivariate analysis is concerned with association between just two random variables and, in particular, with the important concept of correlation [see STATISTICS, DESCRIPTIVE, *article on* ASSOCIATION; MULTIVARIATE ANALYSIS, *article on* CORRELATION METHODS]. These ideas extend to more than two random variables, and then new possibilities enter. An important one is that of partial association: how are spelling and grammar scores associated if reading comprehension is held fixed? The partial association notion is important in survey analysis, where a controlled experiment is often impossible [see SURVEY ANALYSIS; EXPERIMENTAL DESIGN, *article on* QUASI-EXPERIMENTAL DESIGN].

Multivariate analysis also considers statistical methods bearing on the joint structure of the means that correspond to the several kinds of observations, and on the whole correlation structure.

[4]Factor analysis falls in the multivariate area, but it has a special history and a special relationship with psychology [see FACTOR ANALYSIS AND PRINCIPAL COMPONENTS]. Factor-analytic methods try to replace a number of measurements by a few basic ones, together with residuals having a simple probability structure. For example, one might hope that spelling, grammar, and reading-comprehension abilities are all proportional to some quantity not directly observable, perhaps dubbed "linguistic skill," that varies from person to person, plus residuals or deviations that are statistically independent.

The standard factor analysis model is one of a class of models generated by a process called mixing of probability distributions [see DISTRIBUTIONS, STATISTICAL, *article on* MIXTURES OF DISTRIBUTIONS]. An interesting model of this general sort, but for discrete, rather than continuous, observations, is that of latent structure [see LATENT STRUCTURE].

Another important multivariate topic is classification and discrimination, which is the study of how to assign individuals to two or more groups on the basis of several measurements per individual [see MULTIVARIATE ANALYSIS, *article on* CLASSIFICATION AND DISCRIMINATION]. Less well understood, but related, is the problem of clustering, or numerical taxonomy: what are useful ways for forming groups of individuals on the basis of several measurements on each? [See CLUSTERING.]

Time series. Related to multivariate analysis, because of its stress on modes of statistical dependence, is the large field of time series analysis, sometimes given a title that includes the catchy phrase "stochastic processes." An observed time series may be regarded as a realization of an underlying stochastic process [see TIME SERIES]. The simplest sort of time series problem might arise when for each child in an educational experiment there is available a set of scores on spelling tests given each month during the school year. More difficult problems arise when there is no hope of observing more than a single series, for example, when the observations are on the monthly or yearly prices of wheat. In such cases—so common in economics—stringent structural assumptions are required, and even then analysis is not easy.

This encyclopedia's treatment of time series begins with a general overview, oriented primarily toward economic series. The overview is followed by a discussion of advanced methodology, mainly that of spectral analysis, which treats a time series as something like a radio signal that can be decomposed into subsignals at different frequencies, each with its own amount of energy. Next comes a treatment of cycles, with special discussion of how easy it is to be trapped into concluding that cycles exist when in fact only random variation is present. Finally, there is a discussion of the important technical problem raised by seasonal variation, and of adjustment to remove or mitigate its effect. The article on business cycles should also be consulted [*see* BUSINESS CYCLES: MATHEMATICAL MODELS].

The topic of Markov chains might have been included under the time series category, but it is separate [*see* MARKOV CHAINS]. The concept of a Markov chain is one of the simplest and most useful ways of relaxing the common assumption of independence. Methods based on the Markov chain idea have found application in the study of panels (for public opinion, budget analysis, etc.), of labor mobility, of changes in social class between generations, and so on [*see, for example,* PANEL STUDIES; SOCIAL MOBILITY].

Sample surveys and related topics. The subject of sample surveys is important, both in theory and practice [*see* SAMPLE SURVEYS]. It originated in connection with surveys of economic and social characteristics of human populations, when samples were used rather than attempts at full coverage. But the techniques of sample surveys have been of great use in many other areas, for example in the evaluation of physical inventories of industrial equipment. The study of sample surveys is closely related to most of the other major fields of statistics, in particular to the design of experiments, but it is characterized by its emphasis on finite populations and on complex sampling plans.

Most academically oriented statisticians who think about sample surveys stress the importance of probability sampling—that is, of choosing the units to be observed by a plan that explicitly uses random numbers, so that the probabilities of possible samples are known. On the other hand, many actual sample surveys are not based upon probability sampling [*for a discussion of the central issues of this somewhat ironical discrepancy, see* SAMPLE SURVEYS, *article on* NONPROBABILITY SAMPLING].

Random numbers are important, not only for sample surveys, but for experimental design generally, and for simulation studies of many kinds [*see* RANDOM NUMBERS; SIMULATION].

An important topic in sample surveys (and, for that matter, throughout applied statistics) is that of nonsampling errors [*see* ERRORS, *article on* NONSAMPLING ERRORS]. Such errors stem, for example, from nonresponse in public opinion surveys, from observer and other biases in measurement, and from errors of computation. Interesting discussions of these problems, and of many others related to sampling, are given by Cochran, Mosteller, and Tukey (1954).

Sociologists have long been interested in survey research, but with historically different emphases from those of statisticians [*see* SURVEY ANALYSIS; INTERVIEWING IN SOCIAL RESEARCH]. The sociological stress has been much less on efficient design and sampling variation and much more on complex analyses of highly multivariate data. There is reason to hope that workers in these two streams of research are coming to understand each other's viewpoint.

Nonparametric analysis and related topics. I remarked earlier that an important area of study is robustness, the degree of sensitivity of statistical methods to errors in assumptions. A particular kind of assumption error is that incurred when a special distributional form, for example, normality, is assumed when it does not in fact obtain. To meet this problem, one may seek alternate methods that are insensitive to form of distribution, and the study of such methods is called nonparametric analysis or distribution-free statistics [*see* NONPARAMETRIC STATISTICS]. Such procedures as the sign test and many ranking methods fall into the nonparametric category.

For example, suppose that pairs of students—matched for age, sex, intelligence, and so on—form the experimental material, and that for each pair it is determined entirely at random, as by the throw of a fair coin, which member of the pair is exposed to one teaching method (A) and which to another (B). After exposure to the assigned methods, the students are given an examination; a pair is scored positive if the method A student has the higher score, negative if the method B student has. If the two methods are equally effective, the number of positive scores has a binomial distribution with basic probability $\frac{1}{2}$. If, however, method A is superior, the basic probability is greater than $\frac{1}{2}$; if method B is superior, less than $\frac{1}{2}$. The number of observed positives provides a simple test of the hypothesis of equivalence and a basis for estimating the amount of superiority that one of the teaching methods may have. (The above design is, of course, only sensible if matching is possible for most of the students.)

The topic of order statistics is also discussed in

one of the articles on nonparametric analysis, although order statistics are at least as important for procedures that do make sharp distributional assumptions [see Nonparametric statistics, *article on* order statistics]. There is, of course, no sharp boundary line for distribution-free procedures. First, many procedures based on narrow distributional assumptions turn out in fact to be robust, that is, to maintain some or all of their characteristics even when the assumptions are relaxed. Second, most distribution-free procedures are only partly so; for example, a distribution-free test will typically be independent of distributional form as regards its level of significance but not so as regards power (the probability of rejecting the null hypothesis when it is false). Again, most nonparametric procedures are nonrobust against dependence among the observations.

Nonparametric methods often arise naturally when observational materials are inherently nonmetric, for example, when the results of an experiment or survey provide only rankings of test units by judges.

Sometimes the form of a distribution is worthy of special examination, and goodness-of-fit procedures are used [see Goodness of fit]. For example, a psychological test may be standardized to a particular population so that test scores over the population have very nearly a unit-normal distribution. If the test is then administered to the individuals of a sample from a different population, the question may arise of whether the score distribution for the different population is still unit normal, and a goodness-of-fit test of unit-normality may be performed. More broadly, an analogous test might be framed to test only normality, without specification of a particular normal distribution.

Some goodness-of-fit procedures, the so-called chi-square ones, may be regarded as falling under the counted-data rubric [see Counted data]. Others, especially when modified to provide confidence bands for an entire distribution, are usually studied under the banner of nonparametric analysis.

Dispersion. The study of dispersion, or variability, is a topic that deserves more attention than it often receives [see Variances, statistical study of]. For example, it might be of interest to compare several teaching methods as to the resulting heterogeneity of student scores. A particular method might give rise to a desirable average score by increasing greatly the scores of some students while leaving other students' scores unchanged, thereby giving rise to great heterogeneity. Clearly, such a method has different consequences and applications than one that raises each student's score by about the same amount. (Terminology may be confusing here. The traditional topic of analysis of variance deals in substantial part with means, not variances, although it does so by looking at dispersions among the means.)

Design. Experimental design has already been mentioned. It deals with such problems as how many observations to take for a given level of accuracy, and how to assign the treatments or factors to experimental units. For example, in the study of teaching methods, the experimental units may be school classes, cross-classified by grade, kind of school, type of community, and the like. Experimental design deals with formal aspects of the structure of an experimental layout; a basic principle is that explicit randomization should be used in assigning "treatments" (here methods of teaching) to experimental units (here classes). Sometimes it may be reasonable to suppose that randomization is inherent, supplied, as it were, by nature; but more often it is important to use so-called random numbers. Controversy centers on situations in which randomization is deemed impractical, unethical, or even impossible, although one may sometimes find clever ways to introduce randomization in cases where it seems hopeless at first glance. When randomization is absent, a term like "quasi experiment" may be used to emphasize its absence, and a major problem is that of obtaining as much protection as possible against the sources of bias that would have been largely eliminated by the unused randomization [see Experimental design, *article on* quasi-experimental design].

An important aspect of the design of experiments is the use of devices to ensure both that a (human) subject does not know which experimental treatment he is subjected to, and that the investigator who is measuring or observing effects of treatments does not know which treatments particular observed individuals have had. When proper precautions are taken along these two lines, the experiment is called *double blind*. Many experimental programs have been vitiated by neglect of these precautions. First, a subject who knows that he is taking a drug that it is hoped will improve his memory, or reduce his sensitivity to pain, may well change his behavior in response to the knowledge of what is expected as much as in physiological response to the drug itself. Hence, whenever possible, so-called placebo treatments (neutral but, on the surface, indistinguishable from the real treatment) are administered to members of the control group. Second, an investigator who knows which subjects are having which treatments may easily, and quite unconsciously, have his observations biased by preconceived opinions. Problems may arise even if the investigator knows only which subjects are in the same group. Assignment to treatment by the use of random numbers, and random ordering of individuals for observation, are important devices to ensure impartiality.

The number of observations is traditionally regarded as fixed before sampling. In recent years, however, there have been many investigations of sequential designs in which observations are taken in a series (or in a series of groups of observations), with decisions made at each step whether to take further observations or to stop observing and turn to analysis [see SEQUENTIAL ANALYSIS].

In many contexts a response (or its average value) is a function of several controlled variables. For example, average length of time to relearn the spellings of a list of words may depend on the number of prior learning sessions and the elapsed period since the last learning session. In the study of response surfaces, the structure of the dependence (thought of as the shape of a surface) is investigated by a series of experiments, typically with special interest in the neighborhood of a maximum or minimum [see EXPERIMENTAL DESIGN, *article on* RESPONSE SURFACES].

Philosophy. Statistics has long had a neighborly relation with philosophy of science in the epistemological city, although statistics has usually been more modest in scope and more pragmatic in outlook. In a strict sense, statistics is part of philosophy of science, but in fact the two areas are usually studied separately.

What are some problems that form part of the philosophy of science but are not generally regarded as part of statistics? A central one is that of the formation of scientific theories, their careful statement, and their confirmation or degree of confirmation. This last is to be distinguished from the narrower, but better understood, statistical concept of testing hypotheses. Another problem that many statisticians feel lies outside statistics is that of the gap between sampled and target population.

There are other areas of scientific philosophy that are not ordinarily regarded as part of statistics. Concepts like explanation, causation, operationalism, and free will come to mind.

A classic publication dealing with both statistics and scientific philosophy is Karl Pearson's *Grammar of Science* (1892). Two more recent such publications are Popper's *Logic of Scientific Discovery* ([1935] 1959) and Braithwaite's *Scientific Explanation* (1953). By and large, nowadays, writers calling themselves statisticians and those calling

themselves philosophers of science often refer to each other, but communication is restricted and piecemeal. [*See* SCIENCE, *article on* THE PHILOSOPHY OF SCIENCE; *see also* CAUSATION; PREDICTION; SCIENTIFIC EXPLANATION; Dahl 1968.]

Measurement is an important topic for statistics, and it might well be mentioned here because some aspects of measurement are clearly philosophical. Roughly speaking, measurement is the process of assigning numbers (or categories) to objects on the basis of some operation. A measurement or datum is the resulting number (or category). But what is the epistemological underpinning for this concept? Should it be broadened to include more general kinds of data than numbers and categories? What kind of operations should be considered?

In particular, measurement *scales* are important, both in theory and practice. It is natural to say of one object that it is twice as heavy as another (in pounds, grams, or whatever—the unit is immaterial). But it seems silly to say that one object has twice the temperature of another in any of the everyday scales of temperature (as opposed to the absolute scale), if only because the ratio changes when one shifts, say, from Fahrenheit to Centigrade degrees. On the other hand, it makes sense to say that one object is 100 degrees Fahrenheit hotter than another. Some measurements seem to make sense only insofar as they order units, for example, many subjective rankings; and some measurements are purely nominal or categorical, for example, country of birth. Some measurements are inherently circular, for example, wind direction or time of day. There has been heated discussion of the question of the meaningfulness or legitimacy of arithmetic manipulations of various kinds of measurements; does it make sense, for example, to average measurements of subjective loudness if the individual measurements give information only about ordinal relationships?

The following are some important publications that deal with measurement and that lead to the relevant literature at this date: Churchman and Ratoosh (1959); Coombs (1964); Pfanzagl (1959); Adams, Fagot, and Robinson (1965); Torgerson (1958); Stevens (1946); Suppes and Zinnes (1963). [*See* STATISTICS, DESCRIPTIVE; *also relevant are* PSYCHOMETRICS; Georgescu-Roegen 1968; Torgerson 1968.]

Communication and fallacies. There is an art of communication between statistician and nonstatistician scientist: the statistician must be always aware that the nonstatistician is in general not directly interested in technical minutiae or in the parochial jargon of statistics. In the other direction, consultation with a statistician often loses effectiveness because the nonstatistician fails to mention aspects of his work that are of statistical relevance. Of course, in most cases scientists serve as their own statisticians, in the same sense that people, except for hypochondriacs, serve as their own physicians most of the time.

Statistical fallacies are often subtle and may be committed by the most careful workers. A study of such fallacies has intrinsic interest and also aids in mitigating the communication problem just mentioned [see FALLACIES, STATISTICAL; see also ERRORS, article on NONSAMPLING ERRORS].

Criticisms

If statistics is defined broadly, in terms of the general study of the leap from observations to inference, decision, or whatever, then one can hardly quarrel with the desirability of a study so embracingly characterized. Criticisms of statistics, therefore, are generally in terms of a narrower characterization, often the kind of activity named "statistics" that the critic sees about him. If, for example, a professor in some scientific field sees colleagues publishing clumsy analyses that they call statistical, then the professor may understandably develop a negative attitude toward statistics. He may not have an opportunity to learn that the subject is broader and that it may be used wisely, elegantly, and effectively.

Criticisms of probability in statistics. Some criticisms, in a philosophical vein, relate to the very use of probability models in statistics. For example, some writers have objected to probability because of a strict determinism in their *Weltanschauung.* This view is rare nowadays, with the success of highly probabilistic quantum methods in physics, and with the utility of probability models for clearly deterministic phenomena, for example, the effect of rounding errors in complex digital calculations. The deterministic critic, however, would probably say that quantum mechanics and probabilistic analysis of rounding errors are just temporary expedients, to be replaced later by nonprobabilistic approaches. For example, Einstein wrote in 1947 that

. . . the statistical interpretation [as in quantum mechanics] . . . has a considerable content of truth. Yet I cannot seriously believe it because the theory is inconsistent with the principle that physics has to represent a reality. . . . I am absolutely convinced that one will eventually arrive at a theory in which the objects connected by laws are not probabilities, but conceived facts. . . . However, I cannot provide logical arguments for my conviction, but can only call on my little finger as a witness, which cannot claim any authority to be respected outside my own skin. (Quoted in Born 1949, p. 123)

Other critics find vitiating contradictions and paradoxes in the ideas of probability and randomness. For example, G. Spencer Brown sweepingly wrote that

. . . the concept of probability used in statistical science is meaningless in its own terms [and] . . . , however meaningful it might have been, its meaningfulness would nevertheless have remained fruitless because of the impossibility of gaining information from experimental results. (1957, p. 66)

This rather nihilistic position is unusual and hard to reconcile with the many successful applications of probabilistic ideas. (Indeed, Spencer Brown went on to make constructive qualifications.) A less extreme but related view was expressed by Percy W. Bridgman (1959, pp. 110–111). Both these writers were influenced by statistical uses of tables of random numbers, especially in the context of parapsychology, where explanations of puzzling results were sought in the possible misbehavior of random numbers. [See RANDOM NUMBERS; see also Schmeidler 1968.]

Criticisms about limited utility. A more common criticism, notably among some physical scientists, is that they have little need for statistics because random variability in the problems they study is negligible, at least in comparison with systematic errors or biases. This position has also been taken by some economists, especially in connection with index numbers [see INDEX NUMBERS, article on SAMPLING]. B. F. Skinner, a psychologist, has forcefully expressed a variant of this position: that there are so many important problems in which random variability is negligible that he will restrict his own research to them (see Skinner 1956 for a presentation of this rather extreme position). In fact, he further argues that the important problems in psychology as a field are the identification of variables that can be observed directly with negligible variability.

It often happens, nonetheless, that, upon detailed examination, random variability is more important than had been thought, especially for the design of future experiments. Further, careful experimental design can often reduce, or bring understanding of, systematic errors. I think that the above kind of criticism is sometimes valid—after all, a single satellite successfully orbiting the earth is enough to show that it can be done—but that usually the criticism represents unwillingness to consider statistical methods explicitly, or a semantic confusion about what statistics is.

Related to the above criticism is the view that statistics is fine for applied technology, but not for fundamental science. In his inaugural lecture at Birkbeck College at the University of London, David Cox countered this criticism. He said in his introduction,

. . . there is current a feeling that in some fields of fundamental research, statistical ideas are sometimes not just irrelevant, but may actually be harmful as a symptom of an over-empirical approach. This view, while understandable, seems to me to come from too narrow a concept of what statistical methods are about. (1961)

Cox went on to give examples of the use of statistics in fundamental research in physics, psychology, botany, and other fields.

Another variant of this criticism sometimes seen (Selvin 1957; Walberg 1966) is that such statistical procedures as hypothesis testing are of doubtful validity unless a classically arranged experiment is possible, complete with randomization, control groups, pre-establishment of hypotheses, and other safeguards. Without such an arrangement—which is sometimes not possible or practical—all kinds of bias may enter, mixing any actual effect with bias effects.

This criticism reflects a real problem of reasonable inference when a true experiment is not available [see EXPERIMENTAL DESIGN, *article on* QUASI-EXPERIMENTAL DESIGN], but it is not a criticism unique to special kinds of inference. The problem applies equally to any mode of analysis—formal, informal, or intuitive. A spirited discussion of this topic is given by Kish (1959).

5 Humanistic criticisms. Some criticisms of statistics represent serious misunderstandings or are really criticisms of *poor* statistical method, not of statistics per se. For example, one sometimes hears the argument that statistics is inhuman, that "you can't reduce people to numbers," that statistics (and perhaps science more generally) must be battled by humanists. This is a statistical version of an old complaint, voiced in one form by Horace Walpole, in a letter to H. S. Conway (1778): "This sublime age reduces everything to its quintessence; all periphrases and expletives are so much in disuse, that I suppose soon the only way to [go about] making love will be to say 'Lie down.'"

A modern variation of this was expressed by W. H. Auden in the following lines:

Thou shalt not answer questionnaires
Or quizzes upon World-Affairs,
    Nor with compliance
Take any test. Thou shalt not sit
With statisticians nor commit
    A social science.

From "Under Which Lyre: A Reactionary Tract for the Times." Reprinted from *Collected Poems*, by W. H. Auden, edited by Edward Mendelson, by permission of Random House, Inc. Copyright 1946 by W. H. Auden.

Joseph Wood Krutch (1963) said, "I still think that a familiarity with the best that has been thought and said by men of letters is more helpful than all the sociologists' statistics" ("Through Happiness With Slide Rule and Calipers," p. 14).

There are, of course, quite valid points buried in such captious and charming criticisms. It is easy to forget that things may be more complicated than they seem, that many important characteristics are extraordinarily difficult to measure or count, that scientists (and humanists alike) may lack professional humility, and that any set of measurements excludes others that might in principle have been made. But the humanistic attack is overdefensive and is a particular instance of what might be called the two-culture fallacy: the belief that science and the humanities are inherently different and necessarily in opposition.

6 Criticisms of overconcern with averages. Statisticians are sometimes teased about being interested only in averages, some of which are ludicrous: 2.35 children in an average family; or the rare disease that attacks people aged 40 on the average —two cases, one a child of 2 and the other a man of 78. (Chuckles from the gallery.)

Skinner made the point by observing that "no one goes to the circus to see the average dog jump through a hoop significantly oftener than untrained dogs raised under the same circumstances . . ." (1956, p. 228). Krutch said that "Statistics take no account of those who prefer to hear a different drummer" (1963, p. 15).

In fact, although averages are important, statisticians have long been deeply concerned about dispersions around averages and about other aspects of distributions, for example, in extreme values [see NONPARAMETRIC STATISTICS, *article on* ORDER STATISTICS; *and* STATISTICAL ANALYSIS, SPECIAL PROBLEMS OF, *article on* OUTLIERS].

In 1889 the criticism of averages was poetically made by Galton:

It is difficult to understand why statisticians commonly limit their inquiries to Averages, and do not revel in more comprehensive views. Their souls seem as dull to the charm of variety as that of the native of one of our flat English counties, whose retrospect of Switzerland was that, if its mountains could be thrown into its lakes, two nuisances would be got rid of at once. (p. 62)

Galton's critique was overstated even at its date, but it would be wholly inappropriate today.

Another passage from the same work by Galton refers to the kind of emotional resistance to statistics that was mentioned earlier:

Some people hate the very name of statistics, but I find them full of beauty and interest. Whenever they are not brutalized, but delicately handled by the higher methods, and are warily interpreted, their power of dealing with complicated phenomena is extraordinary. They are the only tools by which an opening can be cut through the formidable thicket of difficulties that bars the path of those who pursue the Science of man. (1889, pp. 62–63)

[7]One basic source of misunderstanding about averages is that an individual may be average in many ways, yet appreciably nonaverage in others. This was the central difficulty with Quetelet's historically important concept of the average man [*see the biography of* QUETELET]; a satirical novel about the point, by Robert A. Aurthur (1953), has appeared. The average number of children per family in a given population is meaningful and sometimes useful to know, for example, in estimating future population. There is, however, no such thing as the average family, if only because a family with an average number of children (assuming this number to be integral) would not be average in terms of the reciprocal of number of children. To put it another way, there is no reason to think that a family with the average number of children also has average income, or average education, or lives at the center of population of the country.

Criticisms of too much mathematics. The criticism is sometimes made—often by statisticians themselves—that statistics is too mathematical. The objection takes various forms, for example:

(1) Statisticians choose research problems because of their mathematical interest or elegance and thus do not work on problems of real statistical concern. (Sometimes the last phrase simply refers to problems of concern to the critic.)

(2) The use of mathematical concepts and language obscures statistical thinking.

(3) Emphasis on mathematical aspects of statistics tends to make statisticians neglect problems of proper goals, meaningfulness of numerical statistics, and accuracy of data.

Critiques along these lines are given by, for example, W. S. Woytinsky (1954) and Corrado Gini (1951; 1959). A similar attack appears in Lancelot Hogben's *Statistical Theory* (1957). What can one say of this kind of criticism, whether it comes from within or without the profession? It has a venerable history that goes back to the early development of statistics. Perhaps the first quarrel of this kind was in the days when the word "statistics" was used, in a different sense than at present, to mean the systematic study of states, a kind of political science. The dispute was between those "statisticians" who provided discursive descriptions of states and those who cultivated the so-called *Tabellenstatistik*, which ranged from typographically convenient arrangements of verbal summaries to actual tables of vital statistics. Descriptions of this quarrel are given by Westergaard (1932, pp. 12–15), Lundberg (1940), and Lazarsfeld (1961, especially p. 293).

The *ad hominem* argument—that someone is primarily a mathematician, and hence incapable of understanding truly statistical problems—has been and continues to be an unfortunately popular rhetorical device. In part it is probably a defensive reaction to the great status and prestige of mathematics.

In my view, a great deal of this kind of discussion has been beside the point, although some charges on all sides have doubtless been correct. If a part of mathematics proves helpful in statistics, then it will be used. As statisticians run onto mathematical problems, they will work on them, borrowing what they can from the store of current mathematical knowledge, and perhaps encouraging or carrying out appropriate mathematical research. To be sure, some statisticians adopt an unnecessarily mathematical manner of exposition. This may seem an irritating affectation to less mathematical colleagues, but who can really tell apart an affectation and a natural mode of communication?

An illuminating discussion about the relationship between mathematics and statistics, as well as about many other matters, is given by Tukey (1961).

[8]**Criticisms of obfuscation.** Next, there is the charge that statistics is a meretricious mechanism to obfuscate or confuse: "Lies, damned lies, and statistics" (the origin of this canard is not entirely clear: see White 1964). A variant is the criticism that statistical analyses are impossible to follow, filled with unreadable charts, formulas, and jargon. These points are often well taken of specific statistical or pseudostatistical writings, but they do not relate to statistics as a discipline. A popular book, *How to Lie With Statistics* (Huff 1954), is

in fact a presentation of horrid errors in statistical description and analysis, although it could, of course, be used as a source for pernicious sophistry. It is somewhat as if there were a book called "How to Counterfeit Money," intended as a guide to bank tellers—or the general public—in protecting themselves against false money.

George A. Lundberg made a cogent defense against one form of this criticism, in the following words:

. . . when we have to reckon with stupidity, incompetence, and illogic, the more specific the terminology and methods employed the more glaring will be the errors in the result. As a result, the errors of quantitative workers lend themselves more easily to detection and derision. An equivalent blunder by a manipulator of rhetoric may not only appear less flagrant, but may actually go unobserved or become a venerated platitude. (1940, p. 138)

Criticisms of sampling per se. One sometimes sees the allegation that it is impossible to make reasonable inferences from a sample to a population, especially if the sample is a small fraction of the population. A variant of this was stated by Joseph Papp: "The methodology . . . was not scientific: they used sampling and you can't draw a complete picture from samplings" (quoted in Kadushin 1966, p. 30).

This criticism has no justification except insofar as it impugns *poor* sampling methods. Samples have always been used, because it is often impractical or impossible to observe a whole population (one cannot test a new drug on every human being, or destructively test all electric fuses) or because it is more informative to make careful measurements on a sample than crude measurements on a whole population. Proper sampling—for which the absolute size of the sample is far more important than the fraction of the population it represents—is informative, and in constant successful use.

Criticisms of intellectual imperialism. The criticism is sometimes made that statistics is not the whole of scientific method and practice. Skinner said:

. . . it is a mistake to identify scientific practice with the *formalized constructions* [italics added] of statistics and scientific method. These disciplines have their place, but it does not coincide with the place of scientific research. They offer *a* method of science but not, as is so often implied, *the* method. As formal disciplines they arose very late in the history of science, and most of the facts of science have been discovered without their aid. (1956, p. 221)

I know of few statisticians so arrogant as to equate their field with scientific method generally. It is, of course, true that most scientific work has been done without the aid of statistics, narrowly construed as certain formal modes of analysis that are currently promulgated. On the other hand, a good deal of scientific writing is concerned, one way or another, with statistics, in the more general sense of asking how to make sensible inferences.

Skinner made another, somewhat related point: that, because of the prestige of statistics, statistical methods have (in psychology) acquired the honorific status of a shibboleth (1956, pp. 221, 231). Statisticians are sorrowfully aware of the shibboleth use of statistics in some areas of scientific research, but the profession can be blamed for this only because of some imperialistic textbooks—many of them not by proper statisticians.

Other areas of statistics

The remainder of this article is devoted to brief discussions of those statistical articles in the encyclopedia that have not been described earlier.

Grouped observations. The question of grouped observations is sometimes of concern: in much theoretical statistics measurements are assumed to be continuous, while in fact measurements are always discrete, so that there is inevitable grouping. In addition, one often wishes to group measurements further, for simplicity of description and analysis. To what extent are discreteness and grouping an advantage, and to what extent a danger? [*See* STATISTICAL ANALYSIS, SPECIAL PROBLEMS OF, *article on* GROUPED OBSERVATIONS.]

Truncation and censorship. Often observations may reasonably follow some standard model except that observations above (or below) certain values are proscribed (truncated or censored). A slightly more complex example occurs in comparing entrance test scores with post-training scores for students in a course; those students with low entrance test scores may not be admitted and hence will not have post-training scores at all. Methods exist for handling such problems. [*See* STATISTICAL ANALYSIS, SPECIAL PROBLEMS OF, *article on* TRUNCATION AND CENSORSHIP.]

Outliers. Very often a few observations in a sample will have unusually large or small values and may be regarded as outliers (or mavericks or wild values). How should one handle them? If they are carried along in an analysis, they may distort it. If they are arbitrarily suppressed, important information may be lost. Even if they are to be suppressed, what rule should be used? [*See* STATISTICAL ANALYSIS, SPECIAL PROBLEMS OF, *article on* OUTLIERS.]

Transformations of data. Transformations of data are often very useful. For example, one may take the logarithm of reaction time, the square root of a test score, and so on. The purposes of such a transformation are (1) to simplify the structure of the data, for example by achieving additivity of two kinds of effects, and (2) to make the data more nearly conform with a well-understood statistical model, for example by achieving near-normality or constancy of variance. A danger of transformations is that one's inferences may be shifted to some other scale than the one of basic interest. [*See* STATISTICAL ANALYSIS, SPECIAL PROBLEMS OF, *article on* TRANSFORMATIONS OF DATA.]

Approximations to distributions. Approximations to distributions are important in probability and statistics. First, one may want to approximate some theoretical distribution in order to have a simple analytic form or to get numerical values. Second, one may want to approximate empirical distributions for both descriptive and inferential purposes. [*See* DISTRIBUTIONS, STATISTICAL, *article on* APPROXIMATIONS TO DISTRIBUTIONS.]

Identifiability—mixtures of distributions. The problem of identification appears whenever a precise model for some phenomenon is specified and parameters of the model are to be estimated from empirical observations [*see* STATISTICAL IDENTIFIABILITY]. What may happen—and may even fail to be recognized—is that the parameters are fundamentally incapable of estimation from the kind of data in question. Consider, for example, a learning theory model in which the proportion of learned material retained after a lapse of time is the ratio of two parameters of the model. Then, even if the proportion could be observed without any sampling fluctuation or measurement error, one would not separately know the two parameters. Of course, the identification problem arises primarily in contexts that are complex enough so that immediate recognition of nonidentifiability is not likely. Sometimes there arises an analogous problem, which might be called identifiability of the model. A classic example appears in the study of accident statistics: some kinds of these statistics are satisfactorily fitted by the negative binomial distribution, but that distribution itself may be obtained as the outcome of several quite different, more fundamental models. Some of these models illustrate the important concept of mixtures. A mixture is an important and useful way of forming a new distribution from two or more statistical distributions. [*See* DISTRIBUTIONS, STATISTICAL, *article on* MIXTURES OF DISTRIBUTIONS.]

Applications. Next described is a set of articles on special topics linked with specific areas of application, although most of these areas have served as motivating sources for general theory.

Quality control. Statistical quality control had its genesis in manufacturing industry, but its applications have since broadened [*see* QUALITY CONTROL, STATISTICAL]. There are three articles under this heading. The first is on acceptance sampling, where the usual context is that of "lots" of manufactured articles. Here there are close relations to hypothesis testing and to sequential analysis. The second is on process control (and so-called control charts), a topic that is sometimes itself called quality control, in a narrower sense than the usage here. The development of control chart concepts and methods relates to basic notions of randomness and stability, for an important normative concept is that of a *process in control*, that is, a process turning out a sequence of numbers that behave like independent, identically distributed random variables. The third topic is reliability and life testing, which also relates to matters more general than immediate engineering contexts. [*The term "reliability" here has quite a different meaning than it has in the area of psychological testing; see* PSYCHOMETRICS.]

⁹*Government statistics.* Government statistics are of great importance for economic, social, and political decisions [*see* GOVERNMENT STATISTICS]. The article on that subject treats such basic issues as the use of government statistics for political propaganda, the problem of confidentiality, and the meaning and accuracy of official statistics. [*Some related entries are* CENSUS; DEMOGRAPHY; PUBLIC POLICY AND STATISTICS; VITAL STATISTICS; *see also* Eldridge 1968; Grauman 1968; Liu 1968; Mayer 1968; Moriyama 1968; Ruggles 1968; Sauvy 1968; Spengler 1968; Spulber 1968; Whitney 1968.]

Index numbers. Economic index numbers form an important part of government statistical programs [*see* INDEX NUMBERS]. The three articles on this topic discuss, respectively, theory, practical aspects of index numbers, and sampling problems.

Statistics as legal evidence. The use of statistical methods, and their results, in judicial proceedings has been growing in recent years. Trademark disputes have been illuminated by sample surveys; questions of paternity have been investigated probabilistically; depreciation and other accounting quantities that arise in quasi-judicial hearings have been estimated statistically. There are conflicts or apparent conflicts between statistical methods and legal concepts like those relating to hearsay evidence. [*See* STATISTICS AS LEGAL EVIDENCE.]

Statistical geography. Statistical geography, the use of statistical and other quantitative methods in geography, is a rapidly growing area [*see* GEOGRAPHY, STATISTICAL]. Somewhat related is the topic of rank–size, in which are studied—empirically and theoretically—patterns of relationship between, for example, the populations of cities and their rankings from most populous down. Another example is the relationship between the frequencies of words and their rankings from most frequent down. [*See* RANK–SIZE RELATIONS.]

Quantal response. Quantal response refers to a body of theory and method that might have been classed with counted data or under linear hypotheses with regression [*see* QUANTAL RESPONSE]. An example of a quantal response problem would be one in which students are given one week, two weeks, and so on, of training (say 100 different students for each training period), and then proportions of students passing a test are observed. Of interest might be that length of training leading to exactly 50 per cent passing. Many traditional psychophysical problems may be regarded from this viewpoint [*see* PSYCHOPHYSICS].

Queues. The study of queues has been of importance in recent years; it is sometimes considered part of operations research, but it may also be considered a branch of the study of stochastic processes [*see* QUEUES; OPERATIONS RESEARCH]. An example of queuing analysis is that of traffic flow at a street-crossing with a traffic light. The study has empirical, theoretical, and normative aspects.

Computation. Always intertwined with applied statistics, although distinct from it, has been computation [*see* COMPUTATION]. The recent advent of high-speed computers has produced a sequence of qualitative changes in the kind of computation that is practicable. This has had, and will continue to have, profound effects on statistics, not only as regards data handling and analysis, but also in theory, since many analytically intractable problems can now be attacked numerically by simulation on a high-speed computer [*see* SIMULATION].

Cybernetics. The currently fashionable term "cybernetics" is applied to a somewhat amorphous body of knowledge and research dealing with information processing and mechanisms, both living and nonliving (see Maron 1968; Stagner 1968). The notions of control and feedback are central, and the influence of the modern high-speed computer has been strong. Sometimes this area is taken to include communication theory and information theory (see Pollack 1968, which stresses applications to psychology).

WILLIAM H. KRUSKAL

BIBLIOGRAPHY

GENERAL ARTICLES

BOEHM, GEORGE A. W. 1964 The Science of Being Almost Certain. *Fortune* 69, no. 2:104–107, 142, 144, 146, 148.

KAC, MARK 1964 Probability. *Scientific American* 211, March: 92–108.

KENDALL, MAURICE G. 1950 The Statistical Approach. *Economica* New Series 17:127–145.

KRUSKAL, WILLIAM H. (1965) 1967 Statistics, Molière, and Henry Adams. *American Scientist* 55:416–428. → Previously published in Volume 9 of *Centennial Review.*

WEAVER, WARREN 1952 Statistics. *Scientific American* 186, Jan.:60–63.

INTRODUCTIONS TO PROBABILITY AND STATISTICS

BOREL, ÉMILE F. E. J. (1943) 1962 *Probabilities and Life.* Translated by M. Baudin. New York: Dover. → First published in French.

GNEDENKO, BORIS V.; and KHINCHIN, ALEKSANDR IA. (1945) 1962 *An Elementary Introduction to the Theory of Probability.* Translated from the 5th Russian edition, by Leo F. Boron, with the editorial collaboration of Sidney F. Mack. New York: Dover. → First published as *Elementarnoe vvedenie v teoriiu veroiatnostei.*

○MORONEY, M. J. (1951) 1958 *Facts From Figures.* 3d ed., rev. Harmondsworth (England) and Baltimore: Penguin. → Reprinted with minor revisions in 1967.

○MOSTELLER, FREDERICK; ROURKE, ROBERT E. K.; and THOMAS, GEORGE B. JR. (1961) 1970 *Probability With Statistical Applications.* 2d ed. Reading, Mass.: Addison-Wesley.

TIPPETT, L. H. C. (1943) 1956 *Statistics.* 2d ed. New York: Oxford Univ. Press.

WALLIS, W. ALLEN; and ROBERTS, HARRY V. 1962 *The Nature of Statistics.* New York: Collier. → Based on material presented in the authors' *Statistics: A New Approach* (1956).

WEAVER, WARREN 1963 *Lady Luck: The Theory of Probability.* Garden City, N.Y.: Doubleday.

YOUDEN, W. J. 1962 *Experimentation and Measurement.* New York: Scholastic Book Services.

ABSTRACTING JOURNALS

▶*Current Index to Statistics: Applications, Methods, and Theory.* → Published since 1975 under the joint sponsorship of the American Statistical Association and the Institute of Mathematical Statistics. Its initial editor is Brian L. Joiner.

Mathematical Reviews. → Published since 1940.

Psychological Abstracts. → Published since 1927. Covers parts of the statistical literature.

○*Quality Control and Applied Statistics: Abstract Service.* → Published from 1956 through 1975. Discontinued.

Referativnyi zhurnal: Matematika. → Published since 1953.

Statistical Theory and Method Abstracts. → Published since 1959.

Zentralblatt für Mathematik und ihre Grenzgebiete. → Published since 1931.

WORKS CITED IN THE TEXT

ADAMS, ERNEST W.; FAGOT, ROBERT F.; and ROBINSON, RICHARD E. 1965 A Theory of Appropriate Statistics. *Psychometrika* 30:99–127.

AURTHUR, ROBERT A. 1953 *The Glorification of Al Toolum.* New York: Rinehart.

►Becker, Howard S. 1968 Observation: I. Social Observation and Social Case Studies. Volume 11, pages 232–238 in *International Encyclopedia of the Social Sciences*. Edited by David L. Sills. New York: Macmillan and Free Press.

Born, Max (1949) 1951 *Natural Philosophy of Cause and Chance*. Oxford: Clarendon.

Bortkiewicz, Ladislaus von 1909 Die statistischen Generalisationen. *Scientia* 5:102–121. → A French translation appears in a supplement to Volume 5, pages 58–75.

Braithwaite, R. B. 1953 *Scientific Explanation: A Study of the Function of Theory, Probability and Law in Science*. Cambridge Univ. Press. → A paperback edition was published in 1960 by Harper.

Bridgman, Percy W. 1959 *The Way Things Are*. Cambridge, Mass.: Harvard Univ. Press.

Brown, G. Spencer. → See under Spencer Brown.

Churchman, C. West; and Ratoosh, Philburn (editors) 1959 *Measurement: Definitions and Theories*. New York: Wiley.

Cochran, William G.; Mosteller, Frederick; and Tukey, John W. 1954 *Statistical Problems of the Kinsey Report on Sexual Behavior in the Human Male*. Washington: American Statistical Association.

Coombs, Clyde H. 1964 *A Theory of Data*. New York: Wiley.

Cox, D. R. 1961 *The Role of Statistical Methods in Science and Technology*. London: Birkbeck College.

►Dahl, Robert A. 1968 Power. Volume 12, pages 405–415 in *International Encyclopedia of the Social Sciences*. Edited by David L. Sills. New York: Macmillan and Free Press.

Deming, W. Edwards 1965 Principles of Professional Statistical Practice. *Annals of Mathematical Statistics* 36:1883–1900.

►Eldridge, Hope T. 1968 Population: VII. Population Policies. Volume 12, pages 381–388 in *International Encyclopedia of the Social Sciences*. Edited by David L. Sills. New York: Macmillan and Free Press.

Galton, Francis 1889 *Natural Inheritance*. London and New York: Macmillan.

►Georgescu-Roegen, Nicholas 1968 Utility. Volume 16, pages 236–267 in *International Encyclopedia of the Social Sciences*. Edited by David L. Sills. New York: Macmillan and Free Press.

Gini, Corrado 1951 Caractère des plus récents développements de la méthodologie statistique. *Statistica* 11: 3–11.

Gini, Corrado 1959 Mathematics in Statistics. *Metron* 19, no. 3/4:1–9.

►Grauman, John V. 1968 Population: VI. Population Growth. Volume 12, pages 376–381 in *International Encyclopedia of the Social Sciences*. Edited by David L. Sills. New York: Macmillan and Free Press.

Hogben, Lancelot T. 1957 *Statistical Theory: The Relationship of Probability, Credibility and Error; An Examination of the Contemporary Crisis in Statistical Theory From a Behaviourist Viewpoint*. London: Allen & Unwin.

Huff, Darrell 1954 *How to Lie With Statistics*. New York: Norton.

Kadushin, Charles 1966 Shakespeare and Sociology. *Columbia University Forum* 9, no. 2:25–31.

Kish, Leslie 1959 Some Statistical Problems in Research Design. *American Sociological Review* 24:328–338.

Krutch, Joseph Wood 1963 Through Happiness With Slide Rule and Calipers. *Saturday Review* 46, no. 44:12–15.

Lazarsfeld, Paul F. 1961 Notes on the History of Quantification in Sociology: Trends, Sources and Problems. *Isis* 52, part 2:277–333. → Also included in Harry Woolf (editor), *Quantification*, published by Bobbs-Merrill in 1961.

►Liu, Ta-Chung 1968 Economic Data: III. Mainland China. Volume 4, pages 373–376 in *International Encyclopedia of the Social Sciences*. Edited by David L. Sills. New York: Macmillan and Free Press.

Lundberg, George A. 1940 Statistics in Modern Social Thought. Pages 110–140 in Harry E. Barnes, Howard Becker, and Frances B. Becker (editors), *Contemporary Social Theory*. New York: Appleton.

►Maron, M. E. 1968 Cybernetics. Volume 4, pages 3–6 in *International Encyclopedia of the Social Sciences*. Edited by David L. Sills. New York: Macmillan and Free Press.

►Mayer, Kurt B. 1968 Population: IV. Population Composition. Volume 12, pages 362–370 in *International Encyclopedia of the Social Sciences*. Edited by David L. Sills. New York: Macmillan and Free Press.

►Moriyama, Iwao M. 1968 Mortality. Volume 10, pages 498–504 in *International Encyclopedia of the Social Sciences*. Edited by David L. Sills. New York: Macmillan and Free Press.

Pearson, Karl (1892) 1957 *The Grammar of Science*. 3d ed., rev. & enl. New York: Meridian. → The first and second editions (1892 and 1900) contain material not in the third edition.

●Pfanzagl, J. (1959) 1971 *Theory of Measurement*. 2d ed. New York: International Publications Service. → First published as *Die axiomatischen Grundlagen einer allgemeinen Theorie des Messens*, Statistical Institute of the University of Vienna, New Series, No. 1.

►Pollack, Irwin 1968 Information Theory. Volume 7, pages 331–337 in *International Encyclopedia of the Social Sciences*. Edited by David L. Sills. New York: Macmillan and Free Press.

Popper, Karl R. (1935) 1959 *The Logic of Scientific Discovery*. New York: Basic Books; London: Hutchinson. → First published as *Logik der Forschung*. A paperback edition was published in 1961 by Harper.

►Ruggles, Richard 1968 Economic Data: I. General. Volume 4, pages 365–369 in *International Encyclopedia of the Social Sciences*. Edited by David L. Sills. New York: Macmillan and Free Press.

►Sauvy, Alfred 1968 Population: II. Population Theories. Volume 12, pages 349–358 in *International Encyclopedia of the Social Sciences*. Edited by David L. Sills. New York: Macmillan and Free Press.

►Schmeidler, Gertrude R. 1968 Parapsychology. Volume 11, pages 386–399 in *International Encyclopedia of the Social Sciences*. Edited by David L. Sills. New York: Macmillan and Free Press.

Selvin, Hanan C. 1957 A Critique of Tests of Significance in Survey Research. *American Sociological Review* 22:519–527. → See Volume 23, pages 85–86 and 199–200, for responses by David Gold and James M. Beshers.

Selvin, Hanan C.; and Stuart, Alan 1966 Data-dredging Procedures in Survey Analysis. *American Statistician* 20, no. 3:20–23.

SKINNER, B. F. 1956 A Case History in Scientific Method. *American Psychologist* 11:221–233.

SPENCER BROWN, G. 1957 *Probability and Scientific Inference*. London: Longmans. → The author's surname is Spencer Brown, but common library practice is to alphabetize his works under Brown.

►SPENGLER, JOSEPH J. 1968 Population: III. Optimum Population Theory. Volume 12, pages 358–362 in *International Encyclopedia of the Social Sciences*. Edited by David L. Sills. New York: Macmillan and Free Press.

►SPULBER, NICOLAS 1968 Economic Data: II. The Soviet Union and Eastern Europe. Volume 4, pages 370–372 in *International Encyclopedia of the Social Sciences*. Edited by David L. Sills. New York: Macmillan and Free Press.

►STAGNER, ROSS 1968 Homeostasis. Volume 6, pages 499–503 in *International Encyclopedia of the Social Sciences*. Edited by David L. Sills. New York: Macmillan and Free Press.

STEVENS, S. S. 1946 On the Theory of Scales of Measurement. *Science* 103:677–680.

SUPPES, PATRICK; and ZINNES, JOSEPH L. 1963 Basic Measurement Theory. Volume 1, pages 1–76 in R. Duncan Luce, Robert R. Bush, and Eugene Galanter (editors), *Handbook of Mathematical Psychology*. New York: Wiley.

TORGERSON, WARREN S. 1958 *Theory and Methods of Scaling*. New York: Wiley.

►TORGERSON, WARREN S. 1968 Scaling. Volume 14, pages 25–39 in *International Encyclopedia of the Social Sciences*. Edited by David L. Sills. New York: Macmillan and Free Press.

TUKEY, JOHN W. 1961 Statistical and Quantitative Methodology. Pages 84–136 in Donald P. Ray (editor), *Trends in Social Science*. New York: Philosophical Library.

TUKEY, JOHN W. 1962 The Future of Data Analysis. *Annals of Mathematical Statistics* 33:1–67, 812.

WALBERG, HERBERT J. 1966 When Are Statistics Appropriate? *Science* 154:330–332. → Follow-up letters by Julian C. Stanley, "Studies of Nonrandom Groups," and by Herbert J. Walberg, "Statistical Randomization in the Behavioral Sciences," were published in Volume 155, on page 953, and Volume 156, on page 314, respectively.

WALPOLE, HORACE (1778) 1904 [Letter] To the Hon. Henry Seymour Conway. Volume 10, pages 337–338 in Horace Walpole, *The Letters of Horace Walpole, Fourth Earl of Orford*. Edited by Paget Toynbee. Oxford: Clarendon Press.

WESTERGAARD, HARALD L. 1932 *Contributions to the History of Statistics*. London: King.

WHITE, COLIN 1964 Unkind Cuts at Statisticians. *American Statistician* 18, no. 5:15–17.

►WHITNEY, VINCENT H. 1968 Population: V. Population Distribution. Volume 12, pages 370–376 in *International Encyclopedia of the Social Sciences*. Edited by David L. Sills. New York: Macmillan and Free Press.

WOYTINSKY, W. S. 1954 Limits of Mathematics in Statistics. *American Statistician* 8, no. 1:6–10, 18.

Postscript

[1]A new and unusually deep approach to descriptive statistics has been taken by Bickel and Lehmann (1975). Their stress is on interpretation valid under broad conditions and on efficiency and robustness of estimation.

The new article on exploratory data analysis written for this encyclopedia describes an aggressive, venturesome kind of descriptive statistics that has been developed, in large part under the leadership of John W. Tukey. [See DATA ANALYSIS, EXPLORATORY; see also Tukey 1977. *Further comments about data analysis appear in the Introduction to the present work.*]

[2]Ambiguities in the plus-or-minus sign notation, and related questions of expressing uncertainty, are discussed by Eisenhart ([1963] 1968).

[3]In connection with case studies, broadly viewed, I cite Margaret Mead's presidential address (1976) to the American Association for the Advancement of Science. I would have cited Dukes (1965) in the main article had I known of it. Still earlier interesting treatments that I wish I had known before are by Redfield (1940), Wallin (1941), and Allport (1942). They speak to us across the decades.

[4]Two important new articles written for this encyclopedia deal with factor analysis and principal components and with multidimensional scaling [see FACTOR ANALYSIS AND PRINCIPAL COMPONENTS, *article on* BILINEAR METHODS; SCALING, MULTIDIMENSIONAL].

[5]Humanistic criticisms continue to be easily found, as in Richard Gilman's review (1975, p. 2) of Saul Bellow's novel *Humboldt's Gift*: "This is not quite the culmination of the book. If it were, its blandness and superficiality, its statistician's wisdom, would do more damage to Bellow than he in fact allows." Another example appeared in a speech (1974) by His Eminence Franziskus Cardinal König:

With statistics one may, however, prove anything and everything—including the contrary. And sociology, it seems, is a perfect solvent. The inroads made by sociology into the sphere of theology have produced but a new variety of the old belief in miracles: The naive belief in the infallibility of questionnaires, of representative surveys and of analyses of cross-sections of public opinion. This encroachment of a sociology not inhibited by any awareness of its limitations has thus had deplorable effects.

There is also a moving version of the humanistic criticism in Elizabeth Barrett Browning's *Aurora Leigh* (1856):

A red-haired child
Sick in a fever, if you touch him once,
Though but so little as with a finger-tip,
Will set you weeping; but a million sick . . .
You could as soon weep for the rule of three
Or compound fractions.

[6]Criticisms of overconcern with averages also abound. An almost-novel (1975) by Paul Wilkes deals with a year in the life of an American family that is almost average in a number of Census-like ways. Needless to say, the family is far from average in many other ways. The canard about averages appears without restraint in a novel (1968, p. 80) by Desmond Bagley: "He was average in every way; not too tall, not too short, not too beefy and not too scrawny. He wore an average suit and looked the perfect average man. He might have been designed by a statistician. He had a more than average brain —but that didn't show."

[7]An elaboration of this paragraph is given in Kruskal (1973*b*).

[8]A variant criticism of obfuscation appeared in a 1976 newspaper column of medical advice by Dr. Robert Mcndelsohn. In response to an inquiry about an article read by a correspondent, Dr. Mendelsohn wrote (I paraphrase for brevity):

> It is statistically possible to prove just about any premise, and its opposite. When you see a study whose conclusions you find impossible to believe, recall the following story.
>
> A well-known statistician traveled frequently, but by train instead of by plane because statistical analysis had revealed a significant chance that a bomb might be on a plane. One day a friend was amazed to see the statistician boarding a plane, and asked him about it. The statistician replied that he had recalculated to find the probability of *two* bombs being on any one plane. "It's so small that it's not worth considering. So I'm carrying my own bomb."

The use of this example as a criticism is particularly ironic, because the two-bomb story is an old favorite used by teachers of statistics to illustrate a *fallacious* argument based on a misconception about conditional probability.

[9]The ncw article on statistics and public policy, written specially for this encyclopedia, and the updated article on government statistics together present a comprehensive picture of the relationships between statistics and a variety of policy problems and issues, both in and out of government. [*See* GOVERNMENT STATISTICS; PUBLIC POLICY AND STATISTICS.] The 1971 report of the U.S. President's Commission on Federal Statistics should also be cited, as should three important books, two on experimental evaluation of social innovations and the third a reader dealing with statistical aspects of public policy. *Social Experimentation* (1974), edited by Riecken and Boruch; *Experimental Testing of Public Policy* (Social Science Research Council Conference . . . 1976), edited by Boruch and Riecken; and *Statistics and Public Policy* (1977) by Fairley and Mosteller. [*Further comments appear in the Introduction to the present work.*]

Remarks on references. A supplementary survey of general publications in probability and statistics would be long and unwieldy; instead I list a selection of articles, textbooks, and monographs that I know and that seem to me of special interest.

Useful general articles, or collections of such articles, are Bartlett (1976), Kendall (1966; 1972), and Kruskal (1973*a*; 1974). Two volumes of interesting reprints are Steger (1971) and Lieberman (1971).

Statistics: A Guide to the Unknown (1972), edited by Tanur et al., is an important and unusual introduction to statistics. Another unusual introductory work is *Statistics by Example* (1973), prepared and edited by the Joint Committee on the Curriculum in Statistics and Probability of the American Statistical Association and the National Council of Teachers of Mathematics.

Among textbooks, I call particular attention to Cox and Hinkley (1974), Hodges and Lehman ([1965] 1970), Mosteller and Tukey (1977), and Tukey (1977).

In the main article I should have cited Morgenstern ([1950] 1963) as a stimulating study of error structure in economics. It is also important to cite the standard encyclopedic work by Kendall and Stuart ([1958–1966] 1973–1977).

There has been a great deal of bibliographic progress in statistics since publication of the main article. A partial list of the more important publications appears as the second section of the bibliography that follows.

WILLIAM H. KRUSKAL

ADDITIONAL BIBLIOGRAPHY

WORKS CITED IN THE POSTSCRIPT

ALLPORT, GORDON W. 1942 *The Use of Personal Documents in Psychological Science.* New York: Social Science Research Council.

BAGLEY, DESMOND 1968 *The Viveso Letter.* London: Collins; Garden City, N.Y.: Doubleday. → A paperback edition was published in 1970 by Fontana.

BARTLETT, M. S. 1976 *Probability, Statistics and Time: A Collection of Essays.* London: Chapman & Hall; New York: Wiley.

BICKEL, P. J.; and LEHMANN, E. L. 1975 Descriptive Statistics for Nonparametric Models. Parts 1 and 2. *Annals of Statistics* 3:1038–1044, 1045–1069. → Part 1: Introduction. Part 2: Location.

COX, D. R.; and HINKLEY, D. V. 1974 *Theoretical Statistics.* London: Chapman & Hall; New York: Wiley.

DUKES, WILLIAM F. 1965 $N = 1$. *Psychological Bulletin* 64:74–79. → Reprinted in Steger (1971).

EISENHART, CHURCHILL (1963) 1968 Expression of the Uncertainties of Final Results. *Science* 160:1201–1204. → First published as Chapter 23 in *Experimental Statistics,* National Bureau of Standards Handbook No. 91.

FAIRLEY, WILLIAM B.; and MOSTELLER, FREDERICK 1977 *Statistics and Public Policy.* Reading, Mass.: Addison-Wesley.

GILMAN, RICHARD 1975 Review of *Humboldt's Gift,* by Saul Bellow. *New York Times Book Review* Aug. 17, pp. 1–3.

HODGES, J. L. JR.; and LEHMANN, E. L. (1965) 1970 *Elements of Finite Probability.* 2d ed. San Francisco: Holden-Day.

JOINT COMMITTEE ON THE CURRICULUM IN STATISTICS AND PROBABILITY OF THE AMERICAN STATISTICAL ASSOCIATION AND THE NATIONAL COUNCIL OF TEACHERS OF MATHEMATICS 1973 *Statistics by Example.* 4 vols. Reading, Mass.: Addison-Wesley. → Volume 1: *Exploring Data.* Volume 2: *Weighing Chances.* Volume 3: *Detecting Patterns.* Volume 4: *Finding Models.* Prepared and edited by a committee chaired by Frederick Mosteller and including William H. Kruskal, Richard F. Link, Richard S. Pieters, and Gerald R. Rising.

KENDALL, MAURICE G. 1966 Statistical Inference in the Light of the Theory of the Electronic Computer. *Review of the International Statistical Institute* 34:1–12.

KENDALL, MAURICE G. 1972 Measurement in the Study of Society. Pages 133–147 in William A. Robson (editor), *Man and the Social Sciences.* London: Allen & Unwin; Beverly Hills, Calif.: Sage.

KENDALL, MAURICE G.; and STUART, ALAN (1958–1966) 1973–1977 *The Advanced Theory of Statistics.* 3 vols. London: Griffin; New York: Macmillan. → Volume 1: *Distribution Theory,* 4th ed., 1977. Volume 2: *Inference and Relationship,* 3d ed., 1973. Volume 3: *Design and Analysis, and Time-series,* 3d ed., 1976. Early editions of Volumes 1 and 2, first published in 1943 and 1946, were written by Kendall alone. Stuart became a joint author on later, renumbered editions in the three-volume set.

KÖNIG, FRANZ 1974 The Future of Religion. *New York Times* Dec. 21, p. 27, col. 2. → Excerpts from a speech at the University of Chicago.

KRUSKAL, WILLIAM H. 1973*a* The Committee on National Statistics. *Science* 180:1256–1258.

KRUSKAL, WILLIAM H. 1973*b* Babies and Averages. Volume 1, pages 49–59 in Joint Committee on the Curriculum in Statistics and Probability of the American Statistical Association and the National Council of Teachers of Mathematics, *Statistics by Example.* Volume 1: *Exploring Data.* Reading, Mass.: Addison-Wesley.

KRUSKAL, WILLIAM H. 1974 The Ubiquity of Statistics. *American Statistician* 28, no. 1:3–6.

LIEBERMAN, BERNHARDT 1971 *Contemporary Problems in Statistics: A Book of Readings for the Behavioral Sciences.* New York: Oxford Univ. Press.

MEAD, MARGARET 1976 Towards a Human Science. *Science* 191:903–909.

MENDELSOHN, ROBERT 1976 A Numbers Game Anyone Can Win. *Chicago Daily News* May 22–23, p. 18, col. 5.

MORGENSTERN, OSKAR (1950) 1963 *On the Accuracy of Economic Observations.* 2d ed., completely rev. Princeton Univ. Press. → Also published in German (Physica-Verlag, 1965), Japanese (Hosei Univ. Press, 1968), Russian (Statistika, 1968), and French (Dunod, 1972).

MOSTELLER, FREDERICK et al. (editors) 1973 *Statistics by Example.* 4 vols. → See under Joint Committee on the Curriculum in Statistics and Probability of the American Statistical Association and the National Council of Teachers of Mathematics.

MOSTELLER, FREDERICK; and TUKEY, JOHN W. 1977 *Data Analysis and Regression: A Second Course in Statistics.* Reading, Mass.: Addison-Wesley.

REDFIELD, ROBERT 1940 The Folk Society and Culture. Pages 39–50 in Louis Wirth (editor), *Eleven Twenty-six: A Decade of Social Science Research.* Univ. of Chicago Press. → See especially pages 47–48.

RIECKEN, HENRY W.; and BORUCH, ROBERT H. (editors) 1974 *Social Experimentation: A Method for Planning and Evaluating Social Intervention.* New York: Academic Press.

SOCIAL SCIENCE RESEARCH COUNCIL CONFERENCE ON SOCIAL EXPERIMENTS, BOULDER, 1974 1976 *Experimental Testing of Public Policy: Proceedings.* Edited by Robert F. Boruch and Henry W. Riecken. Boulder, Colo.: Westview.

STEGER, JOSEPH A. (editor) 1971 *Readings in Statistics for the Behavioral Scientist.* New York: Holt.

TANUR, JUDITH M. et al. (editors) 1972 *Statistics: A Guide to the Unknown.* San Francisco: Holden-Day. → Joint editors were the members of the Joint Committee on the Curriculum in Statistics and Probability of the American Statistical Association and the National Council of Teachers of Mathematics.

TUKEY, JOHN W. (1970) 1977 *Exploratory Data Analysis.* Reading, Mass.: Addison-Wesley. → First published in a mimeographed "limited preliminary edition."

U.S. PRESIDENT'S COMMISSION ON FEDERAL STATISTICS 1971 *Federal Statistics: Report of the President's Commission.* 2 vols. Washington: Government Printing Office.

WALLIN, PAUL 1941 The Prediction of Individual Behavior From Case Studies. Pages 181–249 in Social Science Research Council, Committee on Social Adjustment, *The Prediction of Personal Adjustment.* Edited by Paul Horst. New York: The Council. → A footnote on page 215 says that Samuel A. Stouffer wrote pages 240–249.

WILKES, PAUL 1975 *Trying Out the Dream: A Year in the Life of an American Family.* Philadelphia: Lippincott.

WORKS OF BIBLIOGRAPHIC INTEREST

DOLBY, JAMES L.; and TUKEY, JOHN W. 1973 *The Statistics Cum Index.* Los Altos, Calif.: R & D Press.

GANI, JOSEPH M. 1970 On Coping With New Information in Probability and Statistics. *Journal of the Royal Statistical Society* Series A 133:442–450.

GANI, JOSEPH M. 1972 Some Comments on the Bibliography of Probability and Statistics. *International Statistical Review* 40:201–207.

KENDALL, MAURICE G.; and DOIG, ALISON G. 1962 *Bibliography of Statistical Literature, 1950–1958.* 2 vols. Edinburgh: Oliver & Boyd; New York: Hafner. → A companion volume covering 1940–1949 was published in 1965, and a pre-1940 volume (including supplements to the prior volumes) was published in 1968.

LANCASTER, H. O. 1968 *Bibliography of Statistical Bibliographies.* Edinburgh: Oliver & Boyd. → Supplements have appeared in *Review of the International Statistical Institute* (later titled *International Statistical Review*) 37 [1969]:57–67; 38 [1970]:258–267; 39 [1971]: 64–73; 40 [1972]:73–81; 41 [1973]:375–379; 42 [1974]:67–70; 43 [1975]:345–349.

LANCASTER, H. O. 1970 Problems in the Bibliography of Statistics. *Journal of the Royal Statistical Society* Series A 133:409–441.

MORAN, P. A. P. 1974 How to Find Out in Statistical and Probability Theory. *International Statistical Review* 42:299–303.

ROSS, IAN C.; and TUKEY, JOHN W. 1974 *Index to Statistics and Probability: Locations and Authors.* Los Altos, Calif.: R & D Press.

ROSS, IAN C.; and TUKEY, JOHN W. 1975 *Index to Statistics and Probability: Permuted Titles.* 2 vols. Los Altos, Calif.: R & D Press.

Science Citation Index (SCI). → Published since 1961 by the Institute of Scientific Information, Philadelphia.

Social Science Citation Index (SSCI). → Published since 1972 by the Institute of Scientific Information, Philadelphia.

TUKEY, JOHN W. et al. 1973 *Index to Statistics and Probability: The Citation Index.* Los Altos, Calif.: R & D Press.

II
THE HISTORY OF STATISTICAL METHOD

The broad river of thought that today is known as theoretical statistics cannot be traced back to a single source springing identifiably from the rock. Rather is it the confluence, over two centuries, of a number of tributary streams from many different regions. Probability theory originated at the gaming table; the collection of statistical facts began with state requirements of soldiers and money; marine insurance began with the wrecks and piracy of the ancient Mediterranean; modern studies of mortality have their roots in the plague pits of the seventeenth century; the theory of errors was created in astronomy, the theory of correlation in biology, the theory of experimental design in agriculture, the theory of time series in economics and meteorology, the theories of component analysis and ranking in psychology, and the theory of chi-square methods in sociology. In retrospect it almost seems as if every phase of human life and every science has contributed something of importance to the subject. Its history is accordingly the more interesting, but the more difficult, to write.

Early history

Up to about 1850 the word "statistics" was used in quite a different sense from the present one. It meant information about political states, the kind of material that is nowadays to be found assembled in the *Statesman's Year-book.* Such information was usually, although not necessarily, numerical, and, as it increased in quantity and scope, developed into tabular form. By a natural transfer of meaning, "statistics" came to mean any numerical material that arose in observation of the external world. At the end of the nineteenth century this usage was accepted. Before that time, there were, of course, many problems in statistical methodology considered under other names; but the recognition of their common elements as part of a science of statistics was of relatively late occurrence. The modern theory of statistics (an expression much to be preferred to "mathematical statistics") is the theory of numerical information of almost every kind.

The characteristic feature of such numerical material is that it derives from a set of objects, technically known as a "population," and that any particular variable under measurement has a distribution of frequencies over the members of the set. The height of man, for example, is not identical for every individual but varies from man to man. Nevertheless, we find that the frequency distribution of heights of men in a given population has a definite pattern that can be expressed by a relatively simple mathematical formula. Often the "population" may be conceptual but nonexistent, as for instance when we consider the possible tosses of a penny or the possible measurements that may be made of the transit times of a star. This concept of a distribution of measurements, rather than a single measurement, is fundamental to the whole subject. In consequence, points of statistical interest concern the properties of aggregates, rather than of individuals; and the elementary parts of theoretical statistics are much concerned with summarizing these properties in such measures as averages, index numbers, dispersion measures, and so forth.

The simpler facts concerning aggregates of measurements must, of course, have been known almost from the moment when measurements began to be made. The idea of regularity in the patterning of discrete repeatable chance events, such as dice throwing, emerged relatively early and is found explicitly in Galileo's work. The notion that measurements on natural phenomena should exhibit similar regularities, which are mathematically expressible, seems to have originated in astronomy, in connection with measurements on star transits. After some early false starts it became known that observations of a magnitude were subject to error even when the observer was trained and unbiased. Various hypotheses about the pattern of such errors were propounded. Simpson (1757) was the first to consider a *continuous* distribution, that is to say, a distribution of a variable that could take any values in a continuous range. By the end of the eighteenth century Laplace and Gauss had considered several such mathematically specified distributions and, in particular, had discovered the most famous of them all, the so-called normal distribution [see DISTRIBUTIONS, STATISTICAL, *article on* SPECIAL CONTINUOUS DISTRIBUTIONS].

In these studies there was assumed to be a "true" value underlying the distribution. Departures from this true value were "errors." They were, so to speak, extraneous to the object of the study, which was to estimate this true value. Early in the nineteenth century a major step forward was taken with the recognition (especially by Quetelet) that living material also exhibited frequency distributions of definite pattern. Furthermore, Galton and Karl Pearson, from about 1880, showed that these distributions were often skew or asymmetrical, in the sense that the shape of the frequency curve for values above the mean was not the mirror image of the curve for values below the mean. In particular it became impossible to maintain that the deviations from the mean were "errors" or that there existed a "true" value; the frequency distribution itself was to be recognized as a fundamental property of the aggregate. Immediately, similar patterns of regularity were brought to light in nearly every branch of science—genetics, biology, meteorology, economics, sociology—and even in some of the arts: distributions of weak verse endings were used to date Shakespeare's plays, and the distribution of words has been used to discuss cases of disputed authorship.

Nowadays the concept of frequency distribution is closely bound up with the notion of probability distribution. Some writers of the twentieth century treat the two things as practically synonymous. Historically, however, the two were not always identified and to some extent pursued independent courses for centuries before coming together. We must go back several millenniums if we wish to trace the concept of probability to its source.

From very ancient times man gambled with primitive instruments, such as astragali and dice, and also used chance mechanisms for divinatory purposes. Rather surprisingly, it does not seem that the Greeks, Romans, or the nations of medieval Europe arrived at any clear notion of the laws of chance. Elementary combinatorics appears to have been known to the Arabs and to Renaissance mathematicians, but as a branch of algebra rather than in a probabilistic context. Nevertheless, chance itself was familiar enough, especially in gambling, which was widespread in spite of constant discouragement from church and state. Some primitive ideas of relative frequency of occurrence can hardly have failed to emerge, but a doctrine of chances was extraordinarily late in coming. The first record we have of anything remotely resembling the modern idea of calculating chances occurs in a fifteenth-century poem called *De vetula*. The famous mathematician and physicist Geronimo Cardano was the first to leave a manuscript in which the concept of laws of chance was explicitly set out (Ore 1953). Galileo left a fragment that shows that he clearly understood the method of calculating chances at dice. Not until the work of Huygens (1657), the correspondence between Pascal and Fermat, and the work of Jacques Bernoulli (1713) do we find the beginnings of a calculus of probability.

This remarkable delay in the mathematical formulation of regularity in events that had been observed by gamblers over thousands of years is probably to be explained by the philosophical and religious ideas of the times, at least in the Western world. To the ancients, events were mysterious; they could be influenced by superhuman beings but no being was in control of the universe. On the other hand, to the Christians everything occurred under the will of God, and in a sense there was no chance; it was almost impious to suppose that events happened under the blind laws of probability. Whatever the explanation may be, it was not until Europe had freed itself from the dogma of the medieval theologian that a calculus of probability became possible.

Once the theory of probability had been founded, it developed with great rapidity. Only a hundred years separates the two greatest works in this branch of the subject, Bernoulli's *Ars conjectandi* (1713) and Laplace's *Théorie analytique des probabilités* (1812). Bernoulli exemplified his work mainly in terms of games of chance, and subsequent mathematical work followed the same line. Montmort's work was concerned entirely with gaming, and de Moivre stated most of his results in similar terms, although actuarial applications were always present in his mind (see Todhunter [1865] 1949, pp. 78–134 for Montmort and pp. 135–193 for de Moivre). With Laplace, Condorcet, and rather later, Poisson, we begin to find probabilistic ideas applied to practical problems; for example, Laplace discussed the plausibility of the nebular hypothesis of the solar system in terms of the probability of the planetary orbits lying as nearly in a plane as they do. Condorcet (1785) was concerned with the probability of reaching decisions under various systems of voting, and Poisson (1837) was specifically concerned with the probability of reaching correct conclusions from imperfect evidence. A famous essay of Thomas Bayes (1764) broke new ground by its consideration of probability in inductive reasoning, that is to say, the use of the probabilities of observed events to compare the plausibility of hypotheses that could explain them [see BAYESIAN INFERENCE].

The linkage between classical probability theory and statistics (in the sense of the science of regularity in aggregates of natural phenomena) did not take place at any identifiable point of time. It occurred somewhere along a road with clearly traceable lines of progress but no monumental milestones. The critical point, however, must have been the realization that probabilities were not always to be calculated a priori, as in games of chance, but were measurable constants of the external world. In classical probability theory the probabilities of primitive events were always specified on prior grounds: dice were "fair" in the sense that each side had an equal chance of falling uppermost, cards were shuffled and dealt "at random," and so on. A good deal of probability theory was concerned with the pure mathematics of deriving the probabilities of complicated contingent events from these more primitive events whose probabilities were known. However, when sampling from an observed frequency distribution, the basic probabilities are not known but are parameters to be estimated. It took some time, perhaps fifty years, for the implications of this notion to be fully realized. Once it was, statistics embraced probability and the subject was poised for the immense development that has occurred over the past century.

Once more, however, we must go back to another contributory subject—insurance, and particularly life insurance. Although some mathematicians, notably Edmund Halley, Abraham de Moivre, and Daniel Bernoulli, made important contributions to demography and insurance studies, for the most part actuarial science pursued a course of its own. The founders of the subject were John Graunt and William Petty. Graunt, spurred on by the information contained in the bills of mortality prepared in connection with the great plague (which hit England in 1665), was the first to reason about demographic material in a modern statistical way. Considering the limitations of his data, his work was a beautiful piece of reasoning. Before long, life tables were under construction and formed the basis of the somewhat intricate calculations of the modern actuary [see LIFE TABLES]. In the middle of the eighteenth century, some nations of the Western world began to take systematic censuses of population and to record causes of mortality, an example that was soon followed by all [see CENSUS; VITAL STATISTICS]. Life insurance became an exact science. It not only contributed an observable frequency distribution with a clearly defined associated calculus; it also contributed an idea that was to grow into a dynamic theory of probability—the concept of a population moving through time in an evolutionary way. Here and there, too, we find demographic material stimulating statistical studies, for example, in the study of the mysteries of the sex ratio of human births.

Modern history

1890–1940. If we have to choose a date at which the modern theory of statistics began, we may put it, somewhat arbitrarily, at 1890. Francis Galton was then 68 but still had twenty years of productive life before him. A professor of economics named Francis Ysidro Edgeworth (then age 45) was calling attention to statistical regularities in election results, Greek verse, and the mating of bees and was about to propound a remarkable generalization of the law of error. A young man named Karl Pearson (age 35) had just been joined by the biologist Walter Weldon at University College, London, and was meditating the lectures that ultimately became *The Grammar of Science*. A student named George Udny Yule, at the age of 20, had caught Pearson's eye. And in that year was born the greatest of them all, Ronald Aylmer Fisher. For the next forty years, notwithstanding Russian work in probability theory—notably the work of Andrei Markov and Aleksandr Chuprov—developments in theoretical statistics were predominantly English. At that point, there was something akin to an intellectual explosion in the United States and India. France was already pursuing an individual line in probability theory under the inspiration of Émile Borel and Paul Lévy, and Italy, under the influence of Corrado Gini, was also developing independently. But at the close of World War II the subject transcended all national boundaries and had become one of the accepted disciplines of the scientific, technological, and industrial worlds.

The world of 1890, with its futile power politics, its class struggles, its imperialism, and its primitive educational system, is far away. But it is still possible to recapture the intellectual excitement with which science began to extend its domain into humanitarian subjects. Life was as mysterious as ever, but it was found to obey laws. Human society was seen as subject to statistical inquiry, as an evolutionary entity under human control. It was no accident that Galton founded the science of eugenics and Karl Pearson took a militant part in some of the social conflicts of his time. Statistical science to them was a new instrument for the exploration of the living world, and the behavioral sciences at last showed signs of structure that would admit of mathematical analysis.

In London, Pearson and Weldon soon began to exhibit frequency distributions in all kinds of fields. Carl Charlier in Sweden, Jacobus Kapteyn and

Johan van Uven in Holland, and Vilfredo Pareto in Italy, to mention only a few, contributed results from many different sciences. Pearson developed his system of mathematical curves to fit these observations, and Edgeworth and Charlier began to consider systems based on the sum of terms in a series analogous to a Taylor expansion. It was found that the normal curve did not fit most observed distributions but that it was a fair approximation to many of them.

Relationships between variables. About 1890, Pearson, stimulated by some work of Galton, began to investigate bivariate distributions, that is to say, the distribution in a two-way table of frequencies of members, each of which bore a value of two variables. The patterns, especially in the biological field where data were most plentiful, were equally typical. In much observed material there were relationships between variables, but they were not of a mathematically functional form. The length and breadth of oak leaves, for example, were dependent in the sense that a high value of one tended to occur with a high value of the other. But there was no formula expressing this relationship in the familiar deterministic language of physics. There had to be developed a new kind of relationship to describe this type of connection. In the theory of attributes this led to measures of *association* and *contingency* [see STATISTICS, DESCRIPTIVE]; in the theory of variables it led to *correlation* and *regression* [see LINEAR HYPOTHESES; MULTIVARIATE ANALYSIS, *article on* CORRELATION METHODS].

The theory of statistical relationship, and especially of regression, has been studied continuously and intensively ever since. Most writers on statistics have made contributions at one time or another. The work was still going strong in the middle of the twentieth century. Earlier writers, such as Pearson and Yule, were largely concerned with linear regression, in which the value of one variable is expressed as a linear function of the others plus a random term. Later authors extended the theory to cover several dependent variables and curvilinear cases; and Fisher in particular was instrumental in emphasizing the importance of rendering the explanatory variables independent, so far as possible.

Sampling. It was not long before statisticians were brought up against a problem that is still, in one form or another, basic to most of their work. In the majority of cases the data with which they were presented were only samples from a larger population. The problems then arose as to how reliable the samples were, how to estimate from

them values of parameters describing the parent population, and, in general, what kinds of inference could be based on them.

Some intuitive ideas on the subject occur as far back as the eighteenth century; but the sampling problem, and the possibility of treating it with mathematical precision, was not fully appreciated until the twentieth century.

Classical error theory, especially the work of Carl Friedrich Gauss in the first half of the nineteenth century, had considered sampling distributions of a simple kind. For example, the chi-square distribution arose in 1875 when the German geodesist Friedrich Helmert worked out the distribution of sample variance for sampling from a normal population. The same chi-square distribution was independently rediscovered in 1900 by Karl Pearson in a quite different context, that of testing distributional goodness of fit [see COUNTED DATA; GOODNESS OF FIT]. In another direction, Pearson developed a wide range of asymptotic formulas for standard errors of sample quantities. The mathematics of many so-called small-sample distribution problems presented difficulties with which Pearson was unable to cope, despite valiant attempts. William Gosset, a student of Pearson's, produced in 1908 one of the most important statistical distributions under the pseudonym of "Student"; and this distribution, arising from a basic small sample problem, is known as that of Student's *t* [see DISTRIBUTIONS, STATISTICAL].

It was Student and R. A. Fisher (beginning in 1913) who inaugurated a new era in the study of sampling distributions. Fisher himself made major contributions to the subject over the ensuing thirty years. In rapid succession he found the distribution, in samples from a normal population, of the correlation coefficient, regression coefficients, multiple correlation coefficients, and the ratio of variances known as *F*. Other writers, notably John Wishart in England, Harold Hotelling and Samuel Wilks in the United States, and S. N. Roy and R. C. Bose in India, added a large number of new results, especially in the field of multivariate analysis. More recently, T. W. Anderson has advanced somewhat farther the frontiers of knowledge in this rather difficult mathematical field.

Concurrently with these spectacular mathematical successes in the derivation of sampling distributions, methods were also devised for obtaining approximations. Again R. A. Fisher was in the lead with a paper (1928) introducing the so-called *k*-statistics, functions of sample values that have simplifying mathematical properties.

The question whether a sampling method is random is a subtle one. It does not always trouble an experimental scientist, when he can select his material by a chance mechanism. However, sometimes the data are provided by nature, and whether they are a random selection from the available population is difficult to determine. In the sampling of human beings difficulties are accentuated by the fact that people may react to the sampling process. As sampling methods spread to the social sciences, the problems of obtaining valid samples at low cost from a wide geographical scatter of human beings became increasingly important, and some new problems of respondent bias arose. In consequence, the sampling of humans for social inquiry has almost developed into a separate subject, dependent partly on psychological matters, such as how questions should be framed to avoid bias, and partly on expense. By 1960 sampling errors in social surveys were well under control; but many problems remained for exploration, notably those of drawing samples of individuals with relatively rare and specialized characteristics, such as retail pharmacists or sufferers from lung cancer. Designing a sample was accepted as just as much a matter of expertise as designing a house [see INTERVIEWING IN SOCIAL RESEARCH; SAMPLE SURVEYS; SURVEY ANALYSIS].

The control of the sample and the derivation of sampling distributions were, of course, only means to an end, which was the drawing of accurate inferences from the sample that ultimately resulted. We shall say more about the general question of inference below, but it is convenient to notice here the emergence, between 1925 and 1935, of two branches of the subject: the theory of estimation, under the inspiration of Fisher, and the theory of hypothesis testing, under the inspiration of Karl Pearson's son Egon and Jerzy Neyman [see ESTIMATION; HYPOTHESIS TESTING].

Estimation. Up to 1914 (which, owing to World War I, actually means up to 1920), the then current ideas on estimation from a sample were intuitive and far from clear. For the most part, an estimate was constructed from a sample as though it were being constructed for a population (for example, the sample mean was an "obvious" estimate of the parent population mean). A few writers—Daniel Bernoulli, Laplace, Gauss, Markov, and Edgeworth—had considered the problem, asked the right questions, and sometimes found partial answers. Ideas on the subject were clarified and extended in a notable paper by Fisher (1925). He introduced the concepts of optimal estimators and of efficiency in estimation, and emphasized the importance of the so-called method

of maximum likelihood as providing a very general technique for obtaining "best" estimators. These ideas were propounded to a world that was just about ripe for them, and the theory of estimation developed at a remarkable rate in the ensuing decades.

The related problem of gauging the reliability of an estimate, that is, of surrounding it with a band of error (which has associated with it a designated probability) led to two very different lines of development, the confidence intervals of Egon Pearson and Neyman and the "fiducial intervals" of Fisher, both originating between 1925 and 1930 [see ESTIMATION, *article on* CONFIDENCE INTERVALS AND REGIONS; FIDUCIAL INFERENCE]. The two proceeded fairly amiably side by side for a few years, and at the time it seemed that they were equivalent; they certainly led to the same results in simpler cases. However, it became clear about 1935 that they were conceptually very different, and a great deal of argument developed which had not been resolved even at the time of Fisher's death in 1962. Fortunately the controversy, although embittered, did not impede progress. (Omitted at this point is any discussion of Bayesian methods, which may lead to intervals resembling superficially those of confidence and fiducial approaches; Bayesian methods are mentioned briefly below.) [See BAYESIAN INFERENCE *for a detailed discussion.*]

Hypothesis testing. In a like manner, the work of Neyman and Pearson (beginning in 1928) on the theory of statistical tests gave a very necessary clarity to procedures that had hitherto been vague and unsatisfactory. In probabilistic terms the older type of inference had been of this type: If a certain hypothesis were true, the probability that I should observe the actual sample that I have drawn, or one more extreme, is very small; therefore the hypothesis is probably untrue. Neyman and Pearson pointed out that a hypothesis could not be tested *in vacuo* but only in comparison with other hypotheses. They set up a theory of tests and—as in the case of estimation, with which this subject is intimately linked—discussed power, relative efficiency, and optimality of tests. Here also there was some controversy, but for the most part the Neyman–Pearson theory was generally accepted and had become standard practice by 1950.

Experimental design and analysis. Concurrently with developments in sampling theory, estimation, and hypothesis testing, there was growing rapidly, between 1920 and 1940, a theory of experimental design based again on the work of Fisher. Very early in his career, it had become clear to him that in multivariate situations the "explanation" of one variable in terms of a set of dependent or explanatory variables was rendered difficult, if not impossible, where correlations existed among the explanatory variables themselves; for it then became impossible to say how much of an effect was attributable to a particular cause. This difficulty, which still bedevils the general theory of regression, could be overcome if the explanatory variables could be rendered statistically independent. (This, incidentally, was the genesis of the use of orthogonal polynomials in curvilinear regression analysis.) Fisher recognized that in experimental situations where the design of the experiment was, within limits, at choice, it could be arranged that the effects of different factors were "orthogonal," that is, independent, so that they could be disentangled. From this notion, coupled with probabilistic interpretations of significance and the necessary mathematical tests, he built up a most remarkable system of experimental design. The new methods were tested at the Rothamsted Experimental Station in England but were rapidly spread by an active and able group of disciples into all scientific fields.

Some earlier work, particularly by Wilhelm Lexis in Germany at the close of the nineteenth century, had called attention to the fact that in sampling from nonhomogeneous populations the formulas of classical probability were a poor representation of the observed effects. This led to attempts to split the sampling variation into components; one, for example, representing the inevitable fluctuation of sampling, another representing the differences between the sections or subpopulations from which members were drawn. In Fisher's hands these ideas were extended and given precision in what is known as the analysis of variance, one of the most powerful tools of modern statistics. The methods were later extended to cover the simultaneous variation of several variables in the analysis of covariance. [*See* LINEAR HYPOTHESES, *article on* ANALYSIS OF VARIANCE.]

It may be remarked, incidentally, that the problems brought up by these various developments in theoretical statistics have proved an immense challenge to mathematicians. Many branches of abstract mathematics—invariants, symmetric functions, groups, finite geometries, *n*-dimensional geometry, as well as the whole field of analysis—have been brought effectively into play in solving practical problems. After World War II the advent of the electronic computer was a vital adjunct to the solution of problems where even the resources of modern mathematics failed. Sampling experi-

ments became possible on a scale never dreamed of before.

Recent developments. So much occurred in the statistical domain between 1920 and 1940 that it is not easy to give a clear account of the various currents of development. We may, however, pause at 1940 to look backward. In Europe, and to a smaller extent in the United States, World War II provided an interregnum, during which much was absorbed and a good deal of practical work was done, but, of necessity, theoretical developments had to wait, at least as far as publication was concerned. The theory of statistical distributions and of statistical relationship had been firmly established by 1940. In sampling theory many mathematical problems had been solved, and methods of approach to outstanding problems had been devised. The groundwork of experimental design had been firmly laid. The basic problems of inference had been explicitly set out and solutions reached over a fairly wide area. What is equally important for the development of the subject, there was about to occur a phenomenal increase in the number of statisticians in academic life, in government work, and in business. By 1945 the subject was ready for decades of vigorous and productive exploration.

Much of this work followed in the direct line of earlier work. The pioneers had left sizable areas undeveloped; and in consequence, work on distribution theory, sampling, and regression analysis continued in fair volume without any fundamental change in concept. Among the newer fields of attention we may notice in particular sequential analysis, decision function theory, multivariate analysis, time series and stochastic processes, statistical inference, and distribution-free, or nonparametric, methods [*see* DECISION THEORY; MARKOV CHAINS; MULTIVARIATE ANALYSIS; NONPARAMETRIC STATISTICS; QUEUES; SEQUENTIAL ANALYSIS; TIME SERIES].

Sequential analysis. During World War II it was realized by George Barnard in England and Abraham Wald in the United States that some types of sampling were wasteful in that they involved scrutinizing a sample of fixed size even if the examination of the first few members already indicated the decision to be made. This led to a theory of sequential sampling, in which the sample number is not fixed in advance but at each stage in the sampling a decision is made whether to continue or not. This work was applied with success to the control of the quality of manufactured products, and it was soon also realized that a great deal of scientific inquiry was, in fact, sequential in character. [See QUALITY CONTROL, STATISTICAL.]

Decision functions. Wald was led to consider a more general approach, which linked up with Neyman's ideas on hypothesis testing and developed into a theory of decision functions. The basic idea was that at certain stages decisions have to be made, for example, to accept or reject a hypothesis. The object of the theory is to lay down a set of rules under which these decisions can be intelligently made; and, if it is possible to specify penalties for taking wrong decisions, to optimize the method of choice according to some criterion, such as minimizing the risk of loss. The theory had great intellectual attraction and even led some statisticians to claim that the whole of statistics was a branch of decision-function theory, a claim that was hotly resisted in some quarters and may or may not stand up to deeper examination.

Multivariate problems. By 1950 the mathematical development of some branches of statistical theory had, in certain directions, outrun their practical usefulness. This was true of multivariate analysis based on normal distributions. In the more general theory of multivariate problems, several lines of development were pursued. One part of the theory attempts to reduce the number of effective dimensions, especially by component analysis and, as developed by psychologists, factor analysis [see FACTOR ANALYSIS AND PRINCIPAL COMPONENTS]. Another, known as canonical correlation analysis, attempts to generalize correlation to the relationship between two vector quantities. A third generalizes distribution theory and sampling to multidimensional cases. The difficulties are formidable, but a good deal of progress has been made. One problem has been to find practical data that would bear the weight of the complex analysis that resulted. The high-speed computer may be a valuable tool in further work in this field. [See COMPUTATION.]

Time series and stochastic processes. Perhaps the most extensive developments after World War II were in the field of time series and stochastic processes generally. The problem of analyzing a time series has particular difficulties of its own. The system under examination may have a trend present and may have seasonal fluctuations. The classical method of approach was to dissect the series into trend, seasonal movement, oscillatory effects, and residual; but there is always danger that an analysis of this kind is an artifact that does not correspond to the causal factors at work, so that projection into the future is unreliable. Even where trend is absent, or has been abstracted, the

analysis of oscillatory movements is a treacherous process. Attempts to apply harmonic analysis to economic data, and hence to elicit "cycles," were usually failures, owing to the fact that observed fluctuations were not regular in period, phase, or amplitude [see TIME SERIES, *article on* CYCLES].

The basic work on time series was done by Yule between 1925 and 1930. He introduced what is now known as an autoregressive process, in which the value of the series at any point is a linear function of certain previous values plus a random residual. The behavior of the series is then determined, so to speak, partly by the momentum of past history and partly by unpredictable disturbance. In the course of this work Yule introduced serial correlations, which measure the relationship between terms of the series separated by specified time intervals. It was later realized that these functions are closely allied to the coefficients that arise in the Fourier analysis of the series.

World War II acted as a kind of incubatory period. Immediately afterward it was appreciated that Yule's method of analyzing oscillatory movements in time series was only part of a much larger field, which was not confined to movements through time. Earlier pioneer work by several writers, notably Louis Bachelier, Eugen Slutsky, and Andrei Markov, was brought together and formed the starting point of a new branch of probability theory. Any system that passes through a succession of states falls within its scope, provided that the transition from one state to the next is decided by a schedule of probabilities and is not purely deterministic. Such systems are known as stochastic processes. A very wide variety of situations falls within their scope, among them epidemics, stock control, traffic movements, and queues. They may be regarded as constituting a probability theory of movement, as distinct from the classical systems in which the generating mechanism behind the observations was constant and the successive observations were independent. From 1945 onward there was a continual stream of papers on the subject, many of which were contributed by Russian and French authors [see MARKOV CHAINS; QUEUES].

Some philosophical questions. Common to all this work was a constant re-examination of the logic of the inferential processes involved. The problem of making meaningful statements about the world on the basis of examination of only a small part of it had exercised a series of thinkers from Bacon onward, notably George Boole, John Stuart Mill, and John Venn, but it remained essentially unsolved and was regarded by some as constituting more of a philosophical puzzle than a barrier to scientific advance. The specific procedures proposed by statisticians brought the issue to a head by defining the problem of induction much more exactly, and even by exposing situations where logical minds might well reach different conclusions from the same data. This was intellectually intolerable and necessitated some very searching probing into the rather intuitive arguments by which statisticians drew their conclusions in the earlier stages of development of the subject.

Discussion has been centered on the theory of probability, in which two attitudes may be distinguished: subjective and objective [see PROBABILITY, *article on* INTERPRETATIONS]. Neither approach is free from difficulty. Both lead to the same calculus of probabilities in the deductive sense. The primary problem, first stated explicitly by Thomas Bayes, however, is one of induction, to which the calculus of probabilities makes no contribution except as a tool of analysis. Some authorities reject the Bayesian approach and seek for principles of inferences elsewhere. Others, recognizing that the required prior probabilities necessitate certain assumptions, nevertheless can see no better way of tackling the problem if the relative acceptability of hypotheses is to be quantified at all [see BAYESIAN INFERENCE]. Fortunately for the development of theoretical statistics, the philosophical problems have remained in the background, stimulating argument and a penetrating examination of the inferential process but not holding up development. In practice it is possible for two competent statisticians to differ in the interpretation of data, although if they do, the reliability of the inference is often low enough to justify further experimentation. Such cases are not very frequent, but important instances do occur; a notable one is the interpretation of the undeniable observed relationship between intensity of smoking and cancer of the lung. Differences in interpretation are particularly liable to occur in economic and social investigations because of the difficulty of performing experiments or of isolating causal influences for separate study.

Robustness and nonparametric methods. The precision of inferences in probability is sometimes bought at the expense of rather restrictive assumptions about the population of origin. For example, Student's *t*-test depends on the supposition that the parent population is normal. Various attempts have been made to give the inferential procedures greater generality by freeing them from these restrictions. For example, certain tests can be shown to be "robust" in the sense that they are not very sensitive to deviations from the basic assumptions [see ERRORS]. Another interesting field is concerned with tests that depend on ranks, order statistics, or even signs, and are very largely independent of the form of the parent population. These so-called distribution-free methods, which are usually easy to apply, are often surprisingly efficient [see NONPARAMETRIC STATISTICS].

The frontiers of the subject continue to extend. Problems of statistical relationship, of estimation in complicated models, of quantification and scaling in qualitative material, and of economizing in exploratory effort are as urgent and lively as ever. The theoretical statistician ranges from questions of galactic distribution to the properties of subatomic particles, suspended, like Pascal's man, between the infinitely large and the infinitely small. The greater part of the history of his subject lies in the future.

MAURICE G. KENDALL

[*The following biographies present further details on specific periods in the history of statistical method. Early period:* BABBAGE; BAYES; BERNOULLI FAMILY; BIENAYMÉ; GALTON; GAUSS; GRAUNT; LAPLACE; MOIVRE; PETTY; POISSON; QUETELET; SÜSSMILCH. *Modern Period:* BENINI; BIRNBAUM; BORTKIEWICZ; FISHER; GINI; GIRSHICK; GOSSET; HOTELLING; KEYNES, JOHN MAYNARD: CONTRIBUTIONS TO STATISTICS; KŐRÖSY; LAZARSFELD; LEXIS; MAHALANOBIS; PEARSON; PRICE; RÉNYI; SAVAGE; SHEWHART; SPEARMAN; VAN DANTZIG; VON MISES; VON NEUMANN; WALD; WIENER; WILKS; WILCOXON; WILLCOX; YOUDEN; YULE; *see also* Smith 1968; Spengler 1968.]

BIBLIOGRAPHY

There is no history of theoretical statistics or of statistical methodology. Westergaard 1932 is interesting as an introduction but is largely concerned with descriptive statistics. Walker 1929 has some valuable sketches of the formative period under Karl Pearson. Todhunter 1865 is a comprehensive guide to mathematical work up to Laplace and contains bibliographical information on many of the early works cited in the text of this article. David 1962 is a modern and lively account up to the time of de Moivre. The main sources for further reading are in obituaries and series of articles that appear from time to time in statistical journals, especially the "Studies in the History of Probability and Statistics" in Biometrika *and occasional papers in the* Journal of the American Statistical Association.

BAYES, THOMAS (1764) 1958 An Essay Towards Solving A Problem in the Doctrine of Chances. *Biometrika* 45:296–315. → First published in Volume 53 of the Royal Society of London's *Philosophical Transactions.* A facsimile edition was published in 1963 by Hafner.
BERNOULLI, JACQUES (1713) 1899 *Wahrscheinlichkeitsrechnung (Ars conjectandi).* 2 vols. Leipzig: Engelmann. → First published posthumously in Latin.
CONDORCET, MARIE JEAN ANTOINE NICOLAS CARITAT, DE 1785 *Essai sur l'application de l'analyse à la probabilité des décisions rendues à la pluralité des voix.* Paris: Imprimerie Royale.

CZUBER, EMANUEL 1898 *Die Entwicklung der Wahrscheinlichkeitstheorie und ihrer Anwendungen.* Jahresbericht der Deutschen Mathematikervereinigung, Vol. 7, No. 2. Leipzig: Teubner.

DAVID, F. N. 1962 *Games, Gods and Gambling: The Origins and History of Probability and Statistical Ideas From the Earliest Times to the Newtonian Era.* London: Griffin; New York: Hafner.

FISHER, R. A. 1925 Theory of Statistical Estimation. *Cambridge Philosophical Society, Proceedings* 22:700–725. → Reprinted in Fisher 1950.

FISHER, R. A. 1928 Moments and Product Moments of Sampling Distributions. *London Mathematical Society, Proceedings* 30:199–238. → Reprinted in Fisher 1950.

FISHER, R. A. (1920–1945) 1950 *Contributions to Mathematical Statistics.* New York: Wiley.

HUYGENS, CHRISTIAAN 1657 De rationciniis in ludo aleae. Pages 521–534 in Frans van Schooten, *Exercitationum mathematicarum.* Leiden (Netherlands): Elsevir.

KOTZ, SAMUEL 1965 Statistical Terminology—Russian vs. English—In the Light of the Development of Statistics in the USSR. *American Statistician* 19, no. 3:14–22.

LAPLACE, PIERRE SIMON (1812) 1820 *Théorie analytique des probabilités.* 3d ed., revised. Paris: Courcier.

ORE, ØYSTEIN 1953 *Cardano: The Gambling Scholar.* Princeton Univ. Press; Oxford Univ. Press. → Includes a translation from the Latin of Cardano's *Book on Games of Chance* by Sydney Henry Gould.

PEARSON, KARL (1892) 1911 *The Grammar of Science.* 3d ed., rev. & enl. London: Black. → A paperback edition was published in 1957 by Meridian.

POISSON, SIMÉON DENIS 1837 *Recherches sur la probabilité des jugements en matière criminelle et en matière civile, précédées des règles générales du calcul des probabilités.* Paris: Bachelier.

SIMPSON, THOMAS 1757 *Miscellaneous Tracts on Some Curious and Very Interesting Subjects in Mechanics, Physical-astronomy, and Speculative Mathematics.* London: Nourse.

►SMITH, M. BREWSTER 1968 Stouffer, Samuel A. Volume 15, pages 277–280 in *International Encyclopedia of the Social Sciences.* Edited by David L. Sills. New York: Macmillan and Free Press.

►SPENGLER, JOSEPH J. 1968 Lotka, Alfred J. Volume 9, pages 475–476 in *International Encyclopedia of the Social Sciences.* Edited by David L. Sills. New York: Macmillan and Free Press.

TODHUNTER, ISAAC (1865) 1949 *A History of the Mathematical Theory of Probability From the Time of Pascal to That of Laplace.* New York: Chelsea.

WALKER, HELEN M. 1929 *Studies in the History of Statistical Method, With Special Reference to Certain Educational Problems.* Baltimore: Williams & Wilkins.

WESTERGAARD, HARALD 1932 *Contributions to the History of Statistics.* London: King.

Postscript

About 1955, statisticians began to take a serious interest in the history of their subject, and a number of articles were published, mainly in *Biometrika.* E. S. Pearson and Maurice G. Kendall edited the publication of a collected set of these papers, 29 in all, in 1970. A second set of 32 papers edited by Kendall and R. L. Plackett was published in 1977.

Written by many different authors from many different points of view, the papers in these two volumes provide a good coverage of the history of statistics and probability from ancient times to the present day.

MAURICE G. KENDALL

ADDITIONAL BIBLIOGRAPHY

HACKING, IAN 1975 *The Emergence of Probability: A Philosophical Study of Early Ideas About Probability, Induction and Statistical Inference.* Cambridge Univ. Press.

KENDALL, MAURICE G.; and PLACKETT, R. L. (editors) 1977 *Studies in the History of Statistics and Probability.* Volume 2. London: Griffin; New York: Macmillan.

MAISTROV, LEONID E. (1967) 1974 *Probability Theory: A Historical Sketch.* Translated and edited by Samuel Kotz. New York: Academic Press. → First published in Russian.

PEARSON, KARL (1921–1933) 1978 *The History of Statistics in the Seventeenth and Eighteenth Centuries, Against the Changing Background of Intellectual, Scientific and Religious Thought.* Edited by E. S. Pearson. London: Griffin; New York: Macmillan. → Lectures by Karl Pearson given at University College, London, during the academic sessions 1921–1933.

PEARSON, E. S.; and KENDALL, MAURICE G. (editors) 1970 *Studies in the History of Statistics and Probability.* Volume 1. London: Griffin; Darien, Conn.: Hafner.

STATISTICS, DESCRIPTIVE

I. LOCATION AND DISPERSION	*Hans Kellerer*
II. ASSOCIATION	*Robert H. Somers*

I
LOCATION AND DISPERSION

A basic statistical need is that of describing a set of observations in terms of a few calculated quantities—descriptive statistics—that express compactly the most salient features of the observational material. Some common descriptive statistics are the sample average, the median, the standard deviation, and the correlation coefficient.

Of course, one is also interested in corresponding descriptive quantities for the underlying population from which the sample of observations was drawn; these population descriptive statistics may usually be thought of as sample descriptive statistics for very large hypothetical samples, so large that sampling variability becomes negligible.

The present article deals with descriptive statistics relating to *location* or position (for instance, the sample average or the median) and to *dispersion* or variability (for instance, the standard deviation). The accompanying article deals with descriptive statistics for aspects of association between two or more statistical variates (for instance, the correlation coefficient).

Most, although not all, descriptive statistics for location deal with a generalized mean—that is, some function of the observations satisfying intuitive restrictions of the following kind: (a) a generalized mean must take values between the lowest and highest observations; (b) it must be unchanged under reorderings of the observations; and (c) if all the observations are equal, the generalized mean must have their common value. There are many possible generalized means; those selected for discussion here have useful interpretations, are computationally reasonable, and have a tradition of use.

Descriptive statistics of dispersion supply information about the scatter of individual observations. Such statistics are usually constructed so that they become larger as the sample becomes less homogeneous.

Descriptive statistics of location

Generalized means. An important family of location measures represents the so-called central tendency of a set of observations in one of various senses. Suppose the observations are denoted by $x_1, x_2, \cdots, x_n$. Then the ordinary average or arithmetic mean is

$$(1) \qquad \frac{1}{n}\sum_{i=1}^{n} x_i.$$

If, however, a function, f, is defined and the average of the $f(x_i)$'s is considered, then an associated generalized mean, M, is defined by

$$f(M) = \frac{1}{n}\sum f(x_i).$$

The summation is from 1 to n, and f has the same meaning on both sides of the defining equation. For the arithmetic mean, f is the identity function. For the geometric mean (when all x_i's are positive), f is the logarithmic function—that is, $\log M =$

$(1/n) \sum \log x_i$, so that

$$M = \sqrt[n]{x_1 x_2 \cdots x_n}.$$

For this procedure to make sense, f must provide a one-to-one relationship between possible values of the x_i's and possible values of the $f(x_i)$'s. Sometimes special conventions are necessary.

For any such generalized mean, the three intuitive restrictions listed earlier are clearly satisfied when f is monotone increasing. In addition, a change in any single x_i, with the others fixed, changes the value of M. Four of the many generalized means having these properties are listed in Table 1. Since the quadratic mean is important primarily in measuring dispersion, only the first three means listed in Table 1 will be discussed in detail.

The arithmetic mean. The arithmetic mean is perhaps the most common of all location statistics because of its clear meaning and ease of computation. It is usually denoted by placing a bar above the generic symbol describing the observations, thus $\bar{x} = (1/n) \sum x_i$, $\bar{a} = (1/n) \sum a_i$, etc. The population analogue of the arithmetic mean is simply the expectation of a random variable describing the population—that is, $E(X)$, if X is the random variable. [See PROBABILITY, *article on* FORMAL PROBABILITY.]

If the x_i's represent, for example, the wages received by the ith individual in a group, then $\sum x_i$ represents the total wages received, and $\bar{x}$, the arithmetic mean or ordinary average, represents the wages that would have been received by each person if everyone in the group had received the same amount.

The major formal properties of the arithmetic mean are:

(a) The sum of the deviations of the x_i's from $\bar{x}$ is zero: $\sum(x_i - \bar{x}) = 0$.

(b) The sum of squared deviations of the x_i's

Table 1 — Four important generalized means

	$f(x)$	Equation for the mean value, M	
Arithmetic mean	x	$M = \dfrac{\sum x_i}{n}$	
Geometric mean	$\log x$	$\log M = \dfrac{\sum \log x_i}{n}$	so that $M = \sqrt[n]{x_1 x_2 \cdots x_n}$, where $x_i > 0$
Harmonic mean	$\dfrac{1}{x}$	$\dfrac{1}{M} = \dfrac{\sum(1/x_i)}{n}$	so that $M = \dfrac{n}{\sum(1/x_i)}$, where $x_i \neq 0$
Quadratic mean	x^2	$M^2 = \dfrac{\sum x_i^2}{n}$	so that $M = \sqrt{\dfrac{\sum x_i^2}{n}}$, where $x_i \geqslant 0$

from x, considered as a function of x, is minimized by $\bar{x}$: $\sum (x_i - \bar{x})^2 \leqslant \sum (x_i - x)^2$ for all x.

(c) If a and b are any two numbers, and if one sets $y_i = ax_i + b$, then $\bar{y} = a\bar{x} + b$, and $\bar{x} = (\bar{y} - b)/a$ when $a \neq 0$. This linear invariance of the arithmetic mean is the basis for so-called coding, changing the origin and scale of the observations for computational convenience. For example, instead of finding the average of the x_i's 110, 125, 145, 190, 210, and 300, it is simpler to subtract 100 from each number and multiply by $\frac{1}{5}$: $y_i = \frac{1}{5}(x_i - 100)$. After finding $\bar{y}$ (16 for the x_i's listed here), it is necessary only to reverse the coding process to obtain $\bar{x} = 5\bar{y} + 100$ (180 in this case).

(d) Suppose the x_i's are divided into m subgroups, where the jth subgroup includes n_j of the x_i's ($\sum n_j = n$) and has arithmetic mean $\bar{x}_j$. Then $\bar{x}$, the arithmetic mean for all the x_i's, is

$$(2) \qquad \bar{x} = \frac{n_1 \bar{x}_1 + n_2 \bar{x}_2 + \cdots + n_m \bar{x}_m}{n_1 + n_2 + \cdots + n_m} = \frac{\sum n_j \bar{x}_j}{n}.$$

When n is large, it is often advisable to summarize the original observations in a frequency table. Table 2 shows how $\bar{x}$ is calculated in this case.

The table shows, for example, that 22 of the 100 x_i lie in the interval from 15 to 20. By eq. (2)

$$\bar{x} = \frac{\text{Sum of the statistical values in the 5 classes}}{\sum f_j}.$$

The numerator may be determined approximately by resorting to the "hypothesis of the class mark": the average value of the observations in a class is the class mark—that is, $w_j = \bar{x}_j$ for all j. Up to the accuracy of the approximation this yields

$$(3) \qquad \bar{x} \cong \frac{\sum w_j f_j}{\sum f_j} = \frac{\sum w_j f_j}{n}.$$

The numerator summation is now from 1 to m (m = the number of classes). For the data in Table 2, $\bar{x} \cong 2{,}335/100 = 23.35$.

A coding transformation may be employed for easier computation:

$$(4) \qquad \bar{x} \cong c \cdot \frac{\sum \left(\frac{w_j - x_0}{c} \right) f_j}{n} + x_0.$$

The numbers x_0 and c should be chosen for convenience in computation. In the example, one might take $x_0 = 22.5$ = the class mark of the central class and $c = 5$ = the class width. This transformation yields Table 3, from which $\bar{x} \cong 5(17/100) + 22.5 = 23.35$, as before.

It often happens that the first or last class is "open"—that is, the lowermost or uppermost class boundary is unknown. If there are only a few elements in the open classes, a reasonable choice of class marks may often be made without fear that the arithmetic mean will be much affected. [*Considerations about choice of class width and position, and about the errors incurred in adopting the hypothesis of the class mark, are given in* STATISTICAL ANALYSIS, SPECIAL PROBLEMS OF, *article on* GROUPED OBSERVATIONS.]

Weighted averages, of the form $\sum a_i x_i / \sum a_i$, are often employed, especially in the formation of index numbers. These bear a clear formal relationship to the arithmetic means computed by formula (3) [*see* INDEX NUMBERS].

The geometric mean. One of the subtle problems that concerned thinkers in the Middle Ages was the following: The true, unknown value of a horse is, say, 100 units of money. Two persons, A and B, estimate this unknown value. A's estimate is 10 units, and B's is 1,000. Which is the worse estimate? There are two objections to the answer "B's estimate is worse, because he is off by 900 units, whereas A is only 90 units off": first, the available room for estimating downward is only 100 units, because negative units do not apply, whereas the room for estimating upward (above 100 units) is unlimited; second, since A's estimate is $\frac{1}{10}$ of the true value and B's is 10 times the true value, both estimates are equally poor, relatively speaking. The two errors balance out on the multiplicative average, for $0.1 \cdot 10 = 1$.

An example will illustrate the use of the geometric mean: The number of television viewers in a certain area was 50 on January 1, 1960, and

Table 2 — *Observations summarized in a frequency table*

CLASS	CLASS INTERVAL	CLASS MARK	FREQUENCY	
(j)		w_j	f_j	$w_j f_j$
1	$10 \leqslant x < 15$	12.5	3	37.5
2	$15 \leqslant x < 20$	17.5	22	385.0
3	$20 \leqslant x < 25$	22.5	38	855.0
4	$25 \leqslant x < 30$	27.5	29	797.5
5	$30 \leqslant x < 35$	32.5	8	260.0
Totals			100	2,335.0

Table 3 — *Frequency table with coded observations*

Class	$y_j = (w_j - 22.5)/5$	f_j	$y_j f_j$
1	-2	3	-6
2	-1	22	-22
3	0	38	0
4	1	29	29
5	2	8	16
Totals		100	17

Table 4 — Television viewers in a certain area[a]

Date (Jan. 1)	Observed number	Assumed formula	Observed and imputed[b]
1960	50	50	50
1961		$50r$	(150)
1962		$50r^2$	(450)
1963		$50r^3$	(1,350)
1964	4,050	$50r^4$	4,050

a. Hypothetical data.
b. Imputed values are shown in parentheses.

4,050 on January 1, 1964. It might be in order to assume the same *relative* increase (by a factor of r) for each of the four years involved. Table 4 can then be drawn up. Under the assumption of constant relative increase per year, $50r^4 = 4,050$, so that $r = \sqrt[4]{4,050/50} = 3$. The corresponding imputed values are given in parentheses in the last column of Table 4. Note that the square root of the product of the initial and final numbers, $\sqrt{50 \times 4,050}$, is 450, the imputed value at the middle year.

The example indicates that the geometric mean may be appropriate in instances of relative change. This is the situation, for example, in several index problems [see INDEX NUMBERS]. Here multiplication takes the place of the addition employed in the arithmetic mean; hence, the procedure is to extract roots instead of to divide. One justification for the term "geometric mean" comes from considering the geometric sequence 1, 2, 4, 8, 16, 32, 64; the central element, 8, is the same as the geometric mean. Calculating the geometric mean makes sense only when each original value is greater than zero. If the data is grouped as in Table 2, the corresponding geometric mean is

$$(5) \qquad M = \sqrt[n]{w_1^{f_1} w_2^{f_2} \cdots w_m^{f_m}}.$$

The geometric mean of n fractions, x_i/y_i, $i = 1, \cdots, n$, is equal to the quotient of the geometric mean of the x_i's divided by the geometric mean of the y_i's. If X is a positive random variable that describes the underlying population, then the population geometric mean is $10^{E(\log X)}$, where logarithms are taken to base 10.

The harmonic mean. The feature common to the following examples is the employment of reciprocals of the original values.

Example 1: A man travels 2 kilometers, traveling the first at the speed $x_1 = 10$ km/hr and the second at the speed $x_2 = 20$ km/hr. What is the "average" speed? The answer $(10 + 20)/2 = 15$ would usually be misleading, for the man travels a longer time at the lower speed, 10 km/hr, than at the higher speed, 20 km/hr—that is, 1/10 as

against 1/20 of an hour. Since speed is defined as distance/time, a better average might be the harmonic mean

$$H = \frac{2}{\dfrac{1}{10} + \dfrac{1}{20}} = 13\tfrac{1}{3} \text{ km/hr.}$$

The point is that speed averaged over time must satisfy the relationship

average speed × total time = total distance,

and only the harmonic average does this (when the individual distances are equal).

The example may be generalized as follows: At first a man travels 60 km at a speed of 15 km/hr and then 90 km at a speed of 45 km/hr. The corresponding weighted harmonic mean is

$$\frac{60 + 90}{60 \cdot \dfrac{1}{15} + 90 \cdot \dfrac{1}{45}} = 25 \text{ km/hr.}$$

Example 2: A market research institute wants to determine the average daily consumption of razor blades by polling 500 persons. Experience has shown that it is better to ask, "How many days does a razor blade last you?" because that is how people usually think. The results of the poll are shown in Table 5, where x_i denotes the number of days a razor blade lasts. Column 3 indicates that the 100 persons in the first group use $\frac{1}{2} \cdot 100 = 50$ blades a day. The average consumption per person per day is

$$\frac{1}{H} = \frac{\sum(1/x_i)f_i}{\sum f_i} = \frac{160}{500} = .32.$$

Other descriptive statistics of location. There are several location measures that are not conveniently described in the form $f(M) = (1/n)\sum f(x_i)$. Most of these other location measures are based on the order statistics of the sample [see NONPARAMETRIC STATISTICS, *article on* ORDER STATISTICS].

The median. If a set of n observations (n odd) with no ties is arranged in algebraic order, the median of the observations is defined as the middle

Table 5 — Useful life of razor blades*

Number of days $= x_i$	Number of persons $= f_i$	Total consumption per day $= (1/x_i)f_i$
2	100	50
3	150	50
4	200	50
5	50	10
Totals	500	160

* Hypothetical data.

member of the ordering. For example, the median, $Me(x)$, of the (ordered) x_i's

$$2, 17, 19, 23, 38, 47, 98$$

is 23. If n is even, it is conventional, although arbitrary, to take as median the average of the two middle observations; for example, the conventional median of

$$2, 17, 19, 23, 38, 47$$

is $(19 + 23)/2 = 21$. The same definitions apply even when ties are present. Thus, the median cuts the set of observations in half according to order.

Unlike the descriptive statistics discussed earlier, the median is unaffected by changes in most individual observations, provided that the changes are not too large. For example, in the first set of numbers above, 2 could be changed to any other number not greater than 23 without affecting the median. Thus, the median is less sensitive to outliers than is the arithmetic mean [see STATISTICAL ANALYSIS, SPECIAL PROBLEMS OF, *article on* OUTLIERS]. On the other hand, the median is still a symmetric function of all the observations, and not, as is sometimes said, a function of only one observation.

The median minimizes the sum of absolute residuals—that is, $\sum |x_i - x|$ is minimum for $x = Me(x)$.

A disadvantage of the median is that if the x_i's are divided into subgroups, one cannot in general compute $Me(x)$ from the medians and sizes of the subgroups.

For a population described by a random variable X the median is any number, Med, such that

$$Pr\{X < \text{Med}\} \leqslant \tfrac{1}{2} \leqslant Pr\{X \leqslant \text{Med}\}.$$

In general, Med is not uniquely defined, although it is when X has a probability density function that is positive near its middle [see PROBABILITY]; then Med has a clear interpretation via

$$\begin{aligned} Pr\{X < \text{Med}\} &= Pr\{X \leqslant \text{Med}\} \\ &= Pr\{X > \text{Med}\} = Pr\{X \geqslant \text{Med}\} \\ &= \tfrac{1}{2}. \end{aligned}$$

(See Figure 1.)

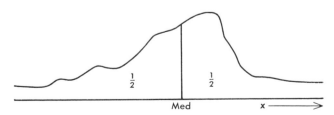

Figure 1 — *Probability density function with median, Med, indicated*

Table 6 — *Frequencies and cumulative frequencies for approximate determination of median*

Class interval	Frequency	Cumulative frequency
$x < 20$	4	4
$20 \leqslant x < 30$	17	21
$30 \leqslant x < 40$	38	59
$40 \leqslant x < 60$	49	108
$60 \leqslant x < 80$	29	137
$80 \leqslant x$	4	141

Table 6 shows how the value of Me is determined *approximately* for grouped observations. The first two columns might represent the frequency distribution of the test scores achieved by 141 subjects. With this as a basis column 3 is derived. The figures in column 3 are found by continued addition (accumulation) of the frequencies given in column 2. The number 59, for example, means that 59 subjects achieved scores of less than 40. In accordance with the definition, the median, Me, is the score achieved by the 71st subject. Since the 59 smallest scores are all less than 40 and since each of the other scores is greater than that, $\text{Me} = 40 + a$, where $0 \leqslant a < 20$, because the fourth class, within which the median lies, has an interval of 20.

The hypothesis that the 49 elements of the fourth class are uniformly distributed [see DISTRIBUTIONS, STATISTICAL] over the interval yields the following relationship for a (linear interpolation): $20 : a = 49 : (71 - 59)$, whence $a = 4.9$ and $\text{Me} = 44.9$. The value $71 - 59 = 12$ enters into the equation because the median is the twelfth score in the median class.

The frequency distribution is a good starting point for computing Me because it already involves an arrangement in successive size classes. The assumption of uniform distribution within the median class provides the necessary supplement; whether this hypothesis is valid must be decided in each individual case. Only a small portion of the cumulative sum table is required for the actual calculation.

In the calculation it was presupposed that the variate was continuous—for instance, that a score of 39.9 could be achieved. If only integral values are involved, on the other hand, the upper boundaries of the third and fourth classes are 39 and 59 respectively. In that case we obtain $\text{Me} = 39 + 5 = 44$.

Quartiles, deciles, percentiles, midrange. The concept that led to the median may be generalized. If there are, say, $n = 801$ observations, which are again arranged in algebraic order, starting with the smallest, it may be of interest to specify the variate values of the 201st, 401st, and 601st observations —that is, the first, second, and third *quartiles*, Q_1, Q_2, and Q_3. Obviously, Q_2 is the same as Me. From the associated cumulative distribution, Q_1 and Q_3 may be obtained as follows (see Figure 2, where, for convenience, a smoothed curve has been used to represent a step function): Draw parallels to the x-axis at the heights 201 and 601; let their intersections with the cumulative distribution be P_1 and P_3. The x-coordinates of P_1 and P_3 then yield Q_1 and Q_3 respectively.

If $Q_1 = \$800$, $Q_2 = \$1,400$, and $Q_3 = \$3,000$ in an income distribution, the maximum income of the families in the lower fourth of income recipients is $800 and that of the bottom half is $1,400, while 25 per cent of the population has an income in excess of $3,000.

The variate values of the 81st, 161st, 241st, $\cdots$, 721st observations in the examples where $n = 801$ determine the 1st, 2d, 3d, $\cdots$, 9th *deciles*. Similarly, the *percentiles* in the example are given by the 9th, 17th, 24th, $\cdots$, 793d ordered observations. In general, if the sample size is n, the ith percentile is given by the sample element with ordinal number $[ni/100] + 1$ when $ni/100$ is not an integer and by the simple average of the two elements with ordinal numbers $ni/100$ and $(ni/100) + 1$ when $ni/100$ is an integer. Here "$[x]$" stands for the largest integer in x. [*For further discussion, see* NONPARAMETRIC STATISTICS, *article on* ORDER STATISTICS.]

If the smallest observation is denoted by $x_{(1)}$ and the largest by $x_{(n)}$, the midrange is $\frac{1}{2}(x_{(1)} + x_{(n)})$ and the quartile average is $\frac{1}{2}(Q_1 + Q_3)$.

Mode. If a variate can take on only discrete values, another useful measure of location is the variate value that occurs most frequently, the so-called mode (abbreviated as Mo). If, for instance, the number of children born of 1,000 marriages, each of which has lasted 20 years, is counted, Mo is the most frequent number of children.

The mode is often encountered in daily life: the most frequent mark in a test, the shoe size called for most often, the range of prices most frequently requested. Generally speaking, the concept of the typical is associated with the mode; consider, for instance, wage distribution in a large factory. The mode is very vivid. However, it possesses concrete significance only when there is a tendency toward concentration about it and when enough observations are available.

In considering a frequency distribution with class intervals that are all of equal width, one convention is to take as the mode the midpoint of the class containing the largest number of elements. This procedure is open to the objection that it does

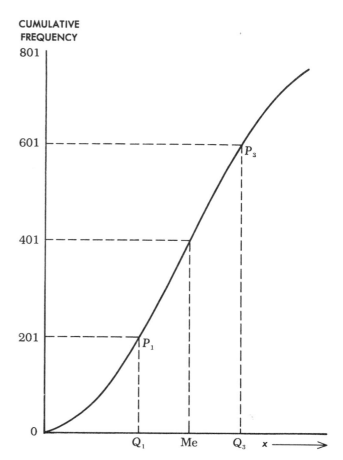

CUMULATIVE FREQUENCY

Figure 2 — Hypothetical cumulative distribution (smoothed from a step function) showing the graphic method of obtaining quartiles

not take into consideration which of the adjoining classes contains a larger number of elements; the mode would be better calculated in such a way that it is closer to the adjacent class that contains the larger number of elements. Several more or less arbitrary ways of doing this have been suggested— for example, by Croxton and Cowden (1939, p. 213) and by Hagood and Price ([1941] 1952, p. 113). On the other hand, for a continuous distribution it makes sense to take the value of x that yields a (relative) maximum as the mode. There may be two or more relative maxima, as may be readily seen from a graph of the distribution.

Further comments on mean values. There exist some quantitative relationships between the mean values.

(*a*) Harmonic mean $\leqslant$ geometric mean $\leqslant$ arithmetic mean. The equal signs apply only to the trivial case where all the observations have the same variate value. This is a purely mathematical result.

(*b*) For a symmetrical distribution (one whose density or frequency function is unchanged by reflection through a vertical line) with exactly one mode, $\bar{x}$, Me, and Mo coincide. For a distribution that is slightly positively skew (that is, one that slopes more gently to the right than to the left—this is not a precise concept), an approximate relationship is Me − Mo $\cong$ 3($\bar{x}$ − Me). In a positively skew distribution the sequence of the three parameters is Mo, Me, and $\bar{x}$; in negatively skew ones, on the other hand, it is $\bar{x}$, Me, and Mo. Kendall and Stuart write as follows about this: "It is a useful mnemonic to observe that the mean, median, and mode occur in the same order (or reverse order) as in the dictionary; and that the median is nearer to the mean than to the mode, just as the corresponding words are nearer together in the dictionary" (1958, p. 40).

Scales of measurement. There is usually a distinction made between four types of scales on which observations may be made; the distinction is briefly characterized by Senders as follows:

If all you can say is that one object is different from another, you have a *nominal scale.*

If you can say that one object is bigger or better or more of anything than another, you have an *ordinal scale.*

If you can say that one object is so many units (inches, degrees, etc.) more than another, you have an *interval scale.*

If you can say that one object is so many times as big or bright or tall or heavy as another, you have a *ratio scale.*   (1958, p. 51)

Scales of measurement are also discussed by Gini (1958), Pfanzagl (1959), Siegel (1956), and Stevens (1959).

For an interval scale there is a meaningful *unit* (foot, second, etc.) but not necessarily an intrinsically meaningful origin. A ratio scale has both a meaningful unit and a meaningful origin.

Some examples of these four types of scales are:

(*a*) *Nominal scale:* classifying marital status into four categories—single, married, widowed, and divorced; a listing of diseases.

(*b*) *Ordinal scale:* answers to the question "How do you like the new method of teaching?" with the alternative replies "Very good," "Good," "Fair," "Poor," "Very poor."

(*c*) *Interval scale:* measuring temperature on the centigrade, Réaumur, or Fahrenheit scales; measuring calendar time (7:09 A.M., January 3, 1968).

(*d*) *Ratio scale:* measuring a person's height in centimeters; measuring temperature on the absolute (Kelvin) scale; measuring age or duration of time.

It is apparent that the four scales are arranged in our listing according to an increasing degree of

Table 7 — Scales of measurement required for use of location measures

Location measure	Scale required
Mode	at least a nominal scale
Median, quartile, and percentile	at least an ordinal scale
Arithmetic mean	at least an interval scale
Geometric and harmonic means	ratio scale

stringency. The nominal scale, in particular, is the weakest method of measurement.

The applicability of a given location measure depends upon the scale utilized, as is shown in Table 7. Only the most frequent value can be calculated for every scale, although Gini (1958) and his school have endeavored to introduce other averages appropriate for a nominal scale.

Table 7 indicates that the median, the quartiles, and the percentiles are relatively invariant for any strictly monotonic transformation in the sense that, for example,

$$\phi(\text{median } X) = \text{median } (\phi(X)),$$

where ϕ is a strictly monotone transformation. Similarly, the arithmetic mean is relatively invariant for any linear transformation $y = ax + b$, and the geometric and harmonic means are relatively invariant in the transformation $y = ax$ (with $a > 0$ for the geometric mean).

Descriptive statistics of dispersion

If there were no dispersion among the observations of a sample, much of statistical methodology would be unnecessary. In fact, there almost always is dispersion, arising from two general kinds of sources: (*a*) inherent variability among individuals (for instance, households have different annual savings) and (*b*) measurement error (for instance, reported savings may differ erratically from actual savings).

Absolute measures of dispersion. A first intuitive method of measuring dispersion might be in terms of $\sum(x_i - \bar{x})$, the sum of residuals about the arithmetic mean. This attempt immediately fails since the sum is always zero. A useful measure of dispersion, however, is obtained by introducing absolute values, yielding the *mean deviation* $(1/n)\sum|x_i - \bar{x}|$.

Variance and standard deviation.   An even more useful measure of dispersion is the *variance,*

$$(6) \qquad s^2 = \frac{1}{n}\sum(x_i - \bar{x})^2,$$

the average squared deviation between an observation and the arithmetic mean. The positive square root of s^2 is called the *standard deviation.* (Often n, in the above definition, is replaced by $n-1$, primarily as a convention to simplify formulas that arise in the theory of sampling, although in some contexts more fundamental arguments for the use of $n-1$ may be given.) [*See* SAMPLE SURVEYS.]

It is of interpretative interest to rewrite s^2 as

$$(7) \qquad s^2 = \frac{1}{2n^2}\sum_i\sum_j(x_i - x_j)^2$$

without mention of $\bar{x}$.

The population variance is $E(X-\mu)^2$, where $\mu = E(X)$ [*see* PROBABILITY, *article on* FORMAL PROBABILITY].

The variance and standard deviation may be computed from a table of grouped observations in a manner analogous to that for the arithmetic mean; coding is particularly useful. Suggestions have been made toward compensating for the class width in a grouping [*see* STATISTICAL ANALYSIS, SPECIAL PROBLEMS OF, *article on* GROUPED OBSERVATIONS].

Variance has an additivity property that is akin to additivity for the arithmetic mean over subgroups. Suppose that the x_i's are divided into two subgroups of sizes n_1, n_2, with arithmetic means $\bar{x}_1$, $\bar{x}_2$, and with variances s_1^2, s_2^2. For convenience, let $p_1 = n_1/n$, $p_2 = n_2/n$. Then, not only does

$$(8) \qquad \bar{x} = p_1\bar{x}_1 + p_2\bar{x}_2,$$

but also

$$(9) \qquad s^2 = p_1s_1^2 + p_2s_2^2 + p_1p_2(\bar{x}_1 - \bar{x}_2)^2.$$

This relationship may be extended to more than two subgroups, and, in fact, it corresponds to a classic decomposition of s^2 into within and between subgroup terms. [*See* LINEAR HYPOTHESES, *article on* ANALYSIS OF VARIANCE.]

Perhaps a more important additivity property of variance refers to sums of observations. [*This is discussed in* PROBABILITY, *article on* FORMAL PROBABILITY, *after consideration of correlation and independence.*]

Other deviation measures of dispersion. The *probable error* is a measure of dispersion that has lost its former popularity. It is most simply defined as $.6745 \times s$, although there has been controversy about the most useful mode of definition. If the x_i's appear roughly normal and n is large, then about half the x_i's will lie in the interval $\bar{x} \pm$ probable error.

Gini has proposed a dispersion measure based on absolute deviations,

$$\Delta = \frac{1}{n(n-1)}\sum_i\sum_{\substack{j \\ i \neq j}}|x_i - x_j|,$$

that has attracted some attention, although it is not widely used.

When the observations are ordered in time or in some other exogenous variate, the mean square successive difference,

$$\delta^2 = \frac{1}{n-1}\sum_1^{n-1}(x_{i+1} - x_i)^2,$$

has been useful. [*See* VARIANCES, STATISTICAL STUDY OF.]

Dispersion measures using order statistics. The simplest measure of dispersion based on the order statistics of a sample is the difference between the largest and smallest x_i: the range [*see* NONPARAMETRIC STATISTICS, *article on* ORDER STATISTICS]. As against its advantage of vividness and ease of calculation there is the disadvantage that the range is fixed by the two extreme values alone. Outliers often occur, especially for large n, so that the range is of limited usefulness precisely in such cases. This is not so serious in small samples, where the range is readily employed. There is a simple relationship connecting the sample size, n, the expected range, and the standard deviation under certain conditions.

The basic shortcoming of the range is avoided by using, say, the 10–90 interpercentile range. This is determined by eliminating the top and bottom 10 per cent of the elements that are ordered according to size and determining the range of the remaining 80 per cent. The 5–95 percentile range can be determined similarly. The semi-interquartile range is $\frac{1}{2}(Q_3 - Q_1)$, in which the bottom and top 25 per cent of all the elements are ignored and half the range of the remaining 50 per cent is calculated.

Relative measures of dispersion. The expressions set forth above are absolute measures of dispersion. They measure the dispersion in the same unit chosen for the several variate values or in the square of that unit. The following example shows that there is also need for relative measures of dispersion, however: A study is to be made of the annual savings (a) of 1,000 pensioners and (b) of 1,000 corporation executives. The average income in (a) might be \$1,800, as against \$18,000 in (b). The absolute dispersion in (b) will presumably be considerably greater than that in (a), and a measure of dispersion that allows for the gross differences in magnitude may be desired. Two of the numerous possibilities available are (1) the coefficient of variation, $s/\bar{x}$ or $(s/\bar{x}) \cdot 100$, which measures s as a fraction or a percentage of $\bar{x}$ (this parameter loses its significance when $\bar{x}$ equals 0 or is close to 0), and (2) the quartile dispersion coefficient $(Q_3 - Q_1)/\mathrm{Me}$, which indicates the interquartile range as a fraction of the median. Such measures are dimensionless numbers.

Concentration curves. Another way of describing dispersion is the construction of a concentration (or Lorenz) curve. [*These techniques are discussed in* GRAPHIC PRESENTATION.]

Populations and samples. In the above discussion a number of descriptive statistics have been defined, both for sets of observations and for populations. A fundamental statistical question is the inductive query "What can be said about the population mean, variances, etc., on the basis of a sample's mean, variance, etc.?" [*Many of the statistical articles in this encyclopedia deal with aspects of this question, particularly* SAMPLE SURVEYS; ESTIMATION; *and* HYPOTHESIS TESTING.]

HANS KELLERER

BIBLIOGRAPHY

Discussions of the statistics of location and dispersion are found in every statistics text. The following are only a few selected references.

CROXTON, FREDERICK E.; and COWDEN, DUDLEY J. (1939) 1955 *Applied General Statistics.* 2d ed. Englewood Cliffs, N.J.: Prentice-Hall. → Sidney Klein became a joint author on the third edition, published in 1967.

DALENIUS, T. 1965 The Mode: A Neglected Statistical Parameter. *Journal of the Royal Statistical Society* Series A 128:110–117.

FECHNER, GUSTAV T. 1897 *Kollektivmasslehre.* Leipzig: Engelmann.

GINI, CORRADO 1958 *Le medie.* Turin (Italy): Unione Tipografico.

HAGOOD, MARGARET J.; and PRICE, DAVID O. (1941) 1952 *Statistics for Sociologists.* Rev. ed. New York: Holt.

KENDALL, MAURICE G.; and BUCKLAND, WILLIAM R. (1957) 1960 *A Dictionary of Statistical Terms.* 2d ed. Published for the International Statistical Institute with the assistance of UNESCO. Edinburgh and London: Oliver & Boyd.

○KENDALL, MAURICE G.; and STUART, ALAN (1958) 1977 *The Advanced Theory of Statistics.* Volume 1: *Distribution Theory.* 4th ed. London: Griffin; New York: Macmillan. → Early editions of Volume 1, first published in 1943, were written by Kendall alone. Stuart became a joint author on later, renumbered editions in the three-volume set.

PFANZAGL, J. 1959 *Die axiomatischen Grundlagen einer allgemeinen Theorie des Messens.* Schriftenreihe des Statistischen Instituts der Universität Wien, New Series, No. 1. Würzburg (Germany): Physica-Verlag.

SENDERS, VIRGINIA L. 1958 *Measurement and Statistics: A Basic Text Emphasizing Behavioral Science Application.* New York: Oxford Univ. Press.

SIEGEL, SIDNEY 1956 *Nonparametric Statistics for the Behavioral Sciences.* New York: McGraw-Hill.

STEVENS, S. S. 1959 Measurement, Psychophysics, and Utility. Pages 18–63 in Charles W. Churchman and Philburn Ratoosh (editors), *Measurement: Definitions and Theories.* New York: Wiley.

VERGOTTINI, MARIO DE 1957 *Medie, variabilità, rapporti.* Turin (Italy): Einaudi.

○YULE, G. UDNY; and KENDALL, MAURICE G. 1950 *An Introduction to the Theory of Statistics.* 14th ed., rev. & enl. London: Griffin; New York: Hafner. → Yule was the sole author of the first edition (1911). Kendall, who became a joint author on the eleventh edition (1937), revised the current edition and added new material to the 1965 printing.

ŽIŽEC, FRANZ (1921) 1923 *Grundriss der Statistik.* 2d ed., rev. Munich: Duncker & Humblot.

Postscript

A new and unusually deep approach to descriptive statistics has been taken by Bickel and Lehmann (1975). Their stress is on interpretation valid under broad conditions and on efficiency and robustness of estimation.

[*See also* DATA ANALYSIS, EXPLORATORY.]

ADDITIONAL BIBLIOGRAPHY

BICKEL, P. J.; and LEHMANN, E. L. 1975 Descriptive Statistics for Nonparametric Models. Parts 1 and 2. *Annals of Statistics* 3:1038–1044, 1045–1069. → Part 1: Introduction. Part 2: Location.

II
ASSOCIATION

● *This article was substantially revised for this volume.*

When two or more variables or attributes are observed for each individual of a group, statistical description is often based on cross-tabulations showing the number of individuals having each combination of score values on the variables. Further summarization is often desired, and, in particular, a need is commonly felt for measures (or indices, or coefficients) that show how strongly one variable is associated with another. In the bivariate case, the primary topic of this article, well-known indices of this kind include the correlation coefficient, the Spearman rank correlation measure, and the mean square contingency coefficient. In this article, the motivations for such measures are discussed, criteria are explained, and some important specific measures are treated, with emphasis throughout on the cross-tabulations of nominal and ordinal measurements. [*For further material on the metric case, see* MULTIVARIATE ANALYSIS, *article on* CORRELATION METHODS.]

Current work in this area is generally careful to make an explicit distinction between a *parameter*—that is, a summary of a population of values (often taken as infinitely large)—and a *statistic*, which is a function of sample observations and most sensibly regarded as a means of making an estimate of some clearly defined population parameter. It is of doubtful value to have a quantity defined in terms of an observed sample unless one is clear about the meaning of the population quantity that it serves to estimate. Recent work on sampling theory presenting, among other things, methods for computing a confidence interval for their γ (illustrated below) is contained in Goodman and Kruskal (1954–1972, part 3).

The correlation coefficient and related measures. In a bivariate normal distribution that has been standardized so that both marginal standard deviations are equal, the correlation coefficient nicely describes the extent to which the elliptical pattern associated with the distribution is elongated. At one extreme, when the distribution is nearly concentrated along a straight line, the correlation coefficient is nearly 1 in absolute value, and each variate is nearly a linear function of the other. At the other extreme, the elliptical pattern is a circle if and only if the correlation coefficient, whose population value is often designated by ρ, is 0. In this case, knowledge of one variate provides no information about the other.

In this bivariate normal case, with equal marginal standard deviations, the regression (conditional expectation) of either variate on the other is a straight line with slope ρ. For the above reasons, it is not unreasonable to adopt ρ as a measure of degree of association in a bivariate normal distribution. The essential idea behind the correlation coefficient was Francis Galton's, but Karl Pearson defined it carefully and was the first to study it in detail. [*See the biographies of* GALTON *and* PEARSON.]

It was pointed out by G. Udny Yule, a student of Pearson's, that ρ also has a useful interpretation even if the bivariate distribution is *not* normal. This interpretation is actually of ρ^2, since $1 - \rho^2$ measures how closely one variate clusters about the least squares linear prediction of it from the other. More generally, it remains true that the two variates have a linear functional relation if and only if $\rho = \pm 1$, but ρ may be 0 without the existence of stochastic independence; in fact, the variates may be functionally related by a properly shaped function at the same time that $\rho = 0$.

More recently, Kruskal (1958, pp. 816–818) has reviewed attempts to provide an interpretation of both Pearson's ρ and the correlation ratios whether or not normality obtains. Both of these measures may be interpreted in terms of expected or average squared deviations, but Kruskal feels that interpretations in these terms are not always appropriate.

In Pearson's early work the assumption of normality was basic, but he soon recognized that one often deals with observations in the form of discrete categories, such as "pass" or "fail" on a test of mental ability. Consequently, he introduced additional coefficients to deal with a bivariate distribution in which one or both of the variables are normal and continuous in principle but become grouped into two or more classes in the process of making the observations. Thus, his biserial correlation coefficient is designed to estimate the value of ρ when one of the normal variables is grouped into two classes, and his tetrachoric coefficient is designed for the same purpose when *both* variables are grouped into two classes, yielding in this latter case a bivariate distribution in the form of a fourfold table (see Kendall & Stuart [1961] 1973, pp. 316–324). More recently, Tate (1955) has pointed out the value of having a measure of association between a discrete random variable that takes the values 0 and 1, and a continuous random variable, especially in the sort of problems that often arise in psychology, where one wishes to correlate a dichotomous trait and a numerically measurable characteristic, such as an ability. He has reviewed two alternative models and both asymptotic and exact distribution theory, all based on normality, and has applied these models to illustrative data.

In addition, Janet D. Elashoff (1971) has proposed several measures of association for the same 0,1-*vs.*-continuous-variable situation without the normality assumption. Employing already tabulated sampling distributions, she examines two models, one in which the two distributions differ only in their mean and another in which variances also differ. Neter and Maynes (1970) have examined the usefulness of a kind of correlation coefficient in regression analysis where the dependent variable is dichotomous; in such instances special analytic methods are often used (see, for example, Goldberger 1964, p. 248).

Yule's "attribute statistics" and recent continuations. Yule's reluctance to accept the assumption of normality in order to measure relationships soon led him to an interest in the relationship between discrete variables, which he referred to as attributes. Most of this work concentrated on dichotomous attributes and, thus, the measurement of relationships in fourfold tables. For this purpose his first proposal was a measure designated Q (for Quetelet, the Belgian astronomer–statesman–statistician); the aspect of the relationship that this represented in quantitative form he designated the "association" of the two attributes (see Yule 1912). The attempt

Table 1 — Illustrative fourfold table, relating race to an attitude of alienation from one's social environment*

		RACE (X)				ASSOCIATION MEASURE
		Negro	White	Total		
ALIENATION (Y)	High	$a = 19$	$b = 55$	74		$Q = \dfrac{ad - bc}{ad + bc} = .66$
	Low	$c = 6$	$d = 85$	91		$d_{yx} = \dfrac{ad - bc}{(a + c)(b + d)} = .37$
	Total	25	140	165		
Per cent highly alienated		76%	39%			

* Cell entries are frequencies.

Source: Templeton 1966.

to coin a special term for the aspect of a relationship measured by a particular coefficient has since been largely discarded, and the term "association" is now used more broadly. [*See the biographies of* QUETELET; YULE.]

The computation of Yule's Q is illustrated in Table 1, where it is adapted to a sociological study of the relationship between race and an attitude of alienation from one's social environment. Alienation was measured by a version of Srole's Anomia Scale (1956); the sample is of adults in Berkeley, California, in 1960. The quantification of this relationship in this way, which in the present case gives $Q = (ad - bc)/(ad + bc) = .66$, might be especially useful if the purpose of the study were to compare different years or communities with regard to the extent to which race may condition an attitude of alienation. Of course, neither an association nor a correlation is in itself evidence of causation. Although the relationship shown here is an objective fact, and therefore a descriptive reflection of social reality with possible consequences for social relations in the United States, it is possible that some other variable, such as socioeconomic status, accounts for at least part of the relationship, and to that extent the relation

observed in Table 1 would be considered "spurious" in the Lazarsfeld ([1955] 1962, p. 123) sense.

It might seem more natural to measure the relation between race and alienation by a comparison of rates of alienation in the two racial groups. Thus, there was a difference of $76 - 39 = 37$ percentage points (see bottom row Table 1) in the measured alienation for these two groups at that time and place. This difference might be considered a measure of association, since it, too, would be 0 if the tabulated frequencies showed statistical independence. Theoretically, such a coefficient (sometimes called the "percentage difference") could achieve a maximum of 100 if the distribution were as in Table 2. For comparability, it is customary to divide by 100, producing a coefficient sometimes referred to as d_{yx} (for X an independent variable and Y a dependent one), which has a maximum of 1.0, as do most other measures of association.

A problem in the use of d_{yx}, the difference between rates of alienation, arises when it is noted that the maximum absolute value of d_{yx} (that is, unity) is achieved only in the special situation illustrated in Table 2, where the row and column marginal distributions must necessarily be equal. This dependence of maximum value on equality of the

Table 2 — Hypothetical fourfold table showing $d_{yx} = 1$

		RACE (X)				ASSOCIATION MEASURE
		Negro	White	Total		
ALIENATION (Y)	High	25	0	25		
	Low	0	140	140		$Q = 1$
	Total	25	140	165		$d_{yx} = 1$
Per cent highly alienated		100%	0%			

marginal distributions can be misleading—for example, in comparisons of different communities where the number of Negroes remains about the same but the average level of alienation varies. In Table 3, Q is 1.0, while d_{yx} is less than 1.0. Table 3 represents the maximum discrepancy that could be observed in the rates of alienation, given the marginal distributions of the obtained sample of Table 1, and for this situation the maximum value of d_{yx} is only .65.

In addition to this independence of its maximum as marginals vary, Q has another desirable property, which is shared by d_{yx}: it has an operational interpretation. Both of these coefficients may be interpreted in a way that involves paired comparisons. Thus, $Q = ad/(ad + bc) - bc/(ad + bc)$ represents the probability of a pair having a "concordant ordering of the variables" less the probability of a pair having a "discordant ordering" when the pair is chosen at random from the set of pairs of individuals constructed by pairing each Negro member of the population with, in turn, each white member of the population and disregarding all pairs except those for which the alienation levels of the two individuals differ. By a "concordant ordering" is meant, in this illustration, a situation in which the Negro member of the pair has high alienation and the white has low alienation.

Closely related to this is the interpretation of $d_{yx} = ad/[(a + c)(b + d)] - bc/[(a + c)(b + d)]$, which is also the difference between the probability of a concordant pair and the probability of a discordant pair. In this case, however, the pair is chosen at random from a somewhat extended set of pairs: those in which the two individuals are of different races (here race is taken as X, the independent variable), whether or not they have different levels of alienation. Thus, d_{yx} asks directly: To what extent can one predict, in this population, the ordering of alienation levels of two persons, one of whom is Negro, the other white? On the other hand, Q asks: To what extent can one predict the ordering of alienation levels of persons of different races among those whose alienation levels are not equal? Thus, Q does not distinguish beween the situation of Table 2, in which no white persons are highly alienated, and that of Table 3, in which about 35 per cent of the whites are alienated.

There is, of course, no reason why one should expect to be able to summarize more than one aspect of a bivariate distribution in a single coefficient. Indeed, some investigators assume that several quantities will be necessary to describe a cross-tabulation adequately, and some go so far as to question the value of summary statistics at all, preferring to analyze and present the whole tabulation with r rows and c columns. This latter approach, however, leads to difficulties when two or more cross-classifications are being compared. The measures of association Q and d_{yx} may be extended to cross-tabulations with more than two categories per variable, as described below. The problems mentioned above extend to such cross-tabulations.

Although correlation theory developed historically from the theory of probability of continuous variables, one may consider the antecedents of "association theory" as lying more in the realm of the logic of classes. Yule's early paper on association traces the origin of his ideas to De Morgan, Boole, and Jevons (in addition to Quetelet); more recently it has been recalled that the American logician Peirce also worked on this problem toward the close of the nineteenth century [see *the biography of* PEIRCE].

The fact that most of the detailed remarks in this article refer to the work of British and American statisticians does not imply that developments have been confined to these countries. In Germany important early work was done by Lipps and Deuchler, and Gini in Italy has developed a series of statistical measures for various types of situations. [*See* GINI.] A comprehensive review of the

Table 3 — *Hypothetical fourfold table showing the maximum possible value of d_{yx} given the observed marginal distribution*

		RACE (X)			ASSOCIATION MEASURE
		Negro	White	Total	
ALIENATION (Y)	High	25	49	74	
	Low	0	91	91	$Q = 1$
					$d_{yx} = .65$
	Total	25	140	165	
Per cent highly alienated		100%	35%		

scattered history of this work is contained in Goodman and Kruskal (1954–1972, parts 1 and 2) and Kruskal (1958). Characteristically, investigators in one country have developed *ad hoc* measures without a knowledge of earlier work, and some coefficients have been "rediscovered" several times in recent decades.

Measurement characteristics and association

In recent years the distinction between the continuous and the discrete—between quantity and quality—often running through the work of logicians has become less sharp because of the addition of the idea of ordinal scales (rankings), which may be either discrete or continuous. Indeed, a whole series of "levels of measurement" has been introduced in the past twenty-five years (see Torgerson 1958, chapters 1–3), and because of their relevance to the measurement of relationships the matter warrants a brief statement here. [*Other discussions of scales of measurement can be found in* STATISTICS, DESCRIPTIVE, *article on* LOCATION AND DISPERSION. *See also* Torgerson 1968.]

As described above, in the past the term "correlation" has referred to a relation between quantities, while the term "association" has been reserved for the relation between qualities. For present purposes it is sufficient to elaborate this distinction only slightly by defining the three categories "metrics," "ordinals," and "nominals." Metrics (for example, height and weight), roughly corresponding to the earlier quantitative variables, are characterized by having an unambiguous and relevant unit of measurement. (Metrics are often further classified according to whether a meaningful zero point exists.) Ordinals (for example, amount of subjective agreement or appreciation) lack a unit of measurement but permit comparisons so that of two objects one can always say that one is less than the other, or that the two are tied, on the relevant dimension. Ordinal scales are for this reason designated "comparative concepts" by Hempel (1952, pp. 58–62). Nominals (for example, types of ideology, geographical regions) are simply classificatory categories lacking any relevant ordering but tied together by some more generic conception. Each of these scales may be either continuous or discrete, except nominals, which are inherently discrete.

Present usage generally retains the term "correlation" for the relationship between metrics (sometimes "correlation" is restricted to ρ and the correlation ratio), and "rank correlation" for the relationship between continuous ordinals, although the term "association" has also been used here. The terms "order association" and "monotonic correlation" have been used for the relationship between discrete ordinals, and the terms "association" and "contingency" are used to refer to the relationship between nominals. (Some statisticians prefer to reserve the term "contingency" for data arising from a particular sampling scheme.)

With ordinal data of a social or psychological origin the number of ordered categories is often small relative to the number of objects classified, in which case the bivariate distribution is most conveniently represented in the form of a cross-classification that is identical in format to the joint distribution over two nominals. Because of the identity of format, however, the different cases are sometimes confused, and as a result a measure of association is used that ignores the ordering of the categories, an aspect of the data that may be crucial to the purposes of the analysis. For this reason it is helpful to conceive of a cross-classification of two ordinals as rankings of a set of objects on two ordinals simultaneously, with ties among the rankings leading to a number of objects falling in each cell of the cross-classification.

Illustrative data, of a type very common in contemporary survey research, are presented in Table 4. In this instance the investigator utilized an additive index of income, education, and occupational prestige to obtain a socioeconomic ranking, which was then grouped into three ordered classes, with observations distributed uniformly over these classes insofar as possible. The alienation scale used in this study is derived from questionnaire responses indicating agreement with 24 such statements as "We are just so many cogs in the machinery of life"; "With so many religions around, one doesn't really know which to believe"; and "Sometimes I feel all alone in the world." Again the results are grouped into three nearly equal classes.

An appropriate coefficient for the measurement of the relation between socioeconomic status and alienation in Table 4 would be gamma (γ), a generalization of Yule's Q, introduced by Goodman and Kruskal (1954–1972, part 1, p. 747). Omitting details of computation, this measure provides in-

Table 4 — Cross-classification of socioeconomic status and alienation*

| | | SOCIOECONOMIC STATUS | | | |
		High	Medium	Low	Total
	High	23	62	107	192
ALIENATION	Medium	61	65	61	187
	Low	112	60	23	195
	Total	196	187	191	574

* Cell entries are frequencies.

Source: Reconstructed from Erbe 1964, table 4(B), p. 207.

formation of the following sort: Suppose one is presented with a randomly chosen pair of individuals from a population distributed in the form of Table 4, the pair being chosen with the restriction that the members be located in different status categories as well as different alienation categories. What are the chances that the individual in the higher status category, say, will also be more alienated? The probability of this event for the population of Table 4 is .215. Similarly, one may ask for the chance that in such a pair of individuals the person with higher status has *less* alienation—this complementary event has probability $1 - .215 = .785$. The difference between these probabilities, $-.570$, is the value of γ, a measure of association with many convenient properties, most of which have been noted in the comments on Q, which is a special case of γ.

The example in Table 4 cross-tabulates two variables in which the categories have a natural ordering. In contrast, Table 5 illustrates the crossing of two nominal or unordered classifications. In this connection, Table 5 is also useful for illustrating an issue that is sometimes confusing in practical applications. The columns of Table 5 identify occupational categories. Often social investigators will take those categories as representing an ordering along a dimension of "occupational status." In the present instance, the investigators did not make that choice. They wished, instead, to summarize the association between occupation and type of district in such a way that their conclusion was not influenced by the manner in which either set of categories was ordered. Accurate measurement of association requires that such a choice be properly made in the light of the purposes underlying the investigation. If those purposes do not provide a clear indication, it may be appropriate to report more than one measure, including a summary of the association that, as above, is sensitive to order and one that, as below, is not.

In summarizing Table 5, the investigators wished to quantify, for purposes of comparison, the extent of occupational segregation within school districts. They accomplished this by a measure of association also introduced by Goodman and Kruskal (1954–1972, part 1, p. 759) designated τ_b (not to be confused with the very different τ_b of Kendall, below). As a measure of association, the former τ_b is 0 when the frequencies are statistically independent and is $+1$ when the frequencies are so distributed that a knowledge of location of an individual on one variable—here, school district—enables perfect prediction of its location on the other variable. In Table 5 the value of τ_b is .075, indicating that, in the language of Goodman and Kruskal's interpretation, the

error of prediction (employing a "proportional" prediction) would be reduced only 7.5 per cent in making predictions of occupational status of father given a knowledge of school district (and the joint frequency distribution), as opposed to making that prediction with a knowledge only of the occupational distribution in the margin. Changing the order of rows or columns does not affect τ_b.

If the frequencies had been distributed as in Table 6, then τ_b would have been 1.0, since knowledge of school district would permit perfect prediction of occupational category.

Because τ_b is intended for nominal variables, the idea of a "negative" association has no meaning, and the measure therefore varies only between 0 and $+1$.

Whereas Goodman and Kruskal interpreted τ_b in prediction terms as described above, Rhodes and others (1965) in their application have presented an interpretation that motivates this measure as an index of spatial segregation, based on a model of chance interaction. Their work thus provides a good example of the derivation of a summary statistic that is appropriate to, and interpretable in the light of, the specific research hypothesis that is the investigator's concern. It is, for their research purposes, a largely irrelevant coincidence that their coefficient happens to be Goodman and Kruskal's τ_b, interpretable in another way.

A slight modification of Goodman and Kruskal's γ, retaining its sensitivity to ordering but yielding an asymmetric coefficient, d_{yx}, proposed by Somers (1962), is also illustrated above in the special case of a fourfold table. (In some references the population value of this measure is designated Δ_{yx}, with d_{yx} reserved for the sample value.) It is asymmetric in that, unlike γ, it will in general have a different value depending on which variable is taken as independent. If d_{xy} is its asymmetric mate, the product $d_{yx}d_{xy}$ yields the square of Kendall's τ_b ([1948] 1975, chapter 3), another useful measure of association for ordinals. This relation between τ_b and d_{yx} is analogous to that between the correlation coefficient and the regression coefficient. More abstractly, τ_b may be seen as a special case of a generalized correlation coefficient (cf. Kendall [1948] 1975, chapter 2), of which both Spearman's rho and Pearson's ρ are also special cases.

A number of other approaches to the measurement of association have been suggested. Basing their work on a criterion of proportional reduction in error, Leik and Gove (1971) introduced a number of measures for ordinal and nominal variables, including the mixed nominal-by-ordinal case. The proportional-reduction-in-error criterion (Costner 1965; see also Rockwell 1974) is analogous to the interpretation of the square of the correlation co-

Table 5 — Distribution of occupations of fathers of children in three school districts*

		FATHER'S OCCUPATION				
		Profes-sional	White-collar	Self-employed	Manual	Total
SCHOOL DISTRICT	A	92	174	68	39	373
	B	39	138	90	140	407
	C	11	111	37	221	380
	Total	142	423	195	400	1,160

* Cell entries are frequencies.

Source: Data from Wilson 1959, p. 839, as presented and analyzed in Rhodes et al. 1965, pp. 687–688.

Table 6 — Hypothetical distribution of frequencies enabling perfect prediction of occupation of father, given knowledge of school district

		FATHER'S OCCUPATION				
		Profes-sional	White-collar	Self-employed	Manual	Total
SCHOOL DISTRICT	A	0	373	0	0	373
	B	407	0	0	0	407
	C	0	0	0	380	380
	Total	407	373	0	380	1,160

efficient in terms of the proportion of explained variance. Hernes (1970) introduced unusual interpretations for a number of common measures of association in the 2×2 table, all based on continuous-time Markov processes in aggregate equilibrium. Some authors have stressed the usefulness of concepts derived from Shannon's information theory for defining measures of association and multiple association (see below) for nominal variables. See, for example, Theil (1967, chapters 1 and 4).

Of the measures of association in current use for tables larger than 2×2 as recently as the early 1950s, only Kendall's τ_b and τ_c were specifically intended for ordered categories, and none of them provided a way of interpreting an association that was intermediate in value between statistical independence and the concept of perfect association that was implicit in the particular measure. Among the measures in use at that time were C (the "coefficient of contingency"), Cramér's V, and Tschuprow's T (cf. Kendall & Stuart [1961] 1973, p. 577), the last two being efforts to obtain chi-square-based measures that achieved a maximum of 1 regardless of the number of columns and rows in the bivariate tabulation. In these last three measures, as in the concept of statistical independence on which the usual chi-square test is based, the distinction between ordered and unordered categories is irrelevant.

The work of Goodman and Kruskal (1954–1972) represented an important breakthrough, since these investigators provided operational interpretations for their measures, yielding an interpretable quantity regardless of the degree of association. Further, these interpretations clearly reflected, when appropriate, the fact that categories had a natural ordering. Progress continues to be made in the development of sampling theory for some of the newer measures of association; Goodman and Kruskal

(1954–1972, parts 3 and 4) also provided ways to calculate approximate confidence intervals for γ, λ, their τ_b and d_{yx}. Consequently, there now exist a number of useful tools in a field that did not exist before 1950.

There has been surprisingly little effort to relate interpretations of measures of association to the form of a bivariate frequency distribution for variables not measured on an interval scale. Reference has almost always been made simply to the presence or absence of statistical independence, or to the fact that one or another form of perfect association is defined by a particular measure of association. Perfect association defined by these measures often lacks a clear motivation. With variables measured on an interval scale, as noted above, bivariate normality has been a useful standard of reference for interpreting the correlation coefficient regardless of the degree of association between the variables. Another illustration of the way in which there may be a natural link between aspects of the form of a distribution and the interpretation of a measure of association is illustrated in the suggestion by Somers (1962, p. 808) that d_{yx} may be provided with a certain interpretation only when the distribution exhibits a regularity of form closely related to Yule's notion of isotropy.

The concept of isotropy (Yule & Kendall 1950, p. 57) is an early and largely ignored effort to distinguish formally one pattern of association or dependence between attributes from another. Isotropy refers to a distribution that is unequivocal regarding the direction of association between two variables. Yule noted that investigators often combine, particularly when categories are ordered, adjacent rows and columns to increase the number of observations in a category, and that this process is sometimes continued until each variable contains only two categories. Under such a transformation, he found that some distributions are invariant in the sense that the sign of the association in the resulting 2×2 table is not influenced by the manner of combining adjacent rows and columns. Such distributions he called isotropic. For example, in Table 4 one investigator might combine the first two columns and the last two rows, while another might make a different choice. Since the distribution is isotropic, both investigators would draw the same conclusion about the negative sign of the 2×2 association; in other situations they might draw opposite conclusions. Since the precise manner of combining categories is often arbitrary in social and psychological investigations, especially when the categories are ordered, this criterion of Yule's has considerable practical import.

In the systematic study of these distributional forms, the work of Lehmann (1966) may represent another breakthrough. Lehmann introduced several concepts of dependence for ordered categories in a bivariate distribution which, in effect, characterize with increasing stringency particular regularity patterns. These patterns are short of the various notions of "complete dependence" that have sometimes been discussed (cf. Kendall & Stuart 1961, p. 540), and are similar to Yule's notion of isotropy. Although Lehmann's purpose in formalizing such regularity was principally to evaluate the effectiveness of statistical tests of ordinal covariation, the usefulness of the concepts he introduced in providing interpretations for measures of association may be an important consequence of his work.

A different approach to the interpretation and evaluation of measures of association that also utilizes specified distributional forms is that of Agresti (1976) who examined the stability of a number of ordinal measures of association when various grids are placed on a bivariate normal distribution. He finds that Kendall's τ_b usually fares better in approximating the correlation in the underlying distribution than τ_c and certain other measures.

Another special kind of problem may arise in certain analyses where association is of interest. In many types of sociological analysis data are not available for cross-tabulating the *individual* observations but rather are available only on the "ecological" level, in the form of rates. Thus, for example, one might have available only the proportion of persons of high income and the proportion of persons voting for each political party for collectivities such as election districts. The investigator, however, may still be interested in the relation at the individual level. Duncan and others (1961) have discussed this problem in detail and have presented methods for establishing upper and lower bounds for the value of the individual association, given ecological data. Goodman (1959) has discussed and elaborated upon these procedures,

and he has shown how it is possible, with minimal assumptions, to make estimates of the value of the individual association from the ecological data.

Multiple and partial association

As should be clear from this brief review, the notion of association between two variables has been operationalized in numerous ways. This is also true, to a lesser extent, for the notion of partial association, which refers to the residual association between two variables remaining after removing the effects of a third variable, often called a control variable. As noted above, if an apparent association between two variables is effectively eliminated by the introduction of a third variable, the original relationship is termed spurious. Sometimes several control variables are employed.

The term "partial association" was introduced by Yule (1900) as an analogue to the notion of partial correlation and is often used whether or not the variables have a natural ordering. Some examples of the measurement of partial association are given in Goodman and Kruskal (1954–1972, part 1, p. 760). Later investigators who emphasized the distinction between ordinal and nominal categories sometimes replaced the term "partial association" with "partial order," "partial rank correlation," or "covariation for measures sensitive to order." (For examples, see Blalock 1974, part 3, and Rockwell 1974.) From a different point of view, several methods are developed in Coleman (1964, chapter 6).

When emphasis shifts to predicting a dependent variable from several independent variables, the term "multiple association" is often employed. A coefficient of multiple association is thus analogous to a multiple correlation coefficient defined for several interval scales. Most measures of multiple associations are insensitive to ordering (cf. Goodman & Kruskal 1954–1972, part 1, p. 761), but Ploch (1974, pp. 356–357) has suggested an approach for ordinal variables.

In this rapidly developing field, a new approach is represented by the work of Goodman (1970) and Bishop et al. (1975), who utilize a strategy for discrete, unordered variables resembling analysis of variance but based on chi-square. It is likely that this approach will be emphasized in the future. [*See* SURVEY ANALYSIS; COUNTED DATA.]

ROBERT H. SOMERS

[*See also* MULTIVARIATE ANALYSIS, *article on* CORRELATION METHODS; SURVEY ANALYSIS, *article on* THE ANALYSIS OF ATTRIBUTE DATA; TABULAR PRESENTATION.]

BIBLIOGRAPHY

AGRESTI, ALAN 1976 The Effect of Category Choice on Some Ordinal Measures of Association. *Journal of the American Statistical Association* 71:49–55.

BISHOP, YVONNE M. M.; FIENBERG, STEPHEN E.; and HOLLAND, PAUL W. 1975 *Discrete Multivariate Analysis.* Cambridge, Mass.: M.I.T. Press.

BLALOCK, HUBERT M. JR. (editor) 1974 *Measurement in the Social Sciences: Theories and Strategies.* Chicago: Aldine.

COLEMAN, JAMES S. 1964 *Introduction to Mathematical Sociology.* New York: Free Press.

COSTNER, HERBERT L. 1965 Criteria for Measures of Association. *American Sociological Review* 30:341–353.

DUNCAN, OTIS DUDLEY; CUZZORT, RAY P.; and DUNCAN, BEVERLY 1961 *Statistical Geography: Problems in Analyzing Areal Data.* New York: Free Press.

ELASHOFF, JANET D. 1971 Measures of Association Between a Dichotomous and a Continuous Variable. Pages 298–300 in American Statistical Association, Social Statistics Section, *Proceedings.* Washington: The Association.

ERBE, WILLIAM 1964 Social Involvement and Political Activity: A Replication and Elaboration. *American Sociological Review* 29:198–215.

GOLDBERGER, ARTHUR S. 1964 *Econometric Theory.* New York: Wiley.

GOODMAN, LEO A. 1959 Some Alternatives to Ecological Correlation. *American Journal of Sociology* 64:610–625.

GOODMAN, LEO A. 1963 On Methods for Comparing Contingency Tables. *Journal of the Royal Statistical Society* Series A 126:94–108.

GOODMAN, LEO A. 1970 The Multivariate Analysis of Qualitative Data: Interactions Among Multiple Classifications. *Journal of the American Statistical Association* 65:226–256.

GOODMAN, LEO A.; and KRUSKAL, WILLIAM H. 1954–1972 Measures of Association for Cross-classifications. Parts 1–4. *Journal of the American Statistical Association* 49 [1954]:732–764; 54 [1959]:123–163; 58 [1963]:310–364; 67 [1972]:415–421. → Part 2: Further Discussion and References. Part 3: Approximate Sampling Theory. Part 4: Simplification of Asymptotic Variances.

GUTTMAN, LOUIS 1941 An Outline of the Statistical Theory of Prediction: Supplementary Study B-1. Pages 253–318 in Social Science Research Council, Committee on Social Adjustment, *The Prediction of Personal Adjustment,* by Paul Horst et al. New York: The Council.

HEMPEL, CARL G. 1952 Fundamentals of Concept Formation in Empirical Science. Volume 2, number 7, in *International Encyclopedia of Unified Science.* Univ. of Chicago Press.

HERNES, GUDMUND 1970 A Markovian Approach to Measures of Association. *American Journal of Sociology* 75:992–1011.

KENDALL, MAURICE G. (1948) 1975 *Rank Correlation Methods.* 4th ed. London: Griffin; New York: Hafner.

KENDALL, MAURICE G.; and STUART, ALAN (1961) 1973 *The Advanced Theory of Statistics.* Volume 2: *Inference and Relationship.* 3d ed. London: Griffin; New York: Hafner. → Early editions of Volume 2, first published in 1946, were written by Kendall alone.

Stuart became a joint author on later, renumbered editions in the three-volume set.

KRUSKAL, WILLIAM H. 1958 Ordinal Measures of Association. *Journal of the American Statistical Association* 53:814–861.

LAZARSFELD, PAUL F. (1955) 1962 Interpretation of Statistical Relations as a Research Operation. Pages 115–125 in Paul F. Lazarsfeld and Morris Rosenberg (editors), *The Language of Social Research: A Reader in the Methodology of Social Research.* New York: Free Press.

LEHMANN, E. L. 1966 Some Concepts of Dependence. *Annals of Mathematical Statistics* 37:1137–1153.

LEIK, ROBERT; and GOVE, WALTER 1971 Integrated Approach to Measuring Association. Pages 279–301 in Herbert L. Costner (editor), *Sociological Methodology, 1971.* San Francisco: Jossey-Bass.

LEWIS, B. N. 1962 On the Analysis of Interaction in Multi-dimensional Contingency Tables. *Journal of the Royal Statistical Society* Series A 125:88–117.

NETER, JOHN; and MAYNES, E. SCOTT 1970 On the Appropriateness of the Correlation Coefficient with a 0,1 Dependent Variable. *Journal of the American Statistical Association* 65:501–509.

PLOCH, DONALD R. 1974 Ordinal Measures of Association and the General Linear Model. Pages 343–368 in Hubert M. Blalock, Jr. (editor), *Measurement in the Social Sciences: Theories and Strategies.* Chicago: Aldine.

RHODES, ALBERT L.; REISS, ALBERT J. JR.; and DUNCAN, OTIS DUDLEY 1965 Occupational Segregation in a Metropolitan School System. *American Journal of Sociology* 70:682–694.

ROCKWELL, RICHARD C. 1974 Continuities in Methodological Research. Part 1. *Social Forces* 53:165–253.

SOMERS, ROBERT H. 1962 A New Asymmetric Measure of Association for Ordinal Variables. *American Sociological Review* 27:799–811.

SOMERS, ROBERT H. 1964 Simple Measures of Association for the Triple Dichotomy. *Journal of the Royal Statistical Society* Series A 127:409–415.

SROLE, LEO 1956 Social Integration and Certain Corollaries: An Exploratory Study. *American Sociological Review* 21:709–716.

TATE, ROBERT F. 1955 Applications of Correlation Models for Biserial Data. *Journal of the American Statistical Association* 50:1078–1095.

TEMPLETON, FREDRIC 1966 Alienation and Political Participation: Some Research Findings. *Public Opinion Quarterly* 30:249–261.

THEIL, HENRI 1967 *Economics and Information Theory.* Chicago: Rand McNally.

TORGERSON, WARREN S. 1958 *Theory and Methods of Scaling.* New York: Wiley.

TORGERSON, WARREN S. 1968 Scaling. Volume 14, pages 25–39 in *International Encyclopedia of the Social Sciences.* Edited by David L. Sills. New York: Macmillan and Free Press.

WILSON, ALAN B. 1959 Residential Segregation of Social Classes and Aspirations of High School Boys. *American Sociological Review* 24:836–845.

YULE, G. UDNY 1900 On the Association of Attributes in Statistics, With Illustrations From the Material of the Childhood Society.... Royal Society of London, *Philosophical Transactions* Series A 194:257–319.

YULE, G. UDNY 1912 On the Methods of Measuring Association Between Two Attributes. *Journal of the Royal Statistical Society* 75:579–652. → Includes ten pages of discussion.

YULE, G. UDNY; and KENDALL, MAURICE G. 1950 *An Introduction to the Theory of Statistics.* 14th ed., rev. & enl. London: Griffin; New York: Hafner. → Yule was the sole author of the first edition (1911). Kendall, who became a joint author on the eleventh edition (1937), revised the current edition and added new material to the 1965 printing.

STATISTICS AS LEGAL EVIDENCE

● *This article was completely revised for this volume.*

A major function of judicial procedures is to resolve issues according to some canon of probability; in criminal cases by evidence that establishes guilt "beyond reasonable doubt," in civil cases by "preponderance of evidence." In spite of such probabilistic language, the law has been slow in accepting statistical methods to make these terms more precise. On the whole, the law appreciates descriptive statistics, but it has been reluctant to acknowledge the evidentiary value of statistical inference. In this article three types of statistical evidence will be discussed: descriptive statistics, inferential statistics, and the evaluation of specific proofs.

Descriptive statistics

Official publications of census-like counts that describe statistical aggregates, such as those published by the Bureau of the Census, are as a rule admissible evidence, in some jurisdictions as a matter of law. Admissibility, of course, does not shield such statistics from critical examination. Although counts based on sampling have encountered hurdles, such sample counts have gained general acceptance, partly because the election polls have established the power of sampling and partly because in some litigations essential knowledge for large aggregates can be secured only through sampling. Market shares based on store samples, mineral deposits from a sample of test drillings, and depreciation estimates from a sample of plant facilities are examples that have come before courts and administrative agencies.

The power of sampling was documented in a law case in which the size of a tax refund was to be determined from the sum of several hundred thousand sales slips. The plaintiff offered to accept an estimate from a sample, minus two-thirds of the estimated standard deviation of that estimate. The court insisted on a complete count—which was 1 per cent off the sample value.

A new path was broken when, in a law case, the activities of the House Un-American Activities Committee over the course of two decades were described through content analysis of the question-and-answer record of a probability sample of all the witnesses who had appeared before the committee. (See Sullivan et al. 1976; Zeisel & Stamler 1976.)

The hearsay objection. The legal hurdle for sample estimates is somewhat higher if they are secured through interviewing. To the law, this is technically hearsay evidence and some jurisdictions will not admit surveys for that reason. A Massachusetts court, for example, rejected an offer of proof (through interviews of a probability sample of eligible persons) that members of the grand jury had not been randomly selected from the population, as the statute required.

Circumventing hearsay objections is sometimes possible. In the Massachusetts case, for instance, the proposition was subsequently proved, albeit with less precision, by sampling and tabulating data from the city directory. When the geographic range of patrons of a drive-in movie theater was at issue, the data were obtained not by interviews with the patrons but by recording and subsequently tracing the license numbers of the parked automobiles. The field workers who jotted down the numbers were competent witnesses under the hearsay rule.

A court may occasionally try to remedy the hearsay defect by testing the survey evidence through testimony of some of the original interviewees. This is a dangerous road to take because, in contrast to interviews, examination in court is a much more artificial interrogation. Moreover, the procedure raises a problem of interviewing ethics, since survey interviewees are, at least implicitly, assured of the full anonymity of their answers. If it became known that such protection cannot be guaranteed, resistance to interviewing might increase prohibitively. One possible way of ensuring privacy to survey respondents is to detach their names from the questionnaires, thus making it possible to present their identity to the court but making it impossible to connect any individual with a set of specific answers. Thus it may be verified that the respondents were in fact interviewed.

State of mind surveys. One of the law's specific exceptions to the hearsay rule has spawned a great amount of sample-survey evidence. That exception allows surveys if the interview response does not concern the truth of the matter asserted but only the interviewee's state of mind. If a respondent is asked, for instance, whether two trademarks represent the same or different manufacturers, the survey maker knows the true facts; the object is to find out whether the interviewee knows them. The following are some of the topics that have been covered by such surveys.

Consumer awareness.   Sometimes the law provides that a trademark or advertising slogan can be protected only so long as it is in sufficient use, that is, sufficiently established in the consumer's mind (*Verkehrsgeltung* in German trademark law).

Confusion of trademarks.   A mark so similar to an existing trademark that the two are likely to be confused may not be registered. The similarity might be created by words, graphic design, or color, or by any combination of the three.

Meaning of trademarks.   The requirement of truth in trademark labeling occasionally imposes the burden of finding out what certain words mean; the issue may be, for instance, whether a term such as "English lavender" or "farmer bread" denotes true origin or merely a type of product.

Proprietary or generic name.   Names, originally protected as brand designations, lose their proprietary character if they have in fact become generic terms, designating the type of product rather than one of its brands. In the United States, "thermos"—to name only one of many—has become a generic term in this fashion. "Vaseline" has lost its proprietary character in some European countries but not in the United States, where it is a specific brand of petroleum jelly.

Misleading advertising.   The Federal Trade Commission and the Food and Drug Administration in the United States have the duty to prohibit misleading advertising. In such procedures two issues arise: one factual—what the product actually does —and one psychological—what the public, guided by the advertising, believes it does.

Community prejudice.   If, for one reason or another, prejudice in a community against a defendant in a criminal proceeding has reached a stage where he can no longer receive a fair trial, the court may postpone the trial or order a "change of venue," that is, move the trial to another community. Proof of such prejudice may be attempted through a public opinion poll among potential jurors.

Design of experiments.   Surveys designed to test a particular legal issue often go beyond mere descriptive recording. In a confusion test the interviewee may be shown a sequence of competing brand products and asked to recall them; in a test of advertising effectiveness, the interviewee may be asked to rate two competing brands on certain claimed properties. The design may involve randomly arranged control groups, to whom a different arrangement is presented—designed, for instance, to determine the amount of unavoidable confusion

between clearly dissimilar brands. The difference between the confusion found in the control group and that found in the experimental groups is then the amount chargeable to brand similarity. [*See* EXPERIMENTAL DESIGN.]

Preparation of legal surveys.   In addition to the general professional ground rules for preparing surveys [*see* SAMPLE SURVEYS], there are problems peculiar to legal surveys. They derive from a variety of sources: from the peculiarly strict requirements of legal proof; from issues of law that cannot be anticipated with precision because they may be decided only during the very trial for which the evidence is prepared; and, unless it is a "state of mind" survey, from the hearsay rule, as discussed above.

Requirements of proof.   The peculiar requirements of legal evidence affect the preparation of sample surveys in several ways. An important one is the prospect of double scrutiny by opposing counsel. The first scrutiny occurs when the admissibility of the particular piece of evidence is debated; at this stage, efforts may be made to prove that the survey is, on its face, irrelevant to the litigated issue, or that is has such technical flaws that the court would be well advised to refuse it. If the offered evidence overcomes this hurdle, its probative value is later explored in even greater detail through cross-examination.

This double scrutiny is often overly exacting. The discovery of but one serious flaw may endanger the entire piece of evidence; the doctrine of *falsus in uno, falsus in omnibus* is sometimes the ground for disbelieving a witness's entire testimony if it is found to be untrue in a single instance, and such a flaw may also hurt the expert witness who presents the survey evidence. His task will be facilitated if every step in the survey procedure is meticulously documented: the definition of the universe to be sampled, the details of the sample design, the process of sample selection, the communications to the interviewers, their control in the field, and finally, the analysis of their reports—all should be documented by the respective research instruments.

To avoid potential bias, the nature of the issue— and, if possible, its very existence—should not be divulged to them. If they must learn the purpose of the survey, they should not learn which side the sponsor is on.

Preparing for legal contingencies.   One of the major difficulties in preparing survey evidence is that some legal uncertainties are likely to be decided only during the trial for which the evidence was prepared. In such situations it is good practice to conduct the survey in a flexible manner.

One of these uncertainties is the definition of the

relevant universe to be sampled. In a trademark confusion case, for instance, is the population those who were purchasers of the particular brands, or of *any* brand of this type of product, or simply all potential customers? One way of solving this problem is to sample the widest universe and to tabulate the results for each subgroup separately.

Another uncertainty is the level of precision at which the survey answers will be relevant. It is sometimes impossible to focus the interviewee's attention on a particular issue without giving him information about the issues in litigation. Thus, a balance must be struck between a more precise answer and possible bias. As an example, consider the following sequence of narrowing focus from a questionnaire designed to explore the respondent's knowledge of a certain merger:

Question: Do you recall any mergers of cement companies in this area during the last two or three years?

[*If no reference is made to the litigated merger, ask:*]

Question: Did you know that the XX Corporation merged with another company?

[*If the answer is yes, ask:*]

Question: Do you happen to know the name of that other company?

In most cases it is preferable to begin with uncontaminated, unaided questions and to delay any contamination as long as possible. At the very end of the interview, even leading questions may be proper, provided their character is openly admitted, just as such questions in cross-examination in court sometimes have their justification.

Statistical inference

Statistical proofs involving the probability of an event, the identification of an individual, or causation pose more difficult problems to the law. The statistical proof could arise from a defense attempt to show that the underlying event had a reasonable probability of occurrence by chance, or from an accuser's attempt to show that that probability was so unreasonably small that it amounted to proof of the opposite.

Probabilistic evidence for the occurrence of an event was explicitly handled by a Swedish court. In a trial for overtime parking, a boundary was put to what constitutes insufficient probability for a finding of guilty. The police constable had marked the position of the valves on two tires on a standard sketch, accurate to the nearest "hour," one valve at the "one o'clock" position and the other at the "twelve o'clock" position. Returning after a time lapse greater than the permitted length of parking, the constable found both valves in the same posi-

tions they had been in earlier. The defendant claimed he had left the place and returned. The Court of Appeals considered odds of $(12 \times 12 =)$ 144:1 insufficient and declared that registering the position of all four tire valves would have been sufficient—odds of $(12^4 =)$ 20,736:1 ("Parkeringsfrägor" 1962, pp. 24, 25). The court may be pardoned for having overstated the odds by calculating them under the assumption that the positions of the valve markings are statistically independent of each other, which almost certainly is not true.

Alleged cheating has evoked similar statistical proofs. In a trial for income tax evasion involving an illegal lottery wheel, the prosecution offered proof that by the rules of the lottery wheel, and the agreed amount of betting, the profits expected according to probability were a multiple of what the defendant had reported. The trial court (in the 1960s) rejected this offer of proof as irrelevant.

The probabilistic evidence in another alleged cheating case did reach the court. An entrance examination attracted attention because the candidate on his second try had improved his score by a large margin. The score was disallowed when it was discovered that the candidate shared with a neighbor 100 correct answers, and 43 identical answers to 45 incorrectly answered questions, although each question offered a choice of 4 incorrect answers. The candidate sued but withdrew when confronted with an affidavit that the probability of this similarity (or a more extreme one) occurring by chance is 1 in a figure with 23 zeros.

In an internal investigation of alleged rigging of a civil service examination, a chi-square test for goodness of fit showed that the distribution of the obtained grades, or a more extreme one, was highly improbable under the hypothesis of no cheating. There was, in the otherwise normal distribution of the scores a suspicious elevation just above, and a complementing depression just below, the passing mark. This led to further investigation and eventual proof of chicanery, albeit not proof in court (McCann Associates 1966, p. 16).

Proof of discrimination (by race, income, sex, and so on) is often attempted by comparing actual with expected distributions. When Dr. Benjamin Spock, whose books have helped millions of women to raise their children, was tried in 1968 for violation of the military service laws, no women sat on his jury. An investigation of 45 jury selections in that court revealed that Dr. Spock's trial judge had found a private way of reducing the proportion of women in all his juries from the normal 29 per cent to an average of 15 per cent. The probability that this or a more extreme reduction could occur by chance was calculated as 1 in a figure with 18 zeros.

In *Castaneda* v. *Partida* (97 S.Ct. 1272, 1977) a Texas prisoner claimed he had been indicted by an improperly constituted grand jury that included only 39 per cent Mexican–Americans in a county where they represented 79 per cent of the population. The U.S. Supreme Court found the petition justified and explains in footnote 17 to the majority opinion the technical rationale and result of the relevant calculations.

In a California case (*Collins*), calculations came under special scrutiny. Two defendants had been charged with robbery, primarily because they fitted an unusual combination of circumstances established through independent testimony. The robbery had been committed by a white woman with blond hair in a ponytail and a black man with mustache and beard driving a partly yellow automobile. A couple fitting that description and owning such an automobile was arrested. During the trial, the prosecutor sought to establish, through expert testimony, that the likelihood of another pair fitting this unusual set of circumstances was infinitesimally small. The jury convicted, but the California Supreme Court, based on a detailed statistical criticism of the argument, acquitted the defendants.

Causal inferences.   As long as cause–effect relationships are inferred from experimentally controlled data, the causal inference in the particular experiment should not be at issue (though the correspondence between the experiment and reality may be). But if causal inferences are drawn from observational, naturally occurring data, new problems arise. Issues such as the following have occupied the courts: Were certain price increases caused by monopolistic arrangements? Was the decline of a brand caused by the "comparing" advertisements of a competitor? Do executions of murderers reduce the number of willful homicides? Answers to these questions must include a demonstration that plausible factors other than the proposed ones had in fact not caused the effect. This is at times difficult to prove. The application of a complicated analytic apparatus, such as regression analysis, does not provide automatic proof. In such situations the courts will do what good scientists do, namely, consider such offers of proof together with other evidence. If all the evidence points in the same direction, our confidence is increased, unless some of the alleged proofs can be shown to be faulty. The courts, however, must be encouraged to weigh the magnitude of faults, because hardly any research operation is free of faults and some

affect the weight of evidence less heavily than others. [*See* CAUSATION.]

Evaluation of specific proofs

Some statistical studies are not in themselves evidence but are, rather, evaluations of types of evidence, some of them more or less standardized testing procedures.

For example, psychologists, beginning with Münsterberg (1908), have been occupied with the problem of reliability of observation and testimony. They have accumulated a great deal of statistics on the difficulties of correctly observing moving objects or quickly developing scenes, of correctly identifying voices, on the reliability of children's testimony, or even on the reliability of psychiatric diagnosis (Marston 1924; Hutchins & Slesinger 1928; Gardner 1933, pp. 391, 407; Messerschmidt 1933, p. 422; McGehee 1937, p. 249).

Of more immediate value is the statistical evaluation of the evidentiary power of specific proofs, such as blood tests for the establishment of paternity (Ross 1958, p. 466; Steinhaus 1954; Łukaszewicz 1955), lie detector tests (Levitt 1955, pp. 632, 637), or with respect to psychiatric diagnosis introduced as evidence (Schmidt & Fonda 1956, p. 262; Ash 1949, p. 272).

In 1978, statistical evidence reached a new level of acceptance when the U.S. Supreme Court, in the case of *Ballew* v. *Georgia*, declared a jury of less than six members as unconstitutional; the Court's opinion relied almost entirely on statistical studies and experiments.

HANS ZEISEL

BIBLIOGRAPHY

ASH, PHILIP 1949 The Reliability of Psychiatric Diagnoses. *Journal of Abnormal and Social Psychology* 44:272–276.

BAADE, HANS W. 1961 Social Science Evidence and the Federal Constitutional Court of West Germany. *Journal of Politics* 23:421–461.

BARKSDALE, HIRAM 1957 *The Use of Survey Research Findings as Legal Evidence.* Pleasantville, N.Y.: Printers' Ink Books.

BLUM, WALTER J.; and KALVEN, HARRY JR. 1956 The Art of Opinion Research: A Lawyer's Appraisal of an Emerging Science. *University of Chicago Law Review* 24:1–69.

FAIRLEY, WILLIAM B. 1973 Probabilistic Analysis of Identification Evidence. *Journal of Legal Studies* 2:493–513.

FAIRLEY, WILLIAM B.; and MOSTELLER, FREDERICK 1974 A Conversation About *Collins. University of Chicago Law Review* 41:242–253.

FEDERAL JUDICIAL CENTER 1973 *Manual for Complex Litigation, With Amendments to January 1, 1973.* Chicago: Commerce Clearing House.

FINKELSTEIN, MICHAEL O. 1973 Regression Models in Administrative Proceedings. *Harvard Law Review* 86: 1442–1475.

GARDNER, DILLARD S. 1933 The Perception and Memory of Witnesses. *Cornell Law Quarterly* 18:391–409.

HUTCHINS, ROBERT M.; and SLESINGER, DONALD 1928 Some Observations on the Law of Evidence-memory. *Harvard Law Review* 41:860–873.

KLEIN, LAWRENCE R. et al. 1976 Review of the Effectiveness of the Death Penalty. Report to the Panel on Deterrence, National Academy of Sciences.

LEVIN, A. LEO 1956 Authentication and Content of Writings. *Rutgers Law Review* 10:632–646.

LEVITT, EUGENE E. 1955 Scientific Evaluation of the "Lie Detector." *Iowa Law Review* 40:440–458.

ŁUKASZEWICZ, J. 1955 O dochodzeniu ojcostwa (On Establishing Paternity). *Zastosowania matematyki* 2: 349–379. → Includes a summary in English.

McCANN ASSOCIATES 1966 Chicago Metropolitan Sanitary Distribution, First Report. Unpublished manuscript.

McGEHEE, EUGENE E. 1937 The Reliability of the Identification of the Human Voice. *Journal of General Psychology* 17: 249–271.

MARSTON, WILLIAM M. 1924 Studies in Testimony. *Journal of Criminal Law and Criminology* 15:5–31.

MESSERSCHMIDT, RAMONA 1933 The Suggestibility of Boys and Girls Between the Ages of Six and Sixteen Years. *Journal of Genetic Psychology* 43:422–437.

MÜNSTERBERG, HUGO (1908) 1933 *On the Witness Stand: Essays on Psychology and Crime.* New York: Broadman. → For a slashing retort, see Wigmore (1909).

NOELLE-NEUMANN, ELISABETH; and SCHRAMM, CARL 1961 *Umfrageforschung in der Rechtspraxis.* Weinheim (Germany): Chemie.

Parkeringsfrägor: II. Tillförlitligheten av det S. K. klocksystemet för parkeringskontroll. 1962 *Svensk juristidining* 47:17–32. → Summarized in Hans Zeisel and Harry Kalven, Jr., "Parking Tickets and Missing Women: Statistics and the Law," pages 102–111 in Judith M. Tanur et al. (editors), *Statistics: A Guide to the Unknown* (San Francisco: Holden-Day, 1972).

PENNSYLVANIA, UNIVERSITY OF, LAW SCHOOL, INSTITUTE OF LEGAL RESEARCH 1956 *Evidence and the Behavioral Sciences.* Edited by A. Leo Levin. Philadelphia: The School.

RICHARDSON, JAMES R. 1961 *Modern Scientific Evidence: Civil and Criminal.* Cincinnati: Anderson.

ROSS, ALF 1958 The Value of Blood Tests as Evidence in Paternity Cases. *Harvard Law Review* 71:466–484.

SCHMIDT, HERMANN O.; and FONDA, CHARLES P. 1956 The Reliability of Psychiatric Diagnosis: A New Look. *Journal of Abnormal and Social Psychology* 52:262–267.

STEINHAUS, HUGO 1954 The Establishment of Paternity. Wrocławskie Towarzystwa Naukowe, *Prace wrocławskiego towarzystwa naukowego* Series A No. 32.

SULLIVAN, THOMAS P.; KAMIN, CHESTER T.; and SUSSMAN, ARTHUR M. 1976 The Case Against HUAC: The Stamler Litigation. *Harvard Civil Rights–Civil Liberties Law Review* 11, no. 2:243–262.

TRIBE, LAURENCE H. 1971 Trial by Mathematics: Precision and Ritual in the Legal Process. *Harvard Law Review* 84:1329–1393. → A comment by Michael O. Finkelstein and William B. Fairley appears on pages 1801–1809.

WEBB, EUGENE J.; CAMPBELL, DONALD T.; SCHWARTZ, RICHARD; and SECHREST, LEE 1966 *Unobtrusive Measures: Nonreactive Research in the Social Sciences.* Chicago: Rand McNally.

WIGMORE, JOHN H. 1909 Professor Muensterberg and the Psychology of Testimony: Being a Report of the Case of Cokestone v. Muensterberg. *Illinois Law Review* 3:399–434.

ZEISEL, HANS 1960 The Uniqueness of Survey Evidence. *Cornell Law Quarterly* 45:322–346.

ZEISEL, HANS 1969 Dr. Spock and the Case of the Vanishing Women Jurors. *University of Chicago Law Review* 37:1–18. → Summarized in Hans Zeisel and Harry Kalven, Jr., "Parking Tickets and Missing Women: Statistics and the Law," pages 102–111 in Judith M. Tanur et al. (editors), *Statistics: A Guide to the Unknown* (San Francisco: Holden-Day, 1972).

ZEISEL, HANS; and STAMLER, ROSE 1976 The Case Against HUAC: The Evidence; A Content Analysis of the HUAC Record. *Harvard Civil Rights–Civil Liberties Law Review* 11, no. 2:263–298.

STEPWISE REGRESSION
See LINEAR HYPOTHESES, *article on* REGRESSION.

STOCHASTIC PROCESSES
See MARKOV CHAINS; QUEUES; TIME SERIES.

STRUCTURAL EQUATIONS
See LINEAR HYPOTHESES, *article on* REGRESSION; MULTIVARIATE ANALYSIS, *article on* CORRELATION METHODS; PREDICTION; SIMULTANEOUS EQUATION ESTIMATION; SOCIAL MOBILITY; SURVEY ANALYSIS.

"STUDENT"
See GOSSET, WILLIAM SEALY.

SUBJECTIVIST METHODS
See BAYESIAN INFERENCE; PROBABILITY, *article on* INTERPRETATIONS.

SUFFICIENCY

Sufficiency is a term that was introduced by R. A. Fisher in 1922 to denote a concept in his theory of point estimation [*see* FISHER, R. A.]. As subsequently extended and sharpened, the concept is used to simplify theoretical statistical problems of all kinds. It is also used, sometimes questionably, in applied statistics to justify certain summarizations of the data, for example, reporting only sample means and standard deviations for metric data or reporting only proportions for counted data.

The sufficiency concept may be explained as follows. Suppose that the probabilities of two given samples have a ratio that does not depend on the unknown parameters of the underlying statistical

Table 1 — Sample points in coin-tossing experiment

NUMBER OF HEADS

0	1	2	3	4
TTTT	HTTT	HHTT	THHH	HHHH
	THTT	HTHT	HTHH	
	TTHT	HTTH	HHTH	
	TTTH	THHT	HHHT	
		THTH		
		TTHH		

model. Then it will be seen that nothing is gained by distinguishing between the two samples; that is, nothing is lost by agreeing to make the same inference for both of the samples. To put it another way, the two samples may be consolidated for inference purposes without losing information about the unknown parameters. When such consolidation can be carried out for many possible samples, the statistical problem becomes greatly simplified.

The argument can best be given in the context of a simple example. Consider tossing a coin four times, with "heads" or "tails" observed on each toss. There are $2^4 = 16$ possible results of this experiment, so the sample space has 16 points, which are represented in Table 1 [see PROBABILITY, *article on* FORMAL PROBABILITY]. For example, the point *THHT* represents the experimental result: tosses 1 and 4 gave tails, while tosses 2 and 3 gave heads. For later convenience, the 16 points are arranged in columns according to the number of occurrences of *H*.

If one makes the usual assumptions that the four tosses are independent and that the probability of heads is the same (say, p) on each toss, then it is easy to work out the probability of each point as a function of the unknown parameter, p: for example, $Pr(THHT) = p^2(1 - p)^2$ [see DISTRIBUTIONS, STATISTICAL, *article on* SPECIAL DISCRETE DISTRIBUTIONS]. In fact, each of the 6 points in column 3 has this same probability. Therefore, the ratio of the probabilities of any 2 points in column 3 has a fixed value, in fact the value 1, whatever the value of p may be, and, as stated earlier, it is not necessary to distinguish between the points in column 3. A similar argument shows that the 4 points in column 2 need not be distinguished from each other; the same is true for the 4 points in column 4. Thus, the sample space may be reduced from the original 16 points to merely the 5 columns, corresponding to the number of *H*'s. No further reductions are justified, since any 2 points in different columns have a probability ratio that depends on p.

To see intuitively why the consolidations do not cost any useful information, consider a statistician who knows that the experiment resulted in one of the 6 points in column 3 but who does not know just which of the 6 points occurred. Is it worth his while to inquire? Since the 6 points all have the same probability, $p^2(1 - p)^2$, the *conditional* probability of each of the 6 points, given that the point is one of those in column 3, is the known number $\frac{1}{6}$ [see PROBABILITY, *article on* FORMAL PROBABILITY]. Once the statistician knows that the sample point is in column 3, for him to ask "which point?" would be like asking for the performance of a random experiment with known probabilities of outcome. Such an experiment can scarcely produce useful information about the value of p, or indeed about anything else.

Another argument has been advanced by Halmos and Savage (1949). Our statistician, who knows that the observed sample point is in column 3 but who does not know which one of the 6 points was observed, may try to reconstruct the original data by selecting one of the 6 points at random (for example, by throwing a fair die or by consulting a table of random numbers [see RANDOM NUMBERS]). The point he gets in this way is not likely to be the point actually observed, but it is easy to verify that the "reconstructed" point has exactly the same distribution as the original point. If the statistician now uses the reconstructed point for inference about p in the same way he would have used the original point, the inference will perform exactly as if the original point had been used. If it is agreed that an inference procedure should be judged by its performance, the statistician who knows only the column, and who has access to a table of random numbers, can do as well as if he knew the actual point. In this sense, the consolidation of the points in each column has cost him nothing.

When a (sample) space is simplified by consolidations restricted to points with fixed probability ratio, the simplified space is called *sufficient:* the term is a natural one, in that the simplified space is "sufficient" for any inference for which the original space could have been used. The original space is itself always sufficient, but one wants to simplify it as much as possible. When all permitted consolidations have been made, the resulting space is called *minimally* sufficient (Lehmann & Scheffé 1950–1955). In the example the 5-point space consisting of the five columns is minimally sufficient; if only the points of column 3 had been consolidated, the resulting 11-point space would be sufficient, but not minimally so.

It is often convenient to define or describe a consolidation by means of a statistic, that is, a function defined on the sample space. For example, let *B* denote the number of heads obtained in the four tosses. Then *B* has the value 0, 1, 2, 3, 4 for the points in columns 1, 2, 3, 4, 5, respectively. Knowledge of the value of B is equivalent to knowledge of the column. It is then reasonable to call B a (*minimal*) *sufficient statistic*. ($B + 2$, B^3, and $\sqrt{B}$, for example, would also be minimal sufficient statistics.) More generally, a statistic is sufficient if it assigns the same value to 2 points only if they have a fixed probability ratio. In Fisher's expressive phrase, a sufficient statistic "contains all the information" that is in the original data.

The discussion above is, strictly speaking, correct only for discrete sample spaces. The concepts extend to the continuous case, but there are technical difficulties in a rigorous treatment because of the nonuniqueness of conditional distributions in that case. These technical problems will not be discussed here. (For a general treatment, see Volume 2, chapter 17 of Kendall & Stuart 1958–1966, and chapters 1 and 2 of Lehmann 1959, where further references to the literature may be found.)

The discovery of sufficient statistics is often facilitated by the Fisher–Neyman factorization theorem. If the probability of the sample point (or the probability density) may be written as the product of two factors, one of which does not involve the parameters and the other of which depends on the sample point only through certain statistics, then those statistics are sufficient. This theorem may be used to verify these examples: (*i*) If B is the number of "successes" in n Bernoulli trials (n independent trials on each of which the unknown probability, p, of success is the same), then B is sufficient. (*ii*) If $X_1, X_2, \cdots, X_n$ is a random sample from a normal population of known variance but unknown expectation μ, then the sample mean, $\bar{X}$, is sufficient. (*iii*) If, instead, the expectation is known but the variance is unknown, then the sample variance (computed around the known mean) is sufficient. (*iv*) If both parameters are unknown, then the sample mean and variance together are sufficient. (In all four cases, the sufficient statistics are minimal.) In all these examples, the families of distributions are of a kind called *exponential*. [*For an outline of the relationship between families of exponential distributions and sufficient statistics, see* DISTRIBUTIONS, STATISTICAL, *article on* SPECIAL CONTINUOUS DISTRIBUTIONS.]

In the theory of statistics, sufficiency is useful in reducing the complexity of inference problems and thereby facilitating their solution. Consider, for example, the approach to point estimation in which estimators are judged in terms of bias and variance [*see* ESTIMATION]. For any estimator, T, and any sufficient statistic, S, the estimator $E(T|S)$—

formed by calculating the conditional expectation of T, given S—is a function of S alone, has the same bias as T, and has a variance no larger than that of T (Rao 1945; Blackwell 1947). Hence, nothing is lost if attention is restricted to estimators that are functions of a sufficient statistic. Thus, in example (*ii*) it is not necessary to consider all functions of all n observations but only functions of the sample mean. It can be shown that $\bar{X}$ itself is the only function of $\bar{X}$ which is an unbiased estimator of μ (Lehmann & Scheffé 1950–1955) and that $\bar{X}$ has a smaller variance than any other unbiased estimator for μ.

Sufficiency in applied statistics. In applied statistical work, the concept of sufficiency is often used to justify the reduction, especially for publication, of large bodies of experimental or observational data to a few numbers, the values of the sufficient statistics of a model devised for the data [*see* STATISTICS, DESCRIPTIVE]. For example, the full data may be 500 observations of a population. If the population is normal and the observations are independent, example (*iv*) justifies reducing the record to two numbers, the sample mean and variance; no information is lost thereby.

Although such reductions are very attractive, particularly to editors, the practice is a dangerous one. The sufficiency simplification is only as valid as the model on which it is based, and sufficiency may be quite "nonrobust": reduction to statistics sufficient according to a certain model may entail drastic loss of information if the model is false, even if the model is in some sense "nearly" correct. A striking instance is provided by the frequently occurring example (*iv*). Suppose that the population from which the observations are drawn is indeed symmetrically distributed about its expected value μ, and that the distribution is quite like the normal except that there is a little more probability in the tails of the distribution than normal theory would allow. (This extra weight in the tails is usually a realistic modification of the normal, allowing for the occurrence of an occasional "wild value," or "outlier.") [*See* STATISTICAL ANALYSIS, SPECIAL PROBLEMS OF, *article on* OUTLIERS.] In this case the reduction to sample mean and variance may involve the loss of much or even most of the information about the value of μ: there are estimators for μ, computable from the original data but not from the reduced data, considerably more precise than $\bar{X}$ when the altered model holds.

Another reason for publication of the original data is that the information suppressed when reducing the data to sufficient statistics is precisely what is required to test the model itself. Thus, the reader of a report whose analysis is based on example (*i*) may wonder if there were dependences among the *n* trials or if perhaps there was a secular trend in the success probability during the course of the observations. It is possible to investigate such questions if the original record is available, but the statistic *B* throws no light on them.

J. L. Hodges, Jr.

BIBLIOGRAPHY

Blackwell, David 1947 Conditional Expectation and Unbiased Sequential Estimation. *Annals of Mathematical Statistics* 18:105–110.

Fisher, R. A. (1922) 1950 On the Mathematical Foundations of Theoretical Statistics. Pages 10.308*a*–10.368 in R. A. Fisher, *Contributions to Mathematical Statistics.* New York: Wiley. → First published in Volume 222 of the *Philosophical Transactions*, Series A, of the Royal Society of London.

Halmos, Paul R.; and Savage, L. J. 1949 Application of the Radon–Nikodym Theorem to the Theory of Sufficient Statistics. *Annals of Mathematical Statistics* 20:225–241.

◯Kendall, Maurice G.; and Stuart, Alan (1958–1966) 1973–1977 *The Advanced Theory of Statistics.* 3 vols. London: Griffin; New York: Macmillan. → Volume 1: *Distribution Theory*, 4th ed., 1977. Volume 2: *Inference and Relationship*, 3d ed., 1973. Volume 3: *Design and Analysis, and Time-series*, 3d ed., 1976. Early editions of Volumes 1 and 2, first published in 1943 and 1946, were written by Kendall alone. Stuart became a joint author on later, renumbered editions in the three-volume set.

Lehmann, E. L. 1959 *Testing Statistical Hypotheses.* New York: Wiley.

Lehmann, E. L.; and Scheffé, Henry 1950–1955 Completeness, Similar Regions, and Unbiased Estimation. *Sankhyā: The Indian Journal of Statistics* 10:305–340; 15:219–236.

Rao, C. Radhakrishna 1945 Information and the Accuracy Attainable in the Estimations of Statistical Parameters. Calcutta Mathematical Society, *Bulletin* 27:81–91.

Postscript

An article by Stigler (1973) presents an interesting discussion of the history of the sufficiency concept and its use.

J. L. Hodges, Jr.

ADDITIONAL BIBLIOGRAPHY

Stigler, Stephen M. (1973) 1977 Laplace, Fisher, and the Discovery of the Concept of Sufficiency. Volume 2, pages 271–277 in Maurice G. Kendall and R. L. Plackett (editors), *Studies in the History of Statistics and Probability.* London: Griffin; New York: Macmillan. → First published in *Biometrika* 60:439–445.

SURVEY ANALYSIS

The articles under this heading were first published in IESS *with a companion article less relevant to statistics.*

I

METHODS OF SURVEY ANALYSIS

● *This article was completely revised for this volume.*

The ways in which causal inferences are drawn from quantitative data depend on the design of the study that produced the data. In experimental studies the investigator, by using one or another kind of experimental control, can remove the effects of the major extraneous causal factors on the dependent variable. The remaining extraneous causal factors can be turned into a chance variable if subjects are assigned randomly to the experimental "treatments." In principle, then, there should be only two sources of variation in the dependent variable: (1) the effects of the independent variables being studied; (2) the effects of the random assignment and of other random phenomena, especially measurement error. By using the procedures of statistical inference, it is possible to arrive at relatively clear statements about the effects of the independent variables. But in survey research (or "observational research," as it is usually called by statisticians), neither experimental control nor random assignment is available to any significant degree. The task of survey analysis is therefore to manipulate such observational data after they have been gathered, in order to separate the effects of the independent variables from the effects of the extraneous causal factors associated with them (Wold 1956).

In the survey, the association of the independent and extraneous variables occurs naturally; in the field experiment, or quasi-experimental design, the extraneous variables usually result from the experimenter's deliberate introduction of a stimulus or his modification of some condition, both of which result in a set of problems different from those considered here [see Experimental design, *article on* quasi-experimental design].

Among the classics of survey analysis are Émile Durkheim's attempt to explain variations in suicide rates by differences in social structure (1897); the studies of soldiers' attitudes conducted by the research branch of the U.S. Army during World War II and reanalyzed afterward in *The American Soldier* (Stouffer et al. 1949); and the series of voting studies that began with *The People's Choice* (Lazarsfeld et al. 1944).

As these examples suggest, survey analysis differs from other nonexperimental procedures for analyzing and presenting quantitative data, notably from demographic analysis. Further, survey analysis is *more* than techniques of probability sampling, and in fact sometimes does not use probability samples. In contrast to the statistical analysis of sample surveys, survey analysis often deals with total populations; even when the data of survey analysis come from a probability sample, the conventional statistical problems of estimating parameters and testing hypotheses are secondary concerns (Tukey 1962). Although survey analysis has historical roots that go back to the earliest work in demography, it differs from demography in the source of its data and, therefore, in the operations it performs on these data. Until recently, demographic analysis largely relied on reworking the published tables of censuses and registers of vital statistics, whereas survey analysts usually constructed their own tables from individual questionnaires or interviews. Although these differences are still important, survey analysts have begun to use some demographic techniques, and demographers have resorted to survey analysis of specially gathered interview data in such areas as labor mobility and family planning. Perhaps the most striking evidence of the convergence of these two lines of inquiry is in the widespread use of the one-in-a-thousand and one-in-ten-thousand samples of the decennial U.S. Census of Population. These samples allow analysts of census data to prepare whatever tables are desired. At this writing, the only other country that offers a public-use sample comparable to that of the U.S. Bureau of the Census is Canada. As other national censuses make their data available in this form, demographic analysis will more closely resemble survey analysis.

The causal emphasis of survey analysis also serves to distinguish it from more narrowly descriptive procedures. It differs from the "social survey," which, at least in Great Britain, has usually been a statistical account of urban life, especially among the poor. And although it has shared with census reports, market research, and opinion polling a reliance on tabular presentation, survey analysis is unlike these fields in seeking to link its data to some body of theory. The theory may be as simple as the proposition that a set of communications has changed certain attitudes, or it may involve an explicit structure of variables, as in analyses of the reasons that people give for having done one thing rather than another.

1. The background of survey analysis

Two basic elements in survey analysis are the use of rates as dependent variables and the explanation of differences in rates by means of their statistical associations with other social phenomena. Both these features first appeared in John Graunt's *Natural and Political Observations Made Upon the Bills of Mortality* (1662), which includes the first data on urban and rural death rates. This one small book thus makes Graunt a major figure in the history not only of survey analysis but also of statistics and demography. With the exception of the life table, which Graunt invented but which was improved significantly a generation later by the astronomer Edmund Halley, Graunt's methods set the pattern for statistical analysis until the middle of the nineteenth century [*see* VITAL STATISTICS; LIFE TABLES; *and the biography of* GRAUNT].

Although Graunt had already noted the approximate constancy of certain rates over time in different areas (for example, the suicide rate and the excess of male births over female), he did not try to provide an explanation of the differences between areas. The German pastor Johann Peter Süssmilch was the first to attempt to explain these differences [Süssmilch 1741; *see also the biography of* SUSSMILCH]. He thus initiated the field of "moral statistics," which was to make much of nineteenth-century statistics, especially in France, resemble modern sociology. But the major figure in the development of moral statistics was the Belgian astronomer Adolphe Quetelet, who made three important contributions to survey analysis: (1) he used multivariate tables to explore the relations between the rates of crime or marriage and such demographic factors as age and sex; (2) he applied the calculus of probability to explaining the constancy of social rates over time; and (3) he helped to establish organized bodies of statisticians, including the Statistical Society of London (later the Royal Statistical Society), and he organized several international statistical congresses. [*See the biography of* QUETELET.]

Whether or not Quetelet was right in trying to explain the stability of rates by drawing on the theory of errors of observation is still a subject of controversy (Hogben 1957; Lazarsfeld 1961). There can be no question, however, that the organization of the statistical societies in England during the 1830s and 1840s was followed by increased application of statistical data to social problems, notably in the statistical demonstrations by John Snow and Thomas Farr of the relation between polluted water and cholera, and in several studies on the differences in the rates of mortality in large and small hospitals.

By the last decade of the nineteenth century, the use of tables for causal analysis had reached a high stage of development, both in England and on the continent of Europe. This was also the period when Charles Booth, disturbed by a socialist claim that a third of the people of London were living in poverty, was conducting his monumental study of the London poor, a study initially intended to uncover the cause of poverty. In France, at about the same time, Émile Durkheim drew on the accumulated work in moral statistics to produce the first truly sociological explanation of differences in suicide rates. The two men and their studies could hardly have been more different. Booth, the successful businessman and dedicated conservative, primarily sought accurate data on the poor of London; his original hope for causal analysis was never realized. Durkheim, the brilliant and ascetic university professor, saw in his analysis of official statistics the opportunity to make sociology a truly autonomous discipline. And yet the two men were alike in one important error of omission: both failed to recognize their need for the statistical tools being developed at the same time in the work of Francis Galton, Karl Pearson, and G. Udny Yule (Selvin 1976).

By 1888, Galton's research on heredity and his acquaintance with Quetelet's use of the normal distribution had led him to the basic ideas of correlation and regression, which were taken up and developed further by Pearson, starting in 1892 (the year Booth was president of the Royal Statistical Society). Three years later, in his first paper on statistics, Yule (1895) called attention to Booth's misinterpretation of some tabular data. Where Booth had claimed to find no association between two sets of variables, Yule computed correlation coefficients that ranged from .28 to .40; a similar table in which Durkheim ([1897] 1951, p. 87) saw "not the least trace of a relation" yields an even higher correlation [Selvin 1965; *see also the biographies of* GALTON; PEARSON; YULE.] According to the then current theory, the coefficient of correlation had meaning only as a parameter of a bivariate normal distribution; Yule had no "right" to make such a computation, but he made it anyway, apparently believing that an illegitimate computation was better than no computation at all.

Two further papers by Yule (1897; 1899) showed the wisdom of this judgment and laid the foundations for much of modern survey analysis. In 1897 he proved that the use of the correlation coefficient to measure association does not depend on the form of the underlying distribution; in particular, he showed that the distribution need not be normal. In the same paper he also gave the formulas of multiple and partial regression and correlation. Two years later he applied these ideas to a survey analysis of "panel" data on poverty—a multiple regression of changes in poverty rates on three independent variables. In four years, 1895–1899, Yule showed how the statistical part of survey analysis could be made truly quantitative. But although some economists and psychologists were early users of multiple regression, it was not until the 1960s that other survey analysts, notably sociologists and students of public opinion, came to see its importance [*see* LINEAR HYPOTHESES, *article on* REGRESSION].

Not content to deal only with continuous variables, Yule also took in hand the analysis of tabular data and set forth its algebraic basis (1900). Although this material appeared in every edition of his *Introduction to the Theory of Statistics* ([1911] Yule & Kendall 1950, chapters 1–3), it had even less effect on survey analysis than did his work on continuous variables. It was finally brought to the attention of social scientists by Paul F. Lazarsfeld in 1948 (see Lazarsfeld & Rosenberg 1955, pp. 115–125), as the basis for his codification of survey analysis.

A new area of application for survey analysis began to appear in the 1920s, first in the form of market research and later in opinion polling, communications research, and election research. Among the factors that promoted these new developments were the change in American social psychology and sociology from speculation to empirical research, the wide availability of punched-card machines, and a new interest in the use of formal statistical procedures, notably at Chicago, where William F. Ogburn and Samuel A. Stouffer taught both correlational and tabular techniques in the 1930s. By the time of World War II, survey analysis had advanced to the point where Stouffer was able, as already mentioned, to organize a group of experienced survey analysts to conduct hundreds of attitude surveys among American soldiers.

Three major developments have shaped survey analysis since the 1940s. The emphasis on closer relations between theory and research has led to greater concern with conceptualization and index formation, as well as with the causal interpretation of statistical relations. The rise of university research bureaus has increased both the quantity and the quality of survey analysis. And the advent of the large computer has brought survey analysts to contemplate once again the vision that Yule had conjured up in 1899—the possibility of replacing the crude assessment of percentaged tables with the more powerful methods of multiple regression and other multivariate procedures (see sections 4 and 5).

2. The structure of survey analysis

Analysis is the study of variation. Beginning with the variation in a dependent variable, the analyst seeks to account for this variation by examining the covariation of the dependent variable with one or more independent variables. For instance, in a sample where 57 per cent prefer the conservative party and 43 per cent the liberal party, the analytic question is why people divide between the two parties. If everyone preferred the same party, there would be no variation and, within this sample, no problem for analysis. The answer to this question comes from examining how the distribution of preferences is affected by a set of independent variables, such as the sex of the individual, the social class of his family, and the size of his community. This combination of a single dependent variable and a set of independent variables is the most common structure examined in survey analysis; it also serves as the building block for other, more complex structures—for example, a study with several dependent variables.

The sequence of steps in the analysis of an ideal experiment is determined largely by the design of the study. In real life, of course, an experimenter almost always confronts new problems in his analysis. The survey analyst, however, has so many more options open to him at each step that he cannot specify in advance all the steps through which an analysis should go. Nevertheless, it is useful to conceive of analysis as a series of cycles, in which the analyst goes through the same formal sequence of steps in each cycle, each time changing some essential aspect of his problem. A typical cycle can be described as follows.

(1) *Measuring the parameters of some distribution.* The concrete form of this step may be as simple as computing percentages in a two-variable table or as complicated as fitting a regression plane to a large set of points. Indeed, the parameters may not even be expressed numerically; in conventional survey analysis (that is, analysis using percentaged contingency tables), two-variable relations may be classified simply as "large" or "small."

(2) *Assessing the criteria for an adequate analysis.* The reasons survey analysts give for stopping one line of investigation and starting another often

appear superficial: they have run out of time, cases, or interest. On further investigation, however, it usually appears that they have stopped for one or more of the following reasons: (*a*) statistical completeness—that is, a sufficiently high proportion of the variation in the dependent variable has been accounted for by the variation in the independent variables; (*b*) theoretical clarity—that is, the meanings of the relations already found and the nature of the causal structure are sufficiently clear not to need further analysis; (*c*) unimportance of error—that is, there is good reason to believe that the apparent findings are genuine, that they are not the result of one or another kind of error. These three reasons, then, can be regarded as criteria for an adequate analysis.

(3) *Changing the analytic model.* With these criteria in mind, the analyst decides whether to stop the analysis or to continue it, either by adding more variables or by changing the basic form of the analysis (for example, from linear to curvilinear regression).

In practice, there are two major sets of procedures by means of which the steps involved in each cycle are taken. One rests on the construction of percentaged contingency tables, the other on the different kinds of multivariate statistical analysis. This neat distinction is bridged by the modern approach to contingency tables. Despite the extensive use of correlational techniques by psychologists, survey analysis has until recently been dominated by percentaged contingency tables. One reason for this dominance was economic: with punched-card machines, running several tables took less time than computing a single correlation coefficient on a desk calculator. The advent of the large electronic computer has brought a revolutionary change in this situation, and statistical computations that might have taken months, if they were done at all, can now be done in minutes. This change has led to new interest in multivariate statistical techniques and to some questioning of the place of tabular methods in survey analysis. Since many textbooks explain the basic ideas of multivariate statistical procedures, it will here be necessary to consider in detail only the logic of tabular analysis.

3. Percentaged contingency tables

Let us recall the illustration of a sample in which 57 per cent prefer the conservative party and 43 per cent the liberal party. Further, let us call the dependent variable (party preference) A and the three independent variables (sex, social class, and size of community) B, C, and D. For simplicity, let each variable take only two values: A_1 (conserva-

Table 1 — Percentage distribution of A*

A_1	57%
A_2	43
Total	100%
	(n = 1,000)

* Hypothetical data.

tive) and A_2 (liberal), B_1 and B_2, and so on. Tabular analysis then begins by considering the distribution of the dependent variable, as in Table 1.

In this simple distribution, 57 per cent of the 1,000 sample cases have the value A_1 and 43 per cent have the value A_2. Analysis proper begins when a second variable is introduced, and the association between them is examined by means of a two-variable table. However, instead of looking at a two-variable table, it is intuitively more appealing to consider once again the distribution of A—or, rather, *two* distributions of A, one for those people classified as B_1 and one for those who are B_2, as in Table 2.

Perhaps the simplest measurement of association is to compare these two univariate distributions— that is, to note the 21 percentage-point difference in the proportion of B_1's and B_2's who respond A_1. These two distributions are, of course, the two columns of a 2×2 table. The point of separating them is to stress the concept of two-variable association as the comparison of univariate distributions.

The same sort of link between different levels appears when a third variable is introduced. This three-variable "elaboration," or multivariate analysis, as Lazarsfeld has called it (not to be confused with the statistical concept of the same name, which is here called multivariate statistical analysis), involves the reexamination of a two-variable association for the two separate subgroups of people classified according to the values of the third variable, C, as in Table 3.

It is necessary to give a numerical measure of the association in each "partial" table, but the above reasoning applies no matter what measure is chosen. Elaboration, then, is simply the comparison

of these two measures of association. That is, when the third variable, C, is introduced, the association between A and B becomes a composite dependent variable; elaboration is the relation between C and some measure of the association between A and B.

This approach to elaboration is another way of describing *statistical interaction*, or the extent to which the association between A and B depends on the value of C. In the usual discussion of elaboration, however, it also appears as something new in the treatment of association. The simple formalization presented here emphasizes the common thread that runs through the treatment of one, two, three, or more variables: the measure of the degree of relation for a given number of variables becomes the dependent variable when a new independent variable is introduced.

Lazarsfeld distinguished three ideal types of configurations that result when the relation between A and B is examined separately for C_1 and C_2. The disappearance of the original relation is called "explanation" when C is causally prior to both A and B, and "interpretation" when C intervenes between A and B. The third type of elaboration, "specification," is essentially an extreme form of interaction, in which at least one partial relation is larger than the original relation or is of opposite sign. A full discussion of elaboration and many examples are given in Hyman (1955, chapters 6 and 7), Rosenberg (1968), and Hirschi and Selvin ([1967] 1973).

Lazarsfeld's several discussions of elaboration have clarified much of what the survey analyst does in practice; they have also led Simon, Blalock, Boudon, and Duncan (whose work is discussed below, in section 7) to mathematize the idea of a causal structure and to extend it beyond the three-variable level. The fundamental ideas of elaboration have thus stood the test of time. However, the percentaged contingency table—the tool that most survey analysts have used to carry out the ideas of elaboration—now appears less satisfactory than it formerly did. It still appears useful as an introduction to analysis for the beginning student and as

Table 2 — Percentage distribution of A for those who are B_1 and B_2*

	B_1	B_2
A_1	45%	66%
A_2	55	34
Total	100%	100%
	(n = 420)	(n = 580)

* Hypothetical data.

Table 3 — Associations of A and B for those who are C_1 and C_2*

	C_1		C_2	
	B_1	B_2	B_1	B_2
A_1	25%	38%	72%	76%
A_2	75	62	28	24
Total	100%	100%	100%	100%
	(n = 240)	(n = 160)	(n = 180)	(n = 420)

* Hypothetical data.

a device for presenting the important findings of a survey to lay readers.

4. Alternatives to tabular analysis

The value of a technique should be judged against the available alternatives to it. Before the development of the log-linear model, the best alternatives to percentaged-contingency-table analysis seemed to be multiple regression (for quantitative dependent variables) and multiple discriminant analysis (for qualitative dependent variables). Two procedures were necessary because conventional tabular analysis, of course, treated quantitative and qualitative variables in essentially the same way. Compared with these techniques, conventional tabular analysis appears to have three principal shortcomings, as follow.

(1) *Lack of a measure of statistical completeness.* The square of the multiple correlation coefficient is the proportion of the variation in the dependent variable that is accounted for by the regression on the independent variables. In regression, the analyst always knows how far he has gone toward the goal of complete explanation (usually linear). By the middle 1960s, no comparable statistic adequate for a large number of independent variables had yet become available in tabular analysis; the analyst therefore did not know whether he had gone far enough, or indeed, whether the introduction of additional variables made an appreciable addition to the explanatory power of those already included.

(2) *Ambiguity of causal inferences.* Even with very large samples, the number of independent variables that can be considered jointly is usually no more than four or five; percentage comparisons involving many variables are usually based on too few cases to be statistically stable. It is often possible, however, to find many more variables that have appreciable effects on the dependent variable. This inability to examine the joint effects of *all* the apparently important independent variables makes the interpretation of any relation between the independent and dependent variables inherently ambiguous. Suppose, for example, that one can examine the effects of only three variables at once; that variables B, C, D, E, and F all have appreciable associations with the dependent variable A; and that, as is usually the case, all these independent variables are intercorrelated. Then it is impossible to draw clear conclusions from any three-variable table, since what appears to be the effect of, say, B, C, and D is also the effect of E and F, but in some unknown and inherently unknowable degree. Maurice Halbwachs (1930) apparently had this

kind of argument in mind when he said that Durkheim's attempt to discern the effects of religion by areal comparisons was fundamentally impossible; any comparison between Catholic areas and Protestant areas involves differences in income, social norms, industrialization, and many other factors. In contrast, the procedures of multivariate statistics can handle dozens or even hundreds of variables at the same time, so that it is possible to represent the relation between a dependent variable and a large set of independent variables. The problem of measuring the separate effects of the independent variables is as yet unsolved (if indeed it is solvable). Such problems as multicollinearity and the high sampling variability of the coefficients in a large equation make it difficult to impute unequivocal meanings to the coefficients of the independent variables. Finally, adding and subtracting independent variables may change the original causal structure in ways that are far from trivial [*see* LINEAR HYPOTHESES, *article on* REGRESSION].

(3) *Lack of a systematic search procedure.* At the beginning of an analysis, the main task is to find the independent variables that are the best predictors of the dependent variables; this task is hampered by the complex intercorrelations among the independent variables. A secondary task is to find the dependent variables that are most predictable from the independent variables; this task is similarly handicapped by the correlations among the dependent variables. The complex intercorrelations make these tedious tasks in tabular analysis. In contrast, "stepwise" regression and discriminant programs rapidly arrange the independent variables in the order of their incremental predictive power, and modern programs allow such analyses to be repeated for other dependent variables in a few seconds. Sonquist and Morgan (1964) have devised a computer program that simulates some aspects of the search behavior of a tabular analyst (see also Sterling et al. 1966).

5. The log-linear model

The difficulty of using simple relative frequencies, or percentages, to study large contingency tables, especially those involving more than two variables or variables that are more than dichotomies, was apparent to Yule from the beginning of his systematic work on the statistics of attributes. In his seminal paper on that subject (1900), Yule discussed the use of summary measures of association and set forth a complex equation relating the various relative frequencies in a system of three dichotomous attributes, which he apparently hoped would prove as useful to analysts of attribute data as the well-known algebra of correlation and regression had proved in the analysis of systems of continuous variables.

Although Yule himself was always careful to make the computation of measures of association contingent on a careful examination of the relative frequencies, writers in the first half of the twentieth century seemed to have lost sight of this important precaution and to have concentrated on a search for an ideal measure of association for all two-variable tables. This line of inquiry is described in the now classic papers of Goodman and Kruskal (1954–1972), in which the choice of measures of association, it is argued, should depend on the investigator's *substantive* interests. Since different investigators have different interests, there can be no one ideal measure of association.

Yule's other line of approach toward simplifying systems of contingency tables, his algebra of dichotomous systems, was ignored by statisticians and survey analysts until the work of Lazarsfeld on "elaboration" (Lazarsfeld & Rosenberg 1955). As remarked above, however, Lazarsfeld's approach to the classification of operations in survey analysis was incomplete. It required that partial relations vanish (rather than the more usual outcome of their less-than-total disappearance) and left the idea of *causal order* outside the system of equations. Lazarsfeld later went on to generalize the algebra of dichotomous systems to many variables and to apply it to empirical data in ways that go beyond Yule and presage Goodman's work, described below [see SURVEY ANALYSIS, *article on* THE ANALYSIS OF ATTRIBUTE DATA].

The idea of analyzing contingency tables by use of linear models like those of the analysis of variance seems to have occurred independently to several sets of investigators. Among them were Nathan Mantel and his fellow researchers in epidemiological applications (Mantel & Haenszel 1959); Alan B. Wilson, of the University of California, Berkeley, who termed his system "least-squares analysis" (Wilson 1964); Frank M. Andrews, James N. Morgan, and John A. Sonquist, of the Survey Research Center, University of Michigan, who called their system "multiple classification analysis" (Andrews et al. 1967); and Leo A. Goodman, of the University of Chicago, who described the "log-linear model" (Goodman 1964; 1970; 1971; 1973).

Whether because Goodman published in journals read by more survey analysts than the monographic publications of the earlier authors or because he had worked out the statistical aspects of his procedures in more detail, his model is now the only one in general use among survey analysts and is, therefore, the only one that we shall discuss here. A more detailed introduction to the log-linear model can be found in Davis (1974), and Bishop, Fienberg, and Holland (1975) includes a complete discussion of the log-linear model and related developments. [See also LINEAR HYPOTHESES, *article on* ANALYSIS OF VARIANCE.]

Both ordinary analysis of variance and the log-linear model use linear combinations of parameters to account for deviations of observed values from the grand mean μ; the log-linear model uses as the observed values the logarithms to the base e (denoted by ln) of the cell frequencies. Other bases could be used as well, but the base e is conventional. When a cell frequency is zero, the logarithm of the frequency is not defined. To avoid this problem, some statisticians add ½ to all cell frequencies; others add ½ only to those frequencies that are actually zero. The table of frequencies being analyzed is usually assumed to be the result of multinomial sampling over the whole table.

Consider a two-way table with I rows for factor A and with J columns for factor B. Let the unknown probability that an observation falls in the (i,j) cell be P_{ij}; here all P_{ij} are assumed to be positive and $\sum P_{ij} = 1$. On the average, the observed count f_{ij} is proportional to P_{ij}. In this two-way table, let $\nu_{ij} = \ln P_{ij}$. The motivation for the logarithmic transformation is that, in the important special case of independence, the model becomes additive rather than multiplicative.

The log-linear model decomposes ν_{ij} into the sum of four terms;

$$\nu_{ij} = \mu + \lambda_i^A + \lambda_j^B + \lambda_{ij}^{AB}.$$

The notation used for the effects includes both subscripts and superscripts; the superscripts identify the factors and the subscripts the corresponding levels. The term λ_2^A refers to the average effect of the second level of factor A; similarly, λ_{14}^{AC} refers to the effect of the first level of factor A and the fourth level of factor C.

As in ordinary analysis of variance, the log-linear model imposes constraints on the parameters to give them well-defined values. The usual constraints are

$$\sum_i \lambda_i^A = 0,$$

$$\sum_j \lambda_j^B = 0,$$

$$\sum_i \lambda_{ij}^{AB} = \sum_j \lambda_{ij}^{AB} = 0;$$

further, since $P_{ij} = \exp(\nu_{ij})$ and $\sum\sum P_{ij} = 1$, there must be a further nonlinear constraint

$$\sum\sum \exp(\mu + \lambda_i^A + \lambda_j^B + \lambda_{ij}^{AB}) = 1.$$

Of course, other constraints are possible. With the conventional constraints, the parameters have the following interpretations:

μ is the average log probability of the cells,

λ_i^A is a row effect that tells how much the average of the ith level of factor A differs from the over-all average,

λ_j^B is the column effect for the jth level of factor B,

λ_{ij}^{AB} is an interaction effect that tells how much ν_{ij} differs from $\mu + \lambda_i^A + \lambda_j^B$.

Note that the two-variable table where the cell entries are the joint probabilities has been converted into a two-way table satisfying traditional ANOVA constraints and another as well. This transformation paves the way for an analysis of contingency tables that is analogous to the traditional analysis-of-variance techniques. The analogy must not be pressed too hard; for example, in the present case the quantity modeled is the logarithm of a probability, while in the ANOVA case what is modeled is the expected value of a separate dependent variable.

The generalization to three or more factors is immediate. Analogously to the two-factor case, let $\nu_{ijk} = \ln P_{ijk}$; then the corresponding log-linear model is

$$\nu_{ijk} = \mu + \lambda_i^A + \lambda_j^B + \lambda_k^C + \lambda_{ij}^{AB} + \lambda_{ik}^{AC} + \lambda_{jk}^{BC} + \lambda_{ijk}^{ABC},$$

where

$$\sum_i \lambda_i^A = \sum_j \lambda_j^B = \sum_k \lambda_k^C = 0,$$

$$\sum_i \lambda_{ij}^{AB} = \sum_j \lambda_{ij}^{AB} = \sum_i \lambda_{ik}^{AC} = \sum_k \lambda_{ik}^{AC}$$

$$= \sum_j \lambda_{jk}^{BC} = \sum_k \lambda_{jk}^{BC} = 0,$$

$$\sum_i \lambda_{ijk}^{ABC} = \sum_j \lambda_{ijk}^{ABC} = \sum_k \lambda_{ijk}^{ABC} = 0,$$

and

$$\sum\sum\sum \exp(\nu_{ijk}) = 1.$$

As before, the main effects, λ_i^A, λ_j^B, and λ_k^C, are the differences between the average of the ν_{ijk} for the level of the factor and the over-all average. The λ_{ij}^{AB} effects are the differences of the average of the ν_{ijk} over factor C and the sum of the over-all average, the λ_i^A effect, and the λ_j^B effect. Analogous relations hold for the other two-factor interactions. The λ_{ijk}^{ABC} parameters are the differences between ν_{ijk} and the grand mean, the sum of the single-factor effects, λ_i^A, λ_j^B, and λ_k^C and the sum of the two-factor interaction effects λ_{ij}^{AB}, λ_{ik}^{AC}, λ_{jk}^{BC}.

The general log-linear model, of course, contains as special cases many of the common contingency-table models. For example, the model of stochastically independent factors, which in the classical Yule formulation is the basis for measuring and testing hypotheses about association (Yule & Kendall 1950, chapter 3), can be expressed as a log-linear model with all interaction terms equal to zero. Let P_i^A be the marginal probability that an observation has level i in factor A, and let P_j^B and P_k^C be the corresponding probabilities for factors B and C, respectively. In the model of stochastically independent factors

$$P_{ijk} = P_i^A P_j^B P_k^C,$$

where $\sum\sum\sum P_{ijk} = 1$ and $\sum P_i^A = \sum P_j^B = \sum P_k^C = 1$. Note that the marginal probability of an observation in level i for factor A is just

$$\sum_j \sum_k P_{ijk} = P_i^A.$$

In general,

$$\lambda_i^A = \frac{\sum_j \sum_k \nu_{ijk}}{JK} - \frac{\sum_i \sum_j \sum_k \nu_{ijk}}{IJK};$$

and under independence,

$$\lambda_i^A = \ln P_i^A + \frac{\sum_j \ln P_j^B}{J} + \frac{\sum_k \ln P_k^C}{K}$$

$$- \frac{\sum_{i'} \ln P_{i'}^A}{I} - \frac{\sum_j \ln P_j^B}{J} - \frac{\sum_k \ln P_k^C}{K}$$

$$= \ln P_i^A - \frac{\sum_{i'=1}^I \ln P_{i'}^A}{I}.$$

Similar relations hold for λ_j^B and λ_k^C. Now, still under independence,

$$\nu_{ijk} = \ln P_{ijk} = \ln P_i^A + \ln P_j^B + \ln P_k^C$$

$$= \mu + \lambda_i^A + \lambda_j^B + \lambda_k^C.$$

The terms λ_i^A, λ_j^B, and λ_k^C are as above and

$$\mu = \frac{\sum_i \ln P_i^A}{I} + \frac{\sum_j \ln P_j^B}{J} + \frac{\sum_k \ln P_k^C}{K}.$$

More complex models, such as Lazarsfeld's elaboration, where factors A and B are independent when factor C is held constant, are also subsumed by the log-linear model.

These various cases are part of a complex hierarchy of models. At the highest tier, the most elaborate model—called a saturated model—includes all possible interaction terms. The model at the second tier excludes the highest-order interactions by setting them to zero but includes all other terms. The third tier comprises a group of models. For a three-way table with factors A, B, and C, the hierarchy is as follows. The model at the first tier has interaction parameters taking any value whatsoever. The model at the second tier has all three factor interactions set at zero. The third tier comprises three models; in each, the three factor interactions are set to zero: the first has AB interactions zero; the second as AC interactions zero; and the third has BC interactions zero.

Thus the hierarchy consists of those linear models in which zero interaction of a given order among a set of factors implies zero interactions of all higher orders for sets of factors including the original one. Restriction to hierarchical models simplifies the interpretation of parameters for procedures in which one starts with the full (that is, saturated) model and moves toward increasingly simple models in the hierarchy, provided at each step that the move is consistent with the data. On the other hand, nonhierarchical models may also be appropriate and useful, as when two or more synergistic factors have to be present simultaneously to obtain interesting responses. An example might be the presence of adequate nutrients in soil to permit plant growth at all. These matters are discussed in Bishop, Fienberg, and Holland (1975, pp. 34, 38–39, et passim).

A central term in the log-linear approach to tables is the "odds ratio." The odds that an event, E, will occur are expressed by the ratio of the probability that it occurs, P_E, to the probability that it will *not* occur, $1 - P_E$. For instance, the probability of throwing a three with a fair die is 1/6; the odds of throwing a three are $(1/6)/(1 - 1/6)$, or 1/5.

More generally, consider the 2×2 array of probabilities for events A and B in Table 4. Here we use P_{1+} and P_{2+} instead of P_1^A and P_2^B for the

marginal probabilities of A, because this notation is simpler in a 2×2 array. The (conditional) odds that B will occur, given that A has occurred, are

$$(P_{11}/P_{1+})/(P_{12}/P_{1+}) = P_{11}/P_{12}$$

and, given that A has not occurred,

$$(P_{21}/P_{2+})/(P_{22}/P_{2+}) = P_{21}/P_{22}.$$

The odds ratio is the ratio of these odds:

$$(P_{11}P_{22})/(P_{12}P_{21}).$$

Note that interchange of rows and columns (that is, of B and A) leaves the odds ratio unchanged. It is also unchanged if the conditional probabilities are changed as shown in Table 5. Similar operations on the columns also leave the odds ratio unchanged; these invariances provide a major motivation for working with the odds ratio. For a fuller discussion of odds ratios, see Fleiss (1973).

An odds ratio equal to 1 means that the conditional probability that B will occur if A has occurred is the same as the conditional probability that B will occur if A has not occurred. Odds ratios not equal to 1 imply a dependence between B and A. For an $R \times C$ table, the odds ratio is defined for two given levels of B and two given levels of A. Since the subtable probabilities do not add to 1, formally we must calculate the rescaled probabilities; in practice, the rescaling factors cancel. Then the odds ratio for the two given levels of B and A is defined to be the odds ratio of the 2×2 subtable corresponding to the indicated levels of the variables.

Returning to dichotomies, suppose that a third event, C, is introduced. Then the three-way table can be split into two two-way tables, as in Table 6.

Table 4 – Probabilities in a 2 × 2 table*

PROBABILITIES

	B	Not B	Sum
A	P_{11}	P_{12}	P_{1+}
Not A	P_{21}	P_{22}	P_{2+}
			1

CONDITIONAL PROBABILITIES

	B	Not B
Given A	P_{11}/P_{1+}	P_{12}/P_{1+}
Given not A	P_{21}/P_{2+}	P_{22}/P_{2+}

* $P_{11} + P_{12} + P_{21} + P_{22} = 1$.

Table 5 – New conditional probabilities, for any ρ between 0 and 1 (exclusive)

	B	Not B
Given A	$P_{11}\rho/P_{1+}$	$P_{12}\rho/P_{1+}$
Given not A	$P_{21}(1-\rho)/P_{2+}$	$P_{22}(1-\rho)/P_{2+}$

Table 6 – Probabilities in a 2 × 2 × 2 table*

		B	Not B
A	C	P_{111}	P_{121}
	Not C	P_{112}	P_{122}
Not A	C	P_{211}	P_{221}
	Not C	P_{212}	P_{222}

* $\sum_i \sum_j \sum_k P_{ijk} = 1$.

Thus there is an odds ratio for each of the two conditional tables, called a second-order odds ratio because there are two variables conditioned. The ratio of these two second-order odds ratios is the third-order odds ratio of *A*, *B*, and *C*. If a fourth event were introduced, then the fourth-order odds ratio would be the ratio of the two conditional third-order odds ratios. A second-order odds ratio of 1 indicates independence of the two factors. A third-order odds ratio of 1 indicates that the degree of dependence does not change. Just as the value of the second-order odds ratio is the same no matter what order the variables enter, the value of the third-order odds ratio is independent of the order of consideration of the variables.

For example, consider Table 7. It shows foster-care placement of Catholic and Protestant (Christian but non-Catholic) children over 12 in New York City in 1973, classified by religion of the child, sex, and status, that is, whether the child was cared for by an agency of its own religion (in-religion) or not (out-religion). Any one of the variables could be used to split the table, but, in view of the enormous interaction between religion and status, let us use sex. The conditional odds ratios are

Male: $\quad(3{,}233/281)/(1{,}120/470) = 4.83$
Female: $\quad(2{,}230/161)/(878/337) = 5.32.$

Both conditional odds ratios are very large and reflect the greater tendency of Catholic children to be cared for in-religion. The value of the third-order odds ratio is

Male/Female: $\quad 4.83/5.32 = .91.$

The third-order odds ratio is so close to 1 that it suggests that there is no need for a third-order interaction model. When status is used as the conditioning variable, the first- and second-order odds ratios indicate an apparent interaction between religion and sex. It is the possibility of simultaneously representing two or more two-factor interactions and testing odds ratios that makes the log-linear model so attractive.

Table 7 — *Foster-care placement (status) by religion and sex of children over 12, New York City, 1973*

		STATUS	SEX Male	Female
RELIGION OF CHILD	Catholic	In-religion	3,233	2,230
		Out-religion	281	161
	Protestant*	In-religion	1,120	878
		Out-religion	470	337

* Christian but non-Catholic.

Source: Data from Young & Finch 1977, p. 202.

The odds are fundamental quantities in the log-linear model because, under the constraints noted above, the λ parameters are simple functions of the odds. For example, in a 2×2 table, λ_1^A is the difference of the average of the ν_{1j}, $(\nu_{11} + \nu_{12})/2$, and the average of all four ν_{ij}; that is,

$$(1) \qquad \lambda_1^A = (\nu_{11} - \nu_{21} + \nu_{12} - \nu_{22})/4.$$

But, $\nu_{11} - \nu_{21} = \ln P_{11} - \ln P_{21} = \ln (P_{11}/P_{21})$, the log of the odds conditioned on the first column. Thus,

$$\lambda_1^A = (\ln[(P_{11}/P_{21})/(P_{12}/P_{22})])/4;$$

and so λ_1^A is a simple one-to-one function of the odds ratio. So are the other λ's.

Two approaches to the statistical analysis of contingency tables using the log-linear model are simultaneous confidence intervals for the λ's and tests of the null hypotheses that certain sets of the λ's are zero. The calculation of a confidence interval for a parameter λ often is based on an estimate $\hat{\lambda}$ of the parameter and an estimated standard deviation of $\hat{\lambda}$. The estimate $\hat{\lambda}$ is based on the expression of λ in terms of the ν_{ij}. In equation (1), λ_1^A was expressed as a linear combination of the ν_{ij} (with the sum of the coefficients zero). This can be generalized for any contingency table; that is, the λ parameters are linear functions of the ν_{ij},

$$\lambda = \sum\sum a_{ij}\nu_{ij},$$

with $\sum\sum a_{ij} = 0$. The a_{ij} are the coefficients of ν_{ij} when one solves for the λ's in terms of the ν_{ij}. An estimator of λ is this expression with an estimate of ν_{ij} $(= \ln P_{ij})$ used for ν_{ij}. One possibility is the log of the observed frequency, $\ln f_{ij}$; however, since the f_{ij} can be zero, the use of

$$y_{ij} = \ln (f_{ij} + \tfrac{1}{2})$$

is preferable. This particular choice of $\frac{1}{2}$ also reduces the asymptotic bias of the estimators of the λ parameters. An estimator of the variance of $\hat{\lambda}$, the estimator of λ, is just

$$\sum\sum a_{ij}^2/(f_{ij} + \tfrac{1}{2}), \qquad \text{where } \hat{\lambda} = \sum\sum a_{ij}y_{ij}.$$

Since increasing the number of variables decreases the average of the counts f_{ij} in any one cell, the estimated variance, $\sum\sum a_{ij}^2/(f_{ij} + \tfrac{1}{2})$, of a $\hat{\lambda}$ can become quite large for complex tables. In particular, estimates of interaction may have so great an estimated variance that they are of little value.

There is a chi-square test of the improvement in fit by using a more complex model in the hierarchy rather than a given model. Goodman's results specify the degrees of freedom in this test. In sifting through a table, some analysts start with the most complex model possible and reduce the complexity downward. This is the opposite of the usual stepwise regression procedure. Typically, the stepwise regression analysis is performed on a large number of variables. The analyst is on a fishing expedition, and his question is whether there are any variables related to his criterion. On the other hand, the variables in a contingency table are usually interrelated in a moderately complex way, and the question is how complex a model is necessary to fit the data.

Table 8 gives the results of fitting the more complex models in the hierarchy to the data of our foster-care example, using a standard program, ECTA (*Everyman's Contingency Table Analysis*). The independence model clearly fails with a chi-square value of 663.7 on 4 degrees of freedom. Fitting the religion–status interaction is clearly the most helpful addition. This reflects the tendency of Catholic children to be cared for in-religion. This model has a P value of .01 and so may be inadequate (see our caveats in section 8 about data dredging). The model using both the religion–status and the religion–sex interactions has an acceptable P value and appears to be a real improvement over the model using only the religion–status interaction. The model using all three two-factor interactions is definitely acceptable and appears to be an improvement over the previous model. Here, as in all model fitting, the investigator must decide whether the loss in simplicity of the model outweighs the gain in goodness of fit. Note that with the log-linear model the various two-factor interactions could be considered simultaneously, and the importance of each could be judged quickly and accurately.

Three-factor tables that have R rows, C columns, and L layers are easily handled by the log-linear model and the standard computer programs associated with it. The interpretation of the output proceeds in exactly the same way as in the $2 \times 2 \times 2$ case. The model is general and can handle any number of factors. The difficulties are that high-order interactions are difficult to interpret practically and that there are substantial data-dredging risks, as there always are in testing multiple hypotheses with the same data. Computationally, the difficulties are minor, but the user who insists upon looking at every model in the hierarchy will have to specify large numbers of models and examine the resulting output.

Problems of multivariate statistics. In emphasizing the defects of tabular analysis and the virtues of multivariate statistics, the above section presents a somewhat one-sided picture. Such procedures as regression and discriminant analysis have serious problems of their own. For example, to treat a nominally scaled variable such as race or geographical region as an independent variable, one must first transform it into a set of "dummy variables" (for instances of this procedure, see Draper & Smith 1966, section 5.3).

Another problem arises in detecting and representing statistical interaction. In their standard forms, regression and discriminant analysis assume

Table 8 — *Results of fitting models to data of Table 7*

MODEL FITTED	CHI-SQUARE STATISTIC (MAXIMUM LIKELIHOOD)	DEGREES OF FREEDOM	P VALUE
Saturated model	0.0	0	—
All two-way interactions and all main effects	0.5	1	.47
Religion–status and religion–sex interactions and all main effects	4.4	2	.11
Religion–status and status–sex interactions and all main effects	9.3	2	.01
Religion–sex and status–sex interactions and all main effects	656.1	2	.00
Religion–status interaction and all main effects	10.7	3	.01
Religion–sex interaction and all main effects	657.5	3	.00
Status–sex interaction and all main effects	662.3	3	.00
Religion, status, and sex main effects (independence)	663.7	4	.00

that there is no interaction. However, several ancillary procedures for detecting interaction are available—for example, analysis of variance, the examination of residuals from the regression, the Sonquist–Morgan "Automatic Interaction Detector" (Sonquist & Morgan 1964), and the stratification of the sample by one or more of the interacting variables, with separate regressions in each part. Similarly, it is possible to represent interaction by appropriate modifications of the standard equations (see Draper & Smith 1966, chapter 5).

On balance, the method of choice appears to be multivariate statistical procedures and contingency table procedures for the early and middle phases of a survey analysis, and tables for presenting the final results. An examination of current journals suggests that this judgment is increasingly shared by social scientists who engage in survey analysis.

6. The meaning of statistical results

In every cycle of analysis, there are problems of imputing or verifying the meanings of variables and relations—that is, there are problems of conceptualization and validation. Much of the meaning that one imputes to variables and relations comes from sources other than the statistical data—from the precipitates of past research and theory, from the history of the phenomena studied, and from a wide range of qualitative procedures. Indeed, it is the skillful interweaving of survey and ancillary data that often distinguishes insightful survey analysis from routine manipulation.

Although conceptualization and validation are partly matters of judgment, the wide range of questions in the typical survey provides an objective basis for imputing meanings to observed variables. At one extreme are such simple procedures as examining the association between a variable of uncertain meaning and one or more additional variables whose meaning is less in question. For example, a self-estimate of "interest in politics" may be validated by seeing how strongly this interest is associated with reading political news, discussing politics with friends, and voting in elections. At the other extreme, the common meanings of large numbers of variables may be extracted by some "rational" scaling procedure, such as Guttman scaling, latent structure analysis, or factor analysis. Again, computer programs have made such procedures much less expensive than they once were, and therefore more desirable than arbitrarily constructed scales. [See FACTOR ANALYSIS AND PRINCIPAL COMPONENTS; LATENT STRUCTURE.]

7. Causal structures

Although survey analysts have always aimed at uncovering causes, the idea that the independent and dependent variables should all be located in a determinate structure of relations—usually represented by a set of boxes and arrows—has only recently been generally accepted. [See CAUSATION.] The first significant methodological advance on Lazarsfeld's formalization was made by Simon (1954); drawing on a large body of work in econometrics, Simon showed how Lazarsfeld's idea of time order, which had been treated separately from the statistical configuration of his three variables, could be combined with them in a system of simultaneous equations. For an autobiographical account by one of the major participants in the econometric study of causality, see Wold (1959). In sociology, Blalock (1964) took up Simon's suggestions and showed how they can be applied to empirical data. Boudon (1965; 1967) has developed a theory of "dependence analysis," which, besides extending the models of Simon and Blalock, shows the relations between these models, ordinary least-squares regression, and the work of Sewall Wright on "path coefficients" (a line of inquiry independently pursued in biology since 1918). Finally, the careful study of causal relations and of structural equations, of which path analysis is a special case, has been a feature of econometric theory since its inception in the 1930s (Goldberger & Duncan 1973; Duncan 1975). It is interesting to note the length of time these methods take to cross the boundaries of substantive disciplines.

The assumptions of path analysis. Path analysis divides the variables studied into two groups: the *endogenous variables*, whose variations are to be explained, and the *exogenous variables*, whose variations are taken as given and, if they were to be explained, would require the introduction of additional antecedent variables. Although it is apparently possible to deal with systems of *feedback* or of *simultaneous reciprocal causation* (Heise 1975) by the procedures of path analysis, most elementary treatments begin with a fully recursive path model. In this model, each of the "later" endogenous variables depends linearly on the "earlier" endogenous variables and on an exogenous residual variable; the "earliest" endogenous variable depends only on exogenous variables. In the models of this discussion and most empirical path models, the exogenous variables are assumed to be uncorrelated. Thus, in Figure 1, the fully recursive model has the equations

$$
\begin{aligned}
X_1 &= p_1 U_1, \\
X_2 &= p_{21}X_1 + p_2 U_2, \\
X_3 &= p_{32}X_2 + p_3 U_3.
\end{aligned}
\tag{2}
$$

In this model, U_1, U_2, U_3, X_1, X_2, and X_3 are standardized random variables (that is, they have mean

0 and variance 1). The variable X_1 is solely determined by U_1 (note that U_1 is logically redundant but is added for notational symmetry); X_2 is a linear combination of the endogenous X_1 and exogenous residual U_2; similarly, X_3 is a combination of X_2 and U_3. The coefficients p_{ij} and p_k are called the path coefficients and are to be estimated from the observations on X_1, X_2, and X_3. The U_i's are unobserved hypothetical variables.

This causal model is the same as Lazarfeld's model of "interpretation"; in econometrics, it is called a causal-sequence model. The variation in each endogenous variable (the X variables) is partitioned between prior endogenous variables and its residual. The path coefficients measure the influence of the associated variables. Note from Figure 1 that the first subscript in the path coefficient relating pairs of endogenous variables refers to the variable at the point of the arrow and the second to the variable at the tail. The notation for the path coefficients involving residual variables is simpler: only the index of the endogenous variable is necessary (for example, p_i), rather than the more complex p_{iU_i}.

A path model is either a diagram, such as Figure 1, or an isomorphic system of equations like (2). The explicit representation of assumed causal order in the system is one of the fundamental advances that linear causal modeling has made over the techniques of elaboration (though, of course, the advance is made at the cost of stronger a priori assumptions). In elaboration, the idea of causal ordering is always expressed outside the elaboration equation, whereas in path analysis it is an essential aspect of the system of equations. Moreover, as can be seen in Figure 2, path analysis allows for the measurement of causal influence in other than ideal-type circumstances, but elaboration requires the ideal-type circumstance of the partial associations vanishing.

Before going on to solve the system of equations

(2) for the unknown path coefficients, note that it provides for the *complete determination* of each endogenous variable. Each residual variable is defined as accounting for all the variation in its "attached" endogenous variable that is not accounted for by the other variables in the system. This accounting for variation is not provided by the Lazarsfeld model.

Calculating the path coefficients from the correlation coefficients. There are five parameters to be estimated from the sample data for the model given by equations (2): p_1, p_2, p_3, p_{21}, and p_{32}. Since the variables in equations (2) have been put in standard score form, both the sample means and their theoretical counterparts, $E(X_i)$, are zero. Thus the first moments are of no use in estimating the parameters, but the path coefficients can be shown to depend on the theoretical second moments: $E(X_i^2)$ and $E(X_iX_j)$. The estimates of the $E(X_iX_j)$ are just the sample correlations r_{ij}, and the $E(X_i^2)$ are equal to 1, again because all variables are standardized. The procedure is to multiply pairs of equations in (2), obtaining such terms as X_i^2 and X_iX_j, and to take expectations. To simplify the calculations, first express the X_i in terms of the U_i:

$$X_1 = p_1U_1,$$
$$X_2 = p_{21}p_1U_1 + p_2U_2,$$
$$X_3 = p_{32}p_{21}p_1U_1 + p_{32}p_2U_2 + p_3U_3.$$

Let $\rho_{ij} = E(X_iX_j)$ be the population correlation between X_i and X_j. Then

$$p_1^2 = 1,$$
$$p_1^2p_{21} = \rho_{12},$$
$$p_{32}p_{21}p_1^2 = \rho_{13},$$
(3)
$$p_{21}^2p_1^2 + p_2^2 = 1,$$
$$p_{32}p_{21}^2p_1^2 + p_{32}p_2^2 = \rho_{23},$$
$$p_{32}^2p_{21}^2p_1^2 + p_{32}^2p_2^2 + p_3^2 = 1.$$

Since there are six equations for five parameters, there must be a condition on the correlations to guarantee the existence of a solution. The solution to system (3) is

$$p_1^2 = 1,$$
$$p_{21} = \rho_{12},$$
$$p_{32} = \rho_{13}/\rho_{12} \quad (\text{or } p_{32} = \rho_{23}),$$
$$p_2 = (1 - \rho_{12}^2)^{1/2},$$
$$p_3 = (1 - \rho_{23}^2)^{1/2}.$$

Estimates of the path coefficients are found by substituting sample correlation coefficients, r_{ij}, for their population counterparts, ρ_{ij}.

Note that the model has a property that we have stated in passing: the path coefficient p_{32} must be

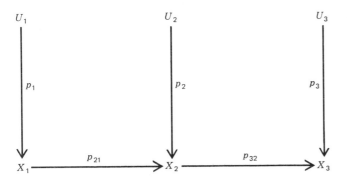

Figure 1 — Path diagram for Lazarsfeld's model of "interpretation"

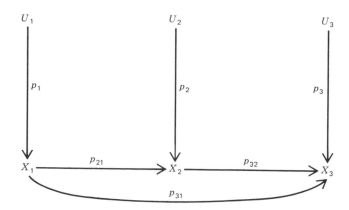

Figure 2 — Path diagram for a nonideal interpretation model

equal to both ρ_{23} and ρ_{13}/ρ_{12}. This is equivalent to $\rho_{13} = \rho_{12}\rho_{23}$, which is Lazarsfeld's criterion for a causal chain. Failure of this equality in the sample casts doubt on the postulated causal structure.

The nonideal interpretation model. Figure 2 differs from Figure 1 in having a direct path from X_1 to X_3 as well as the compound path through X_2. Just as Figure 1 and equation system (2) are equivalent, so are Figure 2 and the following system of equations:

$$\begin{aligned}
X_1 &= p_1 U_1, \\
X_2 &= p_{21}X_1 + p_2 U_2 = p_{21}p_1 U_1 + p_2 U_2, \\
X_3 &= p_{31}X_1 + p_{32}X_2 + p_3 U_3 \\
&= (p_{31}p_1 + p_{32}p_{21}p_1) U_1 + p_{32}p_2 U_2 + p_3 U_3.
\end{aligned} \tag{4}$$

As before, we find all the expected values of the X_i^2 and $X_i X_j$ ($i, j = 1,2,3$ and $i \neq j$) and solve for the path coefficients in terms of the ρ_{ij}. In this case, we obtain

$$\begin{aligned}
p_1^2 &= 1, \\
p_{21}p_1^2 &= \rho_{12}, \\
(p_{31} + p_{32}p_{21})\, p_1^2 &= \rho_{13}, \\
p_{21}^2 p_1^2 + p_2^2 &= 1, \\
(p_{31}p_{21} + p_{32}p_{21}^2)\, p_1^2 + p_{32}p_2^2 &= \rho_{23}, \\
(p_{31} + p_{32}p_{21})^2 p_1^2 + p_{32}^2 p_2^2 + p_3^2 &= 1.
\end{aligned} \tag{5}$$

Solving, we obtain

$$\begin{aligned}
p_{21} &= \rho_{12}, \\
p_{32} &= (\rho_{23} - \rho_{13}\rho_{12})/(1 - \rho_{12}^2), \\
p_{31} &= (\rho_{13} - \rho_{23}\rho_{12})/(1 - \rho_{12}^2), \\
p_1^2 &= 1, \\
p_2^2 &= 1 - \rho_{12}^2, \\
p_3^2 &= (1 - \rho_{12}^2 - \rho_{13}^2 - \rho_{23}^2 + 2\rho_{12}\rho_{13}\rho_{23})/(1 - \rho_{12}^2).
\end{aligned}$$

The path coefficient p_{32} is the population analogue of the sample standardized regression coefficient $\beta_{32.1}$, which is used to estimate it.

Wright's rules describe how to generate the set of equations relating the path coefficients and the correlation coefficients. If there is a direct path from X_i to X_j, the value of the path is just p_{ji}. If there is a path from X_i to X_j through X_k, the value of this compound path is the product of the path coefficients, p_{ki} and p_{jk}. This rule extends to a path with any number of connecting variables. For example, in Figure 2, the value of the path from U_1 to X_3 via X_1 and X_2 is $p_1 p_{21} p_{32}$. If there are two or more paths between X_i and X_j, the value of the total path between X_i and X_j is the sum of the values of the paths between X_i and X_j. In Figure 2, there is a second path between U_1 and X_3: namely, U_1 to X_1 to X_3 with the value $p_1 p_{31}$. The value of the total path is the sum of the two paths, $p_1 p_{31} + p_1 p_{21} p_{32}$. Note that this is just the coefficient of U_1 in the equation for X_3 in (4). Thus, this value represents the increase in X_3 (which is standardized), associated with an increase in U_1 (which is also standardized and here is just X_1), assuming that all other exogenous variables are held constant and that the covariance structure is as described in the model. Note that an individual path coefficient and the total path value may be negative and need not be bounded by any value whatsoever. For a complete statement of Wright's rules, see Li (1975).

The explicit solution of such a system of equations, while didactically useful, is necessary only when the available data is the correlation matrix. In actual research using raw data, it is not necessary to solve such a system of equations, for, when the system of equations is *identifiable* (as it always is in fully recursive models; see Boudon 1965), the path coefficients are standardized *partial regression coefficients*. These quantities are routinely computed in most packages of statistical computer programs such as *SPSS: Statistical Package for the Social Sciences* (Nie et al. 1975). In the model of Figure 2, the researcher would regress X_2 on X_1 and X_3 on X_2 and X_1. The estimated path coefficient p_{21} would be the estimated standardized regression coefficient of X_2 on X_1; and the estimated p_{32} and p_{31} would be the estimated standardized partial regression coefficients $\beta_{32.1}$ and $\beta_{31.2}$, respectively.

Specht (1975) proposes a test of the contribution of an endogenous variable to the model based on the generalized variance and generalized multiple correlation coefficient. The technique of selecting those variables that are most helpful in increasing a statistical measure of completeness is another example of dredging (see section 8; see also McPherson 1976). This use of statistical procedures is an example of model building, and the interpretation of the results of the "tests" must take this into account.

8. Statistical inference in survey analysis

The statistical theory of sample surveys has dealt almost entirely with sampling for descriptive studies, in which the usual problem is to estimate a few predesignated parameters or to test a few predesignated null hypotheses. When survey analysts have tried to apply this theory, they have often ignored or argued away two important assumptions: (1) that the *non*random errors of sampling, interviewing, and coding are negligible, and (2) that all the hypotheses tested were stated before examining the data.

Few survey investigators know the direction and magnitude of the nonrandom errors in their work with any accuracy, for measurement of these errors requires a specially designed study. Stephan and McCarthy (1958) reported the results of several empirical studies of sampling procedures. Such studies provide a rough guide to the survey investigator: precise data on the actual operation of survey procedures are probably available only in large research organizations that frequently repeat the same kind of survey. Without such knowledge, the interpretation of statistical tests and estimates in precise probabilistic terms may be misleading. A "statistically significant" relation may result from nonrandom error rather than from the independent variable, and a relation that apparently is not significant may stem from a nonrandom error opposite in sign and approximately equal in magnitude to the effect of the independent variable.

Lack of information about the nonrandom errors casts doubt on *any* inference from data, not simply on the inferences of formal statistics. However, the practices of many survey analysts justify emphasizing the effects of this lack of knowledge on the procedures of statistical inference, especially on tests of significance. All too often the words "significant at the .05 level" are thought to provide information on the random variables without there being any knowledge of the nonrandom variables— as if this phrase were a certificate of over-all methodological quality.

Even when there is no problem of nonrandom error, the history of the hypothesis being tested affects the validity of probability computations. A survey analyst seldom begins a study with a specific hypothesis in mind. What hypotheses he has are usually diffuse and ill formulated, and the data of the typical survey are so rich and suggestive that he almost always formulates many more hypotheses after looking at the results. Indeed, a survey ana-

lyst usually examines such a small proportion of the hypotheses that can be studied with his data that libraries of survey data have been established to facilitate "secondary analysis," or the restudy of survey data for purposes that may not have been intended by the original investigator (see Hyman 1972).

In the typical survey conducted for scientific purposes, the analyst alternates between examining the data and formulating hypotheses that are then explored further in the same body of data. This kind of "data dredging" is a necessary and desirable part of survey analysis; to limit survey analysis to hypotheses precisely stated in advance of seeing the data would be to make uneconomical use of a powerful and expensive tool. However, the analyst pays a price for this flexibility: although it is legitimate to explore the implications of hypotheses on the same body of data that suggested them, it is dangerous to attach probability statements to these hypotheses (see Selvin & Stuart 1966; for a contrary view, see Finifter 1972).

The situation is analogous to testing a table of random numbers for randomness: the a priori probability of finding six consecutive fives in the first six digits examined is $(.1)^6$, a quantity small enough to cast doubt on the randomness of the process by which the table was generated. If, however, one hunts through thousands of digits to find such a sequence (and the longer one looks, the greater the likelihood of finding it), this probability computation becomes meaningless. Similarly, the survey analyst who leafs through a pile of tables until he finds an "interesting" result that he then "tests for significance" is deceiving himself; the procedure designed to guard against the acceptance of chance results actually promotes their acceptance. The use of computers has exacerbated this problem. Many programs routinely test every relation for significance, usually with procedures that were intended for a single relation tested alone, and many analysts seem unable to resist dressing their dredged-up hypotheses in ill-fitting probabilistic clothes.

The analyst who wants to perform a statistical test of dredged-up hypotheses does not have to wait for a new study. If he has foreseen this problem, he can reserve a random subsample of his data at the outset of the analysis, dredge the remainder of his data for hypotheses, and then test a small number of these hypotheses on the reserved subsample. The analyst who was not so foresighted or who has dredged up too many hypotheses to test on one subsample may be able to use a large data library to provide an approximate test. For example, an analyst who wants to test a dreged-up relation involving variables *A*, *B*, and *C* would look through such a library to find another, comparable study with these same variables. (One rich source of such replications is the General Social Survey conducted annually by the National Opinion Research Center at the University of Chicago and available, along with thousands of other surveys, from the Roper Opinion Research Center, Williamstown, Massachusetts.) The analyst would then divide the sample of this study into a number of random subsamples and see how the dredged-up relation fares in these independent replications. If all or most of the subsamples yield relations in the same direction as his dredged-up relation, he can be reasonably confident that it was not the result of chance.

This procedure is simple, and it will become even simpler when the questions in the data libraries are put on magnetic tape, as the responses are now. However, this "backward replication" does raise some methodological problems, especially concerning the comparability of studies and questions. Neither this procedure nor the use of reserved subsamples has yet been studied in any detail by survey methodologists or statisticians.

HANAN C. SELVIN AND
STEPHEN J. FINCH

[*See also the biography of* LAZARSFELD.]

BIBLIOGRAPHY

ANDREWS, FRANK M.; MORGAN, JAMES N.; and SONQUIST, JOHN A. (1967) 1973 *Multiple Classification Analysis: A Report on a Computer Program for Multiple Regression Using Categorical Predictors.* Rev. ed. Ann Arbor: Survey Research Center, Univ. of Michigan.

BISHOP, YVONNE M. M.; FIENBERG, STEPHEN E.; and HOLLAND, PAUL W. 1975 *Discrete Multivariate Analysis.* Cambridge, Mass.: M.I.T. Press.

BLALOCK, HUBERT M. JR. 1964 *Causal Inferences in Nonexperimental Research.* Chapel Hill: Univ. of North Carolina Press.

BOUDON, RAYMOND 1965 A Method of Linear Causal Analysis: Dependence Analysis. *American Sociological Review* 30:365–374.

BOUDON, RAYMOND 1967 *L'analyse mathématique des faits sociaux.* Paris: Plon.

BOUDON, RAYMOND; and LAZARSFELD, PAUL F. (editors) 1966 *L'analyse empirique de la causalité.* Paris: Mouton.

DAVIS, JAMES A. 1974 Hierarchical Models for Significance Tests in Multivariate Contingency Tables: An Exegesis of Goodman's Recent Papers. Pages 189–231 in Herbert L. Costner (editor), *Sociological Methodology, 1973–74.* San Francisco: Jossey-Bass.

DRAPER, NORMAN R.; and SMITH, H. 1966 *Applied Regression Analysis.* New York: Wiley.

DUNCAN, OTIS DUDLEY 1966 Path Analysis: Sociological Examples. *American Journal of Sociology* 72: 1–16.

DUNCAN, OTIS DUDLEY 1975 *Introduction to Structural Equation Models.* New York: Academic Press.

DURKHEIM, ÉMILE (1897) 1951 *Suicide: A Study in Sociology.* Glencoe, Ill.: Free Press. → First published in French.

FINIFTER, BERNARD M. 1972 The Generation of Confidence: Evaluating Research Findings by Random Subsample Replication. Pages 112–175 in Herbert L. Costner (editor), *Sociological Methodology, 1972.* San Francisco: Jossey-Bass.

FLEISS, JOSEPH L. 1973 *Statistical Methods for Rates and Proportions.* New York: Wiley.

GOLDBERGER, ARTHUR S.; and DUNCAN, OTIS DUDLEY (editors) 1973 *Structural Equation Models in the Social Sciences.* New York: Academic Press.

GOODMAN, LEO A. 1964 Interactions in Multidimensional Contingency Tables. *Annals of Mathematical Statistics* 35:632–646.

GOODMAN, LEO A. 1970 The Multivariate Analysis of Qualitative Data: Interactions Among Multiple Classifications. *Journal of the American Statistical Association* 65: 226–256.

GOODMAN, LEO A. 1971 The Analysis of Multidimensional Contingency Tables: Stepwise Procedures and Direct Estimation Methods for Building Models for Multiple Classifications. *Technometrics* 13:33–61.

GOODMAN, LEO A. 1973 Guided and Unguided Methods for the Selection of Models for a Set of *T* Multidimensional Contingency Tables. *Journal of the American Statistical Association* 68:165–175.

GOODMAN, LEO A.; and KRUSKAL, WILLIAM H. 1954–1972 Measures of Association for Cross-classifications. Parts 1–4. *Journal of the American Statistical Association* 49 [1954]:732–764; 54 [1959]:123–163; 58 [1963]: 310–364; 67 [1972]:415–421. → Part 2: Further Discussion and References. Part 3: Approximate Sampling Theory. Part 4: Simplification of Asymptotic Variances.

GRAUNT, JOHN (1662) 1939 *Natural and Political Observations Made Upon the Bills of Mortality.* Edited with an introduction by Walter F. Willcox. Baltimore: Johns Hopkins Press.

HALBWACHS, MAURICE 1930 *Les causes du suicide.* Paris: Alcan.

HEISE, DAVID R. 1975 *Causal Analysis.* New York: Wiley.

HIRSCHI, TRAVIS; and SELVIN, HANAN C. (1967) 1973 *Principles of Survey Analysis.* New York: Free Press. → First published as *Delinquency Research.*

HOGBEN, LANCELOT T. 1957 *Statistical Theory: The Relationship of Probability, Credibility and Error; An Examination of the Contemporary Crisis in Statistical Theory From a Behaviourist Viewpoint.* London: Allen & Unwin.

HYMAN, HERBERT H. 1955 *Survey Design and Analysis: Principles, Cases, and Procedures.* New York: Wiley.

HYMAN, HERBERT H. 1972 *Secondary Analysis of Sample Surveys: Principles, Procedures, and Potentialities.* New York: Wiley.

LAZARSFELD, PAUL F. 1955 Interpretation of Statistical Relations as a Research Operation. Pages 116–125 in Paul F. Lazarsfeld and Morris Rosenberg (editors), *The Language of Social Research: A Reader in the Methodology of Social Research.* New York: Free Press. → An address given to the American Sociological Society in 1946.

LAZARSFELD, PAUL F. 1961 Notes on the History of Quantification in Sociology: Trends, Sources and Problems. Pages 147–203 in Harry Woolf (editor), *Quantification: A History of the Meaning of Measurement in the Natural and Social Sciences.* Indianapolis, Ind.: Bobbs-Merrill.

LAZARSFELD, PAUL F.; BERELSON, BERNARD; and GAUDET, HAZEL (1944) 1960 *The People's Choice: How the Voter Makes Up His Mind in a Presidential Campaign.* 2d ed. New York: Columbia Univ. Press.

LAZARSFELD, PAUL F.; PASANELLA, ANN K.; and ROSENBERG, MORRIS (editors) 1972 *Continuities in the Language of Social Research.* New York: Free Press. → A companion volume to Lazarsfeld and Rosenberg (1955).

LAZARSFELD, PAUL F.; and ROSENBERG, MORRIS (editors) 1955 *The Language of Social Research: A Reader in the Methodology of Social Research.* New York: Free Press.

LI, CHING CHUN 1975 *Path Analysis: A Primer.* Pacific Grove, Calif.: Boxwood.

McPHERSON, J. MILLER 1976 Theory Trimming. *Social Science Research* 5:95–106.

MANTEL, NATHAN; and HAENSZEL, W. 1959 Statistical Aspects of the Analysis of Data From Retrospective Studies of Disease. *Journal of the National Cancer Institute* 22:719–748.

MOSER, CLAUS A. 1958 *Survey Methods in Social Investigation.* New York: Macmillan.

NIE, NORMAN H. et al. (1970) 1975 *SPSS: Statistical Package for the Social Sciences.* 2d ed. New York: McGraw-Hill.

ROSENBERG, MORRIS 1968 *The Logic of Survey Analysis.* New York: Basic Books.

SELVIN, HANAN C. 1965 Durkheim's *Suicide:* Further Thoughts on a Methodological Classic. Pages 113–116 in Robert A. Nisbet (editor), *Émile Durkheim.* Englewood Cliffs, N.J.: Prentice-Hall.

SELVIN, HANAN C. 1976 Durkheim, Booth and Yule: The Nondiffusion of an Intellectual Innovation. *European Journal of Sociology* 17:39–51.

SELVIN, HANAN C.; and STUART, ALAN 1966 Data-dredging Procedures in Survey Analysis. *American Statistician* 20, no. 3:20–23.

SIMON, HERBERT A. 1954 Spurious Correlation: A Causal Interpretation. *Journal of the American Statistical Association* 49:467–479.

SONQUIST, JOHN A.; and MORGAN, JAMES N. 1964 *The Detection of Interaction Effects: A Report on a Computer Program for the Selection of Optimal Combinations of Explanatory Variables.* Monograph No. 35. Ann Arbor: Survey Research Center, Univ. of Michigan.

SPECHT, DAVID A. 1975 On the Evaluation of Causal Models. *Social Science Research* 4:113–133.

STEPHAN, FREDERICK F.; and McCARTHY, PHILIP J. 1958 *Sampling Opinions: An Analysis of Survey Procedure.* New York: Wiley.

STERLING, T. et al. 1966 Robot Data Screening: A Solution to Multivariate Type Problems in the Biological and Social Sciences. *Communications of the Association for Computing Machinery* 9:529–532.

STOUFFER, SAMUEL A. et al. 1949 *The American Soldier.* 2 vols. Studies in Social Psychology in World War II. Princeton Univ. Press. → Volume 1: *Adjustment During Army Life.* Volume 2: *Combat and Its Aftermath.*

SÜSSMILCH, JOHANN PETER (1741) 1788 *Die göttliche Ordnung in den Veränderungen des menschlichen Geschlechts, aus der Geburt, dem Tode und der Fortpflanzung desselben erwiesen.* 3 vols. Berlin: Verlag der Buchhandlung der Realschule.

TUKEY, JOHN W. 1962 The Future of Data Analysis. *Annals of Mathematical Statistics* 33:1–67.

U.S. BUREAU OF THE CENSUS 1972 *Public Use Samples of Basic Records From the 1970 Census: Description and Technical Documentation.* Washington: The Bureau.

WILSON, ALAN B. 1964 Analysis of Multiple Cross-classifications in Cross-sectional Designs. Berkeley: Survey Research Center.

WOLD, HERMAN 1956 Casual Inference From Observational Data: A Review of Ends and Means. *Journal of the Royal Statistical Society* Series A 119:28–60. → Includes discussion.

WOLD, HERMAN 1959 Ends and Means in Econometric Model Building: Basic Considerations Reviewed. Pages 355–434 in Uif Grenander (editor), *Probability and Statistics: The Harald Cramér Volume.* Stockholm: Almqvist & Wiksell; New York: Wiley.

WOOLF, HARRY (editor) 1961 *Quantification: A History of the Meaning of Measurement in the Natural and Social Sciences.* Indianapolis, Ind.; Bobbs-Merrill.

YOUNG, DENNIS; and FINCH, STEPHEN J. 1977 *Public Needs, Private Service: The Delivery of Foster Care Through a System of Nonprofit Agencies.* Lexington, Mass.: Heath.

YULE, G. UDNY 1895 On the Correlation of Total Pauperism With Proportion of Out-relief. *Economic Journal* 5:603–611.

YULE, G. UDNY 1897 On the Significance of Bravais' Formulae for Regression etc., in the Case of Skew Correlation. Royal Society of London, *Proceedings* 60:477–489.

YULE, G. UDNY 1899 An Investigation Into the Causes of Changes in Pauperism in England, Chiefly During the Last Two Intercensal Decades. Part 1. *Journal of the Royal Statistical Society* 62:249–286.

YULE, G. UDNY 1900 On the Association of Attributes in Statistics, With Illustrations From the Material of the Childhood Society.... Royal Society of London, *Philosophical Transactions* Series A 194:257–319.

YULE, G. UDNY; and KENDALL, MAURICE G. 1950 *An Introduction to the Theory of Statistics.* 14th ed., rev. & enl. London: Griffin; New York: Hafner. → Yule was the sole author of the first edition (1911). Kendall, who became a joint author on the eleventh edition (1937), revised the current edition and added new material to the 1965 printing.

II

THE ANALYSIS OF ATTRIBUTE DATA

Modern social research requires the study of the interrelations between characteristics that are themselves not quantified. Has someone completed high school? Is he native born? Male or female? Is there a relation between all such characteristics? Do they in turn affect people's attitudes and behavior? How complex are these connections: do men carry out their intentions more persistently than women? If so, is this sex difference related to level of education?

Some answers to some of these questions have been made possible by two developments in modern social research: improved techniques of collecting data through questionnaires and observations, and sampling techniques that make such collecting less costly. Empirical generalizations in answer to questions like those above proceed essentially in two steps. First, people or collectives have to be measured on the characteristics of interest. Such characteristics are now often called *variates*, to include simple dichotomies as well as "natural" quantitative variables, like age, or "artificial" indices, like a measure of anxiety. The second step consists in establishing *connections* between such variates. The connections may be purely descriptive, or they may be causal chains based on established theories or intuitive guesswork.

In order to take the second step and study the connection between variates, certain procedures have been developed in *survey analysis*, although the procedures apply, of course, to any kind of data, for instance, census data. Attitude surveys have greatly increased the number of variates that may be connected, and thus the problems of studying connections have become especially visible. The term "survey" ordinarily excludes studies in which there are observations repeated over time on the same individuals or collectives; such studies, usually called panel studies, are not discussed in this article [*see* PANEL STUDIES].

It makes a difference whether one deals with quantitative variables or with classifications allowing only a few categories, which often may not even be ordered. Connections between quantitative variables have long been studied in the form of correlation analysis and its many derivatives [*see* MULTIVARIATE ANALYSIS, *article on* CORRELATION METHODS]. Correlation techniques can sometimes also be applied to qualitative characteristics, by assigning arbitrary numbers to their subdivisions. But some of the most interesting ideas in survey analysis emerge if one concentrates on data where the characterization of objects cannot be quantified and where only the frequencies with which the objects fall into different categories are known. As a matter of fact, the main ideas can be developed by considering only dichotomies, and that will be done here. The two terms "dichotomy" and "attribute" will be used interchangeably.

Dichotomous algebra

Early statisticians (see Yule & Kendall 1950) showed some interest in attribute statistics, but such interest was long submerged by the study of quantitative variables with which the economists and psychologists were concerned. Attribute statistics is best introduced by an example which a few decades ago one might have found as easily in a textbook of logic as in a text on descriptive statis-

tics. The example will also permit introduction of the symbolism needed for this exposition.

Suppose there is a set of 1,000 people who are characterized according to sex (attribute 1), according to whether they did or did not vote in the last election (attribute 2), and according to whether they are of high or low socioeconomic status (SES, attribute 3). There are 200 high-status men who voted and 150 low-status women who did not vote. The set consists of an equal number of men and women. One hundred high-status people did not vote, and 250 low-status people did vote. There are a total of 650 voters in the whole set; 100 low-status women did vote. How many low-status men voted? How is the whole set divided according to socioeconomic status?

There are obviously three attributes involved in this problem. It is convenient to give them arbitrary numbers and, for each attribute, to assign arbitrarily a positive sign to one possibility and a negative sign to the other. The classification of Table 1 is then obtained.

The problem assigns to some of the combinations of attribute possibilities a numerical value, the proportion of people who belong to the "cell." The information supplied in the statement of the problem can then be summarized in the following way:

$$p_{123} = .20,$$
$$p_{1\overline{23}} = .15,$$
$$p_1 = p_{\overline{1}},$$
$$p_{\overline{23}} = .10,$$
$$p_{2\overline{3}} = .25,$$
$$p_2 = .65,$$
$$p_{\overline{123}} = .10.$$

Here the original raw figures are presented as proportions of the whole set; a bar over an index indicates that in this special subset the category of that attribute to which a negative sign has been assigned applies. For example, $p_{\overline{2}3}$ is the number of people who did not vote and are of high status; this proportion has the numerical value $100/1{,}000 = .10$. The number of indices that are attached to a proportion is called its *stratification level*. The problem under discussion ends with two questions. Their answers require the computation of two proportions, $p_{12\overline{3}}$ and p_3.

The derivation of the missing frequencies can

Table 1

Number	Attribute	Sign Assignment	
		+	−
1	Sex	Men	Women
2	Vote	Yes	No
3	SES	High	Low

be done rather simply in algebraic form. For the present purpose it is more useful to derive them by introduction of so-called fourfold tables. The point of departure is, arbitrarily, the fourfold table between attributes 1 and 2. This table is then stratified by introducing attribute 3, giving the scheme of Table 2, in which the information presented in the original problem is underscored. The principal findings are starred.

The tabular form makes it quite easy to fill in the missing cells by addition and subtraction, and this also permits answers to the two questions of the original problem in symbolic, as well as in numerical, form:

The proportion of low-status men voting is given by

$$p_{12\bar{3}} = p_{2\bar{3}} - p_{\bar{1}2\bar{3}} = .15,$$

and the proportion of high-status people is

$$p_3 = p_{\bar{2}3} + (p_2 - p_{23}) = .5.$$

If it were only a task of dividing and recombining sets, the problem would now be solved. But in survey analysis a new element enters which goes beyond the tradition of a calculus of classes. One is interested in the *relation* between attributes, and to express that relation an additional notion and an additional symbol are required. Taking the left side of Table 2, it is reasonable to ask how many men would have voted if sex had no influence on political behavior. The proportion of joint occurrence of two independent events is the product of the two separate proportions ($p_1 \times p_2$), and therefore, under independence, $.65 \times .5 \times 1,000 = 325$ male voters would have been expected. Actually there are 350, indicating that men have a slightly higher tendency to vote than women. If the difference between the empirical and the "independent" figures for all four cells of the fourfold table had been computed, the same result—25 cases, or a difference of 2.5 per cent of the total set—would have been obtained. (Problems of statistical significance are not relevant for the present discussion.)

Here it is useful to introduce the symbolic abbreviation $|12|$,

$$
\begin{aligned}
.025 = |12| &= p_{12} - p_1 p_2 \\
&= -(p_{\bar{1}2} - p_{\bar{1}}p_2) \\
&= -(p_{1\bar{2}} - p_1 p_{\bar{2}}) \\
&= p_{\bar{1}\bar{2}} - p_{\bar{1}} p_{\bar{2}},
\end{aligned}
\tag{1}
$$

which may be called the *cross product* or the *symmetric parameter of second level*. The quantity $|12|$ is the basis of many measures of association in 2×2 tables [see STATISTICS, DESCRIPTIVE, *article on* ASSOCIATION]. Note that $|12| = -|1\bar{2}|$ and, in general, $|ij| = -|i\bar{j}| = |\bar{i}\bar{j}|$.

Cross products can also be computed for the stratified fourfold tables on the right side of Table 2, and by mere inspection new factual information emerges. In the high-status stratum, men and women have the same tendency to vote, and the stratified cross product vanishes. In the low-status stratum, the relation between sex and voting is very marked. The basic question of survey analysis is whether the relation between such cross products of different stratification levels can be put into a general algorithm that can be used to draw substantive inferences. The rest of this presentation is given to the development of the relevant procedures.

In the context of this presentation, with the exception of one case to be mentioned presently, interest is confined to whether the cross product is positive, is negative, or vanishes. This last case occurs when the two attributes are independent. If men and women furnish the same proportion of voters, the cross product will be zero. The same fact can, of course, also be expressed in a different way: the proportion of men is the same among voters and among nonvoters. In empirical studies the cross product will rarely vanish perfectly; it will just be very small. It is a problem in the theory of statistical hypothesis testing to decide when an empirical cross product differs enough from a hypothesized value (often zero) so that the difference is statistically significant. Concern here, however, is

Table 2ᵃ

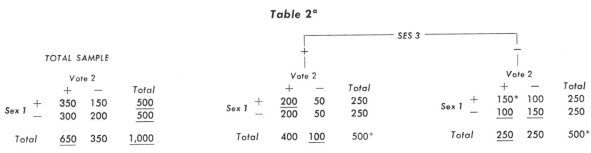

a. Asterisk represents principal finding. Underscoring indicates the original information.

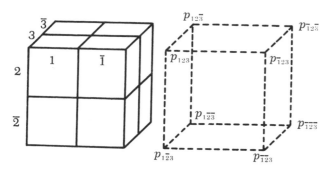

Figure 1 — The dichotomous cube and the relative position of third-level proportions

Table 3 — Front sheet of dichotomous cube

	1	$\bar{1}$	Total
2	p_{123}	$p_{\bar{1}23}$	p_{23}
$\bar{2}$	$p_{1\bar{2}3}$	$p_{\bar{1}\bar{2}3}$	$p_{\bar{2}3}$
Total	p_{13}	$p_{\bar{1}3}$	p_3

Table 4 — Back sheet of dichotomous cube

	1	$\bar{1}$	Total
2	$p_{12\bar{3}}$	$p_{\bar{1}2\bar{3}}$	$p_{2\bar{3}}$
$\bar{2}$	$p_{1\bar{2}\bar{3}}$	$p_{\bar{1}\bar{2}\bar{3}}$	$p_{\bar{2}\bar{3}}$
Total	$p_{1\bar{3}}$	$p_{\bar{1}\bar{3}}$	$p_{\bar{3}}$

only with general logical relations between cross products, and discussion of sampling and measurement error is excluded.

The eight third-level class proportions of a three-attribute dichotomous system can be arranged in the form of a cube. Such a dichotomous cube consists of eight smaller cubes, each corresponding to one of the third-level proportions. The dichotomous cube and the relative position of the third-level proportions are shown in Figure 1.

A second-level proportion can be *expanded* in terms of its third-level components. Thus, for example, $p_{1\bar{2}} = p_{1\bar{2}3} + p_{1\bar{2}\bar{3}}$. No proportion can be negative. Therefore, *if a second-level proportion vanishes, so do its components*; for example, if $p_{\bar{1}2} = 0$, then it follows that $p_{\bar{1}23} = p_{\bar{1}2\bar{3}} = 0$.

Consider now the proportions that lie in the front sheet of the dichotomous cube. Keeping them in their same relative position, they are as shown in Table 3, where the entries are the proportions corresponding to the frequencies in the middle part of Table 2. Table 3 is a fourfold table that summarizes the relation between attributes 1 and 2 within only a part of the complete set of individuals—those who possess attribute 3. Such a table is called a *stratified fourfold table*. The stratified table is bordered by marginal entries of the second level which are the sums of the respective rows and columns, as indicated in the margins of Table 3, as well as by the first-level proportion p_3, the sum of all the entries. The question of dependence or independence of two attributes can also be raised for a stratified table. A conditional cross product could be constructed from the stratified table by first dividing each entry by p_3. But because it makes the whole discussion more consistent and avoids repeated computation of proportions, it is preferable to remain with the original proportions, computed on the base of the total set.

An obvious question, then, would concern the proportion of the high-status men who would have voted if sex had no influence on political behavior.

Under this hypothesized independence the proportion would be given by $p_{13}p_{23}/p_3^2$. The actual proportion of the high-status men who voted is given by p_{123}/p_3. An obvious measure of the relationship between sex and voting within the high-status group is supplied by the difference between this actual proportion and the theoretical one. As an alternative development, if within the subset of individuals who possess attribute 3 there is no relation between attributes 1 and 2, then one would expect to find that there is the same proportion of individuals with attribute 1 in the entire subset as there is in that part of the subset which also possess attribute 2. That is,

$$\frac{p_{13}}{p_3} = \frac{p_{123}}{p_{23}}$$

or

$$\begin{vmatrix} p_{123} & p_{23} \\ p_{13} & p_3 \end{vmatrix} = 0.$$

This determinant will be taken as the definition of the *stratified cross product* between attributes 1 and 2 among the subset possessing attribute 3. The symbol $|12; 3|$ is used for this cross product. In general, then, $|ij; k|$ is defined by

$$(2) \qquad |ij; k| = \begin{vmatrix} p_{ijk} & p_{jk} \\ p_{ik} & p_k \end{vmatrix}.$$

Note that $|ij|$ can be represented as

$$\begin{vmatrix} p_{ij} & p_j \\ p_i & 1 \end{vmatrix}.$$

The elements in the back sheet of the dichotomous cube make up the stratified fourfold table shown in Table 4, which summarizes the relation between attributes 1 and 2 within that subset of individuals who *lack* attribute 3. The cross product of such a stratified fourfold table is defined by

$$\left| ij; \bar{k} \right| = \begin{vmatrix} p_{ij\bar{k}} & p_{j\bar{k}} \\ p_{i\bar{k}} & p_{\bar{k}} \end{vmatrix}.$$

It should be noted that eq. (2) suffices to define both $|ij; k|$ and $|ij; \bar{k}|$ if the index k is permitted to range through both barred and unbarred integers designating particular attributes.

Six fourfold tables can be formed from the elements of the dichotomous cube, one for each of the six sheets. Each of these stratified tables can be characterized by its cross product. Thus, six conditional cross products, $|12; 3|$, $|12; \bar{3}|$, $|23; 1|$, $|23; \bar{1}|$, $|13; 2|$, and $|13; \bar{2}|$, can be formed from the elements of a dichotomous cube.

As can be seen from Figure 1, a dichotomous cube is completely known if the absolute frequency for each of its eight "cells" is known. These eight ultimate frequencies form what is called a *fundamental set;* if they are known, one can compute any of the many other combinations which may be of interest, some of which were used in the introductory example. The number of these possible combinations is large if one thinks of all stratification levels and of all combinations of absence and presence of the three attributes. The search for other fundamental sets, therefore, has always been of interest. Yule (see Yule & Kendall 1950) investigated one which consists of the so-called positive joint frequencies—the frequencies on all stratification levels which have no barred indices, together with the size of the sample (obviously, if the sample size is known, the remaining seven terms in a fundamental set can also be given in the form of proportions).

If one is looking for something like a calculus of associations, the question arises whether a fundamental set whose terms include cross products could be developed. One might then start with the following elements: N, the total number of individuals; p_1, p_2, p_3, the first-level proportions; and $|12|$, $|23|$, $|13|$, the three possible second-level cross products. But these are only seven elements, and so far no third-level data have been utilized. The eighth element must be of the third level, somehow characterizing the dichotomous cube as a whole (that is, depending explicitly on the third-level proportions).

One might choose as the eighth element one of the six stratified cross products, but there is no good reason for choosing one in preference to another. Also, any one of these would lack the symmetry which one can reasonably require of a parameter representing the whole cube.

The choice of the eighth parameter can be determined by three criteria: (*a*) The parameter should be *symmetric;* that is, its value should not be affected if the numbering of the attributes is changed around. For example, $|ij|$ is symmetric, since $|ij| = |ji|$. (*b*) The parameter should be *homogeneous*, in the sense that each of its terms should involve the same number of subscripts. (*c*) The parameter should be such that it can be used, together with lower-level class proportions, to *evaluate any third-level proportion.*

The second-level cross products $|ij| = p_{ij} - p_i p_j$ obviously satisfy the first two conditions. That condition (*a*), symmetry ($|12| = |21|$), and condition (*b*), homogeneity, are satisfied can be seen by inspection. That a condition analogous (for second-level proportions) to the third one is satisfied can be demonstrated by equations like

$$p_{ij} = p_i p_j - |ij|,$$

which can be verified for any combination of bars and indices.

A homogeneous, symmetric parameter of third level can be built up from lower-level proportions and parameters as follows:

$$(3) \quad p_{ijk} = p_i p_j p_k + p_i |jk| + p_j |ik| + p_k |ij| + |ijk|.$$

The introduction of a new mathematical quantity often, at first sight, seems arbitrary, and indeed it is, in the sense that any combination of old symbols can be used to define a new one. The utility of an innovation and its special meaning can only evolve slowly, in the course of its use in the derivation of theorems and their applications. It will be seen that most of the subsequent material in this presentation centers on symmetric parameters of level higher than that of the cross product.

The quantity $|ijk|$, implicitly defined by eq. (3), quite evidently satisfies the criteria of homogeneity and symmetry; that it can be used together with lower-level class proportions to compute any third-level class proportion will now be shown.

If the indices i, j, and k are allowed to range through any three different numbers, barred or unbarred, it is easily shown that the quantities $|ij\bar{k}|$, $|i\bar{j}k|$, and so forth thus defined are not independent but are related to one another and to $|ijk|$ as follows:

$$|ijk| = -|ij\bar{k}| = -|i\bar{j}k| = -|\bar{i}jk| = |i\bar{j}\bar{k}|$$
$$= |\bar{i}j\bar{k}| = |\bar{i}\bar{j}k| = -|\bar{i}\bar{j}\bar{k}|.$$

As a mnemonic, note that if in $|ijk|$ an odd number of indices is barred, the symmetric parameter changes its sign; if an even number is barred, the value of $|ijk|$ remains unchanged. This is easily proved by showing that

$$|ijk| + |ij\bar{k}| = 0.$$

Once the third-level symmetric parameter is defined and computed by eq. (3), the computation of any desired third-level class proportion is possible. For example,

$$p_{\bar{1}2\bar{3}} = p_{\bar{1}}p_2p_{\bar{3}} + p_{\bar{1}}|2\bar{3}| + p_2|\bar{1}\bar{3}| + p_{\bar{3}}|\bar{1}2| + |\bar{1}2\bar{3}|$$
$$= p_{\bar{1}}p_2p_{\bar{3}} - p_{\bar{1}}|23| + p_2|13| - p_{\bar{3}}|12| + |123|$$

With the introduction of the third-level symmetric parameter, a three-attribute dichotomous system can now be completely summarized by *a new fundamental set* of eight data: $N, p_1, p_2, p_3, |12|, |13|, |23|, |123|$.

Through the symmetric parameter, cross products of all levels can be connected. To keep a concrete example in mind, refer to Table 2, where the relation between sex and voting is reported for the whole sample and for two SES strata. It will be seen presently that the following formulas form the core of survey analysis.

Symmetric parameters are substituted into the form

$$|ij; k| = \begin{vmatrix} p_{ijk} & p_{ik} \\ p_{jk} & p_k \end{vmatrix},$$

yielding

$$|ij; k| =$$

(4)
$$\begin{vmatrix} p_ip_jp_k + p_i|jk| + p_j|ik| + p_k|ij| + |ijk| & p_ip_k + |ik| \\ p_jp_k + |jk| & p_k \end{vmatrix}$$

In the last determinant the second row multiplied by p_i is subtracted from the first row, and then the second column multiplied by p_j is subtracted from the first column. This leaves the right side of (4) as

$$\begin{vmatrix} |ijk| + p_k|ij| & |ik| \\ |jk| & p_k \end{vmatrix}.$$

Thus,

(5) $$|ij; k| = p_k|ijk| + p_k^2|ij| - |ik||jk|.$$

By a similar computation,

(6) $$|ij; \bar{k}| = -p_{\bar{k}}|ijk| + p_{\bar{k}}^2|ij| - |ik||jk|.$$

Eqs. (5) and (6) are, of course, related to each other by the general rule of barred indices expressed above.

It is worthwhile to give intuitive meaning to eqs. (5) and (6). Suppose the relation between sex (i) and voting (j) is studied separately among people of high and low SES (k). Then $|ij; k|$ is the

cross product for sex and voting as it prevails in the high SES group. Eq. (5) says that this stratified interrelation is essentially the cross product $|ij|$ as it prevails in the total population corrected for the relation that SES has with both voting and sex, given by the product $|ik||jk|$. But an additional correction has to be considered: the "triple interaction" between all three attributes, $|ijk|$.

Subtracting eq. (6) from eq. (5), using the fact that $p_k + p_{\bar{k}} = 1$, and rearranging terms, one obtains

$$|ijk| = |ij; k| - |ij; \bar{k}| + |ij| (p_k - p_{\bar{k}}).$$

The symmetric parameter thus, in a sense, "measures" the *difference in the association* between i and j under the condition of k as compared with the condition of $\bar{k}$. This is especially true if $p_k = p_{\bar{k}}$, that is, if the two conditions are represented equally often—and often it is possible to manipulate marginals to produce such an equal cut (either by choosing appropriate sample sizes or by dichotomizing at the median).

By dividing eq. (5) and eq. (6) by p_k and $p_{\bar{k}}$, respectively, and adding the two, one obtains

(7) $$\frac{|ij; k|}{p_k} + \frac{|ij; \bar{k}|}{p_{\bar{k}}} = |ij| - \frac{|ik||jk|}{p_kp_{\bar{k}}}.$$

The formula on the left side is analogous to the traditional notion of partial correlation: a weighted average of the two stratified cross products. It may be called the partial association between i and j with k partialed out.

Relation to measures of association. It is very important to keep in mind the difference between a partial and a stratified association. It can happen, for instance, that a partial association is zero, while the two stratified ones have nonzero values, one positive and one negative.

It was mentioned before that the cross products are not "measures of association." They form, however, the core of most of the measures which have been proposed for fourfold tables (Goodman & Kruskal 1954). A typical case is the "per cent difference," which may be exemplified by a well-known paradox [*see* STATISTICS, DESCRIPTIVE, *article on* ASSOCIATION]. Suppose that the three attributes are i, physical prowess; j, intelligence; and k, SES. Designate f_{ij} as the difference between the percentage of intelligent people in the physically strong group and in the physically weak group. It can easily be verified that $f_{ij} = |ij|/p_ip_{\bar{i}}$. The corresponding relation for the subset of high SES people would be

$$f_{ij:k} = \frac{|ij; k|}{p_{ik}p_{\bar{i}k}}.$$

Substituting this expression into eq. (7), one obtains

$$(8)\quad f_{ij}p_ip_{\bar{\imath}} = f_{ij;k}\frac{p_{ik}p_{\bar{\imath}k}}{p_k} + f_{ij;\bar{k}}\frac{p_{i\bar{k}}p_{\bar{\imath}\bar{k}}}{p_{\bar{k}}} + f_{ki}f_{kj}p_kp_{\bar{k}}.$$

The *f*-coefficients are asymmetric. The first subscript corresponds to the item which forms the basis for the percenting; thus, f_{kj} is the difference between the per cents of intelligent people in the high SES subset and in the low SES subset. Now eq. (8) permits the following interpretation: Suppose that the high SES subset has a higher percentage of intelligent people but a lower percentage of physically strong people than the low SES subset ($f_{ki} < 0$ and $f_{kj} > 0$). Then it can happen that *in each SES subset the physically stronger people are more often relatively intelligent, while in the total group*—high and low SES combined—*the opposite is true:*

$$f_{ij;k} > 0, \quad f_{ij;\bar{k}} > 0, \quad \text{but } f_{ij} < 0.$$

Consider one more example of introducing a traditional measure into eq. (7). Suppose someone wants to use the well-known phi-coefficient, which is defined by

$$\phi_{ij} = \frac{|ij|}{\sqrt{p_ip_{\bar{\imath}}p_jp_{\bar{\jmath}}}}.$$

With the use of the obvious symbol, the phi-coefficient applied to a stratified fourfold table would be

$$\phi_{ij;k} = \frac{|ij;k|}{\sqrt{p_{ik}p_{\bar{\imath}k}p_{jk}p_{\bar{\jmath}k}}}.$$

By introducing the last two expressions into eq. (7), one would obtain a relation between stratified and unstratified phi-coefficients.

This is a good place to say a word about the relation between the Yule tradition and the tenor of this presentation. In a rather late edition Yule included one page on "relations between partial associations." He attached little importance to that approach: "In practice the existence of these relations is of little or no value. They are so complex that lengthy algebraic manipulation is necessary to express those which are not known in terms of those which are. It is usually better to evaluate the class frequencies and calculate the desired results directly from them" (Yule & Kendall 1950; p. 59 in the 1940 edition). The few computations Yule presented were indeed rather clumsy. It is easy to see what brought about improvement: the use of determinants, an index notation, and, most of all, the symmetric parameters. Still, it has to be acknowledged that Yule drew attention to the approach which was later developed. Incidentally, Yule re-

ported the theorem on the weighted sum of stratified cross products (eq. 7). It appeared in earlier editions only as an exercise, but in later editions it was called "the one result which has important theoretical consequences." The consequences he had in mind were studies of spurious factors in causal analysis, which he discussed under the title "illusory associations." It will presently be seen what he had in mind.

Modes of explanation in survey analysis

Eq. (7) is undoubtedly the most important for survey analysis. To bring out its implication, it is preferable to change the notation. Assume two original attributes, x and y. Their content makes it obvious that x precedes y in time sequence. If $|xy| > 0$, then this relation requires explanation. The explanation is sought by the introduction of a test factor, t, as a stratification variable. The possible combinations form a dichotomous cube, and eq. (7) now takes the form

$$(9)\quad |xy| = \frac{|xy; t|}{p_t} + \frac{|xy; \bar{t}|}{p_{\bar{t}}} + \frac{|xt|\,|ty|}{p_tp_{\bar{t}}}.$$

This *elaboration* leads to two extreme forms, which are, as will be seen, of major interest. In the first case, the two stratified relations vanish; then eq. (9) reduces to

$$(10)\quad |xy| = \frac{|xt|\,|ty|}{p_tp_{\bar{t}}}.$$

In the second, the test factor, t, is unrelated to x (so that $|xt| = 0$), and then

$$(11)\quad |xy| = \frac{|xy; t|}{p_t} + \frac{|xy; \bar{t}|}{p_{\bar{t}}},$$

which also results if $|ty| = 0$. This form will turn out to be of interest only if one of the two stratified relations is markedly stronger than the other. Call eq. (11) the S form (emphasis on the stratified cross products) and eq. (10) the M form (emphasis on the "marginals").

To this formal distinction a substantive one, the time order of the three attributes, must be added. If x is prior to y in time, then t either can be located *between* x and y in time or can *precede* both. In the former case one speaks of an *intervening* test variable, in the latter of an *antecedent* one. Thus, there are four major possibilities. It is, of course, possible that t is subsequent in time to both x and y. But this is a case which very rarely occurs in survey analysis and is therefore omitted in this presentation.

Given the two forms of eq. (9) and the two relevant time positions of t, essentially four modes arise with two original variables and one test variable, as shown in Table 5. If a relation between two variables is analyzed in the light of a third, either with real data or theoretically, only these four modes or combinations thereof are sought, irrespective of whether they are called interpretation, understanding, theory, or anything else.

Table 5

		STATISTICAL FORM	
		S	M
POSITION OF t	Antecedent	SA	MA
	Intervening	SI	MI

Before this whole scheme is applied to concrete examples, the restriction put on the paradigm of Table 5 should be re-emphasized. Only one test variable is assumed. In actual survey practice it is highly unlikely that in eq. (9) the stratified cross products would disappear after one step. Instead of reaching eq. (10), one is likely to notice just a lowering of the first two terms on the right side of eq. (9). As a result, additional test variables must be introduced. But these further steps do not introduce new ideas. One stops the analysis at the point where one is satisfied with a combination of the four modes summarized in Table 5.

The notion of sequence in time also can be more complicated than appears in this schematic presentation. Sometimes there is not enough information available to establish a time sequence. Thus, when a positive association between owning a product and viewing a television program that advertises it is found, it is not necessarily known whether ownership preceded listening or whether the time sequence is the other way around. Additional information is then needed, if one is to proceed with the present kind of analysis. In other cases a time sequence might be of no interest, because the problem is of a type for which latent structure analysis or, in the case of quantitative variables, factor analysis is more appropriate [see LATENT STRUCTURE; FACTOR ANALYSIS AND PRINCIPAL COMPONENTS]. The problem at hand is to "explain" an empirically found association between x and y. But "explain" is a vague term. The procedures which lead to the main paradigm show that there exist four basic modes of explanation, the combinations of which form the basis for survey analysis. It is also reasonable to relate each type to a terminology which might most frequently be found in pertinent literature.

But although the basic types or modes of analysis are precisely defined, the allocation of a name to each of them is somewhat arbitrary and could be changed without affecting the main distinctions. Now each of the four types in the paradigm will be taken up and exemplified.

Specification. In cases of the type SA, the test variable, t, is usually called a condition. General examples easily come to mind, although in practice they are fairly rare and are a great joy to the investigator when they are found. For example: the propaganda effect of a film is greater among less-educated than among highly educated people; the depression of the 1930s had worse effects on authoritarian families than on other types.

Three general remarks can be made about this type of finding or reasoning: First, it corresponds to the usual stimulus–disposition–response sequence, with x as the stimulus and the antecedent t as the disposition. Second, the whole type might best be called one of *specification*. One of the two stratified associations will necessarily be larger than the original relationship. The investigator specifies, so to speak, the circumstances under which the original relationship holds true more strongly. Third, usually one will go on from the specification and ask why the relationship is stronger on one side of the test dichotomy than it is in the total group. This might then lead to one of the other types of analysis. Durkheim (1897) used type SA in discussing why relatively fewer married people commit suicide than unmarried people. He introduced as a test variable "a nervous tendency to suicide, which the family, by its influence, neutralizes or keeps from development." This is type SA exactly. It does not appear to be a convincing explanation, because the introduction of the hypothetical test variable (tendency to suicide) sounds rather tautological. A more important question is why the family keeps this tendency from development, which leads to type MI, as will be seen later.

Contingency. The type SI is also easily exemplified. In a study of the relationship between job success (y) and whether children did or did not go to progressive schools (x), it is found that if the progressively educated children come into an authoritarian situation (t), they do less well in their work than others; on the other hand, if they come into a democratic atmosphere, their job success is greater.

The relation between type of education and job success is elaborated by an intervening test factor, the work atmosphere. This is a "contingency." In many prediction studies the predicted value depends upon subsequent circumstances that are not related to the predictor. An example is the relation between occupational status and participation in community activities. White-collar people participate more if they are dissatisfied with their jobs, whereas manual workers participate more if they are satisfied.

Correcting spurious relationships. Type *MA* is used mainly in rectifying what is usually called a *spurious relationship*. It has been found that the more fire engines that come to a fire (x), the larger the damage (y). Because fire engines are used to reduce damage, the relationship is startling and requires elaboration. As a test factor, the size of the fire (t) is introduced. It might then be found that fire engines are not very successful; in large, as well as small, fires the stratified relation between x and y vanishes. But at least the original positive relation now appears as the product of two marginal relations: the larger the fire, the more engines called out, on the one hand, and the greater the damage, on the other hand.

Interpretation. Type *MI* corresponds to what is usually called *interpretation*. The difference between the discovery of a spurious relationship and interpretation in this context is related to the time sequence between x and t. In an interpretation, t is an intervening variable situated between x and y in the time sequence.

Examples of type *MI* are numerous. Living in a rural community rather than a city (x) is related to a lower suicide rate (y). The greater intimacy of rural life (t) is introduced as an intervening variable. If there were a good test of cohesion, it would undoubtedly be found that a community's being a rural rather than an urban one (x) is positively correlated with its degree of cohesion (t) and that greater cohesion (t) is correlated with lower suicide rates (y). But obviously some rural communities will have less cohesion than some urban communities. If cohesion is kept constant as a statistical device, then the partial relationship between the rural–urban variable and the suicide rate would have to disappear.

Differences between the modes. It might be useful to illustrate the difference between type *MA* and type *MI* in one more example. During the war married women working in factories had a higher rate of absence from work than single women. There are a number of possible elaborations, including the following:

The married women have more responsibilities at home. This is an intervening variable. If it is introduced and the two stratified relations—between marital status and absenteeism—disappear, the elaboration is of type *MI*. The relation is interpreted by showing what intervening variable connects the original two variables.

The married women are prone to physical infirmity, as crudely measured by age. The older women are more likely to be married and to have less physical strength, both of these as a result of their age. Age is an antecedent variable. If it turns out that when age is kept constant the relation between marital status and absenteeism disappears, a spurious effect of type *MA* is the explanation. Older people are more likely to be married and more likely to need rest at home.

The latter case suggests, again, an important point. After the original relationship is explained, attention might shift to $|ty|$, the fact that older people show a higher absentee rate. This, in turn, might lead to new elaborations: Is it really the case that older women have less physical resistance, be they married or single? Or is it that older women were born in a time when work was not as yet important for women and therefore have a lower work morale? In other words, after one elaboration is completed, the conscientious investigator will immediately turn to a new one; the basic analytical processes, however, will always be the same.

Causal relations. One final point can be cleared up, at least to a degree, by this analysis. It suggests a clear-cut definition of the *causal* relation between two attributes. If there is a relation between x and y, and if for every conceivable *antecedent* test factor the partial relations between x and y do *not* disappear, then the original relation should be called a causal one. It makes no difference here whether the necessary operations are actually carried through or made plausible by general reasoning. In a controlled experiment there may be two matched groups: the experimental exposure corresponds to the variable x, the observed effect to y. Randomization of treatments between groups makes sure that $|xt| = 0$ for any antecedent t. Then if $|xy| \neq 0$ and there have been no slip-ups in the experimental procedure, the preceding analysis always guarantees that there is a causal relation between exposure, x, and effect, y. There are other concepts of causal relations, differing from the one suggested here [*see* CAUSATION].

This has special bearing on the following kinds of discussion. It is found that the crime rate is higher in densely populated areas than in sparsely populated areas. Some authors state that this could not be considered a true causal relation, but such a remark is often intended in two very different ways. Assume an intervening variable—for instance, the increased irritation which is the result of crowded conditions. Such an interpretation does not detract from the causal character of the original relationship. On the other hand, the argument might go this way: crowded areas have cheaper rents and therefore attract poorer, partly demoralized people, who are also more likely to *be* criminals to begin with. Here the character of the inhabitants is antecedent to the characteristics of the area. In this case the original relationship is indeed explained as a spurious one and should not be called causal.

Variables not ordered in time. Explanation consists of the formal aspect of elaboration and some substantive ordering of variables. Ordering by time sequence has been the focus here, but not all variables can be ordered this way. One can distinguish orders of complexity, for example, variables characterizing persons, collectives, and sets of collectives. Other ordering principles could be introduced, for instance, degree of generality, exemplified by the instance of a specific opinion, a broader attitude, and a basic value system. What is needed is to combine the formalism of elaboration with a classification of variables according to different ordering principles. This covers a great part of what needs to be known about the logic of explanation and inference in contemporary survey analysis.

Higher-level parameters

This presentation has been restricted to the case of three attributes, but symmetric parameters can be developed for any level of stratification. Their structure becomes obvious if the parameter of fourth level is spelled out as an example:

$$^1(12) \quad \begin{aligned} p_{1234} = {}& p_1 p_2 p_3 p_4 \\ & + p_1|234| + p_2|134| + p_3|124| + p_4|123| \\ & + p_{12}|34| + p_{13}|24| + p_{14}|23| \\ & + p_{23}|14| + p_{24}|13| + p_{34}|12| \\ & + |1234|. \end{aligned}$$

It is possible for lower-level symmetric parameters to vanish while some higher-level ones do not, and the other way around. The addition of $|1234|$ to the fundamental set would permit the analysis of questions like these: We already know that economic status affects the relation between sex and voting; is this contextual effect greater for whites or for Negroes? If an antecedent attribute (3) only lowers $|12; 3|$, would a fourth intervening attribute explain the residual relation by making $|12; 34| = 0$?

The theorems needed to cope with a larger number of attributes in a survey analysis are often interesting by themselves but are too complex for this summary.

A substantive procedure can always be put into a variety of mathematical forms. Thus, for example, attributes can be treated as random variables, x_i, that can assume the values zero and one. Then the symmetric parameters play the role of covariances, and the difference between two stratified cross products corresponds to what is called interaction. Such translations, however, obscure rather than clarify the points essential for survey analysis. Starting from the notion of spurious correlation, Simon (1954) has translated the dichotomous cube into a system of linear equations that also permits the formalization of the distinction between the *MA* and the *MI* types. In such terminology, however, specifications (*SA* and *SI*) cannot be handled. In spite of this restriction, Blalock (1964) has productively applied this variation of survey analysis to problems for which the symmetric parameters of higher than second level may be negligible.

Polytomous systems have also been analyzed from a purely statistical point of view. A fourfold table is the simplest case of a contingency table that cross-tabulates two polytomous variates against each other. Thus, some of its statistical properties fall under the general heading of nonparametric statistics (Lindley 1964). An additional possibility is to start out with another way to characterize a fourfold table. Instead of a difference, $p_{12}p_{\overline{12}} - p_{1\overline{2}}p_{\overline{1}2}$, that is identical with the cross product, one can build formulas on the so-called odds ratio (Goodman 1965) $p_{12}p_{\overline{12}}/p_{1\overline{2}}p_{\overline{1}2}$. This leads to interesting comparisons between two stratified tables but has not been generalized to the more complex systems that come up in actual survey analysis. Eq. (12) forms the basis of this extension (Lazarsfeld 1961).

PAUL F. LAZARSFELD

[*See also* COUNTED DATA *and the biography of* LAZARSFELD.]

BIBLIOGRAPHY

BLALOCK, HUBERT M. JR. 1964 *Causal Inferences in Nonexperimental Research.* Chapel Hill: Univ. of North Carolina Press.

BOUDON, RAYMOND 1965 Méthodes d'analyse causale. *Revue française de sociologie* 6:24–43.

CAPECCHI, VITTORIO 1967 Linear Causal Models and Typologies. *Quality and Quantity—European Journal of Methodology* 1:116–152.

DURKHEIM, ÉMILE (1897) 1951 *Suicide: A Study in So-*

ciology. Glencoe, Ill.: Free Press. → First published in French.

GOODMAN, LEO A. 1965 On the Multivariate Analysis of Three Dichotomous Variables. *American Journal of Sociology* 71:290–301.

○GOODMAN, LEO A.; and KRUSKAL, WILLIAM H. 1954–1972 Measures of Association for Cross-classifications. Parts 1–4. *Journal of the American Statistical Association* 49 [1954]:732–764; 54 [1959]:123–163; 58 [1963]:310–364; 67 [1972]:415–421. → Part 2: Further Discussion and References. Part 3: Approximate Sampling Theory. Part 4: Simplification of Asymptotic Variances.

HYMAN, HERBERT H. 1955 *Survey Design and Analysis: Principles, Cases and Procedures.* Glencoe, Ill.: Free Press. → See especially Chapter 7.

LAZARSFELD, PAUL F. 1961 The Algebra of Dichotomous Systems. Pages 111–157 in Herbert Solomon (editor), *Item Analysis and Prediction.* Stanford Univ. Press.

LINDLEY, DENNIS V. 1964 The Bayesian Analysis of Contingency Tables. *Annals of Mathematical Statistics* 35:1622–1643.

NOWAK, STEFAN 1967 Causal Interpretation of Statistical Relationships in Social Research. *Quality and Quantity—European Journal of Methodology* 1:53–89.

SELVIN, HANAN C. 1958 Durkheim's *Suicide* and Problems of Empirical Research. *American Journal of Sociology* 63:607–619.

SIMON, HERBERT A. (1954) 1957 Spurious Correlation: A Causal Interpretation. Pages 37–49 in Herbert A. Simon, *Models of Man: Social and Rational.* New York: Wiley. → First published in Volume 49 of the *Journal of the American Statistical Association.*

○YULE, G. UDNY; and KENDALL, MAURICE G. 1950 *An Introduction to the Theory of Statistics.* 14th ed., rev. & enl. London: Griffin; New York: Hafner. → Yule was the sole author of the first edition (1911). Kendall, who became a joint author on the eleventh edition (1937), revised the current edition and added new material to the 1965 printing.

Postscript

[1]In equation (12) the two-subscript coefficients should be products of one-subscript quantities. For example, p_{12} should read $p_1 p_2$, and similarly in the five other cases, so that equation (12) should read

$$
\begin{aligned}
p_{1234} = {} & p_1 p_2 p_3 p_4 \\
& + p_1|234| + p_2|134| + p_3|124| + p_4|123| \\
& + p_1 p_2|34| + p_1 p_3|24| + p_1 p_4|23| \\
& + p_2 p_3|14| + p_2 p_4|13| + p_3 p_4|12| \\
& + |1234|.
\end{aligned}
$$

Closely related to the material discussed in the main article is the work on log-linear models being carried out by Leo Goodman and others. [*For a detailed discussion of these developments, see* SURVEY ANALYSIS, *article on* METHODS OF SURVEY ANALYSIS.]

SÜSSMILCH, JOHANN PETER

Johann Peter Süssmilch (1707–1767), German demographer, was born in Berlin, the son of a corn merchant. He was interested in medicine at an early age, but his parents did not want him to become a physician and sent him to the university at Halle to study Latin and jurisprudence. After he had been there for a while, he decided instead that he would study Protestant theology and enter the ministry. In 1728 he went to the University of Jena to study philosophy, mathematics, and physics.

In 1736 Süssmilch was appointed chaplain to Marshal von Kalckstein's regiment, and he accompanied the regiment during the First Silesian War. The foreword to his famous book *Die göttliche Ordnung in den Veränderungen des menschlichen Geschlechts, aus der Geburt, dem Tode und der Fortpflanzung desselben erwiesen* ("The Divine Order in the Changes in the Human Race . . .") was signed in 1741, "advancing on Schweidnitz." After the war was over he performed pastoral duties, primarily in Berlin, while he carried on his demographic studies. In 1745 he was elected a member of the Akademie der Wissenschaften.

Although Süssmilch wrote on a wide variety of subjects—philosophy, religion, politics, science, and even linguistics—all of his work is profoundly connected with his analytical theory of population. His book on population, *Die göttliche Ordnung,* is the first complete and systematic treatise on the subject. The work was first published in 1741 and was revised in 1761, but it is best known in the fourth, posthumous edition of 1775.

Süssmilch's theory of population was influenced by William Derham and by John Graunt and William Petty. From Derham's *Physico-theology,* published in London in 1713, Süssmilch took the idea that divine providence has established a balance between the size of the population and the supply of food required for subsistence; from Graunt and Petty he learned that it may be possible to discern an underlying order in vital statistics. Süssmilch's assertion of a divine order in population trends reflects the desire, shared by most eighteenth-century scientists, to detect the pattern of the "natural order." He was convinced that if he succeeded in measuring fecundity and mortality, the vital statistics he discovered would agree with the eternal laws of God, whom he compared to an "infinite and exact Arithmeticus . . . who has determined for all things in their temporal state their score, weight, and proportion."

Although works in political arithmetic exerted

a certain influence on Süssmilch, he did not use their rather speculative methods of estimating from faulty or inadequate data. He was one of the first to perceive that the consistency and stability of estimates depends on the number of observations. "One must collect a mass of particular cases over many years and sum up the data for whole provinces before it is possible to detect the concealed rules. Only then does the conformity of these rules to the natural order become apparent" (1741, vol. 1, p. 64).

To obtain reliable estimates, Süssmilch extracted demographic data from many Protestant parish registers. As population mobility was limited and the number of non-Protestants in Prussia was small, the data were fairly representative of the total population. He established the absolute frequencies of births and of deaths and measured the relative growth of population by comparing these quantities, deriving what he called a "rate of special mortality." Arranging the deaths by age groups and comparing the rates for these groups with the rate for a stationary population, he developed a life table. He measured fertility by comparing the number of christened children to the number of married people; as this proportion proved to be relatively constant, he estimated the "general fecundity" of a state by counting the number of married couples.

○ Süssmilch's rate of special mortality averaged 125/100, that is, he found about 125 births per 100 deaths. He used this rate to predict the *Verdoppelungszeit* (that is, the number of years required for doubling the population), referring to calculations of Leonhard Euler. This gave the basis for an estimation of the "divine order" of population growth, which might, however, be disturbed by plagues and wars. Climatic or social obstacles might prevent or defer marriage, thus lengthening the period required for doubling the population. Having established the stability of over-all birth rates and mortality ratios, Süssmilch observed that the rate of growth in urban districts was significantly lower than in rural districts. He therefore concluded that urbanization was a social factor restraining growth. He did not see any definite limits to the tendency of population to increase. Estimating the population capacity of the world at 14,000 million and the population of the world of his time at 1,000 million, he judged that population could grow without any difficulties for the next several centuries.

In order to follow the natural order and to obey divine providence, political measures should support the tendency of population to increase. This interpretation of natural and divine order by

Süssmilch conformed well with the mercantilist theories of his time, which emphasized the advantages of a large population for the wealth and military power of the state.

Süssmilch was highly respected as a learned man at the court of Frederick II of Prussia. Christian Wolf, in his Preface to the first edition, called *Die göttliche Ordnung* "a proof that the theories of probability may be utilized for the comprehension of human life." Süssmilch's work, nevertheless, did not have much influence. An abbreviated Dutch translation is the only foreign edition, and the last German reprint dates from 1798. There are at least two reasons for this neglect of Süssmilch's ideas and methods. First, the Achenwall school of statistics rejected Süssmilch's concern with the philosophical implications of population growth, centering its attention more exclusively on the numerical frequency and stability of vital processes. Second, the pessimistic Malthusian theory of population ran counter to Süssmilch's optimistic views. It was not until modern statistics had to deal with actual demographic problems that the abundance of ideas and the methodological achievements of *Die göttliche Ordnung* were finally acknowledged.

I. ESENWEIN-ROTHE

[*For the historical context of Süssmilch's work, see* DEMOGRAPHY, *article on* THE FIELD, *and the biographies of* GRAUNT *and* PETTY. *See also* Grauman 1968; Sauvy 1968; Spengler 1968.]

WORKS BY SÜSSMILCH

(1741) 1788 *Die göttliche Ordnung in den Veränderungen des menschlichen Geschlechts, aus der Geburt, dem Tode und der Fortpflanzung desselben erwiesen.* 3 vols. Berlin: Verlag der Buchhandlung der Realschule. → Translations of extracts in the text were provided by I. Esenwein-Rothe.

1752 *Der königlichen Residenz Berlin schneller Wachsthum und Erbauung.* Berlin: Haude.

1758 *Gedanken von den epidemischen Krankheiten und dem grösseren Sterben des 1757ten Jahres.* Berlin: Haude.

1766 *Versuch eines Beweises, dass die erste Sprache ihren Ursprung nicht von Menschen, sondern allein vom Schöpfer erhalten habe.* Berlin: Haude.

SUPPLEMENTARY BIBLIOGRAPHY

BONAR, JAMES (1931) 1966 *Theories of Population From Raleigh to Arthur Young.* New York: Kelley.

CRUM, FREDERICK S. 1901 The Statistical Works of Süssmilch. *Journal of the American Statistical Association* 7:335–380.

DERHAM, WILLIAM (1713) 1742 *Physico-theology: Or, a Demonstration of the Being and Attributes of God, From His Works of Creation.* 10th ed. London: Innys.

ELSTER, LUDWIG 1924 Bevölkerungswesen: III. Bevölkerungslehre und Bevölkerungspolitik. Volume 2, pages 735–812 in *Handwörterbuch der Staatswissenschaften*. 4th ed. Jena (Germany): Fischer.

►GRAUMAN, JOHN V. 1968 Population: VI. Population Growth. Volume 12, pages 376–381 in *International Encyclopedia of the Social Sciences*. Edited by David L. Sills. New York: Macmillan and Free Press.

HORVATH, ROBERT 1962 *L'ordre divin* de Süssmilch: Bicentenaire du premier traité spécifique de démographie (1741–1761). *Population: Revue trimestrielle* 17:267–288.

JOHN, VINCENZ 1884 *Geschichte der Statistik: Ein quellenmässiges Handbuch für den akademischen Gebrauch wie für den Selbstunterricht*. Volume 1: Von dem Ursprung der Statistik bis auf Quetelet (1835). Stuttgart (Germany): Enke.

JOHN, VINCENZ 1894 J. P. Süssmilch. Volume 37, pages 188–195 in *Allgemeine deutsche Biographie*. Leipzig: Duncker & Humblot.

KARLSSON, OSKAR 1925 Die Bedeutung Johann Peter Süssmilchs für die Entwicklung der modernen Bevölkerungs-statistik. Dissertation, University of Frankfurt.

KNAPP, GEORG F. 1874 *Theorie des Bevölkerungs-wechsels: Abhandlungen zur angewandten Mathematik*. Brunswick (Germany): Vieweg.

KNORS, HERMANN 1925 Johann Peter Süssmilch: Sein Werk und seine Bedeutung. Dissertation, University of Erlangen.

LANDRY, ADOLPHE (1945) 1949 *Traité de démographie*. 2d ed., rev. Paris: Payot.

LAZARSFELD, PAUL F. 1961 Notes on the History of Quantification in Sociology—Trends, Sources and Problems. *Isis* 52:277–333. → Also published in 1961 on pages 147–203 in Henry Woolf (editor), *Quantification: A History of the Meaning of Measurement in the Natural and Social Sciences*. Indianapolis, Ind.: Bobbs-Merrill.

MEITZEL, S. 1926 Johann Peter Süssmilch. Volume 7, pages 1172–1173 in *Handwörterbuch der Staatswissenschaften*. 4th ed. Jena (Germany): Fischer.

MOHL, ROBERT VON (1855–1858) 1960 *Die Geschichte und Literatur der Staatswissenschaften in Monographien dargestellt*. 3 vols. Graz (Austria): Akademische Druck- und Verlagsanstalt.

REICHARDT, HELMUT 1959 Süssmilch, Johann Peter. Volume 10, pages 267–268 in *Handwörterbuch der Sozialwissenschaften*. Stuttgart (Germany): Fischer.

REIMER, KARL F. 1932 Johann Peter Süssmilch: Seine Abstammung und Biographie. *Archiv für soziale Hygiene und Demographie* New Series 7:820–827.

ROSCHER, WILHELM G. F. (1874) 1924 *Geschichte der National-oekonomik in Deutschland*. 2d ed. Munich and Berlin: Oldenbourg.

►SAUVY, ALFRED 1968 Population: II. Population Theories. Volume 12, pages 349–358 in *International Encyclopedia of the Social Sciences*. Edited by David L. Sills. New York: Macmillan and Free Press.

SCHULZE, KARL 1922 Süssmilch's Anschauungen über die Bevölkerung. Dissertation, University of Halle.

►SPENGLER, JOSEPH J. 1968 Population: III. Optimum Population Theory. Volume 12, pages 358–362 in *International Encyclopedia of the Social Sciences*. Edited by David L. Sills. New York: Macmillan and Free Press.

Die Statistik in der Wirtschaftsforschung: Festgabe für Rolf Wagenführ zum 60. Geburtstag. Edited by Heinrich Strecker and Willi R. Bihn. 1967 Berlin: Duncker & Humblot. → See especially "Johann Peter Süssmilch als Statistiker," by I. Esenwein-Rothe.

TRIPPENSEE, GOTTFRIED G. 1925 Staat und Gesellschaft bei Bielfeld, Süssmilch und Darjes: Ein Beitrag zur Ideengeschichte des preussischen Staates. Dissertation, University of Giessen.

WAPPÄUS, JOHANN E. 1859–1861 *Allgemeine Bevölkerungsstatistik: Vorlesungen*. 2 vols. Leipzig: Hinrichs.

WESTERGAARD, HARALD L. 1932 *Contributions to the History of Statistics*. London: King.

WILLCOX, W. F.; and CRUM, F. S. 1897 A Trial Bibliography of the Writings of Johann Peter Süssmilch, 1707–1767. *Journal of the American Statistical Association* 5:310–314.

T

TABULAR PRESENTATION

Statistical tables are the most common form of documentation used by the quantitative social scientist, and he should cultivate skill in table construction just as the historian learns to evaluate and cite documents or the geographer learns cartography. Table making is an art (as is table reading), and one should never forget that a table is a form of communication—a way to convey information to a reader. The principles of table making involve matters of taste, convention, typography, aesthetics, and honesty, in addition to the principles of quantification.

It is useful to distinguish between raw data tables and analytic tables, although the line between them is somewhat arbitrary. Raw data tables, for example, in census reports, serve a library function: they arrange and explain the figures in such a way as to make it easier for the user to find what he wants. Here principles of accuracy, completeness, and editorial style are important. The reader can find full treatment of these topics in a number of standard sources (see, for example, U.S. Bureau of the Census 1949). It will be sufficient here to stress the importance of showing raw data whenever this is possible. In a research report, basic data can sometimes be presented in appendix tables, and sometimes they can be presented graphically. Sometimes basic data can be deposited with the American Documentation Institute or a similar organization to facilitate public access. As a last, and obviously least satisfactory, resort, a statement that basic data may be obtained from the author should be appended to a research report.

In analytic tables, on the other hand, the data are organized to support some assertion of the author of the research report. Since numbers do not speak for themselves, analytic tables require careful planning and oblige the table maker to steer a course between art and artifice. Without art, he may fail to convey his evidence to the reader, but technique can also be used to deceive. Thus, the most important rule of table making is this: Arrange the table so the reader may both *see* and *test* the inferences drawn in the text.

Among analytic tables, the most common in many of the social sciences are percentage tables. The percentage is an extremely useful statistic in that it is familiar and meaningful to even relatively naive readers; it is highly analogous to the slightly more technical statistical concept of probability; and it is close enough to the raw data so that they can (if the denominator upon which the percentage is based is given) be reconstructed by a critical reader. Percentage tables thus meet the fundamental criterion: the reader may easily see and test the inferences of the author. Percentage tables, however, suffer from the drawback that they become cluttered and confusing unless they are well constructed. Occasionally a very large percentage table is required to present data that might better be summarized by two or three descriptive coefficients. Accordingly, it is important to make a considered decision between the use of percentage tables and the use of descriptive coefficients. If percentage tables (particularly large and complicated ones) are used, consideration must be given to the main principles and strategies for constructing them.

This article deals primarily with percentage tables. Most of the principles described, however, are also applicable to other analytic tables, in which the entries may be means, medians, sums of money, descriptive indexes, and so on. In particular, when dealing with these other sorts of tables, as well as when dealing with percentage tables, one should be sensitive to the importance of using clear, meaningful, and consistent units, of including an informative title and indication of the source of the data, and of making some arrangement to indicate the accuracy of the figures.

Most of these principles apply also to the presentation of tables of empirical distributions by absolute or relative frequencies; in these cases, decisions must be made on such matters as the width of class intervals and the location of class marks. [See STATISTICAL ANALYSIS, SPECIAL PROBLEMS OF, *article on* GROUPED OBSERVATIONS; STATISTICS, DESCRIPTIVE, *article on* LOCATION AND DISPERSION; *see also* Wallis & Roberts 1956, chapter 6; Yule & Kendall 1950, chapter 4.]

Basic principles

The simplest percentage table is one that presents the distribution of answers to a single question or observations on a single variable. Table 1 is a typical, although hypothetical, example that serves to illustrate a number of basic principles.

Table 1 — Attitudes toward job*

	Per cent
I like it very much	58%
I like it somewhat	40
I dislike it	2
	100%
Number of respondents (N) =	2,834
No answer =	8
Not applicable (housewife or unemployed) =	314
Total	3,156

* Hypothetical responses to the question "In general, how do you feel about your job?"

Very few people read a research report word by word from beginning to end. Many are interested in a particular chapter or sections; others are looking for one or two specific tables to answer some limited questions. Since it is a nuisance for these readers to have to read many pages of text in order to understand a table, each table presented should provide sufficient information to be meaningful by itself. This information should, of course, include the source or sources of the data presented in the table, if they originate elsewhere than in the present research.

Headings and footnotes. A table should usually have both a title (caption) and a number; the latter should consistently be used for reference, and in a lengthy monograph, it is convenient to articulate it with the chapter number (for example, Table 3.2 refers to the second table in chapter 3) or even with the page (for example, Table 277a refers to the first table on page 277).

The main title should be short and concrete; subtitles or footnotes should indicate clearly what the table describes. Table 1, for example, gives the wording of the question that was asked. If the data in a table are based upon an index, it should be described, usually in a footnote. Well-known indexes or scales need only be described by their names—for example, "Scores on the Stanford–Binet Intelligence Test, Form B."

If the same items appear in a series of tables, the full information need not be given each time, although if an item reappears after a long gap, there should be a reference to the table containing the full explanation.

Some items, like the dichotomy male–female, are virtually unambiguous. Most questions, however, do have variant forms. Age may be reported to the last birthday or the next birthday, and educational attainment may be in terms of degrees acquired, years of schooling completed, and so on.

Percentages. The figures that appear in percentage tables are of two kinds: the percentages themselves and certain absolute numbers indicating frequencies (N's). It is generally crucial to include both kinds of figures; it is also crucial to avoid confusion between the two. Many of the problems in interpreting percentage tables stem from a failure of the author to make this distinction clear. The following practices should be followed in presenting percentages in tables.

Per cent signs. A per cent sign (%) should be placed after the first percentage in a column of percentages that adds to 100 per cent, as shown in Table 1. This is literally redundant, since the column is labeled "per cent," but in this situation redundancy serves the purpose of reminding the reader that he is examining percentages.

Totals. If the total of the percentages in an additive column is not exactly 100 per cent because of rounding, it is good practice to point this out in a footnote. (If the rounding produces a deviation of more than 1 per cent from 100, the arithmetic should be rechecked. By the same token, the last percentage in a column should be obtained by calculation rather than by subtraction; otherwise, a valuable check on the calculations will be destroyed.)

Multiple responses. If a percentage table presents responses to a question to which multiple responses are permitted (for example, "What brands of coffee did you purchase in the past year?"), the percentages should not be presented as if they were additive. The frequently found notation that "percentages add to more than 100 per cent because of multiple answers" is misleading. Rather, each percentage is better treated as half a dichotomous response (for example, 40 per cent of the sample purchased a particular brand of coffee in the past year, and 60 per cent did not). If there are only a few possible responses, the percentages of respondents falling in the various patterns of response may be of interest.

Decimals.   The number of decimals to retain in presenting percentages (should 38 per cent, 37.8 per cent, or 37.8342 per cent be reported?) must be determined after careful thought. The case for the use of several decimals is based upon the following arguments: (1) a reader who wants to recalculate the data himself can be more accurate if more decimals are given, and (2) one's own check will be more precise. The case against the use of several decimals is that they often give a spurious air of precision (69.231 per cent of 13 cases means simply nine cases), and that they usually add little information, since only in extraordinarily large samples do differences of 1 per cent or less have any meaning. The policy for retaining decimals should, in most cases, remain the same for all of the entries within an individual table, and usually the same within an entire report, except that relatively raw data tables in an appendix are usually more useful if the data are given to more decimals than they are in analytic tables in the body of the report.

Rounding.   Rounding should be to the nearest number. If one decimal place is to be retained, and the original calculation is 76.42 per cent, 76.4 per cent should be used in the table; if the original calculation is 76.48 per cent, 76.5 per cent should be reported. If the original calculation is 76.45000 . . . per cent, an arbitrary convention must be established to guide the rounding. The usual convention is to always round to the *even* possibility. That is, 76.45000 . . . per cent becomes 76.4 per cent, and 76.750000 . . . per cent becomes 76.8 per cent. (See Croxton, Cowden & Klein [1939] 1967, for a discussion of rounding procedures.)

In some investigations, it is desirable to distinguish between a true zero (no observations or responses in a category) and a per cent rounded down to zero (for example, if only one decimal place is being retained, 0.04 per cent will be presented as 0.0 per cent). One convention is to use 0.0e if a true (e = exact) zero is intended. In other cases, the difference may not be important, especially if sampling fluctuation could easily change a true zero to some small nonzero per cent.

Sampling variability.   If the data reported in a table are derived from a probability sample, it is usually very desirable to indicate the extent of sampling variability. Sometimes this is done by adding an indication of standard deviation (for example, 76.5 ± 3.2), but this tends to clutter tables and has the danger of creating misunderstandings. (Is 3.2 the estimated standard deviation, the half length of a 95 per cent confidence interval, or what?) A device favored by the U.S. Bureau of the Census is to give approximate confidence interval widths for per cents in various brackets in a footnote or in a small auxiliary table. (In a particular case, for example, such a footnote might indicate that if a per cent lies between 40 and 60, its 95 per cent confidence interval half width is about 5.) If the tables come from a sample that is not based upon probability sampling, the basic problem of sampling fluctuation becomes much more difficult [see SAMPLE SURVEYS, *article on* NONPROBABILITY SAMPLING]. Errors other than sampling ones are often very important but are typically discussed in the text rather than in a table [see ERRORS, *article on* NONSAMPLING ERRORS]. Of course, if there exists a bias whose magnitude might seriously affect a table's interpretation, it is good practice to refer to the relevant text discussion in a footnote to the table.

Frequencies.   The absolute frequencies that appear in a percentage table are of three kinds: (1) the total number of individuals or cases in the study, (2) the total number of individuals or cases upon which percentages are based (often called N, or base N), and (3) the cases that are excluded from the percentaging for one reason or another.

Some important reasons for excluding cases are inapplicability (for example, in Table 1, an unemployed person cannot have an opinion about his present employment), the refusal of the respondent to answer the question (or the failure of the interviewer to ask or to record it), the inability of the interviewer to locate the respondent (the persistent not-at-homes), and the inconsistency of a response. Cases of this type are often thrown together into a single No Answer or Not Applicable category, but for some purposes (for example, in considering the magnitudes of possible biases) it is important to separate the reasons. The following practices are recommended in reporting Totals, N's, No Answers, and Not Applicables.

Totals.   Unless all cases are accounted for in every table in a series, the total number of cases in the study should be reported in every table. This enables the reader to detect (and perhaps the analyst to avoid) the all-too-common situation in which a result that appears to apply to the entire population studied is really based upon a small fraction of the total. It also forces the analyst to maintain a careful accounting of his cases.

N's or base N's. The number of cases upon which percentages are based should also appear in every table. This is necessary so that the reader may evaluate the reliability of the percentages and, if he desires, reconstruct the original frequencies from the percentages that are presented.

Discrepancies beween Totals and N's. Any discrepancies between Totals and N's should be accounted for by specifying the number of No Answer and Not Applicable cases, as at the bottom of Table 1. It is poor practice to obtain No Answer figures by subtracting the N from the Total, since this destroys a practical check upon the calculations.

The percentaging of No Answers. Some survey analysts include No Answers as a category to be percentaged. This seems inadvisable, since it introduces a logically separate dimension into the classification. The proportion of No Answers depends more strongly on the research procedure than it does on the population being studied. It should be noted that in opinion research the category No Opinion is not the same as No Answer. Respondents who have neither a positive nor a negative opinion on an issue should usually be included in a category to be percentaged, since, for example, differences over time or between subgroups in the proportion of people with no opinion on an issue constitute substantive findings. [*For a discussion of the dangers of a large number of No Opinions and of the effects of No Answers, see* ERRORS, *article on* NONSAMPLING ERRORS.]

Although few analysts follow the convention, in the analysis of sample surveys, a case can be made for including in the No Answer category respondents who were selected for the sample but never interviewed because of refusals, absences from home, and the like. This practice would provide a fairer picture of the coverage of the data than the standard procedure, which is to describe the completed interviews as the Total and to count as No Answers only instances of skipped questions, illegible answers, refusals to answer a question, and so forth.

The reporting of frequencies. As a general rule it is inadvisable to present the absolute frequencies for the various categories being percentaged unless there is some compelling reason to do so (for example, if the frequencies are the basic data for a

Table 2 — Usually wrong!				Table 3 — Right!	
Category	*Per cent*	*N*		*Category*	*Per cent*
High	65%	49		High	65%
Medium	11	8		Medium	11
Low	24	18		Low	24
Total	100%	75		*Total*	100%
				N = 75	

statistical test that will be discussed). In general, frequencies should be reported only if they represent 100 per cent (tables 2 and 3).

There are two reasons for preferring Table 3 over Table 2. First, an absolute frequency that is not presented cannot be confused with a percentage (for example, it is not difficult to read 49 per cent instead of 65 per cent for the top line of Table 2, and the confusion is even more likely to arise if the table is more complicated). Second, no additional information is provided by the individual *N*'s, since the interested reader can calculate the individual *N*'s if he is given the base *N* and the percentages.

Two-variable tables

Two-variable tables present a number of choices for arrangement and layout that do not appear in one-variable tables; Table 4 is a typical example. The following practices are suggested for constructing tables of this type.

Arrangement and layout. There is no consistent opinion on the question of whether percentages should run down the columns (as in Table 4) or across the rows. The former is generally preferable, both because it is consistent with the normal pattern of columns of additive figures and because it places the independent variable (the so-called "causal" variable) at the head of the table and the dependent variable (the variable the analyst is seeking to explain) at the side of the table. (See

Table 4 — Age and cumulative grade-point averages*

Cumulative grade-point average (grouped)	AGE	
	20 or under	Over 20
High	21%	17%
Medium	42	33
Low	37	50
Totals	100%	100%

$$N = 2,705 + 3,110 = 5,815$$

No answer on:		
Grades only	75	
Age only	3	
Both	2	
	80	
Total		5,895

* Hypothetical data.

Zeisel [1947] 1957, pp. 24–41, for a discussion of this preference.)

Which way to percentage. The construction of a two-variable table requires a decision as to which way the percentages are to be computed. For example, in constructing a table from the data used in constructing Table 4, should the table present the distribution of grade scores by age (as in Table 4), or the distribution by age of students in the various grade-point groups, or the distribution of the entire sample by the six age by grade-point groups?

A number of considerations enter into the decision. If, for example, the two samples of students (of differing ages) were drawn according to separate sampling plans, that is, if they were considered as separate populations for sampling purposes, then ordinarily one would compute percentages separately for each age group.

Under the assumption that this sampling issue does not arise, the choice of method of presentation depends on which is most useful for the purposes of the analysis. If the analyst is primarily concerned with the question of how grade-point averages differ between younger and older students, then the method used in Table 4 is most appropriate. If, on the other hand, interest is centered on the age distributions of separate grade-point groups (as might be the case, for example, if the age composition of projected remedial classes is of interest), then percentages should be computed the other way.

Thus no general rule can be given, and indeed it is often desirable to compute the percentages both ways. (A detailed discussion of the issue is given in Zeisel [1947] 1957, pp. 24–41.) Consistency within a report is often a consideration. For example, in analysis of political party preferences, it might be better always to use "per cent Democratic," that is, to let party preference always be the dependent variable, rather than to shift back and forth.

Ordering and placing categories. If the independent variable has no intrinsic order among its categories, the results will typically be clearer if the entries are arranged according to increasing or decreasing values of the dependent variable. Table 5 demonstrates this point; an alternate plan would be to group the fields of study by category (physical sciences, social sciences, humanities) and then order the fields within categories by the frequency of the percentages reporting the dependent variable. Although alphabetical ordering is sometimes desirable, it is the least efficient method in a table containing as few groups as Table 5.

Dichotomous dependent variables. When a de-

Table 5 — Early consideration of graduate study, by field

Field of graduate study	Percentage of graduate students reporting having considered graduate study in field before junior year in college	N*
Geology	77%	107
Chemistry	75	317
Physics	74	289
Botany	61	53
English	58	273
History	47	305
Clinical psychology	41	153
Sociology	26	85
Total N		1,582

* NA not presented here, because data are part of a larger tabulation in the original.

Source: National Opinion Research Center 1962, p. 222.

pendent variable is dichotomous, as in Table 5, a decision has to be made as to whether to present both percentages. The style of Table 5 is generally preferable; the percentage of students who reported that they had *not* considered graduate study in the field is not presented, on the grounds that it is easily computable by the reader through subtraction, and its inclusion would have cluttered the table with unnecessary information. As a general rule, a table should contain as few numbers as possible without excluding vital information.

Per cent signs. It is important that the use of per cent signs in tables be consistent throughout a report. Conventions differ widely, but it is recommended that if a two-variable table reports only one half a dichotomy, a per cent sign should be placed after each percentage if there are only a few (as in Table 6) or after the first percentage in the column if there are many (as in Table 5). If there are many columns or rows, it may be best to eliminate per cent signs altogether, provided that the caption of the table or the headings on the columns or rows indicate clearly that the numbers in the table are percentages. An important consideration is whether or not actual numbers of cases (or amounts such as dollars) should also appear in the table; if they do, per cent signs serve an important role in preventing confusion. In any event, every total percentage should have a per cent sign.

Table 6 — Percentage of students expecting professional employment after graduation, by field of study*

Field of study	Percentage expecting employment	N
Chemistry	45%	650
English	46%	375
Astrophysics	100%	1

* Hypothetical data.

Reporting *N*'s. A two-variable table distributed by one variable contains more than a single base *N* (as in Table 6), and it may contain a large number of *N*'s (as in Table 5). It is essential that an *N* be presented for each row or column of additive percentages; in extremely complex tables, these may be provided in footnotes.

A special problem arises if some of the percentages in a table are based upon a large number of cases while one or more are based upon very few. Table 6 illustrates the problem.

The "100 per cent" figure for astrophysics in Table 6 is unreliable because it is based on only one case, and the unwary reader who does not read the *N*'s upon which percentages are based may mistakenly conclude that a strong difference has been found. One is tempted to protect the reader against this sort of error by omitting categories for which the data are unreliable. If information on one or more categories is excluded from a table, however, the reader is unable to regroup the results and make his own calculations. Furthermore, some readers with a strong interest in a particular category would prefer that any available data be presented, on the grounds that some information is preferable to none at all.

How should this problem be handled? A frequent procedure is to select a value of *N* below which results are considered unreliable. The value selected is, of course, arbitrary: 10 seems to be the lowest that is ever used, and 20 is much more common. There is one good argument for using at least 20; this is the lowest value at which a single case would make no more than a 5 per cent difference. Beyond selecting a value of 20 or more, three styles of presentation are possible: (1) percentages based upon lower values of *N* can be reported but placed in brackets (with a footnote explaining their meaning); (2) the *N* can be reported but the percentage replaced by a dash (again with an explanatory footnote); and (3) procedure (2) can be followed, but in addition to a dash signifying an unreliable percentage, the actual number of cases that would have been the numerator of the percentage can be given adjacent to the *N*, properly labeled. Each style has advantages and disadvantages; the use of brackets is probably the best compromise. (Note that if the data are from a probability sample, and if confidence intervals are reported for each percentage, the great width of the confidence interval when *N* is small serves as a warning of this problem.)

Reporting No Answers. In a cross tabulation of two variables there are three types of No Answers to consider: (1) No Answers on the dependent variable, (2) No Answers on the independent variable, and (3) No Answers on both variables. As a check on calculations, it is necessary to *count* all three types to determine that the totals for each variable are correct. Whether to *report* the No Answers separately or collectively is a matter of individual taste. Two rules of thumb are (1) a total of 10 per cent or more of No Answers of all types is large enough to raise questions in the mind of a critical reader, and (2) No Answers should often be reported separately if one variable contributes a disproportionate share of the total. Table 4 illustrates how No Answers may be reported separately; it also shows that while the total proportion of No Answers is less than 10 per cent, the dependent variable (grade-point average) is the major contributor.

Three-variable tables

Many analyses of survey data require the introduction of a second independent variable, usually an "intervening" variable that specifies or elaborates the relationship between the independent and the dependent variable. Table 7 is a typical example.

If Table 7 is read across the rows, it shows the relationship between combat experience and anxiety at any particular educational level; if it is read down the columns, it shows the relationship between educational level and anxiety at any particular level of combat experience. (One could also look at the row and column totals—these are discussed below.) Since a frequent problem in survey research is to ascertain the relative influence of traits (educational achievement) and exposure to experience (combat), tables having the format of Table 7 are frequently necessary. (See the tables in Kendall & Lazarsfeld 1950; Berelson et al. 1954; Sills 1957; Hyman 1955; Stouffer [1935–1960] 1962; Lazarsfeld & Thielens 1958; Davis 1964.)

Table 7 — Percentage of soldiers with critical scores on the anxiety index, by educational level and combat experience*

EDUCATIONAL ATTAINMENT	COMBAT EXPERIENCE		
	None	*Under fire*	*Actual combat*
Grade school or less	40% (81)	47% (111)	57% (213)
Attended high school	34% (152)	42% (178)	47% (351)
High school graduate	20% (246)	29% (355)	36% (500)

N = 2,187

* Numbers in parentheses are base *N*'s for the percentages.

Source: Adapted from Stouffer et al. 1949, p. 447.

Dichotomizing the dependent variable. The most important principle underlying the construction of tables with two or more independent variables is that in spite of the fact that these tables are mathematically three-, four-, or five-dimensional, they must somehow be presented on two-dimensional paper. (The classic presentation of this idea is in Zeisel [1947] 1957, pp. 67–90.) This can always be done clearly, provided that one of the variables is dependent and that it can be expressed as a dichotomy.

Many variables used in survey research are natural dichotomies, such as voted–did not vote and employed–not employed. Ordered classes, such as low–medium–high, level of education, and military rank can always be dichotomized by combining categories. Items that consist of true qualitative (unordered) categories (such as religious affiliation, field of study, ethnic origin) present difficult problems that are discussed below.

When categories are combined, considerable information is necessarily lost. The analyst who constructs a table is thus placed in a dilemma. If he does not dichotomize the dependent variable, the table may become so complicated that the reader cannot follow it; if he does dichotomize it, the reader may be unable to determine the full relationship between the independent and the dependent variables. Indeed, thoughtless dichotomization may conceal important nonmonotonic relationships— more precisely known as nonisotropic relationships (see Yule & Kendall 1950, pp. 57–59).

There is no clear-cut resolution of the problem of dichotomization, although a few general guidelines can be set down. First, the analyst should inspect the raw data tables; if he is satisfied that the relationship is *essentially* monotonic, the percentage table should present the dependent variable as a dichotomy. Second, if a more complex form of relationship is found, the data are probably better presented by a graph or by a coefficient of some kind. Third, in cases of doubt, a reference should be made to the raw data tables in the appendix, so that the reader may draw his own conclusions about the wisdom of the dichotomization. The analyst should remember, however, that rather large samples are required to detect complex relationships reliably and that all the little "jiggles and bounces" in the data are not grounds for excitement.

Ordered classes should be dichotomized as closely as possible to the median (the cutting point that splits the sample into two groups of equal size). This rule is useful, but it has exceptions. For example, if a study concerns the political participation of young people, the age variable should probably be dichotomized at age 21 (or whatever is the local legal age for voting), even if it is not the median, because it provides a particularly meaningful cutting point. Cutting points that leave only a handful of cases in one group should be avoided (for example, it would usually be foolish to dichotomize economic status into "millionaires" and "all others").

Arrangement and symbols. In presenting three-variable tables, the data should be arranged so that the most important comparisons are those between adjacent numbers, because such comparisons are easier for the reader to make than are comparisons between nonadjacent numbers (see the first part of Table 8). Thus, in planning a complex percentage table, it is often useful to place the base N's in parentheses, below and to the right of the percentages. Table 8 presents the preferred and two less preferable ways of presenting base N's. In the second example in Table 8, the base N (745) intervenes between the two comparisons in the first column; in the third example it intervenes between the two comparisons in the first row; in the first example, neither comparison is obstructed. Furthermore, the reader interested in examining the raw numbers can, if the first example is followed, compute the correlation between the two independent variables by comparing bases that are adjacent to each other in rows and columns. For example, the base N's given in Table 7 make it possible to compute the raw-data (marginal) association between educational level and combat experience.

In general, one independent variable should be displayed in the columns and the other in the rows (see Table 7). This makes for fewer intervening numbers than an all-column or all-row arrangement.

In three-variable tables, it is recommended that a per cent sign be placed adjacent to every percentage in the table. If no per cent signs are used, confusion with the base N's can result; if only the top row of percentages has per cent signs, the non-

Table 8 — Examples of preferred and nonpreferred methods of displaying base N's

Preferred	56%		88%	
		(745)		(635)
	17%		54%	
		(849)		(586)
Not preferred	56%		88%	
	(745)		(635)	
	17%		54%	
	(849)		(586)	
Not preferred	56% (745)		88% (635)	
	17% (849)		54% (586)	

Table 9 — Three-variable table with a trichotomized dependent variable

		INDEPENDENT VARIABLE B					
		Yes			No		
	Yes	Natural sciences	%		Natural sciences	%
		Social sciences	%		Social sciences	%
		Humanities	%		Humanities	%
			100%			100%
INDEPENDENT VARIABLE A			(N)			(N)
	No	Natural sciences	%		Natural sciences	%
		Social sciences	%		Social sciences	%
		Humanities	%		Humanities	%
			100%			100%
			(N)			(N)

additive nature of the percentages may not be apparent.

The use of separate tables. Table 7 presents two partial relationships (between education and anxiety, controlling for combat experience; and between combat experience and anxiety, controlling for education). It does not, however, present the zero-order (two-variable) relationships between the independent and the dependent variables. It is tempting to include these by adding a Totals column giving anxiety percentages by educational level, regardless of combat experience, and a Totals row giving anxiety percentages by amount of combat experience, regardless of educational attainment.

The resulting table would be compact, but it would have drawbacks. First, it would present two mathematically distinct statistical relationships (zero-order and partial relationships) without making a clear visual distinction between them. Second, the natural way of reporting a research finding is to begin with zero-order relationships and then discuss the partials. If Total columns and rows are included, the reader is, in effect, asked to read them while ignoring the partials in the interior of the table.

For these reasons it is generally preferable to present a series of discrete tables. In the present instance, the following sequence is suggested: Table *a*—Education and anxiety; Table *b*—Combat experience and anxiety; Table *c*—Education and combat experience; Table *d*—Table 7. This procedure would be space-consuming, but each table would tell a simple and distinct story. In fact, if this sequence is used, the text almost writes itself:

There is an association between education and anxiety (Table *a*). But combat experience is also associated with anxiety (Table *b*), and the less-educated soldiers are a little more likely to have been in actual combat (Table *c*). However, the educational difference is not a spurious effect of differentials in combat, for when combat experience is controlled, the educational difference remains (Table 7).

In practice, one would provide a fuller description of the tables, explaining the variables and the relationships in more detail, but these four tables provide the basic skeleton for a verbal description of the findings.

It should be remembered that even highly educated people are often poor readers of tables. Accordingly, important tables should be inserted in the text, not at the end of a report or in an appendix; furthermore, only essential tables should appear at all.

Dependent variables that cannot be dichotomized. There remains the problem of how to present in tabular form dependent variables that are true trichotomies, as well as other qualitative classifications that cannot reasonably be dichotomized. If the style of the dichotomy table (for example, Table 8) is followed, the result would be in the form of Table 9, in which major field of study is the dependent variable.

A table format such as that of Table 9 means that column comparisons have a large number of "intervening" figures, and if one is concerned with the effect of variable *A* as well as of variable *B*, the table is hard to read. Although more space is consumed, it is often better to break the table down into subtables, one for each category of the dependent variables, particularly if the nature of the difference varies between the categories of the dependent variable (for example, social science majors vary with *A*, natural science majors vary with *B*, and humanities majors vary with neither). Table 10 shows the format of a table of this kind.

Finally, a trichotomized dependent variable can be presented in two dimensions, without collapsing,

Table 10 — Showing a trichotomized dependent variable with subtables

		PERCENTAGE MAJORING IN THE NATURAL SCIENCES		PERCENTAGE MAJORING IN THE SOCIAL SCIENCES		PERCENTAGE MAJORING IN THE HUMANITIES	
		Independent variable B		Independent variable B		Independent variable B	
		Yes	No	Yes	No	Yes	No
	Yes	%	%	%	%	%	%
		(N)	(N)	(N)	(N)	(N)	(N)
Independent variable A							
	No	%	%	%	%	%	%
		(N)	(N)	(N)	(N)	(N)	(N)

by the use of triangular coordinate graph paper (see Coleman 1961, p. 29; Davis 1964, especially pp. 95–97). This is a very useful procedure, but it must be carefully explained to readers. Other graphical devices have been suggested; for example, see Anderson 1957.

Four-or-more variable tables

If there are four, five, or six independent variables in one table, no new problems arise, but the old ones become intensified. Such tables have more headings, and they usually have smaller case bases and greater problems of dichotomization.

Reading a four-variable table is a difficult task. To assist the reader, a number of subtables should be used (as in Table 10). Also, the analyst must be sure to explain the table and the findings it presents very carefully in the text. The technique of "talking through" a table should be employed, as in this example (referring to Table 11):

Beginning in the upper left-hand cell of the table, observe the entry 58 per cent, which is the percentage employed among fathers under 27 years of age attending private schools. Following across that row, the percentages increase from 58 to 69. A similar trend is found in each row, which means that when family role and type of school attended are held constant, the likelihood of employment increases with age. Turning now to the effect of family role. . . .

Note that the above hypothetical text matter is not an interpretation of the results but, rather, a translation of them from percentages into words. Such a description may justifiably run to several hundred words. (An extreme example is Davis 1964, which consists entirely of the discussion of a single table that cross tabulates nine variables.)

Focusing on one independent variable. One variable among the three or more independent variables is often singled out for attention. Given the fact that variables A, B, and C are all correlated with variable D, the real interest may be in whether C is still correlated with D when A and B are held constant rather than in the independent effects of A and B. Accordingly, if the analyst wants to focus attention on one of the independent variables, he

Table 11 — Percentage of students employed full-time, by age, family role, and type of school*

			AGE	
			Under 27	27 or older
	Father	Private school	58% (40)	69% (226)
		Other school	12% (104)	21% (321)
FAMILY ROLE		Private school	20% (340)	35% (315)
	Other	Other school	4% (889)	9% (544)

Number of respondents (N) =	2,779	
No answer =	63	
Total	2,842	

* Numbers in parentheses are base N's for the percentages.

Source: Adapted from National Opinion Research Center 1962, p. 222.

should present it alone in the columns of the table and use the rows for combinations of the other variables, as in Table 11.

Each row in Table 11 shows an age contrast, controlling for family role and type of school. In order to examine other effects, the reader must shift his eyes between the columns and rows.

The percentage-difference table. An extension of the above presentational strategy is the *percentage difference table*, which presents the effects of a given variable for various combinations of control variables. This table consists of rows and columns laid out according to the control variables, with the entries consisting of the percentage difference in the dependent variable produced by the test variable. Table 12 is an example.

Table 12 — Percentage differences in full-time employment for the age dichotomy in Table 11*

| | | TYPE OF SCHOOL | |
		Private	Other
FAMILY ROLE	Father	+11%	+9%
	Other	+15%	+5%

* Percentage employed among those students who are 27 or older minus the percentage employed among those under 27.

Table 12 demonstrates that age has a positive effect on each of the four control categories. In practice, one would probably not present a percentage difference in a situation where there are so few control categories, but when there are many categories, a difference table can make clear what otherwise appears to be an obscure pattern. In particular, a difference table can reveal complex patterns of interactions. Table 13 is a hypothetical example.

Table 13 summarizes the following complex relationship: "Cramming the night before an examination is associated with better final grades only among the students with high IQs who had done fairly well at midterm; the relationship is the same

for men and women." A complicated pattern such as this might well have been lost if a standard five-variable table format had been used.

The passive role played by sex in Table 13 raises the question of whether or not data should be presented when there are neither effects nor interactions. In general the answer is No, since these no-effect variables add to the complexity of multi-variable tables without adding to their content. In particular, one should avoid the not uncommon practice of maintaining a variable throughout a report simply because it was introduced at an early stage in the analysis and there is reluctance to take the time to recalculate the data. This will avoid the fallacy of pseudo rigor, a tendency to make research appear more meticulous than it is by controlling for irrelevant variables. Nevertheless, there are situations when it is interesting and important to present negative results in tabular form. By presenting such results, the analyst may answer questions that may be in the reader's mind and may prevent others from making unnecessary calculations. In the case of Table 13, for example, it might be argued that there is substantial evidence from other research about sex differences in academic achievement and, accordingly, that it is worthwhile to show that this finding holds for both men and women.

Two important limitations of percentage-difference tables stem from the inherent properties of percentages. First, these tables are appropriate only if both the dependent variable and the independent variable are dichotomized (the range in percentages of the dependent variable produced by variation of the independent variable over multiple levels is not an acceptable measure). Second, the absolute values of the difference must be interpreted with extreme care. When the two percentages are either very large or very small, a slight difference between them may represent as strong an effect as a larger difference when they both are of medium size. Thus, for example, if one calculates the coefficient of association Q for the data

Table 13 — Percentage differences between crammers and moviegoers in final course grades*

| | | MIDTERM GRADES | | | | | |
| | | High | | Medium | | Low | |
		Men	Women	Men	Women	Men	Women
IQ	High	+19%	+21%	+20%	+18%	0%	0%
	Medium	+16%	+15%	+19%	+20%	−1%	+1%
	Low	−2%	−1%	0%	+1%	+1%	−1%

* Hypothetical data showing percentage receiving high grades among those who crammed the night before the examination minus the percentage among those who went to the movies.

Table 14 — Percentage of respondents scored positively on a dependent variable, by age, sex, and education*

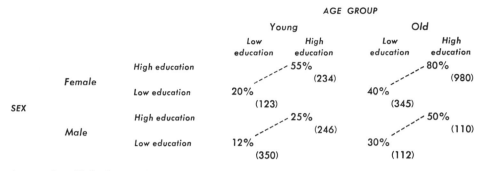

* Numbers in parentheses are base N's for the percentages.

Source: Personal communication from Peter Rossi.

in Table 12, one finds that the Q value associated with the 5 per cent in the lower right-hand corner (Q = .40) is a little higher than for the 15 per cent in the lower left-hand corner (Q = .36). The reason for this difference is that the former is based on two small percentages (9 per cent and 4 per cent), while the latter is based on moderate percentages (35 per cent and 20 per cent). If the size of the effects is important, it is better to use other coefficients of association than the percentage difference. [See STATISTICS, DESCRIPTIVE, *article on* ASSOCIATION.]

Comparisons between independent variables. A different situation obtains when the purpose of a table is to show that each of several independent variables makes a difference in the dependent variable, a situation in survey analysis that is somewhat akin to multiple correlation.

The Rossi stratagem. Given the limitations of geometry, as more comparisons are desired, it becomes increasingly difficult to present all the relevant comparisons adjacent to each other. Peter H. Rossi has developed a way to do this for three independent variables (Table 14). The table has an unorthodox appearance, but it has several advantages over any other form of presentation. First, each vertical comparison involves a sex difference, controlling for the other two variables. Second, each horizontal comparison involves an age difference, controlling for the other two variables. Third, each diagonal comparison (represented by broken lines) involves an educational difference, controlling for the other two variables. Fourth, there are no intervening percentages between any two percentages that are to be compared.

Weaker effects "inside" stronger ones. When there are four or more independent variables, even the stratagem developed by Rossi and illustrated in Table 14 breaks down. In such instances, it is

impossible to place percentages that are to be compared adjacent to each other. When the variables differ considerably and consistently in their percentage effects, the weaker effects should be placed "inside" the stronger effects.

An example using hypothetical data will make the idea clear. Consider four dichotomous independent variables, A, B, C, and D, that produce the following consistent percentage differences in dependent variable E:

$$A = 5\% ; B = 5\% ; C = 25\% ; D = 25\% .$$

It is desired to construct a table showing the simultaneous effects of the four independent variables upon the dependent variable, E. Following the rule of placing the weaker effects "inside" the stronger effects, A and C are paired, as are B and D. The table is then designed so that variable A is nearer the cell entries than variable C, and variable B is "inside" variable D (Table 15, where $\bar{A}$, for example, means "not A.")

The meaning of Table 15 is clear at a glance. The percentages increase steadily up each column and across each row. Therefore, it is apparent that each independent variable adds to the percentage scoring positively on dependent variable, E. Suppose, however, that the rule for the placing of variables was violated, and the weaker effects were placed "outside" and the stronger effects "inside" (Table 16).

Tables 15 and 16 present identical information, but the meaning of the data as they are displayed in Table 16 is not at all clear at a glance. The reader must take the time to make his own specific percentage comparisons if he wishes to verify the statements made in the text of the report.

With actual data the percentage effects are seldom as consistent as those in Table 15, and such

Table 15 — Percentage of respondents scored positively on variable E, by variables A, B, C, and D*

		C̄		C	
		Ā	A	Ā	A
D	B	30% (N)	35% (N)	55% (N)	60% (N)
	B̄	25% (N)	30% (N)	50% (N)	55% (N)
D̄	B	5% (N)	10% (N)	30% (N)	35% (N)
	B̄	0% (N)	5% (N)	25% (N)	30% (N)

* Hypothetical data.

Table 16 — Percentage of respondents scored positively on variable E, by variables A, B, C, and D*

		Ā		A	
		C̄	C	C̄	C
B	D	30% (N)	55% (N)	35% (N)	60% (N)
	D̄	5% (N)	30% (N)	10% (N)	35% (N)
B̄	D	25% (N)	50% (N)	30% (N)	55% (N)
	D̄	0% (N)	25% (N)	5% (N)	30% (N)

* Hypothetical data.

a perfect progression cannot always be displayed. However, if there are four or more variables whose independent effects are to be shown, it is always worth the time to seek the most lucid arrangement of the variables. Note that even in Table 11 the control variables are displayed so that there is a smooth progression up and down the columns, a style that adds considerably to ease of comprehension.

There is no method of tabular presentation that will make small differences any larger or trivial findings of substantive importance. Nevertheless, the presentation of percentage data in a form that will enable the reader to read them and test them is an important aspect of communicating the findings of survey research.

JAMES A. DAVIS AND ANN M. JACOBS

[*See also* GRAPHIC PRESENTATION.]

BIBLIOGRAPHY

○AMERICAN PSYCHOLOGICAL ASSOCIATION (1952) 1974 *Publication Manual of the American Psychological Association.* 2d ed. Washington: The Association. → A style manual giving instructions chiefly for the journals of the American Psychological Association, but containing much generally useful material. Tabular presentation is described on pages 43–50.

ANDERSON, EDGAR 1957 A Semigraphical Method for the Analysis of Complex Problems. National Academy of Sciences *Proceedings* 43:923–927.

BERELSON, BERNARD; LAZARSFELD, PAUL F.; and McPHEE, WILLIAM N. 1954 *Voting: A Study of Opinion Formation in a Presidential Campaign.* Univ. of Chicago Press.

CHAUNDY, THEODOR W.; BARRETT, P. R.; and BATEY, CHARLES 1954 *The Printing of Mathematics: Aids for Authors and Editors and Rules for Compositors and Readers at the University Press.* Oxford Univ. Press. → Described by the authors as the successor to G. H. Hardy's pamphlet, *Notes on the Preparation of Mathematical Papers,* published in 1932. Deals chiefly with the setting of equations; pages 68–69 treat tables. Good exposition of mechanics of typesetting for statistical and technical authors.

○CHICAGO, UNIVERSITY OF, PRESS (1908) 1969 *A Manual of Style.* 12th ed., rev. Univ. of Chicago Press. → A standard manual of typographic and editorial practice. The chapter on tables, pages 273–294, gives a variety of examples from scholarly disciplines.

COLEMAN, JAMES S. 1961 *The Adolescent Society: The Social Life of the Teenager and Its Impact on Education.* New York: Free Press. → See especially pages 29 ff. for examples of trichotomous data laid out on triangular coordinate graph paper.

CROXTON, FREDERICK E.; COWDEN, DUDLEY J.; and KLEIN, SIDNEY (1939) 1967 *Applied General Statistics.* 3d ed. Englewood Cliffs, N.J.: Prentice-Hall. → A standard text. Pages 45–59 of the 1967 edition cover table construction for the presentation of classified statistical data. Earlier editions were by Croxton and Cowden.

DAVIS, JAMES A. 1964 *Great Aspirations: The Graduate School Plans of America's College Seniors.* Chicago: Aldine. → See especially pages 53 ff. for examples of trichotomous data laid out on triangular coordinate graph paper and two-, three-, and four-variable tables of the kind discussed in this article.

DAVIS, JAMES A. et al. 1961 *Great Books and Small Groups.* New York: Free Press.

DIEXEL, KARL 1936 Normung statistischer Tabellen. Institut International de Statistique, *Revue* 4:232–237.

HALL, RAY O. (1943) 1946 *Handbook of Tabular Presentation; How to Design and Edit Statistical Tables: A Style Manual and Case Book.* New York: Ronald Press.

HYMAN, HERBERT H. 1955 *Survey Design and Analysis: Principles, Cases, and Procedures.* Glencoe, Ill.: Free Press.

KENDALL, PATRICIA L.; and LAZARSFELD, PAUL F. 1950 Problems of Survey Analysis. Pages 133–196 in Robert K. Merton and Paul F. Lazarsfeld (editors), *Continuities in Social Research: Studies in the Scope and Method of* The American Soldier. Glencoe, Ill.: Free Press.

LAZARSFELD, PAUL F.; and THIELENS, WAGNER JR. 1958 *The Academic Mind: Social Scientists in a Time of Crisis.* A report of the Bureau of Applied Social Research, Columbia University. Glencoe, Ill.: Free Press.

MYERS, JOHN H. 1950 *Statistical Presentation.* Ames, Iowa: Littlefield.

NATIONAL OPINION RESEARCH CENTER 1958 Survey of Graduate Students. Unpublished manuscript.

NATIONAL OPINION RESEARCH CENTER 1962 *Stipends and Spouses: The Finances of American Arts and Science Graduate Students,* by James A. Davis et al. Univ. of Chicago Press.

SILLS, DAVID L. 1957 *The Volunteers: Means and Ends in a National Organization.* Glencoe, Ill.: Free Press.

STOUFFER, SAMUEL A. (1935–1960) 1962 *Social Research to Test Ideas: Selected Writings.* New York: Free Press.

STOUFFER, SAMUEL A. et al. 1949 *The American Soldier.* Volume 2: *Combat and Its Aftermath.* Studies in Social Psychology in World War II, Vol. 2. Princeton Univ. Press.

U.S. BUREAU OF AGRICULTURAL ECONOMICS (1937) 1942 *The Preparation of Statistical Tables: A Handbook.* Washington: Government Printing Office.

U.S. BUREAU OF THE BUDGET, OFFICE OF STATISTICAL STANDARDS 1963 *Statistical Services of the United States Government.* Rev. ed. Washington: Government Printing Office. → See especially "Presentation of the Data."

U.S. BUREAU OF THE CENSUS 1949 *Manual of Tabular Presentation: An Outline of Theory and Practice,* by Bruce L. Jenkinson. Washington: Government Printing Office. → See the review by Hall in the June 1950 issue of the *Journal of the American Statistical Association.*

WALKER, HELEN M.; and BUROST, WALTER N. 1936 *Statistical Tables: Their Structure and Use.* New York: Columbia Univ. Press.

WALLIS, W. ALLEN; and ROBERTS, HARRY V. 1956 *Statistics: A New Approach.* Glencoe, Ill.: Free Press.

WATKINS, GEORGE P. 1915 Theory of Statistical Tabulation. *Journal of the American Statistical Association* 14:742–757.

○YULE, G. UDNY; and KENDALL, MAURICE G. 1950 *An Introduction to the Theory of Statistics.* 14th ed., rev. & enl. London: Griffin; New York: Hafner. → Yule was the sole author of the first edition (1911). Kendall, who became a joint author on the eleventh edition (1937), revised the current edition and added new material to the 1965 printing.

ZEISEL, HANS (1947) 1957 *Say It With Figures.* 4th ed., rev. New York: Harper. → Designed to initiate the nonstatistical reader into survey analysis. Covers (*inter alia*) multidimensional tables, indexes, analysis of data by cross-tabulation. Copiously illustrated with tables and charts.

TAXONOMY, NUMERICAL

See CLUSTERING; MULTIVARIATE ANALYSIS, *article on* CLASSIFICATION AND DISCRIMINATION; SCALING, MULTIDIMENSIONAL.

TESTING HYPOTHESES

See EXPERIMENTAL DESIGN; HYPOTHESIS TESTING; SIGNIFICANCE, TESTS OF.

TESTS OF SIGNIFICANCE

See SIGNIFICANCE, TESTS OF.

TIME SERIES

I. GENERAL	*Christopher Bingham*
II. ADVANCED PROBLEMS	*P. Whittle*
III. CYCLES	*Herman Wold*
IV. SEASONAL ADJUSTMENT	*Julius Shiskin*

I

GENERAL

► *This article was specially written for this volume. It replaces the article with the same title in* IESS.

A *time series* is a sequence of numerical data in which each item is associated with a particular instant in time. Examples are the daily closing prices of a security traded on an exchange, the noon temperature in Central Park, the hourly number of vehicles crossing a bridge, or the monthly crime index for Minneapolis. In such cases the observation time of each numerical item is an essential piece of information. More generally, if several variables are observed at each time point (air temperature and humidity; investment, income, and consumption; pulse rate, respiration rate, and body temperature), the vector of observations is a *multivariate time series* or multiple time series. The essential point is that the data are linearly ordered according to a known quantity, time. Data linearly ordered with respect to variables other than time, such as position on a line (for example, the thickness of a thread at various distances from its end), may often be suitable for the techniques of time series analysis. Some of these techniques can even be extended to two-dimensional (or more) "time," when each measurement is ordered with respect to two linear quantities such as latitude and longitude (Bartlett 1975). Such generalizations will not be considered here.

From a theoretical point of view, a time series is a collection of random variables, $\{X_t\}$, where the nonrandom index t ranges over some set of instants, such as the trading days in an exchange or all times between the start and the end of an EEG recording. A *continuous parameter* time series is one for which this set is a continuum of instants, as when X_t represents the voltage at time t as displayed on the strip chart produced by an EEG recording device. When the set consists of discrete time points, say January 1, January 2, January 3, and so on, $\{X_t\}$ is said to be a *discrete parameter* time series. Most commonly, but not necessarily, discrete observation times are equally spaced, that is, every second, every day, every quarter, and so on. In the case of multivariate time series, each $\mathbf{X}_t$ is a multivariate random quantity, that is, a random vector.

The definition just given actually defines the more general concept of *stochastic process* (Bartlett 1966; Cox & Miller 1965). Stochastic processes are used as models for a great variety of random phenomena, from the spread of epidemics to the failure of complex machinery, from the behavior of gambling devices to the transmission of nerve impulses. The phrase "stochastic processes" suggests emphasis on the probabilistic, mathematical structure of the phenomenon rather than on statistical analysis of data, although many studies of stochastic processes deal with special analytic techniques for inference. The methods used in the stochastic process literature often differ from those discussed here.

A partial exception to the above distinction is the analysis of *point processes*—stochastic models for data that are the times of occurrence of such events as machine failures, changes of government, or initiations of telephone calls. In the framework given above, X_t for a point process has only two values, say, 0 and 1, with the randomness entering via the times at which X_t is 1. Both regression and spectrum-type methods (discussed below) are used in the analysis of point processes (Bartlett 1966; Cox & Lewis 1966; Lewis 1972). Point processes will not be further discussed in this article.

Many discrete parameter time series are derived by sampling (observing) values of an underlying continuous parameter time series at equally spaced times, Δt time units apart, or by averaging or summing such a series over periods of length Δt. For example, consumption is, in principle, a virtually continuous economic time series, but in practice, available measures of consumption are sums over an interval, seldom shorter than a month. Intuition correctly tells us that, provided the time between observations is short enough, no appreciable information is lost by either the sampling or the summing observational method. Increasing research attention is being given to the question of how well such discrete observations on continuous time series in fact allow the recovery of information when conventional predetermined values of Δt (such as a week, month, or year) are employed.

I shall assume from now on that data are associated with equally spaced times, Δt units apart. For simplicity (it amounts to redefining the basic time unit but has no other effect), I shall usually assume that Δt is 1.

Time series data and the need to analyze them arise in many diverse fields. Among them are aeronautics (Press 1955), astronomy (Moran 1954), biology (Fowler et al. 1972), communications engineering (Blackman & Tukey 1958), de-

mography (Lee 1975), ecology (Holgate 1966), economics (Cootner 1967; Granger & Hatanaka 1964), hydrology (Rao & Kashyap 1974), managerial forecasting (Nelson 1973), medicine (Barlow 1959), meteorology and climatology (Chatfield & Pepper 1971; Hays et al. 1976), oceanography (Groves & Hannan 1968), seismology (Robinson 1967), and traffic control (Miller 1962). Time series analysis aids description, prediction, estimation, control, simulation, and, indeed, virtually all the tasks for which statistical methods are called upon.

General considerations

There are at least two distinguishing features of time series analysis: first, explicit allowance for statistical dependence among observations at differing times, and second, the usual unavailability, even in principle, of independent replicate series. There is only one series for the price of wheat in London; and any hypothetical universe from which it may be considered to have been chosen *cannot* be sampled again.

The possible dependence among data at different times tells us we must beware of analyses that assume the independence of observations to be combined. [*See* PROBABILITY, *article on* FORMAL PROBABILITY *for a discussion of independence.*] Thus, most time series analysis problems involve extracting information from a *sample of size one*, a single observation on a highly multivariate random variable. The statistical yardstick of variability between replicates is not generally available as it is, say, in the analysis of the results of a battery of ability tests, administered to individuals randomly chosen from a well-defined population [*see* CROSS-SECTION ANALYSIS].

These two features—dependence and lack of replication—compel a heavy reliance on more or less restricted models for the statistical structure of time series. Indeed, most models underlying the various forms of time series analysis reintroduce independence, or at least uncorrelatedness, at a deeper structural level than simple replication, and impose some form of temporal homogeneity. These models and methods would have no chance of usefulness unless time series that arise in practice appeared to have some regularities.

Stationarity

One form of regularity apparent in many, but by no means all, time series is *stationarity*. In observational terms this means simply that a graph of the series against time looks qualitatively about the same near one time as near another. Figure 1 gives examples of three time series exhibiting apparent stationarity, two of them computer simulations and one empirical. Although no pattern in any of them exactly repeats, there is clear temporal homogeneity of behavior. More formally, all statistical properties of stationary time series remain unchanged when the period of observation is shifted forward or backward in time. For example, if the time series consisting of the number of personal bankruptcies per month were stationary, the joint probability distribution of the numbers of bankruptcies in November 1971 and January 1972 would be the same as that of the numbers of bankruptcies in June 1976 and August 1976. In particular, the mean or expected value $\mu_t = E(X_t)$ and the variance $E(X_t - \mu_t)^2$ of a stationary time series do not change with time. Furthermore, the *autocovariance* between two X's separated by s time units, $E[(X_t - \mu_t)(X_{t-s} - \mu_{t-s})]$, depends only on time difference or *lag*, s, not on time instant t.

A weaker form of regularity, and one that is sufficient for the development of a rich theory and associated analytic techniques, is *wide-sense stationarity* or stationarity of the first- and second-

moment structures. That is, although the probability distributions may change over time, their means and variances are assumed constant and the autocovariance (and hence *autocorrelation*) is assumed to depend only on the lag. Wide-sense stationarity is sometimes called *weak stationarity* or *second-order stationarity*.

Nonstationarity

Most useful nonstationary time series models are only a step or two removed from stationarity. Perhaps the simplest form of nonstationarity is for only the expectations, the μ_t's, to change with time. We can express this by writing $X_t = \mu_t + e_t$ where e_t is (wide-sense) stationary, and μ_t is sufficiently restricted to permit a useful analysis. Figure 2 illustrates three nonstationary series of this type: the computer-generated Series 2A represents the simple and commonly assumed case in which μ_t varies linearly with time; the computer-generated Series 2B has a weak exactly periodic so-called *seasonal* component, so that $\mu_t = 15 + 3.0 \cos(2\pi t/12) + 1.5 \sin(2\pi t/12)$ (2π radians equals 360° or a full cycle; thus $\cos(2\pi t/12)$ and $\sin(2\pi t/12)$ repeat exactly every 12 time units);

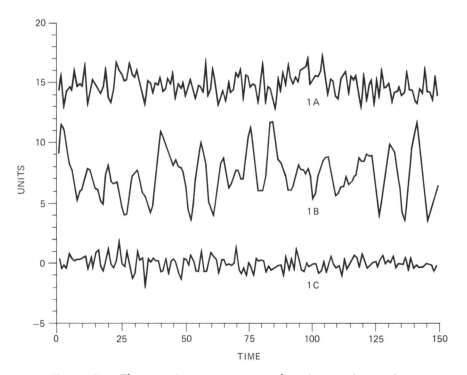

Figure 1 — Three stationary or presumed stationary time series

1A: X_t Independent normal, $\mu = 15$, $\sigma^2 = 1$.

1B: $X_t = 7.5 + 1.42(X_{t-1} - 7.5) - 0.73(X_{t-2} - 7.5) + a_t$, a_t independent normal, $\mu = 0$, $\sigma^2 = .64$.

1C: $X_t =$ daily gain or loss of Control Data common stock in dollars, Dec. 24, 1975, through July 28, 1976 (from the *Wall Street Journal*).

For Series 1A and 1B the time units are arbitrary. For Series 1C, t measures trading days after Dec. 23, 1975.

Series 2C represents an empirical time series with a clear seasonal component superimposed on an irregularly increasing trend. For nonstationary models such as these, the primary interest may be in the expectation function, μ_t, or it may lie in the residual series e_t. In both cases an important tool is *regression analysis* [*see* LINEAR HYPOTHESES, *article on* REGRESSION], modified so as to take proper account of the correlation between the residuals e_t at different times (Hannan 1963). Primarily for tractibility, μ_t is often assumed to be a polynomial in t, or a superposition of trigonometric curves, or a combination of these. In other cases, one time series will be regressed on one or more additional time series considered as independent variables.

The scope of stationary models may be extended in another way by assuming that a series is generated by accumulating or summing values of an underlying stationary series. That is, the sequence $\{X_t\}$ may be such that the first differences, $Y_t = X_t - X_{t-1}$, form a stationary series. Many series of prices or log prices on stock exchanges nearly follow this model, with the further property that the first difference series—the series of gains and

losses—is not only stationary but almost uncorrelated from day to day. This model (stationary independent gains and losses) is the so-called *random walk hypothesis* of price movements. In fact, the deviations from the random walk model are often slight and extremely subtle (Cootner 1967). Series 3A in Figure 3 represents the closing prices of Control Data common stock from December 24, 1975, through July 28, 1976, as reported in the *Wall Street Journal*. The first difference series of gains and losses is displayed as an apparently stationary series in Figure 1 (Series 1C). This approach can be extended by summing series that are themselves sums of stationary series; the second differences, $(X_t - X_{t-1}) - (X_{t-1} - X_{t-2})$, of such series are stationary. Figure 3 also displays examples of computer-generated time series whose first differences (3B) and second differences (3C) are independently identically distributed random variables with zero means (the variances differ). It is noteworthy how much apparently systematic structure is introduced by these simple operations on independent data. Also note the qualitative similarity of the price Series 3A and the artificial Series 3B.

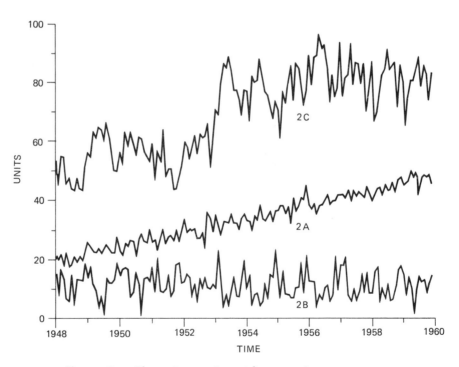

Figure 2 — Three time series with nonstationary means

2A: Linear trend: $X_t = 23 + 2.4(t - 1948) + e_t$; e_t independent normal, $\mu = 0$, $\sigma^2 = 4$.

2B: Seasonal trend: $X_t = 15 + 3\ \cos(2\pi t/12) + 1.5\ \sin(2\pi t/12) + e_t$; e_t independent normal, $\mu = 0$, $\sigma^2 = 16$.

2C: Seasonal and irregular trend: $X_t =$ cattle slaughtered under federal inspection in 10,000's (from *Livestock and Meat Statistics 1962*, U.S.D.A. Statistical Bulletin No. 333, July 1963, table 106).

For all three series, t is in years with $\Delta t = \frac{1}{12}$ year.

Elementary methodology

The development of methods for the statistical analysis of time series has followed the same broad stages as has the development of methods for simpler data. In the first stage, intuitively useful methods were used in arbitrary ways. During a second stage, probabilistic mathematical models that might plausibly be assumed to underlie the generation of the data were developed, together with statistical procedures specifically adapted to these models. In a third stage, the characteristics of various procedures were examined under a variety of models, with the hope that some procedures might be found that behave reasonably under a wider range of models.

The development of methods to estimate some physical quantity (the weight of a bar, the transit time of a star, the temperature of the air) from several independent observations subject to measurement errors may be considered as typical of these stages. The arithmetic average or mean was an early, widely used estimator. Laplace, Gauss, and others later showed that the mean had some optimal properties under various plausible assumptions. It eventually became generally understood, however, that the arithmetic average had undesirable properties under some other, also plausible, models, for which there were alternative better estimators. With this has come the development of estimators that behave well, if not optimally, for a wide range of models. Investigation at this third level continues vigorously today [*see* ERRORS, *article on* EFFECTS OF ERRORS IN STATISTICAL ASSUMPTIONS*; ESTIMATION].

In the case of time series, perhaps the earliest method, and one that still should form part of any analysis, was simply to plot the series and examine its graph intuitively. Without objective confirmatory procedures, however, this can lead easily to the apparent discovery of trends and cycles that do not exist except in the mind of the beholder [*see* TIME SERIES, *article on* CYCLES].

Linear regression and correlogram analysis. Somewhere between the first and second stage, statistical methods adapted from other contexts were applied to time series. Two important such adaptations have been (1) *linear regression* on functions of time and (2) *correlogram analysis*.

The regression approach focused on systematic changes in the level of a time series. Each observa-

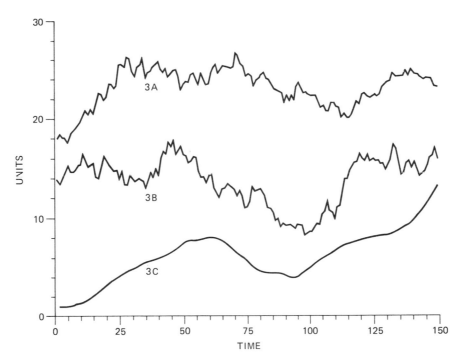

Figure 3 – Three nonstationary series with stationary first or second differences

3A: Daily closing price of Control Data common stock in dollars, Dec. 24, 1975, through July 28, 1976 (from the *Wall Street Journal*).

3B: $X_t - X_{t-1} = e_t$; e_t independent normal, $\mu = 0$, $\sigma^2 = .49$.

3C: $X_t - 2X_{t-1} + X_{t-2} = e_t$; e_t independent normal, $\mu = 0$, $\sigma^2 = 0.0025$.

For Series 3A, *t* measures trading days after Dec. 23, 1975. For 3B and 3C, *t* is in arbitrary units.

tion X_t of a time series was taken to be the sum of a polynomial function in time, one or more trigonometric functions of known period, and a random error with zero expectation, uncorrelated from time to time, and perhaps normally distributed. The estimation of unknown parameters was ordinarily by least squares.

The correlogram approach was concerned with the dependence among observations at different times. The correlogram of a time series is a graph of the correlation coefficients (the autocorrelations) between observations X_t and their lagged neighbors, X_{t-1}, X_{t-2}, $\cdots$, X_{t-s}, $\cdots$, one time unit behind, two units behind, and so on, plotted against the lag s. If a correlogram has sharp distinct maxima or other systematic pattern, a discovery may be asserted.

Both these approaches still have value in particular circumstances and form important parts of more modern techniques. The regression model, however, often requires more information than can reasonably be assumed known, especially in respect to the period of the trigonometric parts. Moreover, the use of least squares for estimation overlooks the need for explicit consideration of statistical dependence between time points. The method does, however, often provide a reasonably objective summary of certain features of time series. An observed correlogram may, because of the correlation between calculated autocorrelations, display spurious, apparently systematic, patterns that are likely to mislead the investigator. Moreover, when the level of a time series is changing, the correlogram may suggest apparent long-term dependence that has no basis in fact. The computation of the correlogram is still common, especially as a first step in choosing an appropriate time-domain model (see below).

Decomposition of time series. A concept underlying a number of methods and problems in time series analysis, especially but not exclusively in economics, is that a time series (or its logarithm) can be additively decomposed into up to four distinct parts—a relatively slowly changing long-term *trend,* a more or less regularly oscillating *cyclic* component, a *seasonal* component with known period (often a year, but sometimes some other period such as a lunar month, or $\frac{1}{60}$ second) [*see* TIME SERIES, *article on* SEASONAL ADJUSTMENT], and an *irregular* remainder representing shorter-term fluctuations or random shocks (Kendall 1973). Depending on emphasis and circumstances, each component has independent interest.

This decomposition of a time series is probably to some degree illusory, although it underlies the development of much important methodology. Models for time series whose realizations display patterns that appear to contain these components

can be constructed *without* their explicit inclusion. Indeed, the Series 3B in Figure 3 might well be seen as having a cyclic pattern with period about 75, with a clearly irregular fluctuation on top of that. Yet, each observation in the series (except with first) was generated simply by adding an independent normal random variable with zero mean to the preceding value. Autoregressive and moving-average models (see below) with no explicit periodic term can display even more clearly marked apparent periodicities. Series 1B in Figure 1 is such a series. In fact, Slutsky (1937) showed that by applying a suitable choice of repeated moving averages, a purely random error series could be made to result in a cosine curve as closely as desired. Even when the decomposition is artificial, it can be useful. Maxwell unified many aspects of the theories of magnetism and light, and predicted the existence of radio and X-rays, in terms of electromagnetic waves of differing frequencies. Yet it is still often useful to treat microwaves and X-rays as different phenomena, following different laws, observable by different devices, and of interest to different groups of scientists. Similarly, fluctuations in time series on different time scales, although perhaps in particular cases describable as all arising from a single mechanism, may still be of independent interest and relevant to differing problems.

In many economic time series there is an obvious, sometimes dominating, *seasonal* component [*see* TIME SERIES, *article on* SEASONAL ADJUSTMENT]. The isolation and prediction of this component may even be of special interest, but more often the analyst desires to disentangle the other components of the series from the seasonal part; for example, a downturn in the seasonal component of employment may mask the fact that employment is up, for that *particular time of year.* Thus an important task in time series analysis is *seasonal adjustment,* the removal of the seasonal component from a time series (Hannan et al. 1970).

For similar reasons, it is sometimes desired to remove the long-term trend from a series (Kendall 1973). Perhaps more important, the long-term trend may represent nonstationarity of the mean, which, if its structure can be estimated by regression or other methods, can be extrapolated, thus providing a forecast or projection of things to come. As Series 3C in Figure 3 shows, however, an apparently slowly changing trend can arise from nothing more than repeated summation of independent disturbances, making extrapolation based on curve fitting risky. Demographers and investors are particularly aware of this and ignore it at their peril.

The cyclic component of time series has probably

attracted most attention historically in the analysis of economic data [*see* TIME SERIES, *article on* CYCLES]. There is no question that many historical series exhibit semiregular oscillations with periods that range from 3 months to 60 years. Many natural phenomena, such as animal populations or sunspot counts, also display apparent cyclic behavior. The most predictable phenomena are those exhibiting true periodicity, exactly repeating themselves after a fixed time period, or at least composed, as are the tides, of the superposition of a small set of periodic phenomena. Thus it was natural to hope that the cyclic component of a time series might be made up of one or more exactly periodic components whose frequencies or periods could be determined from data. The search for *hidden periodicities* has been a fruitful one for time series methodology, if less successful at actually finding cycles on which there could be broad objective agreement (Allais 1962).

The mathematically nicest form for a perfectly periodic phenomenon is a cosine (equivalently, a sine) curve, although other elementary wave forms —such as saw-toothed-shaped ones—have been proposed. The general form of a cosine curve is $A \cos(2\pi f t + \phi)$; the *period* $1/f$ is the length of time before repetition and the *frequency* f is the number of repetitions or cycles per time unit. (The factor 2π is conventionally used as a mathematical device to simplify notation and terminology.) The remaining notation is mnemonic: A for amplitude and ϕ for phase. A sine curve may be obtained from a cosine curve by subtracting $\pi/2$ radians ($= 90°$) from ϕ.

For a long time an important tool in studying cycles was the *Schuster periodogram* (Granger & Hatanaka 1964; Schuster 1897). Any n consecutive observations on a time series can be mathematically (but not necessarily meaningfully) expressed exactly as a superposition of cosine curves (assuming for simplicity that $\Delta t = 1$ and n is odd):

$$(1) \qquad X_t = A_0 + \sum_{k=1}^{(n-1)/2} A_k \cos(2\pi f_k t + \phi_k).$$

Each term is associated with the *frequency* $f_k = k/n$ cycles per unit time. The *amplitude* A_k and *phase* ϕ_k of each term can be calculated from the data, with A_k indicating how important oscillations of period $1/f_k = n/k$ are in the observed time series. The Schuster periodogram is usually defined as the graph of $P_k = \frac{1}{2} n A_k^2$ (or sometimes twice this) as a function of k (or sometimes $1/k$). If the nonstationary mean of X_t contains an exactly cyclic component with period S, then P_k can be expected to have a sharp peak at the integer k nearest to T/S, thus, as was said, revealing the presence, and the approximate period, of the regular oscillation. Figure 4 displays the periodogram of the seasonal Series 2B in Figure 2. Note the peak at $k = 12$, the integer nearest $150/12 = 12.5$.

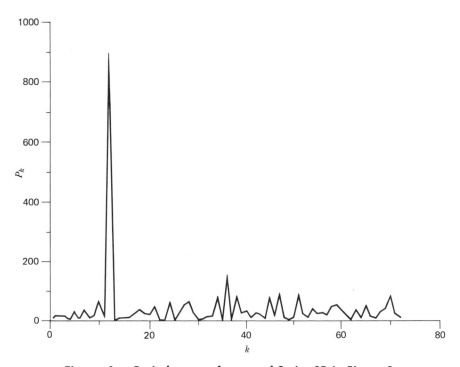

Figure 4 — Periodogram of seasonal Series 2B in Figure 2

The abscissa is in units of $k = Tf = 150f$, where f is frequency in cycles per month.

It was found in practice, however, and later verified theoretically, that P_k has great statistical variability and that its plot against k is extremely irregular. Apparent peaks are easy to find, even in the absence of genuine cycles, because one or more local maxima will often appear substantially larger than neighboring values. Fisher (1929) provided a significance test, appropriate for independent errors, that protects against the selection bias inherent in any procedure paying special attention to the largest of a set of statistics. This test has been modified to allow for correlated series and has been extended to multiple time series (Hannan 1970; MacNeill 1974).

Contemporary time series analysis

Time series analysis today can be roughly divided into two types of methods, *frequency-domain* methods and *time-domain* methods.

In models underlying frequency-domain analysis of a time series, $\{X_t\}$ is expressed as a superposition of independently varying cosine and sine curves with random amplitudes, each contributing only a very small part of the total variability. Such a model can be approximated by

$$(2) \qquad X_t = \mu + \sum_j [U_j \cos(2\pi f_j t) + V_j \sin(2\pi f_j t)],$$

where the U's and V's are uncorrelated random variables with zero expectation and variances $\sigma^2(f_j)$ equal in pairs and functionally dependent on f_j. The frequencies $f_1 = \Delta f, f_2 = 2\Delta f, \cdots, f_j = j\Delta f, \cdots$, are equally spaced, separated by Δf, where Δf is very small, and correspond to periods $1/f_1, 1/f_2, \cdots$. The principal aim of an analysis based on such a model is the estimation of the variances $\sigma^2(f_j)$ of the U's and V's, and the determination of how $\sigma^2(f_j)$ depends on f_j. Sometimes a parametric model is postulated for this dependence, that is, the $\sigma^2(f_j)$ are assumed to be expressible using a small number of unknown constants. More commonly, a frequency-domain analysis is basically nonparametric, assuming only that the $\sigma^2(f_j)$ vary "smoothly" with f_j.

Time-domain methods are, in constrast, usually explicitly parametric, and are based on direct modeling of the lagged relationships between a series and its past. A typical time-domain model would be the second-order autoregressive series

$$(3) \qquad X_t = \phi_1 X_{t-1} + \phi_2 X_{t-2} + a_t,$$

where ϕ_1 and ϕ_2 are constants (unrelated to the phases ϕ_k in equation (1)) and $\{a_t\}$ is a series of uncorrelated random errors with constant variance σ^2. Once such a model has been postulated, the usual aim of analysis is to estimate the ϕ's and σ^2 and other quantities depending on them.

Frequency-domain methods. As can be seen from a comparison of equations (1) and (2), computing the periodogram is an elementary form of frequency-domain time series analysis. It is perhaps most suitable when one is truly searching for exact periodicities. *Spectrum analysis* (alternatively called *spectral analysis*) is the prototype of more modern frequency-domain analysis and is applicable in the much more common situation in which no exact periodicities are expected. Spectrum analysis is based on a generalization of the model given by equation (2), obtained by replacing the summation by a suitably defined integral. The discrete set of frequencies f_j become the continuum of all frequencies between 0 and $1/(2\Delta t)$, the so-called Nyquist frequency, where Δt is the interval between observations (Granger & Hatanaka 1964; Koopmans 1974).

(No frequencies above the Nyquist frequency need be included in the model. An oscillation at a frequency greater than $1/(2\Delta t)$ is observationally indistinguishable from an oscillation at a frequency less than $1/(2\Delta t)$. As an example, consider *daily* noontime observations on sea level. The tides cause the level to fluctuate periodically with a period of 1 cycle per 12.5 hours or about 1.92 cycles per day, a frequency above the Nyquist frequency, 0.5 cycles per day. However, from our noontime observations, we will observe a cycle with frequency about 12.5 days or frequency 0.08 cycles per day, below the Nyquist frequency. This phenomenon is often described by saying that frequencies above $1/(2\Delta t)$ are *aliased* with frequencies below $1/(2\Delta t)$.)

A principal goal is now to estimate the *spectrum density* or simply the *spectrum* of X_t [see TIME SERIES, *article on* ADVANCED PROBLEMS]. The spectrum $S(f)$ is (usually) a continuous function of frequency f that summarizes, in a way somewhat analogous to variance component analysis (Tukey 1961), how the variance of X_t is distributed among oscillations of various frequencies. Except for mathematical niceties having to do with replacing the sum in equation (2) by an integral, a plot of $S(f)$ against f corresponds exactly to a plot of $\sigma^2(f_j)$ against f_j. Usually, spectrum estimation is approached nonparametrically without assuming a particular form for $S(f)$, although sometimes parametric forms, especially ratios of two polynomials in $\cos(2\pi f)$, are assumed, as they are in equation (4) below.

The spectrum of an uncorrelated (that is, one in which all lagged correlations are zero) stationary time series is constant (flat). By analogy with the constant physical spectrum of (ideal) white light, such a completely random series is often called *white noise*. Systematic structure in a time series is indicated by departure from flatness of the spectrum. A peak in $S(f)$ indicates a tendency of the series to oscillate at frequencies near the peak.

A spectacular application of spectrum analysis provided strong evidence that long-term variations in the earth's climate were linked with variations in the earth's orbit. This was accomplished by comparing spectra of climate indicators derived from deep-seabed cores with computed spectra of quantities depending directly on the orbit, such as winter insolation. The comparison showed that peaks in the spectra appeared at essentially the same frequencies (Hays et al. 1976).

Virtually any stationary time series has a spectrum, and for many time-domain models, the spectrum can be given analytically in a form depending directly on the parameters of the model. For instance, the spectrum corresponding to the series whose structure is given by equation (3) is

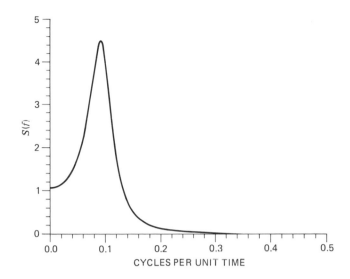

Figure 5 — Spectrum density S(f) of Series 1B in Figure 1

$$S(f) = \frac{.64}{[2\pi(3.5493 - 4.9132\cos 2\pi f + 1.46\cos 4\pi f)]}$$

The abscissa is in units of frequency f cycles per unit time.

$$(4) \qquad S(f) = \frac{\sigma^2}{2\pi[(1 + \phi_1^2 + \phi_2^2) - 2(\phi_1 - \phi_1\phi_2)\cos 2\pi f - 2\phi_2\cos 4\pi f]},$$

where σ^2 = variance (a_t). Series 1B in Figure 1 is a series computer generated according to equation (3), with $\sigma^2 = 0.64$, $\phi_1 = 1.42$, and $\phi_2 = -0.73$. Its spectrum is thus

$$S(f) =$$
$$\frac{.64}{[2\pi(3.5493 - 4.9132\cos 2\pi f + 1.46\cos 4\pi f)]},$$

which is graphed in Figure 5. Note the distinct mode at frequency 0.0909 cycles per unit time. This translates to a period of 11 years (if $\Delta t = 1$ year). A large part of the over-all variability in this series can thus be ascribed to random oscillations in the neighborhood of this frequency, say from 0.06 to 0.15 cycles (period 8.7 to 16.7), the so-called half power points of the spectrum.

Wiener (1949) and Kolmogorov (1939) independently demonstrated the importance of the spectrum in time series models. Practical methods for spectrum analysis were developed in the late 1940s, but were little applied until high-speed computers became widely available. The techniques of *spectrum* (often *spectral*) *analysis* have now become standard statistical methods (Blackman & Tukey 1958; Brillinger 1975; Fishman 1969; Granger & Hatanaka 1964; Hannan 1970; Jenkins & Watts 1968; Kendall & Stuart 1966; Koopmans 1974;

Robinson 1967). Estimation of spectra has to a large degree been revolutionized by the so-called fast Fourier transform (FFT) (Cooley & Tukey 1965; Singleton 1969). Somewhat ironically, the periodogram, whose use has been largely superseded by the estimated spectrum, is now routinely computed, using the FFT, as a preliminary step in computing the spectrum (Bingham et al. 1967).

Time-domain methods. For a single time series, the most widely applied time-domain methods are based on so-called ARIMA (AutoRegressive, Integrated, Moving Average) models (Box & Jenkins 1976; Nelson 1973; Quenouille 1957). These are also widely referred to as Box–Jenkins models, and are constructed by linking three separate types of underlying models.

First and most important are the *autoregressive* (AR) models, a particular example of which was given in equation (3). These can be expressed, using the now standard notation of Box and Jenkins, as (assuming $\Delta t = 1$ and $E(X_t) = 0$)

$$(5) \quad X_t = \phi_1 X_{t-1} + \phi_2 X_{t-2} + \cdots + \phi_p X_{t-p} + e_t,$$

where the ϕ's are constant coefficients and $\{e_t\}$ is a random error series. This expresses the model in the form of a linear regression of X_t on its own past.

The AR model of order 2 (regressing on two past values) was introduced by Yule (1927) to model the variation over time of Wolfer's series of the intensity of sunspot activity. Although the number of sunspots varies remarkably regularly with a period near 11 years, this behavior is not, in fact, well described as the superposition of a small number of oscillators. Yule showed that a second order AR model, as in equation (3), with appropriate parameter values could, to a considerable degree, mimic the behavior of Wolfer's series without invoking an underlying deterministic periodic component. The coefficients $\phi_1 = 1.42$ and $\phi_2 = -0.73$ used in generating Series 1B in Figure 1 represent such a model fit to the sunspot series (Box & Jenkins 1976, p. 239). As remarked above, its spectrum in Figure 5 has a peak centered at 11 years, in substantial conformity with the observed behavior. It has since been shown that an AR model is inadequate for the sunspot data, although as yet no satisfactory alternative has been put forward (Moran 1954).

The general AR model expresses the current observation as a linear combination of past observations plus a residual disturbance term e_t. In a pure AR model, the error e_t is taken to be an uncorrelated stationary series (white noise). In many applications, however, there is evidence that, although the series has some features of an AR series, the disturbance series is not white, but is itself autocorrelated. It is often plausible that the e_t are produced by a superposition of further *uncorrelated* random disturbances a_t with lagged effects, say $e_t = a_t - \theta_1 a_{t-1} - \cdots - \theta_q a_{t-q}$, where now a_t is white noise. That is, $\{e_t\}$ is generated by a so-called *moving-average* (MA) model. The constant coefficients θ_r reflect the influence of the lagged disturbances a_{t-r} on e_t (the use of minus signs follows the convention in Box & Jenkins 1976). The combination of these two models, AR[X_t] with MA disturbances [e_t], can thus be expressed (introducing a mean μ) as

$$(6) \quad X_t = \mu + \sum_{s=1}^{p} \phi_s (X_{t-s} - \mu) + a_t - \sum_{r=1}^{q} \phi_r a_{t-r},$$

and is referred to as a mixed *autoregressive–moving-average* (ARMA) series.

Many stationary time series can apparently be well represented by ARMA models with a relatively small number of nonzero coefficients ϕ_s and θ_r. These coefficients can be estimated by a variety of techniques, including maximum likelihood, a modification of nonlinear least squares, and techniques that transfer a time-domain estimation problem to a frequency-domain estimation problem by estimating a parametric form such as equation (4)

for the spectrum (Aigner 1971; Box & Jenkins 1976; Nelson 1973). One obtains nonstationary series, the ARIMA (I = integrated) models, whose dth differences are stationary ARMA models, by assuming $\{X_t\}$ is a partial sum of an ARMA stationary series, or more generally such a sum iterated d times. Even series with marked seasonalities can often be brought within the domain of these methods by assuming that, say, $X_t - X_{t-12}$ is an ARIMA time series.

An important theoretical result is that a wide class of ARMA models can also be expressed as purely AR models of infinite order (assuming zero mean),

$$(7) \quad X_t = \sum_{s=1}^{\infty} \psi_s X_{t-s} + a_t,$$

where as before the series $\{a_t\}$ is uncorrelated with zero mean. The ψ's are constant coefficients whose values depend on the ϕ's and θ's. For instance, the simple MA model $X_t = a_t + 0.5\, a_{t-1}$ is equivalent to $X_t = a_t + 0.5\, X_{t-1} - (0.5)^2 X_{t-2} + (0.5)^3 X_{t-3} - \cdots$. The summation term in equation (7) is the best prediction of X_t based on the entire past $X_{t-1}, X_{t-2}, \cdots$. The disturbance $\{a_t\}$ represents the new, completely unpredictable, component of the observation X_t at time t, being uncorrelated with the past observations $X_{t-1}, X_{t-2}, \cdots$. For this reason $\{a_t\}$ is sometimes called the *innovation* series.

Multiple time-series models

Relationships between two or more time series are usually more interesting and consequential than the properties of a single series. Although multivariate models are intrinsically more complex than univariate models, analytic techniques based on them are in wide use.

Regression analysis. Regression analysis is probably the most widely applied tool in the analysis of multiple time series. Whenever the values of a dependent variable Y and independent variables $X^{(j)}$ are observed sequentially in time (and the time ordering is known), one is dealing with a time series problem, whether recognized or not. In such a situation, a primary assumption underlying standard least-squares regression analysis, the statistical independence of residuals, is often violated. Lack of independence not only can grossly distort the usual standard errors of regression coefficients but can rob ordinary least squares of optimality properties.

An important advance in regression methodology was the development of tests that can reveal the presence of autocorrelation between residuals. The best known of these is the Durbin–Watson statistic

(Durbin & Watson 1950–1971). This has become a standard option in many computer programs for regression analysis. Its use is equivalent to testing whether the first order autocorrelation of the residual series is zero. When the residuals are autocorrelated, a number of estimation methods have been proposed as more efficient replacements for least squares (Nicholls et al. 1975). Among the more interesting are those that transfer the estimation problem to the frequency domain and apply techniques borrowed from spectrum analysis (Hannan 1963).

Cross-spectrum analysis. The chief tool in the frequency-domain analysis of multiple time series is the *cross-spectrum*, a function of frequency that represents a generalized covariance between oscillations at each frequency in one time series and oscillations at the same frequencies in another. The cross-spectrum is a complex function (in the sense of having both real and imaginary parts). Information concerning the strength of association between series and the amount, if any, one series lags or leads another, can be derived from the cross-spectrum and spectra. A substantial part of standard multivariate analysis (regression, factor analysis, discriminant functions, principal components, MANOVA) can be extended to study interseries relationships in the frequency domain [Hannan 1970, Shumway & Unger 1974; *see also* MULTIVARIATE ANALYSIS; FACTOR ANALYSIS AND PRINCIPAL COMPONENTS].

Multivariate ARMA models. The direct generalization of equation (6) to multivariate series is one useful basis for multivariate time-domain methods. Equation (6) becomes

$$(8) \quad \mathbf{X}_t = \mu + \sum_{s=1}^{p} \Phi_s(\mathbf{X}_{t-s} - \mu) + \mathbf{a}_t - \sum_{r=1}^{q} \Theta_r \mathbf{a}_{t-r},$$

where $\mathbf{X}_t$, μ, and $\mathbf{a}_t$ are now vectors and Φ_s and Θ_r are matrices of constant coefficients. The multivariate disturbance series $\{\mathbf{a}_t\}$ is again uncorrelated from time period to time period and has zero mean. Its separate components, however, may be simultaneously correlated. Under some conditions, equation (8) is equivalent to a multivariate AR model of infinite order that is analogous to equation (7) (Hannan 1970; Quenouille 1957). In that form $\{\mathbf{a}_t\}$ represents the *innovation* vector at time t.

Distributed lags. The *distributed* lags model is another popular way to describe relationships among time series [Dhrymes 1971; Griliches 1967; Nerlove 1972; Sims 1974; *see also* DISTRIBUTED LAGS]. In its simplest form, a "dependent" time series $\{Y_t\}$ is assumed to depend linearly on lagged (occasionally leading) values of an "independent" time series $\{X_t\}$ except for a random error,

$$(9) \quad Y_t = \mu + \sum_s \beta_s X_{t-s} + e_t,$$

where it is assumed that $\{e_t\}$ is uncorrelated with X_t, an important restriction. Ordinarily it is assumed that the β's depend on a relatively small number of underlying parameters. For example, one simple form that has been proposed is that $\beta_s = \gamma \theta^s$ for some constants γ and θ with $|\theta| < 1$ (Koyck 1954). Other, less restricted proposals suggest that β_s be represented as proportional to negative binomial probabilities (Solow 1960), as a polynomial in the lags (Almon 1965; Schmidt 1974), or the coefficients of a rational generating function (Jorgenson 1966). Both theory and practice have shown, however, that uncorrelatedness of the disturbance series $\{e_t\}$ is often an unrealistic assumption. By assuming that $\{e_t\}$ is itself an ARMA series, autocorrelation can be introduced directly into the model (Fishman 1969; Hannan 1965). Useful generalizations allow $\{Y_t\}$ and/or $\{X_t\}$ to be multivariate series and replace the scalar coefficients β_s by matrices of coefficients.

Simultaneous equations models. One of the reasons for the importance and apparent usefulness of distributed lag models in time series analysis is their close relationship with a general class of models that purport to describe in directly interpretable terms the dynamic relationships between simultaneously evolving time series. These models constitute the important class of *simultaneous equations models*. They are expressed as a set of simultaneous linear equations that are satisfied, except for error, by the values of the series whose structure is being described [*see* SIMULTANEOUS EQUATION ESTIMATION]. Each equation is ordinarily associated with some identifiable entity or system in the universe under study. Thus one equation will in some sense describe the behavior of consumers, another the behavior of producers. A standard form for simultaneous equations models is given by the following *structural equations:*

$$(10) \quad \sum_{j=1}^{p} \gamma_{ij} Y_t^{(j)} + \sum_{k=1}^{q} \beta_{ik} X_t^{(k)} = e_t^{(i)},$$
$$i = 1, 2, \cdots, p.$$

Here the $\{Y_t^{(j)}\}$ series are p interrelated so-called *endogenous* (or dependent) time series, and the $\{X_t^{(k)}\}$ series are q so-called *exogenous* series, uncorrelated with the p error series $\{e_t^{(i)}\}$. One or more of the X_t series may be lagged values of the Y_t series. Some of the error series may be identically zero, the associated equations expressing an exact relationship, true by definition, between the variables.

As with distributed lag models, realism has required that the random error series have autocorrelation structure (Nicholls et al. 1975).

Simultaneous equations models can ordinarily be put in a form (the so-called *reduced* form) that is superficially identical to a set of standard multiple linear regression equations in which each $Y_t^{(j)}$ depends on the X series, but with the following essential difference: the components of the error or disturbance time series are correlated with the independent variables $X_t^{(k)}$. As a consequence, ordinary least squares does not properly estimate the unknown coefficients, even for very large samples. Various methods, such as two- and three-stage least squares and partial- and full-information maximum likelihood, are used to obtain estimates of the unknown coefficients in simultaneous equations (Theil 1971). The link with distributed lag models lies in the fact that simultaneous equations models can often be expressed as distributed lag models (the so-called *final form* of the equations) with autocorrelated errors. Although the final form is less informative than the structural equations, as equations (10), for understanding the dynamics of relationships or for predicting the result of manipulating one or more of the component series, it may be used for prediction under unchanging conditions. Important hypotheses concerning the structural equations can, however, impose restrictions on the form of the associated distributed lag models. When these restrictions are observationally refutable, an objective test of the correctness of the original simultaneous equations model can sometimes be performed (Zellner & Palm 1974).

Prediction

One of the most important and fundamental problems in time series analysis, as in statistics generally, is *prediction*. From observations on the present and past of one or more time series, it is desired to estimate a future, as yet unobserved, value of a time series $\{X_t\}$. Clearly, success in prediction can be fruitful in many diverse fields. Knowledge of future interest rates can make money for the investor in bonds; an accurate prediction of Christmas sales can allow the retailer to stock an optimal inventory; in warfare, if one can predict the position of an aircraft by several seconds, the probability of an antiaircraft hit is increased; by anticipating deviations from the norm of output parameters of a chemical process, an engineer can make adjustments to maintain stability. Even when a time series is analyzed primarily to understand the internal and external dynamics of the system generating it, the only completely sound way of validating the conclusions reached is by using them to predict future observations and comparing those predictions with what actually occurs (Christ 1975; Newbold & Granger 1974).

The fundamentals of so-called optimal prediction theory are due to Kolmogorov (1939) and Wiener (1949). For stationary or nearly stationary time series, prediction techniques draw heavily on both the time- and frequency-domain approaches to analysis. Optimal prediction equations are often most elegantly expressed in terms of spectra and cross-spectra. They are often practically implemented by formulae expressing the linear regression of future values on present and past values. The systematic part (expectation) and the random part of a series are treated quite differently. The former is ordinarily forecast by extrapolation of an estimated deterministic function, often a risky venture unless the form of that function has a sound theoretical basis. The random part is predicted by exploiting its correlation, if any, with the random part of the present and past of the series on which the prediction is based (Box & Jenkins 1976). Much actual forecasting of time series, especially in situations such as inventory control when there are many series to predict, is done by relatively ad hoc methods with little attempt at optimality. These include simple exponential smoothing, Holt–Winters forecasts, and seat-of-the-pants guesses (Chatfield 1975; Kendall 1973).

CHRISTOPHER BINGHAM

BIBLIOGRAPHY

Until the 1970s, there were relatively few books on time series analysis readable by nonspecialists. This situation is rapidly changing with new additions to the textbook and monograph literature appearing frequently. Chatfield 1975, Kendall 1973, *and* Nelson 1973 *provide good introductions to time series analysis at not too high a technical level, especially concerning time-domain methods. At a slightly higher level, dealing entirely with ARIMA series, see* Box & Jenkins 1976 *and, for multiple time series,* Quenouille 1957. *The frequency-domain analysis of time series is well presented by* Bloomfield 1976, Brillinger 1975, Jenkins & Watts 1968, *and* Koopmans 1974. *Although obsolete from a computational point of view, the discussion of applied spectrum analysis in* Granger & Hatanaka 1964 *is still among the most readable. More thorough coverage at a considerably higher technical level can be found in* Anderson 1971 *and* Hannan 1970. *The latter is particularly good in its use of frequency-domain techniques and ideas in handling time-domain problems.* Theil 1971 *has good coverage of many of the important topics in the application of distributed lag and simultaneous equations models. The most comprehensive source of bibliographic information,* International Statistical Institute 1965, *unfortunately does not go beyond 1959.* Makridakis 1976 *has a good bibliography; however, its discussion of time series methods is unbalanced and has some misconceptions.*

AIGNER, DENNIS J. 1971 A Compendium on Estimation of the Autoregressive–Moving Average Model From Time Series Data. *International Economic Review* 12: 348–371.

ALLAIS, MAURICE 1962 Test de périodicité; Généralisation du test de Schuster au cas de séries temporelles autocorrelées dans l'hypothèse d'un processus de perturbations aléatoires d'un système stable. International Statistical Institute, *Bulletin* 39, part 2:143–193.

ALMON, SHIRLEY 1965 The Distributed Lag Between Capital Appropriations and Expenditures. *Econometrica* 33:206–224.

ANDERSON, T. W. 1971 *The Statistical Analysis of Time Series.* New York: Wiley.

BARLOW, J. S. 1959 Autocorrelation and Cross-correlation in Electroencephalography. Institute of Radio Engineers, *Transactions on Medical Electronics* 6: 179–183.

BARTLETT, M. S. (1955) 1966 *An Introduction to Stochastic Processes With Special Reference to Methods and Applications.* 2d ed. New York: Cambridge Univ. Press.

BARTLETT, M. S. 1975 *The Statistical Analysis of Spatial Pattern.* London: Chapman & Hall.

BINGHAM, CHRISTOPHER; GODFREY, M. D.; and TUKEY, JOHN W. 1967 Modern Techniques of Power Spectrum Estimation. Institute of Electrical and Electronics Engineers, *Transactions on Audio and Electroacoustics* 15:56–66.

BLACKMAN, R. B.; and TUKEY, JOHN W. 1958 *The Measurement of Power Spectra From the Point of View of Communications Engineering.* New York: Dover.

BLOOMFIELD, PETER 1976 *Fourier Analysis of Time Series: An Introduction.* New York: Wiley.

BOX, GEORGE E. P.; and JENKINS, G. M. (1970) 1976 *Time Series Analysis: Forecasting and Control.* Rev. ed. San Francisco: Holden-Day.

BRILLINGER, DAVID R. 1975 *Time Series: Data Analysis and Theory.* New York: Holt.

CHATFIELD, CHRISTOPHER 1975 *The Analysis of Time Series: Theory and Practice.* London: Chapman & Hall.

CHATFIELD, CHRISTOPHER; and PEPPER, M. P. G. 1971 Time-series Analysis: An Example From Geophysical Data. *Applied Statistics* 20:217–238.

CHRIST, CARL F. 1975 Judging the Performance of Econometric Models of the U.S. Economy. *International Economic Review* 16:54–74.

COOLEY, JAMES W.; and TUKEY, JOHN W. 1965 An Algorithm for the Machine Calculation of Complex Fourier Series. *Mathematics of Computation* 19:297–301.

COOTNER, PAUL H. (editor) (1964) 1967 *The Random Character of Stock Market Prices.* Rev. ed. Cambridge, Mass.: M.I.T. Press.

COX, D. R.; and LEWIS, P. A. W. 1966 *The Statistical Analysis of Series of Events.* London: Methuen; New York: Wiley.

COX, D. R.; and MILLER, H. D. 1965 *The Theory of Stochastic Processes.* New York: Wiley.

DHRYMES, PHOEBUS J. 1971 *Distributed Lags: Problems of Estimation and Formulation.* San Francisco: Holden-Day.

DURBIN, J. R.; and WATSON, G. S. 1950–1971 Testing for Serial Correlation in Least Squares Regression. Parts 1–3. *Biometrika* 37 [1950]:409–428; 38 [1951]: 159–178; 58 [1971]:1–20.

FISHER, R. A. 1929 Tests of Significance in Harmonic Analysis. Royal Society of London, *Proceedings* Series A 125:54–59.

FISHMAN, GEORGE S. 1969 *Spectral Methods in Econometrics.* Cambridge, Mass.: Harvard Univ. Press.

FOWLER, STEPHEN C.; MORGENSTERN, CARL; and NOTTER-

MAN, JOSEPH M. 1972 Spectral Analysis of Variations in Force During a Bar-pressing Time Discrimination. *Science* 176:1126–1127.

GRANGER, CLIVE W. J.; and HATANAKA, M. 1964 *Spectral Analysis of Economic Time Series.* Princeton Univ. Press.

GRENANDER, ULF; and ROSENBLATT, MURRAY 1957 *Statistical Analysis of Stationary Time Series.* New York: Wiley.

GRILICHES, ZVI 1967 Distributed Lags: A Survey. *Econometrica* 35:16–49.

GROVES, G. W.; and HANNAN, EDWARD J. 1968 Time Series Regression of Sea Level on Weather. *Reviews of Geophysics* 6:129–174.

HANNAN, EDWARD J. 1963 Regression for Time Series. Pages 17–37 in Symposium on Time Series Analysis, Brown University, 1962, *Proceedings.* Edited by Murray Rosenblatt. New York: Wiley.

HANNAN, EDWARD J. 1965 The Estimation of Relationships Involving Distributed Lags. *Econometrica* 33: 206–224.

HANNAN, EDWARD J. 1970 *Multiple Time Series.* New York: Wiley.

HANNAN, EDWARD J.; TERRELL, R. D.; and TUCKWELL, N. E. 1970 The Seasonal Adjustment of Economic Time Series. *International Economic Review* 11:24–52.

HAYS, J. D.; IMBRIE, JOHN; and SHACKLETON, N. J. 1976 Variations in the Earth's Orbit: Pacemaker of the Ice Ages. *Science* 194:1121–1132.

HOLGATE, PHILIP 1966 Time-series Analysis Applied to Wildfowl Counts. *Applied Statistics* 15:15–23.

INTERNATIONAL STATISTICAL INSTITUTE 1965 *Bibliography on Time Series and Stochastic Processes.* Edited by Herman Wold. Edinburgh: Oliver & Boyd; Cambridge, Mass.: M.I.T. Press.

JENKINS, G. M.; and WATTS, DONALD G. 1968 *Spectral Analysis and Its Applications.* San Francisco: Holden-Day.

JORGENSEN, D. W. 1966 Rational Distributed Lag Functions. *Econometrica* 34:135–149.

KENDALL, MAURICE G. (1973) 1976 *Time-series.* 2d ed. London: Griffin; New York: Hafner.

KENDALL, MAURICE G.; and STUART, ALAN (1966) 1973 Time-series: General. Time-series: Trend and Seasonality. Stationary Time-series. The Sampling Theory of Serial Correlations. Spectrum Theory. Time-series: Some Further Topics. Volume 3, chapters 45–50 in Maurice G. Kendall and Alan Stuart, *The Advanced Theory of Statistics.* Volume 3: *Design and Analysis, and Time-series.* 3d ed. London: Griffin; New York: Hafner.

KOLMOGOROV, A. N. 1939 Sur l'interpolation et extrapolation des suites stationnaires. Académie des Sciences, Paris, *Comptes rendus* 208, part 2:2043–2045.

KOOPMANS, LAMBERT H. 1974 *The Spectral Analysis of Time Series.* New York: Academic Press.

KOYCK, L. M. 1954 *Distributed Lags and Investment Analysis.* Amsterdam: North-Holland.

LEE, RONALD D. 1975 Natural Fertility, Population Cycles and the Spectral Analysis of Births and Marriages. *Journal of the American Statistical Association* 70:295–304.

LEWIS, P. A. W. (editor) 1972 *Stochastic Point Processes: Statistical Analysis, Theory, and Applications.* New York: Wiley.

MACNEILL, I. B. 1974 Tests for Periodic Components in Multiple Time Series. *Biometrika* 61:57–70.

MAKRIDAKIS, SPYROS G. 1976 A Survey of Time Series Analysis. *International Statistical Review* 44:29–70.

MILLER, A. J. 1962 Road Traffic Flow Considered as a Stochastic Process. Cambridge Philosophical Society, *Proceedings* 58:312–325.

MORAN, P. A. P. 1954 Some Experiments in the Prediction of Sunspot Numbers. *Journal of the Royal Statistical Society* Series B 16:112–117.

NELSON, CHARLES R. 1973 *Applied Time Series Analysis for Managerial Forecasting.* San Francisco: Holden-Day.

NERLOVE, MARC 1972 Lags in Economic Behavior. *Econometrica* 40:221–251.

NEWBOLD, P.; and GRANGER, CLIVE W. J. 1974 Experience With Forecasting Univariate Time Series and the Combination of Forecasts. *Journal of the Royal Statistical Society* Series A 137:131–164. → Includes discussion.

NICHOLLS, D. F.; PAGAN, A. R.; and TERRELL; R. D. 1975 The Estimation and Use of Models With Moving Average Disturbance Terms: A Survey. *International Economic Review* 16:113–134.

PRESS, HARRY 1955 Time Series Problems in Aeronautics. *Journal of the American Statistical Association* 50:1022–1039.

QUENOUILLE, M. H. 1957 *The Analysis of Multiple Time-series.* London: Griffin; New York: Hafner. → Reprinted in 1968 with minor corrections.

RAO, R. A.; and KASHYAP, R. L. 1974 Stochastic Modeling of River Flows. Institute of Electrical and Electronics Engineers, *Transactions on Automatic Control* 19:874–881.

ROBINSON, ENDERS A. 1967 *Multichannel Time Series Analysis With Digital Computer Programs.* San Francisco: Holden-Day.

SCHMIDT, PETER 1974 A Modification of the Almon Distributed Lag. *Journal of the American Statistical Association* 69:679–681.

SCHUSTER, ARTHUR 1898 On the Investigation of Hidden Periodicities With Application to a Supposed 26 Day Period of Meteorological Phenomena. *Terrestrial Magnetism* 3:13–41.

SHUMWAY, R. H.; and UNGER, A. N. 1974 Linear Discriminant Functions for Stationary Time Series. *Journal of the American Statistical Association* 69:948–956.

SIMS, CHRISTOPHER A. 1974 Distributed Lags. Volume 2, pages 289–338 in Michael D. Intriligator and D. A. Kendrick (editors), *Frontiers of Quantitative Economics.* Amsterdam: North-Holland.

SINGLETON, R. C. 1969 A Short Bibliography on the Fast Fourier Transform. Institute of Electrical and Electronics Engineers, *Transactions on Audio and Electroacoustics* 17:166–169.

SLUTSKY, EUGEN E. 1937 The Summation of Random Causes as the Source of Cyclic Processes. *Econometrica* 5:105–146.

SOLOW, ROBERT M. 1960 On a Family of Lag Distributions. *Econometrica* 28:393–406.

THEIL, HENRI 1971 *Principles of Econometrics.* New York: Wiley.

TUKEY, JOHN W. 1961 Discussion, Emphasizing the Connection Between Analysis of Variance and Spectrum Analysis. *Technometrics* 3:191–219.

WIENER, NORBERT 1949 *Extrapolation, Interpolation, and Smoothing of Stationary Time Series, With Engineering Applications.* Cambridge, Mass.: M.I.T. Press.

WOLD, HERMAN (1938) 1954 *A Study in the Analysis of Stationary Time Series.* 2d ed. Stockholm: Almqvist & Wiksell.

WOLD, HERMAN (editor) 1965 *Bibliography on Time Series and Stochastic Processes.* → See under International Statistical Institute.

YULE, G. UDNY 1927 On a Method of Investigating Periodicities in Disturbed Series, With Special Reference to Wolfer's Sunspot Numbers. Royal Society of London, *Philosophical Transactions* Series A 226:267–298.

ZELLNER, ARNOLD, and PALM, FRANZ 1974 Time Series Analysis and Simultaneous Equations Econometric Models. *Journal of Econometrics* 2:17–54.

II
ADVANCED PROBLEMS

Numerical data often occur in the form of a *time series*, that is, a sequence of observations on a variable taken either continuously or at regular intervals of time. As examples consider records of economic variables (prices, interest rates, sales, unemployment), meteorological records, electroencephalograms, and population and public health records. In contrast to much experimental data, the consecutive observations of a time series are not in general independent, and the fascination of time series analysis lies in the utilization of observed dependence to deduce the way the value of a variable (say, steel production) can be shown to be in part determined by the past values of the same variable, or of other variables (say, demand for automobiles).

In the so-called discrete time case, observations are taken at regular intervals of time. The common interval of time can be taken as the unit, so that the value of a variable x at time t can be denoted by x_t, where t takes the values $\cdots, -2, -1, 0, 1, 2, \cdots$. Of course, in practice one observes only a finite set of values, say $x_1, x_2, \cdots, x_n$, but it is useful to imagine that the series can in principle extend indefinitely far back and forward in time. For this reason t is allowed to run to infinity in both directions. (Of course, sometimes one observes a phenomenon continuously, so that x is measured for all t rather than just at intervals—such a process is referred to as continuous. However, since the discrete case is by far the more frequent in the social sciences, this discussion will be limited to that case.)

Suppose that one has a model which explains how the x_t series should develop. The model is termed a *process* and is denoted by $\{x_t\}$; if some of the rules it specifies are probabilistic ones, it is called a *stochastic process*.

Definition of a stationary process

One class of stochastic processes is of particular importance, both in practice and in theory: this is the class of *stationary processes*. A stationary process is one that is in a state of statistical equilibrium, so that its statistical pattern of behavior does not change with time. Formally, the requirement is that for any set of instants of time, $t_1, t_2, \cdots, t_n$, and any time lag, s, the joint distribution of $x_{t_1}, x_{t_2}, \cdots, x_{t_n}$ must be the same as that of $x_{t_1+s}, x_{t_2+s}, \cdots, x_{t_n+s}$. Thus, x_1 and x_5 must have the same univariate distribution, (x_1, x_2) and (x_5, x_6) must have the same bivariate distribution, and so on.

The assumption of stationarity is a strong one, but when it can be made it greatly simplifies understanding and analysis of a process. An intuitive reason for the simplification is that a stationary process provides a kind of hidden replication, a structure that does not deviate too far from the still more special assumptions of independence and identical distribution, assumptions that are ubiquitous in statistical theory. Whether the stationarity assumption is realistic for a particular process depends on how near the process is indeed to statistical equilibrium. For example, because most economies are evolving, economic series can seldom be regarded as stationary, but sometimes a transformation of the variable produces a more nearly stationary series (see the section on "smoothing" a series, below).

Stationarity implies that if x_t has an expectation, then this expectation must be independent of t, so that

$$(1) \qquad E(x_t) = \mu,$$

say, for all t. Furthermore, if x_t and x_{t-s} have a covariance, then this covariance can depend only on the relative time lag, s, so that

$$(2) \qquad \mathrm{cov}(x_t, x_{t-s}) = \Gamma_s.$$

The important function Γ_s is known as the *autocovariance function*.

Processes subject only to the restrictions (1) and (2), and not to any other of the restrictions that stationarity implies, are known as *wide-sense stationary processes*. They are important theoretically, but the idea of wide-sense stationarity is important also because in practice one is often content to work with first-order and second-order moments alone, if for no other reason than to keep computation manageable. This survey will be restricted to stationary processes in the strict sense, unless otherwise indicated.

Note that t need not necessarily mean time. One might, for example, be considering variations in thickness along a thread or in vehicle density along a highway; then t would be a spatial coordinate.

Some particular processes. One of the simplest processes of all is a sequence of independent random variables, $\{\epsilon_t\}$. If the ϵ_t have a common distribution, then the process is strictly stationary—this is the kind of sequence often postulated for the "residuals" of a regression or of a model in experimental design. If one requires of $\{\epsilon_t\}$ merely that its elements have constant mean and variance, m and σ^2, and be uncorrelated, then the process is a wide-sense stationary process. From now on $\{\epsilon_t\}$ will denote a process of just this latter type. Often such a process of "residuals" is presumed to have zero mean (that is, $m = 0$); however, this will not always be assumed here.

What is of interest in most series is just that the observations are *not* independent or even uncorrelated. A model such as

$$(3) \qquad x_t = \alpha x_{t-1} + \epsilon_t$$

(a *first-order autoregression*) takes one by a very natural first step from an uncorrelated sequence, $\{\epsilon_t\}$, to an *autocorrelated* sequence, $\{x_t\}$. Here α is a numerical constant whose value may or may not be known, and the term αx_{t-1} introduces a dependence between observations. Such a model is physically plausible in many situations; it might, for example, crudely represent the level of a lake year by year, αx_{t-1} representing the amount of water retained from the previous year and ϵ_t a random inflow. A common type of econometric model is a vector version of (3), in which x_t, ϵ_t are vectors and α is a matrix.

If observations begin at time T, then the series starts with x_T, x_{T+1} is $\alpha x_T + \epsilon_{T+1}$, x_{T+2} is $\alpha^2 x_T + \alpha \epsilon_{T+1} + \epsilon_{T+2}$, and in general, for $t \geq T$,

$$(4) \qquad x_t = \sum_{k=0}^{t-T-1} \alpha^k \epsilon_{t-k} + \alpha^{t-T} x_T.$$

If $|\alpha| < 1$ and the model has been operative from the indefinitely distant past, then one can let T tend to $-\infty$ in (4) and obtain a solution for x_t in terms of the "disturbing variables" ϵ_t:

$$(5) \qquad x_t = \sum_{k=0}^{\infty} \alpha^k \epsilon_{t-k}.$$

The condition $|\alpha| < 1$ is a necessary one if the infinite sum (5) is not to diverge and if model (3) is to be stable. (By "divergence" one understands in this case that the random variable

$$\xi_T = \sum_{k=0}^{t-T-1} \alpha^k \epsilon_{t-k}$$

does not converge in mean square as T tends to $-\infty$; that is, there does not exist a random variable ξ such that $E(\xi_T - \xi)^2 \to 0$.)

The series $\{x_t\}$ generated by (5) is stationary, and one verifies that

(6) $$\mu = E(x_t) = \frac{m}{1-\alpha},$$

(7) $$\Gamma_s = \Gamma_{-s} = \mathrm{cov}(x_t, x_{t-s}) = \frac{\sigma^2 \alpha^s}{1-\alpha^2}, \qquad s \geqslant 0,$$

where m and σ^2 are, respectively, the mean and the variance of ϵ_t. Note from (7) the exponential decay of autocorrelation with lag.

A useful generalization of (3) is the *pth-order autoregression*,

(8) $$\sum_{k=0}^{p} a_k x_{t-k} = \epsilon_t,$$

expressing x_t in terms of its own immediate past and a stationary residual, ϵ_t. When $p = 1$, (8) and (3) are the same except for trivia of notation: a_0 and a_1 in (8) correspond to 1 and $-\alpha$ in (3). When $p > 1$, a process of type (8) can generate the quasi-periodic variations so often seen in time series. Of course, this is not the only model that can generate such quasi-periodic oscillations (one might, for example, consider a nonlinear process or a Markov chain), but it is probably the simplest type of model that does so.

Corresponding to the passage from (3) to (5), process (8) can be given the moving-average representation

(9) $$x_t = \sum_{k=0}^{\infty} b_k \epsilon_{t-k},$$

which represents x_t as a linear superposition of past disturbances. The sequence b_k is the *transient response* of the system to a single unit disturbance.

The relation between the coefficients a_k and b_k can be expressed neatly in generating function form:

(10) $$B(z) = \sum_{k=0}^{\infty} b_k z^k = \frac{1}{A(z)} = \frac{1}{\sum_{k=0}^{p} a_k z^k}.$$

For example, if z is set equal to 0, then $b_0 = 1/a_0$; if the functions are differentiated once and z is again set equal to 0, then $b_1 = -a_1/a_0^2$. (Discussions of time series sooner or later require some knowledge of complex variables. An introduction to the subject is given in MacRobert 1917.)

The necessary and sufficient condition for series (9) to converge to a proper random variable and for the resulting $\{x_t\}$ series to be stationary is that

$A(z)$, considered as a function of a complex variable z, have all its zeros outside the unit circle; this is again a stability condition on the relation (8). If, however, the stability condition is not fulfilled, relations such as (8) can still provide realistic models for some of the nonstationary series encountered in practice.

A relation such as (9) is said to express $\{x_t\}$ as a *moving average* of $\{\epsilon_t\}$. There are, of course, many other important types of process, particularly the general Markov process [*see* MARKOV CHAINS] and the point processes (see Bartlett 1963), but the simple linear processes described in this section are typical of those that are useful for many time series analyses.

Autocovariance function

The autocovariance function, Γ_s, defined in (2) gives a qualitative idea of the decay of statistical dependence in the process with increasing time lag; a more detailed examination of it can tell a good deal about the structure of the process.

A key result is the following: Suppose that $\{x_t\}$ is a moving average of a process $\{y_t\}$,

(11) $$x_t = \sum_k b_k y_{t-k},$$

where the summation is not necessarily restricted to nonnegative values of k, although in most physical applications it will be. Denote the autocovariances of the two processes by $\Gamma_s^{(x)}$ $\Gamma_s^{(y)}$. Then

(12) $$\Gamma_s^{(x)} = \sum_j \sum_k b_j b_k \Gamma_{j-k+s}^{(y)},$$

so that

(13) $$\sum_s \Gamma_s^{(x)} z^s = \left(\sum_j b_j z^j\right)\left(\sum_j b_j z^{-j}\right)\sum_s \Gamma_s^{(y)} z^s,$$

a generating function relation that will be written in the form

(14) $$g_x(z) = B(z)B(z^{-1})g_y(z).$$

If relation (11) defines a stationary process of finite variance, then (14) is valid for $|z| = 1$ at least.

A deduction from (14) and (10) is that for the autoregression (8) the autocovariance generating function is

(15) $$g_x(z) = \frac{\sigma^2}{A(z)A(z^{-1})}.$$

By calculating the coefficient of z^s in the expansion of this function on the circle $|z| = 1$ one obtains Γ_s; for process (3) one obtains the result (7) as before; for the second-order process with $\alpha_0 = 1$ one obtains

(16)

$$\Gamma_s = \frac{\sigma^2}{(1 - \alpha\beta)(\alpha - \beta)}\left(\frac{\alpha^{s+1}}{1 - \alpha^2} - \frac{\beta^{s+1}}{1 - \beta^2}\right), \quad s \geqslant 0,$$

where α^{-1}, β^{-1} are the zeros of $A(z)$. If these zeros are complex, say, $\alpha, \beta = \rho \exp(\pm i\theta)$, then (16) has the oscillatory form

(17) $\Gamma_s = \dfrac{\sigma^2\{\sin[(s+1)\theta] - \rho^4\sin[(s-1)\theta]\}}{(1 - \rho^2)(1 - 2\rho^2\cos 2\theta + \rho^4)\sin\theta}.$

The autocovariance, Γ_s, has a peak near a lag, s, of approximately $2\pi/\theta$, indicating strong positive correlation between values of x_t and x_{t-s} for this value of lag. The nearer the damping factor, ρ, lies to unity, the stronger the correlation. This is an indication of what one might call a quasi-periodicity in the x_t series, of "period" $2\pi/\theta$, the kind of irregular periodicity that produces the "trade cycles" of economic series. Such disturbed periodicities are no less real for not being strict.

Either from (15) or from the fact that

$$\mathrm{cov}\,(x_{t-j}, \epsilon_t) = \begin{cases} \sigma^2, & j = 0, \\ 0, & j > 0, \end{cases}$$

it can be shown that for the autoregression (8)

(18) $\displaystyle\sum_{k=0}^{p} a_k\Gamma_{j-k} = \begin{cases} \sigma^2, & j = 0, \\ 0, & j > 0. \end{cases}$

These are the Yule–Walker relations, which provide a convenient way of calculating the Γ_s from the coefficients, a_k. This procedure will be reversed below, and (18) will be used to estimate the a_k from estimates of the autocovariances.

Spectral theory

Some of the first attempts at time series analysis concerned the prediction of tidal variation of coastal waters, a problem for which it was natural to consider a model of the type

(19) $\begin{aligned} x_t &= \sum A_j \sin(\omega_j t + \alpha_j) + \epsilon_t \\ &= \sum B_j \cos(\omega_j t) + \sum C_j \sin(\omega_j t) + \epsilon_t. \end{aligned}$

That is, the series is represented as the sum of a number of harmonic components and an uncorrelated residual. If the frequencies, ω_j (corresponding to lunar and diurnal variations and so forth), are known, so that the A_j and α_j are to be estimated, then on the basis of an observed series $x_1, x_2, \cdots, x_n$ the least square estimators of the

coefficients B_j and C_j are approximately

$$\hat{B}_j = \frac{2}{n}\sum_{t=1}^{n} x_t \cos(\omega_j t),$$

$$\hat{C}_j = \frac{2}{n}\sum_{t=1}^{n} x_t \sin(\omega_j t).$$

The approximation lies in the use of

$$\sum_{t=1}^{n} P_t^2 \cong \frac{n}{2},$$

$$\sum_{t=1}^{n} P_t Q_t \cong 0,$$

where P_t and Q_t are any two of the functions of time $\cos(\omega_j t)$, $\sin(\omega_j t)$ $(j = 1, 2, \cdots)$. In this approximation, terms of relative order n^{-1} are neglected. The squared amplitude, $A_j^2 = B_j^2 + C_j^2$, is thus estimated approximately by

(20)

$$\hat{A}_j^2 = \frac{4}{n^2}\left\{\left[\sum x_t \cos(\omega_j t)\right]^2 + \left[\sum x_t \sin(\omega_j t)\right]^2\right\}$$

$$= \frac{4}{n^2}\sum_{s=1}^{n}\sum_{t=1}^{n} x_s x_t \cos[\omega_j(s - t)].$$

This can also be written in the form

$$\hat{A}_j^2 = \frac{4}{n^2}\left|\sum_{t=1}^{n} x_t \exp(-i\omega_j t)\right|^2,$$

which is mathematically (although not computationally) convenient.

The importance of $\hat{A}_j^2$ is that it measures the decrease in residual sum of squares (that is, the improvement in fit of model (19)) obtained by fitting terms in $\cos(\omega_j t)$ and $\sin(\omega_j t)$. The larger this quantity, the greater the contribution that the harmonic component of frequency, ω_j, makes to the variation of x_t. For this reason, if the ω_j are unknown, one can search for periodicities (see below) by calculating a quantity analogous to (20) for variable ω: the periodogram

(21)

$$f_n(\omega) = \frac{1}{n}\left|\sum_{t=1}^{n} x_t \exp(-i\omega t)\right|^2$$

$$= \frac{1}{n}\sum_{s=1}^{n}\sum_{t=1}^{n} x_s x_t \cos[\omega(s - t)]$$

$$= \frac{1}{n}\left\{\left[\sum x_t \cos(\omega t)\right]^2 + \left[\sum x_t \sin(\omega t)\right]^2\right\}.$$

An unusually large value of $f_n(\omega)$ at a particular frequency suggests the presence of a harmonic component at that frequency.

It is an empirical fact that few series are of the type (19): in general, one achieves much greater success by fitting structural models such as an autoregressive one. Even for the autoregressive model, however, or, indeed, for any stationary process, an analogue of representation (19) called the spectral representation holds. Here the sum is replaced by an integral. This integral gives an analysis of x_t into different frequency components; for stationary series the amplitudes of different components are uncorrelated. In recent work the spectral representation turns out to be of central importance: the amplitudes of frequency components have simple statistical properties and transform in a particularly simple fashion if the process is subjected to a moving-average transformation (see eq. (26)); the frequency components themselves are often of physical significance.

So even in the general case the periodogram $f_n(\omega)$ provides an empirical measure of the amount of variation in the series around frequency ω. Its expected value for large n, the spectral density function (s.d.f.),

$$\phi(\omega) = \lim_{n\to\infty} E f_n(\omega)$$

$$(22) \quad = \lim_{n\to\infty} \sum_{s=-n}^{n} \left(1 - \frac{|s|}{n}\right) \exp(-i\omega s) E(x_t x_{t-s})$$

$$= \sum_{s=-\infty}^{\infty} E(x_t x_{t-s}) \exp(-i\omega s),$$

provides the corresponding theoretical measure for a given process.

If the x_t have been reduced to zero mean (which, in fact, affects $\phi(\omega)$ only for $\omega = 0$), then the spectral density function becomes

$$(23) \quad \phi(\omega) = \sum_{s=-\infty}^{\infty} \Gamma_s \exp(-i\omega s),$$

and, as can be seen from (13) and (14), this is only a trivial modification of the autocovariance generating function, $g(z)$, already encountered. In fact,

$$(24) \quad \phi(\omega) = g[\exp(-i\omega)].$$

There is a relation reciprocal to (23), the spectral representation of the autocovariance,

$$(25) \quad \Gamma_s = \frac{1}{2\pi} \int_{-\pi}^{\pi} \exp(i\omega s)\, \phi(\omega)\, d\omega.$$

In more general cases $\phi(\omega)$ may fail to exist for certain values, and $\phi(\omega)\, d\omega$ must be replaced by $dF(\omega)$ in (25), where $F(\omega)$ is the nondecreasing spectral distribution function.

An important property of spectral representations is the simplicity of their transformation under moving-average transformations of the process; relation (14) can be rewritten

$$(26) \quad \phi_x(\omega) = B[\exp(i\omega)]B[\exp(-i\omega)]\phi_y(\omega),$$

showing that the effect of the moving-average operation (11) is to scale each frequency component up or down individually, by a factor $|B[\exp(i\omega)]|^2$.

So, if for the autoregression with spectral density function determined by (15) the polynomial $A(z)$ has zeros at $\rho^{-1} \exp(\pm\theta)$, and ρ is near unity, then $\phi_x(\omega)$ will have peaks near $\omega = \pm\theta$, indicating a quasiperiodicity of "period" $2\pi/\theta$.

Note that for an uncorrelated series $\phi(\omega)$ is constant—all frequencies are equally represented on the average. For an autoregressive series $\phi(\omega)$ is variable but finite—this is an example of a process with continuous spectrum. A process of type (19) has a constant background term in ϕ owing to the "noise," ϵ_t, but also has infinite peaks at the values $\omega = \pm\omega_j$, these constituting a line spectrum component.

For a discrete series one need only consider frequencies in the range $-\pi < \omega \leqslant \pi$, since with observations at unit intervals of time the frequencies $2\pi s + \omega$ (s integral) cannot be distinguished one from the other. This is the aliasing effect, which can occasionally confuse an analysis. If, however, the series has little variation of frequency greater than π (that is, of period less than two time units), then the effect is not serious, for the higher frequencies that could cause confusion hardly occur.

Effect of "smoothing" a series—a caution. In order to isolate the "trend" in a series, $\{x_t\}$, it has sometimes been common to derive a smoothed series, $\{\bar{x}_t\}$, by an averaging operation such as

$$(27) \quad \bar{x}_t = \frac{1}{2m+1} \sum_{k=-m}^{m} x_{t-k},$$

although more elaborate and more desirable types of average are often used.

In terms of frequency, the effect of the operation (27) is to multiply the spectral density function by a factor $B[\exp(i\omega)]B[\exp(-i\omega)]$, where

$$(28) \quad B[\exp(i\omega)] = \frac{\sin[(m+\frac{1}{2})\omega]}{(2m+1)\sin\frac{1}{2}\omega}.$$

This function is graphed as the dotted curve (1) in Figure 1 and is known as the gain function of the transformation (27).

Now, if the purpose of "trend extraction" is to eliminate the high frequencies from a series—that is, to act as a "low-pass filter"—then the ideal gain factor would correspond to the square-shouldered solid curve (2) in Figure 1. (A gain factor is sometimes referred to as a "window.") The function (28) obviously departs considerably from this ideal.

To obtain a moving-average transformation

$$(29) \qquad \bar{x}_t = \sum_{k=-\infty}^{\infty} b_k x_{t-k}$$

which acts as a perfect low-pass filter for the range $-\omega_0 < \omega < \omega_0$ one must choose

$$(30) \qquad b_k = \begin{cases} \dfrac{\omega_0}{\pi}, & k = 0, \\[2mm] \dfrac{\sin(\omega_0 k)}{\pi k}, & k \neq 0. \end{cases}$$

The fact that these coefficients decrease rather slowly means that appreciable truncation of the sum will be necessary (probably at a k-value equal to a multiple of $2\pi/\omega_0$), but the resultant operation will still be a considerable improvement over (27). The gain function of a truncated smoothing oper-

ator is illustrated as the dashed curve (3) in Figure 1.

As was first pointed out by Slutsky (1927), injudicious smoothing procedures can actually have the effect of introducing periodicities—just because the gain function of the averaging operation has a peak at a frequency where it should not. One should always be quite clear about the effect of one's "smoothing" operations, and the way to do this is to graph the corresponding gain factor as a function of frequency.

Attempts are sometimes made to "eliminate" trend in a series by the method of *variate difference*—that is, by calculating series such as

$$y_t^{(1)} = \Delta x_t = x_t - x_{t-1},$$
$$y_t^{(2)} = \Delta^2 x_t = x_t - 2x_{t-1} + x_{t-2}.$$

This measure can have a rough success, in that it largely eliminates deviations from stationarity, although a more fundamental approach would be to fit a model which would actually generate the observed nonstationarity, such as an unstable autoregression. In any case, in evaluating the series obtained after differencing, one must remember that the application of a p-fold difference, Δ^p, has the effect of multiplying the spectral density function of a stationary series by $(2 \sin \tfrac{1}{2}\omega)^{2p}$.

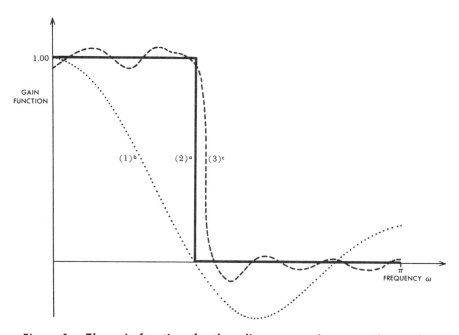

Figure 1 — The gain functions for three linear operations on a time series

a. Curve (2) represents the gain function for the ideal filter passing periods greater than five time units (frequencies less than $2\pi/5$).

b. Curve (1) represents the gain function for a five-term uniform average (formula (27) with $m = 2$).

c. Curve (3) represents the gain function for a finite moving average approximating the ideal filter (the average with weights given by (30) using $\omega_0 = 2\pi/5$, but truncated at $k = \pm 10$).

Sample analogues

Consider now the problem of inference from a sample of n consecutive observations, $x_1, x_2, \cdots, x_n$.

Autocovariance function. Define the uncorrected lagged product-sum,

$$(31) \qquad S_s = \sum_{t=s+1}^{n} x_t x_{t-s} .$$

If $E(x_t) = 0$, then

$$(32) \qquad C_s' = \frac{1}{n-s} S_s$$

certainly provides an unbiased estimate of Γ_s and, under wide conditions, also a consistent one. However, in general the mean will be nonzero and unknown. The sample autocovariance is in such cases naturally measured by

$$(33) \qquad C_s = \frac{1}{n-s} \left[\sum_{t=s+1}^{n} x_t x_{t-s} - \frac{\sum_{t=s+1}^{n} x_t \sum_{t=1}^{n-s} x_t}{n-s} \right],$$

and Ex is estimated by

$$(34) \qquad \bar{x} = \frac{1}{n} \sum_{t=1}^{n} x_t .$$

Minor modifications of (33) will be found in the literature. Expression (33) will in general provide a biased but consistent estimate of Γ_s with a sampling variance of the order of $(n-s)^{-1}$. For a given n the variability of C_s thus increases with s; fortunately, the earlier autocovariances generally contain most of the information.

In order to eliminate problems of scale, investigators sometimes work with the autocorrelation coefficient

$$(35) \qquad r_s = C_s / C_0$$

rather than with C_s, but this is not essential.

Spectral density function. The sample analogue of spectral density function (the periodogram, formula (21)) was introduced before the spectral density function itself. Note from (21) that one can write

$$
(36) \qquad
\begin{aligned}
f_n(\omega) &= \frac{1}{n} \sum_{s=-n}^{n} S_s \cos(\omega s) \\
&= \sum_{s=-n}^{n} \left(1 - \frac{|s|}{n} \right) C_s' \cos(\omega s).
\end{aligned}
$$

If the series has already been corrected for the mean (so that one works with $x_t - \bar{x}$ rather than

x_t), then (36) will become

$$(37) \qquad f_n(\omega) = \sum_{s=-n}^{n} \left(1 - \frac{|s|}{n} \right) C_s \cos(\omega s).$$

Whether one uses formula (36) or formula (37) is not of great consequence. A constant nonzero mean can be regarded as a harmonic component of zero frequency, so the two functions (36) and (37) will differ only near the origin.

The sampling variability of $f_n(\omega)$ does not decrease with increasing n, and $f_n(\omega)$ is not a consistent estimator of $\phi(\omega)$. The problem of finding a consistent estimator will be discussed below.

Fitting and testing autoregressive models

The autoregressive model (8) is a useful trial model, since it usually explains much of the variation and often has some physical foundation. Furthermore, its test theory is typical of a much more general case. The first problem is that of the actual fitting, the estimation of the parameters a_k and σ^2; the second problem is that of testing the fit of the model.

If the ϵ_t and x_t have means 0 and μ, respectively, then the model (8) must be modified slightly to

$$(38) \qquad \sum_{k=0}^{p} a_k (x_t - \mu) = \epsilon_t .$$

One usually assumes the ϵ_t normally distributed—not such a restrictive assumption as it appears. To a first approximation the means and variances of autocorrelations are unaffected by nonnormality (see Whittle 1954, p. 210), and estimates of parameters such as the autoregressive coefficients, a_k, should be similarly robust. For normal processes the log-likelihood of the sample $x_1, x_2, \cdots, x_n$ is, for large n,

(39)

$$L = \text{const.} - \frac{n}{2} \log \sigma^2 - \frac{1}{2\sigma^2} \sum_{t=1}^{n} \left[\sum_{k} a_k (x_{t-k} - \mu) \right]^2 .$$

Maximizing this expression with respect to μ, one obtains the estimator

$$(40) \qquad \hat{\mu} = \frac{\sum_t \sum_k a_k x_{t-k}}{n \sum_k a_k} \cong \frac{1}{n} \sum_{t=1}^{n} x_t .$$

The second approximate equality follows if one neglects the difference between the various averages $(1/n) \sum_{t=1}^{n} x_{t-k}$ ($k = 0, 1, 2, \cdots, p$), that is, if, as is often done, an end effect is neglected. Thus, the maximum likelihood estimator of μ is approximately the usual sample arithmetic mean, despite the dependence between observations. Inserting this estimator in (39), one finds

(41)
$$L \cong \text{const.} - \frac{n}{2}\left(\log \sigma^2 + \frac{1}{\sigma^2}\sum_j \sum_k a_j a_k C_{j-k}\right),$$

so that the maximum likelihood estimators of the remaining parameters are determined approximately by the relations

(42) $$\sum_{k=0}^{p} \hat{a}_k C_{j-k} = 0, \qquad j = 1, 2, \cdots, p,$$

(43) $$\hat{\sigma}^2 = \frac{1}{n}\sum\sum \hat{a}_j \hat{a}_k C_{j-k}$$
$$= \frac{1}{n}\sum_{k=0}^{p} \hat{a}_k C_k.$$

Note the analogue between (42) and (43) and the Yule–Walker relations (18).

To test prescribed values of the a_k one can use the fact that the estimators $\hat{a}_k$ are asymptotically normally distributed with means a_k (respectively) and a covariance matrix

(44) $$[\text{cov}(\hat{a}_j, \hat{a}_k)] = \frac{\sigma^2}{n}[\Gamma_{j-k}]^{-1}.$$

(Here $[\alpha_{jk}]$ denotes a $p \times p$ matrix with typical element α_{jk}, $j, k = 1, 2, \cdots, p$.) This result holds if the ϵ_t are independently and identically distributed, with a finite fourth moment. (See Whittle 1954, p. 214.)

However, a more satisfactory and more versatile approach to the testing problem is provided by use of the Wilks λ-ratio. This will be described in a more general setting below; for the present, note the following uses.

To test whether a given set of coefficients a_1, $a_2, \cdots, a_p$ are zero, treat

(45) $$\psi^2 = (n - p)\log\left(\frac{\sum_{j=0}^{p}\sum_{k=0}^{p} a_j a_k C_{j-k}}{\hat{\sigma}_p^2}\right)$$

as a χ^2 variable with p degrees of freedom (df). Here $\hat{\sigma}_p^2$ has been used to denote the estimator (43), emphasizing the order p assumed for the autoregression.

To test whether an autoregression of order p gives essentially as good a fit as one of order $p + q$, treat

(46) $$\psi^2 = (n - p - q)\log\left(\frac{\hat{\sigma}_p^2}{\hat{\sigma}_{p+q}^2}\right)$$

as a χ^2 variable with q df. In both cases large values of the test statistic are critical.

Fitting and testing more general models

The approximate expression (41) for the log-likelihood (maximized with respect to the mean) can be generalized to any process for which the reciprocal of the spectral density function can be expanded in a Fourier series,

(47) $$\phi(\omega)^{-1} = \sigma^{-2}\sum_{s=-\infty}^{\infty}\gamma_s \exp(i\omega s).$$

The generalized expression is

(48) $$L \cong \text{const.} - \frac{n}{2}\left(\log\sigma^2 + \frac{1}{\sigma^2}\sum_{s=-\infty}^{\infty}\gamma_s C_s\right),$$

where σ^2 is the "prediction variance," the conditional variance of x_t given the values of x_{t-1}, $x_{t-2}, \cdots$. (See Whittle 1954.)

The sum in (48) cannot really be taken as infinite; in most practical cases the coefficients γ_s converge reasonably fast to zero as s increases, and the sum can be truncated.

Another way of writing (48) is

(49)
$$L \cong \text{const.} - \frac{n}{4\pi}\int_{-\pi}^{\pi}\left[\log\phi(\omega) + \frac{f_n(\omega)}{\phi(\omega)}\right]d\omega.$$

In general it will be easier to calculate the sum over autocovariances in (48) than to calculate the integral over the periodogram in (49), but sometimes the second approach is taken.

If the model depends on a number of parameters, $\theta_1, \theta_2, \cdots, \theta_p$ (of which σ^2 will usually be one), then $\phi(\omega)$ will also depend on these, and the maximum likelihood estimators, $\hat{\theta}_j$, are obtained approximately by maximizing either of the expressions (48) and (49). The covariance matrix of the estimators is given asymptotically by

(50)
$$[\text{cov}(\hat{\theta}_j, \hat{\theta}_k)] \cong \frac{2}{n}\left[\frac{1}{2\pi}\int_{-\pi}^{\pi}\frac{\partial\log\phi}{\partial\theta_j}\frac{\partial\log\phi}{\partial\theta_k}d\omega\right]^{-1}.$$

(See Whittle 1954.) Thus, for the moving-average process

(51) $$x_t = \epsilon_t - \beta\epsilon_{t-1},$$

with $|\beta| < 1$, one finds that

(52) $$\sum \gamma_s C_s = \frac{1}{1-\beta^2}\sum_{s=-\infty}^{\infty}\beta^{|s|}C_s.$$

The maximum likelihood estimator of β is obtained by minimizing (52), and expression (52) with $\hat{\beta}$

substituted for β provides the maximum likelihood estimator of $\sigma^2 = \text{var}(\epsilon)$. One finds from (50) that the two estimators are asymptotically uncorrelated, with

$$(53) \qquad \text{var}(\hat{\beta}) \cong \frac{1 - \beta^2}{n}, \qquad \text{var}(\hat{\sigma}^2) \cong \frac{2\sigma^4}{n}.$$

Practical techniques for the calculation of the maximum likelihood estimators in more general cases have been worked out by Durbin (1959) and Walker (1962).

Tests of fit can be based upon the λ-ratio criterion. Let $\hat{\sigma}_p^2$ denote the maximum likelihood estimator of σ^2 when parameters $\theta_1, \theta_2, \cdots, \theta_p$ (one of these being σ^2 itself) are fitted and the values of parameters $\theta_{p+1}, \theta_{p+2}, \cdots, \theta_{p+q}$ are prescribed. Thus, $\hat{\sigma}_{p+q}^2$ will be the maximum likelihood estimator of σ^2 when all $p + q$ parameters are fitted. A test of the prescribed values of $\theta_{p+1}, \cdots, \theta_{p+q}$ is obtained by treating

$$(54) \qquad \psi^2 = (n - p - q)\log\left(\frac{\hat{\sigma}_p^2}{\hat{\sigma}_{p+q}^2}\right)$$

as a χ^2 variable with q df.

Multivariate processes

In few realistic analyses is one concerned with a single variable; in general, one has several, so that $\boldsymbol{x}_t$ must be considered a vector of m jointly stationary variables, $(x_{1t}, x_{2t}, \cdots, x_{mt})$.

There is a generalization (Whittle 1954) of expression (49) for the log-likelihood in such cases, but the only case considered here is that of a multivariate autoregression,

$$(55) \qquad \sum_{k=0}^{p} \boldsymbol{a}_k(\boldsymbol{x}_{t-k} - \boldsymbol{\mu}) = \boldsymbol{\varepsilon}_t,$$

where the $\boldsymbol{a}_k$ are $m \times m$ matrices with $\boldsymbol{a}_0 = \boldsymbol{I}$, and

$$(56) \qquad \begin{aligned} E(\boldsymbol{\varepsilon}_t) &= \boldsymbol{0}, \\ E(\boldsymbol{\varepsilon}_s\boldsymbol{\varepsilon}_t') &= \begin{cases} \boldsymbol{V}, & s = t, \\ \boldsymbol{0}, & s \neq t. \end{cases} \end{aligned}$$

This last assumption states that the vector residuals are mutually uncorrelated but that the covariance matrix of a single vector residual is $\boldsymbol{V}$. As before, the maximum likelihood estimator of the mean vector, $\boldsymbol{\mu}$, is approximately $\bar{\boldsymbol{x}}$, and with this inserted the following generalization of (41) results:

$$(57)$$
$$L \cong \text{const.} - \frac{n}{2}[\log|\boldsymbol{V}| + \sum_j\sum_k \text{tr}\,(\boldsymbol{a}_j'\boldsymbol{V}^{-1}\boldsymbol{a}_k\boldsymbol{C}_{j-k})].$$

Here $\boldsymbol{C}_s$ is the $m \times m$ matrix whose (jk)th element is the sample covariance of x_{jt} and $x_{k,t-s}$, that is, the (jk)th element of $\boldsymbol{C}_s$ is

$$(58)$$
$$C_{sjk} = \frac{1}{n-s}\left[\sum_{t=s+1}^{n} x_{jt}x_{k,t-s} - \frac{\left(\sum_{t=s+1}^{n} x_{jt}\right)\left(\sum_{t=1}^{n-s} x_{kt}\right)}{n-s}\right],$$

and $\text{tr}\,\boldsymbol{A}$ denotes the sum of the diagonal elements of a matrix $\boldsymbol{A}$. The maximum likelihood estimators are given by

$$(59) \qquad \sum_{k=0}^{p} \hat{\boldsymbol{a}}_k\boldsymbol{C}_{j-k} = \boldsymbol{0}, \qquad j = 1, 2, \cdots, p,$$

$$(60) \qquad \hat{\boldsymbol{V}} = \sum_k \hat{\boldsymbol{a}}_k\boldsymbol{C}_{-k}.$$

For the important case $p = 1$, which can be written

$$(61) \qquad \boldsymbol{x}_t = \boldsymbol{\alpha}\boldsymbol{x}_{t-1} + \boldsymbol{\varepsilon}_t,$$

these become

$$(62) \qquad \hat{\boldsymbol{\alpha}} = \boldsymbol{C}_1\boldsymbol{C}_0^{-1},$$

$$(63) \qquad \hat{\boldsymbol{V}} = \boldsymbol{C}_0 - \boldsymbol{C}_1\boldsymbol{C}_0^{-1}\boldsymbol{C}_{-1}.$$

Estimator (62) will, of course, be modified if certain elements of $\boldsymbol{\alpha}$ are known and need not be estimated. Tests of fit can be based on the λ-ratio criterion as before, with expression (57) used for the log-likelihood.

In econometric work, models of type (61) are particularly important. One minor complication is that exogenous variables may also occur in the right-hand side (see (73), below); exogenous variables are variables which are regarded as external to the system and which need not be explained—for example, in a model of a national economy, variables such as overseas prices and technical progress might be regarded as exogenous. A much more severe complication is that of simultaneity: that $\boldsymbol{x}_t$ may be represented as a regression upon some of its own elements as well as upon $\boldsymbol{x}_{t-1}$ and exogenous variables. This latter difficulty has led to an extensive literature, which will not be discussed here (for a general reference, see Johnston 1963).

Regression. An important special type of multivariate process is the regression

$$(64) \qquad x_t = \sum_{j=1}^{r} \beta_j u_{jt} + \eta_t,$$

where x_t (now assumed to be scalar) is regarded as linearly dependent upon a number of variables, u_{jt}, with a stationary residual, η_t (in general autocorrelated). The processes $\{u_{jt}\}$ may be other stationary processes, even lagged versions of the $\{x_t\}$ process itself, or deterministic sequences such as t^ν or $\sin(\omega t)$.

Simple and unbiased estimators of the β_j are the least square estimators, b_j, obtained by minimizing $\sum_{t=1}^{n} \eta_t^2$ and determined by the linear equations

(65) $\displaystyle\sum_{k=1}^{r} b_k(u_{jt}, u_{kt}) = (x_t, u_{jt}), \quad j = 1, 2, \cdots, r,$

where the notation

(66) $\displaystyle (x_t, y_t) = \sum_{t=1}^{n} x_t y_t$

has been used for the simple product-sum. The covariance matrix of these estimators is

(67)

$$\boldsymbol{V}_b = [(u_{jt}, u_{kt})]^{-1} \left[\sum_{s=1}^{n} \sum_{t=1}^{n} u_{js} u_{kt} \Gamma_{s-t} \right] [(u_{jt}, u_{kt})]^{-1},$$

where $[a_{jk}]$ indicates an $r \times r$ matrix with typical element a_{jk} and Γ_s the autocovariance function of $\{\eta_t\}$. If the processes $\{u_{jt}\}$ are jointly stationary, then one can write

(68) $\displaystyle (u_{jt}, u_{k, t\,s}) \cong \frac{n}{2\pi} \int_{-\pi}^{\pi} \exp(i\omega s)\, dM_{jk}(\omega),$

and (67) can be rewritten

(69) $\displaystyle \boldsymbol{V}_b \cong \frac{2\pi}{n} \boldsymbol{M}^{-1} \left[\int_{-\pi}^{\pi} \phi(\omega)\, d\boldsymbol{M}(\omega) \right] \boldsymbol{M}^{-1},$

where $\boldsymbol{M}(\omega) = [M_{jk}(\omega)]$, $\boldsymbol{M} = \boldsymbol{M}(\pi) - \boldsymbol{M}(-\pi)$, and $\phi(\omega)$ is the spectral density function of $\{\eta_t\}$.

This principle can be extended even to mildly nonstationary processes, such as polynomials in t (Grenander & Rosenblatt 1957).

The maximum likelihood estimators, $\hat{\beta}_j$, will in general have smaller variances; these estimators are obtained by maximizing the log-likelihood,

(70) $\displaystyle L \cong \text{const.} - \frac{n}{2} \left[\log \sigma^2 + \frac{1}{\sigma^2} \sum_s \gamma_s \left(x_t - \sum \beta_j u_{jt}, \, x_{t-s} - \sum \beta_j u_{j, t-s} \right) \right].$

They obey the equation system

(71) $\displaystyle \sum_s \sum_k \gamma_s \hat{\beta}_k (u_{kt}, u_{j, t-s}) = \sum_s \gamma_s (x_t, u_{j, t-s}),$

$$j = 1, 2, \cdots, r.$$

Their covariance matrix is

(72) $\displaystyle \boldsymbol{V}_{\hat{\beta}} \cong \frac{2\pi}{n} \left[\int_{-\pi}^{\pi} \phi(\omega)^{-1} d\boldsymbol{M}(\omega) \right]^{-1}.$

If $\phi(\omega)$ has the same value at all ω's for which any of the $M_{jk}(\omega)$ change value, then the variances of the least square estimators will be asymptotically equal to those of the maximum likelihood estimators, and, indeed, the two sets of estimators may themselves be asymptotically equal. For example, (40) above shows that the maximum likelihood estimator of a mean, μ, reduces asymptotically to the least square estimator $\bar{x}$, a great simplification.

If the residual spectral density function, $\phi(\omega)$, involves unknown parameters, then expression (70) must be maximized with respect to these as well as to the β_j. This maximization can be complicated, and a simpler way of allowing for some autocorrelation in $\{\eta_t\}$ is to fit, instead of (64), a model with some autoregression and with uncorrelated residuals,

(73) $\displaystyle \sum_{k=0}^{p} a_k x_{t-k} = \sum_{j=1}^{r} \beta_j^* u_{jt} + \epsilon_t.$

The β_j^* of (73) cannot be identified with the β_j of (64), but they do indicate to what extent the u_{jt} can explain x_t variation.

Spectral (periodogram) analysis

In fitting a parametric model one obtains an estimate of the spectral density function, $\phi(\omega)$, but one often wishes to obtain a direct estimate without the assumptions implied in a model, just as one uses C_s to estimate the autocovariance function, Γ_s.

The periodogram ordinate, $f_n(\omega)$, cannot be used to this end, for although

(74) $\displaystyle E f_n(\omega) \cong \phi(\omega),$

one has also, for normal processes,

(75) $\displaystyle \text{var} f_n(\omega) \cong \phi^2(\omega).$

This variance does not tend to zero with increasing n, and $f_n(\omega)$ is not a consistent estimator of $\phi(\omega)$. That explains the very irregular appearance of the periodogram when it is graphed and the reason one

conceives the idea of estimating $\phi(\omega)$ by smoothing the periodogram.

What one can say concerning distributions is that the quantities

(76) $\displaystyle I_j = \frac{f_n(2\pi j/n)}{\phi(2\pi j/n)}$

are approximately independent standard exponential variables for $j = 1, 2, \cdots, N$, where $N < n/2$. That is, the I_j have approximately a joint probability density function $\exp(-\sum_{j=1}^{N} I_j)$.

Suppose now that one attempts to estimate $\phi(\omega)$ at $\omega = \lambda$ by an estimator of the form

(77)
$$\hat{\phi}(\lambda) = \frac{1}{2\pi} \int_{-\pi}^{\pi} K(\lambda - \omega) f_n(\omega)\, d\omega$$
$$= \sum_{s=-\infty}^{\infty} k_s \Gamma_s \cos(\omega s),$$

where $K(\omega)$ is a symmetric function of period 2π represented by a Fourier series,

$$(78) \qquad K(\omega) = \sum_s k_s \exp(i\omega s).$$

From (76) it follows that

$$(79) \qquad E\hat{\phi}(\lambda) \cong \frac{1}{2\pi} \int K(\lambda - \omega)\phi(\omega)\, d\omega,$$

$$(80) \qquad \operatorname{var} \hat{\phi}(\lambda) \cong \frac{1}{n\pi} \int K^2(\lambda - \omega)\phi^2(\omega)\, d\omega.$$

The Fourier series $K(\omega)$ will be chosen as a function with a peak at the origin; as this peak grows sharper, the bias of $\hat{\phi}(\lambda)$ (determined from (79)) decreases, but the variance (determined by (80)) increases. The best choice will be a compromise between these two considerations. The question of the optimal choice of weight function has been much studied. The choice is partly a matter of convenience, depending upon whether or not one works from the periodogram, that is, upon which of the two formulas (77) is used. If one is calculating digitally, then a simple and useful smoothing formula is Bartlett's, for which

$$(81) \qquad k_s = \begin{cases} 1 - \dfrac{|s|}{m}, & |s| \leqslant m, \\ 0, & |s| > m, \end{cases}$$

$$(82) \qquad K(\omega) = \frac{1}{m}\left[\frac{\sin(\frac{1}{2}m\omega)}{\sin\frac{1}{2}\omega}\right]^2.$$

To test whether a strong peak in the periodogram indicates the presence of a harmonic component (when a delta-function would be superimposed on the spectral density function), one can use the statistic

$$(83) \qquad g = \frac{\max I_j}{\sum\limits_1^N I_j},$$

for which

$$(84) \qquad p(g > u) = \sum_j \frac{N!\,(-1)^{j-1}}{(N-j)!\,j!}\,(1 - ju)^{N-1}.$$

The sum in (84) is taken for all values of j not greater than $1/u$. In constructing the I_j one can use a formula of type (77) to estimate $\phi(\omega)$, although this leads to some underestimate of the relative size of a peak.

A spectral analysis of a multivariate process can lead to some interesting types of investigation, which can only be mentioned briefly here. If

$$(85) \qquad \phi_{xy}(\omega) = \sum_s E(x_t y_{t-s}) \exp(-i\omega s),$$

then for a bivariate stationary process $\{x_t, y_t\}$ the idea of a spectral density function is replaced by that of a spectral density matrix

$$(86) \qquad \mathbf{\Phi}(\omega) = \begin{bmatrix} \phi_{xx}(\omega) & \phi_{xy}(\omega) \\ \phi_{yx}(\omega) & \phi_{yy}(\omega) \end{bmatrix}.$$

Suppose one wishes to investigate the dependence of $\{y_t\}$ upon $\{x_t\}$. As an alternative to low-lag linear models such as

$$(87) \qquad \sum_{k=0}^p a_k y_{t-k} = \sum_{k=0}^q b_k x_{t-k} + \epsilon_t$$

one can apply multivariate techniques to the Fourier components of the processes, at individual values of ω. Thus, $\phi_{yy}(\omega)$ is the variance of the Fourier component of $\{y_t\}$ at frequency ω; one can take a linear regression of this component onto the corresponding Fourier component of $\{x_t\}$ and find a "residual variance,"

$$(88)$$
$$\phi_{yy}(\omega)\left[1 - \frac{\phi_{xy}(\omega)\phi_{yx}(\omega)}{\phi_{xx}(\omega)\phi_{yy}(\omega)}\right] = \phi_{yy}(\omega)[1 - |C_{xy}(\omega)|^2].$$

The quantity $C_{xy}(\omega)$ is the type of correlation coefficient known as the coherency; if it approaches unity in magnitude, then the frequency components at ω of the two series are closely related. It may well happen, for example, that low frequency components of two series keep well in step, although the short-term variation corresponding to high frequency components may be almost uncorrelated between the two series.

Note that (88) is the spectral density function of the process

$$\{\eta_t\} = \{y_t - \sum_{k=-\infty}^{\infty} b_k x_{t-k}\},$$

where the b's are chosen to minimize $\operatorname{var}(\eta)$.

In practice the elements of the spectral density matrix must be estimated by formulas analogous to (77).

A sometimes illuminating alternative to a periodogram analysis is to decompose a series into components corresponding to different frequency bands by using a number of band-pass operators of type (29) and to examine these components individually.

P. WHITTLE

[See also LINEAR HYPOTHESES, *article on* REGRESSION; MARKOV CHAINS.]

BIBLIOGRAPHY

¹*There are no completely satisfactory texts, principally because virtually nobody has a satisfactorily broad grasp of both the theory and the application of time series analysis. A person with a mathematical background may do*

best to read the first four chapters of Grenander & Rosenblatt 1957 *and to follow this with a study of* Hannan 1960. *For someone with less mathematics, Chapters 29 and 30 of* Kendall & Stuart 1966 *provide an introduction that, although not deep, is nevertheless sound and well illustrated.* Jenkins 1965 *surveys some of the more recent work on spectral analysis.*

BARTLETT, M. S. 1963 The Spectral Analysis of Point Processes. *Journal of the Royal Statistical Society* Series B 25:264–296.

DURBIN, J. 1959 Efficient Estimation of Parameters in Moving-average Models. *Biometrika* 46:306–316.

GRANGER, CLIVE W. J.; and HATANAKA, M. 1964 *Spectral Analysis of Economic Time Series.* Princeton Univ. Press.

GRENANDER, ULF; and ROSENBLATT, MURRAY 1957 *Statistical Analysis of Stationary Time Series.* New York: Wiley.

HANNAN, EDWARD J. (1960) 1962 *Time Series Analysis.* London: Methuen.

INTERNATIONAL STATISTICAL INSTITUTE 1965 *Bibliography on Time Series and Stochastic Processes.* Edited by Herman Wold. Edinburgh: Oliver & Boyd.

JENKINS, G. M. 1965 A Survey of Spectral Analysis. *Applied Statistics* 14:2–32.

JOHNSTON, JOHN 1963 *Econometric Methods.* New York: McGraw-Hill.

KENDALL, MAURICE G.; and STUART, ALAN (1966) 1976 *The Advanced Theory of Statistics.* Volume 3: *Design and Analysis, and Time-series.* 3d ed. London: Griffin; New York: Hafner. → The coverage of time series in Chapters 29 and 30 of the first edition (1966) was expanded to Chapters 45–50 of the third edition (1976).

MACROBERT, THOMAS M. (1917) 1947 *Functions of a Complex Variable.* 3d ed. New York: Macmillan.

SLUTSKY, EUGEN E. (1927) 1937 The Summation of Random Causes as the Source of Cyclic Processes. *Econometrica* 5:105–146. → First published in Russian. Reprinted in 1960 in Slutsky's *Izbrannye trudy.*

SYMPOSIUM ON TIME SERIES ANALYSIS, BROWN UNIVERSITY, *1962* 1963 *Proceedings.* Edited by Murray Rosenblatt. New York: Wiley.

WALKER, A. M. 1962 Large-sample Estimation of Parameters for Autoregressive Processes With Moving-average Residuals. *Biometrika* 49:117–131.

WHITTLE, P. 1953 The Analysis of Multiple Stationary Time Series. *Journal of the Royal Statistical Society* Series B 15:125–139.

WHITTLE, P. 1954 Appendix. In Herman Wold, *A Study in the Analysis of Stationary Time Series.* 2d ed. Stockholm: Almqvist & Wiksell.

WHITTLE, P. 1963 *Prediction and Regulation by Linear Least-square Methods.* London: English Universities Press.

WOLD, HERMAN (1938) 1954 *A Study in the Analysis of Stationary Time Series.* 2d ed. With an Appendix by P. Whittle. Stockholm: Almqvist & Wiksell.

Postscript

A trend toward increasing use of spectral methods—typified by Jenkins and Watts (1968), Koopmans (1974), Brillinger (1975), and Bloomfield (1976)—has been followed by a return to the fitting of finite parameter models in the time domain. The new tools for spectral analysis have been the computational use of the fast Fourier transform (see Chatfield 1975, p. 145, or any of the texts cited above) and the development of appropriate smoothing filters (see Koopmans 1974; Brillinger 1975; Bloomfield 1976).

The return to favor of finite parameter models, and to the direct fitting of linear relations (auto- and cross-regressions) in the time domain, has been influenced by the suitability of spectral methods only for long samples from strictly stationary processes, and by the consistent influence exerted by Box and Jenkins in favor of simple time-domain models. The Box–Jenkins program is distinguished for coherency rather than originality or elegance; it takes a simple but integrated approach to the estimation, testing, forecasting, and control of dynamic systems, supplemented by an extensive set of computer programs.

In their treatment of a single time series, Box and Jenkins typically difference the series until it is deemed stationary, and then represent its behavior by a hybrid autoregressive–moving-average (ARMA) model:

$$(89) \qquad \sum_{s=0}^{p} a_s x_{t-s} = \sum_{s=0}^{q} b_s \epsilon_{t-s}.$$

For given values of the coefficients a and b one uses equation (89) to calculate the residuals ϵ_t recursively from the observations x_t; the sum of squares $\sum_1^n \epsilon_t^2$ then supplies a direct evaluation of the quantity $n\sum \gamma_s C_s$ occurring in expression (48). This is then minimized with respect to a and b, either by direct search over a grid of evaluations in parameter space or by hill-climbing methods.

The contrast between the two approaches is typified in treatments of the important problem last referred to in the main article, the determination of the form of dependence of an "output" variable, y_t, upon an "input" variable, x_t. The idea of regressing spectral components of y on those of x, associated with equation (88), can be made into a workable procedure, once one has obtained direct estimates of the spectral density matrix $\phi(\omega)$ using appropriate smoothing filters on the data. In contrast, one can use Box–Jenkins techniques to fit a temporal relation such as equation (87) directly. For a clear discussion of this specific topic, see Chatfield (1975, pp. 213–222).

P. WHITTLE

ADDITIONAL BIBLIOGRAPHY
[1]*The interested reader is very much better served for texts than formerly, even though most continue to be addressed to specialists in the field.* Chatfield 1975, *an excellent text,*

is simple and explicit, yet both inclusive and up to date. Jenkins & Watts 1968, Bloomfield 1976, Brillinger 1975, and Koopmans 1974 are wholly or predominantly devoted to spectral analysis. All give good accounts of this area, but Brillinger's treatment is perhaps somewhat more directed to the specialist. The authors who have concentrated most consistently upon the fitting of finite parameter models in the time domain are undoubtedly Box and Jenkins. Because they have maintained this approach so consistently, and have produced a series of definite recommendations and techniques, they have achieved quite a following. Their approach is presented at length in Box & Jenkins 1970. Hannan 1970 is encyclopedic, and indispensable for the specialist, but more suitable as a reference than as an introduction. Cox & Lewis 1966 still has its niche as the outstanding text on the statistical analysis of point processes.

BLOOMFIELD, PETER 1976 *Fourier Analysis of Time Series: An Introduction.* New York: Wiley.

BOX, GEORGE E. P.; and JENKINS, G. M. (1970) 1976 *Time Series Analysis: Forecasting and Control.* Rev. ed. San Francisco: Holden-Day.

BRILLINGER, DAVID R. 1975 *Time Series: Data Analysis and Theory.* New York: Holt.

CHATFIELD, CHRISTOPHER 1975 *The Analysis of Time Series: Theory and Practice.* London: Chapman & Hall.

COX, D. R.; and LEWIS, P. A. W. 1966 *The Statistical Analysis of Series of Events.* London: Methuen; New York: Wiley.

HANNAN, EDWARD J. 1970 *Multiple Time Series.* New York: Wiley.

JENKINS, G. M.; and WATTS, DONALD G. 1968 *Spectral Analysis and Its Applications.* San Francisco: Holden-Day.

KOOPMANS, LAMBERT H. 1974 *The Spectral Analysis of Time Series.* New York: Academic Press.

III

CYCLES

Cycles, waves, pulsations, rhythmic phenomena, regularity in return, periodicity—these notions reflect a broad category of natural, human, and social phenomena where cycles are the dominating feature. The daily and yearly cycles in sunlight, temperature, and other geophysical phenomena are among the simplest and most obvious instances. Regular periodicity provides a basis for prediction and for extracting other useful information about the observed phenomena. Nautical almanacs with their tidal forecasts are a typical example. Medical examples are pulse rate as an indicator of cardiovascular status and the electrocardiograph as a basis for analysis of the condition of the heart.

The study of cyclic phenomena dates from prehistoric times, and so does the experience that the area has dangerous pitfalls. From the dawn of Chinese history comes the story that the astronomers Hi and Ho lost their heads because they failed to forecast a solar eclipse (perhaps 2137 B.C.). In 1929, after some twelve years of promising existence, the Harvard Business Barometer (or Business Index) disappeared because it failed to predict the precipitous drop in the New York stock market.

Cyclic phenomena are recorded in terms of time series. A key aspect of cycles is the *degree of predictability* they give to the time series generated. Three basic situations should be distinguished:

(*a*) The cycles are fixed, so that the series is predictable over the indefinite future.

(*b*) The cycles are partly random, so that the series is predictable only over a limited future.

(*c*) The cycles are spurious—that is, there are no real cycles—and the series is not predictable.

For the purposes of this article the term "cycle" is used in a somewhat broader sense than the strict cyclic periodicity of case (*a*).

Limited and unlimited predictability

The fundamental difference between situations (*a*) and (*b*) can be illustrated by two simple cases.

The scheme of "hidden periodicities." Suppose that an observed time series is generated by two components. The first is strictly periodic, with period length p, so that its value at time $t + p$ is equal to its value at time t. The second component, superimposed upon the first, is a sequence of random (independent, identically distributed) elements. Thus, each term of the observed series can be represented as the sum of a periodic term and a random one.

Tidal water is a cyclic phenomenon where this model applies quite well (see Figure 1). Here the observed series is the measured water level at Dover, the strictly periodic component represents the lunar cycle, 12 hours and 50 minutes in length (two maxima in one lunar day), and the random elements are the irregular deviations caused by storms, random variations in air pressure, earthquakes, etc.

The periodic component provides a prediction—an unbiased predicted value for a future time with expectation equal to that future value of the periodic component, and with prediction error equal to the random element. The difficulty is that the periodic component is not known and must be estimated empirically. A simple and obvious method is that of Buys Ballot's table; each point on the periodic component is estimated by the average of several points on the observed series, separated in time by the length of the period, p, where p either is known or is assessed by trial and error. The larger is the residual as compared to the cyclic component, the longer is the series needed to estimate with confidence the cyclic component.

The approach of hidden periodicities may be extended, with two or more periodic components being considered. Tidal water again provides a typical illustration. In addition to the dominating lunar component, a closer fit to the data is obtained by considering a solar component with period 183 days.

In view of its simplicity and its many important applications, it is only natural that the approach involving strictly periodic components is of long standing. A distinction must be made, however, between formal representation of a series (which is always possible), on the one hand, and prediction, on the other. Under general conditions, any series, even a completely random one, can be represented by a sum of periodic components plus a residual, and if the number of periodic components is increased indefinitely, the residual can be made as small as desired. In particular, if each of the periodic components is a sine or a cosine curve (a sinusoid), then the representation of the observed series is called a spectral representation. Such a representation, it is well to note, may be of only limited use for prediction outside the observed range, because if the observed range is widened, the terms of the representation may change appreciably. In the extreme case when the observations are all stochastically independent, the spectral representation of the series is an infinite sum of sinusoids; in this case neither the spectral representation nor alternative forecasting devices provide any predictive information.

Irregular cycles. Until rather recently (about 1930), the analysis of oscillatory time series was almost equivalent to the assessment of periodicities. For a long time, however, it had been clear that important phenomena existed that refused to adhere to the forecasts based on the scheme of hidden periodicities. The most obvious and challenging of these was the sequence of some twenty business cycles, each of duration five to ten years, between 1800 and 1914. Phenomena with irregular cycles require radically different methods of analysis.

The scheme of "disturbed periodicity." The breakthrough in the area of limited predictability came with Yule's model (1927) for the irregular 11-year cycle of sunspot intensity (see Figure 2). Yule interpreted the sunspot cycle as similar to the movement of a damped pendulum that is kept in motion by an unending stream of random shocks. [*See the biography of* YULE.]

The sharp contrast between the scheme of hidden periodicities and the scheme of disturbed periodicity can now be seen. In the hidden periodicities model the random elements are superimposed upon the cyclic component(s) without affecting or disturbing their strict periodicity. In Yule's model the series may be regarded as generated by the random elements, and there is no room for strict periodicity. (Of course, the two types can be combined, as will be seen.)

The deep difference between the two types of

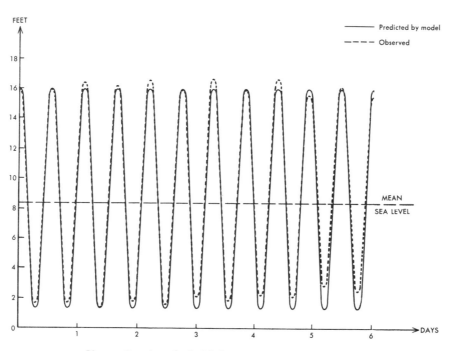

Figure 1 — *Level of tidal water at Dover, 6 days**

* Hypothetical data.

model is reflected in their forecasting properties (see Figure 3). The time scales for the two forecasts have here been adjusted so as to give the same period. In the hidden-periodicities model the forecast over the future time span has the form of an undamped sinusoid, thus permitting an effective forecast over indefinitely long spans when the model is correct. In Yule's model the forecast is a damped sinusoid, which provides effective information over limited spans, but beyond that it gives only the trivial forecast that the value of the series is expected to equal the unconditional over-all mean of the series.

Generalizations. The distinction between limited and unlimited predictability of an observed times series goes to the core of the probability structure of the series.

In the modern development of time series analysis on the basis of the theory of stochastic processes, the notions of predictability are brought to full significance. It can be shown that the series y_t under very general conditions allows a unique

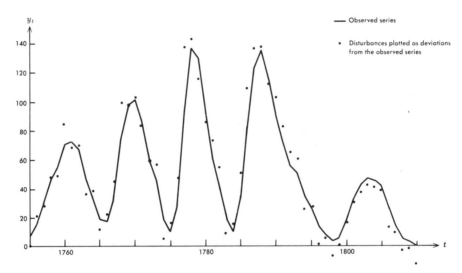

Figure 2 — Sunspot intensity

Source: Adapted from Yule 1927.

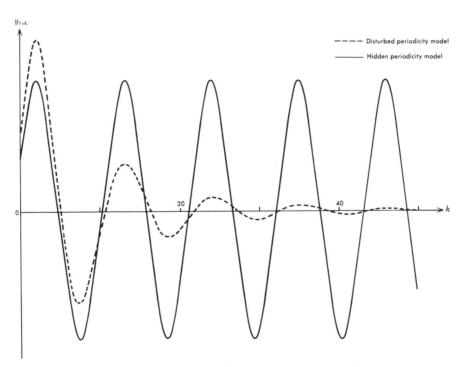

Figure 3 — Forecasting the future on the basis of the past

representation,

$$(1) \qquad y_t = \Phi_t + \Psi_t,$$

known as predictive decomposition, where (a) the two components are uncorrelated, (b) Φ_t is deterministic and Ψ_t is nondeterministic, and (c) the nondeterministic component allows a representation of the Yule type. In Yule's model no Φ_t component is present. In the hidden-periodicities model Φ_t is a sum of sinusoids, while Ψ_t is the random residual. Generally, however, Φ_t, although deterministic in the prediction sense, is random.

The statistical treatment of mixed models like (1) involves a variety of important and challenging problems. Speaking broadly, the valid assessment of the structure requires observations that extend over a substantial number of cycles, and even then the task is difficult. A basic problem is to test for and estimate a periodic component on the supplementary hypothesis that the ensuing residual allows a nondeterministic representation, or, more generally, to perform a simultaneous estimation of the two components. A general method for dealing with these problems has been provided by Whittle (1954); for a related approach, see Allais (1962).

Other problems with a background in this decomposition occur in the analysis of seasonal variation [*see* TIME SERIES, *article on* SEASONAL ADJUSTMENT].

Other stochastic models. Since a tendency to cyclic variation is a conspicuous feature of many phenomena, stochastic models for their analysis have used a variety of mechanisms for generating apparent or genuine cyclicity. Brief reference will be made to the dynamic models for (a) predator–prey populations and (b) epidemic diseases. In both cases the pioneering approaches were deterministic, the models having the form of differential equation systems. The stochastic models developed at a later stage are more general, and they cover features of irregularity that cannot be explained by deterministic methods. What is of special interest in the present context is that the cycles produced in the simplest deterministic models are strictly periodic, whereas the stochastic models produce irregular cycles that allow prediction only over a limited future.

Figure 4 refers to a stochastic model given by M. S. Bartlett (1957) for the dynamic balance between the populations of a predator—for example, the lynx—and its prey—for example, the hare. The data of the graph are artificial, being constructed from the model by a Monte Carlo experiment. The classic models of A. J. Lotka and V. Volterra are deterministic, and the ensuing cycles take the form of sinusoids. The cyclic tendency is quite pronounced in Figure 4, but at the same time the development is affected by random features. After three peaks in both populations, the prey remains at a rather low level that turns out to be critical for the predator, and the predator population dies out.

The peaks in Figure 5 mark the severe spells of poliomyelitis in Sweden from 1905 onward. The cyclic tendency is explained, on the one hand, by the contagious nature of the disease and, on the other, by the fact that slight infections provide immunity, so that after a nationwide epidemic it takes some time before a new group of susceptibles emerges. The foundations for a mathematical theory of the dynamics of epidemic diseases were laid by Kermack and McKendrick (1927), who used a deterministic approach in terms of differ-

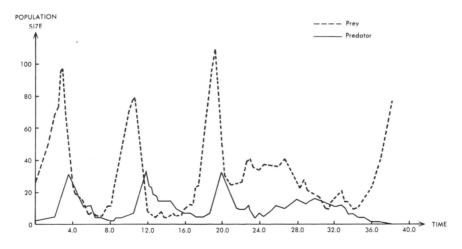

Figure 4 — Predator–prey populations according to a stochastic model

Source: Bartlett 1957, p. 37.

ential equations. Their famous threshold theorem states that only if the infection rate, ρ, is above a certain critical value, ρ_0, will the disease flare up in epidemics. Bartlett (1957) and others have developed the theory in terms of stochastic models; a stochastic counterpart to the threshold theorem has been provided by Whittle (1955).

Bartlett's predator–prey model provides an example of how a cyclic deterministic model may become evolutive (nonstationary) when stochasticized, while Whittle's epidemic model shows how an evolutive deterministic model may become stationary. Both of the stochastic models are completely nondeterministic; note that the predictive decomposition (1) extends to nonstationary processes.

The above examples have been selected so as to emphasize that there is no sharp demarcation between cycles with limited predictability and the *spurious periodicity* of phenomena ruled by randomness, where by pure chance the variation may take wavelike forms, but which provides no basis even for limited predictions. Thus, if a recurrent phenomenon has a low rate of incidence, say λ per year, and the incidences are mutually independent (perhaps a rare epidemic disease that has no aftereffect of immunity), the record of observations might evoke the idea that the recurrences have some degree of periodicity. It is true that in such cases there is an *average period* of length $1/\lambda$ between the recurrences, but the distance from one recurrence to the next is a random variable that cannot be forecast, since it is independent of past observations.

A related situation occurs in the summation of mutually independent variables. Figure 6 shows a case in point as observed in a Monte Carlo experiment with summation of independent variables (Wold 1965). The similarity between the three waves, each representing the consecutive additions of some 100,000 variables, is rather striking. Is it really due to pure chance? Or is the computer simulation of the "randomness" marred by some slip that has opened the door to a cyclic tendency in the ensuing sums? (For an amusing discussion of related cases, see Cole's "Biological Clock in the Unicorn" 1957.)

Figure 6 also gives, in the series of wholesale prices in Great Britain, an example of "Kondratieff waves"—the much discussed interpretation of economic phenomena as moving slowly up and down in spells of some fifty years. Do the waves embody genuine tendencies to long cycles, or are they of a spurious nature? The question is easy to pose but difficult or impossible to answer on the basis of available data. The argument that the "Kondratieff waves" are to a large extent parallel in the main industrialized countries carries little weight, in view of international economic connections. The two graphs have been combined in Figure 6 in order to emphasize that with regard to waves of long duration it is always difficult to sift the wheat of

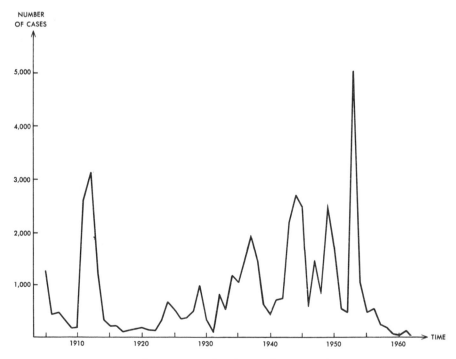

Figure 5 — Paralytic cases of poliomyelitis, Sweden 1905–1962

Source: Unpublished data.

genuine cycles from the chaff of spurious periodicity. (See Garvy 1968.)

Genuine versus spurious cycles

Hypothesis testing.   Cycles are a specific feature in many scientific models, and their statistical assessment usually includes (a) parameter estimation for purposes of quantitative specification of the model, and (b) hypothesis testing for purposes of establishing the validity of the model and thereby of the cycles. In modern statistics it is often (sometimes tacitly) specified that any method under (a) should be supplemented by an appropriate device under (b). Now, this principle is easy to state, but it is sometimes difficult to fulfill, particularly with regard to cycles and related problems of time series analysis. The argument behind this view may be summed up as follows, although not everyone would take the same position:

(i) Most of the available methods for hypothesis testing are designed for use in controlled experiments—the supreme tool of scientific model building—whereas the assessment of cycles typically refers to nonexperimental situations.

(ii) The standard methods for both estimation and hypothesis testing are based on the assumption of independent replications. Independence is on the whole a realistic and appropriate assumption in experimental situations, but usually not for nonexperimental data.

(iii) Problems of point estimation often require less stringent assumptions than those of interval estimation and hypothesis testing. This is frequently overlooked by the methods designed for experimental applications, because the assumption of independence is usually introduced jointly for point estimation, where it is not always needed, and

for hypothesis testing, where it is always consequential.

(iv) It is therefore a frequent situation in the analysis of nonexperimental data that adequate methods are available for estimation, but further assumptions must be introduced to conduct tests of hypotheses. It is even a question whether such tests can be performed at all in a manner corresponding to the standard methods in experimental analysis, because of the danger of specification errors that mar the analysis of nonexperimental data.

(v) Standard methods of hypothesis testing in controlled experiments are thus of limited scope in nonexperimental situations. Here other approaches come to the fore. It will be sufficient to mention *predictive testing*—the model at issue is taken as a basis for forecasts, and in due course the forecasts are compared with the actual developments. Reporting of nonexperimental models should always include a predictive test.

The following example is on the cranky side, but it does illustrate that the builder of a nonexperimental model should have *le courage de son modèle* to report a predictive test, albeit in this case the quality of the model does not come up to the model builder's courage. The paper (Lewin 1958) refers to two remarkable events—the first permanent American settlement at Jamestown, Virginia, in 1607 and the Declaration of Independence in 1776 —and takes the 169 years in between as the basic "cycle." After another 84½ years (½ of the basic cycle) there is the remarkable event of the Civil War, in 1861; after 56 more years (⅓ of the cycle) there is the beginning of the era of world wars in 1917; after 28 more years (⅙ of the cycle) there is the atomic era with the first bomb exploded in

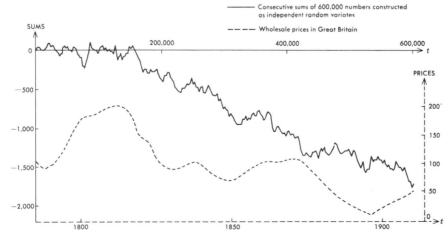

Figure 6 — Spurious cycles and Kondratieff waves

Sources: Consecutive sums, Wold 1965, p. 25; wholesale prices, Piatier 1961.

1945. The paper, published in 1958, ends with the following predictive statement: "The above relation to the basic 169 year cycle of $\frac{1}{1}$, $\frac{1}{2}$, $\frac{1}{3}$, $\frac{1}{6}$ is a definite decreasing arithmetic progression where the sum of all previous denominators becomes the denominator of the next fraction. To continue this pattern and project, we have the 6th cycle—1959, next U.S. Epochal Event—14 year lapse—$\frac{1}{12}$ of 169 years" (Lewin 1958, pp. 11–12). The 1959 event should have been some major catastrophe like an atomic war, if I have correctly understood what the author intimates between the lines in the first part of his article.

It is well to note that this paper, singled out here as an example, is far from unique. Cycles have an intrinsic fascination for the human mind. A cursory scanning of the literature, particularly *Cycles*, the journal of the Foundation for the Study of Cycles, will suffice to show that in addition to the strictly scientific contributions, there is a colorful subvegetation where in quality and motivation the papers and books display all shades of quasi-scientific and pseudoscientific method, down to number mysticism and other forms of dilettantism and crankiness, and where the search for truth is sometimes superseded by drives of self-realization and self-suggestion, not to speak of unscrupulous money-making. The crucial distinction here is not between professional scientists and amateurs. It is all to the good if the search for truth is strengthened by many modes of motivation. The sole valid criterion is given by the general standards of scientific method. Professionals are not immune to self-suggestion and other human weaknesses, and the devoted work of amateurs guided by an uncompromising search for truth is as valuable here as in any other scientific area.

Further remarks

Cycles are of key relevance in the theory and application of time series analysis; their difficulty is clear from the fact that it is only recently that scientific tools appropriate for dealing with cycles and their problems have been developed. The fundamental distinction between the hidden-periodicity model, with its strict periodicity and unlimited predictability, and Yule's model, with its disturbed periodicity and limited predictability, could be brought to full significance only after 1933, by the powerful methods of the modern theory of stochastic processes. On the applied side, the difficulty of the problems has been revealed in significant shifts in the very way of viewing and posing the problems. Thus, up to the failure of the Harvard Business Barometer the analysis of business cycles was

essentially a unirelational approach, the cycle being interpreted as generated by a leading series by way of a system of lagged relationships with other series. The pioneering works of Jan Tinbergen in the late 1930s broke away from the unirelational approach. The models of Tinbergen and his followers are multirelational, the business cycles being seen as the resultant of a complex system of economic relationships. [*See* BUSINESS CYCLES; DISTRIBUTED LAGS.]

The term "cycle," when used without further specification, primarily refers to periodicities in time series, and that is how the term is taken in this article. The notion of "life cycle" as the path from birth to death of living organisms is outside the scope of this presentation. So are the historical theories of Spengler and Toynbee that make a grandiose combination of time series and life cycle concepts, seeing human history as a succession of cultures that are born, flourish, and die. Even the shortest treatment of these broad issues would carry us far beyond the realm of time series analysis; this omission, however, must not be construed as a criticism. (For a discussion of these issues, see Nadel 1968.)

Cycles vs. innovations. The history of human knowledge suggests that belief in cycles has been a stumbling block in the evolution of science. The philosophy of the cosmic cycle was part of Stoic and Epicurean philosophy: every occurrence is a recurrence; history repeats itself in cycles, cosmic cycles; all things, persons, and phenomena return exactly as before in cycle after cycle. What is it in this strange theory that is of such appeal that it should have been incorporated into the foundations of leading philosophical schools and should occur in less extreme forms again and again in philosophical thinking through the centuries, at least up to Herbert Spencer, although it later lost its vogue? Part of the answer seems to be that philosophy has had difficulties with the notion of innovation, having, as it were, a *horror innovationum*. If our philosophy leaves no room for innovations, we must conclude that every occurrence is a recurrence, and from there it is psychologically a short step to the cosmic cycle. This argument being a blind alley, the way out has led to the notions of innovation and limited predictability and to other key concepts in modern theories of cyclic phenomena. Thus, in Yule's model (Figure 2) the random shocks are innovations that reduce the regularity of the sunspot cycles so as to make them predictable only over a limited future. More generally, in the predictive decomposition (1) the nondeterministic component is generated by random elements, innovations, and the component is therefore only of limited predictability. Here there is a close affinity to certain aspects of the general theory of knowledge. We note that prediction always has its cognitive basis in regularities observed in the past, cyclic or not, and that innovations set a ceiling to prediction by scientific methods. [*See* TIME SERIES, *article on* ADVANCED PROBLEMS.]

Mathematical analysis

The verbal exposition will now, in all brevity, be linked up with the theory of stochastic processes. The focus will be on (*a*) the comparison between the schemes of "hidden periodicities" and "disturbed harmonics" and (*b*) spectral representation versus predictive decomposition.

Write the observed series

$$(2) \qquad \cdots, y_{t-1}, y_t, y_{t+1}, \cdots,$$

taking the observations as deviations from the mean and letting the distance between two consecutive observations serve as time unit. Unless otherwise specified, the series (2) is assumed to be of finite length, ranging from $t = 1$ to $t = n$.

Hidden periodicities. With reference to Figure 1, consider first the case of one hidden periodicity. The observed series y_t is assumed to be generated by the model

$$(3) \qquad y_t = x_t + \epsilon_t, \qquad t = 0, \pm 1, \pm 2, \cdots,$$

where x_t, the "hidden periodicity," is a sinusoid,

$$(4) \qquad x_t = \lambda \cos \omega t + \mu \sin \omega t,$$

while

$$(5) \qquad \cdots, \epsilon_{t-1}, \epsilon_t, \epsilon_{t+1}, \cdots$$

is a sequence of random variables, independent of one another and of x_t, and identically distributed with zero mean, $E(\epsilon) = 0$, and standard deviation $\sigma(\epsilon)$. For any λ and μ the sinusoid (4) is periodic, $x_{t+p} = x_t$, with period $p = 2\pi/\omega$, and satisfies the difference equation

$$(6) \qquad x_t - 2\rho x_{t-1} + x_{t-2} = 0, \qquad -1 < \rho < 1,$$

where $\rho = \cos \omega$.

The sinusoid x_t makes a forecast of y_{t+k} over any prediction span k (of course, for real prediction the values of λ, μ, and ρ must be known or assumed),

$$(7) \qquad \mathrm{pred}\, y_{t+k} = x_{t+k}, \qquad k = 1, 2, \cdots,$$

giving the prediction error

$$\Delta(t, k) = y_{t+k} - \mathrm{pred}\, y_{t+k} = \epsilon_{t+k}.$$

Hence the forecast (7) is unbiased and has the same mean-square deviation for all t and k,

$$(8a) \qquad E(\Delta) = 0,$$

$$(8b) \qquad [E(\Delta^2)]^{\frac{1}{2}} = \sigma(\Delta) = \sigma(\epsilon).$$

Further light can be cast on the rationale of the forecast (7) by considering the coefficients λ, μ as limiting regression coefficients of y_t on $\cos \omega t$ and $\sin \omega t$:

$$\lambda = \lim_{n \to \infty} \frac{2}{n} \sum_{t=1}^{n} y_t \cos \omega t;$$

$$\mu = \lim_{n \to \infty} \frac{2}{n} \sum_{t=1}^{n} y_t \sin \omega t.$$

Disturbed periodicity. Yule's model as illustrated in Figure 2 is

$$(9) \qquad y_t - 2\rho y_{t-1} + \gamma^2 y_{t-2} = \epsilon_t, \qquad 0 < \gamma < 1,$$

where the notation makes for easy comparison with model (3):

(*a*) In (3) the disturbances, ϵ_t, are superimposed on the periodic component, x_t, while y_t in (9) is entirely generated by current and past disturbances,

$$(10a) \qquad y_t = \epsilon_t + \alpha_1 \epsilon_{t-1} + \alpha_2 \epsilon_{t-2} + \cdots,$$

where $\alpha_1 = 2\rho$ ($\alpha_0 = 1$) and

$$(10b) \qquad \alpha_k = 2\rho \alpha_{k-1} - \gamma^2 \alpha_{k-2}, \qquad k = 2, 3, \cdots.$$

Hence in (3) the correlation coefficient

$$(11) \qquad r(\epsilon_{t+k}, y_t) = 0$$

for all $k \neq 0$ and t, while (9) gives (11) for all $k > 0$ and t.

(*b*) (See Figure 3.) If the future disturbances, ϵ_{t+k}, were absent, y_{t+k} in (3) would reduce to x_{t+k} and thus make an undamped sinusoid (4), while y_{t+k} in (9), say, y^*_{t+k}, would satisfy the difference equation

$$(12)$$
$$y^*_{t+k} - 2\rho y^*_{t+k-1} + \gamma^2 y^*_{t+k-2} = 0, \qquad k = 1, 2, 3, \cdots,$$

with initial values $y^*_t = y_t$, $y^*_{t-1} = y_{t-1}$, giving

$$(13a) \qquad y^*_{t+k} = \alpha_k \epsilon_t + \alpha_{k+1} \epsilon_{t-1} + \cdots,$$

$$(13b) \qquad y^*_{t+k} = \gamma^k [\lambda^* \cos \omega(t + k) + \mu^* \sin \omega(t + k)].$$

Hence, y^*_{t+k} would make a damped sinusoid with damping factor γ, frequency ω given by $\cos \omega = \rho/\gamma$, and period $2\pi/\omega$, and where the two initial values y_t, y_{t-1} determine the parameters λ^*, μ^*. Since the difference equations in (10b) and (12) are the same except for the initial values, the form (13b) of a damped sinusoid extends to α_k, except that λ^*, μ^* will be different.

(*c*) In (3) the undamped sinusoid, x_t, provides a forecast of y_{t+k} that is unbiased in the sense of (8*a*). In (9) the damped sinusoid, y_{t+k}^*, provides a forecast of y_{t+k},

$$(14) \qquad \operatorname{pred} y_{t+k} = y_{t+k}^*,$$

where

$$(15a) \qquad y_{t+k}^* = E(y_{t+k}|y_t, y_{t-1}),$$

$$(15b) \qquad y_{t+k}^* = E(y_{t+k}|y_t, y_{t-1}, y_{t-2}, \cdots),$$

showing that the forecast (14) is unbiased in the sense of the conditional expectation of y_{t+k} as conditioned by the current and past observations $y_t, y_{t-1}, \cdots$.

(*d*) Figure 7 illustrates that in model (3) the prediction error has constant mean square deviation for all spans k. In (9) it has mean square deviation

$$(16) \qquad [E(\Delta^2)]^{\frac{1}{2}} = (1 + \alpha_1^2 + \cdots + \alpha_{k-1}^2)\sigma(\epsilon)$$

and thus is increasing with k. Formulas (8*b*) and (16) show that in (3) the disturbances, ϵ_t, do not interfere with the sinusoid component, x_t, while in (9) they build up the entire process, y_t.

(*e*) The fundamental difference between models (3) and (9) is further reflected·in the correlogram of y_t,

$$\rho_k = E(y_t y_{t+k})/E(y_t^2), \qquad k = 0, 1, 2, \cdots.$$

In (3) the correlogram is an undamped sinusoid (4), in (9) a damped sinusoid (13*b*). Hence, the two correlograms are curves of the same types as those shown in Figure 3. The graph actually shows the two correlograms, not any two forecasts.

Generalizations. The scheme (3) extends to several hidden periodicities,

$$(17) \qquad x_t = \sum_{i=1}^{h} (\lambda_i \cos \omega_i t + \mu_i \sin \omega_i t),$$

giving the same prediction formulas (7)–(8). Yule's model (9) extends to the *general scheme of autoregression*,

$$(18) \qquad y_t + \beta_1 y_{t-1} + \cdots + \beta_{2h} y_{t-2h} = \epsilon_t,$$

giving expansions of type (10*a*) and (13*a*) and a prediction like (15*a*). Note that formula (17) is a composite undamped swinging. The difference equations (6) and (12) extend from order 2 to order $2h$. The extension of (13*b*) gives y_{t+k}^* as a composite damped swinging.

Stationary stochastic processes. The above models are fundamental cases of stationary stochastic processes. The observed series (2) is seen as a *realization* of the process. Stationarity means that for any fixed n the random variables $\eta_{t+1}, \cdots, \eta_{t+n}$ that generate the observed values $y_{t+1}, \cdots, y_{t+n}$ have a joint probability distribution that is independent of t. In this interpretation a realization corresponds to a sampling point in an n-dimensional distribution, and the parameters λ, μ in model (3) are random variables that vary from one realization (2) of the process to another.

Two general representation theorems for stationary stochastic processes will be quoted briefly.

Spectral representation. The basic reference for spectral representation is Cramér (1940). Any real-valued stationary process η_t allows the representation

$$(19)$$

$$\eta_t - E(\eta) = \frac{2}{\pi} \int_0^{\pi} [\cos \omega t \, d\lambda(\omega) - \sin \omega t \, d\mu(\omega)],$$

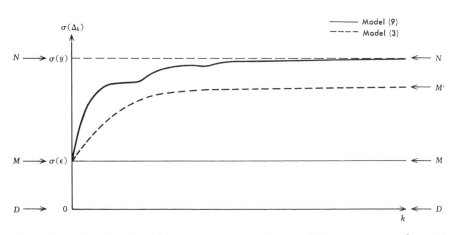

Figure 7 — *The standard deviation, σ(Δ), of the prediction error as a function of the range, k, spanned by the forecast*

where $\lambda(\omega), \mu(\omega)$ are real processes with zero means and zero intercorrelations, with increments $d\lambda(\omega), d\mu(\omega)$ which have zero means and zero intercorrelations, and with variances

$$E\{[d\lambda(\omega)]^2\} = E\{[d\mu(\omega)]^2\} = dV(\omega), \quad 0 \leqslant \omega \leqslant \pi,$$

where $V(\omega)$ is the cumulative spectrum of the process.

Conversely, $\lambda(\omega)$ and $\mu(\omega)$ can be represented in terms of η_t,

$$\lambda(\omega) = \text{l.i.m.}_{T \to \infty} \sum_{t=-T}^{T} \frac{\sin \omega t}{t} [\eta_t - E(\eta)], \quad 0 \leqslant \omega \leqslant \pi,$$

and correspondingly for $\mu(\omega)$. (Here l.i.m. signifies limit in the mean; l.i.m. may exist even if the ordinary limit does not.)

Applying the representation (19) to model (17), the spectrum $V(\omega)$ has discontinuities at the points $\omega = \omega_i$, while the component ϵ_t corresponds to the continuous part of the spectrum. As applied to models (9) and (18), the representation (19) gives a spectrum $V(\omega)$ that is everywhere continuous. Broadly speaking, the spectral representation (19) is useful for analyzing the cyclical properties of the series (2) inside the range of observations, while it is of operative use for prediction outside the observation range only in the case when $V(\omega)$ presents one or more discontinuities.

Predictive decomposition. The basic reference for predictive decomposition is Wold (1938). Any stationary process η_t with finite variance allows the decomposition (1). The deterministic component, Φ_t, can be linearly predicted to any prescribed accuracy over any given span k on the basis of the past observations $y_{t-1}, y_{t-2}, \cdots$. The nondeterministic component, Ψ_t, allows a representation of type (10a) with $\sum \alpha_i^2 < \infty$, correlation properties in accordance with (11) for all $k > 0$ and t, and hence a prediction of type (13a). The ensuing prediction for η_{t+k},

$$\text{pred } \eta_{t+k} = \Phi_{t+k} + \text{pred } \Psi_{t+k},$$

has least square properties in accordance with (16), and if all joint probability distributions are normal (or have linear regressions), the prediction will be unbiased in the sense of (15b).

In models (3) and (17), x_t is the deterministic component and ϵ_t the nondeterministic component. Models (9) and (18) are completely nondeterministic. Levels D, M, N in Figure 7 refer to models that are completely deterministic, mixed, and completely nondeterministic, respectively, each level indicating the standard deviation of the prediction

error for indefinitely large spans k. Making use of the analytical methods of spectral analysis, Kolmogorov (1941) and Wiener (1942) have developed the theory of the decomposition (1) and the nondeterministic expansion (10).

This article aims at a brief orientation to the portrayal of cycles as a broad topic in transition. Up to the 1930s the cyclical aspects of time series were dealt with by a variety of approaches, in which nonscientific and prescientific views were interspersed with the sound methods of some few forerunners and pioneers. The mathematical foundations of probability theory as laid by Kolmogorov in 1933 gave rise to forceful developments in time series analysis and stochastic processes, bringing the problems about cycles within the reach of rigorous treatment. In the course of the transition, interest in cycles has been superseded by other aspects of time series analysis, notably prediction and hypothesis testing. For that reason, and also because cyclical features appear in time series of very different probability structures, it is only natural that cycles have not (or not as yet) been taken as a subject for a monograph.

HERMAN WOLD

[*See also* BUSINESS CYCLES: MATHEMATICAL MODELS; PREDICTION AND FORECASTING, ECONOMIC; Burns 1968.]

BIBLIOGRAPHY

ALLAIS, MAURICE 1962 Test de périodicité: Généralisation du test de Schuster au cas de séries temporelles autocorrelées dans l'hypothèse d'un processus de perturbations aléatoires d'un système stable. Institut International de Statistique, *Bulletin* 39, no. 2:143–193.

BARTLETT, M. S. 1957 On Theoretical Models for Competitive and Predatory Biological Systems. *Biometrika* 44:27–42.

►BROWN, FRANK A. JR. 1972 The "Clocks" Timing Biological Rhythms. *American Scientist* 60:756–766.

BURKHARDT, H. 1904 Trigonometrische Interpolation: Mathematische Behandlung periodischer Naturerscheinungen mit Einschluss ihrer Anwendungen. Volume 2, pages 643–693 in *Enzyklopädie der mathematischen Wissenschaften*. Leipzig: Teubner. → The encyclopedia was also published in French.

►BURNS, ARTHUR F. 1968 Business Cycles: I. General. Volume 2, pages 226–245 in *International Encyclopedia of the Social Sciences*. Edited by David L. Sills. New York: Macmillan and Free Press.

BUYS BALLOT, CHRISTOPHER H. D. 1847 *Les changemens périodiques de température dépendants de la nature du soleil et de la lune mis en rapport avec le prognostic du temps déduits d'observations Neerlandaises de 1729 à 1846.* Utrecht (Netherlands): Kemink.

COLE, LAMONT C. 1957 Biological Clock in the Unicorn. *Science* 125:874–876.

CRAMÉR, HARALD 1940 On the Theory of Stationary Random Processes. *Annals of Mathematics* 2d Series 41:215–230.

Cycles. → Published since 1950 by the Foundation for the Study of Cycles. See especially Volume 15.

►GARVY, GEORGE 1968 Kondratieff, N. D. Volume 8, pages 443–444 in *International Encyclopedia of the Social Sciences.* Edited by David L. Sills. New York: Macmillan and Free Press.

KERMACK, W. O.; and MCKENDRICK, A. G. 1927 A Contribution to the Mathematical Theory of Epidemics. Royal Society of London, *Proceedings* Series A 113: 700–721.

KERMACK, W. O.; and MCKENDRICK, A. G. 1932 Contributions to the Mathematical Theory of Epidemics. Part 2: The Problem of Endemicity. Royal Society of London, *Proceedings* Series A 138:55–83.

KERMACK, W. O.; and MCKENDRICK, A. G. 1933 Contributions to the Mathematical Theory of Epidemics. Part 3: Further Studies of the Problem of Endemicity. Royal Society of London, *Proceedings* Series A 141: 94–122.

KEYSER, CASSIUS J. (1922) 1956 The Group Concept. Volume 3, pages 1538–1557 in James R. Newman, *The World of Mathematics: A Small Library of the Literature of Mathematics From A'h-Mosé the Scribe to Albert Einstein.* New York: Simon & Schuster. → A paperback edition was published in 1962.

KOLMOGOROV, A. N. (1941) 1953 Sucesiones estacionarias en espacios de Hilbert (Stationary Sequences in Hilbert Space). *Trabajos de estadística* 4:55–73, 243–270. → First published in Russian in Volume 2 of the *Biulleten Moskovskogo Universiteta.*

LEWIN, EDWARD A. 1958 1959 and a Cyclical Theory of History. *Cycles* 9:11–12.

MITCHELL, WESLEY C. 1913 *Business Cycles.* Berkeley: Univ. of California Press. → Part 3 was reprinted by University of California Press in 1959 as *Business Cycles and Their Causes.*

►NADEL, GEORGE H. 1968 Periodization. Volume 11, pages 581–585 in *International Encyclopedia of the Social Sciences.* Edited by David L. Sills. New York: Macmillan and Free Press.

PIATIER, ANDRÉ 1961 *Statistique et observation économique.* Volume 2. Paris: Presses Universitaires de France.

SCHUMPETER, JOSEPH A. 1939 *Business Cycles: A Theoretical, Historical, and Statistical Analysis of the Capitalist Process.* 2 vols. New York and London: McGraw-Hill. → An abridged version was published in 1964.

SCHUSTER, ARTHUR 1898 On the Investigation of Hidden Periodicities With Application to a Supposed 26 Day Period of Meteorological Phenomena. *Terrestrial Magnetism* 3:13–41.

TINBERGEN, J. 1940 Econometric Business Cycle Research. *Review of Economic Studies* 7:73–90.

WHITTAKER, E. T.; and ROBINSON, G. (1924) 1944 *The Calculus of Observations: A Treatise on Numerical Mathematics.* 4th ed. Princeton, N.J.: Van Nostrand.

WHITTLE, P. 1954 The Simultaneous Estimation of a Time Series: Harmonic Components and Covariance Structure. *Trabajos de estadística* 3:43–57.

WHITTLE, P. 1955 The Outcome of a Stochastic Epidemic—A Note on Bailey's Paper. *Biometrika* 42: 116–122.

WIENER, NORBERT (1942) 1964 *Extrapolation, Interpolation and Smoothing of a Stationary Time Series, With Engineering Applications.* Cambridge, Mass.: Technology Press of M.I.T. → First published during World War II as a classified report to Section D2, National Defense Research Committee. A paperback edition was published in 1964.

WOLD, HERMAN (1938) 1954 *A Study in the Analysis of Stationary Time Series.* 2d ed. Stockholm: Almqvist & Wiksell.

WOLD, HERMAN 1965 A Graphic Introduction to Stochastic Processes. Pages 7–76 in International Statistical Institute, *Bibliography on Time Series and Stochastic Processes.* Edited by Herman Wold. Edinburgh: Oliver & Boyd.

WOLD, HERMAN 1967 Time as the Realm of Forecasting. Pages 525–560 in New York Academy of Sciences, *Interdisciplinary Perspectives on Time.* New York: The Academy.

YULE, G. UDNY 1927 On a Method of Investigating Periodicities in Disturbed Series, With Special Reference to Wolfer's Sunspot Numbers. Royal Society, *Philosophical Transactions* Series A 226:267–298.

IV
SEASONAL ADJUSTMENT

The objective of economic time series analysis is to separate underlying systematic movements in such series from irregular fluctuations. The systematic movements in the economy—the signals— reveal seasonal patterns, cyclical movements, and long-term trends. The irregular fluctuations—the noise—are a composite of erratic real world occurrences and measurement errors. There are definite advantages in breaking down these two major factors into their respective components. Separation of the systematic components provides a better basis for studying causal factors and forecasting changes in economic activity. Separation of the irregular components provides a basis for balancing the costs of reducing statistical errors against the resultant gains in accuracy. This article is concerned with one of the systematic components, seasonal variations, especially how to measure and eliminate it from economic time series. The relationships between seasonal variations and the other components are also described, with special reference to economic time series in the United States. Characteristics of seasonal variations in different countries, regions, and industries are not discussed.

The seasonal factor. The seasonal factor is the composite effect of climatic and institutional factors and captures fluctuations that are repeated more or less regularly each year. For example, the aggregate income of farmers in the United States displays a definite seasonal pattern, rising steadily each year from early spring until fall, then dropping sharply. Most economic series contain significant seasonal fluctuations, but some (stock prices, for example) contain virtually none.

Changing weather conditions from one season to another significantly affect activities in such industries as construction and agriculture. Movements in a series resulting from this factor are referred to as climatic variations. Differences from year to year in the intensity of weather conditions during each season introduce an irregular element in the pattern of these movements. For example, a very cold winter will have a greater effect on some industries than a winter with average temperature and precipitation.

Intermingled with the effects of variations in climatic conditions are the effects of institutional factors. Thus, the scheduling of the school year from September to June influences the seasonal pattern of industries associated with education, and the designation of tax dates by federal and state authorities affects retail sales and interest rates.

Holidays also help to shape the pattern of activities over 12-month periods. The effects of Christmas and Easter upon the volume of business are widespread, but most direct and largest upon retail sales. Other holidays, such as July 4, Memorial Day, and Labor Day, have a like but generally lesser effect. The number of shopping days between Thanksgiving and Christmas may have some effect upon the volume of Christmas shopping. The effects of certain of these holidays (Easter, Labor Day, and Thanksgiving Day) upon the activities of certain months is uneven, because they do not fall on the same day of the month each year; the dates upon which they fall affect the distribution of activity between two months. The movements resulting from this factor are referred to as holiday variations (and are illustrated by curve 2 of Figure 1, below).

The use of the Gregorian calendar, which provides for months of different lengths and calendar composition, has a special effect upon monthly fluctuations. This effect is due mainly to differences in the character and volume of business activity on Saturdays and Sundays and the variations in the number of these days in the same month in different years. From this point of view there are more than 12 types of months; for example, there are seven different types of months with 31 days, one starting with each different day of the week. The movements resulting from this factor are referred to as calendar, or trading day, variations (and are illustrated by curve 3 of Figure 1, below).

Another type of variation that occurs regularly each year arises from the introduction of new models, particularly in the automobile industry. Although new models are introduced at about the same time each year, the exact date is not predetermined and is separately decided upon by the various companies in the industry. To some extent, these decisions are based upon economic conditions rather than climatic and institutional factors. The movements resulting from this factor are referred to as model year variations.

This complex of factors yields an annual cycle in many economic series, a cycle that is recurrent and periodic. The pattern varies over time, partly because calendar variations are not the same from year to year, but mainly because of changes in the relative importance of firms, industries, and geographic areas. Thus, the seasonal pattern of construction in the United States has been changing as a result of the increasing importance of the south as compared with the north. The annual cycle is not divisible into shorter periods, because any period less than a year will not contain all the factors that determine the annual cycle; for example, holidays are spaced unevenly over the full 12 months, the school schedule spans most of the year, and the tax collection program has a different impact in the various quarters. Unlike business cycle fluctuations, the timing and pattern of seasonal movements in various economic processes, such as production, investment, and financial markets, are not highly correlated.

The role of the seasonal factor. The pattern and amplitude of the seasonal factor are of considerable interest to economists and businessmen. Reducing the waste of resources that are left idle during seasonal low months is one of the targets of economists concerned with accelerating economic growth. Knowledge of the seasonal pattern in the sales of their products (as well as in the materials they purchase) is helpful to companies in determining the level of production that is most efficient in the light of storage facilities, insurance costs, and the risks of forced selling. It can be used to reduce overordering, overproduction, and overstocking.

Some companies forecast only their annual total sales. Then, on the basis of this single forecast, they plan their production schedules, determine their inventory and price policies, and establish quotas for their salesmen. For the companies in this group that also experience large seasonal fluctuations, a good first approximation of the monthly pattern of sales can be obtained by prorating the estimated annual total over the months according to the pattern shown by the seasonal factors. A more refined method involves forecasting the cyclical and trend movements for each of the 12 months ahead and applying the seasonal factors to the forecasts. The seasonal factors can be of further value in making shorter-term forecasts as the year progresses. To keep the forecasts current, the original estimates of the cyclical and trend movements can be revised each month in the light of experience to date, and the seasonal factors can be applied to the revised forecasts.

But the principal interest in economic time series is usually the longer-term cyclical and trend movements. The cycle consists of cumulative and reversible movements characterized by alternating periods of expansion and contraction. It lasts three to four years, on the average. The trend reflects the still longer-run movements, lasting many years.

The nature of the interest in these longer-term movements can be illustrated by the situation in the spring of 1961. About a year earlier a recession had begun in the United States. Although the March and April 1961 data for most economic time series were below the levels reached in March and April 1960, they were higher than in the immediately preceding months. The question was whether the recent improvements were larger or smaller than normal seasonal changes. In forecasting the pattern in the months ahead, it was crucial to know whether the economy had entered a new cyclical phase—whether the economy had been rising or declining to the levels of March and April 1961.

An accurate answer to this question was required to determine the economic programs appropriate at the time. If the underlying movements of the economy were continuing downward, anti-recession measures were in order. But if a reversal had taken place, a different policy was needed. A mistaken reading of statistical trends at such a critical juncture could be costly to the economy; one kind of error could lead to increased unemployment; the other, to eventual inflation.

Cyclical movements are shown more accurately and stand out more clearly in data that are seasonally adjusted. Seasonally adjusted data not only avoid some of the biases to which the widely used same-month-year-ago comparisons are subject but also reveal cyclical changes several months earlier than such comparisons do. Seasonally adjusted series, therefore, help the economic statistician to make more accurate and more prompt diagnoses of current cyclical trends.

Figure 1, computed by the ratio-to-moving-average method (discussed below) by the Bureau of the Census computer program, illustrates various fluctuations discussed above and the resulting seasonally adjusted series. In addition, the figure shows the months for cyclical dominance (MCD) curve. This MCD measure provides an estimate of the appropriate time span over which to observe cyclical movements in a monthly series. In deriving MCD, the average (without regard to sign) percentage change in the irregular component and cyclical component are computed for one-month spans (January–February, February–March, etc.), two-month spans (January–March, February–April, etc.), up to five-month spans. Then MCD is the shortest span for which the average change (without regard to sign) in the cyclical component is larger than the average change (without regard to sign) in the irregular component. That is, it indicates the point at which fluctuations begin to be more attributable to cyclical than to irregular movements, and the MCD curve is a moving average of this many months. (This procedure is explained in full detail in Shiskin 1957a.)

Seasonal adjustment methods. There are many different methods of adjusting time series for seasonal variations. All are, however, based on the fundamental idea that seasonal fluctuations can be measured and separated from the trend, cyclical, and irregular fluctuations. The task is to estimate the seasonal factor and to eliminate it from the original observations by either subtraction or division, or some combination of the two.

All familiar methods of seasonal adjustment, including the well-known link-relative and ratio-to-moving-average methods, follow this simple logic. The link-relative method was introduced in 1919 by Warren M. Persons (1919a; 1919b) of Harvard University. The ratio-to-moving-average method was developed in 1922 by Frederick R. Macaulay (1931) of the National Bureau of Economic Research in a study done at the request of the Federal Reserve Board. The ratio-to-moving-average method has the advantages of more precise measurement of the components and greater flexibility. In addition, it permits analysis of each of the successive stages in the seasonal adjustment process. For these reasons, it was adopted by almost all groups engaged in large-scale seasonal adjustment work, despite the fact that it is relatively laborious.

The ratio-to-moving-average method. The first step in the ratio-to-moving-average method is to obtain an estimate of the trend and cyclical factors by the use of a simple moving average that combines 12 successive monthly figures, thereby eliminating the seasonal fluctuations. Such a moving average is known as a "trend-cycle curve" (see curve 6 of Figure 1), since it contains virtually all the trend and cycle movements and few or none of the seasonal and irregular movements in the

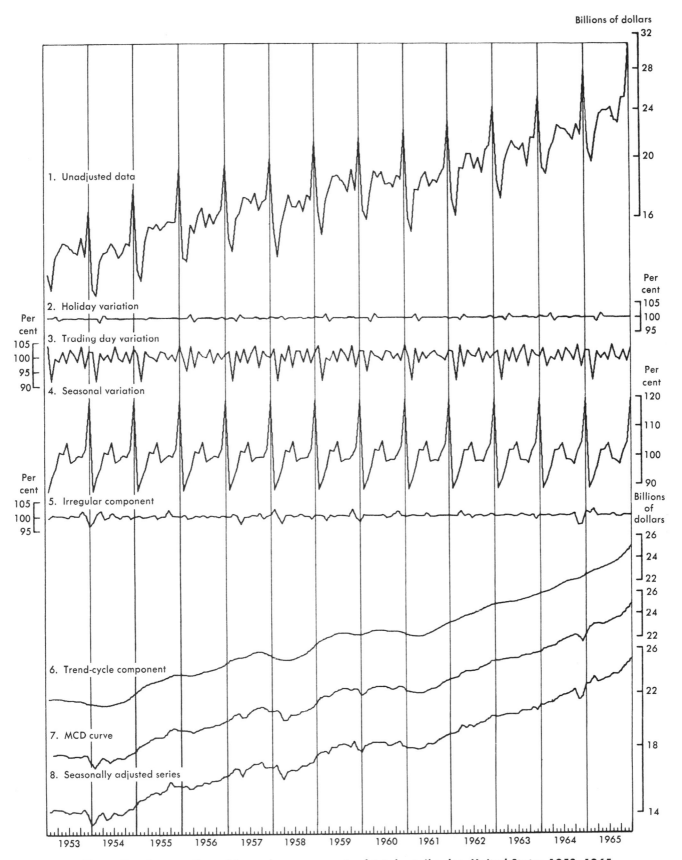

Figure 1 — Systematic and irregular components of total retail sales, United States 1953–1965

Source: U.S. Bureau of the Census 1966, p. 33.

data. Division of the raw data by the moving aver-age yields a series of "seasonal-irregular" ratios. An estimate of the seasonal adjustment factor for a given month is then secured by averaging the seasonal-irregular ratios for that month over a number of years (see curve 4 of Figure 1). It is assumed that the irregular factor will be canceled out in the averaging process. Finally, the original observations are seasonally adjusted by dividing each monthly observation by the seasonal adjustment factor for the corresponding month (see curve 8 of Figure 1). This method yields a multiplicative seasonal adjustment; an additive adjustment can be made by an analogous procedure. At present there is no way of making a simultaneous additive and multiplicative adjustment by this method.

The ratio-to-moving-average method has been programmed for electronic computers and is in widespread use throughout the world. The first seasonal adjustment computer program was developed at the U.S. Bureau of the Census in the summer of 1954. Shortly thereafter, it was used extensively for national series for the United States, Canada, the Organization for Economic Cooperation and Development countries, Japan, and other countries. It has also been utilized by many private concerns to adjust their own data. The U.S. Bureau of Labor Statistics adopted a similar method in 1960, and other adaptations were introduced at about the same time in several other countries.

These programs take advantage of the electronic computer's high-speed, low-cost computations by utilizing more powerful and refined techniques than clerical methods had used in the past. Thus, weighted moving averages are used to represent the trend-cycle factor and to measure changing seasonal patterns. As a result, the computer programs are likely to produce satisfactory results more frequently. They also produce more information about each series—for example, estimates of the trend-cycle and irregular components and of the relations between them. This information can be used for checking the adequacy of the results, for forecasting seasonal and other movements, and for studying the relations among different types of economic fluctuations. For example, for the data graphed in Figure 1 the Bureau of the Census program also gives an indication of the relative importance of the components of the retail sales series by calculating the per cent of the total variation that is contributed by these components over one-month and longer spans. These data are shown in Table 1.

The Bureau of the Census method was designed to analyze a large variety of series equally well.

Table 1 — Per cent of total variation contributed by components of total retail sales, United States, 1953–1965

COMPONENT	PER CENT OF VARIATION	
	Month-to-month	Twelve-month spans
Holiday	0.4	0.5
Trading day	33.1	7.1
Seasonal	64.8	0.0
Irregular	1.2	2.4
Trend-cycle	0.5	90.0
Total	100.0	100.0

Source: U.S. Bureau of the Census 1966, p. 33.

To this end, alternative routines to handle different kinds of series were built into the program, along with techniques for automatically selecting the most appropriate routine for each series. The completeness, versatility, and economy of this method have stimulated broad interest in economic time series analysis in recent years.

This program adjusts for changes in average climatic conditions and institutional arrangements during the year. Adjustments for variations in the number of trading days are also made for some series—for example, new building permits. Further adjustments for variable holidays, such as Easter, are made for certain series, such as retail sales of apparel. Similar adjustments for Labor Day and Thanksgiving Day help bring out the underlying trends. Studies of the effects of unusual weather upon some series have also been started. It is important to note, however, that conventional methods adjust for average weather conditions, and not for the dispersion about this average. For this reason many seasonally adjusted series, such as housing starts, will tend to be low in months when the weather is unusually bad and high in months when the weather is unusually good.

The variants of the ratio-to-moving-average method all give about the same results, and there is considerable evidence that this method adjusts a large proportion of historical series very well. There are, however, some series that cannot be satisfactorily adjusted in this way—for example, those with abrupt changes in seasonal patterns or with constant patterns of varying amplitudes, or those which are highly irregular. Another problem concerns the appropriate seasonal adjustment of an aggregate that can be broken down into different sets of components, each with a different seasonal pattern. However, the principal problem remaining now appears to be obtaining satisfactory seasonal adjustment factors for the current year and the year ahead. These are less accurate than those for previous years, but they play a more important role in the analysis of current economic trends and prospects.

Regression methods. Attempts to use regression methods to analyze time series have been intensified since electronic computers have become available. The basic principle is to represent each of the systematic components by explicit mathematical expressions, usually in the functional form of a linear model. This can be accomplished in a simple form, for example, by regressing the difference between the unadjusted series and the trend-cycle component for each month on the trend-cycle values for that month. The constant term in the regression equation is the additive part of the seasonal component, and the product of the regression coefficient and the trend-cycle value is the multiplicative part of the seasonal component. Thus, this approach has the advantage over the ratio-to-moving-average method that it is not committed to a single type of relationship (e.g., additive or multiplicative) among the seasonal, cyclical, and irregular components of the series.

Another advantage is that the different types of fluctuations can be related to the forces causing them by representing these forces as appropriate variables in the mathematical expressions. Thus, in measuring the seasonal factor, direct allowance can be made, say, for the level of the series or for temperature and precipitation. In certain series, special factors could be taken into account; for example, in measuring the seasonal factor in unemployment, allowance could be made for the number of students in the labor force, or in the case of automobile sales, the level of automobile dealers' inventories could be taken into account. Finally, the mathematical expressions for the estimates of the systematic components provide the basis for deriving measures of variance and significance tests to evaluate the reliability of the estimates; this applies, for example, to estimates of the seasonally adjusted series and to the seasonal component, or to the differences in either series over time.

The principal doubt about the regression approach is whether fairly simple functional forms can adequately measure the implicit economic patterns. Or, to consider the matter from another point of view, do the complex mathematical forms required to represent the systematic movements of historical series constitute a plausible theory of economic fluctuations? A related question is whether either fairly simple functional forms, which only crudely measure historical patterns, or the more complex forms, which fit the past more closely, can provide the basis for accurate forecasts of future patterns.

Thus far, regression methods have been applied to only a small number of series, and their powers to decompose series into the various systematic components and to forecast seasonal factors for future years have not yet been fully tested. While the ratio-to-moving-average method does not have the advantages provided by the mathematical properties of the regression method, extensive tests have demonstrated that it gives good results in practice. Tests completed at the Bureau of the Census show that regression methods yield historical seasonal factors very similar to those yielded by the ratio-to-moving-average method, but that the regression "year-ahead" factors and the trend-cycle curves are less accurate.

Criteria for judging a seasonal adjustment. Although it is not now possible to draw a set of hard-and-fast rules for judging the success of a seasonal adjustment, five guidelines have proved useful.

(*a*) Any repetitive intrayear pattern present in a series before seasonal adjustment should be eliminated and thus should not appear in the seasonally adjusted series, in the trend-cycle component, or in the irregular component. This implies that the seasonal factors are not correlated with the seasonally adjusted series, or with the trend-cycle or irregular components. (The correlations should be computed year by year, because residual seasonality sometimes shows up with inverse patterns in different years.)

(*b*) The underlying cyclical movements should not be distorted. Seasonally adjusted series that in unadjusted form had a large seasonal factor should be consistent in terms of cyclical amplitude, pattern, and timing with other related economic series that either had no seasonal factor at all or had a small seasonal factor compared with the cyclical factor. Similarly, changes in a seasonally adjusted series such as new orders for machinery and equipment should be followed by like changes in a corresponding series such as sales.

(*c*) The irregular fluctuations should behave like a random series when autocorrelations of lags of about 12 months are considered. Autocorrelations of smaller lags need not necessarily behave like the similar autocorrelations of a random series because some irregular influences, such as a long strike, spread their effects over several months. A seasonally adjusted artificial series containing a random component should produce a random series as the irregular component.

(*d*) The sum of the seasonally adjusted series should be equal to the sum of the unadjusted series. For most series, sums are meaningful in economic terms, and the preservation of sums meets the common-sense requirement that the number of units produced, traded, or exported in a year should not be altered by the seasonal adjustment.

(*e*) Revisions in the seasonal factors that take place when data for additional years become available should be relatively small.

Tests of seasonal adjustments. With the massive increase in the number of series seasonally adjusted in recent years, due largely to the increasing use of electronic computers for this purpose, the need for routine objective tests of the quality of the adjustments has grown.

A general type of test involves examining the results of applying a seasonal adjustment procedure to artificial series. One method of constructing suitable artificial series is to combine the irregular, cyclical, and seasonal factors from different real economic series into artificial aggregates; that is, the seasonal factor from one economic series, the trend-cycle factor from another, and the irregular factor from a third are multiplied together to form a new series. A test of the Bureau of the Census method, using 15 different types of such artificial series, revealed that in most instances the "estimated" components trace a course similar to that of the "true" components (Shiskin 1958). Although some limitations were evident, this test showed that the Census method has considerable power to rediscover the different types of fluctuations that were built into the series and does not generate arbitrary fluctuations that have no relationship to the original observations.

A statistical test for the presence of a stable seasonal adjustment component may be made by using the analysis of variance and the associated F-test. This is a test of the null hypothesis that monthly means are equal. Here, the variance estimated from the sum of squares of the differences between the average for each month and the average for all months (between-months variance) is compared with the variance estimated from the sum of squares over all months of the differences between the values for each month and the average for that month (within-months variance). If the between-months variance of the "seasonal-irregular" ratios (computed by dividing the original observations by an estimate of the trend-cycle component) is significantly greater than the within-months variance, it can usually be assumed that there is a true seasonal factor in the series. If the between-months variance is not significantly greater than the within-months variance of the irregular series (computed by dividing the seasonally adjusted series by an estimate of the trend-cycle component), then it can usually be assumed that a complete seasonal adjustment has been made. This test must, however, be used cautiously because differences between months can also appear as a result of differences in the behavior of the irregular component from month to month, because differences between months may be hidden when changes in seasonality in one month are offset by changes in another month, and because the assumptions of the test may not be well satisfied. Nevertheless, the F-test has proved to be a useful test of stable seasonality in practice. [*See* LINEAR HYPOTHESES, *article on* ANALYSIS OF VARIANCE.]

Experience in applying spectral analysis to physical science data has encouraged researchers to explore its use in economics, and this technique is now being used to test for seasonality. Spectral analysis distributes the total variance of a series according to the proportion that is accounted for by each of the cycles of all possible periodicities, in intervals for 2-month and longer cycles. If there is a seasonal pattern in a series, a large proportion of the variance will be accounted for by the 12-month cycle and its harmonics (cycles of 6, 4, 3, 2.4, and 2 months). A significant proportion of the total variance of a seasonally adjusted economic series should be accounted for by a cycle of 45 to 50 months, the average duration of the business cycle, but not by the 12-month cycle or its harmonics. A random series would not be expected to show a significant cycle at any periodicity. While quantitative statistical methods based on suitable assumptions for economic time series have not yet been developed for determining from a spectrum whether seasonality exists, such judgments can often be made from inspection of charts of the spectra.

A question sometimes raised about spectral analysis is whether it is appropriate to consider an economic time series from the viewpoint of the frequency domain, as spectral analysis does, rather than the time domain, as most other methods do. This question comes up mainly because economic series are available for relatively short periods and economic cycles, other than the seasonal, are irregular in length and amplitude. However, the prospect that mathematical representation of a time series in this way may reveal relationships not otherwise apparent would appear to make this alternative view worth further exploration.

These tests do not provide enough information to determine whether all the criteria listed above are satisfied. To this end, comparisons of the sums of seasonally adjusted and unadjusted data are also made, often for all fiscal years in addition to calendar years. The magnitude of revisions resulting from different methods of seasonal adjustment is usually appraised by seasonally adjusting series which cover periods successively longer by one year (e.g., 1948–1954, 1948–1955, 1948–1956, and so forth) and comparing the seasonal factors for the terminal years with the "ultimate" seasonal factors.

Relations of seasonal to other fluctuations. An analysis has been made of the cyclical, seasonal, and irregular amplitudes of a sample of about 150 series considered broadly representative of the different activities of the U.S. economy. This study revealed that, for the post-World War II period, seasonal movements dominate other kinds of month-to-month movements in most current economic series. Seasonal movements are almost always larger than either the irregular or the cyclical movements, and they are often larger than both of the other types combined. More specifically, the average monthly amplitude of the seasonal fluctuations exceeds that of the cyclical factor in 78 per cent of the series, exceeds the irregular factor in 65 per cent of the series, and exceeds the cycle–trend and irregular factors in combination in 45 per cent of the series. Furthermore, where the seasonal factor is larger, it is often much larger. The seasonal factor is three or more times as large as the cyclical factor in 45 per cent of the series, three or more times as large as the irregular factor in 16 per cent of the series, and three or more times as large as the cyclical and irregular fluctuations together in 11 per cent of the series. (See Shiskin 1958.) These results apply to observations of change over intervals of one month; over longer spans the relative importance of the several components would, of course, be different. Table 1 shows how seasonal and trading day fluctuations, which dominate the short-term movements, give way in relative importance to the trend-cycle factor when comparisons are made over longer periods.

These findings emphasize the advantages of seasonally adjusted series over those not so adjusted for studying cyclical movements. Where the seasonal fluctuations are large, a difference in the unadjusted data for two months may be due largely or solely to normal seasonal fluctuations; if the data are seasonally adjusted, the difference can be assumed to be caused chiefly by cyclical or irregular factors.

JULIUS SHISKIN

BIBLIOGRAPHY

Much of the material in this article is discussed in greater detail in Shiskin 1957a; 1958; *and the Shiskin paper in* Organization for European Economic Cooperation 1961. *References that deal with the problem of seasonal adjustment as it relates to current economic conditions are* Organization for European Economic Cooperation 1961 *and*

Shiskin 1957a; 1957b. *Early works on seasonality and seasonal adjustment methods are* Barton 1941; Burns & Mitchell 1946; Hotelling & Hotelling 1931; Kuznets 1933; Macaulay 1931; *and* Persons 1919a; 1919b. *Works dealing with the history of the Bureau of the Census method and its variants are* Organization for European Economic Cooperation 1961; Shiskin & Eisenpress 1957; Shiskin et al. 1965; *and* Young 1965. *Alternative methods are described in* Hannan 1963; Rosenblatt 1963; *and* U.S. Bureau of Labor Statistics 1964. *Tests of seasonal adjustment methods are discussed in* Burns & Mitchell 1946; Granger & Hatanaka 1964; Hannan 1963; Kuznets 1933; Rosenblatt 1963; *and* Shiskin 1957b.

BARTON, H. C. JR. 1941 Adjustment for Seasonal Variation. *Federal Reserve Bulletin* 27:518–528.

BURNS, ARTHUR F.; and MITCHELL, WESLEY C. 1946 *Measuring Business Cycles.* National Bureau of Economic Research, Studies in Business Cycles, No. 2. New York: The Bureau. → See especially pages 43–55, "Treatment of Seasonal Variations" and "Notes on the Elimination of Seasonal Variations."

GRANGER, CLIVE W. J.; and HATANAKA, M. 1964 *Spectral Analysis of Economic Time Series.* Princeton Univ. Press. → See also "Review" by H. O. Wold in *Annals of Mathematical Statistics*, February 1967, pages 288–293.

▶GRETHER, D. M.; and NERLOVE, MARC 1970 Some Properties of "Optimal" Seasonal Adjustment. *Econometrica* 38:682–703.

HANNAN, EDWARD J. 1963 The Estimation of Seasonal Variations in Economic Time Series. *Journal of the American Statistical Association* 58:31–44.

HOTELLING, HAROLD; and HOTELLING, FLOY 1931 Causes of Birth Rate Fluctuations. *Journal of the American Statistical Association* 26:135–149.

KUZNETS, SIMON 1933 *Seasonal Variations in Industry and Trade.* New York: National Bureau of Economic Research.

MACAULAY, FREDERICK R. 1931 *The Smoothing of Time Series.* New York: National Bureau of Economic Research.

ORGANIZATION FOR EUROPEAN ECONOMIC COOPERATION 1961 *Seasonal Adjustment on Electronic Computers.* Report and proceedings of an international conference held in November 1960. Paris: Organization for Economic Cooperation and Development.

PERSONS, WARREN M. 1919a Indices of Business Conditions. *Review of Economics and Statistics* 1:5–107.

PERSONS, WARREN M. 1919b An Index of General Business Conditions. *Review of Economics and Statistics* 1:111–205.

ROSENBLATT, HARRY M. (1963) 1965 *Spectral Analysis and Parametric Methods for Seasonal Adjustment of Economic Time Series.* U.S. Bureau of the Census, Working Paper No. 23. Washington: Government Printing Office. → First published in American Statistical Association, Business and Economics Section, *Proceedings*, pages 94–133.

SHISKIN, JULIUS 1957a *Electronic Computers and Business Indicators.* National Bureau of Economic Research, Occasional Paper No. 57. New York: The Bureau. → First published in Volume 30 of the *Journal of Business.*

SHISKIN, JULIUS 1957b Seasonal Adjustments of Economic Indicators: A Progress Report. Pages 39–63 in American Statistical Association, Business and Economics Section, *Proceedings.* Washington: The Association.

SHISKIN, JULIUS 1958 *Decomposition of Economic Time Series. Science* 128:1539–1546.

SHISKIN, JULIUS; and EISENPRESS, HARRY (1957) 1958 *Seasonal Adjustments by Electronic Computer Methods.* National Bureau of Economic Research, Technical Paper No. 12. New York: The Bureau. → First published in Volume 52 of the *Journal of the American Statistical Association.*

SHISKIN, JULIUS et al. 1965 *The X-11 Variant of the Census Method. II: Seasonal Adjustment Program.* U.S. Bureau of the Census, Technical Paper No. 15. Washington: The Bureau.

U.S. BUREAU OF LABOR STATISTICS 1964 *The BLS Seasonal Factor Method (1964).* Washington: The Bureau.

U.S. BUREAU OF THE CENSUS *Monthly Retail Trade Report* [1966]: August.

YOUNG, ALLAN H. 1965 *Estimating Trading-day Variation in Monthly Economic Time Series.* U.S. Bureau of the Census, Technical Paper No. 12. Washington: The Bureau.

TOLERANCE SETS

See ESTIMATION, *article on* CONFIDENCE INTERVALS AND REGIONS.

TRADE CYCLES

See BUSINESS CYCLES: MATHEMATICAL MODELS.

TRANSFORMATIONS OF DATA

See under STATISTICAL ANALYSIS, SPECIAL PROBLEMS OF.

TRUNCATION AND CENSORSHIP

See under STATISTICAL ANALYSIS, SPECIAL PROBLEMS OF.

V

VALIDITY

See ERRORS, *article on* EFFECTS OF ERRORS IN STATISTICAL ASSUMPTIONS; EXPERIMENTAL DESIGN, *article on* QUASI-EXPERIMENTAL DESIGN; INTERVIEWING IN SOCIAL RESEARCH; PSYCHOMETRICS.

VAN DANTZIG, DAVID

▶ *This article was specially written for this volume.*

When David van Dantzig (1900–1959) died, Holland lost its foremost mathematical statistician. He had begun his career as a pure mathematician, but about 1940, while he was a professor at the Technical University at Delft, he became interested in the field of probability. His first publication on probability theory appeared in 1941. Dismissed by the Germans from his professorship during World War II, van Dantzig continued to study probability and statistics, for he fully understood the importance of this branch of applied mathematics for other sciences, industry, and society in general. His conscientious nature drove him to further study and thought about ways to propagate and develop research and application of mathematics in general, and statistics in particular, after the war.

Before World War II, the field of mathematical statistics had been largely undeveloped in Holland. Although good work had been done by several pioneers in the application of statistics, particularly at the Agricultural University of Wageningen, these activities were isolated and uncoordinated. What was really needed was a bridge between mathematicians and other scientists, engineers, and technicians. Study and research in the domain of applied mathematics (including physical mathematics and statistics) needed to be stimulated and coordinated with similar activities in the field of pure mathematics and numerical methods. Such efforts, and especially the possibility of consultation on a large scale, could not be realized by the universities alone. Thus arose the idea of establishing a mathematical organization that would harbor all these activities—and more—in one institution. This idea, shared by van Dantzig, J. G. van der Corput, and J. F. Koksma, came to fruition in February 1946, when the triumvirate founded the Mathematical Center in Amsterdam.

Having studied probability and statistics during the war years, van Dantzig was in 1946—as he was when he died—a pioneer in this field. With the founding of the Mathematical Center, he determined to realize his ideas on consultation and research as head of its Department of Mathematical Statistics, a task that could be combined with his new professorship (in the "theory of collective phenomena," a chair created especially for him) at the University of Amsterdam. From then on, he was more than busy, planning and teaching courses, carrying out statistical consultation, working with pupils, visiting congresses, delivering lectures, serving on committees and boards, and, besides, producing papers and working out new theories until the very day of his untimely death.

Van Dantzig's theoretical work on probability and statistics won international recognition. His personal relations with outstanding mathematicians and statisticians in many countries were excellent. He was a well-known speaker at international congresses, and in 1950 he was visiting professor for half a year at the University of California at Berkeley. He was elected a member of the International Statistical Institute and a fellow of the Royal

Statistical Society, the Institute of Mathematical Statistics, and the American Statistical Association. He was also a member of the Dutch Royal Academy of Sciences (Koninklijke Nederlandse Akademie van Wetenschappen) and one of the first members of the Dutch Statistical Society (Vereniging voor Statistiek).

Despite all these activities, van Dantzig never lost a strong nostalgia for pure mathematics. "Applied mathematics seems to be like wine," he said (1955d), "it becomes pure just in course of time." This is but one of many expressions of his nostalgia; his (unfinished) endeavor to find a new basis for probability in topology (one of his "old loves") is another.

Van Dantzig did not succeed, however, in disentangling himself from his applied work. Some of the subjects on which he worked were too important. The most important was the statistical and hydrodynamic research on the high water levels at the Dutch coast and the movements of the North Sea, which he worked on for years after the disastrous flood of 1953. The final version of the report of the Mathematical Center's Department of Mathematical Statistics and Department of Applied Mathematics, written under his supervision, and in large part by himself, for the Delta Commission of the Dutch government, was lying on his desk on the day of his death. For weeks he had been putting finishing touches to this voluminous document. Later published as a book, it constitutes a monument to his tireless efforts to further the interests of mathematics and its application to subjects important to society.

Van Dantzig's lively interest in logic, and his extensive knowledge of it, destined him to take part in discussions of the foundations of probability. Although the theory of probability may be seen—following Kolmogorov—as an axiomatic mathematical theory, the application in the form of statistics and the practical interpretation of numerical probability have given and still do give rise to controversies among scientists. [See PROBABILITY, article on INTERPRETATIONS.] Two main branches of interpretation may be distinguished: the objective (frequentist) and the subjective (personal) interpretation. Frequentists see a probability as an approximation of a frequency in the long run, and this was van Dantzig's standpoint. He complemented this interpretation—as have others—by the observation that the practical behavior of mankind indicates that possible events with a very small probability are not expected to occur in isolated observations, and that these probabilities are therefore treated in the same way as zero probability. His further analysis of this interpretation,

of the role of the mathematical model in a scientific investigation, and of the method of its choice, its "switching on" and "switching off," constitute a real contribution to a better comprehension of the statistical method, leading to a sharper and more critical distinction of its different phases and improving the exactness of its applications.

Although not fully satisfied with the frequency interpretation of probability, van Dantzig could never agree with the ideas of the subjectivists, who interpret probability as a degree of belief, or a degree of confidence. His criticism of these ideas was sharp and to the point. In one of his papers (1957b), he launched a heavy attack on Leonard J. Savage's book *The Foundations of Statistics* (1954). [*See the biography of* SAVAGE.] In summary, he wrote:

(1) Statistical work has value only insofar as its results are independent of the preferences of the individual statistician who performs it. Although such an independence in any absolute sense cannot be reached, it can be obtained to a practically sufficient degree, which is not essentially less than one obtainable in other sciences.

(2) Strictly speaking statistics needs as a mathematical tool no calculus of probabilities, but only a calculus of (finite) frequency quotients. The concepts of probability and of infinity are introduced for mathematical convenience only.

(3) Statistics uses the empirical hypothesis that apparatus ('lotteries') exist, admitting random choices of one among any given number of elements. Such apparatus do not exist in absolute perfection and their degree of perfection can only be defined after development of their theory. Their rôle is analogous to that of rigid bodies in euclidean geometry and of perfect clocks in dynamics. Empirical interpretation of probability statements is only possible with reference to such random apparatus or to natural phenomena empirically found to behave statistically sufficiently like these.

(4) Because of imperfections of random apparatus and of simplifying mathematical assumptions probability statements of very great precision have no empirical correlate. In particular the distinction between very small probabilities and zero has none. In accordance with this, actual human behaviour is only understandable on the assumption that possible events having theoretically extremely small probabilities are actually neglected.

(5) Subjective expectations, valuations and preferences and their changes from person to person or in the course of time can and should be investigated by means of 'objective' statistical methods. Trying to use them as a basis of statistics is like trying to gauge a fever thermometer by means of the patient's shivers.

Another paper (1957c) is a critique of R. A. Fisher's book *Statistical Methods and Scientific Inference* (1956). This paper also combines sharp criticism with excellent wit, and it contains an analysis of Fisher's method of fiducial inference. [*See* FIDUCIAL INFERENCE *and the biography of* FISHER.] Although searching for positive points in this method, van Dantzig finally finds that he must reject it. The method of his analysis is well illustrated by a very simple example in the first part of the paper.

Let x be a random variable, normally distributed with unknown mean μ. Then everybody agrees that

$$(1) \qquad P(x < \mu) = \frac{1}{2}.$$

But Fisher would write

$$(2) \qquad P(\mu > x) = \frac{1}{2},$$

which is equivalent to (1), and if an observation of x yields the value 1.37, he would substitute this value into (2), getting

$$(3) \qquad P(\mu > 1.37) = \frac{1}{2}.$$

This statement is—for a frequentist—meaningless, since μ is not a random variable but an unknown constant. If $\mu > 1.37$ is in fact true, then the value of $P(\mu > 1.37)$ is 1, and if $\mu < 1.37$, then that value is 0. Fisher himself adhered to the frequency interpretation and he did not consider (3) as a probability statement but as a new kind of statement, derived from a "fiducial distribution" attributed to μ; this fiducial distribution is also normal, with the same variance as the probability distribution of x and with x itself as mean. Considering (2) as a fiducial statement derived from this fiducial distribution and then substituting $x = 1.37$, one obtains (3).

It is of course a quite legitimate procedure to introduce a new notion called "fiducial distribution," with substitution of a number for x admitted. It is, on the other hand, very confusing to use the same symbol P for this distribution as is used for probability statements. Furthermore, increasing the confusion to the point of incorrectness, Fisher insisted that (2) allows of a frequency interpretation. This is true as long as P means "probability" and as long as no substitution for x is allowed. Van Dantzig makes this clear by rewriting (2) in the form

$$(2a) \qquad \frac{1}{\sigma\sqrt{2\pi}} \int_{-\infty}^{\mu} \exp\left\{ -\frac{1}{2} \frac{(x-\mu)^2}{\sigma^2} \right\} dx = \frac{1}{2}.$$

It is clear that in this formula (which is equivalent to (2) if x is normally distributed) no substitution of 1.37 (or any other number) for x can be allowed. Therefore (2), considered as a fiducial statement with substitution admitted, cannot be equivalent to (2a). But then the meaning of the fiducial statement is totally obscure.

Van Dantzig himself never wrote x for a random variable, but he indicated the random character of a variable by *underlining* its symbol. The distinction between random variables and algebraic ones—neglected by Fisher—was, of course, maintained consistently in the 1950s by several authors, but they used different notations, for example, capital letters for random variables and lowercase letters for algebraic ones, or Greek versus Latin letters. The method of underlining is, however, a very practical one. In this notation, (1) becomes

$$(1') \qquad P(\underline{x} < \mu) = \frac{1}{2}$$

and (2), in its probability sense, becomes

$$(2') \qquad P(\mu > \underline{x}) = \frac{1}{2},$$

clearly equivalent to (1′). But in its fiducial sense it seems to be something like

$$(2'') \qquad P(\underline{\mu} > x) = \frac{1}{2},$$

at least if one insists on a frequency interpretation with substitution for x admitted. For, in this form, x is an algebraic variable and substitution *is* possible. However, (2″) does not in any way follow from (2′) by means of the axioms of probability theory. It should therefore be based on a definition of μ, which Fisher does not give explicitly. Defining

$$(3') \qquad \underline{\mu} \equiv \mu - \underline{x} + x,$$

where μ is the unknown value of the mean of $\underline{x}$ and x is the value assumed by $\underline{x}$ in the observation, (2″) can indeed be proved.

It is not clear whether Fisher would have agreed with this view of his method. Further analysis, along these lines, of more complicated cases led to rather disappointing conclusions, the most important of which was that the Behrens–Fisher two-sample test cannot be justified in this way and is incorrect from the frequentist point of view.

This analysis plainly shows the importance and the power of systematic and clear notation.

During the first years of the Mathematical Center, van Dantzig did most of the consulting work himself. His talks with research workers in other fields of science always compelled them to define sharply the methods and the aim of their investigations. He invariably started a consultation by asking for a formulation of the aim of the experiment; discussion of this point was generally of great value to the experimenter and often led to a change in the experimental design. This made van Dantzig an ardent propagandist for consultation before the actual start of an experiment. He also insisted on building up with a consultee an adequate mathematical model and, if necessary, on adapting the experimental design to bring it closer to the model. Another point that he stressed—which in experiments is too often overlooked—is the essential need to investigate the accuracy (or lack of it) of observations. Because the statistical methods to be used in an analysis must be chosen beforehand, van Dantzig firmly advocated that time and money be allotted to preliminary planning of any project, a phase that he emphatically included in the investigation as a whole. No analysis, he believed, should be considered trifling work to be executed by statistical sorcerers after an experiment has been completed.

Finally, van Dantzig advised that conclusions drawn from an analysis be obtained by "switching off" the mathematical model, and be painstakingly formulated in exact statistical language. He always discussed practical or scientific interpretations of findings with consultees and helped them to "translate" their conclusions into the language of the particular domain of their experiment, but the responsibility of final interpretations always remained with the experimenters themselves.

Thus, van Dantzig not only succeeded in forming a group of statistical pupils who were well schooled in statistical consultation, but he also taught many scientists in other fields to think more clearly and to realize the importance of paying due attention to the theoretical preparation and the possibility of well-founded statistical analysis that are necessary for experiments on a high scientific level.

Van Dantzig's contributions to the mathematical theory of statistics (and to other branches of mathematics) were considerable. His most original contribution to statistics is his "theory of collective marks," an extension of the theory of characteristic functions, with an unusual probabilistic interpretation of the "collective mark" itself. (See van Dantzig 1949; 1955c; van Dantzig & Scheffer 1954; van Dantzig & Zoutendijk 1959.) A popular version with some applications can be found in Hemelrijk (1959; 1960).

J. HEMELRIJK

WORKS BY VAN DANTZIG

For a complete bibliography of van Dantzig's mathematical and statistical papers, see Statistica neerlandica 13:415–432.

1941 Mathematische en empirische grondslagen van waarschijnlijkheidstheorie (Mathematical and Empirical Foundations of the Calculus of Probability). *Nederlands tijkschrift voor natuurkunde* 8:70–91. → Discussion on pages 91–93.

1949 Sur la méthode des fonctions génératices. Pages 29–46 in Centre National de la Recherche Scientifique (France), *Le calcul des probabilités et ses applications.* Colloques Internationaux, No. 13. Paris: The Center.

1951a Sur l'analyse logique des relations entre le calcul des probabilités et ses applications. Volume 4, pages 49–66 in Congrès International de Philosophie des Sciences, Paris, 1949, *Actes.* Volume 4: *Calcul des probabilités.* Paris: Hermann.

1951b On the Consistency and the Power of Wilcoxon's Two-sample Test. Akademie van Wetenschappen, Amsterdam, *Proceedings* Series A (Mathematical Sciences) 54:1–8. → Also in *Indagationes mathematicae* 13:1–8.

1952a Les problèmes que pose l'application du calcul des probabilités. Pages 53–65 in Paul P. Gillis (editor), *Théorie des probabilités: Exposés sur ses fondaments et ses applications.* Louvain (Belgium): Nauwelaerts.

1952b Utilité d'une distribution de probabilités: Distribution des probabilités des utilités. Pages 242–244 in Colloque d'Econométrie, Paris, 1952 [*Actes*]. Centre National de la Recherche Scientifique (France), Colloques Internationaux, No. 40. Paris: The Center.

1954 VAN DANTZIG, DAVID; and HEMELRIJK, J. Statistical Methods Based on Few Assumptions. *Bulletin of the International Statistical Institute* 24, no. 2:239–267.

1954 VAN DANTZIG, DAVID; and SCHEFFER, C. On Arbitrary Hereditary Time-discrete Stochastic Processes, Considered as Stationary Markov Chains, and the Corresponding General Form of Wald's Fundamental Identity. Akademie van Wetenschappen, Amsterdam, *Proceedings* Series A (Mathematical Sciences) 57: 377–388. → Also in *Indagationes mathematicae* 16: 377–388.

1955a Laplace, probabiliste et statisticien, et ses précurseurs. *Archives internationales d'histoire des sciences* 8, no. 30:27–37.

1955b Sur les ensembles de confiance généraux et les méthodes dites non paramétriques. Pages 73–91 in Colloque sur l'Analyse Statistique, Brussels, 1954 [*Actes*]. Organized by the Centre Belge de Recherches Mathématiques. Liège: Thone; Paris: Masson.

1955c Chaînes de Markof dans les ensembles abstraits et applications aux processus avec régions absorbantes et au problème des boucles. Université de Paris, Institut Henri Poincaré, *Annales* 14, fascicle 3:145–199.

1955d The Function of Mathematics in Modern Society and Its Consequence for the Teaching of Mathematics. *Euclides* 31:88–102. → Also in *Enseignement mathématique* 2, no. 1:159–178.

1956 Econometric Decision Problems for Flood Prevention. *Econometrica* 24:276–287.

1957a Mathematical Problems Raised by the Flood Disaster, 1953. Volume 1, pages 218–239 in International Congress of Mathematicians (New Series), Seventh, Amsterdam, 1954, *Proceedings.* Amsterdam: North-Holland.

1957b Statistical Priesthood: Savage on Personal Probabilities. *Statistica neerlandica* 11:1–16.

1957c Statistical Priesthood: II. Sir Ronald on Scientific Inference. *Statistica neerlandica* 11:185–200.

1959 VAN DANTZIG, DAVID; and ZOUTENDIJK, G. Itérations markoviennes dans les ensembles abstraits. *Journal de mathématiques pures et appliquées* New Series 38:183–200.

SUPPLEMENTARY BIBLIOGRAPHY

FISHER, R. A. (1956) 1973 *Statistical Methods and Scientific Inference.* 3d ed., rev. & enl. New York: Hafner; Edinburgh: Oliver & Boyd.

HEMELRIJK, J. 1959 David van Dantzig's Statistical Work. *Synthese* 11:335–351.

HEMELRIJK, J. 1960 The Statistical Work of David van Dantzig (1900–1959). *Annals of Mathematical Statistics* 31:269–275.

SAVAGE, LEONARD J. (1954) 1972 *The Foundations of Statistics.* Rev. ed. New York: Dover.

VARIANCE, ANALYSIS OF
See under LINEAR HYPOTHESES.

VARIANCE COMPONENTS
See LINEAR HYPOTHESES, *article on* ANALYSIS OF VARIANCE.

VARIANCES, STATISTICAL STUDY OF

This article discusses statistical procedures related to the dispersion, or variability, of observations. Many such procedures center on the variance as a measure of dispersion, but there are other parameters measuring dispersion, and the most important of these are also considered here. This article treats motivation for studying dispersion, parameters describing dispersion, and estimation and testing methods for these parameters.

Some synonyms or near synonyms for "variability" or "dispersion" are "diversity," "spread," "heterogeneity," and "variation." "Entropy" is often classed with these.

Why study variability?

In many contexts interest is focused on variability, with questions of central tendency of secondary importance—or of no importance at all. The following are examples from several disciplines illustrating the interest in variability.

Economics. The inequality in wealth and income has long been a subject of study. Yntema (1933) uses eight different parameters to describe this particular variability; Bowman (1956) emphasizes curves as a tool of description [*cf.* Wold 1935 *for a discussion of Gini's concentration curve and* Kolmogorov 1958 *for Lévy's function of concentration; see also* Kravis 1968; Lebergott 1968].

Industry. The variability of industrial products usually must be small, if only in order that the products may fit as components into a larger system or that they may meet the consumer's demands; the methods of quality control serve to keep this variability (and possible trends with time) in check. [*An elementary survey is* Dudding 1952; *more modern methods are presented in* Keen & Page 1953; Page 1962; 1963; *see also* QUALITY CONTROL, STATISTICAL.]

Psychology. Two groups of children, selected at random from a given grade, were given a reasoning test under different amounts of competitive stress; the group under higher stress had the larger variation in performance. (The competitive atmosphere stimulated the brighter children, stunted the not-so-bright ones: see Hays 1963, p. 351; for other examples, see Siegel 1956, p. 148; Maxwell 1960; Hirsch 1961, p. 478.)

General approaches to the study of variability

The simplest approach to the statistical study of variability consists in the computation of the sample value of some statistic relating to dispersion [*see* STATISTICS, DESCRIPTIVE, *article on* LOCATION AND DISPERSION]. Conclusions as to the statistical significance or scientific interpretation of the resulting value, however, usually require selection of a specified family, $\mathcal{F}$, of probability distributions to represent the phenomenon under study. The choice of this family will reflect the theoretical framework within which the investigator performs his experiment(s). In particular, one or several of the parameters of the distributions of $\mathcal{F}$ will correspond to the notion of variability that is most relevant to the investigator's special problem.

The need for the selection of a specified underlying family, $\mathcal{F}$, is typical for statistical methodology in general and has the customary consequences: ideally speaking, each specified underlying family, $\mathcal{F}$, should have a corresponding statistic (or statistical procedure) adapted to it; even if a standard statistic (for example, variance) can be used, its significance and interpretation may vary widely with the underlying family. Unfortunately, the choice of such a family is not always self-evident, and hence the interpretation of statistical results is sometimes subject to considerable "specification error." [*See* ERRORS, *article on* EFFECTS OF ERRORS IN STATISTICAL ASSUMPTIONS.]

Two of the special families of probability distributions that will not be discussed in this article are connected with the methods of factor analysis and of variance components in the analysis of variance.

The factor analysis method analyzes a sample of N observations on an n-dimensional vector $(X_1, X_2, \cdots, X_n)$ by assuming that the X_i ($i = 1, \cdots, n$) are linear combinations of a random error term, a (hopefully small) number of "common factors," and possibly a number of "specific factors." (These assumptions determine a family, $\mathcal{F}$.) Interest focuses on the coefficients in the linear combinations (factor loadings). Unfortunately, the method lacks uniqueness in principle. [*See* FACTOR ANALYSIS AND PRINCIPAL COMPONENTS; *see also the survey by* Henrysson 1957.]

The variance components method, in one of its simpler instances, analyzes scalar-valued observations, x_{ijk} ($k = 1, \cdots, n_{ij}$), on n_{ij} individuals, observed under conditions $\mathcal{C}_{ij}$ ($i = 1, \cdots, r$; $j = 1, \cdots, s$), starting from the assumption that $x_{ijk} = \mu + a_i + b_j + c_{ij} + e_{ijk}$, where the a_i, b_j, c_{ij}, e_{ijk} are independent normal random variables with mean 0 and variances σ_a^2, σ_b^2, σ_c^2, σ_e^2, respectively. The objective is inference regarding these four variances, in order to evaluate variability from different sources. [*See* LINEAR HYPOTHESES, *article on* ANALYSIS OF VARIANCE.]

Parameters describing dispersion

Scales of measurement. Observations may be of different kinds, depending on the scale of measurement used: classificatory (or nominal), partially ordered, ordered (or ordinal), metric (defined below), and so forth. [*See* PSYCHOMETRICS *and* STATISTICS, DESCRIPTIVE, *for further discussion of scales of measurement.*] With each scale are associated transformations that may be applied to the observations and that leave the essential results of the measurement process intact. It is generally felt that parameters and statistical methods should in some sense be invariant under these associated transformations. (For dissenting opinions, see Lubin 1962, pp. 358–359.)

As an example, consider a classificatory scale. Measurement in this case means putting an observed unit into one of several unordered qualitative categories (for instance, never married, currently married, divorced, widowed). Whether these categories are named by English words, by the numbers 1, 2, 3, 4, by the numbers 3, 2, 1, 4, by the numbers 100, 101, 250, 261, or by the letters A, B, C, D does not change the classification as such. Hence, whatever it is that statistical methods extract from classification data should not depend on the names (or the transformations of the names) of the categories. Thus, even if the categories have numbers for their names, as above, it would be meaningless to compute the sample variance from a sample.

Parameters in general. Given a family, $\mathcal{G}$, of probability distributions, an identifiable parameter is a numerical-valued function defined on $\mathcal{G}$. Let P be a generic member of $\mathcal{G}$, and let m be a positive integer. Most of the parameters for describing dispersion discussed in this article can be defined as

$$E_P g(X_1, X_2, \cdots, X_m),$$

where $X_1, X_2, \cdots, X_m$ are independently and iden-

tically distributed according to P and g is an appropriate real-valued function. For example, the variance may be defined as $E_P[\frac{1}{2}(X_1 - X_2)^2]$. Given a family, $\mathcal{G}$, of probability distributions, one evidently has a wide choice of parameters (choosing a different g will usually yield a different parameter).

Different parameters will characterize (slightly or vastly) different aspects of $\mathcal{G}$. For instance, part of the disagreement between the eight methods of assessing variability described by Yntema (1933) stems from the fact that they represent different parameters. Of course, it is sometimes very useful to have more than one measure of dispersion available.

Dispersion parameters. A listing and comparison of various dispersion parameters for some of the scales mentioned above will now be given.

Parameters for classificatory scales. In a classificatory scale let there be q categories, with probabilities θ_i ($i = 1, 2, \cdots, q$); $\sum_i \theta_i = 1$. The dispersion parameter chosen should be invariant under name change of the categories, so it should depend on the θ_i only. If all θ_i are equal, diversity (variability) is a maximum, and the parameter should have a large value. If one θ_i is 1, so that the others are 0, diversity is 0, and the parameter should conventionally have the value 0. A family of parameters having these and other gratifying properties (for example, a weak form of additivity; see Rényi 1961, eqs. (1.20) and (1.21)) is given by

$$H_\alpha(P) = (1 - \alpha)^{-1} \log_2 \left(\sum_{i=1}^{q} \theta_i^\alpha \right) \geq 0, \qquad \begin{array}{l} \alpha > 0, \\ \alpha \neq 1, \end{array}$$

the amount of information of order α (entropy of order α). Note that

$$\lim_{\alpha \to 1} H_\alpha(P) = -\sum_i \theta_i \log_2 \theta_i = -E_P \log_2 P(X)$$

is Shannon's amount of information (see Pollack 1968). This information measure has a stronger additivity property—Blyth (1959) points out that if the values of X are divided into groups, then the dispersion of X = between group dispersion + expected within group dispersion. Miller and Chomsky (1963) discuss linguistic applications.

There are other measures of dispersion for classificatory scales besides the information-like ones. (For example, see Greenberg 1956.)

Parameters for metric scales. On a metric scale observations are real numbers, and all properties of real numbers may be used.

(*a*) For probability distributions with a density f, there is the information-like parameter

$$H_1(f) = -E_f \log_2 f(X) = -\int f(x) \log_2 f(x)\, dx \geq 0,$$

whenever the integral exists. This parameter is not

invariant under arbitrary transformations of the X-line, although it is under translations. (For interesting maximum properties in connection with rectangular, exponential, normal distributions, see Rényi 1962, appendix, sec. 11, exercises 12, 17.) For a normal distribution with standard deviation σ, $H_1(f) = \frac{1}{2}\log_2(\sigma^2 2\pi e)$.

(*b*) Traditional measures of dispersion for metric scales are the standard deviation, $\sigma \geqslant 0$, and the variance, $\sigma^2 = E_P[(X - \mu)^2]$, where $\mu = E_P X$. As mentioned above, an alternative definition is

$$\sigma^2 = \tfrac{1}{2}E_P[(X_1 - X_2)^2],$$

half the expectation of the square of the difference of two random variables, X_1 and X_2, independently and identically distributed. This definition of σ^2 suggests a whole string of so-called mean difference parameters, listed below under (*c*), (*d*), and (*e*), all of which, like σ and σ^2, are invariant under translations only.

(*c*) Gini's mean difference is given by

$$\delta_1^1 = E_P|X_1 - X_2| = \int\int |x_1 - x_2|\, P(dx_1)\, P(dx_2).$$

The integral at the right is in general form; if X_1, X_2 have the density function f, the integral is

$$\delta_1^1 = \int\int |x_1 - x_2|f(x_1)f(x_2)\,dx_1 dx_2.$$

Wold (1935, pp. 48–49) points out the relationship between this parameter and Cramér's ω^2 method for testing goodness of fit. As can be seen in Table 1, below, Gini's mean difference is a distribution-dependent function of σ.

There are variate difference parameters that involve the square of "higher-order differences"; they are distribution-free functions of σ. An example is

$$E_P[(X_3 - 2X_2 + X_1)^2] = 6\sigma^2.$$

There are also variate difference parameters involving the absolute value of higher-order differences; they are distribution-dependent functions of σ. An

example is

$$\delta_2^1 = E_P|X_3 - 2X_2 + X_1|.$$

(*d*) By analogy with the first definition of the variance, there are dispersion parameters reflecting absolute variation around some measure of central tendency. Examples are the mean deviation from the mean, μ,

$$\delta_\mu = E_P|X - \mu|,$$

and from the median, Med X,

$$\delta_{\mathrm{Med}} = E_P|X - \mathrm{Med}\,X|.$$

These are distribution-dependent functions of σ.

(*e*) There are dispersion parameters based on other differences. Two examples are the expected value of range of samples of size n,

$$E_P W_n = E_P[X_{(n)} - X_{(1)}],$$

where $X_{(1)} = \min(X_1, X_2, \cdots, X_n)$ and $X_{(n)} = \max(X_1, X_2, \cdots, X_n)$; and the difference of symmetric quantile points,

$$\xi_{1-\alpha} - \xi_\alpha,$$

where

$$\int_{-\infty}^{\xi_\alpha} f = \alpha = \int_{\xi_{1-\alpha}}^{\infty} f,$$

and f is the density of the probability distribution P. Both these parameters are distribution-dependent functions of σ. Note that this last parameter, the difference of symmetric quantile points, is not based on expected values of random variables.

(*f*) Another dispersion parameter is the coefficient of variation, σ/μ (given either as a ratio or in per cent), invented to eliminate the influence of absolute size on variability (for example, to compare the variation in size of elephants and of mice). Sometimes it does exactly that (Banerjee 1962); sometimes it does nothing of the sort (Cramér 1945, table 31.3.5). (For further discussion, see Pearson 1897.)

Table 1 — A comparison of dispersion parameters

Distributional form	Parameters of distribution			Coefficient of variation	Ratios of mean difference to σ				
	σ	μ	Med X	σ/μ	δ_1^1/σ	δ_2^1/σ	δ_μ/σ	$\delta_{\mathrm{Med}}/\sigma$	EW_n/σ
Normal	σ	μ	μ	σ/μ	$2/\sqrt{\pi}$	$2\sqrt{3/\pi}$	$\sqrt{2/\pi}$	$\sqrt{2/\pi}$	$\sim 2\sqrt{2}\log_e n$
Exponential	θ	θ	$\theta\log_e 2$	1	1	16/9	$2/e$	$\log_e 2$	$\sim \gamma + \log_e n$ [a]
Double exponential	$\theta\sqrt{2}$	λ	λ	$\theta\sqrt{2}/\lambda$	$3\sqrt{2}/16$	b	$\tfrac{1}{2}\sqrt{2}$	$\tfrac{1}{2}\sqrt{2}$	$\sim \sqrt{2}\,[\gamma + \log_e(n/\sqrt{2})]$ [a]
Rectangular	$\dfrac{b-a}{2\sqrt{3}}$	$\dfrac{b+a}{2}$	$\dfrac{b+a}{2}$	$\dfrac{b-a}{\sqrt{3}(b+a)}$	$2/\sqrt{3}$	b	$\tfrac{1}{2}\sqrt{3}$	$\tfrac{1}{2}\sqrt{3}$	$2\sqrt{3}\left(\dfrac{n-1}{n+1}\right)$

a. Here γ is Euler's constant: $\gamma = 0.5772157\ldots$

b. Not known from the literature.

Because they are distribution-dependent functions of σ, the parameters cited under (a), (c), (d), and (e) are undesirable for a study of the variance, σ^2, unless one is fairly sure about the underlying family of probability distributions. This will be illustrated below. Despite this drawback, these parameters are, of course, quite satisfactory as measures of dispersion in their own right.

Comparison of dispersion measures. Table 1 lists the quotient of several of the above-mentioned parameters divided by σ, together with other relevant quantities. It gives these comparisons for the distributions of the types listed in the first column, with parameter specifications as indicated in the next two columns. (The parameterization is the same as that in DISTRIBUTIONS, STATISTICAL, *article on* SPECIAL CONTINUOUS DISTRIBUTIONS.) The sign "~" before an entry denotes an asymptotic result (for large n or large μ). Table 1 illustrates how tribution dependence of these parameters can really be. [*See* ERRORS, *article on* NONSAMPLING ERRORS, *for further discussion.*]

Multivariate distributions. Most of the parameters discussed for univariate distributions can be generalized to multivariate distributions, usually in more than one fashion. The variance, for instance, is the expected value of one-half the square of the distance between two random points on the real line. Generalization may be attained by taking the distance between two random points in k-space or by taking the content of a polyhedron spanned by $k + 1$ points in k-space. Thus, a rather great variety of multivariate dispersion parameters are possible. [*See* MULTIVARIATE ANALYSIS *and, for example,* van der Vaart 1965.]

Statistical inference

Shannon's amount of information. Consider, first, point estimation of Shannon's amount of information for discrete distributions. Suppose a sample of size n is drawn from the probability distribution, with q categories and probabilities θ_i, described earlier. Suppose n_i observations fall in the ith category; $\sum_{i=1}^q n_i = n$. Then

$$\hat{H} = -\sum_{i=1}^q \frac{n_i}{n} \log_2 \frac{n_i}{n}$$

suggests itself as the natural estimator for $H_1(P) = -\sum_i \theta_i \log_2 \theta_i$. The properties of this estimator have been studied by Miller and Madow (1954) and (by a simpler method) by Basharin (1959). The sampling distribution of $\hat{H}$ has mean

$$E\hat{H} = H_1(P) - \frac{q-1}{2n} \log_2 e + O\left(\frac{1}{n^2}\right).$$

(The term $O(1/n^2)$ denotes a function of n and the θ_i such that for some positive constant, c, the absolute value of the function is less than c/n^2.) So for "small" n the bias, the difference between $E\hat{H}$ and $H_1(P)$, is substantial. [*For low-bias estimators, see* Blyth 1959; *for a general discussion of point estimation, see* ESTIMATION, *article on* POINT ESTIMATION.]

The variance of one population. Procedures for estimating the variance of a single population and for testing hypotheses about such a variance will now be described.

Point estimation for general $\mathcal{G}$. Let the underlying family, $\mathcal{G}_\mu$, consist of all probability distributions with density functions and known mean, μ, or of all discrete distributions with known mean, μ. In both cases the theory of U-statistics (see, for example, Fraser 1957, pp. 135–147) shows that the minimum variance unbiased estimator of σ^2, given a sample of size n, is

$$s_1^2 = \frac{1}{n} \sum_{i=1}^n (x_i - \mu)^2.$$

Note that the sampling variance, $\mathrm{var}_P(s_1^2)$, which measures the precision of the estimator relative to the underlying distribution, P (a member of $\mathcal{G}_\mu$), is definitely distribution dependent. If $\mathcal{G}$ is, again, the family of all absolutely continuous (or discrete) distributions now with *unknown mean*, then the uniformly minimum variance unbiased estimator of σ^2 is

$$s_2^2 = \frac{1}{n-1} \sum_{i=1}^n (x_i - \bar{x})^2,$$

where $\bar{x}$ is the sample mean. Again, $\mathrm{var}_P(s_2^2)$ is very much distribution dependent.

For more restricted families of distributions it is sometimes possible to find other estimators, with smaller sampling variances. Also, if the unbiasedness requirement is dropped, one may find estimators that, although biased, are, on the average, closer to the true parameter value than a minimum variance unbiased estimator: for the family of normal distributions,

$$s_3^2 = \frac{1}{n+1} \sum_{i=1}^n (x_i - \bar{x})^2$$

is such an estimator of σ^2.

Distribution dependence. To illustrate the dependence of the quality of point estimators upon the underlying family of probability distributions, Table 2 lists the sampling variance of s_2^2 for random samples from 5 different distribution families. It is seen that the quotient $\mathrm{var}_P(s_2^2)/\mathrm{var}_N(s_2^2)$

Table 2 — The sampling variance of s_2^2

Distribution	var(s_2^2)[a]
Normal	$\dfrac{1}{n}2\sigma^4$
Exponential	$\dfrac{1}{n}8\sigma^4$
Double exponential	$\dfrac{1}{n}32\sigma^4$
Rectangular	$\dfrac{1}{n}(4/5)\sigma^2$
Pearson type VII[b] $(f(x) = \kappa(1 + x^2 a^{-2})^{-p};$ $p > 5/2)$	$\dfrac{1}{n}\left(1 + \dfrac{3}{2p - 5}\right)2\sigma^4$

a. The reader should add a term $O(1/n^2)$ to each entry.
b. Example is due to Hotelling 1961, p. 350.

(where P indicates some nonnormal underlying distribution and N a normal one) may vary from $2/5$ to ∞. Hence, unless $\mathcal{G}$ can be chosen in a responsible way, little can be said about the precision of s_2^2 as an estimator of σ^2 (although for large samples the higher sample moments will be of some assistance in evaluating the precision of this estimator).

Normal distributions. Tests and confidence intervals on dispersion parameters for the case of normal distributions will now be discussed. In order to decide whether a sample of n observations may have come from a population with known variance, σ_0^2, or from a more heterogeneous one, test the hypothesis H_0: $\sigma^2 = \sigma_0^2$ against the one-sided alternative $H_A^{(1)}$: $\sigma^2 > \sigma_0^2$, where σ^2 is the (unknown) variance characterizing the sample (Rao 1952, sec. 6a.1, gives a concrete example of the use of this one-sided alternative). In order to investigate only whether the sample fits into the given population in terms of homogeneity, test H_0: $\sigma^2 = \sigma_0^2$ against $H_A^{(2)}$: $\sigma^2 \neq \sigma_0^2$, where $H_A^{(2)}$ is a two-sided alternative. If the underlying family is normal, the most powerful level-α test for the one-sided alternative rejects H_0 whenever

$$\sum_{i=1}^{n}(x_i - \bar{x})^2 \equiv S^2 > \sigma_0^2\chi_{1-\alpha,n-1}^2,$$

where $\chi_{\delta,n-1}^2$ is the 100δ per cent point of the chi-square distribution for $n - 1$ degrees of freedom (so that $\chi_{1-\alpha,n-1}^2$ is the upper 100α per cent point of the same distribution). [*For further discussion of these techniques and the terminology, see* HYPOTHESIS TESTING.]

The most powerful unbiased level-α test for the two-sided alternative rejects H_0 whenever

$$S^2 < \sigma_0^2 C_1 \quad \text{or} \quad S^2 > \sigma_0^2 C_2.$$

Here $C_1 = \chi_{\gamma,n-1}^2 = \chi_{\lambda,n+1}^2$, and $C_2 = \chi_{1-\beta,n-1}^2 = \chi_{1-\nu,n+1}^2$,

with $\beta + \gamma = \alpha = \lambda + \nu$ (see Lehmann 1959, chapter 5, sec. 5, example 5, and pp. 165 and 129; for tables, see Lindley et al. 1960: $\alpha = 0.05, 0.01, 0.001$). In practice the nonoptimal equal-tail test is also used, where $C_1 = \chi_{\gamma,n-1}^2$ and $C_2 = \chi_{1-\beta,n-1}^2$, with $\beta = \gamma = \frac{1}{2}\alpha$. For the latter test the standard chi-square tables suffice, and the two tests differ only slightly unless the sample size is very small.

The one-sided and two-sided confidence intervals follow immediately from the above inequalities; for example, a two-sided confidence interval for σ^2 at level α is

$$S^2/C_1 < \sigma^2 < S^2/C_2.$$

Nonnormal distributions. The above discussion of the distribution dependence of point estimators of dispersion parameters should have prepared the reader to learn that the tests and confidence interval procedures discussed above are not robust against nonnormality. Little has been done in developing tests or confidence intervals for σ^2 when $\mathcal{G}$ is unknown or broad. Hotelling (1961, p. 356) recommends using all available knowledge to narrow $\mathcal{G}$ down to a workable family of distributions, then adapting statistical methods to the resulting family.

Mean square differences. For a large family of absolutely continuous distributions with unknown mean, the minimum variance unbiased estimator, s_2^2, was introduced above. An alternative formula is

$$s_2^2 = \frac{1}{2n(n - 1)}\sum_i\sum_j(x_i - x_j)^2.$$

This formula suggests another estimator of $2\sigma^2$, unbiased, but not with minimum variance:

$$d_1^2 = \frac{1}{n - 1}\sum_{i=1}^{n-1}(x_i - x_{i+1})^2.$$

If the indices $1, 2, \cdots, n$ in the sample $x_1, x_2, \cdots, x_n$ indicate an ordering of some kind (for example, the order of arrival in a time series), then d_1^2 is called the first mean square successive difference. Similarly,

$$d_2^2 = \frac{1}{n - 2}\sum_{i=1}^{n-2}(x_i - 2x_{i+1} + x_{i+2})^2,$$

the second mean square successive difference, is an unbiased estimator of $6\sigma^2$.

If the underlying family, $\mathcal{G}$, is normal, then asymptotically (for large n)

$$\text{var}\left(\frac{1}{2}d_1^2\right) \sim \frac{3\sigma^4}{n}; \quad \text{var}\left(\frac{1}{6}d_2^2\right) \sim \frac{3.89\sigma^4}{n}$$

(see Kamat 1958). These estimators, although

clearly less precise than s_2^2, are of interest because they possess a special kind of robustness—against trend. Suppose the observations $x_1, \cdots, x_n$ have been taken at times $t_1, \cdots, t_n$ from a time process, $X(t) = \phi(t) + Y$, where ϕ is a smoothly varying function (trend) of t, and the distribution of the random variable Y is independent of t (for example, ϕ might describe an expanding economy and Y the fluctuations in it). Let an estimator be sought for var(Y). Most of the trend is then eliminated by considering only the successive differences $x_i - x_{i+1} = \phi(t_i) - \phi(t_{i+1}) + y_i - y_{i+1}$, thus making for an estimator of var(Y) with much less bias. These methods have been applied to control and record charts by Keen and Page (1953), for example.

Little work has been done on studying the sampling distributions of successive difference estimators in cases where the underlying distribution is nonnormal. Moore (1955) gave moments and approximations of d_1^2 for four types of distributions.

The standard deviation of one population. Since $\sigma = \sqrt{\sigma^2}$, one might feel that the standard deviation, σ, should be estimated by the square root of a reasonable estimator of σ^2. This is, indeed, often done, and for large sample sizes the results are quite acceptable. For smaller sample sizes, however, the suboptimality of such estimators is more marked (specifically, $Es_2 \neq \sigma$; if the underlying family is normal, an unbiased estimator is

$$s_2^* = \frac{s_2 \sqrt{\tfrac{1}{2}(n-1)}\ \Gamma\left[\tfrac{1}{2}(n-1)\right]}{\Gamma\left(\tfrac{1}{2}n\right)},$$

where Γ is the gamma function). Therefore, there has been some interest in alternative estimators, like those now to be described.

Estimation via alternative parameters. In Table 1 it was pointed out that, depending on the underlying family, $\mathcal{G}$, of distributions, certain relations exist between σ and other dispersion parameters, θ, of the form $\theta = v_{\mathcal{G}}\sigma$. So if one knows $\mathcal{G}$, one may estimate θ by, say, $T(x)$, apply the conversion factor $1/v_{\mathcal{G}}$, and find an unbiased estimator of σ.

Thus, the mean successive differences,

$$d_1^1 = \frac{1}{n-1} \sum_{i=1}^{n-1} |x_i - x_{i+1}|,$$

$$d_2^1 = \frac{1}{n-2} \sum_{i=1}^{n-2} |x_i - 2x_{i+1} + x_{i+2}|,$$

are, if $\mathcal{G}$ is normal, unbiased estimators of $2\sigma/\sqrt{\pi}$ and $2\sigma\sqrt{3/\pi}$, respectively, with sampling variances

(see Kamat 1958) given by

$$\mathrm{var}\left(\frac{\sqrt{\pi}}{2}d_1^1\right) = \frac{0.826\sigma^2}{n} + \mathrm{o}\left(\frac{1}{n}\right),$$

$$\mathrm{var}\left(\frac{\sqrt{\pi}}{2\sqrt{3}}d_2^1\right) = \frac{1.062\sigma^2}{n} + \mathrm{o}\left(\frac{1}{n}\right).$$

(Here the term o($1/n$) denotes a function that, after multiplication by n, goes to zero as n becomes large.) See Lomnicki (1952) for the sampling variance of $[n(n-1)]^{-1}\sum_i\sum_j|x_i - x_j|$, Gini's mean difference, for normal, exponential, and rectangular $\mathcal{G}$.

Again, if

$$d_m = \frac{1}{n} \sum_{i=1}^{n} |x_i - \bar{x}|$$

and $\mathcal{G}$ is normal, then $d_m\sqrt{(\pi/2)/[(n-1)/n]}$ is an unbiased estimator of σ; its sampling variance is

$$\mathrm{var}(d_m\sqrt{(\pi/2)/[(n-1)/n]})$$

$$= \frac{\sigma^2}{n}\left[\left(\frac{\pi-2}{2}\right) + \frac{1}{2n} + \mathrm{o}\left(\frac{1}{n}\right)\right],$$

which is close to the absolute lower bound, $\sigma^2/(2n)$. The properties of

$$d_{\mathrm{Me}(x)} = \frac{1}{n} \sum_{i=1}^{n} |x_i - \mathrm{Me}(x)|,$$

where Me(x) is the sample median, differ slightly, yet favorably, from those of d_m. The literature on these and similar statistics is quite extensive.

The last column of Table 1 suggests the use of the sample range, $W_n = X_{(n)} - X_{(1)}$, to estimate σ; the conversion factor now depends on both the underlying distribution and the sample size, n (for normal distributions, see David 1962, p. 113, table 7A.1). With increasing n, the precision of converted sample ranges as estimators of σ decreases rapidly. One may then shift to quasi ranges $(X_{(n-r+1)} - X_{(r)})$ or, better still, to linear combinations of quasi ranges (see David 1962, p. 107). The use of quasi ranges to obtain confidence intervals for interquantile distances $(\xi_{1-\alpha} - \xi_\alpha)$ was also discussed by Chu (1957). This type of estimator employs order statistics. A more efficient use of order statistics is made by the so-called best unbiased linear systematic statistics and by approximations to these [for more information, see NONPARAMETRIC STATISTICS, *article on* ORDER STATISTICS]. These linear systematic statistics are especially useful in case the data are censored [see STATISTICAL ANALYSIS, SPECIAL PROBLEMS OF, *article on* TRUNCATION AND CENSORSHIP]. It should also be mentioned that grouping of data poses special problems for the use of estimators based on order statistics [*see* STATISTICAL ANALYSIS, SPECIAL PROBLEMS OF, *article on* GROUPED OBSERVATIONS].

Comparing variances of several populations. As in the example of increased variation on a reasoning test with competitive stress, discussed above, it appears that situations will occur in which interest is focused on differences in variability as the response to differences in conditions. Two groups were compared in the example, but the situation can easily be generalized to more than two groups. Thus, one may want to apply more than two levels of competitive stress, and one may even bring in a second factor of the environment, such as different economic backgrounds (in which case one would have a two-way classification).

Bartlett's test and the F-test. Consider k populations and k samples, one from each population (in the reasoning-test example, each group of children under a given level of stress would constitute one sample). Let the observations $x_{r1}, x_{r2}, \cdots, x_{rn_r}$ be a random sample from the rth population ($r = 1, \cdots, k$). Let

$$\bar{x}_r = \frac{1}{n_r} \sum_{i=1}^{n_r} x_{ri}.$$

Define $\quad S_r^2 = \sum_{i=1}^{n_r} (x_{ri} - \bar{x}_r)^2 \quad$ and $\quad \nu_r = n_r - 1,$

where $\quad n = \sum_{r=1}^{k} n_r \quad$ and $\quad \sum_{r=1}^{k} \nu_r = n - k.$

Bartlett's 1937 test of the hypothesis H_0: the k variances are equal, against H_A: not all variances are equal, assumes all samples to be drawn from normal distributions and rejects H_0 if and only if the statistic L is too large, where

$$L = (n - k) \log_e \frac{\sum\limits_{r=1}^{k} S_r^2}{\sum\limits_{r=1}^{k} \nu_r} - \sum_{r=1}^{k} \nu_r \log_e \frac{S_r^2}{\nu_r}.$$

The true values of the means of the k populations do not influence the outcome of this test. The distribution of L is known to be chi-square, with $k - 1$ degrees of freedom, for large samples; for samples of intermediate size, it is desirable to use, as a closer approximation, the fact that $L/(1 + c)$, where

$$c = \frac{1}{3(k - 1)} \left(\sum \frac{1}{\nu_r} - \frac{1}{n - k} \right),$$

has approximately the same chi-square distribution. Bartlett's test is unbiased. Against these virtues there is one outstanding weakness: the test has

total lack of robustness against nonnormality [*see* ERRORS, *article on* EFFECTS OF ERRORS IN STATISTICAL ASSUMPTIONS].

For $k = 2$, Bartlett's test reduces to a variant of the *F*-test: reject H_0: $\sigma_1 = \sigma_2$ in favor of $H_A^{(2)}$: $\sigma_1 \neq \sigma_2$, if S_2^2/S_1^2 is either too large or too small. The one-sided *F*-test rejects H_0 in favor of $H_A^{(1)}$: $\sigma_1 > \sigma_2$, if $S_1^2/S_2^2 > (\nu_1/\nu_2) F_{1-\alpha;\nu_1,\nu_2}$, where $F_{1-\alpha;\nu_1,\nu_2}$ is the upper 100α per cent point of the *F*-distribution for ν_1 and ν_2 degrees of freedom. The *F*-test *in this context* naturally has the same lack of robustness against nonnormality as Bartlett's test.

Alternate test for variance heterogeneity. Bartlett and Kendall (1946) proposed an alternative approach: apply analysis of variance techniques to the logarithms of the k sample variances. The virtue of this suggestion is that the procedure can be generalized immediately to a test of variances in a two-way classification. Box (1953, p. 330) showed that this test, too, is nonrobust against nonnormality. More robust procedures are described below.

Variances of two correlated samples. McHugh (1953) quotes a study of the effect of age on dispersion of mental abilities; the same group of persons was measured at two different ages. Naturally the two samples are correlated, and the *F*-test does not apply. Under the assumption that the pairs $(x_{11}, x_{21}), \cdots, (x_{1i}, x_{2i}), \cdots, (x_{1n}, x_{2n})$ constitute a sample from a bivariate normal distribution with variances σ_1^2 and σ_2^2 and correlation coefficient ρ, the hypothesis H_θ: $\sigma_1^2/\sigma_2^2 = \theta$ is tested by the statistic

$$T_\theta = \frac{(S_1^2 - \theta S_2^2) \sqrt{n - 2}}{\sqrt{4 S_1^2 S_2^2 \theta (1 - r^2)}},$$

where S_1^2 and S_2^2 are as defined in the discussion of Bartlett's test, above, and r is the sample correlation coefficient. The statistic T_θ is distributed under the null hypothesis as Student's t with $n - 2$ degrees of freedom. One-sided tests, two-sided tests, and confidence intervals follow in the customary manner. Specifically, the hypothesis H_1: $\sigma_1^2 = \sigma_2^2$ is tested by means of the statistic

$$T_1 = \frac{(S_1^2 - S_2^2) \sqrt{n - 2}}{\sqrt{4 S_1^2 S_2^2 (1 - r^2)}}.$$

This method, which was proposed by Morgan (1939) and Pitman (1939), is based on the easily derived fact that the covariance between $X + Y$ and $X - Y$ is the difference between the variances of X and Y, so that the correlation between the sum and difference of the random variables is zero if and only if the variances are equal.

Testing for variance-heterogeneity preliminary to ANOVA. The analysis of variance assumes equality of variance from cell to cell. Hence, it is sometimes proposed that the data be run through a preliminary test to check this assumption, also called that of homoscedasticity; variance heterogeneity is also called heteroscedasticity.

There are two objections to this procedure. First, the same data are subjected to two different statistical procedures, so the two results are not independent. Hence, a special theoretical investigation is needed to find out what properties such a double procedure has (see Kitagawa 1963). Second (see Box 1953, p. 333), one should not use Bartlett's test for such a preliminary analysis, because of its extreme lack of robustness against nonnormality: one might discard as heteroscedastic data that are merely nonnormal, whereas the analysis of variance is rather robust against nonnormality. (An additional important point is that analysis of variance is fairly robust against variance heterogeneity, at least with equal numbers in the various cells.) [*See* SIGNIFICANCE, TESTS OF, *for further discussion of preliminary tests.*]

In view of the relative robustness of range methods, Hartley's suggestion (1950, pp. 277–279) of testing for variance heterogeneity by means of range statistics is quite attractive.

Robust tests against variance heterogeneity. Box (1953, sec. 8) offers a more robust *k*-sample test against variance heterogeneity: each of the *k* samples is broken up into small, equal, exclusive, and exhaustive random subsets, a dispersion statistic is computed for each subset, and the within-sample variation of these statistics is compared with the between-sample variation. (Box 1953 applies an analysis of variance to the logarithms of these statistics; Moses 1963, p. 980, applies a rank test to the statistics themselves.)

Another approach applies a permutation test, which amounts to a kurtosis-dependent correction of Bartlett's test (Box & Andersen 1955, p. 23). The results are good, although they are better in the case of known means than in the case of unknown means.

Still another procedure uses rank tests [*see* NONPARAMETRIC STATISTICS, *articles on* THE FIELD *and on* RANKING METHODS; *see also a survey by* van Eeden 1964]. Moses (1963, secs. 3, 4) makes some enlightening remarks about things a rank test for dispersion can and cannot be expected to do.

H. ROBERT VAN DER VAART

[*See also* STATISTICS, DESCRIPTIVE, *article on* LOCATION AND DISPERSION.]

BIBLIOGRAPHY

BANERJEE, V. 1962 Experimentelle Untersuchungen zur Gültigkeit des Variationskoeffizienten V in der Natur, untersucht an zwei erbreinen Populationen einer Wasserläuferart. *Biometrische Zeitschrift* 4:121–125.

BARTLETT, M. S. 1937 Properties of Sufficiency and Statistical Tests. Royal Society of London, *Proceedings* Series A 160:268–282.

BARTLETT, M. S.; and KENDALL, D. G. 1946 The Statistical Analysis of Variance-heterogeneity and the Logarithmic Transformation. *Journal of the Royal Statistical Society,* Series B 8:128–138.

BASHARIN, G. P. 1959 On a Statistical Estimate for the Entropy of a Sequence of Independent Random Variables. *Theory of Probability and Its Applications* 4:333–337. → First published in Russian in the same year, in *Teoriia veroiatnostei i ee primeneniia,* of which the English edition is a translation, published by the Society for Industrial and Applied Mathematics.

BLYTH, COLIN R. 1959 Note on Estimating Information. *Annals of Mathematical Statistics* 30:71–79.

BOWMAN, MARY J. 1956 The Analysis of Inequality Patterns: A Methodological Contribution. *Metron* 18, no. 1/2:189–206.

BOX, G. E. P. 1953 Non-normality and Tests on Variances. *Biometrika* 40:318–335.

BOX, G. E. P.; and ANDERSEN, S. L. 1955 Permutation Theory in the Derivation of Robust Criteria and the Study of Departures From Assumptions. *Journal of the Royal Statistical Society* Series B 17:1–34.

CHU, J. T. 1957 Some Uses of Quasi-ranges. *Annals of Mathematical Statistics* 28:173–180.

CRAMÉR, HARALD (1945) 1951 *Mathematical Methods of Statistics.* Princeton Mathematical Series, No. 9. Princeton Univ. Press.

DAVID, H. A. 1962 Order Statistics in Short-cut Tests. Pages 94–128 in Ahmed E. Sarhan and Bernard G. Greenberg (editors), *Contributions to Order Statistics.* New York: Wiley.

DUDDING, BERNARD P. 1952 The Introduction of Statistical Methods to Industry. *Applied Statistics* 1:3–20.

FRASER, DONALD A. S. 1957 *Nonparametric Methods in Statistics.* New York: Wiley.

GREENBERG, JOSEPH H. 1956 The Measurement of Linguistic Diversity. *Language* 32:109–115.

HARTLEY, H. O. 1950 The Use of Range in Analysis of Variance. *Biometrika* 37:271–280.

HAYS, WILLIAM L. 1963 *Statistics for Psychologists.* New York: Holt.

HENRYSSON, STEN 1957 *Applicability of Factor Analysis in the Behavioral Sciences: A Methodological Study.* Stockholm Studies in Educational Psychology, No. 1. Stockholm: Almqvist & Wiksell.

HIRSCH, JERRY 1961 The Role of Assumptions in the Analysis and Interpretation of Data. *American Journal of Orthopsychiatry* 31:474–480. → Discussion paper in a symposium on the genetics of mental disease.

HOTELLING, HAROLD 1961 The Behavior of Some Standard Statistical Tests Under Non-standard Conditions. Volume 1, pages 319–359 in Berkeley Symposium on Mathematical Statistics and Probability, Fourth, 1960, *Proceedings.* Edited by Jerzy Neyman. Berkeley and Los Angeles: Univ. of California Press.

KAMAT, A. R. 1958 Contributions to the Theory of Statistics Based on the First and Second Successive Differences. *Metron* 19, no. 1/2:97–118.

KEEN, JOAN; and PAGE, DENYS J. 1953 Estimating Variability From the Differences Between Successive Readings. *Applied Statistics* 2:13–23.

KITAGAWA, TOSIO 1963 Estimation After Preliminary Tests of Significance. *University of California Publications in Statistics* 3:147–186.

KOLMOGOROV, A. N. 1958 Sur les propriétés des fonctions de concentrations de M. P. Lévy. Paris, Université, Institut Henri Poincaré, *Annales* 16:27–34.

►KRAVIS, IRVING B. 1968 Income Distribution: I. Functional Share. Volume 7, pages 132–145 in *International Encyclopedia of the Social Sciences*. Edited by David L. Sills. New York: Macmillan and Free Press.

►LEBERGOTT, STANLEY 1968 Income Distribution: II. Size. Volume 7, pages 145–154 in *International Encyclopedia of the Social Sciences*. Edited by David L. Sills. New York: Macmillan and Free Press.

LEHMANN, E. L. 1959 *Testing Statistical Hypotheses.* New York: Wiley.

LINDLEY, D. V.; EAST, D. A.; and HAMILTON, P. A. 1960 Tables for Making Inferences About the Variance of a Normal Distribution. *Biometrika* 47:433–437.

LOMNICKI, Z. A. 1952 The Standard Error of Gini's Mean Difference. *Annals of Mathematical Statistics* 23:635–637.

LUBIN, ARDIE 1962 Statistics. *Annual Review of Psychology* 13:345–370.

McHUGH, RICHARD B. 1953 The Comparison of Two Correlated Sample Variances. *American Journal of Psychology* 66:314–315.

MAXWELL, A. E. 1960 Discrepancies in the Variances of Test Results for Normal and Neurotic Children. *British Journal of Statistical Psychology* 13:165–172.

MILLER, GEORGE A.; and CHOMSKY, NOAM 1963 Finitary Models of Language Users. Volume 2, pages 419–491 in R. Duncan Luce, Robert R. Bush, and Eugene Galanter (editors), *Handbook of Mathematical Psychology.* New York: Wiley.

MILLER, GEORGE A.; and MADOW, WILLIAM G. (1954) 1963 On the Maximum Likelihood Estimate of the Shannon–Wiener Measure of Information. Volume 1, pages 448–469 in R. Duncan Luce, Robert R. Bush, and Eugene Galanter (editors), *Readings in Mathematical Psychology.* New York: Wiley.

MOORE, P. G. 1955 The Properties of the Mean Square Successive Difference in Samples From Various Populations. *Journal of the American Statistical Association* 50:434–456.

MORGAN, W. A. 1939 A Test for the Significance of the Difference Between the Two Variances in a Sample From a Normal Bivariate Population. *Biometrika* 31:13–19.

MOSES, LINCOLN E. 1963 Rank Tests of Dispersion. *Annals of Mathematical Statistics* 34:973–983.

PAGE, E. S. 1962 Modified Control Chart With Warning Lines. *Biometrika* 49:171–176.

PAGE, E. S. 1963 Controlling the Standard Deviation by Cusums and Warning Lines. *Technometrics* 5:307–315.

PEARSON, KARL 1897 On the Scientific Measure of Variability. *Natural Science* 11:115–118.

PITMAN, E. J. G. 1939 A Note on Normal Correlation. *Biometrika* 31:9–12.

►POLLACK, IRWIN 1968 Information Theory. Volume 7, pages 331–337 in *International Encyclopedia of the Social Sciences*. Edited by David L. Sills. New York: Macmillan and Free Press.

○RAO, C. RADHAKRISHNA 1952 *Advanced Statistical Methods in Biometric Research.* New York: Wiley. → Reprinted with corrections in 1970 and 1974 by Hafner.

RÉNYI, ALFRÉD 1961 On Measures of Entropy and Information. Volume 1, pages 547–561 in Berkeley Symposium on Mathematical Statistics and Probability, Fourth, 1960, *Proceedings.* Edited by Jerzy Neyman. Berkeley and Los Angeles: Univ. of California Press.

¹RÉNYI, ALFRÉD 1962 *Wahrscheinlichkeitsrechnung, mit einem Anhang über Informationstheorie.* Berlin: Deutscher Verlag der Wissenschaften. → The German edition of Rényi's textbook, first published in Hungarian in 1954 and revised in each of several successive editions in various languages.

SIEGEL, SIDNEY 1956 *Nonparametric Statistics for the Behavioral Sciences.* New York: McGraw-Hill.

VAN DER VAART, H. ROBERT 1965 A Note on Wilks' Internal Scatter. *Annals of Mathematical Statistics* 36:1308–1312.

VAN EEDEN, CONSTANCE 1964 Note on the Consistency of Some Distribution-free Tests for Dispersion. *Journal of the American Statistical Association* 59:105–119.

WOLD, HERMAN 1935 A Study on the Mean Difference, Concentration Curves and Concentration Ratio. *Metron* 12, no. 2:39–58.

YNTEMA, DWIGHT B. 1933 Measures of the Inequality in the Personal Distribution of Wealth or Income. *Journal of the American Statistical Association* 28:423–433.

Postscript

Some additional interesting examples of the importance of variability as a phenomenon in its own right are collected here.

Biology. The biologists have a running battle with variability and the concomitant lack of reproducibility of their experiments. For an account of one botanist's attack on this problem by means of careful control of environmental conditions for plant growth, see Went (1957, especially chapter 15). For a discussion of physiological variability of a roughly periodic nature in mammals, see Savage et al. (1962), on circadian variations in susceptibility to harmful agents; the effect of these variations "may be as drastic as the difference between death and survival" (p. 310). The near extinction of many wildlife populations has a consequence that has not received due attention in the popular press but that will be harmful to their future even if protective measures succeed in preventing their extinction, namely, the consequence of sharply reduced genetic variation (see, for example, Bonnell & Selander 1974). Low variability as a condition for reproductive success is discussed by Emlen and Demong (1975), who show that bank swallows, who nest gregariously, do their foraging socially; this makes it important for the birds to have a highly synchronized reproductive cycle (this is the above-mentioned low variability), because this allows the birds to pool information on where the food is; the authors use the reciprocal of the standard deviation of hatch dates as a measure of the degree of synchronization.

Psychology. Two groups of subjects (see Siegel 1956, p. 148) were asked to judge the amount of hostility in the characters of a motion picture; the subjects in one group were known to have difficulties in handling their own aggressive impulses, while the subjects in the other group did not; it turned out that the average values of the scores of the two groups did not differ, but their spreads did, the first group having the larger spread. Again, Maxwell (1960) found that the scores obtained with the WISC battery of tests from children attending a psychiatric clinic showed a greater dispersion (in addition to a lower average) than those obtained from normal children. Also, Hammond and Kern (1959, quoted in Hammond & Householder 1962, p. 115) demonstrated that medical students participating in a particular clinical program showed more variation in appraising the personality characteristics of their patients than other students (a result perhaps indicating that the participating students became more aware of individual differences among their patients). Similarly, Light and Smith (1970) stress the importance of looking at variability, along with average values, when evaluating educational programs. This point is emphasized even more by Klitgaard (1975, in which see his remark on spreads; see also Klitgaard & Hall 1975). He points out that equality is an increasingly voiced goal of education and that many recent writers have emphasized equality of outcomes as a major educational aim. For those who suspect that this aim can only be attained by stultifying the growth of gifted children, Klitgaard adds that every experienced teacher knows that effective teaching will increase the variance of the groups being taught, and usually markedly. Whichever way the variance among students be interpreted, it is clearly an important parameter in the evaluation of educational efforts: according to the second way of thinking, a large variance would signify the development of each child to his fullest potential, and a small variance would measure the degree of success in the attainment of a goal of social engineering —a goal that Russian education proclaims not to pursue, insofar as that education is based on sharp intellectual competition among students, special schools for unusually gifted students, Olympiad contests in mathematics, competitive entrance examinations to universities, and so on (see for example, Boltianskii & Iaglom 1965).

Economics. Fisher and Lorie (1970, p. 100) list several reasons (mostly related to the risk that goes with investments) that variability studies regarding investment returns are important. These authors characterize variability by various statistics, some of which are uncommon: not only standard deviation and coefficient of variation, but also mean deviation and relative mean deviation, as well as Gini's mean difference (δ_1^1, see the main article) and Gini's coefficient of concentration (δ_1^1/μ); the authors contend that Gini's two expressions are nonparametric measures, and are invulnerable to departures from normality, but unfortunately they give no reasons for this statement; in the main article it is shown, however, that the standard deviation can be defined as the square root of $\frac{1}{2}E_P(X_1 - X_2)^2$; this is neither more nor less parametric than Gini's mean difference, $E_P|X_1 - X_2|$. In any case, this paper provides a genuine application of Gini's measures of dispersion.

Another area of progress has been in research concerning *statistical methods* for the comparison of variances of several populations. Two among several papers are by Hall (1972), who for most situations recommends a jackknife procedure, and by Brown and Forsythe (1974), who tend to recommend replacing the sample mean (in existing test criteria for the equality of variances) with some more robust location estimator in order to obtain more robust tests for the equality of variances. Continuing research is expected.

H. ROBERT VAN DER VAART

ADDITIONAL BIBLIOGRAPHY

BOLTIANSKII, V. G.; and IAGLOM, I. M. 1965 Skolnyi matematicheskii kruzhok pri MGU i moskovskie matematischeskie Olimpiady. Pages 3–50 in *Sbornik zadach moskovskikh matematicheskikh Olimpiad.* Moscow: Prosveschenie.

BONNELL, M. L.; and SELANDER, K. K. 1974 Elephant Seals: Genetic Variation and Near Extinction. *Science* 184:908–909.

BROWN, MORTON B.; and FORSYTHE, ALAN B. 1974 Robust Tests for the Equality of Variances. *Journal of the American Statistical Association* 69:364–367.

EMLEN, STEPHEN T.; and DEMONG, NATALIE J. 1975 Adaptive Significance of Synchronized Breeding in a Colonial Bird: A New Hypothesis. *Science* 188:1029–1031.

FISHER, LAWRENCE; and LORIE, JAMES H. 1970 Some Studies of Variability of Returns on Investments in Common Stocks. University of Chicago, Graduate School of Business, *Journal of Business* 43:99–134.

HALL, I. J. 1972 Some Comparisons of Tests for Equality of Variances. *Journal of Statistical Computation and Simulation* 1:183–194.

HAMMOND, KENNETH R.; and HOUSEHOLDER, JAMES E. 1962 *Introduction to the Statistical Method: Foundations and Use in the Behavioral Sciences.* New York: Knopf.

KLITGAARD, ROBERT E. 1975 Going Beyond the Mean in Educational Evaluation. *Public Policy* 23:59–79.

KLITGAARD, ROBERT E.; and HALL, GEORGE R. 1975 Are There Unusually Effective Schools? *Journal of Human Resources* 10:90–106.

LIGHT, RICHARD J.; and SMITH, PAUL V. 1970 Choosing a Future: Strategies for Designing and Evaluating New Programs. *Harvard Educational Review* 40:1–28.

¹RÉNYI, ALFRÉD 1970 *Probability Theory.* Edited by H. A. Lauwerier and W. T. Koiter. Amsterdam: North-Holland; New York: American Elsevier. → The English edition of Rényi's textbook, first published in Hungarian in 1954 and revised in each of several successive editions in various languages.

SAVAGE, I. RICHARD; RAO, M. M.; and HALBERG, FRANZ 1962 Test of Peak Values in Physiopathologic Time Series. *Experimental Medicine and Surgery* 20, no. 4:309–317.

WENT, FRITS W. et al. 1957 *The Experimental Control of Plant Growth, With Special References to the Earhart Plant Research Laboratory at the California Institute of Technology.* Waltham, Mass.: Chronica Botanica.

VITAL STATISTICS

Vital statistics are statistics on principal events in the life of an individual. They usually are gathered at the time of an event such as birth, marriage, the dissolution of a marriage, and death. Vital statistics are commonly compiled from records of vital events registered through offices that are organized as part of a vital registration system.

Vital registration systems are generally organized units of government. They presuppose a well-established civil administrative organization with trained officials and, most usually, local offices as well as a central one. Local offices are primarily responsible for the collection of information, while both local and central offices process the information for statistical purposes. Local offices are generally responsible for maintaining a legally valid record of the vital events. Thus they are useful to the inhabitants of the population when it becomes necessary to prove a vital event such as birth, nationality, descent, or relationship by marriage. The information-processing offices that provide vital statistics in summary form are charged with this responsibility in the interest of the formation of public policy. [*See* GOVERNMENT STATISTICS.]

The most accurate vital statistics are found in countries that are in an advanced state of economic development; in many of the less developed countries vital registration is still rudimentary, partial, or inaccurate. In order to be complete and reliable, vital registration must be compulsory, i.e., the law must place an obligation on defined classes of persons to notify the registering official of the occurrence of a vital event. This is usually easiest in the case of marriage, which in most countries involves a ceremony before an official of the state or the church, who will record the event, which confers a new status on the spouses. In those countries in which consensual unions are common, however, marriage statistics may give an incomplete count of the number of women who are exposed to a relatively high risk of pregnancy. Furthermore, death registration may be easier to enforce than birth registration, since the disposal of a human body is normally subject to police or sanitary regulations, which require a certificate of registration of death to be produced before the body can be disposed of.

History of vital registration. Vital registration was often preceded by parochial registration of baptisms, burials, and marriages. Parochial registration, however, tends to be incomplete, particularly in the case of births, since not all children who are born are baptized; in particular, the practice relating to the registration of babies who die before baptism may vary in different parishes. In Scandinavia, the work of registration is still carried out by the clergy, although they act as agents of the state.

The oldest systems of vital registration are found in the Scandinavian countries: Finland started in 1628 and Denmark in 1646, Norway and Sweden following in 1685 and 1686 respectively. In America, the General Court in Boston enacted a registration law for the colony of Massachusetts in 1639 which stated "that there be records kept . . . of the days of every marriage, birth and death of every person within this jurisdiction." In 1644 an explicit obligation was placed on "all parents, masters of servants, executors and administrators . . . to bring unto the clerk of the writs the names of such belonging to them, as shall either be born or die." The law was tightened in 1692, when penalty clauses for failure to register were reinforced, but the system remained incomplete until the nineteenth century (Gutman 1959).

In England and Wales parochial registration of baptisms and burials began as early as 1538. John Graunt, who is generally considered the father of modern demography, utilized these data in his work *Natural and Political Observations Made Upon the Bills of Mortality,* which was first published in 1662. An act passed in 1694 provided for the registration of births and deaths throughout the country, but it was in force for only ten years and few of the returns made under its provisions have been located. Estimates of population in England and Wales in the eighteenth century have to be based on the parochial registers, since the system of civil registration was not established until 1836. Even then, the Births and Deaths Registration Act did not lay down any penalties for failure to comply with its provision, an omission that was not repaired until 1874.

In other European countries vital registration was gradually introduced throughout the nineteenth century, and was complete in most areas by the beginning of the twentieth century. In some states, however, compulsory and complete registration was introduced very much later; in Poland, for example, it was not introduced until 1946. In Russia, vital registration was in the hands of the ecclesiastical authorities before the revolution and was only transferred to the civil power afterwards. A registration area was built up, and by 1926 it was working with reasonable efficiency in European Russia. It has gradually been extended to cover the rest of the Soviet Union.

Outside Europe, North America, and Australia, registration is more recent. Japan, the most industrialized and developed country of Asia, introduced a modern registration law in 1898, although household registers had been kept before that date. In India and Pakistan no complete and compulsory system of vital registration exists at present, although partial and incomplete systems operate in a number of areas. In Africa the position is even less satisfactory. Birth and death registration in colonial days was applied only to the population of European, and sometimes to that of Asian, origin; for the indigenous African population, registration operated in a few towns at the most and was often of questionable accuracy. In Latin America, although registration became compulsory in most areas in the nineteenth century, the systems were frequently lacking in accuracy and left much to be desired in other respects.

In the United States vital registration developed slowly. As is the case in most federal countries, the responsibility for vital registration lies with the individual states and not with the federal government. By 1859 eight states had established registration systems, and the progress was resumed after the Civil War. The federal government's influence made itself felt after 1902, when the Bureau of the Census was established as a permanent organization. In 1903 Congress passed a law stressing the importance of a unified system of registration, and model registration laws were drafted for the guidance of individual states. A death registration area and later a birth registration area were set up, admission to which depended upon the achievement of a certain degree of completeness of registration.

Administration of vital registration. Systems of vital registration are normally administered through a network of local registration offices, each of which is responsible for a well-defined local area. It is often convenient to have the boundary of the registration district coincide with that of a local government unit. The onus of informing the registrar of the occurrence of a vital event is placed by law on a definite informant or a substitute when the informant is not available. In the case of births, the legal informant is normally the parent, although in a few countries—of which the United States is the outstanding example—responsibility rests with the attendant at the birth. Obviously, this arrangement is possible only when the vast majority of births are medically attended; and, on the whole, registration by the parent is preferred. However, the completeness of birth registration depends on other factors than the identity of the informant.

In the case of a death, the obligation to register again most frequently devolves upon a relative, or, failing him, a person present at the death. In the United States and New Zealand this responsibility, however, devolves upon the undertaker who arranges for the funeral. In many of the more developed countries, the cause of death must also be stated at registration; this responsibility usually has to be carried out by a medical practitioner. Thus, in England and Wales the medical certificate of death is given by the doctor who attended the deceased before death or (in cases of sudden death) by the pathologist who conducted the autopsy. Either of these persons can *notify* the registrar of the death, but the obligation to *register* it rests with the next of kin. In the case of marriages the informants are normally the groom and bride, although in some areas it is the person solemnizing the marriage who actually registers it.

The time allowed for registration in different countries varies; it is normally shorter for a death than a birth. As an extreme example, the Cuban law (as of 1950) required a death to be registered immediately, but a birth only had to be registered within a year of its occurrence. In England and Wales five days are allowed for a death registration, but 42 days for the registration of births.

The form in which vital events are registered varies from country to country. As the registration system serves as the legal record of the vital event, a certificate of registration is normally issued to each informant. This may carry all the information obtained at registration, but more frequently some of the material collected is used for statistical purposes only and does not appear on the certificate. The minimum information collected at a birth is normally the date and place of its occurrence, the sex of the child, and the name of its father (in the case of a legitimate birth). In some vital statistics systems, however, a good deal of additional information is collected, e.g., the age of the mother, the occupation or age of the father, the length of the parents' marriage, how many brothers and sisters the child has, and in some cases its weight at birth. For death registration, the name, age, and sex of the deceased person, together with the date and place of death, constitute the minimum amount of information desirable. In many vital statistics systems information is sought regarding the decedent's marital status, occupation, and cause of death. The minimum information normally required when a marriage is registered is the marital condition of the bride and groom and their ages, although often details about their occupations and sometimes the occupations of their parents are also included.

There are a number of common difficulties connected with vital registration and vital statistics. In the case of death registration, there have been periodical revisions of the International List of Causes of Death. These revisions have affected the comparability of cause-specific death rates over time. Moreover, the treatment of multiple causes of death may differ in different countries, although the World Health Organization has recently made recommendations, endorsed by the Statistical Commission of the UN, for the adoption of a uniform International Medical Certificate of Death.

Another difficulty lies in the definition of a live birth and in the classification of stillbirths or fetal deaths. Thus, in Belgium a child born alive but dying before registration (that is, within three days of birth) is registered as stillborn. In Colombia, stillbirths are not registrable; in Cuba, survival for at least one day is required before a birth can be registered as live. In Great Britain any child born after the twenty-eighth week of pregnancy that at any time after being expelled from its mother drew breath or showed any sign of life is regarded as liveborn. Stillbirth or fetal death rates calculated in accordance with different definitions therefore cannot be comparable.

The uses of vital statistics. The information collected at vital registration is used principally in the study of population movements. Since censuses can only be taken periodically (often at decennial intervals), vital statistics serve as the principal instrument for making intercensal estimates of population. The decomposition of population growth into births, deaths, and migration is essential if its nature and causes are to be fully understood, and a knowledge of mortality and fer-

tility rates is also necessary if reasonable assumptions are to be made for projection of population trends.

Historically, interest first arose in studying mortality statistics. Reference has already been made to John Graunt's pioneer study in the seventeenth century. In the eighteenth and nineteenth centuries, interest in accurate mortality statistics was stimulated by the growth of life insurance, for which adequate data on the variation of mortality with age and sex were necessary, and by the struggle against infectious and other diseases. In this connection, special mention must be made of the work of William Farr, who entered the British General Register Office as compiler of abstracts shortly after its foundation in 1837 and who served in it until his retirement in 1880. He developed the British system of death registration into an instrument for measuring the sanitary condition of the country, and his studies on mortality differences between different occupations contributed to the understanding of industrial hazards. Farr was also one of the prime movers in making mortality statistics internationally comparable and in constructing a statistical nosology of diseases that was to be used in the study of causes of death. The International List of Causes of Death has been revised from time to time, and at present the responsibility for the list lies with the World Health Organization.

Birth registrations form the basis of both fertility and natality statistics. In connection with census data on the structure of the population, they can be used to assess marital fertility and to establish fertility differences between different social groups; they may also be useful in studies on population genetics. In industrial societies, in which mortality is low, population projections will be dependent mainly on the assumptions made with respect to fertility and on the assessment of trends. Complex breakdown of births by parental age, occupation, duration of marriage, birth order, and sometimes interval since preceding birth, are required to make reasonable assumptions; and registration systems have become more complex in order that this information may be made available. Much the same considerations apply to the study of marriage statistics.

EUGENE GREBENIK

[See also CENSUS; DEMOGRAPHY, *article on* THE FIELD; GOVERNMENT STATISTICS; SOCIAL RESEARCH, THE EARLY HISTORY OF; *and the biographies of* GRAUNT *and* Kőrösy. *See also* Freedman 1968; Kirkpatrick 1968; Moriyama 1968; Petersen 1968; Pressat 1968; Reiss 1968; Rosen 1968; Spengler 1968; Thomas 1968.]

BIBLIOGRAPHY

BENJAMIN, BERNARD (1959) 1960 *Elements of Vital Statistics.* London: Allen & Unwin; Chicago: Quadrangle Books.

EDGE, PERCY GRANVILLE 1944 *Vital Statistics and Public Health Work in the Tropics.* London: Baillière, Tindal & Cox.

○FARR, WILLIAM 1885 *Vital Statistics: A Memorial Volume of Selections From the Reports and Writings of William Farr.* Edited for the Sanitary Institute of Great Britain by Noel A. Humphreys. London: The Institute. → Reprinted, with an introduction by M. Susser and A. Adelstein, in 1975 by Scarecrow. Also reprinted, in 1976, by Arno.

►FREEDMAN, RONALD 1968 Fertility. Volume 5, pages 371–382 in *International Encyclopedia of the Social Sciences.* Edited by David L. Sills. New York: Macmillan and Free Press.

GUTMAN, ROBERT 1959 *Birth and Death Registration in Massachusetts, 1639–1900.* New York: Milbank Memorial Fund. → First published in the *Milbank Memorial Fund Quarterly* 36 [1958] and 37 [1959].

►KIRKPATRICK, CLIFFORD 1968 Family: II. Disorganization and Dissolution. Volume 5, pages 313–322 in *International Encyclopedia of the Social Sciences.* Edited by David L. Sills. New York: Macmillan and Free Press.

KOREN, JOHN (editor) 1918 *The History of Statistics: Their Development and Progress in Many Countries.* New York: Macmillan. → Published for the American Statistical Association.

KUCZYNSKI, ROBERT R. 1948–1953 *Demographic Survey of the British Colonial Empire.* 3 vols. Oxford Univ. Press. → Volume 1: *West Africa,* 1948. Volume 2: *South African High Commission Territories: East and Central Africa, Mauritius, and the Seychelles,* 1949. Volume 3: *West Indian and American Territories,* 1953.

LORIMER, FRANK 1961 *Demographic Information on Tropical Africa.* Boston Univ. Press.

►MORIYAMA, IWAO M. 1968 Mortality. Volume 10, pages 498–504 in *International Encyclopedia of the Social Sciences.* Edited by David L. Sills. New York: Macmillan and Free Press.

►PETERSEN, WILLIAM 1968 Migration: I. Social Aspects. Volume 10, pages 286–292 in *International Encyclopedia of the Social Sciences.* Edited by David L. Sills. New York: Macmillan and Free Press.

►PRESSAT, ROLAND 1968 Nuptiality. Volume 11, pages 223–226 in *International Encyclopedia of the Social Sciences.* Edited by David L. Sills. New York: Macmillan and Free Press.

►REISS, ALBERT J. JR. 1968 Sociology: I. The Field. Volume 15, pages 1–23 in *International Encyclopedia of the Social Sciences.* Edited by David L. Sills. New York: Macmillan and Free Press.

►ROSEN, GEORGE 1968 Public Health. Volume 13, pages 164–170 in *International Encyclopedia of the Social Sciences.* Edited by David L. Sills. New York: Macmillan and Free Press.

►SPENGLER, JOSEPH J. 1968 Lotka, Alfred J. Volume 9, pages 475–476 in *International Encyclopedia of the Social Sciences.* Edited by David L. Sills. New York: Macmillan and Free Press.

SPIEGELMAN, MORTIMER 1963 The Organization of the Vital and Health Statistics Monograph Program. Pages 230–249 in Milbank Memorial Fund, *Emerging Techniques in Population Research: Proceedings of a Round Table at the Thirty-ninth Annual Conference . . . September 18–19, 1962.* New York: The Fund.

►THOMAS, BRINLEY 1968 Migration: II. Economic Aspects. Volume 10, pages 292–300 in *International Encyclopedia of the Social Sciences.* Edited by David L. Sills. New York: Macmillan and Free Press.

UNITED NATIONS, STATISTICAL OFFICE 1955 *Handbook of Vital Statistics Methods.* Studies in Methods, Series F, No. 7. New York: United Nations.

WESTERGAARD, HARALD 1932 *Contributions to the History of Statistics.* London: King.

VON BORTKIEWICZ, LADISLAUS

See BORTKIEWICZ, LADISLAUS VON.

VON MISES, RICHARD

Richard von Mises (1883–1953), who contributed notably to the field of applied mathematics, was born in Lemberg, in the Austro–Hungarian Empire. His father, Arthur von Mises, held a doctoral degree from the Institute of Technology in Zurich and was a prominent railroad engineer in the civil service. On his travels all over the empire he was often accompanied by his family, and von Mises was born on one of these journeys. The family home was in Vienna. Von Mises was the second of three brothers; the eldest, Ludwig, is an economist of international reputation; the youngest died while still a boy. His father's family included engineers, physicians, bankers, and civil servants. Among the members of his mother's family were philologists and bibliophiles.

Von Mises attended the Akademische Gymnasium in Vienna and graduated in 1901 with high distinction in Latin and mathematics. He then studied mechanical engineering at the Vienna Technical University. In 1906, immediately after finishing these studies, he became an assistant to Georg Hamel, who had just accepted a professorship of mechanics at the Technical University in Brünn (now Brno). In 1908 von Mises was awarded a doctorate by the Technical University in Vienna and in the same year obtained the *venia legendi* (*Privatdozentur*) at Brünn, with an inaugural dissertation entitled *Theorie der Wasserräder* (1908). But after only one year (at the age of 26), he was called to Strasbourg as associate professor of applied mathematics, the field he made famous.

After five happy and fruitful years in Strasbourg, von Mises joined the newly formed Flying Corps of the Austro–Hungarian Army at the outbreak of World War I (he already had a pilot's license). He was soon recalled from service in the field to act as technical adviser, organizer, and instructor. In the *Fliegerarsenal* in Aspern, he taught the theory of flight to German and Austrian officers; these lectures constituted the first version of his *Fluglehre* (1918), which went through many editions. He was commissioned to design the first large airplane of the empire, the "Grossflugzeug." At the same time he was working on two basic papers on probability (discussed below).

When the war was over, von Mises could not return to Strasbourg, which had become French. After a brief interlude as lecturer in Frankfurt, he was called in 1919 to the Technical University in Dresden as professor, and in 1920 to the University of Berlin as professor of applied mathematics and director of the Institute of Applied Mathematics. This institute was actually founded by him and was a precursor of several similar institutes in Europe and America. In 1921 he founded the *Zeitschrift für angewandte Mathematik und Mechanik*, the first journal of its kind. As its editor until 1933, he exerted a profound influence on applied mathematics all over the world. For von Mises, applied mathematics included mechanics, practical analysis, probability and statistics, and some aspects of geometry and philosophy of science. He educated a generation of young applied mathematicians. His first assistant was Hilda Geiringer, who held a PH.D. in "pure" mathematics but turned, under his influence, to applied mathematics. She became his collaborator and later his wife.

When, in 1933, von Mises recognized that it would be both unwise and undignified to remain in Berlin, he accepted the position of professor of mathematics and director of the mathematical institute in Istanbul, Turkey, at the university that had been revitalized by Kemal Atatürk. He reorganized the institute, lectured in French and in Turkish, maintained close relations with Turkish professors and dignitaries, and became a leading figure at the university. But in 1939, with the approach of World War II, he felt he had to leave Istanbul; and he accepted a position as lecturer in the School of Engineering at Harvard University. There, he was appointed, in rapid succession, associate professor and Gordon McKay professor of aerodynamics and applied mathematics. He continued his own scientific work as well as the education of undergraduates, postgraduates, and research workers.

The fields to which von Mises made distinctive contributions are (1) mechanics and geometry, (2) probability and statistics, (3) philosophy of science, and (4) analysis. Of these, the first two categories occupied him the most. Geometry captivated him all his life, and most of his geometric contributions are closely connected with mechanics.

The outstanding feature of his work is a striving for clarity and complete understanding. In his contributions to mechanics no vague statements, no *ad hoc* engineering theories are tolerated; explanations of observations follow strictly from the principles of mechanics. Particularly important achievements are his *Theorie der Wasserräder* (1908); his wing theory (1917–1920), which is based on conformal mapping; and his celebrated work on plasticity (1913; 1925; 1928a; 1949).

The main directions of von Mises' thought on the theory of probability appeared in his first major papers on the subject, the "Fundamentalsätze der Wahrscheinlichkeitsrechnung" and the "Grundlagen der Wahrscheinlichkeitsrechnung," both of 1919. Von Mises considered probability as a science of the same epistemological type as, say, mechanics. Its mathematical construction is distilled from experience. The main concept, introduced in the "Grundlagen," is the *Kollektiv* (also denoted as "irregular collective"), which, in the simplest case, idealizes the sequence of results of the repeated tossings of a coin under unaltered circumstances. The collective as a mathematical notion is thus an infinite sequence of zeros and ones (heads and tails). If among the first N terms of the sequence there are N_0 zeros and N_1 ones, $N_0 + N_1 = N$, the *frequencies* N_0/N, N_1/N are given. For reasons of mathematical expediency it is then assumed that, in the abstract sequence, the *limits of these frequencies* exist as N tends toward infinity. In addition, the infinite sequence is to have the property of *randomness*; vaguely explained, this means the following: if we consider not all N trials (not all N terms of the sequence) but only the second, fourth, sixth, . . . or only those whose number is a prime number or only those which follow a run of three "ones," we obtain by such a *selection* a frequency N_1'/N' (and N_0'/N'), and it is postulated that for any such selection

$$\lim_{N\to\infty} N_1'/N' = \lim_{N\to\infty} N_1/N = p_1.$$

This p_1 is the *probability* of the result one, and $p_0 = 1 - p_1$ is that of the result zero. Randomness is the mathematical equivalent of the "impossibility of a gambling system" and thus characterizes the sequences which form the subject of probability calculus.

Von Mises then built up probability theory, by means of collectives, in one or more dimensions. In 1938 Abraham Wald proved the "consistency"— i.e., existence in the mathematical sense—of the collective, indicating precise conditions. Von Mises accepted Wald's results as a necessary and valuable complement. He felt that in mathematics, as well as in any other science, the unceasing improvement and refinement of existing concepts must parallel the creation and extension of new concepts.

Von Mises' theory is in contrast with the a prioristic theory of Laplace, whose definition of probability is both logically unsatisfactory and too narrow. Laplace and his followers therefore had to distinguish between a "theoretical" and an "empirical" probability; the mathematical theorems proved with the theoretical definitions were then unhesitatingly applied to problems where Laplace's "equally likely" and "favorable" events often failed to exist. Von Mises showed in a penetrating analysis that for modern probability, as used in physics, biology, and some of the social sciences, Laplace's definition is quite insufficient.

Von Mises' frequency theory also differs from today's abstract measure–theoretical approach, most closely associated with Kolmogorov. The contrast is not between "frequency" and "measure": in von Mises' developments, as well as in Kolmogorov's, both frequencies and measures are essential. Von Mises wanted to lay the conceptual foundations of the science of probability; Kolmogorov, the axiomatic foundations of the calculus of probability. [*See* PROBABILITY.]

The "Fundamentalsätze" deals with two basic general problems. (1) Given n distributions (for example, n dice with given probabilities $p_i^{(\nu)}$, $i = 1,2, \cdots, 6$; $\nu = 1,2, \cdots, n$ for the six faces of the dice), with results X_ν in the νth trial, what is the distribution, as $n \to \infty$, of the sum $X_1 + X_2 + \cdots + X_n$ (equivalently, of the average) of these results? Regarding this group of problems, indicated here by a very special case, von Mises proved in 1919 two basic "local" theorems and studied the most general problem. (2) Perhaps an even more important contribution is his formulation and study of the second fundamental problem. Consider again a very special case. A coin with *unknown* heads-probability is thrown n times, and "heads" turn up n_1 times. What inference can we make from this observed result about the unknown heads-probability of the coin? Obviously, this is the typical problem of inference from an observed *sample* to an unknown "theoretical" value. This problem was considered by von Mises as the crucial problem of theoretical statistics. This "Bayesian" point of view has been widely attacked by R. A. Fisher and his students but seems to be more and more accepted today. For von Mises, statistics was just one (very important and general) application of probability theory.

In the last years of his life von Mises introduced the fundamental concept of a *statistical function* (as important as the concepts earlier introduced by him of *distribution* and of *sample space*), which led to vast generalizations of the two problems of the "Fundamentalsätze." Von Mises' work in probability and statistics is incorporated in many papers and in three books: his *Wahrscheinlichkeitsrechnung* of 1931 (Volume 1 of *Vorlesungen aus dem Gebiete der angewandten Mathematik*), a comprehensive textbook of his theory; his *Probability, Statistics and Truth* (1928b), a lucid presentation in nontechnical language of his foundations of probability and their applications in statistics, biology, and physics; and his lecture notes, *Mathematical Theory of Probability and Statistics* (1964), which restates and extends the foundations of the theory and builds on them a unified theory of probability and statistics, with particularly original contributions to statistics.

Von Mises did not believe that statistical explanations in physics—and other domains of knowledge—are of transient utility while deterministic theories are the definite goal. He thought that a judgment of what constitutes an "explanation" is, like anything else, subject to change and development. The "Laplacean daimon" of complete determinacy is no longer accepted, nor is an immutable law of causality. Philosophers, von Mises thought, are apt to try to "eternalize" the current state of scientific affairs, just as Kant held Euclidean space as an absolute category. In contrast with these "school philosophers," he called himself a "positivist." In an address given shortly before his death he said, "He is a positivist who, when confronted by any problem reacts in the manner in which a typical contemporary scientist deals with his problems of research." Von Mises thought of science in the general sense of the German *Wissenschaft*. In his book *Positivism* (1939) he followed up this conception through the various domains of thought and of life.

Von Mises loved poetry: He could recite long passages from Goethe, as well as from such modern poets as Hofmannsthal, Verlaine, Altenberg, and, in particular, Rilke. In Rilke's esoteric poetry he found a confirmation of his belief that in areas of life not yet explored by science, poetry expresses the experiences of the mind:

> Nicht sind die Leiden erkannt,
> nicht ist die Liebe gelernt,
> und was im Tod uns entfernt,
>
> ist nicht entschleiert.
> Einzig das Lied überm Land
> heiligt und feiert.

Pain we misunderstand,
love we have yet to learn,
and death, from which we turn,

awaits unveiling.
Song alone circles the land
hallowing and hailing.

> *Sonnets to Orpheus*, First Part, XIX
> Frankfurt am Main: Insel, 1923.
> London: Hogarth, 1936.

Von Mises was a recognized authority on the life and work of Rilke. Over a lifetime, he compiled the largest privately owned Rilke collection (now at Harvard's Houghton Library), for which a 400-page catalogue was published in 1966 by the Insel Verlag, Leipzig.

HILDA GEIRINGER

[*For the historical context of von Mises' work, see the biography of* LAPLACE; *for discussion of the subsequent development of von Mises' ideas, see the biographies of* FISHER *and* WALD.]

WORKS BY VON MISES

1908 *Theorie der Wasserräder.* Leipzig: Teubner. → Reprinted from Volume 57 of the *Zeitschrift für Mathematik und Physik.* Partially reprinted in Volume 1 of von Mises' *Selected Papers.*

1913 Mechanik der festen Körper im plastisch-deformablen Zustand. Gesellschaft der Wissenschaften, Göttingen, Mathematisch–Physikalische Klasse *Nachrichten* [1913]:582–592.

1917–1920 Zur Theorie des Tragflächenauftriebs. *Zeitschrift für Flugtechnik und Motorluftschiffahrt* 8:157–163; 11:68–73, 87–89.

(1918) 1957 *Fluglehre: Theorie und Berechnung der Flugzeuge in elementarer Darstellung.* 6th ed. Edited by Kurt Hohenemser. Berlin: Springer.

1919a Fundamentalsätze der Wahrscheinlichkeitsrechnung. *Mathematische Zeitschrift* 4:1–97.

1919b Grundlagen der Wahrscheinlichkeitsrechnung. *Mathematische Zeitschrift* 5:52–100.

1925 Bemerkungen zur Formulierung des mathematischen Problems der Plastizitätstheorie. *Zeitschrift für angewandte Mathematik und Mechanik* 5:147–149.

1928a Mechanik der plastischen Formänderung von Kristallen. *Zeitschrift für angewandte Mathematik und Mechanik* 8:161–185.

(1928b) 1957 *Probability, Statistics and Truth.* 2d rev. English edition. New York: Macmillan. → First published in German. This edition was edited by Hilda Geiringer.

(1931) 1945 *Vorlesungen aus dem Gebiete der angewandten Mathematik.* Volume 1: Wahrscheinlichkeitsrechnung und ihre Anwendung in der Statistik und theoretischen Physik. New York: Rosenberg.

(1939) 1951 *Positivism: A Study in Human Understanding.* Cambridge, Mass.: Harvard Univ. Press. → First published as *Kleines Lehrbuch des Positivismus.*

1949 Three Remarks on the Theory of the Ideal Plastic Body. Pages 415–429 in *Reissner Anniversary Volume: Contributions to Applied Mechanics.* Ann Arbor, Mich.: Edwards.

Mathematical Theory of Probability and Statistics. Edited by Hilda Geiringer. New York: Academic Press, 1964. → Based upon lectures given in 1946.

Selected Papers of Richard von Mises. 2 vols. Providence, R.I.: American Mathematical Society, 1963–1964. → Contains von Mises' writings first published between 1908 and 1954. Volume 1: *Geometry, Mechanics, Analysis.* Volume 2: *Probability and Statistics; General.* Includes a bibliography on pages 555–568 in Volume 2.

SUPPLEMENTARY BIBLIOGRAPHY

CRAMÉR, HARALD 1953 Richard von Mises' Work in Probability and Statistics. *Annals of Mathematical Statistics* 24:657–662. → Includes a selected bibliography of von Mises' work.

WALD, ABRAHAM (1938) 1955 Die Widerspruchsfreiheit des Kollektivbegriffes. Pages 25–45 in Abraham Wald, *Selected Papers in Statistics and Probability.* New York: McGraw-Hill.

Postscript

[1]Only a year after Wald's paper (1938), Ville published an important work (1939) that pointed out that the property of randomness (as defined in the main article) may contradict the law of the iterated logarithm. This led to a setback of von Mises' frequency theory for many years. In 1970 the theory was reshaped in an interesting and promising manner by Schnorr.

LEOPOLD K. SCHMETTERER

ADDITIONAL BIBLIOGRAPHY

SCHNORR, CLAUS P. 1970 *Zufälligkeit und Wahrscheinlichkeit: Eine algorithmische Begründung der Wahrscheinlichkeitstheorie.* Berlin and New York: Springer.

VILLE, JEAN 1939 *Étude critique de la notion de collectif.* Paris: Gauthier-Villars.

VON NEUMANN, JOHN

John von Neumann, mathematician, was born in Budapest in 1903 and died in Washington, D.C., in 1957. He was the first of the great creative mathematicians to devote major effort to the social sciences. After studying in Budapest and Zurich, von Neumann became a *Privatdozent* in Berlin; in 1931 he received an appointment at Princeton University, and in 1933 he joined the Institute for Advanced Study in Princeton, where he remained for the rest of his life. In 1955, on leave from the institute, he was made a member of the U.S. Atomic Energy Commission. For his scientific work and public services he received several honorary doctorates, academy memberships, prizes, medals, and other distinctions.

Von Neumann's genius ranged over many areas of pure mathematics as well as applied fields. He made important contributions to the axiomatics of set theory, mathematical logic, Hilbert space theory, operator theory, group theory, and measure theory. He proved the ergodic theorem, established a continuous geometry without points, introduced almost-periodic functions on groups, and at the end of his life was much concerned with nonlinear differential equations. In addition, he had a consuming interest in numerical applications, ranging from the development of new computing techniques to the study of the mathematical validity of large-scale numerical operations as they are carried out by modern electronic computers.

Von Neumann's work in physics was manifold. In his *Mathematical Foundations of Quantum Mechanics* (1932), a study of enduring significance, he laid a firm basis for this new field by the first comprehensive use and development of Hilbert space. In his study "The Logic of Quantum Mechanics" (see von Neumann & Birkhoff 1936) he revealed the inner logical structure of quantum mechanics and suggested that each science has its own specific logic. Von Neumann's influence was felt in hydrodynamics, mechanics of continua, astrophysics, and meteorology. In statistics he made contributions to trend analysis, and he developed the Monte Carlo method. He established the logical basis for electronic computer design and built the first of the truly modern flexible machines. He was also concerned with the development of a logical theory of automata and proved the possibility of a self-reproducing machine. This work (1966) is closely related to his "Probabilistic Logics" (1956).

Von Neumann's work had great importance for the social sciences. For example, he opened up entirely new avenues in mathematical economics. In 1928 he published a fundamental paper on the theory of games of strategy in which the now famous minimax theorem was proved for the first time. This theorem establishes that, in a two-person zero-sum game with finite numbers of strategies, there always exist optimal strategies for each player. Each player is assumed to choose a strategy independently, and in ignorance, of his opponent's choice. Selection of an optimal strategy is shown to involve the selection of proper probabilities of adopting each of the pure strategies available. [*See* GAME THEORY.]

This work was developed further in *Theory of Games and Economic Behavior* (von Neumann & Morgenstern 1944). The theory was extended to n-persons ($n \geqslant 3$) and to cases where the sum of winnings by all players is a constant different from zero or is variable. The *Theory of Games* also developed a theory of individual choice in situations of risk, which has given rise to an extensive literature on utility. Game theory, besides analyzing games proper, is taken as a model for economic and social phenomena; it applies to all situations where the participants do not control or know the probability distributions of all variables on which the outcome of their acts depends, situations that therefore cannot be described as ordinary maximum or minimum problems (even allowing for side conditions). Since the publication of the *Theory of Games,* hundreds of books and papers by many authors in many countries have furthered and applied the theory.

In 1937 von Neumann wrote on the general equilibrium of a uniformly expanding closed economy under conditions of constant returns to scale in production and unlimited supply of natural resources. Employing the minimax theory, he proved that the economy's expansion factor must equal the interest factor. The linear production relations in the model include linear inequalities and take full account of alternative processes and of indirect production among industries. In these respects, the model is the forerunner of linear programming and activity analysis, both of which are related to game theory by virtue of the minimax theorem. This work, together with that of Abraham Wald, marked the beginning of a new period in mathematical economics. (See Arrow 1968.) Von Neumann showed that the representation of an economic system requires a set of inequalities since, for example, for any good, both the amount produced and the price must necessarily be nonnegative. A solution of the system must satisfy the inequality constraints, and the existence of a solution is not ensured merely by the equality of the number of unknowns and the number of equations.

A fundamental element in von Neumann's mathematical work is the close relation of his thought to the physical and social sciences. He was firmly convinced that the greatest stimulus for mathematics has always come from the mathematician's involvement with empirically given problems; the simultaneous development of calculus and mechanics is the most striking example. He also believed that the mathematical treatment of the social sciences must be quite different from that of the physical sciences. His profound involvement with the social sciences and his very good knowledge of the natural sciences give special weight to his judgment that these two types of science have different mathematical structures. He expected the mathematical study of social phenomena to bring about the development of new mathematical techniques. He took the largely combinatorial approach of game theory as an indication that the time when this would happen might still be remote.

While von Neumann was primarily interested in the mathematical problems of the physical sciences, he nevertheless had a profound concern for the social sciences, which he considered to be in a state comparable to that of physics prior to Newton. This concern expressed itself also in his interest in history and politics, two fields in which he read widely. He had great influence on his contemporaries not only through the large amount of his published work but also through his many contacts with scientists all over the world.

OSKAR MORGENSTERN

[*See also* GAME THEORY; PROGRAMMING.]

WORKS BY VON NEUMANN

(1928) 1959   On the Theory of Games of Strategy. Volume 4, pages 13–42 in A. W. Tucker and R. Duncan Luce (editors), *Contributions to the Theory of Games.* Princeton Univ. Press. → First published in German.

(1932) 1955   *Mathematical Foundations of Quantum Mechanics.* Investigations in Physics, No. 2. Princeton Univ. Press. → First published in German.

(1936) 1962   VON NEUMANN, JOHN; and BIRKHOFF, GARRETT The Logic of Quantum Mechanics. Volume 4, pages 105–125 in John von Neumann, *Collected Works.* Edited by A. H. Taub. New York: Pergamon.

1937   Über ein ökonomisches Gleichungssystem und eine Verallgemeinerung des Brouwerschen Fixpunktsatzes. *Ergebnisse eines mathematischen Kolloquiums* 8:73–83.

○(1944) 1953   VON NEUMANN, JOHN; and MORGENSTERN, OSKAR *Theory of Games and Economic Behavior.* 3d ed., rev. Princeton Univ. Press. → A paperback edition was published in 1964 by Wiley.

1950   *Functional Operators.* 2 vols. Annals of Mathematical Studies, Nos. 21–22. Princeton Univ. Press.

(1956) 1963   Probabilistic Logics and the Synthesis of Reliable Organisms From Unreliable Components. Volume 5, pages 329–378 in John von Neumann, *Collected Works.* Edited by A. H. Taub. New York: Pergamon.

(1958) 1959   *The Computer and the Brain.* New Haven: Yale Univ. Press. → Published posthumously.

1960   *Continuous Geometry.* Princeton Mathematical Series, No. 25. Princeton Univ. Press. → Published posthumously.

1966   *Theory of Self-reproducing Automata.* Edited and completed by A. W. Burks. Urbana: Univ. of Illinois Press. → Published posthumously.

Collected Works. Edited by A. H. Taub. 6 vols. New York: Pergamon, 1961–1963.

SUPPLEMENTARY BIBLIOGRAPHY

►ARROW, KENNETH J.   1968   Economic Equilibrium. Volume 4, pages 376–389 in *International Encyclopedia of the Social Sciences.* Edited by David L. Sills. New York: Macmillan and Free Press.

BEHNKE, HEINRICH; and HERMES, HANS   1957   Johann von Neumann: Ein grosses Mathematikerleben unserer Zeit. *Mathematisch-physikalische Semesterberichte* 5: 186–190.

BOCHNER, S.   1958   John von Neumann, December 28, 1903–February 8, 1957. Volume 32, pages 438–457 in National Academy of Sciences, Washington, D.C., *Biographical Memoirs.* Washington: The Academy. → Includes a ten-page bibliography.

John von Neumann, 1903–1957.   1958   American Mathematical Society, *Bulletin* 64, no. 3, part 2.

KUHN, H. W.; and TUCKER, A. W.   1958   John von Neumann's Work in the Theory of Games and Mathematical Economics. American Mathematical Society, *Bulletin* 64, no. 3, part 2:100–122.

MORGENSTERN, OSKAR   1958   Obituary: John von Neumann, 1903–1957. *Economic Journal* 68:170–174.

ULAM, S.   1958   John von Neumann, 1903–1957. American Mathematical Society, *Bulletin* 64, no. 3, part 2: 1–49. → See especially the bibliography on pages 42–48. See also pages 48–49, "Abstracts of Papers Presented to the American Mathematical Society."

WALD, ABRAHAM

Abraham Wald (1902–1950) was a mathematical statistician and a geometer. Given the fashions of this century, his fame as a statistician is by far the greater.

Mathematical statistics

Wald's interest in mathematical statistics became primary around 1938 and continued without interruption until his death. At ease in mathematical analysis, Wald contributed to the solutions of many of the specialized statistical problems of that period (see Wolfowitz 1952). However, it is with two broad lines of statistical research that his name is always linked: statistical decision theory and sequential analysis.

Statistical decision theory. By 1938 there was available a considerable body of theory dealing with the relationship between observable data and decision making, resulting from work along two closely related lines. One line dealt with the estimation problem—the problem of forming, from observable sample data, estimates, which are in some sense "best," of characteristics of populations described by probability distributions. The other line began with hypotheses concerning these probability distributions and sought "best" tests, based on observable sample data, of these hypotheses. In these two lines of research, R. A. Fisher, J. Neyman, and E. S. Pearson, all working in England, were particularly prominent.

Both of these developments can, of course, be viewed as branches of the more general problem of making decisions in the face of uncertainty, and others must have thought of them as such. But it was Wald who first formally dealt with them in this way. As early as 1939, in one of his first papers (and possibly his finest) in mathematical statistics, Wald introduced a general mathematical structure for (single-sample) decision making, sufficiently general to include both estimation (point and interval) and hypothesis testing. He introduced such fundamental concepts as the multiple decision space and weight and risk functions, and for one of the solutions of the decision problem he introduced the principle of minimization of maximum risk. (There is currently some difference of opinion as to the dependence, in this last area, of Wald's work on von Neumann's great paper of 1928.) Such concepts as the least favorable a priori distribution and admissible regions are also found in this first paper. Wald continued his broad analysis of the decision problem, with a long interruption during the war, and it slowly but steadily gained acceptance. His work culminated in his formal and very general *Statistical Decision Functions* (1950), which incorporates his earlier researches into Bayes' and minimax solutions, as well as his later researches on complete classes of decision functions. The important connection between the decision problem and the zero-sum two-person game is also described at length in this book. Wald continued his work on decision theory in the short time he lived after the publication of his book, his research centering on the role of randomization in the decision process.

His total work in decision theory is probably his most important contribution to mathematical statistics.

Sequential analysis. Wald's second major achievement in mathematical statistics is sequential analysis. The notion that in some sense it is economical to observe and analyze data sequentially, rather than to observe and analyze a single sample of predetermined fixed size, was not a new one. Intuitive support for this notion is immediate; if the evidence shown in sequentially unfolding data is sharply one-sided, it seems reasonable to believe that the inquiry can be terminated early, with lengthier inquiries reserved for those situations in which the issue at hand appears, via the sequentially unfolding data, to be in greater doubt. This notion and the partial mathematical formulation of it were to be found in the statistical literature; among those who dealt with it before Wald was Walter Bartky of Chicago, and among Wald's contemporaries, George Barnard, working in England. But again it was Wald, in 1943, who first formulated mathematically and solved quite generally the problem of sequential tests of statistical hypotheses. He introduced the particular method of the sequential probability ratio test and, with Wolfowitz (1948), showed its optimal properties. He found operating characteristic and average sample number functions; he introduced, if he did not completely solve, the problem of sequential tests of composite hypotheses (utilizing weight functions); and he began vital discussions of such basic topics as multivalued decisions and optimal sequential estimation. All this, plus many special problems, were gathered together in *Sequential Analysis* (1947), a book surprisingly easy to read, less formal and more elementary in structure than his work on decision functions.

Influence on statistical research. Wald's strictly mathematical approach to problems had heavy impact on American research in statistical theory. Up to 1939, one finds excellent researches in statistical theory that nevertheless sometimes lack a firm mathematical basis. Wald's approach was different: his formulations of decision theory and sequential tests of hypotheses were strictly mathematical. Wald can be associated with the beginning of a separation, continuing through the present, of American statistical research from (the parent) British statistical research. With notable exceptions, of course, current issues of *Biometrika* (a leading British statistical journal) and of the *Annals of Mathematical Statistics* (a journal of predominantly American authorship) will show at a glance the difference between the more formal, more mathematical American school—largely inspired and to some extent trained by Wald—and the more intuitive, more applied, less mathematical British school—influenced by such statistical innovators as Fisher, who were less impressed by the value of formal mathematical structure.

A second consequence of Wald's *modus operandi* is notable. Up to 1939, theoretical statisticians were primarily interested in rather limited problems. Wald's formulation of problems was often so broad that his work was difficult to read, but in setting out problems in broad terms, he greatly facilitated later research by others. For example, research in such difficult areas as sequential tests of composite hypotheses was much facilitated by Wald's extensive outline, however incomplete, of this area in his general formulation of sequential theory.

Wald was at heart a mathematician. Although he was not openly opposed to intuitive justification or to popularization, he had no serious interest in either and he asserted that such activities, in the absence of or as substitutes for logical structure, are not permanently useful. Nevertheless, Part 1 of Wald's first full-scale report of his researches in sequential analysis (see Columbia . . . 1943) includes numerous heuristic and intuitive arguments and justifications of the sequential idea and of approximate formulas for risks of error, many of them originated by Wald himself. These surely help the reader understand, in a nonmathematical way, the nature of this new development; but they do not seem quite in the character of Wald. All this changes in Part 2, where Wald introduced cylindric random variables and abruptly tackled the difficult mathematical problem at hand (see Columbia . . . 1945).

Wald's attitude toward specialized application was similar. He was always willing to help practical statisticians; and although his improvisations, approximations, guesses, and *ad hoc* solutions did not generally match the quality of his formal work, he nevertheless offered them freely. Yet his interest in such areas was casual.

With respect to the originality of his contributions to mathematical statistics, Wald is in a class with Fisher and Neyman. But of all workers in this field Wald combined best a profound understanding of the value of the precise formulation of broad and significant areas of statistical inference with the mathematical equipment to handle them. His ability to recognize a major statistical area when he saw one and to do something about it was impressive. Were he alive today, he might well be able to formulate the Bayesian inference problem in such a way that its mathematical structure and its consequences would be clearly set apart from the philosophical and intuitive controversy which no amount of mathematics can ever settle.

Work in geometry and other fields

Wald's other major contribution was in geometry. Far closer than mathematical statistics to the core of mathematics itself, Wald's work here may someday be regarded as his major achievement. At present it is hardly known. Wald went to Vienna briefly in 1927, permanently in 1930; during the period 1931–1936 he worked in geometry with Karl Menger. His major work centered on the problem of the curvature of surfaces. He wrote on many topics in topology and metric spaces, measure and set theory, and lattice theory; and he was the first to prove the existence of a collective in probability theory. His activity in this area had ended by 1943—in fact, there was little after 1936.

Wald also did important work in econometrics and mathematical economics. From 1932 to 1937 and, sporadically, later, he made valuable contributions to such diverse subjects as seasonal corrections to time series, approximate formulas for economic index numbers, indifference surfaces, the existence and uniqueness of solutions of extended forms of the Walrasian system of equations of production, the Cournot duopoly problem, and finally, in his much-used work written with Mann (1943), stochastic difference equations. By all odds, the most important of these were the papers on the existence of a solution to the competitive economic model, written in 1935 and 1936 for Menger's colloquium; an expository version, published in 1936, was translated in the October 1951 issue of *Econometrica*. These papers, along with von Neumann's slightly earlier oral discussion using Brouwer's fixed-point theorem, are the first in which a competitive existence theorem is rigorously proved. Some of Wald's conditions would be deemed overly strong today, but it was a pioneering accomplishment to have provided such a rigorous proof—some 26 years before Uzawa's demonstration of the equivalence of Wald's existence theorem and the fixed-point theorem. This paper alone guarantees Wald's permanent fame in economics.

Intellectual career

Wald was a superb teacher. There were no gimmicks or jokes—only precision and clarity. Sometimes, as Wolfowitz has noted (1952), the precision was labored, for Wald was generally content with any solid proof and seldom went to the trouble of searching for briefer and more elegant proofs. But his lectures were effective. The present author was the only student in a course of Wald's in the early days of sequential analysis, and with

care and skill Wald taught him the content of his "green book" (see Columbia . . . 1943; the contents of this book were classified by the U.S. government until after the war). Wald was not often electrifying, but his admirable teaching during the 1940s still helps to sustain statistical research and teaching. The notebooks (194?; 1941; 1946) created by his students from his lectures are testimony to the quality of Wald's instruction; they are rigorous at the level Wald had in mind, and they remain, some 25 years after their appearance, useful and even provocative for the modern teacher and student.

In Wald's case, more than in the case of most, the work and the man were the same; he lived his work, and his happiest hours were devoted to it. It could have been Wald who said, "Let's go down to the beach and prove some theorems."

Wald was born in 1902, in Cluj, Rumania. After private schooling and self-schooling (the consequence of complications arising from his family's Jewish orthodoxy), Wald, well-trained in mathematics, finally settled in Vienna in 1930. Soon after, he worked for five years in geometry under Menger. In 1932 he began five years of work in econometrics and mathematical economics at the Austrian Institute for Business Cycle Research. In 1938, the year of the *Anschluss*, Wald accepted an invitation—one which probably saved his life—from the Cowles Commission to do econometric research in the United States. Later in 1938 he was brought by Harold Hotelling to Columbia to work in mathematical statistics, and he remained there for the rest of his life. While on a lecture tour of India in 1950, he died in an airplane crash.

Wald was a quiet and gentle man, deeply immersed in his work. He was fairly aloof from small talk, and he had few hobbies. But he was not indifferent to recognition; in the controversies that occasionally developed in the hyperactive and hypersensitive wartime atmosphere of Columbia's Statistical Research Group (of which Hotelling was official investigator and of which W. Allen Wallis was director of research), Wald displayed an entirely normal combination of passive distaste for dispute and active interest in the handling of his work.

Apart from the pleasure he took in his work, Wald had a reasonable share of joy during his life. His marriage to Lucille Lang, who perished with him in India, and his two children, Betty and Robert, were sources of happiness to him. He also had his full share of sorrows, chief among them the death of eight of the nine European members of his immediate family in the gas chambers of Auschwitz.

The scholars whose professional lives were most closely related to Wald's include Harold Hotelling at the University of North Carolina and J. Wolfowitz at Cornell. Hotelling, himself one of the most distinguished figures in American statistical research, brought Wald to Columbia in 1938, securing for him a Carnegie fellowship and an assistant professorship, and helped him through a difficult period of adjustment. However, although Wald's early interest in certain areas of mathematical statistics was initiated by problems brought to his attention by Hotelling, they did not work together; their approach to problems, as well as the kind of problems that interested them, was somewhat different. In particular, Hotelling's interdisciplinary interests contrasted with Wald's strictly statistical interests. But they had great respect for each other, and Hotelling played a major role in Wald's career.

Wolfowitz was Wald's leading student. Oriented mathematically almost exactly as Wald was, Wolfowitz wrote no fewer than 15 papers with Wald and was his closest friend. It is nearer to the truth to say that it was the team of Wald and Wolfowitz —rather than Wald alone—that gave much of American statistical inference the rather severe mathematical character it has today, though this is not to imply that either would be in sympathy with mathematically difficult work divorced from statistical reality.

HAROLD FREEMAN

[*For the context of Wald's work, see* ESTIMATION; GAME THEORY; HYPOTHESIS TESTING. *For discussion of the subsequent development of Wald's ideas, see* DECISION THEORY; SEQUENTIAL ANALYSIS.]

BIBLIOGRAPHY

For an account of Wald's many specialized contributions to mathematical statistics, see Wolfowitz 1952. *For comment on his early contribution to probability, see* Menger 1952; von Mises 1946. *Discussions of his contributions to mathematical economics and econometrics are found in* Morgenstern 1951; Tintner 1952. *His major work in statistical decision theory is in* Wald 1950. *For his major work in sequential analysis, see* Wald 1947. *For a commentary on his work in geometry, see* Menger 1952. *A bibliography of Wald's published work is contained in* Selected Papers . . . 1955. *For lecture notes compiled by his students, see* Wald 194?; 1941; 1946.

WORKS BY WALD

(1936) 1951 On Some Systems of Equations of Mathematical Economics. *Econometrica* 19:368–403. → First published in German in Volume 7 of *Zeitschrift für Nationalökonomie*.

►1936 *Berechnung und Ausschaltung von Saisonschwankungen*. Vienna: Springer.

(1939) 1955 Contributions to the Theory of Statistical Estimation and Testing Hypotheses. Pages 87–114 in Abraham Wald, *Selected Papers in Statistics and Probability*. New York: McGraw-Hill. → First published in Volume 10 of the *Annals of Mathematical Statistics*.

194? Notes on the Theory of Statistical Estimation and of Testing Hypotheses. Notes prepared by Ralph J. Brookner. Unpublished manuscript. → Lectures given in a one-semester course at Columbia University, and available in the Columbia University Libraries.

1941 Lectures on the Analysis of Variance and Covariance. Notes prepared by Ralph J. Brookner. Unpublished manuscript. → Lectures given in 1941 at Columbia University, and available in the Columbia University Libraries.

1943 WALD, ABRAHAM; and MANN, H. B. On the Statistical Treatment of Linear Stochastic Difference Equations. *Econometrica* 11:173–220.

1943 COLUMBIA UNIVERSITY, STATISTICAL RESEARCH GROUP *Sequential Analysis of Statistical Data: Theory*. New York: Columbia Univ. Press.

1945 COLUMBIA UNIVERSITY, STATISTICAL RESEARCH GROUP *Sequential Analysis of Statistical Data: Applications*. New York: Columbia Univ. Press.

1946 Notes on the Efficient Design of Experimental Investigation. Unpublished manuscript. → Lecture notes of a one-semester course given in 1943 at Columbia University, and available in the Columbia University Libraries.

1947 *Sequential Analysis*. New York: Wiley.

(1948) 1955 WALD, ABRAHAM; and WOLFOWITZ, J. Optimum Character of the Sequential Probability Ratio Test. Pages 521–534 in Abraham Wald, *Selected Papers in Statistics and Probability*. New York: McGraw-Hill. → First published in Volume 19 of the *Annals of Mathematical Statistics*.

(1950) 1964 *Statistical Decision Functions*. New York: Wiley.

Selected Papers in Statistics and Probability. New York: McGraw-Hill, 1955. → Published posthumously. Contains writings first published between 1938 and 1952, and a bibliography of Wald's works on pages 20–24.

SUPPLEMENTARY BIBLIOGRAPHY

HOTELLING, HAROLD 1951 Abraham Wald. *American Statistician* 5:18–19.

MENGER, KARL 1952 The Formative Years of Abraham Wald and His Work in Geometry. *Annals of Mathematical Statistics* 23:14–20.

MORGENSTERN, OSKAR 1951 Abraham Wald, 1902–1950. *Econometrica* 19:361–367.

TINTNER, GERHARD 1952 Abraham Wald's Contributions to Econometrics. *Annals of Mathematical Statistics* 23:21–28.

►TINTNER, GERHARD 1976 Abraham Wald, 1902–1950. Volume 14, pages 121–124 in *Dictionary of Scientific Biography*. Edited by Charles C. Gillispie. New York: Scribner's.

VON MISES, RICHARD (1946) 1964 *Mathematical Theory of Probability and Statistics*. Edited and augmented by Hilda Geiringer. New York: Academic Press.

WOLFOWITZ, J. 1952 Abraham Wald, 1902–1950. *Annals of Mathematical Statistics* 23:1–13.

WALKER, FRANCIS A.

Francis Amasa Walker (1840–1897), American economist and statistician, was born in Boston, the offspring of an old New England family. His father, Amasa Walker, a prominent manufacturer, retired from business the year of his son's birth and devoted the remainder of his life to public service and economic studies, becoming the outstanding American economist of his time, a distinction later assumed by his son.

Walker attended Amherst College and subsequently served in the Civil War; less than five years after graduation he was brevetted brigadier general. After the war he was called to Washington as chief of the bureau of statistics in the Treasury Department. He proved to be an administrator of great ability and was appointed superintendent of the censuses of 1870 and 1880. In these positions he acquitted himself with great distinction. Doing such work also provided him with the opportunity of becoming acquainted with a huge mass of statistical data relating to the economy of the United States.

Guided by his father, Walker studied economics and in 1872 was appointed professor of political economy and history at Yale's Sheffield Scientific School. The decade which followed was one of unusual literary productivity. In 1876 Walker published *The Wages Question;* in 1878 a long discussion entitled *Money;* in 1879 a briefer one, *Money in Its Relations to Trade and Industry;* in 1883 both *Land and Its Rent* and a full-length textbook, *Political Economy.* Meanwhile, in 1881, Walker had been appointed president of the Massachusetts Institute of Technology and in this post again proved to be an able administrator. In spite of the pressure of new duties, Walker's interest in economics continued. He became the leader of the profession, serving as president of the long-established American Statistical Association from 1883 to 1896 and helping the new and at the time controversial American Economic Association on its way by serving as its first president from 1886 to 1892. In addition, he held a large number of public offices, ranging from membership on the New Haven Board of Education, where he favored the discontinuation of religious exercises in public schools, to an appointment as U.S. commissioner to the International Monetary Conference, which convened at Paris in 1878. He took a stand on many public issues, including parochial schools and the "new immigration," both of which he opposed. In politics he was a Republican but turned "mugwump" in 1884 and voted for Grover Cleveland.

Walker's economic views differed from those of earlier American economists in a number of important respects. He considered economics a science rather than an art, concerned with principles rather than precepts. Economists, he said, ought "to teach and not to preach" (1899a). Walker also refused to adhere to the opinion dear to many protectionists that economics should be developed in the form of a "national political economy" that would lend itself to immediate application to practical politics. He was not a dogmatic exponent of laissez-faire; rather, he recognized the existence of economic conflicts of various sorts and referred to instances of "imperfect competition" calling for the intervention of the government. In the field of distribution Walker generalized Ricardo's concept of differential rent and applied it to the earnings of entrepreneurs. Wages appeared to him as a residual share left over after the product had been diminished by rent, interest, and profit. This over-all theory of distribution did not win many adherents. It placed a ceiling on the earnings of labor no less effectively than that imposed by the old wages-fund doctrine, according to which wages are the quotient of the employers' "wages fund" divided by the number of workers. This doctrine Walker demolished with lasting effect by making wages a function of the product, and it is for this contribution that he is remembered best in the history of economic thought.

Walker's views of monetary questions were given forceful expression in *International Bimetallism* (1896). In the face of a gold supply that was inadequate relative to the growth of the world economy, Walker strongly urged the international monetization of silver.

In the field of statistics Walker was a pioneer in supplementing tabular presentation with graphic material, sometimes shown in color. The *Statistical Atlas of the United States*, published under his editorship in 1874, set new standards for official statistical publications. The rise of statistics as an increasingly important field of professional specialization was in no small measure due to Walker's influence in the academic world as well as to his efforts aiming at the establishment of a permanent staff for the census. As a publicist who could count on a wide audience he made the public aware of the importance of adequate statistical data. Here as well as in his other pursuits he was also a leading figure in the international field.

HENRY W. SPIEGEL

[*See also* Cartter 1968; Muth 1968.]

WORKS BY WALKER

(1876) 1904 *The Wages Question: A Treatise on Wages and the Wages Class.* New York: Holt.

(1878) 1891 *Money.* New York: Holt.

(1879) 1907 *Money in Its Relations to Trade and Industry.* New York: Holt.

(1883*a*) 1891 *Land and Its Rent.* Boston: Little.

(1883*b*) 1888 *Political Economy.* 3d ed., rev. & enl. New York: Holt.

1896 *International Bimetallism.* New York: Holt.

1899*a* *Discussions in Economics and Statistics.* 2 vols. New York: Holt. → Volume 1: *Finance and Taxation, Money and Bimetallism, Economic Theory.* Volume 2: *Statistics, Natural Growth, Social Economics.* Published posthumously.

1899*b* *Discussions in Education.* New York: Holt. → Published posthumously.

SUPPLEMENTARY BIBLIOGRAPHY

►CARTTER, ALLAN M. 1968 Wages: I. Theory. Volume 16, pages 397–403 in *International Encyclopedia of the Social Sciences.* Edited by David L. Sills. New York: Macmillan and Free Press.

DEWEY, DAVIS R. 1934 Walker, Francis Amasa. Volume 15, pages 323–324 in *Encyclopaedia of the Social Sciences.* New York: Macmillan.

DORFMAN, JOSEPH 1949 General Francis A. Walker: Revisionist. Volume 3, pages 101–110 in Joseph Dorfman, *The Economic Mind in American Civilization.* New York: Viking.

DUNBAR, CHARLES F. 1897 The Career of Francis Amasa Walker. *Quarterly Journal of Economics* 11:436–448. → Originally published in the *Proceedings* of the American Academy of Arts and Sciences.

FITZPATRICK, PAUL J. 1957 Leading American Statisticians in the Nineteenth Century. *Journal of the American Statistical Association* 52:301–321.

FITZPATRICK, PAUL J. 1962 The Development of Graphic Presentation of Statistical Data in the United States. *Social Science* 37:203–214.

HUTCHISON, T. W. (1953) 1962 *A Review of Economic Doctrines, 1870–1929.* Oxford: Clarendon.

LAUGHLIN, J. LAURENCE 1897 Francis Amasa Walker. *Journal of Political Economy* 5:228–236.

MUNROE, JAMES P. 1923 *A Life of Francis Amasa Walker.* New York: Holt. → Includes a bibliography of Walker's writings and addresses.

►MUTH, RICHARD F. 1968 Rent. Volume 13, pages 454–461 in *International Encyclopedia of the Social Sciences.* Edited by David L. Sills. New York: Macmillan and Free Press.

►NEWTON, BERNARD 1968 *The Economics of Francis Amasa Walker, 1840–1897.* Clifton, N.J.: Kelley. → Includes an introduction by Joseph Dorfman.

SPIEGEL, HENRY W. 1960 Francis A. Walker. Pages 143–153 in Henry W. Spiegel (editor), *The Rise of American Economic Thought.* Philadelphia: Chilton.

TAUSSIG, FRANK W. (1896) 1932 *Wages and Capital: An Examination of the Wages Fund Doctrine.* London School of Economics and Political Science.

WRIGHT, CARROLL D. 1897 Francis Amasa Walker. *Journal of the American Statistical Association* 5:245–290. → A bibliography of Walker's writings and addresses appears on pages 276–290.

WESTERGAARD, HARALD

Harald Ludvig Westergaard (1853–1936), Danish statistician, economist, and social reformer, exerted a strong influence on Danish statistics and social research for many years. He had been interested in pure mathematics from his youth, but after receiving an M.SC. degree in mathematics, he went on to study political economy and statistics.

His first major work was an essay on a subject set by the University of Copenhagen: "Summary and Evaluation of the Recent Studies of the Death Rate in Different Classes of Society." The paper was much praised and was soon published in German as *Die Lehre von der Mortalität und Morbilität* (1882). This comprehensive work proved the turning point in Westergaard's career and was one of the factors that secured him an appointment at the University of Copenhagen. The book attracted favorable notice both in Denmark and abroad and was for many years the standard manual for death-rate statistics. It deals with both theoretical and practical aspects of these statistics and contains a wealth of international statistical facts.

Westergaard's characteristic and partly original approach appears for the first time in this work. This approach stresses the application of the law of errors—in connection with calculations of demographic frequencies (for example, death rates)—to the study of the appropriateness of the data and, at the same time, discusses the limitations caused by inadequacies in the data. These inadequacies can be exemplified by such insufficiencies in the census data as double counting, delayed recording of births and of deaths among newborns, and missing data on age of illiterates. Throughout his life Westergaard emphasized that for further development of scientific statistics the improvement of mathematical methods was less important than the attempt to procure better data (see, for example, 1916). The mathematical methods used by Westergaard in his 1882 book—for instance, his use of standard errors and of approximations by the normal distribution—were not new, but he was original in the way he adapted well-known principles of probability theory to his discussion of practical statistics.

In 1883 he joined the University of Copenhagen as a lecturer in political science and the theory of statistics, the first to teach the latter subject at the university. He was made professor of political science in 1886, a position he retained until his retirement in 1924.

Westergaard wrote well, and his textbooks in

statistics, sociology, and political science were widely used in Denmark and abroad. *Die Grundzüge der Theorie der Statistik* (1890) developed the fundamental idea of using formal probability theory to analyze practical statistics. He was attracted by the Gaussian normal distribution, and his research was concentrated on demographic fields where this distribution is relevant and reasonable as a basic assumption. Although he was aware of the limitations of this approach, in his textbook on theory he urged the applied statistician to consider whether the absence of normality in a particular case might not simply reflect such inadequacies of the data as those discussed above. The *Theorie* was severely criticized for being incomplete in its proofs, for implicitly assuming proofs, and for avoiding problems. Westergaard himself recognized the limitations of his principle of making the normal distribution the keystone of all statistical work, but he did little to demonstrate the mathematical reasons for the limitations of this distribution.

In an article published in 1918, "On the Future of Statistics," Westergaard, largely through an intuitive approach to mathematical statistics, demonstrated great clearsightedness about many problems that were not solved until much later. For example, he urged further work on testing statistical hypotheses from samples. However, the article is also permeated by his partiality toward normal distributions and his contempt for skewed distributions. He believed that the appearance of a nonnormal distribution is evidence of a failure to determine causality and that a closer study of single principal causes is profitable.

After his retirement from the university Westergaard published *Contributions to the History of Statistics* (1932) A work of lasting value, it remains unique in its wealth of detail and its historical meticulousness. It shows how much statistical knowledge has increased, from its small beginnings in the seventeenth century to its considerable scope at the end of the nineteenth. Westergaard placed great emphasis on the need for causal analysis. In writing about the future of statistics he said: "The great problem in all science is to find the causality, to enable us to trace the causes of a given phenomenon and to foretell coming events, where the causes in action are known" (1918, p. 499); and he sought this kind of analysis in history no less than in statistics.

Westergaard belongs to the generation of Pontus Fahlbeck, and there are striking similarities between them. Two influences were of primary importance for both: first, the work of Quetelet and, second, the rich flow of demographic data that began in the middle of the nineteenth century. At an early stage of their maturity Westergaard and Fahlbeck witnessed the rapid progress made by Galton, Pearson, and others of the English statistical school, who stressed the development of the theoretical aspects of statistics, but neither of them was able to assimilate these modern trends. Instead, their attitude toward these developments was skeptical and negative—although Westergaard was the better informed and the less negative of the two. Faced with an uncongenial intellectual situation, Westergaard took up the historical studies that led to his masterful history of statistics.

In addition to research and teaching, Westergaard was prominent in insurance, banking, and humanitarian activities. A quotation from one of his articles summarizes his philosophy: "Political economy is not solely based on facts and formal logical conclusions; it is closely connected with human interests, and consequently every theory is stamped by its author's philosophy" (1881, p. 1).

KAI RANDER BUCH

[*Directly related is the entry on* STATISTICS. *See also* Moriyama 1968.]

WORKS BY WESTERGAARD

1881 Spørgsmaalet om alderdomsforsørgelse. *Nationaløkonomisk tidsskrift* 18:1–30. → The extract in the text was translated by Kai Rander Buch.

(1882) 1901 *Die Lehre von der Mortalität und Morbilität.* 2d ed. Jena (Germany): Fischer.

1890 *Die Grundzüge der Theorie der Statistik.* Jena (Germany): Fischer.

1916 Scope and Method of Statistics. American Statistical Association, *Publications* 15:225–276.

1918 On the Future of Statistics. *Journal of the Royal Statistical Society* 81:499–520.

1932 *Contributions to the History of Statistics.* London: King.

SUPPLEMENTARY BIBLIOGRAPHY

Fortegnelse over et udvalg af Professor Harald Westergaards skrifter. 1937 *Nationaløkonomisk tidsskrift* 75:246–261.

Harald Westergaard. 1937 *Journal of the Royal Statistical Society* 100:149–150.

Harald Westergaard. 1943 Volume 25, pages 403–412 in *Dansk biografisk leksikon.* Copenhagen: Schultz.

►MORIYAMA, IWAO M. 1968 Mortality. Volume 10, pages 498–504 in *International Encyclopedia of the Social Sciences.* Edited by David L. Sills. New York: Macmillan and Free Press.

NYBØLLE, HANS C. 1937 Harald Westergaard: 19. April 1853–13. December 1936. Copenhagen, Universitet, *Festskrift udg. af Københavns Universitet i anledning af universitets aarfest.* [1937]: 136–144.

WIENER, NORBERT

Norbert Wiener was born in Columbia, Missouri, in 1894 and died while visiting Stockholm, Sweden, in 1964. A child prodigy, he became a widely respected mathematician and teacher. During the last twenty years of his life, he became known throughout the world as a founder of and spokesman for the new science that he had named "cybernetics."

Cybernetics, in a narrow sense, is the study of the relationship between information processing and purposeful behavior, both in machines and in animals. In a wider sense the concepts of cybernetics apply to social systems as well and suggest new ways to analyze complex social organizations in terms of the flow and processing of information. However, the more fundamental promise of cybernetics lies not in its ability to help explain the behavior of complex systems, but rather in the fact that the explanations are framed in the new language of information and control. Because of this, the real revolution stimulated by Wiener's notions on cybernetics is a conceptual one that reaches deep into the foundation and structure of the sciences.

Wiener's education and intellectual outlook were enormously influenced by his father, Leo Wiener, who himself had been an intellectually precocious child. Leo Wiener, who was born in Russia and educated in Europe, arrived in the United States at the age of 21 and later became a teacher. Eventually he became professor of Slavic languages and literature at Harvard, where he taught for thirty years before his retirement. Leo Wierner, like James Mill, the father of an earlier child prodigy, had his own theories about educating children. Under his father's rigorous tutelage and discipline, Norbert Wiener at the age of seven was reading books on biology and physics that were beyond even his father's scope. He entered high school at the age of nine and graduated three years later. He then entered Tufts College and graduated—*cum laude* in mathematics—at 15. After a false start toward advanced work in biology, Wiener studied philosophy; he received his PH.D. from Harvard in 1913, at the age of 18.

Upon leaving Harvard, Wiener secured a postdoctoral fellowship that allowed him to travel to Cambridge University, where he studied epistemology and logic with Bertrand Russell and mathematics with G. H. Hardy. After his stay at Cambridge, he went to Göttingen, where he studied mathematics with Landau and David Hilbert and philosophy with Husserl. During a brief period following his return to America he was a writer for the *Encyclopedia Americana*, a mathematician computing ballistic tables for the U.S. Army at the Aberdeen proving grounds, and also a journalist for the *Boston Herald*. Then, in the spring of 1919, Wiener accepted a position in the mathematics department of the Massachusetts Institute of Technology, where he remained, eventually to become a full professor and later Institute professor.

His first work at M.I.T. was on the theory of Brownian motion. This work, which was influenced by the notions of J. Willard Gibbs on statistical mechanics and Lebesgue on probability, shaped his subsequent statistical treatment of the problems of information and communication. Wiener's early mathematical work on harmonic analysis had a later impact on his notions about the filtering and predicting of time series. Thus, much of the mathematical work that he developed during the first part of his career later influenced his ideas on cybernetics.

About 1940, when the United States was gearing itself for a possible war, Wiener became involved in the problem of designing fire control equipment. It was around this complex set of problems that his ideas on information processing and control coalesced to form the basis of cybernetics. The problem of fire control is to design a machine that, when fed radar tracking data, will compute how to aim a gun so that its projectile will intersect the path of the moving target at the appropriate time. This involves not only a theory of prediction and a mechanism to embody the theory but also a theory of stability and control.

In the course of this work, Wiener and his colleague Julian Bigelow (who was later to direct the construction of the first von Neumann-type electronic computer at Princeton University) recognized the critical role of feedback in the organization of a control system. This recognition led to the conjecture that the kinds of information processing and feedback loops necessary to control a mechanical system might resemble those in the cerebellum that control purposeful human behavior. If this conjecture was true, then similar kinds of breakdowns in the internal information-processing mechanisms of a man and of a mechanical control system would produce similar pathological behavior. These ideas were recorded in a paper jointly authored by Wiener, Bigelow, and A. Rosenblueth called "Behavior, Purpose and Teleology," which was published in 1943. It makes explicit the thesis that the brain can be viewed, in a mechanical way, as a kind of computing machine, and that the concepts of information and control are adequate to explain purposeful motor behavior. Left implicit, however, is the further conjecture that the concepts of information and control will be adequate to explain the mechanisms and processes underlying the behavioral correlates of so-called "higher mental functions" involved in thinking. Wiener was not able to publish a fuller treatment of these ideas until the end of World War II.

In 1948 Wiener published his book *Cybernetics: Or Control and Communication in the Animal and the Machine,* which became a best seller and was reprinted many times and translated into many languages. In that now-famous book, Wiener attempted to bring together the concepts underlying information processing, communication, and control. He described the relationship of these cybernetic concepts to other disciplines ranging from neurophysiology, mathematical logic, and computer science to psychology and sociology. His book had an impact on many scientists in these fields, stimulating them to take a fresh look at their own work from a cybernetic point of view. It suggested to psychologists that the behavioral correlates of thinking, remembering, learning, and so forth could be analyzed in terms of the underlying information processes. And much work on the computer simulation of behavior has emerged from that suggestion. The notion of viewing the brain as a kind of computing machine stimulated neurophysiologists not merely to make comparisons between components and coding in both systems but also to try to interpret the logical organization of the brain in terms of information processing and control mechanisms.

The diversity of cybernetic applications in different fields sheds light on the unifying aspect of its basic concepts. Traditionally, a chasm has separated work on the psychology of complex behavior from work on those physiological mechanisms that produce behavior. The gap between these two fields is, in fact, a communication gap caused by the semantic mismatch of concepts from the languages of physiology and psychology. The concepts of behavioristic psychology are too gross and elaborate to fit with the more atomistic concepts from the language of physiology. This same kind of gap would make it impossible to explain the behavior of a digital control computer in terms of the basic physics of its switches, wires, and so forth. Because cybernetics deals with a set of concepts intermediate between psychology and physiology it can provide a conceptual bridge to span both disciplines. The deeper meaning of cybernetics, which lies in

the structure of its language and its role in analyzing complex systems in terms of information processing, communication, and control, has yet to be fully unfolded by philosophers of science.

The extent to which Wiener himself saw this philosophical dimension of his work is not clear. However, he did see clearly some of the social-scientific implications of cybernetics. He believed, and others have subsequently developed the notion, that the economy can be viewed as a control system aimed at maintaining certain conditions of economic growth and that economic instability in the form of period booms and slumps is similar to oscillations in a poorly designed mechanical control system. In a similar vein Wiener argued that society can be examined and understood in terms of the flow and processing of information between individuals and social groups.

Wiener was particularly fearful of the expanding role of the computing machine. He recognized very early that machines could and would eventually displace an increasing number of workers both in the factory and in the office, and he thought that if economic incentives pushed automation ahead of our understanding of its consequences, technological unemployment could shatter social and economic stability. He was also concerned about the potential misuse of computers in decision making and feared that as machines became increasingly complex their users would be less aware of the consequences of their instructions to the machines. As a result, a decision maker might cause a machine to initiate some action the consequences of which might, in fact, be contrary to his actual desires. During the last ten years of his life Wiener traveled widely, lecturing and writing about cybernetics and the potential dangers to a society vastly influenced by computers and automation.

Earlier in his life Wiener had received recognition for his contributions to mathematics, and shortly before his death he was awarded the National Medal of Science by the president of the United States.

M. E. MARON

[See also SIMULATION *and the biographies of* BABBAGE *and* VON NEUMANN. *See also* Gochman 1968; Kaplan 1968; Kendler & Kendler 1968; Maron 1968; Mitchell 1968; Parsons 1968; Pollack 1968; Rapoport 1968.]

WORKS BY WIENER

1943 ROSENBLUETH, A.; WIENER, NORBERT; and BIGELOW, J. Behavior, Purpose and Teleology. *Philosophy of Science* 10, no. 1:18–24.
(1948) 1961 *Cybernetics: Or, Control and Communication*

in the Animal and the Machine. 2d ed. Cambridge, Mass.: M.I.T. Press.

(1950) 1954 *The Human Use of Human Beings: Cybernetics and Society.* 2d ed. Boston: Houghton Mifflin.

1953 *Ex-prodigy: My Childhood and Youth.* New York: Simon & Schuster.

1956 *I Am a Mathematician; The Later Life of a Prodigy: An Autobiographical Account of the Mature Years and Career of Norbert Wiener and a Continuation of the Account of His Childhood in* Ex-prodigy. Garden City, N.Y.: Doubleday. → A paperback edition was published in 1964 by M.I.T. Press.

SUPPLEMENTARY BIBLIOGRAPHY

▶GOCHMAN, DAVID S. 1968 Systems Analysis: V. Psychological Systems. Volume 15, pages 486–495 in *International Encyclopedia of the Social Sciences.* Edited by David L. Sills. New York: Macmillan and Free Press.

▶KAPLAN, MORTON A. 1968 Systems Analysis: IV. International Systems. Volume 15, pages 479–486 in *International Encyclopedia of the Social Sciences.* Edited by David L. Sills. New York: Macmillan and Free Press.

▶KENDLER, HOWARD H.; and KENDLER, TRACY S. 1968 Concept Formation. Volume 3, pages 206–211 in *International Encyclopedia of the Social Sciences.* Edited by David L. Sills. New York: Macmillan and Free Press.

▶MARON, M. E. 1968 Cybernetics. Volume 4, pages 3–6 in *International Encyclopedia of the Social Sciences.* Edited by David L. Sills. New York: Macmillan and Free Press.

▶MITCHELL, WILLIAM C. 1968 Systems Analysis: III. Political Systems. Volume 15, pages 473–479 in *International Encyclopedia of the Social Sciences.* Edited by David L. Sills. New York: Macmillan and Free Press.

Norbert Wiener, 1894–1964. 1966 American Mathematical Society, *Bulletin* 72, no. 1, part 2. → A bibliography appears on pages 135–145.

▶PARSONS, TALCOTT 1968 Systems Analysis: II. Social Systems. Volume 15, pages 458–473 in *International Encyclopedia of the Social Sciences.* Edited by David L. Sills. New York: Macmillan and Free Press.

▶POLLACK, IRWIN 1968 Information Theory. Volume 7, pages 331–337 in *International Encyclopedia of the Social Sciences.* Edited by David L. Sills. New York: Macmillan and Free Press.

▶RAPOPORT, ANATOL 1968 Systems Analysis: I. General Systems Theory. Volume 15, pages 452–458 in *International Encyclopedia of the Social Sciences.* Edited by David L. Sills. New York: Macmillan and Free Press.

Postscript

Wiener's ideas about cybernetics have continued to grow and spread. New journals, professional societies, and special conferences dealing with various aspects of cybernetics have emerged. (See, for example, *Journal of Cybernetics* and IEEE *Transactions on Systems, Man, and Cybernetics.*) Many new books and articles (in journals ranging from those on psychology and linguistics to those on engineering and sociology), dealing with different dimensions of cybernetics, have been published. However, it is difficult to fully assess the current state of this field because the name "cybernetics" continues to mean different things to different people. (See Maron 1968.) Some who accept a broad interpretation put the rapidly growing fields of computer science, information theory, control theory, bionics, robotics, artificial intelligence, and brain theory under the heading of cybernetics. (See, for example, Arbib 1972; 1975). Yet there are others working on related problems who simply refuse to describe their research as cybernetics. (See, for example, Cherry 1975.)

Cybernetics as Wiener defined it—the study of information, communication, and control in men and machines—still represents more of a program than a "new science." That is, cybernetics aims at the full study of the relationship between intelligent behavior (and higher-level mental functions) and the underlying information processes correlated with that behavior. Still, none of the theoretical concepts and laws that would be needed to unify the properly bounded cybernetics if (as some have thought) it were to become the "new science" of information has yet been developed.

Looking more closely, we see that researchers have continued in their attempts to relate the behavioral indications of intelligence to the information processes that could generate the behavior in question. But real progress has been painfully slow in most areas and almost nonexistent in others. For example, the cybernetic study of language comprehension continues to encounter major intellectual roadblocks. To understand how language is understood, we need (but as yet we do not have) some general cybernetic theory of how the cognitive system itself is organized and how incoming linguistic information interacts with and eventually modifies an internal model of the external world. (See, for example, Miller 1974.)

With the introduction into society of new and complex machines and associated techniques have come, in the opinion of some observers, unexpected, unplanned, and often undesired side effects and aftereffects. (See, for example, Weizenbaum 1972.) Wiener predicted them in the case of computers and sounded an early warning. Nevertheless, his deep concerns about misuses and unwanted effects of computers and technological systems have materialized. (See, for example, Taviss 1970; Wessel 1974; Rothman & Mossmann 1975.) Computers have continued to become more powerful (in terms of their speed, information-storage capacity, and information-processing capability); they have become much less expensive; they have become much more widely used in all facets of science and society. And the misuses and side effects have multiplied. With automated machines, systems, factories, and refineries have come increased worker alienation and (as Wiener predicted) increased technological unemployment.

There has been an enormous explosion in the growth and use of computerized databanks, which store and process all kinds of information about people—information on health, travel, credit, finances, education, employment, and so on. (See, for example, Warner & Stone 1970; Westin 1971.) Increasingly, decisions about people have been based on this stored information, which often is incomplete, incorrect, out of date, ambiguous, irrelevant, and confidential. This has led to errors that have been difficult to correct; and it has led to an invasion of privacy and an infringement of other basic civil rights. (See, for example, Miller 1971; Westin & Baker 1972; U.S. Department of Health, Education, & Welfare 1973.) Further, some uses of databanks have led to a sense of dehumanization: people have felt that important decisions affecting their lives have been made without consulting them—they have been "measured"; data about them have been manipulated by computer; but *they* have not been consulted. People feel that they have been treated as mere objects (machines).

In fact, for an increasing number of people, the ideas of cybernetics seem to imply that human beings are really "nothing but" computing machines. But cybernetics implies no such conclusion. It is a fallacy to talk as if brains and people are in the same logical category. Brains can be described, as Wiener did, as special-purpose control computers that store and process information so that their owners can behave in adaptive, intelligent ways. But those owners are *people* who think and feel and love. To say that people are machines is not false; it is nonsense.

M. E. MARON

[*See also* COMPUTATION; ETHICAL ISSUES IN THE SOCIAL SCIENCES.]

ADDITIONAL BIBLIOGRAPHY

ARBIB, MICHAEL A. 1972 *The Metaphorical Brain: An Introduction to Cybernetics as Artificial Intelligence and Brain Theory.* New York: Wiley.

ARBIB, MICHAEL A. 1975 Cybernetics After 25 Years: A Personal View of System Theory and Brain Theory. Institute of Electrical and Electronics Engineers, *Transactions on Systems, Man, and Cybernetics* 5: 359–363.

CHERRY, COLIN 1975 Celebration of the 25th Anniversary of Norbert Wiener's Cybernetics. Institute of Electrical and Electronics Engineers, *Transactions on Systems, Man, and Cybernetics* 5: 366–368.

MARON, M. E. 1968 Cybernetics. Volume 4, pages 3–6 in *International Encyclopedia of the Social Sciences.* Edited by David L. Sills. New York: Macmillan and Free Press.

MILLER, ARTHUR R. 1971 *The Assault on Privacy: Computers, Data Banks, and Dossiers.* Ann Arbor: Univ. of Michigan Press.

MILLER, GEORGE A. 1974 Needed: A Better Theory of Cognitive Organization. Institute of Electrical and Electronics Engineers, *Transactions on Systems, Man, and Cybernetics* 4: 95–97.

ROTHMAN, STANLEY; and MOSMANN, CHARLES (1972) 1975 *Computers and Society.* New ed. Edited by Kay Nerode. Chicago: Science Research. → See especially Chapters 9–11.

TAVISS, IRENE (editor) 1970 *The Computer Impact.* Englewood Cliffs, N.J.: Prentice-Hall.

TOULMIN, STEPHEN 1964 The Importance of Norbert Wiener. *New York Review of Books* Sept. 24: 3–5.

U.S. DEPARTMENT OF HEALTH, EDUCATION, AND WELFARE, SECRETARY'S ADVISORY COMMITTEE ON AUTOMATED PERSONNEL DATA SYSTEMS 1973 *Records, Computers, and the Rights of Citizens: Report.* DHEW Publication No. (OS) 7394. Washington: Government Printing Office.

WARNER, MALCOLM; and STONE, MICHAEL 1970 *The Data Bank Society: Organizations, Computers and Social Freedom.* London: Allen & Unwin.

WEIZENBAUM, JOSEPH 1972 On the Impact of the Computer on Society. *Science* 176: 609–614.

WESSEL, MILTON R. 1974 *Freedom's Edge: The Computer Threat to Society.* Reading, Mass.: Addison-Wesley.

WESTIN, ALAN F. (editor) 1971 *Information Technology in a Democracy.* Cambridge, Mass.: Harvard Univ. Press.

WESTIN, ALAN F.; and BAKER, MICHAEL A. 1972 *Databanks in a Free Society: Computers, Record-keeping, and Privacy.* New York: Quadrangle.

WILCOXON, FRANK

▶ *This article was specially written for this volume.*

Frank Wilcoxon (1892–1965) was a renowned statistician, an excellent chemist, and an outstanding human being. Trained as a physical chemist, he made research contributions to physical chemistry, biochemistry, plant pathology, and entomology. He was a significant member of that pioneering group of statisticians who, although trained in other disciplines, did so much for the early development of statistical methodology. His name is best known in association with fundamental, widely used elementary ranking methods [*see* NONPARAMETRIC STATISTICS, *article on* RANKING METHODS], but within statistics, he also contributed to biological assay and sequential analysis. Wilcoxon's work in statistics is marked by the elegance of simplicity and easy application. Frank Wilcoxon had an intense and enthusiastic interest in understanding his world and a joie de vivre that he communicated to all around him.

Wilcoxon was born in Glengarriffe Castle, near Cork, Ireland, to wealthy American parents. His father was a poet, outdoorsman, and hunter. Wilcoxon spent his boyhood at Catskill, New York, on the banks of the Hudson River, and developed his lasting love for nature and the water there. His early instruction was by private tutors. He received the B.S. degree in 1917 from Pennsylvania Military College, after a number of temporary jobs, such as being a merchant sailor for a few days in New York harbor, manning a gas pumping station in an isolated area of West Virginia, and climbing trees as a tree surgeon. Early manhood was a difficult period of Wilcoxon's life. The school's military system did not agree with his ideas of personal freedom.

Following a World War I position with the Atlas Powder Company at Houghton, Michigan, Wilcoxon enrolled at Rutgers University in 1920. He achieved the M.S. degree in chemistry in 1921 and continued graduate study at Cornell University, where he received the PH.D. degree in physical chemistry in 1924. While at Cornell, he held an initial assistantship and during 1924–1925 was a postdoctoral Heckscher fellow. McCallan, in his biography (1966) of Wilcoxon, notes that he was much influenced by the independent thinking and philosophy of Wilder D. Bancroft, Cornell professor of physical chemistry. At Cornell, Frank met Frederica Facius, an undergraduate from Pittsburgh, and they were married on May 27, 1926. Frank and Freddie later became well known and loved in the statistical community, particularly through their regular participation in the Gordon Research Conference on Statistics in Chemistry and Chemical Engineering.

Although trained as a physical chemist, Wilcoxon's first position on leaving Cornell was as a postdoctoral fellow with the Crop Protection Institute to investigate the use of copper compounds as fungicides. The project, sponsored by the Nichols Copper Company, was located at the Boyce Thompson Institute for Plant Research in Yonkers, New York. On termination of the fellowship, Wilcoxon worked with the Nichols Copper Company in Maspeth, in Queens, from 1928 to 1929. In 1929, the Boyce Thompson Institute received a grant from the Hermann Frasch Foundation and brought Wilcoxon back as a group leader of an investigation of the action of insecticides and fungicides. He continued with the Boyce Thompson Institute until World War II, when he took leave. For two years he was in charge of the control laboratory of the Ravenna Ordnance Plant in Ohio, operated by the Atlas Powder Company.

In 1943, Wilcoxon became group leader of the insecticide and fungicide laboratory of the American Cyanamid Company in Stamford, Connecticut. During the latter part of the period with the Boyce Thompson Institute and at the American Cyanamid Company, his interest, study, and knowledge of statistics grew, and in 1950 he transferred to the Lederle Laboratories Division of the company at Pearl River, New York, where he developed a statistical consulting group. Wilcoxon continued in statistical consulting until his retirement in 1957, when he returned to the Boyce Thompson Institute as a part-time consultant on statistics.

Wilcoxon's teaching career included approximately 12 years (1929–1941) at the Brooklyn Polytechnic Institute. Although he held full-time jobs during this period, he also taught physical chemistry to graduate students in the evenings. He found this teaching to be rewarding and stimulating, and many of his students formed long-term attachments to him and corresponded with him for many years. He resigned very reluctantly in 1941 to begin his wartime work for the Ravenna Ordnance Plant.

Teaching did not play a major role again in the life of Wilcoxon until 1960, when he joined the newly created Department of Statistics at Florida State University as a part-time Distinguished Lecturer. Although it had been nearly twenty years since he last taught formally, the courses that he developed on statistical methodology for students of the natural sciences were soon overenrolled, despite his insistence that no grading credit be given for any problem with an incorrect final numerical answer, even though the correct methodology might have been used. Wilcoxon's knowledge and experience contributed greatly to the development of the young department. It was a productive period for him also, for he had eager collaborators who worked with him on ideas for research that he had accumulated through his years of statistical consulting. Wilcoxon died of a heart attack on November 18, 1965, in Tallahassee, Florida, following a canoe trip. He pursued his interests in statistics and the outdoors vigorously and fully to his final days.

Contributions to chemistry. Frank Wilcoxon's contributions to statistics are at the center of this biography, but his contributions to chemistry were also substantial. This work is described in the McCallan biography (1966).

An anecdote relating to this spanning of two fields is of interest. When Florida State University sought approval for a doctoral program in statistics from the Board of Control (now the Board of Regents) in 1961, the only comment before approval was from a board member who said "I see that

Frank Wilcoxon is associated with this proposal and any program with Frank Wilcoxon is good enough for me." A citrus grower, he was familiar with Wilcoxon's contributions to the development of the insecticide malathion.

Contributions to statistics.   Early in his career with the Boyce Thompson Institute, Wilcoxon was intrigued with C. R. Orton's greenhouse studies, with replications, on organic mercurial seed treatments. The first edition of Sir Ronald A. Fisher's book, *Statistical Methods for Research Workers* (1925), had just been published, and a group had been formed to study it. Besides Wilcoxon, the group included F. E. Denny, a biologist, and W. J. Youden, a chemical engineer who later made important contributions to the design of experiments. C. I. Bliss, an entomologist at the Storrs Agricultural Experiment Station, who later contributed to the development of probit analysis, was also a visitor at the institute.

Frank Wilcoxon was largely self-taught in statistics. It is not surprising that his first work in statistics (1945*a*) related to plant pathology and dose–effect curves, nor that another early paper, with John T. Litchfield, Jr. (1949) was on a simplified method of evaluating dose–effect experiments. Wilcoxon's collaboration with Litchfield was a happy one and led to other joint papers (1953; 1955).

Wilcoxon's most significant contribution to statistics unquestionably was the development of two cornerstones of ranking methods: the two-sample rank–sum statistic and the one-sample (or paired sample) signed-rank statistic, both proposed in a paper (1945*b*) that appeared in the first volume of *Biometrics,* then the *Biometrics Bulletin.* These statistics, now widely known as the Wilcoxon rank–sum statistic and the Wilcoxon signed-rank statistic, were an inspirational force in the development of ranking methods, the central theme of nonparametric methods in statistics.

In the two-sample problem, Wilcoxon reasoned that by ranking the observations in the two samples jointly from smallest to largest, the sum of the ranks obtained by one of the samples could be used to make inferences about the difference between the population locations (means, medians). Thus, if W denotes the sum of the ranks assigned to the sample 1 observations, then large (small) values of W would indicate that the mean of population 1 was larger (smaller) than the mean of population 2. Furthermore, Wilcoxon showed how to calculate the probability, under the hypothesis that the two populations were indeed the same, that W would exceed any given value. This calculation

enables one formally to test the hypothesis of equivalent populations.

The motivation for the signed-rank test is similar to that of the rank–sum test. Here the data consist of n (say) paired replicates, and it is convenient to label the two observations in a pair as the control and treatment responses. Let Z denote the amount by which a treatment response exceeds the control response in a matched pair. Wilcoxon's signed-rank statistic is obtained as follows. Rank the absolute values of the Z's from smallest to largest. Then attach to each rank the sign of the Z-difference to which it corresponds. Call the ranks, with (attached) positive signs, the positive signed ranks; T^+, the sum of those positive signed ranks, is the Wilcoxon signed-rank statistic. Wilcoxon reasoned that large (small) values of T^+ indicate that the median of the Z-distribution is greater (less) than 0. Again, he showed how to calculate the probability, under the hypothesis that the Z-distribution is symmetric about 0, that T^+ would exceed any given value. This calculation enables one to obtain the significance probability for the hypothesis test.

There are numerous advantages to the Wilcoxon tests based on W and T^+ when compared to the classical t-tests based on referring suitably standardized differences between sample means to Student's t-distribution. These advantages include (1) ease and rapidity of calculation, (2) availability of exact significant levels without the restrictive normality assumption, (3) relative insensitivity to outlying sample observations, (4) invariance under certain monotonic transformations of the data, (5) applicability to situations where the data are ordinal, (6) excellent power properties for wide classes of alternative distributions, and (7) availability (through inversion) of distribution-free confidence intervals for the location parameters of interest. Wilcoxon did not, of course, delineate each of these advantages. Much of that would come only after years of work by subsequent researchers. Furthermore, Wilcoxon was the first to admit that, at the time of his proposals, he was not aware of all the remarkable properties of the rank–sum and signed-rank statistics. Nevertheless, it was clear from his commitment to the development of ranking methods—as evidenced by the publication of the first edition of his booklet *Some Rapid Approximate Statistical Procedures* (1947; revised several times)—that he knew or sensed that his rank procedures, and other procedures based on ranks, possessed many (yet undiscovered) desirable properties.

There has been discussion about how much

credit is due Wilcoxon for his small but elegant paper (1945b) in the *Biometrics Bulletin*. Critics can point out that, for the two-sample problem, the treatment of unequal sample sizes via the rank–sum statistic was not covered by Wilcoxon, but left for Mann and Whitney (1947). Others cite Kruskal's scholarly notes (1957) on the history of the rank–sum test, which show that there were independent proposals of the two-sample test, or of closely related procedures, and that some of these proposals preceded the one by Wilcoxon.

The above qualifications do not dim the luster of Wilcoxon's contribution. Despite the earlier use of ranks in different contexts by, for example, Hotelling and Pabst (1936) and Friedman (1937), it was not until Wilcoxon's 1945 paper that the great concentration of energy toward solving other nonparametric problems began. Here was a basic idea: to replace the actual sample values by their ranks, thereby reducing relatively complicated distributional theory to simple counting, and simultaneously (as already indicated) gaining a number of statistical benefits. Furthermore, the very fact that Wilcoxon could describe his methods in a concise, nonmathematical manner reduced the possible fears of many potential users, not trained in statistics, and thus hastened the widespread adoption of the ranking methods.

Whereas Wilcoxon focused his statistical research in the 1940s and early 1950s on dose–effect curves and elementary ranking methods, the latter part of the 1950s found him mulling over how the desirable properties of ranking methods could be utilized in more complicated data situations. He began to consider multiple-comparison procedures —procedures that, on the basis of the sample observations, make possible a number of decisions about the parameters of interest. Wilcoxon was interested in the two-way classification (n blocks, k treatments) and, in particular, in the joint distribution of rank totals that arise as follows. In each of n blocks, rank the k responses (one response for each of k treatments) from 1 to k. Then, add up (over the blocks) the ranks for each treatment, to obtain statistics $R_1, R_2, \cdots, R_k$ (say). These rank totals are known as Friedman rank sums. What is the joint distribution of these k rank sums when the hypothesis of equivalent treatments is true? This question can be answered by direct (though tedious, when n and k are large) enumeration. In 1956, in an unpublished notebook, Wilcoxon performed the correct probability computation for the case $n = 3$, $k = 3$, and several other configurations of small sample sizes. This interest in multiple-comparison procedures was still strong in the 1960s when

Wilcoxon suggested and largely directed a dissertation by Peter Dunn-Rankin that formed the basis for their joint paper (1966). Furthermore, Wilcoxon's inclusion of multiple-comparison procedures in the 1964 revision of *Some Rapid Approximate Statistical Procedures* accelerated the acceptance of these procedures as an important part of statistical methodology.

Wilcoxon's major statistical contribution of the 1960s was not, however, in the area of multiple comparisons, but in nonparametric sequential analysis. Wilcoxon reasoned that since methods based on ranks had proven so successful in fixed sample-size situations, they ought to be extended to situations where the data were obtained sequentially and where the number of observations in the sample(s) was itself a random quantity. This idea led to a number of sequential ranking procedures developed jointly by Wilcoxon, Ralph A. Bradley, and others (1963; 1964; 1965; 1966). The methods developed were based on sequential theory derived by Wald (1947) and a tractable class of nonparametric alternatives developed by E. L. Lehmann (1953). Although it is perhaps still too early to assess these sequential methods, these papers have acted as catalysts for other statisticians to formulate and attack a number of new theoretical problems including the difficult termination probability problems associated with nonparametric sequential methods. (See also Govindarajulu 1975 for coverage of nonparametric sequential procedures.)

Wilcoxon was statistically active up to his death. He was still pursuing a long-term interest in a multivariate generalization of his two-sample rank–sum test. Suppose the two samples consisted of p-variate vector observations, rather than univariate observations. What would be a good way to test the hypothesis of equivalent multivariate populations? Bradley (1967) described two of Wilcoxon's proposals for this problem. These proposals were simple and intuitive, but, to date, both lack an adequate theoretical base.

In 1966, Wilcoxon and Cuthbert Daniel devised factorial experiment designs that would be robust against certain linear and quadratic trends. We quote from their paper: "The basic idea, due to Frank Wilcoxon, is that certain of the ordered contrasts appearing in the 2^p system are orthogonal to linear and to quadratic trends" (Daniel & Wilcoxon 1966, p. 261). It is fitting that the basic idea is attributed to Wilcoxon, for he was an "ideas" man. His ideas were simple, elegant, useful, and stimulating.

Brief biographies, by their very nature, may fail to communicate the warmth and personality of a person. The following, somewhat disjointed, comments are designed to bring out the character of the man. Frank Wilcoxon's interests were broad, his intellectual ability great. An accomplished musician with superior training on the classical guitar, he played several other string instruments, owned a rare guitar, and was a discriminating record collector. McCallen (1966) notes that "the Wilcoxons visited Russia in 1934 and 1935 to see at first hand this huge experiment in planned economic life. The future developments were a disappointment to them." Wilcoxon, who studied Russian briefly but intensively to prepare for this visit, retained a proficient reading knowledge of it (and four other languages) throughout his life. (In the penultimate summer of his life, he studied enough Dutch so that he could read a book of interest to him in that language; when his mother lost her sight, he learned Braille.) Frank Wilcoxon's teaching exuded enthusiasm for statistics; he was loved by his students, and he was always available to talk about his favorite subject. He was interested in mathematical games, puzzles, combinatorics, nomographs, and the hatchet planimeter—he very effectively simulated two-sample ranking under randomness through use of a long-necked, stoppered flask containing two sets of colored beads. Late in life, he was introduced to a method of "divining" for buried pipes and water lines using two L-shaped iron rods —it seemed to work, as he located and traced most of the pipes at the Hampton School while attending a Gordon Research Conference. It was characteristic that he had not dismissed the procedure as obvious nonsense but sought an explanation, consulting various geologists and physical scientists. He was proud of the accident-free record established by his control group in the explosives industry. He was proud of his tree-climbing ability but was shaken by a minor fall in his seventies. Through middle age, Frank and Freddie vacationed with made-to-measure bicycles, covering 100 miles in a day; he did not own an automobile until late in life. At Florida State University, he was one of the first owners of a Honda motorcycle and rode it regularly to work.

Frank Wilcoxon was a fellow of the American Statistical Association and the American Association for the Advancement of Science. He was an early chairman of the Gordon Research Conference on Statistics in Chemistry and Chemical Engineering. The Chemical Division of the American Society for Quality Control annually awards the Frank Wilcoxon Prize for the best papers of the year on practical applications published in *Technometrics*. Florida State University has honored Wilcoxon

through designation of the statistics library and reading room as the Frank Wilcoxon Memorial Room.

RALPH A. BRADLEY AND
MYLES HOLLANDER

[*For the historical context of Wilcoxon's work, see the biography of* YOUDEN. *Related material is treated in* NONPARAMETRIC STATISTICS, *articles on* THE FIELD *and* RANKING METHODS; LINEAR HYPOTHESES, *articles on* ANALYSIS OF VARIANCE *and* MULTIPLE COMPARISONS; SEQUENTIAL ANALYSIS; QUANTAL RESPONSE.]

WORKS BY WILCOXON

A complete bibliography of Wilcoxon's publications, including many in chemistry, appears in Karas & Savage 1967.

1945a   Some Uses of Statistics in Plant Pathology. *Biometrics Bulletin* 1, no. 4:41–45.
1945b   Individual Comparisons by Ranking Methods. *Biometrics Bulletin* 1, no. 6:80–83. → Reprinted in the Bobbs-Merrill Reprint Series in the Social Sciences, S-541.
1946   Individual Comparisons of Grouped Data by Ranking Methods. *Journal of Economic Entomolgy* 39: 269–270.
(1947a) 1964   *Some Rapid Approximate Statistical Procedures.* Stamford, Conn.: Stamford Research Laboratories, American Cyanamid Company. → Revised in 1949 and revised, jointly with Roberta A. Wilcox, in 1964.
1947b   Probability Tables for Individual Comparisons by Ranking Methods. *Biometrics* 3:119–122.
1949   LITCHFIELD, J. T. JR.; and WILCOXON, FRANK A Simplified Method of Evaluating Dose–Effect Experiments. *Journal of Pharmacology and Experimental Therapeutics* 96:99–113.
1951   THAYER, F. D. JR.; BIANCO, E. G.; and WILCOXON, FRANK The Application of Statistics in the Tanning Laboratory: I. The Use of a Youden Square. *Journal of the American Leather Chemists Association* 46: 669–684. → Includes discussion.
1953   LITCHFIELD, J. T. JR.; and WILCOXON, FRANK The Reliability of Graphic Estimates of Relative Potency From Dose–Per Cent Effect Curves. *Journal of Pharmacology and Experimental Therapeutics* 108:18–25.
1954   BURCHFIELD, H. P.; and WILCOXON, FRANK Comparison of Dosage-response and Time-response Methods for Assessing the Joint Action of Antimetabolites. Boyce Thompson Institute for Plant Research, *Contributions* 18:79–82.
1955   LITCHFIELD, J. T. JR.; and WILCOXON, FRANK The Rank Correlation Method. *Analytical Chemistry* 27: 299–300.
1963   WILCOXON, FRANK; RHODES, L. J.; and BRADLEY, RALPH A. Two Sequential Two-sample Grouped Rank Tests With Applications to Screening Experiments. *Biometrics* 19:58–84.
1964   WILCOXON, FRANK; and BRADLEY, RALPH A. A Note on the Paper "Two Sequential Two-sample Grouped Rank Tests With Applications to Screening Experiments." *Biometrics* 20:892–895.
1965   BRADLEY, RALPH A.; MARTIN, DONALD C.; and WILCOXON, FRANK Sequential Rank Tests: I. Monte

Carlo Studies of the Two-sample Procedure. *Technometrics* 7:463–483.

1966 BRADLEY, RALPH A.; MERCHANT, SARLA D.; and WILCOXON, FRANK Sequential Rank Tests: II. Modified Two-sample Procedures. *Technometrics* 8:615–623.

1966 DANIEL, CUTHBERT; and WILCOXON, FRANK Factorial 2^{p-q} Plans Robust Against Linear and Quadratic Trends. *Technometrics* 8:259–278.

1966 DUNN-RANKIN, PETER; and WILCOXON, FRANK The True Distribution of the Range of Rank Totals in the Two-way Classification. *Psychometrika* 31:573–580.

1970 WILCOXON, FRANK; KATTI, S. K.; and WILCOX, ROBERTA A. Critical Values and Probability Levels for the Wilcoxon Rank Sum Test and the Wilcoxon Signed Rank Tests. Volume 1, pages 171–259 in H. Leon Harter and Donald B. Owen (editors), *Selected Tables in Mathematical Statistics*. Chicago: Markham. → A revision of tables developed by the same authors and prepared and distributed in 1963 by the Lederle Laboratories Division of the American Cyanamid Company, in cooperation with the Department of Statistics, Florida State University.

SUPPLEMENTARY BIBLIOGRAPHY

BRADLEY, RALPH A. 1966a Obituary: Frank Wilcoxon. *Biometrics* 22:192–194.

BRADLEY, RALPH A. 1966b Frank Wilcoxon. *American Statistician* 20, no. 1:32–33.

BRADLEY, RALPH A. 1967 Topics in Rank–Order Statistics. Volume 1, pages 593–605 in Berkeley Symposium on Mathematical Statistics and Probability, Fifth, *Proceedings*. Volume 1: *General Theory*. Berkeley: Univ. of California Press.

DUNNETT, CHARLES W. 1966 Frank Wilcoxon, 1892–1965. *Technometrics* 8:195–196.

FISHER, R. A. (1925) 1970 *Statistical Methods for Research Workers*. 14th ed., rev. & enl. Edinburgh: Oliver & Boyd; New York: Hafner.

Frank Wilcoxon Spanned Two Fields. 1965 *New York Times* Nov. 19:p. 39, col. 2.

FRIEDMAN, MILTON 1937 The Use of Ranks to Avoid the Assumption of Normality Implicit in the Analysis of Variance. *Journal of the American Statistical Association* 32:675–701.

GOVINDARAJULU, Z. 1975 *Sequential Statistical Procedures*. New York: Academic Press.

HOTELLING, HAROLD; and PABST, MARGARET RICHARDS 1936 Rank Correlation and Tests of Significance Involving No Assumption of Normality. *Annals of Mathematical Statistics* 7:29–43.

KARAS, JAMES; and SAVAGE, I. RICHARD 1967 Publications of Frank Wilcoxon (1892–1965). *Biometrics* 23:1–11.

KRUSKAL, WILLIAM H. 1957 Historical Notes on the Wilcoxon Unpaired Two-sample Test. *Journal of the American Statistical Association* 52:356–360.

LEHMANN, E. L. 1953 The Power of Rank Tests. *Annals of Mathematical Statistics* 24:23–43.

McCALLAN, S. E. A. 1966 Frank Wilcoxon (September 2, 1892—November 18, 1965). Boyce Thompson Institute for Plant Research, *Contributions* 23:143–145.

MANN, H. B.; and WHITNEY, D. R. 1947 On a Test of Whether One of Two Random Variables Is Stochastically Larger Than the Other. *Annals of Mathematical Statistics* 18:50–60.

WALD, ABRAHAM 1947 *Sequential Analysis*. New York: Wiley. → Reprinted by Peter Smith. A paperback edition was published in 1973 by Dover.

WILKS, S. S.

The various professional roles of the statistician Samuel Stanley Wilks (1906–1964) so parallel the development of mathematics in the mid-twentieth century that he seems ready-made for the hero in a sociological novel entitled "The Professional Mathematician." Like most important mathematicians, Wilks early made strong research contributions, and his innovations opened new lines of research for others. But not very many mathematicians succeeded, as he did, in attracting fine students who, in their turn, used their training to specialize in such diverse areas as statistics, mathematical statistics, probability, sociology, governmental service, and defense research. Still fewer mathematicians have advanced their fields by major editorial commitments as Wilks did, both through the *Annals of Mathematical Statistics* and through the Wiley Publications in Statistics. What is rare among mathematicians is the belief, which Wilks held, in the value of organization, distinct from individual achievement, for the future of mathematics. Acting on this belief, he deliberately devoted much of his career to scientific societies, to committees of governmental agencies, and to public and private foundations; his contributions ranged far beyond the limits of even the broadest interpretations of mathematics. The combination of these activities produced a dedicated scholar, a major educator, and a public servant.

Wilks was born to Chance C. and Bertha May Gammon Wilks in Little Elm, Texas. He and his two younger brothers were raised on a small (for Texas), 250-acre farm near Little Elm. During his formal education, he had a notable set of teachers. W. M. Whyburn, later chairman of the department of mathematics at the University of North Carolina, taught Wilks in the seventh grade. In high school, Wilks used to sneak off to take a college mathematics course during study hour. In 1926 he took a bachelor's degree in architecture at North Texas State Teachers College. While earning his M.A. in mathematics at the University of Texas (he received it in 1928), he studied topology with R. L. Moore and statistics with E. L. Dodd, who encouraged him to join Henry L. Rietz at the University of Iowa, then the leading center in the United States for the study of mathematical statistics.

At Iowa, E. F. Lindquist introduced him to a problem which led to Wilks's thesis, entitled "On the Distributions of Statistics in Samples From a Normal Population of Two Variables With Matched Sampling of One Variable" (1932*a*). This began Wilks's long series of contributions to multivariate analysis, his interest in the applications of statistical methods, and his lifelong relation with the fields of education, testing, and the social sciences. He received a PH.D. in mathematical statistics in 1931.

After the National Research Council awarded him a National Research Fellowship, Wilks and his bride, Gena Orr of Denton, Texas, went to Columbia University so that he could work with Harold Hotelling. There Wilks also listened to C. E. Spearman and met Walter Shewhart of Bell Telephone Laboratories, who then and later introduced him to many research problems arising from industrial applications of statistics.

The next year his fellowship was renewed, and Wilks went to work in Karl Pearson's department of applied statistics at University College, London. Wilks's only child, Stanley Neal, was born there in October 1932. In London Wilks met and worked with Egon Pearson, who remained a lifelong friend, and, of course, he met R. A. Fisher and Jerzy Neyman. At midyear he moved to Cambridge, where he worked with John Wishart and got to know M. S. Bartlett and W. G. Cochran. By the end of his two-year fellowship he had published six papers, two of which grew out of his doctoral thesis; another was entitled "Moments and Distributions of Estimates of Population Parameters From Fragmentary Samples" (1932*b*).

In this important paper Wilks dealt with the problem of missing values in multivariate data —in some kinds of investigations two or more characteristics are to be measured for each member of the sample, but occasionally the value of one variable or another may be missing, as when only part of a skeleton is found in an archeological study. Wilks found for bivariate normal distributions the maximum likelihood equations for estimating the parameters and suggested some alternative estimators. He also suggested the determinant of the inverse of the asymptotic covariance matrix of a set of estimators as the appropriate measure of information that estimators jointly contain.

At this period, mathematical statisticians were developing exact and approximate distributions of statistics of more and more complex quantities under idealized assumptions, making it possible to assess evidence offered by bodies of data about more and more complicated questions. Wilks's paper "Certain Generalizations in the Analysis of Variance" (1932*c*) was a major contribution to this development.

Among the multivariate criteria proposed by Wilks, the one most used today is likely the one denoted by W in his 1932 *Biometrika* paper. In 1947 Bartlett used the notation Λ for this statistic as applied in a wider variety of contexts than originally proposed by Wilks, and in 1948 C. R. Rao introduced the frequently heard term "Wilks's Λ criterion." This criterion provided a multivariate generalization of what is now called the analysis of variance F test. If the "among" sum of squares in analysis of variance is generalized to a $p \times p$ matrix $\mathbf{A}$ of sums of squares and products and the "within" sum of squares is similarly generalized to a $p \times p$ matrix $\mathbf{B}$, then the Wilks criterion is a ratio of determinants, namely, $\det(\mathbf{B})/\det(\mathbf{A} + \mathbf{B})$. In 1932 it was a considerable feat to determine, as Wilks did, the null distributions of this and many similar statistics.

In this and other papers through the years, Wilks found the likelihood ratio criterion for testing many hypotheses in multivariate problems. The criteria repeatedly turn out to be powers of ratios of products of determinants of sample covariance matrices essentially like the formula above, and their moments are products of beta functions. He suggested the determinant of the covariance matrix as the generalized variance of a sample of points in a multidimensional space, and he found the distribution of the multiple correlation coefficient.

Wilks also found the likelihood ratio criterion and its moments for testing the null hypothesis that several multivariate populations have equal covariance matrices. Much later, in 1946, he found the likelihood ratio criteria for testing whether in one multivariate population the variances are equal and the covariances are equal, for testing whether all the means are equal if the variances are equal and the covariances are equal, and for testing all these hypotheses simultaneously. These problems arise from studying whether several forms of a test are nearly parallel. Elsewhere he showed that for a test consisting of many items, scorings based upon two different sets of weights for the items would produce pairs of total scores that were highly correlated across the individuals taking the test. The implication is that modest changes in the weights of the items on a long test make only small changes in the relative evaluations of individuals. He also suggested the likelihood test for independence in contingency tables, and he found the large-sample distribution of the likelihood ratio for testing composite hypotheses.

Wilks encouraged his students to develop nonparametric or distribution-free methods and made major contributions to this area himself. In particular, he invented the statistical idea of tolerance limits, by analogy with the term as used in industry in connection with piece parts. Shewhart had asked for ways to make guarantees about mass-produced lots of parts. Wilks found that by using order statistics, for example, the largest and smallest measurements in a sample, one could make confidence statements about the fraction of the true population contained between the sample order statistics. To illustrate, for a sample of size 10 from a continuous population, the probability is 0.62 that at least 80 per cent of the population is contained between the smallest and largest sample values. Wilks and others have extended this idea in many directions.

In 1933 Wilks joined the department of mathematics at Princeton University, which was his base of operations for the rest of his life. At Princeton he gradually developed both undergraduate and graduate courses in mathematical statistics, repeatedly writing up course notes, producing his long-awaited hard-cover *Mathematical Statistics* in 1962, and publishing several soft-cover books that had a substantial influence on the teaching of mathematical statistics. Wilks wrote or coauthored six books and about forty research papers.

Wilks's first doctoral student, Joseph Daly, received his degree in 1939, and thereafter Princeton had a small but strong graduate program in statistics.

Wilks contributed to education in many ways. From the time of his arrival in Princeton, he participated in the work of the College Entrance Examination Board (and the Educational Testing Service), and some of his research problems arose from this source. He served on the Board's commission on mathematics and was a coauthor of their experimental text in probability and statistics for secondary school students. Later he became a member of the Advisory Board of the School Mathematics Study Group (SMSG) and a visiting lecturer for various groups.

Wilks was one of a small group of statisticians who organized the Institute of Mathematical Statistics in 1935 and later negotiated the arrangements transferring the *Annals of Mathematical Statistics* from the private ownership of its founder and first editor, Harry C. Carver, to the Institute. Thereupon Wilks took over the editorship of the *Annals* (serving from 1938 through 1949) and with it, in effect, the shaping of the long-run future of the Institute. To quote John Tukey, "A marginal journal with a small subscription list became an unqualified first-rank journal in its field; a once marginal society grew to adulthood in size and responsibility and contribution. There is no doubt that the wisdom and judgment of Wilks was crucial; some of us suspect it was irreplaceable" (1965, p. 150).

Starting about 1941, Wilks began research for the National Defense Research Committee. During World War II, he became director of the Princeton Statistical Research Group, which had branches both in Princeton and in New York City. And he took an active part in the development and operation of the short courses that introduced statistical quality control to American industry. That such organizational developments persist is illustrated by the American Society for Quality Control, which was 20,000 strong in 1966. In 1947 Wilks was awarded the Presidential Certificate of Merit for his contributions to antisubmarine warfare and to the solution of convoy problems.

After World War II, Wilks devoted his time more and more to national affairs and services to his profession, and less to research.

He served the Social Science Research Council in many capacities, including successively the chairmanship of each of its three major subdivisions, over a period of 18 years. And from 1953 until his death he served the Russell Sage Foundation as a member of the Board of Directors and of the Executive Committee. He served the National Science Foundation both on its Divisional Committee for the Mathematical, Physical, and Engineering Sciences and on that for the Social Sciences.

Over the years he worked on uncounted committees for the Institute of Mathematical Statistics, the American Statistical Association, and the federal government. He served in many capacities the Division of Mathematics of the National Research Council, chairing the division from 1958 to 1960. He helped organize the Conference Board of the Mathematical Sciences and served as its chairman in 1960. He was a member of the U.S. National Commission for UNESCO from 1960 to 1962 (see Anderson 1965*a* pp. 3–7).

He was president of the Institute of Mathematical Statistics in 1940, and of the American Statistical Association in 1950. His other honors include election to the International Statistical Institute, the American Philosophical Society, and the American Academy of Arts and Sciences. In 1947 the

University of Iowa honored him with a Centennial Alumni Award. After his sudden death on March 7, 1964, several memorials were established, including an S. S. Wilks Memorial Fund at Princeton University and an American Statistical Association Samuel S. Wilks Memorial Award.

FREDERICK MOSTELLER

[*For the historical context of Wilks's work, see the biographies of* FISHER, R. A.; PEARSON; SPEARMAN; *for discussion of the subsequent development of his ideas, see* MULTIVARIATE ANALYSIS; NONPARAMETRIC STATISTICS, *article on* ORDER STATISTICS.]

WORKS BY WILKS

1932a   On the Distributions of Statistics in Samples From a Normal Population of Two Variables With Matched Sampling of One Variable. *Metron* 9, no. 3/4:87–126.

1932b   Moments and Distributions of Estimates of Population Parameters From Fragmentary Samples. *Annals of Mathematical Statistics* 3:163–195.

1932c   Certain Generalizations in the Analysis of Variance. *Biometrika* 24:471–494.

1932d   On the Sampling Distribution of the Multiple Correlation Coefficient. *Annals of Mathematical Statistics* 3:196–203.

1935   The Likelihood Test of Independence in Contingency Tables. *Annals of Mathematical Statistics* 6:190–196.

1938a   The Large-sample Distribution of the Likelihood Ratio for Testing Composite Hypotheses. *Annals of Mathematical Statistics* 9:60–62.

1938b   Weighting Systems for Linear Functions of Correlated Variables When There Is No Dependent Variable. *Psychometrika* 3:23–40.

1941   Determination of Sample Sizes for Setting Tolerance Limits. *Annals of Mathematical Statistics* 12:91–96.

1946   Sample Criteria for Testing Equality of Means, Equality of Variances, and Equality of Covariances in a Normal Multivariate Distribution. *Annals of Mathematical Statistics* 17:257–281.

(1957) 1959   COLLEGE ENTRANCE EXAMINATION BOARD, COMMISSION ON MATHEMATICS *Introductory Probability and Statistical Inference: An Experimental Course*, by E. C. Douglas, F. Mosteller . . . , and S. S. Wilks. Princeton, N.J.: The Board.

1962   *Mathematical Statistics*. New York: Wiley.

1965   Statistical Aspects of Experiments in Telepathy. Parts 1–2. *New York Statistician* 16, no. 6:1–3; no. 7:4–6. → Published posthumously.

1965   GUTTMAN, IRWIN; and WILKS, SAMUEL S. *Introductory Engineering Statistics*. New York: Wiley. → Published posthumously.

►*Collected Papers: Contributions to Mathematical Statistics*. Edited by T. W. Anderson. New York: Wiley, 1967. → Facsimile reprints of Anderson 1965a and the 48 articles listed in Anderson 1965b, with the latter rearranged.

SUPPLEMENTARY BIBLIOGRAPHY

●AMERICAN STATISTICAL ASSOCIATION   1965   Memorial to Samuel S. Wilks. *Journal of the American Statistical Association* 60:939–966. → The contributors and their topics are Frederick F. Stephan and John W. Tukey, "Sam Wilks in Princeton"; Frederick Mosteller, "His Writings in Applied Statistics"; Alexander M. Mood, "His Philosophy About His Work"; Morris H. Hansen, "His Contributions to Government"; Leslie E. Simon, "His Stimulus to Army Statistics"; Morris H. Hansen, "His Contributions to the American Statistical Association"; W. J. Dixon, "His Editorship of the *Annals of Mathematical Statistics*"; and (unsigned) "The Wilks Award."

ANDERSON, T. W.   1965a   Samuel Stanley Wilks: 1906–1964. *Annals of Mathematical Statistics* 36:1–23.

ANDERSON, T. W.   1965b   The Publications of S. S. Wilks. *Annals of Mathematical Statistics* 36:24–27.

COCHRAN, W. G.   1964   S. S. Wilks. International Statistical Institute, *Review* 32:189–191.

CRAIG, CECIL C.   1964   Professor Samuel S. Wilks. *Industrial Quality Control* 20, no. 12:41 only.

►EISENHART, CHURCHILL   1975a   A Supplementary List of Publications of S. S. Wilks. *American Statistician* 29:, no. 1:25–27.

►EISENHART, CHURCHILL   1975b   Samuel S. Wilks and the Army Experiment Design Conference Series. Pages 1–48 in Conference on Design of Experiments in Army Research, Development and Testing, Twentieth, *Proceedings*. Army Research Office Report 75–2. Research Triangle Park, N.C.: U.S. Army Research Office.

►EISENHART, CHURCHILL   1976   Samuel Stanley Wilks. Volume 14, pages 381–386 in *Dictionary of Scientific Biography*. Edited by Charles C. Gillispie. New York: Scribner's.

GULLIKSEN, HAROLD   1964   Samuel Stanley Wilks: 1906–1964. *Psychometrika* 29:103–104.

MOSTELLER, FREDERICK   1964   Samuel S. Wilks: Statesman of Statistics. *American Statistician* 18, no. 2:11–17.

PEARSON, E. S.   1964   Samuel Stanley Wilks: 1906–1964. *Journal of the Royal Statistical Society* Series A 127:597–599.

TUKEY, JOHN W.   1965   Samuel Stanley Wilks: 1906–1964. Pages 147–154 in American Philosophical Society, *Yearbook, 1964*. Philadelphia: The Society.

WILLCOX, WALTER F.

Walter Francis Willcox (1861–1964), American statistician, lived 103 years, was a persistent walker, and even after his ninetieth birthday attended meetings of the International Statistical Institute in Rio de Janeiro, New Delhi, and Stockholm. Such durability attracts attention, and Willcox was afraid that he would be remembered more for his feet than for his head. Actually, he deserves an outstanding place among American social scientists and statisticians as a pioneer teacher–scholar in demography.

Willcox played a major role in transforming the academic approach to social questions from one based on law guided by philosophy to one based on factual investigation guided by provisional theory. His work in social science grew out of a revolt against philosophy. Having studied philosophy and law at Amherst College and Columbia University (Amherst awarded him an A.B. in 1884 and an A.M. in 1888; Columbia awarded him an LL.B. in 1887 and a PH.D. in 1891), in 1889 he went to Berlin, where he began a thesis on divorce. He approached the subject in the tradition of natural law, as a study of practical ethics, until he encountered Bertillon's *Étude démographique du divorce* (1883). Its empirical method opened up a new world to him, and its substance completely overturned his former convictions. After returning to the United States, he applied Bertillon's method to the American data that Carroll D. Wright, then commissioner of labor, had published in *A Report on Marriage and Divorce 1867–1906 in the United States* (see U.S. Bureau of the Census 1908–1909). The result was a major work, *The Divorce Problem: A Study in Statistics* (1891).

In 1891 Willcox joined the faculty of Cornell University, at first teaching statistics in a philosophy course entitled "Applied Ethics." He remained at Cornell for forty years, becoming professor of economics and statistics in 1901 and retiring in 1931.

Willcox's writings ranged widely over the field of demography, covering birth, death, marriage, divorce, migration, the composition of population, and problems of method relating both to censuses and to vital statistics. He tended to study topics of practical importance, which he treated carefully with simple, yet imaginative, manipulative facility and presented as lucidly as possible without showmanship. Much of his work is dated, as indeed he asserted it would be, but much of it is remarkably fresh. His studies were valuable in themselves, often illuminating for the first time diverse problems of American society. But perhaps more important, the fact that he brought statistics to bear on sociology made it possible for later scholars, with access to vastly improved data, to probe more deeply into social problems and their interrelationships.

Of Willcox's studies, the two whose relevance is probably the least dated are the one on the population of China (see 1930a) and the one on the measurement of fertility (see 1910–1911). The former study provided the documentation to which most discussions of China's population continue to refer. In the latter, Willcox's measurement was based on the ratio of children under age five to women of childbearing age. The United States did not have complete registration of births until 1933, and Willcox used his ratio, which he obtained from the decennial census, to make the point that the birth rate in the United States had been declining, at least since the early years of the nineteenth cen-

tury. Willcox was probably the first to employ this ratio, which is now widely used as a measure of fertility whenever birth registration data are not available or are incomplete.

Outside the field of demography, Willcox turned his attention to economic productivity, problems of public health and social welfare, the role of social statistics as an aid to the courts, public opinion and Prohibition legislation, the history of statistics and biographical sketches of early statisticians (whom he preferred to call statists, a term that in his time did not have its present meaning), and above all, the perennial problem of apportionment of representation. Willcox first became interested in apportionment in 1900 and retained this interest all his life: as late as June 1959 he testified before a subcommittee of the Committee on the Judiciary of the House of Representatives, and the *New York Times* printed a letter from him on the subject written in his 101st year. He favored apportionment based on the method of major fractions, a method that had once been used by Congress but that had subsequently been replaced by the method of equal proportions. It was probably more through Willcox's efforts than through those of any other single person that the law now provides for automatic apportionment when Congress fails to act.

From 1899 to 1901 Willcox was one of five professional statisticians in charge of the twelfth, or 1900, census of the United States. He served as chief of the division of methods and results and was responsible for providing supplementary analyses of the data gathered in the census. These analytical investigations were the prototypes of what are now called census monographs. He and his staff issued reports on such topics as age statistics, proportions of children in the population, Negroes, illiteracy, and teachers; in 1906 the monumental *Supplementary Analysis and Derivative Tables*, 12th Census (see U.S. Bureau of the Census 1906) appeared.

In addition to serving as a chief statistician for the census, Willcox served on the Board of Health of New York State from 1899 to 1902, as statistical expert for the War Department on the census of Cuba and Puerto Rico during 1899/1900, as dean of the College of Arts and Sciences at Cornell from 1902 to 1907, and as president of the American Statistical Association in 1912 and of the American Economic Association in 1915. He was particularly interested in international statistics and was elected president of the International Statistical Institute in 1947.

FRANK W. NOTESTEIN

WILLCOX, WALTER F. *1255*

[Directly related are the entries CENSUS; DEMOGRAPHY, article on THE FIELD; LIFE TABLES; GOVERNMENT STATISTICS. See also Clark 1968; Grauman 1968; Mayer 1968.]

WORKS BY WILLCOX

(1891) 1897 *The Divorce Problem: A Study in Statistics.* 2d ed. Columbia University, Faculty of Political Science, Studies in History, Economics and Public Law, Vol. 1, No. 1. New York: Columbia Univ. Press.

1894 The Relation of Statistics to Social Science. National Conference of Charities and Correction, *Annual Report* 21:86–93.

1897 *Density and Distribution of Population in the United States at the Eleventh Census.* American Economic Association, Economic Studies, Vol. 2, No. 6. New York: Macmillan.

1910–1911 The Change in the Proportion of Children in the United States and in the Birth Rate in France During the Nineteenth Century. American Statistical Association, *Publications* 12:490–499. → Since 1914 called the *Journal of the American Statistical Association.*

(1930a) 1931 A Westerner's Effort to Estimate the Population of China and Its Increase Since 1650. International Statistical Institute, *Bulletin* 25, Part 3:156–170. → First published in the *Journal of the American Statistical Association.*

1930b Census. Volume 3, pages 295–300 in *Encyclopaedia of the Social Sciences.* New York: Macmillan.

1933 *Introduction to the Vital Statistics of the United States, 1900 to 1930.* Washington: Government Printing Office.

Studies in American Demography. Ithaca, N.Y.: Cornell Univ. Press, 1940. → Contains writings first published in the decades prior to 1940, as well as previously unpublished material. See especially pages 541–547 for a bibliography of the more important writings of Willcox.

SUPPLEMENTARY BIBLIOGRAPHY

BERTILLON, JACQUES 1883 *Étude démographique du divorce et de séparation de corps dans les différents pays de l'Europe.* Paris: Masson.

►CLARK, TERRY N. 1968 Bertillon, Jacques. Volume 2, pages 69–71 in *International Encyclopedia of the Social Sciences.* Edited by David L. Sills. New York: Macmillan and Free Press.

►GRAUMAN, JOHN V. 1968 Population: VI. Population Growth. Volume 12, pages 376–381 in *International Encyclopedia of the Social Sciences.* Edited by David L. Sills. New York: Macmillan and Free Press.

LEONARD, WILLIAM R. 1961 Walter F. Willcox: Statist. *American Statistician* 15:16–19.

►MAYER, KURT B. 1968 Population: IV. Population Composition. Volume 12, pages 362–370 in *International Encyclopedia of the Social Sciences.* Edited by David L. Sills. New York: Macmillan and Free Press.

RICE, STUART A. 1964 Walter Francis Willcox: March 22, 1861–October 30, 1964. *American Statistician* 18:25–26.

U.S. BUREAU OF THE CENSUS 1906 *Supplementary Analysis and Derivative Tables: Twelfth Census of the United States.* Washington: Government Printing Office.

U.S. BUREAU OF THE CENSUS 1908–1909 *A Report on Marriage and Divorce, 1867–1906 in the United States.* By Carroll D. Wright. 2 vols. Washington: Government Printing Office.

YZ

YOUDEN, W. J.

► *This article was specially written for this volume.*

Wiliam John (Jack) Youden was born in Townsville, Australia, on April 12, 1900. His family of five immigrated to America in 1907 after a five-year stay at Dover, England. He obtained his B.S. degree in chemical engineering from the University of Rochester in 1921, an M.A. (1923) and a PH.D. (1924) in chemistry from Columbia University. From 1924 to 1948 he was on the staff of the Boyce Thompson Institute for Plant Research in Yonkers, New York, except for a three-year tour of duty as an operations analyst (1942–1945) with the U.S. Army Air Force. For his important contributions (bombing accuracy) to the allied victory in World War II, he was awarded the Medal of Freedom in 1946. In 1948 Youden joined the Applied Mathematics Division of the National Bureau of Standards. He retired from full-time employment in 1965 but stayed on as a guest worker until his death on March 31, 1971. He was buried at National Cemetery, Gettysburg, Pennsylvania, in deference to his generally expressed wishes to remain in his adopted country.

Youden began his professional life as a chemist. He held two U.S. patents and was instrumental in developing a hydrogen ion concentration apparatus, which was named after him. He also made an excursion into social sciences in his "Science Looks at 'Social Breakdown'" (1940) in which he advised social scientists to follow the path taken by physical scientists and adopt the "operational point of view" to define abstract terms; that is, to define abstractions so that they could be realized repeatedly through measurements or observations.

Youden's transformation from chemist to statistician came early in his career. He read the first edition of R. A. Fisher's *Statistical Methods for Research Workers* (1925) soon after its publication and began to try out the early experimental designs on agricultural experiments. During 1931–1932 he commuted from Yonkers to Columbia University to attend Harold Hotelling's lectures on statistical inference. In 1936 he attended a six-week summer session at the Statistical Laboratory of Iowa State College where Fisher gave a series of lectures on the design of experiments and topics on the theory of statistics. Publication of "Use of Incomplete Block Replications in Estimating Tobacco-Mosaic Virus" (Youden & Zimmerman 1937) earned Youden a Rockefeller fellowship during the academic year 1937–1938 to work under the direction of Fisher at the Galton Laboratory, University College, London. "Youden squares" was the name given by Fisher and Yates to this new class of rectangular experimental designs in the introduction to the first edition of their *Statistical Tables for Biological, Agricultural and Medical Research* (1938, p. 18). [*See the biographies of* FISHER *and* HOTELLING.]

Youden squares represented a class of symmetrical balanced incomplete block designs that possessed the characteristic "double control" of Latin-square designs, without the restriction that the number of replications must equal the number of treatments [*see* LINEAR HYPOTHESES, *article on* ANALYSIS OF VARIANCE; EXPERIMENTAL DESIGN]. In the 1937 paper, Youden showed that by means of his new rectangular designs it was possible to adjust completely for the influence of leaf positions (rows), and make allowance for plant differences (columns), while the number of test solutions (treatments) that could be compared simultaneously could be increased. An example of seven treatments (A, B, ⋯, G) arranged in seven blocks (plants 1, 2, ⋯, 7) of three replications (leaves) is given in Table 1. Note that each block includes exactly three treatments and that each treatment appears three times, once in each of the three leaf positions. In addition, each treatment appears exactly once with each of the other treatments, for example, B appears with A and C in block 1, with D and F in block 2, and with E and G in block 5. This arrangement permits a simple untangling of the effects of plant, leaf, and treatment, providing that the combined long-run effect of these three factors on the number of lesions produced is the sum of the individual factor effects.

The method of eliminating block effects (that is, to account for the fact that the three values on one plant may be all high or all low because of the idiosyncrasies of that plant) is described in Youden and Zimmerman (1937). For brevity, I paraphrase it here.

Blocks 1, 2, and 5, respectively, are used to evaluate the following expressions:

$$2B - (A + C): \quad 154 - (18 + 30) = 106,$$
$$2B - (D + F): \quad 152 - (86 + 83) = -17,$$
$$2B - (E + G): \quad 130 - (47 + 86) = -3.$$

The sum of these three expressions is

$$6B - (A + C + D + F + E + G): \quad 86.$$

This may be written in the alternative form

$$7B - (A + B + C + D + E + F + G): \quad 86.$$

Dividing by 7 yields

$$\hat{B} - \text{mean of all} = 12.3,$$

or

$$\hat{B} = 1626/21 + 12.3 = 89.7.$$

Here $\hat{B}$ stands for the mean value of treatment B, and 1626 is the total of all the data. This procedure may be followed for the other six treatments and the mean values of the treatments before adjustments for block effects compared with the values after adjustments.

Youden had a talent for reducing a complicated idea to its essentials and then expressing the idea through simple language. Moreover, he worked hard at it. His audience were specialists in other fields who had no statistical training, and he was highly successful in convincing his audience of the reasonableness and usefulness of a statistical idea, omitting all unnecessary details, formulas, and jargon.

All his five books were written in this spirit. His first, *Statistical Methods for Chemists* (1951), was written "for those who make measurements and interpret experiments"; his second, *Statistical Design* (1960), is a collection of 36 bimonthly columns addressed to chemists and engineers; his third, *Experimentation and Measurement* (1962), was sponsored by the National Science Teachers Association to acquaint youth with the laws of measurements; his fourth, *Statistical Techniques for Collaborative Tests* (1967), was written for the Association of Official Analytical Chemists; and his fifth, *Risk, Choice, and Prediction* (1974), published posthumously, is his record and interpretation of the experiments with cards, pennies, and dice that he performed while on summer vacation in 1970, at Jade-bu, his cottage high on a mountaintop near Geilo, Norway.

Youden's searching mind responded to all worthwhile challenges arising from practical aspects of experimental problems. He once wrote, "Experiment design, to be successful, must fit the particular requirements of the science, the problem and the experimental environment" (1965). "Youden squares" were created to accommodate more treatments than Latin squares on the limited number of leaves on a plant; the "Youden plot" was designed to give simple but forceful interpretation of results of interlaboratory tests (1959); Youden's "ruggedness test" checks on the sensitivity of a test method to changes of conditions likely to be encountered in practice (1961a; 1961b).

Each of these statistical innovations resulted from Youden's recognition of the particular error structure inherent in a problem and his ingenuity in fitting a powerful statistical tool to its solution.

Table 1 — Seven treatments in seven blocks with three replications*

		PLANT NUMBER						
		1	2	3	4	5	6	7
	Lower leaf	A (18)	B (76)	C	D	E (47)	F	G
REPLICATION	Middle leaf	B (77)	D (86)	F	E	G (81)	A	C
	Upper leaf	C (30)	F (83)	E	A	B (65)	G	D

* Numerical values of lesion counts are shown in parentheses for corresponding treatments in blocks 1, 2, and 5 for use in the illustrative analysis.

Physical scientists and engineers, easily understanding the logic behind these schemes and realizing their potential payoffs, adopted them unhesitatingly, and gave Youden's name to the two-sample interlaboratory test and the ruggedness test. The two-sample technique is currently in use as a routine procedure in many fields (for example, National Conference of Standard Laboratories; cement, paper, and rubber industries; oil residue analysis) where interlaboratory agreement is important. The simple experiment required of each laboratory yields a summary of results in graphical form that pinpoints trouble spots and shows the direction for improvements.

Following the same philosophy in fitting the design to the problem, Youden developed "chain blocks" with W. S. Connor to accommodate spectrographic determination of chemical elements on photographic plates (Youden & Connor 1953*b*); "calibration designs" where only "differences" are observed for two objects in blocks of size two (Youden & Connor 1954); and "partially replicated Latin squares" with J. S. Hunter to check whether the usual requirement of additive effects for rows, columns, and treatments in a Latin square has been met (Youden & Hunter 1955). His extreme rank–sum test was likewise conceived as a device to highlight the messages contained in experimental results (1963).

During his earlier working years at the National Bureau of Standards, Youden spent much of his time serving as mentor to aspiring young statisticians who came to the Statistical Engineering Laboratory to serve apprenticeship after graduation. Often in the morning, the "master" would try out his thoughts and ideas on the novices gathering around him, leaving them to develop the mathematical theory of the topic. Joint papers flourished, and the interest generated by Youden in the design of experiments is reflected by the NBS Applied Mathematical Series of publications on that subject.

His time was always free to high school and college students who were interested in experimentation and wished to talk to him. Two of his books (1962; 1974), cloaking complicated statistical ideas in elementary experiments and simple language, were specifically aimed at this audience. One of the youngsters whose career Youden influenced is a 1976 Nobel laureate in medicine, D. Carleton Gajdusek. In an interview with the *Rochester Times-Union* (October 15, 1976), Gajdusek recalled an incident that had happened to him at the age of 16: The famous statistician William Youden had taken him aside one day, pointed out that the University of Rochester had an excellent physics department, and

advised him to go there. He followed Youden's advice and went on to Harvard Medical School after graduation from Rochester.

The interaction of Youden and his colleagues with physical scientists at the Bureau of Standards generated new families of statistical designs and statistical techniques; moreover, skillful applications of statistically planned experiments to bring about greater precision and economy were also demonstrated in many fields of research: in the road testing of truck tires, in intercomparison of national radioactivity standards, in measuring the freezing temperatures of chemical cells, and so on. It was largely recognition of these early accomplishments of Youden (and of Churchill Eisenhart who was then chief of the Statistical Engineering Laboratory, National Bureau of Standards) that led Fisher to say in his 1951 Bateson lecture (1952), apropos the conception and fruitful application of modern principles and techniques of statistical design and analysis, that "solid progress is indeed being made, but it is in places like Rothamsted in this country [England] or in the Bureau of Standards in America, rather than in places organized primarily for education."

Youden was one of the first to recognize and to capitalize on important differences between experimentation in the biological and agricultural sciences on the one hand and in the physical sciences on the other. Of paramount importance, he noted, is the difference in the magnitude of the errors of measurement in the two areas. Physical measurements can often be made with such high precision that quantities of interest can be determined with acceptably small standard errors from as few as two or three determinations. Youden took full advantage of these better experimental conditions in physical-science experimentation by "making one measurement do the work of two" (Youden & Connor 1953*a*).

The high degree of precision with which physical scientists can repeat their measurements, with the same instrumentation setup and within a short time period, sometimes lures them into claiming an unwarranted confidence in their results and neglecting the vagaries of other components of error, such as day-to-day variations, differences among nominally identical instruments and standards, and possible systematic errors that are not obvious a priori. Youden spared no effort in his lectures and writings to impress upon his audiences that a realistic estimate of experimental error could be obtained only through "hidden replication," that is, that the measure of precision should not be calculated from a series of straightforward replications but from discrepancies between the measured quantities and the corresponding adjusted values under conditions specified by a careful design. In searching for unsuspected sources of systematic error, he advised the experimenter, contrary to the time-honored tradition of "varying one factor at a time," to vary several factors at a time by proper planning, such that the effect of each factor can be measured with better precision than the traditional way using the same number of measurements (1961a).

During the two decades he was with the Bureau of Standards, as part of his "missionary work" in acquainting scientists and engineers with the effective use of modern tools of statistical design and analysis of experiments, Youden gave more than 210 talks around the country. The clarity of his presentations and the interest he could generate among his audiences are reflected in many a statistician's lament that "I am not a Dr. Youden." In 1972, the Chemical Division of the American Society for Quality Control renamed its "outstanding lecture" series the Youden Memorial Lectures to highlight his command of public speaking. In addition, the Jack Youden Prize was established by the same society in 1969 to be awarded annually for the best expository paper published in its journal, *Technometrics*.

Youden was president of the Philosophical Society of Washington in 1967, and delivered his last major address as retiring president in 1968. In it, he set forth schemes for incorporating investigations of systematic errors into experimental determinations of fundamental constants, and he pleaded for explicit effort by scientists to accumulate objective evidence of the precision and accuracy of their work. His paper (1972a) was published in *Technometrics* after his death, together with a special invited paper (1972b) he had delivered at a general session of the Institute of Mathematical Statistics at their annual meeting in 1956.

In addition to the Medal of Freedom, Youden's honors included the Wilks medal (1969) of the American Statistical Association, the Shewhart medal (1969) of the American Society for Quality Control, and the U.S. Department of Commerce Gold Medal for Exceptional Service (1962). Youden was a fellow of the Royal Statistical Society of London, the American Society for the Advancement of Science, the American Statistical Association, and the Institute of Mathematical Statistics, and a member of the Biometric Society, the Mathematical Association of America, the American Society for Quality Control, and the International Statistical Institute.

H. H. Ku

WORKS BY YOUDEN

For a bibliography of 110 papers dating from 1924 to 1972, see Lasater 1972.

1937 Youden, W. J.; and Zimmerman, P. W. Use of Incomplete Block Replications in Estimating Tobacco-Mosaic Virus. Boyce Thompson Institute for Plant Research, *Contributions* 9, no. 1:41–48. → Reprinted in Lasater (1972).

1940 Science Looks at "Social Breakdown." *Community Chests and Councils* 15, no. 8:115–116.

1951 *Statistical Methods for Chemists*. New York. Wiley. → Translated into Italian.

1953a Youden, W. J.; and Connor, W. S. Making One Measurement Do the Work of Two. *Chemical Engineering Progress* 49, no. 10:549–552. → Reprinted in Lasater (1972).

1953b Youden, W. J.; and Connor, W. S. The Chain Block Design. *Biometrics* 9:127–140.

1954 Youden, W. J.; and Connor, W. S. New Experimental Designs for Paired Observations. U.S. National Bureau of Standards, *Journal of Research* 53:191–196.

1955 Youden, W. J.; and Hunter, J. S. Partially Replicated Latin Squares. *Biometrics* 11:399–405.

1959 Graphical Diagnosis of Interlaboratory Test Results. *Industrial Quality Control* 15, no. 11:24–28. → Reprinted in Lasater (1972) and in *Precision Measurement and Calibration* (1969).

1960 *Statistical Design*. Washington: American Chemical Society. → A collection of 36 bimonthly columns from *Industrial and Engineering Chemistry* first published during 1954–1959.

1961a Systematic Errors in Physical Constants. *Physics Today* 14, no. 9:32–42. → Also in *Technometrics* 4, no. 1:111–123. Reprinted in *Precision Measurement and Calibration* (1969).

1961b Experimental Design and ASTM Committees. *Materials Research and Standards* 1, no. 11:862–867. → Reprinted in *Precision Measurement and Calibration* (1969).

1962 *Experimentation and Measurement*. National Science Teachers Association, Vistas of Science Series, No. 2. New York: Scholastic Book. → Translated into Arabic.

1963 Ranking Laboratories by Round-robin Tests. *Materials Research and Standards* 3, no. 1:9–13. → Reprinted in *Precision Measurement and Calibration* (1969).

1965 The Evolution of Designed Experiments. Pages 59–67 in IBM Scientific Computing Symposium on Statistics, Yorktown Heights, N.Y., 1963, *Proceedings*. White Plains, N.Y.: IBM Data Processing Division. → Reprinted in Lasater (1972).

1967 *Statistical Techniques for Collaborative Tests*. Washington: Association of Official Analytical Chemists.

1972a Enduring Values. *Technometrics* 14, no. 1:1–11.

1972b Randomization and Experimentation. *Technometrics* 14, no. 1:13–22. → A paper first presented at the 1956 annual meeting of the Institute of Mathematical Statistics.

1974 *Risk, Choice, and Prediction: An Introduction to Experimentation*. North Scituate, Mass.: Duxbury.

SUPPLEMENTARY BIBLIOGRAPHY

Highlights of Youden's career are summarized in Lasater 1972 *and* Eisenhart 1976; *his contributions to the theory of statistics are reviewed in* Eisenhart & Rosenblatt 1972.

Eisenhart, Churchill 1976 William John Youden. Volume 14, pages 552–557 in *Dictionary of Scientific*

Biography. Edited by Charles C. Gillispie. New York: Scribner's.

EISENHART, CHURCHILL; and ROSENBLATT, JOAN R. 1972 W. J. Youden, 1900–1971. *Annals of Mathematical Statistics* 43, no. 4:1035–1040.

FISHER, R. A. 1952 Statistical Methods in Genetics. *Heredity* 6, part 1:1–12 → The 1951 Bateson lecture.

FISHER, R. A.; and YATES, FRANK (1938) 1963 *Statistical Tables for Biological, Agricultural and Medical Research*. 6th ed., rev. & enl. Edinburgh: Oliver & Boyd; New York: Hafner.

LASATER, H. A. (editor) 1972 W. J. Youden, 1900–1971. *Journal of Quality Technology* 4, no. 1:1–67. → Memorial issue; includes reprints of nine of Youden's papers and one of his bimonthly columns from *Industrial and Engineering Chemistry* as well as an extensive bibliography.

Precision Measurement and Calibration: Selected NBS Papers. 1969 National Bureau of Standards Special Publication No. 300. Volume 1: *Statistical Concepts and Procedures*. Edited by H. H. Ku. Washington: Government Printing Office. → Includes reprints of 14 of Youden's papers on design of experiments and the interlaboratory test.

YULE, G. UDNY

George Udny Yule (1871–1951), British statistician, was the only child of Sir George Udny Yule, an Indian civil servant, and the nephew of Sir Henry Yule, a distinguished Orientalist and traveler. He was educated at Winchester and at University College, London, where he graduated in engineering in 1892. Feeling that engineering was not his métier, he tried experimental physics and spent a year at Bonn in research under Hertz on electric waves in dielectrics. Yule's first four published papers, in fact, dealt with this subject. But physics also failed to hold his interest. In his mature work there is little evidence of his background, except at the end of his life, when the philological gifts of his family became manifest in his studies of literary vocabulary.

When Yule returned to London in 1893, Karl Pearson was beginning to form his famous statistical unit at University College. Having known Yule as a student and divined something of his talents, Pearson offered him a demonstratorship (a kind of junior lectureship), which Yule promptly accepted. He held this post for six years, but the salary was scarcely a living wage, and in 1899 he resigned to earn his bread and butter as secretary to an examining body in London. His separation from University College, however, was more formal than real: in particular he gave, from 1902 to 1909, a series of lectures that formed the basis of his *Introduction to the Theory of Statistics* (1911). This rapidly diffused his reputation throughout the scientific world; 50 years later, after two revisions by M. G. Kendall, it was still a standard text.

In 1912 the University of Cambridge decided to create a lectureship in statistics. Yule was offered the lectureship and accepted it. Apart from the interruption of World War I (during which he acted as director of requirements at the Ministry of Food), he spent the rest of his life at St. John's College, Cambridge. He was promoted from lecturer to reader (the equivalent of the American assistant professor) and continued as reader until his retirement in 1930. At that point he was physically fit enough to learn to fly, but shortly afterward he was grounded and, indeed, virtually confined to college by a partial heart block. After some years of relative inactivity he took a new lease on life, producing in 1944 his *Statistical Study of Literary Vocabulary*; but his health continued to decline, and he died of heart failure in 1951.

Honors came to him in a steady stream. He received the gold medal of the Royal Statistical Society in 1911, was elected a fellow of the Royal Society in 1922, and was president of the Royal Statistical Society from 1926 to 1928. He was also elected to various foreign societies, and his *Introduction to the Theory of Statistics* was translated into several languages.

Yule's contributions to the development of theoretical statistics were extensive and profound. They may be classified into four main groups, corresponding roughly in time to his period in London (1893–1912), the war interval (1913–1919), his heyday at Cambridge (1920–1931), and his final studies (1938–1946).

When Yule joined Pearson in 1893, the science of statistics as it became known in the middle of the twentieth century scarcely existed. Pearson was beginning his series of memoirs on frequency curves and on correlation. The practical applications of this work lay mainly in biology, and he made occasional forays into social medicine. Yule was an ideal complement. He took the theory of correlation, then in a rather elementary state, and in two basic memoirs (1897 and 1907) laid the foundations of the theory of partial correlation and of linear regression for any number of variables [*see* MULTIVARIATE ANALYSIS, *article on* CORRELATION METHODS]. Yule's method almost immediately became standard practice. Characteristically, Yule's theoretical studies of regression were accompanied by practical studies, notably on the relationship between pauperism and outrelief (i.e., relief given outside institutions by local authorities).

From problems of relationships among measurable variables Yule was led to the parallel problems among attributes—i.e., those qualities that form the basis of classification on a nonmeasured basis, such as sex, inoculation against disease, or eye color. This in turn led him to revive Boole's logic of class frequencies (1901) and to develop a theory of association, culminating in a fundamental paper (1912). The work was illustrated by studies of smallpox and vaccination. These interests led to the formation of a lifelong friendship with the epidemiologist Major Greenwood. Their joint work on the interpretation of inoculation statistics (1915), now almost a textbook commonplace, was a landmark in medical statistics.

Yule's second period, 1913–1919, short as it was, produced two basic papers in collaboration with Greenwood. In the first (1917), Yule produced a theoretical scheme to account for the so-called negative binomial distribution, for which fresh applications were still being discovered nearly 50 years later. In the second (1920), the authors discussed compound distributions, with particular reference to industrial accidents. In the light of later elaborations these early attempts seem simple. But the simplicity is that of genius, and if it is the first step that counts, Yule must be credited with a great many first steps. The 1917 paper contains the beginnings of what in later hands became an important class of stochastic processes.

The third period saw the full exercise of Yule's abilities. Some earlier studies in Mendelian inheritance emerged in a mathematical theory of evolution (1924), which attracted no attention from geneticists but introduced some J-shaped frequency distributions that later proved of great interest in other subjects. Likewise, his interests in vital statistics, a subject on which he lectured for many years, culminated in a paper (1925) on the growth of population and the factors that control it. His greatest work, however, undoubtedly lay in his papers on time series (1926; 1927).

In his earlier work on correlation Yule had been puzzled by the high correlations that were noted between unrelated quantities observed over a course of time. For example, suicide rate was highly correlated with membership in the Church of England, and more recently, in the same category, there has been observed in Sweden a remarkable correlation between the fall in birth rate and the decline in the population of storks. Yule called these "nonsense-correlations" and successfully set out to explain them. Incidentally, in so doing, he frightened economic statisticians off correlation analysis for two or three decades. He then proceeded to discuss time series in terms of their internal correlations,

devising the correlogram for the purpose. In his papers on sunspots, he effectively laid the basis of what is known as the theory of "autoregressive" time series [see TIME SERIES]. In later hands this has developed into a large and complicated subject, but no one has ever surpassed Yule's peculiar blend of insight, theoretical analysis, and insistence on practical application.

His illness over the years 1931–1938 prevented the publication of any serious research. Toward the end of that period, however, he became interested in the statistical characteristics of prose style, with particular reference to questions of disputed authorship. His earlier work concerned sentence length, but he later turned to noun frequency. The master had not lost his touch: once more his work formed the basis of extensive further research by others, and his technique was applied in such an unrelated field as bacteriology. But he himself had finished and, as his health steadily failed, set himself to wait for the end, which came in his 81st year.

Yule's outstanding contribution to statistics results not so much from any one quality as from his combination of qualities. He was not a great mathematician, but his mathematics was always equal to the task. He was not trained in economics or sociology, but his wide knowledge of human relationships enabled him to write with insight on both subjects. He had the precision, the persistence, and the patience of a true scientist but never lost sight of the humanities. He was a kindly, genial, highly literate, approachable man who refused to embroil himself in the controversies that mar so much of statistical literature. Above all, he had the flair for handling numerical data that characterizes the truly great statistician.

MAURICE G. KENDALL

[*For the historical context of Yule's work, see the biography of* PEARSON. *See also* LINEAR HYPOTHESES, *article on* REGRESSION; STATISTICS, DESCRIPTIVE, *article on* ASSOCIATION.]

WORKS BY YULE

[1]*There is no collected edition of Yule's works. A complete bibliography is given in* Kendall 1952.

1897 On the Significance of Bravais' Formulae for Regression, &c., in the Case of Skew Correlation. Royal Society of London, *Proceedings* 60:477–489.
1901 On the Theory of Consistence of Logical Class-frequencies, and Its Geometrical Representation. Royal Society of London, *Philosophical Transactions* Series A 197:91–133.
1907 On the Theory of Correlation for Any Number of Variables, Treated by a New System of Notation.

Royal Society of London, *Proceedings* Series A 79: 182–193.

○(1911) 1950   YULE, G. UDNY; and KENDALL, MAURICE G. *An Introduction to the Theory of Statistics.* 14th ed., rev. & enl. London: Griffin; New York: Hafner. → Kendall, who became a joint author on the eleventh edition (1937), revised the current edition and added new material to the 1965 printing.

1912   On the Methods of Measuring Association Between Two Attributes. *Journal of the Royal Statistical Society* 75:579–652. → Contains ten pages of discussion.

1915   GREENWOOD, MAJOR; and YULE, G. UDNY The Statistics of Anti-typhoid and Anti-cholera Inoculations, and the Interpretation of Such Statistics in General. Royal Society of Medicine, Section of Epidemiology and State Medicine, *Proceedings* 8, part 2:113–194.

1917   GREENWOOD, MAJOR; and YULE, G. UDNY On the Statistical Study of Some Bacteriological Methods Used in Water Analysis. *Journal of Hygiene* 16:36–54.

1920   GREENWOOD, MAJOR; and YULE, G. UDNY An Inquiry Into the Nature of Frequency Distributions Representative of Multiple Happenings With Particular Reference to the Occurrence of Multiple Attacks of Disease or of Repeated Accidents. *Journal of the Royal Statistical Society* 83:255–279.

1924   A Mathematical Theory of Evolution, Based on the Conclusions of Dr. J. C. Willis, F. R. S. Royal Society of London, *Philosophical Transactions* Series B 213: 21–87.

1925   The Growth of Population and the Factors Which Control It. *Journal of the Royal Statistical Society* 88:1–58.

1926   Why Do We Sometimes Get Nonsense-correlations Between Time-series?—A Study in Sampling and the Nature of Time-series. *Journal of the Royal Statistical Society* 89:1–64.

1927   On a Method of Investigating Periodicities in Disturbed Series, With Special Reference to Wolfer's Sunspot Numbers. Royal Society of London, *Philosophical Transactions* Series A 226:267–298.

1944   *The Statistical Study of Literary Vocabulary.* Cambridge Univ. Press.

SUPPLEMENTARY BIBLIOGRAPHY

KENDALL, MAURICE G.   1952   G. Udny Yule, 1871–1951. *Journal of the Royal Statistical Society* 115:156–161. → Includes a complete bibliography.

YATES, FRANK   1952   George Udny Yule, 1871–1951. Volume 8, pages 309–323 in Royal Society of London, *Obituary Notices of Fellows.* Cambridge: The Society. → Includes a bibliography on pages 320–323.

Postscript

[1]A selected set of Yule's papers has now been published (Stuart & Kendall 1971).

MAURICE G. KENDALL

ADDITIONAL BIBLIOGRAPHY

STUART, ALAN; and KENDALL, MAURICE G. (editors)   1971   *Statistical Papers of George Udny Yule.* London: Griffin; New York: Hafner.

ZIPF'S LAW

See RANK–SIZE RELATIONS.

CORRIGENDA

Page 191, lefthand column, lines 30–31. The citation of Corry (1968) is unsupported by a reference in the bibliography. On page 197, the following reference should be inserted in alphabetical order: "CORRY, BERNARD 1968 Marshall, Alfred. Volume 10, pages 25–33 in *International Encyclopedia of the Social Sciences*. Edited by David L. Sills, New York: Macmillan and Free Press."

Page 332, lefthand column, line 8 from foot. The citation of Guttman (1953) is unsupported by a reference in the bibliography. On page 335, the following reference should be inserted in alphabetical order: "GUTTMAN, LOUIS 1953 Image Theory for the Structure of Quantitative Variates. *Psychometrika* 18:277–296."

Page 358, righthand column. The following significant biography, published too late for citation, should be added to the additional bibliography: "BOX, JOAN FISHER 1978 *R. A. Fisher: The Life of a Scientist*. New York: Wiley."

Page 589, lefthand column, line 7. The cross-reference to SCALING should read SCALING, MULTIDIMENSIONAL.

Directory
of Contributors

ACKOFF, RUSSELL L.
University of Pennsylvania, The Wharton School
OPERATIONS RESEARCH

ADELMAN, IRMA
University of Maryland, College Park
SIMULATION, *article on* ECONOMIC PROCESSES

ALLEN, R. G. D.
London School of Economics and Political Science
BOWLEY, ARTHUR LYON

ANDERSON, OSKAR A.
University of Munich
Emendation of ANDERSON, OSKAR N.

ANDERSON, T. W.
Stanford University
MULTIVARIATE ANALYSIS, *article on* CLASSIFICATION AND DISCRIMINATION

ANDREWS, DAVID F.
University of Toronto
DATA ANALYSIS, EXPLORATORY

ANSCOMBE, F. J.
Yale University
STATISTICAL ANALYSIS, SPECIAL PROBLEMS OF, *article on* OUTLIERS

ARMITAGE, P.
University of Oxford
SEQUENTIAL ANALYSIS

ARROW, KENNETH J.
Harvard University
GIRSHICK, MEYER A.

ASHENHURST, ROBERT L.
University of Chicago
COMPUTATION

BARTLETT, M. S.
Worthing, England
FISHER, R. A.

BARTON, ALLEN H.
Columbia University
ORGANIZATIONS, *article on* METHODS

BENJAMIN, B.
City University (London)
GRAUNT, JOHN

BERRY, BRIAN J. L.
Harvard University
GEOGRAPHY, STATISTICAL

BILLINGSLEY, PATRICK
University of Chicago
MARKOV CHAINS

BINGHAM, CHRISTOPHER
University of Minnesota, Twin Cities
TIME SERIES, *general article*

BIRNBAUM, ALLAN
LIKELIHOOD

BLISCHKE, WALLACE R.
University of Southern California
DISTRIBUTIONS, STATISTICAL, *article on* MIXTURES OF DISTRIBUTIONS

BORGATTA, EDGAR F.
City University of New York, Queens College
SOCIOMETRY

BOX, GEORGE E. P.
University of Wisconsin, Madison
EXPERIMENTAL DESIGN, *article on* RESPONSE SURFACES

BRADLEY, RALPH A.
Florida State University
MULTIVARIATE ANALYSIS, *overview article;*
WILCOXON, FRANK

BROWN, BYRON W. JR.
Stanford University
QUANTAL RESPONSE

BUCH, KAI RANDER
Technical University of Denmark (Lyngby)
WESTERGAARD, HARALD

BURKHOLDER, DONALD L.
University of Illinois at Urbana–Champaign
ESTIMATION, *article on* POINT ESTIMATION

BUSH, ROBERT R.
MODELS, MATHEMATICAL

CAMPBELL, DONALD T.
Northwestern University
EXPERIMENTAL DESIGN, *article on*
QUASI-EXPERIMENTAL DESIGN

CANNELL, CHARLES F.
University of Michigan, Ann Arbor
INTERVIEWING IN SOCIAL RESEARCH

CARROLL, J. DOUGLAS
Bell Laboratories (Murray Hill, New Jersey)
SCALING, MULTIDIMENSIONAL

CATTELL, RAYMOND B.
Champaign, Illinois
SPEARMAN, C. E.

CHAPMAN, DOUGLAS G.
University of Washington
COUNTED DATA

CHERNOFF, HERMAN
Massachusetts Institute of Technology
DECISION THEORY

CHRIST, CARL F.
Johns Hopkins University
ECONOMETRIC MODELS, AGGREGATE

COCHRAN, WILLIAM G.
South Orleans, Massachusetts
EXPERIMENTAL DESIGN, *article on* THE DESIGN OF
EXPERIMENTS

COLEMAN, JAMES S.
University of Chicago
LAZARSFELD, PAUL F.

COX, D. R.
*Imperial College of Science and Technology
(London)*
QUEUES

CURETON, EDWARD E.
University of Tennessee at Knoxville
PSYCHOMETRICS

DAVID, F. N.
Kensington, California
GALTON, FRANCIS

DAVID, H. T.
Iowa State University of Science and Technology
GOODNESS OF FIT

DAVID, HERBERT A.
Iowa State University of Science and Technology
NONPARAMETRIC STATISTICS, *article on* RANKING
METHODS

DAVIS, JAMES A.
Harvard University
TABULAR PRESENTATION

DEANE, PHYLLIS
Royal Economic Society (London)
KING, GREGORY; PETTY, WILLIAM

DEMING, W. EDWARDS
Washington, D.C.
SAMPLE SURVEYS, *article on* THE FIELD; SHEWHART,
WALTER A.

DE FINETTI, BRUNO
University of Rome
PROBABILITY, *article on* INTERPRETATIONS

DUGUÉ, DANIEL
University of Paris
BIENAYMÉ, JULES

DUNNETT, CHARLES W.
McMaster University
SCREENING AND SELECTION

EDWARDS, WARD
University of Southern California
DECISION MAKING, *article on* PSYCHOLOGICAL
ASPECTS

EISENHART, CHURCHILL
National Bureau of Standards (Washington, D.C.)
GAUSS, CARL FRIEDRICH

ELASHOFF, JANET D.
University of California, Los Angeles
ERRORS, *article on* EFFECTS OF ERRORS IN
STATISTICAL ASSUMPTIONS

ELASHOFF, ROBERT M.
University of California, Los Angeles
ERRORS, *article on* EFFECTS OF ERRORS IN
STATISTICAL ASSUMPTIONS

ESENWEIN-ROTHE, I.
University of Erlangen–Nuremberg
SÜSSMILCH, JOHANN PETER

FAIRLEY, WILLIAM B.
Commonwealth of Massachusetts
PUBLIC POLICY AND STATISTICS

FELS, EBERHARD M.
ANDERSON, OSKAR N.

FÉRON, R.
University of Lyons
POISSON, SIMÉON DENIS

FINCH, STEPHEN J.
State University of New York at Stony Brook
SURVEY ANALYSIS, *article on* METHODS OF SURVEY
ANALYSIS

FISHER, FRANKLIN M.
Massachusetts Institute of Technology
STATISTICAL IDENTIFIABILITY

FRASER, D. A. S.
University of Toronto
FIDUCIAL INFERENCE

FRÉCHET, MAURICE
LAPLACE, PIERRE SIMON DE

FREEMAN, HAROLD
Massachusetts Institute of Technology
WALD, ABRAHAM

GEIRINGER, HILDA
VON MISES, RICHARD

GJEDDEBÆK, N. F.
A/S Ferrosan (Søborg, Denmark)
STATISTICAL ANALYSIS, SPECIAL PROBLEMS OF, *article on* GROUPED OBSERVATIONS

GOLDHAMER, HERBERT
SOCIAL MOBILITY

GOOD, IRVING JOHN
Virginia Polytechnic Institute and State University
FALLACIES, STATISTICAL

GOSNELL, HAROLD F.
Bethesda, Maryland
RICE, STUART

GRANGER, GILLES-GASTON
University of Provence
CONDORCET

GREBENIK, EUGENE
University of Leeds
VITAL STATISTICS

GREENBERG, BERNARD G.
University of North Carolina at Chapel Hill
NONPARAMETRIC STATISTICS, *article on* ORDER STATISTICS

GUITTON, HENRI
University of Paris
COURNOT, ANTOINE AUGUSTIN

GUMBEL, E. J.
BORTKIEWICZ, LADISLAUS VON

HAAVELMO, TRYGVE
University of Oslo
BUSINESS CYCLES: MATHEMATICAL MODELS

HAIGHT, FRANK A.
Pennsylvania State University, University Park
DISTRIBUTIONS, STATISTICAL, *article on* SPECIAL DISCRETE DISTRIBUTIONS

HAMAKER, HUGO C.
Technological University of Eindhoven
QUALITY CONTROL, STATISTICAL, *article on* ACCEPTANCE SAMPLING

HEISS, KLAUS-PETER
ECON, Inc. (Princeton, New Jersey)
BENINI, RODOLFO; LEXIS, WILHELM

HEMELRIJK, J.
University of Amsterdam
VAN DANTZIG, DAVID

HERMANN, CHARLES F.
Ohio State University
SIMULATION, *article on* POLITICAL PROCESSES

HILDRETH, CLIFFORD
University of Minnesota, Twin Cities
EDGEWORTH, FRANCIS YSIDRO

HOBBS, NICHOLAS
Vanderbilt University
ETHICAL ISSUES IN THE SOCIAL SCIENCES

HODGES, J. L. JR.
University of California, Berkeley
SUFFICIENCY

HOEFFDING, WASSILY
University of North Carolina at Chapel Hill
HOTELLING, HAROLD

HOLLANDER, MYLES
Florida State University
WILCOXON, FRANK

HOOVER, ETHEL D.
Bethesda, Maryland
INDEX NUMBERS, *article on* PRACTICAL APPLICATIONS

IRWIN, J. O.
Schaffhausen, Switzerland
GOSSET, WILLIAM SEALY

JACOBS, ANN M.
New York City
TABULAR PRESENTATION

JACQUARD, ALBERT
Institut National d'Études Démographiques (Paris)
DEMOGRAPHY, *postscript to article on* DEMOGRAPHY AND POPULATION GENETICS

JOHNSON, NORMAN L.
University of North Carolina at Chapel Hill
DISTRIBUTIONS, STATISTICAL, *article on* APPROXIMATIONS TO DISTRIBUTIONS

KAHN, ROBERT L.
University of Michigan, Ann Arbor
INTERVIEWING IN SOCIAL RESEARCH

KELLERER, HANS
STATISTICS, DESCRIPTIVE, *article on* LOCATION AND DISPERSION

KENDALL, MAURICE G.
International Statistical Institute (London)
STATISTICS, *article on* THE HISTORY OF STATISTICAL METHOD; YULE, G. UDNY

KEYFITZ, NATHAN
Harvard University
GOVERNMENT STATISTICS

KIRK, DUDLEY
Stanford University
DEMOGRAPHY, *article on* THE FIELD

KLEIN, LAWRENCE R.
University of Pennsylvania
SIMULTANEOUS EQUATION ESTIMATION

KONÜS, A. A.
Institute for Economics and Statistics (Moscow)
SLUTSKY, EUGEN

KRUSKAL, JOSEPH. B.
Bell Laboratories (Murray Hill, New Jersey)
FACTOR ANALYSIS AND PRINCIPAL COMPONENTS, *article on* BILINEAR METHODS; SCALING, MULTIDIMENSIONAL; STATISTICAL ANALYSIS, SPECIAL PROBLEMS OF, *article on* TRANSFORMATIONS OF DATA

KRUSKAL, WILLIAM H.
University of Chicago
PRICE, RICHARD; SAVAGE, LEONARD JIMMIE; SIGNIFICANCE, TESTS OF; STATISTICS, *article on* THE FIELD

KU, H. H.
National Bureau of Standards (Washington, D.C.)
YOUDEN, W. J.

KUHN, THOMAS S.
Princeton University
SCIENCE, *article on* THE HISTORY OF SCIENCE

LANDAU, DAVID
New York City
QUETELET, ADOLPHE

LAZARSFELD, PAUL F.
QUETELET, ADOLPHE; SURVEY ANALYSIS, *article on*
THE ANALYSIS OF ATTRIBUTE DATA

LÉCUYER, BERNARD
*Group d'Étude des Méthodes de l'Analyse
Sociologique (Paris)*
SOCIAL RESEARCH, THE EARLY HISTORY OF

LEHMANN, E. L.
University of California, Berkeley
HYPOTHESIS TESTING

LEVENSON, BERNARD
*City University of New York, Mount Sinai School
of Medicine*
PANEL STUDIES

LINDLEY, DENNIS V.
University College (London)
BIRNBAUM, ALLAN; KEYNES, JOHN MAYNARD:
CONTRIBUTIONS TO STATISTICS

LUCE, R. DUNCAN
Harvard University
MATHEMATICS; MODELS, MATHEMATICAL

McCARTHY, PHILIP J.
Cornell University
INDEX NUMBERS, *article on* SAMPLING

MADANSKY, ALBERT
University of Chicago
LATENT STRUCTURE

MARON, M. E.
University of California, Berkeley
WIENER, NORBERT

MARSCHAK, JACOB
DECISION MAKING, *article on* ECONOMIC ASPECTS

MAXWELL, A. E.
Maudsley Hospital (London)
FACTOR ANALYSIS AND PRINCIPAL COMPONENTS,
article on FACTORING CORRELATION MATRICES

MOORE, PETER G.
London Graduate School of Business Studies
NONPARAMETRIC STATISTICS, *article on* RUNS

MORGENBESSER, SIDNEY
Columbia University
SCIENTIFIC EXPLANATION

MORGENSTERN, OSKAR
GAME THEORY, *article on* THEORETICAL ASPECTS;
VON NEUMANN, JOHN

MORRISON, EMILY
Ithaca, New York
BABBAGE, CHARLES

MORRISON, PHILIP
Massachusetts Institute of Technology
BABBAGE, CHARLES

MOSES, LINCOLN E.
Department of Energy (Washington, D.C.)
STATISTICAL ANALYSIS, SPECIAL PROBLEMS OF,
article on TRUNCATION AND CENSORSHIP

MOSTELLER, FREDERICK
Harvard University
ERRORS, *article on* NONSAMPLING ERRORS;
WILKS, S. S.

MULLER, MERVIN E.
World Bank (Washington, D.C.)
RANDOM NUMBERS

MUNDLAK, YAIR
Hebrew University of Jerusalem
CROSS-SECTION ANALYSIS

NEMENYI, PETER
Virginia State College
LINEAR HYPOTHESES, *article on* MULTIPLE
COMPARISONS

NERLOVE, MARC
Northwestern University
DISTRIBUTED LAGS

NEWELL, ALLEN
Carnegie–Mellon University
SIMULATION, *article on* INDIVIDUAL BEHAVIOR

NOETHER, GOTTFRIED E.
University of Connecticut
PROBABILITY, *article on* FORMAL PROBABILITY

NOTESTEIN, FRANK W.
Princeton University
WILLCOX, WALTER F.

OBERSCHALL, ANTHONY
Vanderbilt University
SOCIAL RESEARCH, THE EARLY HISTORY OF

ORE, ØYSTEIN
BERNOULLI FAMILY

OWEN, DONALD B.
Southern Methodist University
DISTRIBUTIONS, STATISTICAL, *article on* SPECIAL
CONTINUOUS DISTRIBUTIONS

PAGE, E. S.
University of Newcastle Upon Tyne
QUALITY CONTROL, STATISTICAL, *article on*
PROCESS CONTROL

PERRIN, EDWARD B.
Battelle Human Affairs Research Centers (Seattle)
Postscript to LIFE TABLES

PFANZAGL, J.
University of Cologne
ESTIMATION, *article on* CONFIDENCE INTERVALS
AND REGIONS

PULLUM, THOMAS W.
University of Washington
Postscript to SOCIAL MOBILITY

QUANDT, RICHARD E.
Princeton University
PROGRAMMING

RAO, C. RADHAKRISHNA
Indian Statistical Institute (New Delhi)
MAHALANOBIS, P. C.

RAPOPORT, ANATOL
University of Toronto
RANK–SIZE RELATIONS; RICHARDSON, LEWIS FRY

RIKER, WILLIAM H.
University of Rochester
COALITIONS, THE STUDY OF

ROBERTS, HARRY V.
University of Chicago
BAYESIAN INFERENCE

RUIST, ERIK
Stockholm School of Economics
INDEX NUMBERS, *article on* THEORETICAL ASPECTS

SALVEMINI, TOMMASO
University of Rome
GINI, CORRADO

SAVAGE, I. RICHARD
Yale University
NONPARAMETRIC STATISTICS, *article on* THE FIELD

SCHMETTERER, LEOPOLD K.
University of Vienna
Postscript to VON MISES, RICHARD

SCHMID, CALVIN F.
University of Washington
GRAPHIC PRESENTATION

SCHUESSLER, KARL F.
Indiana University at Bloomington
PREDICTION

SCRIVEN, MICHAEL
University of California, Berkeley
SCIENCE, *article on* THE PHILOSOPHY OF SCIENCE

SEAL, HILARY L.
École Polytechnique Fédérale de Lausanne
BAYES, THOMAS; MOIVRE, ABRAHAM DE

SELVIN, HANAN C.
State University of New York at Stony Brook
SURVEY ANALYSIS, *article on* METHODS OF SURVEY ANALYSIS

SHAFER, GLENN
Princeton University
Postscript to BERNOULLI FAMILY

SHEYNIN, O. B.
Moscow
Postscript to BORTKIEWICZ, LADISLAUS VON; KEPLER, JOHANNES

SHISKIN, JULIUS
Bureau of Labor Statistics (Washington, D.C.)
TIME SERIES, *article on* SEASONAL ADJUSTMENT

SHUBIK, MARTIN
Cowles Foundation for Research in Economics (New Haven)
GAME THEORY, *article on* ECONOMIC APPLICATIONS

SIMON, HERBERT A.
Carnegie–Mellon University
CAUSATION; SIMULATION, *article on* INDIVIDUAL BEHAVIOR

SOMERS, ROBERT H.
Pacific Telephone and Telegraph Company (San Francisco)
STATISTICS, DESCRIPTIVE, *article on* ASSOCIATION

SPIEGEL, HENRY W.
Catholic University of America
WALKER, FRANCIS A.

SPIEGELMAN, MORTIMER
LIFE TABLES

STANLEY, JULIAN C.
Johns Hopkins University
LINEAR HYPOTHESES, *article on* ANALYSIS OF VARIANCE

STEINDL, JOSEF
Austrian Institute of Economic Research (Vienna)
SIZE DISTRIBUTIONS IN ECONOMICS

STEVENS, JOSEPH C.
Yale University
PSYCHOPHYSICS

STIGLER, STEPHEN M.
University of Wisconsin, Madison
Postscript to LAPLACE, PIERRE SIMON DE; *postscript to* PEIRCE, CHARLES SANDERS

STROTZ, ROBERT H.
Northwestern University
ECONOMETRICS

STUART, ALAN
University of London
SAMPLE SURVEYS, *article on* NONPROBABILITY SAMPLING

SUCHMAN, EDWARD A.
EPIDEMIOLOGY

SUPPES, PATRICK
Stanford University
MATHEMATICS; MODELS, MATHEMATICAL

SUSSER, MERVYN
Columbia University
EPIDEMIOLOGY

SUTTER, JEAN
DEMOGRAPHY, *article on* DEMOGRAPHY AND POPULATION GENETICS

TAEUBER, CONRAD
Georgetown University
CENSUS

TATE, ROBERT F.
University of Oregon
MULTIVARIATE ANALYSIS, *article on* CORRELATION METHODS

THIRRING, LAJOS
Budapest
KŐRÖSY, JÓZSEF

VAN DER VAART, H. ROBERT
North Carolina State University at Raleigh
VARIANCES, STATISTICAL STUDY OF

VINCZE, ISTVÁN
Hungarian Academy of Sciences (Budapest)
RÉNYI, ALFRÉD

WALKER, HELEN M.
Claremont, California
PEARSON, KARL

WALLACE, DAVID L.
University of Chicago
CLUSTERING

WHITIN, T. M.
Wesleyan University
INVENTORY CONTROL THEORY

WHITTLE, P.
University of Cambridge
TIME SERIES, *article on* ADVANCED PROBLEMS

WIENER, PHILIP P.
Temple University
PEIRCE, CHARLES SANDERS

WILLIAMS, EVAN J.
University of Melbourne
LINEAR HYPOTHESES, *article on* REGRESSION

WOLD, HERMAN
University of Goteborg
TIME SERIES, *article on* CYCLES

WRIGHT, CHARLES R.
University of Pennsylvania
EVALUATION RESEARCH

ZARNOWITZ, VICTOR
University of Chicago
PREDICTION AND FORECASTING, ECONOMIC

ZEISEL, HANS
University of Chicago
STATISTICS AS LEGAL EVIDENCE

ZELEN, MARVIN
Harvard University
QUALITY CONTROL, STATISTICAL, *article on*
RELIABILITY AND LIFE TESTING

Acknowledgments by Contributors

THE FOLLOWING CONTRIBUTORS wish to extend their thanks for the special assistance they received in the preparation of their original articles in the *International Encyclopedia of the Social Sciences:*

DONALD L. BURKHOLDER to R. R. Bahadur, for his valuable assistance.

BRUNO DE FINETTI to Leonard J. Savage, for help in all stages of the preparation of his article.

ROBERT M. ELASHOFF to the U.S. Public Health Service, for research support.

R. DUNCAN LUCE and PATRICK SUPPES to the Carnegie Corporation and the National Science Foundation, for research support.

FREDERICK MOSTELLER to the Office of Naval Research and the National Science Foundation, for research support; to Mrs. Cleo Youtz, for her aid in all parts of the preparation of his article on nonsampling errors; to the following, who helped prepare source materials: Sarah Biss, E. P. Cornelius, Jr., Ralph B. D'Agostino, Elizabeth A. Feigon, Hedda Fish, Chien-pai Han, Richard Jay Light, William S. Mosteller, and Rhett Tsao; to Harvard University's Social Relations 199, class of 1966, who aided with criticisms and suggestions, as did Yvonne M. M. Bishop, William G. Cochran, A. P. Dempster, Jacob J. Feldman, Stephen E. Fienberg, Steven Fosburg, Gudmund Iversen, Leslie Kish, Thomas Lehrer, Richard Link, Lincoln E. Moses, Martin Schatzoff, Seymour Sudman, George B. Thomas, Jr., and John W. Tukey.

HERMAN WOLD to Birgitta Hedman, for research assistance.

The following contributors wish to extend their thanks for the special assistance they received in the preparation of new, revised, or amended materials for the *International Encyclopedia of Statistics:*

H. T. DAVID to Shashikala Sukhatme, for collaboration in preparing the postscript to "Goodness of Fit."

WILLIAM B. FAIRLEY to the National Science Foundation, for research support, and to Frederick Mosteller, Herbert I. Weisberg, Richard J. Light, Edward R. Tufte, and Richard Zeckhauser, for helpful comments.

JOSEPH B. KRUSKAL to Harry H. Harman, whose book on factor analysis was more helpful than any other source in providing a foundation for the viewpoint expressed in his article on bilinear methods.

DENNIS V. LINDLEY to D. R. Cox, B. Norton, and Alan Stuart, for assistance in the preparation of the biography of Allan Birnbaum.

FREDERICK MOSTELLER to John Marden, Morris Hansen, and Joseph Waksberg for valuable assistance.

ANATOL RAPOPORT to Churchill Eisenhart, for calling his attention to several items of relevant material.

WILLIAM H. KRUSKAL, as editor, to Churchill Eisenhart, for assistance in the preparation of the original biography of Laplace.

WE HAVE TRIED to construct a friendly index—that is, one that is self-explanatory, logical, helpfully redundant, and easily entered. Both proper names and topics appear as main entries in a single alphabetical order. Capital letters show article titles. All subentries are article titles (often shortened), showing contexts. Acronyms and abbreviations head each alphabetical listing.

Names of all authors whose work is discussed or cited appear as main entries. Where it has seemed useful, substantive mentions have been distinguished from merely supportive citations: substantive mentions are recorded as subentries; page numbers of supportive citations follow "*also cited*" at the end of the entry. The abbreviation "*t*" or "*f*" after a page number means "table" or "figure."

Cross-references connect related entries, clarify synonymous terms, and explain wording inverted for alphabetical sense. Compound terms ("linear transformations"; "random sampling") should be sought first under the former part ("linear"; "random"). Broad topics ("estimation"; "testing") should be sought under both general and specific rubrics. For example, specific rubrics corresponding to "estimation" include "least squares estimation," "maximum likelihood estimation," and "sufficiency."

Index

THANKS ARE DUE to many people for help in the prepara-
tion of this index. Index cards were compiled by Joel A.
Waksman and were controlled during editing by Shane
Hewitt. Editorial and/or clerical assistance was provided
by Gilda Abramowitz, Jeffrey Apter, Alicia Denefe, Edward
Kinchley Evans, Diane Martella, Donald A. Paradise,
Stephen Pender, Dennis Seetoo, Nancy Simpson, Eugene
Waering, and Donald Ward.
As these concluding pages go to press, we also thank
Stuart David, Yaron Fidler, Arlene Lee, Frank Lee, and
Anthony Paccione, the graphic artists responsible for
page layout from *A* through *Z*. — THE EDITORS

B

F

G

I

J

K

Keats, J. A., 403
Keen, Joan, 1215, 1220
Keeney, Ralph L., 126, 799
Keeping, E. S., 621
Keleti, Károly, 491
Kellerer, Hans

contributor: Statistics, descriptive, *article on* location and dispersion, 1101–1109

Anderson, Oskar N., 3
Kelley, Florence, 1030
Kelley, Truman L.
order statistics, 649
Spearman, C. E., 1037
Kelling, George L., 794, 796, 797
Kellogg, Paul, 1030
Kelman, Herbert C., 863
Kemeny, John G., 590, 924
Kemp, K. W., 811
Kemp, M. C., 423
Kempthorne, Oscar
goodness of fit, 402
also cited: 9, 288
Kendall, D. G., 1221
Kendall, Maurice G.

contributor: Statistics, *article on* the history of statistical method, 1093–1101; Yule, G. Udny, 1261–1263

correlation methods, 616, 622
Edgeworth, Francis Ysidro, 201
geography, 390
random numbers, 841, 842
ranking methods, 664
statistics, descriptive: association, 1114
statistics, descriptive: location and dispersion, 1107
Yule, G. Udny, 1261
also cited: 27, 126, 161, 171, 255, 258, 263, 330, 446, 613, 614, 618, 624, 627, 645, 660, 665, 1091, 1101, 1110, 1115, 1116, 1124, 1127, 1132, 1142, 1145, 1156, 1161, 1172, 1175, 1178, 1191, 1262, 1263
Kendall, Patricia L.
Lazarsfeld, Paul F., 506
also cited: 39, 1160
Kendall's tau, 1114–1116
Kendler, Howard H., 919, 959, 964, 1243
Kendler, Tracy S., 919, 959, 964, 1243
Kendrick, John W., 181
Keniston, K., 475
Kennard, Robert W., 539
Kennedy, Eugene C., 276
Kennedy Institute of Bioethics (Washington, D.C.,), 276
Kenny, D. A., 303, 304
KEPLER, JOHANNES, 487–488

Kermack, W. O., 1195
Kerrich, J. E., 566
Kessler, A., 974
Kettenring, J. J., 1055
Keuls, M.
multiple comparisons, 566
order statistics, 654
Keyfitz, Nathan

contributor: Government statistics, 413–425

sample surveys, 878
KEYNES, JOHN MAYNARD, 488–489
Bortkiewicz, Ladislaus von, 25
Bowley, Arthur Lyon, 29
econometric models, aggregate, 181
econometrics, 192
Edgeworth, Francis Ysidro, 198, 199
government statistics, 419
Lexis, Wilhelm, 508
prediction and forecasting, economic, 718
probability: interpretations, 745, 746
significance, tests of, 954
also cited: 8, 510
Keynesian economics
inventory control theory, 486
Khinchin, Aleksandr Ia. (Khintchine, A. Y.)
queues, 838
Slutsky, Eugen, 1001
also cited: 816
Kiaer, Anders N.
Bowley, Arthur Lyon, 28
sample surveys, 868
Kiefer, J.
goodness of fit, 403, 405, 408
inventory control theory, 483
also cited: 406, 565
Kim, P. J., 239
Kimball, A. W., 82, 338
Kinch, S. H., 204
King, Arnold J., 269
King, E. P., 934
King, Frederick A., 961, 964
KING, GREGORY, 489–491
social research, history of, 1013
King, Leslie J., 390
King, Stanley H., 203, 204
King, Willford, 181, 463
Kinnard, Douglas, 793
Kinostatistics
graphic presentation, 435
Kinsey, Alfred C., 468, 471, 475
Kirby, Bernard C., 710
Kirk, Dudley

contributor: Demography, *article on* the field, 136–144

Kirkpatrick, Clifford, 1227

Kish, Leslie
nonsampling errors, 214, 216
also cited: 215, 870, 1084
Kitagawa, Tosio, 950, 1222
Klahr, David, 965
Klapper, Joseph T., 688
Klein, David, 471, 478
Klein, Lawrence R.

contributor: Simultaneous equation estimation, 979–994

economic models, aggregate, 181, 186, 187
prediction and forecasting, economic, 724, 727
also cited: 31, 154
Klein, Sidney, 627, 1157
Klein–Goldberger model
simulation: economic processes, 967
Klineberg, Otto, 279, 280
Klitgaard, Robert E., 1224
Kloek, T.
index numbers: theoretical aspects, 454
simultaneous equation estimation, 989
Kluckhohn, Clyde, 477
Kluckhohn, Florence R., 212
Knapp, G. F., 1025
Kneebone, G. T., 590
Knibbs, George H., 420
Knight, Frank H., 122
Knowledge, attitudes, and practices, field surveys of
demography, 141
Knuth, Donald E., 845, 846
Koch, Robert, 202
Koehler, David H., 59
Kogan, Marvin, 710, 1012
Kogan, Nathan, 114
Kohlberg, Lawrence, 273, 276
Koksma, J. F., 1211
Koller, Siegfried, 882
Fisher, R. A., 357
goodness of fit, 403
prediction and forecasting, economic, 721
probability: interpretations, 746
random numbers, 845
time series, 1175, 1178
von Mises, Richard, 1230
also cited: 1215
Kolmogorov (Kolmogoroff), A. N.
cycles, 1201
Kolmogorov–Smirnov tests
goodness of fit, 401, 403–404, 408
Kondratieff, N. D., 2
Kondratieff waves
cycles, 1196
Konefsky, Samuel J., 700
König, Franziskus Cardinal, 1090
König, René, 76, 712

M

P

W